Phosphate deposits of the world

VOLUME 3

Neogene to Modern phosphorites

Phosphate deposits of the world

VOLUME 3

Neogene to Modern phosphorites

EDITED BY

WILLIAM C. BURNETT

Department of Oceanography, Florida State University, Tallahassee, Florida

STANLEY R. RIGGS

Department of Geology, East Carolina University, Greenville, North Carolina

International Geological Correlation Programme Project 156: Phosphorites

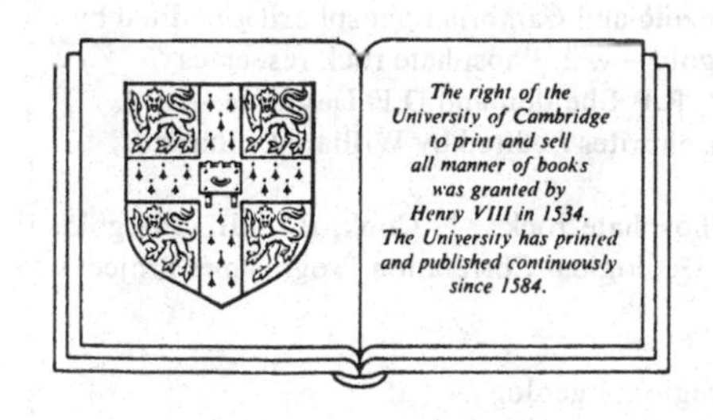

CAMBRIDGE UNIVERSITY PRESS

Cambridge
New York Port Chester
Melbourne Sydney

CAMBRIDGE UNIVERSITY PRESS
Cambridge, New York, Melbourne, Madrid, Cape Town, Singapore, São Paulo

Cambridge University Press
The Edinburgh Building, Cambridge CB2 2RU, UK

Published in the United States of America by Cambridge University Press, New York

www.cambridge.org
Information on this title: www.cambridge.org/9780521333702

First published 1990
This digitally printed first paperback version 2006

A catalogue record for this publication is available from the British Library

Library of Congress Cataloguing in Publication data
Phosphate deposits of the world.

(v. 3: World and regional geology series)
'International Geological Correlation Programme
Project 156: Phosphorites.'
Includes bibliographies.
Contents: v. 1. Proterozoic and Cambrian phosphorites / edited by
P.J. Cook and J.H. Shergold – v. 2. Phosphate rock resources /
edited by A.J.G. Notholt, R.P. Sheldon and D.F. Davidson. – v. 3.
Neogene to modern phosphorites / edited by William C. Burnett,
Stanley R. Riggs.
1. Phosphates. 2. Phosphate rock. I. Cook, P.J. II. Shergold,
J.H. III. International Geological Correlation Programme. Project
156 Phosphorites.

IV. Series: World and regional geology.
TN913.P56 1986 553.6′4 84-21396

ISBN-13 978-0-521-33370-2 hardback
ISBN-10 0-521-33370-9 hardback

ISBN-13 978-0-521-03418-0 paperback
ISBN-10 0-521-03418-3 paperback

Dedication

Vincent Ellis McKelvey (1916–87)

Vincent E. McKelvey, a Rocky Mountain Overthrust belt geologist, mineral resource geologist, policy analyst, and ninth director of the US Geological Survey (USGS) from 1971 to 1978, died of amyotrophic lateral sclerosis (Lou Gehrig's disease) at his home in St Cloud, Florida, on January 23, 1987. He is survived by his wife, Genevieve, and his geologist son, Gregory.

Vince made many fundamental contributions to the geology of phosphorites, but perhaps his most important was his early recognition of the value of the upwelling hypothesis of the origin of phosphorite, advanced by A.V. Kazakov at the Moscow IGC in 1936. Vince applied this model to the Phosphoria Formation of the western US at the IGC in Algiers in 1952 (McKelvey *et al.*, 1953) and in subsequent publications, bringing the Kazakov model squarely to the attention of phosphorite geologists worldwide. In 1969, he published a map of phosphorite and other minerals of possible economic value on the present seafloor, giving the first comprehensive overview of phosphorites on the modern oceanfloor. His publications on vanadium, uranium and organic-rich rocks of the Phosphoria Formation as well as the worldwide relationship between phosphorites, trace metals, oil shales and petroleum source beds showed the immense economic value of these sediments of upwelling origin. He recast these concepts of upwelling sediments to formulate a phosphate exploration model that led to a decade of discoveries from 1955 to 1965 that included, to my own knowledge, Mexico, Peru, Angola, Turkey, Saudi Arabia, Iran, India and Australia, and possibly others as well.

In the 1960s, Vince turned his efforts toward the use of mineral resource knowledge in governmental affairs. In the early 1960s when he was the USGS Assistant Geologist for Economic Geology, he recognized that the Department of Interior needed estimates of the United States petroleum resources, and initiated a USGS program to meet that need. When the energy crisis of the early 1970s arrived, USGS preliminary estimates were available. However, these estimates were not universally accepted, and a major controversy on the size of the national petroleum endowment ensued. Mark Twain once remarked that most disputes arise because people use the same words to refer to different things or different words to refer to the same thing. The United States petroleum argument was no exception to Twain's good sense, and this led Vince, in 1972, to publish a paper on resource estimates and public policy (*American Scientist*, **60**, 32–40), in which he formulated the principles of resource classification and terminology. Vince paid a price for his belief at the time, as he was labeled an unrealistic resource optimist by some who failed to see the vision of these principles. Time has treated Vince fairly, however, as his resource nomenclature system, or variations of it, have been adopted worldwide and have helped what once could have been termed 'resource semantic uproar' evolve into critical analysis.

Those who worked with and for Vince McKelvey quickly came to realize that he was a man of great personal integrity. This was perhaps never shown more forcefuly than in 1971, when he was the US representative to the Economic and Technical Subcommittee of the United Nations Law of the Sea Committee. The Asian–African Legal Consultative Committee, which was organized by Asian and African countries to develop a united policy, was generally suspicious of western delegates, but they recognized Vince as a highly knowledgeable mineral resource scientist who put truth before politics. He was the only westerner asked to participate in their meeting in Colombo, Ceylon, in 1981 to give technical background for the various Law of the Sea proposals. This dedication to objectivity and the desire to put geologic knowledge to good use in national and international affairs made Vince a valuable advisor. Harry Truman once expressed a desire for a one-handed economist to advise him, not one that would say this on the one hand and that on the other. Vince would probably not have satisfied such a politician very well as he was always two-handed on complex issues. However, he always tried, usually successfully, to identify the underlying simplicities of complex issues. Vince's advisory assignments to politically-appointed administrators covered such areas as outer continental shelf oil, gas, and sulfur leasing policy (1968), international seafloor minerals policy (1971–75, 1978), and Project Independence Blueprint Interagency Oil Task Force (Chairman, 1974). Vince's commitment that the USGS should always be a non-advocacy science agency has kept the organization he helped to build at the forefront of scientific credibility.

I was privileged to work for Vince at several stages in his USGS career; first on his Phosphoria project as a sampler-digger, and later as a young geologist just awarded my geologic pick, and then again much later as his chief geologist when he was director of the USGS. He was a wonderful person to work for. His warmth, humor, and calm made difficult times fun, and his insight, leadership, and encouragement made geologic endeavors exciting. That was the way he planned it, so that everybody could give their most, just as he in turn gave his most to the science of geology and to its application in government affairs.

Richard P. Sheldon
Washington, DC

Contents

Contributors

P. Aharon, Department of Geology, Louisiana State University, Baton Rouge, Louisiana 70803, USA

M. Allen, Department of Geology and Geophysics, Yale University, New Haven, Connecticut 06511, USA

P. Baker, Department of Geology, Duke University, Durham, North Carolina 27708, USA

E.J. Barron, Earth System Science Center, 537 Deike Building, The Pennsylvania State University, University Park. Pennsylvania 16802, USA

G.N. Baturin, Institute of Oceanography, Academy of Science, 23 Krasikova, 117 218 Moscow, USSR

H. Belayouni, University of Tunis, Unite de Geologie, Faculte des Sciences, Tunis, Tunisia

R.A. Berner, Department of Geology and Geophysics, Yale University, New Haven, Connecticut 06511, USA

I.I. Bersenev, Institute of Pacific Oceanography, Far Eastern Branch, Academy of Science, Vladivostock, USSR

G. Birch, ESSO Australia (Exploration), PO Box 4047, Sydney N.S.W. 2001, Australia

J.M. Bremner, Department of Geology, Marine Geoscience, University of Cape Town, Rondebosch, Cape Town 7700, South Africa

W.C. Burnett, Department of Oceanography, Florida State University, Tallahassee, Florida 32306, USA

P.J. Cook, Division of Continental Geology, Bureau of Mineral Resources, Canberra 2601, Australia

R. Flicoteaux, Laboratoire de géologie dynamique et de pétrologie de la surface, Faculté des Sciences et Techniques de St-Jérôme, 13 397—Marseille Cedex 13, France

L.A. Frakes, Department of Earth Sciences, University of Adelaide, Adelaide, South Australia 5001, Australia

C. Galli-Olivier, Department of Geology, Universidad Autonoma de Baja California Sur, La Paz, Baja California Sur 23080, Mexico

J. Gamiño, Department of Geology, Universidad Autonoma de Baja California Sur, La Paz, Baja California Sur 23080, Mexico

G. Garduño, Department of Geology, Universidad Autonoma de Baja California Sur, La Paz, Baja California Sur 23080, Mexico

R.E. Garrison, Earth Sciences Board, University of California, Santa Cruz, California 95064, USA

C.R. Glenn, Hawaii Institute of Geophysics, University of Hawaii, Honolulu, Hawaii 96822, USA

V.V. Gusev, Institute of Pacific Oceanography, Far Eastern Branch, Academy of Science, Vladivostock, USSR

A.C. Hine, Department of Marine Sciences, University of South Florida, St Petersburg, Florida 33701, USA

A.V. Ilyin, Institute of the Lithosphere, Academy of Sciences, Staromonetny per. 22, Moscow 109180, USSR

M. Kastner, Scripps Institution of Oceanography, La Jolla, California 92093, USA

Y. Kolodny, Department of Geology, The Hebrew University, Jerusalem, Israel

L.R. Kump, Earth System Science Center, The Pennsylvania State University, 210 Deike Building, University Park, Pennsylvania 16802, USA

J.R. Lappartient, Laboratoire de Géologie Dynamique et de Pétrologie de la Surface, Faculté des Sciences et Techniques de St-Jérôme, 13 397—Marseille Cedex 13, France

E.P. Lelikov, Institute of Pacific Oceanography, Far Eastern Branch, Academy of Science, Vladivostock, USSR

B. Loebner, Pacific and Arctic Marine Geology, US Geological Survey MS 97, Menlo Park, California 94025, USA

P.M. Mallette, Newmont Exploration Ltd, 240 South Rock Blvd, Reno, Nevada 89502, USA

J. M. McArthur, Department of Geology, University College London, Gower St, London, WC1E 6BT, England

G.H. McClellan, Department of Geology, University of Florida, Gainesville, Florida 32611, USA

G.W. O'Brien, Division of Continental Geology, Bureau of Mineral Resources, Canberra 2601, Australia

J.T. Parrish, Department of Geosciences, Gould-Simpson Building, The University of Arizona, Tuscon, Arizona 85721, USA

L. Pietrafesa, Department of Marine, Earth and Atmospheric Sciences, North Carolina State University, Raleigh, North Carolina 27695, USA

D.Z. Piper, Pacific and Arctic Marine Geology, US Geological Survey MS 97, Menlo Park, California 94025, USA

P. Popenoe, Office of Energy and Marine Geology, US Geological Survey, Woods Hole, Massachusetts 02543, USA

J.P. Prian, BRGM, Direction des Activités Minières, BP 6009, 45060 Orleans, France

M. Rachidi, Laboratoire de Géochimie Organique, Universite d'Orleans, F 45067 Orleans, France

G.I. Ratnikova, Institute of the Lithosphere, Academy of Sciences, Staromonetny per. 22, Moscow 109180, USSR

C.E. Reimers, Scripps Institution of Oceanography, La Jolla, California 92093, USA

S.R. Riggs, Department of Geology, East Carolina University, Greenville, North Carolina, 27858, USA

J. Rogers, Department of Geology, Marine Geoscience, University of Cape Town, Rondebosch, Cape Town 7700, South Africa

A.M. Rossfelder, GEOMAREX, PO Box 2244, La Jolla, California 92038, USA

M.W. Sandstrom, Methods Research & Development Program, US Geological Survey, 5293 Ward Road, Arvada, CO 80002, USA

S.M. Savin, Department of Geological Sciences, Case Western Reserve University, Cleveland, Ohio 44106, USA

T. Scott, Florida Geological Survey, Department of Natural Resources, Tallahassee, Florida 32303, USA

R.P. Sheldon, Geologic Consultant, 3816 T. Street NW, Washington, DC 20007, USA

A. Shemesh, Lamont-Doherty Geological Observatory, Columbia University, Palisades, New York, 10964, USA

Scott W. Snyder, Department of Geology, East Carolina University, Greenville, North Carolina, 27858, USA

Stephen W. Snyder, Department of Marine, Earth & Atmospheric Sciences, North Carolina State University, Raleigh, NC 27659 USA

C.P. Summerhayes, Institute of Oceanographic Sciences, Deacon Laboratories, Brook Road, Wormley, Godalming, Surrey, GU8 5UB United Kingdom

G.M. Sustrac, BRGM, Direction Scientifique, BP 6009, 45060 Orleans, France

J.G. Trichet, Laboratoire de Géochimie Organique, Universite d'Orleans, F 45067 Orleans, France

S.J. Van Kauwenbergh, International Fertilizer Development Center, Box 2040, Muscle Shoals, Alabama 35662, USA

F. Woodruff, Center for Earth Sciences, University of Southern California, Los Angeles, California 90089, USA

Preface

History of IGCP Project 156 – Phosphorites

Phosphate deposits of the world is a series of multi-author volumes summarizing the results of an international research program on phosphate deposits (IGCP Project 156). The International Geological Correlation Program (IGCP) was jointly established by UNESCO and the International Union of Geological Sciences (IUGS) in 1972 in order to facilitate the acquisition and distribution of knowledge of mankind's resources and environment. A project to study sedimentary phosphate deposits (phosphorites) was formally accepted as IGCP Project 156 in early 1977.

The original Project Directors, Peter J. Cook and John Shergold (both currently affiliated with the Australian Bureau of Mineral Resources), organized the inaugural meeting in the form of a 'Field Workshop and Symposium'. This workshop, held in Australia in 1978, became the model for over a dozen similar meetings over the next 10 years. Scientists from a variety of relevant disciplines assembled, both in a seminar and field setting, for discussions concerning the origin of marine phosphorite. The success of the first meeting encouraged many others to become involved and eventually the Project had many hundred participants. Project 156 went on to become one of the largest, best known, and successful of all IGCP projects.

It was recognized early in the development of the Project that in order to be successful in addressing the wide range of questions relevant to an understanding of phosphorites, it would be necessary to set up thematic working groups. A total of four such working groups were ultimately established:

Working Group I – 'Proterozoic and Cambrian Phosphorites', (Co-Chairmen: P.J. Cook and J.H. Shergold);

Working Group II – 'International Phosphate Resource Data Base', (Co-Chairmen: A.J.G. Notholt and R.P. Sheldon);

Working Group III – 'Young Phosphogenic Systems', (Co-Chairmen: W.C. Burnett and S.R. Riggs); and

Working Group IV – 'Cretaceous and Tertiary Phosphorites', (Co-Chairmen: K. Al-Bassam, J. Lucas, and S. Sassi).

Each of the volumes in this series *Phosphate Deposits of the World* summarizes the results of one working group, with the working group chairman as editors.

Accomplishments of Project 156

Over the life of the project, there have been significant advances in the dual aims of facilitating an exchange of ideas and information, and encouraging research into all aspects of phosphate deposits. Some of the more significant achievements are:

1. *The holding of meetings:* The Project has convened several very successful international field workshops and seminars, most of them at the site of a major phosphate deposit: Australia (1978), western USA (1979), Mongolian People's Republic (1980), Mexico (1981), India (1981), People's Republic of China (1982), Morocco and Senegal (1983), USSR (1984), southeastern USA (1985), Venezuela (1986), Tunisia (1987), Jordan (1988), Peru (1988), and England (1988). These meetings have provided ample opportunity to visit the deposits at the outcrop and have allowed maximum interaction among the participants in an informal setting.
2. *The encouragement of research:* Although the Project never had the financial resources to directly sponsor research, it had an impressive catalytic role in the research funding of many of its participants. One outstanding example was 'PHOSREP', the *Phos*phate *Re*search *P*roject, held in Australia in 1987. Funding was provided by the Australian and US governments for a multidisciplinary, international team of scientists to carry out a comprehensive study of pre-Cambrian, Neogene, and modern phosphorites in Australia, both on land and offshore. This and other Project-related research has had a major impact on our understanding of the nature and origin of phosphate deposits.
3. *Exchange of information:* A considerable amount of information has been disseminated by participants in the Project through their publications in scientific journals and in special publications prepared by the Project for International Conferences. *Phosphate Deposits of the World* represents the major contribution in this area. In addition, the Project has published 19 issues of the IGCP 156 newsletter, distributed to over 700 scientists in 60 countries.
4. *Training:* The Project has provided training to many scientists from less-developed countries by ensuring that a significant number attend each of our international workshops, and by holding training courses on phosphate geology. This component of our Project has been very successful and has ensured that many geologists from developing countries have obtained firsthand knowledge of phosphate deposits.

Purpose and content of this volume

This volume, *Neogene and Modern phosphorites*, the third in the series of four volumes on phosphate deposits of the world, is a collection of papers put together to present new information pertaining to the origin of Neogene marine phosphorites. However, a glance at the table of contents will show that we have not restricted the volume to only those papers dealing specifically with phosphorites, but have included several papers that deal with associated authigenic sediments (dolomite, clay minerals, etc.) and the prevailing oceanic environment during the Miocene, one of the most important phosphogenic periods in the geologic record.

A major portion of the papers contained in this volume were presented at a 2-day symposium on the 'Genesis of Neogene and Modern Phosphorites' organized by Project 156. The symposium, held in Tallahassee, Florida in 1985, brought together some of the foremost experts in the world in many diverse areas – geochemistry, paleoceanography, sedimentary petrology, and many other fields. The purpose of both the symposium and this volume is to summarize the most current thinking with respect to the origin of marine phosphorites.

This volume has evolved beyond a standard proceedings volume. We have added several components that have either developed since the conference was held or were not possible to present at that time. For example, several of the papers in Part II describe Neogene deposits (Cuba, Sea of Japan, etc.) that have never before appeared in the western literature.

We have organized this volume into four parts: (1) *The Modern setting* – a description of upwelling processes in the modern ocean and examples of phosphorite formation in both upwelling and non-upwelling environments; (2) *Modern and Neogene phosphorites and associated sediments* – contains new descriptions of Neogene deposits from several areas of the world, on land, offshore, and oceanic islands; (3) *The Neogene environment* – contains insights into the environmental setting present in the world's oceans at the time of one of the most important phosphogenic episodes; and (4) *Neogene phosphorites of California and the southeastern US* – contains a detailed investigation of two of the best known Miocene age deposits including a special 4-part section on the latest research accomplished on the North Carolina continental margin. Together, these papers on phosphorites, associated sediments, and prevailing environments should offer the most up-to-date insights available on the origin of marine sedimentary phosphorite.

Acknowledgments

This volume could never have been possible without the assistance of many organizations and individuals. The 1985 symposium was primarily sponsored by the National Science Foundation, the Florida Institute of Phosphate Research, and Florida State and East Carolina Universities. Susan Lampman and the staff at the Florida State University Center for Professional Development provided outstanding conference support. Editorial assistance was provided by Sheila Heseltine and Alison Watkins at FSU and the staff of Cambridge University Press. Finally, we thank all the authors and reviewers who worked on this project – we hope you are pleased with the collective fruits of our labor.

William C. Burnett Stanley R. Riggs
(Co-Project Leaders, 1984–1988)

PART 1

The Modern setting

1

Upwelling processes associated with Western Boundary Currents

L. PIETRAFESA

Abstract

Upwelling processes associated with Western Boundary Currents (WBCs) include not only classical wind driven coastal upwelling but also mechanisms due to buoyancy flux, topography and boundary current frontal instabilities. Sharp breaks in cross-shelf bottom topography can create localized transient pockets of uplifted isopycnals during occasions of wind-forced upwelling events. Longshore variations in bathymetry interacting with a WBC can have the effect of torquing the sheared jet across isobaths and kinematically inducing upwelling, creating regions of persistent or quasi-permanent upwelling. Isolated topographic ridges or bumps, such as the Charleston Bump, a topographic rise sitting atop the continental slope directly in the path of the Gulf Stream can cause WBCs to deflect offshore creating a downstream wave trough which, via bottom Ekman suction, can support a continental slope region of persistent upwelling. WBCs are characterized by frontal instabilities which manifest themselves as shelf-break hugging, downstream-propogating frontal meanders and filaments. These features are found to be responsible for creating shelf-break upwelling domes and ridges. These events can act occasionally in concert with wind and buoyancy stress-forcing to create shelf-wide upwelling. All of the above WBC related upwelling processes are shown to be important to the supply of high concentrations of phosphate to local continental margin and shelf regions. Relatively high values of phosphate are shown to exist where the upwelling processes described above are present in WBC systems, such as the Gulf Stream.

Introduction

In the classic standard entitled *Upwelling* by R.L. Smith (1968), the following description of the 'phosphate' distribution is given:

> Phosphate, or dissolved inorganic phosphorous, generally increases with depth from very low values at the surface to maximum values at some intermediate depth. It is one of the principal nutrients and therefore important to life in the sea. High values occur in upwelling regions where subsurface water comes to the surface. The usual mid-latitude oceanic surface concentration is of the order of 0.2 μg-atom/l. In the upwelling regions of Peru (Posner, 1957) and the Benguela Current (Hart & Currie, 1960) the surface concentration is between 1 and 2 μg-atom/l. Values of over 2 μg-atom/l were found at the surface of the south-east Arabian coast (Royal Society, 1963), and Laurs (1967) has reported values of 2.5 μg-atom/l near the coast during intensive upwelling off Oregon.

It appears clear that values of phosphates which make for the importance of 'upwelling' as a physical process in supplying dissolved inorganic phosphorous to continental margin regions are 1–2.5 μg-atom/l versus ambient values of an order of magnitude less. What is upwelling?

Smith (1968) defines upwelling as 'an ascending motion, of some minimum duration and extent, by which water from subsurface layers is brought into the surface layer'. Smith points out that while the word 'upwelling' has been used extensively throughout oceanography, dating back to Sverdrup (1938), a precise definition had not been coined prior to Smith (1968). Since the Smith summary, much has been written about the physical, biological and chemical oceanography of coastal upwelling zones. The importance of these zones was further emphasized by Ryther (1969). He estimated that fully one-half of the world's fish food supply is derived from upwelling regions.

A recent survey publication entitled *Coastal Upwelling* (Richards, 1981) documents the new knowledge derived from such major programs as the National Science Foundation – International Decade of Ocean Exploration Coastal Upwelling Ecosystems Analysis. These programs have provided much new insight into the interrelationships between the physics and biology of wind-driven upwelling systems such as are found along the Pacific coasts of the United States and South America, and the Atlantic coasts of North and South Africa. These are all eastern boundaries of ocean basins (cf. Fig. 1.1). However, few studies were made of upwelling regions and mechanisms which might exist along the western sides of ocean basins.

Smith (1968) states that 'the major atmospheric systems' indigenous to the eastern sides of ocean basins 'are favorable to upwelling over extended areas and periods'. He also notes that 'upwelling of more limited extent can be found elsewhere, when and where the local winds transport water in the surface layer offshore'. Smith alludes to several reports of upwelling along the northeastern coast of Florida (Green, 1944; Taylor & Stewart, 1959), North Carolina (Wells & Gray, 1960) and on the Caribbean coast of Venezuela (Richards, 1960). So, by 1968 there was some testimony, albeit minimal, concerning the occurrences of

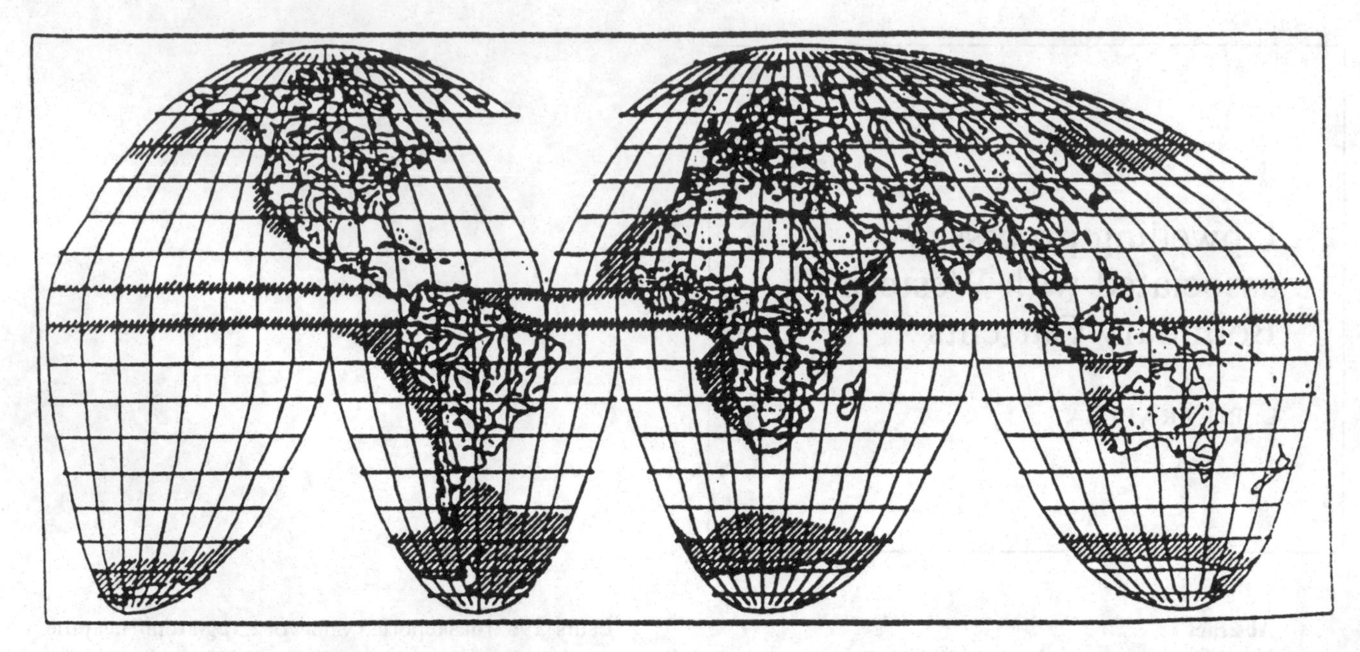

Fig. 1.1. General world areas of upwelling which are also areas of high organic production (Fairbridge, 1966).

upwelling along the western sides of ocean basins. The important note here is that strong, longshore Western Boundary Currents (WBCs) occur on the western sides of ocean basins, and where upwelling is present in these regions, such processes could be associated with the presence of the WBCs.

Even if upwelling were to occur along western sides of ocean basins, is there enough evidence that phosphate is present in the water columns of such regimes to justify an interrogation of the dynamics? In a US Department of the Interior literature survey of the National Oceanographic Data Center's records of the observed distribution of oceanic variables along the continental margin of the southeastern seaboard of the continental United States (ERT, 1979a,b) the following statement is made:

> Most high phosphate concentrations (>0.5 μg-atom/l) occur in the lower part of the water column over the shelf break region... Relatively high mid-shelf and near-shore values are occasionally noted both at the surface and near the bottom in various sections and in various seasons.

Thus, as of 1978, there was evidence of 'relatively high' phosphate occurring in the South Atlantic Bight (Fig. 1.2), the western boundary current region of the southeastern continental USA. Herein, Atkinson, Pietrafesa & Hoffman (1982), provided the pictoral, diagrammatic relationships between water column temperature/nitrate/phosphate values shown in Figure 1.3 for the region between Capes Fear and Hatteras, NC (Fig. 1.4). Clearly, as shown in Figure 1.3, phosphate concentrations of 1–2 μg-atom/l are not uncommon in these waters. Could the occurrences of high phosphate in North Carolina coastal waters be related to upwelling phenomena?

Recently, investigators such as Pietrafesa & Janowitz, (1979), Hofmann *et al.*, (1980), Janowitz & Pietrafesa, (1980), Pietrafesa & Janowitz, (1980), Blanton *et al.*, (1981), Hofmann, Pietrafesa & Atkinson, (1981), Leming & Mooers (1981), Atkinson, Pietrafesa & Hofmann, (1982), Janowitz & Pietrafesa, (1982), McClain, Pietrafesa, & Yoder, (1984) and Pietrafesa, Janowitz & Wittman, (1985) have discussed a variety of upwelling processes which occur in the South Atlantic Bight (SAB) of the USA. The SAB is the continental margin region between Cape Canaveral, Florida and Cape Hatteras, NC (Figs. 1.2, 1.3). It is likely that similar phenomena are present along other western oceanic boundaries, such as the western Pacific and the Mid-Atlantic Bight of the USA, South America and Africa. We shall discuss the upwelling mechanisms which can be present along western boundaries of ocean basins, test their application to a specific locale, the SAB, and make inferences for similar regions of the world's oceans.

Upwelling theory

Consider the classical balance relations in a form conventionally used in oceanographic applications (for a complete derivation refer to Fofonoff, 1962):

(a) (b) (c) (d) (e) (f)

$$\frac{Du}{Dt}=fv-\alpha\frac{\partial P}{\partial x}+\frac{\partial}{\partial x}\left(A_x\frac{\partial u}{\partial x}\right)+\frac{\partial}{\partial y}\left(A_y\frac{\partial u}{\partial y}\right)+\frac{\partial}{\partial z}\left(A_z\frac{\partial u}{\partial z}\right) \quad (1.1)$$

(g) (h) (i) (j) (k) (l)

$$\frac{Dv}{Dt}=fv-\alpha\frac{\partial P}{\partial y}+\frac{\partial}{\partial x}\left(A_x\frac{\partial v}{\partial x}\right)+\frac{\partial}{\partial y}\left(A_y\frac{\partial v}{\partial y}\right)+\frac{\partial}{\partial z}\left(A_z\frac{\partial v}{\partial z}\right) \quad (1.2)$$

(m) (n)

$$\frac{\partial P}{\partial z}=-g\rho \quad (1.3)$$

(o) (p) (q)

$$\frac{\partial u}{\partial x}+\frac{\partial v}{\partial y}+\frac{\partial w}{\partial z}=0 \quad (1.4)$$

(r) (s) (t) (u)

$$\frac{D\rho}{Dt}=\frac{\partial}{\partial x}\left(K_x\frac{\partial\rho}{\partial x}\right)+\frac{\partial}{\partial y}\left(K_y\frac{\partial\rho}{\partial y}\right)+\frac{\partial}{\partial z}\left(K_z\frac{\partial\rho}{\partial z}\right) \quad (1.5)$$

(v) (w) (x) (y)

$$\frac{D}{Dt}=\frac{\partial}{\partial t}+u\frac{\partial}{\partial x}+v\frac{\partial}{\partial y}+w\frac{\partial}{\partial z} \quad (1.6)$$

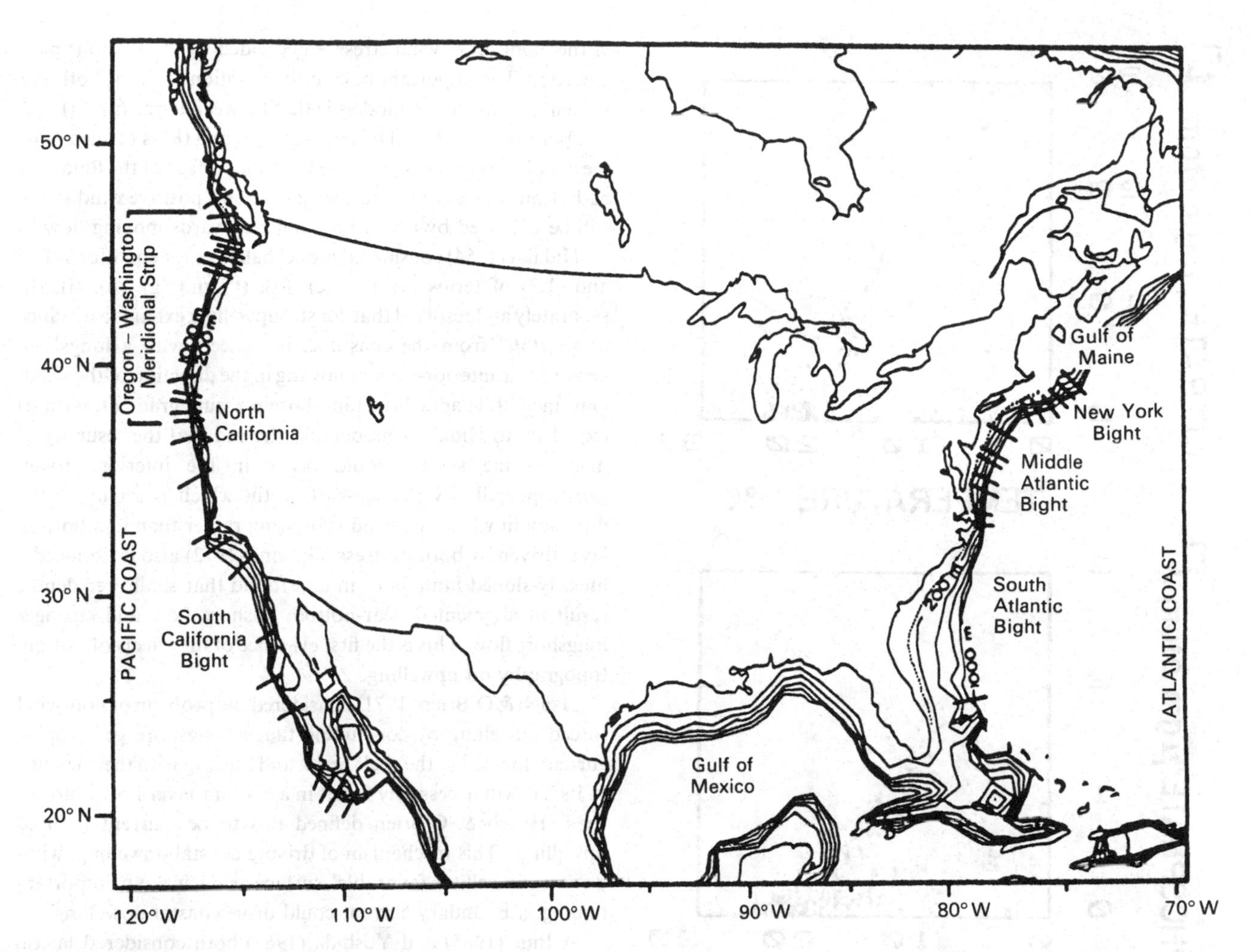

Fig. 1.2. Mainland United States continental margin regions. Submarine (continental slope) canyons are noted as straight-line cuts across the continental margins.

where u, v, w are the offshore (+x), alongshore (+y) and upwards (+z), right-handed Cartesian coordinate components of the Eulerian velocity of a water parcel, t is time, f is the Coriolis parameter ($2\Omega\sin\theta$), α is the specific volume of sea water, ρ is the density of sea water, P is pressure and the A's are eddy viscocity coefficients and the K's, the eddy diffusion counterparts in the x, y and z directions. Ω is the Earth's angular velocity and θ is latitude.

Equations (1.1) and (1.2) relate the relative acceleration of a water parcel, in the cross-shelf (or diabathic) and alongshelf (or parabathic) directions, respectively, as a function of time and space (via definition (b)) to the effect of the Earth's rotation, pressure gradient forces, and horizontal and vertical frictional forces. Relation (1.3) states the hydrostatic balance, while equations (1.4) and (1.5) provide for the conservation of volume and mass and for the distribution of the thermohaline field as well as its disposition as a forcing function.

Take equations (1.1) and (1.2), cross-differentiate them, and subtract the resulting relations, i.e. form the vectoral 'curl', while assuming that $(\partial v/\partial x - \partial u/\partial y) \ll f$ (the low Rossby number assumption). Next, combine this 'curl' relation with equation (1.4) to yield

$$\overset{(a)}{\frac{\partial w}{\partial z}} = \overset{(b)}{\frac{1}{f}\left(\frac{\mathrm{D}}{\mathrm{D}t}\left(\frac{\partial v}{\partial x}\frac{\partial u}{\partial y}\right)\right)} + \overset{(c)}{\beta v} + \overset{(d)}{\frac{\partial \alpha}{\partial x}\frac{\partial P}{\partial y}} - \overset{(e)}{\frac{\partial \alpha}{\partial y}\frac{\partial P}{\partial x}}$$
$$-\frac{\partial}{\partial y}\left(\overset{(f)}{\frac{\partial}{\partial x}\left(A_x\frac{\partial u}{\partial x}\right)} + \overset{(g)}{\frac{\partial}{\partial y}\left(A_y\frac{\partial u}{\partial y}\right)} + \overset{(h)}{\frac{\partial}{\partial z}\left(A_z\frac{\partial u}{\partial z}\right)}\right)$$
$$+\frac{\partial}{\partial x}\left(\overset{(i)}{\frac{\partial}{\partial x}\left(A_x\frac{\partial v}{\partial x}\right)} + \overset{(j)}{\frac{\partial}{\partial y}\left(A_y\frac{\partial v}{\partial y}\right)} + \overset{(k)}{\frac{\partial}{\partial z}\left(A_z\frac{\partial v}{\partial z}\right)}\right) \quad (1.7)$$

In summary, the vertical gradient of the vertical velocity w, is related to the change in the vertical component of the relative vorticity following a water parcel, plus the meridional advection of planetary vorticity, the effects of the solenoidal pressure-density fields (which vanish in the Bousinesq limit), and the effects of the horizontal and vertical diffusion of relative vorticity.

In the classical Ekman (1905) and Sverdrup (1938) pictures of upwelling, an alongshore longshore wind blowing with a coastline to its left in the northern hemisphere would drive surface waters seaward with a transport equal to the magnitude

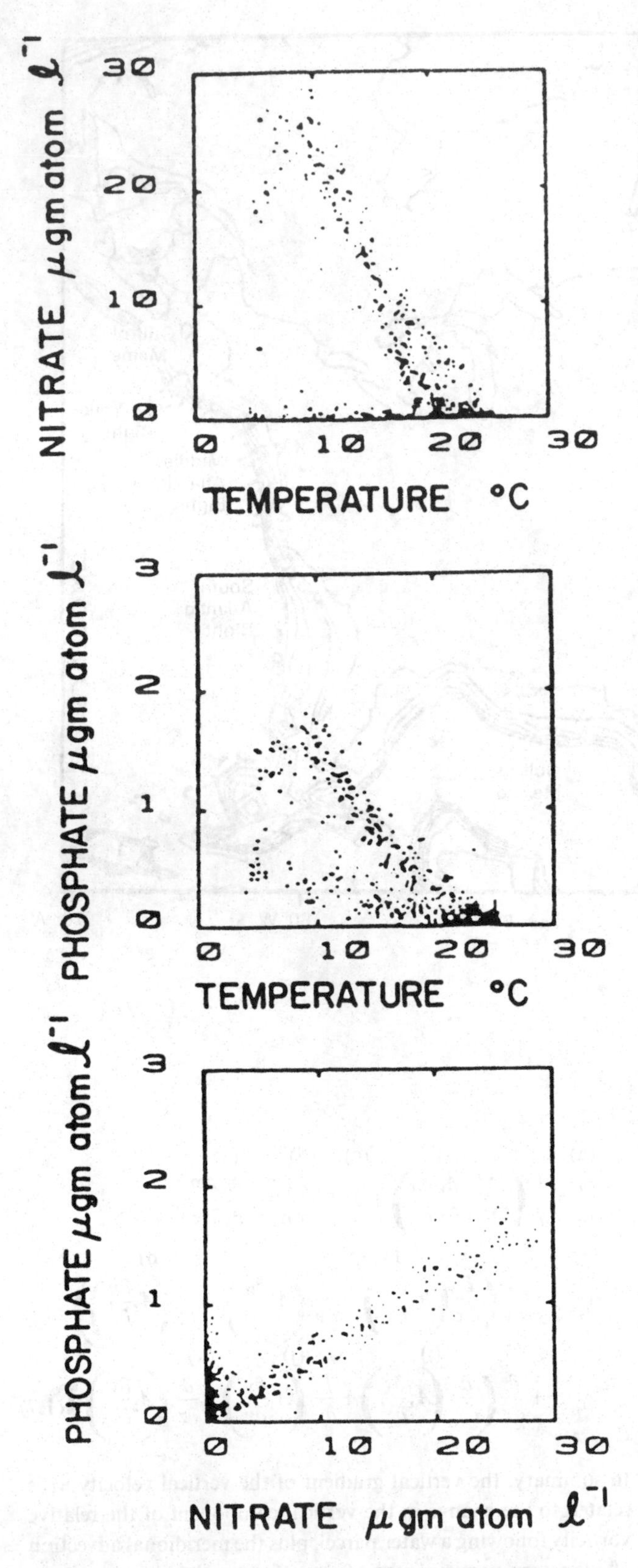

Fig. 1.3. Temperature, nitrate and phosphate relationships for outer Carolina Cape bay waters (Atkinson, Pietrafesa & Hofmann, 1982).

of the alongshore wind stress $\tau(y)$ divided by the Coriolis parameter, f. The important balance in equation (1.7) for both the Ekman and Sverdrup models is that between terms (a), (h) and (k). If a vertical integral of terms (a), (h) and (k) is formed with the conditions that a rigid lid rests on the surface of the fluid and no bottom stress is effected then a non-zero positive wind stress will be balanced by a vertically finite upwards-moving flow.

Hidaka (1954) considered model balances for equations (1.1) and (1.2) of terms (a), (b), (c), (d), (f) and (g), (h), (j), (l), separately and resolved that coastal upwelling extends a distance of $(A_x/f)^{1/2}$ from the coastline, in concert with a longshore geostrophic interior-current flowing in the direction of the wind. Garvine (1971) added the longshore pressure gradient, term (i) (eqn 1.2), to Hidaka's model and showed that the resupply of mass to the system could occur in the interior, driven geostrophically by the sea-surface tilt which is set up in the direction in which the wind is blowing rather than in a bottom layer driven by bottom stress. Garvine (1972) also introduced a linearly-sloped finite bottom and found that shallower depths result in augmented near-bottom onshore flow and stronger longshore flow. This is the first evidence of the effects of bottom topography on upwelling.

Hsueh & O'Brien (1971) considered the problem of non-wind forced upwelling by concluding that a longshore geostrophic current, located at the shelf break and flowing with the coastline to its left will necessarily result in a bottom-layer flow onto the shelf. Hsueh & O'Brien defined this to be 'current induced upwelling'. This mechanism of driving coastal upwelling, without an 'upwelling favorable' surface wind was an important finding; a boundary current could drive coastal upwelling.

Arthur (1965) and Yoshida (1967) both considered lateral topographic effects via balances of terms (a), (b) and (c) in equation (1.7). The argument is that as a longshore current flows by a wall to its left, if a cape appears, i.e. if the wall 'moves away' from the current, then a cyclonic vortex will form in the lee of the cape and upwelling will occur in the center of the vortex. This mechanism must be treated judiciously because, in the inviscid limit, over a flat bottom, under a rigid lid, on an f-plane, the vertical velocity is identically zero, since the material derivative of the potential vorticity is equal to a zero. Terms (h) and (k) can provide for vertical motions driven by both the curl of the wind stress and the curl of bottom stress but these effects can only be estimated given the paucity of atmospheric wind data collected serially from meteorological buoys and given very few bottom stress estimates based on observations.

In all of the aforementioned models, a key ingredient missing from consideration was stratification. As indicated by equations (1.5) and (1.7), upwelling is a non-linear coupled process. The degree of non-linearity or of baroclinic effect is moot; the point is that the system is coupled via the strong stratification associated with upwelling. The studies of Charney (1955), Lineykin (1955), Yoshida (1967), Tomczak (1970), Leetma (1971), Allen (1972 and 1973), Blumsack (1972), Hsueh & Kenney (1972), O'Brien & Hurlburt (1972), Hurlburt & Thompson (1973), McNider & O'Brien (1973), Pietrafesa (1973), Thompson & O'Brien (1973), Pietrafesa & Janowitz (1979), all incorporated the effects of stratification and were able to deduce many of the more salient features of the upwelling process, as discussed by Mooers, Col-

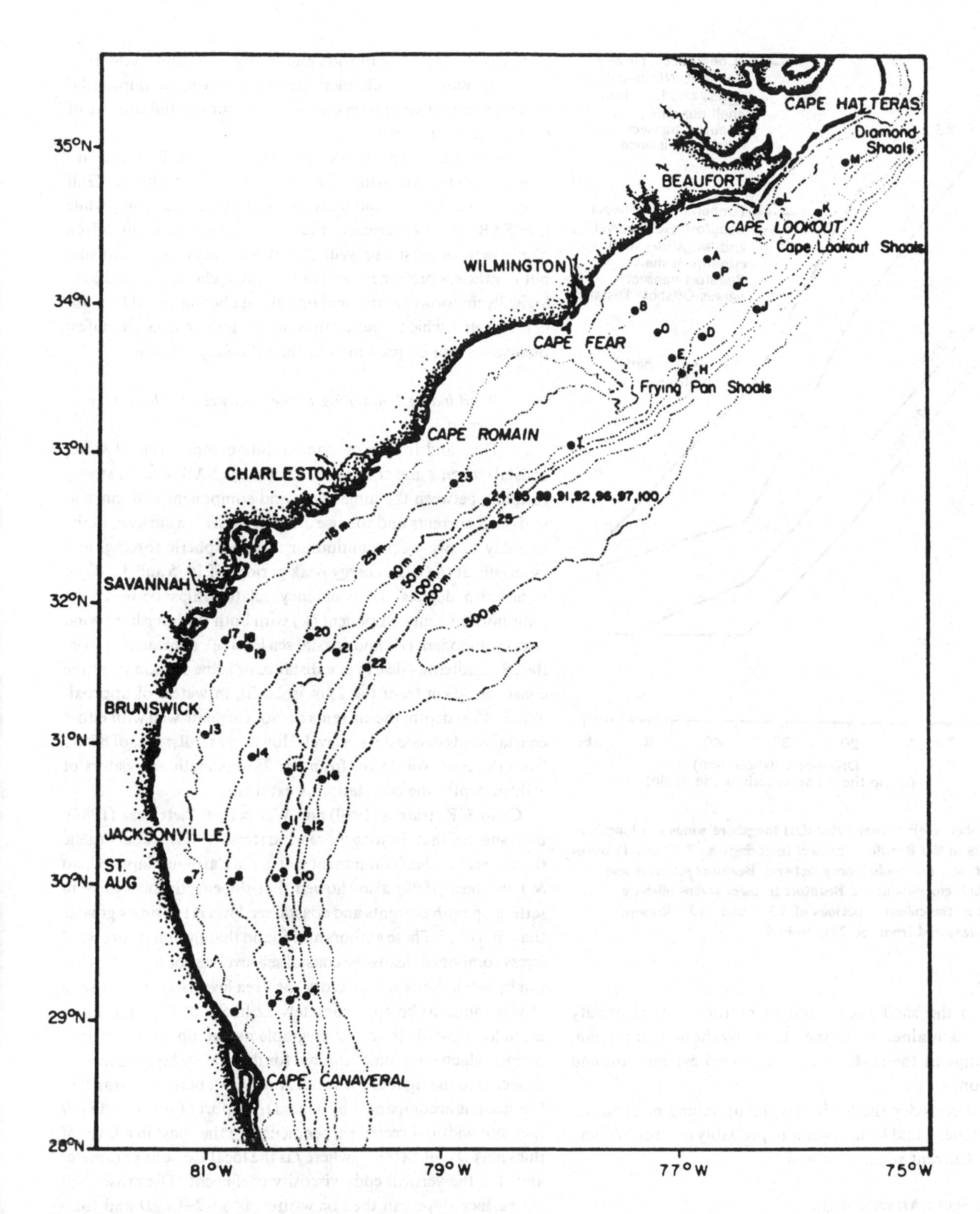

Fig. 1.4. South Atlantic Bight plus North Carolina State University and University of Miami current meter moorings, 1975–81.

lins & Smith (1976). Several important length scales were derived from these studies; they include an horizontal length scale, measured from the coastline seaward, within which 'coastal' upwelling is confined. It is the baroclinic radius of deformation and for a two-layer fluid is equal to $((\rho_2-\rho_1)gh_1h_2/\rho_2f^2(h_1+h_2))^{1/2}$ where h_1, and h_2 are upper and lower layer thicknesses, and ρ_1 and ρ_2 are upper and lower layer densities.

Prior to 1972, only wind forcing was considered to be important as a surface forcing mechanism of continental margin upwelling, but Stommel & Leetma (1972), Pietrafesa (1973) and Pietrafesa & Janowitz (1979) considered the effects of thermohaline processes, via buoyancy flux 'stress' and pressure gradients. Included in these effects are such features as outwelling estuarine plumes, horizontal density differences from

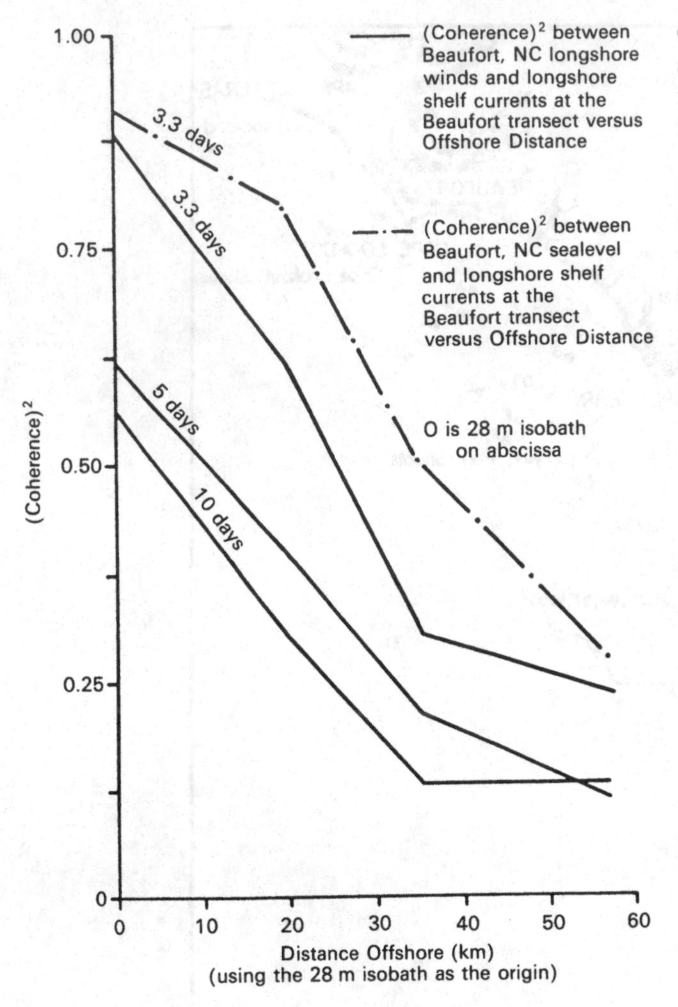

Fig. 1.5. (Coherence)^2 between Beaufort longshore winds and longshore shelf currents at the Beaufort transect (moorings A, P, C and J) versus offshore distance; and (coherence)^2 between Beaufort sea level and longshore shelf currents at the Beaufort transect versus offshore distance; all at the coherent periods of 3.3, 5 and 10 d. Offshore distance is measured from the 28 m isobath.

the coast to the shelf break, top to bottom vertical density differences maintained at the shelf break by the adjacent ocean, heat exchange at the surface, and navifacial evaporation and precipitation.

We now consider the SAB and the upwelling mechanisms which exist there, and by implication, probably in other Western Boundary Current regions as well.

The South Atlantic Bight

The SAB, the southeastern-most coastline of the continental USA (Fig. 1.4), is characterized by a shallow, broad continental shelf. Shelf-break depths generally vary between 50–80 m and while the cross-shelf (or diabathic) width of the shelf proper is only about 25 km off both Cape Canaveral and Hatteras, the shelf is generally of the order of 100–150 km wide throughout the central portion of the SAB. To the north, there are three prominent coastal cusps, the Carolina Capes, which are partially separated from each other by extensive shoals. There are only minor land-derived freshwater sources in the Carolina Capes. In contrast, the Florida and Georgia coastlines comprise a curving land barrier characterized by low-lying swamps, tidal flats and several large rivers, which elicit substantial sources of freshwater to the shelf.

The SAB is graced by several persistent upwelling mechanisms: these include wind forcing, topographic effects, Gulf Stream frontal events and buoyancy flux influences. Thus, while the SAB does experience the more familiar, wind-driven phenomena of coastal upwelling, it also features several unusual but nontheless prominent and biologically, chemically and geologically important additional upwelling phenomena. The manner within which these upwelling phenomena manifest themselves will be presented in the following sections.

Wind-induced upwelling in the presence of a boundary current

Time and frequency domain intercomparisons of wind, coastal currents and coastal sea level in the SAB reveal a strong coupling between the longshore wind component and inner to mid-shelf currents and the rise and fall of the sea surface, in the two-day to two-week continuum of atmospheric forcing, and especially at the wind energy peak periods of 10, 5 and 3.3 days. Figure 1.5 depicts the coherency relationships between the longshore current component (V) with both the longshore wind stress component (τ^y) and coastal sea level (η), as measured from the 28 m isobath. Clearly at a distance of some 100 km from the coast (or 50 km from the 28 m isobath), in waters of approximately 45 m depth, the currents do not correlate well with either coastal winds or coastal sea level. However at a distance of 85 km from the coast (or 35 km from the 28 m isobath), in waters of <40 m depth, the correlation is excellent.

Chao & Pietrafesa (1980) and Weisberg & Pietrafesa (1983), both showed that the longshore wind stress (τ^y) is more energetic than the cross-shelf component (τ^x) in the Carolina Capes. Chao & Pietrafesa (1980) also showed that the effectiveness of τ^y in setting up both currents and coastal sea level is ten times greater than that of τ^x. These authors also found that the longshore wind stress component leads the coastal sea level response at Charleston by 8–9 h, however the cross-shelf sea level slope response to τ^y was found to be approximately 22 h. So, at the Charleston latitude, cross-shelf sea level gradients set up in an inertial period; which is the time required for the surface layer flow to be deflected to the right of the wind. A wind, τ^y, blowing parallel to the coast, is accompanied by a mass transport of magnitude τ^y/f (per unit width) directed perpendicular to the coast in a layer of thickness $D=\pi(2A/f)^{1/2}$, where f is the local Coriolis parameter and A is the vertical eddy viscosity coefficient. The cross-shelf sea surface slope can then be written as $\beta=2\pi\tau^y/\rho gD$ and from geostrophy the alongshore slope current is $v=g\beta/f$. The time required for the slope current to be established is H/fD, where H is the total water depth. This scenario is found to be always true in 28 m of water, partially true in 40 m and occasionally true at 75 m. Inner-shelf waters of depth 28 m or less respond closely to the wind while those at 75 m rarely do so. At 40 m, response to the wind is evident but not primary.

Any scenario of a steady-state response to the longshore component of the wind in inner shelf waters in the SAB is convenient but overly simplistic. Mid- to inner shelf currents and

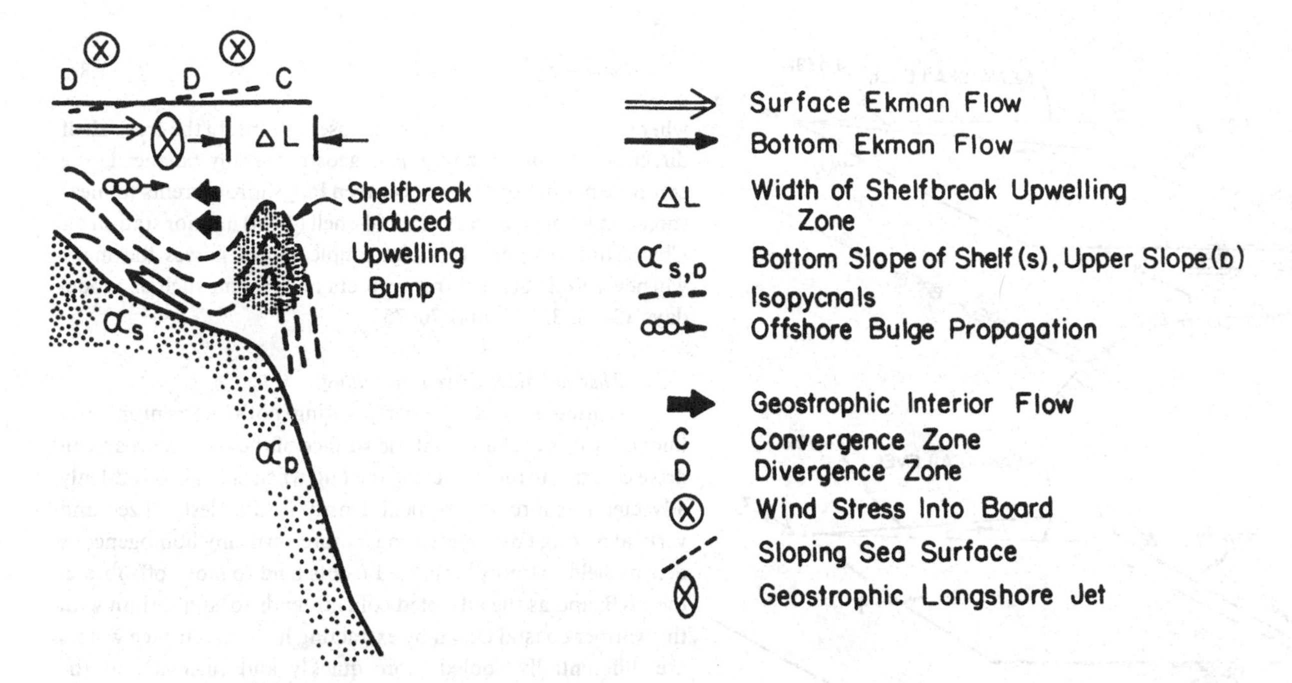

Fig. 1.6. Conceptual schematic of wind-induced upwelling processes in the South Atlantic Bight.

sea-level response to the wind have been described, documented and verified as time dependent responses to the total wind stress vector over variable topography by Chao & Pietrafesa (1980) and Janowitz & Pietrafesa (1980), respectively. Winds from the west to south quadrant cause sea level to fall at the coast while winds from the north to east quadrant cause a rise in sea level at the coast. Thus, a longshore wind directed to the northeast drives a surface layer offshore, which is initially balanced by an interior flow onshore and longshore. Within a day a bottom frictional layer becomes fully established and the flow within is directed onshore. Since this bottom layer is supplying mass, the interior flow drops off somewhat from its initial strength. A longshore jet is set up during this process. Since the shallow waters spin-up more rapidly, isopycnals appear to rise more quickly in shallow water. Actually, they simply have less vertical distance to traverse. The longshore jet has a negative cross-shelf gradient, i.e. its strength decreases with offshore distance (negative relative vorticity) and consequently the isopycnal rise process is suppressed. Since the spin-up process occurs first in shallower waters, the supression of the rise creates the impression of an isopycnal bulge moving from shallow to deeper waters. This process, shown pictorially in Figure 1.6, has been documented both theoretically and observationally by Janowitz & Pietrafesa (1980) to be the fundamental wind-induced upwelling process in the Carolina Capes. The terms in equations (1.1) and (1.2) considered by Janowitz & Pietrafesa (1980) to be of essential importance under conditions favorable for wind-driven upwelling are linearized (a), (b), (c) and (f) and linearized (g), (h), (i), and (l), respectively. This balance was confirmed by Purba (1984) for the Florida/Georgia region (the Georgia Bight) as well.

As shown in Figure 1.4, the mean alignment of the SAB is SSE–NNW from Cape Canaveral, Florida to New Brunswick, Georgia and southwest to northeast from Savannah, Georgia to Cape Hatteras, North Carolina. However there are also three cusps, the Carolina Cape bays, located in the northern half of the SAB. The coastal curvature is such that a favorable wind for upwelling can drive surface waters not only seaward but also against a northern boundary, such that sea level will rise in the direction of the wind and currents will be driven, geostrophically, in an interior cross-shelf flow directed to the left of the wind. This happens frequently in the northern halves of Long, Onslow and Raleigh bays (Askari, 1985).

In the southern end of the SAB, off Florida and Georgia, and in the southern half of the Carolina Cape bays, a northeastward wind which appears on the leading or eastern edge of an eastward moving low pressure weather system, a typical spring storm in the SAB (Weisberg & Pietrafesa, 1983), will drive surface waters offshore where the coastline is first encountered. Since the coastline is curved, the midpoint of the coastline will feel the wind first and consequently, sea level will fall at these midpoints earlier than to the north or south thereby creating a northward, longshore drop in sea level off Florida–Georgia and in the southern halves of the Carolina bays (Fig. 1.7). This drop will induce an offshore interior flow and decelerate the longshore jet. Thus in the southern part of the Georgia Bight and in the southern halves of the Carolina bays, compensatory flows are more confined to a bottom layer while in the northern part of the Georgia Bight and the bays the bottom flow is less intense since the interior onshore flow can supply the necessary mass.

Winds favorable for upwelling can occur any time of year on a transient basis (Fig. 1.8*a*) throughout the SAB and are especially persistent during spring and summer (Fig. 1.8*b*).

The degree to which the subinertial frequency time dependent response of the coastal ocean to atmospheric forcing is predicted by the theory of Janowitz & Pietrafesa (1980) can be tested by

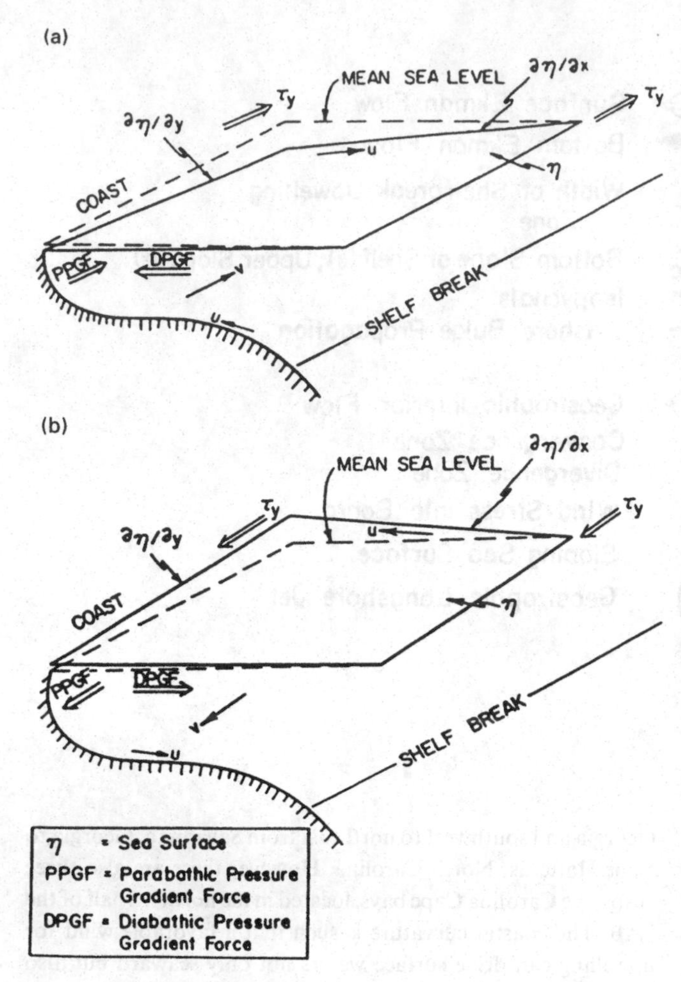

Fig. 1.7. Response of sea level to longshore wind stresses favorable for: (a) upwelling; and (b) downwelling in the southern halves of Raleigh and Onslow Bays, and from Cape Canaveral, Florida, to Charleston, South Carolina.

comparing longshore currents measured at a mooring site to those predicted by the model. Consider data acquired 12 m from the surface in 45 m of water at a mooring site located off Charleston, SC, for the 100-d period 13 April 1977–22 July 1977. The balance suggested by Janowitz & Pietrafesa (1980), states that the longshore wind-driven current is equal to the sum over time of the quantity defined as the difference between the magnitude of the observed longshore wind stress minus the square of the measured near bottom longshore velocity times a drag coefficient, all divided by the depth of the water column. In other words, the longshore wind-driven velocity is equal to the sum over time of the longshore wind stress minus the longshore component of bottom stress, all divided by depth. The cross-shelf velocity component is equivalent to the negative of the time rate of change of the longshore flow divided by the Coriolis parameter and the resulting vertical velocity is equal to the value of the depth multiplied by the offshore variation of the cross-shelf flow. These relationships are;

$$\frac{\partial v}{\partial t}=\frac{\tau^y - C_1 v}{h} \tag{1.8a}$$

$$u=-\frac{1}{f}\frac{\partial v}{\partial t} \tag{1.8b}$$

and

$$w=-z\frac{\partial u}{\partial x} \tag{1.8c}$$

where $C_1 = 0.01$, computed from observations. In the cross-shelf direction, the interior region is geostrophically balanced. We compare predictions of wind-driven longshore currents to measured longshore currents at a mid-shelf (40 m) interior station off Charleston in figure 1.9. The simple model misses the mark whenever Gulf Stream front waters are present such as model days 12–18, 32–46, and 70–76.

Thermohaline-driven upwelling

During late fall to early spring both momentum and buoyancy flux exchanges at the surface of the coastal ocean can drive coastal currents. During the fall, when cold air is suddenly advected offshore, the vertical density field is destabilized and vertical mixing ensues, creating a more vertically homogeneous density field. Atmospheric cold fronts tend to move offshore in the SAB, and as the advected cold air tends to equilibrium with the warmer coastal ocean by extracting heat, the surface waters are differentially cooled more quickly and efficiently at the surface and in shallower waters. This process has a destabilizing effect in the water column and the net result is that the isopycnals tend to slope downward in the offshore direction (ERT, 1979a). This negative slope of the diabathic density gradient effects a southward flow throughout the water column with the flow becoming increasingly southward with increasing depth, at a particular mooring location, relative to the surface flow. This process occurs via the thermal wind balance. It continues until late February to March and then reverses itself.

In the spring, warmer air begins to move offshore and the coastal ocean heats up, beginning in shallower waters. This process, in concert with an increase in fresh coastal water,

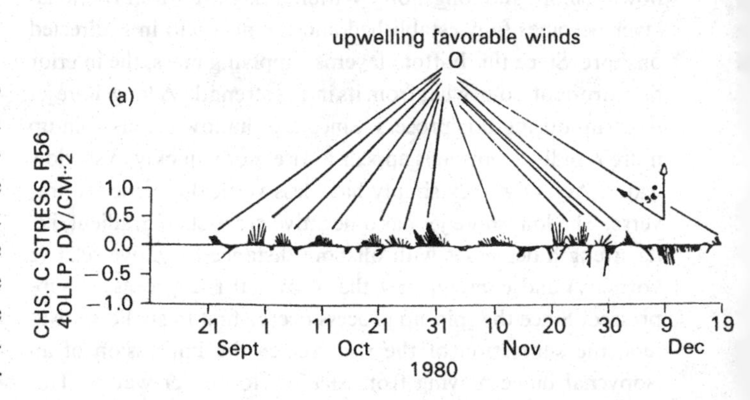

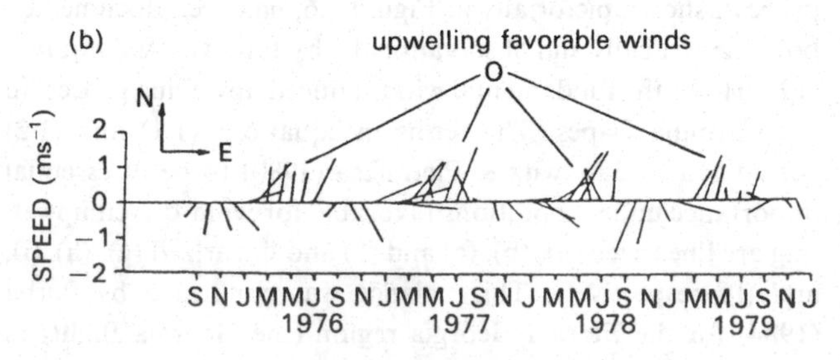

Fig. 1.8. Wind velocity vectors at the Charleston, South Carolina coastal station: (a) 40-hour low pass wind vectors; and (b) 30-day low pass wind vectors (adapted from Weisberg & Pietrafesa, 1983).

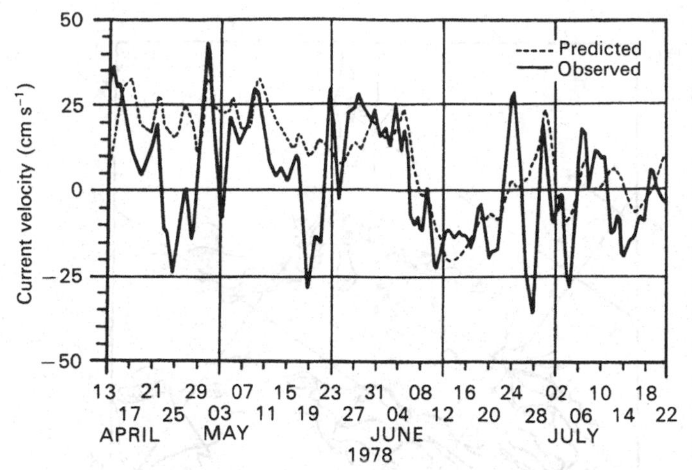

Fig. 1.9. Predicted versus observed mid-depth, longshore currents at a 40 m, mid-depth station, offshore of Charleston using the Janowitz–Pietrafesa (1980) model prediction.

derived from spring runoff, causes the isopycnals to slope downward toward the coast, supporting a northward flow in the water column. A flux across the shelf of heavy water can occur in spring in the SAB due to thermohaline forcing exacted from the positive salinity gradient set up by coastal waters made fresh by spring runoff and by the ever present highly saline Gulf Stream. Stommel & Leetma (1972) presented a theoretical analysis of circulation on a continental shelf under vertically homogeneous, horizontally stratified conditions. They considered a flat-bottomed, semi-infinite vertical plane shelf which was mechanically driven by surface winds and by a net input of freshwater, per unit length of coastline, along the surface. Stommel & Leetma's (1972) paper, the dissertations of Leetma (1969), Pietrafesa (1973) and the work of Pietrafesa & Janowitz (1979) are amongst the few which have dealt with thermohaline shelf forcing. The latter find that during the SAB spring transition period, freshwater runoff is at a maximum and forms a band of low salinity water near the coast, thus producing large cross-shelf salinity gradients. This effect combined with increased surface heating, which minimizes the shelf–Gulf Stream temperature differential, tends to produce maximum positive horizontal density gradients. Since this is also a period when vertical mixing begins to diminish because of a decrease in surface wind stress, a density-driven flow develops, which tends to stratify the shelf. Lighter (fresher) water, moves offshore on the surface and denser (saltier) water intrudes onshore along the bottom. This spring intrusion may also upwell cooler, deeper Gulf Stream waters having higher nutrient content onto the shelf for subsequent biological uptake. Both the drifter results of Bumpus (1973) and the averaged current meter data (fig. 1.10) indicate onshore bottom flows for this period. Summer heating and diminished vertical mixing tend to increase the temperature contrast in the stratified layers.

For the winter period in the Mid-Atlantic Bight shelf region, Stommel & Leetma (1972) found an approximate functional dependence of the salt penetration scale across the shelf (L) on the Ekman number (E) and wind stress τ. Here the salt penetration scale is defined as: $L = So/(\partial S/\partial X)$ and the vertical Ekman number as $E = A/fH^2$, where So is the mean salinity, X is in the cross-shelf direction, A is the vertical eddy viscosity, f is the Coriolis parameter and H is water depth. As E becomes small, L increases and depends mainly on the wind stress. Also, alongshelf winds appear to have more influence on the salt penetration distance than do cross-shelf winds. As the depth becomes shallow, mixing increases. The available salt is thus mixed with the subsurface water, decreasing both the transport of salt onshore and the values of L. We have taken the liberty of recalculating the Stommel & Leetma estimate by deriving exact solutions and find that (Fig. 1.11) density driving, while generally less effective than wind forcing in moving heavy bottom water across the shelf, nontheless is capable of transporting such waters from the shelf break across the width of most of the SAB shelf. For example thermohaline forcing can drive salt across a shelf of width 10^7–10^8 km, more effectively than can a 1 dyne/cm^2 wind stress directed either shoreward or with the coast to its right for vertical Ekman numbers of .02–1.

Climatological distributions of salinity presented by Atkinson *et al.* (1982) suggest that low salinity water is carried northward and offshore in spring and southward against the coast in autumn. During April and May surface water with salinity <36 ppt is found along the shelf break between 32 and 33° N. This water is clearly a mixture of low salinity inner shelf and Gulf Stream water. According to Blanton & Atkinson (1983), the evacuation of low salinity water adjacent to the Georgia coast occurs most rapidly in spring just after river discharge reaches a maximum. The rate of evacuation is very likely related to the strength and persistence of the longshore wind stress. The water in the upper layers of the inner shelf frontal zone is likely carried northward with a strong offshore component such that low salinity water originally present in surface layers adjacent to the Georgia (and probably South Carolina) coast is advected toward the outer shelf between 32 to 33° N. The coupling of two upwelling driving mechanisms, wind stress and buoyancy stress during spring leads to replenishment

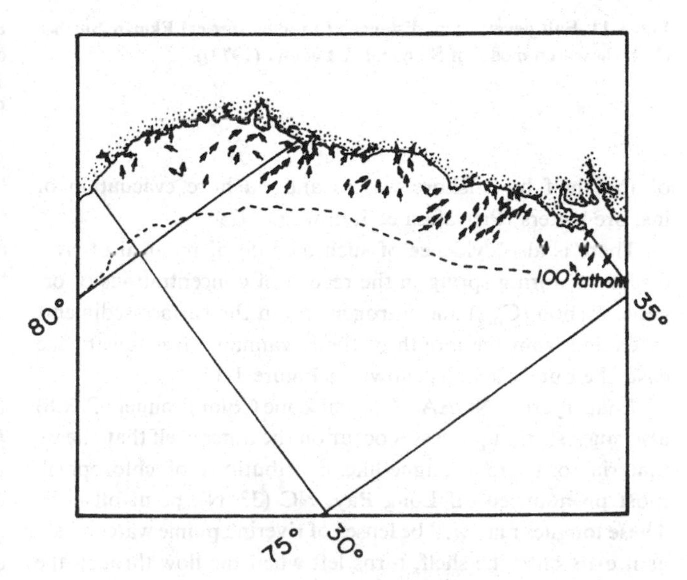

Fig. 1.10. Near-bottom current vector means for late March, April, May and early June as inferred from 10 years of bottom drifter data (Bumpus, 1973) and moored current meter data.

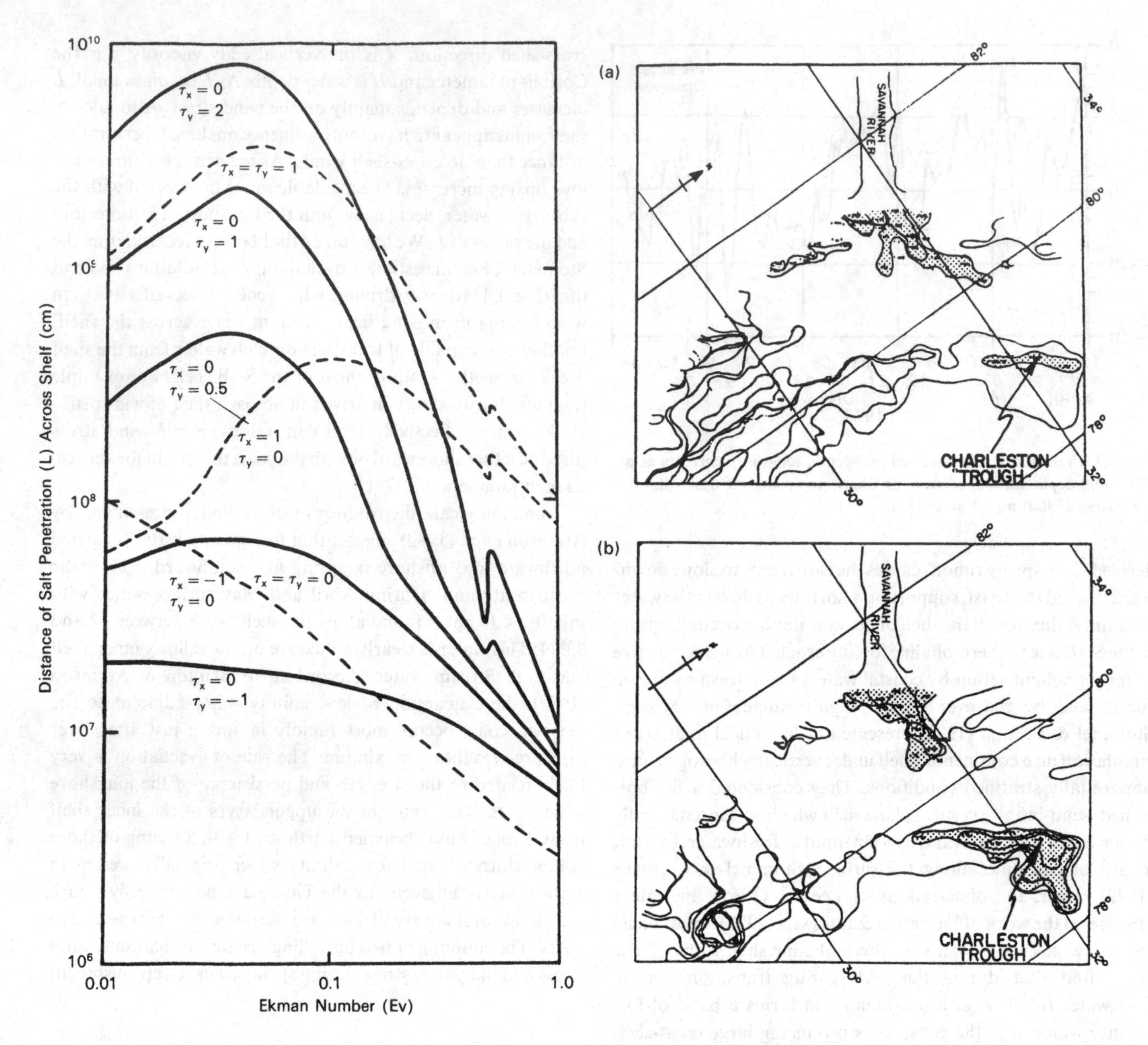

Fig. 1.11. Salt penetration, distance (L) versus vertical Ekman Number (Ev), (based on model of Stommel & Leetma (1972)).

Fig. 1.12. Stippled areas indicate areas of 1% concentrations of: (a) organic carbon; and (b) nitrogen in surface sediments. Regions of high organic carbon and nitrogen are found off the mouth of the Savannah River and in the Charleston Trough (adapted from Premuzic, 1980).

of the shelf by offshore waters and offshore evacuation of inshore waters (Pietrafesa & Janowitz, 1979).

There is also evidence of such a route of removal of river discharge during spring in the record of concentrations of organic carbon (C_{org}) and nitrogen (N) in the surface sediments extending from the mouth of the Savannah River toward the east, the outer shelf, as shown in Figure 1.12.

Imagery from NOAA's Coastal Zone Color Scanner (CZCS) also suggests that processes occur on the inner shelf that release material offshore. Tongue-like distributions of chlorophyll, most pronounced off Long Bay, NC (33° N), point offshore. These tongues may well be lenses of riverine plume water which as it exits onto the shelf, turns left when the flow through the entire water column at the mouth of the estuary is outward. As the river discharge decreases, the bottom layer of the estuary reverses and the net flow is into the mouth, the surface plume will turn south (Zhang, Janowitz & Pietrafesa, 1987). So lenses of fresh water, carrying river nutrients may occasionally turn left, during spring on the Georgian shelf (Fig. 1.13) and when conditions change, the lenses may pinch off from their coastal source.

Topographic effects

Effects of topography on coastal flows have been difficult to quantify. At first glance, the Pacific coast of the continental United States appears to have a fairly regular coastline and topography was historically believed to play a minor role in the overall physics of the coast, save for Arthur's (1965) notion of cape-induced upwelling. However, on the eastern United States continental margin, the topographic aspect ratio of the shelf is an order of magnitude less than its west coast counterpart and the coastal configuration, for example the Carolina Capes shoals, including Frying Pan, Lookout and Diamond (Fig. 1.2) and the

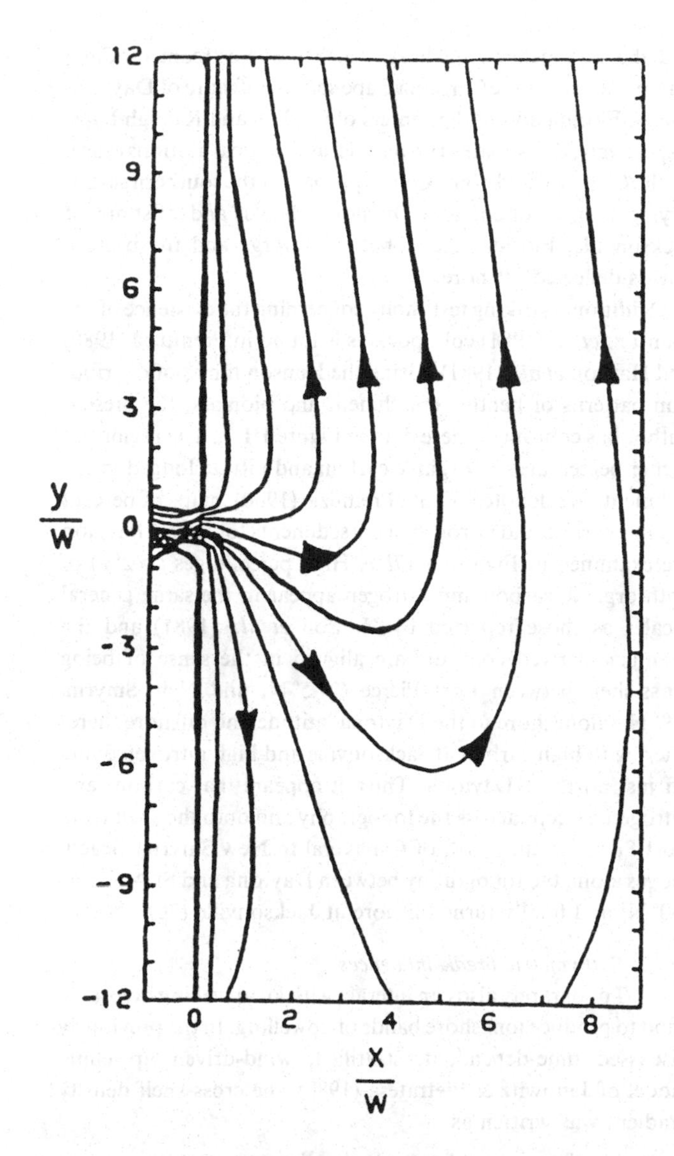

Fig. 1.13. Streamline pattern in the lower layer of the estuarine plume model of Zhang, *et al.* (1987). Horizontal distances are presented in multiples of the estuary mouth width.

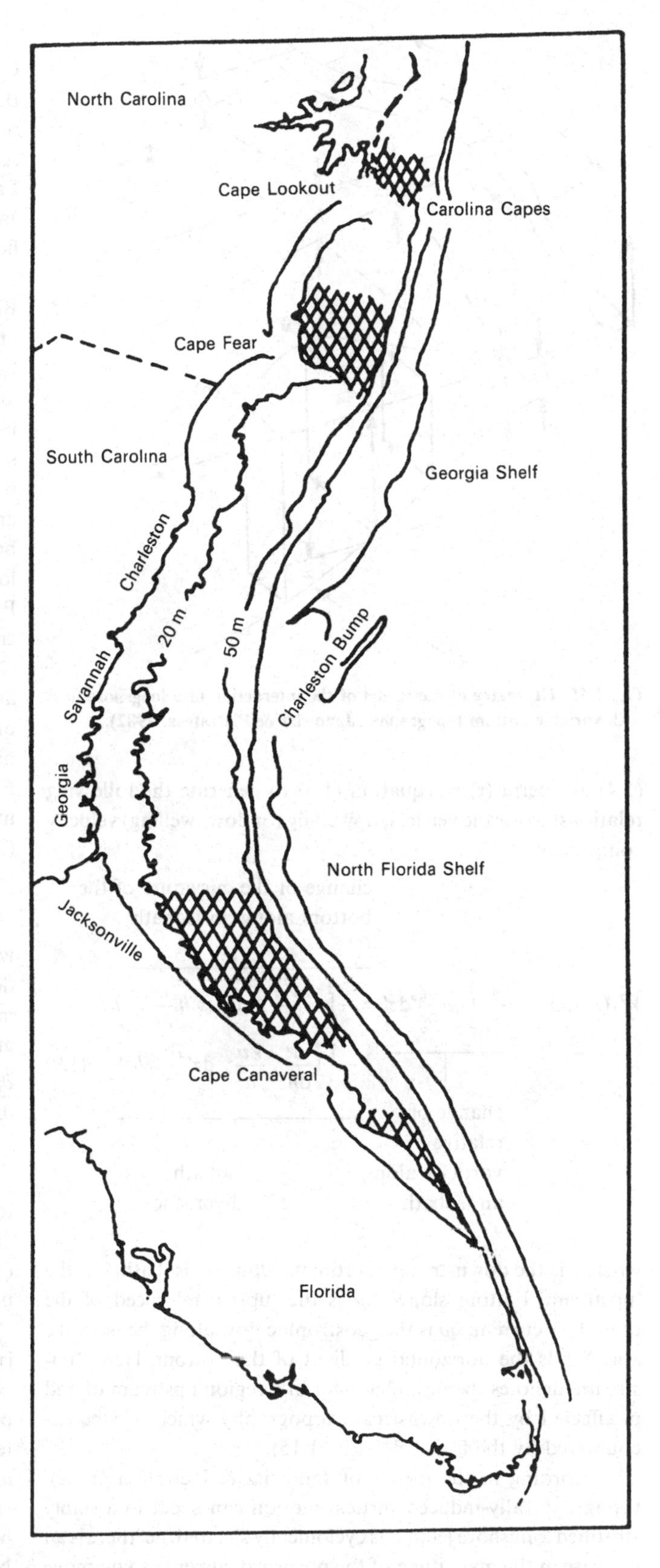

Fig. 1.14. Pockets of relatively cold bottom water (cross-hatch pattern) found in the South Atlantic Bight.

coastal circulation were believed to be linked (Abbe, 1895). As a consequence, while west coast investigators have historically ignored bottom and sidewall effects, these effects were recognized as probably significant in the Carolina Capes.

Blanton *et al.*, (1981) found cold bottom waters to be present on the northern side of the Carolina Cape shoals and invoked Arthur's west coast notion of cape-induced upwelling (Arthur, 1965) to explain the persistence of the cold water pockets. However from equation (1.7) the concept of cape-induced vertical motion fails at both Cape Canaveral and in the Carolina Capes since in the inviscid limit over a flat bottom, on an f-plane, there is no mechanism to provide for vertical motion. Albeit, an alternative and physically consistent theory was presented by Janowitz & Pietrafesa (1982) to explain these cold pockets, shown in Figure 1.14.

Janowitz & Pietrafesa (1982) considered a sheared, variable density boundary current flowing with a wall to its left over variable topography via the balances of terms (a), (b), (c) in equation (1.1), (g) (h), (i) in equation (1.2), equations (1.3) and

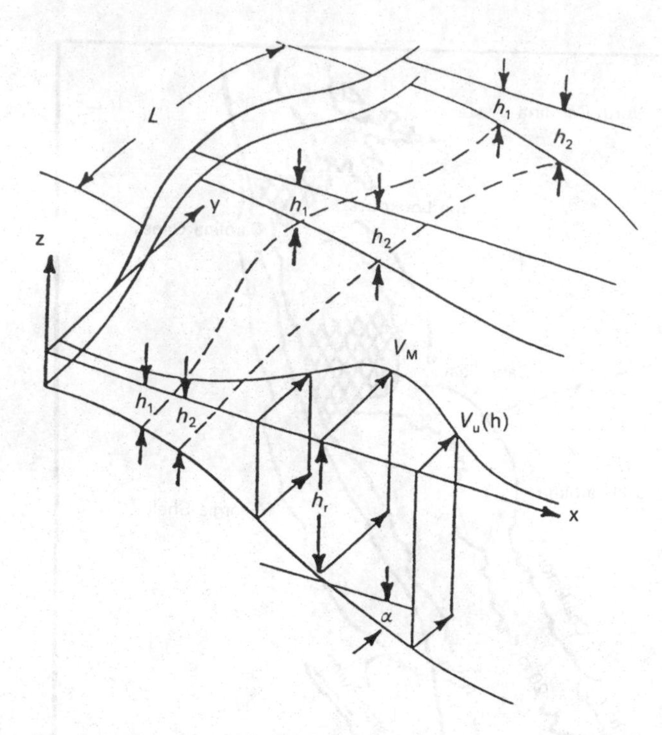

Fig. 1.15. Geometry of the model of the interaction of a longshore flow with variable bottom topography (Janowitz & Pietrafesa, 1982).

(1.4) and term (r) in equation (1.5) to generate the following relationship for the vertical, upwelling (or downwelling) velocity component

change of the curvature of the bottom along an isobath

$$W_1(x,y,z) = -\frac{1}{f_0}\int_z^{\circ} qo\frac{\partial \zeta_\circ}{\partial \zeta}dz' = -\int_z^{\circ} \frac{Vu}{\alpha}^2 dz'|\nabla h|\frac{\partial}{\partial s}\nabla^2 h$$

$$-\frac{1}{3}\int_z^{\circ}\frac{d}{dh}\frac{Vu}{\alpha}dz'\frac{\partial}{\partial s}|\nabla h|^3 \qquad (1.9)$$

change of relative vorticity along an isobath

isobath divergence

where s is the downstream coordinate along an isobath, α is the 'upstream' bottom slope, Vu is the 'upstream' speed of the boundary current, qo is the geostrophic flow along the isobaths and ∇h is the horizontal gradient of the bottom. Here, 'upstream' denotes the flow character in a region upstream of and unaffected by the downstream topography which will be encountered by the flow field (Fig. 1.15).

According to the theory of Janowitz & Pietrafesa (1982), topographically-induced vertical motion can occur in a stably stratified longshore jet that is cyclonically sheared: i.e. there is an increase in the magnitude of the poleward current as you move seaward from an east coast. They found that upwelling (downwelling) and onshore (offshore) flow will occur in a cyclonically sheared current whenever the isobaths diverge (converge) in the downstream flow direction. Hence, as the Gulf Stream flows past Cape Canaveral or either Frying Pan or Lookout shoals (Fig. 1.16), the bathymetry begins to diverge, and the flow deflects onshore. Farther downstream of Cape Canaveral and of the Carolina Cape shoals, offshore of Daytona Beach, Florida and in the centers of Onslow and Raleigh bays, respectively, the isobaths run parallel and vertical motions cease. As the Gulf Stream Front (GSF) approaches the southern side of Frying Pan, Lookout and Diamond shoals, and offshore of Jacksonville, Florida, the isobaths converge and the bottom flow is deflected offshore.

Additional striking testimony concerning the existence of the Blanton, *et al*, (1981) cold pockets is found in Premuzic (1980) and Hanson *et al*., (1981). Using the Hanson tables of distribution patterns of benthic enrichment and biomass, the present author has contoured these data in Figure 1.17*a*. The regions of higher percentages of organic carbon and nitrate lodged in the sediments are denoted. From Premuzic (1980), maps of percent organic carbon and nitrogen in the sediments for the SAB region are examined in Figures 1.17*b*,*c*. High percentages (>2%) of both organic carbon and nitrogen appear in the same general locales as those reported by Hanson *et al*., (1981) and the Premuzic-derived contours are aligned in the sense of being cross-shelf between Fort Pierce (27.5° N) and New Smyrna (29° N), alongshore to the Daytona latitude and offshore thereafter, with high carbon at Jacksonville and high nitrogen dying off just north of Daytona. Thus, it appears that carbon (and nitrogen) sweeps across the topography and onto the shelf from Fort Pierce to the south of Canaveral to New Smyrna Beach, moves along the topography between Daytona and St Augstine (30° N) and finally turns offshore at Jacksonville (30.2° N).

Bathymetric break influences

Topography also can interact with an upwelling favorable wind to produce longshore bands of upwelling. In the previously discussed time-dependent, stratified, wind-driven upwelling model of Janowitz & Pietrafesa (1980), the cross-shelf density gradient was written as

$$\frac{\partial \rho}{\partial x} = -|z \quad \frac{\partial \rho}{\partial z}| \quad \frac{\tau^y}{f^2 d} \quad \frac{1}{h^2} \quad \frac{\partial h}{\partial x}$$

$$(\eta e^{-\eta})(2-\eta-h)\frac{(\partial^2 h/\partial x^2)}{(\partial h/\partial x)^2} \qquad (1.10)$$

for a linear bottom stress law, $\tau_B = \rho_0 c_1 v(x, -h, t)$ where d is $(A/2f)^{1/2}$, A is vertical eddy viscosity coefficient, h is local depth, $\partial h/\partial x$ is local slope of bottom η is ftd/h, local, non-dimensional spin-up time and other variables are as previously defined.

If the bracketed factor, $h(\partial^2 h/\partial x^2)/(\partial h/\partial x)^2$ in equation (1.10) is negative, then the density becomes an increasing function of x, i.e. the density increases as you move seaward. Since η is a positive parameter then the only factor that could make the last term positive is the constant +2. In order for η to exceed 2 in magnitude, at a 30° latitude, a wind would have to blow persistently for the order of a month or so. Thus, if the last factor is to be negative then the quantity $h(\partial^2 h/\partial x^2)/(\partial h/\partial x)^2$ must exceed 2. Now, on shelves where the bottom varies smoothly and increases monotonically seaward, the above inequality is not easily satisfied. Alternatively though, if the bottom slope has localized discontinuities located between montonically increasing regions, then the above criterion can be satisfied. The width of the upwelling band, ΔL, can be approximated as $\Delta L = 2h(\alpha_D - \alpha_S)/(\alpha_D + \alpha_S)^2$ where α is the shelf slope and the subscripts D

(a)

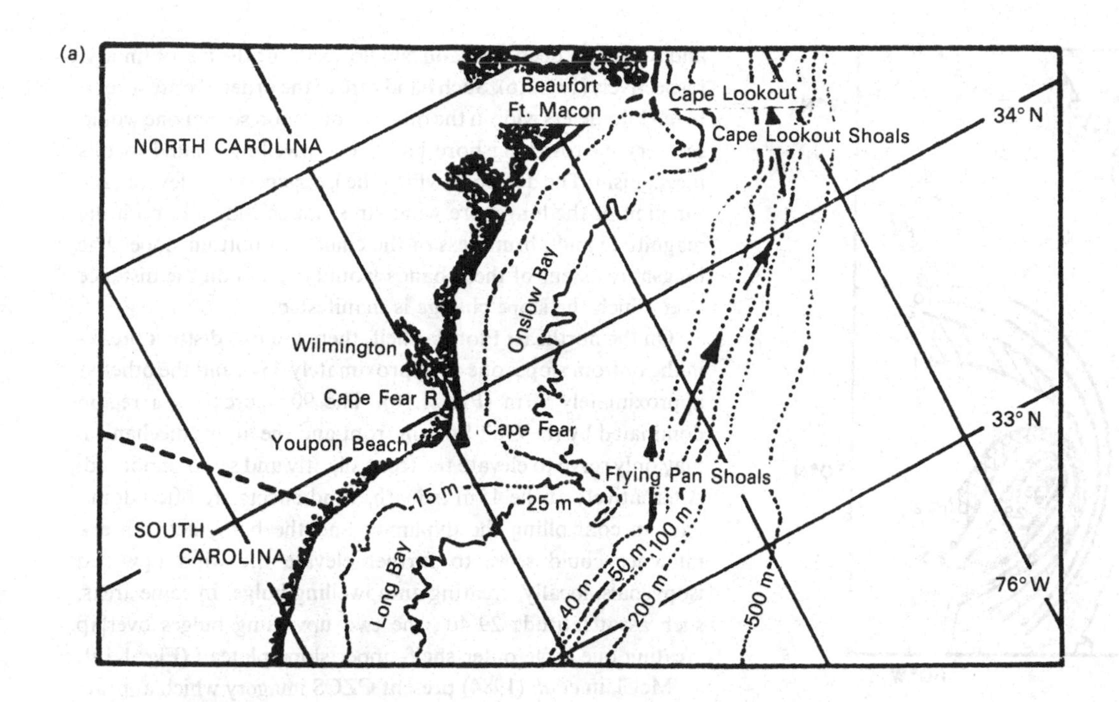

(b)

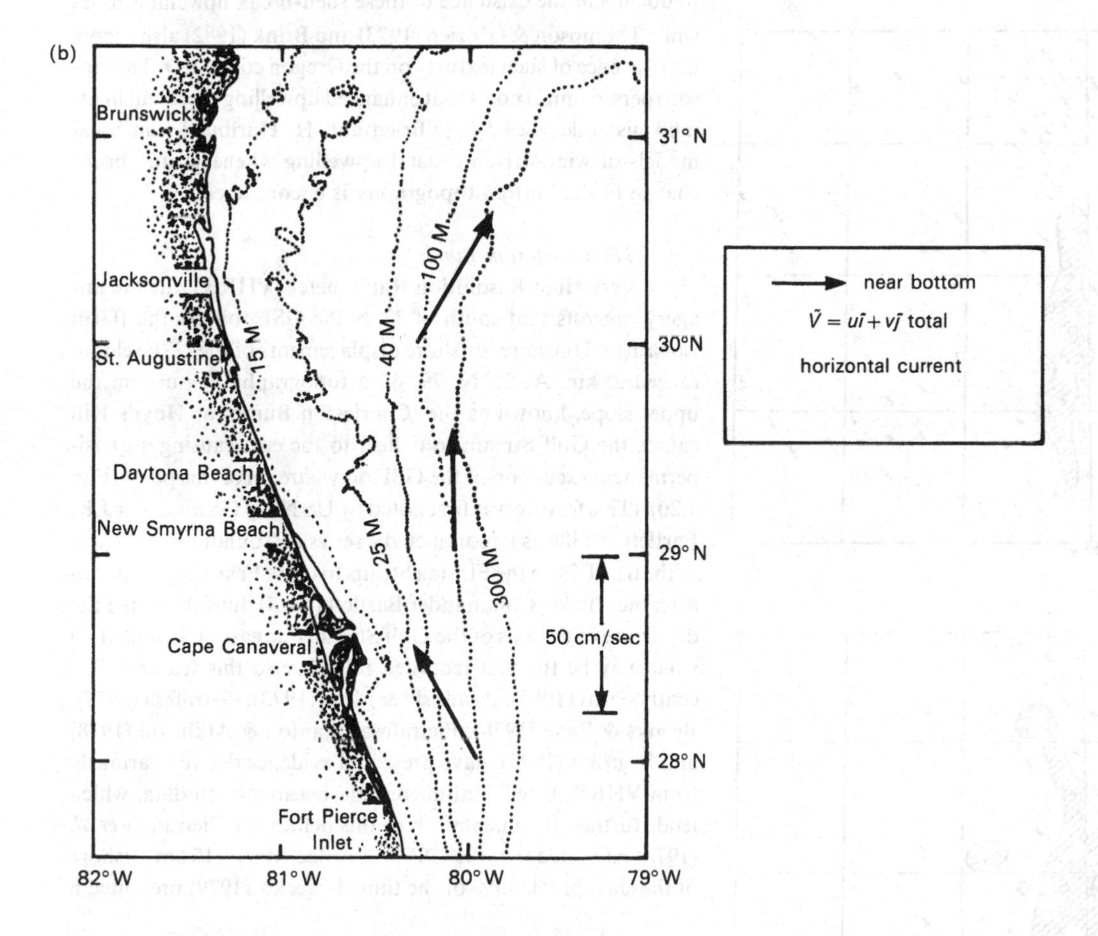

Fig. 1.16. (a) Bottom flow patterns around Frying Pan and Cape Lookout Shoals and in Onslow Bay created by the interaction of the Gulf Stream front and bottom topography via the theory of Janowitz & Pietrafesa, 1982. (b) Near-bottom vectors indicating onshore flow in diverging isobath region and offshore flow in converging isobath regions.

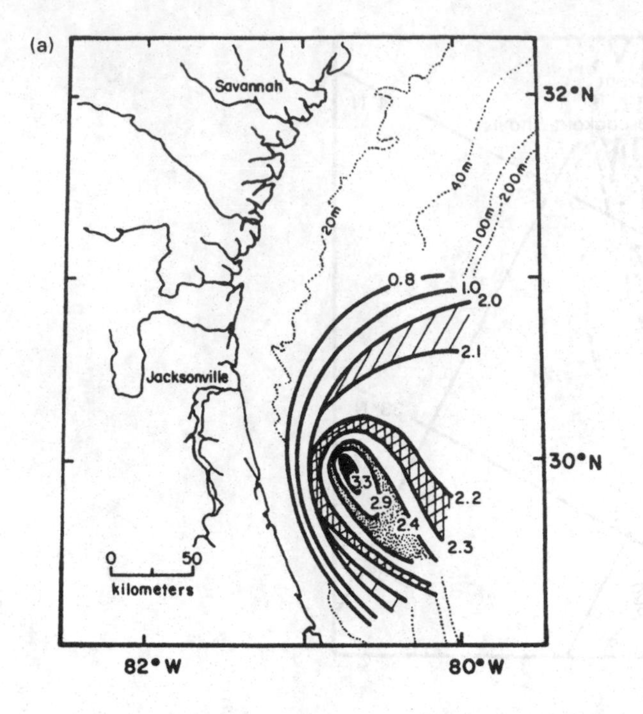

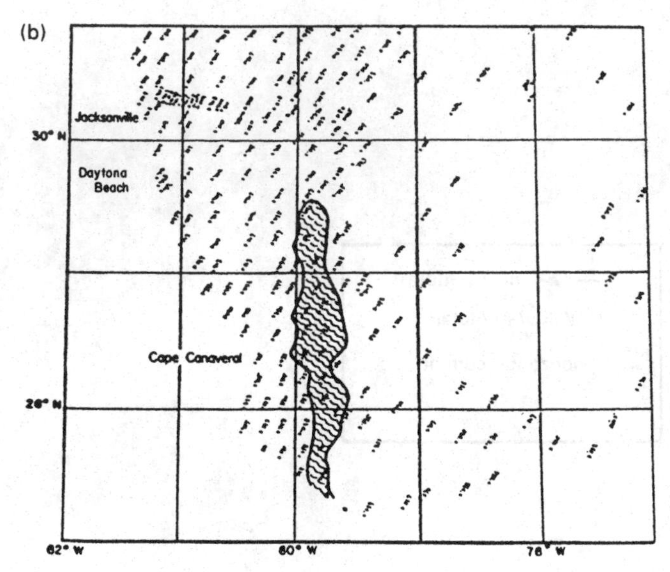

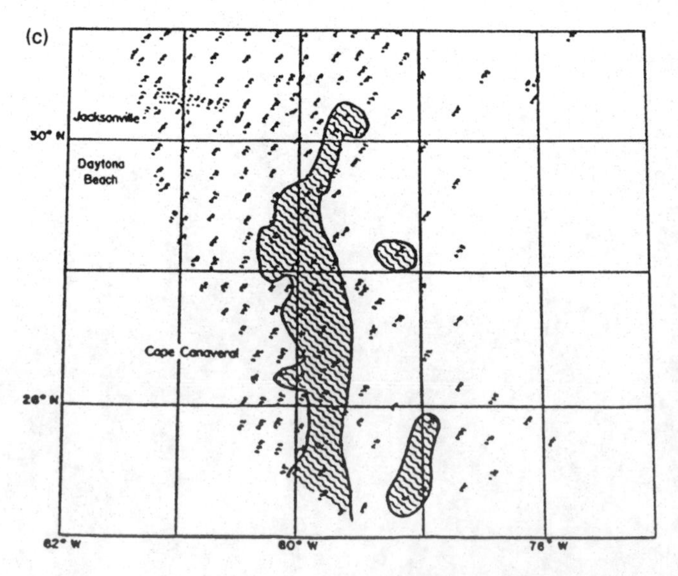

and S refer to 'deep side' and 'shallow side' of the discontinuity, respectively (Fig. 1.6). Such bands are of the order of only several to ten kilometers wide in the offshore direction so that one would see very narrow longshore bands of upwelling related to this mechanism. The degree to which the isopycnals are elevated is a function of the longshore wind stress magnitude and both the magnitude and abruptness of the change in bottom slope. The longshore extent of these bands would depend on the distance over which the slope change is manifested.

On the northeast Florida shelf, there are two distinct breaks in the bottom slope, one at approximately 45 m and the other at approximately 90 m (Fig. 1.18). The 90 m break is a region dominated by the Gulf Stream front and the above mechanism may only serve to elevate the front slightly and so go unnoticed. Alternatively at the 45 m isobath, wind effects are often dominant in controlling the dynamics and the bathymetric break influence could serve to further elevate the wind upwelled isopycnals, locally, creating an upwelling bulge. In some areas, such as at latitude 29°40′, the two upwelling bulges overlap creating one large outer shelf–upper slope plateau (Fig. 1.19).

McClain *et al.* (1984) present CZCS imagery which appears to document the existence of these shelf-break upwelling zones while Thompson & O'Brien (1973) and Brink (1982) also report the existence of such features on the Oregon coast. J.D. Thompson (pers. comm.) notes that enhanced upwelling is present in his (and his colleagues J.J. O'Brien and H. Hurlburt) numerical models of wind-driven coastal upwelling whenever an abrupt change in the bottom topography is encountered.

The Charleston Bump

Very High Resolution Radiometer (VHRR) satellite imagery suggests that south of 32° N the GSF follows the 100 m isobath and onshore–offshore displacements of the GSF seldom exceed 25 km. At 32° N, 79° W, a topographic feature on the upper slope, known as the 'Charleston Bump' or Hoyt's Hill causes the Gulf Stream to deflect to the east causing a quasi-permanent excursion of the GSF downstream of this point (Fig. 1.20). (This feature was first noted by US Navy Commander J.R. Bartlett in 1880 as he conducted a series of 'soundings and hauls of the trawl' from the Florida Straits to Cape Fear aboard the US Streamer Blake. Commander Bartlett (1883) dutifully noted the departure of the axis of the Gulf Stream due east of Savannah in what may be the first recorded reference to this feature). Recently, Pratt (1963), Pashinski & Maul (1973), Pietrafesa (1977), Brooks & Bane (1978), Pietrafesa, Blanton & Atkinson (1978) and Legeckis (1979), have presented evidence derived variously from VHRR, satellite altimeter and ocean station data, which lends further documentation of this deflection. Pietrafesa *et al.* (1978) estimated that the GSF was deflected 60–110 km offshore of the shelf break 70% of the time. Legeckis (1979) presented a

Fig. 1.17. (a) Sediment organic carbon content (%) on the continental shelf off Georgia and Florida in June, 1977 (after Hanson *et al.*, 1981). Only areas where carbon content > 2% are shown. (b) Organic nitrogen in the surface sediments on Florida–Georgia continental margins. Stippling indicates areas where concentrations are > 2% (after Premuzic, 1980). (c) Organic carbon in the surface sediments on Florida–Georgia continental margins. Stippling indicates areas where concentrations are > 2% (after Premuzic, 1980).

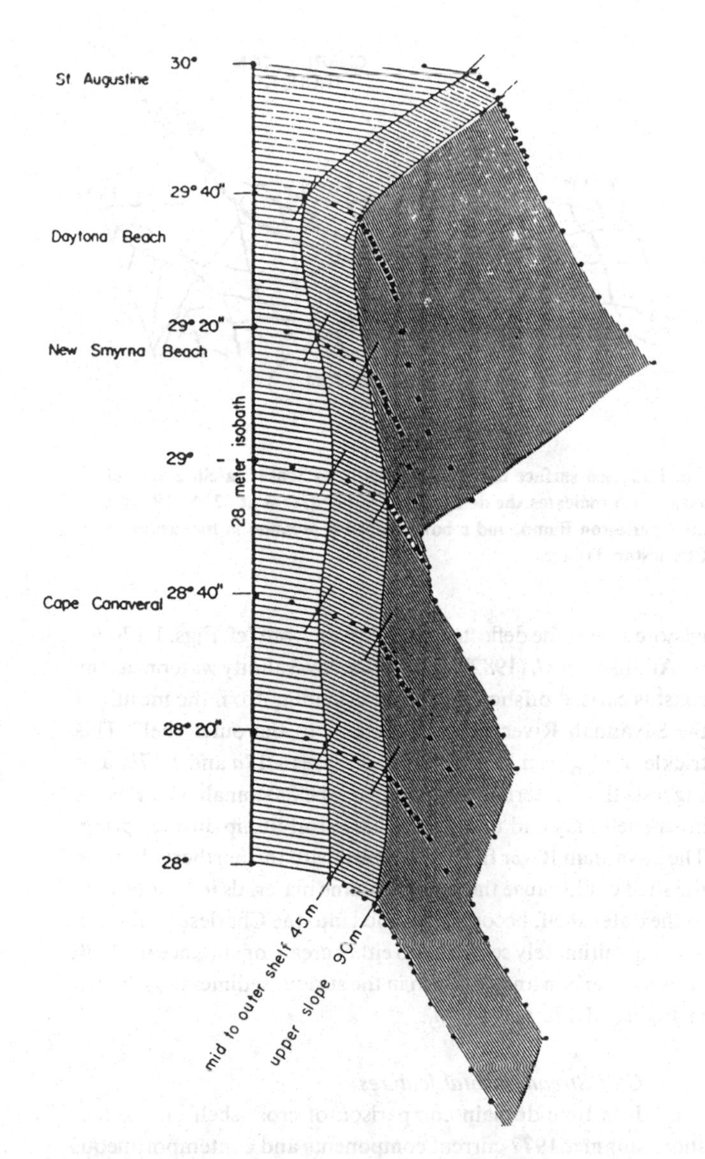

Fig. 1.18. Three-dimensional view of northeastern Florida continental margin seaward of the 28 m isobath. The two distinct breaks in the bottom topography are delineated.

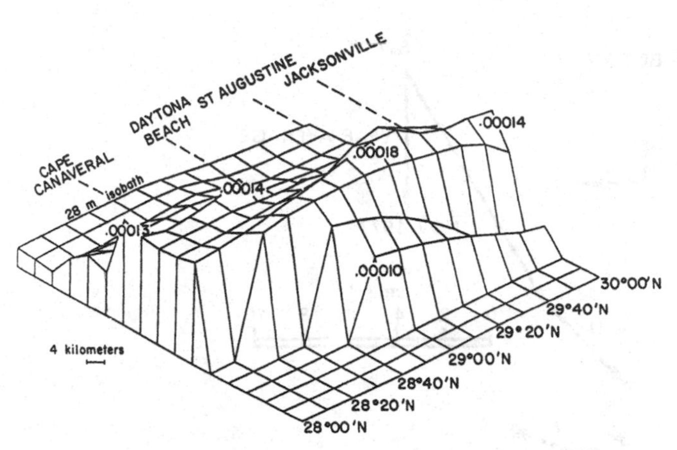

Fig. 1.19. Ridges of localized upwelling appearing on the northeast Florida shelf, as depicted in a surface contour of the density perturbation field 5 m above the bottom. The upwelling ridges were created 1 day after the onset of a 0.5 dyne cm^{-2} upwelling favorable wind as described by Janowitz & Pietrafesa (1980).

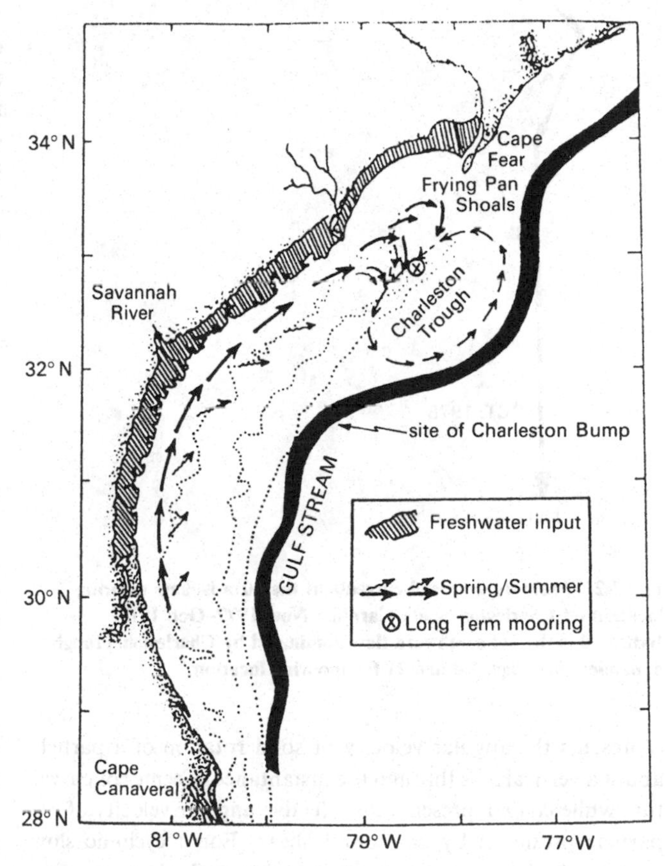

Fig. 1.20. Schematic of transport pathway connecting near coastal waters to the Charleston Trough.

glossary of VHRR imagery which shows that downstream of the 'bump', lateral meanders of the GSF propagate to the north (east) with lateral displacements on the order of 10–100 km.

The interaction of the Gulf Stream with the topographic irregularity of the Charleston Bump, would affect the generation of a quasi-stationary topographic Rossby wave with wavelength on the order of 100–300 km (Rooney, Janowitz & Pietrafesa, 1978). In the lee, i.e. to the north and west of the point of deflection, the trough of the wave would exist. The eastern half of the trough is defined by the Gulf Stream front as observed in VHRR satellite imagery but the western part of the trough is not easily discerned by infra-red techniques. We are aided here by current meter and satellite altimeter data.

Monthly averaged currents on the outer shelf at the 40 m isobath offshore of Cape Romain (east of Charleston, South Carolina) display a net southward motion excepting during the spring when an eastward motion is suggested (Fig. 1.21). Additionally, a sea surface topography map (Fig. 1.22) constructed from SEASAT satellite altimeter data suggests that the trough is a large bowl-like depression about which there might be a cyclonic circulation. The horizontal circulation resulting from such a cyclonic cell would have a vertical component of relative vorticity which can be expressed as

$$\text{vertical component of relative vorticity} = \frac{c}{R} - \frac{\partial c}{\partial n} \quad (1.11)$$

where c is the horizontal flow field, R is the radius of curvature and n is an ordinate perpendicular to c at any point. C/R

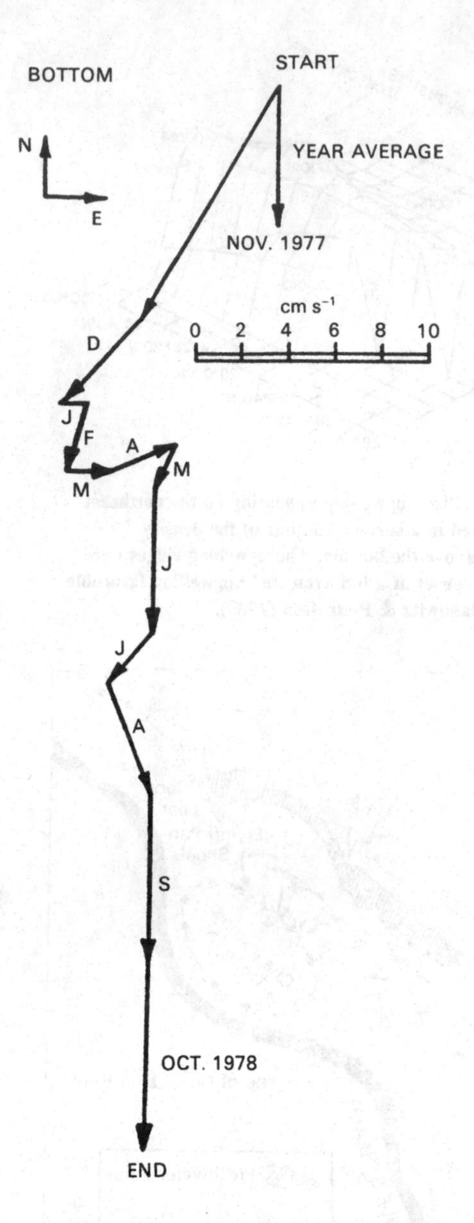

Fig. 1.21. Monthly averaged currents at the 40 m isobath mooring location off Charleston South Carolina Nov. 1977–Oct. 1978. Indicated is the net southward flow dominated by Charleston Trough dynamics. See Figs. 1.4 and 21 for mooring location.

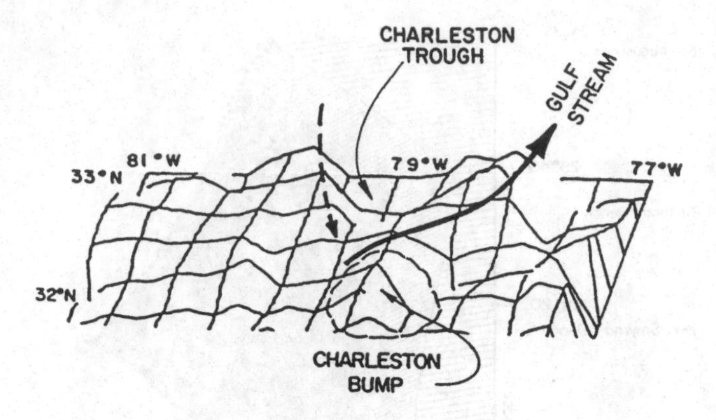

Fig. 1.22. Sea-surface map constructed from Seasat satellite altimeter data which indicates the deflection of Gulf Stream at 32° N, 79° W by the Charleston Bump, and a bowl-shaped depression in the center of Charleston Trough.

represents the angular velocity of solid rotation of a particle about a vertical axis through the instantaneous center of curvature while $\partial c/\partial n$ represents the effective angular velocity of the particle produced by horizontal shear. For a cyclonic slow pattern, the first term is positive and in a cyclonic shear zone, such as the GSF, the second term is also positive so that the vorticity is positive and there would be persistent upwelling within the trough itself. Hydrographic observations (Fig. 1.23) support this notion.

Imagery from NOAA's CZCS (image compliments of C. McClain, pers. comm.) shows a large pool of high chlorophyll- *a* situated near the shelf break at 33° N (Fig. 1.24). We hypothesize that such pools result from the large cyclonic trough north of 32° N.

Furthermore, relatively high concentrations of organic carbon and nitrogen are found in the region to the north and on the inshore side of the deflected Gulf Stream front (cf. Figs. 1.17*a*,*b*).

Atkinson *et al.* (1982) suggest that low salinity water near the coast is carried offshore in spring extending from the mouth of the Savannah River towards the east to the outer shelf. This trickle of C_{org} and N is shown in Figures 1.17*a* and 1.17*b*, and suggests that materials released from the Savannah–Charleston inner shelf may end up on the Charleston Bump during spring. The Savannah River freshet, coupled with the northerly flow on the shelf could cause the riverine plume materials to be advected to the outer shelf, become entrained into the Charleston Trough gyre and ultimately settle out to either create or enhance the high values of carbon and nitrogen in the surface sediments as shown in Figures 1.17*c* and 1.17*d*.

Gulf Stream frontal features

In a time domain comparison of cross-shelf and alongshore summer 1977 current components and contemporaneous temperature (T), Pietrafesa & Janowitz (1980) found that at both the shelf break and mid-shelf locations, a northeastward alongshore current (v) occasionally will become diminished in magnitude or possibly reverse in concert with a decrease in temperature. Typically, this v and T scenario is followed several days hence by a return of a northeastward current and an increase in temperature (Fig. 1.25). The rise in T usually peaks at values characteristic of those associated with the Gulf Stream Front (GSF). The cross-shelf component of current (u) will have gone from being positive (offshore) to negative (onshore) to positive again during this period and preceeds the changes in v. The drop in temperature related to the decrease in positive v could result from the fact that the warm frontal waters have moved offshore, making way for a horizontal advection of cooler shelf–slope waters affecting an upwelling of cool, deep shelf–slope waters as a vertical consequence of a one-sided horizontal divergence. Conductivity data collected at these sites indicate that the converging water masses are of Gulf Stream origin. These types of summertime variations in u, v and T are typical for mid- to outer shelf waters. These events are manifestations of the Gulf Stream lateral meanders observed by Webster (1961a,b). The u, v, T scenario is repeated about every 6–7 d with no apparent preference for season. However, wintertime u, v, T

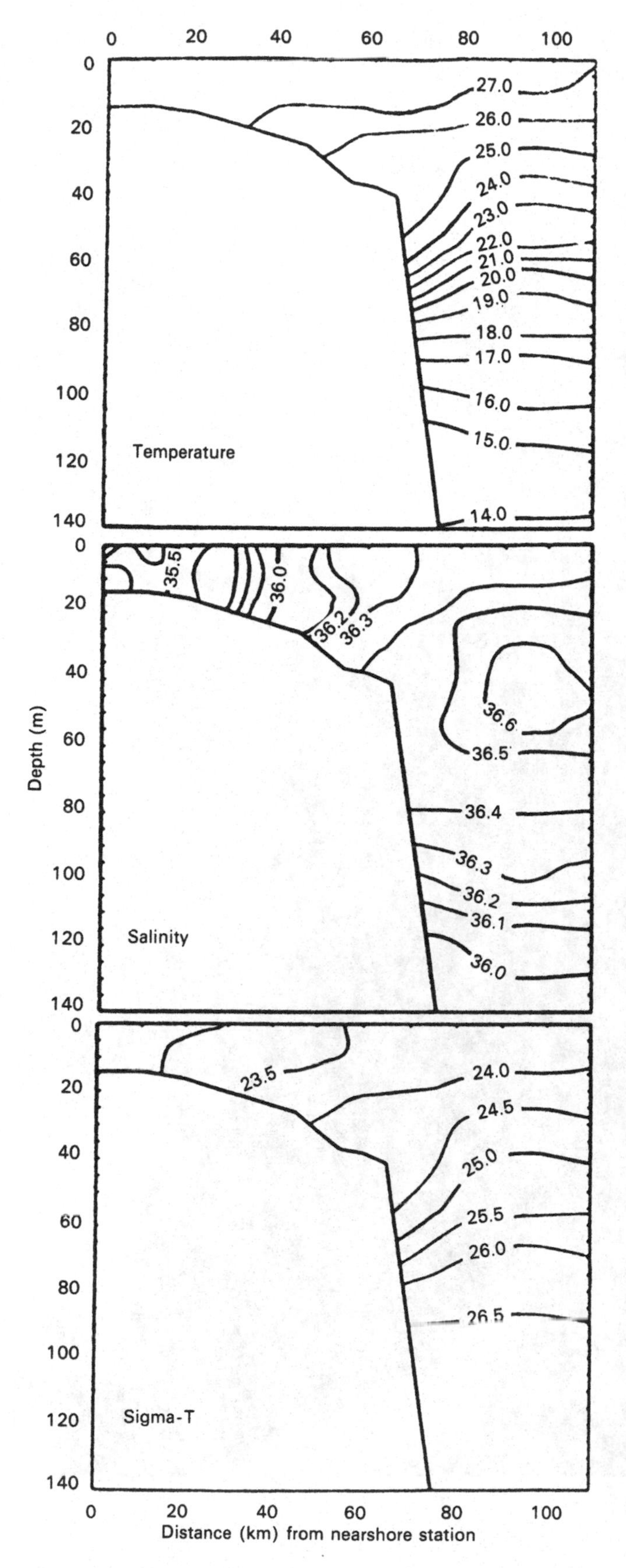

Fig. 1.23. T, S, σ_T dome of upwelled water in a cross-continental margin transect across the Charleston Trough. See Fig. 1.20 for Trough location.

relations do differ somewhat from their summertime counterparts. During the winter, similar types of phenomena are observed on the outer shelf but not in mid-shelf waters. During the months when the shelf waters are cold enough to block the incursion of warmer water masses, these wave-like features are prevented from penetrating across the shelf. In summer, warm shelf waters are more amenable to penetration by offshore waters. Also, during the winter months, mechanical forcing of shelf waters by the atmosphere is greatly increased in concert with the larger wind stresses present (Saunders, 1977; Weisberg & Pietrafesa, 1983). Consequently, the outer portion of the shelf reflects the combined effects of reduced penetration of the GSF onto the shelf and increased wind forcing.

We now consider observational evidence concerning the frequency of occurrence and characterization of these waves which are actually GSF filaments and meanders. We first consider the appearance of features that clearly fit Webster's (1961a,b) description of the frontal meandering phenomenon. At Site K TOP, *u*, *v* current stick and temperature data acquired during the summer of 1977 (Fig. 1.25*b*) provide the following description of the Eulerian manifestation of frontal meanders; the time series of current vectors and temperature is broken into three temporal pieces, 12–16, 16–19 and 19–23 July. It is proposed that three northeastward propagating frontal waves swept by the mooring during the 12–24 July period. The three waves are denoted as A, B and C in Figure 1.25*a* and correspond to lines 1–1¹, 2–2¹ and 3–3¹ and in Figure 1.25*b*. The GSF oscillates as a long wave with crest and trough in the horizontal plane, in keeping with the description provided by Legeckis (1979) based on VHRR imagery. As the crest of wave A approaches the current meter site (Point 1 in Figure 1.25*a* and day 12 in Figure 1.25*b*), the northeastward flow increases and the temperature rises. As wave A moves past the mooring, the trough approaches the site (day 13) and the longshore flow decreases in magnitude and finally reverses in direction, all in concert with a drop in temperature (day 15). Site K moves through the trough, relatively speaking, between days 13.5–16 only to encounter wave B, which proceeds to move by K along the 2–2¹ line between days 16–19. The scenario is repeated for wave C between days 19–22.5 along line 3–3′. The streamlines of flow within the crests and troughs of the waves are as shown in Figure 1.25*b*. Note that as the axis of the GSF shifts farther onshore the current vector simply sways right to left as the temperature rises and then falls, as suggested by wave B. Wave B probably had an amplitude similar to that of wave A. The amplitude of wave C, on the other hand, may have been much larger than those of A or B, or alternatively wave C's passage by site K occurred along a skewed line. Without synoptic ship or other data and/or continuous VHRR imagery, the point is moot. Line 3–3′ is the suggested path taken by wave C by site K. Satellite VHRR imagery interpreted by Mr R. Perchall of the US Naval Oceanography Office shows five meander-like bulges of Gulf Stream frontal features into Raleigh Bay, over site K, during the period 22 June–27 July 1977 encompassing the data set shown in Figure 1.25.

The conceptual model (fig. 1.25*a*) of the events that transpire during the passage of a Gulf Stream frontal meander (Fig. 1.25*b*) by a moored current meter indicates that as the GSF moves offshore, i.e. the crest of the wave passes by a site, and the trough

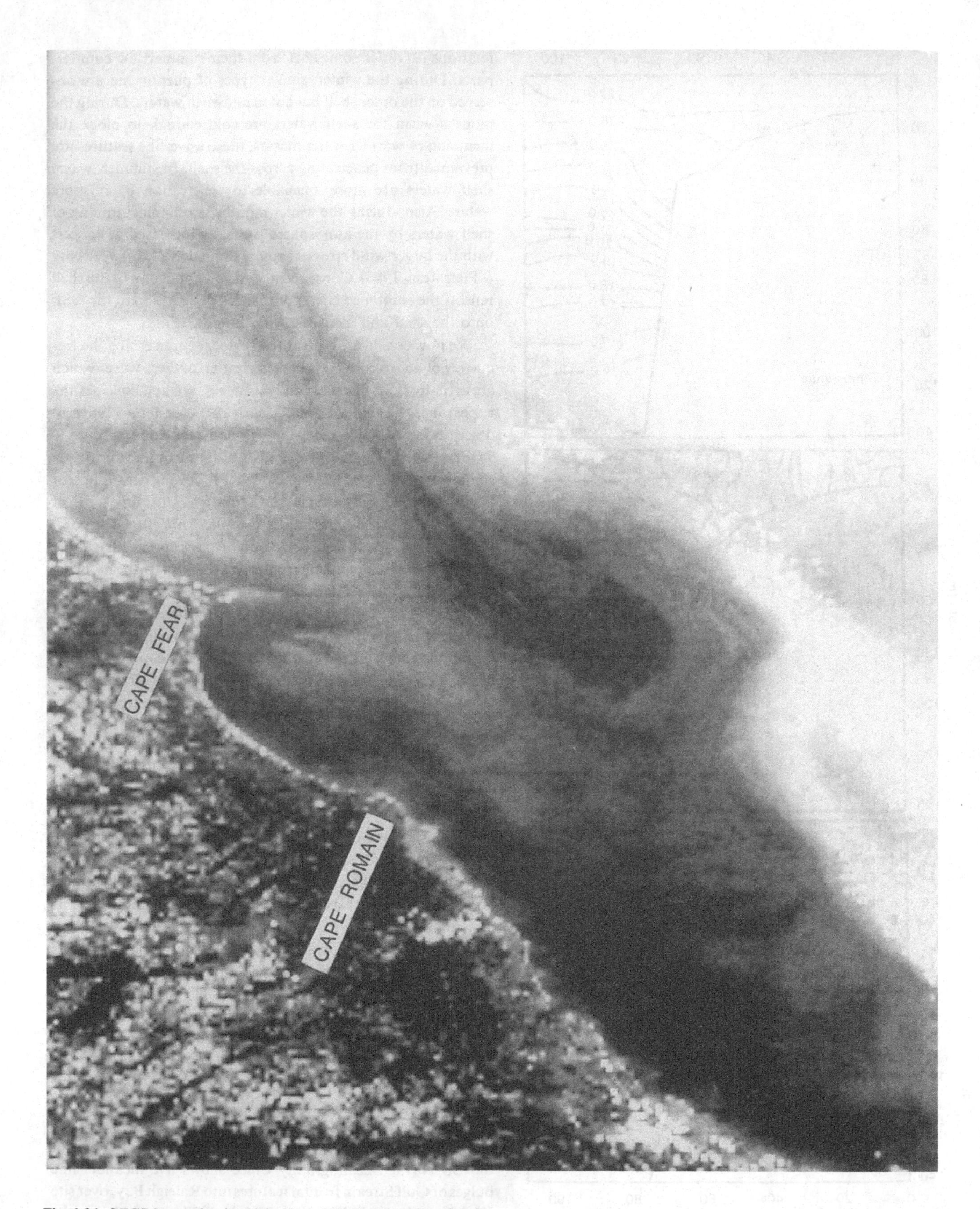

Fig. 1.24. CZCS image showing high chlorophyll concentrations found in the Charleston Trough (C. McClain, pers. comm.).

(a)

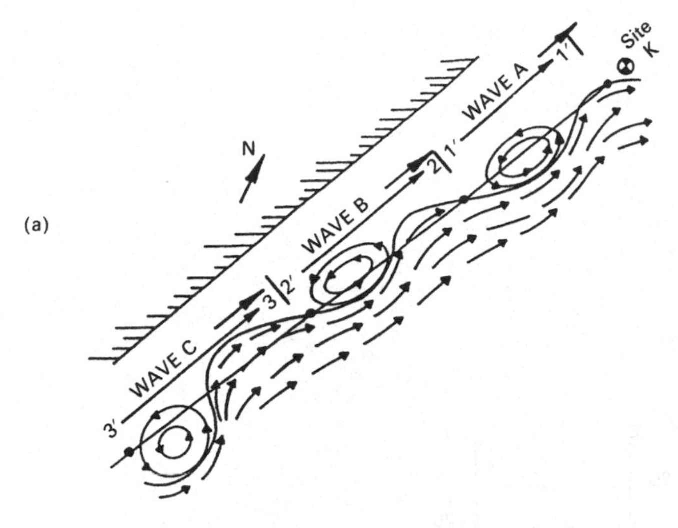

K-TOP 40 HRLP CURRENT VECTORS AND TEMPERATURE

(b)

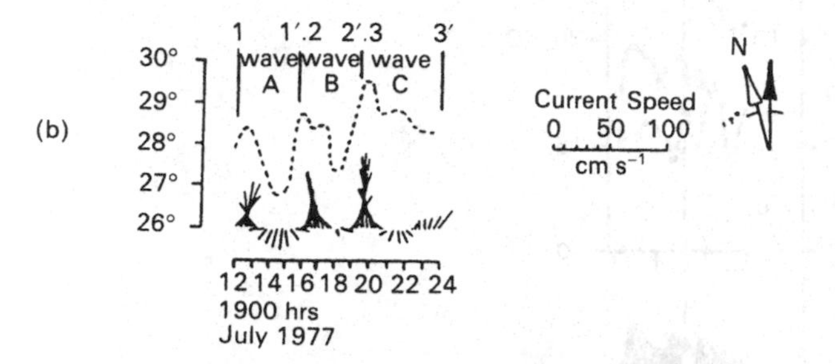

Fig. 1.25. Conceptual streamline flow pattern of three consecutive Gulf Stream frontal meanders moving by: (a) mooring Site K based on; (b) mooring horizontal current and temperature data.

of the wave appears, two upwelling processes occur. The offshore movement of the front obviously effects a one-sided divergence, which can be compensated for by a flow inward from the shoreward side or by an advection of water up from below. It is this latter convergence mechanism, vertical advection, i.e. upwelling, which seems to be principally responsible for resupplying water to the system. In addition, if the trough of the wave is rotating cyclonically, as the current meter data suggest, then upwelling occurs as a natural sequence of the positive vertical vorticity generated. Kim *et al.* (1980), Lee, Atkinson & Legeckis (1981) and Yoder (1981) have found this to be the condition off the coast of Georgia, as well.

Pietrafesa & Janowitz (1980) provide an intercomparison of parabathic, alongshore current components, which indicate a lag of phase (or propagation of event) towards the northeast in nearsurface (< 17 m) mid-shelf to outershelf waters during the summers of 1977 and 1978. Propagation speeds ranged from 21 to 69 cm/s, with wavelengths ranging from 111 to 528 km and periods between 2.8 and 8.0 d. These results encompass the pioneering results of Webster (1961a, b) in the original description of meanders. How about filaments?

GSF filaments are meanders that grow in amplitude in the horizontal plane at their crest and fold back around the cool pool sitting in the wave trough (Pietrafesa & Janowitz, 1980; Sun, 1982; Pietrafesa, 1983). Two examples of the circulation and temperature scenarios are presented in Figures 1.26 and 1.27. A filament was present in Onslow Bay on 23 April 1979 (fig. 1.26*a*). On the shoreward side of the tongue, at mooring sites O and P (Figs. 1.26*b*, *c*) flow is to the north, while at sites H and J (figs. 1.26*d*, *e*) flow is to the south. The warm Gulf Stream waters, which define the filament to a VHRR sensor, deepen due to the downward vertical velocity. To the east, i.e. offshore, of the filament and to the west, i.e. onshore of the GSF, is a colder water mass which is counterclockwise rotating and is, in fact, the trough of the wave. This same scenario is depicted in Figure 1.27 off the coast of northeastern Florida on April 24 1980.

The filament formation occurs in concert, either as cause or effect, and the wave trough realizes an intensified cyclonic torque and spins up more intensely. The net effect is a rapid vertical ascension of isopycnals, a 'doming up' phenomenon. As the filaments advects north(east)ward along the GSF, a series of shelf-break/slope cold pool eddies are spun up, forming an upwelled ridge, which, as a function of the intensity of cyclonic torquing, peaks with varing elevation (Fig. 1.28). It is these features, the Gulf Stream frontal filaments which represent the major upwelling mechanism and provide for the major source of nutrients, such as phosphate, as shown in Figure 1.29*b*.

As shown in Figure 1.3, a neat fairly compact set of relationships exists between PO_4, NO_4 and temperature (T) for newly upwelled GSF water. From the moored temperature data acquired at 25 mooring locations by 55 current meters and thermistor chains (in the array shown in Fig. 1.29*a*), we can solve for PO_4 in terms of T and create phosphate concentration maps throughout the array at a specific height in the water column. For example, in Figure 1.29*b* we see a 2-dimensional representation of phosphate on 24 April 1980, at a depth 3 m above the bottom from just north of Cape Canaveral to Jacksonville of the Gulf Stream frontal filament, cold core eddy event shown previously in Fig. 1.27. Water with phosphate concentrations of 0.5–1.5 μg-atom l^{-1}, are being upwelled at the shelfbreak by the presence and passage of a filament. These numbers are comparable to Smith's (1968) values of 1–2 μg-atom l^{-1} as a measure of the importance of upwelling. While these features are transients, they take 2–5 d to pass a site and a new one appears every 7–14 d over the entire year.

Does this phosphate make it onto the shelf? Yes, if the wind or buoyancy flux or bottom topography happen to be accommodating. The most likely candidate areas for high phosphate concentrations in the water column with the greatest potential for accumulation in the sediments occurs from Ft Pierce to New Smyrna, Florida, to the left and downstream of the Charleston Bump and on the northern side of the Carolina Capes.

Conclusions

Upwelling processes in boundary current regions include not only classical wind-driven mechanisms but also mechanisms due to buoyancy flux, topography and boundary current frontal instabilities. Bottom topography can affect wind-driven upwelling circulation, and, through the interaction of boundary currents with the topography, several mechanisms for upwelling can be created. All of these mechanisms have been found to be present in the South Atlantic Bight and all could contribute in varying degrees to the upwelling and ultimate deposition of phosphate-rich sediments on the continental margin. Although

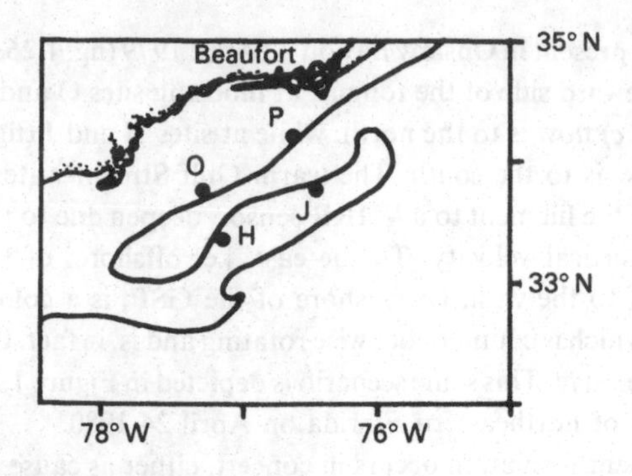

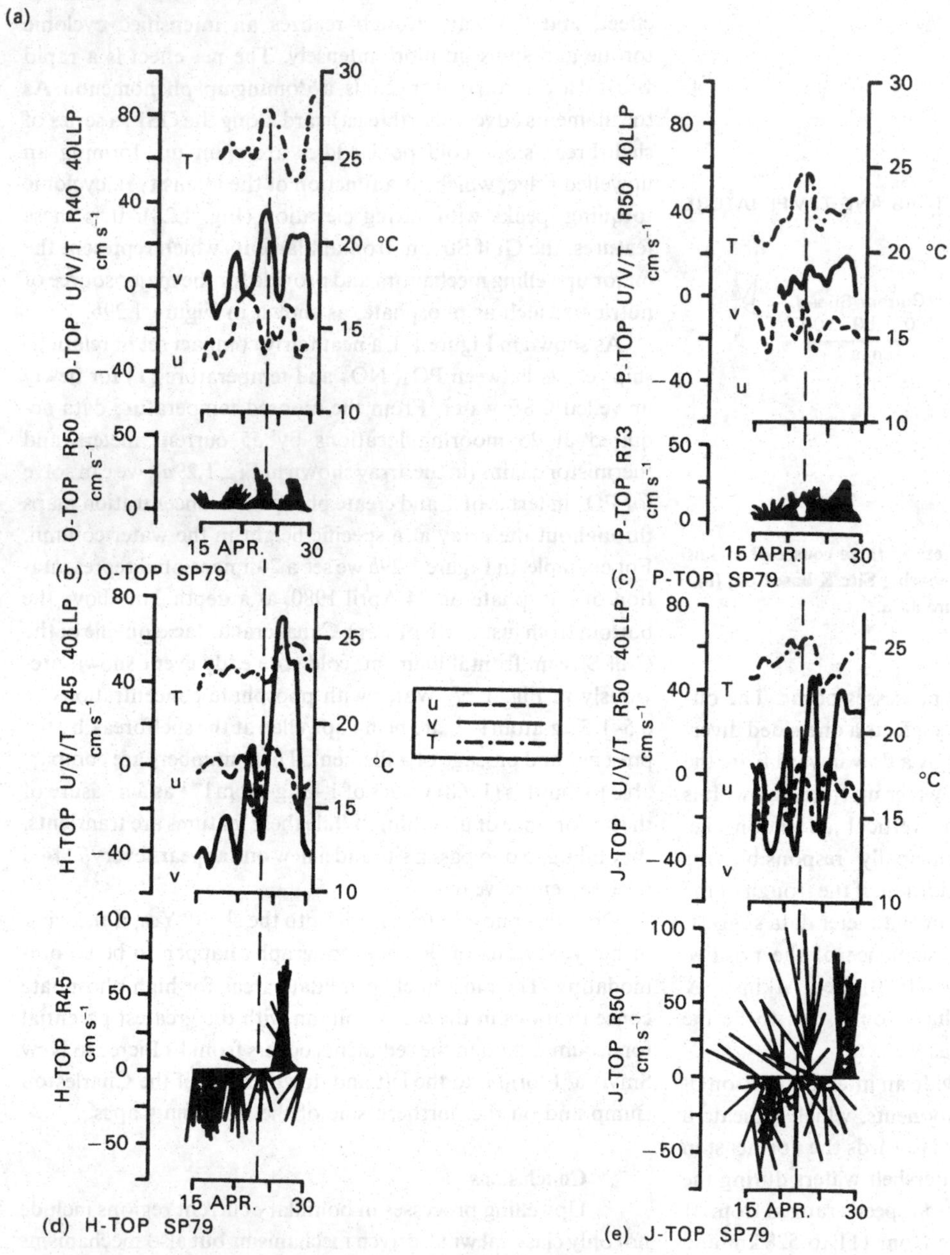

Fig. 1.26. Eulerian horizontal current and temperature signature during presence of frontal filament in Onslow Bay: (a) Gulf Stream frontal filament to Onslow Bay, 23 April, 1979. Courtesy Dr Steven Baig, (NOAA/AOML); (b) horizontal current components, temperature and current stick time series at Site 0 top during filament presence; (c) horizontal current components, temperature and current stick time series at Site P top during filament presence; (d) horizontal current components, temperature and current stick time series at Site H top during filament presence; (e) horizontal current components, temperature and current stick time series at Site J top during filament presence.

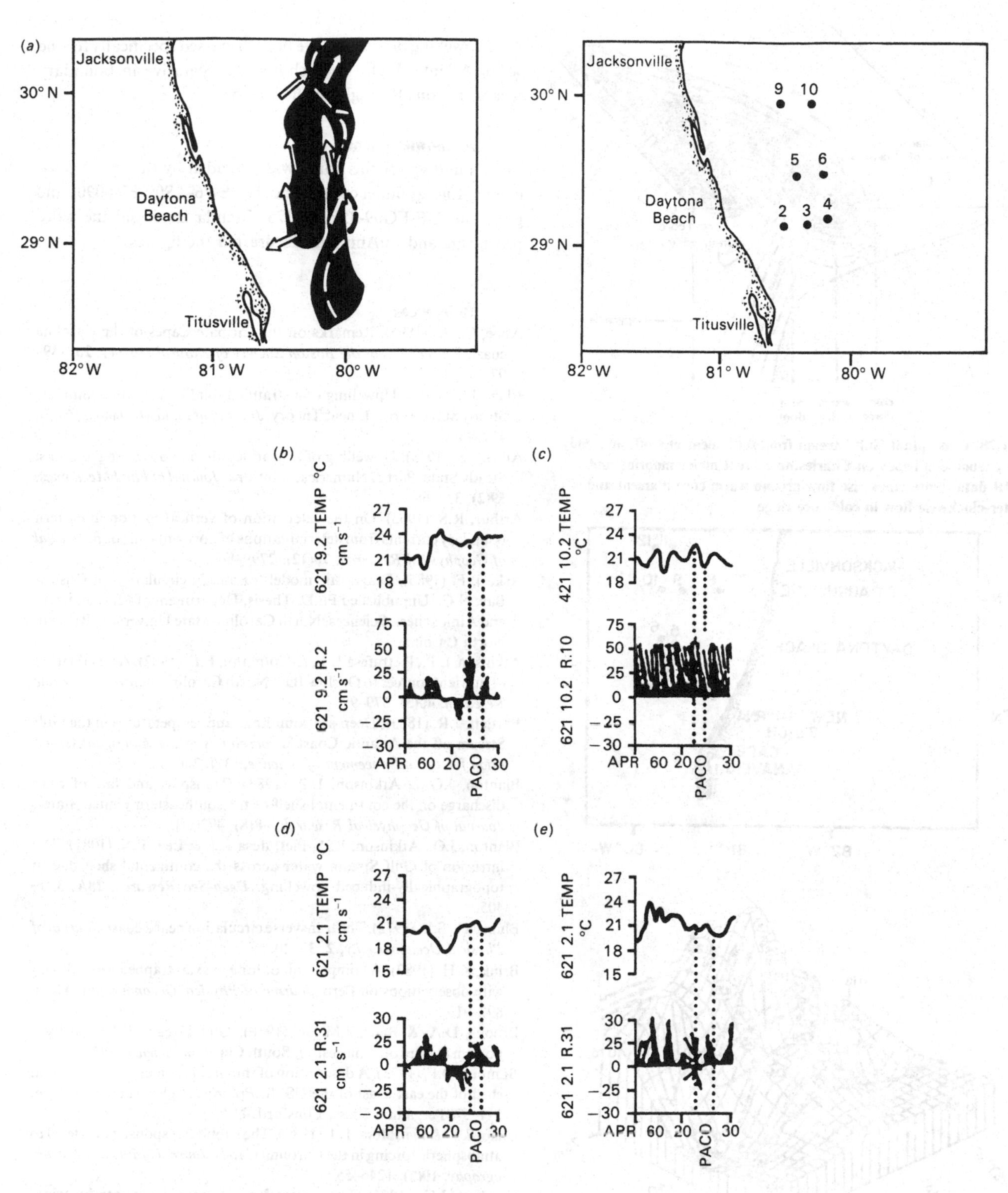

Fig. 1.27. (a) Airborne Radiation Thermometer (ART) image of filament 'Fred', 24 April, 1980. Current vectors from 17 m below surface at mooring sites in the event are shown for (b) Site 9; (c) Site 10; (d) Site 2; and (e) Site 3.

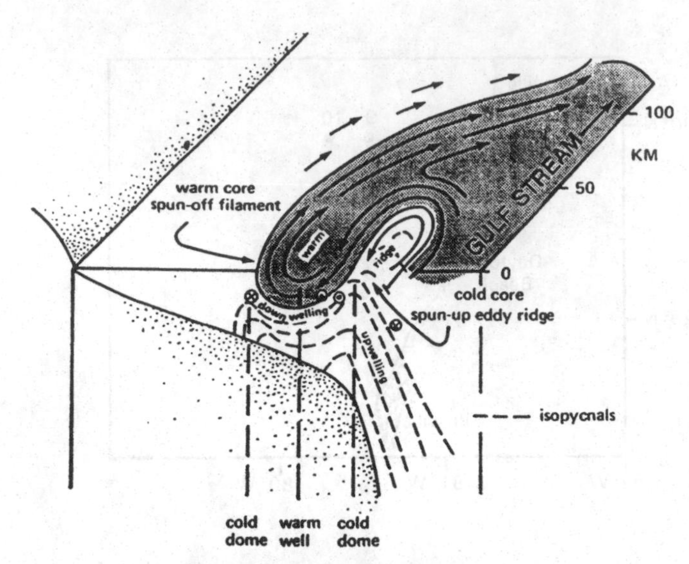

Fig. 1.28. Conceptual Gulf Stream frontal filament and offshore eddy ridge phenomena based on Charleston current meter mooring and VHRR data. Note clockwise flow around warm core filament and counter-clockwise flow in cold core ridge.

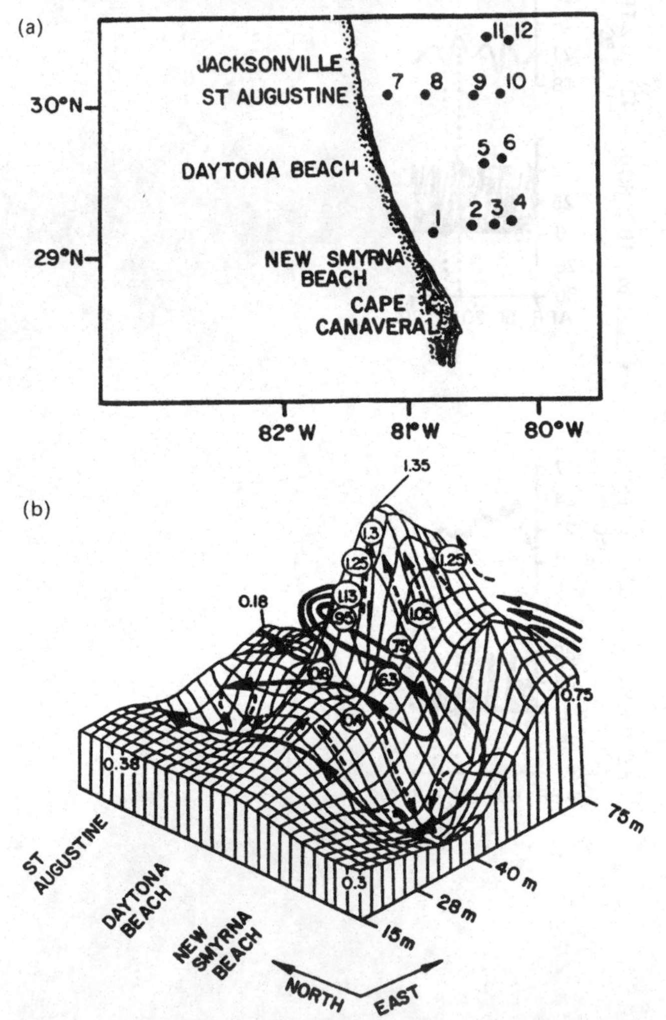

Fig. 1.29. (a) Station locations of fixed temperature sensors at a height of 3 m above the bottom over the region of the northeast Florida shelf during the Gabex I experiment February–May, 1980. (b) Phosphate concentrations (μg atoms l^{-1}) at 3 m above the bottom on 24 April 1980 across and along the instrumented shelf region shown in (a) above. Concentrations were calculated from the observed temperatures, using the relationships between temperature and phosphate shown in Figure 1.3. Arrows depict circulation pattern.

these upwelling processes have been discussed specifically for the South Atlantic Bight, they should be operative in boundary current regions throughout the world.

Acknowledgments

Funding for this study was provided by the US Department of Energy under contract no. DOE-76-AS09-EY00902 and grant no. DE-FG09-85ER60376. Brenda Batts did the word processing and LuAnn Salzillo drafted the figures.

References

Abbe, C., Jr. (1895). Remarks on the Cuspate Capes of the Carolina coast. *Proceedings of the Boston Society of Natural History*, **26**, 489–97.

Allen, J.S. (1972). Upwelling of a stratified fluid in a rotating annulus: Steady State. Part 1. Linear Theory. *Journal of Fluid Mechanics*, **56(3)**, 429–45.

Allen, J.S. (1973). Upwelling of a stratified fluid in a rotating annulus: Steady State. Part 2. Numerical solutions. *Journal of Fluid Mechanics*, **59(2)**, 337–68.

Arthur, R.S. (1965). On the calculation of vertical motion in eastern boundary currents from determinations of horizontal motion. *Journal of Geophysical Research*, **70(12)**, 2799–803.

Askari, F. (1985). Diagnostic model for steady circulation of Onslow Bay, N.C., Unpublished Ph.D. Thesis, Department of Marine, Earth and Atmospheric Sciences, North Carolina State University, Raleigh, North Carolina.

Atkinson, L.P., Pietrafesa L.J., & Hofmann, E.E. (1982). An evaluation of nutrient sources to Onslow Bay, North Carolina. *Journal of Marine Research*, **40(3)**, 679–99.

Bartlett, J.R. (1883). Deep-sea soundings and temperatures in the Gulf Stream off the Atlantic Coast. *Proceedings of the American Association for the Advancement of Science*, (31), 349–52.

Blanton, J.O. & Atkinson, L.P. (1983). Transport and fate of river discharge on the continental shelf of the southeastern United States. *Journal of Geophysical Research*, **88(8)**, 4730–8.

Blanton, J.O., Atkinson, L.P., Pietrafesa L.J. & Lee, T.N. (1981). The intrusion of Gulf Stream water across the continental shelf due to topographically-induced upwelling. *Deep-Sea Research*, **28A**, 393–405.

Blumsack, S.L. (1972). The transverse circulation near a coast. *Journal of Physical Oceanography*, **2**, 32–40.

Brink, K.H. (1982). A comparison of long coastal trapped wave theory with observations off Peru. *Journal of Physical Oceanography*, **12(8)**, 897–913.

Brooks, D.A. & Bane, J.M. Jr. (1978). Gulf Stream deflection by a bottom feature off Charleston, South Carolina. *Science*, **201**, 1225–6.

Bumpus, D.F. (1973). A description of the circulation on the continental shelf of the east coast of the US. In *Progress in Oceanography*, **6**, pp. 111–57. Pergamon Press, Elmsford, N.Y.

Chao, S.Y. & Pietrafesa, L.J. (1980). The subtidal response of sea level to atmospheric forcing in the Carolina Capes. *Journal of Physical Oceanography*, **10(8)**, 1246–55.

Charney, J.G. (1955). The generation of oceanic currents by wind. *Journal of Marine Research*, **14**, 477–98.

Ekman, V.W. (1905). *Arkiv. Matem. Astr. Fysik.*, Stockholm, **2(11)**, 1–52.

ERT (1979a). Summary and analysis of physical oceanographic and meteorological information on the continental shelf and Blake Plateau from Cape Hatteras to Cape Canaveral. *ERT* **2(1–5)**. Environmental Research and Technology, Inc. Prepared for Bureau of Land Management, Department of Interior, Washington, DC.

ERT (1979b). Summary and analysis of physical oceanographic and meteorological information on the Continental Shelf and Blake Plateau from Cape Hatteras to Cape Canaveral. *ERT* **3 (6, 7, 8)**. Environmental Research Technologies. Prepared for Bureau of Land Management, Department of Interior, Washington, DC.

Fairbridge, R.W. (1966). *The Encyclopedia of Oceanography*. Van Nostrand Reinhold, Princeton, New Jersey.

Fofonoff, N.P. (1962). Dynamics of ocean currents. In *The Sea*, I, pp. 325–80. Interscience, New York.

Garvine, R.W. (1971). A simple model of coastal upwelling dynamics. *Journal of Physical Oceanography*, **1(4)**, 169–79.

Garvine, R.W. (1972). The effect of bathymetry on the coastal upwelling of homogeneous water. *Journal of Physical Oceanography*, **3(1)**, 47–56.

Green, C.K. (1944). Summer upwelling-northeast coast of Florida. *Science*, **100 (New Series Vol. C)**, 546–7.

Hanson, R.B., Tenore K.R., Bishop, S., Chamberlain, C., Pamatmat, M.M. & Tiejen, J. (1981). Benthic enrichment in the Georgia Bight related to Gulf Stream intrusions and estuarine outwelling. *Journal of Marine Research*, **39(3)**, 417–41.

Hart, T.J. & Currie, R.I. (1960). *Discovery*, **XXI**, 123–298.

Hidaka, K. (1954). A contribution to the theory of upwelling and coastal currents. *Transactions American Geophysical Union*, **35**, 431–44.

Hofmann, E.E., Pietrafesa, L.J. & Atkinson, L.P. (1981). A bottom water intrusion in Onslow Bay, North Carolina. *Deep Sea Research*, **28A**, 329–45.

Hofmann, E.E., Pietrafesa, L.J., Klinck, J.M. & Atkinson, L.P. (1980). A time dependent model of nutrient distribution in continental shelf waters. *Journal of Ecological Modelling*, **10(3)**, 193–214.

Hurlburt, H.E. & Thompson, J.D. (1973). Coastal upwelling on a β plane. *Journal of Physical Oceanography*, **3(1)**, 16–32.

Hsueh, Y. & Kenney, R.N. (1972). Steady coastal upwelling in a continuously stratified ocean. *Journal of Physical Oceanography*, **2(1)**, 27–33.

Hsueh, Y. & O'Brien, J.J. (1971). Steady coastal upwelling induced by an along-shore current. *Journal of Physical Oceanography*, **1(4)**, 180–6.

Janowitz, G.S., & Pietrafesa, L.J. (1980). A model and observations of time dependent upwelling over the mid-shelf and slope. *Journal of Physical Oceanography*, **10(10)**, 1574–83.

Janowitz, G.S. & Pietrafesa, L.J. (1982). The effects of alongshore variation in bottom topography on a boundary current – (topographically induced upwelling). *Continental Shelf Research*, **1(2)**, 123–41.

Kim, H.H., McClain, C.R., Blaine, L.R., Hart, W.D., Atkinson, L.P. & Yoder, J.A. (1980). Ocean chlorophyll studies from a U-2 aircraft platform. *Journal of Geophysical Research*, **85(7)**, 3982–90.

Laurs, R.M. (1967). Ph.D. thesis, (no title available) Oregon State University, Corvallis.

Lee, T.N., Atkinson, L.P. & Legeckis, R. (1981). Observations of a Gulf Stream frontal eddy on the Georgia Continental Shelf, April 1977. *Deep-Sea Research*, **28A**, 347–78.

Leetma, A. (1969). 'On the theory of coastal upwelling'. Unpublished Ph.D. Thesis. Department of Meteorology, Massachusetts Institute of Technology, Cambridge, Massachusetts.

Leetma, A. (1971). Some effects of a stratification on rotating fluids. *Journal of Atmospheric Science*, **28**, 65–71.

Legeckis, R. (1979). Satellite observations of the influence of bottom topography on the seaward deflection of the Gulf Stream off Charleston, South Carolina. *Journal Physical Oceanography*, **9(3)**, 483–97.

Leming, T.D. & Mooers, C.N.K. (1981). Cold water intrusions and upwelling near Cape Canaveral, Florida. Coastal upwelling. *Coastal and Estuarine Science*, **1**, 63–71.

Lineykin, P.S. (1955). On the determination of the thickness of the baroclinic layer in the ocean. *Koklady Academii Naak USSR*, **101**, 461–4.

McClain, C.R., Pietrafesa, L.J. & Yoder, J.A. (1984). Observations of Gulf Stream-induced and wind-driven in the Georgia Bight using ocean color and infrared imagery. *Journal of Geophysical Research*, **89(c3)**, 3705–23.

McNider, R.T. & O'Brien, J.J. (1973). A multi-layer transient model of coastal upwelling. *Journal of Physical Oceanography*, **3(3)**, 258–73.

Mooers, C.N.K., Collins, C.A. & Smith, R.L. (1976). The dynamic structure of the frontal zone in the coastal upwelling region off Oregon. *Journal of Physical Oceanography*, **6(1)**, 3–21.

O'Brien, J.J. & Hurlburt, H.E. (1972). A numerical model of coastal upwelling. *Journal of Physical Oceanography*, **2(1)**, 14–26.

Pashinski, D.J. & Maul, G.A. (1973). Use of ocean temperature while coasting between the straits of Florida and Cape Hatteras. *Marine Weather Log*, **17(1)**, 1–3.

Pietrafesa, L.J. (1973). 'Steady baroclinic circulation on a continental shelf.' Unpublished Ph.D. thesis, University of Washington, Seattle.

Pietrafesa, L.J. (1977). Winds, sea level, currents and hydrography on the North Carolina Continental Shelf summer–fall, 1975 and 1976. *Transactions American Geophysical Union*, **58(12)**, 1173.

Pietrafesa, L.J. (1983). Survey of a Gulf Stream frontal filament. *Geophysical Research Letters*, **10(3)**, 203–6.

Pietrafesa, L.J., Blanton, J.O. & Atkinson, L.P. (1978). Evidence for deflection of the Gulf Stream at the Charleston Rise. *Gulfstream*, **IV(9)**, 3–7.

Pietrafesa, L.J., & Janowitz, G.S. (1979). On the efforts of buoyancy flux on continental shelf circulation. *Journal of Physical Oceanography*, **9(5)**, 911–18.

Pietrafesa, L.J. & Janowitz, G.S. (1980). On the dynamics of the Gulf Stream front in the Carolina Capes. *Proceedings of the Second International Symposium on Stratified Flows*, 24–27 June 1980, Trondheim, Norway, pp. 184–97. Tapin, Norway.

Pietrafesa, L.J., Janowitz, G.S. & Wittman, P.A. (1985). Physical oceanographic processes in the Carolina Capes. Oceanography of the Southeastern US Continental Shelf. *Coastal and Estuarine Sciences*, **2**, 23–32.

Posner, G.S. (1957). No title available. *Bulletin of Bingham Oceanography College*, **16**, 106–55.

Pratt, R.M. (1963). Bottom currents on Blake Plateau. *Deep-Sea Research*, **10(3)**, 245–9.

Premuzic, E.T. (1980). *Organic Carbon and Nitrogen in the Surface Sediments of World Oceans and Seas: Distribution and Relationship to Bottom Topography*. Brookhaven National Laboratory, under contract with the department of Energy, New York.

Purba, M. (1984). 'Parametic evaluation of physical oceanographic observations in the Georgia Bight.' Ph.D. thesis, Department of Marine, Earth and Atmospheric Sciences, North Carolina State University, Raleigh, North Carolina.

Richards, F.A. (1960). Some chemical and hydrographic observations along the north coast of South America–I. Cabo Tres Puntas to Curacao, including the Cariaco Trench and the Gulf of Cariaco. *Deep Sea Research*, **7**, 163–82.

Richards, F.A. (ed.) (1981). *Coastal Upwelling. Coastal and Estuarine Sciences 1*. American Geophysical Union, Washington, DC.

Rooney, D.M., Janowitz, G.S. & Pietrafesa, L.J. (1978). A simple model of deflection of the Gulf Stream by the Charleston Rise. *Gulfstream*, **IV(11)**, 3–7.

Royal Society (1963). *International Indian Ocean Expediton R.R.S. Discovery Cruise 1*. Cruise Report, London, 24 pp.

Ryther, J.H. (1969). Photosynthesis and fish production in the sea. *Science*, **166**, 72–6.

Saunders, P.M. (1977). Wind stress on the ocean over the eastern continental shelf of North America. *Journal of Physical Oceanography*, **7**, 555–66.

Smith, R.L. (1968). Upwelling. *Oceanographic Marine Biological Annual Review*, **6**, 11–46.

Stefansson, U. & Atkinson, L.P. (1971). Nutrient–density relationships in the western North Atlantic between Cape Lookout and Bermuda. *Limnology and Oceanography*, **16**, 51–9.

Stommel, H. & Leetma, A. (1972). Circulation on the continental shelf. *Proceedings of the National Academy of Sciences, USA*, **69(11)**, 3380–4.

Sun, L.C. (1982). 'On the dynamic variability of the Gulf Stream front.' Ph.D. thesis, North Carolina State University, Raleigh, NC.

Sverdrup, H.U. (1938). On the process of upwelling. *Journal of Marine Research*. **1**, 155–64.

Taylor, C.B. & Stewart, H.B. (1959). Summer upwelling along the east coast of Florida, *Journal of Geophysical Research*, **64**, 33–40.

Thompson, J.D. & O'Brien, J.J. (1973). Time-dependent coastal upwelling. *Journal of Physical Oceanography*. **3(1)**, 33–46.

Tomczak, M., Jr. (1970). Eine linear theorie des stationaren auftriebs im stetig geschichteten meer. *Deutsche Hydrographische Zeitschrift*, **5**, 214–34.

Webster, F. (1961a). A description of Gulf Stream meanders off Onslow

Bay. *Deep-Sea Research*, **8**, 130–43.
Webster, F. (1961b). The effects of meanders on the kinetic energy balance of the Gulf Stream. *Tellus*, **13**, 392–401.
Weisberg, R.H., & Pietrafesa, L.J. (1983). Kinematics and correlation of the surface wind field in the South Atlantic Bight. *Journal of Geophysical Research*, **88(8)**, 4593–610.
Wells, H.W. & Gray, I.E. (1960. Summer upwelling off the northeast coast of North Carolina. *Limnology Oceanography*, **5**, 108–9.
Yoder, J.A. (1981). Environmental control of phytoplankton production on the southeastern US continental shelf. *Oceanography of the Southeastern US Continental Shelf, Coastal and Estuarine Sciences*, vol. 2, pp. 93–103. American Geophysical Union, Washington, DC.
Yoshida, K. (1967). Circulation in the eastern tropical oceans with special references to upwelling and undercurrents. *Japanese Journal of Geophysics*, **4(2)**, 1–75.
Zhang, Q.H., Janowitz, G.S. & Pietrafesa, L.J. (1987). The interaction of estuarine and shelf waters: a model and applications. *Journal of Physical Oceanography*, **17(4)**, 455–69.

2
Diagenesis of phosphorus in sediments from non-upwelling areas

R.A. BERNER

Abstract

Most organic matter burial in marine sediments occurs in non-upwelling areas and it is here where most phosphorous diagenesis also takes place. Interstitial waters of these sediments characteristically indicate increases in dissolved phosphate with depth due predominently to the release of P accompanying the decomposition of organic matter but also to the release of adsorbed P from the surfaces of hydrous ferric oxides accompanying their reductive dissolution. Precipitation of interstitial dissolved P upon further burial is witnessed by concentration profile reversals (decrease in concentration with depth) or, in the case of fine-grained carbonate sediments, by low dissolved concentrations at all depths. Precipitation may result in the formation of carbonate fluorapatite (evidenced by decreases in both dissolved fluoride and phosphate with depth) or vivianite (evidenced, in very highly organic sediments below the zone of sulfate reduction, by a buildup of dissolved $Fe2^{+}$ accompanying reversal in the dissolved P profile.)

Mathematical models can be fitted to interstital water and other sediment chemical data to derive rate constants for authigenic phosphate precipitation and P release accompanying organic matter decomposition. Such models also can be used to deduce the C:P ratio of organic matter undergoing decomposition during early diagenesis. Although some modeling of this sort has been done, much more is needed for sediments of both upwelling and non-upwelling areas.

Introduction

Phosphorus is an important and often limiting nutrient in natural aquatic systems. Thus, it exerts a major control on primary productivity and, consequently, plays an important role in the biogeochemical carbon and other cycles. In the marine environment most phosphorus secreted by planktonic organisms is regenerated to solution within the water column (e.g. Broecker & Peng, 1982). Nevertheless, some is buried with its host organic matter and it is this burial which constitutes a major control on the overall level of phosphorus in the ocean. Froelich *et al.* (1982) have shown that burial of phosphorus associated with organic matter is quantitatively one of the two most important processes of phosphorus removal in the present oceans. (Association with calcareous skeletal debris is the other). Thus, it is instructive to examine the factors that bring about organic-P burial and the effects of various diagenetic processes on the release of phosphate to pore solution and its precipitation as authigenic minerals.

Table 2.1. *Present-day rates of organic carbon burial in marine sediments* (After Berner, 1982)

Sediment type	Organic carbon burial rate[a] (10^{12}g y^{-1})
Deltaic–shelf–estuarine sediments	104
Biogenous sediments of high productivity regions (upwelling, etc.)	11
Shallow water carbonate sediments	6
Remaining pelagic sediments	5
Total	126

[a]Values are corrected for diagenetic loss in the top few meters.

Most work on phosphorus diagenesis has been devoted to organic-rich sediments of classic upwelling regions (e.g. Manheim, Rowe & Jipa, 1975; Burnett, 1977; d'Anglejan, 1979; Kolodny, 1980; Baturin, 1982; Calvert & Price, 1983; Froelich *et al.*, 1983 Jahnke *et al.*, 1983) because of the high concentrations of phosphorite found there. However, there are many other modern sediment areas which exhibit notable phosphorus diagenesis. Most organic matter deposition in the oceans occurs, not in classic upwelling regions (or in 'organic-rich' anoxic basins such as the Black Sea), but in rather ordinary sediments of moderate organic matter content, which are deposited on shelves, in deltas and estuaries, and in nearshore basins (see Table 2.1). It is here that most of the world's sediment is deposited each year along with most of the world's organic matter. Thus, if we are interested in how phosphorus is turned over during organic matter decomposition, it is imortant to look at the early diagenesis of P in the sediments of non-upwelling areas. This is the subject of the present paper.

Interstitial dissolved phosphate as a diagenetic indicator

Figures 2.1, 2.2 and 2.3, illustrate some representative dissolved phosphate profiles for organic-rich sediments of non-upwelling areas. (Additional similar data can be found in Rittenberg, Emery & Orr, 1955; Nissenbaum, Presley & Kaplan, 1972; Hartmann *et al.*, 1973 and 1976; Sholkovitz, 1973; Price,

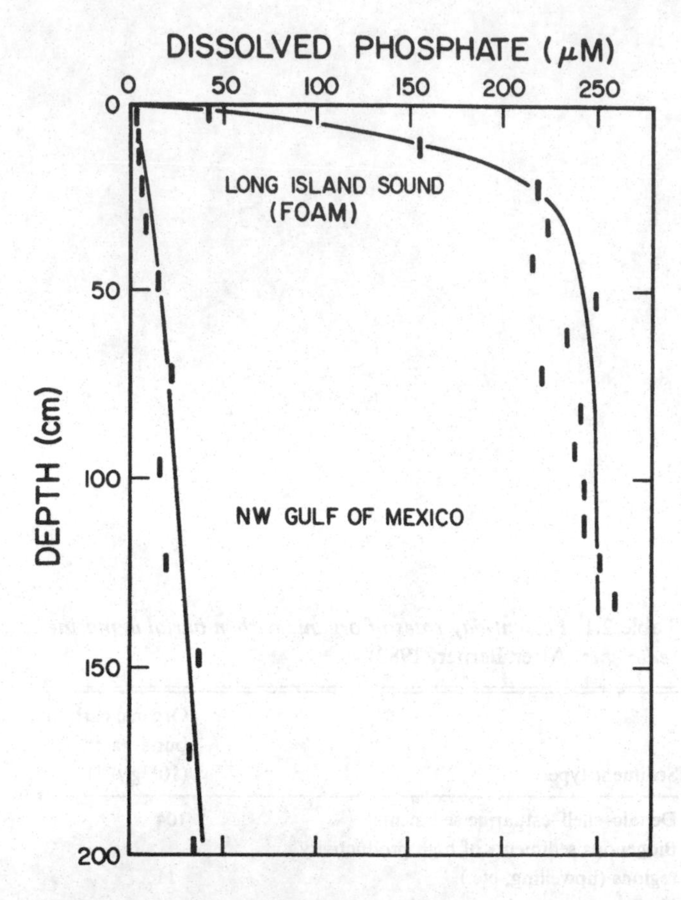

Fig. 2.1. **Plots of interstitial dissolved (molybdate reactive) phosphate versus sediment depth for the FOAM site of Long Island Sound (data from Aller, 1977) and for a shelf sediment from the northwestern Gulf of Mexico (data from Filipek & Owen, 1981).**

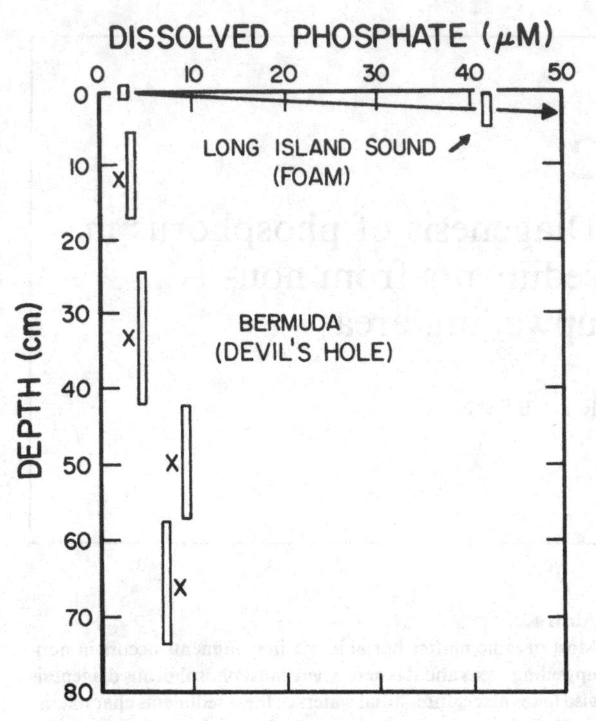

Fig. 2.2. Plot of interstitial dissolved phosphate versus sediment depth for the organic-rich $CaCO_3$ sediments of Devil's Hole, Bermuda. Points marked by an x represent concentrations calculated for each depth interval for phosphorite saturation at measured values of pH and Ca^{2+} and F^- concentrations (data from Berner, 1974). For comparison, data for the FOAM site, a similar organic-rich but siliciclastic sediment, is reproduced from Figure 2.1.

1976; Murray, Grundmanis & Smethie, 1978; and Elderfield *et al.*, 1981.) All profiles exhibit an initial increase in dissolved phosphate downward from the sediment–water interface. At greater depths, however, there is distinctly different behavior. Some sediments, such as the FOAM site of Long Island Sound, NY (Fig. 2.1), show a continued but decelerating rise of dissolved phosphate, with an approach to a limiting asymptotic value. Other sediments from the northwest Gulf of Mexico (Filipek & Owen, 1981) show a constant increase to the bottom of the core (Fig. 2.1). The sediment from Florida Bay (Fig. 2.2) shows constant low phosphate concentrations (but still distinctly higher than that in the overlying seawater) at all depths even though the sediment contains abundant decomposing organic matter. Finally, the Sachem sediment from Long Island Sound (Fig. 2.3) shows a reversal at depth with positive and negative gradients separated by a subsurface concentration maximum.

How can one explain these profiles in terms of diagenetic processes? Rise in dissolved phosphate with depth is usually ascribed to the bacterial breakdown of organic matter with liberation of phosphorus to solution. By this model, continued rise with eventual leveling off, as shown in Figure 2.1, is due presumably to exhaustion of the source of the phosphorus. In other words, successive utilization of less reactive organic compounds with depth (e.g. Westrich & Berner, 1984) results in decreasing organic decomposition rates and, thus, successively slower rates of release of phosphate to pore solution. However, the near-surface, initial deceleration of phosphate release with depth in many sediments, can be explained in a completely different way. For example, the rapid rise in concentration in the uppermost portions of sediments from Long Island Sound (e.g. the top 15 cm for the FOAM site in Fig. 2.1) has been ascribed to the release of adsorbed phosphate from hydrous ferric oxide minerals during reduction of the minerals, by H_2S or by various organic compounds, to dissolved Fe^{2+} (Krom & Berner, 1981). Data backing this idea for another Long Island Sound locality (NWC) are shown in Table 2.2. This process appears to be important in other marine sediments (Filipek & Owen, 1981; Hines & Lyons, 1984) as well as in the sediments of lakes (Mackereth, 1966; Emerson & Widmer, 1978). Thus, the initially rapid rise in the concentration of dissolved phosphate with depth, in many sediments may be due to iron reduction, (or in rarer cases the dissolution of phosphatic biogenic debris – see Suess, 1981) and not only to organic matter decomposition.

Once below the near-surface zone of iron reduction, continued increases in dissolved phosphate concentration must be due to liberation of phosphate from decomposing organic matter. However, apparent deceleration in the rate of liberation may be due, not only to depletion of the more reactive organic compounds, as mentioned above, but also to the precipitation of

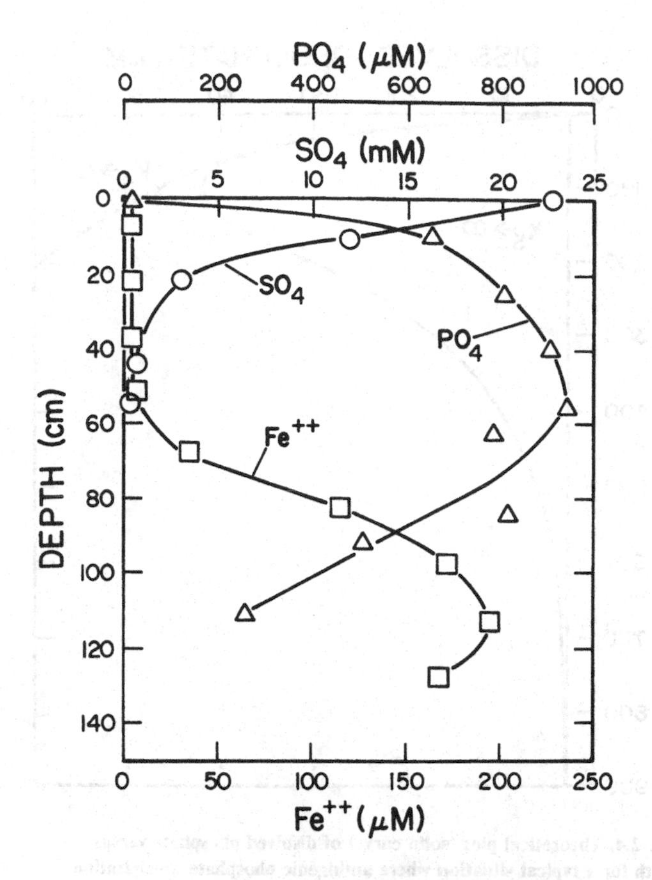

Fig. 2.3. Pore-water chemical data versus sediment depth, for the Sachem site of Long Island Sound (data from Martens, Berner & Rosenfeld, 1978).

authigenic phosphate minerals. Combination of a removal process, such as precipitation, with the continued release from organic matter could bring about a progressive flattening of the curve of dissolved phosphate versus depth. In fact, continuation of this process until the rate of precipitation exceeds that of release from organic matter can result in reversals in gradient.

Precipitation of dissolved phosphate to form authigenic phases is the most likely explanation for observed gradient reversals and concentration decreases with depth. Such decreases have been used, along with decreases in dissolved fluoride, to document the precipitation of phosphorite (authigenic carbonate fluorapatite) in upwelling regions (e.g. Froelich *et al.*, 1983; Jahnke *et al.*, 1983). Similar documentation in non-upwelling regions is relatively lacking but, in the writer's opinion, phosphorite precipitation is probably more widespread than is presently recognized. Interstitial waters in non-carbonate, organic-rich sediments commonly attain high degrees of supersaturation with respect to fluorapatite (e.g. Berner, 1974; Jahnke, 1981) and, thus, phosphorite would be expected to form at many localities. To test this idea, what is needed, among other things, are more analyses of pore waters for dissolved phosphate, and especially dissolved fluoride, as well as a method of detecting the presence of low concentrations of authigenic apatite in sediments.

Documentation of *probable* apatite precipitation has been shown for shallow water, fine-grained carbonate sediments. Levels of dissolved phosphate are very low in such sediments (e.g. Berner, 1974; Morse *et al.*, 1985) and behave as if they were

Table 2.2. *Sediment analyses from the NWC site of Long Island Sound (after Krom & Berner, 1981)*

Depth(cm)	Organic C%	Organic P%[a]	Inorganic P%	H_2S-extractable P%
0–1 (oxic)	2.5	0.043	0.043	0.007
30–40 (anoxic)	2.0	0.026	0.036	—

[a]Note that the drop in inorganic phosphorus with depth is equivalent to the amount at the surface that is extractable by treatment with H_2S. This indicates the presence of appreciable P adsorbed on reducible hydrous ferric oxides.

in equilibrium with a calcium phosphate mineral (see Fig. 2.2). In the absence of precipitation, judging from the degree of organic matter decomposition, one would expect to find much higher concentrations of dissolved phosphate in these sediments, but this is not the case. (In Fig. 2.2 data from the FOAM site are shown for comparison.) Apparently, fine-grained $CaCO_3$ acts as a preferential nucleus for apatite crystallization, as has been demonstrated by a number of laboratory experiments (e.g. Stumm & Leckie, 1970). In fact, a relative lack of fine-grained carbonate in clay-rich, terrigenous sediments is one factor which might help explain why high levels of supersaturation, with respect to phosphorite, are attained in such sediments and why widespread phosphorite formation hasn't been more commonly observed. However, the converse of this cannot be assumed. There is good evidence for phosphorite formation in organic-rich Mexican continental margin sediments which contain *no* $CaCO_3$ (Jahnke *et al.*, 1983). Thus, although $CaCO_3$ may promote phosphorite formation, there must be other, yet to be discovered geochemical factors which are equally important.

Decreases in dissolved phosphate with depth are not all accountable in terms of the precipitation only of phosphorite. In unusually organic-rich sediments, the phosphate may precipitate instead as vivianite, $Fe_3(PO_4)_2\ 8H_2O$. An example, taken from the Sachem site of Long Island Sound is shown in Figure 2.3. In the marine environment, vivianite formation comes about only as a result of a rather unusual set of circumstances. If sedimentation of abundant organic matter is sufficiently rapid, as is the case for the Sachem site, enough highly reactive organic compounds are buried to anoxic depths so that complete removal of interstitial dissolved sulfate takes place via bacterial sulfate reduction. Once this occurs the source of H_2S (in other words, sulfate), is no longer available and H_2S is consequently consumed by reactions with detrital iron minerals. This allows the build-up of dissolved Fe^{2+}, which otherwise would be precipitated by H_2S to form insoluble sulfides. The Fe^{2+}, then, continues to rise until the solubility of vivianite is exceeded and the precipitation of both Fe^{2+} and dissolved phosphate occurs. This helps explain why phosphate precipitation and dissolved Fe^{2+} build up at the Sachem site takes place only below the depth where sulfate disappears (see Fig. 2.3). It also helps to explain why vivianite formation is so common in organic-rich freshwater lake sediments (e.g. Emerson & Widmer, 1978) where, because of very low initial sulfate concentrations, complete reduction and removal of sulfate near the sediment surface is common.

Theoretical modeling of phosphorus diagenesis

The explanation of pore water profiles of dissolved phosphate can also be done in a quantitative manner. Using standard theoretical methods (e.g. see Berner, 1980) one can derive a set of equations, which takes into consideration all of the processes discussed above as well as several additional ones involved in diagenesis in general. Appropriate steady state equations quantifying the effects of organic matter decomposition, authigenic mineral precipitaton, particle bioturbation, ionic diffusion, adsorption on solid surfaces, and sediment burial on the concentrations of interstitial dissolved phosphate, organic matter, and authigenic mineral P, are:

$$\frac{D_s}{1+K_P}\frac{\partial^2 C}{\partial x^2}-\omega\frac{\partial C}{\partial x}+\frac{k a_p G}{1+K_P}-\frac{k_s(C-C_s)^n}{1+K_P}=0 \quad (2.1)$$

$$D_B\frac{\partial^2 G}{\partial x^2}-\omega\frac{\partial G}{\partial x}-kG=0 \quad (2.2)$$

$$-\omega\frac{\partial A}{\partial x}+k_s(C-C_s)^n=0 \quad (2.3)$$

where C is the concentration of dissolved phosphate, G is the concentration of decomposable organic matter, A is the concentration of (solid) authigenic mineral P, x is the depth below the sediment–water interface, ω is the rate of burial due to sedimentation, k_s is rate constant for authigenic mineral precipitation, C_s is the concentration of dissolved phosphate at solubility equilibrium with the authigenic phase, n is the positive integer derived from precipitation rate experiments, k is the first-order rate constant for organic matter decomposition, a_p is the P:C ratio of decomposing organic matter, K_P is the adsorption equilibrium constant for phosphate, D_s is the ionic diffusion coefficient of phosphate in pore water, and D_B is the bioturbation mixing coefficient for sediment particles. (In deriving equations (2.1–2.3) it was assumed that: there is negligible compaction, pore water flow, or biological irrigation; only one reactivity class of organic compounds is undergoing decomposition over the depth range of interest; and liberation of adsorbed phosphate via the reduction of ferric oxide minerals occurs so soon after burial that it can be considered to be above the zone of applicability of the equations.)

These equations can be solved for the appropriate boundary conditions to yield rather complicated expressions relating each concentration to depth in the sediment. This will not be done here. However, to illustrate the types of solution, the results for C as a function of depth, for the situation where $D_B=0$ (no bioturbation) and $n=1$ (first-order preipitation kinetics), as derived by Berner (1980), is shown in Figure 2.4. Note that a reversal in slope and a dissolved phosphate maximum occur as a natural consequence of the model. Also shown by the dashed lines is the result when authigenic precipitation is very rapid ($k_s\rightarrow\infty$) and when it is very slow ($k_s\rightarrow 0$). These two end-member cases, along with the general case (here where k_s is assumed equal to $10k$), mimic the types of dissolved phosphate versus depth profiles actually found in sediments (Figs. 2.1–2.3). The situation of very rapid precipitation (constant low C with depth) is characteristic of fine-grained carbonate sediments and that of continuous increase with leveling off is common in clay-rich terrigenous sediments.

By fitting analytical solutions to the above equations to actual measured profiles, (and independently determining values of the parameters D_s, D_B, K_p, ω, a_p, C_s, and n) one can actually calculate values of k_s and k. In this way diagenetic modeling enables the determination of critical rate parameters which can then be used to predict the diagenetic behavior of phosphorus at other localities. More modeling of this sort, along with determinations of D_s, K_p, etc., is sorely needed, not only for sediments of deltas and estuaries, etc., but also for those from the classic upwelling areas. (Some work on modeling of upwelling areas has been completed recently by Van Cappellen & Berner, 1988).

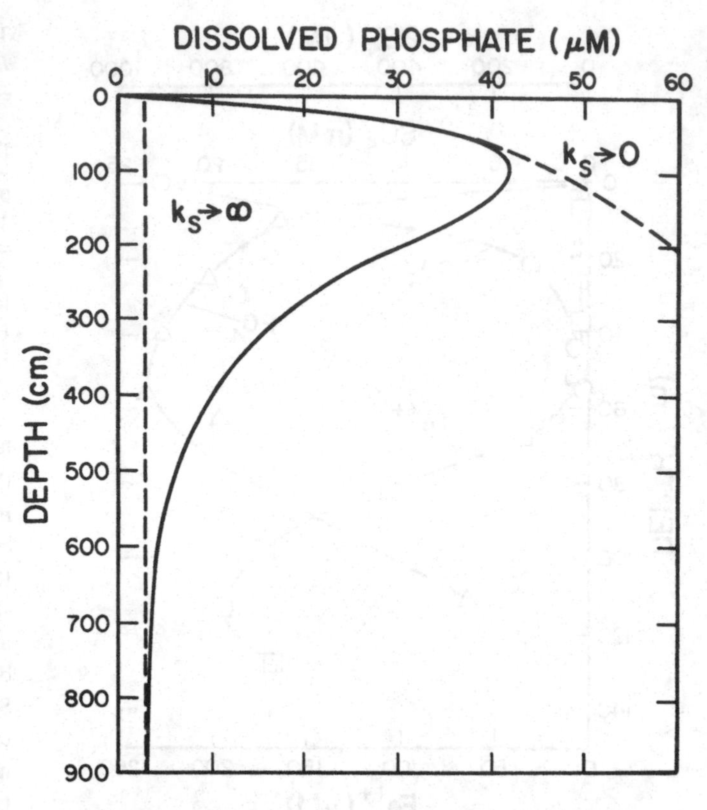

Fig. 2.4. Theoretical plot (solid curve) of disolved phosphate versus depth for a typical situation where authigenic phosphate precipitation takes place at depth. Values assumed: $D_s=100\,cm^2y^{-1}$; $\omega=0.1\,cm\,y^{-1}$; $k=1.2\times(10^3y)^{-1}$; $k_s=10k$; $C_o=1\,\mu M$; $C_{eq}=3\,\mu M$; $K_p=3$; $a_pG_o=1000\,\mu M$ (dashed curves for $k_s=0$ and $k_s>>10k(\rightarrow\infty)$).

P:C ratios in sediments

The average atomic ratio of phosphorus to carbon in marine sedimentary organic matter (P:C) has often been assumed to be the same as that in average marine plankton (e.g. – Broecker, 1982), the value for plankton being approximately 0.01 (Redfield, Ketchum & Richards, 1963). However, there is good evidence (e.g. see Grill & Richards, 1964) that during sedimentation there is preferential removal of P relative to C as a result of aerobic decomposition within the water column and at the sediment–water interface. Consequently the P:C ratio of most organic matter in marine sediments is considerably less than 0.01. Froelich *et al.* (1982) have summarized P:C data for a number of locations and they show that for a variety of continental margin sediments P:C atomic ratios range from 0.001 to 0.005. (Some organic-poor deep-sea pelagic sediments show high P:C ratios but these sediments represent a miniscule proportion of annual organic carbon burial as shown in Table 2.1).

The measured ratio of P:C in sedimentary organic matter may decrease with depth as continued organic decomposition, with preferential P loss, takes place (e.g. see Table 2.2 and Hartmann *et al.*, 1973). One way to spot this trend (or any other trends) from existing data without necessitating numerous determinations of organic P (which is difficult) is through the use of diagenetic modeling and pore-water chemical data. Concentration profiles of dissolved phosphate and HCO_3^- (or SO_4^{2-} where the dominant organic decomposition process is bacterial sulfate reduction) can be used to calculate the P:C ratio, of the decomposing organic matter, a_P (Berner, 1977). The appropriate equation, where sulfate reduction is the dominant organic decay process, is:

$$a_P = \frac{D_{sp}k + (1+K_P)\omega^2}{2\,D_{sS}k}\,\frac{dC_P}{dC_S} \qquad (2.4)$$

Here the subscripts P and S refer to dissolved phosphate and sulfate, respectively, and all other symbols are as defined above. If the calculated a_P value for the decomposing material is different from that for the total organic matter, then there is diagenetic fractionation of organic P and C.

An example of the use of equation (2.4) is its application to a sediment from the FOAM site of Long Island Sound (Krom & Berner, 1981). Here the decomposing and total organic matter were found to have essentially identical P:C ratios within the zone of sulfate reduction (see Table 2.3). However, additional data, on sediments of the northwestern African continental margin (Hartmann *et al.*, 1973), enable calculation of a_P values which decrease with depth and which are essentially equal to the P:C ratio of the total organic matter at some depths but not at others (Table 2.3). Obviously more work along these lines needs to be undertaken before anything applicable to sediments in general can be concluded.

Acknowledgments

Research supported by National Science Foundation Grants EAR 8420334 and EAR 8617600.

Table 2.3. *Calculation of* a_P *(P/C ratio of decomposing organic matter) for sediments of the FOAM site of Long Island Sound (Krom & Berner, 1981) and two cores from the northwestern African continental margin (Hartmann* et al, *1973, data for depths below 100 cm)*

Location	Depth range (cm)	$dC_P:dC_s$ ($\times 0.001$)	a_P ($\times 0.001$)	Measured: organic P:C ($\times 0.001$)
Long Island Sound	20–120	7.4	3.0	3.3 ± 0.6
Northwest African margin	90–500	3.6	1.4	1.8 ± 0.2
Northwest African margin	500–800	1.9	0.7	1.9 ± 0.2

Assumed values: $D_{sS} = 100\,cm^2y^{-1}$; $D_{sp} = 75\,cm^2y^{-1}$ $K_P = 2$ (for the NW African sediments, due to slow burial, the term $(1+K_P)\omega^2$ is assumed to be much less than $D_{s_P}k$).

References

Aller, R.C. (1977). 'The influence of macrobenthos on chemical diagenesis of marine sediments'. Unpublished Ph.D. thesis, Yale University.

Baturin, G.N. (1982). Phosphorites on the sea floor: origin, composition and distribution. In *Developments in Sedimentology*, **33**, 1–343.

Berner, R.A. (1974). Kinetic models for the early diagenesis of nitrogen sulfur, phosphorus, and silicon in anoxic marine sediments. In *The Sea*, 5, ed. E.D. Goldberg, Wiley, New York, pp. 427–50.

Berner, R.A. (1977). Stoichiometric models for nutrient regeneration in anoxic sediments. *Limnology and Oceanography*, **22**, 781–6.

Berner, R.A. (1980). *Early Diagenesis: A Theoretical Approach.* Princeton University Press, Princeton.

Berner, R.A. (1982). Burial of organic carbon and pyrite sulfur in the modern ocean: its geochemical and environmental significance. *American Journal of Science*, **282**, 451–73.

Broecker, W.S. (1982). Ocean chemistry during glacial time. *Geochimica et Cosmochimica Acta*, **46**, 1689–706.

Broecker, W.S. & Peng, T.-H. (1982). *Tracers in the Sea.* Eldigio Press, Madrid.

Burnett, W.C. (1977). Geochemistry and origin of phosphorite deposits from off Peru and Chile. *Geological Society of America Bulletin*, **88**, 813–23.

Calvert, S.E. & Price, N.B. (1983). Geochemistry of Namibian shelf sediments, In *Coastal Upwelling*, Pt. A, ed. E. Suess & J. Thiede, pp. 337–75. Plenum Publishing, New York.

D'Anglejan, B.F. (1979). Phosphorite deposits off Baja, California. In *Report on the Marine Phosphatic Sediments Workshop*, ed. W.C. Burnett, & R.P. Sheldon. East-West Resource Systems Institute, Honolulu 65p.

Elderfield, H., McCaffrey, R.J., Luedtke, N., Bender, M. & Truesdale, V.W. (1981). Chemical diagenesis in Narraganasett Bay sediments. *America Journal of Science*, **281**, 1021–55.

Emerson, S. & Widmer, G. (1978). Early diagenesis in anaerobic lake sediments, II. Thermodynamic and kinetic factors controlling the formation of iron phosphate, *Geochimica Cosmochimica Acta* **42**, 1307–16.

Filipek, L.H. & Owen, R.M. (1981). Diagenetic controls of phosphorus in outer continental shelf sediments from the Gulf of Mexico. *Chemical Geology*, **33**, 181–204.

Froelich, P.N., Bender, M.L., Luedtke, N.A., Heath, G.R. & De Vries, T. (1982). The marine phosphorus cycle. *American Journal of Science*, **282**, 474–511.

Froelich, P.N., Kim, K.H., Jahnke, R., Burnett, W.C., Soutar, A. & Deakin, M. (1983). Pore water fluoride in Peru continental margin sediments: Uptake from seawater. *Geochimica Cosmochimica Acta*, **47**, 1605–12.

Grill, E.V. & Richards, F.A. (1964). Nutrient regeneration from phytoplankton decomposing in seawater. *Journal of Marine Research*, **22**, 51–9.

Hartmann, M.P., Muller, P.J., Suess, E. & Van Der Weijden, C.H. (1973). Oxidation of organic matter in recent sediments. *'Meteor' Forschung-Ergeb*, **C 12**, 74–86.

Hartmann, M.P., Muller, P.J., Suess, E. & Van Der Weijden, C.H. (1976). Chemistry of late Quaternary sediments and their interstitial waters from the northwestern African continental margin. *'Meteor' Forschung-Ergeb*, **C 24**, 1–67.

Hines, M. & Lyons, W.B. (1984). (unpublished data on phosphorus in sediments).

Jahnke, R.A. (1981). 'Current phosphorite formation and the solubility of synthetic fluorapatite.' Unpublished Ph.D. thesis, University of Washington, Seattle.

Jahnke, R.A., Emerson, S.R., Roe, K.K. & Burnett, W.C. (1983). The present day formation of apatite in Mexican Continental margin sediments. *Geochimica Cosmochimica Acta*, **47**, 259–66.

Kolodny, Y. (1980). The origin of phosphorite deposits in light of the occurrence of recent sea floor phosphorites, (extended abstract). *SEPM Special Publications*, **29**, 249.

Krom, M.D. & Berner, R.A. (1981). The diagenesis of phosphorus in a nearshore marine sediment. *Geochimica Cosmochimica Acta*, **45**, 207–16.

Mackereth, F.J.H. (1966). Some chemical observations on post-glacial lake sediments, *Philosophical Transactions of the Royal Society*, **Section B**, **250**, 165–220.

Manheim, F., Rowe, G. & Jipa, D. (1975). Marine phosphorite formation off Peru. *Journal of Sedimentary Petrology*, **45**, 243–51.

Martens, C.S., Berner, R.A. & Rosenfeld, J.K. (1978). Interstitial water chemistry of anoxic Long Island Sound sediments II; Nutrient regeneration and phosphate removal, *Limnology and Oceanography*, **23**, 605–17.

Morse, J.W., Zullig, J.J., Bernstein, L.D., Millero, F.J., Milne, P., Mucci, A. & Choppin, G.R. (1985). Chemistry of calcium carbonate-rich shallow water sediments in the Bahamas, *American Journal of Science*, **285**, 147–85.

Murray, J.W., Grundmanis, V. & Smethie, W.N. (1978). Interstitial water chemistry in the sediments of Saanich Inlet. *Geochimica Cosmochimica Acta*, **42**, 1011–26.

Nissenbaum, A., Presley, B.J. & Kaplan, I.R. (1972). Early diagenesis in a reducing fjord, Saanich Inlet, British Columbia I: chemical and isotopic changes in major components of interstitial waters, *Geochimica Cosmochimica Acta*, **32**, 1007–27.

Price, N.B. (1976). Chemical diagenesis in sediments, In *Chemical Oceanography*, v.6 ed. J.P. Riley & R. Chester, pp. 1–58, Academic Press, New York.

Redfield, A.C., Ketchum, B.H. & Richards, F.A. (1963). The influence of organisms on the composition of seawater. In *The Sea*, v.2 ed. M.N. Hill, pp. 26–77, Wiley, NY.

Rittenberg, S.C., Emery, K.O. & Orr, W.L. (1955). Regeneration of nutrients in sediments of marine basins. *Deep-Sea Research*, **3**, 23–45.

Sholkovitz, E.R. (1973). Interstitial water chemistry of the Santa Barbara Basin sediments. *Geochimica Cosmochimica Acta*, **37**, 2043–73.

Stumm, W. & Leckie, J.O. (1970). Phosphate exchange with sediments: its role in the productivity of surface waters, In *Advances in Water Pollution Research*, **2**, 26/1–16.

Suess, E. (1981). Phosphate regeneration from sediments of the Peru continental margin by dissolution of fish debris. *Geochimica Cosmochimica Acta*, **45**, 577–88.

Van Cappellen, P. & Berner, R.A. (1988). A mathematical model for authigenic apatite formation in modern marine sediments. *American Journal of Science*, **288**, 289–383.

Westrich, J.T. & Berner, R.A. (1984). The role of sedimentary organic matter in bacterial sulfate reduction: the G model tested. *Limnology and Oceanography*, **29**, 236–49.

3

Organic matter in Modern marine phosphatic sediments from the Peruvian continental margin

M.W. SANDSTROM

Abstract

The compositions of aliphatic hydrocarbons, humic acids and protokerogen in a sediment core from the Peru continental margin near 15° S provide detailed information about the diagenetic transformations of marine organic matter in a region where phosphorites form. Aliphatic hydrocarbons were identified by capillary gas chromatography–mass spectrometry and humic acids and protokerogen were characterized by elemental (C,H,N,O,S,P) and carbon isotopic composition.

Reactive, labile organic matter derived from bacteria and phytoplankton, including the C_{25} isoprenoid 2,6,10,15,19-pentamethyleicosane derived from methanogenic bacteria, and a branched C_{20} monoene and suite of branched C_{25} polyunsaturated alkenes of presumed microbial origin, were predominant in the surface (0–3 cm) sediments and decreased rapidly with depth. Long chain *n*-alkanes derived from higher plants were more abundant in the subsurface samples presumably because of greater resistance to microbial degradation. The abundance of C_{27}-C_{29} Δ^2-sterenes and C_{27} and C_{30} hopenes in the subsurface sediments compared to the surface suggests a diagenetic origin rather than a direct microbial source. Humic acid and protokerogen fractions contained less than 6% of the total sediment phosphorus, implying that diagenetic transformations of these geopolymers are unlikely to provide a significant source of phosphorus for the formation of phosphorites. Hence the precipitation of sedimentary carbonate fluorapatite is probably associated with diagenesis of biogenic organic phosphorus compounds, including lipids or nucleic acids.

Introduction

Upwelling regions of the Peru–Chile margin have some of the highest levels of marine primary productivity in the world, with average values as high as 3 g $C m^{-2} d^{-1}$, in contrast to values in the range of 0.5–1.4 g $C m^{-2} d^{-1}$ for other shelf ecosystems (Walsh, 1981). A significant portion of this primary productivity is transported to the sediments, resulting in some sediment organic carbon values as high as 21% (Krisseck, Scheidegger, & Kulm, 1980). The Peru–Chile margin is also notable for being one of the few regions in the world where modern (present to 0.1 Ma) phosphorites occur (Baturin, Merkulova, & Chalov, 1972; Burnett & Veeh, 1977; Burnett, Beers & Roe, 1982), lending support to early hypotheses on the importance of upwelling in the formation of phosphorites (see Bentor, 1980; Kolodny, 1981; Sheldon, 1981 for reviews). The crucial link between upwelling and phosphorite formation appears to be the biological assimilation and regeneration of phosphorus from organic matter in the sediments and an investigation of the diagenesis of organic matter in sediments from the Peru–Chile upwelling system may provide insight into this relationship.

Many recent investigations have documented the organic carbon accumulation rates (Muller & Suess, 1979; Walsh, 1981; Henrichs & Farrington, 1984), composition of bulk organic matter (Pentina & Chetverikova, 1977; Poutanen & Morris, 1983), and various lipid classes (Gagosian, Volkman & Nigrelli, 1983; Smith, Eglinton & Morris, 1983; Wakeham, Farrington, & Volkman, 1983) in surficial sediments and particulate matter in the Peru upwelling area. In general these studies have been concerned with composition of lipids in the surface sediment compared with those produced in the overlying water column. Few data are available concerning the diagenetic transformations of the organic matter in the sediments that may identify important processes related to the origin of the phosphorites. Details of the depth profiles of amino acids (Henrichs, Farrington, & Lee, 1984) and several lipid classes (Volkman *et al.*, 1983) in the sediment have been presented. Cunningham & Burnett (1985) have shown that compositional patterns of amino acids in nodules are similar to those in unlithified surface sediments, suggesting that the nodules and sediments constitute a single genetic series.

In an effort to further document the genetic link between phosphorites and associated organic-rich sediment, the hydrocarbon distributions in a sediment core representing 180 y of deposition in a region of the Peru margin where modern phosphorites occur are presented here. Depth profiles of the hydrocarbons and composition of major classes of organic matter are used to examine the diagenetic sequence from labile organic matter at the sediment surface to more resistant organic matter at depth in the core and to identify possible sources of organic matter. These results are briefly compared to hydrocarbon distributions in modern and ancient phosphorites. Hydrocarbons were chosen because some of these biological markers survive early diagenesis and occur in geologically older phosphorites (Powell, Cook, & McKirdy, 1975; Sandstrom, 1980; Amit & Bein, 1982; Belyouni & Trichet, 1983).

Geological setting

Modern phosphorites from the Peru–Chile continental margin are associated with an upper slope and shelf mud facies which occurs mainly between 10–14° S at water depths between 100 and 600 m (Krisseck *et al.*, 1980). These sediments are within the region of most intense upwelling, and where the oxygen minimum layer impinges on the slope and outer shelf.

The sediments have high concentrations of organic carbon (5–21%), total phosphorus (average 0.3%), and rich diatom assemblages. Texturally, the sediments are anomalously fine-grained, which may be an effect of the rapid sedimentation of fine-grained terrigenous particles incorporated into fecal pellets (Krisseck *et al.*, 1980). Bacterial mats of non-photosynthetic sulfur-oxidizing bacteria of the genera *Beggiatoa* and *Thioploca* occur in the surface sediments (Gallardo, 1977; Reimers, 1982; Henrichs & Farrington, 1984). The mats consist of cell chains and muciliganous sheaths that bind the sediment and give it a highly porous fabric (Reimers, 1982).

A distinctive feature of this upper slope mud facies is the occurrence of phosphorite nodules and concretions within the sediments or on the seabed surface. Radiometric dating of some of these nodules provided the first confirmation of the presence of Holocene phosphorite deposits in areas of coastal upwelling (Baturin *et al.*, 1972) and since then the phosphorites have been the subject of numerous geochemical and geological investigations.

Initial descriptions suggested that lithologic types ranging from friable to compact represented a continuous series of progressively older phosphorites, which has been generally confirmed by uranium series dating of the nodules (Burnett & Veeh, 1977; Burnett, Veeh & Soutar, 1980). Additionally, it has been shown that the nodules grow slowly at rates of millimeters per thousand years, and that nodules at the sediment–water interface apparently grow downward into the soft muds (Burnett *et al.*, 1982) while buried nodules grow upward (Kim & Burnett, 1986). This fits previous models for the origin of the nodules, in which the nodules are thought to be diagenetically precipitated from sediment pore waters and not by direct precipitation from overlying waters (Burnett 1977; Froelich *et al.*, 1983). The source of the phosphorus is presumably from microbially-mediated regeneration of organic matter, although Suess (1981) has suggested that dissolution of phosphatic fish debris may also be important.

Methods

Sample material

Sediment samples are from a 20 cm box core (Station 7408–1904; core #4) collected during August 1974 from the Peru upwelling region (14°38.9′ S; 76°10.3′ W) in a water depth of 183 m. The core samples were placed in a freezer soon after recovery and kept in a frozen state until analysis in 1980. ^{210}Pb activities of the box core samples determined by Koide & Goldberg (1982) are shown in Figure 3.1. They interpreted the change in slope of the ^{210}Pb profile as the result of an apparent increase in sedimentation rate from 0.15 cm y^{-1} below 7 cm to 0.32 cm y^{-1} in the upper 7 cm. Alternatively, the apparent change in slope of the ^{210}Pb profile may be due to bioturbation

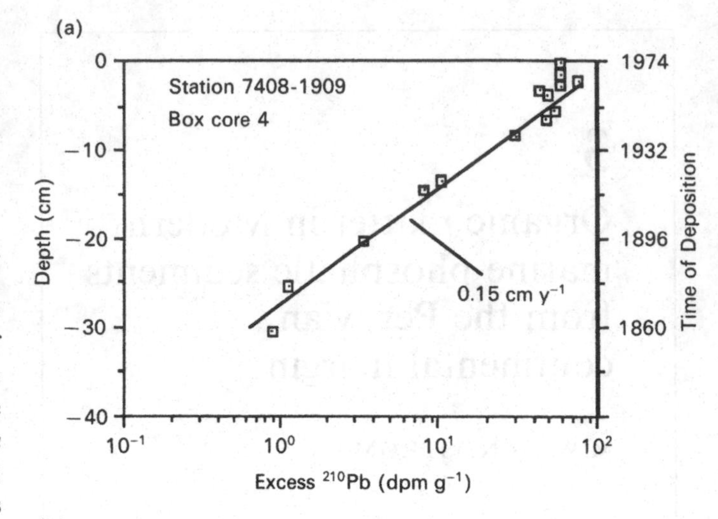

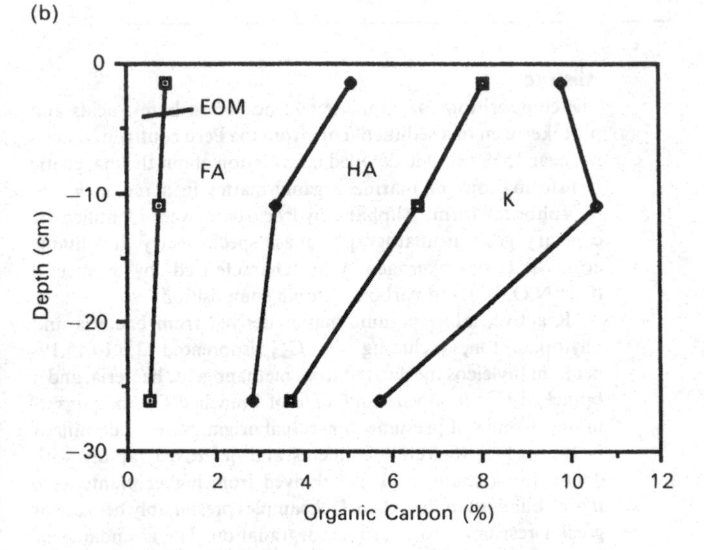

Fig. 3.1. Depth profile of: (a) excess ^{210}Pb (from Koide & Goldberg, 1982); and (b) major organic fractions in Peru sediments. Fractions are calculated as percentage of total organic carbon. EOM: extractable organic matter; FA: fulvic acids; HA: humic acids; K: protokerogen.

to a depth of 7 cm. The calculated times of sedimentation for the depth intervals sampled are presented in Table 3.1.

Semi-quantitative X-ray mineralogical analysis of sediment samples indicated that quartz, potassium feldspar, and plagioclase were the predominant mineral phases present with minor amounts of clays in the lower core sections. Carbonate fluorapatite (CFA) and calcite were not detected.

Analytical procedures

The frozen sediments (5–10 g) were allowed to thaw in Soxhlet thimbles and rinsed with distilled water (100 ml) to remove soluble salts, then Soxhlet extracted with methanol for 24 h followed by dichloromethane–methanol (9:1) for 76 h. The water and methanol fractions were rinsed with hexane, the hexane rinse added to the dichloromethane–methanol, and the combined solvents removed on a rotary evaporator to concentrate the samples to 5 ml. Each extract was passed through a column of activated copper to remove elemental sulfur, and the solvent again removed by evaporation to a known volume (500 μl) on a sand bath under a stream of nitrogen. The weight of

Table 3.1. *Concentrations of phosphorus, organic carbon, extractable organic matter (EOM) and hydrocarbon compound class distribution in Peru box core 7408–1909. Approximate ages were calculated from ^{210}Pb profile*

Depth interval (cm)	Total phosphorus (mg g^{-1})	Organic carbon (% dry wt)	EOM	EOM compound class[a] Aliphatic	Aromatic	Heteroatomic	Age (y)
			μg g^{-1} dry sediment				
0–3	2.66	9.8	11000	340	220	5300	0–9
10–12	4.01	10.6	9400	70	140	3800	42–55
25–27	2.92	5.7	6000	60	40	3100	98–112

[a]Extractable organic matter classes separated by column chromatography into aliphatic and aromatic hydrocarbons and heteroatomic compounds.

extractable organic matter (EOM) was measured by transferring 2 μl aliquots to tared weighing pans and weighing after evaporation of solvent.

The EOM was separated by adsorption chromatography on columns of alumina over silica gel (1:2). Aliphatic and aromatic hydrocarbons were eluted with hexane, aromatic hydrocarbons with benzene, and polar heteroatomic compounds with methanol. Hydrocarbons in the hexane fractions were analyzed by capillary gas chromatography (GC) on a Varian 2440 instrument fitted with a flame ionization detector and SP-2100 fused silica capillary column. Concentrations of individual hydrocarbons were determined by peak area comparison of external standards. Samples from the 0–3 and 25–27 cm depth intervals were analyzed by gas chromatography-mass spectrometry (GC–MS) using a Finnigan 4023 instrument interfaced with a Finnegan 9610 GC. Individual hydrocarbons were identified by comparison of retention times and mass spectra with those of published structural assignments.

Humic acids were isolated from the solvent-extracted sediments by extraction with 2 N NaOH after carbonates had been dissolved with 3 N HCl. The acid and base extractions were conducted under a nitrogen atmosphere and residues centrifuged at 1500 g for 20 min. Humic materials were filtered through 0.45 μm glass fibre filters, acidified to pH 2 with HCl, and the precipitated humic acids allowed to settle. Salts were removed from the humic acids by gel permeation chromatography on columns of Bio-gel P-2. The kerogen residue was treated with 6 N HCl:70% HF (1:1) to remove minerals, followed by rinse with 25% $AlCl_3$- 6HOH to remove fluorides.

Carbon, hydrogen, nitrogen, sulfur content of the humic acid and kerogen fractions were determined using Carlo Erba instruments. Phosphorus was determined by spectrometry (Murphy & Riley, 1962) after dissolution with a nitric–perchloric acid (3:1) mixture. Iron was determined by atomic absorption spectrometry. The equivalent pyrite sulfide sulfur was calculated from these iron analyses and subtracted from the total sulfur to give the corrected organic sulfur content (e.g. Durand & Monin, 1980). Samples for stable carbon isotope analysis were prepared using the method of Sofer (1980).

Carbon and sulfur determination of the bulk sediment were conducted by combustion using a LECO instrument. Organic carbon was determined using the LECO instrument after treatment with 2 N hydrochloric acid. Total phosphorus was determined by spectrophotometry (Shapiro & Brannock, 1962).

Results

Diagenesis and transformation of organic matter

The sediments from the Peru upwelling region contain relatively high amounts of reactive organic matter that is remineralized in the upper few centimeters of sediment. Total organic carbon (TOC) values in the sediments analyzed here range from 10.6 to 5.7% (Table 3.1). These values are very high compared to non-upwelling continental margin sediments, and reflect the high accumulation rate of organic matter in the upwelling region. The pronounced decrease (about two-fold) in TOC with depth in the core could be attributed to changes in accumulation rate of organic matter over time (Walsh, 1981; Henrichs & Farrington, 1984) as well as remineralization of organic matter. However, the ^{210}Pb results do not show any major changes in sedimentation rate in the bottom of the core (Fig. 3.1). Since the changes in composition of organic matter discussed below indicate the loss of labile components, the overall decrease in TOC with depth in the core most likely results from remineralization.

The amount of solvent extractable organic matter (EOM) in the sediments is relatively high, representing about 9–11% of the TOC (Table 3.1). The aliphatic and aromatic hydrocarbons comprise about 0.6% of the TOC in the surface and decrease to about 0.2% over the length of the core. Functionalized polar components (e.g. alcohols, ketones, phospholipids), which elute in the heteroatomic fraction, represent more than 90% of the EOM in all depth intervals. These concentrations of EOM and hydrocarbons are very high in comparison to shelf sediments from non-upwelling areas (Palacas, Love & Gerrild, 1972; Gearing *et al.*, 1976; Venkatesan *et al.*, 1981) and reflect the high accumulation rate of marine lipids in these sediments.

The depth profiles of EOM and hydrocarbons both show significant decreases with depth (Fig. 3.2). The pronounced decrease in hydrocarbons in the surface compared to TOC or EOM suggests that the surface layer is more enriched in bacterial lipids compared to deeper layers or contains a hydrocarbon component that is more reactive than the rest of the organic matter.

However, changes in the amount of humic substances are quantitatively the most important in regard to the decrease in TOC in the sediment core. The humic acid fraction represents approximately 30% of the total organic carbon in the surface and 10–12 cm depth intervals, and decreases to 15% in the 25–27 cm depth interval (Fig. 3.1). The insoluble organic matter or

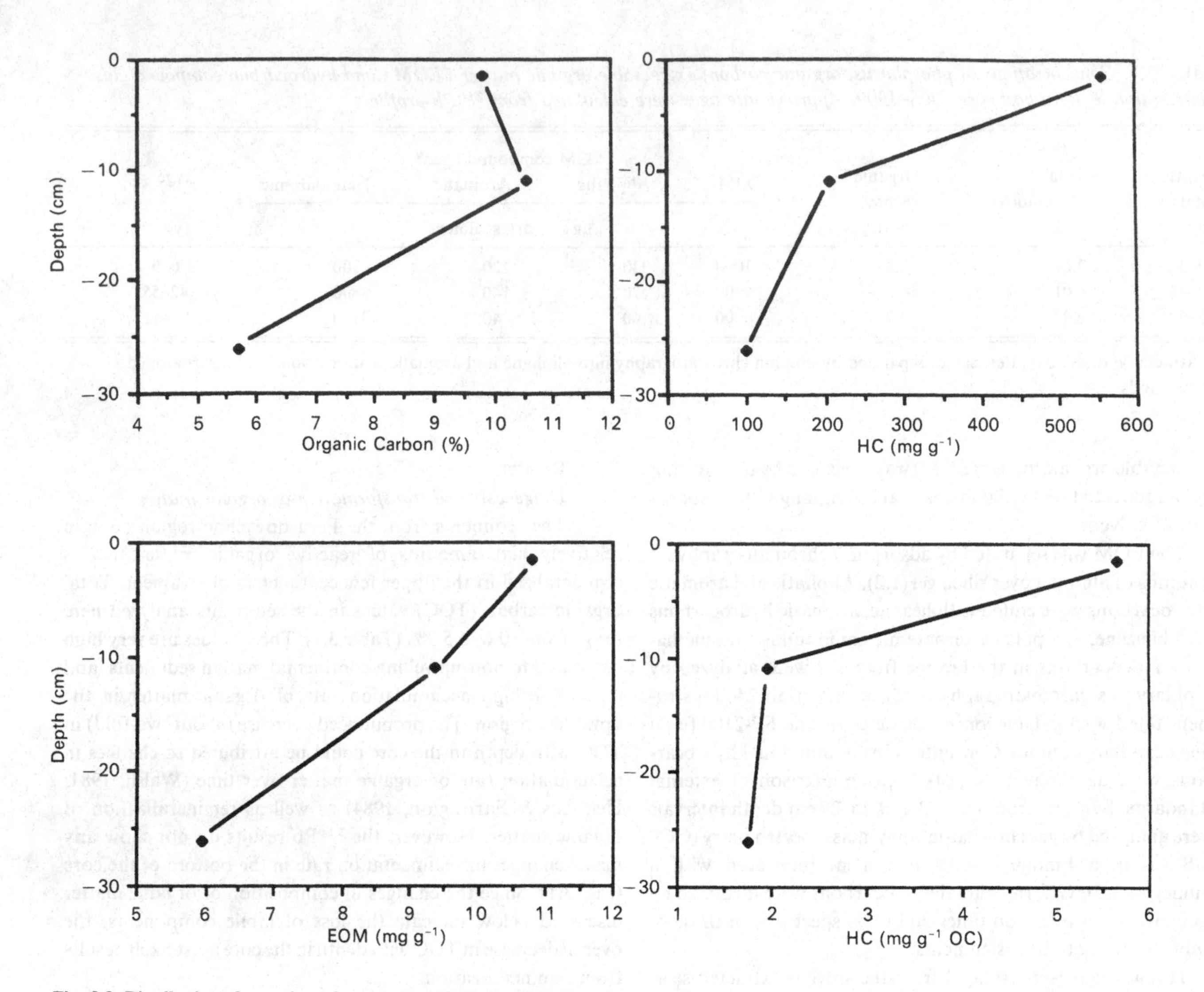

Fig. 3.2. Distribution of organic carbon, extractable lipids (EOM), and hydrocarbons (HC) in box core 7408–1909.

protokerogen, which includes lignin and microbial cell wall components, increases from 18% of the TOC in the surface sediment to about 38% in the subsurface depth intervals. The remainder (41–44%) of the organic matter consists of fulvic acids, amino acids, and other water- and acid-soluble organic substances. For comparison, Henrichs, Farrington & Lee (1984) found that total hydrolyzable amino acids make up about 22% of the TOC in Peru surface sediments, and decreased by about 50% over the depth of the cores.

The abundance of humic substances in the surface sediments may be related to the bacterial mats of *Thioploca* and *Beggiatoa* which are characteristic of these sediments. Reimers (1982) examined the organic matter in the mats using microscopic techniques and found an abundance of cell chains and muciliganous sheaths derived from filamentous bacteria. This type of organic matter consists largely of polysaccharides and other biopolymers that would contribute to the humic fraction of the TOC.

The changes in relative abundance of humic acid and protokerogen fractions indicated by the gradual decrease in the humic: protokerogen ratio with depth (Table 3.2) reflects the gradual polycondensation and insolubilization of organic matter occurring in addition to the remineralization of organic matter during early diagenesis. This is consistent with the change in fabric of Peru sediments with depth that appears to be related to the degradation of the microbial mat structure (Reimers, 1982). These diagenetic reactions are also indicated by changes in the chemical composition of the humic and protokerogen fractions of the Peru sediment (Table 3.2).

The humic acid and protokerogen fractions from the box core sediments have H:C, N:C and O:C ratios characteristic of very immature marine or algal sources of organic matter (Stuermer, Peters & Kaplan, 1978). A predominantly marine source for the organic matter is also indicated by the $\delta^{13}C$ values of the humic acid and protokerogen fractions which are between −20.2 and −22.0 per mil (Table 3.2) for all depth intervals (Nissenbaum & Kaplan, 1972).

The humic acids are characterized by higher H:C, N:C, and O:C ratios than the corresponding protokerogen fractions, which reflects the higher content of nitrogen- and oxygen-con-

Table 3.2. *Chemical composition and carbon isotope data for humic acid and protokerogen fractions isolated from Peru box core 7408–1909 sediments*

Depth interval (cm)	Sample[a] Yield ($mg\,g^{-1}$ DS)	Composition (% Ash-free dry wt) C	H	N	S[b]	O[c]	P	Ash (%)	Atomic Ratio H/C	N/C	S/C	O/C	C/P	$\delta^{13}C$ (ppt)
Humic acid														
0–3	75.64	50.59	7.41	6.39	1.77	33.60	0.24	22.97	1.75	0.108	0.012	0.50	544	−20.4
10–12	110.56	47.43	7.54	5.62	1.61	37.50	0.31	38.56	1.89	0.101	0.012	0.59	395	−21.1
25–27	23.20	48.68	7.03	5.78	1.20	37.07	0.23	25.21	1.72	0.102	0.009	0.57	546	−20.5
Protokerogen														
0–3	17.51	55.97	7.35	6.33	1.17	29.17	0.023	5.81	1.56	0.097	0.007	0.39	6275	−21.7
10–12	39.75	50.37	7.37	6.62	—	21.15	0.050	8.77	1.45	0.094	—	0.26	3114	−20.7
25–27	20.32	58.75	6.76	5.81	2.04	26.59	0.048	19.68	1.37	0.085	0.012	0.34	3157	−22.0

[a]Dry weight of humic acid or protogerogen fraction, $mg\,g^{-1}$ dry sediment.
[b]Determined by difference or total sulfur and equivalent sulfide sulfur from iron analysis.
[c]Determined by difference.

taining functional groups of the humic acids. There was an overall decrease in H:C and N:C ratios of the humic acids and protokerogen with depth in the core resulting from the loss of these components (Fig. 3.3).

The S:C ratios of the humic acid and protokerogen fractions were low in comparison to other anoxic marine sediments (Stuermer *et al.*, 1978). The sulfur content of the insoluble organic matter in marine sediments is thought to be derived in part from sulfide produced by microbial sulfate reduction (Nissenbaum & Kaplan, 1972). In the Peru box core sediment the low sulfur content of the humic acid and protokerogen may in part be the result of oxidation of hydrogen sulfide by the *Thioploca* and *Beggiatoa* bacterial mats which effectively remove hydrogen sulfide that could react with sediment organic matter. It should be noted that microbial or chemical oxidation of sulfur can only occur where suitable oxidizing agents (e.g. nitrate, oxygen) are present, near the sediment surface.

The phosphorus content of the humic acid and protokerogen fractions varied from 0.23 to 0.31% and 0.02–0.05%, respectively. This is about 1–2 orders of magnitude less than the phosphorus content of plankton or bacteria (Redfield, Ketchum & Richards, 1963), and, as in the case of nitrogen, reflects the preferential loss of phosphorus during diagenesis of the organic matter. The humic acid and protokerogen fractions combined represent <6% of the total phosphorus in the sediments, and thus do not appear to be an important source of phosphorus for the precipitation of CFA during further insolubilization and condensation reactions. Other fractions of organic matter such as labile lipid components are probably a more important source of regenerated phosphorus.

Molecular composition of hydrocarbons

Detailed information about the origin and diagenetic changes of organic matter in the sediments is provided by examination of the molecular composition of the aliphatic and olefinic hydrocarbons. Capillary gas chromatograms of the aliphatic hydrocarbon fractions of the Peru sediment are shown in Figure 3.4.

There were distinct changes in hydrocarbon composition with depth in the core, with the most pronounced changes occurring between the surface (0–3 cm) and subsurface samples. The surface sample is characterized by a predominance of components eluting between n-C_{15} and n-C_{23}, including a series of n-alkanes, a narrow unresolved complex mixture (UCM), and a number of branched alkenes. The abundance of these components decreases in the subsurface samples, while a series of n-alkanes and terpenoids with carbon number greater than 25 correspondingly increases. The hydrocarbon distribution is similar to that previously reported for other surface sediments from the Peru upwelling region (Smith, Eglinton & Morris, 1983; Volkman *et al.*, 1983) and so may be considered broadly representative of the region.

Alkanes

The n-alkanes consist of two components which vary in relative abundance over the length of the core (Table 3.3). n-Alkanes with less than 20 carbon atoms (n-C_{14}–n-C_{20}) and having no carbon number preference ($CPI_{14-20} = 1.03$) are relatively abundant in the 0–3 cm depth interval. The subsurface samples are characterized by a bimodal distribution in which the n-alkanes with less than 20 carbon atoms are less abundant than a series of long-chain n-alkanes (n-C_{25}–n-C_{33}) maximizing at n-C_{29} and having strong odd carbon-number predominance ($CPI_{21-33} = 3.6$–4.1).

The source of the latter component is attributed to an input of terrestrial higher plants since the alkane distribution is typical of the epicuticular waxes of higher plants (Eglinton & Hamilton, 1967). These hydrocarbons have been found in many hemipelagic and shelf sediments, and their abundance, relative to n-alkanes derived from marine sources, is commonly attributed to a greater resistance to microbial degradation, perhaps because of an association with lignin or sporopollenin. Thus the increase in n-C_{29} with depth in the core (Fig. 3.4) may be a diagenetic effect which results in a relative enrichment of more refractory components of the organic matter.

In addition, fluctuations in the amount of plantwax n-alkanes

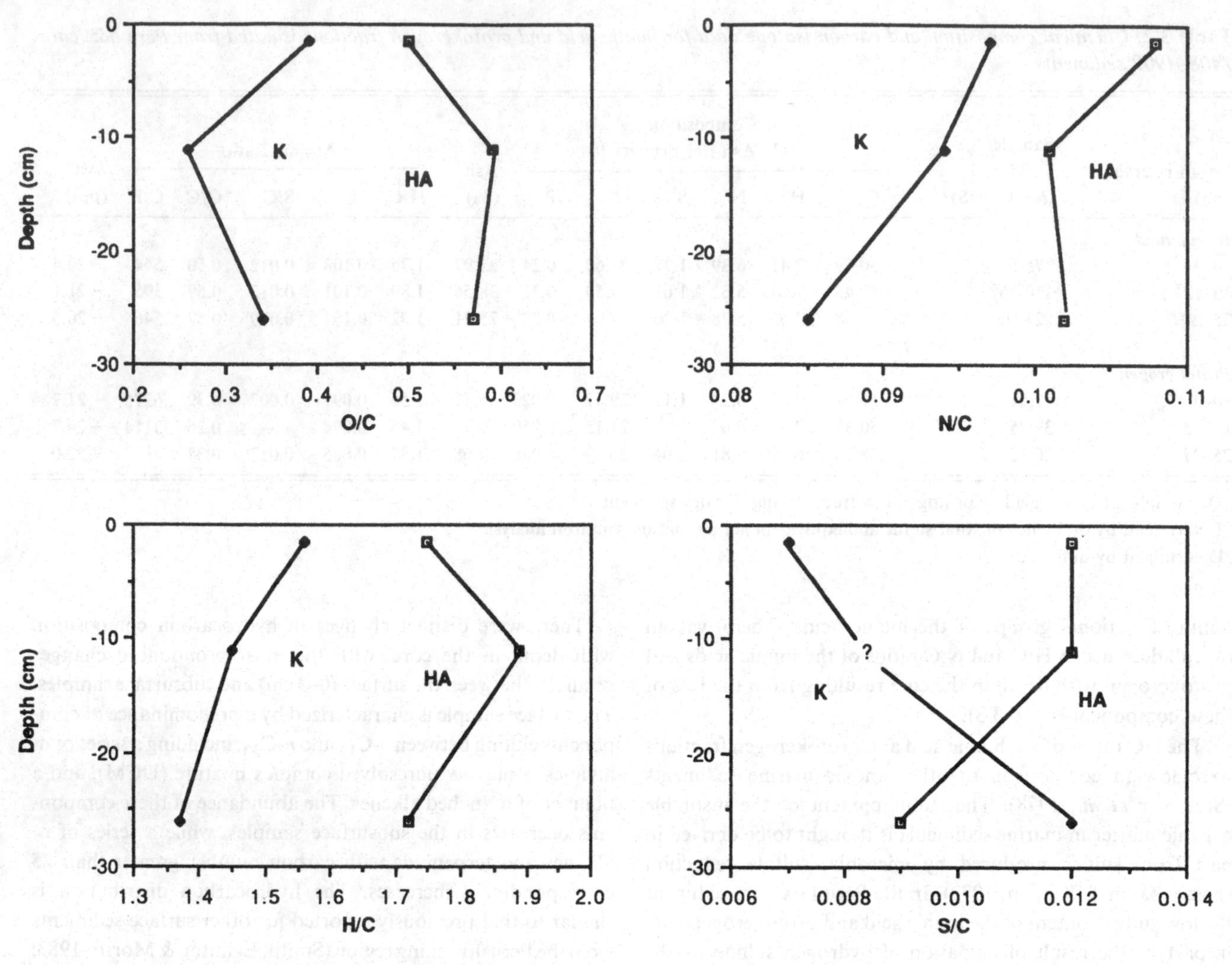

Fig. 3.3. Depth profiles of atomic H/C, O/C, N/C, and S/C of humic acid (HA) and protokerogen (K) fractions in Peru sediment.

deposited in the sediments could be due to climatic changes or minor changes in marine productivity. Volkman *et al.* (1983) reported variations in plantwax alkanes in Peru continental margin sediments which they attributed to fluctuations in accumulation rate of terrigenous material. However, each of the depth intervals samples here include a number of strong El Niño events (Quinn *et al.*, 1978), so major differences in rainfall or runoff cannot easily explain the relative abundance of plantwax alkanes in Core 1905. However, if the ^{210}Pb data are interpreted in terms of increased sedimentation rate in the upper 7 cm of the core rather than bioturbation (Fig. 3.1), the low concentrations of plantwax alkanes in the surface could be explained by higher accumulation of marine organic matter.

The C_{14-20} *n*-alkanes and associated narrow UCM which characterize the surface sediment can be attributed to direct or indirect microbial sources. Generally, similar alkane distributions have been reported from non-photosynthetic bacteria (Han & Calvin, 1969) and anaerobic decomposition products of salt marsh plant detritus (Johnson & Calder, 1973), and have thus been attributed to microbial sources in recent marine and lacustrine sediments (Douglas, Eglington & Maxwell, 1969; Hatcher, Simoneit & Gerchakov, 1977; Venkatesan *et al.*, 1981). Implicit in this interpretation is that *in situ* microbial activity breaks down *n*-alkanes derived from plankton with resynthesis of the C_{14-20} alkane components and associated narrow UCM.

This is consistent with the absence of specific alkane biomarkers derived from phytoplankton. The C_{17} *n*-alkane and 7- and 8-methylheptadecane are predominant components of green algae and cyanobacteria, respectively (Han & Calvin, 1969), and also occur in algal mats (e.g. Philp, 1980). Similarly, the C_{21} alkene cis-3,6,9,12,15,18-heneicosahexaene is a characteristic alkene of diatoms (Blumer, Mullin & Guillard, 1970). These hydrocarbon biomarkers were absent from the sediments despite the fact that the sediments contain large amounts of diatom frustules. Hence, it appears that microbial transformation of plankton biomarkers can occur quite rapidly since the surface sediment represented about 10 y of accumulation.

Although the C_{14-20} alkane distribution is nonspecific in terms of the type of bacteria involved, the predominance of these components in the surface sediments indicates maximum microbial activity or biomass in the surface sediments, in agreement with the reported abundance of bacterial mats in the surface sediments (Reimers, 1982; Henrichs & Farrington, 1984). Furthermore, the depth distribution (Table 4.3) indicates that these components are very reactive and are preferentially degraded relative to C_{21-33} *n*-alkanes and organic carbon.

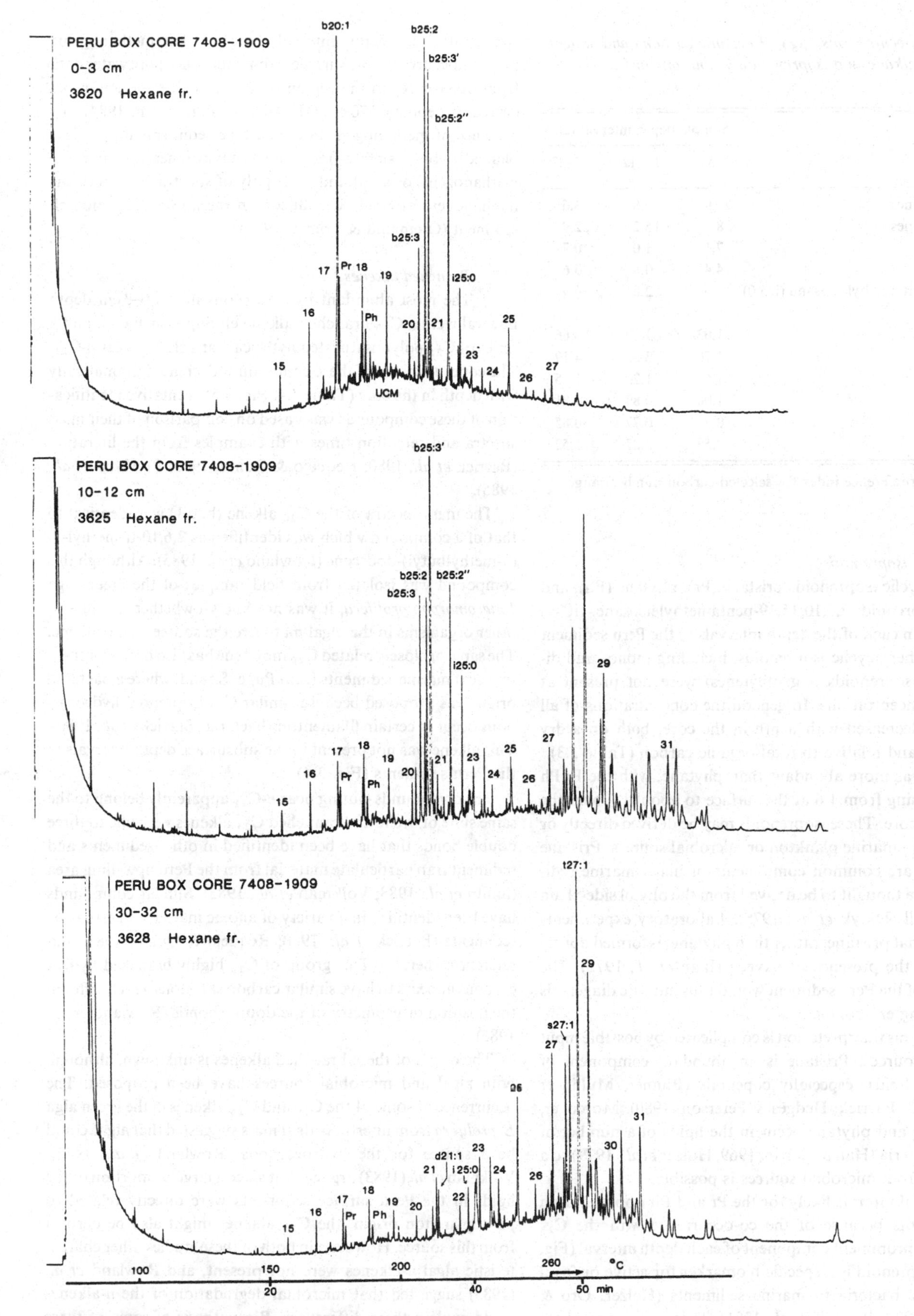

Fig. 3.4. Capillary gas chromatograms of aliphatic hydrocarbons from Peru box core sediment 7408–1909. Numbers refer to carbon number of *n*-alkanes; identifications of other components are given in the text.

Table 3.3. *Concentrations ($\mu g\ g^{-1}$ organic carbon) and salient features of* n-*alkane and isoprenoid distributions in Peru sediments*

Parameter	Sample depth interval (cm)		
	0–3	10–12	25–27
C_{14}–C_{20} *n*-alkanes	29.9	5.8	3.6
C_{21}–C_{33} *n*-alkanes	8.6	15.2	22.9
Pristane (Pr)	7.1	1.0	0.7
Phytane (Ph)	4.4	0.8	0.6
2,6,10,15,19-pentamethyleicosane (i25:0)	7.0	2.8	0.9
CPI_{14-20}[a]	1.03	1.00	0.66
CPI_{21-33}[a]	1.76	3.62	4.19
Pr/Ph	1.62	1.20	1.15
Pr/nC_{17}	1.00	0.89	0.81
Ph/nC_{18}	0.71	0.77	0.85
i25:0/Ph	1.59	3.27	1.52

[a]CPI – carbon preference index for selected carbon number range.

Acyclic isoprenoids

The acyclic isoprenoids pristane (Pr), phytane (Ph), and the C_{25} isoprenoid 2,6,10,15,19-pentamethyleicosane (iC_{25}) were present in each of the depth intervals in the Peru sediment (Fig. 3.4). Other acyclic isoprenoids, including mono- and di-unsaturated isoprenoids (e.g. phytenes) were not present at detectable concentrations. In general the concentrations of all isoprenoids decreased with depth in the core, both on a dry weight basis and relative to total organic carbon (Table 3.3).

Pristane was more abundant than phytane, with the Pr:Ph ratios decreasing from 1.6 at the surface to about 1.2 over the length of the core. These isoprenoids may be derived directly or indirectly from marine plankton or microbial sources. Pristane and phytane are common components of many marine sediments, and are thought to be derived from the phytol side-chain of chlorophyll (Didyk *et al.*, 1978). Laboratory experiments have shown that pristane, rather than phytane, is formed during diagenesis in the presence of oxygen (Ikan *et al.*, 1975). The Pr:Ph ratios of the Peru sediment would thus indicate diagenesis in an oxidizing environment.

However, this interpretation is complicated by possible input from other sources. Pristane is an abundant component of zooplankton lipids, especially copepods (Blumer, Mullin & Thomas, 1963; Barrick, Hedges & Peterson, 1980). Moreover, both pristane and phytane occur in the lipids of a number of species of bacteria (Han & Calvin, 1969; Holzer *et al.*, 1979) so a direct input from microbial sources is possible.

A microbial input is likely for the Pr and Ph isoprenoids in these sediments because of the co-occurrence with the C_{25} isoprenoid, a prominent component of each depth interval (Fig. 3.4). This isoprenoid is a specific biomarker for active or fossil methanogenic bacteria in marine sediments (Holzer, Oro & Tornabene, 1979; Brassell *et al.*, 1981). These microorganisms may be responsible for part of the observed changes in the lipid material; for example, reworking the short-chain alkanes derived from phytoplankton. The presence of this compound in the surface (0–3 cm) depth interval indicates that methanogenic bacteria occurred in the surface sediments. Since pore water data from other cores in the region indicates sulfate reduction occurred to depths of 70 cm (Henrichs & Farrington, 1984), the presence of methanogens in the surface sediments may be explained by sulfate-free microenvironments, inactive methanogens, or an abundant supply of substrate for both the methanogens and sulfate reducers in the anoxic, organic-rich sediment (Oremland & Polcin, 1982).

Branched alkenes

The most abundant hydrocarbons in the 0–3 cm depth interval were a C_{20} branched alkene eluting near *n*-C_{17}, and a series of C_{25} polyunsaturated hydrocarbons eluting near *n*-C_{21}. The concentrations of these compounds decreased dramatically with depth in the core (Table 3.4, Fig. 3.5). Tentative identification of these compounds was based on comparison of their mass spectra and retention times with examples from the literature (Barrick *et al.*, 1980; Requejo & Quinn, 1985; Rowland *et al.*, 1985).

The mass spectra of the C_{20} alkene (b20:1) was identical to that of a compound which was identified as 2,6,10-trimethyl-7-(3-methylbutyl)-dodecene (Rowland *et al.*, 1985). Although this compound was isolated from field samples of the green alga *Enteromorpha prolifera*, it was not known whether the alga or other organisms in the algal mat were the source of the alkene. The same or closely related C_{20} monoene has also been identified in recent marine sediments from Puget Sound, where a bacterial origin was proposed because similar C_{20} isoprenoid hydrocarbons occur in certain filamentous bacteria (Barrick *et al.*, 1980). This alkene was not present in the subsurface depth intervals of the Peru sediments (Fig. 3.4).

The compounds eluting near *n*-C_{21} apparently belong to the same suite of cyclic and branched C_{25} alkenes with one to three double bonds that have been identified in other sediments and sediment trap particulate material from the Peru upwelling area (Smith *et al.*, 1983; Volkman *et al.*, 1983). Similar compounds have been identified in a variety of anoxic marine and estuarine sediments (Barrick *et al.*, 1980; Requejo & Quinn, 1983, and references therein). This group of C_{25} highly branched hydrocarbons appears to have similar carbon skeletons which differ in the position or geometry of the double bonds (Rowland *et al.*, 1985).

The origin of these branched alkenes is unknown, although both algal and microbial sources have been proposed. The occurrence of some of the C_{20} and C_{25} alkenes in the green alga *E. prolifera* from intertidal algal mats suggested that algae could be a source for the hydrocarbons (Rowland *et al.*, 1985). Volkman *et al.* (1983) argued that since a large proportion of the lipids in the Peru surface sediments were directly related to phytoplankton origin, the C_{25} alkenes might also be derived from this source. However, in both of these studies other characteristic algal *n*-alkenes were not present, and Rowland *et al.* (1985) suggested that microbial degradation of the *n*-alkenes could explain these differences. Biosynthesis of some of these alkenes by anaerobic bacteria in the sediments is also probable since anaerobic bacteria contain C_{20}, C_{25}, and C_{30} unsaturated hydrocarbons (Holzer *et al.*, 1979).

Table 3.4. *Structural assignments, retention time, and concentrations of branched cyclic and acyclic, steroid, and triterpenoid, hydrocarbons in aliphatic hydrocarbons fractions isolated from Peru box core 7408–1909*

Peak[a] Compound	Composition	Molecular Weight	KI[b]	Concentration ($ng\,g^{-1}$ dry sediment) sample depth (cm) 0–3	10–12	25–27
Cyclic and acyclic branched hydrocarbons						
b20:1	$C_{20}H_{40}$	280	1705	4590	—	—
b25:3	$C_{25}H_{46}$	346	2045	934	491	—
b25:2	$C_{25}H_{48}$	348	2074	2202	680	—
b25:3	$C_{25}H_{46}$	346	2091	2035	2093	30
b25:2	$C_{25}H_{48}$	348	2187	1981	537	—
Total C_{25} branched hydrocarbons				7152	3801	30
Steroid hydrocarbons						
s27:1 cholest-2-ene	$C_{27}H_{46}$	370	2830	110	790	250
s28:1 24-methylcholest-2-ene	$C_{28}H_{48}$	384	2927	—	108	28
s28:2 24-methylcholestadiene?	$C_{28}H_{46}$	382	2954	—	tr	18
s29:1 24-ethylcholest-2-ene	$C_{29}H_{50}$	398	3022	—	tr	tr
Triterpenoid hydrocarbons						
t27:1 22-29,30-trisnorneohop-13(18)-ene	$C_{27}H_{44}$	368	2816	—	200	200
t27:1 22,29,20-trisnorneohop-17(21)-ene	$C_{27}H_{44}$	368	2886	tr[c]	1680	660
t30:1 hop-17(21)-ene	$C_{30}H_{50}$	410	3059	—	200	170
t30:1 hop-13(18)-ene?	$C_{30}H_{50}$	410	3127	—	200	30
t31:0 homohopane	$C_{31}H_{54}$	426	3342	—	tr	tr

[a]See Figure 3.4.
[b]Kovats retention index.
[c]Trace.

The concentration data for the C_{25} alkenes indicates that individual isomers have different rates of diagenetic transformation but all decrease to relatively low levels on a dry weight basis and relative to organic carbon over the depth of the core.

Steroid hydrocarbons

Steroid hydrocarbons identified in the Peru sediment consist of a simple series of C_{27}–C_{29} monosterenes (steroidal alkenes) which were most abundant in the lower section of the core (Table 3.4). The C_{27} monosterene (cholest-2-ene) was the most abundant sterene in each sample and a major component of the hydrocarbon fraction in the subsurface samples (Fig. 3.4). The C_{28} and C_{29} Δ^2-sterenes were below the detection limit in the surface sediment, and only trace components in the subsurface core sections. Other steroid hydrocarbons, including Δ^4- or Δ^5-sterenes, steradienes, steranes, 4-methyl steranes, diasteranes or diasterenes were not present.

The Δ^2-sterenes are though to be diagenetic intermediates in the transformation of biogenic sterols to the steranes, diasteranes and diasterenes found in geologically older sediments (Dastilung & Allbrecht, 1977; Edmunds, Brassell & Eglington, 1980; Gagosian *et al.*, 1980). Their abundance in other Recent anoxic marine sediments (Gagosian & Farrington, 1978) and suspended particulates from the O_2 minimum zone (Wakeham *et al.*, 1983) has been attributed to microbial transformations of steroidal precursors.

However, the relatively low concentrations of sterenes in the surface sediment of the Peru core compared to deeper depth intervals is not consistent with a direct input from either water column fecal pellets or production within the living bacterial mat. The depth profile of the monosterenes may reflect purely chemical diagenetic reactions rather than direct microbial production of sterenes. Alternatively, the microbial community responsible for the transformation of the steroidal precursors may be most abundant in the subsurface sediments, and not associated with the *Beggiotoa* and *Thioploca* mats. An analogous situation occurs in algal mats, where monosterenes are absent from the living layers of the mat but present below the mat (Philp, 1980).

Triterpenoid hydrocarbons

Triterpenoid hydrocarbons present in the Peru box core sediment consist predominantly of C_{27} $\Delta^{17(21)}$-hopene and $\Delta^{13(18)}$-hopene isomers (Table 4.4). Traces of the homologous C_{30} hopene isomers were also present. These hydrocarbons are derived either directly from microbial sources, or indirectly by dealkylation and isomerization of microbial hopenes. Hop-17(21)-ene has been identified in lipids of methylotrophic bacteria, while another C_{30} hopene, hop-22(29)-ene, is a ubiquitous component of procaryotic organisms in general (Bird *et al.*, 1971; De Rosa *et al.*, 1971). In addition, the lipids of many types of bacteria contain a variety of extended (C_{35}) hopanoids that may be the precursors for either the C_{27} or C_{30} hopenes found in the sediments (Ourisson, Albrecht & Rohmer, 1979).

The predominance of the C_{27} hopenes in the Peru sediments is unusual in that other marine sediments from non-upwelling

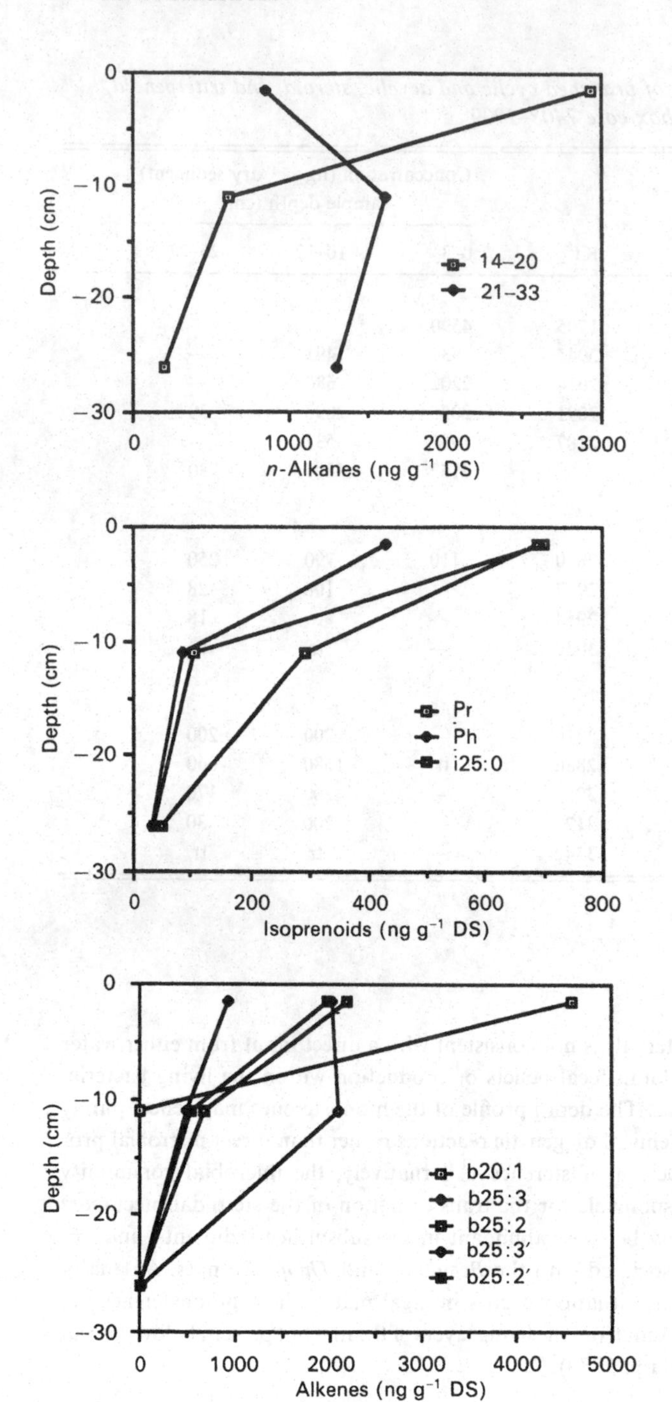

Fig. 3.5. Depth profiles of C_{14-20} and C_{21-33} *n*-alkanes, iC_{25}, and C_{25} branched alkenes.

environments contain a variety of degraded and alkylated (C_{27}–C_{35}) hopanoids (e.g. Simoneit, 1978). This may indicate that in the Peru sediments, there is a direct, but unknown microbial source for the C_{27} hopenes. Alternatively, the hopanoid precursors may be preferentially transformed by microbial hydroxylation and elimination reactions to produce the stable C_{27} hopanoid ring structure. Both processes indicate a high bacterial biomass in the sediments, which is consistent with the other geochemical data. The depth profile of the hopenes, like the sterenes, shows highest concentration in the subsurface rather than the surface depth interval (Table 3.4), thus suggesting that the hopenes are not directly produced in the *Beggiotoa* and *Thioploca* mats.

Discussion

Diagenesis of organic matter and phosphorite formation

Recent studies using uranium series methods to determine the growth rates of phosphorite nodules from the Peru margin have shown that the surface nodules grow downward while buried nodules apparently grow upward in the soft sediment (Burnett *et al.*, 1982; Kim & Burnett, 1986). Additionally, pore-water studies of fluoride display downward decreasing gradients which require diffusion from overlying water into the sediment and removal, presumably by incorporation into phosphorites forming in the upper few tens of centimeters (Froelich *et al.*, 1983). The origin of the phosphorus is unknown, but it is generally thought that regeneration from organic matter is an important source.

The composition of organic matter in the surface (0–3 cm) sediments described here reflects a predominantly marine source from phytoplankton and bacteria. Most of the original algal hydrocarbons have been modified by microbial reworking or modifications. Smith *et al.* (1983) found that highly labile fatty acids derived from phytoplankton were predominant in the surface layer of sediment from Peru. It is reasonable to assume that other labile components derived from phytoplankton, such as phospholipids, nucleic acids, or storage polyphosphates, are transported to the sediments where they are degraded to provide a source of phosphorus for the formation of CFA pellets or nodules.

There are several features of the organic matter in the Peru sediments relevant to an understanding of the mechanism of phosphorite formation in sediments. The relatively low concentrations of phosphorus in the humic acid and protokerogen fractions demonstrate that adsorption of regenerated phosphorus by humic acids is not a significant mechanism for preventing diffusion of regenerated PO_4^{3-} from the sediments. The molecular composition of the hydrocarbons, which indicates a high degree of microbial modification of the organic matter, suggests that a direct or indirect role for microbial processes in phosphorite formation is likely. The C_{20} monoene identified in the surface sediments has structural features similar to C_{20} isoprenoid hydrocarbons in actinomycetes (Barrick *et al.*, 1980). Some of these filamentous bacteria deposit CFA in oral cavities of man and primates (Rizzo, Scott & Blade, 1963), and although the association is tenuous, the possibility that the C_{20} monoene in the surface sediments represents a biomarker for bacteria that can directly precipitate CFA deserves more investigation.

Comparison with organic matter in phosphorites

The distribution of the major fractions of organic matter in the sediment core (Fig. 3.2) shows that phosphorite nodules growing near the sediment surface would incorporate organic matter which consists mainly of humic substances, with smaller amounts of protokerogen and EOM. This is in agreement with the composition of organic matter in geologically older phosphorites, which are characterized by a relatively high proportion of humic acids compared to non-phosphatic sediments (Amit & Bein, 1982; Belyouni & Trichet, 1983; Sandstrom, 1986). This could be due to the preferential adsorption of humic acids by the CFA mineral matrix during growth of the nodules, similar to the preferential enrichment of acidic amino acids in phosphorite nodules from Peru (Cunningham & Burnett, 1985).

Since most of the humic acids appear to become insolubilized with depth in the core, the predominance of humic acids in ancient phosphorites suggests that they formed near the sediment surface where humic acids were more abundant.

At the same time, these results indicate that caution should be exercised in interpretation of the depositional environment of formation of phosphorites from the composition of organic matter. The abundance of humic acids has been attributed to oxidative conditions during deposition of phosphatic sediments (Amit & Bein, 1982; Belyouni & Trichet, 1983; Sandstrom, 1986). However, the co-occurrence of abundant humic acids and the C_{25} isoprenoid derived from methanogenic bacteria in the Peru surface sediments demonstrates that humic acids are not necessarily indicative of oxic sediments. Their abundance in ancient phosphorites may be due to an abundant supply of marine organic matter, formation near the sediment surface, or preferential adsorption rather than oxidative conditions during formation of the phosphorites.

The molecular composition of the surface sediment also shares features in common with geologically older phosphorites. The *n*-alkanes in phosphorites range from C_{15} to C_{24} with no odd/even predominance, have maximum abundance between C_{19} or C_{22}, and are associated with a narrow UCM (Powell *et al.*, 1975; Sandstrom, 1980; Amit & Bein, 1982; Belyouni & Trichet, 1983). This is similar to the suite of C_{14}–C_{20} *n*-alkanes which are abundant in the Peru surface sediment (Fig. 3.4). Phosphorites also contain relatively low levels of extractable steroid and triterpenoid hydrocarbons (Sandstrom, 1980). These hydrocarbons are only abundant in the deeper sections of the Peru sediment, probably because their formation is the result of geochemical as well as microbial processes in the sediments. This implies that the ancient phosphorites without extractable steroid and triterpenoid hydrocarbons formed relatively rapidly near the sediment surface prior to the diagenetic formation of the steroid and triterpenoid hydrocarbons. Alternatively, the steroid and triterpenoid precursors may have been strongly bound by the CFA mineral surface or humic matter in the phosphorites. For example, Belyouni & Trichet (1983) have identified monoaromatic steroids strongly bound to humic acids isolated from Tunisian phosphorites.

Summary and conclusions

Surface sediments from a box core from the Peru upwelling region near 15° S contain relatively high concentrations of extractable lipids and hydrocarbons derived predominantly from phytoplankton and bacteria. These components decrease dramatically with depth in the upper 30 cm of the sediments because of rapid and preferential degradation or insolubilization of these components. Specific molecular indicators of abundant microbial activity, including Δ^2-sterenes, C_{25} alkenes, and the C_{25} isoprenoid 2,6,10,15,19-pentamethyleicosane, indicate the overall decrease in organic carbon with depth in the core is the result of microbial decomposition processes rather than changes in carbon accumulation rate.

These results provide circumstantial evidence for the accumulation of relatively undegraded organic material presumably rich in phosphorus in surface sediments and its subsequent transformation and release in the sediments. The phosphorus content of the humic acid and protokerogen fractions represents < 5% of the total phosphorus in the sediments. Hence the formation of the phosphorites is probably related to early diagenesis of biogenic organophosphorus compounds, e.g. lipids, rather than diagenetic transformation of geopolymers.

Acknowledgments

I am grateful to Peter J. Cook for his support and guidance. Robert Cunningham, William Burnett, and an anonymous reviewer provided valuable comments. Andrew Soutar, Scripps Institution of Oceanography, kindly provided the box core samples. This work was funded by a Post Graduate Scholar award from Research School of Earth Sciences, Australian National University.

References

Amit, O. & Bein, A. (1982). Organic matter in Senonian phosphorites from Israel. Origin and diagenesis. *Chemical Geology*, **37**, 277–87.

Barrick, R.C. & Hedges, J.I. (1981). Hydrocarbon geochemistry of the Puget Sound region. II. Sedimentary diterpenoid, steroid, and triterpenoid hydrocarbons. *Geochimica Cosmochimica Acta*, **45**, 381–92.

Barrick, R.C., Hedges, J.I. & Peterson, M.L. (1980). Hydrocarbon geochemistry of the Puget Sound region. I. Sedimentary acyclic hydrocarbons. *Geochimica Cosmochimica Acta*, **44**, 1349–62.

Baturin, G.N., Merkulova, K.I. & Chalov, P.I. (1972). Radiometric evidence for recent formation of phosphatic nodules in marine shelf sediments. *Marine Geology*, **13**, 37–41.

Belyouni, H. & Trichet, J. (1983). Preliminary data on the origin and diagenesis of the organic matter in the phosphate basin of Gasfa (Tunisia). In *Advances in Organic Geochemistry 1981*, ed. M. Bjoroy *et al.*, pp. 328–35. John Wiley, Chichester.

Bentor, Y.K. (1980). Phosphorites – the unsolved problems. In *Marine Phosphorites: Geochemistry, Occurrence, Genesis*, ed. Y.K. Bentor, pp. 3–17. Society Economic Paleontology Mineralogy Special Publication No. 29, Tulsa.

Bird, C.W., Lynch, T.M., Pirt, S.J. & Reid, W.W. (1971). The identification of hop-22(29)-ene in procaryotic organisms. *Tetrahedron Letters*, **34**, 3189–90.

Blumer, M., Mullin, M.N. & Guillard, R. (1970). A polyunsaturated hydrocarbon (3,6,9,12,15,18-heneicosane) in the marine food web. *Marine Biology*, **6**, 226–35.

Blumer, M., Mullin, M.N. & Thomas, D.W. (1963). Pristane in zooplankton. *Science*, **140**, 974.

Brassell, S.C., Wardroper, A.M.K., Thomson, I.D., Maxwell, I.R. & Eglinton, G. (1981). Specific acyclic isoprenoids as biologic markers of methanogenic bacteria in marine sediments. *Nature*, **290**, 693–6.

Burnett, W.C. (1977). Geochemistry and origin of phosphorite deposits from off Peru and Chile. *Geological Society of America Bulletin*, **88**, 813–23.

Burnett, W.C., Beers, M.J. & Roe, K.K. (1982). Growth rates of phosphate nodules from the continental margin off Peru. *Science*, **215**, 1616–18.

Burnett, W.C. & Veeh, H.H. (1977). Uranium series disequilibrium studies in phosphorite nodules from the west coast of South America. *Geochimica Cosmochimica Acta*, **41**, 755–65.

Burnett, W.C., Veeh, H.H. & Soutar, A. (1980). Uranium series, oceanographic and sedimentary evidence in support of Recent formation of phosphate nodules off Peru. In *Marine Phosphorites*, ed. Y.K. Bentor, Society Economic Paleontologist Mineralogist Special Publication No. 29, pp. 61–71.

Cunningham, R. & Burnett, W.C. (1985). Amino acid biogeochemistry and dating of offshore Peru–Chile phosphorites. *Geochimica Cosmochimica Acta*, **49**, 1413–19.

Dastilung, M. & Allbrecht, P. (1977). Δ^2-Sterenes as diagenetic intermediates in sediments. *Nature*, **269**, 678–9.

De Rosa, M., Gambaciorta, A., Minale, L. & Bullock, J.D. (1971).

Bacterial Triterpanes. *Chemical Communications*, **1971**, 619–20.

Didyk, B.M., Simoneit, B.R.T., Brassell, S.C. & Eglinton, G. (1978). Organic geochemical indicators of paleoenvironmental indicators of sedimentation. *Nature*, **272**, 216–22.

Douglas, A.G., Eglinton, G., & Maxwell, J.R. (1969). The hydrocarbons of coorongite. *Geochimica Cosmochimica Acta*, **33**, 569–77.

Durand, B. & Monin, J.C. (1980). Elemental analysis of kerogens. In *Kerogen* ed. B. Durand, pp. 114–42. Editions Technip, Paris.

Edmunds, K.L.H., Brassell, S.C. & Eglinton, G. (1980). The short term diagenetic fate of 5-cholestan-3-ol: *in situ* radiolabeled incubations in algal mats. In *Advance in Organic Geochemistry 1979* ed. A.G. Douglas & J.R. Maxwell, pp. 427–34. Pergamon Press, Columbus, Ohio.

Eglinton, G. & Hamilton, R.J. (1967). Leaf epicuticular waxes. *Science*, **156**, 1322.

Froelich, P.N., Kim, K.H., Jahnke, R., Burnett, W.C., Soutar, A. & Deakin, M. (1983). Pore water fluoride in Peru continental margin sediments: uptake from seawater. *Geochimica Cosmochimica Acta*, **47**, 1605–12.

Gagosian, R.B. & Farrington, J.W. (1978). Sterenes in surface sediments from southwest African shelf and slope. Geochimica Cosmochimica Acta, **47**, 1091–101.

Gagosian, R.B., Smith, S.O., Lee, C., Farrington, J.W. & Frew, N.M. (1980). Steroid transformations in Recent marine sediments. In *Advances in Organic Geochemistry 1979*, ed. A.G. Douglas & J.R. Maxwell, pp. 407–19. Pergamon Press, Columbus, Ohio.

Gagosian, R.B., Volkman, J.K. & Nigrelli, G.E. (1983). The use of sediment traps to determine sterol sources in coastal sediments off Peru. In *Advances in Organic Geochemistry, 1981*, ed. M. Bjorøy, Wiley & Sons, New York.

Gallardo, V.A. (1977). Large benthic microbial communities in sulfide biota under Peru–Chile subsurface counter current. *Nature*, **268**, 331–2.

Gearing, P., Gearing, J.N., Lytle, T.F. & Lytle J.S. (1976). Aliphatic and olefinic hydrocarbons in 60 northeast Gulf of Mexico shelf sediments. *Geochimica Cosmochimica Acta*, **40**, 1005–17.

Han, J. & Calvin, M. 1969. Hydrocarbon distribution of algae and bacteria and microbiological activity in sediments. *Proceedings National Academy Sciences USA*, **64**, 436–43.

Hatcher, P.G., Simoneit, B.R.T. & Gerchakov, S.M. (1977). The organic geochemistry of a Recent sapropelic environment: Mangrove Lake, Bermuda. In *Advances in Organic Geochemistry 1975*, ed. R. Compos & J. Goni, pp. 469–84. Enadisma, Madrid.

Henrichs, S.M. & Farrington, J.W. (1984). Peru upwelling region sediments near 15° S. Remineralization and accumulation of organic matter. *Limnology and Oceanography*, **29**, 1–19.

Henrichs, S.M., Farrington, J.W. & Lee, C.L. (1984). Peru upwelling region sediments near 15° S. 2. Dissolved free and total hydrolyzable amino acids. *Limnology and Oceanography*, **29**, 20–34.

Holzer, G., Oro, J. & Tornabene, T.G. (1979). Gas chromatographic–mass spectrophotometric analysis of neutral lipids from methanogenic and thermoacidophilic bacteria. *Journal of Chromatography*, **186**, 795–809.

Ikan, R., Aizenshtat, Z., Baedecker, M.J. & Kaplan, I.R. (1975). Thermal alteration experiments on organic matter in recent marine sediments II. Isoprenoids. *Geochimica Cosmochimica Acta*, **39**, 187–94.

Johnson, R.W. & Calder, J.A. (1973). Early diagenesis of fatty acids and hydrocarbons in salt marsh environment. *Geochimica Cosmochimica Acta*, **37**, 1943–55.

Kim, K.H. & Burnett, W.C. (1986). Uranium-series growth history of a quaternary phosphatic crust from the Peruvian continental margin. *Chemical Geology*, **58**, 227–44.

Koide, M. & Goldberg, E.C. (1982). Transuranic nuclides in two coastal marine sediments off Peru. *Earth and Planetary Science Letters*, **57**, 263–77.

Kolodny, Y. (1981). Phosphorites. In *The Sea, Vol. 7: The Oceanic Lithosphere*, ed. C. Emiliani, pp. 981–1023. Wiley & Sons, New York.

Krisseck, L.A., Scheideggar, K.F. & Kulm, L.D. (1980). Surface sediments of the Peru–Chile continental margin and the Nazca Plate. *Geological Society America Bulletin*, **91**, 321–31.

Muller, P.J. & Suess, E. (1979). Productivity, sedimentation rate, and organic matter in the oceans. I. Organic carbon preservation. *Deep Sea Research*, **26**, 1347–62.

Murphy, J. & Riley, J.P. (1962). A modified single solution method for the determination of phosphate in natural waters. *Analytical Chimica Acta*, **27**, 31–6.

Nissenbaum, A. & Kaplan, I.R. (1972). Chemical evidence for the *in situ* origin of marine humic substances. *Limnology and Oceanography*, **17**, 570–82.

Oremland, R.S. & Polcin, S. (1982). Methanogenesis and sulfate reduction: competitive and noncompetitive substrates in estuarine sediments. *Applied and Environmental Microbiology*, **44**, 1270–6.

Ourisson, G., Albrecht, P. & Rohmer, M. (1979). The hopanoids. Paleochemistry and biochemistry of a group of natural products. *Pure and Applied Chemistry*, **51**, 709–29.

Palacas, J.G., Love, A.H. & Gerrild, P.M. (1972). Hydrocarbons in surface sediments of Choctawhatchee Bay, Florida and their implications for genesis of petroleum. *American Association Petroleum Geologists Bulletin*, **56**, 1402–18.

Pentina, T.Y. & Chetverikova D.P. (1977). Structure of the non-bituminous part of organic matter in Recent sediments from the Peruvian region of the Pacific Ocean. *Oceanology*, **17**, 532–5.

Philp, R.P. (1980). Comparative organic geochemical studies of recent algal mats and sediments of algal origin. In *Biogeochemistry of Ancient and Modern Environments*, ed. P.A. Trudinger & M.R. Walter, pp. 173–86. Springer-Verlag, New York.

Poutanen, E.L. & Morris, R.J. (1983). The occurrence of high molecular weight humic compounds in the organic-rich sediments of the Peru continental shelf. *Oceanologica Acta*, **6**, 21–8.

Powell, T.G., Cook, P.J. & McKirdy, D.M. (1975). Organic geochemistry of phosphorites: relevance to petroleum geochemistry. *American Association Petroleum Geologists Bulletin*, **59**, 618–32.

Quinn, W.H., Zopff, D.O., Short, K.S. & Kuo Yang, R.T.W. (1978). Historical trends and statistics of the Southern Oscillation, El Nino, and Indonesian droughts. *Fishery Bulletin*, **76**, 663.

Redfield, A.C., Ketchum, B.H. & Richards, F.A. (1963). The influence of organisms on the composition of seawater. In *The Sea*, ed. M.N. Hill, pp. 26–77. Interscience, New York.

Reimers, C.E. (1982). Organic matter in anoxic sediments off Central Peru: relations of porosity, microbial decomposition and deformation properties. *Marine Geology*, **46**, 175–97.

Requejo, A.G. & Quinn, J.G. (1983). Geochemistry of C_{25} and C_{30} biogenic alkenes in sediments of Narragansett Bay Estuary. *Geochimica Cosmochimica Acta*, **47**, 1075–90.

Requejo, A.G. & Quinn, J.G. (1985). C_{25} and C_{30} biogenic alkene in sediments and detritus of a New England salt marsh. *Estuarine, Coastal, and Shelf Science*, **20**, 281–97.

Rizzo, A.A., Scott, D.B. & Blade, H.A. (1963). Oral calcification in man. *Annals New York Academy Science*, **109**, 14–22.

Rowland, S.J., Yon, D.A., Lewis, C.A. & Maxwell, J.R. (1985). Occurrence of 2,6,10-trimethyl-7-(3-methylbutyl)-dodecane and related hydrocarbons in the green algae *Enteromorpha prolifera* and sediments. *Organic Geochemistry*, **8**, 207–13.

Sandstrom, M.W. (1980). Organic geochemistry of some Cambrian phosphorites. In *Advances in Organic Geochemistry 1979*, ed. A.G. Douglas & J.R. Maxwell, pp. 123–31. Pergamon Press, Oxford.

Sandstrom, M.W. (1986). Geochemistry of organic matter in Middle Cambrian phosphorites from the Georgina Basin, northeastern Australia. In *Phosphate Deposits of the World. Volume 1. Proterozolic and Cambrian phosphorites*, ed. P.J. Cook & J.H. Shergold, pp. 268–79. Cambridge University Press, Cambridge.

Shapiro, L. & Brannock, W.W. (1962). Rapid analysis of silicate, carbonate, and phosphate rocks. *US Geological Survey Bulletin*, **1144-A**, 174–83.

Sheldon, R.P. (1981). Ancient marine phosphorites. *Annual Review Earth Planetary Science*, **9**, 251–84.

Simoneit, B.R.T. (1978). The organic geochemistry of marine sediments. In *Chemical Oceanography, Volume 7*, ed. J.P. Riley & R. Chester, pp. 233–311. Academic Press, New York.

Smith, D.J., Eglinton, G. & Morris, R.J. (1983). The lipid chemistry of an interfacial sediment from the Peru continental shelf: fatty acids, alcohols, aliphatic ketones, and hydrocarbons. *Geochimica Cosmochimica Acta*, **47**, 2225–32.

Sofer, Z. (1980). Preparation of carbon dioxide for stable carbon isotope analysis of petroleum fractions. *Analytical Chemistry*, **52**, 1389–91.

Stuermer, D.H., Peters, K.E. & Kaplan, I.R. (1978). Source indicators of humic substances and protokerogen. *Geochimica Cosmochimica Acta*, **42**, 989–98.

Suess, E. (1981). Phosphate regeneration from sediments of the Peru continental margin by dissolution of fish debris. *Geochimica Cosmochimica Acta*, **45**, 577–88.

Venkatesan, M.I., Sandstrom, M.W., Brenner, S., Ruth, E., Bonilla, J. & Bjoroy *et al.*, pp. 228–40. Wiley, Chichester.

Kaplan, I.R. (1981). Organic geochemistry of surficial sediments from the Eastern Bering Sea. In *The Eastern Bering Sea Shelf: Oceanography and Resources*, ed. D.W. Hood & J.A. Calder, pp. 389–409. University of Alaska Press, Fairbanks.

Volkman, J.K., Farrington, J.W., Gagosiam, R.B. & Wakeham, S.G. (1983). Lipid composition of coastal marine sediments from the Peru upwelling region. In *Advances in Organic Geochemistry, 1981*, ed. M.

Wakeham, S.G., Farrington, J.W. & Volkman, J.K. (1983). Fatty acids, wax esters, triglycerides, and glyceryl ethers associated with particles collected in sediment traps in the Peru upwelling. In *Advances in Organic Geochemistry, 1981*, ed. M. Bjoroy *et al.*, pp. 185–97. Wiley, Chichester.

Walsh, J.J. (1981). A carbon budget for overfishing off Peru. *Nature*, **290**, 300–4.

4

Pore water, petrologic and stable carbon isotopic data bearing on the origin of Modern Peru margin phosphorites and associated authigenic phases

C. R. GLENN

Abstract

Pore water, petrographic, stable-carbon isotopic and lattice-bound CO_2 data from modern Peru margin upper-slope/outer-shelf phosphorites are reviewed which provide insight into their origin and paragenesis relative to co-occurring authigenic minerals (glauconite, pyrite and dolomite). Glauconites are precipitated relatively early following the partial reduction of ferric iron, followed by carbonate fluorapatite, pyrite, and then dolomite precipitation at progressively deeper levels in the sediment. As in many ancient economic phosphorite deposits, the phosphatic facies off Peru consist of nodules, crusts, coatings, and strata dominated by phosphatic peloid grains ('ooids' and structureless 'pellets', intraclasts, biogenic grains etc.) in association with organic carbon-rich biosiliceous sediments. Textural evidence, chemical and isotopic pore-water data and $\delta^{13}C$ values from phosphatic phases suggests phosphate precipitates within a few, to at most a few tens of centimeters below the sediment–water interface in association with suboxic to perhaps anoxic microbial degradation of organic matter. Variations in nodular cement growth rates, birefringence/optical-anisotrophy and crystallinity are suggested to have been produced by variable lattice substitutions that reflect changes in pore-water carbonate ion concentrations and thus the extent of organic matter degradation. Relatively high concentrations of dissolved phosphorus in pore waters that are not related to sulfate reduction processes promote phosphate precipitation directly below the sediment–water interface. High threshold carbonate-ion concentrations at depth apparently poison deeper phosphate precipitation. Below such critical depths, high alkalinity and diminished reactive iron and sulfate favor the development of dolomite while precluding further development of carbonate fluorapatite and pyrite. Periodic sediment reorganization (bioturbation, current winnowing, mass wasting, etc.) plays an important role in mechanically concentrating peloid grains, maintaining phosphatic phases at critical depth levels in the sediment, and in mixing ordered mineral paragenesis into complicated sequences.

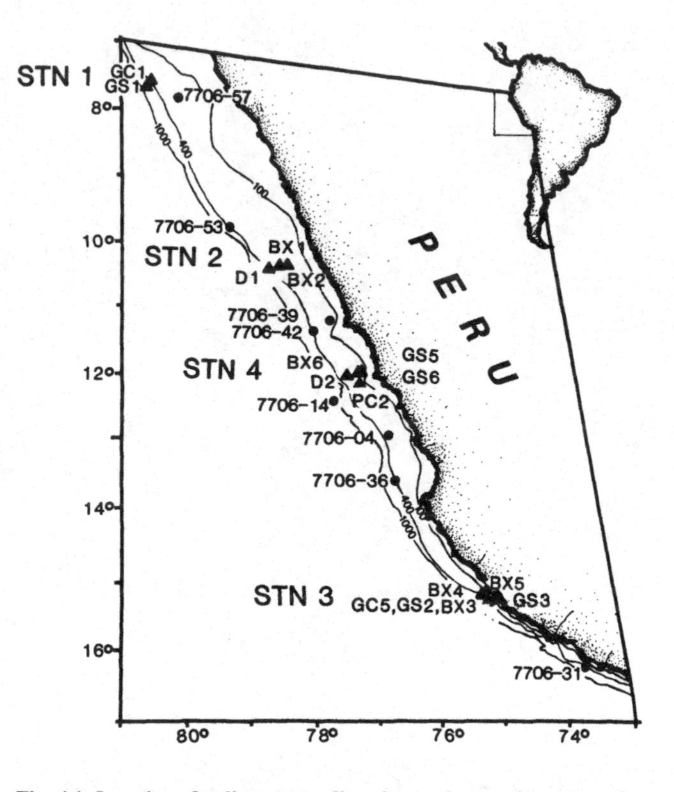

Fig. 4.1. Location of sediment sampling sites at four stations along the outer shelf–upper slope of the Peru margin. (BX) Box core, (D) Dredge haul, (GC) Gravity core, (GS) Grab sample, (PC) Piston core. 7706 sites of Suess (1981) are included for comparison.

Introduction

Similar to many ancient phosphoritic strata cropping out on land, Pleistocene- to Recent-age phosphatic materials occurring along the Peru margin consist of nodules, crusts, pebble- to silt-sized peloid grains ('pellets') and fish debris found in association with organic-rich and biosiliceous sediments. Because of these similarities, modern phosphorites are of great interest because they supply important information which may help to constrain models for the formation of ancient phosphorite occurrences. Although such particulars as facies associations, the size of the deposits, and the depositional or post-depositional mechanics governing the economic upgrading of ancient phosphorites may significantly differ from those associated with modern occurrences, specific pore-water and crystal-chemical parameters controlling phosphate mineralization may be common to both. This paper addresses some of the specifics regulating authigenic phosphate growth and its association with other authigenic phases in modern Peru margin surficial sediments by reviewing the petrology and stable carbon isotopic composition of these components within the light of recent pore-water geochemical results. Details of many of the results summarized here, as well as sampling strategy, geochemical methodology, etc. are discussed in Glenn & Arthur (1988) and Glenn *et al.* (1988). Sampling sites along the margin are shown in figure 4.1.

Geologic setting

Phosphatic minerals of the Peru continental margin chiefly reside within an upper slope/outer shelf mud lens consisting of organic carbon-rich muds and biosiliceous sediments that are a result of low continental fluvial discharge, sustained nutrient upwelling and high biological productivities (cf. Reimers & Suess, 1983a,b; Scheidegger & Krissek, 1983). Oxidative consumption of organic matter in the water column produces a well developed mid-water oxygen-minimum zone which impinges oxygen-deficient waters against outer shelf and upper slope sediments between about 5° S and 20° S (fig. 4.1). Although the character of the sediments is typically finely laminated (to burrow mottled near the outer margins of the oxygen-minimum zone), numerous erosional unconformities in the sediments suggest intermittent slope failure and scouring by bottom currents on scales of thousands of years (cf. DeVries & Pearcy, 1982; Reimers & Suess, 1983a). As discussed below, such variations between sedimentation and sediment reworking/winnowing may have played an important role in concentrating authigenic phosphatic detritus into winnowed lag deposits, as well as serving as a mechanism for maintaining phosphatic phases at shallow burial levels optimum for phosphate precipitation from pore waters.

Petrology

Phosphatic components of the Peru Margin outer shelf/upper slope muds consist of sand- to silt-sized phosphatic grains or peloids ('pellets') and larger accretionary masses such as crusts and nodules. Peloid grains are found randomly disseminated in the sediments or as dense concentrations in friable layers or beds. In a few instances, the majority of the cores were composed of these grains. Phosphatic peloids are also a common component of some carbonate fluorapatite (CFA) nodules and crusts. They are cemented into these accretionary bodies as either randomly disseminated grains or as individually cemented sediment–grain layers.

The chief phosphatic mineral phase in both nodules and peloids is CFA. Although an anisotropic mineral, CFA is frequently cryptocrystalline and thus due to aggregate polarization appears optically isotropic (i.e. psuedo-isotropic) under cross polarization. When the size of CFA crystallites reach a few microns, however, they begin to exhibit low orders of birefringence (first-order greys–pale yellow). This variation in birefringence (or lack of it) with crystallite size is an important point to bear in mind when considering crystal habits of this typically fine-grained material. As discussed below, the variation in CFA crystallite size as observed in freshly deposited sediments may be in part a reflection of variations in pore-water chemistry.

Phosphatic grains

As in many ancient phosphorite deposits (cf. Mabie & Hess, 1964; Cook, 1976; Riggs, 1979), Peru margin phosphatic grains possess a variety of morphologies including structureless peloids, coated grains ('ooids'), phosphatized or originally phosphatic biogenic debris, intraclasts ('polynucleated peloids') erosionally derived from previously cemented substrates, and irregular-shaped phosphatic 'clumps' (Figs. 4.2 and 4.3). The latter are morphologically transitional between structureless peloids and interstitial cements (Fig. 4.4). Subrounded to ovule-shaped cryptocrystalline structureless and coated grains dominate CFA grain abundances here. Whereas structureless peloids generally lack prominent internal morphology, coated grains typically contain a central phosphatic or siliciclastic nucleus grain surrounded by a concentric structure of light yellow brown–dark brown psuedo-isotropic lamina occasionally interlayered with thin (<10 μm) bands of relatively inclusion-free, first-order grey birefringent CFA (Fig. 4.2*b*). Unlike typical carbonate ooids, the shape and internal arrangement of phosphatic coated grains are highly irregular, and their lamina typically thicken and thin, or pinch-out around the grain.

Linear programing and least-squares regression analysis and extended factor analysis of electron microprobe data (for details see Glenn & Arthur, 1988) suggest that the coated grains and structureless peloids are chemically distinct species which reflect differing environments of pore-water precipitation. Structureless peloids are relatively inclusion-free, sodium-substituted phases which probably represent ripped-up and rounded interstitial cements, while coated grains are sodium-poor pore-water precipitates that have incorporated substantial quantities of potassium-bearing siliciclastic detritus, pyrite and organic matter. Dense accumulations of these microinclusions as concentric lamina impart the banded texture to the coated grains. The irregular arrangement of the bands may be caused by restrictions at grain–grain contacts and resumption of precipitation over previously terminated portions of the grains suggests periodic grain rotation and reworking by bottom currents or perhaps by burrowing infauna. Some of the grains have broken edges indicative of such reworking (e.g. Fig. 4.2). The periodic appearance of inclusion-poor birefringent bands may also indicate changes from normal band accretion to perhaps minor episodes of the fringe cement development described below.

The mechanism of band development in Peru margin CFA coated grains and microlaminated crusts (see below) remains unresolved. The presence of organic matter inclusions in the coated grains and crusts suggests that in addition to inorganic CFA precipitation, coating and inclusion-binding may be in some way related to filamentous bacterial or fungal mediation of CFA growth. For example, many microbial-like structures have been observed in a variety of phosphorites (e.g. O'Brien *et al.*, 1981; Soudry & Champetier, 1983; Williams & Reimers, 1983; Lucas & Prévôt, 1984 and 1985; Dahanayake & Krumbein, 1985; Southgate, 1986), and similar CFA-replaced structures have been produced in association with apatite synthesis experiments in the laboratory (Lucas & Prévôt, 1985). Lucas & Prévôt (1985) suggested that such microorganisms locally aid in CFA precipitation by hydrolyzing nucleic acids and thereby liberating

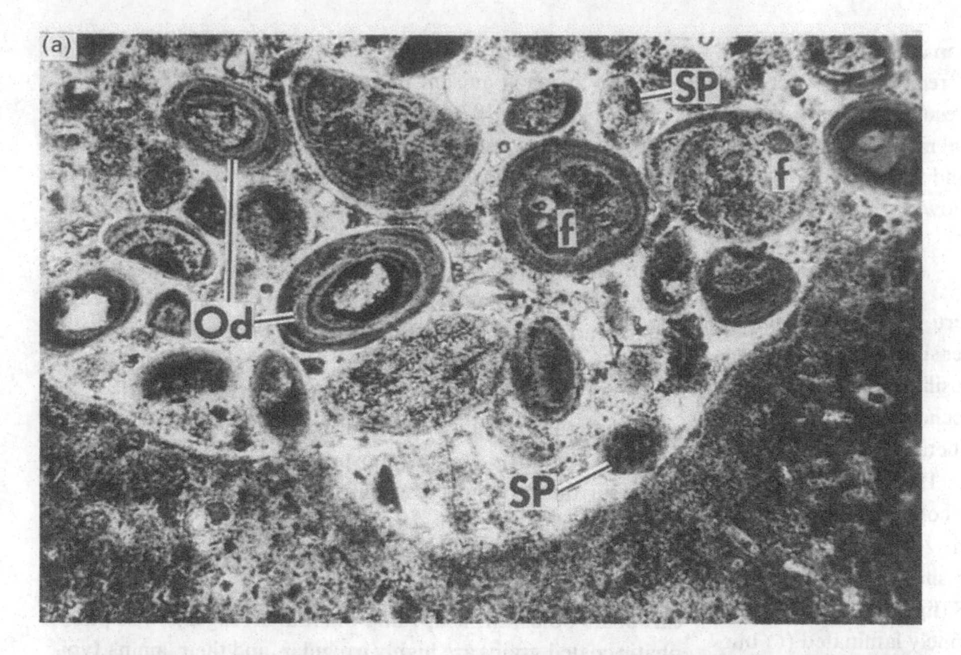

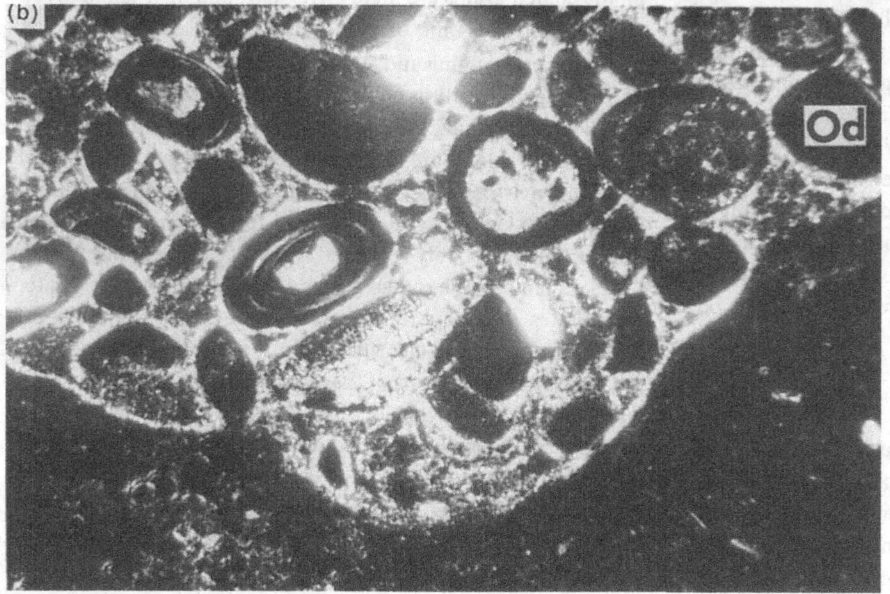

Fig. 4.2. Photomicrographs of phosphatic peloids infilling bored depression in CFA nodule from station 2 (D1): (a) plain light; (b) crossed nicols with maximized light intensity to emphasize various cements. Note birefringent grain-fringing and partially birefringent interstitial CFA cements. (Od) Coated peloids, (SP) Structureless peloids, (f) CFA replaced and coated foraminifera. Note broken peloids and birefringent interlayers in coated grains. Horizontal field = 1.3 mm.

orthophosphate to solution. However, in addition to microbial-binding mechanisms, dissolved and particulate organic compounds may also be electrochemically adsorbed on grain surfaces, and the binding of siliciclastic impurities may be in turn due to the high adsorption capacity of such compounds for oppositely charged clay minerals and dissolved metallic cations in pore waters (cf. Rashid, 1985). Thus, the distinction between abiotic and true biologically-mediated mechanisms of CFA growth and inclusion binding remains an important question in phosphorite research.

Nodules

The internal fabric of Peru margin phosphate nodules is highly variable between samples and especially between stations. At Station 1 (Fig. 4.1), near the outer margin of the oxygen-minimum zone, the nodules are relatively non-layered semi-friable–well-indurated structures containing scattered glauconite and minor phosphate grains and well-preserved burrow structures (Fig. 4.5). South of Station 1, under more oxygen-deficient bottom waters, the nodules are typically well-indurated darkly colored bodies often composed of alternating, lenticular

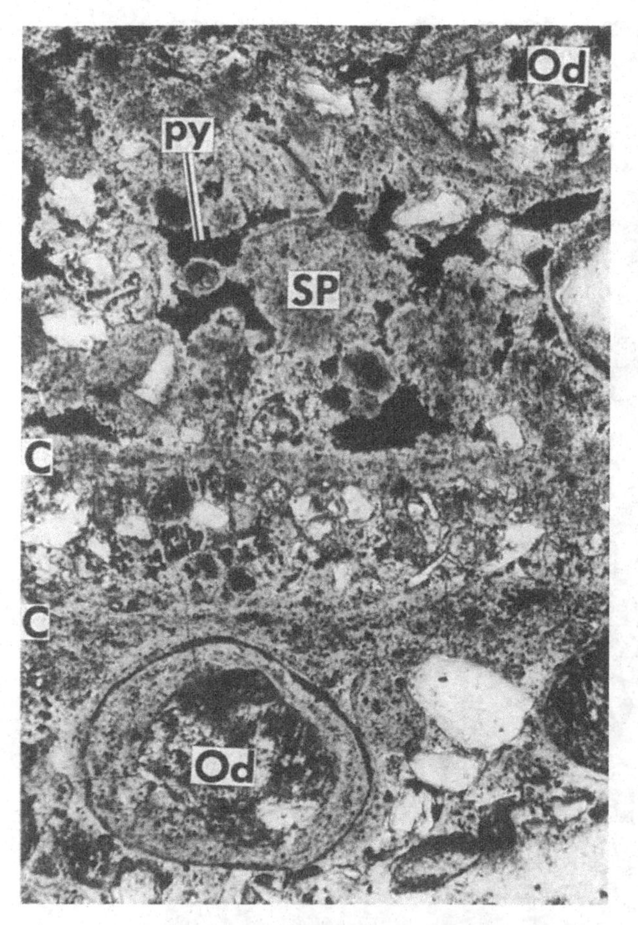

Fig. 4.3. Photomicrograph of three distinctly sorted and cemented CFA and siliciclastic grain layers in nodule from GS5. Layers are consecutively cemented by phosphatic fringe cement, partially developed interstitial phosphatic cement, and finally, opaque (black) interstitial pyrite (py). Microlaminated crusts (C) separate the layers. Structureless (SP) and coated (Od) peloids are also indicated. Plain light; long dimension of photo is 1.3 mm.

layers of phosphatically cemented size- and/or morphologically-sorted phosphatic, biogenic, and siliciclastic grains (Figs. 4.3 and 4.6). At Station 4, where bottom-water oxygen contents are a minimum and organic carbon contents reach a maximum, the phosphatically-cemented layers form phosphatic crusts rinding dolomicritic cores (Fig. 4.6). Individual sediment layers coating the nodules often *each* record the sequential precipitation of: birefringent grain-fringing and pore-lining CFA cements, followed by infilling remaining pore space by more equant and psuedo-isotropic interstitial CFA cements, and finally capping each sediment–cement generation with a thin, completely psuedo-isotropic and relatively inclusion-free microlaminated CFA crust. The growth of such layered nodules is therefore punctuated, with each additional sediment layer being bound by a new triple-generation of phosphatic cement.

Relative to other CFA cements, *fringing cements* (Fig. 4.2) are relatively coarse-grained and inclusion-free anisotropic CFA cements which occur as thin (< 13 μm) rims surrounding grains and lining cavities in a near isopachous manner. A prerequisite for their growth appears to be ample pore space in which to grow or possibly rapid, displacive growth. Crystals show negative elongation (length-fast) and are aligned normal to grain surfaces. These cements are similar to the occasionally interlayered birefringent bands of the coated grains described above, and both may be morphological variants produced by the same basic process. *Interstitial CFA cements* (Figs. 4.2, 4.3 and 4.4) range from psuedo-isotropic to more birefringent varieties transitional with fringe cements, and cements binding individual grain layers are usually distinct from one another in terms of their relative psuedo-isotropism, crystallinity, and microinclusion staining. Increases in interstitial cement psuedo-isotropism (and thus diminishing crystallite size) are usually accompanied by increases in the degree of carbonate microfossil phosphatization and a progressive darkening of the groundmass due to increases in organic matter and possibly pyrite inclusions (cf. Fig. 4.7). Such progressive phosphatization is similar to the 'convergent diagenesis' of Negev peloids described by Soudry & Nathan (1980). Interstitial CFA cements may also display meniscus-like textures when their development has been abruptly halted (Fig. 4.3). Episodically sealing and thus separating individually-cemented sediment layers in the nodules are thin, relatively inclusion-free *microlaminated coatings or crusts* now apparent as yellow–brown anastomosing seams cross-cutting the nodules (Figs. 4.3 and 4.7). Although occasionally containing very thin birefringent layers, these binding cements are usually completely psuedo-isotropic. Unless pitted by biogenic borings, the upper surface of individual crusts or seams are usually planar. Their lower (nodule-inner) surfaces, however, are usually undulose, following the relict microtopography of previous nodule surfaces (Fig. 4.3). As these cements encrust previously cemented depositional packages (i.e. lenses or layers of peloid grains), they represent the last stage of CFA precipitation in the lithification of each package.

Mineral paragenesis

Diagenetic accessory minerals associated with the Peru margin phosphorites include glauconite, pyrite and dolomite, and their occurrence in the phosphatic nodules and peloids supplies important information about various stages of organic matter degradation and mineral paragenesis characterizing Peru margin sediments during CFA precipitation. The distribution of these minerals generally appears to be one reflecting the depth distribution of suboxic to anoxic bacterial populations in marine pore waters (Table 4.1). Glauconite is most abundant in the phosphatc nodules preserving sediment burrow structures recovered from beneath the relatively oxygenated bottom-water environment at Station 1 (see also Burnett, 1980). As glauconite contains iron in both its reduced and oxidized states, its development is probably very early, taking place within suboxic sediments near the sediment–water interface after partial reduction of ferric iron. There, soluble reduced iron may be reoxidized and incorporated along with small amounts of Fe^{2+} during the glauconitization process.

Both CFA and pyrite replace, and thus generally post-date glauconite formation at Station 1 (Glenn & Arthur, 1988). Pyrite is an important earmark for sulfate reduction processes (cf. Berner, 1984 and 1985) and is a ubiquitous phase in the Peru margin phosphorites. Its formation is favored by available reac-

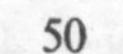

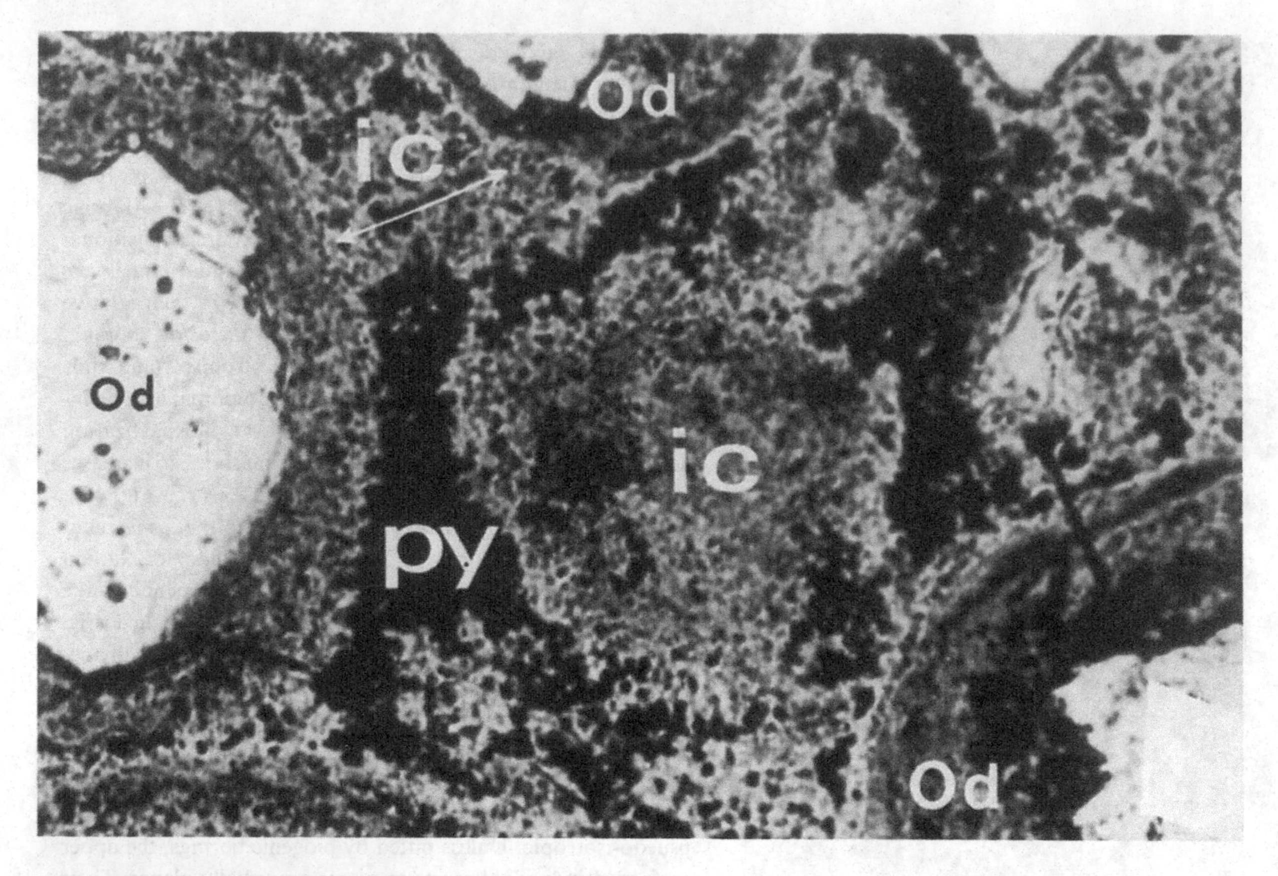

Fig. 4.4. Close-up of interparticle cements as in top layer in Figure 4.3. After reworking, interstitial CFA cements (ic) may serve as starting points in the development of structureless peloids. Note meniscus-like appearance of cement between grains and that although pyrite (py) is void filling, replacement of CFA is also indicated by small pyrite embayments along coated peloid (Od) and cement boundaries. Plain light; long dimension of photo is 125 μm.

tive iron and an abundance of highly reactive marine organic compounds in these sediments. Within Peru CFA nodules it occurs as both a common mineral inclusion in all CFA grain and cement morphologies as well as a replacement (chiefly after glauconite and dolomite) and pore-filling cement. Precipitation of pyrite and CFA appears to be nearly coincident, with pyrite precipitation continuing beyond that of CFA where it may replace the CFA, or infill remaining pore space after partial interstitial CFA cementation (cf. Figs. 4.3 and 4.4; Glenn & Arthur, 1988).

Well-ordered dolomite nodules form the cores of many of the phosphate nodules recovered from Station 4 (Fig. 4.6). Carbon isotopic ratios of these dolomites (see below) indicate precipitation within the zone of microbial sulfate reduction and later methanogenesis. Subsequent biological boring and encrustation of these dolomicrites by multiple generations of phosphatic and pyritic cements (Fig. 4.6), however, indicates their later exhumation and exposure at the sea floor followed by reburial. Results of factor analysis of geochemical data from the phosphatized portions of the dolomite cores of these nodules (cf. Glenn & Arthur, 1988) suggests that as dolomite phosphatization takes place it does so as a dissolution–reprecipitation process with the two minerals remaining as intimately mixed, yet structurally distinct phases.

Stable carbon isotopic results

Carbon isotope and lattice-bound CO_2 data from CFA may be used to help constrain where and when phosphorite formation occurs. The primary assumption is that as CFA phases grow, they do so in isotopic and chemical equilibria with the total dissolved carbon (TDC) in the waters from which they precipitate (see discussions in Glenn *et al.*, 1988). If CFA precipitates directly from seawater at the sediment–water interface, then the carbon isotopic composition of incorporated CO_2 should be characteristic of bottom water (about 0 per mil (‰), relative to PDB). If, on the other hand, CFA precipitates within the sediments, the $\delta^{13}C$ values of lattice-bound CO_2 should mimic those of the pore-water TDC and display relatively extreme negative to positive values characteristic of suboxic to anoxic organic matter degradation processes (Table 4.1).

The isotopic composition and abundance of TDC in marine sediments depends primarily on the relative proportion of CO_2 derived from organic matter degradation in the sediments and that of the TDC of bottom waters overlying the sediments. Degradation of marine organic matter typically contributes CO_2 with a carbon isotopic composition of about −18‰ to 24‰, while the carbon isotopic composition of modern deep water TDC is about +/−0.5‰. Because of high rates of organic matter degradation at most of the Peru margin sites, the TDC–$\delta^{13}C$ in

Table 4.1. *Successive microbial oxidation of organic mater (CH_2O) and change in the isotopic composition of marine pore waters with downward decreasing metabolic free energy yields (after Curtis, 1977; Irwin, Curtis & Coleman, 1977; Froelich* et al., *1979; Berner, 1980; Coleman, 1985; and others).*

TDC-$\delta^{13}C$	Environment	Diagenetic zone	Reaction
FROM:	oxic	aerobic oxidation	$CH_2O+O_2 \rightarrow CO_2+H_2O \rightarrow HCO_3^-+H^+$
+/−0.5‰ (bottom water)			
~	suboxic	manganese reduction	$CH_2O+3CO_2+H_2O+2MnO_2 \rightarrow 2Mn^{2+}+4HCO_3^-$
~		nitrate reduction	$5CH_2O+4NO_3^- \rightarrow 2N_2+4HCO_3^-+CO_2+3H_2O$
~		ferric iron reduction	$CH_2O+7CO_2+4Fe(OH)_3 \rightarrow 4Fe^{2+}+8HCO_3^-+3H_2O$
~	anoxic	sulfate reduction	$2CH_2O+SO_4^{2-} \rightarrow H_2S+2HCO_3^-$
TO:			
−25‰			
		methanogenesis	$2CH_2O \rightarrow CH_4+CO_2$
TO: +25‰			

pore waters decreases sharply with increasing sediment depth as TDC, total alkalinity (TA) and carbonate ion concentration increases during early diagenesis (Fig. 4.8; the carbonate concentration of the pore waters may be approximated by the difference between TA and TDC as TA is dominated by the carbonate alkalinity). The depletion in TDC–$\delta^{13}C$ in organic carbon-rich marine sediments takes place in suboxic to anoxic sulfate-reducing environments up to the point at which all dissolved sulfate has been consumed by sulfate-reducing bacteria. Authigenic carbonate-bearing mineral phases that precipitate within these zones may thus exhibit values ranging from about 0‰ for phases forming near the sediment–water interface, to values as negative as −24‰ for samples forming in association with extensive sulfate reduction (Table 5.1). Concurrent with or following the last sulfate reduction processes methane-producing bacteria begin to utilize organic matter and/or TDC in pore waters whereby extremely light biogenic methane is produced leaving residual TDC enriched in ^{13}C (cf. Claypool & Kaplan, 1974). Therefore, below the zone of sulfate reduction, methane production results in a progressive increase in TDC–$\delta^{13}C$ from about −24‰ to about +25‰, depending on the extent of methanogenesis (Table 5.1). Authigenic mineral phases that precipitate in this zone may exhibit a range in ^{13}C from very depleted to very enriched.

Carbon isotope ratios (relative to PDB) for the structural carbonate of Peru margin CFA and dolomite are shown in Figure 4.9. $\delta^{13}C$ values obtained from CFA nodules and CFA grains are about the same, ranging from about −5‰–0‰ and −3‰–0‰, respectively. In comparison with pore-water $\delta^{13}C$ values from these sediments (e.g. Fig. 4.8), the range of CFA $\delta^{13}C$ suggests that all CFA morphologies have formed at or near the sediment–water interface in largely suboxic pore waters to perhaps those characterized by initial sulfate reduction. These shallow depths are also suggested by dissolved interstitial fluoride uptake by CFA within a few to perhaps a few tens of centimeters of the sediment–water interface (Figs. 4.10, 4.11 and 4.12; for detailed discussion see Froelich *et al.*, 1983 and 1988). The dolomitic cores of CFA nodules recovered from 12° S, however, display both more strongly negative CFA-$\delta^{13}C$ values (to about −11‰), which suggest precipitation in the zone of sulfate reduction, and more positive values (to about +6‰) indicative of precipitation at greater depth in the sediment within the zone of methanogenesis (cf. Fig. 4.9 and Table 4.1). Although ages for the dolomitic (interior) portions of the nodules encrusted by CFA have not been determined, it is apparent that they must have formed first, perhaps meters below the sediment–water interface, and then later have been exhumed and exposed to the seafloor (where they were bored) prior to their reburial and encrustation by CFA and pyrite.

Sources of phosphorus

The petrographic and isotopic considerations discussed above indicate the Peru margin CFA nucleates within a few to at most a few tens of centimeters of the sediment-water interface, largely in association with early microbial suboxic organic matter degradation processes. At first, this result may seem somewhat surprising because in most oceanic settings elevated pore-water phosphorus concentrations are normally encountered with depth in the sediment, chiefly in association with anoxic sulfate-reduction processes (e.g. BX6, Fig. 4.12). However, sulfate reduction is not the only source of interstitial phosphorus in these sediments. Shallow pore-water phosphate maxima of at least 20 μM PO_4, such as those shown in figures 4.10, 4.11 and 4.12, are also commonly present within the upper few centimeters of these sediments, and these maxima show no correlation to decreases in dissolved sulfate. Figure 4.13 illustrates the relationship between dissolved sulfate and phosphate in the pore waters for the variety of the cores and indicates that there is much dissolved phosphate in excess of that attributable to sulfate reduction processes in these sediments (also see Suess, 1981). Similar interfacial phosphate spikes are also found in association with the formation of Recent phosphorite along the Mexican continental margin (Jahnke *et al.*, 1983) and off the eastern coast of Australia (O'Brien & Heggie, 1988). These interfacial phosphate spikes are probably linked to suboxic organic degradation processes at the interface (cf. Froelich *et al.*, 1979 and 1988) and/or the release of adsorbed phosphorus from ferric oxyhydroxides upon encountering reducing conditions in pore waters (cf. Berner, 1973; Krom & Berner, 1980 and 1981; De Lange, 1986; Glenn, 1987; Froelich *et al.*, 1988; O'Brien & Heggie, 1988). Another possibility is that the phosphate spikes are in some way related to the metabolic activity of sulfur-oxidizing bacterial mats (*Thioploca* sp.) commonly present on

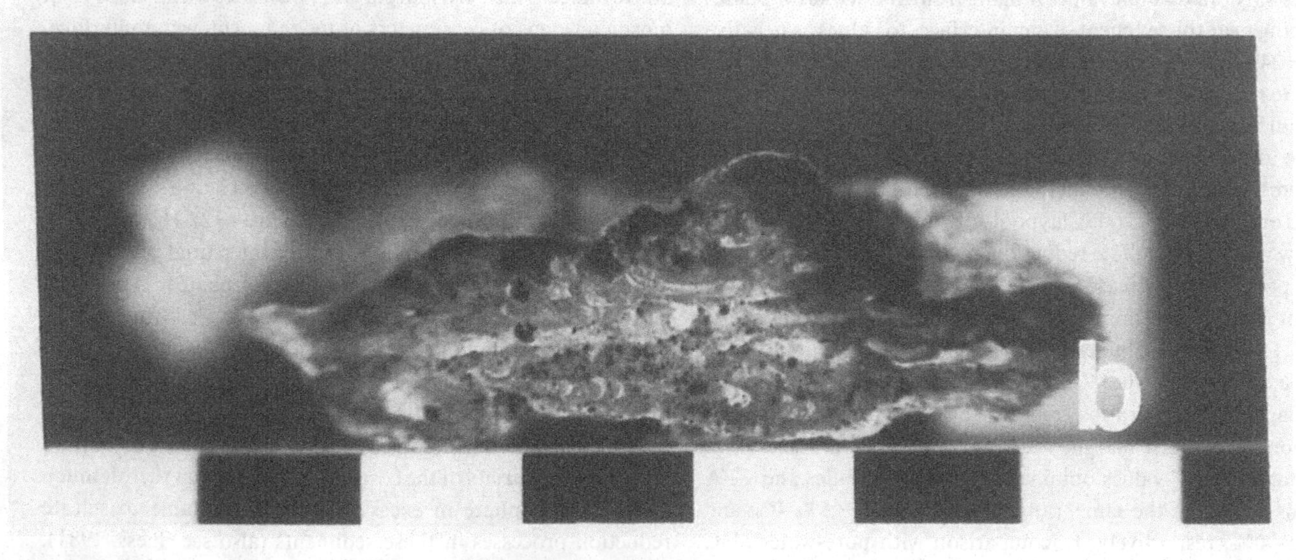

Fig. 4.5. (a) Surface morphology of glauconitic CFA nodule from Station 1 (GS1); light color and poor consolidation is typical of more recent nodules. Note darker features which may be borings of semilithified crust. (b) Cross-section of more indurated glauconitic CFA nodule from Station 1 (GS1). Note well preserved burrows with backfill structures. Scale bars are 1 cm.

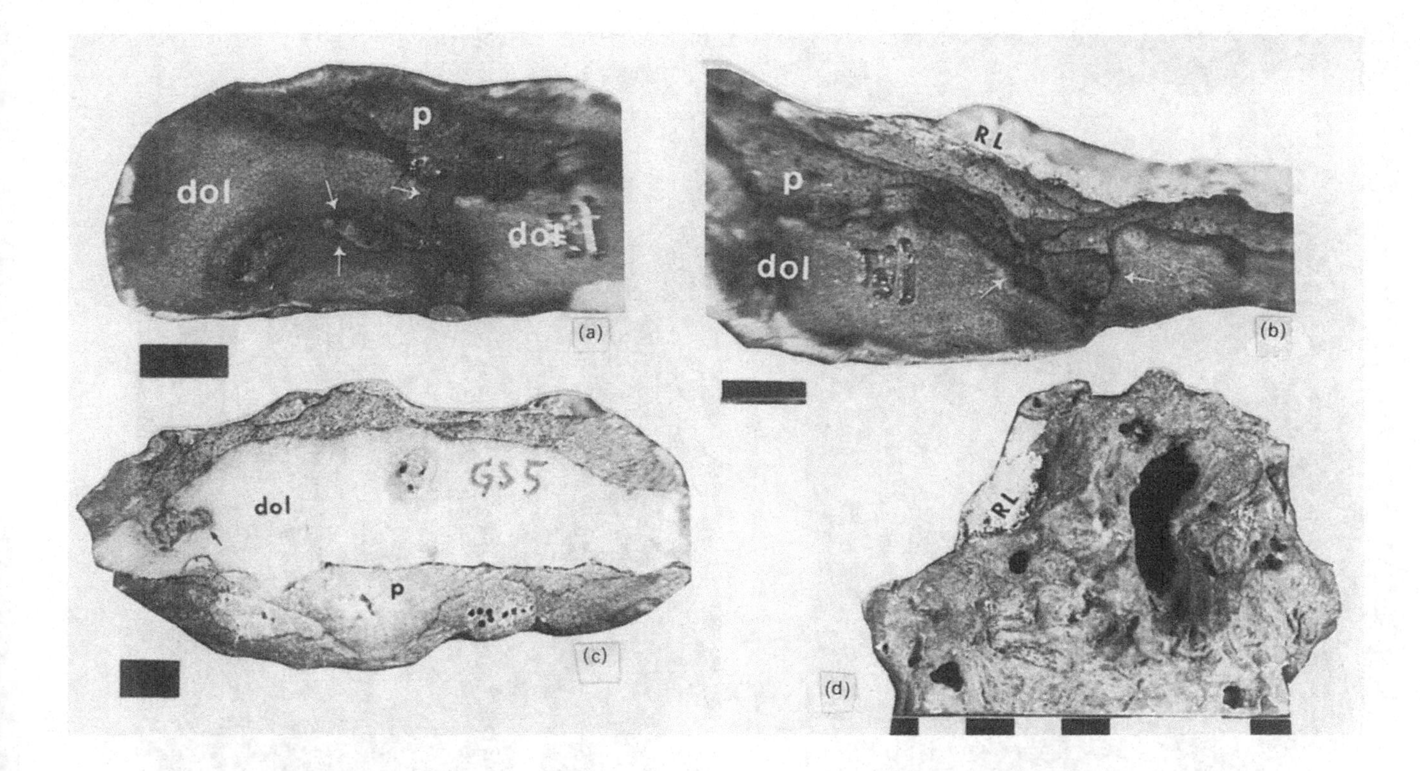

Fig. 4.6. Phosphate nodules from Station 4. The highly bored interior portions of these nodules (arrows) are composed of dolomicrite (dol) and are encased by multiple generations of phosphatic sediment and cement (p), the most recent layers of which (RL) are typically lighter-colored and relatively poorly consolidated (Figs. 4.6B and D). At least six CFA layers are present above bored surface in (B). Outer irregular surfaces of sample shown in (D); convolutions in surface are phosphatized annelid encrustations. Small holes and grooves in slabbed surfaces are areas sampled for isotopic analyses. Scale bars in cm.

the seafloor and in the sediments in this region. The dissolution of fish debris in near-surface sediments is another potential source (cf. Suess, 1981; Froelich *et al.*, 1983 and 1988). Because in most instances interstitial phosphorus appears to be readily available throughout the upper meter or more of Peru margin sediments (Figs. 4.10–4.13) due to a variety of potential sources, an intriguing problem thus becomes explaining why CFA precipitation is limited to within a few centimeters of the sediment–water interface.

Carbonate substitution in CFA

While the availability of dissolved phosphorus and fluoride obviously play important roles in regulating the precipitation of CFA (cf. Froelich *et al.*, 1983 and 1988), perhaps less well appreciated is the potential influence of dissolved carbonate ion in pore waters. Carbonate substitution in CFA has been well studied and quantified for a number of deposits, and it is now accepted among most workers that CO_3^{2-} substitutes for PO_4^{3-} on a 1:1 molar basis with the charge imbalance being chiefly satisfied through additional incorporation of F^- and substitution of Na^+ for Ca^{2+}, in the most simple case (cf. McClellan, 1980; Nathan, 1984; Glenn & Arthur, 1988). Some of the porewater F^- illustrated in Figs. 4.10–4.12 may be being consumed in this manner. The maximum amount of substitution (about 6.3% CO_2 (by weight) as measured by X-ray diffraction (XRD)) may be structurally controlled by the number of unit cell repetitions that are possible before excessive carbonate substitution disrupts the lattice and prevents further growth of CFA crystallites (McClellan & Lehr, 1969). This suggestion is generally supported by a number of independent studies. LeGeros *et al.* (1967) and McClellan & Lehr (1969) demonstrated that carbonate substitution severely limited the size and acicularity of apatite crystallites, and Ames (1959) demonstrated that increased rates of synthetic apatite formation (after carbonates) were correlated with decreases in the ratio of $HCO_3^-:PO_4^{3-}$ bathing the precipitate. More recent experimental studies further indicate that CFA stability may generally increase with decreasing pH (Nathan & Sass, 1981) and thus decreasing carbonate speciation (e.g. Krauskopf, 1979), and that carbonate substitution in synthetic CFA increases CFA solubility (Jahnke, 1984). Thus, it would appear that if CFA growth rates and crystallite size are inversely proportional to the carbonate-ion concentration of formational waters, then there may exist a threshold $CO_3^{2-}:PO_4^{3-}$ activity ratio in pore waters above which CFA cannot precipitate (Glenn & Arthur, 1988; Glenn *et al.*, 1988). This ratio is apparently minimized in uppermost Peru margin sediments where most CFA precipitation is inferred because

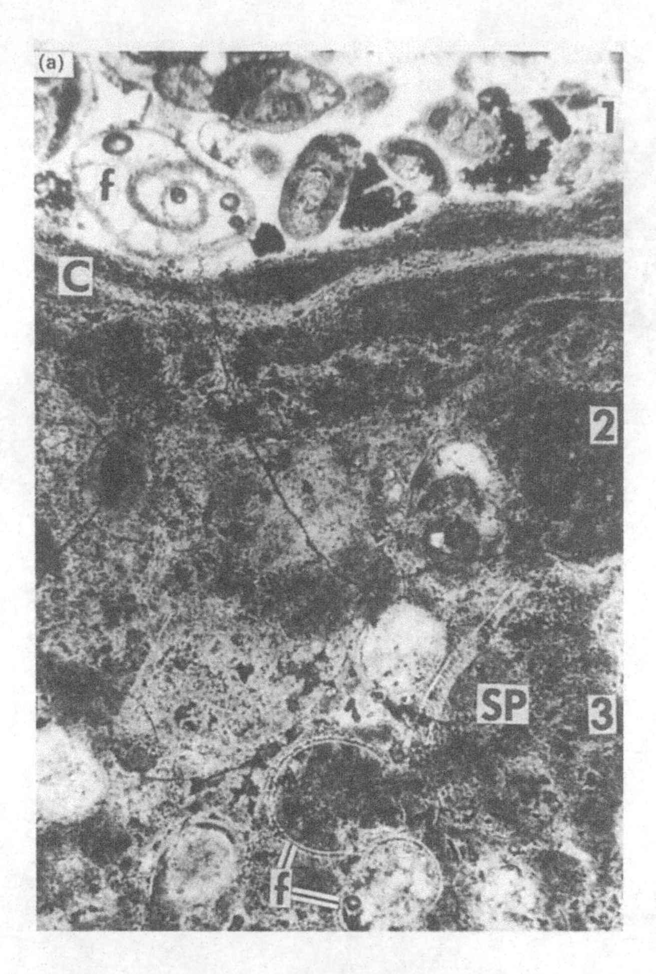

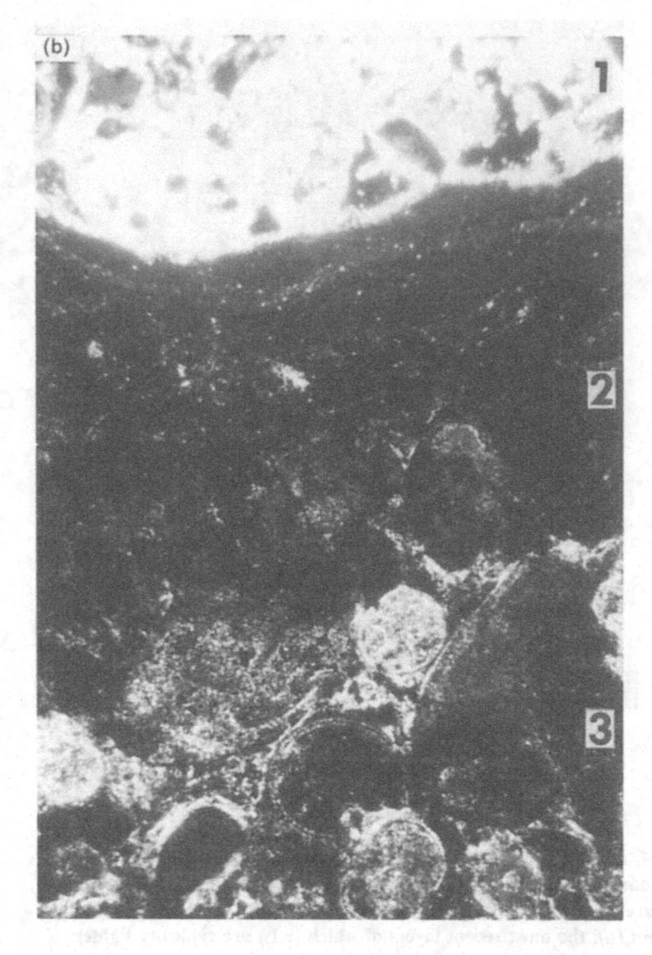

Fig. 4.7. Various degrees of phosphatization (2 > 3 > 1) in separate grain layers encrusting a CFA nodule from D-1. (a) plain light; (b) crossed nicols. Long dimension of view is 1.3 mm. Light intensity maximized to show detail in more heavily phosphatized layer 2. f = foraminifera, C = microlaminated crust, SP = structureless peloid with altered ('phosphomicritized') fringe cement. Cement in layer 1 is grain fringing and foraminifera have not been replaced. In layer 3 foraminifera, CFA peloids and fringe cements are still discernable, but foraminifera have been replaced by birefrigent CFA. In layer 2 both grains and cement are dark brown in plain light and approach complete optical isotropism under crossed nicols. A *minimum* of three discreet depositional–diagenetic events is indicated.

phosphate concentrations are abnormally high (Figs. 4.10–4.12) and carbonate-ion concentrations are relatively low (Fig. 4.8). Below the suboxic interfacial phosphate maxima, however, this ratio increases as a result of continued CO_2 and alkalinity production, phosphate consumption by CFA, and ultimately by decreases in phosphate production as a consequence of organic matter degradation.

Figures 4.14, 4.15 and 4.16 illustrate that there is a general trend for increasing lattice-bound carbonate contents (as determined by XRD and expressed as CO_2) with decreasing $\delta^{13}C$ values in Peru-margin CFA nodules and to a lesser extent in CFA grains. This general trend can be attributed to a progressive increase in carbonate substitution in the CFA lattice in response to increased pore-water carbonate ion concentrations below the sediment–water interface as the result of organic matter decomposition (Glenn *et al.*, 1988; Arthur, Glenn & Froelich, unpublished; cf. Fig. 4.8). The scatter in these figures at least partially reflects the variability in pore-water $\delta^{13}C$ compositions and carbonate ion contents between sites and between individually analyzed components. Figures 4.15 and 4.16 provide two examples of this relationship for specific settings. Figure 4.15 shows the $\delta^{13}C$–CO_2 relationship for peloidal CFA and phosphatized biogenic debris extracted from the peloid-rich layers at Station 4 (GS-6), and Figure 4.16 illustrates the relationship for consecutive radiometrically-dated slices sub-sampled from one CFA nodule (GS-1n) at Station 1 (see below). In both figures note the general trend for increasing carbonate contents with progressive decreases in CFA $\delta^{13}C$.

Figure 4.15 shows the variation in structural CFA–CO_2 as a function of $\delta^{13}C$ for various size classes of CFA peloids and for phosphatized biogenic debris separated from sample GS-6. The data points for bulk peloids and phosphatized biogenic debris in this figure each represent tens of grains collected from differing depth intervals in this highly phosphatic core. These data suggest

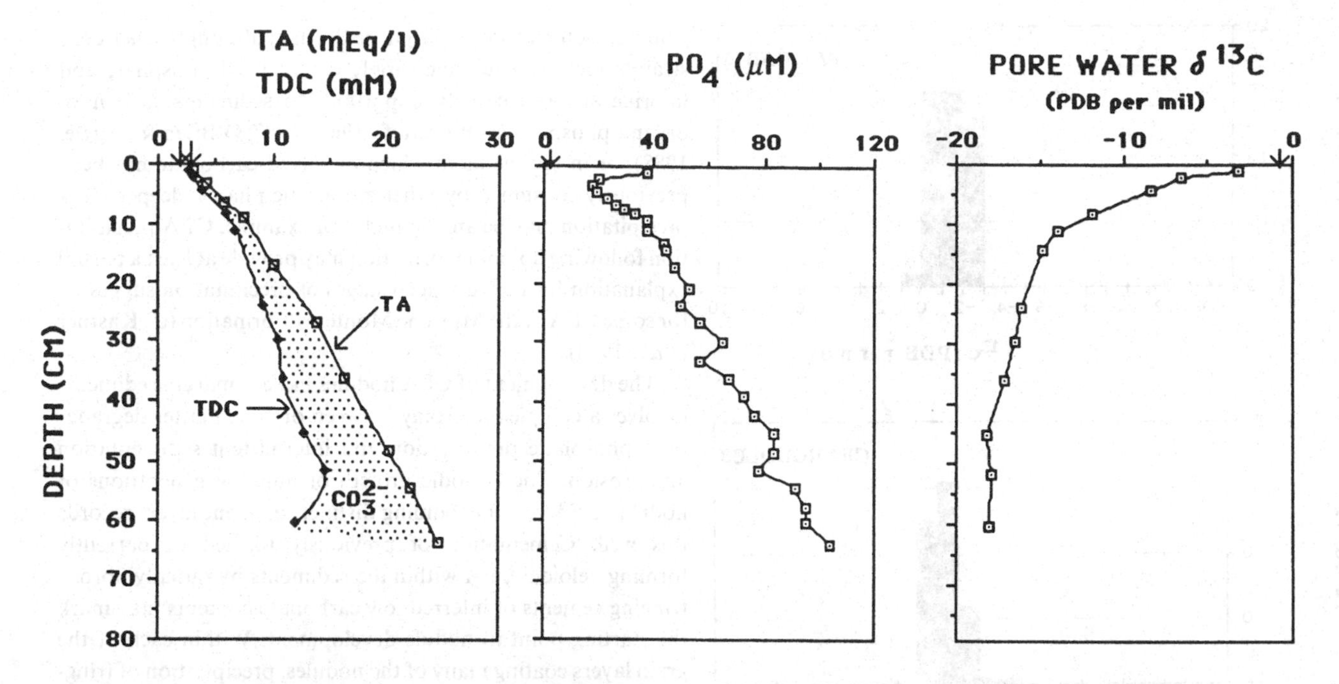

Fig. 4.8. Pore-water profiles of total alkalinity (TA), total dissolved carbon (TDC), dissolved phosphate and TDC-$\delta^{13}C$ for core BX-6. Arrows along axes indicate bottom-water concentrations. Width of stippled area (difference between TA and TDC) approximates total dissolved carbonate-ion in solution (mMl^{-1}). CO_3^{2-}/PO_4^{3-} ratios are low in the zone of inferred CFA precipitation near the sediment–water interface.

that carbonate substitution in CFA decreases in the order of larger peloids > smaller peloids > phosphatized biogenic debris, that CFA-^{13}C is depleted in peloids relative to phosphatized biogenic debris, and that only pelletal grains larger than 600 μm are more depleted than about −2‰. Thus, pellet enlargement and phosphatization of biogenic debris appears to take place as pore-water alkalinity and carbonate ion concentrations increase and TDC-$\delta^{13}C$ values decrease as organic matter is progressively degraded in the sediment. The relatively tight clustering of $\delta^{13}C$ and XRD–CO_2 values for the smaller and most abundant size class of peloids (< 600 μm; cf. Baker & Burnett, 1988) from throughout core GS-6 (Fig. 4.15) reinforces the interpretation that most pellet grains form relatively early and probably within a few centimeters of the sediment–water interface. The spread of $\delta^{13}C$ values for the individual pebble-sized (> 2 mm) peloids extracted from a single sedimentary layer in sample GS-6 (Fig. 4.15), however, may indicate reworking (either by transport or bioturbation) of grains derived from a variety of pore-water environments. Four of these grains show anomalous C^{13} enrichment which may indicate that they formed in association with the dissolution of precursory carbonates, or that these grains were repeatedly phosphatized (and substituted) very close to the sediment–water interface. Other CFA–$\delta^{13}C$/CO_2 variations in peloid grains are discussed in more detail by Glenn *et al.* (1988).

Figures 4.16 and 4.17 illustrate results from an oriented CFA nodule recovered from Station 1 which grew upwards from within the sediments towards the sediment–water interface (cf. Kim & Burnett, 1985). The $\delta^{13}C$ of this nodule increases as a function of decreasing age (Fig. 4.17), starting at −2.1‰ about 6250 BP and increasing about +0.5‰ per thousand years until about 4900 BP. This trend of upwards growth with ^{13}C enrichment (Fig. 4.17) is in agreement with the concept of pore-water ^{13}C depletion with depth (e.g. Fig. 4.8), and suggests that precipitation occurred within a few centimeters of the sediment–water interface. Furthermore, because growth rates for this nodule (1.2–1.3 cm ky^{-1}) are about an order of magnitude lower than the average sedimentation rate at this site, (ca. 10 cm ky^{-1}; Kim & Burnett, 1985), some mechanism of sediment removal (such as winnowing and erosion) is required in order to have maintained the upper surface of the nodule at sediment depths shallow enough to sustain such relatively heavy $\delta^{13}C$ values. Figure 4.16 shows the relationship between CFA XRD–CO_2 and $\delta^{13}C$ for two laterally adjacent portions (A and B) of the nodule. In accord with the results for CFA-peloid grains, both portions of the nodule show similar general trends of decreasing CFA structural carbonate with relative enrichments in ^{13}C. Note in Figure 4.17 that molar accumulation rates of phosphorus into the nodule CFA, and thus the growth rates of this nodule, vary with age, $\delta^{13}C$, and carbonate ion substitution. The trend of the data suggests faster CFA precipitation rates when in contact with pore waters relatively enriched in TDC-$\delta^{13}C$ and improverished in dissolved carbonate. This assertion is in good agreement with the classic experimental work of Ames (1959) which suggested increasing apatite growth rates under decreasing molar HCO_3^-:PO_4^- ratios in solution.

Conclusions

The above discussions collectively suggest that CFA growth in Peru margin sediments is confined to within a few, to

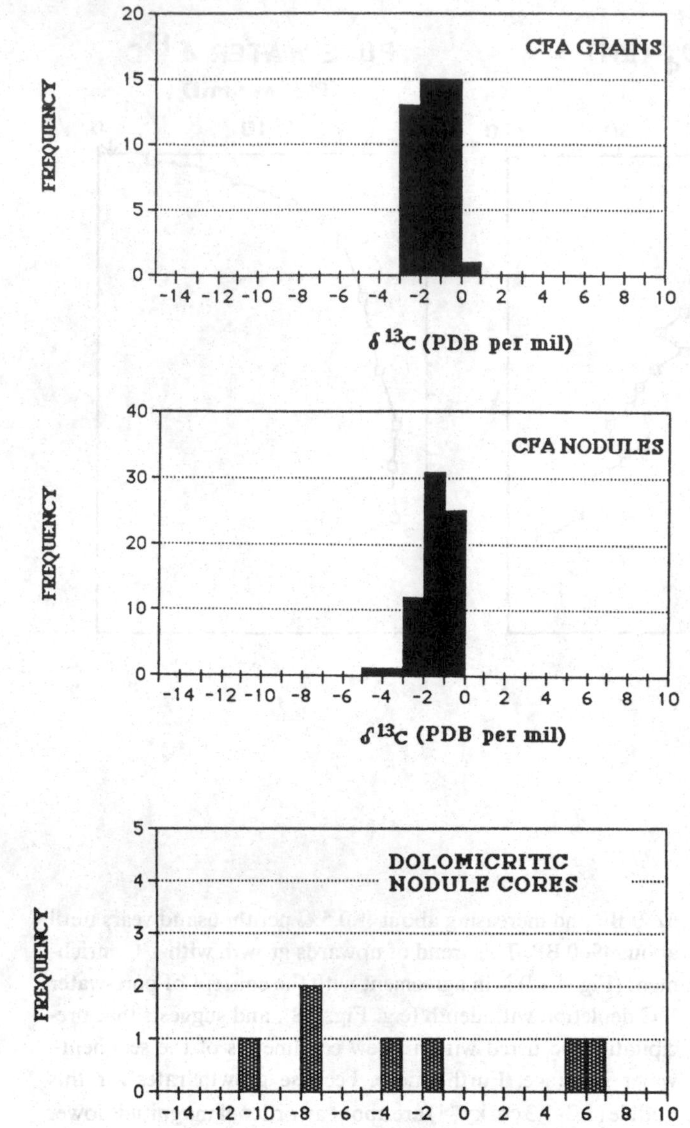

Fig. 4.9. Histogram frequency (number of samples) plots of $\delta^{13}C$ variations in structural carbonate of Peru margin CFA grains and nodules and dolomicritic concretionary cores.

perhaps a few tens of centimeters of the sediment–water interface and that rates of CFA growth in these sediments decrease in association with decreasing pore-water $\delta^{13}C_{TDC}$ compositions and exponentially downward increasing pore-water TDC, alkalinity, and CO_3^{2-} ion concentrations. In addition, the general relationship of increasing carbonate substitution and decreasing $\delta^{13}C$ values for the structural carbonate in Peru margin pelletal and nodular CFA also imply that increases in carbonate and accompanying charge-balancing substitutions in the CFA lattice are forced by increases in the ambient dissolved carbonate content of the pore waters, and that above a certain dissolved CO_3^{2-}:PO_4^{3-} concentration level in pore waters, structural inhomogeneities in the lattice prevent further CFA growth. This interpretation may explain why CFA precipitation is confined to within close proximity of the sediment–water interface in organic carbon-rich Peru margin sediments, although other constraints such as a suitable supply of dissolved phosphate and fluorine are also equally important. In sediments with non-organic phosphorus sources (cf. Glenn, 1987; O'Brien & Heggie, 1988), or in sediments in which dissolved carbonate has been previously consumed by other authigenic phases, deeper CFA precipitation may be anticipated. For example, CFA precipitation following dolomite formation may provide at least a partial explanation for the very deep stages of precipitation suggested for some CFA of the Miocene Monterey Formation (cf. Kastner *et al.*, 1984).

The development of CFA nodules in Peru margin sediments involves a complex interplay between organic matter degradation, phosphate precipitation and intermittent sedimentation and erosion. The episodic growth of multiple generations of nodular CFA cements binding distinct sediment layers records this well. Cementation of previously formed or currently forming peloidal CFA within the sediments by radially fibrous fringing cements of inferred low carbonate contents may mark the starting point in nodule development. Within each of the grain layers coating many of the nodules, precipitation of fringing CFA usually rapidly gives way to infilling of remaining pore space with more equant and optically-isotropic interstitial CFA cements, and ultimately to the capping of each sediment–cement generation by thin, optically-isotropic CFA crusts or seams. The interstitial CFA cements binding each of the individual grain layers are usually distinct from one another in terms of their relative ability to display birefringence, crystallinity, and abundance of microinclusions, factors presumably reflecting variations in growth rates as controlled by dissolved carbonate in pore waters. Early, relatively coarse and birefringent fringe cement development may occur when CO_3^{2-}:PO_4^{3-} activity ratios are minimal and CFA growth rates are maximized.

The paragenesis of authigenic components (glauconite, CFA, pyrite, and dolomite) in these sediments is interpreted to be one generally following the depth distribution of suboxic to anoxic bacterial populations in pore waters (Fig. 4.18). Such depth-stratified mineral zonations may also be transposed in the lateral dimension in response to lateral variations in organic carbon accumulation rates and bottom-water oxygen deficiencies. Glauconites are inferred to form relatively early in these sediments (especially near the margins of the oxygen-minimum zone) following the partial reduction of ferric iron. CFA precipitation may also be related to the reduction of ferric iron in that this reaction may supply an important source of dissolved phosphorus to CO_3^{2-}-deficient pore waters directly below the sediment–water interface. At deeper levels in the sediment CFA precipitation may be poisoned by elevating alkalinity as a consequence of further bacterial degradation of organic matter. Pyrite precipitation overlaps and then continues beyond that of CFA precipitation at progressively deeper levels in the sediment in association with sulfate-reduction processes. Whereas excessive pore-water alkalinities and associated carbonate ion concentrations may inhibit CFA precipitation at depth in these sediments, such conditions may promote the precipitation of dolomite in association with deeper sulfate reduction and methanogenesis (cf. Baker & Burns, 1985; Machel & Mountjoy, 1986). In ad-

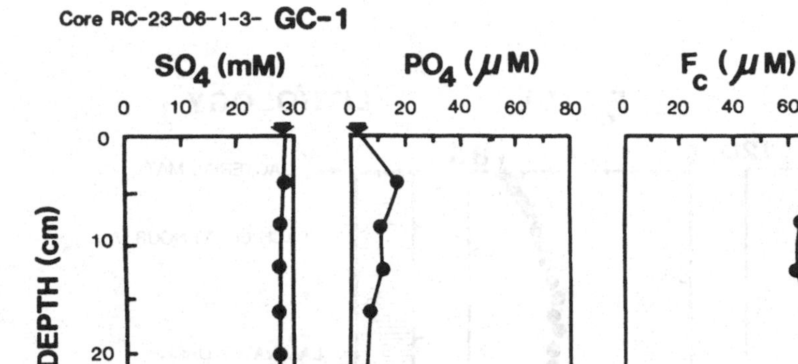

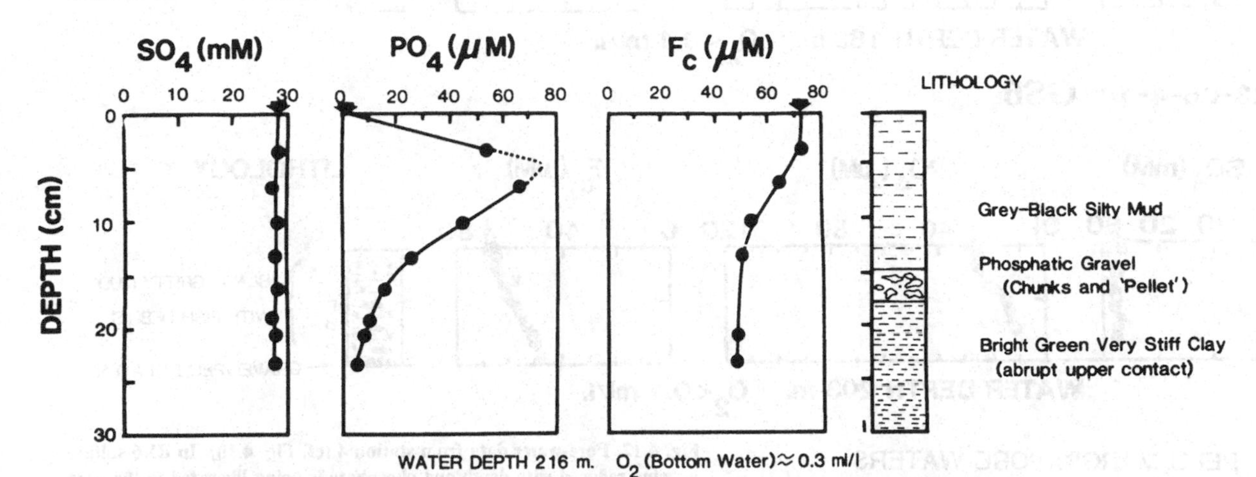

Fig. 4.10. Dissolved sulfate, phosphate and colorimetrically-determined fluoride (F_c) pore-water profiles from stations 1 (GC-1) and 2 (BX2) with shipboard lithologic descriptions. Arrows indicate bottom-water concentrations. Note dissolved phosphate production and uniformity of sulfate valves (lack of sulfate reduction) in top 10 cm of cores. CFA precipitation is suggested below 10 cm by draw-downs in PO_4 and F.

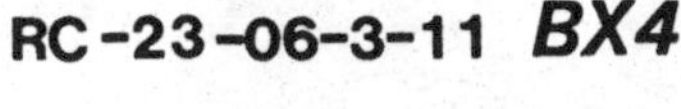

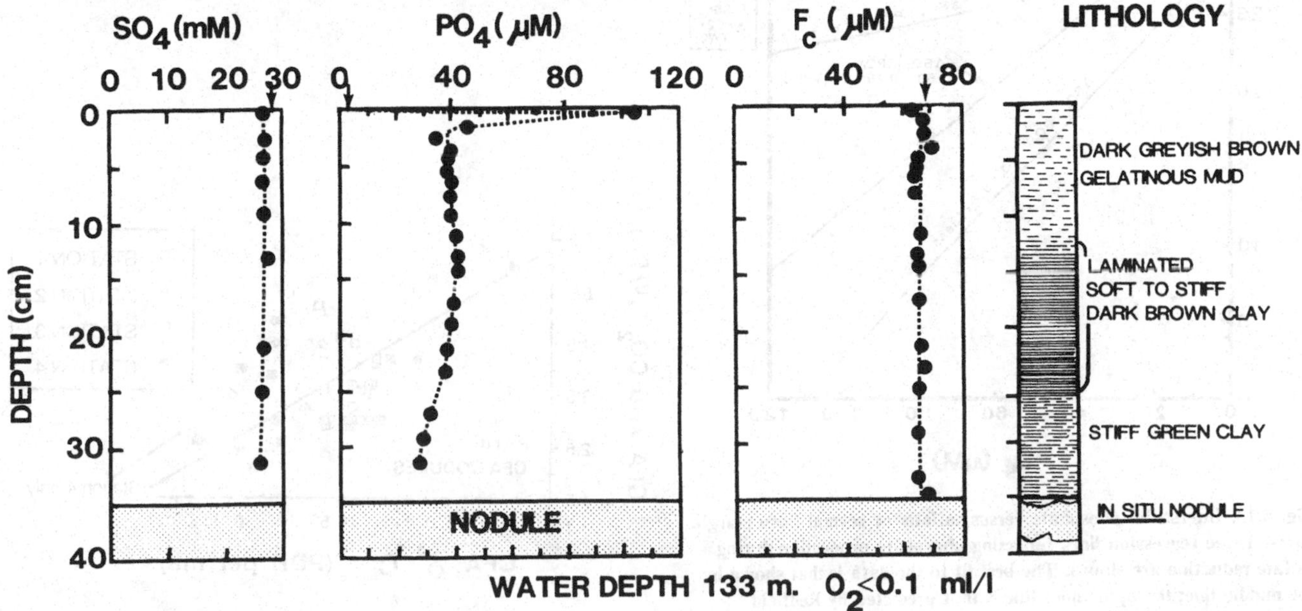

Fig. 4.11. Pore-water data from station 3 (cf. Fig. 4.10). Note lack of evidence for sulfate reduction and strong interfacial phosphate spike. Below slight 4.5 μm drop in F at about 1 cm depth, fluoride displays relatively conservative behavior.

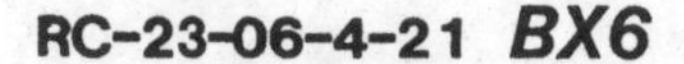

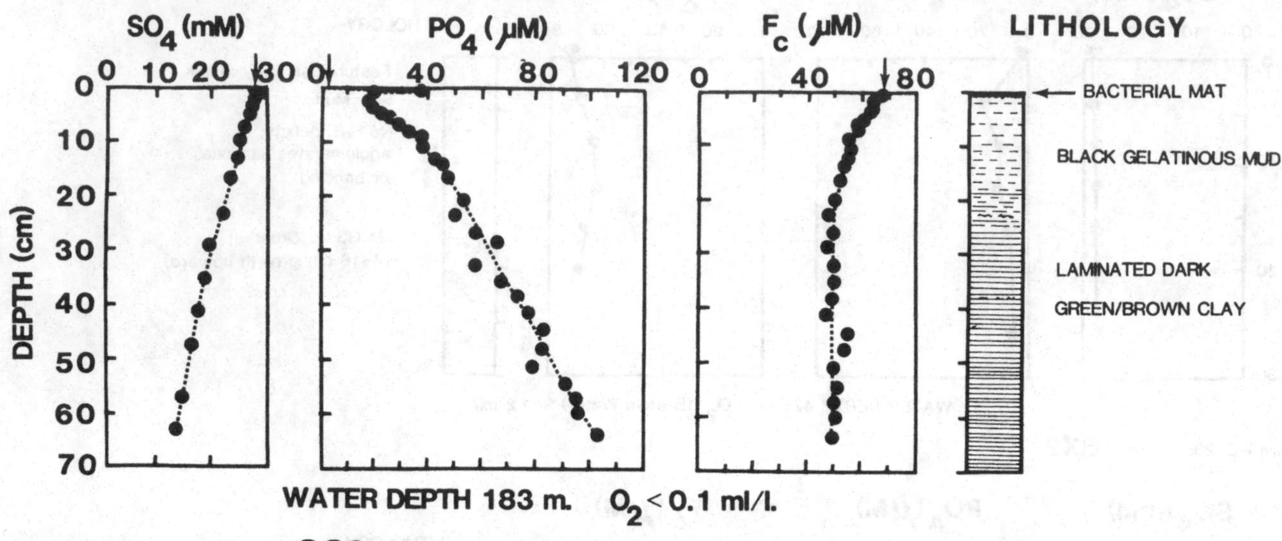

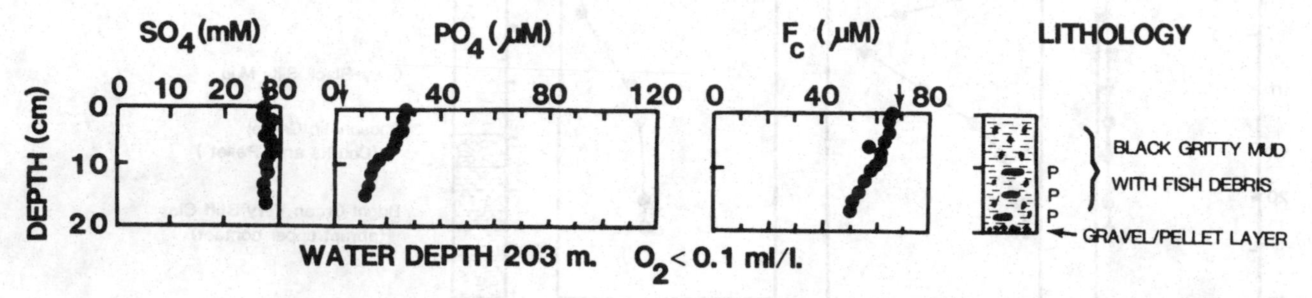

Fig. 4.12. Pore-water data from station 4 (cf. Fig. 4.10). In BX6 sulfate is being reduced with depth and phosphate is being liberated to the pore waters in the process. Interfacial phosphate spikes and fluoride uptake with depth is evident in both cores. In addition to CFA nodules, up to 80% of GS6 is composed of CFA peloids and phosphatized biogenic debris.

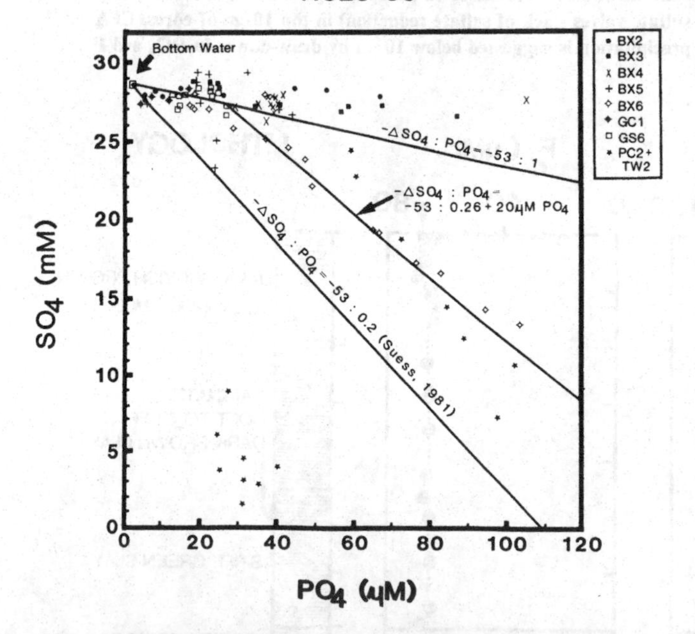

Fig. 4.13. Interstitial phosphate versus sulfate for several Peru margin cores. Three regression lines reflecting phosphate production during sulfate reduction are shown. The best fit to the data is that shown by the middle line; the uppermost line is that predicted by Redfield stoichometry (Richards, 1965) for undegraded marine organic matter. The relationship suggested by the data of Suess (1981) for other sites is included for comparison. Note the 'excess' phosphate hovering above the Redfield relationship at bottom-water SO_4 concentrations.

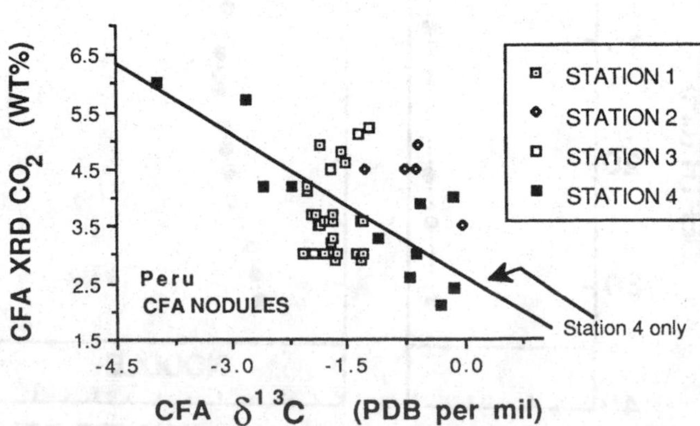

Fig. 4.14. $\delta^{13}C$ and weight % CO_2 as determined by X-ray diffraction for the structural carbonate in Peru margin CFA nodules. The data shows grouping by sample locality. The regression curve is for Station 4 (12° S) nodules only, where $\%CO_2 = 2.62 - 0.84\delta^{13}C$, $r = -0.86$.

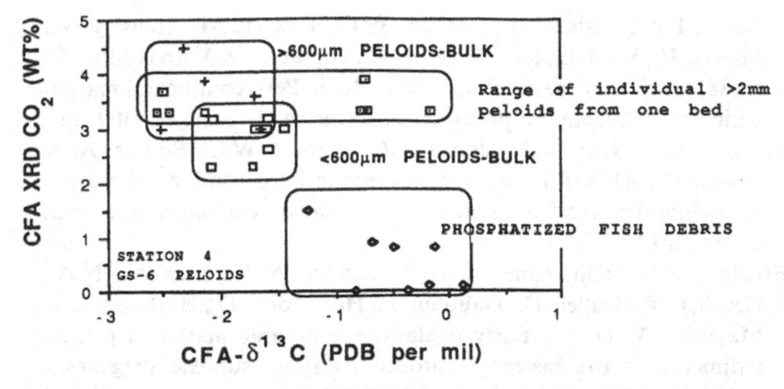

Fig. 4.15. $\delta^{13}C$ versus weight % CO_2 as determined by XRD for the structural carbonate in CFA grains from core GS-6.

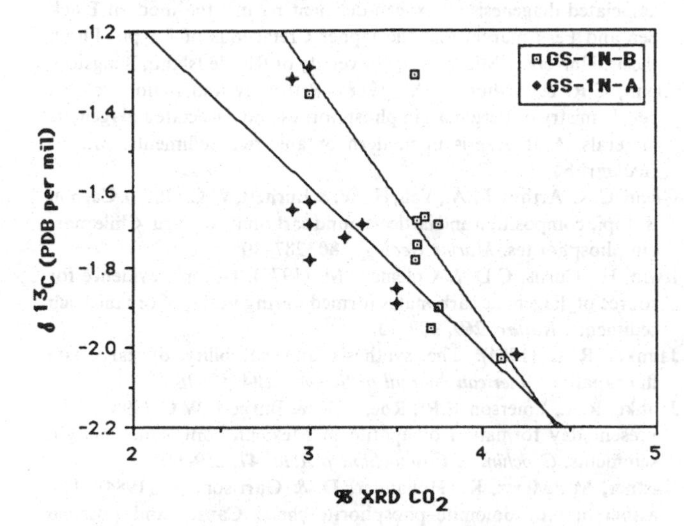

Fig. 4.16. $\delta^{13}C$ versus weight % XRD CO_2 for two laterally adjacent portions of nodule GS-1N.

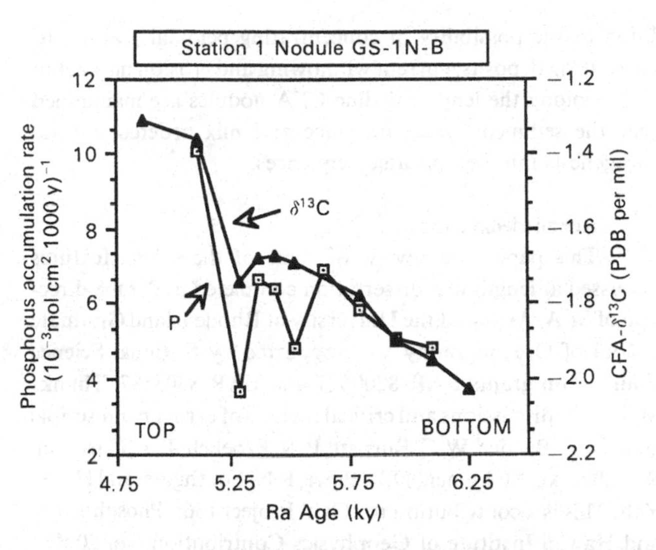

Fig. 4.17. CFA-$\delta^{13}C$ and phosphorus accumulation rate across CFA nodule GS-1N as a function of age. $\delta^{13}C$ increases about 0.5‰ $(10^3y)^{-1}$ as the nodule grew upwards towards the sediment–water interface. The relative depletion in ^{13}C between 5.5 ky and 5.25 ky suggests re-burial into slightly more depleted pore waters. P- and F-accumulation rates display a perfect linear correlation (not shown). Accumulation rates calculated from data from Kim & Burnett (1985).

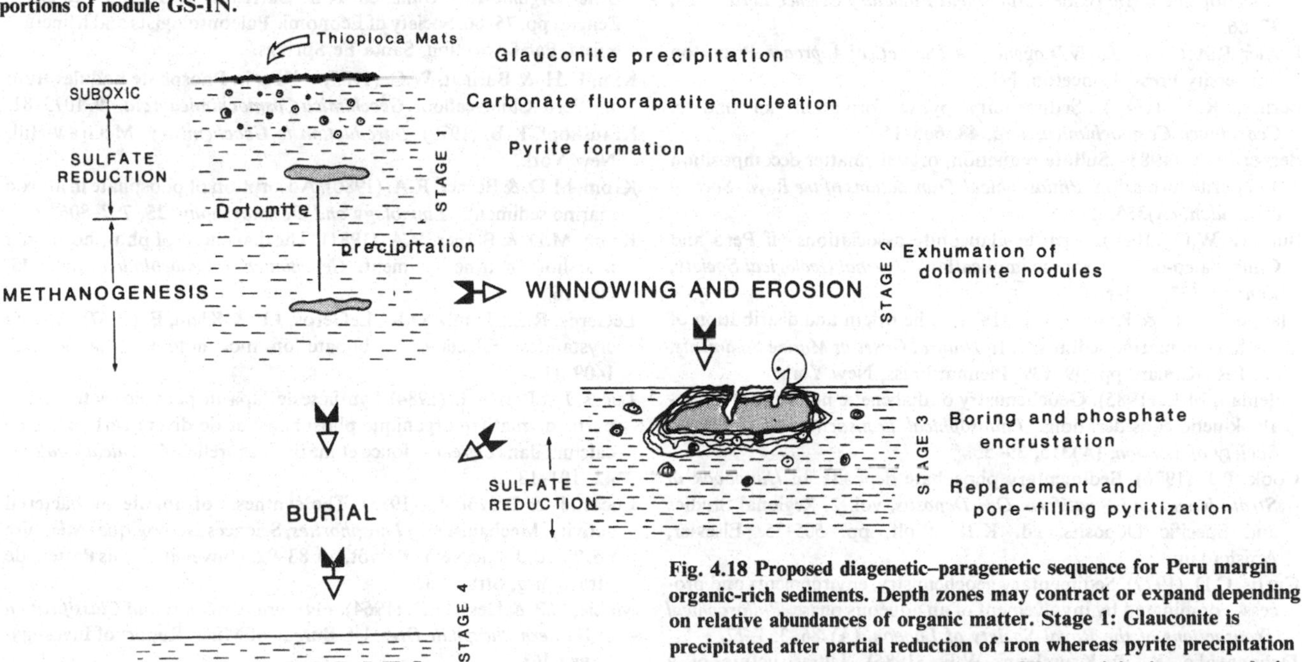

Fig. 4.18 Proposed diagenetic–paragenetic sequence for Peru margin organic-rich sediments. Depth zones may contract or expand depending on relative abundances of organic matter. Stage 1: Glauconite is precipitated after partial reduction of iron whereas pyrite precipitation is tied to the zone of sulfate reduction. CFA precipitation occurs during suboxic diagenesis directly below the sediment–water interface in association with interfacial pore-water phosphate spikes and is depth-limited by increases in carbonate alkalinity. Pyrite is depth-limited by available reactive iron. Dolomite precipitation in aided by increases in carbonate alkalinity (and decreases in dissolved sulfate?) with depth. With winnowing and erosion (stage 2) scattered peloids may be concentrated, dolomite nodules may be brought to the seafloor, and critical sediment depth positions of CFA precipitation may be maintained under high rates of sedimentation.

dition to the possibility of concentrating peloidal grains into surficial lag deposits, current winnowing and erosion may act to both prolong the length of time CFA nodules are maintained near the sediment–water interface and mix ordered mineral paragenesis into complicated sequences.

Acknowledgments

This paper is a review of some of the salient features discussed at length in a dissertation completed under the direction of M.A. Arthur at the University of Rhode Island Graduate School of Oceanography and supported by National Science Foundation grants EAR-8200521 and EAR-8403587. Thanks for helpful discussions and critical reviews of earlier manuscripts go to M.L. Bender, W.C. Burnett, P.N. Froelich, R.E. Garrison, R.A. Jahnke, M. Leinen, D.Z. Piper, P.N. Southgate, and H.-W. Yeh. This is a contribution to IGCP Project 156: 'Phosphorites' and Hawaii Institute of Geophysics Contribution No. 2071.

References

Ames, L.L. (1959). The genesis of carbonate-apatite. *Economic Geology*, **54**, 829–41.

Arthur, M.A., Glenn, C.R. & Froelich, P.N. (1989). Total Dissolved CO_2, alkalinity, and ^{13}C in pore waters of organic carbon rich muds of the Peru margin. *Geochimica Cosmochimica Acta*, (in press).

Baker, K.B. & Burnett, W.C. (1988). Distribution, texture, and composition of modern phosphate pellets in Peru shelf muds. *Marine Geology*, **80**, 195–213.

Baker, P.A., & Burns, S.J. (1985). Occurrence and formation of dolomite in organic-rich continental margin sediments. *American Association of Petroleum Geologists Bulletin*, **69**, 1917–30.

Berner, R.A. (1973). Phosphate removal from seawater by adsorption on volcanogenic ferric oxides. *Earth and Planetary Science Letters*, **18**, 77–86.

Berner, R.A. (1980). *Early Diagenesis A Theoretical Approach*. Princeton University Press, Princeton, NJ.

Berner, R.A. (1984). Sedimentary pyrite formation: an update. *Geochimica Cosmochimica Acta*, **48**, 605–15.

Berner, R.A. (1985). Sulfate reduction, organic matter decomposition and pyrite formation. *Philosophical Transactions of the Royal Society of London*, **(A)315**, 25–38.

Burnett, W.C. (1980). Apatite–glauconite associations off Peru and Chile: paleo-oceanographic implications. *Journal Geological Society, London*, **137**, 754–64.

Claypool, G.E. & Kaplan, I.R. (1974). The origin and distribution of methane in marine sediments. In *Natural Gases in Marine Sediments*, ed. I.R. Kaplan, pp. 99–139. Plenum Press, New York.

Coleman, M.L. (1985). Geochemistry of diagenetic non-silicate minerals: kinetic considerations. *Philosophical Transactions of the Royal Society of London*, **(A)315**, 39–56.

Cook, P.J. (1976). Sedimentary phosphate deposits. In *Handbook of Strata-bound and Stratiform Ore Deposits*, vol. 2, Regional Studies and Specific Deposits, ed. K.H. Wolf, pp. 505–35. Elsevier, Amsterdam.

Curtis, C.D. (1977). Sedimentary geochemistry: environments and processes dominated by involvement of an aqueous phase. *Philosophical Transactions of the Royal Society of London*, **(A)286**, 353–72.

Dahanayake, K. & Krumbein, W.E. (1985). Ultrastructure of a microbial mat-generated phosphorite. *Mineralogica Deposita*, **20**, 260–5.

De Lange, G.J. (1986). Early diagenetic reactions in interbedded pelagic and turbidic sediments in the Nares Abyssal Plain (western North Atlantic): consequences for the composition of sediment and interstitial water. *Geochimica Cosmochimica Acta*, **50**, 2543–61.

DeVries, T.J. & Pearcy, W.G. (1982). Fish debris in sediments of the upwelling zone off central Peru: a late Quaternary record. *Deep-Sea Research*, **28**, 87–109.

Froelich, P.N., Arthur, M., Burnett, W.C., Deakin, M., Hensley, V., Jahnke, R., Kaul, L., Kim, K., Roe, K., Soutar, A. & Vathakanon, C. (1988). Early diagenesis of organic matter in Peru continental margin sediments: phosphorite precipitation. *Marine Geology*, **80**, 309–46.

Froelich, P.N., Kim, K.H., Jahnke, R., Burnett, W.C., Soutar, A. & Deakin, M. (1983). Pore water fluoride in Peru continental margin sediments: Uptake from seawater. *Geochimica Cosmochimica Acta*, **47**, 1605–12.

Froelich, P.N., Klinkhammer, G.P., Bender, M.L., Luedtke, N.A., Heath, G.R., Cullen, D., Dauphin, P., Hammond, D., Hartman, B. & Maynard, V. (1979). Early oxidation of organic matter in pelagic sediments of the eastern equitorial Atlantic: suboxic diagenesis. *Geochimica Cosmochimica Acta*, **43**, 1075–90.

Glenn, C.R. (1987). 'Phosphorus fluxes, phosphorite sedimentation and associated diagenesis in oxygen-deficient basins: the modern Black Sea and Peru margin and the Upper Cretaceous of Egypt.' Ph.D. thesis. University Microfilms, University of Rhode Island, Kingston.

Glenn, C.R. & Arthur, M.A. (1988). Petrology and major element geochemistry of Peru margin phosphorites and associated diagenetic minerals: Authigenesis in modern organic-rich sediments. *Marine Geology*, **80**, 231–67.

Glenn, C.R., Arthur, M.A., Yeh, H.-W. & Burnett, W.C. (1988). Carbon isotopic composition and lattice-bound carbonate of Peru–Chile margin phosphorites. *Marine Geology*, **80**, 287–307.

Irwin, H., Curtis, C.D. & Coleman, M. (1977). Isotopic evidence for source of diagenetic carbonates formed during burial of organic-rich sediments. *Nature*, **269**, 209–13.

Jahnke, R.A. (1984). The synthesis and solubility of carbonate fluorapatite. *American Journal of Science*, **284**, 58–78.

Jahnke, R.A., Emerson, S.R., Roe, K.K. & Burnett, W.C. (1983). The present day formation of apatite in Mexican continental margin sediments. *Geochimica Cosmochimica Acta*, **47**, 259–66.

Kastner, M., Mertz, K., Hollander, D. & Garrison, R. (1984). The association of dolomite–phosphorite–chert: Causes and possible diagenetic sequences. In *Dolomites of the Monterey Formation and Other Organic-Rich Units*, ed. R.E. Garrison, M. Kastner, & D.H. Zenger, pp. 75–86. Society of Economic Paleontologists and Mineralogists, Pacific Section, Santa Fe Springs.

Kim, K.H. & Burnett, W.C. (1985). ^{226}Ra in Phosphate nodules from the Peru–Chile seafloor. *Geochimica Cosmochimica Acta*, **49**, 1073–81.

Krauskopf, K.B. (1979). *Introduction to Geochemistry*. McGraw-Hill, New York.

Krom, M.D. & Berner, R.A. (1980). Adsorption of phosphate in anoxic marine sediments. *Limnology and Oceanography*, **25**, 797–806.

Krom, M.D. & Berner, R.A. (1981). The diagenesis of phosphorus in a nearshore marine sediment. *Geochimica Cosmochimica Acta*, **45**, 207–16.

LeGeros, R.Z., Trautz, O.R., LeGeros, J.P. & Klein, E. (1967). Apatite crystallites: Effects of carbonate on morphology. *Science*, **155**, 1409–11.

Lucas, J. & Prévôt, L. (1984). Synthèse de l'apatite par voie bacteriénne à partir de matière organique phosphatée et de divers carbonates de calcium dans des eaux douce et marine naturelles. *Chemical Geology*, **42**, 101–18.

Lucas, J. & Prévôt, L. (1985). The synthesis of apatite by bacterial activity: Mechanism. In *Phosphorites*, Sciences Geologiques Mémoire no. 77, ed. J. Lucas & L. Prévôt, pp. 83–92. Université Louis Pasteur de Strasbourg, Strasbourg.

Mabie, C.P. & Hess, H.D. (1964). *Petrographic Study and Classification of Western Phosphate Ores*. US Bureau of Mines Report of Investigations 6468.

Machel, H.-G. & Mountjoy, E.W. (1986). Chemistry and environments of dolomitization – a reappraisal. *Earth Science Reviews*, **23**, 175–222.

McClellan, G.H. (1980). Mineralogy of carbonate fluorapatites. *Journal of the Geological Society, London*, **137**, 657–81.

McClellan, G.H. & Lehr, J.R. (1969). Crystal chemical investigation of natural apatites. *American Mineralogist*, **54**, 1374–91.

Nathan, Y. (1984). The mineralogy and geochemistry of phosphorites. In *Phosphate Minerals*, ed. J.O. Nriagu & P.B. Moore, pp. 275–91. Springer-Verlag, Berlin.

Nathan, Y. & Sass, E. (1981). Stability relations of apatites and calcium carbonates. *Chemical Geology*, **34**, 103–11.
O'Brien, G.W., Harris, J.R., Milnes, A.R. & Veeh, H.H. (1981). Bacterial origin of East Australian continental margin phosphorites. *Nature*, **294**, 442–4.
O'Brien, G.W. & Heggie, D. (1988). East Australian continental margin phosphorites. *EOS*, **69**, 2.
Rashid, M.A. (1985). *Geochemistry of Marine Humic Compounds*. Springer-Verlag, New York.
Reimers, C.E. & Suess, E. (1983a). Spatial and temporal patterns of organic matter accumulation on the Peru continental margin. In *Coastal Upwelling – Its Sedimentary Record, Part B: Sedimentary Records of Ancient Coastal Upwelling*, ed. E. Suess & J. Thiede, pp. 311–46. Plenum Press, New York.
Reimers, C.E. & Suess, E. (1983b). Late Quaternary fluctuations in the cycling of organic matter off central Peru: A proto-kerogen record. In *Coastal Upwelling – Its Sedimentary Record, Part A: Responses of the Sedimentary Regime to Present Coastal Upwelling*, ed. J. Thiede & E. Suess, pp. 497–526. Plenum Press, New York.
Richards, F.A. (1965). Anoxic basins and fjords. In *Chemical Oceanography*, 1, ed. J.P. Riley and G. Skirrow, pp. 611–45. Academic Press, London.
Riggs, S.R. (1979). Petrology of the Tertiary phosphorite system of Florida. *Economic Geology*, **74**, 195–220.
Scheidegger, K.F. & Krissek, L.A. (1983). Zooplankton and nekton: Natural barriers to the seaward transport of suspended terrigenous particles off Peru: In *Coastal Upwelling – Its Sedimentary Record, Part A: Responses of the Sedimentary Regime to Present Coastal Upwelling*, ed. J. Thiede & E. Suess, pp. 303–33. Plenum Press, New York.
Soudry, D. & Champetier, Y. (1983). Microbial processes in the Negev phosphorites (southern Israel). *Sedimentology*, **30**, 411–23.
Soudry, D. Nathan, Y. (1980). Phosphate peloids from the Negev phosphorites. *Journal of the Geological Society, London*, **137**, 749–55.
Southgate, P.N. (1986). Cambrian phoscrete profiles, coated grains, and microbial processes in phosphogenesis: Georgina Basin, Australia. *Journal of Sedimentary Petrology*, **56**, 429–41.
Suess, E. (1981). Phosphate regeneration from sediments of the Peru continental margin by dissolution of fish debris: *Geochimica Cosmochimica Acta*, **45**, 577–88.
Williams, L.A. & Reimers, C. (1983). Role of bacterial mats in oxygen-deficient marine basins and coastal upwelling regimes: preliminary report. *Geology*, **11**, 267–9.

5

Phosphorite growth and sediment dynamics in the Modern Peru shelf upwelling system

W.C. BURNETT

Abstract
The Peru shelf has served as a useful modern analogue for the type of environment in which many ancient phosphorite deposits are thought to have formed. Sediment accumulation rates, biological mixing processes, rates of phosphate mineralization, and other processes may be studied on much shorter time scales than possible through study of ancient land deposits. Uranium-series and radiocarbon studies of various types of phosphatic grains and associated sediment with modern phosphorite have provided data which can be used to constrain models of deposition for similar sedimentary phosphorites in the geologic record.

Introduction

Phosphorites and upwelling

The occurrence of phosphorites on the ocean floor has been recognized since the original HMS Challenger expedition (Murray & Renard, 1891). Workers over the past few decades have located and described marine phosphorites along many of the continental margins, often in areas of strong coastal upwelling with associated high biological productivity. A general relationship between phosphorite development and processes related to upwelling has been recognized for many years (Kazakov, 1937; McKelvey, Swanson & Sheldon, 1953; Sheldon, 1964; McKelvey, 1967). It is now known that phosphorite occurrences may favor but are not restricted to areas of intense upwelling. For example, ancient phosphorites are also found in what are today low-productivity environments such as on the summits of many seamounts in the Pacific Ocean (Burnett, Cullen & McMurtry, 1987), off the east coast of New South Wales, Australia (O'Brien & Veeh, 1980; O'Brien *et al.*, 1981) and off the southeast coast of the United States (Manheim, Pratt & McFarlin, 1980; Riggs, 1984). However, the deposits off the US southeast coast may be related to western boundary current upwelling during periods of higher sea level in the Neogene (Riggs, 1984). Thus, there may be an 'upwelling-phosphorite' link in that case after all.

As recently as 1970, it was speculated that phosphorites may not be forming on the modern-day sea floor at all (Kolodny, 1969; Kolodny & Kaplan, 1970). This was based on a rather broadbrush study of the uranium isotopic composition of phosphorites from several areas of the world's oceans. A few years later, it became clear through closer study of a few deposits, that modern formation of phosphorite was occurring in at least two areas: (1) offshore Namibia (Baturin, Merkulova & Chalov, 1972); and (2) on the Peru–Chile continental margin (Veeh, Burnett & Soutar, 1973). These initial findings strengthened the argument of a causative link between upwelling and phosphorite formation as both these areas are characterized by persistent and strong upwelling. Further studies have demonstrated modern phosphorite formation occurring on the continental slope of New South Wales, Australia (O'Brien & Veeh, 1980; O'Brien *et al.*, 1981); off Baja California, Mexico (Jahnke *et al.*, 1983); and most recently off the west coast of India (Borole, Rajagopalau & Somayajulu, 1987).

With the exception of the New South Wales deposits, all Recent phosphorites discovered to date are located in areas of modern upwelling. The phosphorites off east Australia may represent a rather unique set of circumstances where slow sediment accumulation rates, mixing by benthic organisms, and iron cycling have provided an environment where phosphate, although present at low concentrations in the bottom and pore waters, precipitates as carbonate fluorapatite (CFA), just below the sediment mixed layer (O'Brien & Heggie, 1988). Although this general process probably occurs in other low-productivity environments such as Long Island Sound (Ruttenberg & Berner, 1987), the flux of other sediment components is normally sufficiently large to mask the presence of CFA precipitates to the point where recognition is extremely difficult.

Upwelling, therefore, is clearly not the only process which can cause the prerequisite conditions for CFA mineralization. It is, however, a very effective means of producing a suitable environment and probably the single most important marine environment that produces a high flux of well preserved organic matter to the seafloor. Thus, although the Kazakov model of direct, inorganic precipitation of CFA out of nutrient-rich ascending seawater is almost certainly not correct in detail, the link between oceanic upwelling and the occurrence of major phosphorite deposits still holds. The fact that phosphorite may occasionally form in other marine environments does not detract from this association.

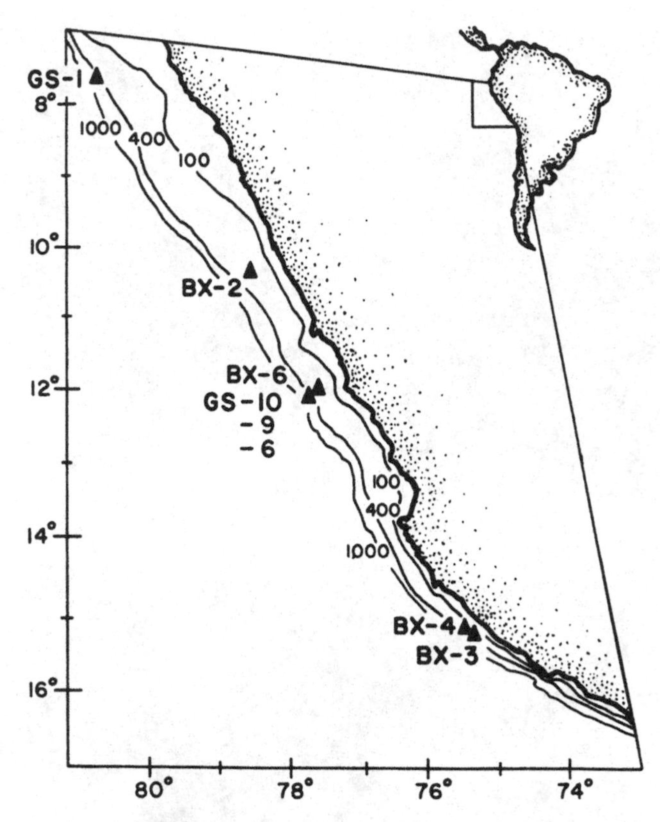

Fig. 5.1. Index map of the Peru margin with locations of all samples discussed in this paper.

Nodules, pellets and other grains

Phosphorites sampled from the modern ocean floor have been described in a variety of ways including: irregular masses, oolites, pellets, grains and intracasts. The literature on ocean-floor phosphorites is dominately concerned with *nodular* phosphorite. This is one reason cited as evidence that modern ocean-floor deposits should not be used as an analogue for the large deposits studied on land, which tend to be primarily *pelletal*. This apparent discrepancy, however, may have been due largely to a sampling bias – nodules are easily recovered and their presence is obvious during dredging and grab sampling. Pellets are not nearly as evident when sampled and may easily wash out of some types of sampling equipment.

In spite of the limited amount of data published on ocean-floor phosphate pellets, this morphology has been reported from all major phosphate occurrences found in the ocean. Baturin (1971) described pellets from the shelf of southwest Africa as granular nodules 0.1–0.3 mm in diameter, yellowish-gray in color, with P_2O_5 contents of 5–11%, and occurring within an organic-rich diatomaceous ooze. Birch (1979) reported the presence of phosphorite pellets from both the western continental margin of South Africa and from the adjacent coastal terrace. D'Anglejan (1967) described black, ovoidal–spherical grains from 0.125–0.250 mm in diameter from the continental margin off Baja California, Mexico. These grains were generally structureless and optically isotropic. Burnett (1977) reported pellets in the sediments of the continental shelf of Peru but did not have sufficient quantities available for detailed study. In more recent studies, Burnett and his colleagues (Baker & Burnett, 1988; Burnett *et al.*, 1988; Kim & Burnett, 1988) have shown that black, spherical phosphatic pellets, very similar to those described from the large deposits of the geologic record, may constitute a significant fraction (up to > 60% in one case) of the 'upwelling suite' of sediments on the Peru shelf.

The Peru shelf environment

The organic-rich muds of the Peru shelf and upper slope contain a coastal upwelling sediment facies that is rich in organic carbon (ca. 2–20%; Froelich *et al.*, 1988) and opaline silica (ca. 6–13%; Froelich *et al.*, 1988) with significant concentrations of the authigenic minerals CFA, glauconite and dolomite, as well as several percent alumino-silicate detrital minerals (Burnett, 1977; Suess, 1981; Scheidegger & Krissek, 1982; Krissek & Scheidegger, 1983; Reimers & Suess, 1983; Kulm, Suess & Thornburg, 1984). An upper slope mud lens between 10.5° S and 13.6° S latitude and 100–600 m in depth has been identified as a very distinctive sediment in this area (Krissek, Scheidegger & Kulm, 1980). An oxygen minimum layer lies at the depositional center of this so-called 'mud lens' (Suess, 1981), resulting in organic carbon contents approaching 20% in the surface sediments (Walsh, 1981). Fish debris, as well as phosphate nodules and pellets, often constitute several weight percent of the sediment (Muller & Suess, 1979; Baker & Burnett, 1988).

An expedition was organized in 1982 to geochemically investigate the sediments, pore waters, and associated phosphorites on the Peru shelf. The results of this cruise on the R/V *Robert Conrad* have recently been reported (Burnett & Froelich, 1988 and other papers in the same issue). Ocean-bottom samples were collected by box coring and other techniques in four separate areas on the Peru margin (Fig. 5.1). This paper will assess some of the findings from the *Conrad* cruise in terms of evaluating the Peru shelf as a suitable analogue for evaluating depositional models for geologic phosphorite deposits.

Sediment texture and composition

Grain-size separation of the detrital, organic-rich, Peru shelf muds revealed that 0.125–0.5 mm phosphate pellets were abundant in some cores (Baker & Burnett, 1988). One core in particular (box core BX-2) contained extremely rich (up to ~60% by weight) layers of phosphate pellets. When individual size fractions of the bulk sediment and the estimated concentrations of phosphorite pellets are plotted against depth in this core, the predominance of phosphate pellets in the 3Φ (125–250 μm) and 4Φ (63–125 μm) size fractions below 5 cm is evident (Fig. 5.2). In all but the uppermost layers, the pelletal concentration is just slightly lower or the same as the total mass in the 2Φ (250–500 μm), 3Φ (125–250 μm) and 4Φ (63–125 μm) size classes. Other cores sampled during the *Conrad* cruise have smaller quantities of phosphate pellets, from below detection limits to approximately 30% by weight.

Examination of pellets separated from sample BX-2 and other cores from the Peru shelf shows them to be typically black, spherical–ovoidal in shape and usually between 125–250 μm in diameter. In many respects, the phosphate pellets are similar to the grains sampled off Baja California, Mexico, described by D'Anglejan (1967) and those from off North Carolina described by Riggs *et al.* (Chapter 29 this volume). Size fractions coarser

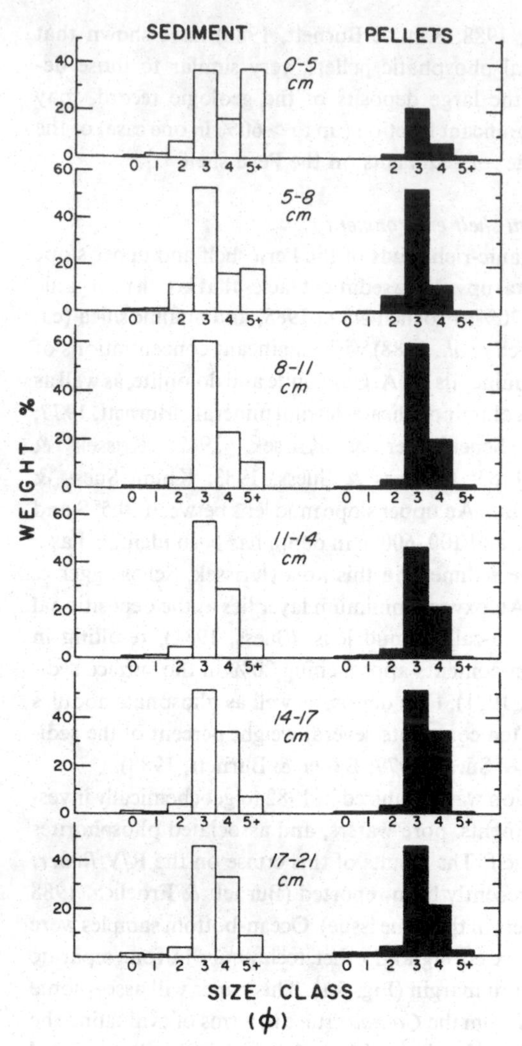

Fig. 5.2. Grain size distribution of total sediment (open boxes) and phosphate pellets (closed boxes) as a function of depth in box core BX-2 (from Baker & Burnett, 1988).

(a)

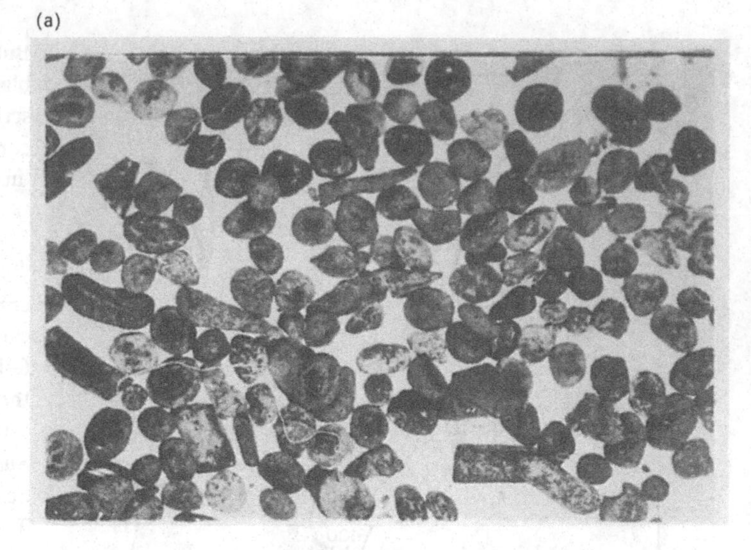

(b)

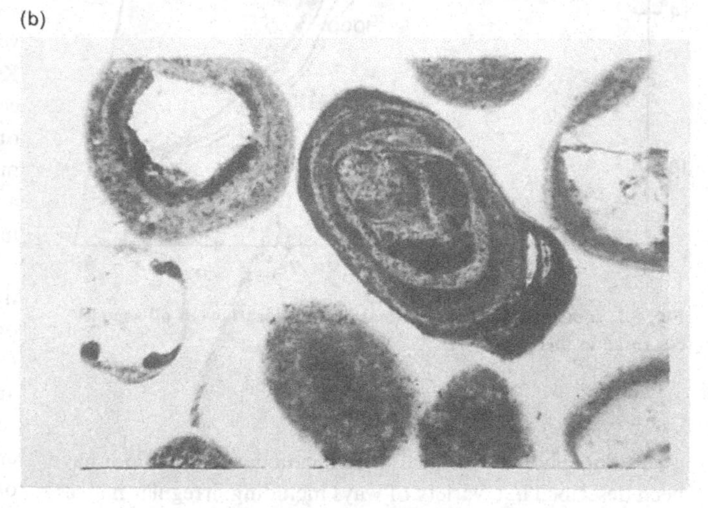

Fig. 5.3. (a) Photomicrograph of 125–250 μm size fraction from core GS-6. Dark, rounded objects are phosphate pellets. Elongated particles are skeletal phosphate grains. (b) Photomicrograph of 250–500 μm pellets from GS-9 as viewed in thin section.

than about 250 μm tend to be dominated by fish bones and other elongated particles of biogenous origin (Fig. 5.3). Sediment grains < 63 μm in diameter contained few phosphate pellets and were dominately detrital silts (quartz, feldspar), clay minerals and organic matter. When the separated phosphate pellets were observed in thin section, concentric oolite-like structures around an inorganic or skeletal nucleus were seen very commonly (Fig. 5.3*b*).

When geochemically analyzed, the phosphate pellets separated from Peru shelf muds have compositions consistent with phosphorites from the geologic record. The main mineral species, based on both X-ray diffraction analysis and chemical analysis appears to be CFA, with F:P_2O_5 weight ratios of close to 0.11 and > 1.0% structural CO_2. In addition, it was observed in two of the cores with high pellet concentrations (BX-2 and GS-9) that the F:P_2O_5 ratio increased systematically downcore within the main pellet size class (3ϕ in BX-2; 2ϕ in GS-9). This implies that the initial solid formed during the precipitation of phosphate pellets was somewhat deficient in fluoride, and fluoridation of these pellets continues until a steady-state composition is obtained.

By compiling all available elemental data on these sediments, together with some assumptions regarding the phase composition of the components, Froelich *et al.* (1988) estimated the overall sedimentary component composition of the cores from the *Conrad* cruise. An examination of the changing composition of BX-2, the most pellet-rich core, with depth displays several features (Fig. 5.4). The top interval of the core is rich in organic matter and silica while the CFA component becomes more prevalent at the point (~ 7 cm) where most of the organic matter has been oxidized. This is also the point in the sediment where bioturbation becomes less important. In fact, there may be a link between biological mixing processes and the distribution of phosphatic grains in this core, as will be explained further in the next section.

Sediment accumulation and mixing

One of the advantages of studying a modern 'analogue' environment is that the rate processes may be directly measured instead of being inferred as is the case when working with the geologic record. Sediment accumulation rates based on excess ^{210}Pb and radiocarbon profiles from six cores of organic-rich

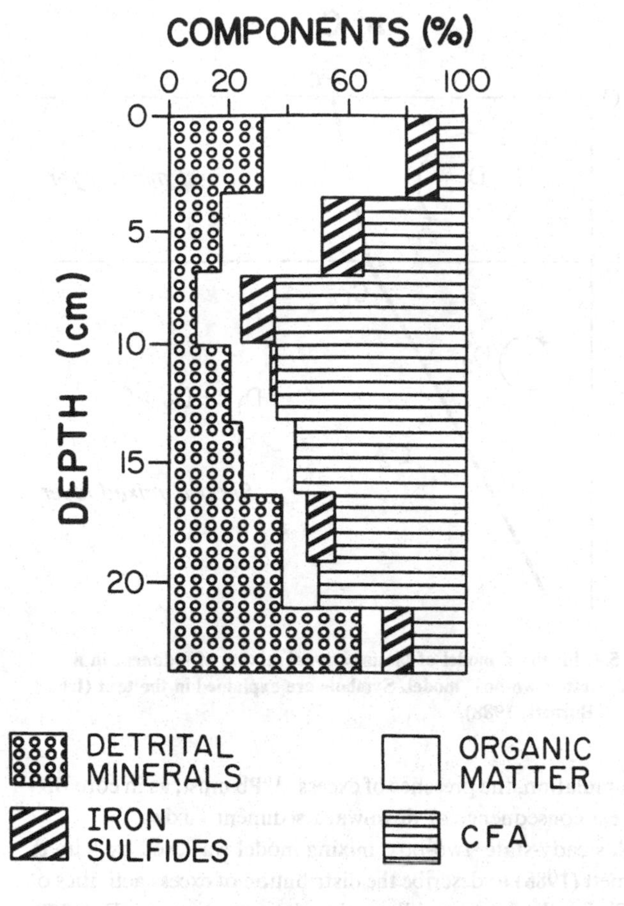

Fig. 5.4. Estimated abundance of sedimentary components of the total sediment as a function of core depth in core BX-2 (from data presented in Froelich *et al.*, 1988).

mud collected from the 12° S latitude (BX-6, GS-8, GS-10) and 15° S latitude (BX-3, BX-4, BX-5) areas of the Peru continental margin, where persistent strong upwelling occurs (Zuta, Riviera & Bustamante, 1975), are in the range of 0.028–0.18 cm y^{-1} (Kim & Burnett, 1988). These rates are reasonably close to values previously reported at 11° S (0.16 cm y^{-1}) and at 13° S (0.05 cm y^{-1}) by DeMaster (1981). Henrichs & Farrington (1984) reported ranges of 0.4–1.3 cm y^{-1} (Sta. 4) and 0.3–1.1 cm y^{-1} (Sta. 5a) for two cores collected very close to the site at 15°. These values are significantly higher than the estimates made by Kim & Burnett (1988). This may be because the downcore ^{210}Pb data was interpreted differently in the two studies – Henrichs & Farrington neglected bioturbation and Kim & Burnett did not include a correction for compaction. Some differences in accumulation rates are to be expected, however, even in cores located in nearly the same area. The Peru shelf is a dynamic sedimentary system and mass movement of sediments is a common feature (DeMaster, 1981). Such sediment movements can easily result in significant differences of accumulation rates in cores sampled only a few kilometers apart. Cores BX-5 and BX-6 contained intervals in the sedimentary record with constant activities in the excess ^{210}Pb profiles, probably caused by slumping. DeMaster (1981) pointed out that on a 1000-y time scale an indeterminate amount of Peru margin sediment is episodically removed from the upper continental margin, possibly through slumping or bottom-current scouring. Core intervals deposited by slumps

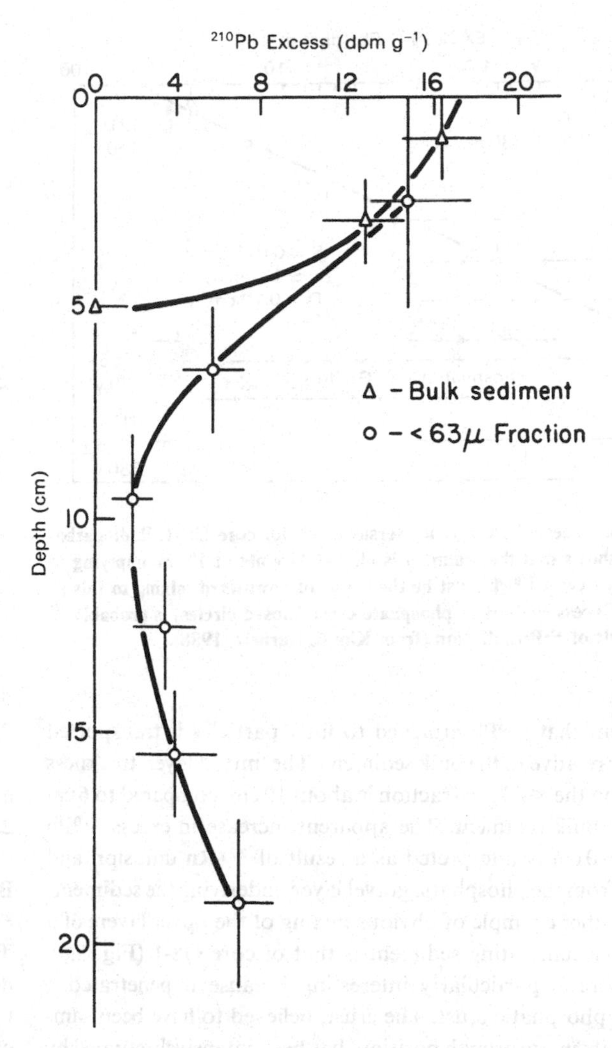

Fig. 5.5. Excess ^{210}Pb activity versus depth for total sediment (triangles) and $<63\ \mu m$ size fraction (circles) in core BX-2 (from Burnett *et al.*, 1988).

(with constant excess ^{210}Pb activity), as well as biological particle mixing signals in the upper parts of cores, could cause an overestimation of sedimentation rates if not resolved by closely spaced sample analysis.

Estimated sedimentation rates for cores collected in the northern study areas of the Peru shelf are much lower. Core GS-ls (7° S latitude) has an accumulation rate of only 0.011 cm y^{-1}. Core BX-2 (10° S latitude), with the highest content of phosphatic pellets (up to ~60% by weight) yielded the lowest sedimentation rate (0.0017 cm y^{-1}) of all cores studied from the Peru continental margin.

Evaluation of the effect of sediment mixing by bioturbation is important for consideration of how sedimentary particles (including phosphate pellets) may be transported within the sediment after deposition. There is some mixing in core BX-2 as evidenced by the presence of excess ^{210}Pb in the upper 6 cm of the bulk sediment (Fig. 5.5). This must be due to mixing since the ^{14}C sediment accumulation rate for this core is only 0.0017 cm y^{-1}. The existence of excess ^{210}Pb down to a depth of 6 cm (corresponding to more than 100 half-lives of ^{210}Pb) must, therefore, be a consequence of bioturbation. When the excess ^{210}Pb for the $<63\ \mu m$ fraction is plotted versus core depth, it is

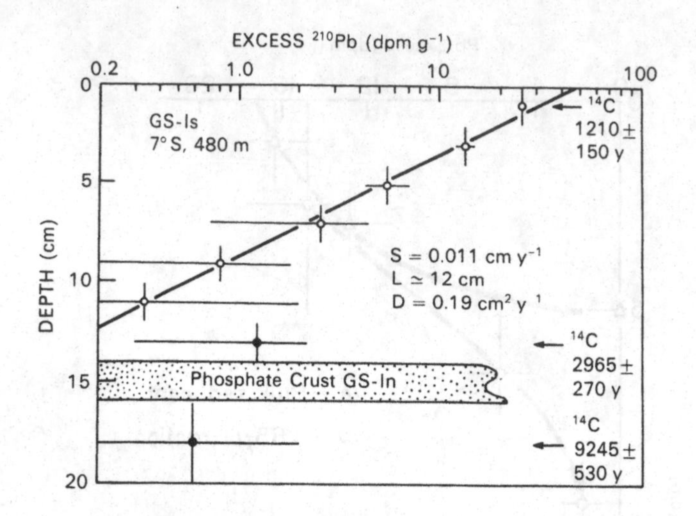

Fig. 5.6. Excess ^{210}Pb activity versus depth for core GS-1. Radiocarbon dating shows that the sediment is almost 3 ky old at 12 cm, implying that any excess ^{210}Pb must be the result of downward mixing to this depth. Excess ^{210}Pb near phosphate crust (closed circles) is probably the result of ^{222}Rn diffusion (from Kim & Burnett, 1988).

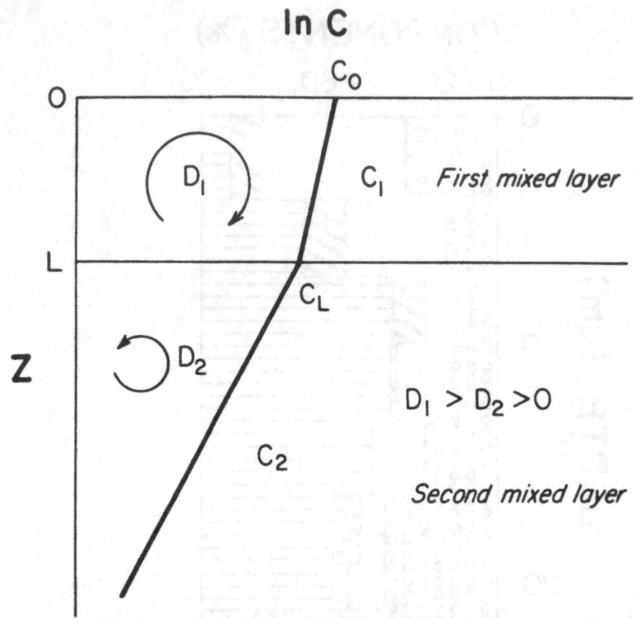

Fig. 5.7. Idealized model of bioturbational mixing of sediment in a steady-state 'two-box' model. Symbols are explained in the text (from Kim & Burnett, 1988).

apparent that ^{210}Pb attached to finer particles is transported further relative to the bulk sediment. The 'mixed layer' thickness based on the <63 μm fraction is about 10 cm, compared to 6 cm for the bulk sediment. The apparent increase in excess ^{210}Pb below 10 cm is interpreted as a result of ^{222}Rn diffusion and decay from the phosphatic gravel layer underlying the sediment.

Another example of obvious mixing of the upper layers of a slowly accumulating sediment is that of core GS-1 (Fig. 5.6). This core is particularly interesting because it penetrated a buried phosphatic crust. The crust, believed to have been sampled in its *in situ* growth position, has been intensively studied by geochemical, mineralogical and radiochemical techniques (Kim & Burnett, 1986).

The ^{210}Pb profile of core GS-1 displays an exponential trend to 12 cm depth which corresponds to an 'apparent' rate of sediment accumulation of 0.071 cm y^{-1} (Fig. 5.6). The significantly higher ^{210}Pb activity at the 12–14 cm interval is probably due to a contribution from decay of diffusive ^{222}Rn emanating from the buried phosphatic crust found 14 cm below the sediment surface. Organic carbon fractions of three intervals of GS-1 sediment were also dated by ^{14}C techniques. Extrapolating the radiocarbon ages from the two intervals above the phosphate crust, Kim & Burnett calculated an apparent core top radiocarbon age of 1064 y. Assuming the ^{14}C age distribution is a reflection of sedimentation and mixing, the sedimentation rate, S (cm y^{-1}), can be calculated by an equation similar to that derived by Peng *et al.* (1977):

$$S=\lambda L/(e^{\lambda t}-1) \tag{5.1}$$

where λ is the decay constant of ^{14}C (1.21×10^{-4} y^{-1}), t is the core top radiocarbon age (1064 y), and L is the thickness of the mixed layer (12 cm based on the excess ^{210}Pb profile). A sedimentation rate for GS-1 of 0.011 cm y^{-1} results from this calculation, significantly lower than the 'apparent' rate of 0.071 cm y^{-1} based on the assumption that the ^{210}Pb profile represents only radioactive decay. With this very slow rate of sediment accumulation, the presence of excess ^{210}Pb must, as in core BX-2, be a consequence of downward sediment mixing.

A steady-state 'two-box' mixing model was used by Kim & Burnett (1988) to describe the distribution of excess activities of ^{210}Pb for the more rapidly accumulating sediments (Fig. 5.7). This approach was preferred over the 'one-box' model used for deep-sea sediments (Goldberg & Koide, 1962; DeMaster & Cochran, 1982; Aller & DeMaster, 1984; Cochran, 1985) because the time period of sediment accumulation under consideration was of the same order as the half-life of the tracer nuclide (^{210}Pb). The model assumes that relatively fast biological particle mixing occurs in the first (upper) box and that mixing also occurs in the second (lower) box, but at a slower rate. The model equation governing the distribution of the radionuclide within the first box, being affected by particle mixing, burial, and radioactive decay, is:

$$dC_1/dt=D_1(\partial^2C_1/\partial z^2)-S(\partial C_1/\partial z)-\lambda C_1=0 \tag{5.2}$$

and within the second box:

$$dC_2/dt=D_2(\partial^2C_2/\partial z^2)-S(\partial C_2/\partial z)-\lambda C_2=0 \tag{5.3}$$

where the subscripts refer to the first (1) and the second box (2), z is the depth below the sediment–water interface (cm), C is the excess activities of the pertinent radionuclide over its parent activity (dpm g^{-1}), D is the biological particle mixing coefficient (cm^2 y^{-1}), for which it is assumed that $D_1>D_2>0$ and S is the sediment accumulation rate (cm y^{-1}), λ is the decay constant of the appropriate radionuclide (^{210}Pb$=0.0311$ y^{-1}; ^{234}Th$=10.5$ y^{-1}). Particle dissolution is assumed to be negligible in this model. General and analytical solutions to these equations may be found in Kim & Burnett (1988). An example of the results for one core for which this model was successfully applied is BX-3 (Fig. 5.8).

Three cores from the 12° S latitude area displayed three distinctly different patterns of mixing and sedimentation: the

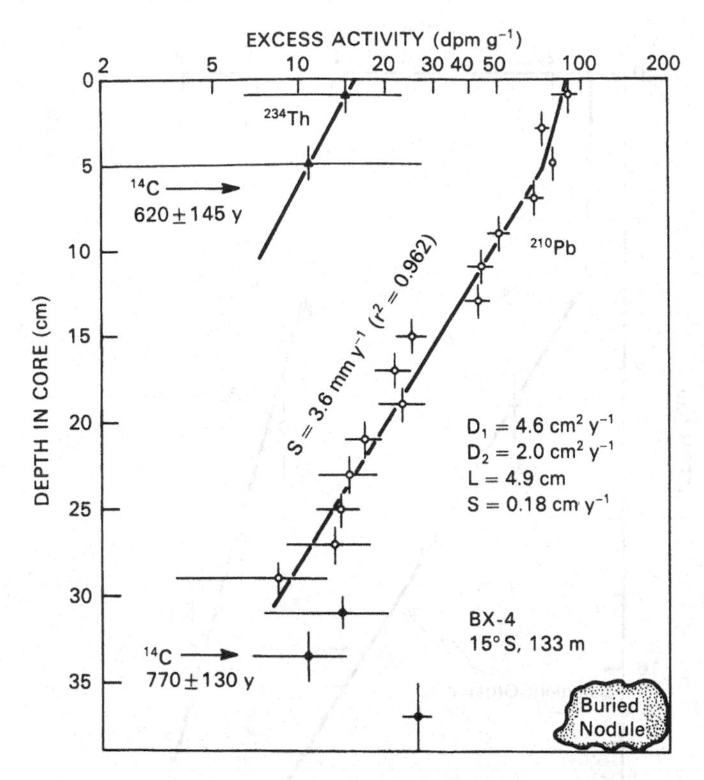

Fig. 5.8. Excess ^{210}Pb and ^{234}Th versus depth in rapidly accumulating sediment BX-4. The distribution of ^{210}Pb within these sediments may be explained in terms of the 'two-box' model (Fig. 5.7). Three data points below 30 cm (solid circles) were assumed to be due to radon diffusion from the buried nodule and not considered in the regression analysis for sediment accumulation and mixing rates (from Kim & Burnett, 1988).

mixed layer was absent in the core from the innermost part of the oxygen-minimum zone (BX-6); a very rapid, intense mixing was observed in the shallowest core GS-8, and the mixed layer was missing in GS-10, possibly due to sediment slumping.

The ^{210}Pb-derived particle mixing coefficients for the six sediment cores from the 12° S and 15° S latitude areas fall in the range of 0.72–2.6 cm² y⁻¹ for the second mixed layer and 4.2–4.6 cm² y⁻¹ for the upper mixed layer. The estimates for the first mixing coefficient (D_1) are an order of magnitude lower than those calculated for hemipelagic cores from the Panama Basin and Atlantic continental rise (Aller & DeMaster, 1984). However, they are on the same order of magnitude as, or up to one order lower than, those calculated from coastal marine sediment from other areas (Santschi *et al.*, 1986). The estimates for D_1 are most similar to values estimated for deep-sea sediments (Nozaki *et al.*, 1979; Peng, Broecker & Berger, 1979). Estimates of the second mixing coefficient (D_2) are on the same order of magnitude as those calculated for normal coastal marine sediments (Santschi *et al.*, 1986).

Phosphorite deposits on the Peru shelf

There are a variety of phosphatic grain types present on the Peru margin. A thorough petrologic description of them is found in Glenn & Arthur (1988) who also worked on materials recovered from the Conrad expedition. Irregular-shaped nodular grains, often longer than about 5 cm, are abundant in the surface sediments of the Peru margin (Burnett, 1977). Of particular interest is the GS-1 core because of the apparently *in situ* phosphatic crust occurring 14 cm below the sediment surface (Fig. 5.6). Fragments of this crust, apparently broken by the impact of the coring device, were recovered in an oriented position and studied in detail (Kim & Burnett, 1986).

Radiochemical results (Fig. 5.9) show that the age of the crust increases downward implying slow, unidirectional upward growth at a rate of approximately 1.3 cm $(10^3 y)^{-1}$. This is consistent with rates derived from nodules dredged off Peru although the direction of growth is opposite to that inferred from other nodule samples (Burnett, Beers & Roe, 1982). Furthermore, significant changes with depth in composition and crystallinity were noted. Unlike the BX-2 and GS-9 pellet samples, however, the compositional change observed was not fluoridation but an increase in structural carbonate. The time scale of these transitions in the GS-1 crust is approximately the same as for the pellet samples. The difference, therefore, may be a reflection of the difference in morphology. Baker & Burnett (1988) have reported that SEM observations reveal that the microcrystallite size of CFA within Peru shelf phosphatic pellets are distinctively smaller (~0.x μm) than those of phosphatic nodules (~x.0 μm). It is not clear what physical and chemical reasons are responsible for the innate distinctions in crystal size between these different morphologic types. Presumably, the compositional and crystal size differences may be related to the formational process.

While study of the growth directions and rates of indurated phosphatic crusts and nodules was relatively straightforward because their size allowed convenient sub-sampling, phosphatic pellet studies had to be approached in an entirely different manner. Pellets were separated into individual size classes from four cores for the purpose of determining ages and growth rates by uranium-series isotopes and accelerator mass spectrometry (AMS). Because of their small grain size (125–250 μm), 'composite' samples had to be analyzed, each one consisting of thousands to tens of thousands of individual pellets. The results were surprising and very interesting. The apparent uranium-series ($^{230}Th/^{234}U$) ages were very old (30–100 ky) yet showed a progressive increase in a downcore direction. Since the organic matter fraction of the sediment was known to be young (< 10 ky) from conventional radiocarbon dating, it appeared as if 'old', possibly reworked, pellets occurred in a 'young' sediment matrix. All radiocarbon sediment ages in BX-2 are < 10 ky and AMS ^{14}C ages of pellet samples are less than 12 ky with both age trends displaying an internally consistent increase in age below an upper mixed zone (Fig. 5.10). Uranium-series results indicate ages apparently tens of thousands of years older than the companion radiocarbon ages, yet both data sets display internally consistent trends of increasing age downcore. In addition, the CFA pellet ^{14}C ages are systematically offset by a few thousand years (older) than the matching sediment organic carbon ages.

If the pellets contained within the core BX-2 are indeed an *in situ* deposit as much evidence suggests, the 'old' uranium-series results must be reconciled with the 'young' radiocarbon ages of the sediment and the 'young' AMS-^{14}C ages of the separated pellets. Although this appears to be an 'open-system' case with discordant ^{230}Th, ^{231}Pa, and ^{14}C ages, the internal consistency within each individual data set argues against this interpretation.

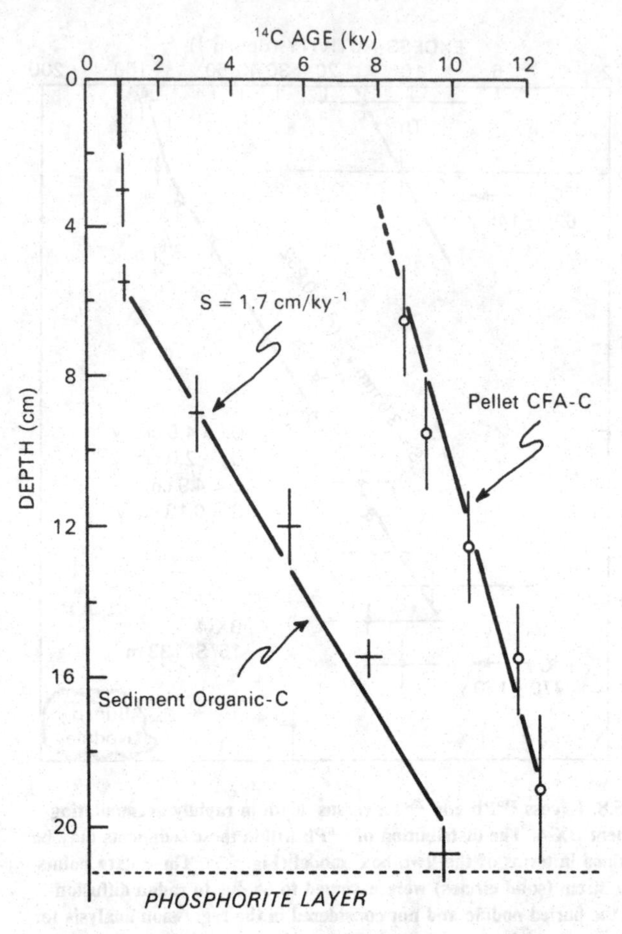

Fig. 5.10. Radiocarbon years versus depth for both the total sediment organic carbon (determined by conventional methods) and CFA-carbon extracted from phosphate pellets (determined by acceleration mass spectrometry) for core BX-2. Thorium ages of the phosphate pellets (not shown) also display internally-consistent results with progressively greater ages downcore but are systematically older than both sets of radiocarbon ages (from Burnett *et al.*, 1988).

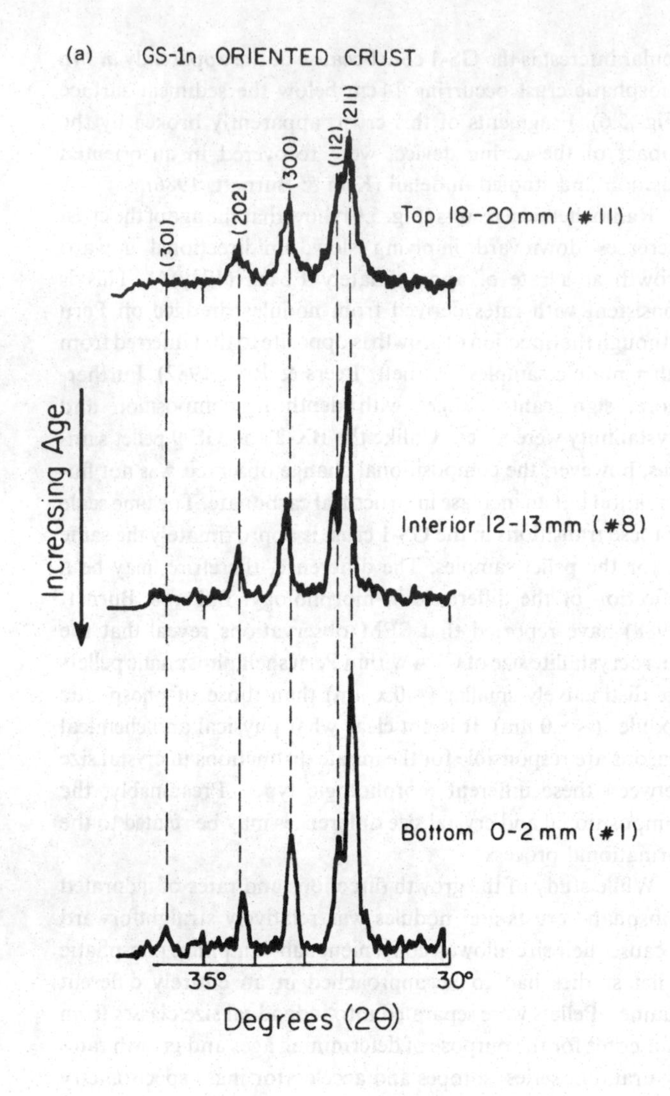

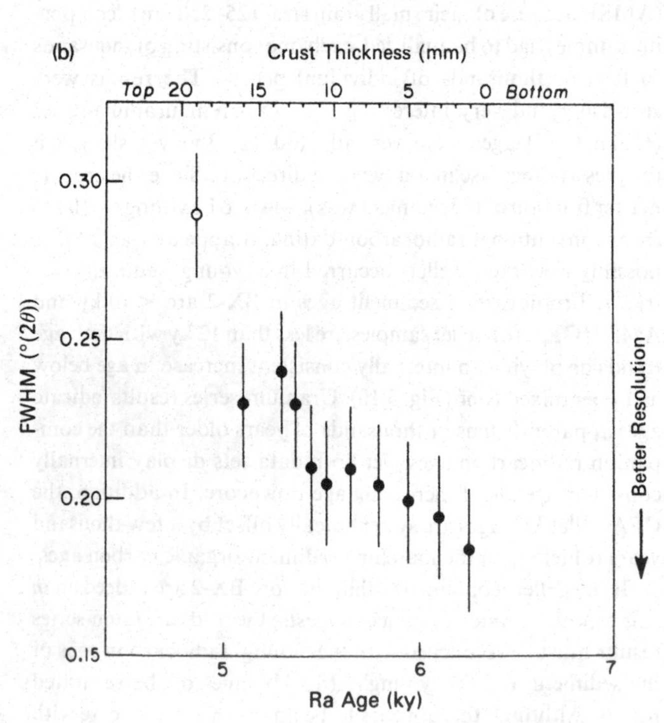

Fig. 5.9. Age and crystallinity trends in phosphate crust GS-1 (from Kim & Burnett, 1986). (a) Portion of X-ray diffractograms from top (youngest), middle, and bottom (oldest) portions of phosphate crust GS-1. There is an obvious improvement in the resolution of these CFA peaks in the older layers. (b) Resolution of the 211 CFA peak, expressed as 'full width at half maximum' (FWHM), as a function of crust thickness and age.

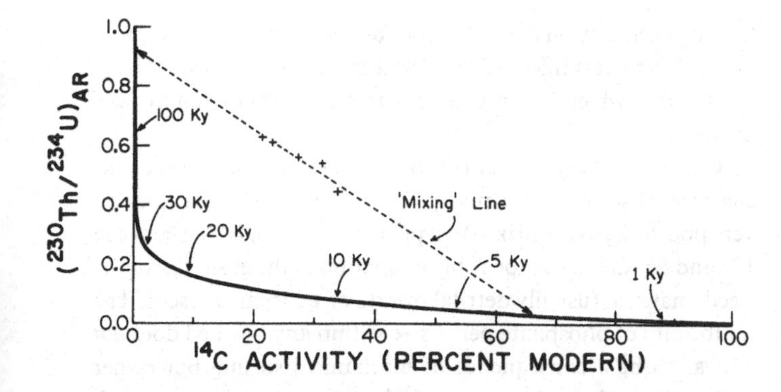

Fig. 5.11. 'Concordia' plot of the activity ratio $^{230}Th/^{234}U$ versus % modern ^{14}C together with data points for BX-2 phosphate pellets. The pellet data fall along a mixing line between 'old' and 'young' end members, implying that the pellet samples analyzed are 'composites' of two generations (from Burnett *et al.*, 1988).

Another possibility, which can explain all these observations, is that all radiochemical measurements are 'composites' resulting from natural physical intermixing of 'old' versus 'young' pellets within these sediments. Should this mixture be relatively simple, say with only two end-members, independent uranium-series and radiocarbon age determinations should constrain the possible mixing pattern.

The locus of points which describes the progression of the $^{230}Th/^{234}U$ activity ratio and the ^{14}C activity (expressed as percent modern) together with actual data points for BX-2 phosphate pellets (Fig. 5.11) shows that the data fall considerably off this concordia. However, the BX-2 data points fall along a fairly well-defined trend line ($r = 0.95$) which may represent a mixing line between two end-members. The intersections of this mixing line with the concordia curve should define the end-member compositions if this interpretation is correct. These results can thus be explained by a physical mixture of 'old' pellets with a $^{230}Th/^{234}U$ activity ratio of about 0.92 and no ^{14}C activity (approximately 254 ky) together with 'young' pellets with a $^{230}Th/^{234}U$ activity ratio of about 0.03 and 67% modern ^{14}C (approximately 3 ky). Assuming this model is correct, the fraction of old pellets within any layer of BX-2 may be calculated based on the 'composite' $^{230}Th/^{234}U$ activity ratios or the percent modern (PM) ^{14}C activities.

The model fractions of 'old' pellets contained within BX-2 show that there is a systematic increase downward of these recycled pellets (Fig. 5.12). This implies that pellets from a previously existing generation contained in an underlying lag deposit (partially preserved in the bottom of the box core) were progressively mixed into the Recent sediments together with modern pellets forming contemporaneously within the sediment. Thus, the measured disequilibrium relationships have apparently been influenced more by mixing of older pellets into Recent sediment than by radioactive decay. Since sufficient sensitivity to measure uranium-series isotopes of single pellets is not available, the 'composite' samples result in ages which represent integrated rather than depositional ages. If this interpretation is correct, the majority (~75% at the core top) of pellets contained in core BX-2 formed *in situ* during the depositional history of the Recent sediment at this site. Further work will be

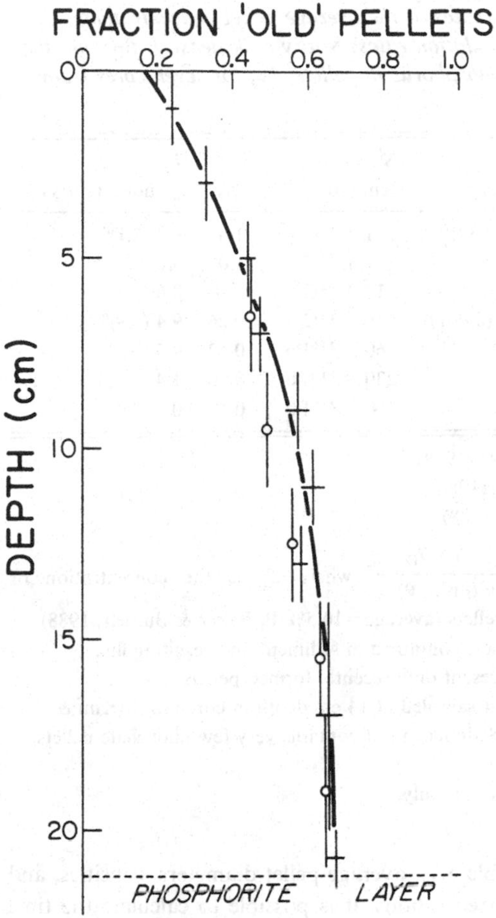

Fig. 5.12. Estimated fraction of 'old' pellets contained within different layers of core BX-2 (from Burnett *et al.*, 1988).

necessary to evaluate if the different grain populations may be identified by petrographic or geochemical criteria.

By scaling the concentration of phosphate pellets within the Peru shelf sediment cores to the net rate of sediment accumulation, it is possible to estimate uptake fluxes of phosphorus by pellet formation. These calculations (Table 5.1) show that the fluxes appear to remain relatively constant over a much wider range of pellet concentrations and sediment accumulation rates. With the exception of core GS-10, which may have 'missing' sediment based on the very low ^{210}Pb concentrations and high extrapolated ^{14}C core top age, and GS-1, which contains the *in situ* crust, all cores have pellet uptake rates in the range of 0.5–4.4 μmol P$(cm^2\ y)^{-1}$. The same cores have a range of net sediment accumulation of greater than two orders of magnitude. Uptake rates of phosphate are maintained at reasonably constant levels as a result of an inverse relationship between pellet concentration and sediment accumulation rates. If pellet formation is controlled by the release and subsequent precipitation of dissolved phosphate at the sediment–seawater interface and diffusion of fluoride from the overlying water, it seems reasonable that these rates should be somewhat constant. For example, fluoride diffusion through uniform sediments should be relatively constant because of its constant concentration in the overlying water column (Froelich *et al.*, 1983).

Based upon the estimated uptake rates of phosphorus in these

Table 5.1. *Estimates of the average pellet concentrations, X_p; sediment accumulation rates, S; dry bulk density, B_D; and the burial flux of phosphorus in pellets, F_p, for seven cores from the Peru shelf*

Core number	X_p[a] fraction	S[b] (cm $(10^3\ y)^{-1}$	B_D[c]	F_p[d] (μmol P $(cm^2y)^{-1}$)
BX–2	0.76 (0.37)[e]	1.7–^{14}C	1.0	4.4 (2.1)[e]
GS–6	0.65	1.8–^{14}C	1.0	4.0
GS–1s[f]	—	11.0–^{14}C	—	8.6
GS–10	0.097 (0.061)[g]	110 –^{14}C	0.26	9.4 (5.9)[g]
BX–3	0.008	160 –^{210}Pb	0.12	0.52
BX–6	0.018	230 –^{210}Pb	0.10	1.4
BX–4	0.005	360 –^{210}Pb	0.12	0.7

[a]Baker & Burnett (1988).
[b]Kim & Burnett (1988).
[c]Froelich *et al.* (1988).
[d]$F_p = \frac{C_p \times X_p \times S \times B_D}{31 \times 10^{-6}\ g\ (\mu mol\ P)^{-1}}$, where C_p is the concentration of phosphorus in pellets (average = 10.5% P, Baker & Burnett, 1988).
[e]Fraction of pellets contained in sediment and resulting flux corrected to represent only recently formed pellets.
[f]Phosphatic crust sampled at 14 cm depth in core GS–1 (Kim & Burnett, 1986). Sediment itself contains very few phosphate pellets (<2%).
[g]125–500 μm fraction only.

sediments (Table 5.1), average pellet diameters, densities, and phosphate concentrations, it is possible to calculate the time required for individual pellets to form. Based on these calculations, individual Peru shelf phosphate pellets apparently grow in < 10 y! Furthermore, based on the sediment accumulation rates for these same cores, the equivalent core length or 'reaction zone' is quite narrow, < 1 cm in every case. Thus, if phosphate pellets begin to form at the sediment–seawater interface, then they will be completely formed within just the first few millimeters or so and are buried afterwards. Some pellets may, of course, cycle through this 'reaction zone' more than once by the action of bioturbational mixing. Several trips through the reaction zone may account for the concentric layering apparent in some pellets. Pore water fluoride (Froelich *et al.*, 1983) and phosphate (Froelich *et al.*, 1988), as well as stable isotope results (Glenn *et al.*, 1988), are all internally consistent with a model of fast pellet formation near the sediment–seawater interface.

Conclusions and future work

Sediments of the Peru shelf upwelling zone display a heterogeneous distribution of thin (few centimeters) phosphatic pellet beds which may be comparable in some respects to the large geologic deposits studied on land. Cores with several tens of percent phosphate pellets lie within kilometers of cores which are essentially organic-rich muds devoid of pellets. This is probably an effect of uneven bathymetry, variability in bottom currents, patchiness of productivity, the distribution of the oxygen-minimum zone and other factors. Heterogeneous distribution of pellet beds also seems to be the case for many laminated phosphorite sequences studied on land. The phosphatic Meade Peak Member of the Phosphoria Formation, for example, has lateral continuity on the order of kilometers only for beds which are several meters thick. Within these major beds are minor beds and laminae which are only correlatable over tens to hundreds of meters.

Grain size analysis of cores containing significant pellet concentrations showed that phosphate pellets reside in a poor to very poorly sorted matrix. Most pellets were found to be between 125 and 500 μm in size, possibly a function of the grain size of the 'seed' material (usually detrital quartz or feldspar) present. The uniformity of phosphate pellet size within any one bed does *not* appear to be a consequence of mechanical sorting but rather uniform depositional processes. In fact, some of our results (such as the distribution of excess ^{210}Pb by size) imply that mixing may result in a sediment with less 'sorting'.

Studies of sediment accumulation and mixing rates show that Peru shelf sediment has an extremely wide range of accumulation rates, from about 1.7 cm $(10^3y)^{-1}$ to 360 cm $(10^3y)^{-1}$. Biological mixing of these sediments on the other hand, is more uniform with mixing coefficients of the upper sediments (D_1) ranging from 4.2 to 4.6 $cm^2\ y^{-1}$, lower than most normal coastal environments but similar to values estimated for deep-sea sediments. Bioturbation is thus a significant process even in these oxygen-depleted waters of the Peru shelf. Uranium-series and AMS ^{14}C studies show that pellet distribution is significantly affected by sediment mixing. Significant redistribution of pellets within sediment core BX-2, for example, has occurred by mixing of an older, underlying generation of phosphate pellets into Recent sediment with contemporaneously formed pellets.

Based upon ^{14}C and ^{210}Pb ages of associated sediment, phosphate pellets appear to form very quickly, probably within less than about 10 y after growth is initiated. Assuming that these pellets form at the sediment–seawater interface, the thickness of the 'reaction' zone is < 1 cm for the cores examined. Estimated authigenic uptake rates of phosphorus into modern Peru shelf phosphate *pellets* agree closely with rates based on pore-water models but are somewhat higher than rates previously calculated for phosphate *nodules*. This may explain the predominance of phosphatic pellets over nodules in the geologic record.

The information we have gained concerning sediment texture, accumulation, mixing rates, phosphorite mineralization rates and redistribution of phosphatic pellets should constrain possible models of 'phosphogenesis' for a geologic deposit deposited within a coastal upwelling environment. It is now clear that not all sedimentary phosphorites are, in fact, formed in that type of environment. Continued study of a modern system, such as the Peru shelf should continue to benefit the long-term goal of understanding the controls and mechanisms of sedimentary phosphorite formation.

Acknowledgments

The author wishes to thank several colleagues who have assisted in his obtaining at least some degree of understanding of marine phosphorites. I particularly would like to thank R.P. Sheldon, S.R. Riggs, P.J. Cook and R.E. Garrison, all 'real' geologists. Essentially all of the contents of this paper were done in collaboration with former students of mine (K.H. Kim, K.B. Baker and P.A. Chin). I thank these guys for their good work. A. Watkins helped on the editing of this paper as well as most papers in this volume. S.R. Riggs provided a good thorough review of an earlier version of this paper. All of the research presented here was supported by grants from the Ocean Sciences Division of the National Science Foundation (OCE8317181 and OCE8520724).

References

Aller, R.C. & DeMaster, D.J. (1984). Estimates of particle flux at the deep-sea floor using $^{234}Th/^{238}U$ disequilibrium. *Earth and Planetary Science Letters*, **67**, 308–18.

Baker, K.B. & Burnett, W.C. (1988). Distribution, texture, and composition of modern phosphate pellets in Peru shelf muds. *Marine Geology*, **80**, 195–213.

Baturin, G.N. (1971). Stages of phosphorite formation on the ocean floor. *Nature*, **29**, 232.

Baturin, G.N., Merkulova, K.I. & Chalov, P.I. (1972). Radiometric evidence for recent formation of phosphatic nodules in marine shelf sediments. *Marine Geology*, **13**, 37–41.

Birch, G.F. (1979). Phosphorite pellets and rock from the western continental margin and adjacent coastal terrace of South Africa. *Marine Geology*, **33**, 91–116.

Borole, D.V., Rajagopalau, G. & Somayajulu, B.L.K. (1987). Radiometric ages of phosphorites from the west coast of India. *Marine Geology*, **78**, 161–5.

Burnett, W.C. (1977). Geochemistry and origin of phosphorite deposits from Peru and Chile. *Geological Society of America Bulletin*, **88**, 813–23.

Burnett, W.C., Baker, K.B., Chin, P.A., McCabe, W. & Ditchburn, R. (1988). Uranium-series and AMS ^{14}C studies of modern phosphatic pellets from Peru shelf muds. *Marine Geology*, **80**, 215–30.

Burnett, W.C., Beers, M.J. & Roe, K.K. (1982). Growth rates of ocean-floor phosphate nodules from the continental margin of Peru. *Science*, **215**, 1616–18.

Burnett, W.C., Cullen, D.J. & McMurtry, G.M. (1987). Open-ocean phosphorites – in a class by themselves? In *Marine Minerals: Resource Assessment Strategies*, ed. P.G. Teleki, M.R. Dobson, J.R. Moore & U. Von Stackelberg, pp. 119–31. D. Reidel, Dordrecht.

Burnett, W.C. & Froelich, P.N. (1988). Preface to special issue on the origin of marine phosphorite. The results of the R.V. Robert D. Conrad cruise 23-06 to the Peru Shelf. *Marine Geology*, **80**, iii–vi.

Cochran, J.K. (1985). Particle mixing rates in sediments of the eastern equatorial Pacific: evidence from ^{210}Pb, $^{239,240}Pu$ and ^{137}Cs distribution of MANOP sites. *Geochimica Cosmochimica Acta*, **99**, 1195–1210.

D'Anglejan, B.F. (1967). Origin of marine phosphorites off Baja California, Mexico. *Marine Geology*, **5**, 15–44.

DeMaster, D.J. (1981). The supply and accumulation of silica in the marine environment. *Geochimica Cosmochimica Acta*, **45**, 1715–32.

DeMaster, D.J. & Cochran, J.K. (1982). Particle mixing rates in deep-sea sediments determined from excess ^{210}Pb and ^{32}Si profiles. *Earth and Planetary Science Letters*, **61**, 257–71.

Froelich, P.N., Arthur, M., Burnett, W.C., Deakin, M., Hensley, V., Jahnke, R., Kaul, R., Kim, K.H., Roe, K., Soutar, A. & Vathakanan, C. (1988). Early diagenesis of organic matter in Peru continental margin sediments: phosphorite precipitation. *Marine Geology*, **80**, 309–46.

Froelich, P.N., Kim, K.H., Jahnke, R., Burnett, W.C., Soutar, A. & Deakin, M. (1983). Pore water fluoride in Peru continental margin sediments: Uptake from seawater. *Geochimica Cosmochimica Acta*, **47**, 1605–12.

Glenn, C.R. & Arthur, M.A. (1988). Petrology and major element geochemistry of Peru margin phosphorites and associated diagenetic minerals: authigenesis in modern organic-rich sediments. *Marine Geology*, **80**, 231–67.

Glenn, C.R., Arthur, M.A., Yeh, H-W. & Burnett, W.C. (1988). Carbon isotopic composition and lattice-bound carbonate of Peru–Chile margin phosphorites. *Marine Geology*, **80**, 287–307.

Goldberg, E.D. & Koide, M. (1962). Geochronological studies of deep-sea sediments by the thorium–ionium method. *Geochimica Cosmochimica Acta*, **26**, 417–50.

Henrichs, S.M. & Farrington, J.W. (1984). Peru upwelling region sediments near 15° S. I. Remineralization and accumulation of organic matter. *Limnology and Oceanography*, **29**, 1–19.

Jahnke, R., Emerson, S.R., Kim, K.H. & Burnett, W.C. (1983). The present day formation of apatite in Mexican continental margin sediments. *Geochimica Cosmochimica Acta*, **47**, 259–66.

Kazakov, A.V. (1937). The phosphorite facies and genesis of phosphorites. *Scientific Institute of Fertilizers and Insecto-Fungicides Transactions*. **142**, 95–113.

Kim, K.H. & Burnett, W.C. (1986). Uranium-series growth history of a Quaternary phosphatic crust from the Peruvian continental margin. *Chemical Geology*, **58**, 227–44.

Kim, K.H. & Burnett, W.C. (1988). Accumulation and biological mixing of Peru margin sediments. *Marine Geology*, **80**, 181–94.

Kolodny, Y. (1969). Are marine phosphorites forming today? *Nature*, **224**, 1017–19.

Kolodny, Y. & Kaplan, I.R. (1970). Uranium isotopes in sea floor phosphorites. *Geochimica Cosmochimica Acta*, **34**, 3–24.

Krissek, L.A. & Scheidegger, K.F. (1983). Environmental controls on sediment texture and composition in low oxygen zones off Peru and Oregon. In *Coastal Upwelling: Its Sediment Record*, Part B, ed. J. Thiede & E. Suess, pp. 163–80. Plenum Press, New York.

Krissek, L.A., Scheidegger, K.F. & Kulm, L.D. (1980). Surface sediments of the Peru–Chile continental margin and the Nazca Plate. *Geological Society of America Bulletin*, **1(91)**, 321–31.

Kulm, L.D., Suess, E. & Thornburg, M. (1984). Dolomites in organic-rich muds of the Peru forearc basins: analogue to the Monterey Formation. In *Dolomites of the Monterey Formation and Other Organic-Rich Units*, 41, ed. R.E. Garrison, M. Kastner & D.H. Zenger, pp. 29–47. Society of Economic Palaeontologists and Mineralogists, Pacific Section, Santa Fe Springs.

Manheim, F.T., Pratt, R.M. & McFarlin, P.F. (1980). Composition and origin of phosphorite deposits of the Blake Plateau. In *Marine Phosphorites*, ed. Y.K. Bentor, pp. 117–38. Society of Economic Palaeontologists and Mineralogists, Special Publications No. 29, Tulsa.

McKelvey, V.E. (1967). Phosphate deposits. *United States Geological Survey Bulletin*, **1252-D**, 1–21.

McKelvey, V.E., Swanson, R.W. & Sheldon, R.P. (1953). The Permian phosphate deposits of the western United States. In *Origine des Gisements de Phosphates de Chaux, 19th International Geology Congress*, **11**, 45–64. Algiers.

Muller, P. & Suess, E. (1979). Productivity, sedimentation rate and sedimentary organic matter in the oceans – I: Organic carbon preservation. *Deep-Sea Research*, **26**, 1347–62.

Murray, J. & Renard, A.F. (1891). *Deep Sea Deposits, Scientific Results of the Exploration Voyage of HMS Challenger, 1872–1876*. Longmans, London.

Nozaki, Y., Cochran, J.K., Turekian, K.K. & Keller, G. (1979). Radiocarbon and ^{210}Pb distribution in submersible-taken deep-sea cores from Project FAMOUS. *Earth and Planetary Science Letters*, **34**, 167–73.

O'Brien, G.W., Harris, J.R., Milnes, A.R. & Veeh, H.H. (1981). Bacterial origin of East Australian continental margin phosphorites. *Nature*, **294**, 442–4.

O'Brien, G.W. & Heggie, D. (1988). East Australian continental margin phosphorites. *EOS*, **69**, 2.

O'Brien, G.W. & Veeh, H.H. (1980). Holocene phosphorite on the East Australian continental margin. *Nature*, **288**, 690–2.

Peng, T.H., Broecker, W.S. & Berger, W.H. (1979). Rates of benthic mixing in deep-sea sediments as determined by radioactive tracers. *Quaternary Research*, **11**, 141–9.

Peng, T.H., Broecker, W.S., Kipphut, G. & Shackleton, N. (1977). Benthic mixing in deep sea cores as determined by ^{14}C dating and its implications regarding climate stratigraphy and the fate of fossil fuel CO_2. In *The Fate of Fossil Fuel CO_2 in the Oceans*, ed. N.R. Anderson & A. Malahoff, pp. 355–73. Plenum, New York.

Reimers, C.E. & Suess, E. (1983). Spatial and temporal patterns of organic matter accumulation on the Peru continental margin. In *Coastal Upwelling: Its Sediment Record, Part B*, ed. J. Thiede & E. Suess, pp. 311–46. Plenum, New York.

Riggs, S.R. (1984). Paleoceanographic model of Neogene phosphorite deposition, US Atlantic continental margin. *Science*, **223**, 123–31.

Ruttenberg, K.C. & Berner, R.A. (1987). Evidence for precipitation of authigenic carbonate fluorapatite in non-upwelling environments (abstract). *Transactions, American Geophysical Union*, **68**, 1711.

Santschi, P.H., Li, Y.H., Bell, J.J., Trier, R.M. & Kawtaluk, K. (1986). Pu in coastal marine environments. *Earth and Planetary Science Letters*, **51**, 248–65.

Scheidegger, K.F. & Krissek, L.A. (1982). Dispersal and deposition of eolian and fluvial sediments off Peru and northern Chile. *Geological Society of America Bulletin*, **93**, 150–62.

Sheldon, R.P. (1964). Paleolatitudinal and paleogeographic distribution of phosphorite. *US Geological Survey Professional Paper*, **501-C**, 106–13.

Suess, E. (1981). Phosphate regeneration from sediments of the Peru continental margin by dissolution of fish debris. *Geochimica Cosmochimica Acta*, **45**, 577–88.

Veeh, H.H., Burnett, W.C. & Soutar, A. (1973). Contemporary phosphorites on the continental margin of Peru. *Science*, **181**, 844–45.

Walsh, J.J. (1981). A carbon budget for overfishing off Peru. *Nature*, **290**, 300–4.

Zuta, S., Riviera, T. & Bustamante, A. (1975). Hydrological aspects of the main upwelling areas off Peru. In *Upwelling Ecosystems*, ed. R. Boje & M. Tomczak, pp. 235–56. Springer-Verlag, New York.

PART 2

Modern and Neogene phosphorites and associated sediments

6
Occurrence of dolomite in Neogene phosphatic sediments

P. BAKER AND M. ALLEN

Abstract
Dolomite is a common mineral in the phosphatic sediments of the Miocene-age Pungo River Formation of North Carolina, Hawthorn Group of Florida, Monterey Formation of California, as well as many other phosphatic sediments around the world. This dolomite forms as a product of early diagenesis in these organic-rich sediments. The dolomite is originally a poorly ordered, Mg-depleted protodolomite, but eventually undergoes a solution–reprecipitation reaction which increases its ordering and alters its chemical and stable isotopic composition. The high organic carbon contents of these sediments are necessary to promote microbial sulfate reduction. Sulfate reduction is the dominant process of organic carbon mineralization in these sediments, liberating the organically-bound phosphorus to solution to allow apatite precipitation. Sulfate reduction also elevates pore-water alkalinity and removes dissolved sulfate from the pore waters enabling dolomite formation.

The overall detrital sedimentation rate is one of the most important controls on the eventual fraction of dolomite (or apatite) forming in the sediments. If the rate of terrigenous sedimentation is too high, dolomite (or apatite) will be masked by detritus. Abundant dolomite formation requires low detrital sedimentation rates. Similarly, phosphorite (*sensu stricto*) formation does not necessarily require reworking or winnowing (suggested by many other workers), but it does require low detrital sedimentation rates.

Dolomite precipitation in phosphatic sediments may take place within the zone of sulfate reduction or in the deeper zone of methanogenesis. High abundances of dolomite in a sediment require that most of this dolomite formed at shallow burial depths. Phosphate precipitation occurs shallower and faster than dolomite precipitation. Most of the apatite in a phosphorite deposit precipitates within a few centimeters of the sediment–water interface in the uppermost zone of sulfate reduction. Small amounts of apatite may precipitate at greater depths. In most cases, dolomite precipitation must post-date apatite precipitation.

The stable isotopic compositions of apatites and dolomites originally reflect the environment of authigenesis. When buried, however, both minerals undergo diagenetic isotopic exchange reactions with the pore waters. This exchange obscures the interpretation of the environment of authigenesis based solely on stable isotopic analysis.

Introduction

Dolomite ($CaMg(CO_3)_2$) is a mineral which commonly occurs in phosphatic sediments of marine origin. Other associated minerals usually include opaline silica (and its diagenetic derivatives, opal-CT and quartz), glauconite, and pyrite, and often include magnesium-bearing silicate minerals such as sepiolite and palygorskite. The existence of each of these mineral species is a direct result of biological productivity in the overlying water column (silica) or subsequent microbially-mediated oxidation of organic matter at or near the seawater–sediment interface (pyrite, dolomite, glauconite, and apatite).

It has previously been proposed that the association between the magnesium-bearing minerals, such as dolomite and apatite, is more than coincidental. Martens & Harris (1970) experimentally synthesized calcium phosphate from phosphate-enriched, seawater-like solutions. They showed that amorphous calcium phosphates could be formed over a large range of solution Mg:Ca ratios, but that only at solution Mg:Ca ratios less than about 0.2 (molar ratio) could they produce crystalline apatite on the time scale of their experiments. Their results were confirmed by the work of Nathan & Lucas (1972 and 1976). These observations led many workers to conclude that, in nature, apatite can only form in the near absence of dissolved magnesium. Bentor (1980) pointed out that marine interstitial waters are frequently depleted of dissolved magnesium, further suggesting the authigenic precipitation of dolomite or magnesium-bearing clay minerals. For this reason he suggested that apatite precipitation could occur in these pore waters. His proposal was strengthened by the close association in some (but certainly not all) deposits of dolomite and apatite and the antithetical relation between calcite and apatite as observed, for example, by Gulbrandsen (1960) in the Phosphoria Formation.

More recently, however, Jahnke *et al.*(1983) have shown that crystalline apatite is presently forming in marine pore waters that have seawater levels of dissolved calcium and magnesium. Jahnke *et al.* (1983) hypothesized that an initial precipitate of amorphous calcium phosphate forms in phosphate-enriched pore waters and recrystallizes to crystalline apatite on a time scale of 10–100 y. The existence of such a mechanism, at least in laboratory experiments, was confirmed by Gulbrandsen, Roberson & Neil (1984) who were able to synthesize crystalline carbonate fluorapatite from phosphate- and fluoride-enriched seawater after nearly ten years.

In this paper we shall first review the geology and geochemistry of dolomites occurring in Neogene phosphatic sediments from three locations: the Pungo River Formation of North Carolina, the Hawthorn Group of Florida, and the Monterey Formation of California. These ancient marine sediments will then be compared with modern marine sediments thought to be presently undergoing dolomite and phosphate authigenesis. We will address the following questions. What are the characteristics of dolomites associated with Neogene phosphates? What is the relative timing of dolomite and apatite authigenesis in these sediments? Why do these minerals commonly occur together? What is the chemical and sedimentary environment of formation of these two minerals?

Geology of dolomite occurrences in the Pungo River Formation

The geology of dolomite occurrences in the Pungo River Formation, North Carolina, was studied by Allen (1985). These occurrences are perhaps of greatest interest here because the dolomite is so closely associated with very large and pure phosphate deposits. In the other two locations under study, phosphate comprises only a minor portion of the total sedimentary section.

The Pungo River Formation is of Middle Miocene age and has been correlated with the Calvert Formation of Maryland and Virginia (Brown, 1958) and the Hawthorn Group of Florida, Georgia and South Carolina (Gibson, 1967; Riggs, 1979). It is characterized by cyclic sedimentation. Each cycle consists of a lower, dominantly terrigenous subunit which grades upward into a phosphate-rich subunit, and is capped by a carbonate subunit (Riggs, 1979 and 1984; Riggs *et al.*, 1982, Scarborough, Riggs & Snyder, 1982). The tops of the carbonate units are often lithified and bored, indicative of submarine exposure (Riggs, 1984).

The terrigenous subunits are composed dominantly of quartz sand, illite, and smectite, with minor kaolinite and clinoptilolite. The phosphate-rich facies is composed of about 30–50% phosphate grains, with dolomite present in minor quantities. Dolomite dominates the mineralogy of the carbonate subunits where it comprises from 10–85% of the total sediment.

The sediments of the carbonate facies are primarily clayey and sandy dolomitized packstones and wackestones, although some samples are clayey, phosphatic and dolomitic quartz sandstones. Non-fossil grains are composed of silt- to sand-sized quartz (2–30%), carbonate fluorapatite pellets and intraclasts (1–25%), clays (5–20%), pyrite (1–5%), glauconite (0–3%) and feldspars (0–1%). The quartz grains are angular–subangular and are fair to poorly sorted.

The phosphate pellets in the carbonate facies contain a variety of disseminated biogenic particles including foraminiferal fragments, diatoms, teeth and bone fragments, silt-sized quartz grains, glauconite and very fine (<10 μm) dolomite crystals. These crystals are much smaller than the dolomite rhombs in the surrounding matrix. Microcrystalline aggregates of pyrite also occur within the pellets. It is likely that the phosphate was formed as an early diagenetic phase soon after pellet formation. The presence of smaller than average dolomite rhombs in the phosphate grains suggests that this included dolomite may have formed prior to pellet formation and was subsequently incorporated into the pellet.

The fossil grains observed in thin section in the dolomitic sediments include barnacle plates, bryozoa, planktonic and benthonic foraminifera, arthropod fragments, worm tubes, rare echinoderm spines and pelecypod fragments, fish teeth and bones and diatoms. Foraminifera, when present, are recrystallized and poorly preserved. Bryozoa are often present as highly dissolved fragments. Pelecypod molds, present at the tops of some of the cycles, are occasionally filled with large crystals of calcite spar, indicative of a phase of late stage freshwater alteration. Barnacle plates, which originally consisted of low-Mg calcite, remain well preserved. Many of the barnacle fragments contain a fine (10–20 μm thick) isopachous calcite cement rind lining their compartments. This cement is the only calcite cement seen in thin section.

The dolomite occurs as rhombs ranging in size from 10–75 μm with the modal size being about 45 μm. In carbonate-rich sediments these rhombs intergrow into larger aggregates of microsucrosic dolomite. In muddier, lower carbonate sediments, the rhombs are dispersed. They often contain opaque and highly birefringent centers or inclusions similar to the rhombs described by Sibley (1980) from the Pliocene Seroe Domi Formation of Bonaire. Some dolomite centers are hollow. Similar grains were also identified by Weaver & Beck (1977) in the Miocene Hawthorn Group of Florida. Textures of the dolomite rhombs, as observed by scanning electron microscopy, vary from totally smooth surfaces with sharp unabraded edges and corners, to surfaces with flame-like structures (see also Weaver & Beck, 1977), to surfaces with some pitting as a result of the dissolution of precursor calcium carbonate.

X-ray diffractograms (XRD) of these dolomites display highly attenuated super-lattice reflections relative to ideally ordered dolomite. The magnesium content estimated from XRD peak shifts range from 42 to 48 mol% $MgCO_3$. These estimates are in reasonable agreement with concentrations determined by bulk chemical analyses and electron microprobe analyses. All the calcite observed in these samples was determined to be low-Mg calcite.

The $\delta^{18}O$ values of the Pungo River dolomites (53 samples, uncorrected for the difference between the dolomite phosphoric acid fractionation factor and the calcite phosphoric acid fractionation factor) range from $+3.14$ to $+5.60$‰ PDB. The oxygen isotope values clearly indicate a marine origin for these dolomites and eliminate any possibility of significant diagenetic re-equilibration with fresh waters. Applying a fractionation factor for dolomite–water oxygen isotopic fractionation (using Epstein *et al.*, 1953, for calcite–water fractionation, adding a constant $+3.5$‰ offset for dolomite–calcite fractionation) and assuming that very shallow marine pore waters have the same oxygen isotopic composition as contemporaneous ocean water, yields a range of 8–18° C for the formation temperature of Pungo River dolomites. These are reasonable temperatures for formation in temperate deep shelf waters.

The $\delta^{13}C$ values of the Pungo River dolomites range from -4.35‰ to $+1.01$‰ PDB with a mean of -2.62‰ PDB. These mostly slightly negative values are probably the result of the mixture of a small amount (about 20%) of isotopically-light carbon produced by organic carbon oxidation during early diagenesis and a large amount (about 80%) of isotopically-heavier carbon derived either from seawater or from a calcium carbonate precursor.

Table 6.1. *Mean chemical composition of dolomites from the Miocene Pungo River Formation (this study), the Miocene Monterey Formation (Burns & Baker, 1987), and the Quaternary of DSDP Site 479 (Baker, unpublished)*

Analyses of the Pungo River Formation were done by atomic absorption spectrophotometry on hand-picked, pure dolomite separates.

Location	Sr (ppm)	Na (ppm)	Fe (ppm)	Mn (ppm)	Zn (ppm)	Mg (mol % $MgCO_3$)
Pungo River Formation	226	987	339	32	15	43.7
Monterey Formation						
Shell Beach	228	—[a]	421	390		47.2
Mussel Rock	302	—	577	513	—	47.4
El Capitan	442	1080	38 000	1065	—	42.3
Arroyo Seco	511	568	61 000	1215	—	38.4
Chico Martinez	716	1009	16 000	1150	—	45.0
DSDP Site 479	565	—	19 000	880	—	45.0

[a]Not analyzed.

The minor element contents of these dolomites are generally lower than those of most marine biogenic carbonates and most other young dolomites (Table 6.1). For example, the average strontium concentration of Pungo River dolomites is 226 ppm. Using the distribution coefficient of strontium in dolomite reported by Baker & Burns (1985), results in a calculated molar dissolved Sr:Ca ratio nearly identical to that of seawater. The relatively high sulfate concentrations (average of about 700 ppm as SO_4, Baker, unpublished data) of these dolomites indicate that some dissolved sulfate was present in the pore waters during the precipitation of the dolomite, but the presence of pyrite suggests that dissolved sulfate concentrations were probably somewhat less than the seawater value. The carbon isotopic composition of the dolomites proves that dolomite precipitation took place in a pore-water environment at least somewhat different from seawater. It seems most likely that the site of authigenesis of the dolomite was within the zone of sulfate reduction, but that not all of the dissolved sulfate had been reduced.

The common lack of clear-cut paragenetic relationships between the apatite and dolomite (an exception was noted earlier) implies that the authigenesis of these two different minerals took place in different diagenetic environments, i.e. under conditions of non-steady-state sedimentation. Sea-level cycles obviously effected important changes in the composition and abundance of the primary components of the sediments (e.g. detrital minerals, primary carbonate minerals, siliceous microorganisms, and organic matter) and these changes altered the course of subsequent diagenesis. There is no evidence to support the contention that dolomite precipitation is necessary to allow phosphate precipitation.

Geology of dolomite occurrences in the Hawthorn Group

The geology of dolomite occurrences in the Hawthorn Group was studied by Prasad (1985) and most of the following discussion is based upon his work. The Hawthorn Group is of Middle Miocene age and, as mentioned previously, is correlated to the Pungo River Formation (Riggs, 1984). It extends from southern Florida to southern South Carolina. The lithology of the Hawthorn is somewhat similar to that of the Pungo River. The former is far more carbonate-rich, but the same cyclic patterns of sedimentation were developed in both regions (Riggs, 1979 and 1984). Regionally, dolomite in the Hawthorn occurred predominantly in the outer-shelf sediments of southern and eastern Florida and lesser amounts were formed to the north and northwest in the shallow coastal regions near the ancient Sanford and Ocala Highs (Riggs, 1984).

The lithology of the carbonate facies of the Hawthorn is similar to that in the Pungo River. By far the most common lithology is wackestone with abundant silt-sized dolomite rhombs comprising from 50% to 98% of the rock (Prasad, 1985). Packstone and grainstone lithologies are rarer and are usually composed of calcite.

Prasad (1985) classified the Hawthorn dolomites into three types: dolosilts, replacement dolomites, and void-lining or overgrowth dolomites. Dolosilts vary from carbonate-poor dolomitic mudstones to nearly pure carbonate, microsucrosic dolomites. Coarse microsucrosic dolomite is by far the most common type of dolomite in the Hawthorn. These dolosilts are nearly identical to the Pungo River Formation wackestones. Usually skeletal components present in these samples are calcitic, although echinoderm plates and red algal fragments may be partially dolomitized (replacement dolomites). In carbonate-poor samples, inclusion-rich dolomites are common and, upon occasion, zoned dolomites can be found.

XRD analyses of the Hawthorn dolomites yield an average composition of 46 mol% $MgCO_3$ (Prasad, 1985). They also display moderately well-developed superstructure reflections. $\delta^{18}O$ values (10 samples, Prasad, 1985) range from +1.97‰ to +3.39‰ relative to PDB. These values are indicative of a marine origin and yield precipitation temperatures ranging from 17° C to 23° C, somewhat warmer than the dolomites of the Pungo River. The $\delta^{13}C$ values range from −2.92% to +3.30‰ relative to PDB, similar to the Pungo River dolomites. The slightly higher $\delta^{13}C$ values of Hawthorn dolomites may be a result of a smaller contribution of organic carbon-derived carbon and a larger fraction of calcium carbonate-derived carbon to the precipitated dolomite.

Geology of dolomite occurrences in the Monterey Formation, California

A great deal of work has been done in recent years on the dolomites of the Miocene Monterey Formation of California. Much of this work is published in a volume edited by Garrison, Kastner & Zenger (1984). More recent papers include those by Burns & Baker (1987), Compton & Siever (1986) and Kastner, Reimers & Garrison (Chapter 25, this volume).

The deeper basinal sections of the Monterey Formation often can be divided into a lower calcareous member, a middle phosphatic or organic shale member, and an upper siliceous member (Pisciotto, 1978; Isaacs, 1980). Of major interest to the present study are the dolomites of the calcareous and phosphatic members. These sediments accumulated at sedimentation rates varying from those typical of pelagic carbonates, 10–20 m my^{-1}, to much higher rates in shalier intervals, up to 400 m my^{-1} (Mertz, 1984).

The lithology of the organic shale member is generally a foraminiferal shale or mudstone (Pisciotto & Garrison, 1981). Interestingly, Isaacs (1980) noted that the organic shales of the Santa Barbara coastal sections contain no dolomite, but do contain more abundant phosphate (averaging an estimated 2% P_2O_5) and organic matter (up to 24%, by weight) than any other members. In the Santa Maria Basin, however, Compton & Siever (1984) report that dolostone (containing 50–90% dolomite, by weight) comprises 10–18% by volume of the phosphatic member. They state that: 'dolostone beds and nodules tend to be associated with stratigraphic intervals that are richer in phosphate and organic matter. Laminated dolomites occur in the phosphatic lithofacies of the Pismo Basin (Surdam & Stanley, 1981a); organic matter content is high (5–20% weight), and phosphate content is also high (0.5–14% P_2O_5 by weight, Surdam & Stanley, 1981b). The organic shales of the Sandholt Member of the Monterey Formation in the Santa Lucia Mountains typically contain 0.5–5% P_2O_5 by weight and 0.5–7% organic carbon by weight (Mertz, 1984). Here the dolomite content is highest (about 1–30% by weight) in organic-rich, phosphate-poor, laminated shales and lowest (about 0.5–2.5% by weight) in faintly laminated, less organic, but more phosphatic, shales (Mertz, 1984). In summary, dolomite is abundant in the Monterey Formation and is closely associated with organic-rich sediments. In some cases, dolomite is abundant in phosphatic lithologies, but, in other cases, it is not intimately associated with the most phosphatic lithologies.

Dolomite occurs either as beds and nodules where dolomite cements or replaces pre-existing grains to form quite pure deposits. Alternatively, dolomite occurs as disseminated rhombs in much less pure deposits (see Isaacs, 1984). Compositional ranges in Monterey dolomites reported by Pisciotto (1981), partly based on the work of Murata, Friedman & Cremer (1972), are: $CaCO_3 = 47.8–58.0$ mol%, $MgCO_3 = 27.9–51.1$ mol%, $FeCO_3 = 0.0–16.6$ mol%, and $MnCO_3 = 0.0–0.5$ mol%. The $CaCO_3$ content of these dolomites averages 52 mol% (Burns & Baker, 1987) with no significant difference of $CaCO_3$ content between low-iron (<2 mol% $FeCO_3$) and high-iron dolomites.

Iron and manganese concentrations are correlated: both are high in dolomites from sections rich in terrigenous detritus and low in dolomites from carbonate- and phosphate-rich sections (Burns & Baker, 1987). This is probably due to the greater abundance of available iron and manganese in the associated detrital sediments. Strontium concentrations are also higher in dolomites from detrital-rich sections and lower in dolomites from carbonate- or phosphate-rich sections. This has been interpreted by Baker & Burns (1985) to be due to higher Sr:Ca ratios in the pore waters of detrital-rich sections, as well as the changing stoichiometry of the dolomite-forming reaction in carbonate-rich sediments. Sulfate concentrations of Monterey dolomites range from 70–250 ppm SO_4 (Baker, unpublished data). These low sulfate concentrations indicate that most Monterey dolomites were precipitated in pore waters with greatly reduced concentrations of dissolved sulfate relative to seawater.

The stable isotopic compositions of the dolomites show facies-dependent relationships similar to those for elemental compositions. This was first noted by Murata *et al.* (1972) and Pisciotto (1981). Dolomites from the detrital-rich intervals tend to have highly positive values of $\delta^{13}C$ (ranging from about 0‰ to +15‰ PDB) and those from calcareous or phosphatic intervals most commonly have negative values of $\delta^{13}C$ (ranging from about 0‰ to −15‰ PDB). The oxygen isotopic ratios of Monterey dolomites, which commonly range from +5‰ to −5‰ PDB (Burns & Baker, 1987), indicate precipitation in marine pore waters, often over a rather large temperature (burial depth) range. As discussed by Baker & Burns (1985), in sections with abundant dolomite, the dolomite must have formed at shallow depths (within a few meters of the seawater–sediment interface) and at an early stage during diagenesis. In the detritus-rich sections, there is far less dolomite and it formed at depth and over a wider temperature range. Some of these conclusions are discussed below in more detail.

The occurrence of dolomite in Quaternary organic-rich marine sediments

Dolomite is presently forming in deep-sea hemipelagic sediments along the continental margins of eastern North America (e.g. DSDP Site 533), northern South America (e.g. DSDP Site 147), western Central America (e.g. DSDP Site 496), the Gulf of California (e.g. DSDP Site 479), western North America (e.g. DSDP Site 467), western Africa (e.g. DSDP Site 532), eastern Asia (e.g. DSDP Site 440) and certainly in many other locations. In none of these locations however, does authigenic dolomite account for more than about 5% by weight of the total sediment. The conditions of dolomite authigenesis in this general setting have been described recently by Baker & Burns (1985). Their findings are summarized in the following paragraphs.

One of the key ingredients for dolomite formation is abundant (greater than about 0.5% by weight of the sediment) organic carbon. Dolomite formation has never been observed in sites which do not have microbial sulfate reduction. Two possible reasons for this association are the removal of dissolved sulfate, an inhibitor of dolomite precipitation (Baker & Kastner, 1981) and the increase in alkalinity which accompanies sulfate reduction.

Sedimentation rate is another control of dolomite abundance. At sedimentation rates greater than about 500 m my^{-1}, the dolomite content of sediments is so diluted by detrital input as to be negligible. Also, much of the calcium content of authigenic dolomite is supplied from precursor pelagic calcite (foraminifera and coccoliths). Since these components have maximum accumulation rates of about 20 m my^{-1}, much higher detrital sedimentation rates also tend to dilute this carbonate precursor (see Burns & Baker, 1987, Table 1). Optimal conditions for dolomite authigenesis include abundant marine-derived organic carbon, abundant calcium carbonate precursor and low rates of detrital sediment accumulation. Intuitively, it seems likely that the most favorable combination of these conditions would be attained during times of high sea level when terrigenous detrital sediments are most efficiently trapped on the shelves and sediments underlying regions of high productivity are less diluted. As we show below, phosphorite formation is favored by similar conditions and Riggs (1984) has shown that Neogene phosphogenesis is coincident with high sea-level stands.

Baker & Burns (1985) and Compton & Siever (1986) have shown on the basis of pore-water mathematical models that most of the magnesium needed to form these dolomites is derived from overlying seawater and transported to the reaction site by molecular diffusion. In sediments with abundant dolomite, this places serious constraints on how deep and how late in the history of the sediment dolomite can form. As mentioned above, highly dolomitic sediments must form very near the seafloor and very soon after sediment deposition.

The formation of phosphate in Quaternary marine sediments

In this section we briefly discuss the authigenesis of phosphate in modern sediments in order to better understand the dolomite–apatite paragenesis. In fact, there are very few published studies documenting the present-day precipitation of phosphate from marine interstitial waters. The most relevant of these are the papers by Jahnke *et al.* (1983) and Froelich *et al.* (1983).

Jahnke and co-workers analyzed interstitial waters recovered from box cores taken from the western Mexican continental shelf. These sites are located within a well developed oxygen-minimum zone at water depths from 200 to 400 m. There was no detectable $CaCO_3$ at any study site. The sediment accumulation rate measured at one site of proven phosphate precipitation was about 100 m my^{-1}. At all sites, microbial sulfate reduction commences almost at the sediment–water interface. In most cases, pore-water phosphate concentrations rapidly increase downward below the sediment–water interface, but within a few centimeters of the interface, the phosphate concentrations reach a maximum and then decrease downcore, presumably due to phosphate precipitation at greater depth in the sediments. These observed downcore pore-water concentration gradients were as large as -0.013 μmol cm^{-4} for phosphate and -0.0025 μmol cm^{-4} for fluoride.

Froelich *et al.* (1983) analyzed interstitial waters recovered from cores taken on the Peruvian continental shelf from water depths of 86–425 m. Many of these cores were observed to contain a *Thioploca* bacterial mat at the sediment–water interface (the significance of which will be discussed below). They observed either constant dissolved fluoride concentrations versus depth (at two sites) or downward-decreasing values of fluoride. Phosphate concentrations were not reported. The two sites with constant gradients had measured sedimentation rates of 820 and 3500 m my^{-1}, and the site with the steepest fluoride concentration gradient (about -0.0015 μmol cm^{-4}) had a sedimentation rate of 180 m my^{-1}. Since the F:P ratio of modern Peru shelf carbonate fluorapatite is about 1/3 (Froelich *et al.*, 1983), the phosphate flux to the site of apatite precipitation is expected to be three times higher than the fluoride flux.

Using the above data, we can calculate the diffusional fluxes (J_i) of phosphate or fluoride ions into modern phosphate-precipitating sediments. These fluxes can be calculated approximately from Fick's First Law (e.g. Berner, 1980):

$$J_i = -\Phi D_i (dc_i/dz)_o \quad (6.1)$$

where subscript i refers to the phosphate or fluoride ion, subscript o refers to the sediment–water interface, Φ is the fractional porosity, D is the diffusion coefficient of the ion in the pore waters (incorporating the effects of adsorption as needed), c is the concentration of the ion and z is the sub-bottom depth. Reasonable values for these parameters are: $\Phi = 0.90$ (Froelich *et al.*, 1983), $D = 2.0 \times 10^{-6}$ $cm^2 s^{-1}$ for phosphate (Krom & Berner, 1980) and $D = 7.6 \times 10^{-6}$ $cm^2 s^{-1}$ for fluoride (applying Stokes-Einstein relationship to the data in Li & Gregory, 1974, and correcting for tortuosity). The concentration gradients noted above thus result in calculated diffusional fluxes into the sediments for fluoride of 1.1×10^{-14} mol $cm^{-2} s^{-1}$ (Froelich *et al.*, unpublished data) and 1.8×10^{-14} mol $cm^{-2} s^{-1}$ (Jahnke *et al.*, unpublished data) and 2.3×10^{-14} mol $cm^{-2} s^{-1}$ for phosphate (Jahnke *et al.*, unpublished data). These calculated fluxes are dependent on the choice of values for porosity, diffusion coefficient and especially the interfacial concentration gradient. The important point of these calculations, as previously pointed out by Suess (1981), is that even the maximum observed fluxes of phosphate (or fluoride) at the sites discussed above or at sites studied by Suess (1981) are insufficient to form a very pure phosphate deposit because the sedimentation rates are so high.

This can be shown by a calculation of the fraction of phosphate forming in a sediment based upon the following expression:

$$\%P_2O_5\text{, by weight} = [(100)(J\text{, mol P}(cm^2\,s)^{-1})(71\text{ g }P_2O_5(\text{mol P})^{-1})(3.15 \times 10^7\,s\,y^{-1})(10^{-4})] / [(S\text{, sedimentation rate, m my}^{-1})(2.5\text{ g sediment cm}^{-3})(1-\Phi)] \quad (6.2)$$

Using a value of $\Phi = 0.9$, this expression becomes approximately:

$$\%P_2O_5 = (J/S) \times 9 \times 10^{15} \quad (6.3)$$

This equation does not account for the porosity changes as phosphate is added to the sediment. In Figure 6.1, values of weight % P_2O_5 in the sediments are plotted as a function of the flux of phosphate and the total sedimentation rate. We have also plotted the Pungo River Formation phosphorites (average sedimentation rate of about 3 m my^{-1} and P_2O_5 content of about 10–30% by weight), sediments from the unpublished studies of Jahnke *et al.* and Froelich *et al.* (utilizing reported sedimentation

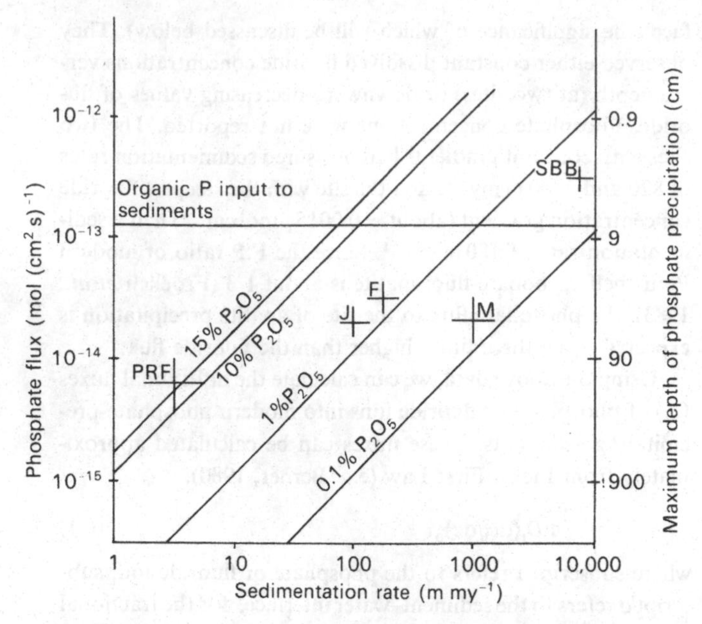

Fig. 6.1. Values of weight % P_2O_5 in sediments calculated as a function of the phosphate flux into the sediments and the sedimentation rate. The phosphate flux can also be converted into an approximate maximum depth of phosphate precipitation, as explained in the text. The horizontal line represents the organically-bound phosphorus flux to the sediments of the Peru upwelling region, likely a maximum phosphorus flux to any sediments. Only a fraction of this flux will ever be available for phosphogenesis. PRF, J, F, M, and SBB, refer to the Pungo River Formation, Mexican margin (Jahnke *et al.*, 1983), Peru margin (Froelich *et al.*, 1983), Monterey Formation (Mertz, 1984), and the Santa Barbara Basin (Sholkovitz, 1973), respectively.

rates and the phosphate diffusional fluxes into the sediments as calculated above), laminated organic-rich sediments from the Monterey Formation (Mertz, 1984, utilizing estimated original sedimentation rates and weight $\%P_2O_5$), and modern laminated sediments from the Santa Barbara Basin (Sholkovitz, 1973, using measured sedimentation rates and measured concentrations of inorganic phosphorus). It is clear from this calculation that phosphorite (*sensu stricto*) formation is dependent on both an adequate source of phosphate *and* a favorably low rate of sedimentation.

The Peruvian shelf is one of the best studied regions of modern phosphate precipitation and is therefore an interesting area in which to construct a tentative phosphate budget (Fig. 6.2). Somewhat different budgets have previously been discussed by Suess (1981) and Froelich *et al.* (1982). It should be noted here that the sediments of the Peruvian shelf rarely constitute a phosphorite deposit, although recently Dr William Burnett and Dr Phillip Froelich (Burnett, Chapter 5 this volume) have collected very rich pelletal phosphate sediments in localized regions of the shelf. More frequently, however, these sediments contain $<2\%$ P_2O_5 by weight (Manheim, Rowe & Jipa, 1975; Suess, 1981). Since the Peruvian upwelling system is one of the most productive marine environments known, it is likely that the estimated fluxes of phosphorus in this system represent upper limits on the phosphorus fluxes in most other marine environments.

Rates of gross primary productivity in the Peru upwelling system have been measured by a number of workers. We use a typical value of 3 g C (m^{-2}) from Barber *et al.* (1978). Much of this organic matter is rapidly regenerated in the euphotic zone and in the underlying water column (Suess, 1980). Measured downward fluxes of particulate organic carbon sinking through the sub-euphotic zone water column average about 10% of the primary productivity (Staresenic, 1978; Von Brokel, 1981; Lee & Cronin, 1982). This organic carbon is estimated to have an average molar C:P ratio of about 250:1 (Sholkovitz, 1973; Bishop *et al.*, 1977; Suess; 1981, Froelich *et al.*, 1982). From this estimate we get an input to the sediments of 1.2×10^{-13} mol $(cm^2\ s)^{-1}$ of organically-bound phosphorus. This input is believed to be the maximum flux of phosphorus potentially available for phosphogenesis in any marine environment.

Since we are mostly interested in phosphate precipitation localized in sediments of the oxygen minimum zone (Burnett, Veeh & Soutar, 1980), we follow the fate of this organic matter in suboxic to anoxic sediments. The role of the bacteria, *Thioploca*, and associated microorganisms in the cycling of phosphorus and organic carbon in Peruvian shelf sediments has not yet been quantitatively assessed. Gallardo (1977) initially described the abundant bacterial flora occurring on the sediment surface in the oxygen-minimum zone off the Chilean margin. More recently, a number of workers have speculated on their potential importance in organic carbon cycling (Henrichs & Farrington, 1984) and phosphate cycling (Reimers, Kastner & Garrison, Chapter 24, this volume). If sulfide oxidation is its major energy source, then *Thioploca* cannot play a dominant role in the energetics of organic carbon cycling relative to the sulfate reducers. If (by analogy with *Beggiatoa*) *Thioploca* is not sulfide-dependent and is therefore not an obligate chemolithotroph (Sieburth, 1979), then it may be capable of utilizing a variety of organic compounds for nutrition and may play a dominant role in the recycling of organic detritus and the release of inorganic phosphorous at the sediment–water interface of anoxic sediments.

If, for lack of better data, we assume that the quantitatively most important mineralization of organic matter takes place by sulfate reduction within the sediments (Jorgensen, 1982), then the amount of organically-bound phosphate released to the pore fluids is dependent on the amount of microbial sulfate reduction. In Peruvian sediments, rates of sulfate reduction have been directly measured on incubated samples (Rowe & Howarth, 1985) and have been calculated by pore-water solute modeling (Henrichs & Farrington, 1984). We have used an average value of 1.4×10^{-14} mol $P(cm^2\ s)^{-1}$ from the calculated rates of organic carbon mineralization reported by Henrichs & Farrington (1984) assuming a 250:1 C:P ratio. Parenthetically, we note that it is not clear that this ratio is at all appropriate for the present calculation. Suess (1981) suggested a much higher value, but Aller (1980) has demonstrated that phosphorus is released preferentially over nitrogen which is released preferentially over carbon during the early diagenesis of organic matter. He has therefore suggested a lower value for the remineralized C:P ratio in Long Island Sound sediments.

The fate of this released phosphate is either to diffuse upward and exit the sediment column (regenerated phosphorus) or to be precipitated *in situ* as some form of solid phosphate. It is also possible that some of the upward-diffusing phosphate can be

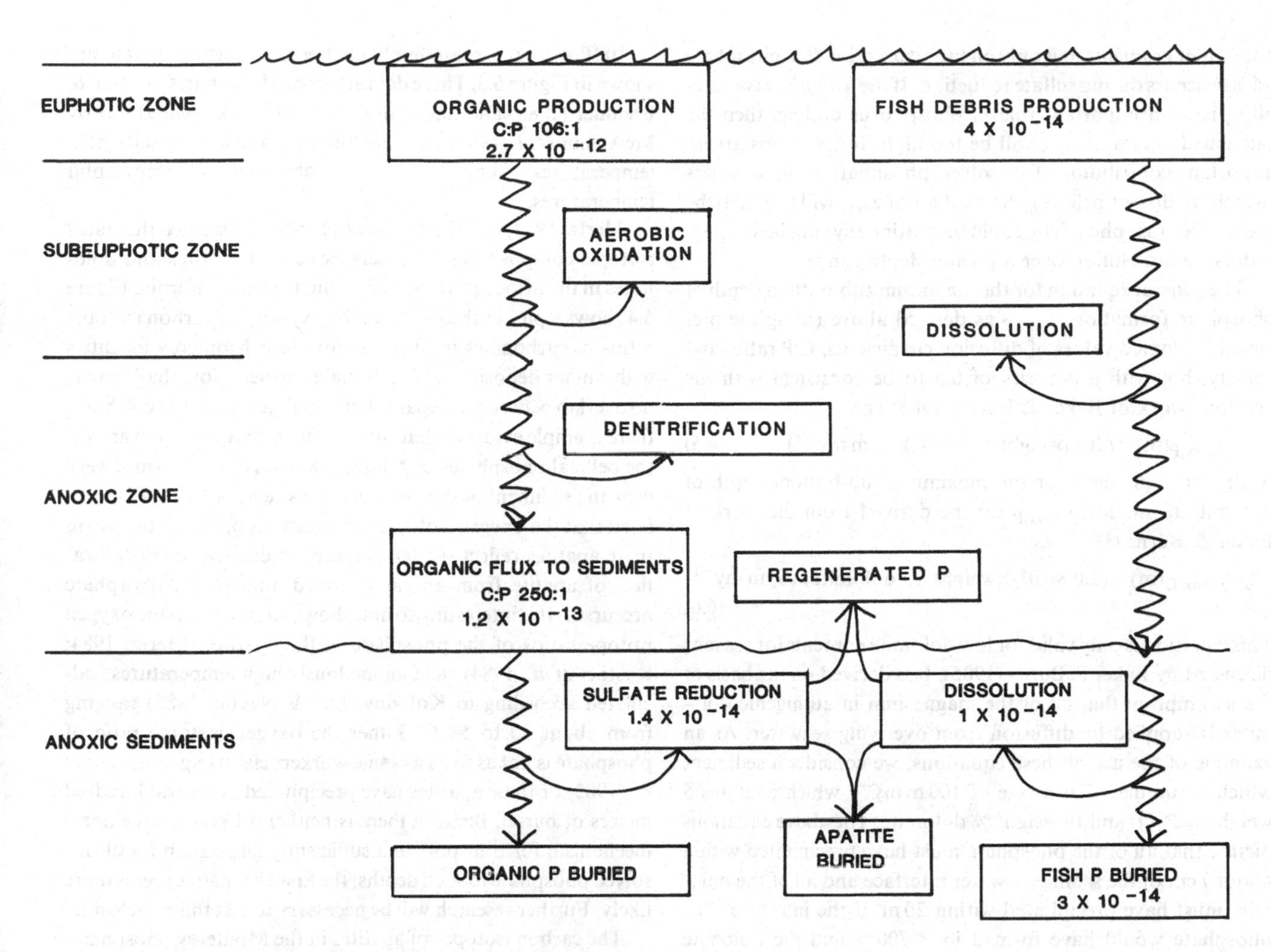

Fig. 6.2. Schematic phosphorus budget for the Peru upwelling region. Estimated fluxes of phosphorus (in units of mol P $cm^{-2}s^{-1}$) are taken from sources given in the text. According to this budget, only a fraction of 2.4×10^{-14} mol P $cm^{-2}s^{-1}$ is available for phosphogenesis in the Peru region.

adsorbed on iron oxide particulates formed at the seawater–sediment interface (Aller, 1980; Baccini, 1985). Organically-bound phosphorus which escapes oxidation during sulfate reduction will be buried. This is the fate of the largest fraction (about 90%, we estimate here) of organic phosphorous entering the Peruvian anoxic sediments.

Another source of phosphorus to the sediments is the incorporation of fish debris. This fish debris is comprised of hydroxyapatite which is thermodynamically unstable in seawater and tends to rapidly dissolve or to convert to carbonate fluorapatite (CFA) (Atlas & Pytkowicz, 1977; Jahnke, 1984). Estimates of the magnitude of fish debris production and sedimentation in the Peruvian upwelling region are highly variable (cf. Suess, 1981; Froelich *et al.*, 1982), and all of the phosphate fluxes associated with fish debris production and dissolution should be regarded as order-of-magnitude estimates. The fish debris upon reaching the seafloor can be dissolved in the sediments or buried. The dissolved phosphate can diffuse out of the sediment column or be precipitated as some form of solid phosphate. We believe, on the basis of sedimentological evidence, that fish debris is not an important source of phosphate in most phosphorites.

If we assume that all of the phosphorus flux into the pore waters is derived from the oxidation of organic matter during sulfate reduction (i.e. no fish debris), then we can calculate the maximum depth below the sediment–water interface of phosphate precipitation. For example, arbitrarily consider a phosphate flux of 1×10^{-14} mol $(cm^2\ s)^{-1}$. This implies an oxidation of 2.5×10^{-12} mol $(cm^2\ s)^{-1}$ of organic carbon (C:P is 250:1) by reduction of 1.25×10^{-12} mol $(cm^2\ s)^{-1}$ of sulfate. The diffusional flux of this sulfate into the sediments is given by:

$$J = -D(dc/dz)_o \quad (6.4)$$

If $D = 4 \times 10^{-6}\ cm^2\ s^{-1}$, $c_o = 28 \times 10^{-6}$ mol cm^{-3} (the seawater value), then a simple linear concentration–depth profile predicts that $c = 0$ at $z_{max} = 90$ cm. Thus, all of this organic matter is oxidized and all of the organically-bound phosphate is released to pore waters within about 90 cm of the seawater–sediment interface. Similar reasoning (i.e. transport calculations based on reasonable phosphate concentration gradients) demonstrates that the phosphate cannot diffuse very far down from this zone without precipitation. Froelich *et al.* (1983) reached a similar conclusion based upon the uptake of dissolved fluoride. Note that the above calculations are not based upon observed sulfate pore water gradients, but only on the assumption that phosphate release occurs during sulfate reduction. If the *Thioploca* community plays an important role in phosphorus cycling, then the calculated values of z_{max} will be too high. If fish debris are an important contributor of dissolved phosphate to pore waters (which we do not believe), then values of z_{max} will probably be too low because phosphate could be continually supplied to pore waters by dissolution over a greater depth range.

The general equation for the maximum sub-bottom depth of phosphate formation, $z_{max,P}$ as derived above (using the previously assumed values of diffusion coefficients, C:P ratio, and density, but with a porosity of 0.8 to be consistent with the previous work of Baker & Burns, 1985) is:

$$z_{max,P}(\text{m}) = (36)/(\text{weight \% } P_2O_5)(S, \text{ m my}^{-1}) \qquad (6.5)$$

A similar expression for the maximum sub-bottom depth of dolomite formation, $z_{max,P}$ can be derived from the work of Baker & Burns (1985) as:

$$z_{max,D}(\text{m}) = (2.0 \times 10^4)/(\text{weight \% dolomite})\ (S, \text{ m my}^{-1}) \qquad (6.6)$$

This equation is only valid for low dolomite contents for reasons discussed by Baker & Burns (1985). It is derived on the basis of the assumption that all of the magnesium in authigenic dolomites is supplied by diffusion from overlying seawater. As an example of the use of these equations, we consider a sediment which accumulated at a rate of 100 m my^{-1} which contains 5 weight % P_2O_5 and 10 weight % dolomite. The above equations dictate that all of the phosphate must have precipitated within about 7 cm of the sediment–water interface and all of the dolomite must have precipitated within 20 m of the interface. The phosphate would have formed in <700 y and the dolomite would have formed in $<2 \times 10^5$ y. A typical phosphorite containing 25 weight % P_2O_5 (at a porosity of 0.9) cannot accumulate faster than about 10 m my^{-1} (Fig. 6.1), thus all of the phosphate must precipitate within about 30 cm of the interface within about 3×10^4 y of deposition.

The timing of dolomite and apatite formation: stable isotopic evidence

The stable isotopic evidence concerning the timing and environment of formation of associated dolomite and apatite is interesting but ambiguous. Kolodny & Kaplan (1970) first demonstrated that the carbonate substituent ions in apatite have stable carbon and oxygen isotopic compositions which are distinctly different from those of coexisting calcite. In most samples apatite carbonate was found to be isotopically lighter than calcite carbonate in both carbon and oxygen. Shemesh, Kolodny & Luz (1983) showed that the carbonate ions of apatites of Holocene age have oxygen isotopic compositions similar to those expected for calcite in equilibrium with the oxygen isotope ratios observed in the phosphate ions of the apatite. In slightly older (even Pleistocene to Miocene) apatites, however, the observed oxygen isotopic ratios of the carbonate are lighter than would be predicted on the basis of the observed P:O isotopic ratios, implying that apatite carbonate may be easily exchanged during diagenesis. This had previously been argued by McArthur (1978) and McClellan (1980) on purely chemical grounds. Stable isotopic ratios of carbon and oxygen in modern ($<100\,000$ y) and presumably unaltered apatite carbonate are shown in Figure 6.3. These data are from McArthur, Coleman & Bremner (1980) and Shemesh *et al.* (1983). As pointed out by McArthur *et al.* (1980 and 1986) the oxygen isotopic results yield temperatures similar to the observed environmental temperatures.

Mertz (1984) and Kastner *et al.* (1984) have studied the stable isotopic compositions of closely associated apatites and dolomites in the Miocene Monterey Formation in California. Figure 6.4 shows a plot of their data on the oxygen and carbon isotopic ratios of carbonates in apatites from four Monterey localities with similar depositional and burial histories. Note that there is no overlap between the apatite fields in Figures 6.3 and 6.4. Since there is ample evidence that some of these apatites (for example, the pelletal phosphates of Mertz, 1984) must have formed very near the sediment–water interface, Kastner *et al.* (1984) speculated that the oxygen isotopic compositions of the carbonate in these apatites reflect the temperature of diagenetic crystallization of apatite from an early-formed amorphous phosphate precursor. It is interesting to note, however, that even the oxygen isotope ratios of the phosphate in the apatites (Mertz, 1984; Kastner *et al.*, 1984) yield anomalously high temperatures (calculated according to Kolodny, Luz & Navon, 1983) ranging from about 30 to 50° C. Either the oxygen isotopic ratio of phosphate is not as fixed as some workers claim (e.g. Shemesh *et al.*, 1983) or these apatites have precipitated at several hundred meters of burial. Because there is neither a likely source nor a mechanism for transport of a sufficiently large quantity of dissolved phosphate to such depths, the first alternative seems more likely. Further research will be necessary to test this conclusion.

The carbon isotopes of apatites in the Monterey phosphates could either reflect origin in the zone of sulfate reduction (very shallow) or in the much deeper zone of thermocatalytic decarboxylation. Mertz (1984) concluded the former. We contend that it is more likely that the very labile $CaCO_3$–C in the original apatite was exchanged during deep burial and heating (similar to the $CaCO_3$–O isotopic exchange) and thus obtained the carbon isotopic signature of the zone of thermocatalytic decarboxylation. The few positive carbon isotopic values of the apatites seen in Figure 6.4 are due to a contribution of carbon from the intermediate depth zone of microbial methanogenesis where any CO_2 remaining in solution is isotopically heavy.

Also shown in Figure 6.4 are the carbon and oxygen isotopic compositions of dolomites associated with the phosphates (Kastner *et al.*, 1984; Mertz, 1984). As Mertz (1984) has previously discussed, these dolomites must have obtained their carbon and oxygen isotopic compositions in the zone of methanogenesis. The data of figure 6.4 span a calculated range of precipitation temperatures from 13 to 65° C and burial depths from perhaps 50 to 900 m (utilizing a paleogeothermal gradient estimate of 60° C km^{-1} from Garrison & Gråham (1984) and a bottom water paleotemperature of 10° C). Such relatively great depths of formation of the dolomite at this location in the Monterey are consistent with the conclusions of Baker & Burns (1985) and Burns & Baker (1987): in sections with abundant terrigenous detrital sediment and only small amounts of dolomite, the dolomite can form at relatively great burial depth and high temperature. In addition, there may be a certain amount of

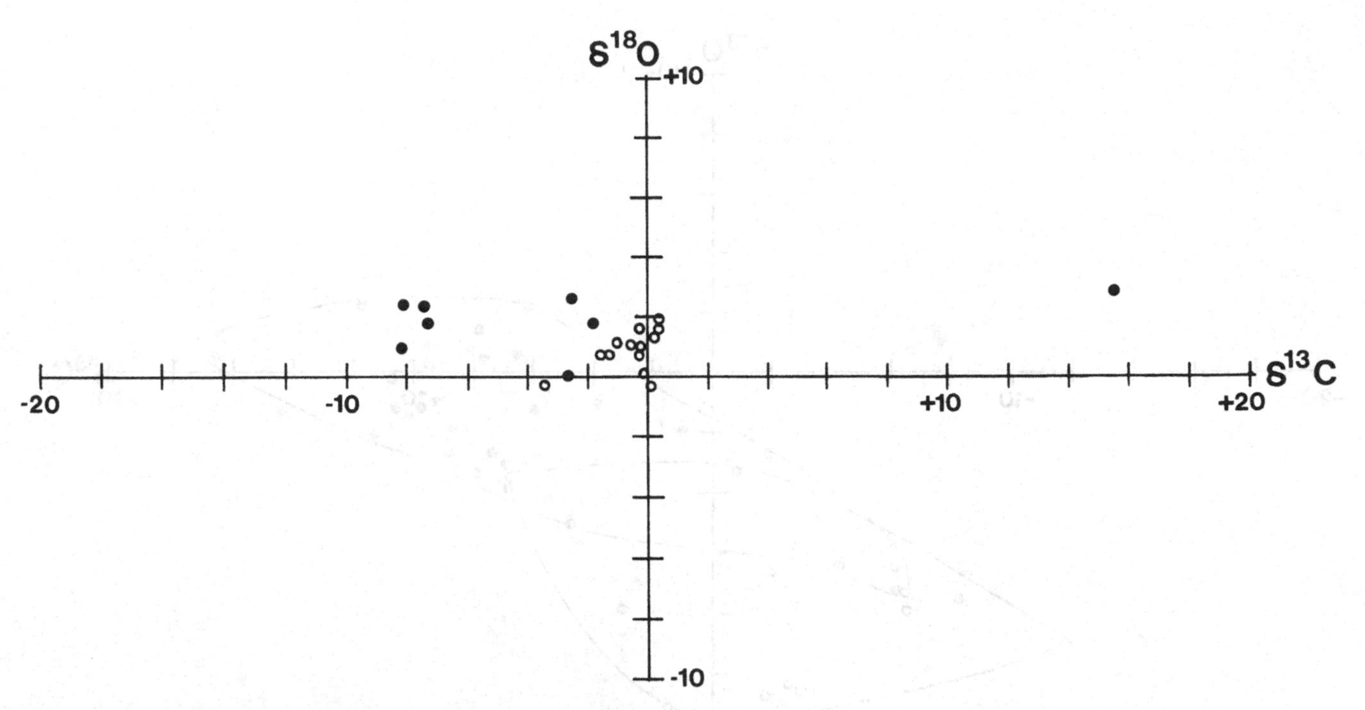

Fig. 6.3. Carbon and oxygen isotopic compositions of carbonate in modern (< 100 000 y) apatites. Open circles are samples from the Peru–Chile margin; closed circles are from the Namibian margin. Data are from McArthur *et al.* (1980) and Shemesh *et al.* (1983). Values are reported as per mil relative to PDB.

diagenetic resetting of the oxygen isotopic compositions of these dolomites.

One of the most interesting lines of evidence regarding the origin of phosphorites is the concentration and isotopic composition of the sulfate incorporated within the apatite structure. Apatites precipitated in the zone of sulfate reduction would be expected to have lower sulfate concentrations than apatites not formed during sulfate reduction and the sulfur isotopic composition of this sulfate should be heavier than that of contemporaneous seawater (e.g. Goldhaber & Kaplan, 1974). McArthur (1985) has reported that there is no consistent difference in the sulfate content between apatites precipitated in oxic environments and those precipitated in anoxic environments. In most of the apatite samples which have been analyzed for sulfur isotopes (Nathan & Nielsen, 1980; Benmore, Coleman & McArthur, 1983), the values of $\delta^{34}S$ are similar to or higher than contemporaneous seawater as predicted. In a couple of samples, however, including a modern Peru shelf apatite (Benmore *et al.*, 1983) and a Miocene-age Peruvian apatite (McArthur *et al.*, 1986), the values of $\delta^{34}S$ are much lighter than seawater. Benmore reasoned that this sulfate must have been derived from the oxidation of H_2S diffusing upward into oxic sediments and subsequently being coprecipitated in the apatite. If they are correct in their interpretation, then we must conclude that the sulfide-oxidizing bacteria, such as *Thioploca*, may play a pivotal role in releasing organically-bound phosphate to the pore waters enabling phosphate precipitation. A possible alternative is that acid-soluble sulfide was oxidized to sulfate during their sample preparation (digestion in 10% HCl for an unspecified amount of time). The only reported case (to our knowledge) of marine pore-water sulfate having a sulfur isotopic composition lighter than seawater was one near-surface sample from a core taken in the Gulf of California (Goldhaber & Kaplan, 1980, $\delta^{34}S = +17.7‰$, $SO_4^{2-} = 32.1$ mM). These values were attributed to sulfide oxidation 'either *in situ* or during handling' (Goldhaber & Kaplan, 1980). In future studies of sulfate isotopes in apatite, it is recommended that apatite samples be rapidly dissolved under a bubbling nitrogen atmosphere to remove acid-volatile sulfides and avoid their oxidation to sulfate. This should eliminate the ambiguity of the above results.

Conclusions

(1) Dolomite and apatite are commonly associated. This is probably not a symbiotic relationship. Instead, both minerals form authigenically in marine sediments under similar, but not identical conditions. The key controls on the formation of abundant quantities of both minerals are: high amounts of marine organic detritus and low overall sedimentation rates. In marine sediments with high organic carbon contents, microbial sulfate reduction is the dominant process of organic carbon mineralization. This process liberates abundant phosphate and alkalinity to shallow pore waters and removes sulfate from these pore waters creating chemical conditions conducive to the precipitation of both apatite and dolomite. At high sedimentation rates, typical of sites with high organic contents, the formation of these minerals is masked by detrital sediments.

(2) Phosphate precipitation occurs shallower and faster than dolomite precipitation. Most of the phosphate in a phosphorite deposit precipitates within the uppermost zone of sulfate reduction within a few centimeters of the sediment–water interface. Small amounts of phosphate may precipitate at greater depth. Both the carbonate (carbon and oxygen) and phosphate (oxy-

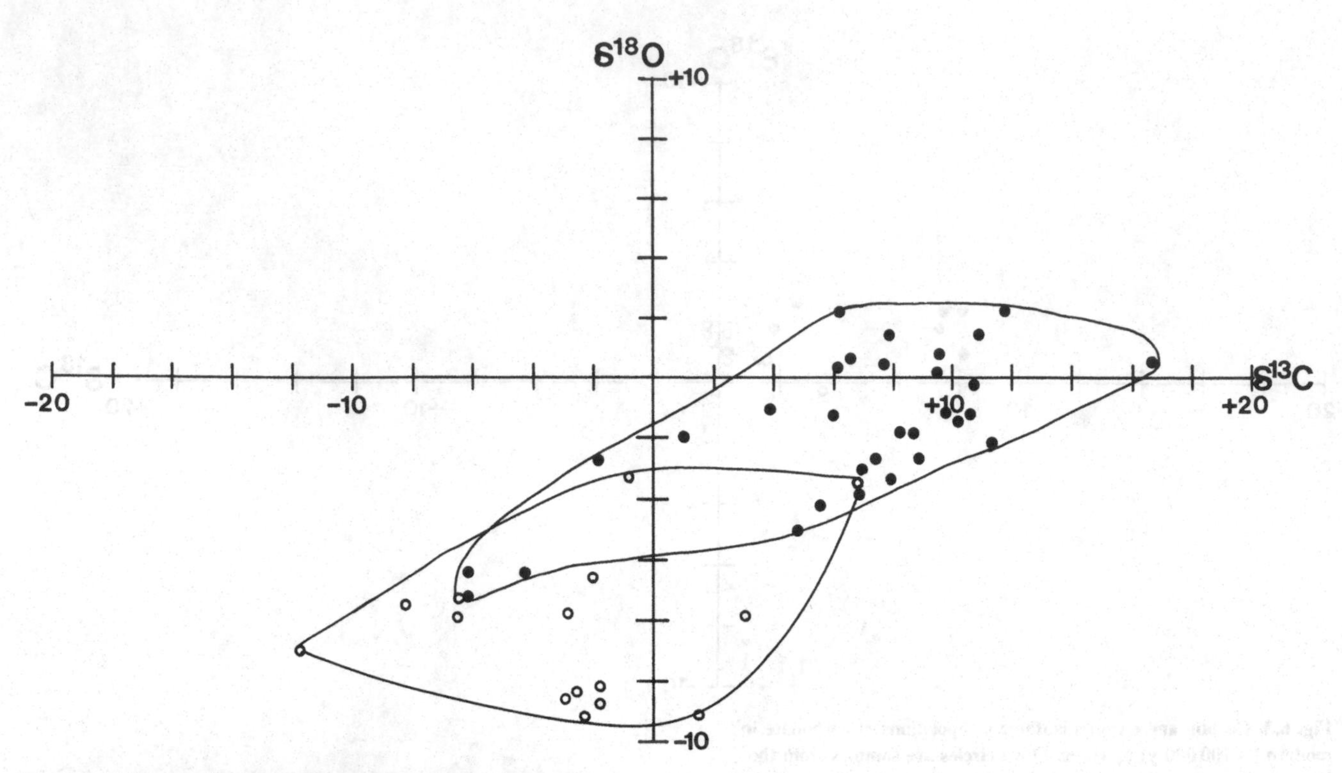

Fig. 6.4. Carbon and oxygen isotopic compositions of carbonate in closely associated apatites (open circles) and dolomites (closed circles) from the Monterey Formation. Data are from Mertz (1984) and Kastner *et al.* (1984). Values are reported as per mil relative to PDB.

gen) stable isotopic ratios of a phosphate can apparently be easily reset during burial diagenesis.

(3) Dolomite precipitation in phosphatic sediments may take place within the zone of sulfate reduction or in the zone of methanogenesis. Baker & Burns (1985) have shown that if there are large amounts of dolomite present (greater than about 10% by weight) then it must form in the zone of sulfate reduction within a few meters of the sediment–water interface. In either case, most of this dolomite forms deeper and more slowly than (or in a paragenetic sequence, it postdates) the initial phosphate precipitate. The stable carbon and oxygen isotopic ratios of Neogene-age dolomite often reflect the environment of authigenesis. Because of material balance relationships between pore waters and authigenic dolomite, carbon isotopic ratios usually remain intact and oxygen isotopic ratios are usually altered during burial diagenesis.

Acknowledgments

We thank Stan Riggs who provided us with all the samples for the study of the Pungo River Formation and who encouraged us throughout this work. We thank both Stan and Bill Burnett who invited us to present this study at the IGCP Project 156 conference. Sach Prasad gave us access to his unpublished MSc thesis on the dolomites of the Hawthorn Group. Stephen Burns contributed to much of the work on the dolomites of the Monterey Formation. Bill Showers and Rob Dunbar both assisted in isotopic analyses. This paper has benefitted from reviews by Yaakov Nathan, Marty Goldhaber and Bill Burnett. Funding for this study was partially provided by awards from the American Association for Petroleum Geologists Grants-in-Aid and the Geological Society of America Research Grants (to Allen), and the American Chemical Society–Petroleum Research Fund and Texaco, Inc. (to Baker).

References

Allen, M.R. (1985). 'The origin of dolomite in the phosphatic sediments of the Miocene Pungo River Formation, North Carolina.' Unpublished M.Sc. thesis, Duke University, North Carolina.

Aller, R.C. (1980). Diagenetic processes near the sediment–water interface of Long Island Sound. I. Decomposition and nutrient element geochemistry (S, N, P). *Advances in Geophysics*, **22**, 237–350.

Atlas, E.L. & Pytkowicz, R.M. (1977). Solubility behavior of apatites in seawater. *Limnology and Oceanography*, **22**, 290–300.

Baccini, P. (1985). Phosphate interactions at the sediment–water interface. In *Chemical Processes in Lakes*, ed. W. Stumm, pp. 189–205. John Wiley, New York.

Baker, P.A. & Burns, S.J. (1985). Occurrence and formation of dolomite in organic-rich continental margin sediments. *American Asociation of Petroleum Geologists Bulletin*, **69**, 1917–30.

Baker, P.A. & Kastner, M. (1981). Constraints on the formation of sedimentary dolomite. *Science*, **213**, 214–16.

Barber, R.T., Huntsman, S.A., Kogelschatz, J.E., Smith, W.O., Jones, B.H. & Paul, J.C. (1978). *Carbon, Chlorophyll and Light Extinction from JOINT-II 1976 and 1977*. CUEA Data Report 49, Corvallis, Oregon.

Benmore, R.A., Coleman, M.L. & McArthur, J.M. (1983). Origin of sedimentary francolite from its sulphur and carbon isotope composition. *Nature*, **302**, 516–18.

Bentor, Y.K. (1980). Phosphorites–the unsolved problems. In *Marine Phosphorites*, ed. Y.K. Bentor, pp. 3–18. Society of Economic Paleontologists and Mineralogists, Tulsa.

Berner, R.A. (1980). *Early Diagenesis*. Princeton University Press, Princeton, NJ.

Bishop, J.K., Edmond, J.M., Ketten, D.R., Bacon, M.P. & Sliker, W.B. (1977). The chemistry, biology and vertical flux of particulate matter

from the upper 400 m of the equatorial Atlantic Ocean. *Deep-Sea Research*, **24**, 511–48.
Brown, P.M. (1958). The relation of phosphorites to ground water in Beaufort County, North Carolina. *Economic Geology*, **53**, 85–101.
Burnett, W.C., Veeh, H.H. & Soutar, A. (1980) U-series, oceanographic and sedimentary evidence in support of recent formation of phosphate off Peru. In *Marine Phosphorites*, ed. Y.K. Bentor, pp. 61–72. Society of Economic Paleontologists and Mineralogists, Tulsa.
Burns, S.J. & Baker, P.A. (1987). Geochemical studies of the origin of dolomite in the Miocene Monterey Formation of California. *Journal of Sedimentary Petrology*, **57**, 128–39.
Compton, J.S. & Siever, R. (1984). Magnesium mass balance calculations for the Monterey Formation, Santa Maria basin, California (abstract). In *Dolomites of the Monterey Formation and Other Organic-rich Units*, ed. R.E. Garrison, M. Kastner & D.H. Zenger, p. 214. Society of Economic Paleontologists and Mineralogists, Pacific Section, Los Angeles.
Compton, J.S. & Siever, R. (1986). Diffusion and mass balance of Mg during early dolomite formation, Monterey Formation. *Geochimica Cosmochimica Acta*, **50**, 125–36.
Epstein, S., Buchsbaum, H.A., Lowenstam, H.A. & Urey, H.C. (1953). Revised carbonate–water isotopic temperature scale. *Geological Society of America Bulletin*, **64**, 1315–26.
Froelich, P.N., Bender, M.L., Luedtke, N.A., Heath, G.R. & DeVries, T. (1982). The marine phosphorous cycle. *American Journal of Science*, **282**, 474–511.
Froelich, P.N., Kim, K.H., Jahnke, R., Burnett, W.C., Soutar, A. & Deakin, M. (1983). Pore water fluoride in Peru continental margin sediments: uptake from seawater. *Geochimica Cosmochimica Acta*, **47**, 1605–12.
Gallardo, V. (1977). Large benthic microbial communities in sulphide biota under Peru–Chile subsurface countercurrent. *Nature*, **268**, 331–2.
Garrison, R.E. & Graham, S.A. (1984). Early diagenetic dolomites and the origin of dolomite-bearing breccias, lower Monterey Formation, Arroyo Seco, Monterey County, California. In *Dolomites of the Monterey Formation and Other Organic-rich Units*, ed. R.E. Garrison, M. Kastner & D.H. Zenger, pp. 87–102. Society of Economic Paleontologists and Mineralogists, Pacific Section, Los Angeles.
Garrison, R.E., Kastner, M. & Zenger, D.H. (1984). *Dolomites of the Monterey Formation and Other Organic-rich Units*. Society of Economic Paleontologists and Mineralogists, Pacific Section, Los Angeles.
Gibson, T.G. (1967). Stratigraphy and paleoenvironment of the phosphatic Miocene strata of North Carolina. *Geological Society of America Bulletin*, **78**, 631–50.
Goldhaber, M. & Kaplan, I. (1974). The sulfur cycle. In *The Sea*, 5, ed. E. Goldberg, pp. 569–655. Wiley-Interscience, New York.
Goldhaber, M.B. & Kaplan, I.R. (1980). Mechanisms of sulfur incorporation and isotope fractionation during early diagenesis in sediments of the Gulf of California. *Marine Chemistry*, **9**, 95–143.
Gulbrandsen, R.A. (1960). Petrology of the Mead Peak phosphatic shale member of the Phosphoria Formation at Coal Canyon, Wyoming. *US Geological Survey Bulletin*, **1111C**, 71–146.
Gulbrandsen, R.A., Roberson, C.E. & Neil, S.T. (1984). Time and the crystallization of apatite in seawater. *Geochimica Cosmochimica Acta*, **48**, 213–18.
Henrichs, S.M. & Farrington, J.W. (1984). Peru upwelling region sediments near 15° S. 1. Remineralization and accumulation of organic matter. *Limnology and Oceanography*, **29**, 1–19.
Isaacs, C.M. (1980). 'Diagenesis in the Monterey Formation examined laterally along the coast near Santa Barbara, California.' unpublished Ph.D. thesis, Stanford University, Stanford.
Isaacs, C.M. (1984). Disseminated dolomite in the Monterey Formation, Santa Maria and Santa Barbara areas, California. In *Dolomites of the Monterey formation and Other Organic-rich Units*, ed. R.E. Garrison, M. Kastner & D.H. Zenger, pp. 155–70. Society of Economic Paleontologists and Mineralogists, Pacific Section, Los Angeles.
Jahnke, R.A. (1984). The synthesis and solubility of carbonate fluorapatite. *American Journal of Science*, **284**, 58–78.
Jahnke, R.A., Emerson, S.R., Roe, K.K. & Burnett, W.C. (1983). The present day formation of apatite in Mexican continental margin sediments. *Geochimica Cosmochimica Acta*, **47**, 259–66.
Jorgensen, B.B. (1982). Mineralization of organic matter in the sea bed–the role of sulphate reduction. *Nature*, **296**, 643–5.
Kastner, M, Mertz, K., Hollander, D. & Garrison, R. (1984). The association of dolomite–phosphorite–chert: causes and possible diagenetic sequences. In *Dolomites of the Monterey Formation and Other Organic-rich Units*, ed. R.E. Garrison, M. Kastner & D.H. Zenger, pp. 75–86. Society of Economic Paleontologists and Mineralogists, Pacific Section, Los Angeles.
Kolodny, Y. & Kaplan, I.R. (1970). Carbon and oxygen isotopes in apatite CO_2 and coexisting calcite from sedimentary phosphorite. *Journal of Sedimentary Petrology*, **40**, 954–9.
Kolodny, Y, Luz, B. & Navon, O. (1983). Oxygen isotopes in phosphates of biogenic apatites, I. Fish bone apatite-rechecking the rules of the game. *Earth and Planetary Sciences Letters*, **64**, 398–404.
Krom, M.D. & Berner, R.A. (1980). The experimental determination of the diffusion coefficients of sulfate, ammonium, and phosphate in anoxic marine sediments. *Limnology and Oceanography*, **25**, 327–37.
Lee, C. & Cronin, C. (1982). The vertical flux of particulate organic nitrogen in the sea: decomposition of amino acids in the Peru upwelling area and the equatorial Atlantic. *Journal of Marine Research*, **40**, 227–51.
Li, Y-H. & Gregory, S. (1974). Diffusion of ions in sea water and in deep-sea sediments. *Geochimica Cosmochimica Acta*, **38**, 703–14.
Manheim, F.T., Rowe, G.T. & Jipa, D. (1975). Marine phosphorite formation off Peru. *Journal of Sedimentary Petrology*, **45**, 243–51.
Martens, C.S. & Harris, R.C. (1970). Inhibition of apatite precipitation in the marine environment by Mg-ions. *Geochimica Cosmochimica Acta*, **34**, 621–25.
McArthur, J.M. (1978). Systematic variations in the contents of Na, Sr, CO_3 and SO_4 in marine carbonate–fluorapatite and their relation to weathering. *Chemical Geology*, **21**, 89–112.
McArthur, J.M. (1985). Francolite geochemistry – compositional controls during formation, diagenesis, metamorphism and weathering. *Geochimica Cosmochimica Acta*, **49**, 23–36.
McArthur, J.M., Benmore, R.A., Coleman, M.L., Soldi, C., Yeh, H.-W. & O'Brien, G.W. (1986). Stable isotopic characteristics of francolite formation. *Earth and Planetary Sciences Letters*, **77**, 20–34.
McArthur, J.M., Coleman, M.L. & Bremner, J.M. (1980). Carbon and oxygen isotopic composition of structural carbonate in sedimentary francolite. *Journal of the Geological Society*, **137**, 669–74.
McClellan, G.H. (1980). Mineralogy of carbonate fluorapatites. *Journal of the Geological Society*, **137**, 675–82.
Mertz, K.A. (1984). Diagenetic aspects, Sandholt Member, Miocene Monterey Formation, Santa Lucia Mountains, California. In *Dolomites of the Monterey Formation and Other Organic-rich Units*, ed. R.E. Garrison, M. Kastner & D.H. Zenger, pp. 49–74. Society of Economic Paleontologists and Mineralogists, Pacific Section, Los Angeles.
Murata, K.J., Friedman, I. & Cremer, M. (1972). Geochemistry of diagenetic dolomites in Miocene marine formations of California and Oregon. *US Geological Survey Professional Paper* **724-C**.
Nathan, Y. & Lucas, J. (1972). Synthese de l'apatite a partir du gypse: application au probleme de la formation des apatites carbonatees par precipitation directe. *Chemical Geology*, **9**, 99–112.
Nathan, Y. & Lucas, J. (1976). Experiences sur la precipitation directe de l'apatite dans l'eau de mer; implications dans la genese des phosphorites. *Chemical Geology*, **18**, 181–6.
Nathan, Y. & Nielsen, H. (1980). Sulfur isotopes in phosphorites. In *Marine Phosphorites*, ed. Y.K. Bentor, pp. 73–8. Society of Economic Paleontologists and Mineralogists, Tulsa.
Pisciotto, K.A. (1978). 'Basinal sediments and diagenetic aspects of the Monterey Shale, California'. Unpublished Ph.D. Thesis, University of California, Santa Cruz.
Pisciotto, K.A. (1981). Review of secondary carbonates in the Monterey Formation, California. In *The Monterey Formation and Related Siliceous Rocks of California*, ed. R.E. Garrison, R.G. Douglas, K.A.

Pisciotto, C.M. Isaacs & J.C. Ingle, pp. 273–83. Society of Economic Paleontologists and Mineralogists, Pacific Section, Los Angeles.
Pisciotto, K.A. & Garrison, R.E. (1981). Lithofacies and depositional environments of the Monterey Formation, California. In *The Monterey Formation and Related Siliceous Rocks of California*, ed. R.E. Garrison, R.G. Douglas, K.A. Pisciotto, C.M. Isaacs & J.C. Ingle, pp. 97–122. Society of Economic Paleontologists and Mineralogists, Pacific Section, Los Angeles.
Prasad, S. (1985). 'Microsucrosic dolomite from the Hawthorn formation (Miocene) of Florida: distribution and development'. Unpublished M.Sc. Thesis, University of Miami, Miami.
Riggs, S.R. (1979). Phosphorite sedimentation in Florida – a model phosphogenic system. *Economic Geology*, **74**, 195–220.
Riggs, S.R. (1984). Paleoceanographic model of Neogene phosphorite deposition, US Atlantic continental margin. *Science*, **223**, 123–31.
Riggs, S.R., Lewis, D.W., Scarborough, A.K. & Snyder, S.W. (1982). Cyclic deposition of Neogene phosphorites in the Aurora area, North Carolina, and their possible relationship to global sea-level fluctuations. *Southeastern Geology*, **23**, 189–204.
Rowe, G.T. & Howarth, R. (1985). Early diagenesis of organic matter in sediments off the coast of Peru. *Deep-Sea Research*, **32**, 43–55.
Scarborough, A.K., Riggs, S.R. & Snyder, S.W. (1982). Stratigraphy and petrology of the Pungo River Formation, central Coastal Plain of North Carolina. *Southeastern Geology*, **23**, 205–16.
Shemesh, A., Kolodny, Y. & Luz, B. (1983). Oxygen isotope variations in phosphate of biogenic apatites, II. Phosphorite rocks. *Earth and Planetary Sciences Letters*, **64**, 405–16.
Sholkovitz, E. (1973). Interstitial water chemistry of the Santa Barbara basin sediments. *Geochimica Cosmochimica Acta*, **37**, 2043–73.
Sibley, D.F. (1980). Climate control of dolomitization, Seroe Domi Formation (Pliocene), Bonaire, NA. In *Concepts and Models of Dolomitization*, ed. D.H. Zenger, J.B. Dunham & R.L. Ethington, pp. 247–58. Society of Economic Paleontologists and Mineralogists, Tulsa.
Sieburth, J.M. (1979). *Sea Microbes*. Oxford University Press, New York.
Staresenic, N. (1978). 'The vertical flux of particulate organic matter in the Peru coastal upwelling as measured with a free-drifting sediment trap.' Unpublished Ph.D. Thesis, Woods Hole Oceanographic Institution–Massachusetts Institute of Technology, Woods Hole.
Suess, E. (1980). Particulate organic carbon flux in the oceans – surface productivity and oxygen utilization. *Nature*, **288**, 260–3.
Suess, E. (1981). Phosphate regeneration from sediments of the Peru continental margin by dissolution of fish debris. *Geochimica Cosmochimica Acta*, **45**, 577–88.
Surdam, R.C. & Stanley, K.O. (1981a). Diagenesis and migration of hydrocarbons in the Monterey Formation, Pismo Syncline, California. In *The Monterey Formation and Related Siliceous Rocks of California*, ed. R.E. Garrison, R.G. Douglas, K.A. Pisciotto, C.M. Isaacs & J.C. Ingle, pp. 317–27. Society of Economic Paleontologists and Mineralogists, Pacific Section, Los Angeles.
Surdam, R.C. & Stanley, K.O. (1981b). Stratigraphic and sedimentologic framework of the Monterey Formation, Pismo Syncline, California. In *Guide to the Monterey Formation in the California Coastal Area, Ventura to San Luis Obispo*, ed. C.M. Isaacs, pp. 83–91. American Association of Petroleum Geologists, Pacific Section, Camarillo, California.
Von Brokel, K. (1981). A note on short-term production and sedimentation in the upwelling region off Peru. In *Coastal Upwelling*, ed. F.A. Richards, pp. 291–7. American Geophysical Union, Washington, DC.
Weaver, C.E. & Beck, K.C. (1977). *Miocene of the Southeastern United States*. Elsevier, Amsterdam.

7

Organic geochemistry of phosphorites: relative behaviors of phosphorus and nitrogen during the formation of humic compounds in phosphate-bearing sequences

J.G. TRICHET, M. RACHIDI
AND H. BELAYOUNI

Abstract

This paper introduces new data concerning the behavior of phosphorus and nitrogen in humic compounds associated with phosphorites. According to Nissenbaum (1979), phosphorus and nitrogen can be released in important quantities in solutions when fulvic acids condense to form higher molecular weight humic acids. This relationship is studied in phosphatic and non- or poorly-phosphatic samples from the Paleocene Gafsa Basin, in Tunisia.

Our results show that the higher the inorganic phosphorus content in the strata the lower the percentage of phosphorus in humic acids as well as lower atomic P:C ratios within the humic acids. According to Rachidi (1983), amino acid nitrogen is depleted by a factor ranging between 6 and 20 when passing from fulvic to humic acids. This is comparable to the highest depletion occurring within the phosphatic beds we studied. These two results support Nissenbaum's theory concerning the process during which high quantities of P and N can be liberated in the sedimentary solutions.

However, some properties of the fulvic acids contained within these sediments, in particular their richness in unstable amino acids (serine and threonine), give evidence that these acids could be merely alteration products of humic acids rather than their precursors.

Introduction

Most data on the composition of organic matter associated with phosphorites and phosphatic sediments have been published since 1972 and relate to a small number of deposits: the phosphatic sediments off Namibia (Romankevich & Baturin, 1972; Morris & Calvert, 1977), various basins in the world (Powell, Cook & McKirdy, 1975), the Middle Cambrian Georgina Basin in north-eastern Queensland, Australia (Sandstrom, 1980, 1982, 1986), the Phosphoria Formation (Maughan, 1980), the Basin of Gafsa, Tunisia (Belayouni & Trichet, 1980; Belayouni, 1983; Belayouni & Trichet, 1983, and 1984; Rachidi, 1983), the Upper Cretaceous sequence of southern Israel (with comparison to some Namibian and Moroccan samples; Amit & Bein, 1982; Bein & Amit, 1982) and the Recent phosphorites off Peru and Chile (Cunningham & Burnett, 1985).

Among the results provided by these studies, two relationships are of particular interest: (1) the nature and origin of the organic matter in the phosphate-rich and interbedded strata; and, (2) the presence and abundance of humic compounds within the phosphatic strata (Fig. 7.1).

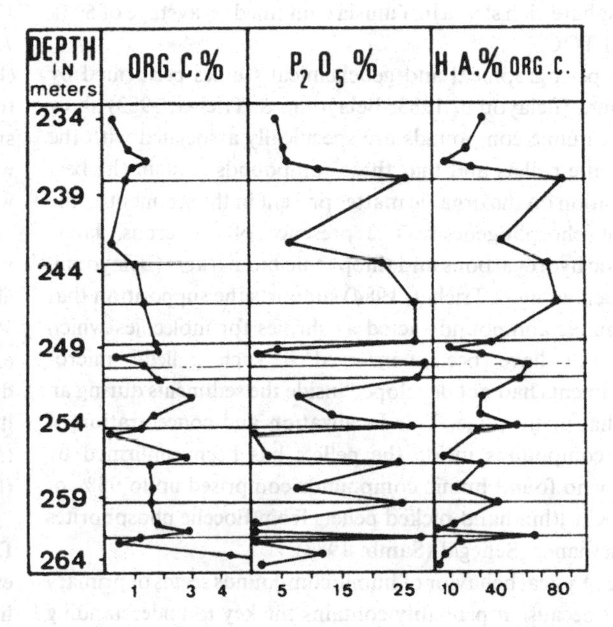

Fig. 7.1. Variation of the total organic carbon and phosphorus content and of the amount of humic acids (HA) (HA carbon % of total organic carbon) in phosphate-rich and phosphate-poor strata of the core S15 of Redeyef (Gafsa Basin, Tunisia) (after Belayouni, 1983).

When the organic geochemical analyses were performed not only on the phosphate-rich strata but also on the interbedded strata (siltstones in the Georgina Basin; shales and cherts in the Israel series; shales, marls, cherts and limestones in the Gafsa Basin), the nature and origin of the organic matter proved to be essentially the same, i.e. planktonic marine in all strata. This observation led some authors to the conclusion that the genesis of the related phosphorites took place in marine epicontinental basins in which variations of oceanographic physico-chemical parameters, particularly those associated with hydrodynamics (Maughan, 1980), were responsible for progressive variations at the water–sediment interface. The result was the deposition of organic-rich, dark sediments (generally shales or cherts) under low Eh conditions, limestones during oxidizing conditions, and phosphate under intermediate conditions (Bein & Amit, 1982; Belayouni, 1983). The sedimentological transitions during such periods must have been continuous, as shown by the progressive changes in the lithological facies of the successive strata.

The presence and abundance of humic compounds within the phosphatic strata was noted as early as 1938 in the Tunisian Basin by Vincent & Boischot, and this relationship has been confirmed by all other authors who have worked more recently on the organic geochemistry of phosphate-rich beds. For example, Romankevich & Baturin (1972) found that in the phosphatic sediments off Namibia approximately 50% of the total organic carbon is in the humic form; Amit & Bein (1982), working in Israel, found that the humic carbon in Senonian phosphorites accounts for more than 70% of the total organic carbon (TOC); Sandstrom (1982) working on the Georgina Basin phosphorites reported that humic compounds represent between 5 and 60% of the TOC; and Belayouni (1983) and Belayouni & Trichet (1983) established that humic compounds in phosphate-rich strata in Tunisia contained an average of 59% of total TOC.

The petrographical and geochemical studies performed by Belayouni (Belayouni, 1983; Belayouni & Trichet, 1983) show that the humic compounds are specifically associated with the phosphatic pellets and that these compounds contain the best information on the organic matter present in the sediment at the time of phosphogenesis. The presence of numerous, labile aromatic hydrocarbons and thiophenic biomarkers (Belayouni, 1983; Belayouni & Trichet, 1984) supports the supposition that these humic compounds acted as shrines for molecules which would not have been conserved if such pelletal micro-environments had not developed inside the sediments during an early diagenetic stage. This localization and concentration of humic compounds inside the pellets has been confirmed by Samb, who found humic compounds comprised up to 95% of the TOC within hand-picked pellets from Eocene phosphorites of Casamance, Senegal (Samb, 1983).

The physical behavior of humic compounds seems of primary interest because it probably contains the key to understanding the incorporation and conservation of phosphate inside the pellets. The very close association between humic compounds and apatite suggests that these two types of compounds precipitated simultaneously, or penecontemporaneously, and that the organic matrix of humic acids may have protected calcium phosphate against diffusion and dissolution in the interstitial solutions.

The general properties of these humic compounds have been presented and published elsewhere (Amit & Bein, 1982; Sandstrom, 1982; Bein & Amit, 1982; Belayouni, 1983; Belayouni & Trichet, 1983; Rachidi, 1983). They are summarized here, in order to discuss the behavior of inorganic phosphorus during the very early stage of its incorporation in the pellets. The main characteristics of humic compounds associated with phosphorites are:

(1) They result from marine, planktonic organic matter. This is indicated both by petrographical studies, which show the absence of any noticeable higher land plant remains and, conversely, the abundance of planktonic fossils (Belayouni & Trichet, 1980; Fauconnier & Slansky, 1980; Belayouni, 1983), as well as by geochemical studies (Sandstrom, 1980 and 1982; Amit & Bein, 1982; Bein & Amit, 1982; Belayouni, 1983; Belayouni & Trichet, 1983 and 1984) which show high H:C, N:C and S:C ratios (Table 7.1, Figs 7.2, 7.3) and the absence of phenolic compounds (Ishiwatari, 1970; Rashid & King, 1970; Brown *et al.*, 1972; Nissenbaum & Kaplan, 1972; Huc, 1973; Huc & Durand, 1974 and 1977; Debyser & Gadel, 1977; Debyser *et al.*, 1978).

If the condensation processes usually have been related to Maillard reactions between nitrogenous compounds and carbohydrates (Maillard, 1912; Enders & Theis, 1938), they also have been accompanied by the incorporation of important amounts of lipids, i.e. hydrocarbons, fatty acids and other polar compounds (Belayouni, 1983; Rachidi, 1983; Belayouni & Trichet, 1984).

(2) The sulfur content of these humic acids is intrinsically very high (Sandstrom, 1982 and 1986); Amit & Bein, 1982; Bein & Amit, 1982; Belayouni, 1983; Belayouni & Trichet, 1983), up to 12% by weight of the humic compounds (Belayouni, 1983; Table 7.1). As shown by electron spectroscopy for chemical analyses (ESCA), the sulfur can have two origins, one part being inherited from molecules of living organic matter in the form of C–SO_2 sulfone bonds, the other part resulting from the reaction of H_2S with the sedimentary organic matter in the anoxic environment where phosphogenesis occurred (Belayouni, 1983).

(3) The N-content of the humic acids is characterized by values of $<2.5\%$, appreciably lower than the 5.5–6.5% generally encountered in humic acids of marine origin (Ishiwatari, 1970; Rashid & King, 1970; Huc, 1973), and by the liberation of ammonia (due to deamination) and amino acids by acid hydrolysis (5.6N HCl, 24h, 110° C); the weaker the condensation of humic compounds, the greater the intensity of deamination (ammonia production) relative to the amino acids release (Belayouni, 1983).

(4) The humic compounds are remarkably rich in lipids. Depending upon the solvent used for the extraction, the total extractable carbon can vary from 12–79% by weight of the humic acids in samples from the Gafsa Basin (Belayouni, 1983; Rachidi, 1983; Belayouni & Trichet, 1984).

The composition of the extract is itself remarkable, as 57–91% of the total extractable carbon is contained in polar compounds, especially asphaltenes. The hydrocarbons contain between 9 and 32% of the TOC of the humic acids and their molecular distribution indicates that they originated in marine planktonic organic matter that had been altered biochemically by bacteria during the first stages of diagenesis (Powell *et al.*, 1975; Maughan, 1980; Amit & Bein, 1982; Bein & Amit, 1982; Sandstrom, 1982; Belayouni, 1983; Belayouni & Trichet, 1984).

The best preserved of the biomarkers, which include C-ring mono-aromatic steroids, benzohopanoids, C35-thiophenic pentacyclic triterpenoid, mono-unsaturated fatty acids and $\beta\beta$-hopanoic acids, as noted above, are found only in the extractable fraction of the humic compounds from the phosphatic strata. Similarly, they occur also in the bulk extract from shale or chert strata (Belayouni, 1983; Belayouni & Trichet, 1984).

Both the origin and the character of these humic compounds provide a good hydrocarbon-generating potential, as documented by Rock-Eval pyrolysis and pyrochromatography

Table 7.1. *Chemical analyses of humic acids extracted from two cores of the Gafsa Basin (core Redeyef S15 'RD'; core M'Dilla S10B 'MD'), after Belayouni (1983)*

Depth (m)	Lithology	C%	H%	N%	O%	S%	Ash	(H/C)	(O/C)	(N/C)	(S/C)
RD 237.20	Marl	55.51	7.10	1.88	18.92	9.37	7.00	1.53	0.26	0.029	0.063
RD 238.10	Phosphate	36.75	5.41	1.18	15.22	8.28	32.00	1.68	0.30	0.027	0.059
RD 242.40	Limestone	33.42	4.99	0.99	18.30	5.97	35.00	1.79	0.41	0.025	0.067
RD 243.80	Phosphate	46.77	6.06	1.42	13.38	10.34	22.00	1.55	0.21	0.026	0.083
RD 246.40	Phosphate	43.53	5.51	1.34	13.81	9.55	26.00	1.52	0.24	0.026	0.082
RD 249.00	Phosphate	38.24	6.22	1.23	18.00	7.96	28.00	1.95	0.35	0.027	0.078
RD 250.00	Phosphate	45.89	6.15	1.61	13.24	9.05	24.00	1.61	0.22	0.030	0.074
RD 251.00	Phosphate	45.65	6.53	1.56	13.98	9.62	22.60	1.71	0.23	0.029	0.079
RD 252.30	Marl	56.70	8.47	1.85	17.68	12.10	3.50	1.79	0.23	0.028	0.000
RD 253.60	Marl	52.99	7.18	2.00	18.71	10.58	8.00	1.62	0.26	0.032	0.075
RD 254.20	Phosphate	52.44	7.00	1.59	17.96	10.00	11.00	1.60	0.26	0.026	0.071
RD 258.30	Marl	43.39	6.97	1.19	26.39	7.38	14.50	1.92	0.46	0.023	0.064
RD 259.40	Phosphate	54.88	8.52	1.97	13.66	9.79	11.00	1.86	0.20	0.030	0.067
RD 262.90	Marl	53.80	5.80	1.78	14.69	10.46	13.50	1.29	0.20	0.028	0.073
RD 263.60	Marl	34.47	4.92	1.06	15.80	6.95	35.00	1.71	0.34	0.026	0.075
MD 149.00	Marl	41.07	7.17	0.92	39.04	3.54	8.00	2.09	0.71	0.020	0.032
MD 176.80	Marl	48.77	6.50	2.30	20.74	6.56	15.00	1.60	0.32	0.040	0.050
MD 185.00	Marl	38.35	7.80	0.63	39.65	3.16	—	2.44	0.77	0.014	0.031
MD 203.00	Marl	44.93	6.29	1.18	25.29	6.19	15.00	1.67	0.42	0.023	0.051
MD 205.00	Phosphate	57.94	6.13	1.96	18.29	9.98	5.00	1.27	0.24	0.029	0.065
MD 207.60	Limestone	49.76	6.09	1.48	20.85	7.40	15.00	1.46	0.31	0.025	0.056
MD 208.30	Phosphate	48.22	5.76	1.72	16.73	7.75	20.00	1.43	0.26	0.030	0.060

of the humic acids (Belayouni, 1983; Belayouni & Trichet, 1984) or by direct pyrochromatography of the asphaltenes previously extracted from the humic acids (Rachidi, 1983).

Initial stages in the incorporation of humic acids in pellets

The behavior of phosphorus and nitrogen during the first stages of the formation of humic acids within the phosphate-rich strata is of particular interest. The general assumption regarding the mutual relationship between humic compounds and phosphorus is that the latter is released in soils and sediments when low molecular weight organic constituents (fulvic acids) condense to the humic state (Nissenbaum, 1979). Working on marine sediments, Nissenbaum found that 0.40–0.80% of organically-bound P was contained in fulvic acids but only 0.10–0.20% was contained in humic acids.

We attempted to determine if such a general relationship could be demonstrated for P during diagenesis of the Tunisian phosphorites. We investigated the comparative behavior of nitrogen during this diagenetic step as well to learn if nitrogen (or certain N-containing compounds) also decreased during the passage of FA to HA and, in particular, to see if the low values of N:C in the humic acids associated with phosphorites (relative to other humic acids of marine origin) could be explained by such a decrease.

Phosphorus behavior during the early stages of incorporation

Analyses were performed on 15 samples from core S15 at Redeyef, 10 samples from core S6 at Metlaoui, and 9 samples from core S10.B at M'Dilla. All cores were from the Paleocene Basin of Gafsa as described geologically and lithologically by Sassi (1974), Belayouni (1983) and Belayouni & Trichet (1984). The cores were chosen because they all displayed an alternation of non- or poorly reworked phosphate-rich strata (Tables 7.1 and 7.2; Figs. 7.4, 7.5, 7.6). Tables 7.1 and 7.2 give basic information on the lithological and geochemical properties of the samples.

Rock samples were treated first with chloroform and then 2N HCl for 16 h. Humic compounds were extracted from 20 g samples with 100 ml 0.1N NaOH under magnetic stirring for 1 h. This operation was repeated four times. The alkaline portion of the solution was separated from the residual fraction by centrifugation (3000 rpm, 5 min). The alkaline extract was acidified by 2N HCl to pH 2 leading to humic acid precipitation. The humic acids were separated from the fulvic acids by centrifugation, redissolved in 0.1N NaOH (50–100 ml) and purified by dialysis against deionized water (2 d). They were then separated by ion-exchange on a Dowex 50W-8X, H^+ column and freeze-dried.

Phosphorus in purified humic acids

We used the Tunisian alternating phosphate-rich and phosphate-poor strata (shales or limestones) in order to determine if the liberation of organic phosphorus initially associated with fulvic acids had been more efficient in phosphate-rich strata than in phosphate-poor strata.

Table 7.2 and figures 7.4, 7.5 and 7.6 show the values of the two ratios HAP (% of HA) and atomic P:C in HA versus the

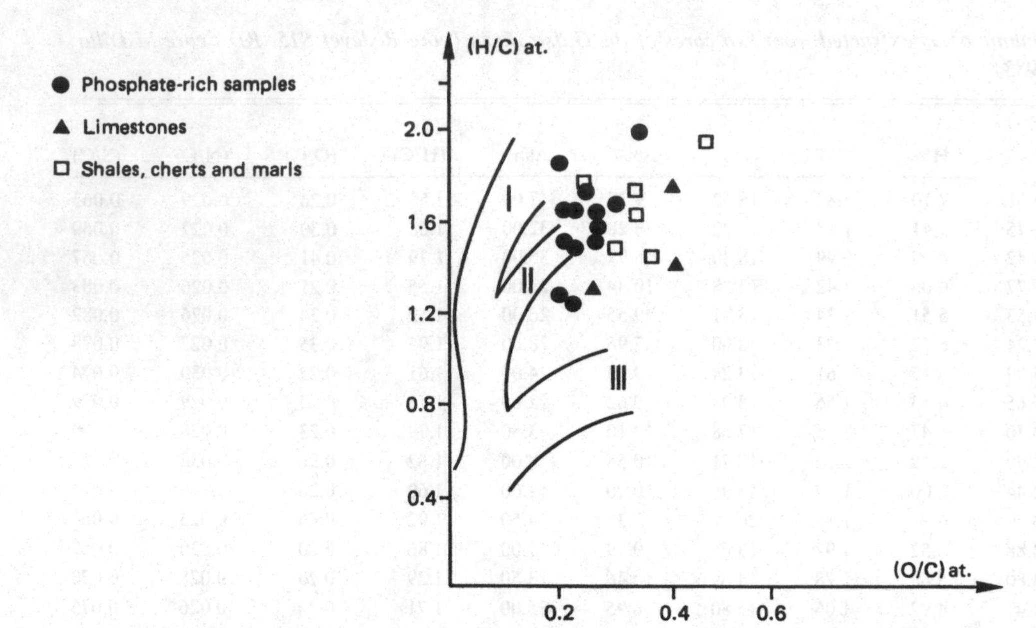

Fig. 7.2. Atomic ratios H/C vs O/C in humic acids extracted from strata differing by their lithological facies in the core S15 from Redeyef (Gafsa Basin, Tunisia) (from Belayouni & Trichet 1983).

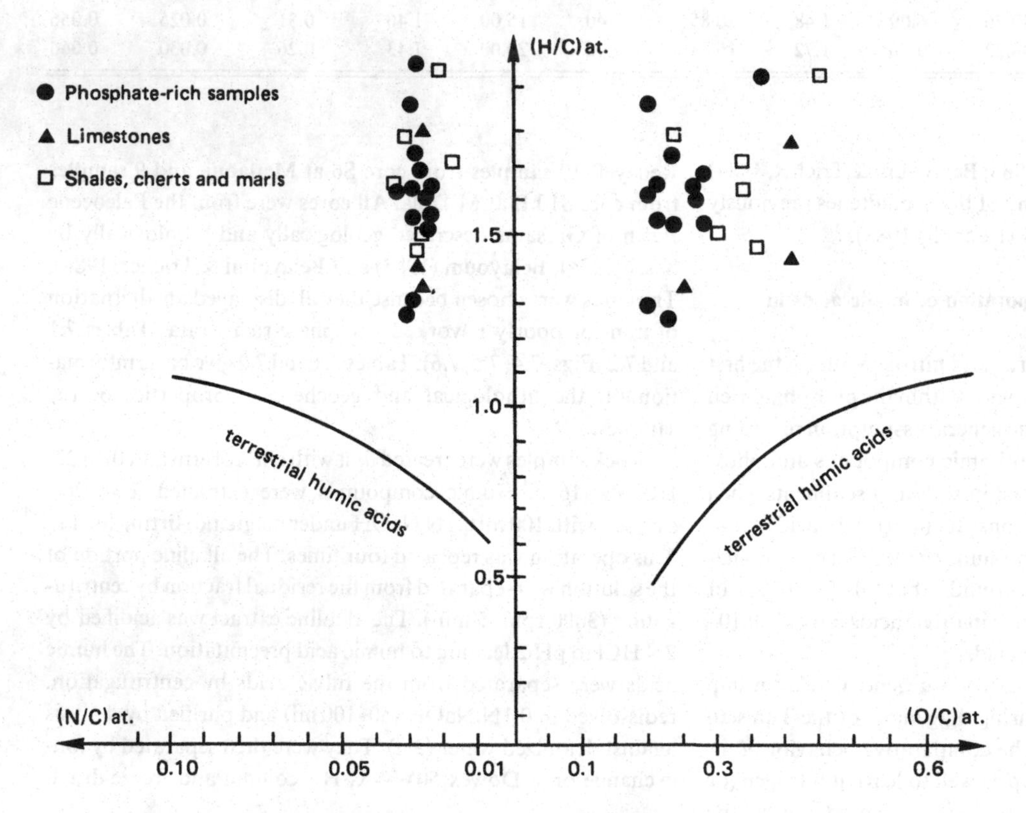

Fig. 7.3. Variations of atomic H:C, N:C and O:C ratios in humic acids extracted from strata differing by their lithological facies in the core S15 from Redeyef (Gafsa Basin, Tunisia). Terrestrial humic acids, after Debyser & Gadel, 1977.

Table 7.2. *Analytical data on carbon and phosphorus contents in bulk sediments and in related humic acids (HA) in samples cored in three places (Redeyef, M'Dilla and Metlaoui) in the Gafsa Basin (after Belayouni, 1983; and Rachidi, 1983)*

Depth (m)	Lithology	Organic carbon %	P %	HAC % of TOC	HAC % of HA	HAP % of HA	(P:C) at. HA $\times 10^{-3}$
RD 237.20	Marl	0.90	4.50	26.00	55.51	0.021	0.15
RD 238.10	Phosphate	0.78	10.96	86.00	36.75	0.040	0.42
RD 242.40	Limestone	0.32	2.82	44.00	34.42	0.050	0.58
RD 243.80	Phosphate	1.04	9.42	75.00	46.77	0.030	0.25
RD 246.40	Phosphate	1.35	11.60	81.00	43.53	0.030	0.27
RD 249.00	Phosphate	1.79	11.60	38.00	38.24	0.025	0.25
RD 250.00	Phosphate	1.36	12.34	63.00	45.89	0.030	0.25
RD 251.00	Phosphate	1.85	11.76	55.00	45.65	0.023	0.20
RD 252.30	Marl	3.04	3.42	33.00	56.70	0.076	0.52
RD 253.60	Marl	1.67	5.80	29.00	52.90	0.251	1.83
RD 254.20	Phosphate	1.27	11.30	54.00	29.67	0.027	0.35
RD 258.30	Marl	1.65	3.27	14.00	43.39	0.132	1.18
RD 259.40	Phosphate	1.62	10.90	41.00	54.88	0.042	0.30
RD 262.90	Marl	4.30	5.07	7.00	53.80	0.030	0.22
RD 263.60	Marl	7.80	0.75	5.00	34.47	0.042	0.47
MD 149.00	Marl	0.39	3.74	62.00	41.07	0.036	0.34
MD 153.00	Limestone	0.71	3.40	68.00	55.42	0.045	0.31
MD 176.00	Marl	1.63	0.65	22.00	44.94	0.180	1.55
MD 176.80	Marl	0.87	4.68	74.00	48.77	0.024	0.19
MD 185.00	Marl	0.68	3.16	21.00	38.55	0.047	0.47
MD 203.00	Marl	1.84	0.55	17.00	44.93	0.054	0.46
MD 205.00	Phosphate	1.38	11.40	67.00	57.94	0.023	0.15
MD 207.60	Limestone	0.80	4.50	38.00	49.76	0.055	0.43
MD 208.30	Phosphate	1.18	9.24	83.00	48.22	0.029	0.23
MT 159.80	Phosphate	1.02	11.13	49.00	34.67	0.066	0.74
MT 161.10	Phosphate	1.23	11.35	39.00	37.36	0.030	0.31
MT 163.10	Phosphate	1.03	11.22	83.00	49.36	0.071	0.55
MT 164.80	Phosphate	0.93	8.90	83.00	35.50	0.029	0.32
MT 165.00	Marl	1.52	5.46	41.00	39.27	0.051	0.50
MT 167.20	Marl	1.97	2.28	37.00	18.69	0.037	0.76
MT 169.50	Phosphate	1.00	10.58	64.00	42.14	0.060	0.55
MT 169.90	Marl	2.03	0.52	31.00	44.96	0.127	1.10
MT 174.30	Marl	3.22	1.63	31.00	35.95	0.015	0.16
MT 176.00	Phosphate	1.11	10.32	59.00	38.04	0.021	0.21

total P-content in the three cores studied. An inverse correlation exists between P-content in the sediment and the values of both ratios. This suggests that the humic acids of the phosphate-rich strata are poorer in organic-P when compared to the humic acids of the phosphate-poor strata. In other words, the humic acids are most abundant but poorer in organic-P in the phosphate-rich strata, and less abundant but richer in organic-P in poorly phosphatized strata.

The atomic C:P ratio varies in phosphate-rich samples between 2100 (mean value for five samples in the Metlaoui core, Table 7.2) and 6600 (for the richest sample in P in the M'Dilla core, Table 7.2, with a mean value of 3300 in seven samples of the Redeyef core, Table 7.2). In the same cores this ratio varies in shales and marls between 1000 (four samples in the Redeyef core, Table 7.2) and 1700 (five samples in the Metlaoui core, Table 7.2), with a value of 1430 in four samples in the M'Dilla core (Table 7.2).

These ratios compare favorably with those given by Nissenbaum (1979) for humic acids of marine origin, which vary from 300 to 400 in recent sediments, up to 1000 in deeper sediments (Saanich Inlet). The values of C:P for the Tunisian samples fall in the range of such values in non-recent sediments, though emphasizing the originality of the composition of humic acids associated with phosphate-rich beds in which intensive and efficient mineralization of P provoked an important decrease of the P-content relative to carbon.

Finally, the P-content of the HA is lower in the Tunisian phosphatic series (0.02–0.04%) than in deep sediments such as those of the Saanich Inlet (3345–3475 m; 0.06% P) in which a maximum amount of P is supposed to have been mineralized (Nissenbaum, 1979).

Unfortunately, due to the small amounts of fulvic acids in the samples (generally <4% of TOC, Amit & Bein, 1982; Belayouni, 1983), and due to the difficulty of removing the acids from inorganic impurities, it has not been possible to prepare a sufficient amount of purified fulvic acid to perform a direct

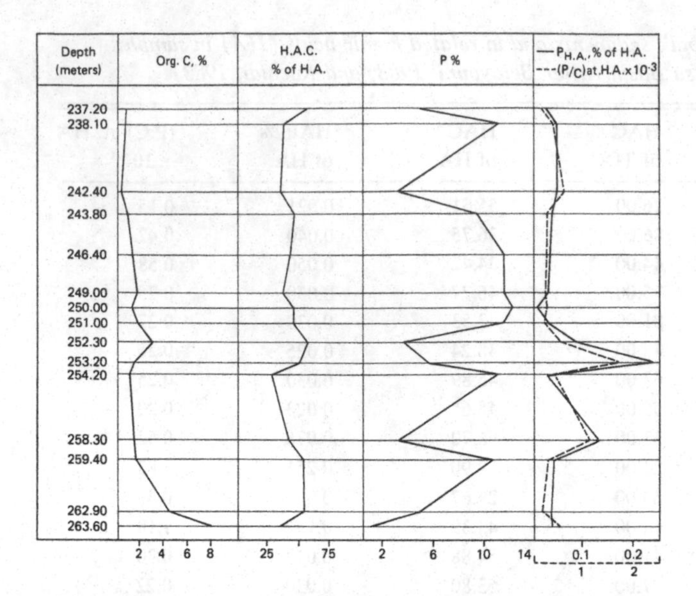

Fig. 7.4. Variations in organic carbon and phosphorus contents in bulk samples and related humic acids (HA) in samples from the core S15 of Redeyef (Gafsa Basin, Tunisia). Lithological data are given in Table 7.2.

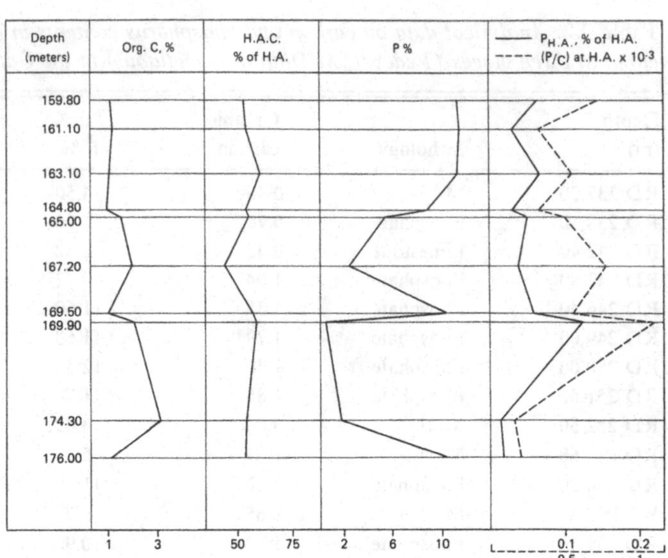

Fig. 7.6. Variations of organic carbon and phosphorus contents in bulk samples and related humic acids (HA) in samples from the core S6 of Metlaoui (Gafsa Basin, Tunisia). Lithological data are given in Table 7.2.

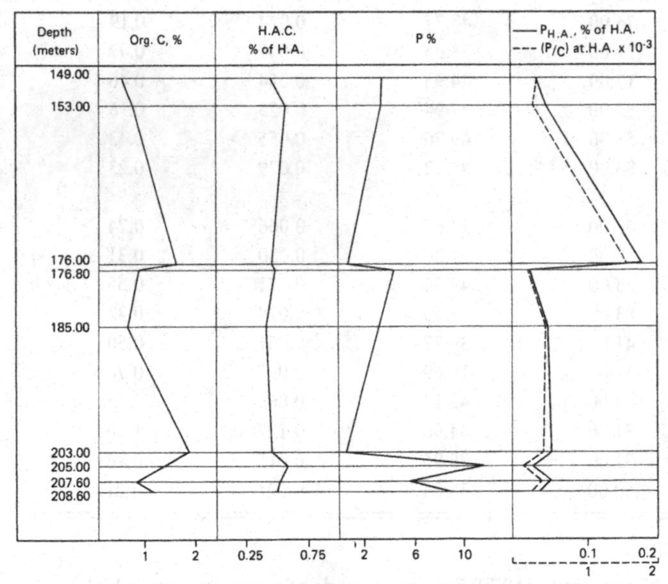

Fig. 7.5. Variations of organic carbon and phosphorus contents in bulk samples and related humic acids (HA) in samples from the core S10B of M'Dilla (Gafsa Basin, Tunisia). Lithological data are given in Table 7.2.

analysis of its organic-P content. Moreover, the nature and significance of the fulvic acid fraction within the bulk organic matter can be controversial, as will be discussed later.

Amino acids and nitrogen in humic compounds

An advantage in studying amino acids (AA) in humic compounds is that they can be analyzed by further acid hydrolysis of the acid solution in which the fulvic acids are separated from the humic acids. Also, they can be compared with the amino acids contained in the humic acids which are obtained by redissolution in 0.1N NaOH of the precipitate of the humic acids formed during addition of HCl in the bulk humic compounds solution.

Analyses have been performed on three samples of different lithology, one being a shale (RD 262.90 m), another a phosphorite (RD 254.20 m) and the third a limestone sample (RD 262.00 m).

The extraction of fulvic and humic acids was performed by a method differing on two points from the one previously described: (1) the bulk humic compounds obtained by NaOH treatment of the samples were added to a 1% wt. NaCl solution to floculate the larger colloids; and (2) after their precipitation by 2N HCl, the humic acids were redissolved in 0.1N NaOH then hydrolyzed by 5.6N HCl without being purified by dialysis, and freeze dried. The organic carbon content of fulvic and humic acids was measured in a Carmhograph apparatus.

Figures 7.7, 7.8 and 7.9 show the relative amounts of different amino acids, respectively, in the fulvic and humic acids extracted from the shale, phosphate and limestone samples in the Redeyef core. These values are related to the organic carbon content of the fulvic and humic acids.

Table 7.3 shows the values of the amount of amino acids in the fulvic and humic acids relative to the organic carbon content of the fraction in each facies (shale, phosphate and limestone), and the decrease of AA content relative to carbon when passing from fulvic acids to humic acids (Rachidi, 1983).

The conclusions that have been drawn from the amino acid analyses in Tunisian samples of different lithological facies (shales, phosphates, and limestones) are as follows (Rachidi, 1983):

(1) the fulvic acids are richer in AA than the humic acids, by a factor that ranges between 6 and 20. A point to be noted regarding the behavior of nitrogen during the genesis of humic acids is that if the relative decrease in the amino acid content between fulvic and humic acids is great (>84%), it is greatest in the phosphate-rich sample (Table 7.3).

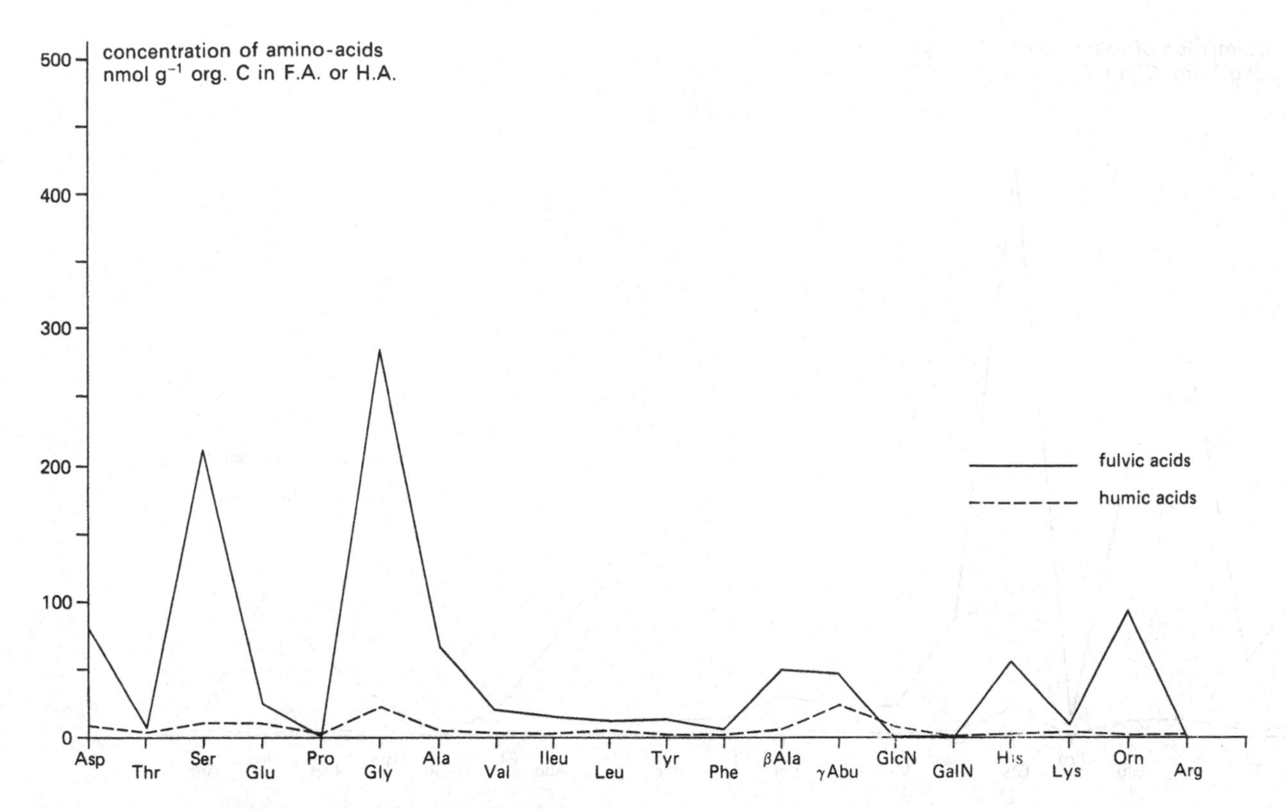

Fig. 7.7. Amino-acid distribution in fulvic and humic acids extracted from a shale sediment (RD 262.90 m, core S15 Redeyef, Gafsa Basin, Tunisia) (after Rachidi, 1983).

This result compares favorably with the previous conclusions regarding phosphorus behavior, and shows that the formation of humic acids is accompanied by a strong mineralization of both P and N from lighter organic precursors. In the case of shales or limestones, the P and N which have been liberated at that time have been delivered to the interstitial waters; whereas in the case of phosphorites, N has been lost (explaining the low N:C values in humic acids), but some P has been preserved inside the apatite minerals and in close association with the contemporaneously formed humic acids.

(2) Regarding the nature of the amino acids, Rachidi (1983) showed that some amino acids display a predominance over others: glycine (Gly), serine (Ser), glutamic acid (Glu), aspartic acid (Asp), alanine (Ala), γ-amino-butyric acid (γ-Abu), β-alanine (β-Ala) and ornithine (Orn) are the dominant species.

A planktonic origin of the organic matter is in good agreement with the abundance of aspartic acid, glutamic acid and glycine, considered as dominant amino acids in organic matter of marine origin (Degens & Mopper, 1976). However, this conclusion must be considered with respect to the alteration processes on the amino acid content occurring during diagenesis (see further points).

The abundance of acidic amino acids (Glu and Asp) is related to the presence of calcium-containing minerals. Aspartic acid is known to be abundant in biogenic calcareous sediments (Trichet, 1968 and 1972; Suess, 1970 and 1973; Mitterer, 1971 and 1972; Muller & Suess, 1977), and the abundance of both amino acids has been noted in calcium phosphate-rich materials (Belayouni, 1978 and 1983). In particular, the importance of diacidic amino acids over neutral ones has been observed by Rachidi (1983) in the stable residue of the organic matter belonging to the different lithological facies of the Tunisian Basin (stable residue is the fraction of organic matter which is resistant to extraction by chloroform, 2N HCl treatment, humic compounds extraction and HCl–HF treatment of the humic residue). In these stable residues, the ratio of diacidic to neutral amino acids shifts from 0.30 in shales and cherts to 0.45 and 0.48, respectively, in phosphate and limestone samples. The relation between the acidic amino acids and the calcium phosphate-minerals is supposed to be structural according to Garcia-Ramos & Carmona (1982).

The abundance of γ-amino-butyric acid is generally considered as an indicator of microbial activity. This amino acid is not utilized by bacteria, and an increase in its relative amount (due to bacterial feeding on other amino acids) has been used to estimate bacterial activity in sediments (Bremner, 1967). The abundance of this organic acid in the fulvic fraction of all sedimentary facies indicates the role played by microbes in the evolution of the organic matter.

The amount of β-alanine has been related to oxidizing conditions within the environment (Degens & Mopper, 1976), this amino acid being considered the result of the oxidation of aspartic acid. The presence of this acid and its greater abundance in the fulvic fractions relative to the humic fractions is understandable in this regard.

The dominance of ornithine over the other basic amino acids, especially arginine, from which it is derived, underlines the important degradation undergone by the organic matter within all the sedimentary facies.

The abundance of serine, associated with noticeable amounts

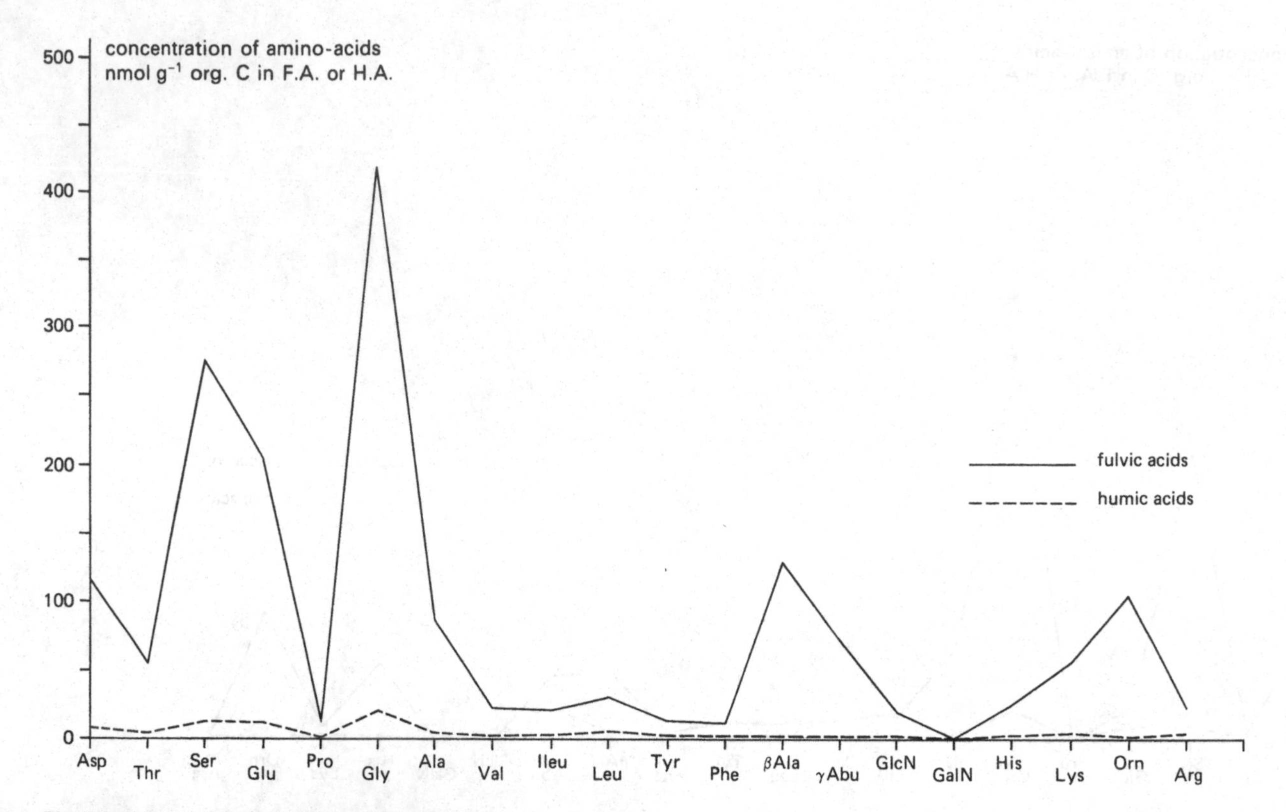

Fig. 7.8. Amino-acid distribution in fulvic and humic acids extracted from a phosphate sediment (RD 254.20 m, core S15 Redeyef, Gafsa Basin, Tunisia) (after Rachidi, 1983).

of threonine, is remarkable in the fulvic and humic samples, and is important, as these amino acids are considered to be unstable molecules in samples older than Pleistocene (Bada & Man, 1980; Miller & Hare, 1980). From this perspective, their abundance in Paleocene samples would be considered the result of contamination by more recent organic matter.

As will be discussed shortly, this abundance may have two explanations: serine and threonine may have been protected efficiently against degradation inside the humic acids, as has been shown for aromatic hydrocarbons and organic sulfur-containing molecules (Belayouni & Trichet, 1984); or contamination did actually occur, or a microbial activity is still active in the strata.

In conclusion, the major factor concerning the amino-nitrogen content in the humic compounds is the considerably higher proportion of such compounds in fulvic acids relative to humic acids. According to the scheme proposed by Nissenbaum (1979), such a disproportion means that there is an important loss of nitrogen during the condensation of fulvic acids into humic acids. This loss appears to have been greatest in the phosphatic samples.

Discussion and conclusions

As evidenced by the close geometrical association between humic acids and calcium phosphate inside pellets, and by the good mathematical correlations existing between P_2O_5 content and HAC:TOC. in phosphorites (Belayouni, 1983), the relation between apatite-P and humic acids composition is considered significant. The simplest model for this relationship is one in which apatite-P has been inherited directly from organic-P that was initially associated with the precursors of the humic acids. This model is supported by all the evidence for the biogenic origin of phosphorus in phosphorites, by geochemical and petrographical data (Belayouni, 1983) and also by the experimental soil studies by Brannon & Sommers (1985a, b). Brannon & Sommers show that the incorporation of organic phosphorus into humic polymers is the most efficient process, and is likely to occur when the organic phosphorus belongs to organic compounds containing both amino acids and phosphate ester functional groups. When such a compound reacts with oxidized polyphenols (humic acid precursors), a heavier humic complex is formed in which the introduced phosphorus is stabilized. This reaction emphasizes the relationship existing between P and N behavior during the formation of high molecular weight humic compounds. If such a reaction is applied to phosphatogenic sediments (in which all geochemical and petrographical data prove the marine planktonic origin of the organic matter), the nitrogen- and phosphorus-bearing molecules can be considered as marine planktonic originated fulvic acids. This relationship, associated with the mechanism of P and N release from fulvic acids proposed by Nissenbaum (1979), suggests that the apatite P can be provided through the mineralization of initially organically-bound P.

However the incorporation of phosphorus in humic compounds as biogenic phosphorus does not exclude the possibility of the addition of some soluble ionic phosphoric species to humic entities, as shown in soil studies by Levesque & Schnitzer (1967), Levesque (1969), Schnitzer & Khan (1972), or as shown through radioisotope chemistry by Lean (1972).

The present paper focuses attention on steps in the very early

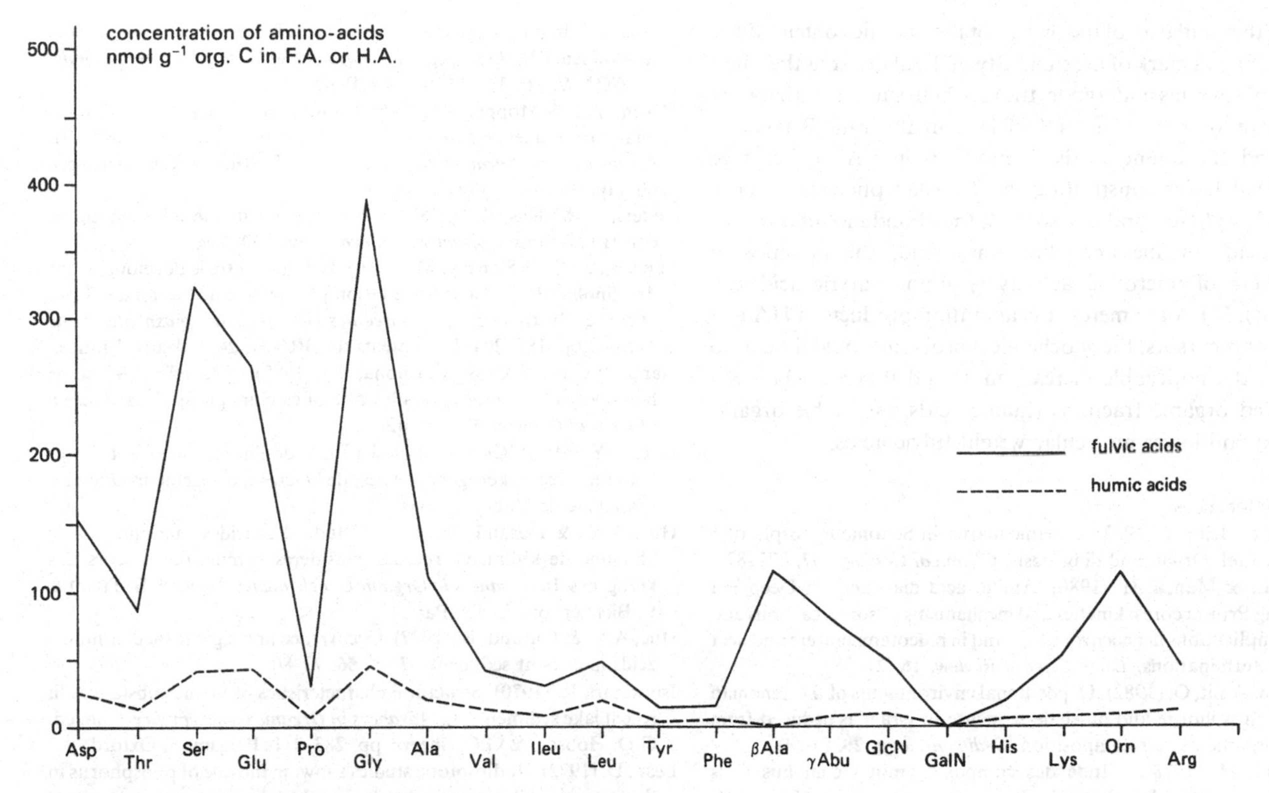

Fig. 7.9. Amino-acid distribution in fulvic and humic acids extracted from a limestone sediment (RD 262.00 m, core S15 Redeyef, Gafsa Basin, Tunisia) (after Rachidi, 1983).

Table 7.3. *Amino acid content in fulvic and humic acids in three samples from the Redeyef S15 core*

Depth (m)	AA in fulvic acids (nmol g^{-1} of FA org.C)	AA in humic acids (nmol g^{-1} of HA org.C)	Relative decrease (%)
RD 262.90 (shale)	1009	131	87.0
RD 254.20 (phosphate)	1706	87	94.9
RD 262.00 (limestone)	1987	303	84.7

mineralization of organic phosphorus. Following the assumption by Nissenbaum (1979) that the most important loss of N and P, relative to carbon, occurred when fulvic acids condensed to form humic acids, it has been shown that:

(1) amino acid nitrogen was actually highly depleted when fulvic acids were transformed into humic acids in all the sedimentary strata of the Gafsa Basin, the largest depletion occurring in the phosphatic strata adjacent to shale or limestone strata:

(2) if the humic acids that are contained in the shales and phosphatic sediments are actually depleted in phosphorus (relative to carbon), as described by Nissenbaum (1979) for marine humic acids (C:P ratio varying from 300 to 1000), the depletion is highest in the phosphatic beds (C:P varying between 2000 and 6600 in phosphate beds and between 1000 and 1700 in shale strata). This result confirms the efficient mineralization of organic phosphorus when apatite forms. Finally, the P content in the HA is lower in the Tunisian phosphatic series than in sediments in which maximum amounts of P are supposed to have been mineralized (Nissenbaum, 1979).

With regard to the comparative geochemistry of phosphorus and nitrogen during phosphatogenesis, these results, along with those by Nissenbaum (1979) and Brannon & Sommers (1985a, b), emphasize the exchange similarities of N and P in the humic acids.

The relation between fulvic and humic acids can, however, be considered from another point of view. According to the usual assumption, and in particular that of Nissenbaum (1979), the FA are supposed to be the precursors of HA, but another possibility is that fulvic acids can result from either microbial or non-microbial alteration of a previous stock of organic matter, as documented in some oceanic sediments by Pelet (1983a, b), or by Poutanen & Morris (1983). From this standpoint, fulvic acids must be considered as chemical entities differing from humic acids, not only by their lower molecular weight, but also by chemical specificity. This form of fulvic acid generation has been documented by Blondeau (1986).

According to this interpretation, the genetic relationship between FA and HA as proposed by Nissenbaum (1979) could be questioned. It is true that in the Tunisian samples there are three details which can be interpreted equivocally, either as a mark of the similarity of the origins of the organic content of FA and HA, or as a mark of the similarity of the alterations that their organic matter has undergone: the very high similarities between the total amount of FA (<4% of TOC in all strata; Belayouni, 1983) and the amino acids distribution in FA in the three lithological facies constituting the Tunisian phosphatic series (shale, phosphates, and limestones); the abundance of unstable amino acids (serine and threonine); and, the presence of biomarkers of microbial activity (γ-amino-butyric acid and ornithine). If FA are merely the alteration products of HA and not their precursors, the geochemical problem would then be to interpret the noticeable increase in N and P content between condensed organic fractions (humic acids and stable organic residues) and lower molecular weight fulvic acids.

References

Amit, O. & Bein, A. (1982). Organic matter in Senonian phosphorites from Israel. Origin and diagenesis. *Chemical Geology*, **37**, 277–87.

Bada, J.L. & Man, E.H. (1980). Amino acid diagenesis in Deep Sea Drilling Project cores: kinetics and mechanisms of some reactions and their implications in geochronology and in paleotemperature and heat flow determinations. *Earth Science Review*, **16**, 21–55.

Bein, A. & Amit, O. (1982). Depositional environments of the Senonian chert, phosphorite and oil shale sequence in Israel as deduced from their organic matter composition. *Sedimentology*, **29**, 81–90.

Belayouni, H. (1978). 'Etude des composés aminés contenus dans quelques séries phosphatées'. Thèse de Doctorat de Spécialité, Université d'Orléans, Orléans, France.

Belayouni, H. (1983). 'Etude de la matière organique dans la série phosphatée du bassin de Gafsa-Metlaoui (Tunisie). Application à la compréhension des mécanismes de la phosphatogenèse'. Thèse de Doctorat ès-Sciences, Université d'Orléans, Orléans, France.

Belayouni, H. & Trichet, J. (1980). Glucosamine as a biochemical marker for dinoflagellates in phosphate sediments. In *Advances in Organic Geochemistry 1979*, ed. A.G. Douglas & J.R. Maxwell, pp. 205–10. Pergamon, Oxford.

Belayouni, H. & Trichet, J. (1983). Preliminary data on the origin and diagenesis of the organic matter in the phosphate basin of Gafsa (Tunisia). *Advances in Organic Geochemistry 1981*, ed. M. Bjorøy *et al.*, pp. 328–35. Pergamon, Oxford.

Belayouni, H. & Trichet, J. (1984). Hydrocarbons in phosphatized and non-phosphatized sediments from the phosphate basin of Gafsa. *Organic Geochemistry*, **6**, 741–54.

Blondeau, R. (1986). Comparison of soil humic and fulvic acids of similar molecular weight. *Organic Geochemistry*, **9**, 47–50.

Brannon, C.A. & Sommers, L.E. (1985a). Preparation and characterization of model humic polymers containing organic phosphorus. *Soil Biology and Biochemistry*, **17**, 213–19.

Brannon, C.A. & Sommers, L.E. (1985b). Stability and mineralization of organic phosphorus incorporated into model humic polymers. *Soil Biology and Biochemistry*, **17**, 221–7.

Bremner, J.M. (1967). Nitrogenous compounds. In *Soil Biochemistry*, ed. A.D. Mc Laren, G.H. Peterson & M. Dekker, pp. 19–58. M. Dekker, New York.

Brown, F.S., Baedecker, M.J., Nissenbaum, A. & Kaplan, I.R. (1972). Early diagenesis in a reducing fjord, Saanich Inlet, British Columbia, III: changes in organic constituents of sediment. *Geochimica Cosmochimica Acta*, **36**, 1185–203.

Cunningham, R. & Burnett, W.C. (1985). Amino acid biogeochemistry and dating of offshore Peru–Chile phosphorites. *Geochimica Cosmochimica Acta*, **49**, 1413–19.

Debyser, Y. & Gadel, F. (1977). Etude géochimique des composés humiques et des Kérogènes. In *Géochimie Organique des Sédiments Marins Profonds, ORGON I, Mer de Norvège*, pp. 247–68. CNRS, Paris.

Debyser, Y., Gadel, F., Leblond, C. & Martinez, M.J. (1978). Etude des composés humiques, des kérogènes et de la fraction hydrolysable dans les sédiments, In *Géochimie Organique des Sédiments Marins Profonds, ORGON II*, pp. 339–56. CNRS, Paris.

Degens, E.T. & Mopper, K. (1976). Factors controlling the distribution and early diagenesis of organic material in marine sediments. In *Chemical Oceanography*, 6, 2nd edn, ed. J.P. Riley & R. Chester, pp. 60–110. Academic Press, New York.

Enders, C. & Theis, K. (1938) Die Melanoidine und ihre Beziehung zu den Huminsäuren, *Brennstoff Chemie*, **19**, 360–439.

Fauconnier, D. & Slansky, M. (1980). Relation entre le développement des dinoflagellés et la sédimentation phosphatée du bassin de Gafsa (Tunisie). In *Géologie Comparée des Gisements de Phosphates et de Pétrole*, pp. 185–204. Documents du BRGM, 24, Orleans, France.

Garcia-Ramos, J.V. & Carmona, P. (1982). The effect of some homopolymers on the crystallization of calcium phosphates. *Journal of Crystal Growth*, **57**, 336–42.

Huc, A.Y. (1973) 'Contribution à l'étude de l'humus marin et de ses relations avec les kérogènes'. Thèse de Doctorat d'Ingénieur-Docteur, Université de Nancy, France.

Huc, A.Y. & Durand, B. (1974). Etude des acides humiques et de l'humine de sédiments récents considérés comme précurseurs des kérogènes. In *Advances in Organic Geochemistry 1973*, ed. B. Tissot & F. Bienner, pp. 53–72. Paris.

Huc, A.Y. & Durand, B. (1977). Occurrence and significance of humic acids in ancient sediments, *Fuel*, **56**, 73–80.

Ishiwatari, R. (1970). Structural characteristics of humic substances in recent lake sediments. In *Advances in Organic Geochemistry 1966*, ed. G.D. Hobson & G.C. Speers, pp. 283–311. Pergamon, Oxford.

Lean, D. (1972). Radiosotope studies showing the role of phosphorus in the formation of a large molecular weight organic compound present in lake water, In *Proceedings of the International Meeting on Humic Substances 1973*, pp. 159–70.

Levesque, M. (1969). Characterization of model and soil organic matter metal-phosphate complexes, *Canadian Journal of Soil Science*, **49**, 365–73.

Levesque, M. & Schnitzer, M. (1967). Organo-metallic interactions in soils, 5, Preparation and properties of fulvic acid-metal-phosphates, *Soil Science*, **103**, 183–90.

Maillard, L.C. (1912). Action des acides aminés sur les sucres: formation des mélanoïdines par voie méthodique. *Comptes Rendus de l'Académie des Sciences, Paris*, **154**, 66–8.

Maughan, E.K. (1980). Relation of phosphorites, organic carbon and hydrocarbons in the Permian Phosphoria formation, western United States of America. In *Géologie Comparée des Gisements de Phosphates et de Pétrole*, pp. 63–91. Documents du BRGM, 24, Orleans, France.

Miller, G.H. & Hare, P.E. (1980). Amino acid geochronology: integrity of the carbonate matrix and potential of molluscan fossils. In *Biogeochemistry of Amino Acids*, ed. P.E. Hare *et al.*, pp. 415–43. John Wiley and Sons, New York.

Mitterer, R.M. (1971). Influence of natural organic matter on $CaCO_3$ precipitation. In *Carbonate Cements*, ed. O.P. Bricker, pp. 252–8. Johns Hopkins University Press, Studies in Geology, 19, Boston.

Mitterer, R.M. (1972). Biogeochemistry of aragonite mud and oolites. *Geochimica Cosmochimica Acta*, **36**, 1407–22.

Morris, R.J. & Calvert, S.E. (1977). Geochemical studies of organic rich sediments from the Namibian shelf. I – The organic fractions. In *A Voyage of Discovery; George Deacon 70th Anniversary Volume*, ed. M. Engel, pp. 647–65. Pergamon, Oxford.

Müller, P.J. & Suess, E. (1977). Interaction of organic compounds with calcium carbonate. III – Amino-acid composition of sorbed layers. *Geochimica Cosmochimica Acta*, **41**, 941–9.

Nissenbaum, A. (1979). Phosphorus in marine and non-marine humic substances. *Geochimica Cosmochimica Acta*, **43**, 1973–8.

Nissenbaum, A. & Kaplan, I.R. (1972). Chemical and isotopic evidence for the *in situ* origin of marine humic substances. *Limnology and Oceanography*, **17**, 570–82.

Pelet, R. (1983a) Preservation and alteration of present day sedimentary organic matter. In *Advances in Organic Geochemistry 1981*, ed. Bjorøy *et al.*, pp. 241–50. Pergamon, Oxford.

Pelet, R. (1983b). Connaissance de la sédimentation organique actuelle et récente: vue d'ensemble sur les missions ORGON. In *Géochimie Organique des Sédiments Marins. D'Orgon à Misedor*, pp. 453–80. CNRS, Paris.

Poutanen, E.L. & Morris, R.J. (1983) The occurrence of high molecular weight humic compounds in the organic-rich sediments of the Peru continental shelf. *Oceanologica Acta*, **6(1)**, 21–8.

Powell, T.G., Cook, P.J. & McKirdy, D.M. (1975). Organic geochemistry of phosphorites, relevance to petroleum geochemistry. *Bulletin of the American Association of Petroleum Geologists*, **59**, 618–32.

Rachidi, M. (1983). Contribution à la connaissance de la matière organique de certaines séries phosphatées'. Thèse de Doctorat de Spécialité, Université d'Orléans, Orleáns, France.

Rashid, M.A. & King, L.H. (1970). Major oxygen containing functional group in humic and fulvic acid fractions isolated from contrasting marine environments, *Geochimica Cosmochimica Acta*, **34**, 193–202.

Romankevich, Ye. A & Baturin, G.N. (1972). Composition of the organic matter in phosphorites from the continental shelf of South West Africa. *Geokhimia*, **6**, 719–26. (English translation, *Geochemistry International*, **9**, 464–70).

Samb, M. (1983). 'Typologie et caractérisation physico-chimique de particules phosphatées. Application à la série phosphatée Eocène de Casamance (Sénégal)'. Thèse de Docteur-Ingénieur, Université d'Orléans, Orléans, France.

Sandstrom, M.W. (1980). Organic geochemistry of a Cambrian phosphorite. *Advances in Organic Geochemistry 1979*, ed. A.G. Douglas & J.R. Maxwell, pp. 123–31. Pergamon, Oxford.

Sandstrom, M.W. (1982). 'Organic geochemistry of phosphorites and associated sediments'. Unpublished. Ph.D. thesis, Australian National University, Canberra.

Sandstrom, M.W. (1986). Proterozoic and Cambrian phosphorites-specialist studies: geochemistry and organic matter in Middle Cambrian phosphorites from the Georgina Basin, north-eastern Australia. In *Phosphate Deposits of the World*, vol. I, ed. P.J. Cook, & J.H. Shergold, pp. 268–79. Cambridge University Press, Cambridge.

Sassi, S. (1974). 'La sédimentation phosphatée au Paléocène dans le sud et le centre-ouest de la Tunisie'. Thèse de Doctorat es-Sciences, Université de Paris-Orsay, Paris.

Schnitzer, M. & Khan, S.U. (1972). *Humic Substances in the Environment*. M. Dekker, New York.

Suess, E. (1970). Interaction of organic compounds with calcium carbonate. I. Association phenomena and geochemical implications. *Geochimica Cosmochimica Acta*, **34**, 157–68.

Suess, E. (1973). Interaction of organic compounds with calcium carbonate. II. Organo-carbonate association in recent sediments. *Geochimica Cosmochimica Acta*, **37**, 2435–47.

Trichet, J. (1968). Etude de la composition de la fraction organique des oolites. Comparaison avec celle des membranes des bactéries et des cyanophycées. *Comptes Rendus des Séances de l'Académie des Sciences, Paris*, **267**, 1492–4.

Trichet, J. (1972). Etude du mécanisme de la nucléation du carbonate de calcium dans des dépôts algaires. Lien avec la biogéochimie comparée de quelques types de dépôts dus à des cyanophycées (Polynésie Française). *Proceedings of the 24th International Geological Congress, Montréal*, **7**, 631–8.

Vincent, V. & Boischot, P. (1938). Nature et composition de la matière organique des phosphates de l'Afrique du Nord. Etude du phosphate de Gafsa. *Comptes Rendus des Séances de l'Académie des Sciences, Paris*, **24**, 1247–9.

8

Neogene to Holocene phosphorites of Australia

P.J. COOK AND G.W. O'BRIEN

Abstract

Numerous minor phosphorite occurrences are present throughout the onshore Neogene sequences of southeastern Australia. In addition, minor phosphorite (probably reworked from the underlying Cretaceous) also occurs within Neogene sediments in Western Australia. The east Australian occurrences are predominantly nodular and have been tentatively assigned ages ranging from Middle Miocene to Early Pliocene.

Many of these phosphorites are associated with a Late Miocene–Early Pliocene hiatus. However, since the ages of these nodules are not well-constrained, it is uncertain whether these phosphorites formed during the Middle Miocene and were subsequently reworked during the hiatus, or whether they actually formed on the hiatal surface during the Late Miocene–Early Pliocene. In general, both the origin and the distribution of the onshore Neogene phosphate occurrences are poorly understood.

Nodular phosphorites and phosphatic sediments have also been documented from several areas around the Australian continental margin. The most significant seafloor occurrence appears to be located off northern New South Wales, where an almost continuous veneer of phosphatic nodules mantles the outer continental shelf and upper slope between 29–32° S. These nodules fall into two distinct categories: ferruginous and non-ferruginous. Ferruginous nodules typically occur on the outer shelf and have been recovered at water depths ranging from 197 to 241 m. They are relict (> 800 ky), well-indurated, goethite-rich (10–50% Fe_2O_3), and are often conglomeratic; many may be as old as Middle Miocene.

In contrast, non-ferruginous nodules are restricted to the upper slope and generally occur in water depths ranging from 365–455 m. These nodules lie within unconsolidated, glauconitic, foraminiferal sands, are usually friable, and contain very little or no goethite. Uranium-series isotopic data show that they range in age from Holocene to Late Pleistocene. Both the non-ferruginous nodules and the associated sediments have low ($\leqslant$0.5%) total organic carbon concentrations. In addition, oceanographic and sedimentological data suggest that neither coastal upwelling nor enhanced biological productivities were important factors in the origin of these Quaternary phosphorites.

Neogene phosphatic nodules occur on the western Tasmanian shelf. These nodules are commonly slightly ferruginous and occur as a lag within bryozoan-rich sands. The distribution of the nodules appears to be very patchy over much of the shelf. Slightly phosphatic (< 3% P_2O_5) sediments also occur on the northwestern Australian shelf. There the phosphate occurs predominantly as phosphate grains, angular fragments and as phosphatic internal molds of organisms, with the zone of P_2O_5 enrichment loosely following the 180 m isobath between 13–18° S. Minor phosphatic nodules and crusts of Neogene age have also been found on seamounts in the Tasman Sea, and on the South Tasman Rise.

While both the onshore Neogene and offshore Neogene–Holocene phosphorite occurrences can be areally extensive, they are characteristically thin and of low grade. In addition, offshore Neogene phosphorites are typically highly ferruginous. As such, the potential for the discovery of economic phosphorite deposits within this part of the sedimentary section in Australia appears to be low.

Introduction

Australia is a major importer of phosphate rock. Christmas Island and Nauru are the major suppliers at this time, with some imports from Florida, North Africa and the Middle East. Because of this dependency on imports, numerous phosphate exploration programs have been carried out within Australia. This resulted in the discovery of major phosphate resources in Cambrian sediments of the Georgina Basin in northern Australia in 1966 (Russell, 1967). This major (though as yet uneconomic) discovery diverted attention from younger sequences in southern Australia, where there are a number of phosphate occurrences. The most significant Neogene occurrences known to date are those of the onshore basins of Victoria. In addition, there are fairly extensive phosphorite occurrences offshore from western Tasmania and northern New South Wales.

In this paper we will briefly review these occurrences of Neogene–Holocene phosphorites. Mention will be made of some of the insular guano deposits, though these are almost certainly of Quaternary rather than Neogene age. The location of these guano deposits may in part be related to areas of high marine productivity during the Quaternary and, as such, are possibly relevant to the occurrence of phosphorites. Onshore Neogene cave deposits are not discussed.

Neogene basins

The Tertiary was a time of deep weathering in Australia. Most of the onshore sedimentary basins contain little more than a thin veneer of Neogene continental sediments. Onshore Neogene marine sediments are found only in the Perth Basin and the southeastern basins (such as the Murray and Otway Basins) (Fig. 8.1). Thin Neogene marine sequences probably occur in

Fig. 8.1. Australian sedimentary basins containing Neogene marine sediments.

most offshore areas, with the exception of the Gulf of Carpentaria region, where lacustrine and continental sediments predominate.

The thickest Australian Neogene marine sequences probably occur in the basins offshore from northwest Western Australia (Fig. 8.1) where calcareous sediments predominate. There are no confirmed reports of Neogene phosphorites, apart from isolated samples dredged from the outer shelf and upper slope. Phosphate rock has been reported from a number of the islands in this area, including several where the phosphate was mined such as Ashmore Reef, the Lacepede Islands and the Houtman–Albrolhos Islands (Fig. 8.2). Whilst there are reports of 'phosphatised rock' (Hutchinson, 1950), there seems little doubt that all of these insular deposits are guano or guano-derived.

The Western Australian Neogene sequence thins considerably, contains many hiatuses, and becomes less calcareous in the Perth Basin. Phosphate nodules occur within the Pliocene Ascot Beds (which unconformably overlie Cretaceous–Eocene sediments) in the Cooljarloo area (Fig. 8.2). The Ascot Beds are a thin (2 m), shallow, marine bioclastic sandy calcarenite. The phosphate occurs in the form of nodules, with phosphate contents of up to 26% P_2O_5 (Baxter & Hamilton, 1980). The nodules contain *Inoceramus* prisms, together with foraminifera and radiolarians of Cretaceous age. As the Ascot Beds are fairly confidently dated as Pliocene (Kendrick, 1980), it is likely that the nodules have been reworked from the underlying Cretaceous sediments which are known to be phosphatic in parts of the Perth Basin (Matheson, 1948). In the south of the continent, in the Eucla and adjacent basins, thin carbonates once more predominate. Very minor occurrences of phosphate rock of guano origin have been reported from some of the islands in this area such as Marum, Bickers, and Brothers Islands east of Port Lincoln and the Recherche Archipelago (Fig. 8.2). Some of these were able to support small-scale mining operations in the last century. There are no reports of phosphorites within the Neogene sequence.

The onshore basins of southeastern Australia (Fig. 8.1) also contain a number of phosphorite occurrences; these occurrences will be discussed in some detail later. Offshore, a number of phosphorite occurrences have been reported off the western and southern coasts of Tasmania, on the South Tasman Rise and in the Tasman Sea. Onshore, there are no reported occurrences of Neogene marine phosphorites in South Australia, Tasmania,

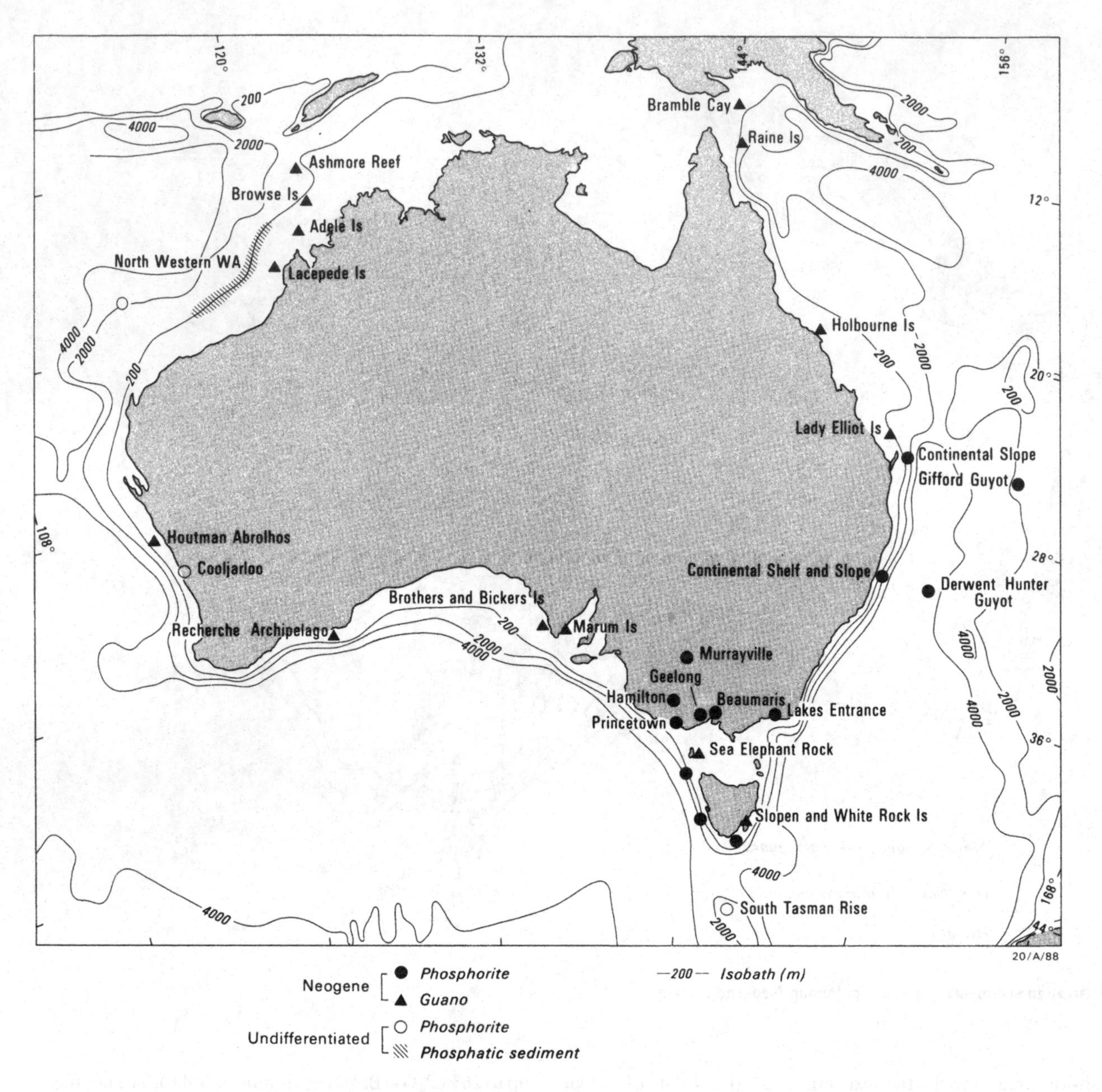

Fig. 8.2. Distribution of Neogene–Modern phosphorites and guano deposits.

New South Wales or Queensland. A few minor guano occurrences have been reported by Twelvetrees (1917) and Hutchinson (1950) off the southeast coast of Tasmania.

Along most of the New South Wales and southern Queensland coast, the offshore Neogene sequence is likely to be thin. Thicker sequences are developed in the basins of north Queensland such as the Capricorn Basin, the Townsville Trough and the Queensland Trough (Fig. 8.1). The only drilling carried out in this area has been in the Capricorn Basin, and no phosphorites were reported. Significant occurrences of insular phosphate rock (including phosphatized coralline limestone), all of which were mined on a small scale in the last century, occur off the Queensland coast (Fig. 8.2). These include the deposits at Raine, Lady Elliot, and Holbourne Islands and Bramble Cay. All of these appear to be accumulations of guano, or rock phosphatized by the remobilization of guano-derived phosphate.

Phosphorites are therefore relatively uncommon in the Neogene basins of Australia. The only significant occurrences are in southeastern Australia (especially Victoria) and in the adjacent offshore areas. These occurrences will now be considered in detail.

Neogene phosphorites of Victoria

In the Late Jurassic–Early Cretaceous, rifting between Australia and Antarctica produced an easterly-trending rift (the Bassian Rift) and a series of extensional basins. Sedimentation in these basins was initially dominated by volcaniclastics but, as break-up advanced, normal marine conditions extended through the basins along the southern margin of the continent. By the Late Mesozoic–Early Tertiary, three major basins – the Otway, Bass and Gippsland Basins – had developed along the southern margin of Victoria. Between the Otway and Gippsland Basins localized subsidence formed a series of small basins, notably the Torquay, Port Phillip and Western Port Sub-Basins

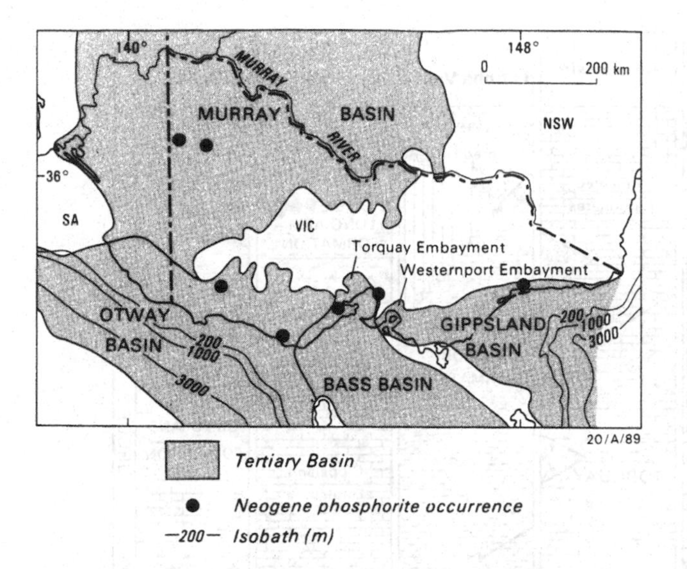

Fig. 8.3. Neogene sedimentary basins of southeastern Australia.

or Embayments (Fig. 8.3). The Torquay Sub-Basin and the Sorrento Graben may represent a failed arm of the Otway rift system (Gunn, 1975). The northwest corner of the State lies within the large, but comparatively shallow, intracratonic Murray Basin. The Murray Basin is structurally distinct from the southern basins.

Phosphorites occur in the Neogene portion of all of these basins (Fig. 8.3). Whilst these occurrences are relatively minor, they extend from the Hamilton area in the west to Lakes Entrance in the east, a distance of over 500 km. Carter (1978 a, b and 1985) has suggested that these extensive occurrences can be related to a Late Miocene–Pliocene eustatic event.

The generalized Neogene stratigraphy of the Victorian basins is shown in Figure 8.4. There are three features to note: The dominance of carbonates in the first half of the Neogene; a widespread hiatus in the Upper Miocene and possibly extending into the Pliocene; and the association of phosphatic sediments with this hiatus.

Murray Basin

The Neogene sediments of the Murray Basin are characterized by a series of complex facies changes (Fig. 8.5) that can be related to eustatic sea-level changes (Brown, 1983). Terrigenous clays are common in the east of the basin whereas carbonates predominate in the western, more marine portion of the basin. The latest Oligocene–Middle Miocene Murray Group consists of several hundred meters of limestones deposited during the Late Oligocene–Miocene sea-level rise documented by Vail, Mitchum & Thompson (1977). To the east, this sequence gives way to a series of predominantly fine-grained terrigenous sediments, notably the Geera clay and the Olney Formation. Carter (1985) reports that the Late Bairnsdalian and Mitchellian Zones of the Upper Miocene appear to be missing and suggests, because of the lack of clear evidence of erosion, that the Late Miocene was a time when there was a marine but non-depositional environment in the Murray Basin.

The limestones of the Murray Group and its correlatives are overlain by the most extensive marine unit of the Basin, the Bookpurnong Formation and its equivalents. This is a thin (ca. 30 m) clayey interval containing Early Pliocene fossils that locally overlies the Murray Group with a well defined disconformity marked by 'an eroded, ferruginised and barnacle-encrusted surface' (Carter, 1985). This description is reminiscent of a hardground surface and, significantly, it is at this stratigraphic level that phosphorites occur. The Bookpurnong Beds are overlain by the Loxton–Parilla Sand (up to 90 m thick). More detailed descriptions of these units and their correlatives are given by Abele *et al.* (1976), Brown (1983) and Carter (1985). The main points to note here are that, in the western part of the basin (where the phosphorites are known to occur), sedimentation during the Early–Middle Miocene consisted mainly of carbonates. This was followed during the Late Miocene–Early Pliocene by a period of little or no sedimentation, apart from the local deposition of minor phosphorites. Hardgrounds may have developed during this time. In the Early Pliocene, the onset of marine sedimentation was marked by the deposition of an extensive marine clay with a slightly phosphatic unit at the base. Whether the phosphatic interval was related to reworking during the Late Miocene–Early Pliocene hiatus, or deposition at the Early Pliocene transgression cannot be determined from the limited information currently available.

In the Murrayville area of the Basin, phosphatic sediments have been intersected in a number of bores. The phosphatic interval is reported by Chapman (1916) to consist of thin (2–9 m) sands composed in part of phosphatic and glauconitic grains, with numerous small fish bones and teeth. Phosphate nodules are also present in places. There is strong evidence that this phosphatic interval is also located at the base of the Bookpurnong Beds. Because of lack of outcrop and poor subsurface information, the extent of this phosphatic unit in this part of the Murray Basin is uncertain.

Otway Basin

The Otway Basin is separated from the Torquay Embayment of the Bass Basin by the Otway Ranges High and from the Murray Basin by Palaeozoic rocks of the Western Highlands. Within the basin is a series of 'highs' and 'embayments' which had a significant effect on sedimentation. The generalized Neogene stratigraphy of the basin is shown in Figure 8.6. The Miocene sequence is composed predominantly of carbonates of the Heytesbury Group, particularly the Port Campbell Limestone and the Gellibrand Marl. At the base of the group, however, the Late Oligocene Clifton Formation comprises 12 m of limonitic sandstones and calcarenites with a bed of limonitic nodular phosphate in the middle of the unit (Baker, 1945; Abele *et al.*, 1976).

The Heytesbury Group is overlain by the Moorabool Viaduct Formation (and its correlatives), with a major hiatus between the two. The Moorabool Viaduct Formation, of Late Miocene–Pliocene age, consists of ferruginous and calcareous sands with minor gravels. The base of the Formation is marked by a phosphatic nodule bed about 5 cm thick; a second nodule bed occurs about 1.5 m above the base. The Pliocene sequence is terminated by a major phase of Late Pliocene volcanism which resulted in the deposition of extensive basalts over much of western Victoria.

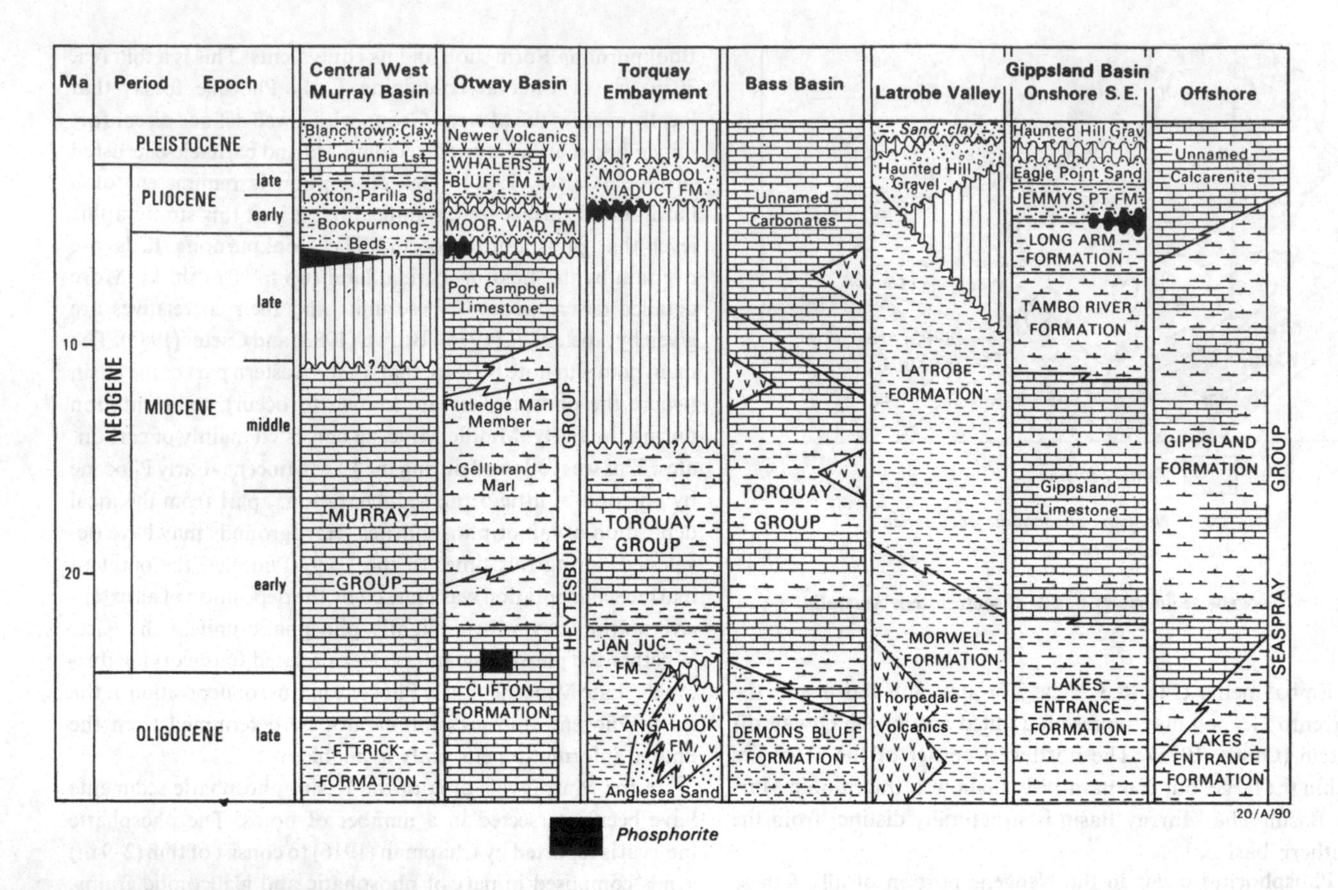

Fig. 8.4. Generalized Neogene stratigraphy of Victoria (after Wilford & Langford, pers. comm.)

Phosphorites are only known from a limited number of localities within the Otway Basin, notably in the Hamilton area near the northern margin of the basin, at Minhamite 40 km southeast of Hamilton, and near Princetown on the eastern margin of the basin. At Hamilton, phosphatic nodules occur within a calcareous clay disconformably overlying limestones and clays of Early and Middle Miocene age (Carter, 1978a, b). The phosphatic unit contains abundant *remanié* foraminifera derived from the underlying Miocene units. At Minhamite, phosphate nodules occur in the basal 40 cm of a calcareous clay which disconformably overlies Middle Miocene limestones and clays. Again, there are Middle Miocene *remanié* foraminifera within the nodule bed.

Only the Princetown occurrences have been documented in any detail (Baker, 1945). The Princetown nodules were first described by Wilkinson (1865) whose studies are amongst the earliest accounts of a phosphorite. In his description of the Princetown occurrences, Baker (1945) documents a nodule bed 30–90 cm thick, composed of nodules ranging in length from 1 to 30 cm. The nodules range from ovate and smooth to highly irregular and are composed mainly of dense collophane but with some included detrital grains and fossil fragments. In a single analysis of a nodule, Baker records a P_2O_5 content of approximately 15%. Fossils are relatively common within the nodule bed and include corals, pelecypods, gastropods, echinoderms, cetacean bones and shark teeth. The nodule bed is located immediately above a diastem and overlying a prominent limestone. However, Baker assigned an identical age (Janjukian–Miocene) to both the nodule bed and the underlying limestone and concluded from this that, unlike the Geelong occurrences of the Torquay Embayment (see later), the Princetown nodules were not reworked from the underlying limestone but had formed *in situ*.

Bass Basin (Torquay Embayment)

The Torquay Embayment and the adjacent Port Philip Basin have been variously regarded as part of the Otway, Gippsland and Bass Basins. Here they are regarded as constituting the onshore northern margin of the predominantly offshore Bass Basin. The Neogene sequence comprises the upper part of the Torquay Group and the Moorabool Viaduct Formation (Fig. 8.6), with a hiatus between these two units. The Late Oligocene–Early Miocene part of the Group is made up of the 25–90 m of carbonates, including the Point Addis Limestone; a poorly-bedded partially-cemented fossiliferous grainstone–packstone and the Jan Juc Marl, a mixed skeletal grainstone–fossiliferous glauconitic mudstone (Reeckmann, 1979; Link & Thompson, 1985). These carbonates are in turn overlain by a series of thin units, including the massive fossiliferous Puebla Clays (up to 30 m thick), which are followed by the Cellepora Beds (> 15 m of carbonate muds and silts), the Zeally Limestone (> 15 m of bedded fossiliferous grainstones) and finally the Miocene Yellow Bluff Beds which consist of > 14 m of bioturbated thinly interbedded packstones and claystones with echinoderms and bryozoans.

This upper unit of the Torquay Group is disconformably

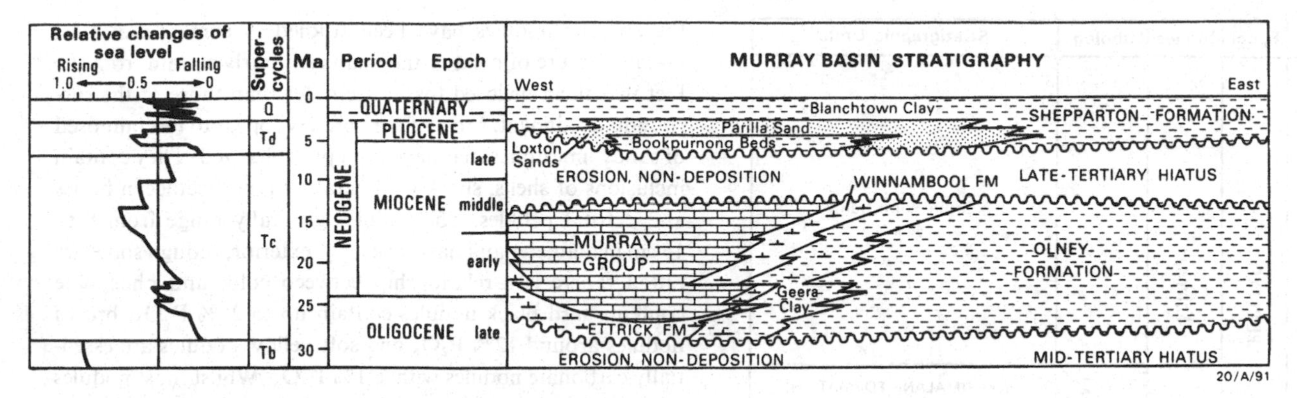

Fig. 8.5. Neogene stratigraphy of the Murray Basin (after Brown, 1983).

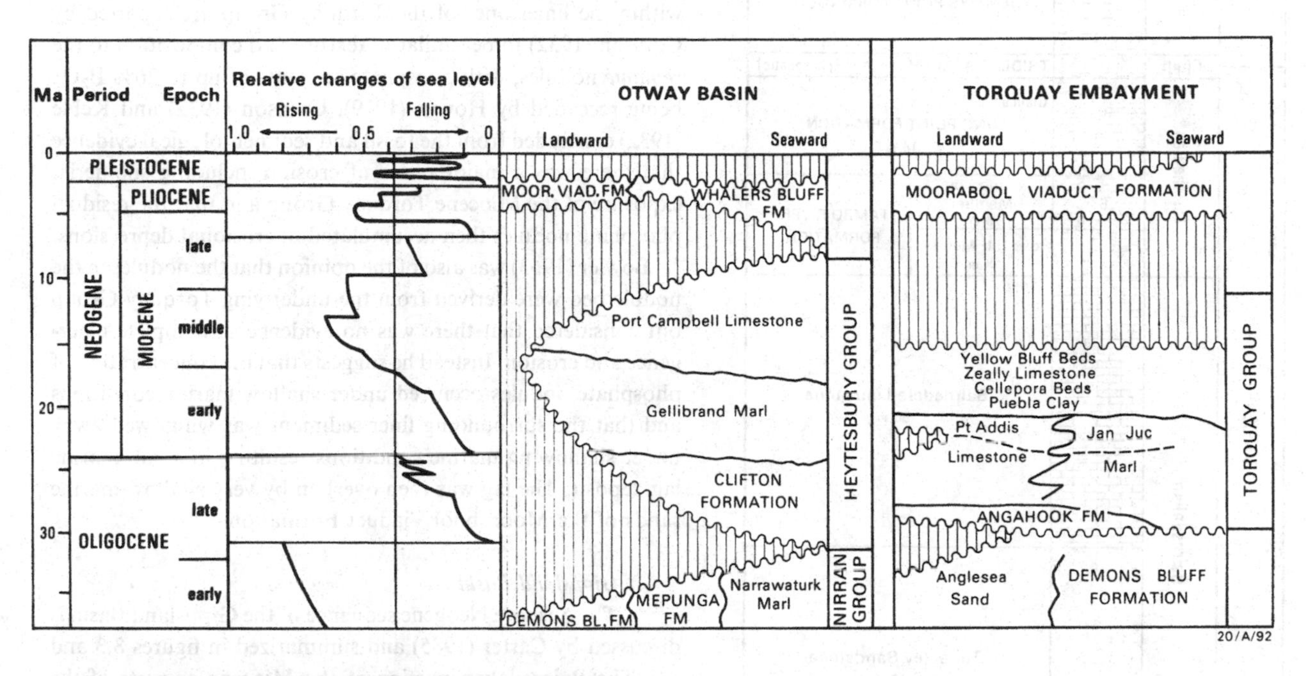

Fig. 8.6. Neogene stratigraphy of the Otway Basin and the Torquay Embayment (after Reeckmann, 1979).

overlain by the Moorabool Viaduct Formation of Late Miocene–Pliocene age, comprising sub-tidal and intertidal calcareous sands and calcarenites (Abele *et al.*, 1976). Phosphatic nodules occur on the disconformity surface at the base of the Moorabool Viaduct Formation.

To the east of Torquay, on the Nepean Peninsula, a somewhat thicker Neogene section is preserved in the Sorrento Graben (Mallett & Holdgate, 1985) and the hiatus at the top of the Torquay Group equivalent (the Fyansford Formation) is much less evident. Within the Sorrento Graben, the upper part of the Neogene is represented by the Brighton Group, followed disconformably by the Wannaeue Formation, with a total thickness of about 120 m. A glauconite horizon at the top of the Fyansford Formation (i.e. equivalent to the top of the Torquay Group) is correlated by Mallett & Holdgate (1985) with a glauconitic shelly sand containing vertebrate remains and phosphate nodules. The microfauna in this unit are taken by Mallett & Holdgate to indicate a Late Miocene age within the *Globorotalia acostaensis* or *G. conomiozea* Zones. At Beaumaris, the boundary between the Fyansford Formation and the overlying Brighton Group is marked by a disconformity and the phosphatic nodule bed. However, no phosphate nodules have been encountered in any of the Nepean Peninsula bores.

In their interpretation of the Nepean Peninsula sequence, Mallett & Holdgate (1985) suggest that

> Throughout the Middle and Late Miocene, lithologies of the Fyansford Formation reflect overall shallowing with changes from clays to shelly sands and coquinas. No distinct changes occur in the Middle Miocene but there is a transition to sands at about the beginning of the Late Miocene. This corresponds with a global sea-level fall of Vail *et al.* (1977). Within the Late Miocene, calcarenites, coquinas nodule beds, quartzose marine sands of the lower Brighton Group and probably the first of the non-marine upper Brighton Group accumulated. These changes are interpreted as due to shallowing resulting first in coarsening of sediment, with glauconite, reworked materials and coquinas. This was succeeded by a phase of very slow deposition (with erosion of some of the section in

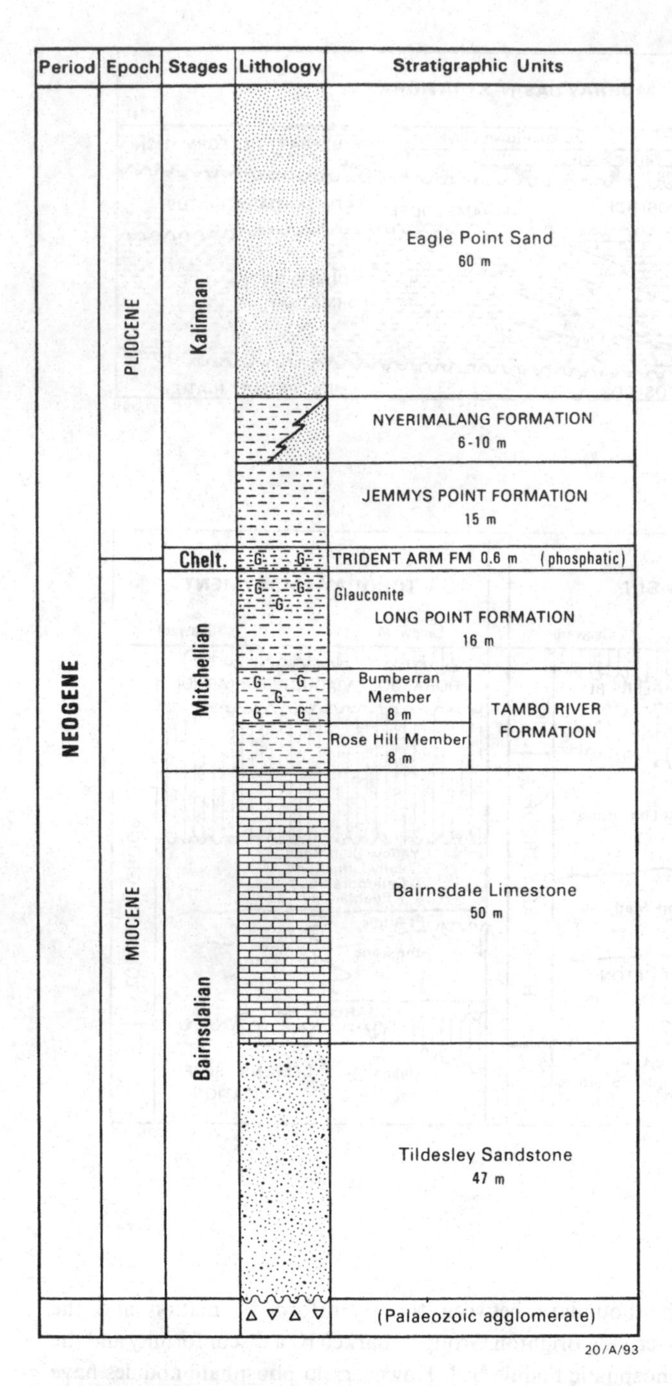

Fig. 8.7. Neogene stratigraphy of the eastern part of the Gippsland Basin (after Carter, 1985).

places) during which nodule and bone beds accumulated. Prograding littoral deposits covered the nodules and were succeeded by non-marine sands and clays.

The phosphatic nodules of the Geeong area have been described by Coulson (1932), Keble (1932) and Bowler (1963). In the first detailed discussion on the nodules, Coulson (1932) recognized two types of nodules – '*remanié* and concretionary'. The concretionary nodules occur within the carbonates of the Torquay Group, the *remanié* nodules in a thin (10 cm) bed on the disconformity surface between the top of the Torquay Group and the base of the Moorabool Viaduct Formation. Little information is available on the concretionary nodules. In contrast, the *remanié* nodules have been studied in some detail. The *remanié* nature of most of the nodules is fairly certain from the fact that their enclosed fossils are older than those of the surrounding sediment. Most of the nodules appear to be composed of sandy limestones that have been phosphatized. They contain inclusions of shells, shark teeth, crab remains, cetacean bones and detrital pebbles. The nodules generally range from 8 to 15 cm in diameter and have a smooth exterior, though some are pitted. There is a relationship between color and phosphate content; hard black nodules contain up to 25% P_2O_5, brown nodules around 12% P_2O_5 and soft yellow nodules are essentially carbonate nodules with $\leqslant 1\%$ P_2O_5. Whilst most nodules appear to have a fairly uniform phosphate content, a few show concentric banding. The concretionary phosphate nodules within the limestones of the Torquay Group are reported by Coulson (1932) to be similar in texture and composition to the *remanié* nodules, with phosphate contents of up to 26% P_2O_5 being recorded by Howitt (1919). Coulson (1932) and Keble (1932) concluded from the fossil and sedimentological evidence that there was a major phase of erosion including sub-aerial exposure of the Miocene Torquay Group and that the residual phosphate nodules then accumulated in erosional depressions.

Bowler (1963) was also of the opinion that the nodules in the nodule bed were derived from the underlying Torquay Group but considered that there was no evidence of complete emergence and erosion. Instead he suggests that the concentration of phosphate nodules occurred under shallow marine conditions and that the surrounding finer sediment was winnowed away under shallowing marine conditions resulting in a sub-marine lag deposit. The lag was then overlain by very shallow marine sands of the Moorabool Viaduct Formation.

Gippsland Basin

The onshore Neogene sequence of the Gippsland Basin is discussed by Carter (1985) and summarized in figures 8.3 and 8.7. The Bairnsdalian portion of the Miocene consists of the Tildesley Sandstone (a coarse ferruginous quartz sandstone of mainly non-marine affinities) followed by the marine Bairnsdale Limestone. Carter (1985) suggests that within the Bairnsdale Limestone there is evidence of rising sea level up to about Mid-Miocene time, followed by a lowering of sea level. The regression continued throughout the remainder of the Miocene, though conditions remained predominantly marine. Glauconite is especially common in the Tambo River Formation and relatively common towards the top of the Long Point Formation. Overlying this unit is a thin (0.6 m) phosphatic nodule bed, the Trident Arm Formation, which is interpreted by Carter as having formed at the time of maximum regression; bryozoan-encrusted nodules in the nodule bed are taken as evidence for a period of non-deposition.

The Trident Arm Formation is overlain by the Early Pliocene Jemmy's Point Formation, a partly marine sandy and argillaceous – partly calcareous unit, which is in turn overlain by probable lagoonal Pliocene sands of the Nyerimalong Formation and the Eagle Point Sand.

Comparison of the Gippsland section with other Neogene sections in southeastern Australia has lead Carter (1978b, 1979 and 1985) to the conclusion that the phosphatic nodule beds throughout this area are essentially synchronous and were deposited during a widespread latest-Miocene–earliest-Pliocene regression. He does, however, qualify this by pointing out that the nodules at some localities are abraded and encrusted by benthic organisms, and takes this as evidence that the period of maximum phosphate deposition was slightly before the time of maximum regression. However, Mallett & Holdgate (1985) query the correlation of the Gippsland and Bass (Beaumaris) nodules, pointing out that the nodules lie within the *G. puncticulata* zone in the Gippsland Basin and are therefore probably younger than those of the Bass Basin (Torquay Embayment), though possibly synchronous with the nodules of the Hamilton area of the Otway Basin.

Offshore phosphorites and phosphatic sediments

Marine sedimentary phosphorites and phosphatic sediments have been discovered in three distinct geographical provinces on the Australian continental shelf and upper continental slope. These provinces are off northern New South Wales, Tasmania and northwestern Western Australia. All of these phosphate occurrences were discovered between 1965 and 1968, a period when several private companies and the Australian Bureau of Mineral Resources (BMR) were actively exploring offshore for economic phosphorite occurrences. Phosphorite exploration was also carried out in the Great Australian Bight, off South Australia, and off the southern part of the west coast of Western Australia, though no phosphatic sediments were recovered (Noakes & Jones, 1973; Johns, 1976). In addition, sedimentary phosphate has been found on the South Tasman Rise, and on seamounts in the Tasman Sea.

The northern New South Wales occurrence has been moderately well documented. In contrast, only one paper (Slater & Goodwin, 1973) has dealt with the phosphate on the Tasman Sea seamounts. Similarly, the offshore Tasmanian phosphorites have only been described in company reports, in spite of the fact that this phosphate occurrence appears to have a moderately large areal extent. Phosphatic sediments from northwestern Western Australia have also received little study, though preliminary investigations (Jones, 1968; Noakes & Jones, 1973) suggest that the province may be extensive, but of very low grade.

Northern New South Wales

The earliest work on phosphorites from this area was by Loughnan & Craig (1962) who described phosphorites recovered from off Coffs Harbour and Port Macquarie as 'goethite and siderite cemented nodules'. In late 1966, Global Marine Inc. carried out a dredging cruise along the outer continental shelf and upper slope between Port Macquarie and Brisbane; phosphorite was recovered from several localities. These samples were subsequently described by von der Borch (1970) who subdivided the nodules into three types on the basis of their morphology. Group A nodules, recovered from a depth of 385 m off Yamba (29°23′ S, 153°50′E) are small (average diameter 3 cm), dull grey, friable, irregular concretions without a surface iron oxide gloss (Fig. 8.8*a*). These concretions are mostly composed of calcareous skeletal material (predominantly planktonic foraminifera), silt-sized angular quartz, glauconite grains, and rare goethite grains, set in a phosphatic matrix. Groups B and C nodules were recovered from shallower water (210–290 m) south of the Group A area, off Coffs Harbour. Group B nodules (Fig. 8.8*b*) are similar to Group A but are well-indurated, sub-rounded and glazed, with a fairly thick (>0.05 mm) coating of goethite. Group C nodules (Fig. 8.8*c*) are composed of cemented agglomerations of Group B nodules. On the basis of enclosed crab remains (*Ommatocarcinus corioensis*) and the assemblage of foraminifera in the ferruginous Groups B and C nodules, the phosphorites in this area were assigned a probable Middle Miocene age. Von der Borch (1970) proposed that the nodules originally formed during a period of relatively high organic productivity during the Miocene, and were subsequently eroded out of the Miocene strata and exposed and reworked on the seafloor. Group A nodules were considered to have experienced minimal reworking and exposure, whereas in the case of Group C nodules, exposure and reworking was prolonged. The P_2O_5 concentrations of the nodules were highly variable, with the highest being obtained for a Group A nodule (21.2% P_2O_5) whereas Groups B and C concretions ranged from 7.8 to 20.7% P_2O_5.

Following Global Marine's initial work, the BMR carried out marine geological surveys of the continental shelf off southern Queensland and northern New South Wales in 1970 and 1972 (Marshall, 1980). Ferruginous Groups B and C nodules were recovered from relict, non-depositional areas on the outer shelf between Coffs Harbour and Port Macquarie, at water depths ranging from 197 to 241 m. The petrology and geochemistry of these ferruginous nodules were subsequently described in detail by Marshall & Cook (1980), Cook & Marshall (1981) and Marshall (1983). No Group A nodules were recovered during either survey, probably because the sampling (which was predominantly at water depths of <300 m) did not extend far enough down the upper slope.

Preliminary uranium-series studies (Kress & Veeh, 1980) on the Global Marine Inc. samples had indicated that the Group A nodules might be predominantly of late Pleistocene age, rather than of Middle Miocene age as suggested by von der Borch (1970). In 1979, scientists from Flinders University collaborated with the Commonwealth of Scientific and Research Organization (CSIRO) (RV 'Sprightly') and the New Zealand Oceanographic Institution (NZOI) (RV 'Tangaroa') in separate cruises to the northern New South Wales continental margin; these cruises concentrated on delineating the occurrence of the Group A nodules on the upper slope. Dredge and grab samples, as well as piston and box cores, were taken at water depths ranging from 105 to 655 m. Subsequently, O'Brien & Veeh (1980) obtained Late Pleistocene–Holocene uranium-series ages for Group A nodules recovered from the upper continental slope between 29–31° S, at water depths ranging from 360 to 420 m. In contrast, the ferruginous Groups B and C nodules, which occur at depths of <300 m, invariably had ages in excess of the uranium-series dating limit (>250 000 y) and were considered to be of probable Middle Miocene age, as previously suggested from palaeontological evidence (von der Borch, 1970).

The depositional environment of the Holocene phosphorites on the East Australian continental margin is strikingly different from those of the modern phosphogenic provinces off Peru–

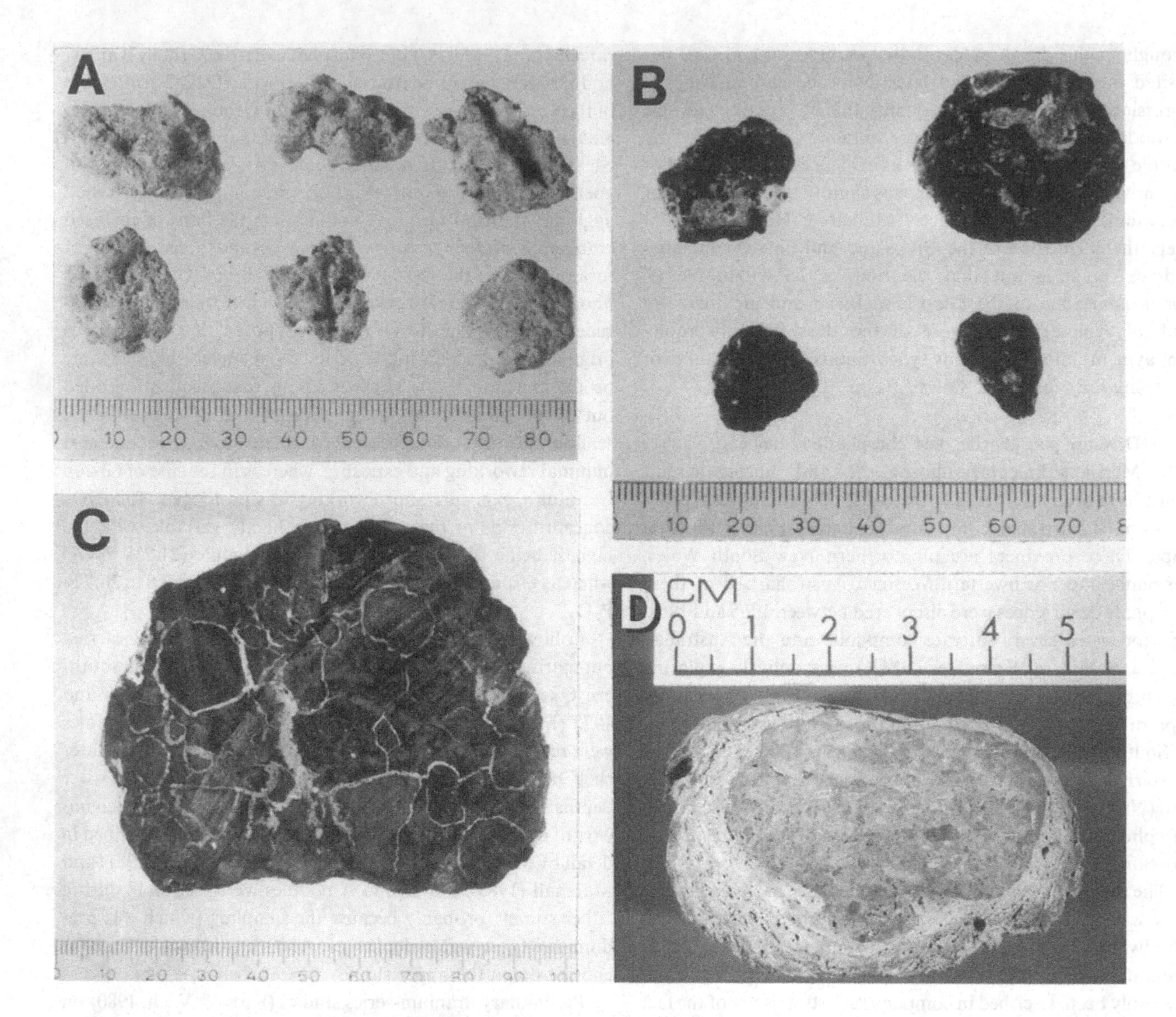

Fig. 8.8. Photographs of Neogene–Modern phosphatic nodules dredged from the continental shelf. (a) Earthy, friable, non-ferruginous phosphatic nodules from location G7 (water depth 385 m, Evans Head–Yamba region). Radiometric ages are Late Pleistocene–Holocene. (b) High ferruginous, polished, well-indurated phosphatic nodules encrusted with assorted calcareous organisms (location P915, water depth 350–420 m, Nambucca Heads region). Possibly Middle Miocene age. (c) Large, highly ferruginous, conglomeratic phosphatic nodule showing cemented agglomeration of smaller ferruginous phosphatic nodules (location G18, water depth 210 m, Coffs Harbour region). Probably of Middle Miocene age. (d) Phosphatic nodule (dark) with an encrusting rind (white) of calcareous algae from the northwest Tasmanian shelf.

Chile and Namibia. Both of these areas are characterized by an eastern boundary current, with strong, largely continuous coastal upwelling and very high associated organic productivity. The phosphorites form within organic-rich diatomaceous oozes, apparently from phosphate released during the bacterial degradation of organic matter (Baturin, 1971a, b; Atlas, 1975; Burnett, 1977). In contrast, the East Australian continental margin appears to be an area of only moderate seasonal upwelling and low biological productivities and, consequently, the organic carbon concentrations of the sediments are much lower. In view of the inferred (genetic) relationship between coastal upwelling–high organic productivity and marine phosphorites (Brongersma-Sanders, 1957; McKelvey, 1967; Baturin, 1971a, b), the formation of contemporary phosphorite would not be expected in a low productivity zone such as off eastern Australia, Indeed, the importance of this deposit lies in the fact that its origin cannot readily be ascribed to coastal upwelling and/or high organic carbon fluxes. It may thus provide clues to the processes operating in other phosphorite deposits that also formed in calcareous, relatively organic-poor sediments. The closest comparison to the offshore east Australia deposits are perhaps those off the southeastern United States. There, too,

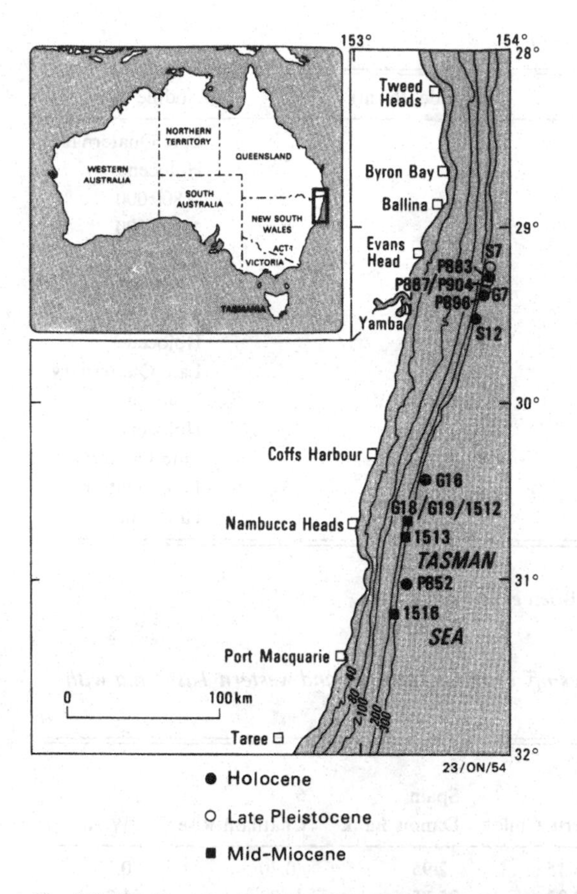

Fig. 8.9. The East Australian continental margin showing location of Holocene, Late Pleistocene and Middle Miocene phosphatic nodules. Ages obtained from uranium-series and palaentological data in von der Borch (1970), O'Brien & Veeh (1980), O'Brien *et al.* (1981), O'Brien & Veeh (1983) and O'Brien *et al.* (1986).

extensive phosphorites occur flanked by a western boundary current (the Gulf Stream) in an area not noted for its high organic productivity.

In the following discussion, phosphorites from the east Australian continental margin will be broadly subdivided into two types: (i) non-ferruginous, and (ii) ferruginous. Non-ferruginous nodules correspond to the Group A classification of von der Borch (1970) and the Type I of O'Brien & Veeh (1980), whereas the ferruginous nodules are equivalent to von der Borch's Groups B and C, and O'Brien & Veeh's (1980) Type II.

Non-ferruginous nodules

The non-ferruginous nodules occur on the upper slope between 29 and 31° S (see Table 8.1, Fig. 8.9) at water depths ranging from 350 to 455 m (O'Brien & Veeh, 1980; O'Brien *et al.*, 1981; O'Brien, 1982; O'Brien & Veeh, 1983; O'Brien *et al.*, 1986). They are typically friable to moderately consolidated, and occur within unconsolidated glauconitic, foraminiferal grainstones and packstones. Petrographic evidence suggests that the nodules, which are predominantly of packstone texture, have formed by the cementation of the associated sediments with cryptocrystalline apatite (O'Brien, 1983).

Average geochemical compositional data for the non-ferruginous nodules are presented in Table 8.2. Glauconite, clay minerals and quartz are relatively abundant in these nodules, and this results in moderately high average SiO_2 (24.0%), Al_2O_3 (3.53%), K_2O (0.89%) and Fe_2O_3 (5.00%) concentrations. The P_2O_5 concentrations are highly variable, but typically range between 7.0 and 15.0%, with an average of 11.0% P_2O_5. The high abundance of biogenic carbonate in the nodules has resulted in high CaO concentrations, and $CaO:P_2O_5$ ratios which are considerably greater than typical marine phosphorite values. Significantly, electron microprobe analysis of small areas (< 10 μm) of phosphatic matrix in the nodules has shown that the $CaO:P_2O_5$ ratios of the matrix average 2.05 (O'Brien & Veeh, 1980), still considerably higher than that of pure apatite (1.62; McClellan, 1980), and implying that fine grained biogenic calcite is intimately admixed with the apatite in the phosphatic matrix. The fluorine concentrations are moderately high (average 2.0%), and the $F:P_2O_5$ ratios are considerably higher than those of pure fluorapatite (0.089) (Rooney & Kerr, 1967). The apatite in the nodules contains a high amount of CO_2 (average 7%) (O'Brien, 1983), and the high $F:P_2O_5$ ratios are probably related to the coupled substitution of F and CO_3^{2-} for PO_4^{3-} (Gulbrandsen, 1966).

Compositionally, the non-ferruginous nodules most closely resemble phosphorites from the Agulhas Bank off South Africa, though the Agulhas Bank samples contain slightly more P_2O_5, but less Al_2O_3 and SiO_2 (i.e. alumino-silicates). They also have some compositional similarities to phosphorites from the Danois Bank off Spain, though the Spanish samples have a much higher Fe_2O_3 content, but lower SiO_2 and P_2O_5 concentrations.

Uranium-series isotopic studies of the non-ferruginous nodules have shown that their uranium concentrations vary between 27 and 303 ppm, with an average of 126 ppm (O'Brien *et al.*, 1986). Similar concentrations have been reported for phosphorites off Peru–Chile (Burnett & Veeh, 1977), Namibia (Veeh, Calvert & Price, 1974), and the Agulhas Bank and the Chatham Rise (Kolodny & Kaplan, 1970). However, because the P_2O_5 concentrations of the East Australian nodules are relatively low compared with other areas, their $U:P_2O_5$ ratios, which range between 4.7 and 23.4 (average 11.9), are typically greater than ratios usually reported for marine phosphorites (O'Brien *et al.*, 1986). Similarly, the tetravalent uranium content (%U(IV) = (U(IV)/UTOTAL) × 100) of the East Australian nodules is high, averaging 80% (O'Brien *et al.*, 1987), and exceeds values reported from phosphorites off Peru–Chile, the sea off California, and the Agulhas Bank (Kolodny & Kaplan, 1970; Burnett, 1974). The amount of reduced uranium (UIV) in marine phosphorites is considered an indicator of the oxidation potential of their depositional environment (Burnett & Veeh, 1977) and yet, paradoxically, the %U(IV) in phosphates found on the apparently well-oxygenated, low carbon flux East Australian upper slope exceeds the %U(IV) in phosphorites known to be forming in largely anaerobic, organic-rich sediments on the continental shelves off Peru–Chile and Namibia.

The radiometric ages of the East Australian non-ferruginous nodules range from 2500 to greater than 250 000 y (O'Brien *et al.*, 1986); Holocene ages were obtained from five different locations between 29° and 31° S (see Fig. 8.9) at water depths ranging from 365 to 450 m. The age data (O'Brien & Veeh, 1980; O'Brien *et al.*, 1986) suggest that phosphogenesis has been almost continuous throughout the late Quaternary in this area rather than being

Table 8.1. *Site data for phosphorites from the East Australian continental margin*

Sample location	Latitude (°S)	Longitude (°E)	Water depth (m)	Nodule age[a]
G7	29°23′	153°50′	385	Late Quaternary
G16	30°25′	153°26′	365	Holocene
G18	30°41′	153°18′	210	>800 000
G19	30°41′	153°19′	230	>800 000
1512	30°39.6′	153°19.8′	197	>800 000
1513	30°43.4′	153°18.6′	200	>800 000
1516	31°10.6′	153°13.9′	241	>800 000
P852	31°01.0′	153°18.7′	450	Holocene
P887	29°18.8′	153°50′	370	Late Quaternary
P883	29°16.9′	153°52.0′	400	Holocene
P896	29°23.9′	153°49.4′	440	Holocene
P904	29°18.0′	153°50.3′	405	Late Quaternary
S7	29°14.3′	153°51.3′	412	Late Quaternary
S12	29°25.1′	153°48.5′	376	Holocene

[a]Radiometric age of nodules from specified location.
Data from O'Brien & Veeh (1980), O'Brien *et al.* (1981), O'Brien & Veeh (1983); O'Brien *et al.* (1986).

Table 8.2. *Comparison of average chemical composition of phosphatic nodules off Eastern Australia and western Tasmania with phosphorites from other offshore areas*

	1a Eastern Australia non-ferruginous	1b Eastern Australia ferruginous	2 Western Tasmania	3 Agulhas Bank	4 Peru–Chile	5 Spain Danois Bank	6 Chatham Rise	7 SW Africa
Al_2O_3	3.53	3.46	1.18	2.13	5.15	2.95	0.30	0.37
CaO	31.9	19.6	46.8	36.87	33.93	25.35	44.3	51.3
F	2.0	2.0	n.d.	2.15	2.22	0.7	1.9	2.39
Fe_2O_3	5.00	29.5	4.07	6.17	2.85	27.82	2.73	0.87
K_2O	0.89	1.04	0.34	1.55	1.30	0.43	n.d.	0.13
MgO	1.31	2.59	1.37	1.38	1.07	1.65	3.07	0.60
MnO	0.04	0.06	0.04	0.01	n.d.	0.32	n.d.	n.d.
Na_2O	0.9	0.6	1.0	0.76	0.85	0.56	n.d.	n.d.
P_2O_5	11.0	9.6	10.2	17.53	22.61	7.82	20.8	32.1
SiO_2	24.0	17.6	4.65	14.79	22.13	14.64	5.7	1.4
SO_3	0.66	0.55	0.72	1.12	0.40	n.d.	n.d.	n.d.
TiO_2	0.32	0.26	0.07	0.10	n.d.	0.17	n.d.	0.06
Loss	19.6	14.7	28.8	16.40	8.78	17.65	n.d.	n.d.
CaO/P_2O_5	2.90	2.04	4.59	2.10	1.50	3.24	2.13	1.60
F/P_2O_5	0.18	0.21	—	0.123	0.098	0.090	0.091	0.074

[a]Total iron reported as Fe_2O_3
n.d. = not determined
1a = 4 analyses (Kress & Veeh, 1980; O'Brien & Veeh, 1980)
1b = 12 analyses (Kress & Veeh, 1980; Cook & Marshall, 1981)
2 = 1 analysis (this study, sample T-1)
3 = 12 analyses (Parker, 1975)
4 = 15 analyses (Burnett, 1977)
5 = 13 analyses (Lucas, Prevot & Lamboy, 1978)
6 = 5 analyses (Pasho, 1972)
7 = 2 analyses (Price & Calvert, 1978)

restricted to high sea-level stands, as reported for phosphorites off Peru–Chile (Burnett & Veeh, 1977).

Because of the steepness of the continental slope in this area (Marshall, 1979), the late Pleistocene and Holocene nodules are restricted to a relatively narrow band on the upper slope (Fig. 8.9), located approximately 35–40 km offshore (O'Brien *et al.*, 1981; O'Brien, 1982; O'Brien & Veeh, 1983). Upwelling on the East Australian continental margin in this area is weak and seasonal and is typically restricted to within 10–15 km of the coast (O'Brien & Veeh, 1983). Moreover, upwelling occurs when the rapidly flowing East Australian Current moves close to the coast; the strong, southward-flowing current interacts with the continental slope, producing an Ekman layer in which the mass transport is directed up into the shallower water (Godfrey, 1973; Rochford, 1975; Garrett, 1979). Upwelling events, while enhancing biological productivities in the nearshore areas, are contemporaneous with periods of high bottom-current velocities on the upper slope (O'Brien, 1982); these strong currents probably limit the transport of organic matter from the inshore upwelling areas. In addition, biological productivities over the upper slope between 29–32° S are low, averaging 0.1 g C $(m^2\ d)^{-1}$ (Jitts, 1965), and the organic carbon concentrations of both the nodules and the associated sediments are typically low ($\leqslant 0.5\%$ weight). Consequently, the formation of these phosphorites cannot be readily related to either coastal upwelling or periods of high organic productivity (O'Brien *et al.*, 1981; O'Brien & Veeh, 1980, 1983; O'Brien, 1982).

Scanning electron microscope (SEM) studies of the Quaternary non-ferruginous nodules have shown that all of the detectable phosphorus is located within what appear to be bacterial cellular structures. The cellular structures are typically 1.5–2.0 μm in length, and 0.75–1.0 μm in width, and resemble non-filamentous, botuliform bacilli (O'Brien *et al.*, 1981). The bacterial cells are now composed of apatite which is apparently an intracellular deposit rather than an extracellular encapsulation. It has been proposed that the nodules have formed via the *post-mortem* transformation of slowly growing, phosphorus-rich bacterial cells to apatite (O'Brien *et al.*, 1981). The bacteria slowly proliferate on the seafloor, binding the unconsolidated surface or near-surface foraminiferal sands and forming a semi-lithified bacterial mat or layer. Phosphogenesis appears to be favored by periods of very low current velocities on the seafloor, as evidenced by the abundance of very fine grained coccoliths, quartz grains, and clay minerals within the phosphatic matrix. It is likely that the phosphatic layer is broken up when bottom current velocities subsequently increase.

Since periods of coastal upwelling are linked to the shoreward movement of the East Australian current, which in turn results in strong bottom currents, it is likely that the phosphatic nodules and associated sediments undergo reworking and erosion at such times (O'Brien, 1982). The genetic model proposed by O'Brien *et al.* (1981) and O'Brien & Veeh (1983) for the East Australian phosphorites envisages that the phosphorites form slowly at, or very near, the sediment–water interface, within low carbon flux, oxic environments on the upper slope. This model is supported by recent stable isotopic data (McArthur *et al.*, 1986). The $\delta^{13}C$ values in the non-ferruginous nodules show no organic carbon contributions, and the $\delta^{34}S$ values are identical to values for contemporaneous evaporite sulphate, thus indicating an oxic phosphogenic depositional environment.

Ferruginous nodules

Ferruginous nodules are typically restricted to the outer shelf and have been recovered from water depths ranging from 197 to 241 m (Kress & Veeh, 1980; Marshall, 1980; Marshall & Cook, 1980; O'Brien & Veeh, 1980; Cook & Marshall, 1981; O'Brien *et al.*, 1981; Marshall, 1983). They are typically well-indurated, with a well developed geothite coating or surface glaze. They range in size from granules to boulders and occur on relict, current-swept areas of the outer shelf, probably as a gravel horizon (Marshall, 1980). It appears from preliminary work (Table 8.1, Fig. 8.9) that the ferruginous nodules are most abundant between 30° and 31° S, off Coffs Harbour.

Average geochemical compositional data for the ferruginous nodules are presented in Table 8.2. The average Fe_2O_3 concentration is 29.5%, approximately six times higher than that of the non-ferruginous nodules. All other elements, with the exception of calcium, are present in broadly similar concentrations in both ferruginous and non-ferruginous nodules. CaO averages only 19.6% in the ferruginous nodules, much less than in the non-ferruginous types (31.9%). This difference is probably due to the widespread dissolution of biogenic $CaCO_3$, as well as replacement of carbonate by goethite, both of which have been observed in the ferruginous nodules (Marshall, 1980; Marshall & Cook, 1980; Cook & Marshall, 1981). As a consequence of the $CaCO_3$ dissolution and replacement, the average $CaO:P_2O_5$ ratio of the ferruginous nodules is much less than that of the non-ferruginous types (2.04 and 2.90, respectively). Compositionally the ferruginous nodules most closely resemble ferruginous phosphorites from the Danois Bank off Spain. The compositional data suggests that the ferruginous nodules have formed by the ferruginization of nodules with similar geochemical compositions to those of the non-ferruginous Quaternary types.

The ferruginous nodules invariably have $^{230}Th/^{234}U$ ages which are in excess of the dating limit of the method (> 250 000 y). In fact, the $^{234}U:^{238}U$ ratios in these nodules are typically in secular equilibrium, indicating that their ages exceed 800 000 y (O'Brien & Veeh, 1980; O'Brien *et al.*, 1981; O'Brien *et al.*, 1986). The uranium-series ages are therefore consistent with the palaeontological data, which indicates that the ferruginous nodules are predominantly of Middle Miocene age (von der Borch, 1970). The uranium concentrations in the ferruginous nodules vary between 18 and 153 ppm, with an average of 67 ppm (O'Brien *et al.*, 1986), considerably less than the average uranium concentration of the non-ferruginous types (126 ppm). The $U:P_2O_5$ ratios range from 3.0–13.3 (average 7.9) and are similar to those of the non-ferruginous nodules, implying that the lower uranium concentrations in the ferruginous nodules are principally due to a lower average P_2O_5 (i.e. apatite) concentration.

Several theories have been proposed for the origin of the ferruginous nodules. Marshall & Cook (1980) and Cook & Marshall (1981) proposed that these nodules, like those from off Peru–Chile and Namibia, formed within organic-rich sediments from phosphate released during the bacterial degradation of organic matter. Moreover, these workers proposed that the precipitation of apatite, and the associated goethite and glauconite, was largely synchronous. Apatite precipitation was considered to have taken place in two stages, an earlier stage consisting of more coarsely crystalline hexagonal apatite platelets, 2–3 μm diameter, and a later stage comprising clusters of acicular hexagonal crystals about 1 μm long. Subsequently, Marshall (1983) also proposed that these nodules formed via apatite precipitation from phosphate-enriched pore waters. However, as an alternative to organically-derived phosphate, Marshall suggested that substantial inorganic phosphate was absorbed onto the ferric hydroxides during deposition of the iron-rich shelf sediments, and that this phosphate was released when the ferric hydroxides were buried, and encountered reducing conditions at depth within the sediment.

Other workers (O'Brien *et al.*, 1981; O'Brien, 1983; O'Brien & Veeh, 1983) have proposed that the ferruginous nodules formed by the same processes as the non-ferruginous Quaternary nodules, but the nodules have subsequently undergone extensive diagenetic modifications. The modifications include apatite recrystallization (to form large, well developed crystals), iron-enrichment, and loss of organic carbon (O'Brien, 1986; O'Brien *et al.*, 1987). This proposal is supported by the work of O'Brien (1983 and 1986) and O'Brien *et al.* (1987), who have shown that the non-ferruginous nodules can gain up to 5% Fe_2O_3 within 60 000–70 000 y of their formation, solely as a result of ferric hydroxide precipitation onto the nodule surfaces. Moreover, the stable isotopic data for the ferruginous nodules are very similar to those of the non-ferruginous nodules (McArthur *et al.*, 1986), again implying that both nodule types formed by similar processes.

In conclusion, phosphorite nodules from the East Australian continental margin fall into two types. Non-ferruginous nodules are friable and occur within unconsolidated, organic-poor, glauconitic, foraminiferal sands on the upper slope, at water depths between 350–455 m. They usually have finite (< 250 000 y) uranium-series ages, and Holocene ages have been obtained from five distinct locations between 29° and 31° S. Ferruginous nodules are well indurated and occur on relict areas of the outer shelf at water depths varying between 197 and 241 m. They are invariably older than the dating limit of the uranium-series method, and have undergone extensive diagenetic modifications. These modifications have virtually destroyed the original depositional fabric of these nodules, though they probably formed by similar mechanisms to the non-ferruginous types.

Tasmania

In 1966, Ocean Mining AG carried out exploration for phosphorite on the continental shelf off western, southern and southeastern Tasmania. Nodules were recovered from 89 of the 180 dredge stations occupied at water depths ranging from 60 to 155 m; preliminary results (OMAG unpublished report) indicate that nodules occur in greatest abundance on the western Tasmanian shelf. The nodules range from 2 to 45 cm in diameter, are angular to subrounded and, while their surfaces are generally rough and pitted, they are occasionally smooth and highly polished. They typically range in color from light grey to dark brown. The majority of nodules are uniformly microcrystalline and structureless, though some are conglomeratic, whereas other rounded nodules possess a well defined concentric structure. The P_2O_5 concentration of most of the nodules varied between < 1 and 13%, though nodules recovered from four different stations off western Tasmania had P_2O_5 concentrations ranging from 13% to a maximum of 26%. Recent work on the Tasmanian nodules (O'Brien, unpublished data) has indicated that many of the nodules off Tasmania are actually carbonate nodules rather than phosphate nodules. For example, some of the nodules described as having 'well-defined concentric structures' are actually carbonate nodules largely composed of encrusting coralline algae (Fig. 8.8*d*). An accurate assessment of the abundance of phosphorites off Tasmania is difficult, though recent extensive sampling off western Tasmania (Jones & Davies, 1983) recovered only two slightly phosphatic (maximum 3.6% P_2O_5), ferruginous limestone nodules from the middle and northwesten shelf. It thus seems likely that the distribution of even moderately phosphate-rich nodules is patchy off western Tasmania.

The geochemical composition of a phosphate nodule (sample T-1) from off western Tasmania is given in Table 8.2. The nodule has low concentrations of Fe_2O_3, Al_2O_3 and K_2O, indicating that clay minerals, glauconite and quartz are present in only minor amounts. CaO is present in high concentration (46.8%) and is mostly present within biogenic carbonate. The P_2O_5 concentration is relatively low (10.2%) and consequently the $CaO:P_2O_5$ ratio is very high (4.59). Nodule T-1 has a much lower silicate–aluminosilicate content than phosphorites from most other areas, with the exception of Chatham Rise and Namibia. The nodule is quite different in composition from the East Australian nodules, though the P_2O_5 concentration is similar. It appears that the Tasmanian nodule is compositionally most similar to phosphorites from the Chatham Rise, though the Chatham rise nodules have much higher P_2O_5 concentrations. X-ray diffraction studies of sample T-1 have shown that the apatite contains 7.3% structurally-bound carbonate, similar to the amount of structural CO_2 in the East Australian nodules. The total organic carbon concentration in T-1 is only 0.27%.

The concentrations and activity ratios of the uranium-series isotopes in sample T-1 are given in Table 8.3. The nodule contains 67 ppm uranium, lower than the average uranium concentration of the non-ferruginous East Australian samples, but broadly similar to the concentrations in nodules from off Peru–Chile, and off California, and also the ferruginous nodules off East Australia. The $U:P_2O_5$ ratio of nodule T-1 (6.6) is also similar to some Peru–Chile and East Australian phosphorites, but much higher than that of nodules from off Namibia (Baturin & Kochenov, 1974; Baturin, Merkulova & Chalov, 1974). The thorium concentration of the Tasmanian nodule (1.6 ppm) is lower than values reported from most other regions, and probably results from T-1's low alumino-silicate content. Both the $^{234}U/^{238}U$ and $^{230}Th/^{234}U$ activity ratios are in secular equilibrium, indicating that the nodule has an age of > 250 000 y and probably > 800 000 y.

Petrographically, the Tasmanian phosphatic nodules consist of abundant, small (< 0.1 mm diameter) planktonic foraminifera, occasional fine sand-sized, partly oxidized glauconite pellets, and rare silt-sized angular quartz, in a mixed phosphatic and calcareous micritic matrix. Collophane also commonly oc-

Table 8.3. *Uranium-series data for a phosphatic nodule from off the west coast of Tasmania. Errors quoted are based on counting statistics ($\pm 1\sigma$)*

Sample no	U (ppm)	Th (ppm)	$^{234}U/^{238}U$	$^{230}Th/^{234}U$	Age (ky)	U/P_2O_5 ($\times 10^{-4}$)
T − 1[a]	67	1.6	0.99 ± 0.015	1.01 ± 0.036	> 250	6.6

[a] Analysis performed at Flinders University by methods described by Burnett & Veeh (1977).

curs as an internal mold within foraminiferal tests. The nodules show evidence of bioturbation, and can best be described as burrowed phosphatic mudstones. In contrast, the carbonate nodules are basically slightly ferruginous, skeletal packstones, composed of abundant, moderately coarse-grained, poorly sorted bryozoal, foraminiferal, echinoid, and molluscan debris, in a micritic carbonate matrix; crustose coralline algae encrust most of the nodules. Quartz grains and glauconite pellets are much more abundant in the carbonate nodules than in the phosphatic mudstones. The carbonate nodules also occasionally contain small (1–4 mm), subrounded, ferruginous phosphatic mudstone lithoclasts which have clearly undergone extensive mechanical reworking.

While the phosphate and carbonate nodules are both obviously relict sediments, the carbonate nodules show strong similarities to the mixed relict–modern bryozoan sands that presently mantle the western shelf of Tasmania (Jones & Davies, 1983), suggesting that they were originally deposited in an open shelf environment. In contrast, the phosphate nodules were probably deposited in a very low energy, deeper water environment, with a restricted benthic fauna and a slow rate of terrigenous sedimentation. The mechanism of phosphate precipitation is, at this stage, unknown. There is no evidence of replacement of the foraminiferal tests by apatite, though the replacement may have only affected the fine-grained micritic matrix, a process which may occur in Agulhas Bank phosphorites (Parker & Siesser, 1972).

Northwestern Western Australia

The Bureau of Mineral Resources conducted a sampling cruise off northwestern Western Australia in late 1967. 163 bottom stations were sampled on the continental shelf and upper slope between latitudes 13° and 18°30′ S at water depths ranging from 18 to 366 m (Jones, 1968). No phosphorite nodules were found, though minor amounts of phosphate were present over large areas of the shelf. The phosphate occurs predominantly as relatively pure phosphate grains and angular fragments (up to 3 mm in diameter), and as phosphatic internal molds within foraminifera, lamellibranchs and gastropods (Jones, 1968). Highly angular lithic fragments consisting of biogenic carbonate, clay minerals and collophane also occur, as do partially phosphatized shell fragments and serpulid tubes.

The P_2O_5 concentrations are generally low, reaching a maximum of 9–10% in the < 2 mm-fraction of fine sediments located beyond the shelf break, in water depths between 247 and 329 m (Jones, 1968). The P_2O_5 concentration of the fine (< 2 mm) fraction varied between 3 and 7% at 11 other stations, and between 'trace' and 2% at another 46 stations. It is likely that the P_2O_5 concentrations of the 'bulk' sediment samples would be approximately half those of the < 2 mm fractions (Jones, 1968), and thus the bulk sediment P_2O_5 concentrations in this area do not exceed 5%. Nevertheless, a zone of P_2O_5 enrichment loosely follows the 180 m isobath almost continuously from 13° to 18° S (Jones, 1973).

One of the principal reasons that northwestern Western Australia was chosen for phosphate exploration was that coastal upwelling, considered by many workers to be a key link in the phosphogenic cycle, could be expected in this area (Jones, 1973; Noakes & Jones, 1973). Subsequent work (Holloway *et al.*, 1985) has shown, however, that upwelling on the North West Shelf is weak and restricted to the summer months. Thus the lack of phosphorite nodules in this area may possibly be attributable to this lack of upwelling. The actual mechanism of phosphate enrichment in this area is not known, though Jones (1973) has noted a close association between biogenic detritus and collophane, and suggested either direct precipitation of apatite or replacement of lime mud as possible mechanisms.

A few phosphatic encrustations have been dredged from upper slope and marginal plateaux locations during a North West Shelf cruise by the RV Sonne. No detailed work has been done on this material but it appears that there has been late diagenetic phosphatization of originally calcareous sediment.

South Tasman Rise

Marine sedimentary phosphate was discovered recently (O'Brien & Davies, 1986) on the South Tasman rise (46°12.7′ S, 127°0.5′ E), south of Tasmania (Fig. 8.2), during a joint BGR–BMR cruise of the research vessel RV Sonne. The phosphate is present mainly within fractures and vesicles in a manganese-encrusted scoriaceous, palagonitic basalt breccia that was dredged from the seafloor at water depths of between 1600 and 1810 m. The sedimentary phosphate consists of angular grains of quartz, K-feldspar, oligoclase, biotite, glauconite and epidote, together with Neogene planktonic foraminifera, in a collophanic matrix. At this stage it is not known whether it represents an isolated occurrence or part of a more extensive deposit. Preliminary results suggest that similarities exist between the phosphate from the South Tasman Rise and that from the Blake Plateau, off the coast of South Carolina.

Tasman Sea seamounts

The Tasmanitid guyots are a series of north–south-trending seamounts located in the Tasman Sea, approximately half-way between the Australian mainland and Lord Howe Rise. Phosphorite has been recovered from the upper flanks and summit platforms of three of these seamounts, namely Taupo

Bank, Barcoo Guyot and Derwent Hunter Guyot, at water depths ranging from 302–375 m (Slater & Goodwin, 1973). The phosphorite occurs as two distinct types: phosphatized ferruginous microfossiliferous limestone, and thin discontinuous phosphatic veneers on basalt. In addition, phosphatic fossiliferous limestone has also been found at a depth of 283 m on Gifford Guyot, near the Lord Howe Rise.

Slater & Goodwin (1973) propose that the apatite in the phosphatic limestones has formed via the replacement of a micritic carbonate precursor, whereas the phosphatic veneers on basalt are the result of direct precipitation. In both cases, phosphorite formation is considered to have been induced via the upwelling of cold, phosphate-rich Tasman Sea bottom water around the margins of the guyots.

Discussion

One of the striking features to come out of this overview of Neogene–Holocene phosphorites of Australia is the comparative lack of information on most of the occurrences. A large amount of geochemical data are now available on the offshore New South Wales phosphorites but many questions remain to be resolved. Even basic information such as their facies relationships, associated microfaunas, their ages (other than 'greater than 800 000 years' or 'mid-Miocene') and their relationship to the onshore occurrences are not known at this time.

There is a similar dearth of information for the onshore deposits, and basic petrological and geochemical data are lacking for most occurrences. Micro-palaeontological information is available but there is no agreement on correlation between the various occurrences. Information on the distribution of nodules is unreliable. This lack of basic data makes it difficult to place these occurrences into a broad depositional model or to relate them to the picture of global Neogene phosphogenesis that is emerging (see, for example, Riggs & Sheldon, Chapter 18, this volume). Nevertheless, the following tentative statements can be made on the nature and origin of Neogene to Holocene phosphorites in Australia.

Distribution

Neogene phosphorite occurrences are distributed sporadically in an arcuate belt almost 2000 km long around the southeastern margin of Australia. They extend from longitude 142° E to 153° E and from approximately 25° S to 46° S, and thus fall within a similar latitudinal range to the Neogene phosphogenic province of the southeastern United States. The Australian phosphorites are isolated occurrences, in contrast to the massive Neogene phosphorite deposits in the southeastern United States.

Age

Most of the Neogene phosphorites are poorly dated, and the offshore deposits particularly so. Without accurate ages, the extent to which these occurrences have been reworked from older deposits cannot be established. In the case of the West Australian occurrences, the phosphatic nodules have been quite clearly reworked from much older (Cretaceous) units. In most instances, however, this distinction is less clear. As in many other phosphatic sequences, vertebrate remains, crustacea, etc. are fairly common in places. As a general rule, the Neogene phosphorites are of Middle Miocene–Early Pliocene age and appear to be associated with a widespread Late Miocene–Early Pliocene disconformity. This disconformity can in turn be related to a well defined low sea-level stand documented by Vail *et al.* (1977).

Lithology

The Neogene–Holocene phosphorite occurrences of Australia include nodular, concretionary and encrusting forms, though nodular forms predominate. There are no reports of grainstone (pelletal) phosphorites.

The abundance of iron in Neogene phosphatic nodules is a feature of the offshore occurrences and, in this respect, the Australian nodules are comparable with some overseas occurrences, notably those of the Danois Bank in the eastern Atlantic. There is very little petrological and geochemical information available on the onshore occurrences and, consequently, it is difficult to compare them with other occurrences.

Depositional model

The sedimentary associations of offshore nodular intervals are unknown at this time. Onshore, the phosphatic nodules are in general overlain by Upper Miocene sediments and almost invariably occur on, or a short distance above, the top of a widespread limestone sequence of Oligocene–Early Miocene age. A similar limestone sequence commonly underlies Neogene phosphorites in other parts of the world. There are some reports of scattered phosphate nodules within these carbonates. As mentioned previously, the phosphate nodules are usually quite closely related to the disconformity surface though in the Torquay Embayment, for example, nodules occur up to several metres above the base of the Moorabool Viaduct Formation, a largely detrital unit. There are some very minor occurrences of glauconite at about the same stratigraphic level as the phosphate nodules, for example, in the Sorrento Graben. Glauconite grains are relatively common in the offshore Neogene–Holocene occurrences.

There is evidence, therefore, of a spatial and temporal link between Neogene phosphorite occurrences and both iron-rich sediments and carbonates. A fall in sea level, resulting in a hiatus and either the reworking of nodules or the formation of nodules at the disconformity surface, is also a feature of most occurrences. What is lacking is the classical black shale–chert–phosphorite assemblage. It can of course be argued that there are no Neogene phosphorite *deposits* known in Australia and that, therefore, the typical upwelling sedimentary assemblage would not be expected. Alternatively, these occurrences could be seen as constituting the most landward edge of a phosphogenic system; in that case any organic-rich sediments would be located seawards in the Bass, Otway or Gippsland Basins.

Origin of Neogene phosphorites

The sparse information presently available on the Neogene phosphorites of Australia indicates that they formed contemporaneously with the much more extensive deposits of the southeastern United States and elsewhere (Carter, 1978b). This suggests that the Australian occurrences are consistent with a pattern of global phosphogenesis which, by analogy with the Cambrian phosphogenic episode (Cook & Shergold, 1985), may be linked to secular changes in the phosphate concentration of the world ocean. This is, however, only one of the factors necessary to produce conditions suitable for the formation of a phosphorite (Cook, 1984; Cook & Shergold, 1986). Other major factors include a relatively low palaeolatitude (around 40° S) and a major change (fall) of sea level, following a sharp eustatic rise. Perhaps more than any other factor, coastal upwelling has been invoked to explain the formation of phosphate deposits (Kazakov, 1937; McKelvey, Swanson & Sheldon, 1952 and many others) though, in reality, it may be only one of several prerequisites for phosphorite formation. Indeed, it can be argued, by analogy with the Holocene phosphorite ocurrences off Eastern Australia, that coastal upwelling plays no part whatsoever in phosphogenesis. However, it could also be argued that the East Australian Current contains gyrals which influence both the distribution of (as yet undiscovered) high productivity areas and also the distribution of Holocene phosphorites, in a similar manner to that described for the Gulf Stream off the southeastern United States. Similarly, during the mid-Miocene high sea-level stand, the East Australian Current would have impinged more directly upon the continental shelf, perhaps producing dynamic upwelling in places. Such a process may have been active off North Carolina during the Miocene high stand (Riggs *et al.*, 1985).

Rather different conditions would have prevailed in the Bass, Otway and Murray Basins. There, easterly-directed currents could have interacted locally with topographic highs to produce dynamic upwelling, or with the north–south-oriented coasts to produce entrainment of coastal waters, resulting in localized phosphate-rich environments in the nearshore zone. This would also have most likely occurred during a high sea-level stand when Bass Strait was less restricted than it is at the present day. As noted earlier, however, coastal upwelling is only one of the potential components necessary for the formation of a phosphorite and, in the case of the Australian Neogene phosphorites occurrences, it may have been relatively unimportant. Conversely, it could be argued that the lack of major upwelling is the reason why the Australian Neogene phosphorites are relatively unimportant!

Any major deposits?

No major Neogene phosphate deposits are known in Australia at the present time. Several previous exploration efforts in the Neogene part of the geologic column have been unsuccessful (Howard, 1966 and 1967; Eddington, 1967; Grasso, 1967;).

Nevertheless, some of the right ingredients for the accumulation of major Neogene deposits do seem to be present; notably the very widespread (though admittedly low-grade) phosphorite occurrences throughout southeastern Australia; the presence of Neogene shallow marine sediments deposited within approximately 40° of the palaeo-Equator; and, relatively low rates of terrigenous sedimentation. The short answer to the question of whether there are any major Australian Neogene phosphorites is, therefore, '*perhaps*'!

Acknowledgments

We thank Mrs D. Christensen and Mrs Y. Grecian for typing this manuscript. Drafting was done by Ms H. Apps and Mr T. Kimber. We thank Dr A.N. Carter and Professor C.C. von der Borch for their helpful comments on the manuscript. G.E. Wilford and R. Langford provided access to unpublished information. This paper is published with permission of the Director of the Bureau of Mineral Resources.

References

Abele, C., Gloe, C.S., Hocking, J.B., Holdgate, G., Kenley, P.R., Lawrence, C.R., Ripper, D. Threlfall, W.F. (1976). 'Geology of Victoria'. *Geological Society Australia, Special Publication*, **5**, 177–274.

Atlas, E.L. (1975). 'Phosphate equilibria in sea water and interstitial waters.' Ph.D. Unpublished thesis, Oregon State University, Corvallis, Oregon.

Baker, G. (1945). Phosphate deposit near Princetown, Victoria, Australia, *Journal of Sedimentary Petrology*, **15(3)**, 88–92.

Baturin, G.N. (1971a). Stages of phosphorite formation on the ocean floor. *Nature*, **232**, 61–62.

Baturin, G.N. (1971b). Formation of phosphate sediments and water dynamics. *Oceanology*, **11**, 373–6.

Baturin, G.N. & Kochenov A.K. (1974). Uranium content of oceanic phosphorites. *Litologiya Poleznye Iskopaemye*, **1**, 124–9.

Baturin, G.N., Merkulova, K.I. & Chalov, P.I. (1974). Absolute dating of oceanic phosphorites by disequilibrium dating. *Geokhimiya*, **5**, 801–7.

Baxter, J.L. & Hamilton, R. (1980). The Yoganup Formation and Ascot Beds as possible facies equivalents. *Geological Survey of West Australia, Annual Report*, **1980**, 42–3.

Birch G.F., Thomson J., McArthur J.M. & Burnett W.C. (1983). Pleistocene phosphorites off the west coast of South Africa. *Nature*, **302**, 601–3.

Bowler, J.M. (1963). Tertiary stratigraphy and sedimentation in the Geelong-Maude area, Victoria. *Proceedings of the Royal Society of Victoria*, **76**, 69–136.

Brongersma-Sanders, M. (1957). Mass mortality in the sea. *In Treatise on Marine Ecology and Palaeoecology, vol. 1*, ed. J.W. Hedgepath, pp. 941–1010. Ecological Society of America Memoir 67.

Brown, C.M. (1983). Discussion: A Cainozoic history of Australia's Southeast Highlands. *Journal of the Geological Society of Australia*, **30**, 483–6.

Burnett, W.C. (1974). Phosphorite deposits from the sea floor off Peru and Chile: Radiochemical and geochemical investigations concerning their origin. *Hawaii Institute of Geophysics Report*, **74–3**, 164 pp.

Burnett, W.C. (1977). Geochemistry and origin of phosphorite deposits from off Peru and Chile. *Geological Society of America Bulletin*, **88**, 813–23.

Burnett, W.C. & Veeh, H.H. (1977). Uranium-series disequilibrium studies in phosphorite nodules from the west coast of South America. *Geochimica Cosmochimica Acta*, **41**, 755–64.

Burnett, W.C., Veeh, H.H. & Soutar. A. (1980). U-series, oceanographic and sedimentary evidence in support of Recent formation of phosphate nodules off Peru. *Society of Economic Paleontologists and Mineralogists, Special Publication* **29**, 61–71.

Carter, A.N. (1978a). The discovery of a phosphatic nodule bed at the base of the Jemmy's Point Formation in East Gippsland, Victoria. *Search*, **9**, 370–2.

Carter, A.N. (1978b). Phosphatic nodule beds in Victoria and the late Miocene–Pliocene eustatic event. *Nature*, **276**, 258–9.

Carter, A.N. (1979). Pliocene eustacy and the onset of sand barrier formation in Gippsland, Victoria. *Nature*, **280**, 131–2.

Carter, A.N. (1985). A model for depositional sequences in the Late Tertiary of south-eastern Australia. *South Australian Department of Mines Special Publication*, **5**, 13–27.

Chapman, F. (1916). Cainozoic geology of the Mallee and other Victorian bores. *Records Geological Survey Victoria*, **3**, 326–430.

Cook, P.J. (1984). Spatial and temporal controls on the formation of phosphate deposits. In *Phosphate Minerals*, ed. J.O. Nriagu & P.B. Moore, pp. 242–74. Springer-Verlag, Berlin.

Cook, P.J. & Marshall, J.F. (1981). Geochemistry of iron and phosphorus-rich nodules from the East Australian continental shelf. *Marine Geology*, **41**, 205–21.

Cook, P.J. & Shergold, J.H. (1985). Late Proterozoic–Cambrian phosphorites and phosphogenesis. *27th International Geological Congress Proceedings*, **15**, 397–444.

Cook, P.J. & Shergold, J.H. (1986). Proterozoic and Cambrian phosphorites. *Phosphate Deposits of the World, Volume 1*, Cambridge University Press, Cambridge.

Coulson, A. (1932). Phosphatic nodules in the Geelong district. *Proceedings of the Royal Society of Victoria*, **44**, 118–27.

Eddington, S.M. (1967). Results of test drilling in the Otway Basin, M.E.L. Nos 64, 65, 72 for Esso Minerals (Australia). Open file Mines Department, Victoria.

Garrett, C. (1979). Topographic Rossby waves off East Australia: identification and role in shelf circulation. *Journal of Physical Oceanography*, **9**, 244–53.

Godfrey, J.S. (1973). Comparison of the East Australian Current with the western boundary flow in Bryan and Cox's 1968 numerical model ocean. *Deep-Sea Research*, **20**, 1059–76.

Grasso, R. (1967). Report on drilling operations for phosphate, Princetown area, Otway Basin. M.E.L. 56 to 60, for Continental Oil Coy (Australia). Open file Mines Department, Victoria.

Gulbrandsen, R.A. (1966). Chemical composition of phosphorites of the Phosphoria Formation. *Geochimica Cosmochimica Acta*, **30**, 769–78.

Gunn, P.J. (1975). Mesozoic–Cainozoic tectonics and igneous activity–south-eastern Australia. *Journal of the Geological Society of Australia*, **22(2)**, 218–22.

Holloway, P.E., Humphries, S.E., Atkinson, M. & Imberger, J. (1985). Mechanism for nitrogen supply to the Australian north-west shelf. Australian Journal of Marine and Freshwater Resources, **36**, 753–64.

Howard, P.F. (1966). Completion report Killawarra area, for IMC Dev. Corp. Nov. 1966, Open file, Mines Department, Victoria.

Howard, P.F. (1967). Completion report of E.L. 40 Otway Basin, for IMC Dev. Corp., Open file, Mines Department, Victoria.

Howitt, A.M. (1919). Phosphate nodules in limestone, Thompson's Creek, near Mordialloc. *Records of the Geological Survey of Victoria*, **4(3)**, 262.

Hutchinson, G.E. (1950). The biochemistry of vertebrate excretion. *American Museum of Natural History Bulletin*, **96**, 253–9, 264–74.

Jitts, H.R. (1965). The summer characteristics of primary productivity in the Tasman and Coral Seas. *Australian Journal of Marine Freshwater Research*, **16**, 151–62.

Johns, R.K. (1976). Phosphate – South Australia. In *Economic Geology of Australia and Papua New Guinea. 4. Industrial Minerals and Rocks*, 4, ed. C.L. Knight, pp. 282–5.

Jones, H.A. (1968). A preliminary account of the sediments and morphology of part of the North West Australian continental shelf and upper continental slope. *Bureau of Mineral Resources, Division of Geology and Geophysics (Australia)*, **Record 1968/84**.

Jones, H.A. (1973). Marine geology of the north west Australian continental shelf. *Bureau of Mineral Resources of Australia Bulletin*, **136**, 102 pp.

Jones, H.A. & Davies, P.J. (1983). Superficial sediments of the Tasmanian continental shelf and part of Bass Strait. *Bureau of Mineral Resources of Australia Bulletin*, **218**, 1–10.

Kazakov, A.V. (1937). The phosphorite facies and the genesis of phosphorites. *Geological Investigations of Agricultural Ores. Transactions Scientific Institute of Fertilizers and Insecto-fungicides* (USSR), **142**, 93–113.

Keble, R.A. (1932). Notes on the faunas of the Geelong nodule beds. *Proceedings of the Royal Society of Victoria*, **44(2)**, 129–33.

Kendrick, G.W. (1980). Molluscs from the Ascot Beds from the Cooljarloo heavy mineral deposit, Western Australia. *Geological Survey of Western Australia*, **1980**, 44.

Kolodny, Y. & Kaplan, I.R. (1970). Uranium isotopes in sea floor phosphorites. *Geochimica Cosmochimica Acta*, **34**, 3–24.

Kress, A.G. & Veeh, H.H. (1980). Geochemistry and radiometric ages of phosphatic nodules from the continental margin of northern New South Wales, Australia. *Marine Geology*, **36**, 143–57.

Link, A. & Thompson, B.P. (1985). Notes for the Torquay Embayment workshop. *Petroleum Exploration Society of Australia*, Field Notes, 12 pp.

Loughnan, F.C. & Craig, D.C. (1962). A preliminary investigation of the recent sediments off the east coast of Australia. *Australian Journal of Marine Freshwater Research*, **31**, 48–56.

Lucas, J., Prevot, J. & Lamboy, M. (1978). Les phosphorites de la marge norde de l'Espagne. Chimie, mineralogie, genese. *Oceanologica Acta*, **1(1)**, 55–72.

Mallett, C.W. & Holdgate, G.R. (1985). Subsurface Neogene stratigraphy of Nepean Peninsula, Victoria. *South Australia Department of Mines Special Publication*, **5**, 233–45.

Marshall, J.F. (1971). Phosphatic sediments on the eastern Australian upper continental slope. *Bureau of Mineral Resources of Australia, Records* 1971/59.

Marshall, J.F. (1979). The development of the continental shelf of northern New South Wales. *Bureau of Mineral Resources, Journal of Australian Geology & Geophysics*, **4**, 281–8.

Marshall, J.F. (1980). Continental shelf sediments: southern Queensland and northern New South Wales. *Bureau of Mineral Resources of Australia, Bulletin*, **207**, 1–39.

Marshall, J.F. (1983). Geochemistry of iron-rich sediments on the outer continental shelf off northern New South Wales. *Marine Geology*, **51**, 163–75.

Marshall, J.F. & Cook, P.J. (1980). Petrology of iron- and phosphorus-rich nodules from the Eastern Australian continental shelf. *Journal of the Geological Society, London*, **137**, 765–71.

Matheson, R.S. (1948). The Dandaragan phosphate deposits. *Geological Survey of Western Australia, Bulletin*, **124**.

McArthur, J.M., Benmore, R.A., Coleman, M.L., Soldi, C., Yeh, H.W. & O'Brien, G.W. (1986). Stable isotopic characterisation of francolite formation. *Earth and Planetary Sciences Letters*, **77**, 20–34.

McClellan, G.H. (1980). Mineralogy of carbonate fluorapatites. *Journal of the Geological Society, London*, **137**, 675–81.

McKelvey, V.E., Swanson, R.W. & Sheldon, R.P. (1952). The Permian phosphorite deposits of Western United States. In *Origine des gisements de phosphates de chaux 19th International Geological Congress*, **11**, 45–64. Algiers.

McKelvey, V.E. (1967). Phosphate deposits. *US Geological Survey Bulletin*, **1252–D**, 1–21.

Noakes, L.C. & Jones, H.A. (1973). Mineral Resources Offshore. *Bur. Miner. Resour. Geol. Geophys. Aust. Rec.* 1973/55.

O'Brien, G.W. (1982). Origin of east Australian continental margin phosphorites. In *Fifth International Field Workshop and Seminar on Phosphorites*, pp. 435–54. Kunming, China.

O'Brien, G.W. (1983). 'Geochemistry and origin of phosphatic nodules and associated sediment from the East Australian continental margin.' Ph.D. thesis, Flinders University of South Australia, Australia.

O'Brien, G.W. (1986). Reworking: a major control on the uranium-series and major element chemistry of phosphorites from the East Australian continental margin. (abstract). *12th International Sedimentological Congress*, Canberra, Australia.

O'Brien, G.W. & Davies, H. (1986). Sedimentary phosphate discovered on the South Tasman Rise. *Bureau of Mineral Resources Newsletter* **4**, 1.

O'Brien, G.W., Harris, J.R., Milnes, A.R. & Veeh, H.H. (1981). Bacterial origin of East Australian continental margin phosphorites. *Nature*, **294**, 442–4.

O'Brien, G.W. & Veeh, H.H. (1980). Holocene phosphorite on the East Australian continental margin. *Nature*, **288**, 690–2.

O'Brien, G.W. & Veeh, H.H. (1983). Are phosphorites reliable indicators of upwelling? In *Coastal Upwelling and its Sediment Record, Part*

A, ed. E. Suess & J. Thiede, pp. 399–419. Plenum Press. New York.

O'Brien, G.W., Veeh, H.H., Cullen, D.J. Milnes, A.R. (1986). Uranium-series isotopic studies of marine phosphorites and associated sediments from the East Australian continental margin. *Earth and Planetary Sciences Letters*, **80**, 19–35.

O'Brien, G.W., Veeh, H.H., Milnes, A.R. & Cullen, D.J. (1987). Seafloor weathering of phosphate nodules off East Australia: its effect on uranium oxidation state and isotopic composition. *Geochimica Cosmochimica Acta*, **51**, 2051–64.

Ocean Mining AG (OMAG), (1966). Phosphates off Tasmania. Unpublished Company Report, 13 pp.

Parker, R.J. (1975). The petrology and origin of some glauconitic and glauco-conglomeratic phosphorites from the South African continental margin. *Journal of Sedimentary Petrology*, **45**, 230–42.

Parker, R.J. & Siesser, W.G. (1972). Petrology and origin of some phosphorites from the South African continental margin. *Journal of Sedimentary Petrology*, **42**, 434–40.

Pasho, D.W. (1972). Character and origin of marine phosphorites. Office of Marine Geology. *US Geological Survey Report*, USC-GEOL, 72–5.

Price, N.B. & Calvert, S.E. (1978). The geochemistry of phosphorites from the Namibian shelf. *Chemical Geology*, **23**, 151–70.

Reeckmann, S.A. (1979). 'Detailed stratigraphy of the Tertiary sequence Torquay, Victoria – facies, environment and diagenesis'. Unpublished Ph.D. thesis, Melbourne University, Australia.

Riggs, S.R., Lewis, D.W., Scarborough, A.K., Snyder, S.W. (1985). Cyclic deposition of Neogene phosphorites in the Aurora area, North Carolina, and their possible relationships to global sea level fluctuations. *Southeastern Geology*, **23(4)**, 189–204.

Rochford, D.J. (1975). Nutrient enrichment of East Australian coastal waters II. Laurieton upwelling. *Australian Journal of Marine Freshwater Resources*, **26**, 223–43.

Rooney, T.P. & Kerr, P.F. (1967). Mineralogic nature and origin of phosphorite, Beaufort County, North Carolina. *Geological Society of America Bulletin*, **78**, 731–48.

Russell, R.T. (1967). Discovery of major phosphate deposits in northwest Queensland. *Queensland Government Mining Journal*, **68**, 153–7.

Slater, R.A. & Goodwin, R.H. (1973). Tasman Sea Guyots. *Marine Geology*, **14**, 81–99.

Twelvetrees, W.H. (1917). Phosphate deposits in Tasmania. *Tasmania Geological Survey Mineral Research*, **3**.

Vail, P.R., Mitchum, R.M. & Thompson, S. (1977). Global cycles of relative changes of sea level. *AAPG Memoir*, **26**, 83–97.

Veeh, H.H., Burnett, W.C. & Soutar, A. (1973). Contemporary phosphorites on the continental margin of Peru. *Science*, **181**, 844–5.

Veeh, H.H., Calvert, S.E. & Price, N.B. (1974). Accumulation of uranium in sediments and phosphorites on the South West African Shelf. *Marine Chemistry*, **2**, 188–202.

Von der Borch, C.C. (1970). Phosphatic concretions and nodules from the upper continental slope, northern New South Wales. *Journal of the Geological Society of Australia*, **16**, 755–9.

Wilkinson, C.S. (1865). Report on the Cape Otway country (from Parliamentary Papers, 1864–1865). *Report Geological Survey of Victoria*, **1863–64**, 21–8.

9

Miocene phosphorites of Cuba

A.V. ILYIN AND G.I. RATNIKOVA

Abstract

Occurrences of Miocene phosphorites in Cuba are significant to understanding the genesis of Neogene phosphorites for two reasons. First, this extends the known distribution of a major phosphogenic event from the extensive deposits of the southeastern United States into the Caribbean Basin, where additional deposits might occur. Second, it helps to define critical paleoceanographic conditions and paleogeography associated with development of the Miocene episode that lead to a unique global sedimentologic event. The following review describes the character of known examples of Miocene phosphorite occurring within the Caribbean Basin. It is based upon geologic studies of the phosphate deposits by Pokryshkin (1969) and Mederos & Krasilnikova (1984) and investigations of the tectonics of Cuba by Pusharovsky (1966) and Mossakovsky & Chekhovitch (pers. comm.).

Introduction

Miocene phosphorites were first found and described in the Havana Province by Pokryshkin in the early sixties (1969). By the late seventies, several phosphate deposits in the southern part of the Province had been explored in detail by Mederos & Krasilnikova (1984). The Loma Candela, Pipian and Meseta Rokha phosphate deposits occur on a plateau, up to 200 m above sea level, in the Guines Pipian district (Fig. 9.1).

Pre-Miocene stratigraphy

Western Cuba is probably underlain by crystalline Precambrian rocks; however, they are not exposed (Pusharovsky, 1966). These basement rocks are overthrusted by folded and faulted rocks of Cretaceous and Paleocene age that consist of andesitic to basaltic volcanics, ophiolite suites, and deep-water pelagic sediments. The Cretaceous–Paleocene rocks form the tectonic framework of western Cuba and crop out mainly along the northern portion of the country (Fig. 9.1) (Pusharovsky, 1966).

The post-Paleocene sedimentary cover is gently folded and overlies the severely folded Cretaceous–Paleocene units in sharp angular unconformity. The west–east-trending and west-plunging Madruga Anticline is a prominent structural feature that occurs south of Havana (Fig. 9.1). This large-scale anticline has a core of Cretaceous and Paleocene rocks with upper Eocene sediments occurring on northern and southern limbs. Both limbs are inclined at 10–20°. The uplift of the anticlinal core is believed to have produced the basins for the post-Paleocene sedimentation which began with Middle–Upper Eocene micritic limestones intercalated with marls and marly clays up to 20 cm thick. The 25 m-thick Eocene sediments contain siliceous and siliceous–carbonate nodular inclusions in the upper portion.

Oligocene sediments are composed of interbedded sequences of white–pinkish, fine-grained, clayey limestone, with scattered iron and manganese oxide inclusions and green-gray, calcareous clay. The sequence is capped by an erosional surface characterized by common burrow structures filled with fragments of overlying Neogene calcarenite and phosphorite gravels.

Miocene stratigraphy

Oligocene sediments are unconformably overlain by middle Miocene glauconitic and phosphatic sandstone. The latter sediments grade upsection into thinly laminated, yellow-gray to green, phosphatic calcarenites. Phosphorites occur around both limbs and near the western closure of the Loma Candela Anticline (Fig. 9.2), a smaller-scale structure than the Madruga Anticline in Figure 9.1. Seaward into the depositional basins and away from the Loma Candela Anticline, the Miocene sediments grade rapidly from phosphorites into slightly phosphatic calcarenite and limestones. Middle Miocene phosphate-bearing sediments are up to 20 m thick and increase in thickness rapidly into the depositional basins and away from the anticline (Fig. 9.2). Middle–Upper Miocene sediments consist of irregularly distributed facies of slightly phosphatic to nonphosphatic, detrital reef limestones and dolomites.

The Middle Miocene phosphate deposits of Meseta Rokha and Loma Candela are located near the core of the Loma Candela Anticline (Fig. 9.2). Outward from the axis of the anticline, phosphorites grade into non-phosphatic limestone. The poorly phosphatic sediments of the Pipian deposits are located on the southern limb and rather far from the axis of the anticline. The Katalina deposit occurs on the northern limb, inclined at 20–25°, and is in close proximity to the axis of the anticline (Fig. 9.2). All of the deposits are small; Loma Candela (Figs. 9.3, 9.4) and Meseta Rokha (Fig. 9.5) are relatively the largest deposits.

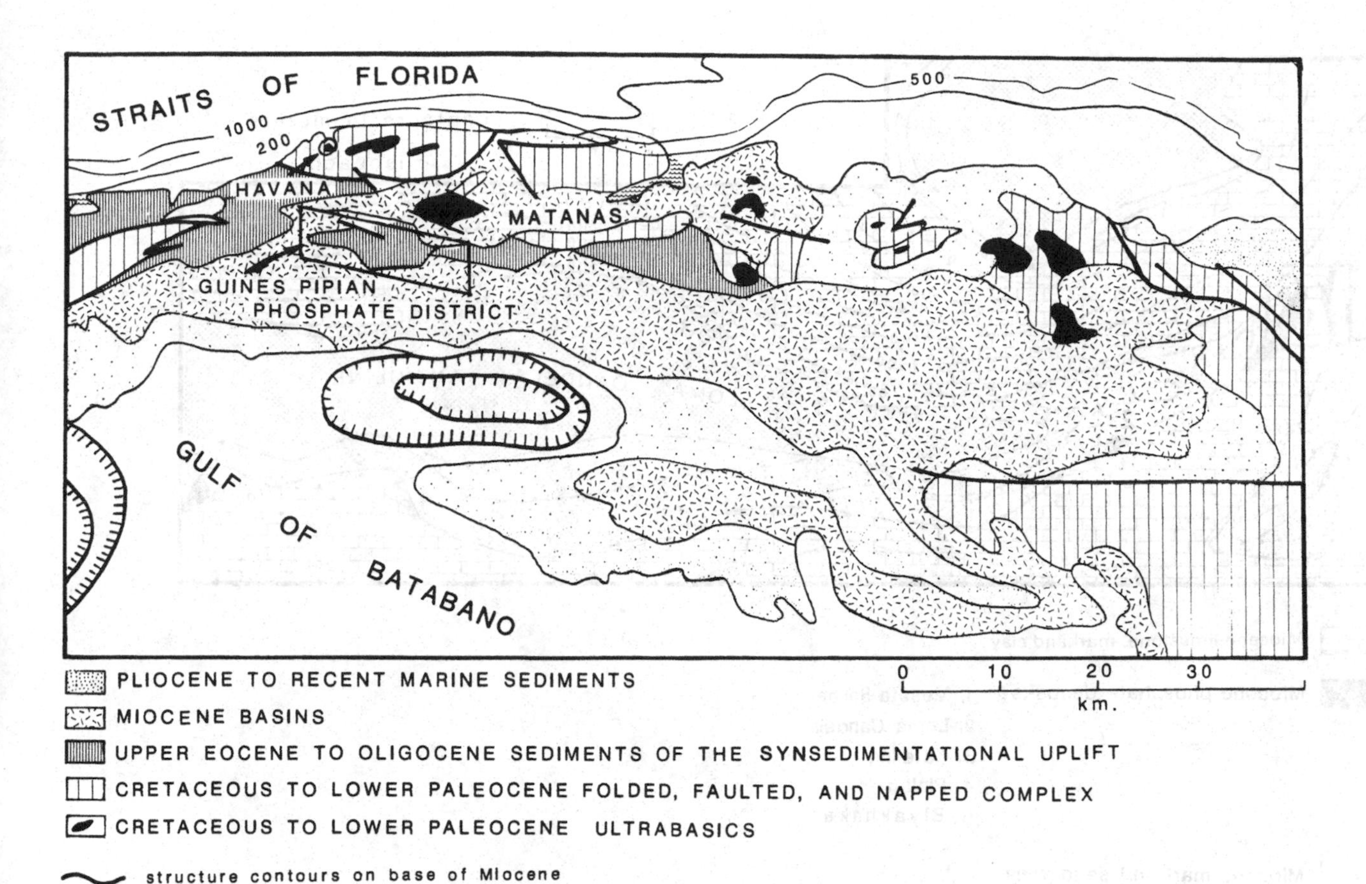

Fig. 9.1. Schematic geological map of the western Cuba (after Pusharovsky, 1966) showing the location of the Guines Pipian Phosphate District. Contours in the Gulf of Batabano are on the base of the Miocene and delineate the deep depositional basins.

Miocene phosphorites

Phosphorites of the Guines Pipian District are pelletal and consist essentially of carbonate fluorapatite, calcite and glauconite. The phosphorites have a bedded geometry which ranges from 0.2–3.5 m in thickness with a mean thickness of about 1.5 m. The bed can be traced in various places from 2.5–10 km and is lithologically non-uniform. The lower part of the bed consists of pelletal phosphorites with some phosphate pebbles and grades upward into nodular or conglomeratic phosphorite and then to phosphatic calcarenites.

Pelletal phosphorites occur locally within the Meseta Rokha deposit, are rare in Loma Candela, and completely absent in other places. Pelletal phosphorites contain up to 9.5% P_2O_5 (Table 9.1) and consist of 20–25% phosphate grains, 30–35% glauconite grains and 40–45% detrital carbonate grains, all in a slightly indurated calcareous and argilaceous mud matrix. Iron hydroxides constitute up to 5% of the sediment. The phosphate pellets are from 0.2 to 1.0 mm in diameter and have variable composition (Table 9.2).

Phosphatic calcarenites are present throughout the Guines Pipian District and average 3% P_2O_5 (Table 9.1). They range from 0.7 m to 35 m in thickness with an average thickness of about 8 m. In the Meseta Rokha deposit, they are from 0.7 m to 18 m thick. These calcarenites consist of 80–85% calcite sand grains, 5–10% spherical apatite grains from 0.05 mm to 0.1 mm in diameter, and are cemented with calcareous mud. Subordinant components include iron hydroxides, glauconite, and quartz.

Pelletal phosphorites and phosphatic calcarenites are locally weathered to various extents. The final product of this weathering is a phosphatic-rich clay which contains the highest phosphate content (average 14.5% P_2O_5; Table 9.1).

Phosphatic conglomerates are localized mainly in the southern part of the Loma Candela deposit, where they range from 0.2 to 0.8 m thick. They also occur in the Katalina deposit. The conglomerates are composed of semirounded fragments of phosphate that are 3–5 cm in diameter, contain an average of 9.36% P_2O_5 (Table 9.1) and are cemented with carbonate.

Petrography

Two varieties of phosphate pellets occur in the Miocene sediments. First are light brown, rounded grains 0.3–0.8 mm in diameter with abundant argillaceous microinclusions. They have a specific density of about 3 $g\,cm^{-3}$ and a refractive index of 1.594 ± 0.003. Second are transparent, light yellow, isotropic, and irregularly shaped grains of 0.2–0.5 mm in diameter. They have a specific density of 3 $g\,cm^{-3}$ and a refractive index of 1.608 ± 0.003. Cell parameters are: $a = 9.315 \pm 0.002$ A° and $c = 6.893 \pm 0.01$ A°. Mineralogically, both varieties of phosphate pellets are carbonate fluorapatite with considerable carbon sub-

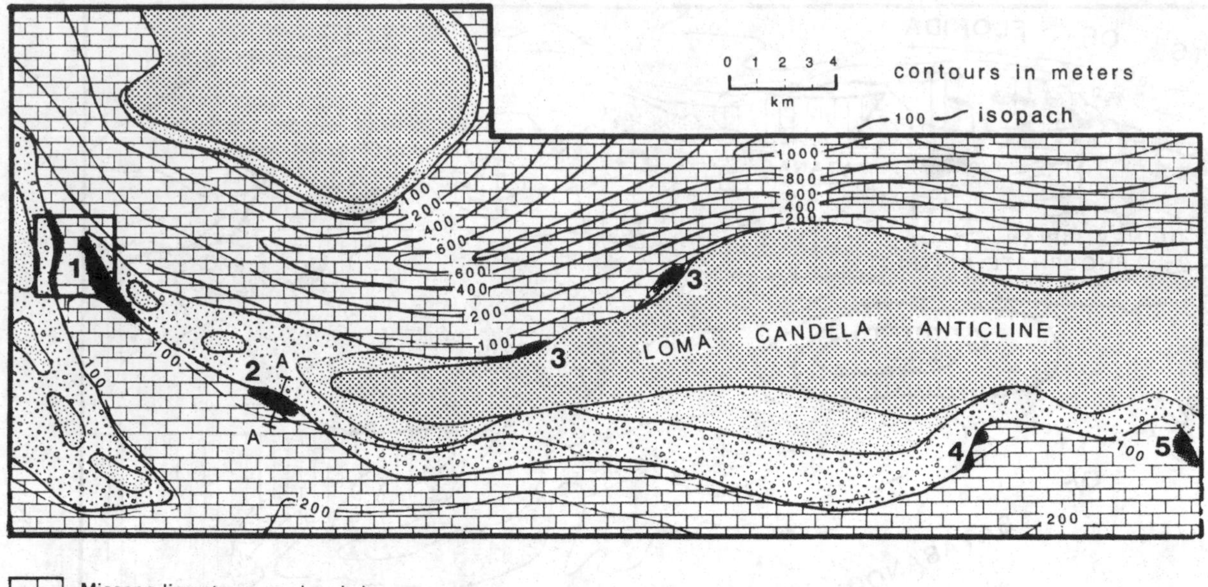

Miocene limestone, marl and clay

Miocene phosphate deposits

1. Meseta Rokha
2. Loma Candela
3. Katalina
4. Pipian
5. Biyakhaka

Miocene marl and sandstone

Miocene calcareous sandstone

Upper Eocene to Oligocene sediments

Fig. 9.2. Geologic map of the Guines Pipian Phosphate District outlined in Figure 9.1 and showing (a) areas of pre-Miocene sediments that form the Miocene uplifts and associated sedimentary basins; (b) location of the Middle Miocene phosphate deposits, (c) associated Miocene lithologies; (d) location of the geologic cross-section A–A′ in Figure 9.4; (e) location of the geologic map of the Meseta Rokha phosphate deposit in Figure 9.5, and (f) isopach contours of the Miocene sediments in meters.

stitution for phosphorus. Sieve analyses show that the −0.2 mm and +0.074 mm and the −0.074 mm +0.044 mm size fractions are most enriched in phosphate grains. These size fractions contained 15.2% P_2O_5 and 12.8% P_2O_5, respectively.

Two varieties of glauconite grains also occur in the sediments. First are light green–yellow green, spherical grains 0.1–0.3 mm in diameter. These grains are internally uniform with weak pleochroism with a refractive index of 1.590 and a specific density of $2.2\,g\,cm^{-3}$. Second are bright green, irregular–semirounded grains 0.3–0.5 mm in diameter, with strong pleochroism.

Phosphate grains of the first variety, glauconite and calcite were formed by sedimentation and early diagenesis at the sediment–seawater interface.

Geologic history

In the Early Miocene, western Cuba was differentiated into synsedimentational uplifts and subsiding basins; the low uplifts did not supply much clastic material. Phosphate accumulated in shallow water areas close to the uplifts. The region of phosphate deposition occurs in the northern portion of a large Miocene basin, the Batabano Basin (Fig. 9.1). The base of Miocene sediments on the southern shelf of Cuba occurs at depths up to 3000 m below present sea level. Another basin to the west and occuring between Pilar del Rio, St Filippe and Pinos Islands has even thicker Eocene, Oligocene, Miocene, Pliocene and Recent sediments. The base of the Miocene is about 4000 m below present sea level.

Phosphogenesis occurred on the Atlantic shelf in conjunction with the Early Miocene transgression. It was followed by regression and erosion of previously formed sediments. By the beginning of the Middle Miocene, the area experienced another transgression causing erosion, redistribution of sediments, and forming the second generation of glauconite and phosphate grains. In the Meseta Rokha deposit, glauconitic phosphorite and irregularly scattered phosphatic pebbles were deposited. Down the depositional slope, in the Loma Candela deposit, phosphatic calcarenites accumulated with rounded phosphatic

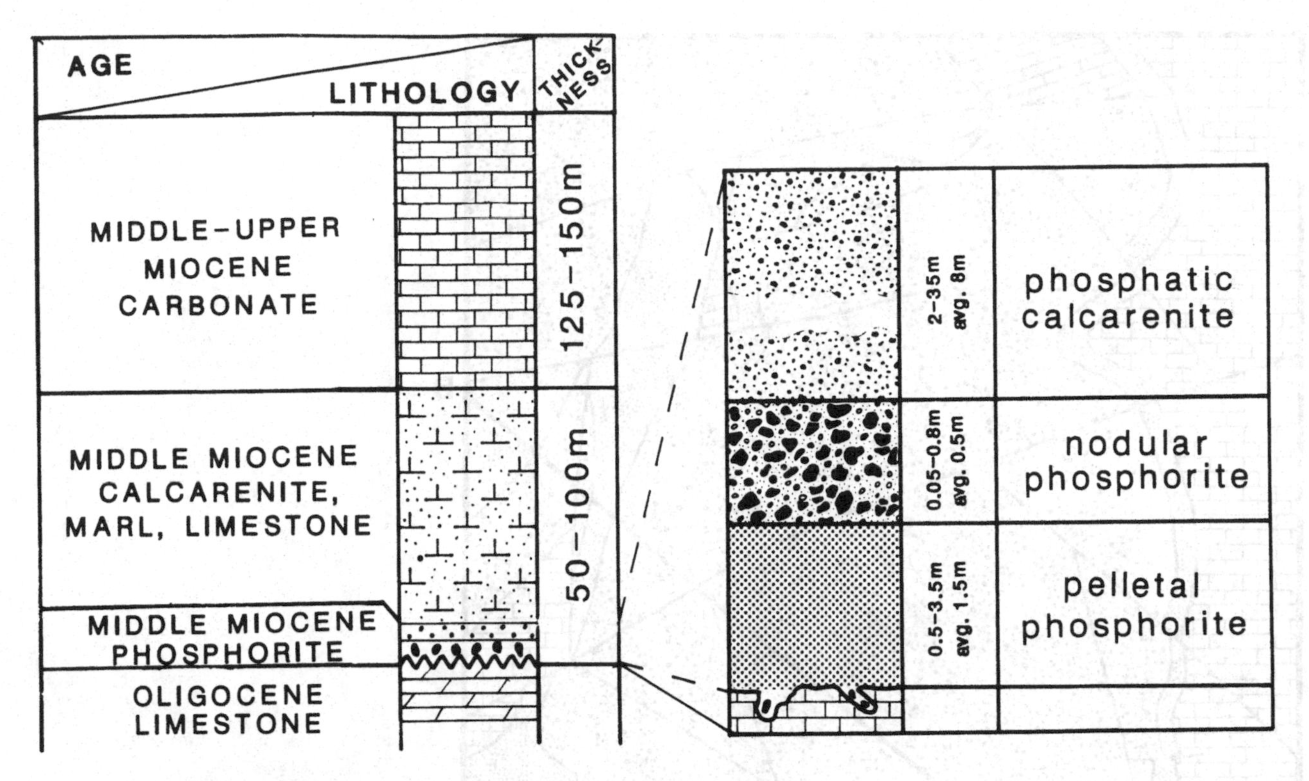

Fig. 9.3. Stratigraphic column of the Oligocene–Miocene sediments in the Loma Candela phosphate deposit with an expanded view of the phosphorite portion of the section.

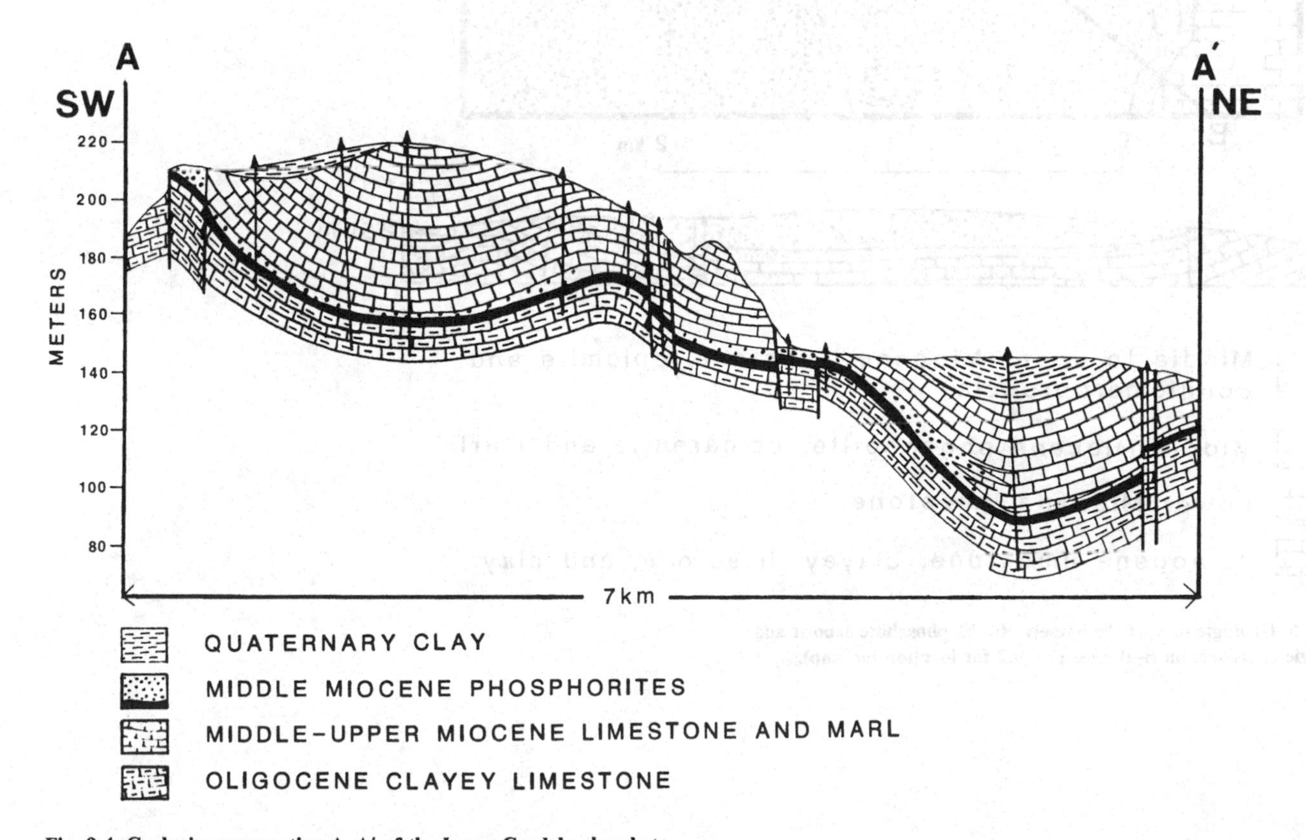

Fig. 9.4. Geologic cross-section A–A′ of the Loma Candela phosphate deposit (see Fig. 9.2 for location of section).

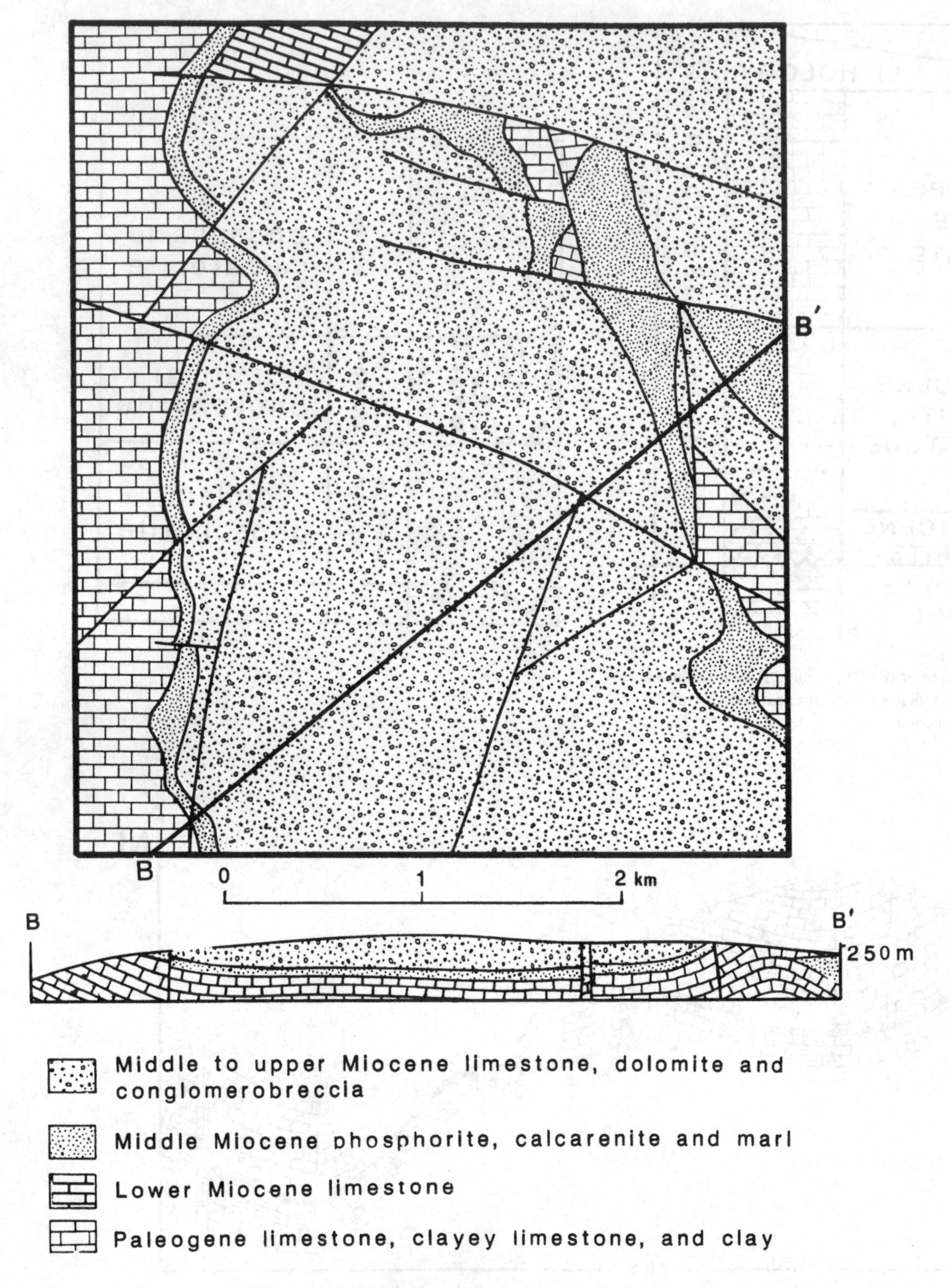

Fig. 9.5. Geologic map of the Meseta Rokha phosphate deposit and geologic cross-section B–B′ (see Fig. 9.2 for location for map).

Table 9.1. *Mean chemical composition in weight percent of pelletal phosphorites (PP), phosphatic calcarenites (PCA), phosphatic conglomerate (PC) and phosphatized clay (PCL) from the Meseta Rokha phosphate deposit*

	PP	PCA	PC	PCL
Al_2O_3	1.36	0.60	1.06	17.08
CaO	49.47	51.27	51.05	40.91
CO_2	26.90	37.91	28.49	11.22
SiO_2	4.18	1.06	1.42	14.66
F	0.85	0.30	0.60	1.90
FeO	0.22	0.28	0.50	0.28
Fe_2O_3	2.45	0.37	2.57	9.36
H_2O	0.64	0.13	0.32	—
K_2O	0.35	0.23	0.20	0.31
MgO	1.0	0.40	0.66	0.67
Mno	—	—	—	—
Na_2O	0.28	—	0.30	0.1
P_2O_5	9.53	3.05	9.36	14.5
SO_3	0.41	0.11	0.39	0.36
TiO_2	0.14	0.06	—	0.21
Insolube residue	4.50	1.61	2.08	29.90
Loss on ignition	3.13	2.15	3.39	9.70
Organic carbon	0.17	0.12	0.16	0.19

pebbles at the base. With further subsidence, thin layers of calcarenites were deposited in Meseta Rokha and thick, slightly- to non-phosphatic calcarenites accumulated in Loma Candela. Further down the depositional slope, foraminiferal limestones were deposited.

By the end of the Middle Miocene, previously deposited sediments were partly eroded, particularly nearshore sediments which were reworked down the depositional slope beyond the Loma Candela area. Another transgression began by the beginning of the Upper Miocene.

Table 9.2. *Chemical composition by weight percent of phosphate pellets and phosphate concentrate from the Meseta Rokha phosphate deposit*

	Pellets	Concentrate
Al_2O_3	1.40	6.00
CaO	48.05	41.64
CO_2	12.54	3.04
SiO_2	1.52	1.18
F	2.02	2.81
Fe_2O_3	3.78	5.20
MgO	1.06	0.44
P_2O_5	22.26	29.08
SO_3	1.42	3.37

Pelletal phosphorites generally formed at the base of the Middle Miocene in the Guines Pipian District and are included in the glauconite–argillaceous–carbonate series. P_2O_5 contents decrease upwards through the sequence. Phosphate accumulation was controlled by the contemporaneous uplifts associated with the Madruga and Loma Candela Anticlines.

The Miocene of Cuba has great promise for the occurrence of major phosphorite deposits, especially in the western part of the island. However, throughout much of this area the Miocene is deeply buried, such as on the southern shelf of Cuba (Fig. 9.1).

References

Mederos, P.H. & Krasilnikova, I.G. (1984). Phosphoritovye mestorozhdenya yuga provincyi Gavana; Phosphorite deposits of the southern part of the Habana Province. *Geology and Exploration*, **5**, 131–3.

Pokryshkin, V.I. (1969). Phosphority respubliki Cuba; Phosphorites of Cuba. *Industry of Raw Chemical Materials*, **2**, 17–23.

Pusharovsky, Yu.M. (1966). Tectonicheskaya karta Cuby; *Tectonic Map of Cuba*. GUGK Publishing Office.

10
Phosphorite deposits in the Upper Oligocene, San Gregorio Formation at San Juan de la Costa, Baja California Sur, Mexico

C. GALLI-OLIVIER, G. GARDUÑO AND J. GAMIÑO

Abstract
In southern Baja California Sur, the San Gregorio Formation, of Late Oligocene age, was deposited in the shelfal and upper slope region of a forearc basin and consists of two members. The lower member is 39 m thick at San Juan de la Costa and contains 15 beds of high-grade phosphorite, with a total aggregate thickness of 3.9 m and P_2O_5 values averaging 17.8%. The phosphate grains at San Juan de la Costa were concentrated by intermittent high energy processes. These grains formed initially in a shallow water marine environment. Later they were deposited on the upper slope and became interbedded with laminated diatomaceous mudrocks. Distinct differences in the character of phosphate grains in the phosphorites and in the mudrocks indicate conclusively that the former were not derived by winnowing of the latter.

Introduction

The San Juan de la Costa phosphorite deposits of Baja California Sur are of Late Oligocene age and are among the largest in Mexico. They are also significant because phosphorites of this age are very sparse along the eastern margin of the Pacific. These deposits and their associated siliceous (diatomaceous) rocks, assigned to the San Gregorio Formation, provide a rare record of paleoceanographic conditions in the eastern Pacific during Late Oligocene time. This paper reports on field and petrologic studies of the San Gregorio Formation in the vicinity of the San Juan de la Costa mine north of La Paz (Fig. 10.1).

Geologic setting

The San Juan de la Costa region was located in a forearc basin along the Pacific continental margin in Late Oligocene time. It was part of a westward-facing open shelf and slope, a setting favorable for upwelling, and a factor perhaps conducive to the formation of phosphorite (Loughman, 1984) in the marine lower member of the San Gregorio Formation. Kim (1987) has noted the presence of abundant diatoms in the San Gregorio formation, a fact supporting Late Oligocene upwelling in this region.

West of this forarc basin the Pacific Farallón Plate was being subducted under the North American Plate, which included the future Baja California granitic crust, in Oligocene time. As a result of subduction, an andesitic-rhyolitic volcanic arc built up during the Oligocene and Miocene, forming the ancestral Sierra Madre Occidental. The explosive activity of the arc produced ashfalls on the western marine continental margin, contributing volcanic as well as terrigenous clasts to the San Gregorio Formation.

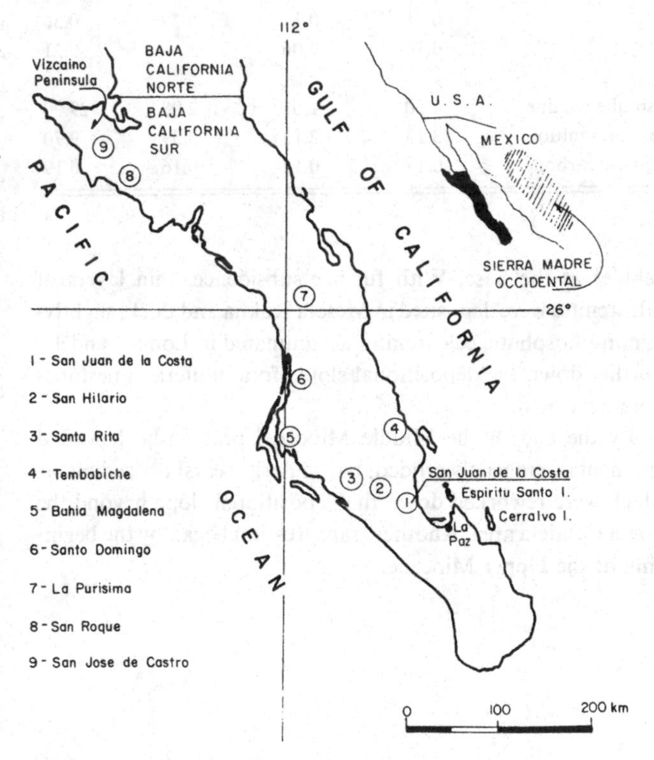

Fig. 10.1. Location of the San Juan de la Costa study area. Numbers correspond to other known phosphorite occurrences in Baja California Sur.

A probable trough-like topography, caused by intra-continental rifting in Late Oligocene time, redirected rivers from their former westerly paths to the Pacific Ocean, consequently sedimentation rates in much of the borderland dropped markedly (Dickinson, 1982); this in turn may have promoted phosphogenesis in sediments of the San Gregorio Formation. By early Miocene time, 40 m of calc–alkaline zeolitic volcanic gravels, green volcaniclastic arenites and pink ashes, derived from

Sierra Madre sources, were deposited together with large *Ostrea* sp. and *Arca* sp. populations in a shallow marine environment, producing the Isidro Formation.

Overlying the Isidro Formation are 1500 m or more of andesitic lahars, tuffs and flows, rhyolitic ignimbrites, and fluvial deposits of the Comondú Formation, the products of volcanic activity and associated sedimentary processes along the eastern margin of the present Baja California Peninsula. This activity culminated in the uplift of the peninsula above sea level (Hausback, 1984) and lasted until 12 Ma. Thus the slow uplift of Western Mexico caused the emergence of the phosphorite-bearing continental margin and the subsequent offlap of the volcanogenic deposits of the Isidro and Comondú Formations. The Gulf of California began to form 5.5 Ma ago after a long period of intra-continental rifting. Extension due to seafloor spreading produced listric normal faulting and the tilting of marginal blocks at the peninsular edge. Many of those blocks subsided and were completely inundated by the ocean; other tilted blocks (e.g. Espíritu Santo and Cerralvo islands) are presently above sea level. Parts of the San Juan de la Costa gulf margin have large landslide deposits possibly associated with the opening of the Gulf of California.

Earlier investigations of phosphorites in Baja California Sur

The phosphate deposits of Baja California Sur were first explored by FORNAS SA in 1956 (Salas, 1978) in the Recent beach sands along the Bahia Magdalena. This led to the discovery of low grade phosphate (3–5% P_2O_5). D'Anglejan (1963 and 1967) studied a pelletiferous–biogenic submarine deposit off the Santo Domingo–Vizcaíno Peninsula coasts and estimated 1.5–3.10^9 tons of P_2O_5. A preliminary exploration by means of an aerial radiometric survey on land discovered phosphate occurrences in San Hilario, San Juan de la Costa, and Santa Rita (Fig. 10.1). Roca Fosfórica Mexicana (ROFOMEX) started an exploration and evaluation program in San Juan de la Costa in 1977. Escandón (1977), Salas (1978), Anonymous (1980) and Mayoral (1982) reported on the geologic and mining engineering characteristics of the deposit. Hausback (1984) described the stratigraphy of the region with particular emphasis on the volcanic rocks.

ROFOMEX is exploiting the Humboldt phosphorite bed of the San Gregorio Formation which varies in thickness from 8 cm to 1.4 m. The average grade of the mined phosphorites is 20.0% P_2O_5, the proven reserves are close to 34 million tons, and the annual production is 0.7 million metric tons. Marginal localities in the San Juan de la Costa area are presently being investigated by means of geologic mapping and drilling.

San Gregorio Formation

The phosphate-bearing San Gregorio Formation was defined by Beal (1948) in the La Purísima area, 260 km northwest of San Juan de la Costa. The rocks of San Gregorio were named Monterey Formation by Heim (1922) due to the similarity of the Miocene Monterey Formation of California although the distance between the aforementioned localities and Southern California is approximately 1100 km. Paleontologic (VanderHoof, 1942), radiometric (Hausback, 1984; McLean, Barron & Hausback, 1984) and diatom biostratigraphic (Kim & Barron, 1986; Kim, 1987) evidence indicates a Late Oligocene age for the San Gregorio Formation, which Applegate (1986) included as a member of his El Cien Formation. The San Gregorio Formation unconformably overlies the Eocene Tepetate Formation although the latter is not exposed in the San Juan de la Costa area. The Isidro Formation (Lower Miocene?) rests unconformably on the San Gregorio Formation (Fig. 10.2).

The San Gregorio Formation is a sequence of marine fossiliferous yellow-brown sediments that are well stratified and have a minimum thickness of 78 m (Escandón (1977) reports 127 m). The formation is almost horizontal and thickens and coarsens upwards (Fig. 10.3). At San Juan de la Costa, the formation can be divided into two members.

Lower member

The lower member of the San Juan de la Costa section is composed of an upper slope facies assemblage of phosphatic sandstones (49%), mudrocks (41%) and phosphorites (10%). The most characteristic sedimentary primary structures are laminations in the mudrocks and hummocky cross-stratification and graded bedding in the phosphorites. Because this member contains the economic phosphorite layers, our research has been focused on it.

Upper member

The upper member is 39 m thick and consists of a marine shallow-water facies assemblage of low-grade phosphatic arenites (53%), coquinites (24%), conglomerates (15%) and siliceous mudrocks (8%). The most characteristic primary sedimentary structures are unidirectional and herringbone cross-stratification. The upper member does not contain phosphorite-rich sediments.

Field characteristics and petrology of the lower member

Phosphorites

Phosphorites of the lower San Gregorio are high-grade phosphate ooidal–peloidal sandstones, which, based on thin-section estimates, consist of > 50% phosphate grains and cements. In weathered outcrops the phosphorite is yellow-brown; if unweathered it is gray. Thickness of the phosphorite beds varies over distances of several hundred meters from a few centimeters to 1.4 m; the latter is the thickness of the Humboldt bed, the thickest and richest at San Juan de la Costa (Fig. 10.3). A total of 15 high-grade phosphorite beds were recognized in the measured section (Fig. 10.3), with an aggregate thickness of 3.87 m. The P_2O_5 contents of these 15 phosphorite beds averages 17.8%; the highest value is 25.8% and the lowest is 15.1%. Many phosphorite beds have graded bedding, others are massive (and probably burrowed), and a few show hummocky cross-stratification. Several phosphorite beds seem to have been deposited by a single high-energy event (e.g. a turbidity current), but a thin persistent bentonite layer within the Humboldt bed indicates at least two separate depositional events.

Most of the phosphorites are grain-supported packstones, but in thin sections the packing of grains varies widely. In some samples, point contacts dominate and suggest early

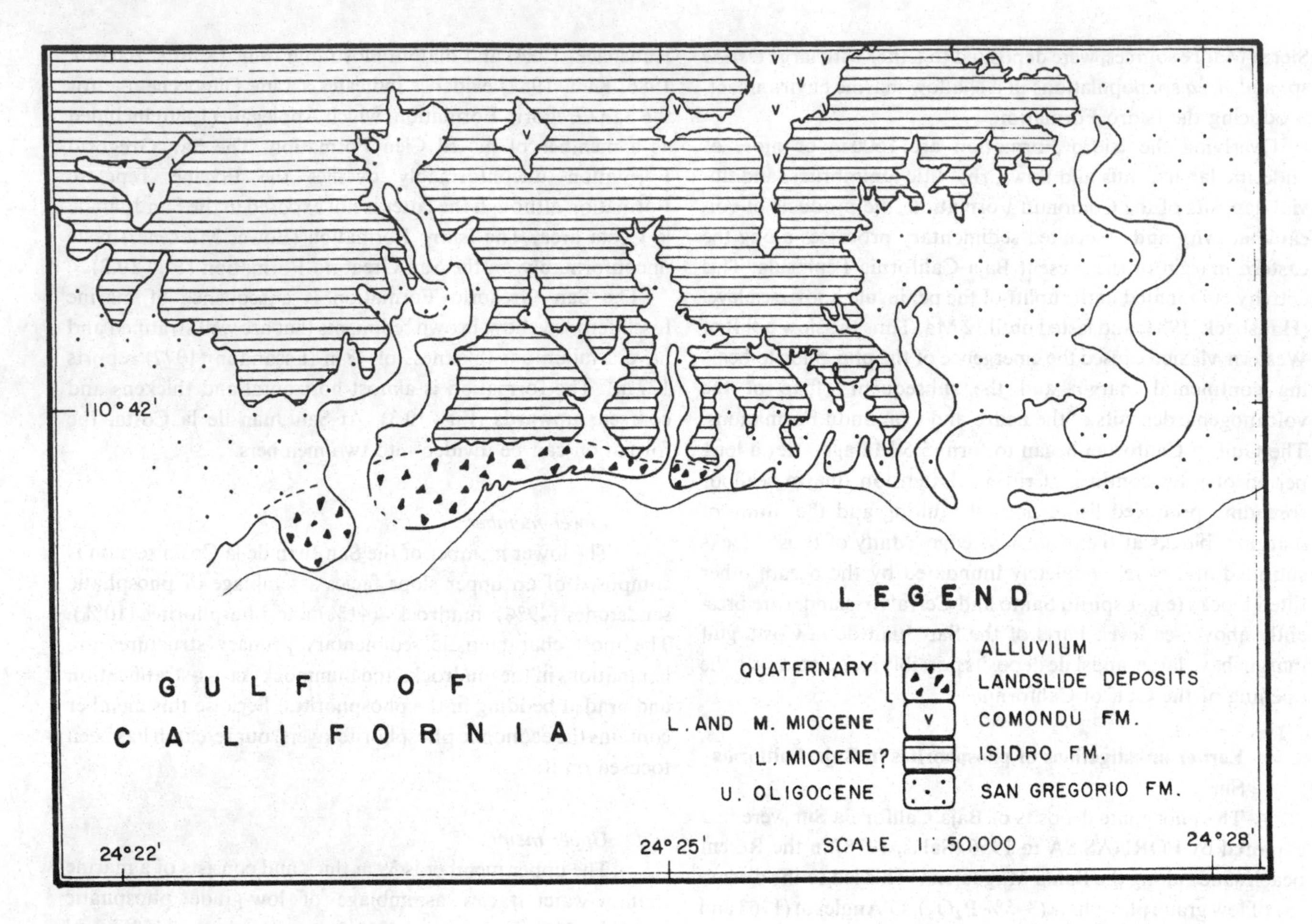

Fig. 10.2. Generalized geologic map of the San Juan de la Costa area.

cementation. In other samples grain contacts are concave–convex and fused, with common distorted micas. Sorting varies from moderate to good.

Framework grains constitute between 70 and 90% of the rock. Grain components are phosphate ooids, peloids, and intraclasts, siliciclastic grains and fossils – many of the latter are phosphatized. On the average ooids are 250 μm in diameter and have nuclei of terrigenous grains or fossils surrounded by concentric laminae which suggest one or more accretion events. Some laminae are eroded or bored. Tangential crystallites within the concentric laminae are common and suggest the ooids may have formed in an agitated environment. Broken ooids and scoured surfaces between laminae are additional evidence of high-energy processes.

Phosphate peloids are spherical–prolate grains lacking laminae or any other discernible internal structure. Algal debris, small terrigenous particles and unidentified crystallites are major components of these massive grains.

Phosphate intraclasts are spherical–irregular grains up to several millimeters in diameter. Composition of the intraclasts is extremely variable and includes algal limestone clasts, bone fragments, and clasts of cemented phosphate ooids and peloids, microsphorite and terrigenous grains.

Terrigenous components are angular, up to granule size, extrabasinal grains of feldspar, quartz, mica, volcanic rock fragments, and glass, and less common hornblende, detrital hematite

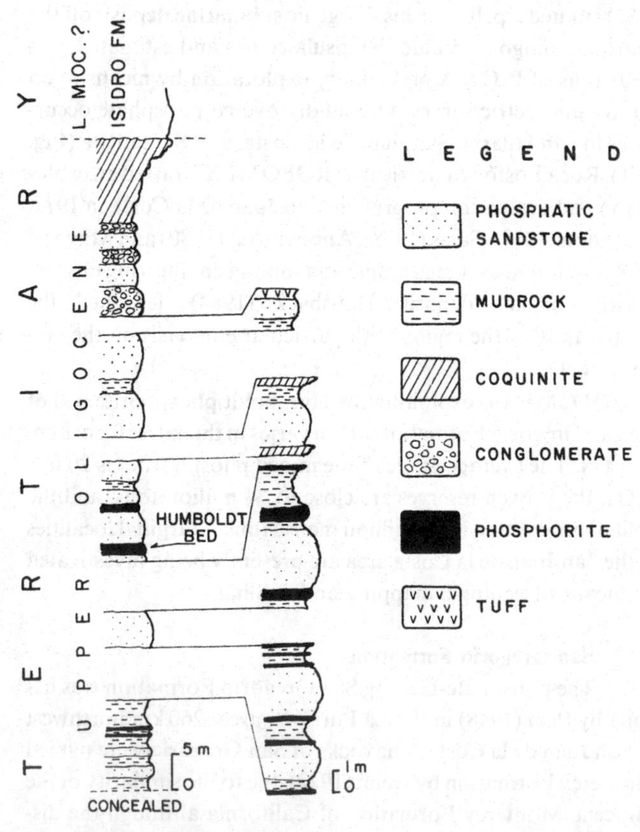

Fig. 10.3. Measured section of the San Gregorio Formation in San Juan de la Costa area.

and zircon. Plagioclase grains, many of which are zoned, are the most common terrigenous component. The abundance of plagioclase in the phosphorite, as well as in other facies of the lower member, suggest that the source area for the sediments was primarily the ancestral Sierra Madre Occidental, then an active volcanic arc located to the east (Fig. 10.1). Authigenic glauconite is rare.

Fossils are foraminifers, mollusks, mammal and fish bones, shark teeth, ostracods, calcareous algae and echinoderm spines. Some of these occur within phosphate ooids, peloids or intraclasts. Burrows are common in some phosphorite beds.

Cements in the phosphorites include carbonate, cryptocrystalline apatite, chalcedony, hematite and zeolite. A few phosphorites have clay or micrite matrices. In many phosphorites the framework grains are rimmed by a 20–25 μm thick layer of phosphatic cement, which fills pores partially or totally, and forms bridges between adjacent grains. The remaining open spaces were later filled by chalcedony. Other phosphorite have cements of sparry calcite and almost isotropic cryptocrystalline apatite.

Phosphatic sandstones

Phosphatic sandstones are low-grade phosphatic ooidal–peloidal sandstones which contain an average of about 20% phosphate grains and cement, as estimated in thin section. The sandstones share many of the macroscopic and microscopic characteristics of high-grade phosphorites with which they form a gradational series. In the lower member, low-grade phosphatic ooidal–peloidal sandstones occur in 19 beds with a total aggregate thickness of 19 m. The average ooid diameter is 163 μm and average P_2O_5 content is 6.8%.

Siliceous mudrocks

Most mudrocks are laminated, gray if unweathered, mixtures of silt, clay, and very fine sand in which framework grains (generally <40%) are embedded in matrix and cement. Kim (1987) has shown that the mudrocks of the San Gregorio Formation contain abundant diatoms. Framework grains are dominantly very angular plagioclase and possibly other feldspars, quartz and mica, with grain sizes between 20 and 100 μm. Glauconite grains, 25–100 μm in diameter, compose <1%. The glauconite is angular and varied in grain size in some beds, and is of uniform size in other beds, suggesting that the glauconite is both authigenic and detrital.

Phosphatic fish scales and bones, sharkteeth and large mammal bones are common in certain strata. Also common are phosphatized mollusks, echinoderms, ostracods, algae and particularly foraminifers. Some beds are bioturbated (*Nereites*) and the burrows are generally filled with phosphate ooids and peloids from overlying phosphorite beds.

Phosphate grains 20–250 μm in diameter occur scattered in the mudrocks. These grains are generally massive rounded peloids, irregular fossil fragments, or diagenetic micronodules and peloids. In thin sections they rarely form >5% of the rock. P_2O_5 analyses of 23 mudrock beds in the lower member average 3.3%, but two samples rich in peloids of probable fecal origin have P_2O_5 contents of 26.6 and 22.9%, respectively.

Origin of the San Juan de la Costa phosphorite

Field and petrological characteristics of the phosphorite show that the phosphate grains are polygenetic and were concentrated during one or more cycles of reworking on the seafloor. The high concentration of phosphate grains, their good sorting, laminae of the cortex surrounding the nuclei of ooids with their tangential crystallites and erosional features present in the laminae suggest the following. The ooidal grains had a long history of accretion, interruptions of growth and repeated erosion that posibly took place in shallow phosphorus-saturated waters that were intermittently agitated. Actual phosphatization of these coated grains may have been microbially-mediated (cf. Soudry & Champetier, 1983). Subsequently many of the grains, mixed with phosphate fecal pellets and intraclasts, and other ooids and fossils, were transported as clastic particles to somewhat greater depths in response to high-energy storms, mainly as turbidity currents. On the upper slope these allochthonous sediments were deposited intercalated with autochthonous siliceous muds, the precursors of the mudrocks.

In contrast, nearly all grains in siliceous mudrocks are either phosphatized fossils or early diagenetic phosphate micronodules and peloids. The latter were probably related to bacterial concentration near the sediment–water interface or in interstitial pore waters. In any event, they differ markedly from the phosphate grains present in the phosphorites.

In summary, the characteristics of the phosphate grains of the phosphorites are different from and much more complex than the phosphate grains of mudrocks. Clearly, the phosphate grains of the granular phosphorites were not derived from winnowing and reworking of phosphate grains in the upper slope mudrocks; they were derived, instead, from a shallow-water environment and transported to an upper slope environment.

Conclusions

(1) The San Gregorio Formation of Baja California Sur is of Late Oligocene age. It accumulated in a shelfal to upper slope part of a forearc basin, in a probable upwelling setting. Only the lower member of San Gregorio contains high-grade beds of phosphate ooidal–peloidal sandstone. At San Juan de la Costa, ROFOMEX exploits of the Humboldt bed, including the phosphate-bearing bioturbated top of the underlying mudrock, revealed averages of 20.0% P_2O_5.

(2) The San Gregorio Formation at San Juan de la Costa has a minimum thickness of 78 m. The measured section is divided into two 39 m members, the upper and lower members. The phosphate-rich lower member is an upper slope facies assemblage of low-grade phosphatic sandstone (49%) siliceous mudrock (41%) and high-grade phosphorite (10%). The high-grade phosphorite occurs in 15 beds with a total aggregate thickness of 3.9 m, and P_2O_5 values that average 17.8%.

(3) Phosphorite components are polygenetic phosphate grains concentrated during one or more cycles of reworking. Ooidal features indicate that they have a long history of accretion, interruptions of growth and repeated erosion. Mixed with phosphate peloids, intraclasts, and fossils, the phosphate ooids were transported by turbidity currents and other high-energy processes from shallower to deeper water environments, and

deposited in an area of siliceous mud accumulation on the upper slope.

(4) Characteristics of the phosphate grains in the phosphorite beds are different from and more complex than the late diagenetic phosphate peloids and micronodules of the siliceous mudrocks. It is concluded that the phosphorites were not derived from winnowing of the phosphate grains in the upper slope siliceous muds, but instead were formed initially on the seafloor in a shallow marine environment.

Acknowledgments

We thank ROFOMEX and particularly A. Castelo, S. Pantoja, M. Martinez and L. Aguilar, who were involved in various stages of the fieldwork. I (C.G.O.) thank the Deutscher Akademischer Austauschdienst (DAAD) for partial support in the preparation of this report. We express our appreciation to Rudolf Fischer (Universität Hannover) and Luis Segura (UABCS) for their advice in the field and laboratory.

References

Anonymous, (1980). Mexico – phosphate rock producer of the future. *Phosphorous and Potassium*, **109**, 20–21, 24.

Applegate, S.P. (1986). The El Cien Formation, strata of Oligocene and early Miocene age in Baja California Sur. *Revista*, **6(2)**, 145–62.

Beal, C.H. (1948). Reconnaissance of the geology and oil possibilities of Baja California, Mexico, *Geological Society of America Memoir*, **31**, 138.

D'Anglejan, B.F. (1963). Sobre la presencia de fosforitas marinas frente a Baja California, México. *Boletin de la Sociedad Geológica Mexicana*, **26(2)**, 95–101.

D'Anglejan, B.F. (1967). Origin of marine phosphorites of Baja California, Mexico, *Marine Geology*, **5**, 15–44.

Dickinson, W.R. (1982). Space-time evolution of Cretaceous–Tertiary tectonomagmatic provinces in the southwestern United States, GSA *Cordilleran Section Meeting*, abstracts with programs, **14(4)**, 160.

Escandón, V.F.J. (1977). Bosquejo geológico de los depósitos de fosforita de San Juan de la Costa, Baja California Sur. *Roca Fosfórica Mexicana*, (unpublished).

Hausback, B.P. (1984). Cenozoic volcanic and tectonic evolution of Baja California Sur, Mexico. In *Geology of the Baja California Peninsula*, ed. V.A. Frizzell, Jr, pp. 219–36. *Society of Economic Palaeontologists and Mineralogists, Pacific Section*, Los Angeles.

Heim, A. (1922). The Tertiary of southern lower California. *Geological Magazine*, **59**, 529–47.

Kim, W.H. (1987). 'Biostratigraphy and depositional history of the San Gregorio and Isidro formations, Baja California Sur, Mexico'. Unpublished Ph.D. thesis, Stanford University, California.

Kim, W.H. & Barron, J.A. (1986). Diatom biostratigraphy of the upper Oligocene to lowermost Miocene San Gregorio Formation, Baja California Sur, Mexico. *Diatom Research*, **1**, 169–87.

Loughman, D.L. (1984). Phosphate authigenesis in the Aramachay Formation (Lower Jurassic) of Peru. *Journal of Sedimentary Petrology*, **54(4)**, 1147–56.

Mayoral, M.L. (1982) 'Geología y yacimientos minerales de fosforita del área de San Juan de la Costa, Municipio de La Paz'. B.C.S. Unpublished Universidad Nacional Autónoma de México, Facultad de Ingeniería.

McLean, H., Barron, J.A. & Hausback, B.P. (1984). The San Gregorio Formation of Baja California, Mexico is Late Oligocene. *Society of Economic Palaeontologists and Mineralogists, Pacific Section*, (abstracts with programs).

Salas, G.P. (1978). Sedimentary phosphate deposits in Baja California, Mexico. *1978 AIME Annual Meeting*, **78-H-75**, 1–22.

Soudry, D. & Champetier, Y. (1983). Microbial processes in the Negev phosphorites (southern Israel). *Sedimentology*, **30**, 411–23.

Van der Hoof, V.L. (1942). An occurrence of the Tertiary marine mammal, Cornwallis, in Lower California. *American Journal of Science*, **240(4)**, 298–301.

11

Phosphates in West and Central Africa – the problem of Neogene and Recent formations

G. SUSTRAC, R. FLICOTEAUX,
J.R. LAPPARTIENT AND J.P. PRIAN

Abstract
The principal phosphate concentrations discovered in West and Central Africa are assigned to the Eocene and to the passage from Eocene to Oligocene. The age of the calcium phosphates at Taiba and Thiès in West Senegal and at Farim in Guinea Bissau is Middle Eocene and locally Upper Eocene–Oligocene. The calcium phosphate deposits at Ouali Diala, located on the Senegal River south of Matam, and at Bofal-Loubboïra in Mauritania are reported to be Lower Eocene (Fig. 11.1). The phosphate occurrences in the People's Republic of Congo are assigned to the Upper Cretaceous, and the deposits in Cabinda to the Upper Cretaceous and Eocene. The aluminum phosphate weathering on the Thiès plateau in West Senegal is polyphase, the various phases identified ranging from Oligocene to Lower or Middle Pleistocene.

As far as is known, the Neogene represented a period of: relatively discrete phosphate genesis, having given rise to the phosphate occurrences of Casamance (southern Senegal) (Gorodiski, 1958) and to some of the phosphates offshore of the People's Republic of Congo and of Gabon[1]; and dismantling or reworking of earlier phosphate formations having given rise to new concentrations such as in Gabon and the Pire Goureye-type deposits in West Senegal. The latter are generally assigned to continental phenomena post-dating the Middle Eocene (Upper Lutetian) and provisionally to the Oligocene (Tessier, 1952). Some uncertainty remains as to the extension of the Miocene marine formations in Senegal, as discussed in work by J.R. Lappartient (1985).

This paper does not give an overview of the geodynamics of the western margin of Africa, but concentrates on stratigraphic and sedimentologic results of exploration.

[1] Dismantling of both the older phosphatic formations and of Neogene to Recent phosphate deposits is effectively involved in this case.

The Neogene as related to phosphates in West Senegal

The stratigraphy of the Tertiary formations in West Senegal has been established by Tessier (1952) and Lappartient (1970 and 1985) on the basis of the macrofauna, and by Chino (1963), Monciardini (1966), Brancart & Flicoteaux (1971), Brancart (1977), Ducasse, Dufaure & Flicoteaux (1978) and Flicoteaux (1980) on the basis of the microfauna. The reconstruction illustrated in Table 11.1 may be drawn from work by these authors, and in particular by Flicoteaux who has studied sections of the three main phosphate deposits in West Senegal (Lam Lam, Pallo and Taïba).

Above the sandy–argillaceous Campanian and Maastrichtian facies which crop out in the Ndiass dome (Fig. 11.2), the Paleocene is sandy at the base and otherwise essentially calcareous. The cause of the intensely karstified aspect of the Paleocene is still subject to discussion. Clayey and marly deposits predominate in the Eocene units. The basal marly–calcareous or sandy layers containing flints, phosphate and glauconite fossilize the 'karstic' Paleocene facies of the Ndiass dome. They are overlain by paper clays containing attapulgite and by marls, calcareous marls, and limestones which predominate towards the top of the Lower Eocene. The Lutetian deposits overlie the earlier deposits in continuity, being distinguished by the microfauna. Clayey–marly and calcareous facies predominate. The top of the series is commonly truncated due to erosion which occurred from the Upper Eocene onwards. The calcium phosphates at Lam Lam and Taïba, which occur within an extensive condensed sequence over the Thiès plateau, are assigned to the Middle Eocene, to part of the Upper Eocene and to the basal Oligocene. A comparative log of the Lam Lam and Taïba sequences is shown in Figure 11.3. At Taïba, the quartzose phosphate sequence which represents the main ore mass abuts to the east against a unit of Middle Eocene phosphatic limestone containing nummulites, or is interdigitated within this unit. This is a lateral facies variation. The phosphate formation disappears to the west, mainly because it is channelled by more recent units assigned to the 'Late Continental' and Ogolian. The Late Continental is a term referring to the Tertiary continental layers of the Sahara and later extended to other basins. It is used for those sediments that result from continental weathering of Neogene marine sequences (Lappartient, 1985). The Ogolian corresponds to an arid episode during the Upper Pleistocene (Elouard, 1962).

The aluminum phosphates which cover the greater part of the Thiès plateau were formed at the expense of the calcium phosphates and are assigned to the period between the Oligocene and the Lower or Middle Pleistocene (Flicoteaux, 1982).

The sequence with which the present study is particularly concerned overlies the Middle Eocene phosphatic limestone. This limestone not only acts as a regional control for the principal calcium phosphate formation, but also represents the bedrock of the phosphate concentrations encountered at the base of the overlying clayey–sandy sequence.

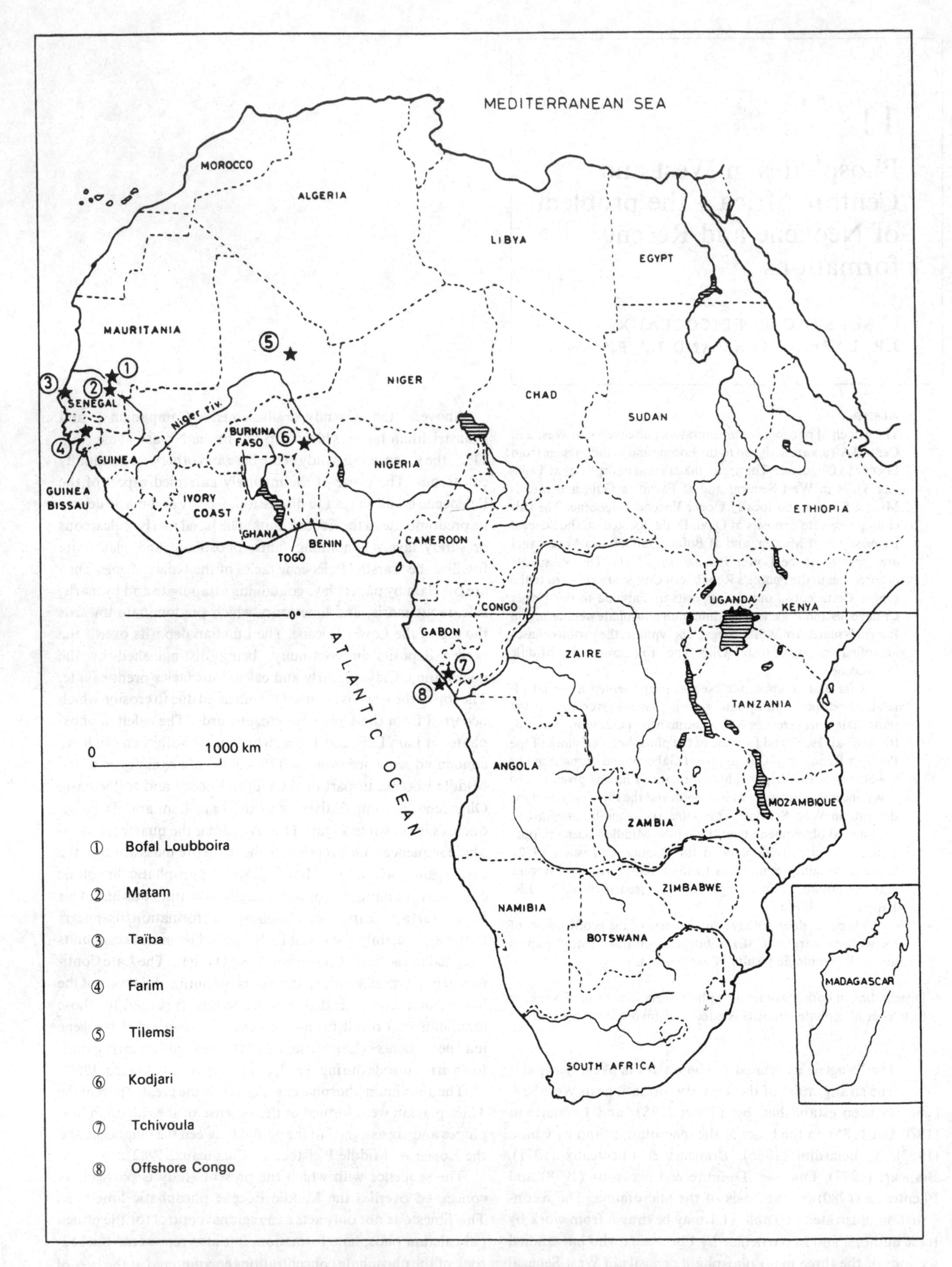

Fig. 11.1. Location of phosphate deposits of west and central Africa in which BRGM has conducted prospecting and assessment work.

Table 11.1. *General correlation between Lam Lam, Taïba and Pallo phosphate deposits in west Senegal (Flicoteaux, 1982; Flicoteaux & Hameh, 1989)*

Age of Deposition[a]	Lam Lam	Taïba		Pallo (Thiès Plateau)
Quaternary	Superficial formations	Dune sand (Ogolian)		Surficial formation
Oligocene or Miopliocene	Iron crust and phosphatic sandstone	Phosphatic sandstone		Ferruginous, kaolinitic and phosphatic sandstone
Upper Eocene–Oligocene	Bedded aluminum phosphate Platy flints and variegated clays	Silico-ferruginous layers	Bedded aluminum phosphate Alternating beds of phosphate, clays and flints	Phosphatic lateroïds
		Variegated clays		
Middle Eocene		Calcium Phosphate	homogeneous facies heterogeneous facies phosphate gravel	Flints
	Lam Lam marl Shelly limestone	Laminated clay		Clay with phosphate beds Shelly limestone

[a]The age proposed in the stratigraphic column is that of the unaltered rocks and not that of the alteration.

Another phosphate occurrence is present 20 km east of the Lam Lam and Taïba phosphates and the aluminum phosphates of the Thiès plateau. Termed 'Gueoul' (Tessier, 1952) from the name of the region where it has been identified, it occurs between the Middle Eocene nummulite-bearing phosphatic limestones and the Late Continental and Ogolian overburden. Previous surveys showed a relatively wide extension, called Pire Goureye phosphate, between Lam Lam, Baba Garage and Mèkhé (Fig. 11.2). This extension is known by the name of the location near which the main mineral concentrations of this type were found.

The Pire Goureye deposit: prospecting and type sequence

Over a vast area stretching from Pire Goureye to Kébémer and Sagata, the Middle Eocene nummulite-bearing phosphate limestones are overlain by a formation consisting of a melange of blocks of varied composition within a clayey–sandy matrix. These rounded blocks of up to 0.40 m in size contain various types of sandstone and phosphate, in places bearing distinguishable nummulites. The phosphate blocks, which form thick accumulations of one to several metres, show relatively low P_2O_5 grades and are aluminum- and iron-rich.

Prospecting was mainly conducted from 1948 onwards by BRGM's predecessor, the Bureau Minier de la France d'Outre Mer (BUMIFOM). Between 1948 and 1950, the Pire Goureye deposit was explored by pits and Platyx drill holes on a 400 m grid, in places tightened to 200 m. The characteristics of the deposit are as follows:

Total area: 428 Ha (4.3 km^2)
Reserves: 25 Mt grading 34.3% P_2O_5 (for scrubbed ore)
Average thickness of barren overburden: 25.6 m
Average thickness of phosphate layer: 6.15 m
Average run-of-mine: 22.8% P_2O_5
Fe–Al grade: commonly 5%

From 1952 onwards, BUMIFOM's work was continued by the Société d'Etudes et de Recherches Minières du Sénégal (SERMIS). Apart from the work completed in the Taïba area which is not discussed in detail here, SERMIS also explored the area covered by Permis Général de Recherches (PGRA) no. 10, including the Pire Goureye deposit in its south part. Work was comprised of almost 200 drill holes in the Mekhé–Kebemer–Sagatta area, but no concentrations equivalent to that at Pire Goureye were discovered. The overburden is at least 40 m thick, overlying a slightly phosphatic formation with a maximum thickness of 1 m. Figure 11.4 shows three typical logs for the area covered by the exploration permit. None of the drill holes penetrated deep into the footwall limestone because the method employed (Banka-type holes) was technically incapable of penetrating the limestones more than superficially.

The presence of Pire Goureye-type phosphates further north was confirmed during prospecting by H. Heetveld in 1964–5 (Fig. 11.2). The phosphates are irregularly distributed, being more-or-less continuous in the south where they are up to 2 m thick, as opposed to 0.5 to 1 m in the center of the permit area. They take the form of gray-yellow clayey sand grading 0.1–8.5% P_2O_5, and are either in direct contact with the Lutetian limestone bedrock or separated from it by 0.5–1.4 m of sand and clay. Further northeast and in the Ferlo valley (Fig. 11.2), the Lower Eocene layers stratigraphically below the Middle Eocene fossiliferous limestones are visible (Trenous, 1970).

The phosphate occurrences of Casamance (S. Senegal) and the problem of the marine Neogene in Senegal

The rare phosphate occurrences assigned to the clearly marine Neogene have been described in the lower part of Casamance (Fig. 11.5). The first occurrence was discovered in 1958 from a petroleum drill hole at Ziguinchor, in which A. Gorodiski describes the following succession in the upper part of the hole:

Recent Quaternary: 0–12 m; beige more-or-less clayey littoral and fluviatile sand containing rare fragments of *Arca* sp.

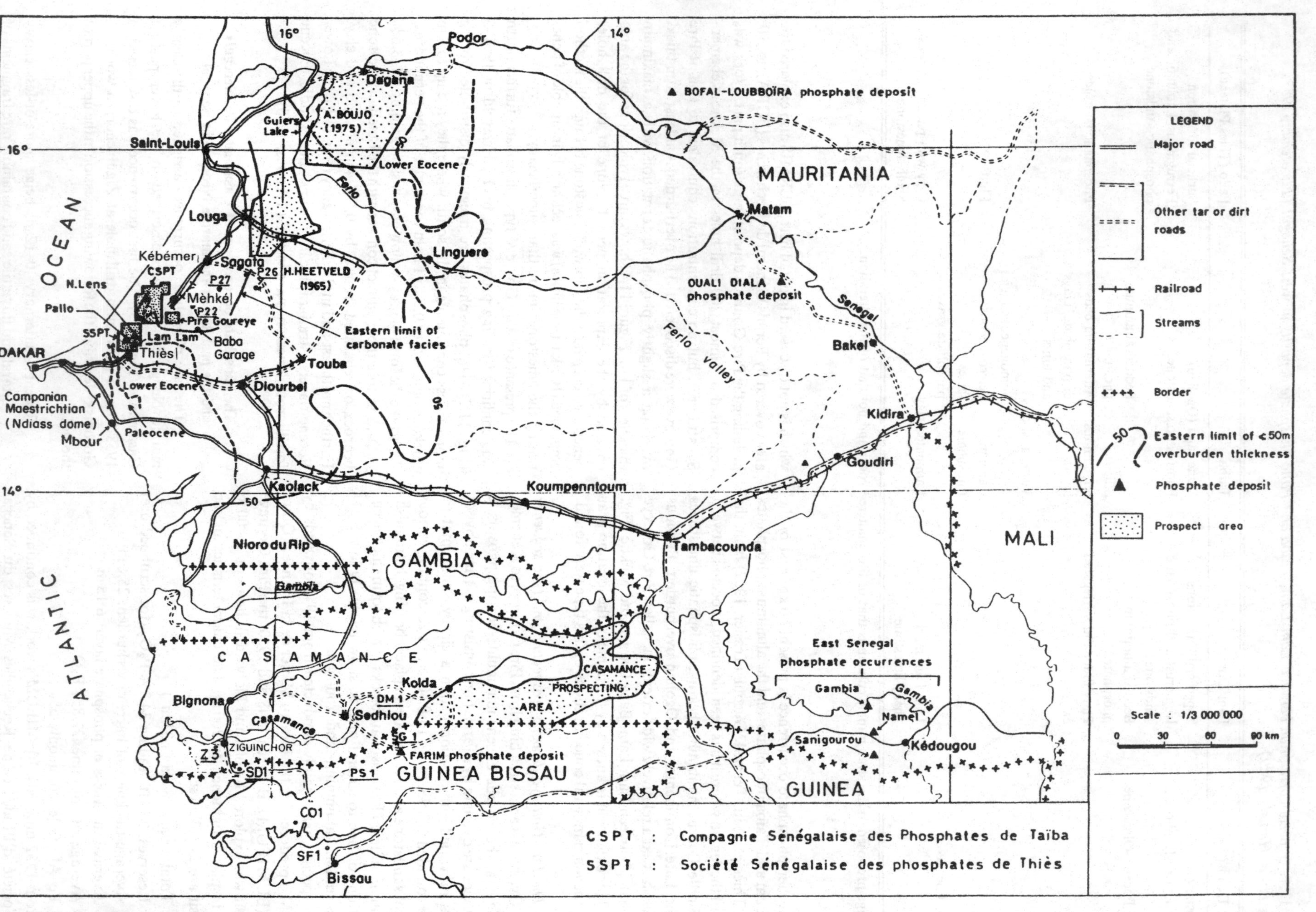

Fig. 11.2. Senegal. Location of phosphate deposits, occurrences and prospect areas.

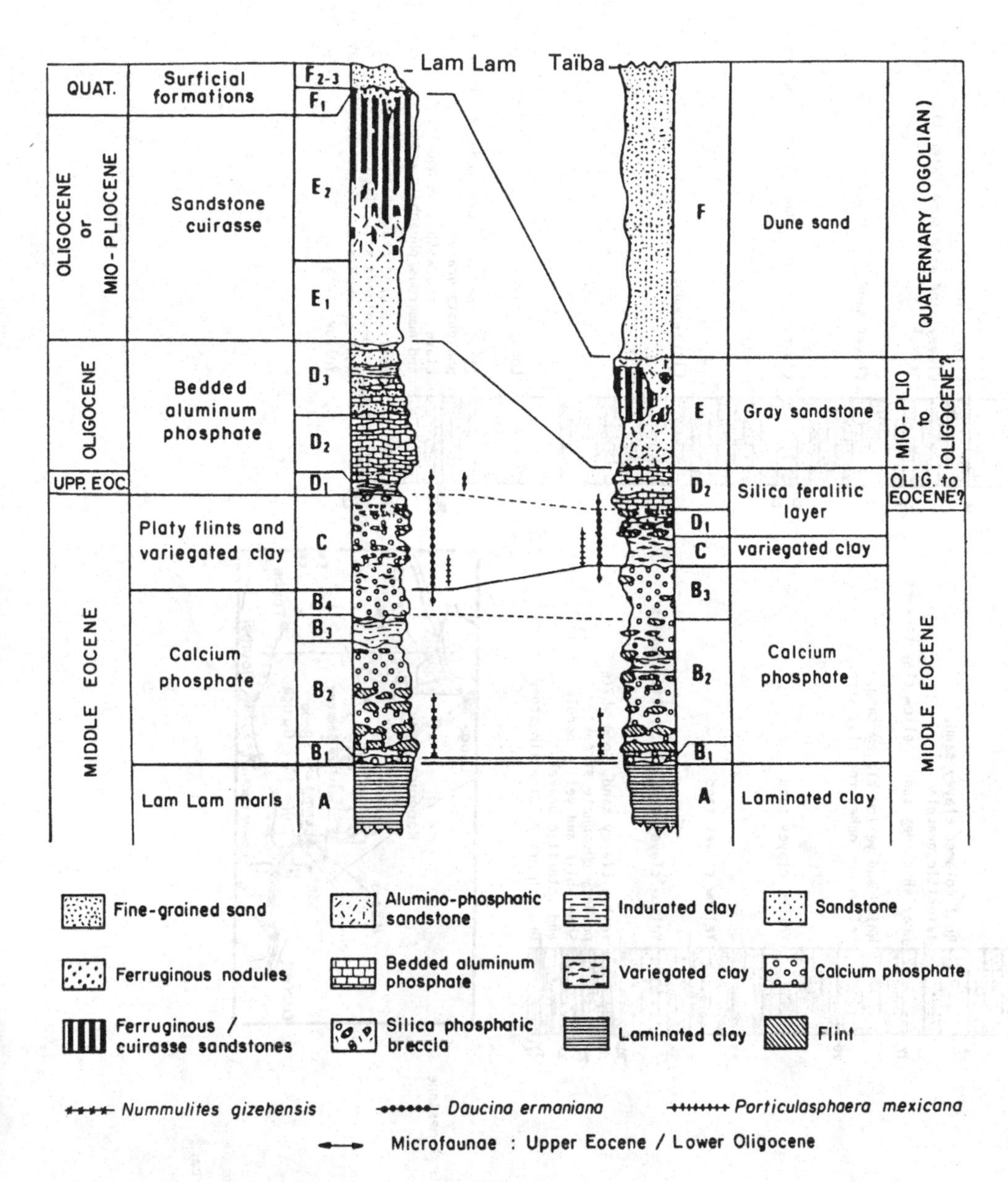

Fig. 11.3. Representative lithostratigraphic sequences of the Lam Lam and Taïba phosphate deposits (from Flicoteaux, 1982, Fig. 39).

Pliocene or Quaternary: 12–21 m; laterite and continental azoic sand
Middle Miocene: 21–42.5 m; clayey sand containing phosphatic grains and fragments
42.5–52.5 m; pale greenish clay
52.5–67 m: dark blue clay
Lower Miocene: 67–96 m; marl and limestone

P_2O_5 values are very low: 0.15–0.45% P_2O_5 between 13 and 21 m; 1–1.8% P_2O_5 between 22 and 42.5 m, and 0.5–0.9% P_2O_5 between 42.5 and 61 m.

The same lithologic succession has been encountered in water exploration boreholes at Bignona (Tessier *et al.*, 1975; Flicoteaux & Medus, 1980), Djibelor and Tanaf (Ly, 1985), in petroleum drill holes at Kafountine and in phosphate exploration drill holes at Farim (Prian & Flicoteaux, unpublished). This succession is underlain at the base by sands and dark gray – black lignitic clays overlying carbonaceous or phosphatic layers of different age (Middle Eocene–Upper Oligocene or basal Miocene). Phosphatic grains, coprolites, teeth and bone fragments are encountered at various levels within this succession, particularly in the blue or green clays and in the clayey-sands located above the limestone horizon. At Bignona, the green clays and clayey-sands have been weathered, and aluminum phosphate is locally observed (Tessier *et al.*, 1975). As at Ziguinchor, the greater part of this predominantly clayey–sandy sequence is assigned to the Miocene, the uppermost weathered and azoic portion, dated as Pliocene or Quaternary by Gorodiski, being assigned to the Late Continental. The exact position of the sequence within the Miocene remains open to discussion: Flicoteaux & Medus (1980) mention the Lower and Middle Miocene and Ly (1985) refers to the Miocene *sensu largo*, although he demonstrates that there may be a hiatus in the Aquitanian at the base and that the sequence does not extend beyond the Upper (not Late) Miocene.

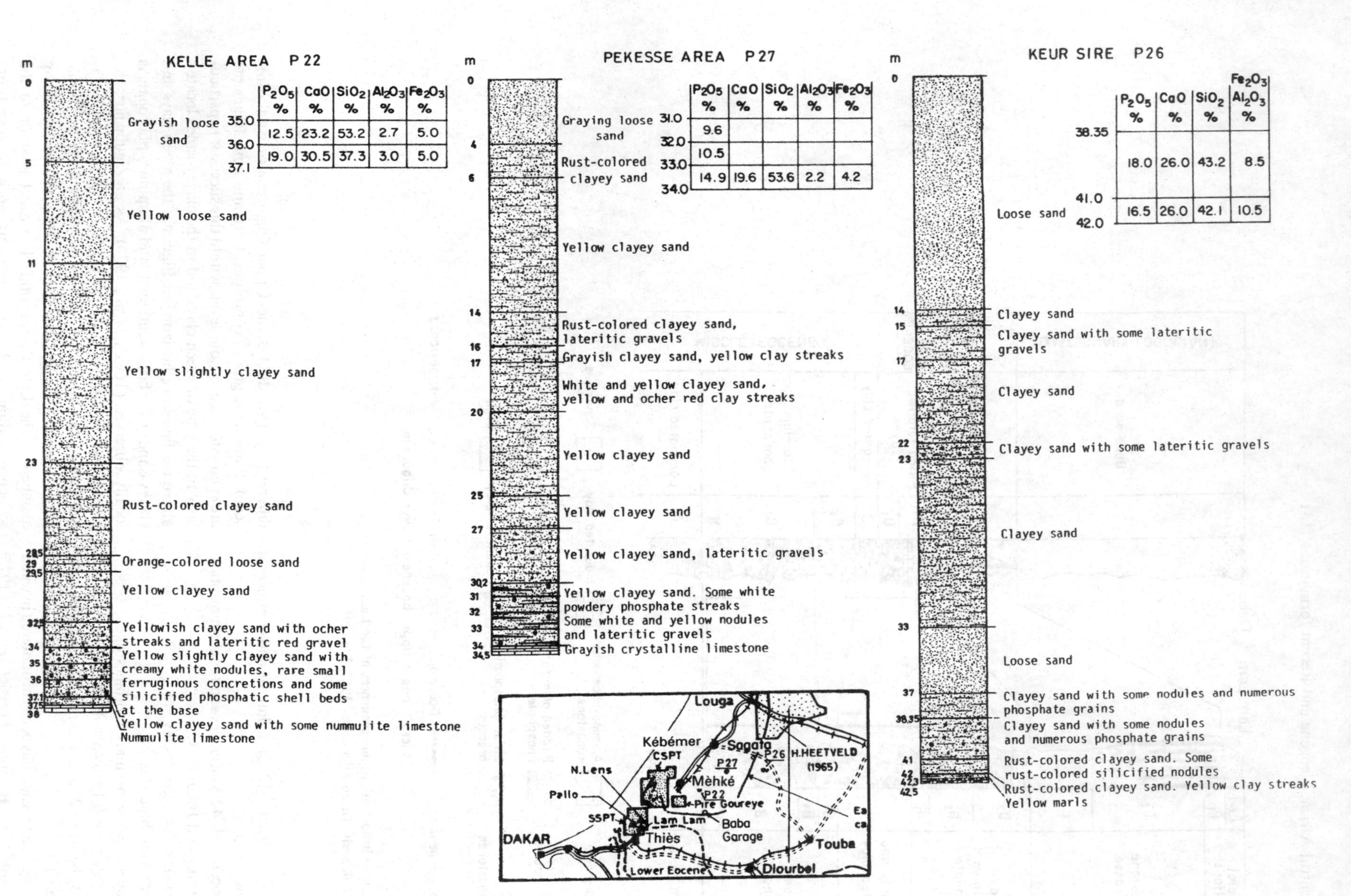

Fig. 11.4. Drilling in the Mèkhé–Kébémer–Sagatta area.

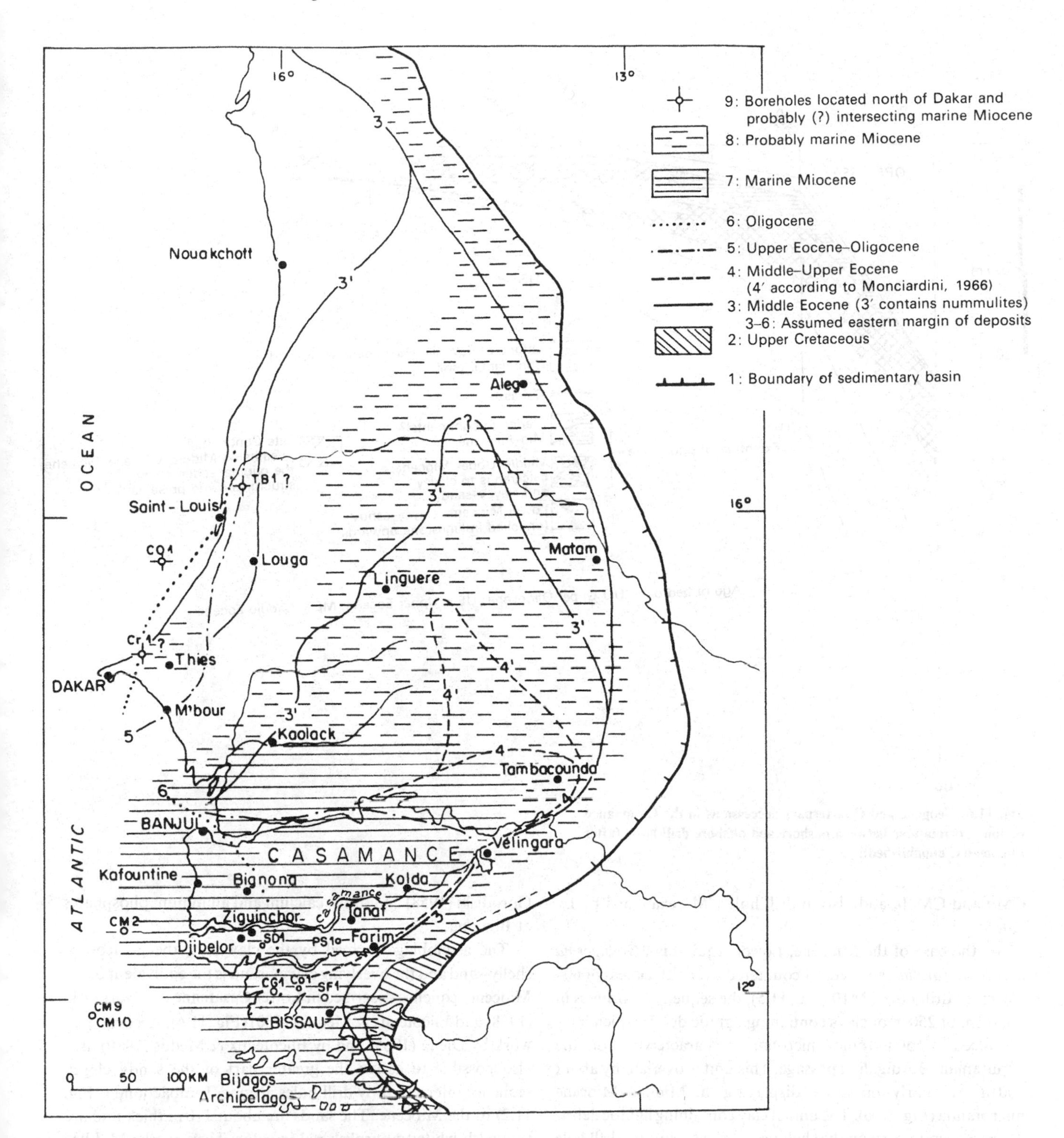

Fig. 11.5. Distribution of Paleogene and Miocene marine deposits (modified after Flicoteaux & Medus, 1980, Lappartient, 1983; and completed by J.P. Prian for Guinea Bissau).

According to data from drill holes in lower part of Casamance (Figs. 11.2 and 11.5), the Miocene and Late Continental cover is too thick (40–200 m) for phosphate concentrations formed during the principal regional episode of phosphate genesis (the Lutetian) to be accessible. For this reason, work between 1980 and 1982 was focussed on the Kolda and Velingara areas (Figs. 11.2 and 11.5) where the chances of encountering a lesser overburden accompanied by condensation of the Eocene sequence are better. As far as the Neogene period is concerned, drilling confirmed the presence of Pire Goureye-type occurrences at the base of the uppermost clayey–sandy sequence assigned in this region to the Late Continental (Bense, 1962). Here again, the occurrences are rich in iron and aluminum.

The marine Miocene is most accurately dated offshore of lower Casamance, at the edge of the continental plateau. The passage from Oligocene to Miocene is continuous in drill holes

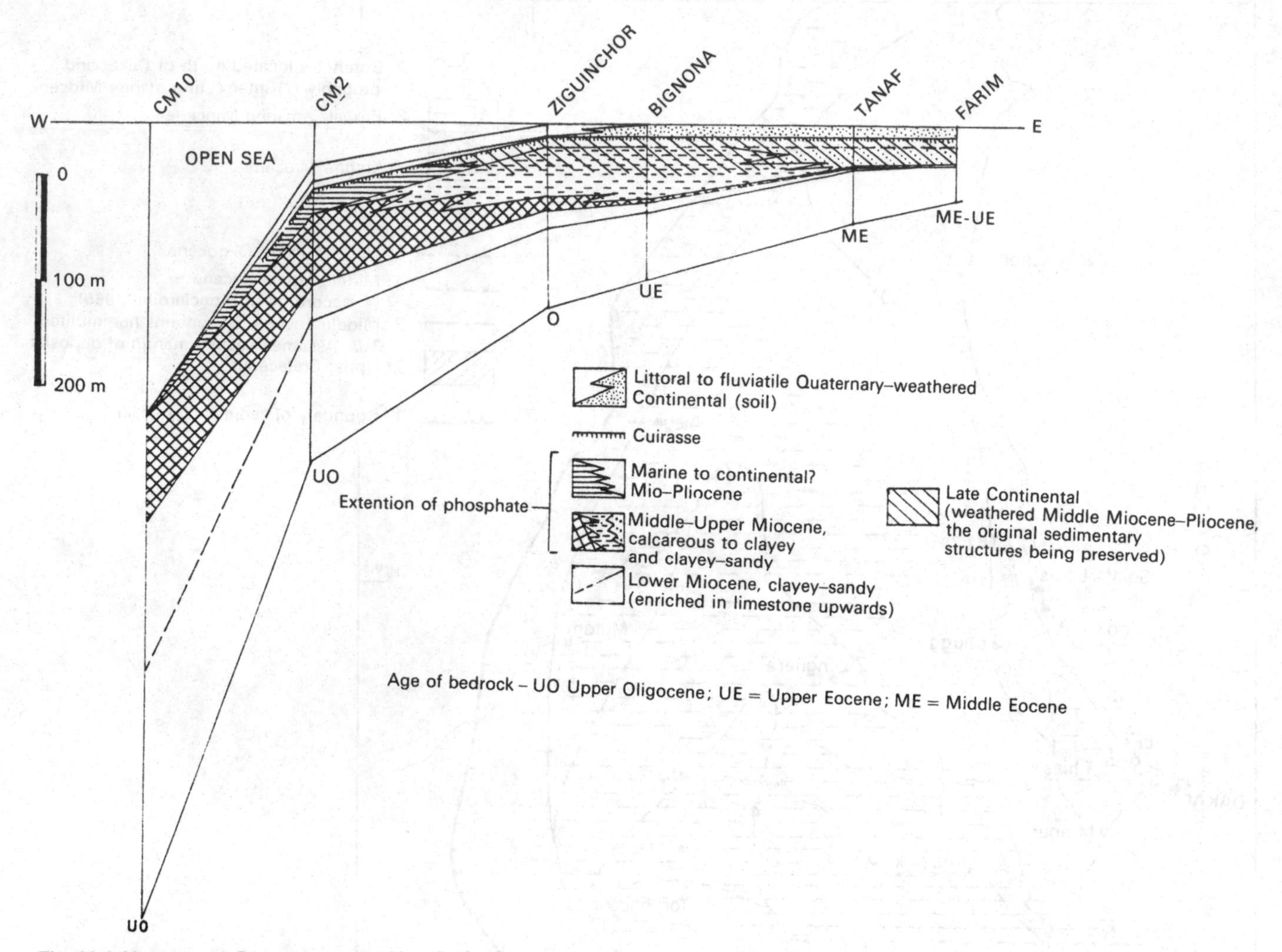

Fig. 11.6. Neogene and Quarternary successions in the Casamance region. Correlations between onshore and offshore drill holes (after Flicoteaux, unpublished).

CM9 and CM10, and also in drill hole CM2 examined by Ly (1985).

At the base of the Miocene, Lower Aquitanian *Globigerina* and *Globoratalia* have been encountered over a thickness of 60–100 m. In drill hole CM10 (Fig. 11.5), the sequence continues in the form of 250 m of clays containing lignitic debris which have produced a planktonic microfauna characteristic of the Aquitanian–Burdigalian passage. This unit is overlain by about 100 m of marly-limestone displaying a Middle Miocene microfauna (Fig. 11.6). The unit of clay containing lignitic debris is again encountered on the shelf near the shoreline (in drill hole CM2), where it is enriched in coarse-grained components. It is assigned to the Lower Miocene by Ly (1985), who distinguishes Aquitanian and Burdigalian above the transgressive Late Oligocene. The determination of Aquitanian and Burdigalian by Ly confirms the opinion advanced by Bellion and Guiraud in the Mineral Plan for the Republic of Senegal (Sustrac, 1984).

The clay unit is overlain by 30–60 m of glauconitic shelly-sand containing a Middle Miocene microfauna (Langhian–Tortonian). The sequence ends with 30–40 m of sand, clay and rare limestone containing glauconitic or phosphatic granules. This is the same zone as that of the occurrence described by Gorodiski (1958) and of the calcium and aluminum phosphates at Bignona.

The assemblage formed by the clay unit, the glauconitic shelly-sand and the overlying clayey-sand is the equivalent of the Miocene sequence described in Lower Casamance by Gorodiski (1958) and Flicoteaux & Medus (1980) (Fig. 11.6). According to work by Dieng (1965) and by Flicoteaux & Medus (1980), it is also possible to assign the greater part of the sandy–clayey sequence intersected by drill holes around Tambacounda (Fig. 11.2) to the Miocene. The sands are about 110 m thick and are yellowish-white to purplish-red in color. They overlie Middle–Upper Eocene calcareous clays themselves overlain by about 10 m of sand capped by a cuirasse and assigned to the Plio-Pleistocene. These sands contain hystrichosphaerids and display a palynologic association identical to that of the shelly facies in drill hole 3 at Ziguinchor (Fig. 11.5). The depositional environment was lagoonal to marine; Dieng (1965) and Flicoteaux & Medus (1980) assign a Middle Miocene age to the sequence, which would tend to indicate that the Oligocene and Lower Miocene are absent in this region. All these indications demonstrate that the marine Miocene extends considerably towards the interior in Lower and Upper Casamance.

In the region between Gambia and the Saloum River, the Late Continental clayey-sandstones, together with the ferruginous cuirasse which caps them, have given rise to marine sediments. They contain Miocene marine fossil fauna (Lappartient, 1978b and 1985), most of the mollusks mentioned by Lappartient (1985) having existed in a marine environment. One typically marine form of echinid is encountered in two fossiliferous deposits. The iron in these sandstones originates from glauconite. The fossil forms recorded make it possible to assign an Upper Burdigalian age to this ferruginous sandstone assemblage.

Further north, between the Saloum River and the Senegal River, i.e. in the area which contains the largest phosphate occurrences of the Pire Goureye-type, the Late Continental covers a very large area, occurring as facies similar to those in the area south of Kaolack. The thickness is variable, being up to 130 m in the south of Ferlo. Despite the absence of formal proof, it is possible to adopt Lappartient's interpretation that these are again, at least partially, weathered marine Miocene formations (Fig. 11.5).

Some units of sand, marl and clay encountered in petroleum drill holes onshore and offshore of the Senegal coast between Cape Verde and the Senegal delta (drill holes Cr1, Co1, and TB1, see Fig. 11.5) may also be Miocene in age. Study of seismic data recorded along the Senegal continental margin (Gomez & Barusseau, 1984) demonstrates that the maximum thickness of the Neogene in this area is effectively encountered on the continental slope and to the south of Dakar.

Whatever the case, and despite the fact that precise micropaleontologic data for the Pire Goureye–Kébémer area are lacking, it is possible that part of the sequence overlying the Lutetian phosphatic limestones represents a weathered or eroded marine facies. Study of the drill hole at Ziguinchor has revealed the presence of very low-grade Miocene marine phosphates. It is possible that the Pire Goureye-type phosphates originate partially from reworking of the Lutetian phosphatic limestones and partially from a regional episode of phosphate genesis during the Miocene. Unfortunately, all the prospecting work completed on such occurrences was carried out between 1948 and 1952 using Banka-type equipment unsuitable for the collection of appropriate samples. Later phosphate prospecting was mainly concentrated on the Lutetian phosphates, neglecting the Pire Goureye-type occurrences which form part of the overburden. The same applies to water boreholes which were concerned with other objectives. The problem of accessibility to samples which might confirm or otherwise the ideas discussed here remains to be solved.

The Miocene between Senegal and Congo

Guinea Bissau

The Miocene is extensively represented over about 8000 km² in Guinea Bissau and crops out locally around the town of Bissau and the Bijagos archipelago. The thickness as explored by drilling is at the most 130 m at the Bijagos islands, becoming progressively thinner eastwards and northeastwards towards the edges of the Mesozoic–Cenozoic sedimentary basin. Oil and water boreholes have shown that the Miocene is discordant over the Maastrichtian sands at the edge of the Silurian–Devonian basement and over the Paleocene and Eocene away from the basin border (Andreini, 1975; Mamedov & Petrosiants, 1980).

Lower Miocene It is difficult to distinguish the lower Miocene from the Oligocene. The lower Miocene is composed of gray sandy-clay containing lignitic fragments. The age of this dark colored clayey-sandy formation overlying the Upper Eocene carbonate and underlying the Middle Miocene limestone in Guinea Bissau is subject to controversy. Freudenthal (1968) has assigned the upper part of the unit to the Lower Miocene but has indicated that the drill holes put down at Cagongue (CG1) and Saô Domingos (SD1) have identified micropaleontologic markers of the Oligocene. However these markers have not been encountered in the drill holes at Safim (SF1) and at Cô (Co1) (Fig. 11.2).

By analogy with the lignite unit known in Casamance (referred to as the 'lignite group') and encountered in drill hole Z3 at Ziguinchor (Fig. 11.2) between 106 and 165 m, this formation probably ranges in age from Lower Oligocene with Lepidocyclina to Middle Miocene with Heterostegina. Oil borehole DM1 at Diana Malari in Upper Casamance clearly shows the presence in this gray clayey–sandy unit of Lower Miocene with *Glogigerina trilocularis*, of Upper Oligocene with *Globigerina ciperoensis* (confirmed by Ly, 1984) and of Lower Oligocene with *Globigerina ouachitaensis* (Drouhin & Couppey, 1961).

In the Farim area, in Guinea Bissau, the gray lignitic clay unit, 10–20 m thick, overlying the Upper Eocene dolomitic limestone and underlying the Middle Miocene limestone, contains abundant tricolporate pollen grains including *Zonocostites ramonae*, *Psilatricolporites minutus* and *Verrutricolporites rotundiporus*, known in the Lower–Middle Oligo-Miocene in Nigeria (Fauconnier, 1983). A thin microconglomeratic layer composed of milky quartz pebbles, silica-aluminous phosphate pebbles and fine–coarse quartz grains occurs at the base of the black sandy-clays, northwest of Farim. This conglomerate has reworked the thin grit-like silica-aluminous phosphate layers occurring at the Upper Eocene–Lower Oligocene interface and marking emersion of the Farim area at the end of the Upper Eocene (Prian & Gama *et al.*, 1987).

Middle Miocene After the mangrove lacustrine–fluvial lacustrine environment of the sandy lignitic clays, a final marine transgression took place, extending eastwards well beyond the boundaries of the Eocene transgression. The sequence consists mainly of yellowish marly-limestone with abundant fauna such as Elphidium, Rotalia, Heterostegina and Amphistegina (Perry, 1967), and glauconitic, phosphatic sand.

The limestone cropping out close to Bissau contains low grade phosphate (Torres, 1948) assaying 5% P_2O_5. These grades are confirmed in the phosphatic limestone outcrops of the Bijagos islands at Bubaque.

In the Farim area, two cored drill holes (PS1, SG1) (Fig. 11.2) put down in 1977 and 1983 to prospect for Eocene phosphate intersected a 7 m layer of Middle Eocene age composed of a bioclastic limestone band 1 m thick overlain by rust-colored, weathered, glauconitic and slightly-phosphatic sands assaying 5% P_2O_5. Phosphate occurs in the form of bone fragments.

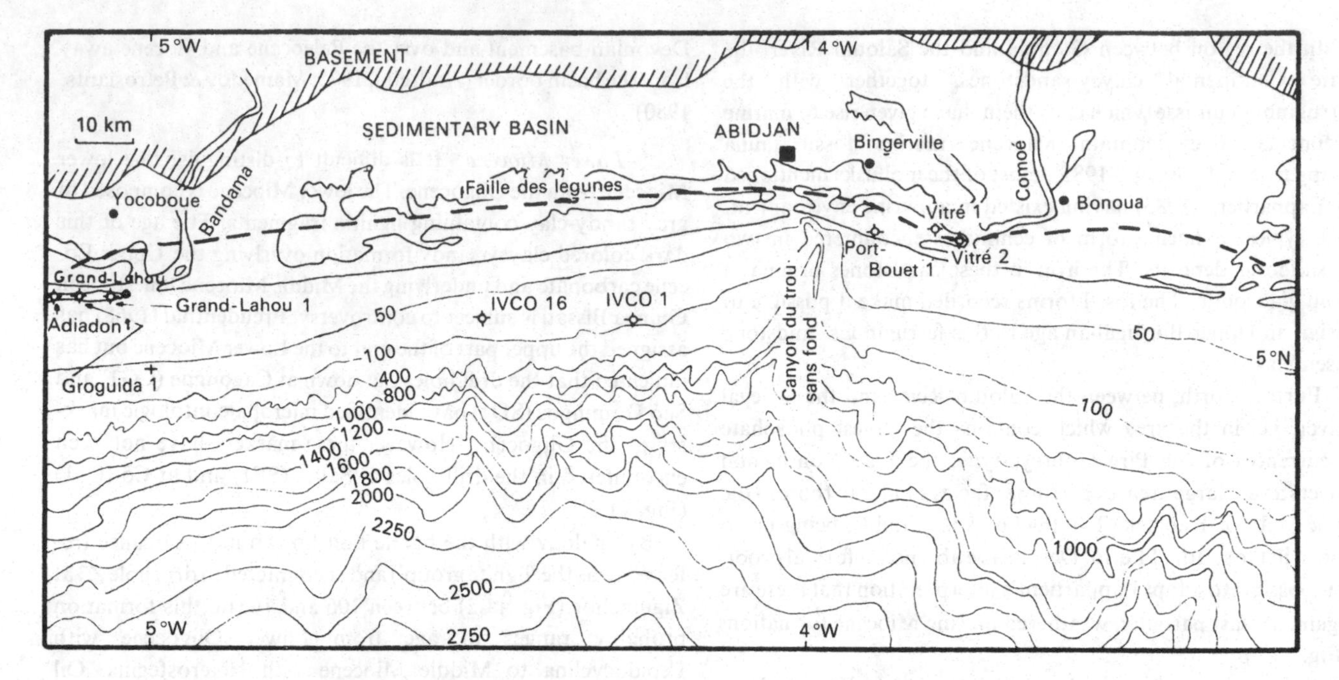

Fig. 11.7. Offshore Ivory Coast. Bathimetry and location of petroleum drill holes.

These layers, probably of major lateral extension, were intersected by a few drill holes put down on outliers located at approximately sea level. However, due to very intensive pedogenetic weathering they are no longer observable in the low lying zones of the Farim area near Rio Cacheu.

Upper Miocene Perry (1967) has assigned an Upper Miocene age to a fairly thin blue-gray clayey sequence displaying *Rotalia*, *Elphidium crispium* and *Nonion dollfusi* intersected in oil boreholes in the Bijagos islands as well as in drill holes sunk at Cagongue (CG1) and Saô Domingo (SD1). However this 10–30 m sequence overlying the Middle Eocene carbonates is generally intensively altered and undatable. It comprises the Upper Miocene and Plio-Quaternary.

Ivory Coast

In Ivory Coast (Simon & Amakou, 1984), work conducted onshore and offshore by the Centre d'Etudes Marines and by oil companies has served to reconstitute the major stages of development of the coastal sedimentary basins.

A major unconformity of the Oligocene was identified, overlain onshore by sandy–clayey deposits attributed, as in East Senegal, to the Upper Continental (UC) and probably Mio-Pliocene in age. The question as to whether three deposits partially represent marine facies has already been raised (Bacchiana *et al.*, 1982).

According to the Direction des Hydrocarbures de Côte d'Ivoire (of which mention is made in Simon & Amakou, 1984), the Upper Continental comprises the following succession from base to top:

(1) a sandy–clayey unit corresponding to fluvial (channel) and lagoonal-deposits;
(2) a gravel complex composed of quartz and ferruginized fragments such as sandstone, siltstone and clay deposited in channels;
(3) a uniform red clayey–sandy unit, known as 'terre de barre'.

Offshore of Abidjan, drill holes Port Bouet 1, Vitre 1 and Vitre 2 (Fig. 11.7) have intersected typically marine clayey formations, Miocene in age and deposited over Cretaceous sediments. In other locations, Miocene sequences overlie Eocene formations.

The Neogene–Holocene sequence is thicker to the west than to the east of Abidjan, i.e. between the Abidjan–Bingerville channel in the east and the Grand Lahou channel in the west. In both these channels the Miocene transgression has progressed further towards the continent.

The Miocene fill is mainly composed of green-gray–olive plastic clay grading into mudstone towards the base of the unit. The clay is slightly calcareous, locally silty and micaceous and contains glauconite (in places berthierite?) commonly disposed in layers and more abundant towards the base of the sequence. The clay also contains fine–coarse rounded quartz grains. Fossils and microfossils are abundant especially in the upper part of the sequence, a large part being affected by pyritization or limonitization. Fine interbeds of brown–gray hard, generally cryptocrystalline–microcrystalline, slightly sandy–silty limestone are present, containing pyrite and dolomite layers generally towards the base of the sequence. To the north and east, the formations become more detrital (sandy limestone, sandstone and siltstone) and contain vegetal organic matter.

Togo–Benin

According to Slansky (1962), the Upper Continental of Benin in particular is divided into two members: (1) a lower member composed of white, pink, red and purple fine-grained sandy, clayey facies, uncomformably overlying various formations, mainly Middle Eocene in age. This member is azoic and probably Upper Eocene to Oligocene in age; and (2) an upper probably Oligocene–Quaternary, coarser-grained, sandy, sandy–clayey and clayey member becoming brown–yellow at the top where it grades into 'terre de barre'. This member is very distinctive where it contains a layer of gravel and rounded quartz pebbles. According to the available data, no phosphate facies of the Pire Goureye-type have been identified in Togo and Benin.

Nigeria

The same formations as in Benin are present in West Nigeria: the poorly-dated Benin and Ogwashi-Asaba Formations, Oligocene to Holocene in age (Whiteman, 1982) and overlying the Eocene formations (Fig. 11.8).

The Ogwashi-Asaba Formation is represented by alternating lignite beds of the 'lignite group' and clay beds. It occurs from West to East Niger, passing through the Niger Delta, and is Oligocene–Miocene in age. The term 'Benin Formation' refers to a sequence of sand and white clay cropping out in the coastal area as a whole.

The sandstone and shale sequences of the Agbada Formation, Eocene–Paleocene in age according to subsurface data, are overlain partly by the Afam clay member, but only in the lower Imo area, and mainly by the Benin Formation which occurs further inland. The Afam clay is represented by a sandy–clayey formation which may correspond to a valley deposit towards the base of the Benin formation. The Afam clay is one of the main clay fill structures of the Niger delta and covers a surface area of 3496 km^2. This Middle–Upper Eocene formation thickens from north to south, reaching a maximum of 762 m at Lubara Creek and disappearing laterally away from the channel axis.

The Benin Formation is mainly sandy with shale intercalations. It varies in thickness from 305 m in the south, i.e. 10–50 km south of the present day coast, to 1830 m and even 3500 m in the center of the delta in the Port Harcourt area. To the south, there is a progressive transition from the Benin Formation to the Holocene Agbada Formation. Miocene–Holocene in age, the Benin Formation consists of a complex unit of marine, deltaic, estuarine, lagoonal and fluvial-lacustrine deposits. Like the Agbada Formation, it is diachronous.

Cameroon and equatorial Guinea

In Cameroon and equatorial Guinca, the Miocene–Holocene formations extensively erode the older Cenozoic formations, cropping out onshore but disappearing offshore. This phenomenon is more widely expressed than in Ivory Coast where Miocene erosion has only affected specific areas.

Unlike the underlying Cretaceous sandstone, siltstone and shale, little attention has been paid to the Miocene formations in drill holes. These formations mainly consist of shales with sandy calcareous interbeds and contain shell fragments and some pyrite aggregates. Glauconite is not recorded.

In West Cameroon, the Miocene succession is fairly similar to the Agbada and Benin formations observed in Nigeria. In a drill hole sunk in equatorial Guinea near the Gabon border, the Miocene–Holocene formations contain 75% of brown-green–olive clays with abundant fossils and about 25% sand. Traces of pyrite and glauconite are noted. This formation unconformably overlies the clayey Maastrichtian.

Gabon

Offshore of Gabon, erosion of the Eocene formations by the Miocene transgression is restricted to smaller areas, as in Ivory Coast. The Paleocene Ikando Formation and the Eocene Ozouri Formation are characterized by extensive silicification (porcelanite and chert) indicating a shallow depositional environment. The Eocene Animba Formation begins with a marine regression which marked a new stage of evolution of the continental flexure. Erosion channels in the Ozouri Formation were subsequently filled with fluvial sand and estuarine and marine clay indicative of a further marine transgression. In some cases, the Ozouri Formation is capped with a hardground or is reworked at the base of the Animba Formation.

The Animba Formation is mainly composed of silt and marly-clay and may reach a thickness of several hundred meters. The Upper Eocene is generally eroded and occurs at the top of the Ngola diapir as Hantkenina clay. The flexuring of the Animba episode increased during the Neogene. The Oligocene–Pliocene N'Tchengue, M'Bega and Mandorove Formations are up to 3000 m thick at the edge of the present day continental slope.

The base of this sequence may be Paleocene in age, but preliminary datings place it in the Lower Miocene (*Globorotalia fonsi*). Offshore, the Miocene is composed of clay and sand, but closer to the continent at the latitude of Libreville, for instance, Senonian calcareous facies are also present, as in Guinea Bissau. The Mio-Pliocene formations become more and more sandy towards the top, first as channel fillings and then in the form of a continuous sand and gravel bed overlain by Pleistocene deposits.

In Gabon, this succession as a whole, together with the preceding members down to the Aptian evaporite of the Ezanga Formation, were intensely disturbed by the evolution of the Atlantic margin: various episodes of flexuring are marked by downthrown blocks, by channeling, extensive in places, and by diapirism of the Ezanga evaporite.

Neogene to Recent phosphates offshore of the People's Republic of Congo and of Gabon

The discovery of concentrations of phosphate and phosphatic nodules in the sediments offshore of Congo and Gabon dates back to work undertaken by the Office de la Recherche Scientifique et Technique Outre Mer (ORSTOM) and by the Département de Géologie de l'Université Marien Ngouabi in Brazzaville (Giresse *et al.*, 1981). Between 1970 and 1978, eight offshore programs were carried out, representing 450 dredgings, 150 cored holes in rock, and 50 Kullenberg cores of recent sediments. The samples collected were used for numerous sedimentologic and stratigraphic studies, particularly of the Holocene sediments and of the bathymetry of the continental shelf (Giresse, 1980b). This exploration led to the discovery of a phosphate concentration offshore of Pointe Noire in the Congo (Fig. 11.9) and of phosphatic gravels dispersed in the sediments in many sectors.

Reconnaissance of these phosphate occurrences began in Gabon under the PHOSCAP operation, (Horn *et al.*, 1979) financed by French public and private funds. The operation was jointly conducted by CNEXO (now IFREMER, the Institut

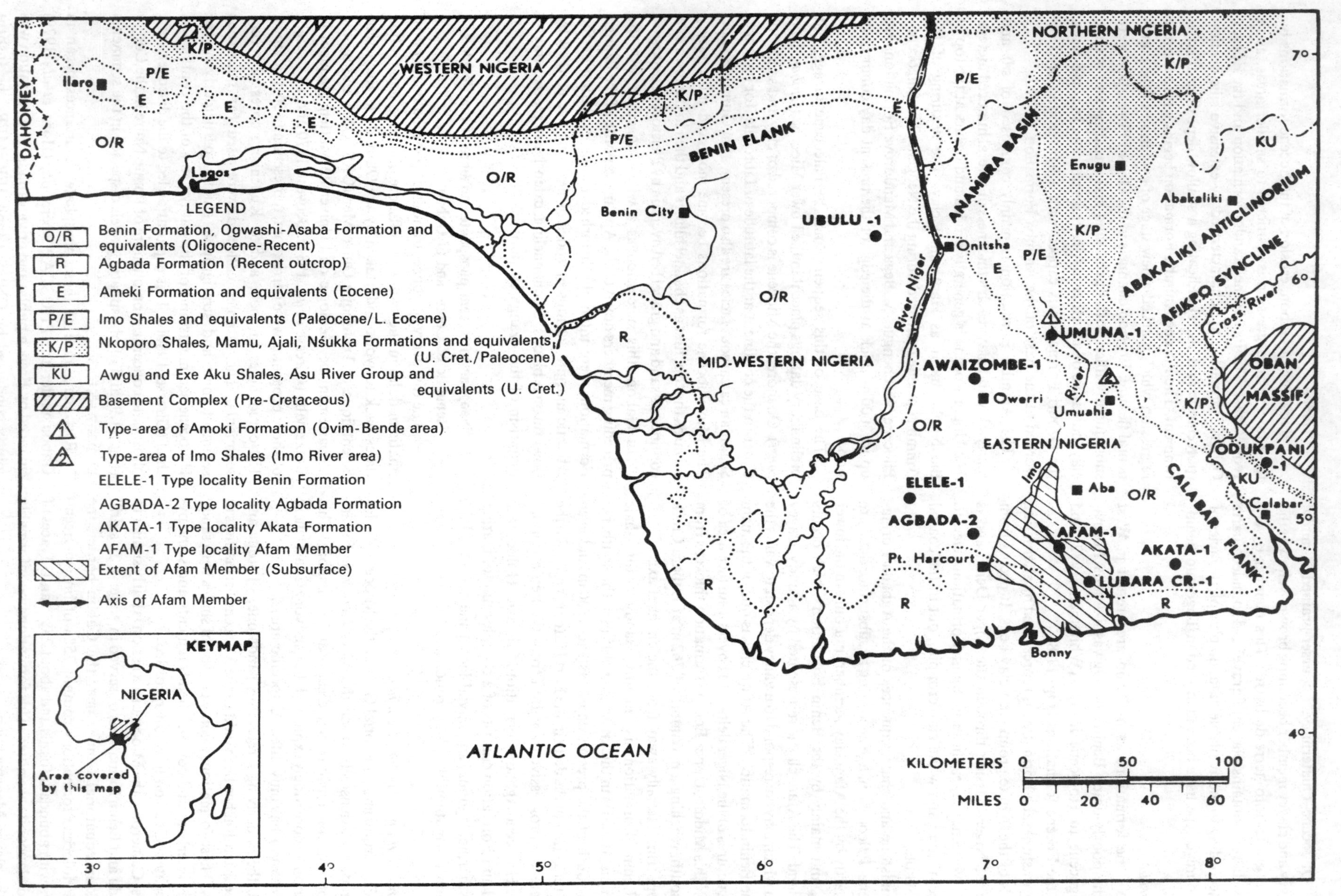

Fig. 11.8. Location of main geological units of western Nigeria and Niger delta (from Whiteman, 1982).

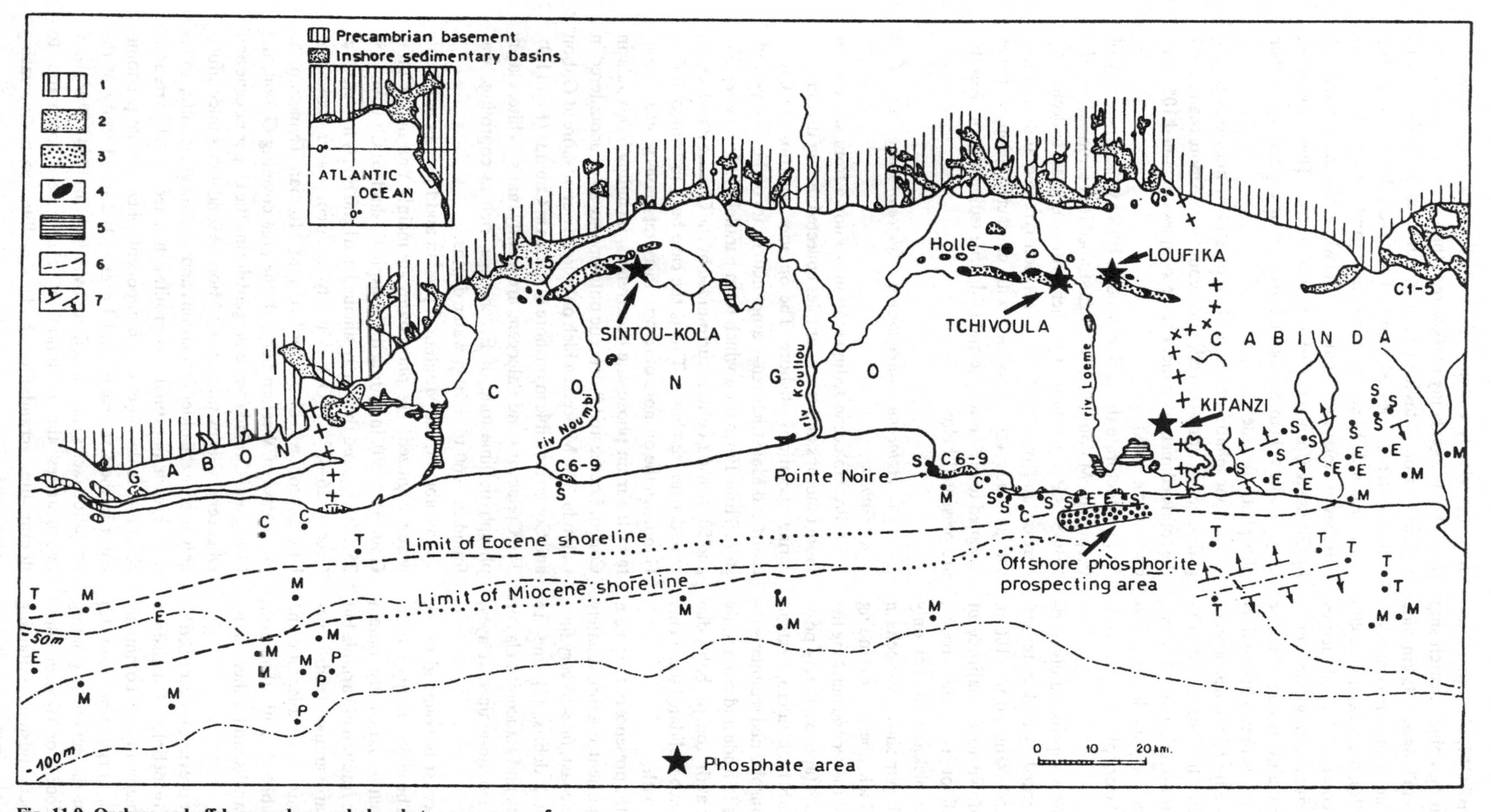

Fig. 11.9. Onshore and offshore geology and phosphate occurrences of Congo (adapted from P. Giresse *et al.*, 1981). 1 – Precambrian base; 2 – Lower Cretaceous (C_{1-5}); 3 – Upper Cretaceous (C_{6-9}); 4 – Maestrichtian phosphatic deposits; 5 – lakes and swamps; 6 – old shorelines; 7 – anticlinal axis. Cenomanian (C), Turonian (T), Senonian (S), Paleocene (P), Eocene (E) and Miocene (M) outcrops.

Français pour la Recherche en Mer), BRGM, and other companies, with the assistance of the oceanographic research ship Capricorne. Work covered an area 150 km long and 35 km wide inshore of the 100–110 m isobath. The program comprised 1350 km of high resolution seismic profiles, the collection of 400 surface samples using CNEXO's Ville corer, 82 3-m vibracores using the BRGM's V2 S3 equipment (with a casing diameter of 100 mm and a length of 3.2 m) and 65 dredgings (using a circular Rallier du Baty-type dredge). Geophysics were conducted during a preliminary phase of study comprising 16 profiles perpendicular to the coast and 5–10 km apart. The distance between profiles was tightened in the south, demonstrating that this area contains the thickest unconsolidated sedimentary deposits. Exploration undertaken during the second work phase was accordingly concentrated on this area.

Over the 150 km studied, the coastline is approximately rectilinear, with a few rare prominences marked by rock outcrops. Two major knickpoints are visible at − 50–60 m and at − 100 m. In the north of the study area and down to the latitude of Mayumba bay (Fig. 11.8), the seafloor is mainly rocky, unconsolidated sediments reaching a thickness of 2 m only locally. South of Mayumba, the isopach map shows two belts in which the sediments are about 10 m thick, the first extending from the coast to the − 40 m isobath and the second marking the depression between the two knickpoints (approximately positioned at − 60 m and − 90 m). In the south, rock outcrops extend seawards from Point Toshibobo. Unconsolidated sediments increase substantially in thickness at this latitude and commonly display a cover of fine grained sediment at the top, probably due to influx from rivers. The thickness of unconsolidated sediments is limited seawards of the − 100 m isobath.

Study of the sediments has revealed the presence of four main facies which form the following bands oriented approximately parallel to the coastline: (1) nearest the coast, fine- to very fine-grained clayey-sand containing vegetal debris; (2) fine- to medium-grained quartzose sand; (3) silty and glauconitic clay or mud; and (4) furthest offshore, medium-grained clayey, glauconitic and shelly sand.

The fine-grained sand nearest the coast is bounded to the south by the Point Toshibobo zone. The clays occupy a band delineated by the − 60 and − 10 m isobaths and locally contain gravels and a quartzose and glauconitic fraction of variable but not inconsiderable importance. The coarse grained sands represent the most-seaward deposits. They are very glauconitic and contain small gravels and abundant 'pebbles' of lithothamniate algae (Melobesiae) and of lamellibranchs and other fauna debris.

Two forms of phosphatic gravel were previously described by Giresse (1980a,b): (1) rock fragments or pebbles of sandstone and limestone, in places up to a decimeter in size, commonly encrusted, and perforated by boring organisms; and (2) coprolites 1–3 cm long and 0.2–0.5 cm in section originating from the disintegration of essentially Eocene or Miocene phosphatic rocks. These latter rocks are either sandstone containing phosphatic gravel, puddingstone containing large coprolites, or decalcified clay, clayey limestone, or clay that contain white coprolites.

These indurated rocks have only been encountered *in situ* seaward of the − 80 m isobath. Between the − 80 m isobath and the coast, clearly circumscribed Miocene coatings remain in places but are generally masked by a thick cover of present day sediments. Coprolites have been encountered in the shelly-sand around outcrops located west of Point Toshibobo, and in the most seaward band of glauconitic and shelly-sand. Two varieties of glossy-brown and friable white coprolites described by Giresse & Cornen (1976) were encountered. These coprolites are generally > 1 cm. Shark teeth are present, together with other phosphatic fragments.

The phosphatic components represent at most 5% of the grain-size fraction assessed, but account for no more than 2% of the total sediment volume. (Three samples contained 10% and one sample 20%).

In a doctorate thesis published in 1983 by the University of Perpignan, Malounguila N'ganga studied four samples from the PHOSCAP operation, together with samples collected by Giresse and by ORSTOM. This petrographic and sedimentologic study was concerned with characterization of facies assigned to pre-Holocene and Holocene sedimentation. Results are presented below.

Pre-Holocene sedimentation offshore of Congo and of Gabon

Pre-Holocene sedimentation is known from several onshore outcrops and from samples collected during dredging or coring operations offshore. The onshore outcrops are Cretaceous and Maastrichtian–Campanian facies of the clayey-sandstone Holle Formation which form parallel bands between 300 and 500 m wide, entirely surrounded by the Cirques Plio-Pleistocene detrital sequence. This is the case both in Congo and in Gabon. In the onshore outcrops, this Cretaceous sequence contains the main phosphate occurrences, including Tchivoula in Congo. Eocene and Paleocene outcrops are also encountered in Cabinda. The Miocene which only crops out offshore at Gabon and Congo is visible in onshore outcrops in Cabinda (Fig. 11.9). The Cretaceous and Paleocene are mainly marls. White, clayey phosphatic limestone of Eocene age, rich in coprolites, was found in − 60 to − 70 m water depths.

In Cabinda, the Montian, Ypresian and Lutetian sediments were recognized. The phosphates are mainly Ypresian (Cunha Gouveia, 1960), including the phosphatic sands at Chivovo (15–25% P_2O_5), and the partly silicified marl and marly limestone at Cacata (26–42% P_2O_5). A marly calcareous layer containing large coprolites is also recorded in the Montian–Thanetian. As in the case of Western Senegal, the area covering Gabon, the Congo and Cabinda was emergent during the Upper Eocene and Oligocene. Clayey phosphatic Miocene sandstones uncomfortably overlie the older formations and contain glauconite, abundant brown phosphatic coprolites, fish teeth and vertebrae. Kaolinite is the main clay component whereas illite predominates in the Cretaceous and Eocene. A detailed study of the alteration facies was carried out by N'landou (1984). This study demonstrates the evolution of carbonate fluorapatite to fluorapatite, crandallite and wavellite with increasing weathering.

According to Giresse & Cornen (1976), the phosphatic Miocene deposits offshore of the Congo–Gabon border occur as three main facies: (1) ocher limestone containing a phosphatic and microcrystalline matrix bearing 20–40% phosphatic nodules, some displaying an oolitic structure; numerous foraminifers and some glauconite gravel are also present; (2) brown sandstone containing a clay or calcite matrix; large Selachian coprolites and quartz grains are present; and (3) ferruginous micro-sandstones, rich in organic matter and quartz. The gray or brown cylindrical coprolites are often separated from the matrix and have accumulated as placers. Average coprolite size is 1–3 cm long and 0.2–0.5 cm wide. Giresse & Cornen (1976) record no phosphate grains in the sand fraction. At two points near the − 105 m isobath, the corer recovered identical white coprolites from undated gray–beige limestone at a depth of 50 m.

Offshore of the Congo–Cabinda border, dredging conducted around the − 35 m isobath about 25 km south of Pointe Noire recovered shelly-sand extremely rich in brown coprolites and urchin spines. The same facies as those recorded further north are commonly encountered; ocher limestone containing gravels or phosphatic pseudo-ooliths and true puddingstone containing large coprolites. According to Giresse & Cornen (1976), the type-section of this phosphate-rich zone is as follows:

0–20 cm: coarse-grained sand containing abundant very brown coprolites;

20–40 cm: a heterogeneous assemblage of dark ocher-colored Miocene sandstone or limestone rich in brown coprolites;

40–60 cm: white decalcified clay containing both brown and white coprolites identical to those in the north;

60–80 cm: white clayey limestone containing white phosphatic coprolities.

The white limestone at the base displays the same facies as in the Ypresian in Cabinda; the two overlying units are assigned to the Miocene. According to Giresse & Cornen, the Miocene sediments contain reworked earlier white coprolites, although deposition of such coprolites continued during this Miocene period. X-ray diffractometry analysis of the coprolites revealed only fluorapatite within both the white (Eocene) coprolites and brown (Miocene) coprolites.

The phosphate concentration discovered south of Pointe Noire was subsequently prospected through the United Nations Development program. About 50 dredge and core samples were collected from an area of 80 km². Although rich in bone debris, the Miocene of Cabinda does not appear to contain other types of phosphate grains, whereas there are occurrences in the same Miocene units in the Cuanza Basin.

Angola

In this area the Upper Cretaceous marls, limestones, silts and sands are overlain by: (1) Paleocene–Eocene limestone, siltstone and clay; and (2) a mainly clayey Oligo–Miocene sequence (the Quifangondo and Luanda Formations) containing fine dolomitic intercalations in the upper part. Pyrite and glauconite are commonly observed in the clay. Burdigalian foraminifera were recorded in the Quifangondo Formation (50 m thick) intersected in a drill hole put down by TOTAL in 1973.

Conclusions

Despite the distances which separate West Senegal–Casamance and Gabon–Congo–Cabinda, the two examples described in this paper display a number of similarities: (1) the main episode of phosphate genesis was pre-Oligocene in both cases. It was Lower and mainly Middle Eocene in West Senegal–Casamance and Upper Cretaceous and partly Eocene in Gabon–Congo–Cabinda; (2) an episode of Late- or post-Eocene emersion occurred in both cases. In Senegal and in Congo, this Upper Eocene–Oligocene emersion was followed by the formation of aluminum phosphates. A similar situation is encountered in the Farim deposit of Guinea Bissau; and (3) The Miocene transgression was associated with reworking of older phosphate deposits (e.g. the Pire Goureye-type occurrences in West Senegal) and with minor phosphate precipitation of the same age (e.g. the occurrences in Casamance and offshore of Congo).

In Senegal, it is possible that some of the deposits previously assigned to the Late Continental are in fact weathered Middle–Upper Miocene. The low grade phosphate deposits virtually ubiquitously present in the Late Continental are thus considered to be Miocene phosphate concentrations.

References

Andreini, J.C. (1975). *Le bassin sédimentaire de Guinée Bissau et le problème des phosphates.* Rapport interne DMG, Bissau.

Bacchiana, C., Brancart, R., Klasz, I. de, Legoux, O. and Paradis, G. (1982). Présence de Miocène inférieur marin dans le 'Continental terminal' de la Basse Côte d'Ivoire. *Revue de micropaléontologie,* **25(3)**, 145–9.

Bense, C. (1962). *Carte Géologique du Sénégal,* échelle 1/500 000, 4 feuilles. BRGM, Dakar.

Brancart, R. (1977). 'Etude micropaléontologique et stratigraphique du Paléogène sur le flanc occidental du horst de Ndias et dans la région de Taïba'. Thèse 3ème cycle Université de Aix-Marseille.

Brancart, R. and Flicoteaux, R. (1971). Age des formations phosphatées de Lam Lam et de Taïba (Sénégal occidental). Données micropaléontologiques, conséquences stratigraphiques et paléogéographiques. *Bulletin de la Société Géologiques de France,* **(7)XIII**, 399–408.

Chino, A. (1963). Quelques précisions sur la série stratigraphique tertiaire du bassin sénégalais. *Rapport du Bureau de Recherches Géologiques et Minières,* **DAK 63-A5.**

Cunha Gouveia, J.A. (1960). Notas sobre of fosfatos sedimentares de Cabinda. *Serv. Geol. y Minas Angola, Bull.* **1**, 49–65.

Dieng, M. (1965). *Contribution à l'étude géologique du Continental terminal du Sénégal* (travaux effectués de 1962 à 1965). *Rapport du Bureau de Recherches Géologiques et Minières,* **DAK 65A 27.**

Drouhin, J.P. & Couppey, Cl. (1961). *Rapport de fin de sondage de Diana-Malari 1 (DM 1).* Rapport Comp. Pétroles Total (Afr. Ouest), RE 10.

Ducasse, O., Dufaure, P. & Flicoteaux, R. (1978). Le passage de l'Eocène inférieur à l'Eocène moyen dans la presqu'île du Cap Vert (Sénégal occidental). Révision micropaléontologique et synthèse stratigraphique. In: Contribution à la connaissance géologique et micropaléontologique de l'Ouest africain. *Cahiers de Micropaléontologie,* **1**, ed. CNRS, pp. 3–28. Paris.

Elouard, P. (1962). Etude géologique et hydrogéologique des formations sédimentaires du Gueba mauritanien et de la vallée du Sénégal. Thèse Sciences, Paris, 1959, et *Mémoire du Bureau de Recherches Géologiques et Minières,* **7.**

Fauconnier, D. (1983). Etude palynologique de 59 échantillons de la mission phosphate Farim (Guinée Bissau). *Etude du Bureau de Recherches Géologiques et Minières,* **83 GEO EM 166.**

Flicoteaux, R. (1980). 'Genèse des phosphate alumineux du Sénégal occidental. Etapes et guides de l'altération'. Thèse Universite de Aix–Marseille, 229pp.

Flicoteaux, R. (1982). Genèse des phosphates alumineux du Sénégal occidental. Etapes et guides de l'altération. Thèse Sciences Aix-

Marseille III, 1980 & *Mémoires Sciences Géologiques Université de Strasbourg*, **67**.

Flicoteaux, R. & Hameh, P. (1989). The aluminous phosphate deposits of Thiès, Sénégal. In *Phosphate Deposits of the World*, vol. 2, A. Notholt, D. Sheldon & D. Davidson, Cambridge University Press, Cambridge, (in press).

Flicoteaux, R. & Medus, J. (1980). Existence d'une lacune entre les termes marins du Paléogène et du Néogène du Sénégal méridional démontrée par les microfaunes et les microflores. *Actes VIè Coll. Afr. Micropaléont.*, Tunis (1974) et *Ann. Mines Géol.*, Tunis, **28, III**, 193–207.

Freudenthal. (1968). *Tertiary Paleontology of Portuguese Guinea*. Rapport Esso, Bordeaux, France.

Giresse, P. (1980a). The Maastrichtian phosphate sequence of the Congo. In *Econ. & Miner. (SEPM)*. Spec. Pub. 29, 193–205.

Giresse, P. (1980b). *Carte sédimentologique du Plateau continental du Congo*. 3 pl. 1/200 000. Notice explicative 85, ORSTOM, Bondy (Fr.).

Giresse, P. & Cornen, G. (1976). Distribution, nature et origine des phosphates miocènes et éocènes sous-marins des plateformes du Congo et du Gabon. *Bulletin du Bureau de Recherches Géologiques et Minières*, **IV, 1**, 5–15.

Giresse, P. & Kouyoumontzakis, G. (1973). Cartographie sédimentologique des plateaux continentaux du Sud du Gabon, du Congo, du Cabinda et du Zaïre. *Cah. ORSTOM, (Bondy) sér. Géol.*, **V, 2**, 235–57.

Giresse, P., Jansen, F., Kouyoumontzakis, G. & Moguedet, G. (1981). Les fonds de la plateforme congolaise, le delta sous-marin du fleuve Congo. Bilan de huit ans de recherches sédimentologiques, paléontologiques, géochimiques et géophysiques. *Travaux et Documents de l'ORSTOM (Bondy)*, **138**, 13–44.

Giresse, P., Barusseau, J.P., Malounguila-N'ganga, D. & Wiber, M. (1984). *Les phosphates au large du Congo et du Gabon. Nature géochimique et conditions mécaniques d'accumulation*, pp. 315–26. Comptes rendus du 2ème séminaire international sur les ressources minérales sous-marines, 19–23 mars 1984, GERMINAL (Groupe d'étude et de recherche de minéralisations au large).

Gomez, R. & Barusseau, J.P. (1984). Disposition des formations post-éocènes de la marge continentale sénégalaise. *Bulletin de la Société Géologiques de France*, **(7), XXVI, 6**, 1107–16.

Gorodiski, A. (1958). Miocène et indices phosphatés de Casamance (Sénégal). *C.R. Somm. Soc. Geol. Fr.*, **13**, 293–7.

Heetveld, H. (1965). *Syndicat du Lac de Guiers. Rapport final. Campagne 1964–65. Rapport du Bureau de Recherche Géologiques et Minières*, DAK, non numéroté.

Horn, R. *et al.* (1979). *Recherche de gravelle phosphatée au droit des côtes du Gabon (opération PHOSCAP). Rapport du Bureau de Recherche Géologiques et Minières*, **79 SGN 318** MAR.

Lappartient, J.R. (1970). L'Eocène inférieur de la Cuesta de Thiès (République du Sénégal). *Trav. Lab. Sci. Terre St-Jérôme*, Marseille (A), **1**, 45p.

Lappartient, J.R. (1978a). *Le Continental terminal du Sénégal. Une formation marine néogène continentale*. Proc. 2d. Working. Conf. Project n° 227, IUGS-IGCP. Revision of the concept of Continental terminal in Africa. Vol. 1. Ahmadu Bello. Univ. Dép. Geol. Oceans. Publ. 7.

Lappartient, J.R. (1978b). *Extension vers le Nord du Golfe miocène casamançais (Sénégal)*. 6ème Réunion Ann. Sci. Terre, Orsay, France.

Lappartient, J.R. (1983). Evolution du bassin sénégalo-mauritanien pendant le Cénozoïque (bordure et partie méridionale). In: Res. commun. séance spec. Soc. Géol. Fr., Marseille, 7–8 mars, 'Bassins sédimentaires en Afrique'. *Trav. Lab. Sci. Terre St-Jérôme, Marseille, (A)*, **15**, 63–4.

Lappartient, J.R. (1985). *Le 'Continental terminal' et le Pleistocène ancien du bassin sénégalo-mauritanien. Stratigraphie, sédimentation, diagénèse, altération. Reconstitution des paléorivages au travers des cuirasses*. Thèse Sciences, Aix-Marseille III.

Lappartient, J.R. & Monteillet, J. (1981). Le gisement fossilifère sénonien supérieur des carrières de Paki (Sénégal). *Bull. Inst. Fond. Afr noire*, **43, 3**, 431–9.

Ly, A. (1984). Biozonation des formations sédimentaires tertiaires de Casamance, au Sénégal. *Journal of African Earth Sciences*, **2, 1**, 57–60.

Ly, A. (1985). *Les sondages tertiaires de Casamance. Approche litho-stratigraphique (Foraminifères) et sédimentologie*. Thèse Universite Aix-Marseille III.

Malounguila-N'Ganga, D. (1983). *Les environnements sédimentaires des plateformes du Nord-Congo et du Sud-Congo au Quaternaire supérieur d'après les données de vibro-carottages*. Thèse 3ème C. Univ. Paul Sabatier Toulouse.

Mamedov, V. & Petrosiants, A. (1980). *Sobre os resultados das pesquiras de fosforites realizadas nas areas occidentais de la Republica da Guinée Bissau nos anos de 1977–1980*. Rapport inédit, DGM Bissau.

Meloux, J., Bigot, M. & Viland, J.C. (1983). *Plan minéral de la République Populaire du Congo*. 2 vol.

Monciardini, C. (1966). La sédimentation éocène au Sénégal. Mém., *Bureau de Recherches Géologiques et Minières*, **43**.

N'landou, J. de Dien (1984). *Le gisement maestrichtien de phosphates de Tchivoula (R.P. Congo), étapes de sédimentogenèse marine, de diagénèse et d'altération*. Thèse 3ème cycle. Universite P. Sabatier. Centre de Rech. Sédim. marine de Perpignan.

Perry, L.J. (1967). *Progress report on the geology of the Bijagos Archipelago*. Esso East Atlantic Study Group.

Prian, J.P. (1986). Geologie de la bordure méridionale du golfe tertiaire de Casamance (Sénégal et Guinée Bissau); partie I: présentation de la géologie régionale. *Bureau de Recherches Géologiques et Minières*, **101**.

Prian, J.P. & P. Gama *et al.* (1987). Le gisement de phosphate éocène de Farim-Saliquinhé (République de Guinée-Bissau). *Chron. Rech. Min.*, **486**, 25–54.

Simon, P. & Amakou, B. (1984). La discordance oligocène et les dépôts postérieurs à la discordance dans le bassin sédimentaire ivoirien. *Bulletin de la Société Géologiques de France*, **(7), XXVI 6**, 1117–25.

Slansky, M. (1962). Contribution à l'étude géologique du bassin sédimentaire côtier du Dahomey et du Togo. *Memoires du Bureau de Recherches Géologiques et Minières*, **11**.

Sustrac, G. (Coord.). (1984). *Plan Minéral de la République du Sénégal*. Bureau de Recherches Géologiques et Minières, Direction des Mines et de la Géologie du Sénégal, Département de Géologie de l'Université de Dakar, Compagnie Générale des Matières Nucléaires, COGEMA. Ed. BRGM, 3 vol.

Tessier, F. (1952). Contribution à la stratigraphie et à la paléontologie de la partie ouest du Sénégal (Crétacé et Tertiaire). Th. Sci. Univ. Marseille, 1950 et *Bull. Dir. Mines AOF* 14, I et II.

Tessier, F. *et al.* (1975). Reform of the concept of 'Continental terminal' in the coastal sedimentation basins of West Africa. *Trav. Lab. Sci. Terre St-Jérôme*, Marseille, (A), 8.

Texeira, J.E. (1968). *Geologia da Guiné Portuguesa* vol. 1, 53–105. Curso de geol. da Ultramar, Junta Invest., Ultram., Fac. Sci., Lisboa.

Torres, A. (1948). *Une formation phosphatée à Bissau (Guinée Portugaise)*. Rep. 18th Sess. Internac. Geol. Cong., Londres.

Trenous, J.Y. (1970). Etude géologique dans le Sénégal nord-occidental et le Ferlo. *Trav. Lab. Sci. Terre St-Jérôme, Marseille, (A)*, **4**.

Whiteman, A. (1982). *Nigeria: its petroleum geology, resources and potential*, 2 vol. Graham & Trotman, London.

Woolsey, R., Ferrante, M. & Le Lann, F. (1984). *Exploration for phosphate in the offshore territories of the Congo*. Comptes rendus du 2ème séminaire international sur les ressources minérales sous-marines, 19–23 mars 1984, GERMINAL (Groupe d'étude et de recherche de minéralisations au large), 329–38.

COLL. Offshore well sections: courtesy of Compagnie Française des Pétroles.

12

Phosphorite deposits on the Namibian continental shelf

J.M. BREMNER AND J. ROGERS

Abstract

On the Namibian shelf we find three types of phosphorite; phosphorite sand, rock phosphorite and concretionary phosphorite. The phosphorite sand is further subdivided into pelletal phosphorite and glauconitized pelletal phosphorite which has a glauconitized outer rim. Both the pelletal and glauconitized pelletal phosphorite varieties are found on the outer shelf and have been dated radiometrically as pre-Quaternary, probably Miocene in age. A model is proposed that has pelletal phosphorite forming as intraclasts on intertidal mudflats in subtropical estuaries, which are inundated daily by phosphate-rich, upwelled, oceanic water. The glauconitized pelletal phosphorite formed in the upper reaches of the estuaries where river-borne iron was available from introduced clay minerals. Both types were distributed laterally during Late Neogene transgressions and regressions as the surf zone passed over the earlier estuarine environments.

Rock phosphorite of authigenic and diagenetic origin is present on the Namibian shelf, but in contrast to the South African situation, it occurs as a very minor constituent.

Concretionary phosphorite forms today by slow authigenic growth in the interstitial waters of a Holocene diatomaceous mudbelt centred off Walvis Bay. The phosphate content of the interstitial water is raised by bacterial decomposition of abundant organic matter in the mud, and by the release of phosphate scavenged by adsorption onto illite–mica settling through the water column. In addition, the pH is increased by decomposing organic matter, which promotes the precipitation of apatite around nuclei such as fish debris.

Francolite is the chief phosphate mineral of each phosphorite type and only the dark, indurated concretionary phosphorite is free from allogenic impurities. Organic matter is particularly abundant (2.21–2.91%) in pelletal phosphorite. Strontium and yttrium are the only trace elements enriched in the phosphorite relative to the average crustal rock and sea water.

Within the upper 10 cm of sediment draping the continental shelf, the total resource of phosphorite sand and concretionary phosphorite amounts to 3020 million tons. Three separate deposits of phosphorite sand and one deposit of concretionary phosphorite have been identified.

Introduction

Three main phosphorite types have been identified on the Namibian continental shelf. The deposits are named here by type because they occur in geographically distinct areas. They are: Phosphorite sand, of which there are two kinds, pelletal phosphorite (pP) and glauconitized pelletal phosphorite (gpP) (described here for the first time), rock phosphorite (rP) and concretionary phosphorite (cP).

The Namibian phosphorites differ markedly from those off South Africa, the former being mainly authigenic in origin, and the latter being mainly diagenetic. A transition zone where phosphorites of both varieties are found is situated between Sylvia Hill and Hondeklip Bay (Fig. 12.1). Diagenetic rP has been found at one locality north of the transition zone (Bremner, 1978) and five small occurrences of authigenic pP are known about south of the transition zone (Birch, Chapter 13, this volume).

Phosphorite sand occurs in three principal deposits centered at 19° S, latitude 12° E longitude (off Rocky Point), 24° S, 14° E (between Walvis Bay and Sylvia Hill) and 29° S, 15° E (off the Orange River) (Fig. 12.2). The northern deposit measures about 400 km in length, 40 km in width, lies mainly on the landward side of the 200 m isobath, and contains higher concentrations of gpP than pP. The central deposit is the largest, measuring 550 km in length, 70 km in width, is situated between the 100 m and 500 m isobaths, and contains pP in greater abundance than gpP. The southern deposit is perhaps 100 km long, 30 km wide, consists entirely of pP, and lies mainly between the 100 m and 200 m isobaths.

Due to minimal rock outcrop on a shelf largely blanketed by unconsolidated sediment, few rP samples have been recovered seaward of the 200 m isobath (Fig. 12.2).

The cP samples are confined to a 740 km-long diatomaceous mud belt, which is situated on the inner shelf, and is widest (76 km) off Walvis Bay (Bremner, 1980a) (Fig. 12.2). The landward flank of the mudbelt usually contains the greatest abundance of concretions, with maximum values being found north of Walvis Bay.

General geology

In southern Africa, onshore phosphorite deposits are known in the Varswater Formation near Langebaanweg (Fig. 12.1), (Tankard, 1974) and in the Whitehill Formation of the Dwyka Group in the northern Cape (Haughton, 1969). One of the constituents of the authigenic phosphorite-sand of the Varswater Formation is pP, whereas authigenic cP is found in

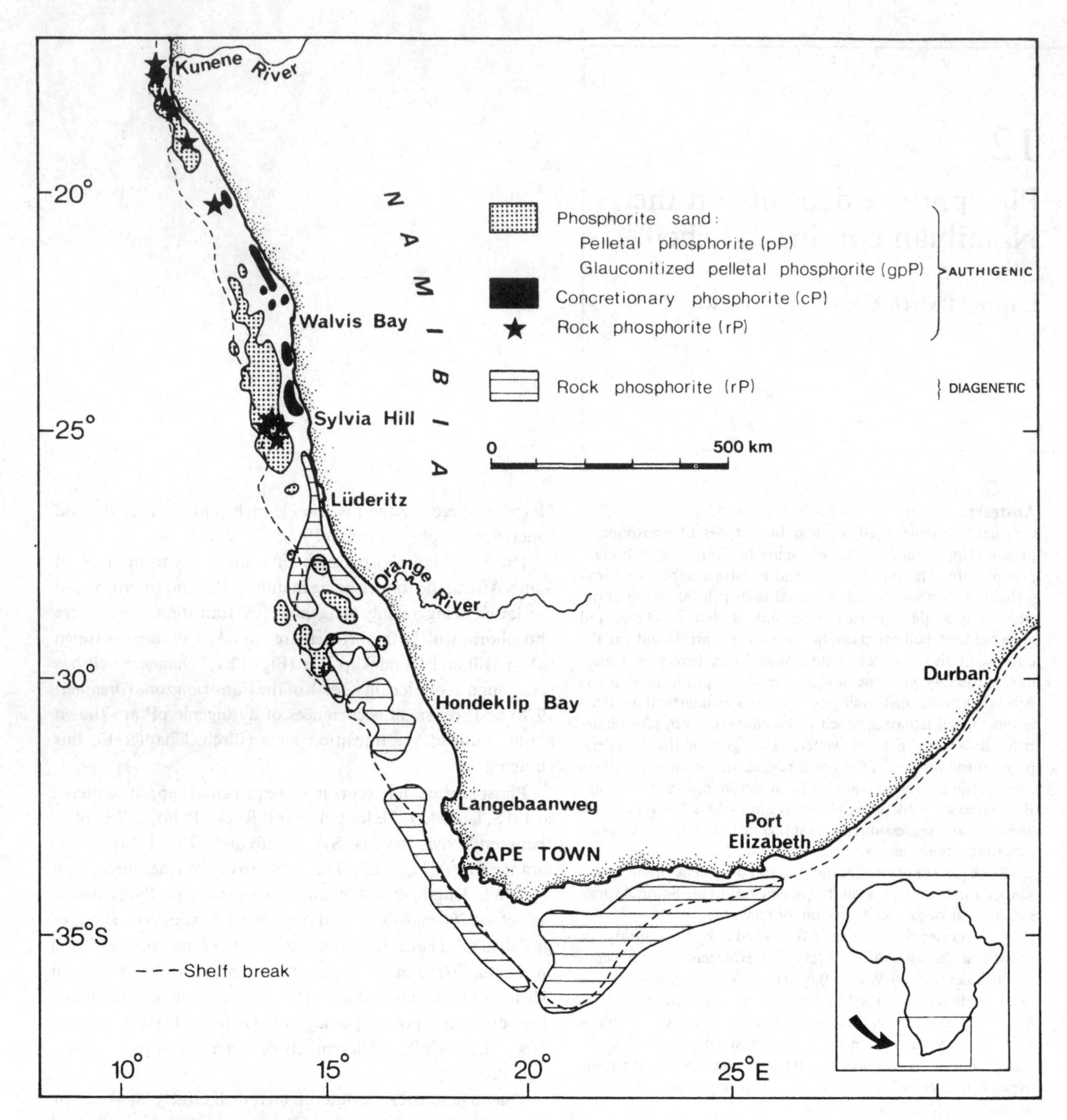

Fig. 12.1. Location of phosphorite types off southern Africa.

carbonaceous shales of the Whitehill Formation.

Offshore, the authigenic phosphorite sand off Namibia is similar to that off Baja California, where the deposit occurs in depths of 100 m (D'Anglejan, 1967), and also resembles the onshore deposits of North Carolina and Florida (Riggs, 1984).

Little is known at this stage about the authigenic rP due to limited sample recovery, but its origin is sometimes indicated by 2 mm wide mudcracks formed by syneresis (Selley, 1976) of an original phosphorite gel. An essential component of authigenic rP is pP (Rogers, 1977; Bremner, 1978).

Diagenetic rP occurs on the outer shelf as a thin phosphatized crust on Neogene limestones and is best developed on the Agulhas Bank (Parker, 1971; Parker & Siesser, 1972; Parker, 1975). Samples dredged from the seafloor are slablike and not 'nodular' as described by their discoverers (Murray & Renard, 1891). Three arenaceous and two conglomeratic types are recognized, which grade from nearshore (quartzose) to mid-shelf (glauconitic) to outer shelf (microfossiliferous) facies (Birch, 1979). In contrast to the phosphorite sand and the cP, diagenetic rP is commonly found worldwide as off Southern California (Emery, 1960), off Florida on the Blake Plateau (Emery & Uchupi, 1972), and off northwest Africa (Summerhayes, Nutter & Tooms, 1972).

Concretionary phosphorite (cP) has been located in

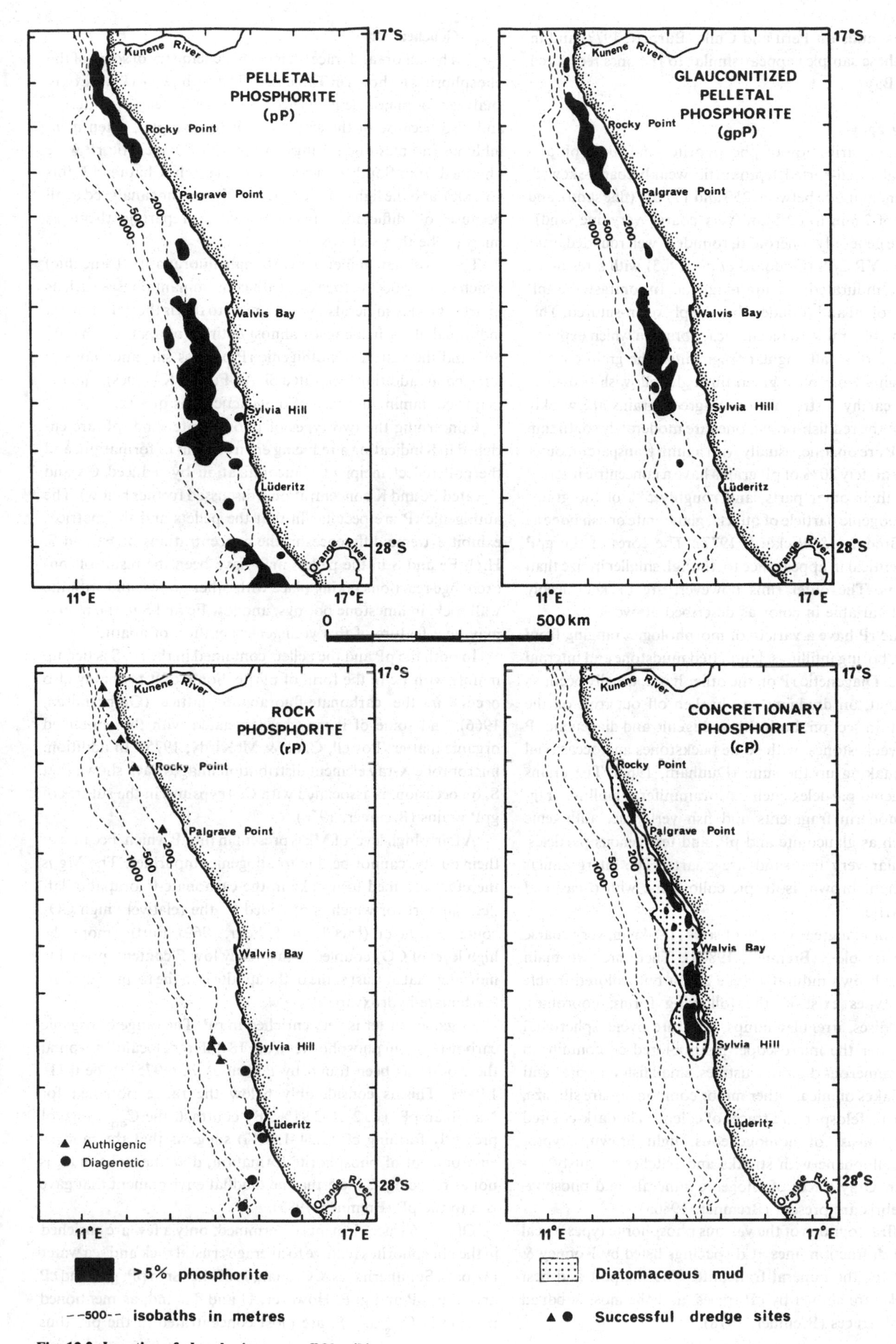

Fig. 12.2. Location of phosphorite types off Namibia.

diatomaceous muds off Peru and Chile (Burnett, 1977) and in some cases, these samples appear similar to the ones recovered near Walvis Bay.

Petrography

The size distribution of phosphorite sand (both pP and gpP) is generally well-sorted, leptokurtic, weakly coarse-skewed, with a prominent mode between 250 and 177 μm (fine sand), and a size-range of 2 mm to 62.5 μm (very coarse–very fine sand).

The pP are generally spheroidal, rounded–well rounded, very dark brown (5YR 2/2) (Goddard *et al.*, 1963) with a resinous, reddish lustre, indurated and non-magnetic. In contrast, the gpP are also spheroidal and rounded, but deeply open-sutured. This renders them vulnerable to mechanical abrasion which explains the abundance of small angular fragments. The grain color is variable ranging from olive green through yellowish-brown to red with an earthy lustre. Yellowish-green grains are weakly magnetic, but the reddish-brown ones are moderately so. In thin section the pP are opaque, usually with a thin, transparent, outer rim. Approximately 20% of pP grains have a concentric internal structure in their outer parts, and roughly 5% of the grains contain an allogenic particle of quartz, glauconite or fish bone as a nucleus (Bremner & Rickard, 1977). The cores of the gpP grains are identical in appearance to, though smaller in size than the pP grains. The outer rims however, are thicker, deeply sutured, and variable in color as described above.

Authigenic rP have a variety of morphologies ranging from whale bones, boring infillings, laminated mudstone and internal mollusc casts. Diagenetic rP on the other hand, usually occur as large slabs that, on dredging, are broken off outcrops on the seafloor. In thin section, both the authigenic and diagenetic rP are mainly wackestones, with some packstones and occasional mudstones making up the suite (Dunham, 1962). The grains include biogenic particles such as foraminifera, mollusc fragments, echinoderm fragments and fish vertebrae, authigenic particles such as glauconite and pP, and terrigenous particles, chiefly angular very fine sand-size quartz. The intergranular material is light brown, isotropic collophane with occasional specks of pyrite.

The cP exhibit a great variety of sizes (1–64 mm; very coarse sand–coarse pebbles) (Bremner, 1980a). There are two main types: a dark brown indurated type and a buff-colored friable type. Both types exist in the following forms: coprolites, dreikanters, discs, irregular lumps, and rare ovoid–spheroidal particles. Under the microscope, light-colored cP contain, in addition to numerous diatom frustules, small fish vertebrae and scales, and flakes of mica. Other minor components are silt-size, angular quartz, feldspar and traces of calcite. The dark-colored lithified cP consist of homogeneous, light brown, cryptocrystalline collophane with streaks and patches of 'dusty' organic matter. Only traces of allogenic minerals, and phosphatized fish debris are present (Bremner, 1980a).

X-ray diffractograms of the various phosphorite types reveal well defined diffraction lines at d-spacings listed by Rooney & Kerr (1967) for the mineral francolite. The sharpest and best defined peaks are shown by cP traces, and the most subdued peaks by gpP traces (Bremner, 1978).

Geochemistry

The major- and trace-element geochemistry of some of the phosphorites is shown in Table 12.1. The authigenic rP were only analyzed for major elements because of the scarcity of material, and also because of the small size of the samples, often mere infillings in limestone borings. Analyses of diagenetic rP were obtained from South African material (Birch, Chapter 13, this volume), and the light colored, friable cP were not analyzed at all because of difficulties involved with preparing them as microprobe thin sections.

The phosphate mineral is carbonate fluorapatite (francolite) which usually occurs together with contaminant phases such as quartz and clay minerals. An exception to this is the dark brown, indurated cP, which consist almost entirely of apatite. The pP, gpP and the matrix of authigenic rP possess silt- and clay-size terrigenous additions (elevated Si, Al, Fe and K values), and the major contaminant phase of diagenetic rP is quartz.

Concerning the two types of phosphorite sand, pP are enriched in S indicating a reducing environment of formation, and the gpP reflect incipient glauconitization by reduced Ca and elevated Fe and K concentrations (discussed further below). The authigenic rP are peculiar in that the pellets and the matrices exhibit extreme differences in the concentrations of Fe and S. High Fe and S in the pellets may have been the result of ion-exchange reactions taking place with other grains, and with the wall rock, in limestone borings, and low Fe and S in the matrix may be a feature of this younger generation of apatite.

In both the pP and the pellets contained in the rP, S is tied up mainly with Fe in the form of pyrite. Some of it probably also occurs in the carbonate-fluorapatite lattice (Gulbrandsen, 1966), and some of it may be associated with the contained organic matter (Powell, Cook & McKirdy, 1975). In addition, microprobe X-ray element distribution images have shown that S, on occasion, is associated with Ca (gypsum) in the sutures of gpP grains (Bremner, 1978).

A fairly high level of Mg is present in the cP which, because of their purity, cannot be due to allogenic impurities. The Mg is therefore ascribed to uptake in the carbonate-fluorapatite lattice, support for which is provided by the relatively high CO_2 content of the cP (McClellan & Lehr, 1969). Furthermore, the high level of CO_2 coupled with a fairly low F content, probably indicates that at least some of the apatite is in the form of dahllite (carbonate hydroxyapatite).

Organic matter is very enriched in pP. The range of organic carbon (C_{org}) in phosphorites from 18 different localities around the world has been found by Powell *et al.* (1975) to be 0.11–1.98%. This is considerably below the range obtained for Namibian pP, i.e. 2.21–2.91%. By contrast, the C_{org} range of presently forming cP (1.24–1.38%) suggests that the modern environment of phosphorite formation, diatomaceous mud, is not as reducing as the estuarine mudflat environment that gave rise to the pP (Bremner, 1983).

Of the 16 trace elements determined, only a few are enriched in the phosphorites relative to average crustal rock and sea water (Tooms, Summerhayes & Cronan, 1969) i.e. Sr in pP, gpP and cP and Y in pP and gpP. However, U and Th and, as mentioned previously C_{org} and S, are most concentrated in the pP, thus

Table 12.1. *Geochemistry of phosphorites*

	pP	gpP			rP(authigenic)			rp(diagenetic)[b]	cP (dark-colored)
		core	rim	average	pellets	matrix[a]	average		
Number of analyses (electron microprobe)	29	24	35	59	14	45	59	46	8
SiO_2	3.09	3.99	4.74	4.43	3.30	5.61	5.06	19.39	.07
Al_2O_3	1.10	1.25	1.04	1.13	.93	1.78	1.58	1.95	.3
FeO*	5.21	5.98	5.82	5.89	8.48	1.10	2.85	6.59	—
MgO	.90	.66	.80	.74	.80	1.76	1.53	1.41	1.42
CaO	45.11	41.72	38.85	40.02	43.26	45.98	45.33	37.28	51.61
Na_2O	1.09	.74	.60	.66	1.06	.95	.98	.90	.79
K_2O	.36	.40	.48	.45	.34	.48	.45	1.38	.04
P_2O_5	27.55	27.99	24.80	26.10	25.05	25.96	25.74	17.42	33.76
S	5.42	2.67	1.55	2.01	8.16	2.35	3.73	.40	.64
CO_2[c]	3.6			3.9			ND	11.3	4.9
F[d]	3.3			2.5			ND	2.3	3.2
C_{org}[e]	2.6			1.4			ND	.4	1.3
F/P_2O_5	.12			.10			ND	.13	.09
Cao/P_2O_5	1.64	1.49	1.57	1.53	1.73	1.77	1.76	2.14	1.53
No. of analyses (X-ray fluorescence)	4			4			—	—	3
Ba	66			62			ND	ND	79
Co	< 10			< 12			ND	ND	<5
Cr	195			300			ND	ND	14
Cu	28			48			ND	ND	4
Mn	30			90			ND	ND	12
Nb	<3			<3			ND	ND	<3
Ni	60			80			ND	ND	7
Rb	7			20			ND	ND	<3
Sr	2260			1630			ND	ND	2350
Th[f]	335			291			ND	ND	110
Ti	463			890			ND	ND	43
U[f]	117			102			ND	ND	37
V	40			67			ND	ND	20
Y	206			284			ND	ND	7
Zn	27			46			ND	ND	84
Zr	29			28			ND	ND	<3

Notes:

*Total iron reported as FeO

[a]Analysed by R.C. Wallace, Geological Survey, Pretoria

[b]Whole rock analyses (XRF) (Birch, this volume)

[c]Analysed using gasometry (Hülsemann, 1966) and X-ray diffraction (LeGeros, 1965)

[d]Analysed using a specific ion electrode by FOSKOR, Phalaborwa

[e]Analysed using a wet chemical technique (Morgans, 1956)

[f]Analysed by the Atomic Energy Board, Pelindaba

suggesting that the four components are intimately associated with each other in these grains. Cu, Ni and Y have the highest concentrations in gpP and correlate strongly with glauconite; their uptake in phosphorite may therefore be a function of lattice distortions created by major-element substitutions. Ba, Sr and Zn are richest in cP and although Ba^{2+} has a much larger ionic radius than Ca^{2+}, it appears to replace Ca in the carbonate-fluorapatite lattice. Zn correlates strongly with CO_2 which implies that it substitutes for Ca as CO_3^{2-} substitutes for PO_4^{3-}.

Age

Phosphorite sand (pP) has been dated radiochemically using equilibrium values of the activity ratios of $^{234}U/^{238}U$ and $^{230}Th/^{234}U$. The system is regarded as radiochemically closed, and a pre-mid-Pleistocene age (> 700 Ky) has been assigned (Thomson *et al.*, 1984). Bremner (1978) postulated a Miocene age on the grounds that stiff muds seaward of the phosphorite have been dated on nannofossil evidence as Miocene–Pliocene (Siesser, 1978). In addition, similar pP found at Langebaanweg

(Fig. 12.1), north of Cape Town, have been repeatedly reworked and recemented, and are associated with Late Miocene–Early Pliocene shark teeth and horse teeth in the Gravel Member of the Varswater Formation (Tankard, 1974).

The age of authigenic rP is considered to be Pliocene (Bremner, 1978) by comparison with numerous Pliocene dates of phosphorite off South Africa (Siesser, 1978). The earlier suggestion by Summerhayes *et al.* (1973) and Rogers (1977) that authigenic rP off Sylvia Hill predated, and was the source rock for pP, has been discounted by Bremner (1978) because the light brown organic matter-poor matrix of authigenic rP is not found as pP. Although mid-Miocene benthic foraminifera have been recovered in phosphatized nummulitic limestones (diagenetic rP) on the southern edge of the study area (Rogers, 1977), this merely provides a maximum age for the subsequent phosphatization episode. Newly available radiochemical work on diagenetic rP off Cape Town however, has indicated a Late Pleistocene age (61.5–75.9 ky) using $^{230}Th/^{234}U$ and $^{231}Pa/^{235}U$ methods (Birch *et al.*, 1983).

The most confidently dated phosphorite is cP, which occurs at depths that were flooded by the Flandrian transgression (Bremner, 1978). Consequently, uranium-series ages of 3900–660 y (Thomson *et al.*, 1984) were not unexpected. These Recent ages for both the dark indurated cP, and the light colored friable cP, conflict with the earlier findings of Baturin, Merkulova & Chalov (1972) and Veeh, Calvert & Price (1974) that the dark cP are older than 700 ky.

Sedimentation

Baturin (1982) is of the opinion that winnowing of cP out of diatomaceous muds gave rise to the phosphorite sand on the midshelf. Bremner (1980a) has pointed out however, that the cP differ morphologically, petrographically and geochemically from the phosphorite sand. Following Kazakov (1937), we suggest that pP and gpP were formed by rapid, inorganic precipitation of apatite from sea water in subtropical estuaries beside an arid west coast. The apatite gel is conceived as being deposited on organic-rich, intertidal mudflats in response to the following set of environmental conditions:

(1) Strong, steady, southerly winds drove low salinity surface water northwards from the mouths of small rivers situated at the heads of northward-opening estuarine lagoons (Fig. 12.3a). Simultaneously, phosphate- and plankton-enriched, upwelled, oceanic water was drawn southwards into the lagoons as a compensatory undercurrent. In other words, water exchange in the lagoons was enhanced by wind-driven, estuarine-like circulation as described by Brongersma-Sanders & Groen (1970).
(2) The pH and Eh of the water in the sheltered lagoons were raised and lowered, respectively, due to the bacterial decomposition of organic matter, particularly on the intertidal mudflats (Berner, 1969; Moshiri & Crumpton, 1978).
(3) Because of high solar insolation in these subtropical latitudes the water temperature was raised, which encouraged bacterial activity and suppressed the solution of newly precipitated apatite (Kramer, 1964; Atlas, 1975).
(4) Small rivers at the heads of the estuaries introduced clay minerals that adsorbed Mg ions from the sea water, and at the same time released adsorbed Fe ions into solution (Martens & Harris, 1970; Drever, 1971; Bachra, Trautz & Simon, 1965).

The precise mechanism of apatite precipitation is not known with certainty but McConnell (1965) found experimentally that the common enzyme carbonic anhydrase (Veitch & Blakenship, 1963) causes precipitation of carbonate hydroxyapatite, and mentions in addition that certain bacteria are capable of initiating the same reaction. Stumm & Morgan (1970) state that microorganisms are probably instrumental in determining such factors as pH, Eh, the type of residual organic matter and the proportions of particulate phosphate to dissolved inorganic phosphate. These microorganisms are therefore assumed to have played an essential role in stimulating the precipitation of apatite from the phosphate-saturated estuarine waters of the proposed model.

The apatite precipitated as a gel on the estuarine intertidal mudflats, and was deposited as a layer a millimetre-or-so thick along with organic matter, and minor amounts of terrigenous silt and clay. During low tide, the mudflats were exposed, desiccated, and the phosphatic surface-layer disintegrated into small sand-size intraclasts (cf. Riggs, 1979 and 1984). These intraclasts were subsequently modified in size during high tide, eventually attaining equilibrium with the prevailing hydraulic regime in the estuaries. That is, the smaller intraclasts physically accreted additional phosphate by the 'snowball' mechanism of Sorby (1879), and the larger intraclasts were eroded down to this equilibrium grain size. The former thus developed a pseudo-oolitic texture around an unstructured center whereas the latter are completely unstructured (Bremner & Rickard, 1977).

Pellets of gpP originated as pP but were located close to the river mouths in the upper reaches of the estuaries, where adsorbed Fe^{3+} ions were released from the clay minerals (Drever, 1971). If the upper reaches still possessed a negative Eh (reducing), Fe^{2+} ions, which would be highly soluble, could readily replace calcium in the apatite lattice (R.E. Loewenthal, pers. comm.). Due to the smaller ionic radius of Fe^{2+} ions compared to Ca^{2+} ions, lattice distortions may have been responsible for the formation of sutures in the rims of the gpP.

The present, extensive, lateral distribution of high concentrations of both pP and gpP is explained by a Late Miocene transgression (Siesser & Dingle, 1981) advancing over the estuaries, winnowing the dense phosphorite grains, and transporting them alongshore northwards. As the transgression proceeded, conditions favorable to phosphogenesis would have been enhanced and the zone over which phosphorite was being formed would have been broadened (cf. Riggs, 1984). Repeated winnowing and northward spreading continued during the Late Pliocene and during the numerous Quaternary fluctuations of sea level.

The matrices of authigenic rP are typically organic matter-

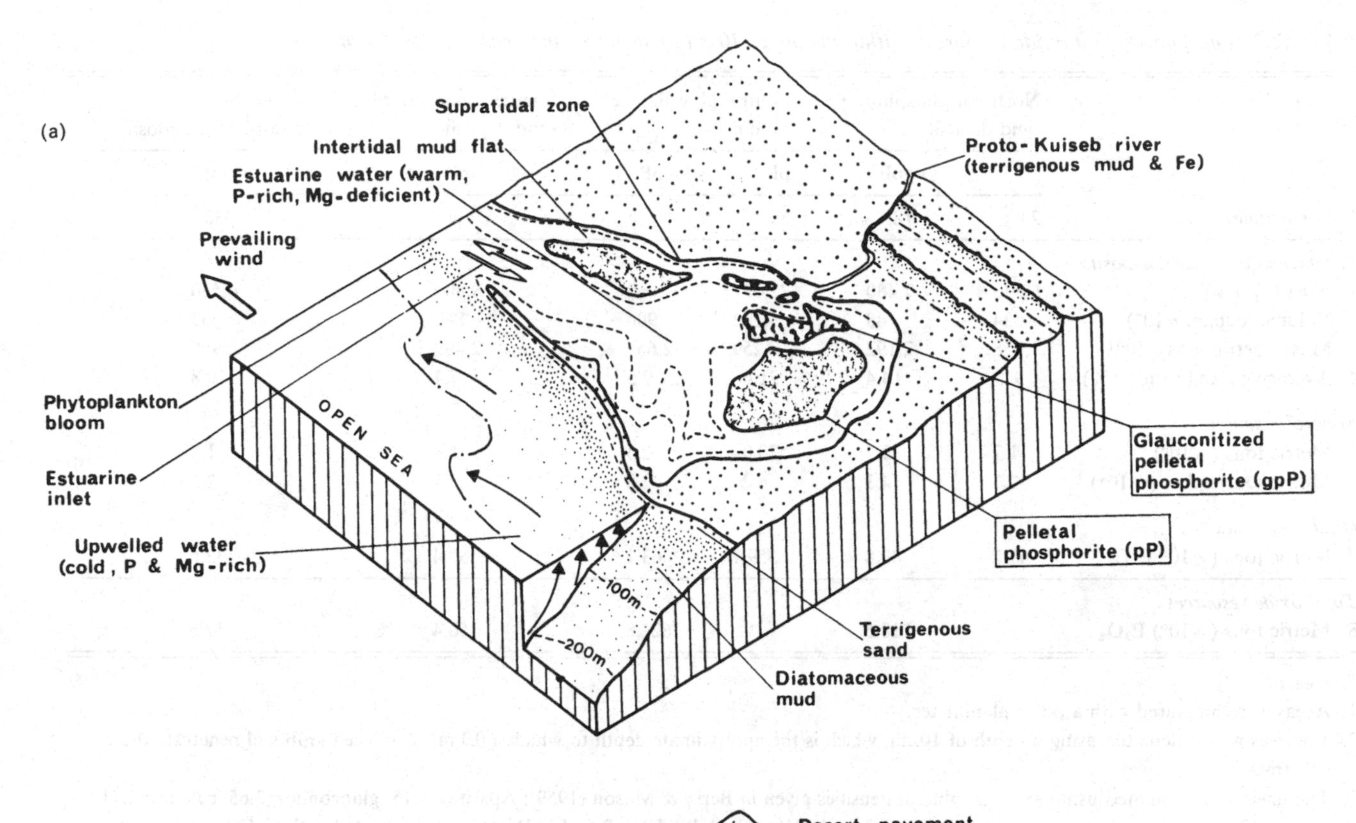

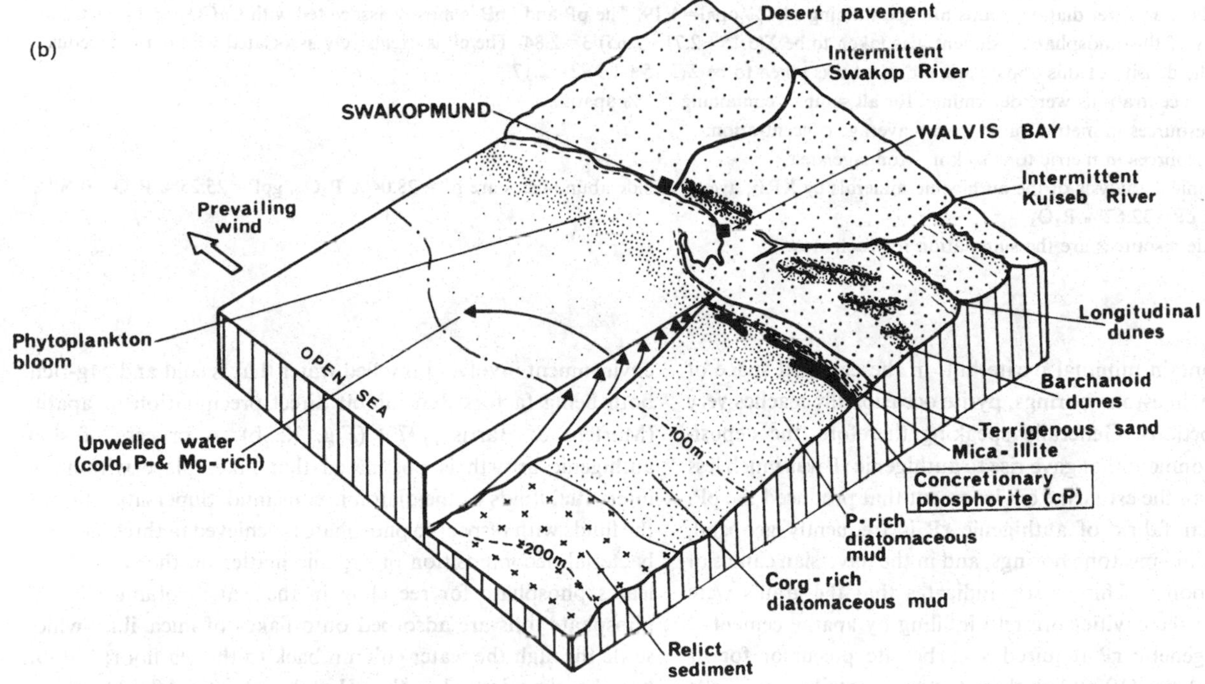

Fig. 12.3. Models of phosphorite formation on the Namibian continental shelf: (a) Pelletal phosphorite (pP) and glauconitized pelletal phosphorite (gpP); (b) Concretionary phosphorite (cP).

Table 12.2. *Phosphorite and oxide resources within the upper 10 cm of sediment in areas with > 5% apatite*

	Northern phosphorite sand deposit		Central phosphorite sand deposit		Southern phosphorite sand deposit	Concretionary-phosphorite deposit
	pP	gpP	pP	gpP	pP	cP
No. of samples	24	18	80	15	24	30
Parameters of mineral deposits						
1. Area (sq. km)	6,340	7,480	32,021	9,060	8,800	4,530
2. Volume (cub.m. × 10^6)	634	748	3,202	906	880	453
3. Mass (metric tons × 10^6)	1,858	2,192	9,259	2,655	2,499	969
4. Average concentration (%)	9.1	14.4	21.6	9.5	7.4	10.8
Mineral resources						
5. Metric tons (× 10^8)	1.7	3.2	20.0	2.5	1.8	1.0
6. Metric tons/sq.km (× 10^4)	2.7	4.3	6.2	2.8	2.1	2.2
Oxide resources						
7. Metric tons (× 10^6 P_2O_5	47.7	80.8	559.1	63.1	50.4	32.6
Total oxide resources						
8. Metric tons (× 10^6) P_2O_5		128.5		622.8	50.4	32.6

**Notes:*
1. Areas were measured with a polar planimeter.
2. Volumes were calculated using a depth of 10 cm, which is the approximate depth to which a 0.1 m^3 Van Veen grab will penetrate the substrate.
3. The mass was calculated using average mineral densities given in Berry & Mason (1959): Apatite = 3.15, glauconite = 2.65, calcite = 2.71, quartz = 2.65, and wet diatomaceous mud containing 54.4% opal = 1.19. The pP and gpP is mostly associated with $CaCO_3$ and quartz, so the density of this phosphatic sediment was taken to be $\Sigma(3.15+2.71+2.65)/3=2.84$. The cP is exclusively associated with diatomaceous mud, so the density of this phosphatic sediment was taken to be $\Sigma(3.15+1.9)/2=2.17$.
4. Average concentrations were determined for all samples containing > 5% apatite.
5. Mineral resources in metric tons = mass × average concentration.
6. Mineral resources in metric tons/sq.km. = tonnage/area.
7. From chemical analysis of the authigenic minerals by XRF, average oxide abundances are pP = 28.04% P_2O_5, gpP = 25.25% P_2O_5, 0.60% K_2O, and cP = 32.62% P_2O_5.
8. Total oxide resources are the summation for each area.

free and contain minimal pyrite but, in the confined space of some of the limestone borings, pyrite occasionally reaches very high proportions. Generally speaking therefore, the depositional environment that gave rise to authigenic rP was much less reducing than the estuarine environment that produced the pP. The granular fabric of authigenic rP is frequently geopetal, particularly in limestone borings, and in the haversian canals of vertebrate bones. This clearly indicates that the grains were washed into the cavities prior to infilling by apatite cement.

The diagenetic rP required a carbonate precursor for its formation. Ames (1959) has shown experimentally that calcite can be replaced by apatite at PO_4^{3-} concentrations as low as 0.1 ppm. Since the PO_4^{3-} content of upwelled oceanic water off Namibia is at least 0.2–0.3 ppm (2–3 μmol/l^{-1}) (Chapman & Shannon, 1985), it is highly probable that the surfaces of limestones on the southern Namibian outer shelf underwent slow phosphatization. Diagenetic rP off the Orange River are similar to the phosphatized limestones of the Agulhas Bank, where classic surface-to-center replacement is observed (Parker, 1971; Birch, 1979).

Bremner (1980a) has detailed a model for cP formation that differs from the model proposed for phosphorite sand. The cP environment involves upwelled water that is cold and Mg-rich, both being factors that inhibit direct precipitation of apatite (Martens & Harris, 1970) (Fig. 12.3b). A process of slow authigenic growth is postulated that takes place within the interstitial fluids of the diatomaceous mud. Supersaturation of the fluids with respect to phosphate is achieved in three steps: (i) bacterial decomposition of organic matter on the seafloor releases phosphate for recycling in the water column; (ii) the phosphate ions are adsorbed onto flakes of mica–illite which settle through the water column back to the sea floor; (iii) On burial of the mica flakes, the pH of the interstitial fluids is raised by protein-decomposing bacteria which releases phosphate to the interstitial fluids. Eventually the fluids become supersaturated with respect to phosphate and slow authigenic growth of apatite commences on nuclei such as fish debris or diatom frustules (Weaver & Wampler, 1972; Edzwald, Toensing & Leung, 1976; Bremner, 1980a).

Resources

Our systematic sampling has allowed us to attempt the first crude estimate of the surficial phosphate resources of the Namibian continental shelf (Table 12.2). We emphasize that sample spacing over the landward part of the diatomaceous mudbelt was about 4 km (2 nm) along lines 19 km apart whereas farther seaward, the spacing was only about 19 km (10 nautical miles). The sampling device used was a 0.1 m^3 modified Van Veen grab, which was designed to penetrate about 10 cm into the substrate. The resources have been calculated for three deposits of phosphorite sand, and the cP deposit, but no estimate is given for authigenic and diagenetic rP because of small sample size and poor sample recovery.

The northern and central deposits of phosphorite sand are flanked on their seaward sides by 'stiff' Neogene and Quaternary non-phosphatic mud, and their landward sides are covered by terrigenous or diatomaceous sediments. The southern deposit of phosphorite sand is open ended to the south, where pP gives way to phosphatic foraminiferal infillings (Rogers, 1977). The cP deposit is richest along the landward edge of the diatomaceous mudbelt (Bremner, 1980a) but nothing is yet known about the size and abundance of cP with depth in the deposit. The physical dimensions of the mudbelt have been described in some detail by Meyer (1973) and Bremner (1980b).

Recovery of phosphorite sand from seafloor depths of < 500 m, and cP from depths of < 200 m, are well within the scope of modern technology as equipment now exists for the recovery of manganese nodules by airlift dredging from the abyssal plains (Mero, 1965). Ports capable of handling small- to medium-size ore carriers are Walvis Bay and Luderitz (Fig. 12.2).

Phosphorite sand can easily be beneficiated by wet-screening (to remove mud and gravel), and by dissolution of the carbonate sand fraction in dilute acetic acid. Quartz can then be separated by floatation methods to produce the final concentrate. Beneficiation of cP would involve gentle wet-screening to concentrate both friable and indurated coarse sand–gravel-size cP and fish debris (Bremner, 1980a). Acetic acid and floatation techniques would then remove the shell and quartz, respectively.

Acknowledgments

We would like to thank the Director of the Geological Survey of South Africa, Mr L.N.J. Engelbrecht, for permission to use and publish these data. Drs R.V. Dingle and I.L. Van Heerden, and Prof W.C. Burnett, critically reviewed the manuscript, and Mesdames S.N. Smith and E.G. Krummeck meticulously attended to the draughting and typing, respectively.

References

Ames, L.L. (1959). The genesis of carbonate apatites. *Economic Geology*, **54**, 829–41.

Atlas, E.L. (1975). 'Phosphate equilibria in seawater and interstitial waters'. Unpublished Ph.D thesis, Oregon State University, Oregon.

Bachra, B.N., Trautz, D.R. & Simon, S.L. (1965). Precipitation of calcium carbonates and phosphates – III. The effect of magnesium and fluoride ions on the spontaneous precipitation of calcium carbonates and phosphates. *Arch. Oral. Biol*, **10**, 731–8.

Baturin, G.N. (1982). Phosphorites on the sea floor. Origin, composition and distribution. *Developments in Sedimentology*, **33**, 1–343.

Baturin, G.N., Merkulova, K.I. & Chalov, P.I. (1972). Radiometric evidence for recent formation of phosphatic nodules in marine shelf sediments. *Marine Geology*, **13**, 37–41.

Berner, R.A. (1969). The synthesis of framboidal pyrite. *Economic Geology*, **64**, 383–4.

Berry, L.G. & Mason, B. (1959). *Mineralogy: Concepts, Descriptions and Determinations*. W.H. Freeman, San Francisco.

Birch, G.F. (1979). A model of penecontemporaneous phosphatization by diagenetic and authigenic mechanisms from the western margin of southern Africa. *Society of Economic Paleontologists Mineralogists Special Publicaion*, **29**, 79–100.

Birch, G.F., Thomson, J., McArthur, J. & Burnett, W.C.A. (1983). Pleistocene phosphorites off the west coast of South Africa. *Nature*, **302**, 601–3.

Bremner, J.M. (1978). 'Sediments on the continental margin off South West Africa between latitudes 17° and 25° S.' Unpublished Ph.D thesis, University of Cape Town, South Africa.

Bremner, J.M. (1980a). Concretionary phosphorite from SW Africa. *Journal Geological Society, London*, **137**, 773–86.

Bremner, J.M. (1980b). Physical parameters of the diatomaceous mud belt off South West Africa. *Marine Geology*, **34**, 67–76.

Bremner, J.M. (1983). Biogenic sediments on the South West African (Namibian) continental margin. In *Coastal Upwelling: Its Sediment Record, part B*, ed. J. Thiede & E. Suess, pp. 73–104. Plenum Press, New York.

Bremner, J.M. & Rickard, R.S. (1977). On the formation of phosphorite pellets. *Proceedings Electron Microscopy Society South Africa*, **7**, 83–4.

Brongersma-Sanders, M. & Groen, P. (1970). Wind and water depth and their bearing on the circulation in evaporite basins. Third Symposium on Salt. *Northern Ohio Geological Society*, pp. 3–7. Cleveland, Ohio.

Burnett, W.C. (1977). Geochemistry and origin of phosphorite deposits from off Peru and Chile. *Bulletin Geological Society America*, **88**, 813–23.

Chapman, P. & Shannon, L.V. (1985). The Benguela ecosystem. Part II. Chemistry and related processes. In *Oceanography Marine Biology Annual Review*, ed. M. Barnes, **23**, 183–251.

D'Anglejan, B.F. (1967). Origin of marine phosphorites off Baja California, Mexico. *Marine Geology*, **5**, 15–44.

Drever, R.V. (1971). Magnesium–iron replacement in clay minerals in anoxic marine sediments. *Science*, **172**, 1334–6.

Dunham, R.J. (1962). Classification of carbonate rocks according to depositional texture. In *Classification of carbonate rocks – a symposium*, 1, ed. W.E. Ham, pp. 108–21. Memoir American Association Petroleum Geologists, Tulsa, Oklahoma.

Edzwald, J.K., Toensing, D.C. & Leung, M.C-Y. (1976). Phosphate adsorption reactions with clay minerals. *Environment Science Technology*, **10**, 485–90.

Emery, K.O. (1960). *The Sea off Southern California*. Wiley, New York.

Emery, K.O. & Uchupi, E. (1972). *Western North Atlantic Ocean: Topography, Rocks, Structure, Water, Life and Sediments*. American Association Petroleum Geologists, Tulsa.

Goddard, E.N., Trask, P.D., De Ford, E.K., Rove, O.N., Singewald, J.T.Jr & Overbeck, R.M. (1963). *Rock-Colour Chart*. Geological Society America, New York.

Gulbrandsen, R.A. (1966). Chemical composition of phosphorites of the Phosphoria Formation. *Geochimica Cosmochimica Acta*, **30**, 769–78.

Gulbrandsen, R.A. (1970). Relation of carbon dioxide content of apatite of the Phosphoria Formation to regional facies. *US Geological Survey Professional Paper*, **700–B**, B9–13.

Haughton, S.H. (1969). *Geological History of Southern Africa*. Geological Society of South Africa, Johannesburg.

Hüsemann, J. (1966). On the routine analyses of carbonates in unconsolidated sediments. *Journal Sedimentary Petrology*, **36**, 622–5.

Kazakov, A.V. (1937). The phosphorite facies and the genesis of phosphorites. In Geological Investigations of Agricultural ores. *Scientific Institute Fertilizers Insecto-Fungicides*, **142**, 95–113.

Kramer, J.R. (1964). Sea water: saturation with apatites and carbonates. *Science*, **206**, 403–4.

LeGeros, R.Z. (1965). Effect of carbonate on the lattice parameters of apatite. *Nature*, **206**, 403–4.

Martens, C.S. & Harris, R.C. (1970). Inhibition of apatite precipitation in the marine environment by magnesium ions. *Geochimica*

Cosmochimica Acta, **34**, 621–5.
McClellan, G.H. & Lehr, J.R. (1969). Crystal–chemical investigation of natural apatites. *American Mineralogist*, **54**, 1374–91.
McConnell, D. (1965). Precipitation of phosphates in sea water. *Economic Geology*, **60**, 1059–62.
Mero, J.L. (1965). *The Mineral Resources of the Sea.* Elsevier, London.
Meyer, K. (1973). Uran-Prospektion vor Sudwestafrika. *Erzmetall*, **26**, 313–17.
Morgans, J.F.C. (1956). Notes on the analysis of shallow water soft substrata. *Journal Animal Ecology*, **25**, 367–87.
Moshiri, G.A. & Crumpton, W.G. (1978). Certain mechanisms affecting water column-to-sediment phosphate exchange in a bayou estuary. *Journal Water Pollution Control Federation*, **50**, 392–4.
Murray, J. & Renard, A.F. (1891). *Deep-sea deposits. Reports Scientific Research. HMS Challenger 1873–1876.* HMSO, London.
Parker, R.J. (1971). The petrography and major element geochemistry of phosphorite nodule deposits on the Agulhas Bank, South Africa. *Bulletin South African National Committee Oceanographic Research Marine Geology Programme*, **2**, 1–94.
Parker, R.J. (1975). The petrology and origin of some glauconitic and glauco-conglomeratic phosphorites from the South African continental margin. *Journal Sedimentary Petrology*, **45**, 230–42.
Parker, R.J. & Siesser, W.G. (1972). Petrology and origin of some phosphorites from the South African continental margin. *Journal Sedimentary Petrology*, **42**, 434–40.
Powell, T.G., Cook, P.J. & McKirdy, D.M. (1975). Organic geochemistry of phosphorites: Relevance to petroleum genesis. *Bulletin American Association Petroleum Geologists*, **59**, 618–32.
Riggs, S.R. (1979). Petrology of the Tertiary phosphorite system of Florida. *Economic Geology*, **74**, 195–220.
Riggs, S.R. (1984). Paleoceanographic model of Neogene phosphorite deposition, US. Atlantic continental margin. *Science*, **223**, 123–31.
Rogers, J. (1977). 'Sedimentation on the continental margin off the Orange River and the Namib Desert'. Unpublished Ph.D thesis, Geology Department, University of Cape Town, South Africa.
Rogers, J. (1980). First report on the Cenozoic sediments between Cape Town and Elands Bay. *Report Geological Survey South Africa*, **1980–165**, 1–64. (Open File).
Rooney, T.P. & Kerr, P.F. (1967). Mineralogic nature and origin of phosphorite, Beaufort County, North Carolina. *Bulletin Geological Society America*, **78**, 731–48.
Selley, R.C. (1976). *An Introduction to Sedimentology*. London, Academic.
Siesser, W.G. (1978). Age of phosphorites on the South African continental margin. *Marine Geology*, **26**, 17–28.
Seisser, W.G. & Dingle, R.V. (1981). Tertiary sea-level movements around southern Africa. *Journal Geology*, **89**, 83–96.
Sorby, H.C. (1879). On the structure and origin of limestones (Presidential address). *Proceedings Geological Society London*, **35**, 56–95.
South African Committee for Stratigraphy (SACS) (1980). Stratigraphy of South Africa. Part 1: Lithostratigraphy of the Republic of South Africa, South West Africa/Namibia and the Republics of Bophuthatswana, Transkei and Venda. *Handbook Geological Survey South Africa*, **8**, 1–690.
Stumm, W. & Morgan, J.J. (1970). *Aquatic chemistry. An Introduction Emphasizing Chemical Equilibria in Natural Waters.* Wiley, New York.
Summerhayes, C.P., Birch, G.F., Rogers, J. & Dingle, R.V. (1973). Phosphate in sediments off southwestern Africa. *Nature*, **243**, 509–11.
Summerhayes, C.P., Nutter, A.H. & Tooms, J.S. (1972). The distribution and origin of phosphate in sediments off northwest Africa. *Sedimentary Geology*, **8**, 3–28.
Tankard, A.J. (1974). Petrology and origin of the phosphorite and aluminium phosphate rock of the Langebaanweg–Saldanha area south-western Cape Province. *Annals South African Museum*, **65**, 217–49.
Thomson, J., Calvert, S.E., Mukherjee, S., Burnett, W.C. & Bremner, J.M. (1984). Further studies of the nature, composition and ages of contemporary phosphorite from the Namibian shelf. *Earth Planetary Science Letters*, **69**, 341–53.
Tooms, J.S., Summerhayes, C.P. & Cronan, P.S. (1969). Geochemistry of marine phosphate and manganese deposits. In *Oceanography Marine Biology Annual Review*, ed. H.H. Barnes, **7**, 49–100.
Veeh, H.H., Calvert, S.E. & Price, N.B. (1974). Accumulation of uranium in sediments and phosphorites on the South West African shelf. *Marine Chemistry*, **2**, 189–202.
Veitch, F.P. & Blakenship, L.C. (1963). Carbonic anhydrase in bacteria. *Nature*, **197**, 76–7.
Weaver, C.E. & Wampler, J.M. (1972). The illite–phosphate association. *Geochimica Cosmochimica Acta*, **36**, 1–13.

13

Phosphorite deposits on the South African continental margin and coastal terrace

G. BIRCH

Abstract

A thin capping of diagenetic (replacement) phosphorite covers extensive areas of the western and southern margins of Southern Africa, whereas minor, isolated occurrences of fragmented diagenetic phosphorite are located on the eastern margin. Minor authigenic (precipitated) phosphorite pellets and rock are found on the western margin of South Africa and more abundantly on the adjacent coastal terrace on land.

Authigenic phosphorite forms in shallow estuaries or embayments in an environment of intense, wind-generated upwelling and high biological productivity. The seafloor sediments become phosphate-enriched by decay of siliceous phytoplankton facilitating interstitial precipitation of apatite to form phosphatic packstones and oolitic pellets at the water–sediment interface. Diffuse, subsurface upwelling at the zone of divergence on the middle and outer shelf is accompanied by moderately enhanced nutrient concentration in the water column and increased zooplankton productivity. Although insufficient phosphate is released at the seafloor for precipitation to occur, lime mud produced through the disaggregation of calcareous exoskeletons is replaced by calcium phosphate. All the components of the bottom sediment are lithified into a near-continuous capping of phosphate rock.

The phosphorite deposits on the Agulhas Bank and western margin of South Africa contain approximately 9000×10^6 metric tons P_2O_5 and the western onland coastal terrace contains about 37.5×10^6 metric tons at 3% P_2O_5.

Diagenetic phosphorites comprise three arenaceous packstone varieties and two conglomeratic types. Arenaceous microfossiliferous limestone occurs at various stages of phosphatization and contains planktonic foraminifera, quartz, and glauconite with minor bryozoan, echinoderm fragments and a poorly sorted terrigenous component. Cement is dominantly apatite with subordinate micrite and/or glauconite. The second arenaceous variety contains abundant glauconite with minor foraminifera and quartz in an apatitic cement. The third subgroup is made up of subordinate quartz, foraminifera and glauconite in a goethite-rich cement. One conglomerate class is similar in composition to the second arenaceous group and the enclosed pebbles are of either of the other two arenaceous varieties. The second conglomerate class is distinguished by abundant macrofossils, bone and coraline algae in an apatite cement. Enclosed pebbles contain minor glauconite and quartz and the cement is goethite.

The composition of phosphatic rocks form part of a series ranging from inner shelf samples, rich in quartz and rock fragments, through glauconite-rich middle shelf phosphorites to deep-water microfossiliferous varieties. Phosphorite is typically tabloid or lenticular and between 5 and 50 cm thick. Extensive reworking is indicated by rounded grain morphology.

Introduction

Extensive areas of the western and southern margins (Murray & Renard, 1891; Collet, 1905; Murray & Philippi, 1908; Cayeux, 1934; Parker, 1971 and 1975) of South Africa are mantled by rich deposits of phosphatic material (Summerhayes *et al.*, 1973), whereas only minor, isolated occurrences are present on the eastern margin (Fig. 13.1). Sedimentary phosphorite is also located on the coastal terrace between the Olifants River and Cape Town (Visser & Schoch, 1973; Tankard, 1974a, b and 1975; Visser & Toerien, 1971; Hendey & Dingle, 1989).

South African phosphorite deposits are comprised of authigenic and diagenetic types (Summerhayes, 1973; Parker, 1975; Birch, 1979a). Authigenic phosphorite forms by precipitation (Kazakov, 1937), interstitial growth or pelletal accretion, and diagenetic phosphorite originates by replacement of a calcareous precursor (Ames, 1959).

Distribution

Authigenic and diagenetic phosphorite are separated regionally. Authigenic quartz packstone and pelloids are located onland on the coastal terrace (Tankard, 1974a, b and 1975; Birch, 1977a and 1979c) and minor, isolated concentrations of pelloids on the continental shelf north of Cape Town (Birch, 1979c) are the extreme southerly extension of similar rich deposits to the north off Namibia (Bremner, 1975 and 1978; Rogers, 1977; Bremner & Rogers, Chapter 12, this volume). Diagenetic rocks are the dominant phosphorite variety on the South African continental margin. Due to surficial sediment cover and an irregular dredging density, it is difficult to determine the exact extent of phosphorite occurrence. A specially designed investigation using close-spaced dredging with seismic and side-scan sonar control (Du Plessis & Birch, 1977) has established that diagenetic phosphorite exists as a near-continuous 'pavement'

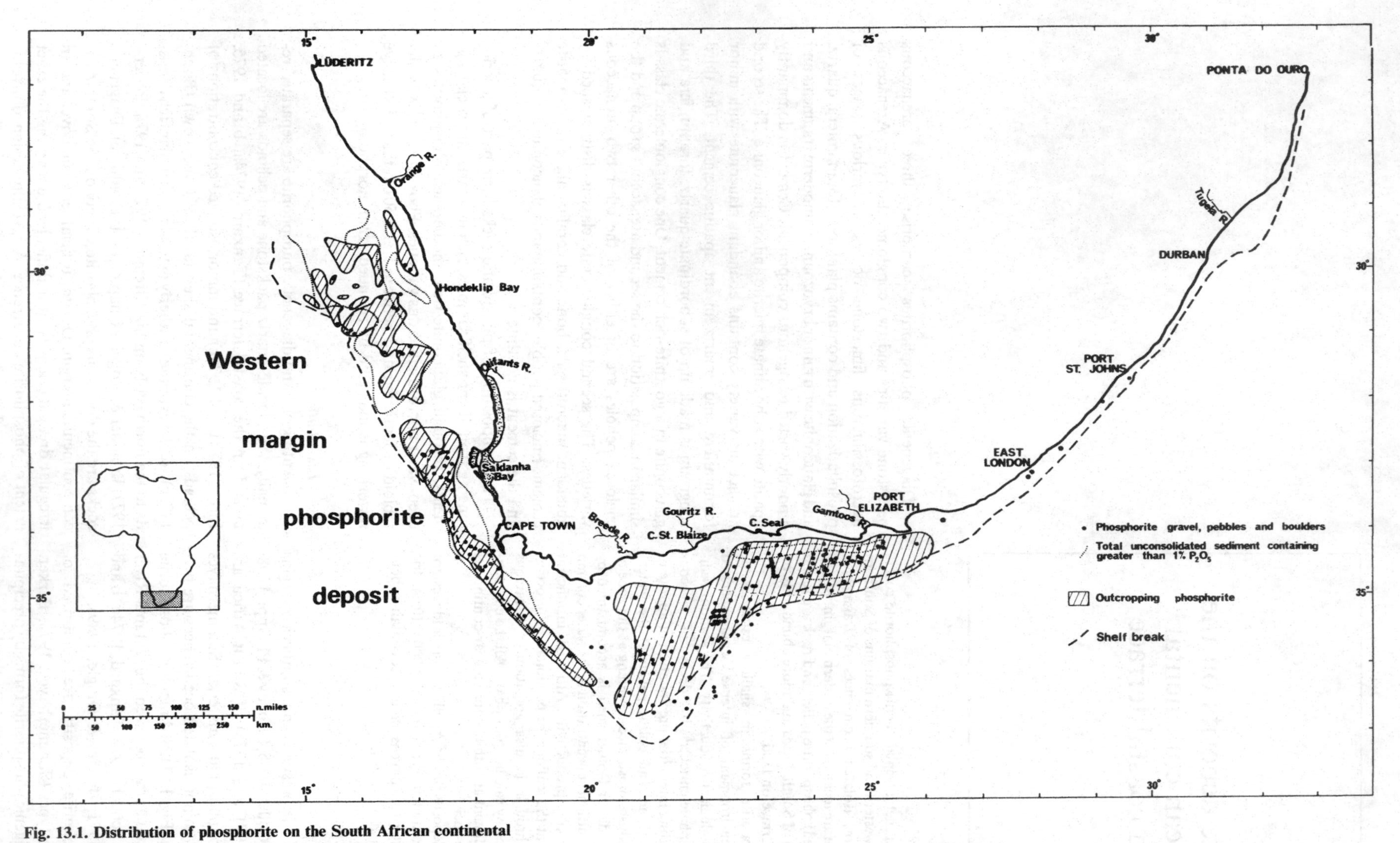

Fig. 13.1. Distribution of phosphorite on the South African continental margin and coastal terrace. Hatched areas represent offshore diagenetic phosphorite. Stippled areas depict onland authigenic phosphorite.

Table 13.1. *X-ray fluorescence analyses of phosphorite*

		Authigenic material from Saldanha Bay		Moroccan phosphorites[a]	
	Diagenetic phosphate rock	Rock	Pellets	'Diagenetic' massive rock	'Authigenic' pelletal
Number of analysis	46	2	1	9	5
Al_2O_3	1.95(2.42)[e]	1.25(2.51)	1.73(1.80)	0.37(0.38)	0.37(0.38)
CaO	37.28(46.25)	24.49(49.10)	46.97(48.99)	49.81(51.03)	49.00(50.25)
Fe_2O_3[b]	6.59(8.18)[c]	0.57(1.14)	1.77(1.85)[c]	0.73(0.75)	0.79(0.81)
K_2O	1.38(1.71)	0.28(0.56)	0.26(0.27)	0.14(0.14)	0.13(0.13)
MgO	1.41(1.75)	0.20(0.40)	0.28(0.29)	0.96(0.98)	0.89(0.91)
Na_2O	0.90(1.13)	0.18(0.36)	0.68(0.71)	1.02(1.04)	1.37(1.40)
P_2O_5	17.42(21.61)	17.45(34.98)	33.16(34.59)	18.58(19.03)	26.16(26.83)
S	0.40(0.49)[d]	1.82(3.65)	0.17(0.18)	1.07(1.10)	1.31(1.34)
SiO_2	19.39	50.12	4.15	2.39	2.48
CO_2	11.30(13.83)[d]	2.60(5.21)	3.81(3.97)	8.42(8.63)	8.30(8.51)
C_{org}	0.43(0.53)[d]	1.28(2.57)	N.D.	0.68(0.70)	0.78(0.80)
F	2.25(2.76)[d]	1.80(3.61)	3.44(3.59)	ND	ND
CaO/P_2O_5	2.14	1.40	1.42	2.68	1.87
F/P_2O_5	0.13[d]	0.10	0.10	—	—

[a]McArthur (1978).
[b]Total Fe as Fe_2O_3.
[c]Total Fe as FeO.
[d]For 36 analyses.
ND Not determined.
[e]Figures in brackets are values expressed on a quartz-free basis.

or capping over an extensive area of the Agulhas Bank and western margin (Fig. 13.1).

Subgroups of diagenetic phosphorite are also regionally separated. Goethite-rich phosphatic rocks are located on the eastern Agulhas Bank and on the western margin north of the Berg River. This distribution pattern is related to the input of fluvially-derived iron onto the continental margin (Birch, 1980). The distribution of glauconite-rich phosphorite and glauconite in the unconsolidated sediment coincide closely because the formation of apatite and glauconite is environmentally related (Rooney & Kerr, 1967; Bailey & Atherton, 1969; Birch, 1979d) and a considerable proportion of glauconite on this margin is formed by the replacement of apatite pellets (Birch, 1977b and 1979a). Arenaceous phosphatic packstones are abundant on the western margin and on the outer shelf of the Agulhas Bank in water too deep to be affected by eustatic events, whereas conglomeratic phosphorite is concentrated on the inner shelf where reworking during transgressive–regressive cycles was most intense.

Mineralogy and geochemistry

Major constituents of authigenic and diagenetic phosphorite are calcite, apatite, glauconite and quartz, and minor minerals include feldspar, illite, chlorite–kaolinite and mixed layer clays. The apatite mineral is francolite, a fluorine-rich (> 1% F) carbonate apatite.

The average CO_2 contents of authigenic phosphorite from the coastal terrace at Saldanha Bay and of pellets from the continental margin is 2.6% and 3.8%, respectively. These values are markedly lower than those of diagenetic phosphorite from the continental shelf which average 5.5% (Parker, 1971; Birch, 1975). However, intermediate values from adjacent localities for the two phosphorite types (Rogers, 1977), make any genetic significance of these different CO_2 contents doubtful (McArthur, 1978).

Because of the great variation in the quartz content of phosphorite rocks, bulk geochemistry is compared on a silica-free basis (Table 13.1). Variations in the average whole-rock geochemical composition of phosphatic material from the South African continental margin and coastal terrace with respect to phosphorite from other localities are mainly a reflection of the dilution caused by non-phosphatic components of deposition. The enrichment of SiO_2, Al_2O_3, K_2O and Fe_2O_3 in phosphorite from South Africa compared to that from Morocco, indicates increased quartz and glauconite in the former deposit. Higher S and C_{org} in the Moroccan material with respect to both the authigenic and diagenetic phosphorite from South Africa probably indicates less reducing conditions and lower biogenic productivity in the latter phosphogenic province.

Authigenic and diagenetic phosphorite from South Africa exhibit a clear divergence in bulk geochemistry. Rich calcareous material and glauconite result in higher Fe_2O_3, MgO and K_2O values and a higher $CaO:P_2O_5$ ratio for diagenetic phosphorites, whereas higher P_2O_5, C_{org} and S (excluding pellets) for authigenic material typify the reducing environment under which they form. Authigenic rocks from Saldanha Bay exhibit higher Al_2O_3, K_2O and S relative to pellets due to the presence of terrigenous minerals (feldspar and clay minerals) in the sediment prior to lithification.

Qualitative X-ray imagery shows the matrix of diagenetic phosphorite to be comprised of a heterogenous mixture of silt-size quartz, micrite and finely-divided glauconite (Birch, 1979d). Pelletal material is either internally structureless or it exhibits a well defined concentric layering. Layering is defined by concentric rings of pyrite concordant with the pellet rim and never extends to the central part of the pellet. Pellets which lack internal structure are composed of a heterogeneous mixture of quartz, clay minerals, pyrite and apatite. Incipient glauconitization of some pellets is indicated by increased Fe, K and S at the rim.

Phosphorite ages

Dating of phosphate mineralization has been attempted by association of rock types (Dingle, 1971, 1973 and 1975) and by foraminifera and calcareous nannoplankton (Siesser, 1978). For diagenetic material, two main stages of phosphatization are possibly the Early Eocene and Late Miocene–Early Pliocene. Authigenic quartzitic packstone at Saldanha Bay is dated as Miocene on shark teeth, mollusca, and penguin bones (Hendey, 1974, 1976 and 1978; Tankard, 1975; Dingle, Lord & Hendey, 1979). The age of authigenic pellets on the continental margin is unknown, but by association with adjacent onland occurrences, they are considered to be Miocene (Birch, 1977b and 1979c).

Contrary to early uranium isotope work (Kolodny & Kaplan, 1970) and electron microscope evidence (Baturin & Dubinshuk, 1974), at least some authigenic and diagenetic material on the western margin is of Pleistocene age (Birch *et al.*, 1983). Two arenaceous phosphorites dated on nannofossil evidence as Early Oligocene–mid-Eocene and a quartzitic authigenic packstone from the western margin have uranium-series disequilibrium ages between 61.5 and 75.9 ky. Detailed radiometric studies, using both $^{230}Th/^{234}U$ and $^{231}Pa/^{235}U$ methods, give consistent results. Exhaustive geochemical tests have established that the assumptions on which radiometric age dating are based are valid and that these new data are meaningful. It is shown that uranium entered the apatite phase rapidly at the time of mineralization, and that overprinting by either secondary enrichment or leaching has not taken place.

Phosphorite genesis

A model has been proposed (Birch, 1980) to explain the penecontemporaneous formation of authigenic and diagenetic phosphorite. The regional distribution of both phosphorite types indicates an association with upwelling of cold, nutrient-rich South Atlantic Central Water (Clowes, 1954) and attendant high biological productivity (Parker & Siesser, 1972; Birch, 1975; Parker, 1975).

The prime genetic process in the formation of diagenetic phosphorite is believed to be the replacement of carbonate by phosphate (Ames, 1959; Simpson, 1964). Diffuse, subsurface upwelling in the zone of divergence between the South East Trade Wind Drift and the Benguela Current system results in increased nutrient content and enhanced zooplankton productivity over the outer shelf and slope (Hart & Curry, 1960; Shannon, 1966; Bang, 1971 and 1973; Calvert & Price, 1971). The source of the phosphate is probably the phosphorus liberated by decaying protoplasm of marine organisms, whereas the calcareous exoskeleton of zooplankton provides calcium carbonate to the seafloor. The western and southern margins of South Africa are wide and flat and extensive areas of the shelf would be submerged in shallow water during sea-level fluctuations. Shallow, phosphate-rich upwelled water lying on the shelf could become heated by solar radiation during regressions, thereby facilitating the replacement process (Ames, 1959). With continuous regression, successively deeper zones of the shelf would become available for mineralization. Phosphatization of micrite in the upper sedimentary layers (0.5 m) would lithify the heterogeneous components of the surficial sediment into a near-continuous phosphorite capping similar to that observed on the Agulhas Bank and western margin today.

Conglomeratic textures originate when the phosphatic surface capping is ripped up during succeeding transgressions and incorporated into a new calcareous surficial sediment which is subsequently phosphatized during the following regression in a period of renewed mineralization.

Studies by O'Brien *et al.*, (1981) in areas of limited upwelling suggest that apatite may be formed within bacterial cellular structures through slow assimulation of phosphorus in seawater not anomalously enriched in phosphate. The regional distribution of South African phosphorite clearly implies a process associated with upwelling and high biological productivity, but the degree of control bacteria may have on the phosphogenic process in both time and space is presently unknown.

It is suggested (Pevear, 1966; Birch, 1979c) that the authigenic phosphorite of Saldanha Bay originated in an estuary adjacent to a region of intense, wind-generated upwelling and high biological productivity (De Decker, 1970). The decay of siliceous phytoplankton would supply abundant phosphorus to the seafloor. High water temperatures during summer, associated with increased upwelling, would facilitate optimum conditions for interstitial precipitation of apatite in grain-supported quartzose sands. Oolitic structure, nuclei and microsurface textures exhibited by pelletal phosphorite are indisputable evidence for accretionary growth, possibly in shallow waters of the estuary. Internally structureless pellets probably formed by periodic fragmentation of phosphatic quartzose packstone during local storms.

In proposing a process for the formation of authigenic phosphorite several physico-chemical constraints must be overcome. These are discussed in detail elsewhere (Brooks, Presley & Kaplan, 1968; Berner, Scott & Tomlinson, 1970; Birch, 1980).

Estimated resources

The near continuous cover of diagenetic phosphorite on the middle and outer shelf of the Agulhas Bank (20–26.5° E longitude) and western margin south of the Orange River (29–37° S latitude) cover an estimated $13\,570 \times 10^6$ and $21\,500 \times 10^6$ m^2, respectively. Assuming an average P_2O_5 content of 16% (Birch, 1975; Parker, 1971) and that the deposits are 0.5 m thick, the western margin deposit contains approximately 3500×10^6

metric tons P_2O_5 and the Agulhas Bank deposit comprises about 5500×10^6 metric tons P_2O_5 (Birch, 1979b).

The authigenic pelletal phosphorite on the western margin is too small for consideration, but the Saldanha Bay deposits have been conservatively estimated at 30×10^6 tons P_2O_5 assuming a P_2O_5 content of between 4.5 and 11% (Roux, 1973). Resources of 37.5×10^6 metric tons at 3% P_2O_5 and 20×10^6 tons at 10% P_2O_5 are estimated by Hendey & Dingle (1989). The latter deposit is mined intermittantly, dependant on market forces and is sold as super-phosphate. It is used in the raw form as land fertilizer and as an animal fodder additive.

Acknowledgments

Drs P.J. Cook and J.M. Bremner and Professor R.V. Dingle are thanked for reading a draft of the paper.

References

Ames, L.L. (1959). The genesis of carbonate apatites. *Economic Geology*, **54**, 829–41.

Bailey, R.L. & Atherton, M.P. (1969). The petrology of glauconitic sandy chalk. *Journal of Sedimentary Petrology*, **39(1)**, 1420–31.

Bang, N.D. (1971). The Southern Benguela Current region in February, 1966. Part II. Bathythermography and air–sea interactions. *Deep-Sea Research*, **18**, 209–24.

Bang, N.D. (1973). Characteristics of an intense ocean frontal system in the upwelled regime west of Cape Town. *Tellus*, **25(3)**, 256–65.

Baturin, G.N. & Dubinshuk, V.T. (1974). Microstructures of Agulhas Bank phosphorites. *Marine Geology*, **16**, 63–70.

Berner, R.A., Scott, M.R. & Thomlinson, C. (1970). Carbonate alkalinity in the pore waters of anoxic marine sediments. *Limnology & Oceanography*, **15**, 544–9.

Birch, G.F. (1975). 'Sediments on the continental margin off the west coast of South Africa.' Unpublished Ph.D. thesis, University of Cape Town, South Africa.

Birch, G.F. (1977a). Phosphorites from the Saldanha Bay region of South Africa. *Royal Society of South Africa*, **42**, 223–40.

Birch, G.F. (1977b). Surficial sediments on the continental margin off the west coast of South Africa. *Marine Geology*, **23**, 305–37.

Birch, G.F. (1979a). The nature and origin of mixed glauconite/apatite pellets from the continental margin off South Africa. *Marine Geology*, **29**, 313–34.

Birch, G.F. (1979b). Phosphatic rocks on the western margin of South Africa. *Journal of Sedimentary Petrology*, **49**, 93–110.

Birch, G.F. (1979c). Phosphorite pellets and rock from the western continental margin and adjacent coastal terrace of South Africa. *Marine Geology*, **33(1/2)**, 91–117.

Birch, G.F. (1979d). The association of glauconite and apatite minerals in phosphatic rocks from the South African continental margin. *Transactions Geological Society of South Africa*, **82(1)**, 43–53.

Birch, G.F. (1980). A model of penecontemporaneous phosphatization by diagenetic and authigenic mechanisms from the western margin of southern Africa. *Society of Economic Palaeontologists and Mineralogists Special Publication No. 29*, 79–100.

Birch, G.F., Thomson, J., McArthur, J.M. & Burnett, W.C. (1983). Pleistocene phosphorites off the west coast of South Africa. *Nature*, **302**, 601–3.

Bremner, J.M. (1975). Faecal pellets, glauconite, phosphorite and bedrock from the Kunene–Walvis continental margin. *GSO/UCT Marine Geology Program Technical Report, Geology Department University of Cape Town*, **7**, 59–68.

Bremner, J.M. (1978). 'Sediments on the continental margin off South West Africa between Sylvia Hill and the Kunene River.' Unpublished Ph.D. thesis, University of Cape Town, South Africa.

Brooks, R.R. Presley, B.J. & Kaplan, I.R. (1968). Trace elements in the interstitial waters of marine sediments. *Geochimica Cosmochimica Acta*, **32**, 397–414.

Calvert, S.E. & Price, N.B. (1971). Upwelling and nutrient regeneration in the Benguela Current, October, 1968. *Deep-Sea Research*, **18**, 505–23.

Cayeux, L. (1934). The phosphatic nodules of the Agulhas Bank. *Annals of the South African Museum*, **31**, 105–36.

Clowes, A.J. (1954). The South African pilchard (Sardinops ocellata); the temperature, salinity and inorganic phosphate content of the surface layer near St. Helena Bay, 1950–1952. *Investigative Reports of the Division of Fisheries, South Africa*, **16**, 1–47.

Collet, L.W. (1905). Les concretions phosphatees de l'Agulhas Bank. *Proceedings of the Royal Society of Edinburgh*, **25**, 862–93.

De Decker, A.H.B. (1970). Notes on the oxygen-depleted sub-surface current off the west coast of South Africa. *Investigative Reports of the Division of Fisheries, South Africa*, **84**, 1–24.

Dingle, R.V. (1971). Tertiary sedimentary history of the continental shelf off southern Cape Province, South Africa. *Transactions, Geological Society of South Africa*, **74**, 173–86.

Dingle, R.V. (1973). Post-Palaeozoic stratigraphy of the eastern Agulhas Bank, South African continental margin. *Marine Geology*, **15**, 1–23.

Dingle, R.V. (1975). Agulhas Bank phosphorites: a review of 100 years of investigation. *Transactions, Geological Society of South Africa*, **77**, 261–4.

Dingle, R.V., Lord, A.R. & Hendey, Q.B., (1979). New Sections in the Vorswater Formation (Neogene) of Langebaon Road, Southwestern Cape, South Africa. Annals *South African Museum*, **78**, 81–92.

Du Plessis, A. & Birch, G.F. (1977). The nature of the sea floor south of Cape Seal in a block bound by the longitudes 23° 15′ E and 23° 35′ E and latitudes 34° 10′ S and 34° 25′ S. *GSO/UCT Marine Geology Program, Bulletin*, **9**, 75–85. Geology Department, University of Cape Town.

Hart, T.J. & Curry, R.I. (1960). The Benguela Current. *Discovery Report*, **31**, 123–298.

Hendey, Q.B. (1974). The Late Cenozoic carnivora of the southwestern Cape Province. *Annals South African Museum*, **63**, 1–369.

Hendey, Q.B. (1976). The Pliocene fossil occurrences in 'E' Quarry, Langebaanweg, South Africa. Annals S. Afr. Museum, **69**, 215–47.

Hendey, Q.B. (1978). The age of the fossils from Baard's Quarry, Langebaanweg South Africa. Annals S. Afr. Museum, **75**, 1–24.

Hendey, Q.B. & Dingle, R.V. (1989). Onshore sedimentary phosphate deposits in southwestern Africa. In *Phosphate Rock Resources*, vol. 2., ed. A.J.G. Notholt, R.P. Sheldon & D.F. Davidson. Cambridge University Press, Cambridge.

Kazakov, A.V. (1937). The phosphorite facies and the genesis of phosphorites, in geological investigations of agricultural ores, USSR. *Institute of Fertilizers and Insecto-Fungicides Transactions*, **142**, 95–115.

Kolodny, Y. & Kaplan, I.R. (1970). Uranium isotopes in sea-floor phosphorite *Geochimica Cosmochimica Acta*, **34**, 3–24.

McArthur, J.M. (1978). Systematic variations in the contents of Na, Sr, CO_3 and SO_4 in marine carbonate–fluorapatite and their relation to weathering. *Chemical Geology*, **21**, 89–112.

Murray, J. & Philippi, E. (1908). Die Grundproben der deutschen Tiefsee Expedition 1898–1899 auf dem Dampfer 'Valdivia'. *Wiss. Ergeb. Dtsch. Tiefsee-Exped.*, IO, Jena, 181–7.

Murray, J. & Renard, A.F. (1891). Deep-sea Deposits. Report Scientific Results HMS Challenger 1873–1876. HMSO, London.

O'Brien, G.W., Harris, J.R., Milnes, A.R. & Veeh, H.H. (1981). Bacterial origin of East Australian continental margin phosphorites. *Nature*, **294(5840)**, 442–4.

Parker, R.J. (1971). The petrography and major element geochemistry of phosphorite nodule deposits on the Agulhas Bank, South Africa. *SANCOR Marine Geology Program Bulletin* **2**, Department of Geology, University of Cape Town.

Parker, R.J. (1975). The petrology and origin of some glauconitic and glauco-conglomeratic phosphorites from the South African continental margin. *Journal of Sedimentary Petrology*, **45(1)**, 230–42.

Parker, R.J. & Siesser, W.G. (1972). Petrology and origin of some phosphorites from the South African continental margin. *Journal of*

Sedimentary Petrology, **42**, 434–40.
Pevear, D.R. (1966). The estuarine formation of United States Atlantic coastal plain phosphorites. *Economic Geology*, **61**, 251–6.
Rogers, J. (1977). 'Sedimentation on the continental margin off the Orange River and the Namib Desert'. Unpublished Ph.D. thesis, University of Cape Town, South Africa.
Rooney, T.P. & Kerr, P.F. (1967). Mineralogic nature and origin of phosphorite, Beaufort County, North Carolina. *Geological Society of America, Bulletin*, **78**, 731–48.
Roux, E.H. (1973). Reserwes, outqinning en verweking van fosfate. *Fertilizer Society of South African Journal* **2**, 45–55.
Seisser, W.G. (1978). Age of phosphorites on the South African continental margin. *Marine Geology* **26**, M17–M28.
Shannon, L.V. (1966). Hydrology of the south and west coasts of South Africa. *Investigative Report of the Division of Sea Fisheries, South Africa*, **58**, 1–62.
Simpson, D.R. (1964). The nature of the alkali carbonate apatites. *American Mineralogist*, **49**, 363–76.
Summerhayes, C.P. (1973). Distribution, origin and economic potential of phosphatic sediments from the Agulhas Bank, South Africa. *Transactions, Geological Society of South Africa*, **76**, 271–7.
Summerhayes, C.P., Birch, G.F., Rogers, J. & Dingle, R.V. (1973). Phosphate in the sediments off Southwestern Africa. *Nature*, **243**, 509–11.
Tankard, A.J.T. (1974a). Petrology of the phosphorite and aluminum phosphate rock of the Langebaanweg–Saldanha area, southwestern Cape Province. *Annals of the South African Museum*, **65(8)**, 217–49.
Tankard, A.J.T. (1974b). Chemical composition of the phosphorites from the Langebaanweg–Saldanha area, Cape Province. *Transactions, Geological Society of South Africa*, **77**, 185–90.
Tankard, A.J.T. (1975). The marine Neogene Saldanha Formation. *Transactions, Geological Society of South Africa*, **78**, 257–64.
Visser, H.N. & Schoch, A.E. (1973). The geology and mineral resources of the Saldanha Bay area. *Marine Geology Survey of South Africa*, **63**, 1–150.
Visser, H.N. & Toerien, D.K. (1971). *Die Geologie von die Gebied tussen Vredendal en Elandsbaai*. Geological Survey of South Africa, Pretoria.

14

Moroccan offshore phosphorite deposits

C.P. SUMMERHAYES AND J.M. MCARTHUR

Abstract

Phosphorites and phosphatic rocks of Upper Cretaceous, Eocene, and Miocene age crop out on the continental margin off Morocco between 31° N and 33° 45′ N latitude, from where they have been recovered in water depths between 80 and 900 m (Tooms & Summerhayes, 1968; Summerhayes, 1970); Summerhayes *et al.*, 1971 and 1972; Bee, 1974; McArthur, 1974, 1978 and 1980). These rocks are most abundant off Cap Blanc, where they have been recovered from the entire shelf width, and off Cap Sim, where they have been recovered mainly from the outer shelf and slope. A few isolated phosphorites have been recovered from the outermost shelf and slope between these two areas. Four types of phosphorite have been recovered from the Moroccan margin: (1) glauconitic phosphatic conglomerates; (2) pelletal conglomeratic phosphorites; (3) ferruginous dolomitic phosphatic limestone; and (4) three types of non-ferruginous phosphatic limestone.

Geology and stratigraphy

Upper Cretacous and Eocene phosphatic rocks have been recovered near Cap Blanc, Morocco. They appear to originate from three distinct horizons in gently-folded concordant strata which strike approximately parallel to the coast and which dip gently seaward beneath a relatively smooth seafloor (Summerhayes, 1970; Summerhayes, Nutter & Tooms, 1971 and 1972; Bee, 1974) (Fig. 14.1). Identification of the specific phosphorite horizons becomes uncertain as they are traced southwards, because folding becomes more pronounced and the bottom topography shows more relief. These changes occur in response to deformation associated with the High Atlas orogeny, and because salt tectonics are, or have been, active in bringing phosphorite to the surface on the slope off Cap Sim. Miocene phosphorites are separated from these older rocks by a major Oligocene unconformity that defines a large part of the present day shelf, especially off Cap Sim. The Miocene phosphorites formed and are now found on this unconformity (Fig. 14.1).

The distribution of offshore phosphorites suggests that they were deposited in two gulfs, one off Cap Sim and the other off Cap Blanc, in the landward extensions of which the present onshore phosphorites were formed (Choubert & Faure-Muret, 1962; Bee, 1974) (Fig. 14.1). The onshore phosphorites are Upper Cretaceous (Maastrichtian) and Eocene in age.

Seismic cross-sections of the shelf may be found in Summerhayes (1970), Summerhayes *et al.* (1971), Bee (1974) and Von Rad *et al.* (1982), who also discuss the regional geology of the Moroccan continental margin.

Morphology and classification

Four principal types of phosphorite have been recovered from the Moroccan margin: (1) glauconitic phosphatic conglomerates, which occur as rounded, tabular, cobbles and boulders weighing up to 50 kg (Fig. 14.2); (2) pelletal conglomeratic phosphorites (Fig. 14.3); (3) ferruginous dolomitic phosphatic limestones; and (4) various types of non-ferruginous phosphatic limestone. The last three phosphorite types are most common as rounded, spheroidal–discoidal, cobbles and pebbles. The samples described here were recovered by dredging and appeared to be loose on the seafloor. They probably represent the products of erosion that took place during lower stands of sea level. All are covered with a shiny glauconite veneer, and many are extensively bored and covered with epifauna, especially worm tubes.

The glauconitic conglomerates and ferruginous phosphatic limestones were recovered almost exclusively south of Safi, whilst the pelletal phosphorites were found only to the north of Safi (Fig. 14.1).

Age

The microfossil ages of datable phosphatic foraminiferal limestones and pelletal phosphorites are mainly Upper Cretacous–Eocene. Two pelletal conglomerates (134 and 966(1), Fig. 14.1) yielded Miocene microfossil ages. Glauconite separated from the matrix of two conglomerates from off Cap Sim gave K/Ar ages of 11–12 Ma (Late Miocene) (Fig. 14.1). Datable pebbles within the conglomerates yielded mainly Eocene microfossil ages but from three sites off El Jadida pebbles found within pelletal conglomerates contained Upper Cretaceous microfossils (Summerhayes, 1970; Bee, 1974). The dates of phosphatization are unknown, and may well not coincide with the microfossil ages. For instance, Birch *et al.* (1983) show that phosphatized limestone from the western continental margin off South Africa contains Miocene microfossils but was phosphatized in the Late Pleistocene. Off Morocco, Upper Cretaceous phosphatic limestone pebbles occur in Eocene

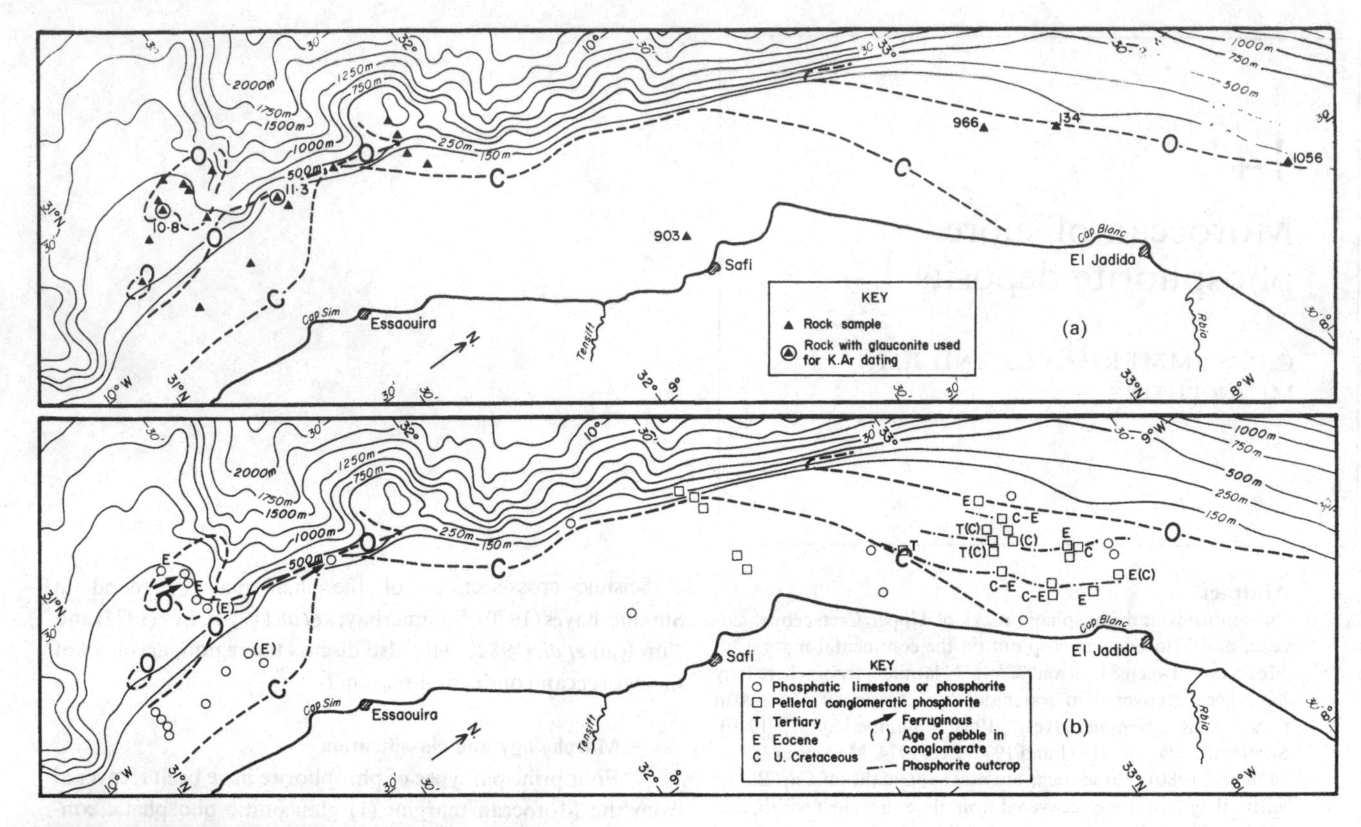

Fig. 14.1. Location where offshore phosphorites were recovered, showing bathymetry (m) and outcrops of major seawardly-dipping unconformities – Oligocene (O) and Mid-Cretaceous (C); (based on Summerhayes, 1970; Summerhayes *et al.*, 1971; Bee, 1974). (a) Distribution of glauconitic facies, mostly glauconitic phosphatic conglomerates except for 134, 903, 966(1), and 1056 (see text for details). Available age determinations suggest these rocks are Miocene. 10.8 and 11.3 are K/Ar dates on glauconites from conglomerates. (b) Distribution of non-glauconitic facies, mostly phosphatic limestones, pelletal phosphatic conglomerates, and ferruginous phosphorites. Available age determinations suggest these rocks are Eocene or Cretaceous. In the northern area we find Cretaceous pebbles in Eocene conglomerates; in the south we find Eocene pebbles in Miocene conglomerates. Cretaceous and Eocene strata appear to crop out in three zones parallel to the coast off Cap Blanc (Bee, 1974). Note: a great many of these samples were mixed with other non-phosphatic rock types that were also found at many other sites on this generally rocky shelf. Age determinations on these other samples, along with geophysical data, helped to define the positions and ages of the unconformities (Summerhayes, 1970; Summerhayes *et al.*, 1971; Bee, 1974).

conglomeratic phosphorite, and Eocene phosphatic pebbles occur in conglomerates with a phosphatic matrix, which contains Miocene glauconite. These facts suggest that at least three episodes of phosphatization have occurred and that all of them are pre-Recent.

Petrography

Glauconitic conglomerates

These rocks are strikingly similar to the glauconitic conglomerates recovered from the Agulhas Bank (Parker, 1970 and 1975; Parker & Siesser, 1972) and the Spanish continental margin (Lucas, Prevot & Lamboy, 1978).

The Moroccan conglomerates contain clasts of various types of limestone and phosphatic limestones set in a collophane–micrite matrix which contains abundant glauconite and quartz (Fig. 14.2).

The clasts are poorly sorted, angular–rounded, and of granule to large cobble size. The most abundant lithologies are calcareous siltstone and foraminiferal limestone, with lesser amounts of aphanitic limestone, coarse shelly limestone, phosphatic limestone, ferruginous phosphatic limestone and ferruginous dolomitic phosphatic limestone. On the continental slope some samples contain glauconitic phosphatic limestone pebbles which are petrographically similar to the matrix. The indented margins of some of these clasts suggest that they were deposited whilst soft and that these conglomerates are partly intra-formational.

Dark green, medium-sand size glauconite comprises between

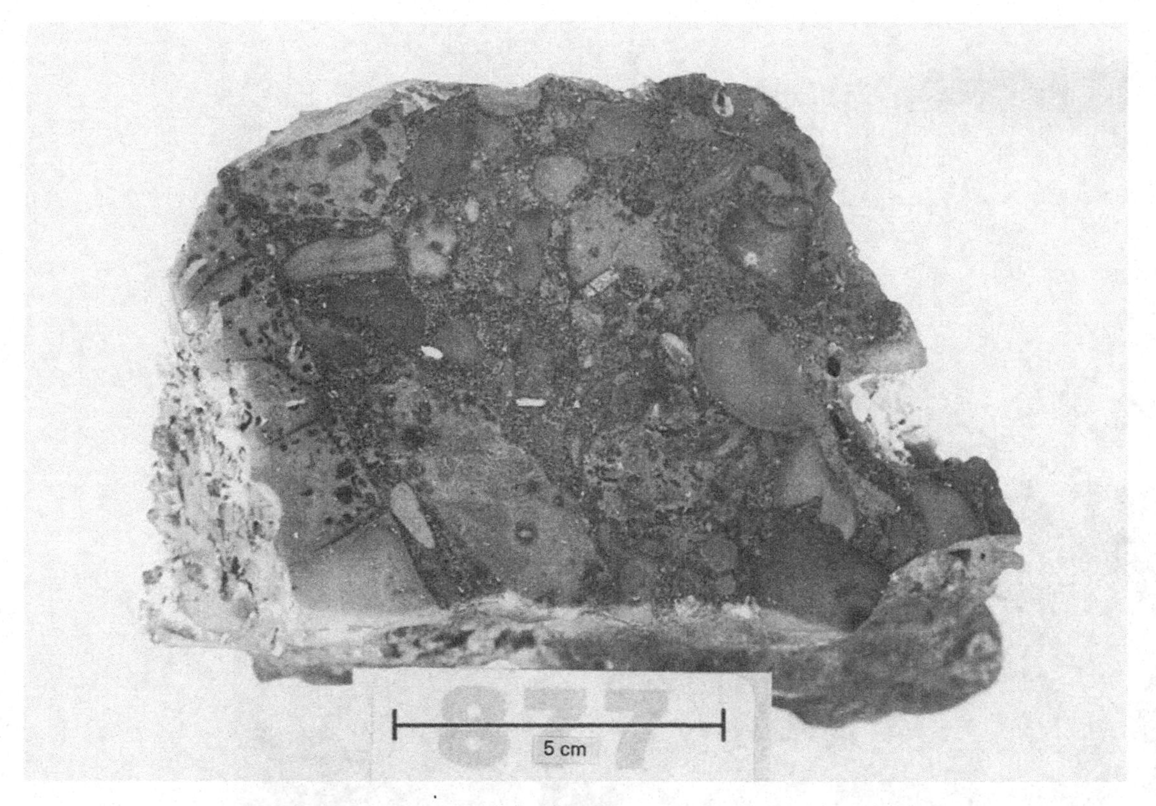

Fig. 14.2. Glauconitic phosphatic conglomerate from the Moroccan continental slope at 31°31′ N 10°12.3′ W, 151 m water depth, off Cap Sim (see Fig. 14.1a). Scale bar is 5 cm.

10 and 60% of the matrix, which can be either grain- or matrix-supported. Some glauconite grains show color changes resulting from oxidation before their incorporation into the matrix. Silt-size quartz comprises up to 30% of the matrix and is most abundant in nearshore samples. Planktonic foraminifera, usually unbroken, are most abundant in slope samples.

The matrix is a fine grained, pale yellow-brown, collophane–micrite mixture clouded with fine grained, carbonaceous and argillaceous material and iron oxide. The degree of phosphatization is variable and patchy with fine-grained components apparently being replaced before coarser grains, many of which have not been phosphatized.

Non-conglomeratic glauconitic phosphatic limestones occur at sites 903 and 1056 (Fig. 14.1). At site 1056 the glauconite fills cavities in a rugose coral complex.

Pelletal conglomeratic phosphorite

These rocks comprise pebbles and granules of phosphatic limestone and phosphorite in a fine-grained, collophane–micrite matrix containing abundant sand-sized phosphorite pellets, some foraminiferal tests (mostly broken), and silt-sized quartz (Fig. 14.3). These seem to be intraformational conglomerates, the pebbles having the same composition as the matrix. Pebble margins are commonly indented, showing that most were semi-indurated when deposited. Other pebbles are partly phosphatized foraminiferal limestones, dolomitic phosphorites, and, rarely, silty limestones.

Between 10 and 70% of the matrix consists of subrounded, medium sand to coarse-silt sized, pale yellow–dark brown pellets of phosphorite that are structureless and usually lack nucleii. They closely resemble the collophane micrite matrix. All pellets and most pebbles have a completely phosphatized rim of clear collophane: some pellets have two or more such zones but are pseud-ooids rather than ooids (e.g. Salvan, 1952). Pseud-ooids form up to 20% of all pellets. Some have composite nucleii that include older pellets, indicating more than one period of reworking. Some pellets contain fish bones.

The matrix of these samples has patches of flaser bedding which fills gaps between the pebbles and consists of layers of graded pelletal sand overlain by muddy sand which is in turn overlain by mud. Individual graded layers are as thin as 0.5 mm. Phosphatization in these layers increases upwards. Pellets are less common ($<10\%$ of matrix) in the unbedded conglomerates.

Ferruginous dolomitic phosphatic limestone

They are essentially fine-grained, ferruginous limestones containing about 14% P_2O_5. Typically they are finely bedded or laminated, with iron staining and phosphatization increasing towards bed surfaces. Planktonic foraminifera make up 10–20% of the rock: they are filled with calcite. The surrounding matrix consists of a mixture of collophane, micrite, and finely disseminated iron oxide, liberally studded with dolomite rhombohedra. Some pebbles of this facies in the conglomerates have been completely phosphatized.

Non-ferruginous phosphatic limestones

Non-conglomeratic pelletal phosphorites: Most of these are like the matrix of the pelletal conglomerates, and most are of the same age (Cretaceous or Eocene); two are Miocene (134, 966(1): see Fig. 14.1).

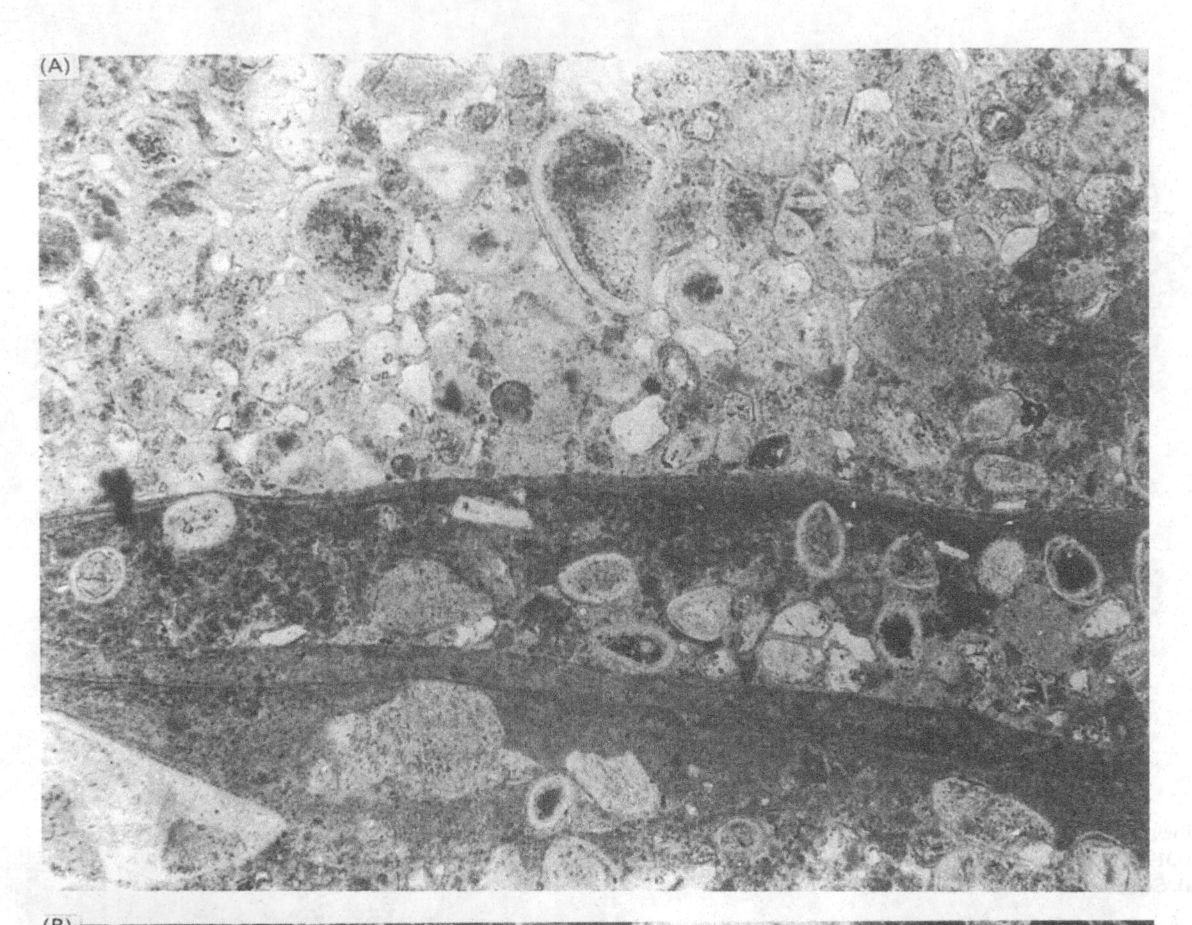

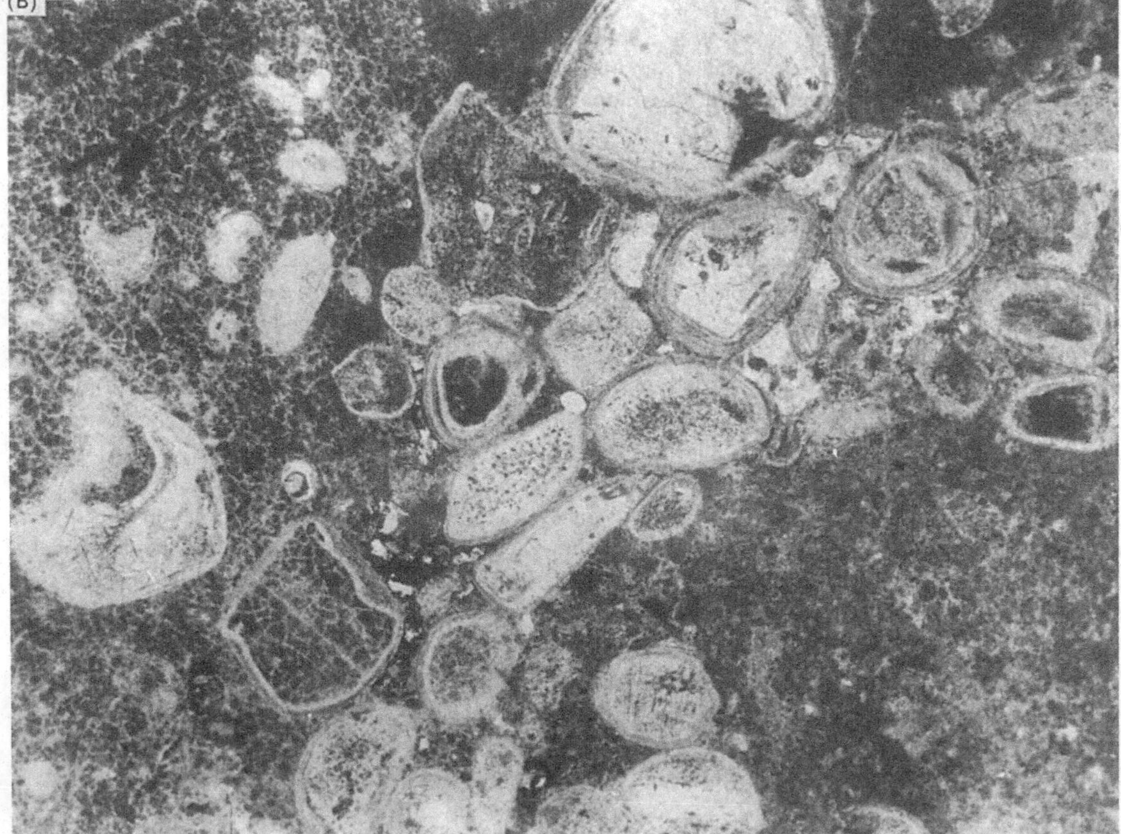

Fig. 14.3. Photomicrograph of pelletal conglomeratic phosphorites showing: (a) Typical pseud-ooids with clear francolite rims, graded bedding, and phosphatized bed surfaces, as described in text (sample 982(1) from 33°06.4′ N, 9°00.1′ W, 109 m water depth off Cap Blanc (see Fig. 14.1a). (b) Grain supported pseud-ooids between pebbles. Note clear calcite filling of pore spaces between grains, also network structures in cloudy matrix of pebbles, as described in text (sample 1004(2) from 33°07.1′ N, 9°04.4′ W, 119 m water depth off Cap Blanc; see Fig. 14.1a).

Phosphatized limestones: Most of these are foraminiferal biomicrites consisting of foraminiferal tests cemented by a fine-grained mixture of collophane and micrite phosphatized to varying degrees. These are like the phosphatized limestones of the Agulhas Bank (Parker & Siesser, 1972). Pebbles of this type occur in the pelletal and glauconitic conglomerates.

Phosphatized shelly biomicrite: These rocks, mostly from the shelf off Safi, consist of large bivalve fragments in a phosphatized micrite matrix containing silt-sized phosphorite pellets.

Mineralogy

The pelletal, pelletal conglomeratic, and non-ferruginous phosphatic limestones from off Cap Blanc contain variable proportions of low-Mg calcite ($\leqslant 50\%$) and francolite ($\leqslant 90\%$), with minor pyrite ($\leqslant 2\%$), organic matter ($\leqslant 2\%$), quartz ($\leqslant 4\%$), iron oxides ($\leqslant 1\%$), and alumino-silicate detrital material ($\leqslant 2\%$), mainly illite (McArthur, 1974 and 1978). Pyrite is usually present as foraminiferal infillings, although stringers of framboidal pyrite occur within larger shell fragments.

Phosphorites from off Cap Sim have a highly variable mineralogy which is governed by the type and abundance of clasts in the conglomerates. Clast mineralogy ranges from relatively pure low-Mg calcite to over 85% francolite, with up to 34% quartz, 23% Fe_2O_3 (as well crystallized goethite), 40% dolomite, and 5% alumino-silicate clastics, mostly illitic in composition. Pyrite is absent; Fe_2O_3 blebs in non-ferruginous limestone clasts may represent original sulphide framboids which have been oxidized by weathering associated with clast formation (McArthur, 1974 and 1978). The conglomerate matrix contains up to 60% well-crystallized glauconite, 30–60% francolite, and up to 30% low-Mg calcite as major components, with quartz ($\leqslant 30\%$) and goethite ($\leqslant 12\%$) as important accessory minerals. Non-conglomeratic phosphorites from off Cap Sim show a range of composition which is similar to that found in conglomerate clasts.

Geochemistry

The chemical composition of the Moroccan offshore phosphorites (Tables 14.1–14.3) is largely a reflection of their mineralogical composition. Ti, Rb, K_2O and Al_2O_3 are largely found within alumino-silicate minerals, of which glauconite is the most easily identifiable. K:Al molar ratios of 0.2–0.4 in many non-conglomeratic samples (McArthur, 1974 and 1978) show that illites contribute markedly to the alumino-silicate fraction, whilst remaining difficult to identify petrographically or by X-ray diffraction. Iron resides largely in pyrite ($FeS_{1.9\pm0.1}$) in samples from off Cap Blanc, and in glauconite and goethite in samples from south of Safi. Strontium is wholly associated with francolite irrespective of the accessory calcite content of the samples, whilst sulfur is partitioned between francolite (as SO_4^{2-}) and pyrite, where present. Where sulfide is present it controls the abundance of As, Cu, Ni and Zn, and contains up to 5% Ni in spot concentrations. Where goethite is present it controls the abundance of As, Mn and V (McArthur, 1978).

Uranium: P_2O_5 ratios range from 2×10^{-4} to 3.6×10^{-3}. In samples from north of Safi, uranium concentrations correlate, though poorly, with the abundance of pelletal collophane, which suggests that secondary reworking has enriched these pellets in U prior to their incorporation in the samples.

The concentration of Pb, I, Cr, and Y, is not related in any clear way to any single mineral or element, although concentrations of Pb and Ba are much higher in samples from the south of Safi than in samples from off Cap Blanc (McArthur, 1974 and 1978).

The francolite in all offshore samples is a highly substituted variety, with a composition very close to that of sample 966(2) (Table 14.1) which contains only 7% impurities. A small fraction (10–20% depending on age and degree of lithification) of the Na in many phosphorites is water-soluble, and is probably relict sea salt, so the structural-Na in sample 966(2) is probably slightly less than the 1.4% Na_2O given in Table 14.1.

The conglomerate clasts within the phosphorites off Cap Sim have clearly been produced by weathering and transportation. This weathering has reduced the concentrations of the structural substituents within their francolite component to levels below those measured in the relatively unweathered samples found off Cap Blanc, with cation substituents being about 17% lower, and anion substituents about 8% lower (McArthur, 1980).

Rare Earth element abundances are presented in Table 14.3; fractionation patterns are very similar to seawater patterns, with pronounced negative Ce anomalies and heavy Rare Earth enrichments (McArthur & Walsh, 1985).

The stable isotopic composition of francolite-CO_2 (Table 14.3) shows that it is close to that of coexisting calcite (Benmore, McArthur & Coleman, 1983): it probably represents relict CO_3^{2-} remaining after phosphatization of carbonate (McArthur, Coleman & Bremner, 1980; Benmore *et al.*, 1983). Petrographic studies also suggest that the francolite originated by the phosphatization of carbonates (Summerhayes, 1970).

Sedimentary history

Glauconitic phosphatic conglomerates

We assume that the Oligocene erosion left an irregular surface carpeted with pebbles and cobbles on beaches and scree slopes. During the Miocene, carbonate muds accumulated in the depressions on this surface and silt-sized quartz was blown in by the Trade Winds. Glauconite grew in the sediment. Periodically, storms agitated the seabed – especially over topographic highs – redistributing pebbles and glauconite grains into the depressions to be mixed with the silty lime muds by bioturbation. Lack of graded bedding suggests that these deposits are not true tempestites: some may be diamictites. Lack of laminations suggests that the bottom waters were oxidizing. On the slope the glauconitic muds were rather unstable, and small-scale slides formed intraformational conglomerates.

Pelletal phosphorite conglomerates

The variable sorting suggests exposure to highly variable energy levels. The combination of structural and textural features suggests that deposition took place in an offshore bar–lagoon complex, with pellets forming on bars and lime muds accumulating in lagoons. Disaggregation of the semi-indurated lime muds during storms, or by migrating tidal channels, provided the pebbles and raw material for pellets. Summerhayes (1970) suggested that landward movement of the pellets, perhaps

Table 14.1. *Chemical composition of offshore Moroccan phosphorite; air-dried 25°C, unwashed (data from McArthur, 1974 and 1978)*

Element	Pelletal phosphorite		Phosphatic limestone	Phosphatic ferruginous limestone	Quartzitic phosphatic ferruginous limestone	Matrix, glauconitic conglomerate
(%)	982	966(2)	135	157	865	898(C2)
Al_2O_3	0.30	0.45	0.30	1.65	2.20	1.80
CaO	48.6	50.3	47.8	34.6	37.9	36.8
Organic C	0.59	0.85	0.35	0.21	0.17	0.26
CO_2	11.3	7.6	18.8	19.8	10.2	7.9
CO_3^{2-} [a]	8,3	8.2	8.5	8.6	7.2	8.3
Fe_2O_3	0.57	0.78	0.57	14.3	6.85	8.10
K_2O	0.10	0.12	0.17	0.25	0.71	2.37
MgO	0.90	0.90	0.85	7.65	4.85	2.30
Na_2O	1.22	1.67	1.11	0.63	1.01	1.06
P_2O_5	25.8	29.3	18.6	12.1	18.7	19.1
Total S	1.23	1.65	0.86	0.30	0.56	—
S^{2-} [b]	0.49	0.81	0.33		—	0.61
SiO_2	2.3	2.30	3.9	6.4	10.3	13.7
Sr	0.20	0.21	0.12	0.078	0.11	0.13
(ppm)						
As	9	25	15	100	47	22
Ba	<4	<4	<4	<4	<4	83
Cr	34	58	29	43	31	84
Cu	17	24	10	14	11	17
I	59	<5	26	50	19	28
Mn	74	65	104	486	125	83
Ni	30	46	20	51	6	12
Pb	7	10	5	20	9	18
Rb	6	8	8	13	29	55
Ti	94	94	178	233	747	380
U	71	129	507	38	61	142
V	19	34	79	184	51	86
Y	237	134	30	29	27	50
Zn	40	82	43	77	37	53

[a] CO_3^{2-}, structural by Gulbrandsen's method (1970)
[b] Sulphur present as sulphide-sulphur.

Table 14.2. *Mineralogical composition of offshore Moroccan phosphorite (data from McArthur, 1974 and 1978)*

% Mineral	Pelletal phosphorite		Phosphatic limestone	Phosphatic ferruginous limestone	Quartzitic phosphatic ferruginous limestone conglomerate clast	Matrix, glauconitic conglomerate
Sample	982	966(2)	135	157	865	898(2)
Calcite	8	4	30	9	<2	<2
Clay Minerals (as Al_2O_3)	(0.30)	(0.45)	(0.30)	(1.65)	(2.20)	(1.8)
Dolomite	<2	<2	<3	34	21	<3
Francolite	88	93	65	36	60	70
Glauconite	Nil	Nil	Nil	Nil	Nil	8
Goethite	Nil	Nil	Nil	14	6	6
Pyrite	1	1.5	0.5	Nil	Nil	Nil
Quartz	<2	<2	<4	5	10	10

Table 14.3. *Rare earth element and stable isotopic composition of offshore Moroccan phosphorite (data from Benmore* et al., *1983; McArthur & Walsh, 1985)*

Sample		Structural CO_2[a]		Element ($\mu g\ g^{-1}$)											
		$\delta^{13}C$‰	$\delta^{18}O$‰	La	Ce	Pr	Nd	Sm	Eu	Gd	Dy	Ho	Er	Yb	Lu
135	Phosphatic	−2.64	−0.66												
136(1)	limestone	−3.44	−1.43												
1016		−3.25	−1.10												
1022(1)		−3.79	−0.92												
1022(2)		—	—	4.7	3.1	0.7	3.5	0.8	0.21	0.9	0.8	0.19	0.58	0.51	0.10
139(1)	Pelletal	−3.08	−0.59												
966(2)	Phosphorite	−3.34	−1.02	57	44	11	45	11	2.6	11	11	2.5	6.9	6.3	1.1
136(3)	Limestone	+0.13	−1.44												
966(3)		+2.23	−1.22	2.0	2.7	0.5	1.5	0.5	0.08	—	0.24	0.05	0.2	0.16	0.03

[a]Relative to PDB.

during transgressions, might account for the accumulation in Morocco of economically valuable pelletal phosphate deposits (Salvan, 1952; Visse, 1953). This hypothesis must be rejected because the onshore and offshore pelletal deposits have different petrographic characteristics (Bee, 1974) and isotopic compositions.

Phosphatized foraminiferal limestones

These rocks probably formed in quiet and moderately deep water, either far from shore or off a low-lying coast with little runoff.

Mechanism of phosphatization

Whilst there is little evidence for direct precipitation of carbonate-apatite from seawater, post-depositional precipitates of fibrous, radially-oriented francolite line pores where the texture is very sandy in both the pelletal and glauconitic facies. The pore spaces here are filled with clear collophane or secondary calcite and, in some glauconitic samples, with hematite. Presumably deposition took place from pore waters.

Phosphatization affected the fine grained components first. As it proceeded in the pelletal samples, the matrix became penetrated by a honeycomb-like network of clear collophane veinlets 2–5 μm across. Phosphatization is usually greatest at the rims of pebbles and pellets, and took place before mixing with the matrix. While the matrix was phosphatized by interstitial solution, pebbles and pellets could have been phosphatized in seawater. The highly phosphatized flaser bedding and erosion surfaces could also have been phosphatized by seawater. Alternatively the phosphatizaion of these surfaces could have taken place by the downward movement of phosphate out of ephemeral, transient deposits of organic-rich mud.

Paleoenvironments

Upwelling currents driven by the Trade Winds bring nutrient-rich water to the surface stimulating productivity off northwest Africa today, and probably did so through much of the history of this continental margin. It seems likely that upwelling was the source of the phosphate in these deposits (e.g. Tooms, Summerhayes & Cronan, 1969; Summerhayes, 1970; Summerhayes *et al.*, 1971).

Similar phosphate deposits, of Maastrichtian age, occur beneath the Saharan shelf (Arthur *et al.*, 1979), and Miocene phosphorites occur on the Saharan slope (Von Rad *et al.*, 1979) as well as further south off Liberia (Schlee, Behrendt & Robb, 1974). An enhancement of productivity occurred off northwest Africa during the Miocene in response to climatic changes that increased upwelling and thereby led to widespread deposition of siliceous planktonic remains (Arthur *et al.*, 1979; Diester-Haass & Schrader, 1979; Von Rad *et al.* 1982), as well as the formation of organic-rich muds on the continental slope (Rullkötter, Cornford & Welte, 1982).

It is not clear why phosphatization was focused in the Cretaceous, Eocene and Miocene. Many different factors govern both upwelling and phosphorite formation (cf. Summerhayes, 1983). Various palaeoceanographic explanations have been proposed to explain the age distribution of these phosphorites, but all are speculative (Summerhayes, 1970; Bee, 1974). One previously unconsidered factor is sampling bias; another is stratigraphy rather than chemistry.

The history of phosphorite deposition off Morocco involves erosion of the Lower Cretaceous during Middle Cretaceous times; deposition of Late Cretaceous phosphatic sediments; deposition of Eocene phosphatic sediments containing pebbles of Cretaceous phosphorites; formation of an erosion surface in the Oligocene; deposition of Miocene phosphatic sediments containing Eocene (and possibly Cretaceous) phosphatic pebbles on the Oligocene erosion surface; and Pleistocene erosion to produce the present shelf surface, with partial removal of the Miocene sequence and re-exposure of the Cretaceous–Eocene sequence. Salt tectonics has caused extensive folding off Cap Sim, and brought older sediments to the surface on the upper slope (the Tafelney Plateau) near Cap Sim (Summerhayes *et al.*, 1971) (Fig. 14.1).

Resources

Evaluation of reserves is difficult in the absence of drill cores and in the recovery of so few rock samples from such a large area. We have no direct measurement of the thickness of the deposit, nor of its areal extent beneath the shelf. As well as these unknowns, anyone making a resource estimate for the deposit would need to consider the known facts, which are that phosphatic rocks crop out continuously along the 500 km of shelf over which they were sampled, that the width of outcrop is about 10 km on average, and that the mean P_2O_5 content of these rocks is about 20%.

Although this must be a large deposit of phosphatic rock, the economic potential of these offshore deposits is low because of their low grade, necessitating considerable beneficiation, their situation in water depths greater than 100 m, and the difficulty in mining hard rock from the seabed.

Acknowledgments

The majority of the work summarized here was undertaken between 1967 and 1973 under a series of NERC research grants awarded to J.S. Tooms. The support of NERC via a Research Assistantship (to CPS) and a Research Studentship (to JMM) is gratefully acknowledged. JMM acknowledges the further support of NERC (Grant GR3/3819) for continued work on these deposits. We are indebted to A.G. Bee for access to his unpublished study of the stratigraphy of the offshore phosphate deposits. This paper is a contribution to IGCP Project 156. The contribution of CPS is published with the approval of BP.

References

Arthur, M.A., Von Rad, U., Cornford, C., McCoy, F.W. & Sarnthein, M. (1979). Evolution and sedimentary history of the Cape Bojador continental margin, northwestern Africa. *Initial Reports of the Deep Sea Drilling Project*, **47(1)**, 773–816.

Bee, A.G. (1974). 'The marine geochemistry and geology of the Atlantic continental shelf of central Morocco'. Unpublished Ph.D thesis, Imperial College, London University, England.

Benmore, R.A., McArthur, J.M. & Coleman, M.L. (1983). Stable isotopic composition of structural carbonate in sedimentary francolite. *Proceedings of the 4th International Field Workshop and Seminar on Phosphorites*, Udaipur, India, 1981.

Birch, G.F., Thomson, J.J., McArthur, J.M. & Burnett, W.C. (1983). Pleistocene phosphorites off the west coast of South Africa. *Nature*, **302**, 601–3.

Choubert, G. & Faure-Muret, A. (1962). Evolution du Domaine Atlasique Marocain Depuis les Temps Paleozoiques. In *Livre de Paul Fallot* vol. 1, pp. 447–527. Société Géologique de France, Paris.

Diester-Haass, L. & Schrader, H-J. (1979). Neogene coastal upwelling history off northwest and southwest Africa. *Marine Geology*, **29**, 39–53.

Gulbrandsen, R.A. (1970). Relation of carbon dioxide content of apatite of the Phosphoria Formation to regional facies. *US Geological Survey Professional Paper*, **700-B**, 9–13.

Lucas, J., Prevot, L., & Lamboy, M. (1978). Les phosphorites de la marge nord de l'Espagne: chimie, mineralogie, genese. *Oceanologica Acta*, **1**, 55–72.

McArthur, J.M. (1974). 'The geochemistry of phosphorite concretions from the continental shelf off Morocco.' Unpublished Ph.D. thesis, Imperial College, London University, England.

McArthur, J.M. (1978). Element partitioning in ferruginous and pyritic phosphorite from the continental margin off Morocco. *Mineralogical Magazine*, 221–8.

McArthur, J.M. (1980). Post-depositional alteration of the carbonate–fluorapatite phase of Moroccan phosphate. Society of Economic Geologists and Palaeontologists, Special Publication 29, pp. 53–60.

McArthur, J.M., Coleman, M.L. & Bremner, J.M. (1980). Carbon and oxygen isotopic composition of structural carbonate in sedimentary francolite. *Journal of Geological Society*, **137**, 669–73.

McArthur, J.M. & Walsh, J.N. (1985). Rare Earth element geochemistry of phosphorites. *Chemical Geology*, **47(3/4)**, 191–220.

Parker, R.J. (1970). Agulhas Bank phosphorite deposits. *South African National Committee Oceanographic Research, Technical Report*, **2**. 95pp.

Parker, R.J. (1975). The petrology and origin of some glauconitic and glauco-conglomeratic phosphorites from the South African continental margin. *Journal of Sedimentary Petrology*, **45**, 230–42.

Parker, R.J. & Siesser, W.G. (1972). Petrology and origin of some phosphorites from the South African continental margin. *Journal of Sedimentary Petrology*, **42**, 434–40.

Rullkötter, J., Cornford, C. & Welte, D.H. (1982). Geochemistry and petrography of organic matter in northwest African continental margin sediments: quantity, provenance, depositional environment, and temperature history. In *Geology of the Northwest African Continental Margin*, ed. U. Von Rad *et al.*, pp. 686–703. Springer-Verlag, Heidelberg.

Salvan, H.M. (1952). Phosphates. In *Geologie des Gites Mineraux Marocains. 19th International Geological Congress* (Algiers), Monographs Regionales Series 3, Maroc No. 1.

Schlee, J., Behrendt, J.C. & Robb, J.M. (1974). Shallow structure and stratigraphy of the Liberian continental margin. *Bulletin of the American Association of Petroleum Geologists*, **58(4)**, 708–28.

Summerhayes, C.P. (1970). 'Phosphate deposits on the Northwest African continental shelf and slope.' Unpublished Ph.D thesis, Imperial College, London University, England.

Summerhayes, C.P. (1983). Sedimentation of organic matter in upwelling regimes. In *Coastal Upwelling – Its Sediment Record*, Part B, ed. J. Thiede & E. Suess, pp. 29–72. Plenum Press, London.

Summerhayes, C.P., Nutter, A.H. & Tooms, J.S. (1971). Geological structure and development of the continental margin of northwest Africa. *Marine Geology*, **11**, 1–25.

Summerhayes, C.P., Nutter, A.H., & Tooms, J.S. (1972). The distribution and origin of phosphate in sediments off northwest Africa. *Sedimentary Geology*, **8**, 3–28.

Tooms, J.S., & Summerhayes, C.P. (1968). Phosphatic rocks from the north-west African continental shelf. *Nature*, **218**, 1241–2.

Tooms, J.S., Summerhayes, C.P., & Cronan, D.S. (1969). Geochemistry of marine phosphate and manganese deposits. *Oceanography and Marine Biology Review*, **7**, 49–100.

Visse, L. (1953). Les Facies Phosphates. *Rev. Inst. Franc. Petr. Ann. Combust, Ligu.*, **7**, 87–99.

Von Rad, U., Cepek, P., Von Stackleberg, U., Wissman, G. & Zobel, B. (1979). Cretaceous and Tertiary Sediments from the Northwest African slope (dredges and cores supplementing DSDP results). *Marine Geology*, **29**, 273–312.

Von Rad, U., Hinz, K., Sarnthein, M. & Seibold, E. (1982). *Geology of the Northwest African Continental Margin*. Springer Verlag, Heidelberg.

15

Neogene phosphorites of the Sea of Japan

I.I. BERSENEV, G.N. BATURIN, E.P. LELIKOV AND V.V. GUSEV

Abstract
Occurrences of Neogene phosphorite in the Sea of Japan, never before reported in the western literature, are described. Phosphorite has been located at over 50 stations throughout the Sea of Japan area, mainly on the slopes of submarine highs. Diagenetic precipitation within unconsolidated clay–diatomaceous oozes apparently produced much of this phosphorite although some formed by phosphatization of Late Miocene carbonate oozes. The geomorphological setting of the Miocene Sea of Japan may have had similarities to the southeastern US continental margin, where phosphorites were also forming at that time.

Introduction

The existence of phosphorites on the floor of the Sea of Japan was first established in 1977 during laboratory chemical assays of specimens sampled in 1973–74. In 1978, phosphorites were discovered on seamounts off the east coast of the Korean peninsula and on the rise of the ancient Chentsov volcano. In the course of expeditions to this area on the *R/V Pervenets*, additional samples of phosphorite have been recovered, together with other rocks from basal horizons of marine Neogene deposits. The results of recent Soviet and Japanese marine geological work in the Sea of Japan have been summarized in several maps and other publications (Geological Survey of Japan, 1979a, b; 1981; Bersenev *et al.*, 1983a, b).

General structure, geology and stratigraphy

The major physiographic features of the Sea of Japan are continental and island shelves, continental slopes and rises, the borderland of Japanese Islands, deep-sea basins (Central and Tsushima), troughs and seamounts. The crust is continental beneath shelf areas, suboceanic beneath the basins, subcontinental beneath the borderland and seamounts, and presumably intermediate beneath the troughs. All of these structural elements contain numerous volcanic mountains and ridges.

The geologic structure of the bottom of the Sea of Japan is determined by pre-Cenozoic consolidated basement and Cenozoic volcanic and sedimentary cover sequences. These features are underlain by continental and subcontinental crust built of formations typical of the surrounding land. The oldest basement rocks are metamorphic, approximately 2–2.7 billion years old. The Paleozoic rocks are represented by volcanic–terrigenous eugeosyncline sequences and terrigenous subplatform deposits and large granitic batholiths of Mid- and Late Paleozoic age, 220–350 million years old (Lelikov, Sédin & Fershtatter, 1984). The Mesozoic sequences are represented by Jurassic marine and Early Cretaceous continental deposits, Late Cretaceous volcanics, and Early and Late Cretaceous granites. The Cenozoic cover consists of volcanic and sedimentary deposits.

Mountains and ridges in basins and troughs, as well as plateaus and mountains on subsea heights, are volcanic. Their acoustic properties do not differ significantly from those of pre-Cenozoic formations so that they are combined and both considered part of the acoustic basement. The time of their formation ranges from Paleocene to Pliocene. The volcanics are represented by sequences of basalt–andesite–rhyolite, alkali–olivine–basalt and alkali–basalt series.

The sedimentary deposits of the lower part of the Cenozoic section are built of continental terrigenous formations divided into Paleocene and Oligocene–Lower Miocene sequences, the former belonging to the acoustic basement. The major part of the sedimentary cover consists of marine Middle–Upper Miocene and Pliocene deposits filling basins and troughs and covering transgressively older rocks within other structures. On the adjacent part of the Asian continental slope, these deposits are named the Valentinov Suite, and are subdivided into Middle and Upper Miocene subsuites. The thickness of the former is 150 m on the shelf and 200–300 m on the slope. Its basal layer consists of sandstones with gravels, pebbles and rare conglomerates. This subsuite includes diatom siltstones and clayey diatom siltstones with layers or lenses of sandstones, siltstones, tuffs and rocks of mixed composition. The upper subsuite lies conformingly on the lower, but transgressively on the basement rocks. The thickness of this subsuite on the shelf and slope is from 100 to 300 m. It contains more diatomaceous rocks and calcareous sandstones (with occasional limestone lenses) than the lower suite.

The Middle and Upper Miocene deposits on plateau areas of the inner part of the Sea of Japan transgressively overlay the rocks of the acoustic basement and in some localities are in contact with them along the faults. These deposits fill grabens and depressions in pre-Neogene relief, smoothing the surfaces of the highs. Their total thickness reaches 500 m or more, but they are absent on steep slopes of tectonic origin. Their composition is similar to these deposits found on the continental slope.

Pliocene deposits on the shelf and continental slope are represented by the Gamov Suite which transgressively overlays the Valentinov Suite or basement rocks near the shore. The thickness of these sediments is about 100–300 m. Its lithology differs from that of the Valentinov Suite by its higher content of pyroclastics and carbonate rocks. The Pliocene deposits on submarine highs conformingly overlay the Upper Miocene deposits or the basement rocks with a thin horizon of poorly lithified basal conglomerates.

Quaternary sediments, together with Pleistocene deposits, have a thickness of about 100 m and cover nearly the whole shelf and the bottoms of deep basins and troughs. On the flat portions of the submarine highs, their thickness is no more than 20 m and they are practically absent on steep slopes.

Phosphorites were found in the basal layers of the Upper Miocene deposits represented mainly by gravels, sandstones, siltstones, diatomites and rocks of intermediate types (Fig. 15.1).

Paleogeography of the Sea of Japan during the Miocene

Reconstruction of the paleogeographic situation in the Sea of Japan during the Miocene epoch is difficult because of the lack of data and differential tectonic movements of separate basement rocks. In the Early Miocene, much of what is today considered the Sea of Japan was above sea level (Fig. 15.2). A relatively deep marine basin was evidently situated in the western part of the Tatar Strait and in the Central and Tsushima Deeps. Shallow shelf seas of longitudinal and sublongitudinal directions were present west of the Sakhalin, Hokkaido and Honshu Islands. These seas correspond to the Okushiri and Mogami grabens and others. Similar basins were presumably situated in the western part of the central Sea of Japan between the mainland and the East Korean Plateau and in the Krishtofovitch Trough.

Present submarine highs were exposed during the Early Miocene, with irregular relief and volcanic plateaus. Alluvial and lake sediments were deposited in depressions and are found now on the Southern Yamato Rise at a depth of 1500 m and on Takuye at the depth of 2000 m. The Honshu Deep was probably occupied by a lake and alluvial plain. These seas were connected with the ocean through the Sakhalin and Korean Straits among others.

In the Middle Miocene, an overall transgression began with differential depression of the crust and a global sea-level rise. Toward the end of that period, the shoreline in the inner part of the sea of East Korean, northern Yamato and Kita Oki Plateaus, was situated on the modern 2000 m isobath (Fig. 15.2). At the same time, the sea flooded most of the Asian continental shelf, although its level was several hundred meters lower compared to Recent time. Vast land territories were left on horst structures of the borderland in the eastern and southeastern parts of the Sea of Japan. In the Middle Miocene, these seas were, for the most part, connected with the ocean.

In the Late Miocene, the rise of the Japanese islands began, followed by regression of the covering seas. The shore of the mainland and the submarine highs, however, display no signs of such regression because the general transgression continued during the Late Miocene. On the continental shelf, the shoreline approached its Recent position. On the East Korean, Yamato and other plateaus, an erosional–depositional abrasion-terrace, 10–20 m wide was formed fixing the position of the insular shelves. This terrace is now situated at 1000–1200 m water depth. In the Late Miocene, the Sea of Japan was connected with the open ocean through the Tatar, Le Perouse, Korean and other straits crossing the present-day Japanese islands.

In the Late Miocene–Early Pliocene, the regression on the continental shore began stretching over the shelf and the upper part of the continental slope, as well as some inner parts of the sea. The Japanese islands continued to rise, whereas the plateaus continued to subside, so that the sea covered them by the end of the Pliocene, except for some of the highest mountains which remained as islands (e.g. Oki, and others). The connection of the sea with the open ocean was thus developed to its present state.

Phosphorites

Phosphorites have been discovered at more than 50 stations within the Sea of Japan, mainly on submarine highs: East Korean, Krishtofovitch, Northern and Southern Yamato Plateaus, and the Oki Ridge. Some isolated samples were also found on the adjacent part of the Asian continental slope. On the submarine highs, phosphorites lie predominantly on slopes and less often on summits covered by Miocene–Pliocene marine deposits.

Lithology

Based on the morphology of the samples, as well as the content and character of the non-phosphatic inclusions, phosphorites are divided into three major groups: (1) massive phosphorites; (2) nodular phosphorites and (3) coarse–fine-grained rocks with phosphatic cement (Lipkina & Shkolnik, 1981; Bersenev, Shkolnhik & Gusev, 1983c and 1984).

Most common are phosphorites of the first group which have been recovered by dredges and grabs as fragments of various form, dimension, density, and color. Their shape is typically equidimensional, angular or plate-like from 1 to 15–20 cm in cross section. Most of these samples are dense and hard, covered by thin black films, light brown–black in color. However, some are not wholly consolidated, but are friable, without black films, and are light yellow in color.

These phosphorites consist of a phosphatic matrix (70–80%) in which silt or fine-grained non-phosphatic material is randomly dispersed, including terrigenous grains of quartz, feldspars, glauconite and its relicts, glauconitized sedimentary rocks and, rarely, granites. The phosphatic component is greenish-yellow–colorless, mainly isotropic, of carbonate fluorapatite (CFA) composition. The phosphatic matrix contains dispersed relics of diatoms, sponge spicules and radiolaria replaced by phosphate, showing that phosphorites had been formed by phosphatization of oozes during diagenesis. Based on our sampling, it is assumed that phosphorites orginally formed layers and lenses up to 20 m thick, predominantly in Late Miocene deposits which had been destroyed on steep slopes, leaving phosphorites behind.

The second group of phosphorites is represented by friable

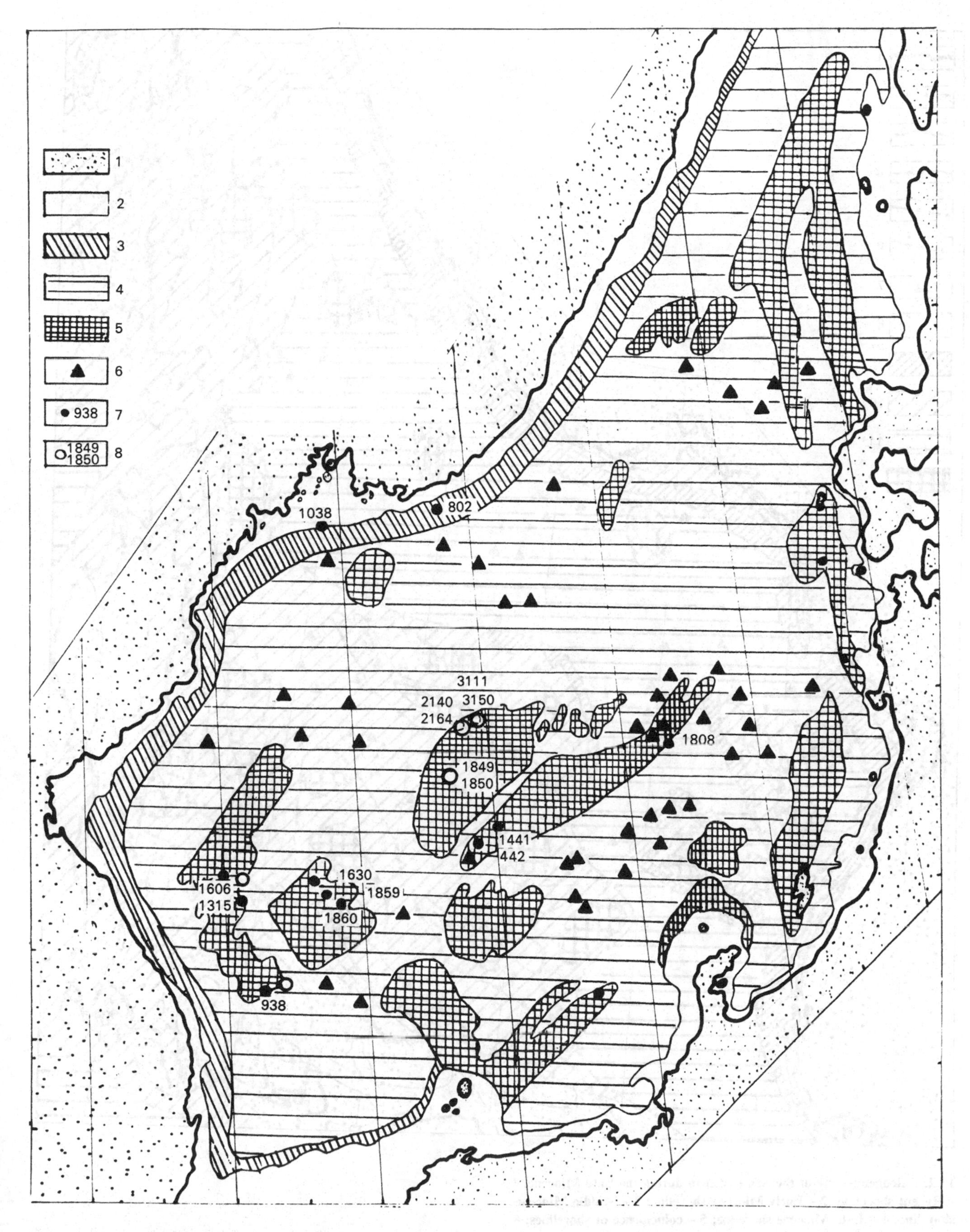

Fig. 15.1. Physiographic structures and phosphorite occurrences of the Sea of Japan. 1 – land; 2 – shelf; 3 – continental slope and rise; 4 – deep-sea basins, troughs and depressions; 5 – sub-sea highs and banks; 6 – volcanic mountains and ridges; 7 – phosphorite occurrences and station numbers; 8 – the same group of stations.

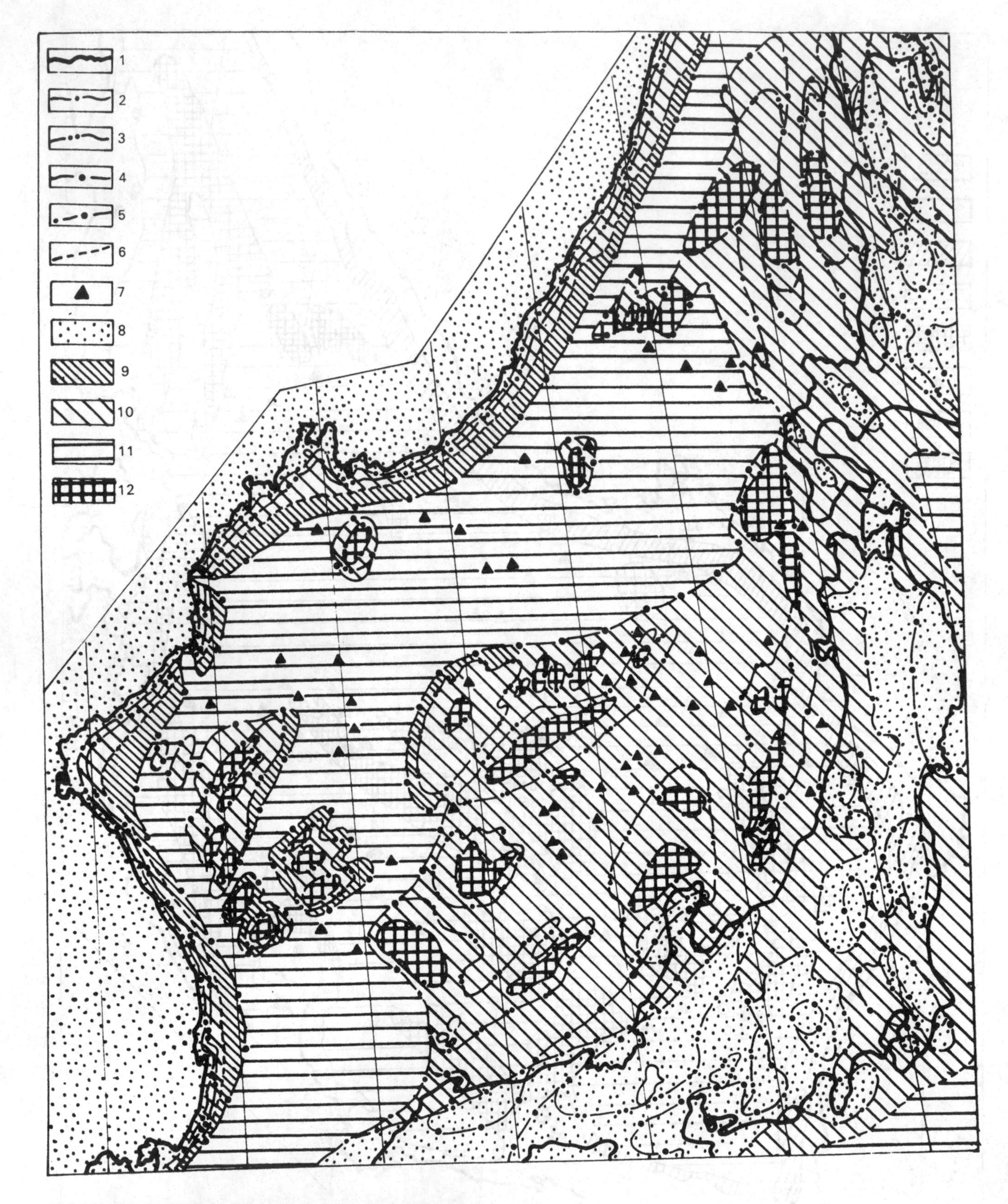

15.2. Paleogeography of the Sea of Japan during the Late Miocene. 1 – Recent shoreline; 2 – Early Miocene shoreline; 3 – Middle Miocene shoreline; 4 – Late Miocene shoreline; 5 – coincidence of shorelines; 6 – Late Miocene shelf break; 7 – volcanoes; 8 – Late Miocene land; 9 – continental slope; 10 – shelves and shallow seas; 11 – deep sea; 12 – emerged parts of sub-sea highs.

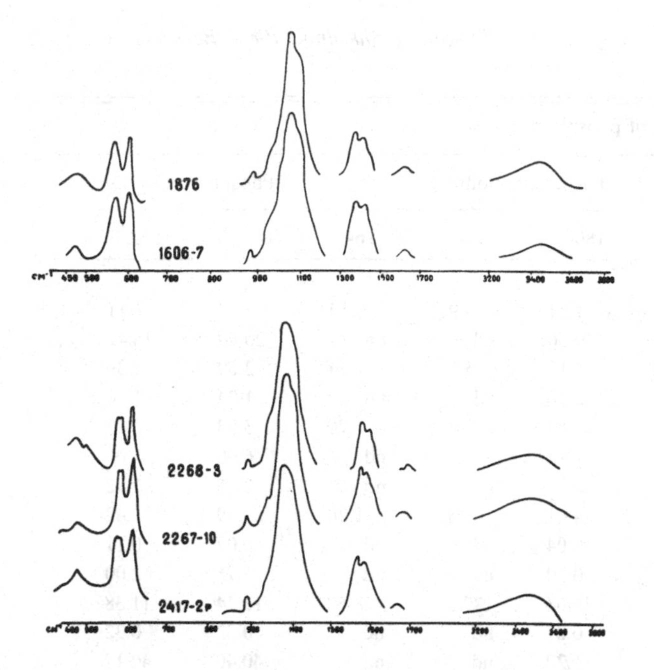

Fig. 15.3. Infrared absorption spectra of phosphorites from the Sea of Japan.

nodules of flattened form, 5–10 cm in diameter, 2–4 cm thick, light grey on the surface and almost black on fractures. In the center, they usually contain nuclei consisting of fish bones, mollusc shells or foraminiferal tests. Their phosphatic matrix contains numerous diatom frustules, some terrigenous material, and relics of glauconite grains. Occasionally glauconite occurs in the form of crossing veins.

The third group of phosphorites is represented by slightly lithified rocks consisting of poorly-sorted terrigenous clastic material of fine-grained–pebble-size, including granitoids, granitporphyrs, acidic pryoclastics, quartz, feldspar, and glauconite representing several generations. The phosphatic matrix is the same as in phosphorites of the first group, with numerous inclusions of relics of diatom frustules.

Nature of phosphatic material

The phosphate of all phosphorites is of CFA composition, light yellow in color, and isotropic, or slightly anisotropic in rare cases. The average refraction index measured in immersion liquids is 1.60. X-ray diffraction analysis showed that CFA is crystallized in the hexagonal class within the P6 3/m space group, and with cell parameters $a = 9.335–9.40 \pm 0.002$Å, and $c = 6.89 \pm 0.003$Å.

Infrared analysis reveals two types of absorption spectra with either two or three lines of absorption from the phosphatic anion within four different fields of oscillation (Fig. 15.3). This is caused by differences of local symmetry of the phosphatic anion. X-ray amorphous phosphate produces a spectra corresponding to a lower local symmetry of the phosphatic ion, whereas crystalline phosphate gives a result more characteristic of CFA (Gusev, Baturin & Pliss, 1985).

Chemical composition of phosphorites

The P_2O_5 content in the Sea of Japan phosphorites of the first group is usually 23–30% and that of the second group 29–31% (Table 15.1). The phosphatic nodules from the mainland continental shelf, on the other hand, contain much less P_2O_5 (14%) because of weak phosphatization of their external shell. The phosphatized coarse and fine-grained rocks usually contain only about 2–4%, and rarely up to 11–13% P_2O_5. The amount of citric acid-soluble phosphorus relative to the total quantity in these rocks reaches 40% or more, an important consideration for their possible utilization as a raw material for phosphatic fertilizer.

The content of major lithogenic components in phosphorites of the first two groups is as follows: SiO_2 1.2–8.8%, Al_2O_3 0.7–2.6% and in phosphorites of the third group 40–46% and 5.7–6.1%, respectively. The CO_2 content of highly phosphatized specimens is in the range of 3.55–6.58%, typical of other oceanic phosphorites (Baturin, 1978 and 1982).

The content of total sulfur in high-grade phosphorites is rather constant, 0.65–0.81%. The fluorine content is 2.00–3.22% with F/P_2O_5 ratios ranging from very low values of 0.066 to more typical values for sedimentary phosphorite of 0.128.

Ultramicroscopic structures of phosphorites

Ultramicroscopic structures reported in oceanic phosphorites include colloform, granular, globular, crystalline and biomorphic textures (Baturin *et al.*, 1985). All of these are found within the phosphorites of the Sea of Japan.

Colloform phosphate is represented by amorphous aggregates of irregular form. It may have either a fluid-like appearance or take a form of fibrous masses and films. The colloform phosphatic matter which gives no diffraction pattern is most common in the massive, as well as in nodular, phosphorites from the Sea of Japan (Fig. 15.4*a*).

Granular phosphate consists of grains from 0.0*x*–0.*x* μm in diameter. Grain morphology varies from equidimensional to irregular. Grains of different size and appearance form larger aggregates up to several micrometers. The grains occasionally grow among amorphous masses of phosphate, in syneresis cracks, and in cavities left after dissolution of diatom frustules (Fig. 15.4*b*).

Globular phosphate is represented by globules from one to several micrometers in diameter, often associated with amorphous masses. The surface of globules may be smooth, rough or cracked, which may indicate the beginning of crystallization. Some of the globules have a fibrous structure. Microdiffraction patterns indicate weak crystallization.

Crystalline phosphate assumes several variations. One shows crystallization in the form of aggregates of ultramicroscopic rods (Fig. 15.4*c*) and globules which reveal radial structure on fractures. The crystallomorphic rods or their aggregates often accumulate on biogenic components, most commonly on relics of diatom frustules. In some places, aggregates of tabular or prismatic hexagonal crystals of CFA are being formed. Microdiffraction investigations show that the CFA crystal structure in this kind of material is well developed. Crystalline phosphate is occasionally found in massive phosphorites and is rare in nodular ones.

Biomorphic structures of phosphatic matter, caused by the presence of abundant biogenic remains, are common in all types

Table 15.1. *Chemical composition (%) of phosphorites from the Sea of Japan (after Lipkina & Shkolnik, 1981; Bersenev* et al., *1984)*

	Type of phosphorite									
	Phosphatized diatomites					Phosphatic nodules			Phosphatized rocks	
Station number	1441	1606	1620	1876	1630	1808	2140	2164	1333A	1333B
Components										
Al_2O_3	1.62	1.99	1.33	1.94	2.62	1.21	0.91	0.67	5.73	6.11
CaO	43.29	39.83	46.90	40.02	42.52	46.56	nd	nd	20.69	18.42
CO_2	4.10	5.03	4.35	3.55	6.58	5.12	5.31	4.74	2.59	2.30
F	2.15	3.22	2.00	2.88	2.42	2.26	nd	nd	1.24	1.16
Fe_2O_3	1.93	2.80	1.61	3.68	2.66	1.53	1.40	1.70	3.33	3.13
Ignition loss	11.68	14.29	10.04	10.48	9.06	11.12	nd	nd	6.90	6.06
K_2O	0.31	0.52	1.94	0.60	nd	0.27	nd	nd	2.75	2.92
MgO	3.84	1.42	1.35	2.87	2.81	4.61	1.35	1.26	3.19	2.42
MnO	0.87	1.22	0.02	0.02	0.09	0.04	nd	nd	0.07	0.11
Na_2O	1.36	1.52	0.61	0.90	nd	0.90	nd	nd	1.75	2.00
P_2O_5	29.05	25.15	30.10	27.27	28.84	30.64	29.66	28.68	13.14	11.38
Total S	0.68	0.66	0.81	0.68	0.65	0.87	nd	nd	0.32	0.32
SiO_2	4.33	8.00	3.60	8.60	8.25	1.22	nd	nd	40.40	40.17
TiO_2	0.06	0.09	0.07	0.08	0.03	nd	nd	0.19	0.19	
Total	101.17	100.71	100.99	101.17	99.62	101.26	—	—	100.24	101.34
$O_2 = F$	0.91	1.36	0.84	0.91	—	0.95	—	—	0.52	0.49
$O_2 = S$	0.34	nd	0.40	0.34	—	0.43	—	—	0.16	0.14
	99.92	99.39	99.75	99.58	—	99.88	—	—	99.56	100.71
F/P_2O_5	0.074	0.128	0.066	0.069	0.084	0.074	—	—	0.094	0.102

nd – not determined

of Sea of Japan phosphorites. Diatom frustules are the most common together with occasional radiolarian and foraminifera tests. Among diatoms, several species have been identified, including *Coscinodiscus marginatus* and *Stephanopyxis spp.*, characteristic of neritic and upwelling conditions.

The degree of phosphatization of organic remains is variable but, in most cases, the phosphatization is nearly complete (Figs. 15.5 and 15.6). The opal of radiolarians is also replaced by finely crystalline phosphate consisting of crystallites 0.5–1 μm in size (Fig. 15.6). The initial stage of phosphatization of diatoms consists of the penetration of gelatinous-type amorphous phosphate inside the cavities and pores of frustules and the formation of rod-shaped crystallites on their surface. Complete substitution of biogenic opal by phosphate is also commonly accompanied by its partial crystallization and formation of phosphatic pseudomorphs after diatoms. Examination under a scanning electron microscope reveals that the cells of diatom frustules are filled by aggregates of parallel, needle-like bodies about 0.1 μm in diameter (Fig. 15.5*a*,b). Sometimes, the diatom frustules are wholly dissolved and the remaining cavities are filled by aggregates of elongated crystallite-like grains of CFA approximately $0.6 \times 1.7\ \mu$m in size (Fig. 15.5*d*). The rare foraminifera tests are filled by amorphous fine-grained phosphate. The non-phosphatic components of phosphorites observed include glauconite (Fig. 15.7*a*), quartz, feldspar and pyrite (Fig. 15.7*b*).

Origin of Sea of Japan phosphorites

As shown on the paleogeographic scheme (Fig. 15.2), the Yamato Rise and many of the subsea mountains were islands during the Miocene. The configuration of the Sea was such that it occurred as a longitudinal channel for a southern current passing across those island shelves. This current presumably produced coastal upwelling which stimulated high biological productivity. The common occurrences of shallow-water diatomaceous oozes among the Miocene deposits of the Sea of Japan and local high abundances of fish remains in the Miocene basement horizon support this interpretation.

The structure and microstructure of these phosphorites imply a diagenetic nature of the phosphatization process took place in unconsolidated clayey-diatomaceous oozes enriched in phosphorus and organic matter.

The general geomorphological setting of the Miocene Sea of Japan is comparable in some ways to the modern California borderland which also consists of a system of submarine highs and depressions (Dietz, Emery & Shepard, 1942; Emery, 1960). The general paleogeographical situation in the Miocene Sea of Japan, however, was probably more similar to that of the southeastern US continental margin where phosphorites of the same age formed in upwelling conditions caused by the action of the Gulf Stream which was flowing closer to the shore during the Miocene high sea-level stand (Riggs, 1984).

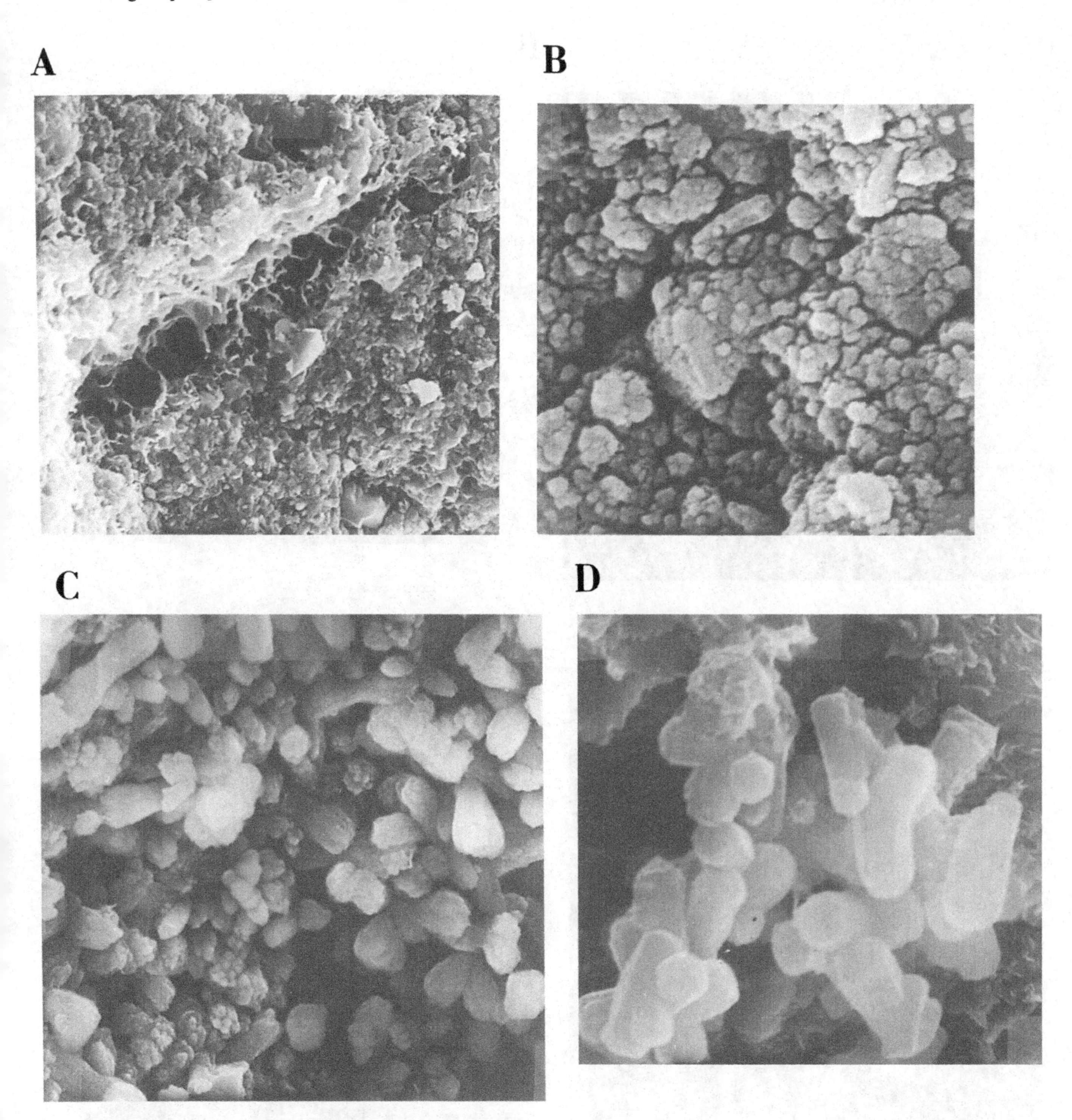

Fig. 15.4. Microstructure of massive phosphorites. (A) Amorphous and fibrous phosphate, × 3200. (B) Granulated phosphate, possibly due to dehydration, × 14400. (C) Formation of aggregates of rod-shaped microcrystals, × 5100. (D) The same, with more crystalline appearance, × 10200.

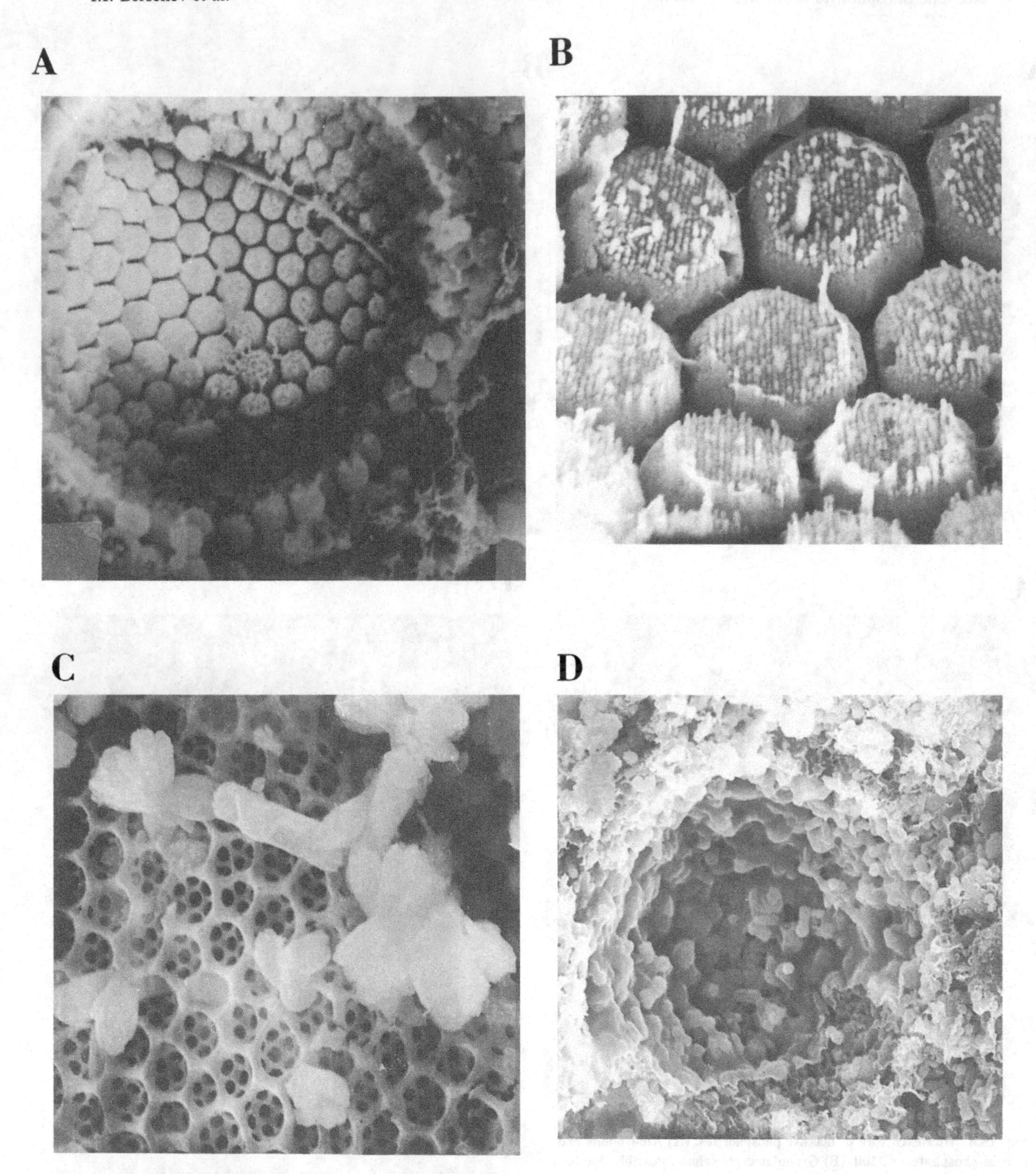

Fig. 15.5. Phosphatization of diatoms in phosphorites. (a) Phosphatized diatom, × 2400. (b) The same, × 9600. (c) Isolated and clustered rod-shaped phosphate crystals on the surface a of diatom frustule, × 7200. (d) Void left after dissolution of diatom and filled by microcrystalline phosphatic material, × 2400.

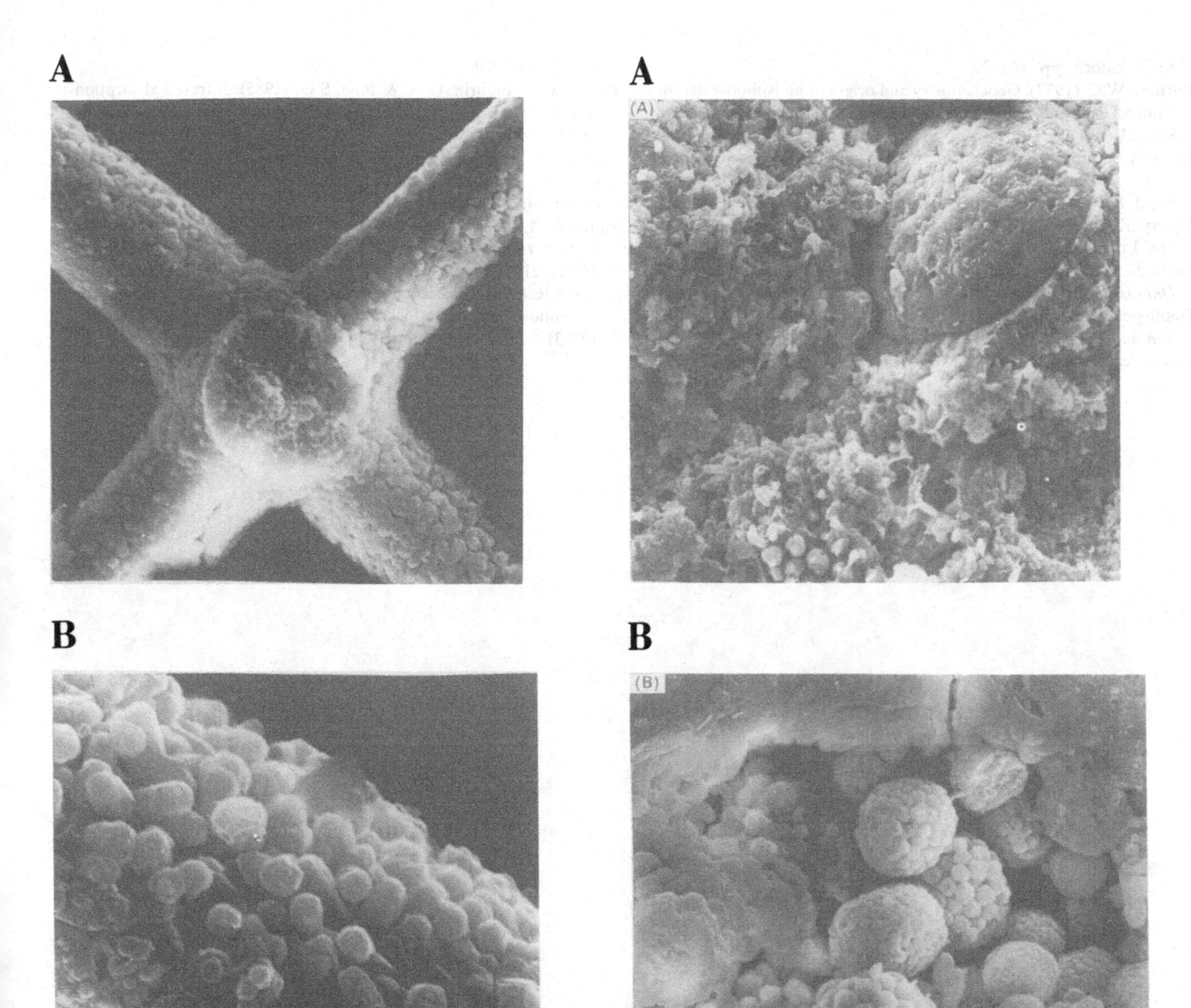

Fig. 15.6. Phosphatization of radiolarians. (a) Phosphatized radiolarian skeleton, × 2700. (b) Crystalline structure of phosphatized radiolarian, × 9000.

Fig. 15.7. Minerals associated with phosphorite from the Sea of Japan. (a) Corroded rounded grain of glauconite, × 2000. (b) Framboidal pyrite, × 2400.

The microstructures of the phosphorites of the Sea of Japan are similar to those of Recent and Late Quaternary phosphorites of Namibian and Chile–Peru shelves (Burnett, 1977; Baturin & Dubinchuk, 1979). These observations imply that a phosphogenic Miocene event in the Sea of Japan may be considered as one of many modifications of diagenetic phosphorite formation connected with biogenic sedimentation under coastal upwelling conditions.

References

Baturin, G.N. (1978). *Phosphorites on the Sea Floor*. Nauka, Moscow.

Baturin, G.N. (1982). *Phosphorites on the Sea Floor. Origin, Composition and Distribution*. Elsevier, Amsterdam.

Baturin, G.N., Bersenev, I.I., Gusev, V.V., Lelikov, E.P. & Shkolnik, E.L. (1985). Electron-microscopic study of phosphorites from the bottom of the Sea of Japan. *Doklady Academii Nauk USSR*, **281(5)**, 1169–72.

Baturin, G.N. & Dubinchuk, V.T. (1979). *Microstructures of Oceanic Phosphorites*. Nauka, Moscow.

Bersenev, I.I., Bezverkhniy, V.L., Lelikov, E.P. & Terekhov, E.P. (1983b). *Geological Structure of the Sea of Japan. Stratigraphy of pre-Cenozoic Deposits*, (preprint). Nauka, Moscow, 28p.

Bersenev, I.I., Bezverkhniy, V.L., Vashchenkova, N.G. *et al.* (1983a). *Geological Structure of the Sea of Japan*, (preprint). Nauka, Moscow, 54p.

Bersenev, I.I., Shkolnik, E.L. and Gusev, V.V., (1983c). Phosphorites from the bottom of the Sea of Japan. *Doklady Academii Nauk USSR*, **271(2)**, 397–401.

Bersenev, I.I., Shkolnhik, E.L. & Gusev, V.V. (1984). Phosphorites of the Sea of Japan. In *Phosphates of Eastern Asia and Bordering Seas*,

Vladivostock, pp. 162–79.
Burnett, W.C. (1977). Geochemistry and origin of phosphorite deposits from off Peru and Chile. *Bulletin Geological Society of America*, **88(6)**, 813–23.
Dietz, R.S., Emery, K.O. & Shepard, F.F. (1942). Phosphorite deposits on the sea floor off California. *Bulletin Geological Society of America*, **53(6)**, 815–48.
Emery, K.O. (1960). *The Sea off Southern California*. Wiley, New York and London.
Geological Survey of Japan (1979a). *Geological Map of Japan and Okhotsk Seas around Hokkaido*, 1:1,000,000.
Geological Survey of Japan (1979b). *Geological Map of Southern Japan Sea and Tsushima Straits*, 1:1,000,000.
Geological Survey of Japan (1981). *Geological Map of the Central Japan Sea*, 1:1,000,000.
Gusev, V.V., Baturin, G.N. & Pliss, S.G. (1985). Infrared absorption spectra of oceanic phosphorites. *Doklady Academii Nauk USSR*, **283(3)**, 694–8.
Lelikov, E.P., Sédin, V.G. & Fershtatter, G.B. (1984). Rubidium and strontium in magmatic rocks from the bottom of the Sea of Japan. *Geokhimiya*, **8**, 1209–17.
Lipkina, M.I. & Shkolnik, E.L. (1981). Phosphorites from the subsea Chentsov volcano in the Sea of Japan. *Doklady Academii Nauk USSR*, **257(1)**, 217–22.
Riggs, S.R. (1984). Paleoceanographic models of Neogene phosphorite deposition, US Atlantic continental margin. *Science*, **223(4632)**, 123–31.

16

Physical and chemical properties of the phosphate deposit on Nauru, western equatorial Pacific Ocean

D.Z. PIPER, B. LOEBNER AND P. AHARON

Abstract

The phosphate deposit on the Island of Nauru is one of the largest insular deposits known. The original reserves were estimated at 90 million tons (Hutchinson, 1950), of which possibly 10% remains. It consists of two units draped over a dolomitized karrenfeld, the pinnacles of which are 3–5 m high. A pelletal unit, 10–25 cm thick, lies against the pinnacles; this unit grades into a conglomeratic unit, the cobbles of which are also composed of pellets. The composition of these two units is quite uniform; the only mineral detected is a carbonate fluorapatite that also contains hydroxyl ion. The major difference between the two units is in their CO_2 content. The pelletal ore contains twice as much CO_2 as the conglomeratic ore. The concentrations of most minor elements in the apatite are uniformly low, as exemplified by the rare-earth elements. The La content is 1 ppm in one sample only, in comparison with a content of 50 ppm in marine deposits. The uranium content of the Nauru deposit, however, is comparable to, though slightly lower than, that of the major sedimentary deposits, and the zinc content exceeds that of the sedimentary deposits.

The low concentrations of the rare-earth elements, as well as of sulphate and fluoride, support the earlier conclusion of other workers, based on the physical characteristics of Nauru, that the source of the phosphate was bird guano. The $\delta^{13}C$ and $\delta^{18}O$ values of the apatite indicate that the guano reacted with rainwater to produce a phosphate-charged acidic solution. This solution was buffered by reaction with the carbonate rocks and sediment of the reef and lagoon to produce supersaturation with respect to, and precipitation of, apatite within the fresh groundwater zone.

Introduction

We have studied the phosphate deposit on the Island of Nauru, in the western equatorial Pacific Ocean (Fig. 16.1), to ascertain its physical and chemical relations to other members of the group of sedimentary pelletal phosphorites. One advantage presented by this insular deposit is its freedom from diluting phases, such as clay minerals, other alumino-silicates, and biogenic silica, which commonly constitute a large part of marine deposits (Sheldon, 1963). Also, it is a relatively young deposit that has likely existed entirely within the meteoric-water zone since its formation. Thus, it may have escaped significant alteration (McArthur, 1985).

Insular deposits provided much of the fertilizer phosphate during the nineteenth century, and Nauru continues to supply the western Pacific with a large fraction of its present requirements. Despite its economic importance, very little is known about the Nauru deposit beyond its bulk composition and appearance. It has been examined twice (Power, 1905; Elschner, 1913) since its discovery in 1899 by Ellis (1935). More recent discussions about this deposit are based almost totally on these two investigations. A few radiometric and chemical analyses of the deposit have been reported quite recently, but nothing was known about the samples other than that they were collected on Nauru. Because most of the deposit has been exploited since mining commenced in 1906, we considered that Nauru warranted one last examination and sampling before its demise, which will possibly come within the next decade. One of us (D.Z.P.) visited Nauru briefly, for 10 days in each of the years 1982 and 1983.

The earlier work on insular deposits documented their distribution, size, composition, and the source of the phosphate. Hutchinson (1950) compiled the available information on these deposits. Space does not permit an adequate review here of these deposits, nor does the completeness of Hutchinson's monumental work require it. The studies on which his account was based established that the source of the phosphorus was bird guano, but they often left unanswered many questions about the development of the deposits after initial accumulation of the guano. This report represents but one step toward an improved understanding of their geochemistry.

Among the insular deposits, that on Nauru belongs to the group of deposits on elevated atolls on which the guano has altered to apatite. These deposits are composed of an earthy to massive carbonate fluorapatite that contains hydroxyl ion (McClellan & Lehr, 1969) substituting for fluoride. The finer grained ore is pelletal, but even the massive ore contains cemented pellets. Deposits on other islands (e.g. Christmas Island) include iron and aluminum phosphates when the underlying substrate is rich in alumino-silicates. These deposits have been mined on many islands, such as Banaba (Ocean), Kita Daito Jima, Anguar, and Makatea in the Pacific Ocean and Aldabra and Christmas Island in the Indian Ocean. Mining continues on Nauru and Christmas Island.

At the other end of the spectrum are the relatively young

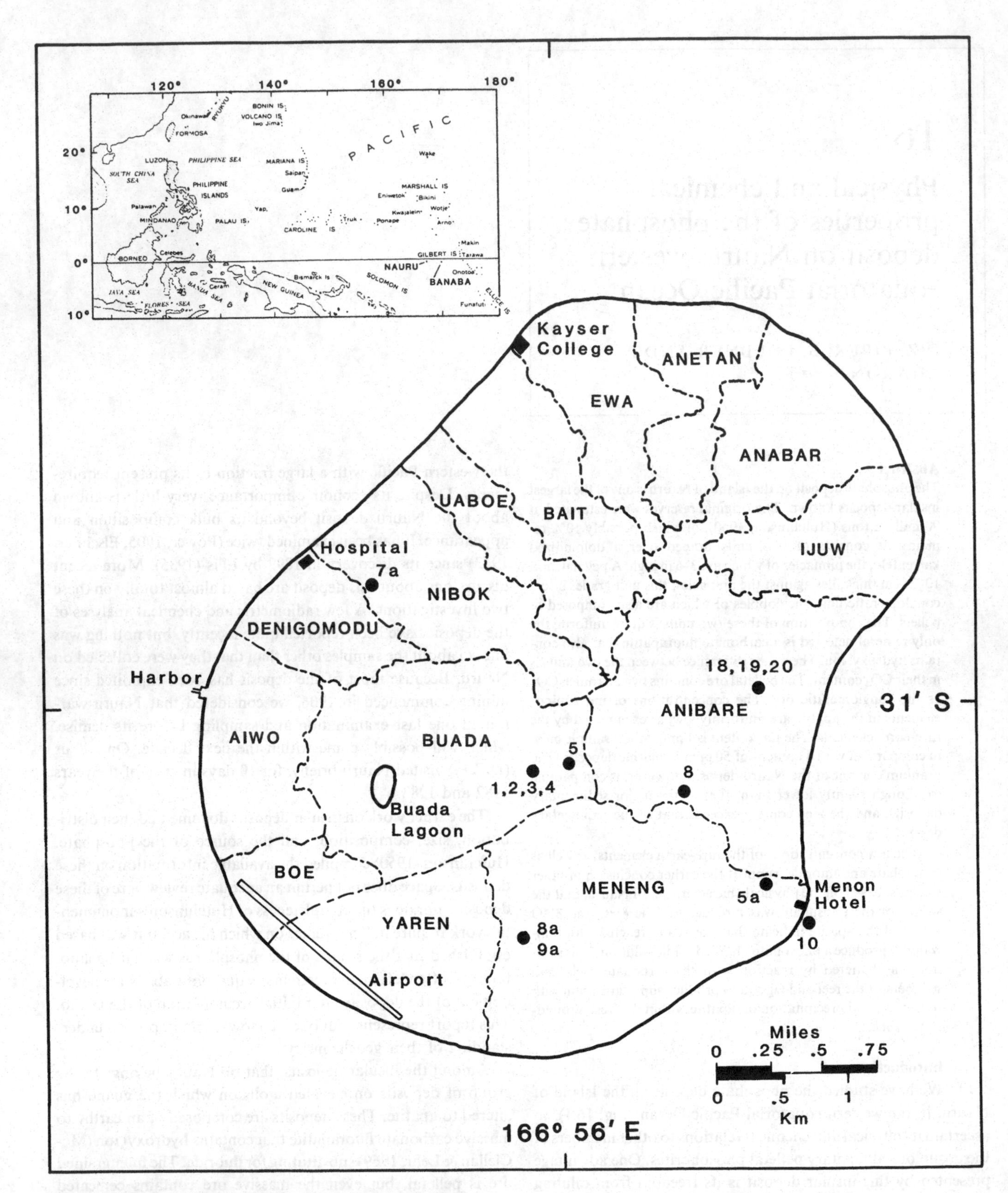

Fig. 16.1. Plan view of Nauru, showing districts and sample locations. Inset map of the western Pacific Ocean shows location of Nauru.

Table 16.1. *Compositions of Nauru phosphate ores, subvitreous phosphate (nauruite), and dolomite. Analyses are reported in weight %, and represent averages of more than 100 microprobe analyses of each sample type. CO_2 and loss on ignition (LOI) at 540°C represent averages of four replicate analyses of each type of ore. Dashes indicate no analysis made*

	Incoherent ore[a]	Coherent ore[b]	Coherent ore[c]	Nauruite	Dolomite	Phosphate[d]	Marine[g] phosphorite
Al_2O_3	0.01	0.24	0.23	0.00	0.05	0.30[d,f]	1.7
CaO	52.3	49.4	52.7	48.9	32.6	54.4	44.0
CO_2	5.8	3.4	2.6	3.4	—	2.0	2.2
F	2.97	2.88	2.97	1.20	0.05	2.62	3.1
FeO	0.08	0.06	0.13	0.03	0.00	—	0.50
	(0.08)[e]		(0.10)[e]		(0.03)[e]		
K_2O	0.00	0.02	0.01	0.00	0.00	tr	0.5
LOI							
(less CO_2)	2.5	2.8	0.8	—	—	—	—
MgO	0.18	0.12	0.10	0.44	19.5	0.00	0.3
Na_2O	0.10	0.19	0.18	0.39	0.02	0.45	0.6
P_2O_5	37.0	37.2	39.0	34.2	0.31	38.9	30.5
SiO_2	0.03	0.01	0.01	0.03	0.00	0.20	11.9
SO_3	—	0.16	0.15	—	—	0.00	1.8

[a]Red ore surrounding pinnacles.
[b]Grey ore adjacent to incoherent ore, but not present at all outcrops.
[c]Yellow ore.
[d]Hutchinson (1950).
[e]FeO by neutron activation analysis, in weight %.
[f]Reported as $Al_2O_3+Fe_2O_3$.
[g]Average composition of phosphorite from the Phosphoria Formation (Gulbrandsen, 1966).

deposits on many coastal as well as oceanic islands. They are composed of hydrated phosphates and oxalates of mostly calcium, sodium, and ammonium. As noted by Hutchinson (1950), the source of these constituents is clearly bird guano, an interpretation further supported by their association with enormous bird colonies. Notable among these deposits were those on the coastal islands of Peru and Chile. The most impressive deposit, if not the largest, was on the Chincha Islands in the Bay of Pisco, where the guano attained a thickness of more than 45 m. Hutchinson employed several lines of evidence to show that it accumulated at the rather phenomenal rate of several meters per 100 y. Because these deposits were largely exploited by the end of the nineteenth century, little can now be ascertained about their chemical and physical properties beyond the brief reports published before their exhaustion at the turn of the twentieth century.

Between these two extremes are small deposits and mere occurrences of phosphate on numerous islands. These deposits can consist of relatively young guano at the surface and of apatite at depth. Some phosphate material may remain on many of these islands. Examples are the deposits that have been mined on Baker, Jarvis, and Malden Islands in the equatorial Pacific, atolls that may be uplifted only a few meters (Menard, 1973) and on which the lagoon is now almost completely filled with $CaCO_3$ sand and rubble and evaporites, such as on Malden and Jarvis Islands (Schlanger & Tracey, 1970). Many islands, such as Clipperton and the Chincha Islands, are sites of current accumulation of bird guano.

Our initial view of the phosphate deposit on Nauru and of insular guano deposits in general was that they formed by a process, or processes, quite different from those that contributed to the formation of the much larger marine pelletal deposits. The information gained during this investigation, however, suggests that the same basic conditions govern the formation of this and all other sedimentary phosphate deposits; differences are in details rather than in the major aspects of phosphogenesis. Expressed simply, phosphate, along with CO_2, is put into solution during sediment diagenesis (Arthur & Jenkyns, 1980) by the oxidation of organic matter. The source of the organic matter, of which phosphate is a minor but integral part, can be bird guano that accumulates on a coral island, or the fecal debris of both vertebrates and invertebrates that accumulates on the seafloor. On the seafloor the solution is saline and the most important oxidizing agents are O_2, NO_3^-, and SO_4^{2-}. On coral islands the solution is fresh to brackish and the oxidizing agent is predominantly O_2. Under conditions of O_2 respiration, the resulting phosphate-charged solution is acidified by CO_2. Reaction with solid $CaCO_3$ brings about supersaturation with respect to, and precipitation of, carbonate fluorapatite. The $\delta^{13}C$ values of marine organic matter, $CaCO_3$, and apatite suggest a stoichiometry for this reaction that requires the organic matter and $CaCO_3$ to contribute equal amounts of carbon to the solution from which the apatite precipitates. The minor-element chemistry of the Nauru deposit reflects a unique source of the phosphate, but its mineralogy, rock association, and texture resemble those of the major sedimentary deposits.

Analytical techniques

The major element chemistry of the phosphate and carbonates were measured by microprobe. Several hundred analyses of each rock type were averaged to give bulk compositions (Table 16.1). Analyses of individual grains were periodically

Table 16.2. *Instrumental neutron activation analyses of Nauru phosphate ores and dolomite, in parts per million (ppm). Incoherent and coherent ores are the same as those in Table 16.1. Dash indicates no analysis reported. Phosphoria analysis (Marine phosphorite) from Gulbrandsen (1966)*

	Incoherent ore	Coherent ore	Coherent ore	Dolomite	Marine phosphorite
As	<9.0	<9.0	<8.0	<2.0	40
Ba	<800	<800	<900	<300	100
Br	8.44	8.82	4.60	3.25	—
Co	2.5	2.3	.3	7.3	<10
Cr	63.9	56.7	59.4	<8.0	1000
Sb	.05	.18	.07	.08	7
Se	1.06	1.02	1.25	<.7	13
Sr	663.0	629.0	690.0	290	1000
Th	<.40	<.40	.14	<.20	—
U	65.2	50.9	68.7	2.5	90
Zn	1155	727	983	18	300
Ce	0.6[a]	0.4[a]	3.3[a]	<3.0[a]	—
Eu	<2.00	<.90	<2.00	.60	—
La	0.8[a]	0.4[a]	1.8[a]	2[a]	300
Lu	<.90	<.90	<1.00	<.30	—
Nd	<3[a]	<5[a]	9[a]	7	300
Sc	<.20	.17	.15	<.07	10
Sm	<.40	<.30	<.30	<.20	—
Tm	<.30	<.30	.30	<.08	—
Yb	—	<.9	—	<.40	10

[a]Correction for fission-product interference exceeds 20% of the reported value. These values agree closely with recently obtained, as-yet unpublished measurements made by induction-coupled-plasma (ICP) spectroscopy except for Nd, for which the ICP measurements were reported as <0.5 ppm.

repeated to ascertain precision and an apatite standard was run during each analytical session; precision is better than ±4%. CO_2 was measured gasometrically using a Leco induction furnace; precision is ±5%. The rare-earth elements and U were measured by instrumental neutron-activation analysis. After correcting for the contribution from U fission (Baedecker & McKown, 1985), the rare-earth element contents are reduced to a level near their detection limit in almost every case (Table 16.2). The precision for U, determined by duplicate analyses of the samples and of a standard irradiated with the samples, is ±10%. Mineral identification was by X-ray diffraction and petrographic microscope.

Experimental procedures for the determination of $^{18}O/^{16}O$ and $^{13}C/^{12}C$ isotopic ratios were designed to permit separate analyses of the apatite and dolomite components. Trace amounts of $CaCO_3$ were removed in 10% acetic acid according to the method of Sullivan & Krueger (1981). The purity of the apatite from both calcite and dolomite was verified by X-ray diffraction techniques.

Samples of apatite and dolomite were dissolved at 25°C in H_3PO_4 for 72 h; the liberated CO_2 was purified in a vacuum extraction line according to the method of McCrea (1950). The CO_2 evolved was analyzed for $^{18}O/^{16}O$ and $^{13}C/^{12}C$ isotopic ratios by using the triple-collector 6/60 gas-source mass spectrometer at Louisiana State University. The measured isotopic ratios were corrected for instrumental characteristics and nominal mass ratios according to the method of Craig (1957). Fractionation factors of 1.01025 and 1.01109 were applied to the measured isotopic ratios of CO_2 from apatite (Kolodny & Kaplan, 1970) and dolomite (Friedman & O'Neil, 1977), respectively, using the conventional assumptions that the apatite CO_2 fractionation is identical to that of calcite, and the calcian dolomite CO_2 fractionation is identical to that of the stoichiometric dolomites. δ notations are relative to the Peedee Belemnite (PDB) standard (Craig, 1957). The overall error (δ) of the isotopic determinations, based on analyses of standards and replicate analyses of three samples, is estimated to be ±0.1‰.

Physiography of Nauru

Nauru (Fig. 16.1) is an uplifted atoll with a maximum elevation of 65 m (Fig. 16.2). It is elongated in a NE–SW direction and has an approximate circumference of 20 km. The island can be divided into two physiographic regions – a coastal plain that encircles the island, and an elevated interior region. The coastal plain consists of a fringing reef and wave-deposited terrace; an approximately 15 m wide beach marks their boundary. The central elevated region of the island rises abruptly from the coastal plain; its highest elevations are at its margin (Fig. 16.2). Several basins and intervening ridges constitute this central region. One of these basins (Buada Lagoon) contains a few tens of centimetres of water, but Buada Lagoon is possibly unique among these basins in other aspects as well, as noted below.

The fringing reef extends unbroken around the island. It has an average width of approximately 100 m and a maximum width in the north of 300 m; it is exposed at low tide and is covered by about a meter of water at high tide. Although living coral may inhabit the outer reef below low tide, on the easily accessible part

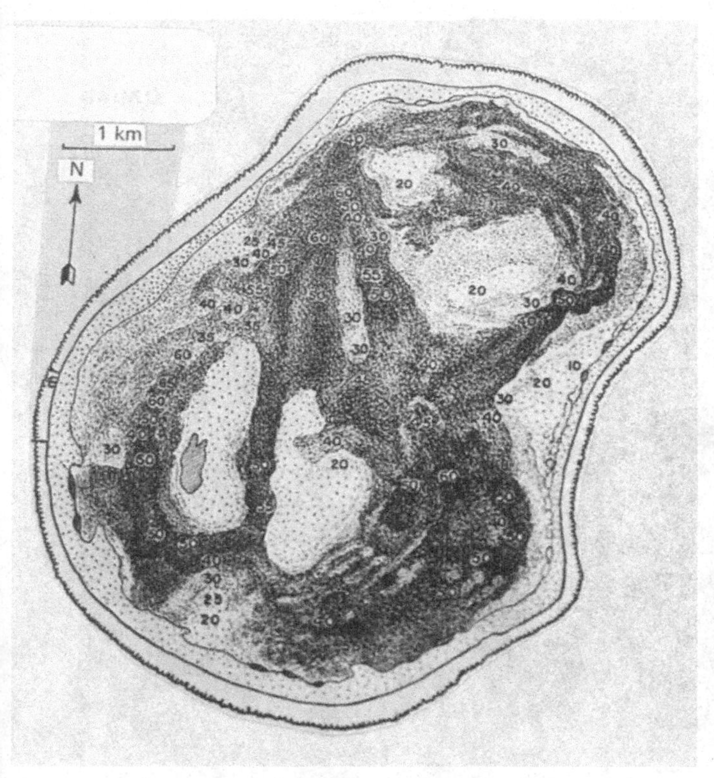

Fig. 16.2. Relief of Nauru. Individual elevations and locations of ridges and basins are given in Hutchinson (1950).

it is restricted to the outer edge of the reef. The outer-reef flat is covered by red algae, whereas the inner reef is covered by green algae.

The reef consists of a dolomite platform around approximately four-fifths of the island and of aragonite along the remaining fifth, along the southwestern margin. The dolomite portion of the reef is punctuated throughout by pinnacles (Fig. 16.3), also composed of dolomite, that reach a maximum height of about 3 m along the eastern part of the island but vary significantly in height from one section of the reef to another. They are black on the exposed surface and grayish-yellow on freshly broken surfaces. Throughout, they exhibit strong evidence of dissolution; their surface texture is strongly pitted and angular (Fig. 16.3). The aragonite portion of the reef has no pinnacles.

The terrace rises from the beach 2–4 m. It is virtually absent at about 0° 32′ S latitude along the east shore, where a 5°–10° slope extends essentially from the reef to the central region. Elsewhere, the terrace extends in a continuous line around the island; it reaches a maximum width of about 500 m in the most heavily developed area. A few pinnacles are present on the terrace, extending as much as 3 m above its otherwise flat surface; they have a texture similar to those on the reef.

Examination of a 4 m deep cistern, excavated into the terrace and within 50 m of the beach, showed the terrace to be composed of angular fragments of carbonate sand and cobbles. The upper 90 cm are grayish in color, beneath which are 2 m of brown conglomeratic debris extending to the water line at the bottom of the excavation. Ellis (1935) described material recovered from a core drilled in 1929 in the terrace landward of the coral reef. The upper 2 m was composed of rubble, similar to that seen in the cistern; below this layer, however, were 2–3 m of pelletal phosphate. This material would be as much as 2 m below the present sea level and well below the level of the reef.

The juncture between the terrace and the slope, extending up to the elevated interior, is marked in many places by a series of depressions, as much as 1.5 m deep (Fig. 16.2). Several of these depressions contain water, and many more apparently held water during the early twentieth century (K. Jacobs, pers. comm., 1983). Many of them have pinnacles within or at their landward margin.

The slope leading to the elevated central part of the island is in places greater than 25°, more typically between 5° and 10°. It is characterized by dolomite, commonly forming pinnacles. Solution is prevalent throughout the slope, as evidenced by the angular and pitted surfaces of the pinnacles and the occurrence of caves. One cave, immediately north of the airfield, extends more than 15 m into the cliff and contains a permanent stream. The walls of the cave are dolomite. No conspicuous drip-stone formation was observed, although minor amounts of low-Mg calcite encrust the cave walls. Smaller caves are also present.

On the Islands of Makatea (Montaggioni *et al.*, 1985) and Ocean (Elschner, 1913), a series of terraces, commonly with caves and conspicuous notchlines, are clearly related to periods of high stands of sea level. The slope on Nauru, however, does not seem to exhibit terracing. Also, too few caves were visited to ascertain whether they exhibit any continuity relative to sea level, as they do so conspicuously on Makatea.

The elevated central area represents about 85% of the island. The outer margin is generally marked by a ridge, slightly elevated above an interior plain (Fig. 16.4). Before mining commenced in 1906, this plain was relatively smooth (Fig. 16.4*a*), with several basins separated by ridges and knolls. A few dolomite pinnacles extended above the surface. They are easily identified even today because of the similarity of their surface texture to that of the pinnacles exposed on the coastal plain (Fig. 16.3). They may occur in a line, as along a ridge, or as single isolated pinnacles.

Mining has revealed the full extent of this dolomitized and pinnacled surface, known as a *karrenfeld* (Fig. 16.4*b*). The karrenfeld extends over the entire central part of the island. The presence of pinnacles throughout most of the coastal plain, however, suggests that the karrenfeld once extended over the entire island. The only area where pinnacles were not observed was in Buada Lagoon (Fig. 16.2) and on the basin floor immediately surrounding the lake. This area was not investigated other than by visual examination from the road that circles the lake; no pinnacles were noted within the basin itself during this cursory inspection. Because this area has not been mined, we could not ascertain whether the karrenfeld extends across the basin beneath a cover of unconsolidated material.

The pinnacles within the karrenfeld average appoximately 3–4 m in height, but can be as high as 7–8 m. They have a smooth surface texture (Fig. 16.4*b*), in contrast to those exposed on the coastal plain (Fig. 16.3); they generally are higher in the basins and over level areas than on slopes, where typically they are no more than 1 m high. The largest pinnacle observed on the island rises 10 m above the surrounding karrenfeld to an elevation of

Fig. 16.3. Photograph of beach and reef. Note angular and pitted texture of dolomite pinnacles.

59 m; it has an angular and pitted surface texture, indicating that it was exposed before mining. Pelletal apatite was present in cracks over its surface, suggesting that it had once been buried and subsequently exposed by erosion. The absence of any undercutting, as commonly occurs for pinnacles on the coastal plain, suggests that erosion was by surface runoff of rainwater rather than by wave action. Within the areas that were being mined in 1982–3, however, the tops of the highest pinnacles almost invariably attain the same height over an area of several thousand square metres (Fig. 16.4). They rise to within a few centimetres of the surface, which is well marked by a 1–5 cm-thick black soil horizon (Fig. 16.5*a*). Earlier sketches (Elschner, 1913; Ellis, 1935) show this same relation between pinnacle height and the surface (Fig. 16.5*b*).

In one basin in the northwest of the island that was mined some 25 years ago, however, as much as 10 m of phosphate was removed before reaching the level of the pinnacles (E. Kertchner, pers. comm., 1983). We have no way of ascertaining whether this material might have represented material transported from around the few pinnacles that extend above the interior plain before mining. Evidence of such reworking, however, was seen in one recent mine-cut face. The black soil horizon in this area reached a maximum thickness of 30 cm but decreased continuously in thickness over a distance of 100 m to typically < 5 cm. This horizon was composed of pelletal phosphate and angular cobbles, up to 6 cm in diameter. The local topography requires that transport distance must have been no more than a few tens of metres.

The surface texture of pinnacles in the mined area of the interior suggests that erosion in the karrenfeld did not reach below the average maximum height of the karrenfeld except on relatively steep slopes and in the few areas where isolated pinnacles extended well above the surrounding karrenfeld. The close proximity of the pinnacles to each other and their relatively uniform height have surely been major factors limiting the depth to which surface erosion has been able to incise the karrenfeld. The elevated 'rim' of the karrenfeld surrounding the interior area (Fig. 16.2) has very likely prevented further loss of material onto the coastal plain and across the reef.

Support for this interpretation of very limited erosion is seen in areas currently being mined in the district of Anabar (Fig. 16.1); the very outer rim of the karrenfeld and even the slope some 10–15 m toward the coastal plain contain pelletal phosphate. The amount of ore in the area is small, possibly owing to the small size of the pinnacles (approx 1 m high), but the ore covers pinnacles of the karrenfeld that extend down the outer slope.

Carbonate rocks on Nauru

Dolomite is the dominant carbonate rock on Nauru, whereas aragonitic coral is exposed along about one-fifth of the reef. The maximum measured $CaCO_3$:$MgCO_3$ molar ratio of 1.8 (Table 16.3, Fig. 16.6) is similar to that of dolomites from other atolls and islands. The dolomite is composed of a reef fauna (W. Sliter, pers. comm., 1984); corals can be seen on pinnacle surfaces throughout the island (Fig. 16.7). Bryozoans, spines of

Fig. 16.4. Interior views of Nauru: (a) Prior to mining. Photograph is from Ellis (1935). (b) The karrenfeld on Nauru, exposed by mining. Pinnacles in the foreground have a relatively uniform height. The pinnacles have a ring structure, representing a dissolution nip. The 'rim' of the interior is seen in the background.

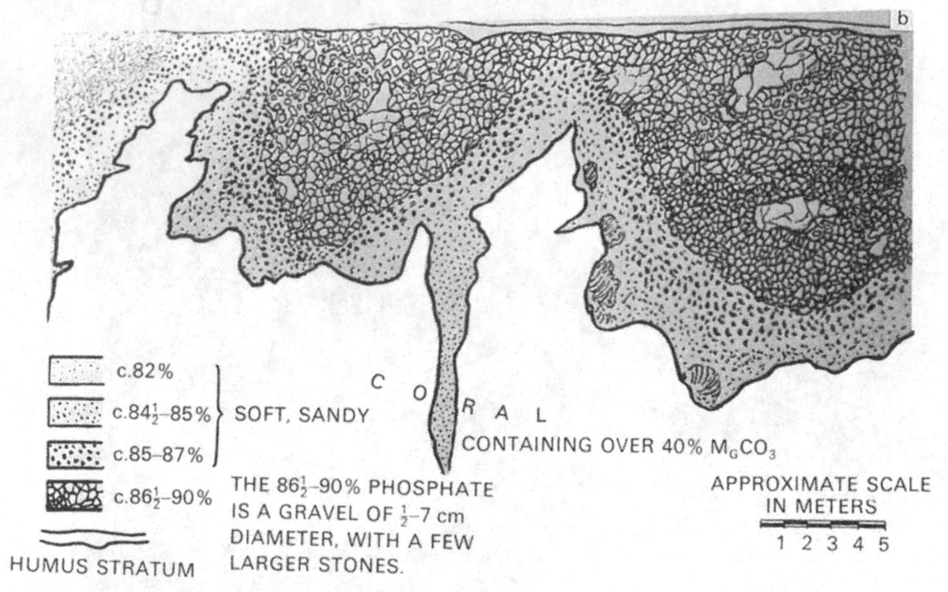

Fig. 16.5. (a) A recently exposed mine surface. The pelletal ore is reddish in color and occurs adjacent to the pinnacle. The conglomeratic ore is yellow in color. The soil horizon is black and about 5 cm thick. The pinnacles extend to near the soil horizon. (b) Cross-section of Nauru phosphate deposit, sketched by Elschner (1913). Compare this with Figure 16.5*a*.

echinoderms, and thick-walled tests of foraminifers are seen in thin section. The mollusc *Tridacna*, represented by specimens up to 30 cm in diameter, has been recovered adjacent to the pinnacles.

Dolomite with strikingly different properties was observed in one small outcrop (Fig. 16.8) on the slope inland from the coastal plain (Fig. 16.2), where it consists of stratified micritic dolomite in alternating beds of white and pale red. The red units have a rich fauna typical of a lagoon environment (E. Moore, pers. comm., 1984); they include ostracods, foraminifers and micromolluscs. The foraminifer *Hastigerina pelagicum*, of Late Miocene–Holocene age (G. Keller, pers. comm., 1984), was identified in one thin section. The white beds contain few or no fossils. Contacts between the top of a white unit and the bottom of a red unit are sharp, whereas the red units grade upward into white units.

A large rounded fragment of dolomite, similar in texture and color to that making up the pinnacles, is embedded in this

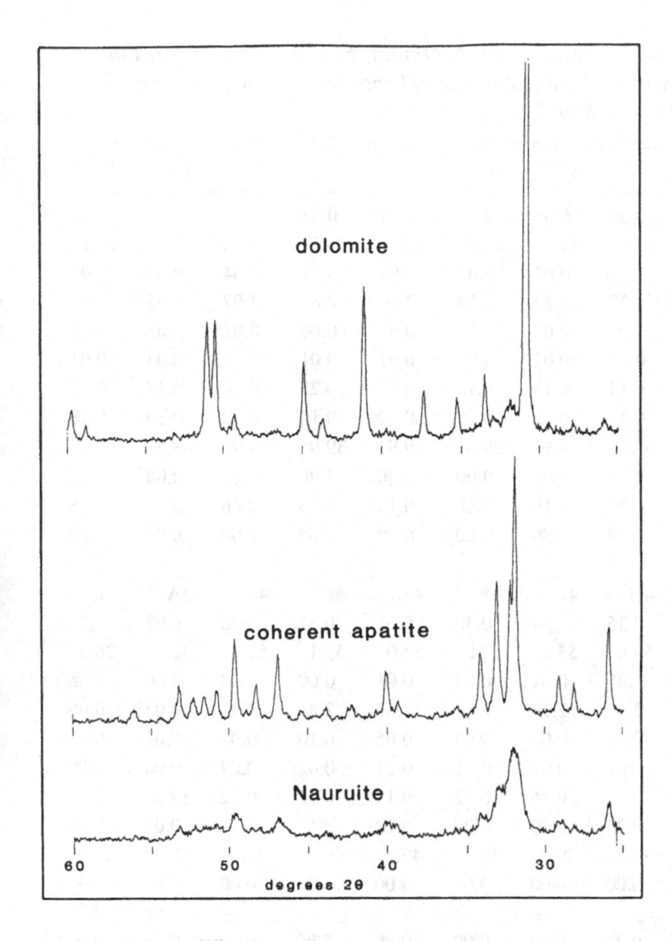

Fig. 16.6. X-ray diffraction patterns of dolomite, coherent apatite, and 'Nauruite'.

deposit (Fig. 16.8) and impregnated by it. This fragment was possibly carried into the lagoon from the reef during a storm. The lagoon sediment appears to be disrupted at this same horizon, but only slightly. The apparent absence of compaction features around the boulder and the sharp contact between the two types of dolomite suggest that lithification and, presumably, dolomitization occurred soon after, or even contemporaneously with, deposition.

The ranges in oxygen and carbon isotopic composition of all the dolomites from Nauru, reef material from the pinnacles and both fossiliferous and non-fossiliferous laminae of the lagoon deposit, are limited (Table 16.4); the $\delta^{18}O$ value averages $+1.3 \pm 0.2‰$ at $\pm 1\sigma$, and the $\delta^{13}C$ value is $+2.3 \pm 0.12‰$. The white units of the lagoon deposit, (NRU 83 5) generally are carbon-light and oxygen-heavy, relative to the highly fossiliferous red units. The results, however, show several exceptions to this trend such that the two groups of analyses overlap. All analyses are enriched in both ^{18}O and ^{13}C relative to marine aragonite (Fig. 16.9).

Phosphate on Nauru

Earlier descriptions of the phosphate deposit suggest that it is quite uniform (Hutchinson, 1950), similar to that remaining in 1982. Major differences have been in nomenclature rather than observations. The cross-section (Fig. 16.5) after Elschner (1913) corresponds closely to recent photographs of freshly exposed mine-cut surfaces, although the pinnacles themselves seem to be smoother than the idealized drawing by Elschner would suggest. The phosphate is essentially pelletal throughout. A red fine-grained pelletal ore (termed 'incoherent ore' by Owen, 1923) lies adjacent to the pinnacles; it contains few subrounded fragments or cemented pellets. This unit grades into a grey ore made up of a mixture of individual pellets and subrounded–subangular pebbles and cobbles ('coherent ore'). This unit grades into a yellow ore that likewise contains abundant cobbles (also referred to as coherent ore). The outer surface of the cobbles is slightly smooth, suggestive of an outer dull veneer. The textures of these two ores are quite similar, the only clear difference is their color. In fact, a grey ore cannot be identified in all outcrops; instead, the incoherent (pelletal) ore grades into the coherent ore, which makes up the major fraction of the Nauru deposit.

In thin section, individual pellets exhibit an oolitic structure (Fig. 16.10) that is faint in the incoherent ore but more strongly developed in some pellets of the coherent ore. Clear concentric layers are present in many pellets in the coherent ore, surrounding a nucleus of either transparent or opaque apatite. Clear and dark layers can show more than one alternating pair but in most samples the pellets contain a single well developed clear layer. In reflected light, the boundary between these layers is pitted, possibly suggesting a somewhat higher concentration of non-phosphatic material. Microprobe analyses, however, showed the composition of layers in individual pellets (Table 16.3, Fig. 16.10) to be uniform. Other pellets show no structure. The pellets in cobbles can be whole or fragmented, with varying internal textures. Some pellets have nucleii of possibly bone material. The cobbles in both ores have an isopachous cement (Fig. 16.10); its composition also is quite uniform and similar to that of the enclosed pellets.

A finely laminated vitreous phosphate has also been described. It is quantitatively of minor importance on Nauru but was apparently much more important on Ocean Island (Ellis, 1935). It was initially given the name 'nauruite', but its mineralogy is the same as that of the pelletal ores (Fig. 16.6). The broader aspect of X-ray diffraction peaks suggests that it is poorly crystalline. A conspicuous occurrence collected here was in direct contact with dolomite, on the top of pinnacles that were in an area of maximum elevation of the karrenfeld.

The major and minor elemental compositions of the ores are all quite similar (Tables 16.1, 16.2). All three ores contain approximately 60 ppm U. Their rare-earth element contents are uniformly low relative to sedimentary phosphorites (Altschuler, 1980). The La content is above 1 ppm in only one sample. The Zn content, however, is extremely high. The SO_4^{2-} content (as SO_3, Table 16.3) is generally less than approximately 0.2%, similar to the concentrations reported by Hutchinson (1950).

The most significant difference is in their CO_2 content. The incoherent ore contains twice as much CO_2 as the coherent ores. The incoherent ore also failed to show any $CaCO_3$ under petrographic or microprobe analysis.

X-ray diffraction analyses show the ores to be a carbonate fluorapatite (Fig. 16.6), although McClellan & Lehr (1969) reported that Nauru apatite has a significant hydroxy ion concentration. This composition accounts for the lower $F:P_2O_5$ ratio of 0.075 than that of marine carbonate fluorapatite of

Table 16.3. *Microprobe analyses of Nauru phosphate ores and associated dolomite samples. Oxides, F, and Cl are reported in weight %. Dashes indicate no analysis made. Each composition represents the average of several spot analyses within a well-defined area. Locations of microprobe analyses are shown in figures 16.7, 16.8 and 16.10*

	2A	2B	2C	2D	2E	2F	2G	2H	3A	3B	3C	3D	3E	3F	3G	3H
Al_2O_3	0.00	0.00	0.00	0.00	0.00	0.00	0.02	0.41	0.36	0.53	0.28	0.21	0.43	0.40	0.28	0.20
CaO	52.4	52.5	51.6	52.6	52.6	51.6	51.8	51.4	49.9	53.8	54.0	53.7	53.7	54.0	53.8	54.1
Cl	—	—	—	—	—	—	—	—	0.07	0.02	0.03	0.01	0.01	0.04	0.03	0.02
F	2.54	3.03	3.00	3.10	3.15	3.02	3.13	2.97	2.37	2.82	2.59	2.60	2.82	3.07	2.95	3.01
FeO	0.07	0.08	0.08	0.07	0.09	0.09	0.11	0.10	0.02	0.02	0.02	0.02	0.00	0.02	0.03	0.01
K_2O	0.00	0.01	0.00	0.00	0.00	0.01	0.00	0.00	0.01	0.02	0.02	0.01	0.01	0.03	0.03	0.01
MgO	0.17	0.16	0.25	0.19	0.17	0.18	0.20	0.23	0.13	0.17	0.16	0.17	0.23	0.16	0.18	0.18
Na_2O	0.15	0.11	0.06	0.10	0.12	0.07	0.08	0.11	0.12	0.17	0.17	0.12	0.10	0.32	0.30	0.18
P_2O_5	36.6	37.1	36.9	38.2	37.3	36.3	36.5	36.4	37.2	39.9	39.6	39.9	39.9	39.9	39.5	39.7
SiO_2	0.04	0.04	0.03	0.03	0.01	0.03	0.04	0.03	0.00	0.00	0.00	0.00	0.00	0.00	0.00	0.00
SO_3	—	—	—	—	—	—	—	—	0.20	0.19	0.20	0.11	0.13	0.46	0.39	0.18
SrO	0.00	0.00	0.00	0.00	0.03	0.00	0.00	0.00	0.00	0.00	0.00	0.00	0.00	0.00	0.00	0.00

	3I	4A	4B	4C	4D	4E	4F	4G	4H	4I	4J	4K	4L	4M	5A	5B
Al_2O_3	0.00	0.19	0.10	0.05	0.00	0.18	0.13	0.10	0.35	0.54	0.30	0.31	0.53	0.26	0.07	4.52
CaO	52.9	53.0	52.4	53.2	53.7	53.0	53.5	52.2	53.6	53.2	52.8	54.0	53.1	52.6	32.7	26.8
Cl	0.01	—	—	—	—	—	—	—	0.05	0.01	0.03	0.01	0.00	0.00	0.00	0.08
F	2.91	2.94	3.09	3.05	2.90	3.14	3.21	3.25	2.55	2.90	2.56	2.67	2.41	2.93	0.03	0.08
FeO	0.00	0.12	0.12	0.13	0.14	0.14	0.14	0.20	0.07	0.07	0.07	0.05	0.10	0.05	0.02	0.03
K_2O	0.01	0.00	0.01	0.01	0.01	0.02	0.01	0.01	0.01	0.02	0.01	0.01	0.02	0.02	0.00	0.00
MgO	0.29	0.08	0.10	0.07	0.05	0.09	0.06	0.17	0.13	0.16	0.22	0.11	0.15	0.22	19.3	12.5
Na_2O	0.13	0.24	0.27	0.11	0.14	0.14	0.11	0.12	0.24	0.20	0.29	0.12	0.12	0.14	0.04	0.03
P_2O_5	38.8	39.1	38.9	39.1	40.3	39.2	38.7	38.1	39.5	39.6	39.4	40.3	40.4	38.9	0.32	1.52
SiO_2	0.00	0.04	0.03	0.02	0.03	0.03	0.02	0.03	0.00	0.00	0.00	0.00	0.00	0.00	0.00	0.32
SO_3	0.13	—	—	—	—	—	—	—	—	—	—	—	—	—	0.08	3.75
SrO	0.00	0.02	0.01	0.00	0.00	0.00	0.00	0.00	0.00	0.00	0.00	0.00	0.00	0.00	0.00	0.00

	5C	5D	5E	5F	5G	5H	8A	8B	8C	8D	8E	8F	17A	17B	17C
Al_2O_3	0.14	0.07	3.91	0.06	3.42	0.61	0.00	0.00	0.00	0.05	0.05	0.05	0.00	0.00	0.00
CaO	32.5	30.7	31.6	30.7	19.8	22.8	30.7	29.3	30.6	31.4	32.7	32.2	49.1	49.3	49.2
Cl	0.01	0.00	0.01	0.01	0.11	0.11	—	—	—	0.00	0.02	0.00	—	—	—
F	0.06	0.00	0.12	0.05	0.08	0.20	0.06	0.02	0.07	0.03	0.03	0.03	1.33	1.38	0.80
FeO	0.00	0.00	0.11	0.00	0.01	0.00	0.00	0.00	0.00	0.00	0.00	0.00	0.04	0.03	0.03
K_2O	0.00	0.00	0.00	0.00	0.00	0.00	0.00	0.00	0.00	0.00	0.00	0.00	0.00	0.00	0.01
MgO	19.5	21.0	14.3	21.1	9.09	9.05	18.8	20.0	18.3	19.9	19.2	19.8	0.49	0.41	0.47
Na_2O	0.04	0.02	0.04	0.02	0.03	0.04	0.04	0.00	0.05	0.04	0.04	0.02	0.38	0.44	0.29
P_2O_5	0.31	0.07	1.65	0.05	1.88	3.74	0.04	0.00	0.03	0.11	0.07	0.05	34.2	34.3	34.7
SiO_2	0.00	0.00	1.61	0.00	0.22	0.17	0.01	0.01	0.02	0.00	0.00	0.00	0.02	0.03	0.02
SO_3	0.12	0.03	0.35	0.03	8.45	7.80	—	—	—	0.06	0.10	0.06	—	—	—
SrO	0.00	0.00	0.00	0.00	0.00	0.00	0.00	0.00	0.00	0.00	0.00	0.00	0.01	0.00	0.00

about 0.10 (Table 16.1). A shift in the spacing between the (004) and (410) X-ray peaks suggests that the incoherent ore has about 2% higher CO_3^{2-} content than the coherent ore (Gulbrandsen, 1970), in agreement with the bulk chemical analyses (Table 16.1). The occurrence of OH^- in the Nauru samples, however, can make direct comparison of the X-ray diffraction traces with those of samples of the carbonate fluorapatites, as analyzed by Gulbrandsen, somewhat uncertain. The broader X-ray peak widths of the incoherent ore than those of the coherent ore suggest possible greater crystallinity of the coherent ore. The nauruite has even broader peaks.

The $\delta^{13}C$ values of the carbonate within the apatite of all three ore types vary only slightly (Table 16.4); they are intermediate between that of marine organic matter of about −20‰ (Aharon & Veeh, 1984; Kroopnick, 1985) and the aragonite–dolomite values of +2.0‰ (Fig. 16.9). Their $\delta^{18}O$:$\delta^{13}C$ ratios plot within the field of freshwater-altered carbonates of western Pacific islands or are slightly enriched in $\delta^{18}O$ (Aharon, Piper & Socki, 1985); i.e. they have light $\delta^{18}O$ values relative to both aragonite and dolomite.

The age of the apatite has not yet been determined. Radiometric measurements establish a minimum age of 0.3 my (Roe & Burnett, 1985), and the foraminifers give a maximum age of approximately 11 my.

Discussion

Studies of Nauru and neighbouring Banaba Island have suggested several episodes of emergence and submergence

a

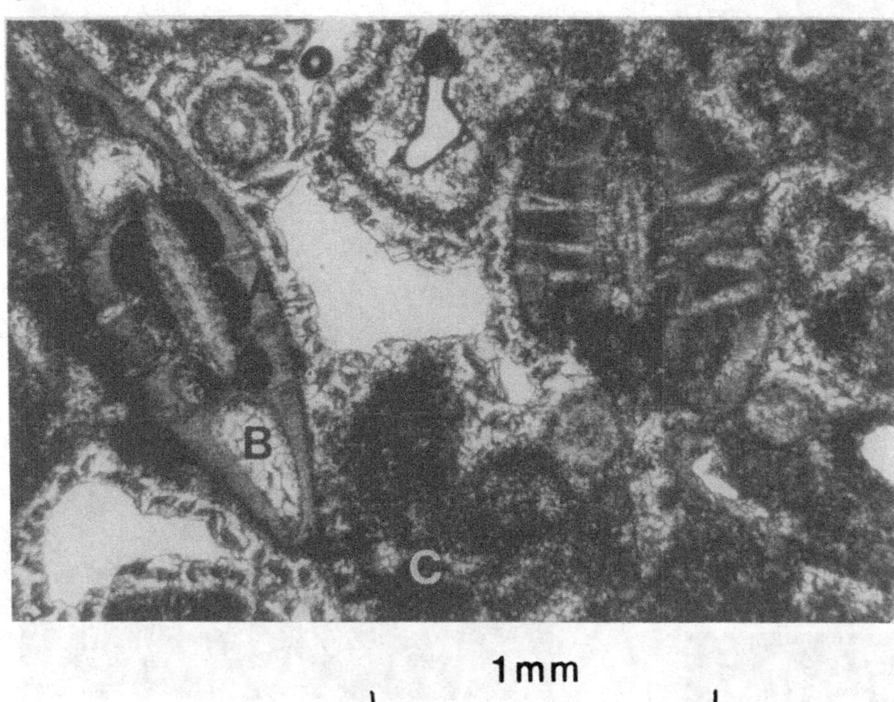

b

Fig. 16.7. (a) Thin section of dolomite from a pinnacle in the island's interior (NRU-8). Labelled spots correspond to analyses 8A–F in Table 16.3. (b) Surface of dolomite pinnacles showing coral, in apparent growth position. Hand lens in center of figure gives scale.

a

c

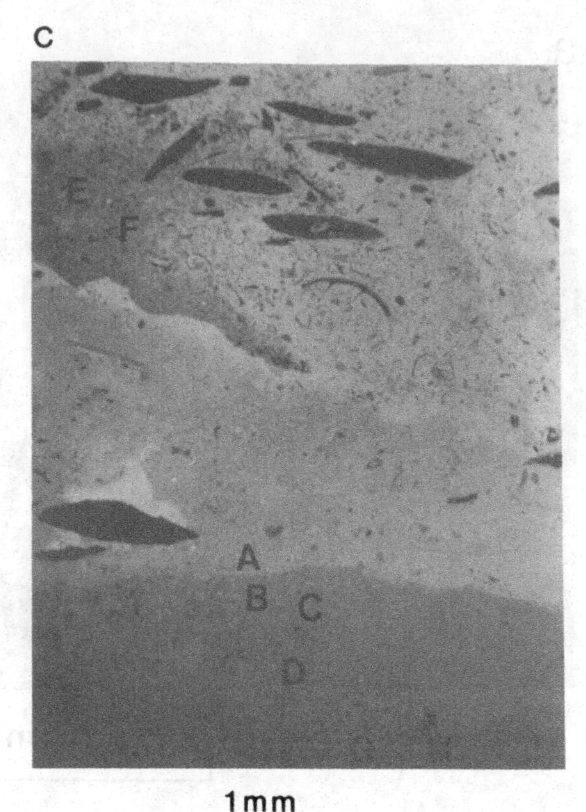

1 mm

b

1 cm

Fig. 16.8. Lagoon sediment (NRU-5). (a) outcrop; location given on Figure 16.1. The thin beds are seen on the fresh surface. (b) Hand specimen; dark layers are reddish with slightly higher Fe contents. (c) Microscope-view with white unit at bottom, and fossiliferous red unit at the top. Labelled spots correspond to analyses 5A–H in Table 16.3.

(Power, 1925), although Owen (1923) and Ellis (1935) argued against submergence after deposition of the guano. In agreement with their interpretation, the history of Nauru can be explained by invoking a single, possibly sporadic episode of emergence following its development as an atoll during a period of submergence. Charles Darwin and many other scientists since his time have outlined the development of an atoll through its various stages, beginning with construction of a volcanic island, followed by submergence forced by isostatic adjustment. For Nauru, the period of submergence has been interrupted by one of emergence, a history recorded by several other uplifted atolls in the Pacific. The uplift of such islands may be in response to flexure of the lithosphere caused by volcanic activity on a neighbouring island (McNutt & Menard, 1978: McNutt, 1984). Nauru and Banaba, however, have no such neighbours. Menard (1973) suggested that their emergence may be related to movement of the lithosphere over a 'bumpy' asthenosphere. If so, a single period of emergence might be expected, given the maximum age of the deposit, the proposed wavelength of such 'bumps' on the asthenosphere, and the rate of lithospheric movement. The remoteness of Nauru (and Banaba) from volcanic islands and tectonically-active areas, however, reduces explanations of this aspect of its origin to considerable speculation. Regardless of the mechanism of emergence, there is no evidence in the present relief to suggest that Nauru has undergone more than a single episode of emergence, although it may be undergoing submergence currently. A study of the slope beyond the outer edge of the reef might elucidate this aspect of the island's history.

The three aspects of the Nauru phosphate deposit addressed in this report are the episodes of dolomitization, phosphatization, and sculpturing of the karrenfeld, all of which likely occurred during the period of emergence. We consider first the process of dolomitization because it may be independent of the other two and may have preceded them. One possible interpretation of the lagoon dolomite deposit (Fig. 16.8) is that it represents alternating periods of normal salinity (the red fossiliferous layers), separated by periods of gradual increase toward hypersalinity and anoxia (the white non-fossiliferous layers), which were abruptly terminated by a return to normal

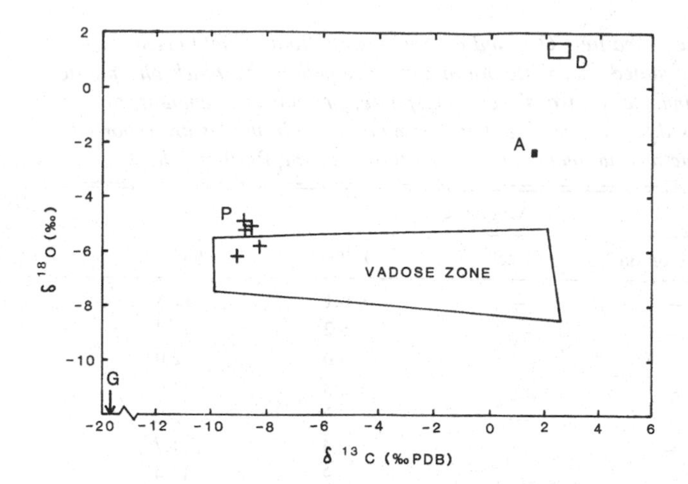

Fig. 16.9. Oxygen and carbon isotope relationship in the phosphate and carbonate components of the sedimentary sequence on Nauru. The isotopic field of recrystallized limestone in the Western Pacific (labelled vadose zone) was defined by Aharon & Veeh (1984). The arrow labelled G approximates $\delta^{13}C$ of marine organic matter. Field A represents aragonite, D dolomite, and P (crosses) apatite.

salinity. During the very early stage of emergence, the lagoon would have become isolated from the open ocean at a time when the area of the lagoon greatly exceeded that of the surrounding island. Under extreme aridity, as suggested by Hutchinson (1950), evaporation may have been sufficient to maintain a high salt content in the lagoon (Schlanger, 1965), in excess of that due to groundwater recharge from the ocean. This process could have been interrupted by storm overwash and a return to normal salinities. The possible duration of hypersaline conditions cannot be ascertained. Clipperton Island may be approaching, or already within, this stage, but a sufficient rainfall prevents its lagoon from becoming hypersaline. As emergence continued, the land area would eventually greatly exceed that of the lagoon. This condition would have been achieved by infilling of the lagoon, as well as by uplift. There are many examples of atolls that are elevated only a few meters and have had their lagoons almost totally filled; these atolls include Baker, Malden, and Christmas Island (Pacific Ocean), to name but three. The remnant of the lagoon would become relatively fresh, as it is today (Hutchinson, 1950); that is, the lagoon had normal salinities initially, became hypersaline during an early stage of uplift, and eventually became brackish–fresh during a late period of uplift. It remains brackish today.

Dolomitization of the coral could have occurred during the hypersaline stage, by reflux of the lagoon water through the carbonate rock (Adams & Rhodes, 1960; Schlanger, 1965; Schlanger & Tracey, 1970). Reflux occurs even today, as suggested by a few millimeters of variation in the lagoon surface, lagging at a constant interval behind the tide on the open reef (Hutchinson, 1950).

The absence of gypsum within the dolomite possibly argues against increasing the Mg:Ca ratio of hypersaline water through precipitation of gypsum (Berner, 1971). One possible explanation for the demise of organisms in the lagoon, as suggested by the non-fossiliferous white dolomitic layers of this deposit, is that the lagoon became anoxic at the same time it became hypersaline. The SO_4^{2-} content could have been lowered by bacterial sulfate reduction, possibly precluding the extensive precipitation of gypsum during dolomitization, i.e. other than in the lagoon. Alternatively, any gypsum that might have precipitated within the dolomite could have been removed by the flux of acidic water when these rocks passed through the freshwater phreatic zone during emergence, as discussed below.

Although we have no evidence for sulfate reduction, it certainly would not be unexpected in a hypersaline (thus, strongly stratified) lagoon (Aharon, Kolodny & Sass, 1977). The influx of nutrients leached from the guano on the surrounding island would have supported a high productivity of organic matter in the lagoon. Degradation of this organic matter on the lagoon floor would then have produced an anoxic condition in the water column. Sulfate reduction occurs in the sediment of the enclosed lagoon on the atoll of Clipperton Island, even though the flux of nutrients from the island should be insignificant by comparison with the conditions that must have existed on Nauru. Rainfall on Clipperton (Niaussat, 1978) is apparently sufficient to prevent hypersaline conditions in the lagoon, and recharge through the reef prevents a major build-up of H_2S.

An alternative mechanism for dolomitization is by the dorag process (Badiozamani, 1973; Rogers, Easton & Downes, 1982). Dolomite stability is enhanced over that of calcite by a mixing of marine- and freshwater in an approximate ratio of between 5:95 and 30:70 (Badiozamani, 1973). Both this and the hypersaline reflux mechanism have been invoked to explain the features of dolomitic rocks (Sears & Lucia, 1980).

A third alternative is dolomitization under conditions of low SO_4^{2-} content but otherwise normal salinities (Baker & Kastner, 1981). Again, this low SO_4^{2-} content could have been achieved by sulfate reduction.

The uniform ^{18}O enrichments in the dolomites of Nauru seem incompatible with a mixed zone dolomitization process (i.e. dorag) as recently demonstrated by Aharon *et al.* (1985). We favor, therefore, dolomitization under conditions of normal marine salinities or by reflux of hypersaline water. In favor of reflux, a hypersaline process represents the more dynamic system. Loss of water from the lagoon would have been by both evaporation and refluxion of the high density hypersaline water. Recharge from the ocean must have maintained the water balance in the lagoon. Recharge into a lagoon of normal salinity could be considerably less, and a freshwater lagoon would be relatively stagnant. This is particularly true at the base of the freshwater lens, advocated as the site of dolomitization in the dorag process (Rogers *et al.*, 1982).

Uplift had the additional consequence of exposing more of the atoll to bird colonization. The small amount of rain that may have fallen on Nauru during its occupancy by this colony (Hutchinson, 1950), nonetheless, was responsible for dissolution and oxidation of the guano. Because NO_3^-, the transition metals, and SO_4^{2-} all have much too low concentrations in freshwater to have contributed substantially to oxidation of the guano, oxygen must have been the major oxidizing agent. An idealized stoichiometry for this reaction is (Froelich *et al.*, 1979):

$$C_{106}H_{263}O_{110}N_{16}P + 106O_2 = 106H_2CO_3 + 16NH_3 + H_3PO_4 \quad (16.1)$$

Table 16.4. *Oxygen- and carbon-isotopic compositions of phosphates (apatite-CO_2) and carbonates on Nauru. Field occurrences of samples are described by the following abbreviations: sp, unconsolidated phosphate ore of sand-size pellets; bs, black phosphatic soil; rc, cast of reef coral; bl, banded phosphate layers on top of pinnacle; tp, translucent phosphate; pr, pinnacle remnant; br, modern beachrock; fl, nfl – alternating fossiliferous red and non-fossiliferous white laminae. Numbers of subsamples are reported in parentheses. ± values give the full range of the analyses. Isotopic measurements are in ‰, relative to the PDB standard*

	Apatite		Dolomite		Aragonite		
Sample	Phosphate ore	Nauruite	Atoll reef	Lagoon	Reef	$\delta^{18}O$	$\delta^{13}C$
NRU-1	bs	—	—	—	—	−6.1	−11.3
NRU-2	sp	—	—	—	—	−6.2	−11.8
NRU-3	sp	—	—	—	—	−6.0	−12.0
NRU-4	sp	—	—	—	—	−5.8	−11.0
NRU-8	—	—	pr	—	—	1.3	3.2
NRU-10	—	—	—	—	br	−2.4	1.7
NRU-18	sp	—	—	—	—	−6.9	−12.4
NRU-19	sp	—	—	—	—	−6.4	−11.3
NRU-20	sp	—	—	—	—	−6.5	−13.5
NRU-8A	—	tp	—	—	—	−6.3	−10.9
NRU-3/1	rc	—	—	—	—	−6.0	−8.5
NRU-5	sp	—	—	—	—	−6.0	−11.4
NRU-9A	bl	—	—	—	—	−5.3	−8.5
NRU-7A	—	—	pr(n=4)	—	—	1.2±0.2	1.5±0.6
NRU-5A	—	—	—	fl(n=11)	—	1.3±0.4	2.3±0.2
NRU-5A	—	—	—	nfl(n=9)	—	1.3±0.2	2.2±0.2

The enormous amount of acid generated by the dissociation of H_2CO_3 in this reaction relative to PO_4^{3-} should have substantially increased the solubility of apatite (Kazakov, 1950; Manheim & Gulbrandsen, 1979; Jahnke, 1984). On Nauru, this acid could be neutralized by reaction with the carbonate rocks (i.e. dolomite and/or $CaCO_3$) by the familiar reaction

$$H_2CO_3 + CaCO_3 = 2HCO_3^- + Ca^{++}. \quad (16.2)$$

Combining reactions (16.1) and (16.2) and allowing phosphate to precipitate as $Ca_3(PO_4)_2$ rather than to react with the acid, we have:

$$106O_2 + C_{106}H_{263}O_{110}N_{16}P + 93Ca(CO_3) = 199HCO_3^- + 91.5Ca^{2+} + 16NH_4^+ + 0.5Ca_3(PO_4)_2 \quad (16.3)$$

Equilibrium constants define the exact partitioning of carbon between the various species H_2CO_3, HCO_3^-, and CO_3^{2-} and the apatite precipitates as $Ca_{10}(PO_4,CO_3)_6(F,OH)_2$ rather than as $Ca_3(PO_4)_2$. The main points of the reaction, however, are: 1) the stoichiometry of this reaction controls the acidity of the water, i.e. carbonate rock acts as a buffer that dampens the effect of the enormous C:P ratio of the organic matter of 106:1; and 2) the carbonate rocks contribute calcium and carbon to the reaction, the latter in approximately a 1:1 ratio with that contributed by the organic matter. This ratio is reflected in the $\delta^{13}C$ value of the apatite of −11‰, almost exactly intermediate between those of the two carbon reservoirs (Table 16.4). That is, the CO_3^{2-} content of the solution from which apatite precipitated reflected equal contributions of CO_2 from carbonate rocks and bird guano.

The maximum amount of carbonate that should have dissolved can be ascertained from the amount of apatite now present and from the stoichiometry of equation (16.3). The phosphate is approximately 4 m thick in the central part of Nauru. Assuming a specific gravity of 2, it required about 80 m of guano of equal specific gravity, assuming that the pinnacles occupy no volume. The amount of aragonite (or dolomite) dissolved would be about 200 m. An ideal stoichiometry for marine plankton (Redfield, 1958; Richards, 1965) has been assumed, but it may not be appropriate. Any CO_2 lost at the surface, by oxidation in air, would have decreased the demand for carbonate dissolution.

A more realistic dissolving potential of the guano may be estimated by considering the composition of 'mature' guano from the Chincha Islands (Hutchinson, 1950), which had a P_2O_5 content of about 12%, a ten-fold increase in its concentration over plankton that required a comparable loss of CO_2. Much of this CO_2 probably was lost to the atmosphere. If a comparable loss of CO_2 is assumed to have been lost on Nauru to the atmosphere, the 'mature' guano still would have had the potential to dissolve 20 m of carbonate rock, giving a ratio of phosphate to dissolved rock of 1:5. Even this reduced amount of dissolution can easily account for the sculpturing of the karrenfeld.

The F and U contents allow for another estimate of the amount of carbonate rock that might have been dissolved. These two elements are present in negligible amounts in the organic matter; their contents in the apatite are approximately 3% and 60 ppm, respectively (Tables 16.1 and 16.2). The dolomite has a U content of 2.5 ppm. Although the F content of the dolomite is unknown, that of coral is about 0.1% (Carpenter, 1969). If these two carbonate rocks acted as the main buffer for the precipitation of apatite and the source of F and U in the apatite, then the amount of carbonate dissolved was approximately 25 (using U)

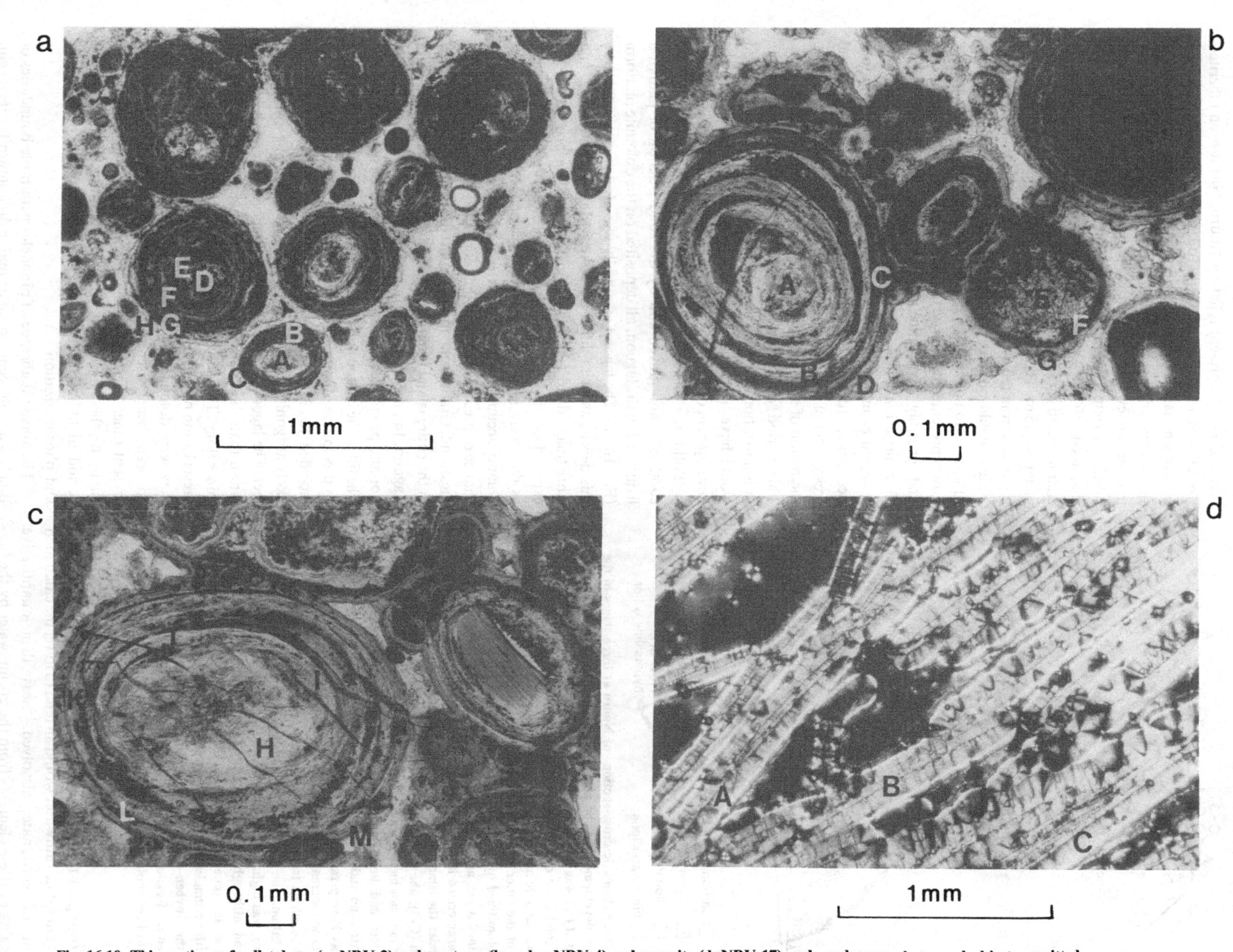

Fig. 16.10. Thin sections of pelletal ore (a, NRU-2), coherent ore (b and c, NRU-4) and nauruite (d, NRU-17). a, b, and c are photographed in transmitted light; d is in crossed nichols. Labelled spots correspond to analyses 2A–H, 4A–M, and 17A–C, respectively, in Table 16.3.

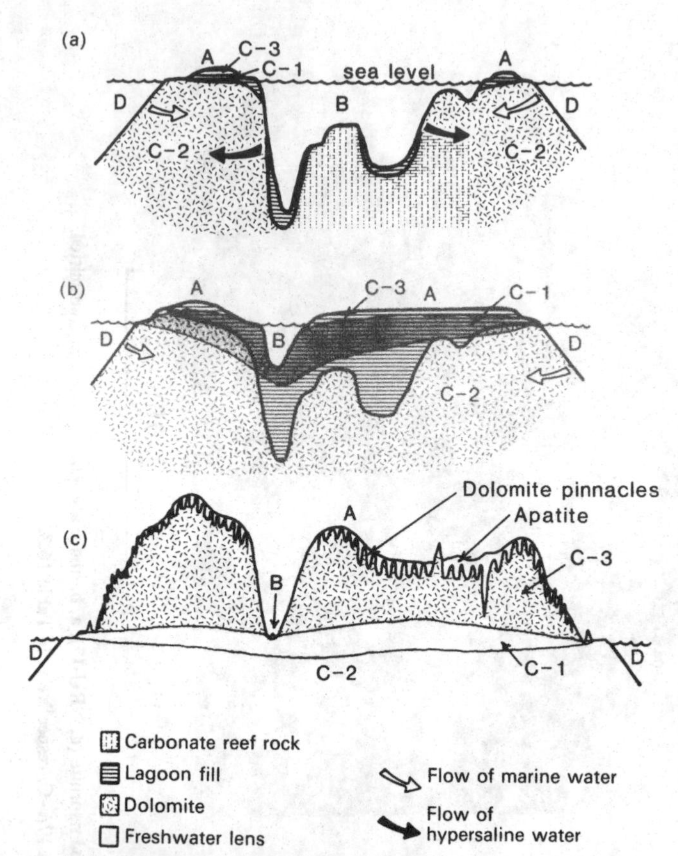

Fig. 16.11. Schematic cross-sections of Nauru at three stages in the historical development of the phosphate deposit and karrenfeld. (a). The island (A) was low-lying and totally encircled the lagoon (B). Within the lagoon evaporation greatly exceeded precipitation. The water was anoxic at times and possibly hypersaline. Aragonite and calcite of the outer reef and of coral ridges and fine muds of the lagoon were dolomitized by reflux of the lagoon water. Seawater recharge maintained a salt and water balance. The phreatic, or saturated, zone was characterized by a thin freshwater lens, the Ghyben–Herzberg lens, under the island (stippled area labelled C-1) and a saline zone beneath (C-2); above was the vadoze zone (C-3). Ocean (D) upwelling may or may not have been high. (b). The atoll was uplifted, at most a few meters, and the lagoon was filled (horizontal lines) mostly with carbonate sand and gravel derived from the reef and island. A small lake may have persisted, and it contained freshwater. It is possible that precipitation increased significantly between the first stage and this time although no information is available to substantiate this interpretation. The Ghyben–Herzberg lens was extensive. Oceanic upwelling was high. Guano (not shown) accumulated on the surface of the island, although a large amount of guano might not have been present at any time. Oxidation and dissolution might have merely kept pace with accumulation. Pelletal formation of apatite was greatest within the Ghyben–Herzberg lens. Initially, apatite precipitation was triggered by dissolution of lagoon fill, but continued as the karrenfeld was etched into dolomite. (c). Emergence continued to the present level.

to 30 (using F) times the amount of apatite formed. Low but significant saline rainwater would all tend to lower the calculated amount of carbonate dissolved. Less than quantitative scavenging of these elements from the groundwater by the precipitating apatite would increase the required amount of carbonate dissolved; the values, however, fall between the maximum and minimum values calculated above. Thus, the carbonates could have provided these elements to the apatite.

The amount of carbonate dissolution raises the question of the origin of the present relief on Nauru. Has it been inherited from the shape of the ancestral atoll and the ridges and basins merely emphasized by dissolution, or has it been etched from a surface quite different from what we see today? The actual case may lie somewhere between these two extremes. The present relief is strikingly similar to that on Clipperton Atoll today (Niaussat, 1978). This lagoon, surrounded by the island, is characterized by several basins that range in depth from a few meters to 45 m. The deepest is sometimes anoxic. Each basin is separated from the others by prominent ridges. Submergence of Nauru would produce a quite similar lagoon (Fig. 16.2). Our interpretation of Nauru requires that, at the stage of its history now represented by Clipperton Atoll, the lagoon became anoxic and possibly hypersaline, initiating the major episode of dolomitization (Fig. 16.11). Clipperton Atoll is not hypersaline. It would be interesting to determine whether dolomitization is occurring, either in the zone of mixing of marine water and fresh water of the Ghyben–Herzberg lens under the island, or in the region of sulfate reduction in the lower part of the lagoon. This period on Nauru might have provided for complete dolomitization of the surface and near-surface rocks. The island would have been under periodic wave attack, precluding the development of a large aviary and preservation of major guano.

Following this stage was an episode of lagoon infilling by coral sand and gravel, and by the solid fraction of plankton living within the lagoon, all deposited on the dolomite platform (Fig. 16.11). The lagoon eventually was almost totally filled by this debris, as exemplified by Baker, Malden, Enderbury, and Christmas Island (Pacific Ocean). As infilling proceeded, the exposed surface was soon inhabited by birds, the inner areas first as they were removed from the wave action of major storms. Guano accumulating within the interior had a much better chance of preservation than that deposited along the margins of the island. Gradual and modest uplift during this period resulted in the entire island eventually being inhabited except for Buada Lagoon. Intense colonization most probably began in the interior and progressed outward as uplift exposed more of the reef. The fill of the lagoon and reef material reacted with groundwater to precipitate apatite. The carbonates that were dissolved thus included the lagoon deposit, which was unconsolidated and likely aragonitic, and several meters of underlying dolomitized reef. The lagoon sediment was surely more susceptible to dissolution than the dolomitized material, owing to its texture and mineralogy. This susceptibility would account for the apparent preservation of the basins and ridges of the ancestral lagoon, as they were dolomitized before infilling of the lagoon, and the apparent near absence of lagoon sediment preserved to the present time. The rather phenomenal rate of accumulation of guano on the Chincha Islands demonstrates that a very short period might have been required for accumulation of the guano and phosphatization.

The apparent absence of pinnacles within the boundaries of Buada Lagoon seems to favor major dissolution for its origin. Although it was considered by early investigators to be the ancestral lagoon, Agassiz (1903) considered that it formed by chemical denudation. Both explanations may be correct. As infilling proceeded, the lagoon became progressively smaller and more shallow, approaching its present size. Surface- and groundwater runoff into the lagoon would have made it fresh, as suggested above. As uplift progressed, the underlying carbonates dissolved away at a rate equal to that of uplift. The present basin was simply cored at a rate approximately equal to that of uplift.

The $\delta^{18}O$ value of the apatite suggests that reaction (16.3) proceeded in the freshwater zone above the zone of saltwater intrusion, either in the vadose zone or the freshwater-saturated zone of the Ghyben–Herzberg lens. Infilling and uplift that exposed the island to colonization by birds would naturally have carried this reaction above the zone of saltwater intrusion. The first ore formed could have been similar to that now in contact with the dolomite, i.e. the incoherent ore. The 'red, deep guano' described on Baker Island (Hutchinson, 1950) closely resembles this incoherent ore on Nauru. Continued uplift and oxidation of guano, at and near the surface, would have carried the boundary between carbonate dissolution and apatite precipitation to progressively greater depth. The youngest, or most recently formed ore is thus that presently in contact with the dolomite; it defines the most recent, or final, reaction boundary between the carbonate and the acidic, phosphate-charged freshwater at the time the island was depleted of guano.

The initially formed apatite, the ore eventually removed to some height above the downward-migrating reaction zone, would have been flushed by the downward-fluxing acidic water. Under this condition, and out of the zone of carbonate sediment, it would have been transformed to a less soluble apatite, as suggested by the lower CO_2 content (Jahnke, 1984) of the overlying coherent ore.

The extremely low rare-earth element contents of Nauru apatite and the strong propensity of apatite to adsorb seawater rare-earth elements (Arrhenius & Bonatti, 1965) further suggest that this deposit has not been exposed to seawater. This interpretation of a freshwater origin is also supported by the occurrence of OH^- in crystal sites (McClellan & Lehr, 1969), and the near absence of SO_4^{2-}.

Our interpretation of the phosphatic oolitic pellets is that they formed *in situ*, possibly in the saturated zone rather than in the vadose zone of the freshwater environment. The isopachous cement of the conglomeratic pebbles that have been examined, which has been interpreted as a feature of the phreatic zone (Longman, 1980), is our only evidence. In contrast, the texture of nauruite and its occurrence on the very top of the karrenfeld would suggest deposition in the water-unsaturated environment of the vadose zone.

Conclusions

The history of Nauru is surely more complex than we have outlined here. Dolomitization occurred during an early stage of uplift in an anoxic normal marine, or hypersaline, environment; and phosphatization occurred at a later stage under freshwater conditions. Dolomitization, however, might have occurred by more than one mechanism, and changes in sea level during the stage of phosphatization may have pushed the zone of apatite precipitation into the vadose zone at the one extreme, or into the saltwater phreatic zone at the other extreme.

Many of the physical and chemical properties of the Nauru phosphate deposit differ considerably from those of the major sedimentary phosphate deposits of the world. The similarities, however, are striking: their association with carbonates, their pelletal texture, and their carbon isotopic compositions relative to that of the associated carbonate rocks and organic matter. The carbonate association is evident from a comparison of the Nauru deposit with the Phosphoria Formation of Idaho (Sheldon, 1963), the Bone Valley Formation of Florida (Riggs, 1984), the Karatau Formation of Kazakhstan (Eganov *et al.*, 1984), and the Monterey Formation of California (Isaacs, 1981). Although few $\delta^{13}C$ measurements are available for these ancient deposits, the apatite of the Phosphoria has a $\delta^{13}C$ value of −3.5‰ – −9.1‰ (Murata, Friedman & Gulbrandsen, 1972), intermediate between that of average Permian organic matter and average carbonate rocks. Although this relation is similar to that for the Nauru deposit (Fig. 16.9), the broad range of $\delta^{13}C$ values for the Phosphoria Formation suggests a far more complex chemistry. The similarities, however, indicate that the chemical reactions which formed the deposit on Nauru may be quite similar to those that formed the major sedimentary phosphate deposits. The major differences are: the source of the phosphate on Nauru was bird guano; and, the solution from which it precipitated was freshwater.

Several questions that remain to be addressed are crucial to a full understanding of Nauru. Does the apparent absence of gypsum within the lagoon sediment require dolomitization under anoxic conditions? What is the nature of the contact between the coral and dolomite along the reef? Do the cliffs around Buada Lagoon show horizons of major dissolution, suggestive of a permanent lake and of still stands of sea level? What mechanisms of dolomitization and island tectonics can explain the occurrence of pinnacles across the present day reef, to its very edge, i.e. is Nauru now submerging? Are significant amounts of phosphate present under the rubble on the wave-deposited terrace and within Buada Lagoon? Many of these problems can be investigated by an examination of the carbonate rocks, which will remain long after the phosphate has been exploited.

Acknowledgments

The assistance of Klaus Jacobs and Brad Elliot on Nauru was invaluable. Conversations with several members of the mining company also proved extremely helpful. These and many other people of Nauru contributed immeasurably to this study. Several palaeontologists identified fossils in hand specimens and thin sections. S. Schlanger, R. Gulbrandsen, and W. Burnett made many helpful suggestions. Despite their tremendous assistance, the manuscript may have wandered from the simplest and most plausible explanation, and that is our responsibility.

References

Adams, J.E. & Rhodes, M.L. (1960). Dolomitization by seepage refluxion. *American Association of Petroleum Geologists Bulletin*, **44**, 1912–21.

Agassiz, A. (1903). Reports on the scientific results of the expedition to the tropical Pacific by the US Fish Commission steamer ALBATROSS, August 1899 to March 1900, IV, the coral reefs to the tropical Pacific. *Museum of Comparative Zoology Memoirs*, **28**, 410pp.

Aharon, P., Kolodny, Y. & Sass, E. (1977). Recent hot brine

dolomitization in the Solar Lake, Gulf of Elat: isotopic, chemical and mineralogical study. *Journal of Geology*, **85**, 27–48.

Aharon, P., Piper, D. & Socki, R.A. (1985). An appraisal of dolomitization models in the context of uplifted atolls. *Society of Economic Palaeontologists and Mineralogists, Annual Meeting*, **2**, 3.

Aharon, P. & Veeh, H.H. (1984). Isotope studies of insular phosphates explain atoll phosphatization. *Nature*, **309**, 614–17.

Altschuler, Z.S. (1980). The geochemistry of trace elements in marine phosphorites. Part I. Characteristic abundances and enrichment. In *Marine Phosphorites*, ed. Y.K. Bentor, pp. 19–30. Society of Economic Palaeontologists and Mineralogists, Special Publication No. 29.

Arrhenius, G. & Bonatti. E. (1965). Neptunism and volcanism in the oceans. *Progress in Oceanography*, **3**, 7–22.

Arthur, M.A. & Jenkyns, H.C. (1980). Phosphorites and paleoceanography. *Oceanologica Acta*, **4 (Supplement)**, 83–96.

Badiozamani, K. (1973). The dorag dolomitization model–application to the Middle Ordovician of Wisconsin. *Journal of Sedimentary Petrology*, **43**, 965–84.

Baedecker, P.A. & McKown, D.M. (1985). *Instrumental Neutron Activation Analyses of Geologic Samples* Unpublished manuscript, 26pp.

Baker, P.A. & Kastner, M. (1981). Constraints on the formation of sedimentary dolomite. *Science*, **213**, 214–16.

Berner, R.A. (1971). *Principles of Chemical Sedimentology*. McGraw-Hill, New York.

Carpenter, R. (1969). Factors controlling the marine geochemistry of fluorine. *Geochimica Cosmochimica Acta*, **33**, 1153–67.

Craig, H. (1957). Isotopic standards for carbon and oxygen and correction factors for mass spectrometric analysis of carbon dioxide. *Geochimica Cosmochimica Acta*, **12**, 133–49.

Eganov, E.A., Ergaliev, G.H., Ilyin, A.V. & Krasnov, A.A. (1984). Karatau phosphate Basin. *27th International Geological Congress, Guidebook*, **45**, 72pp.

Ellis, A.F. (1935). *Ocean Island and Nauru: Their Story*. Angus & Robertson, Sydney.

Elschner, C. (1913). *Corallogene Phosphat-Inseln Austral-Oceaniens und ihre Produkte*. Max Schmidt, Lubeck.

Friedman, I. & O'Neil, J.R. (1977). Compilation of stable isotope fractionation factors of geochemical interest US *Geological Survey Professional Paper*, **440–KK**.

Froelich, P.N., Klinkhammer, G.P., Bender, M.L., Luedke, N.A., Heath, G.R., Cullen, D. Dauphin, P., Hammond, D., Hartman, B. & Maynard, V. (1979). Early oxidation of organic matter in pelagic sediments of the eastern equatorial Atlantic: suboxic diagenesis. *Geochimica Cosmochimica Acta*, **43**, 1020–75.

Gulbrandsen, R.A. (1966). Composition of phosphorites of the Phosphoria Formation. *Geochimica Cosmochimica Acta*, **30**, 769–78.

Gulbrandsen, R.A. (1970). Relation of CO_2 content of apatite of the Phosphoria Formation to regional facies. *US Geological Survey Professional Paper*, **700-B**, 9–13.

Hutchinson, G.E. (1950). The Biogeochemistry of Vertebrate Excretion. *Bulletin of the American Museum of Natural History Bulletin*, **96**, 570pp.

Isaacs, C.M. (1981). Porosity reduction during diagenesis of the Monterey Formation, Santa Barbara coastal area, California. In *The Monterey Formation and Related Siliceous Rocks of California*, ed. R.E. Garrison, *et al*., pp. 257–71. Society of Economic Palaeontologists and Mineralogists, Pacific Section, Los Angeles.

Jahnke, R.A. (1984). The synthesis and solubility of carbonate fluorapatite. *American Journal of Science*, **284**, 58–78.

Kazakov, A.V. (1950). Geotectonics and the formation of phosphorites (English translation). Izves. Acad. Nauk SSSR, **5**, 42–68.

Kolodny, Y. & Kaplan, I.E. (1970). Carbon and oxygen isotopes in apatite CO_2 and co-existing calcite from sedimentary phosphorite. *Journal of Sedimentary Petrology*, **40**, 954–9.

Kroopnick, P. (1985). The distribution of ^{13}C of CO_2 in the world oceans. *Deep-Sea Research*, **32**, 57–84.

Longman, M.W. (1980). Carbonate diagenetic textures from nearsurface diagenetic environments. *American Association of Petroleum Geologists*, **64**, 461–87.

Manheim, F.T. & Gulbrandsen, R.A. (1979). Marine phosphorites. In *Marine Minerals*, ed. R.G. Burns, pp. 151–73. *Mineralogical Society of America*, **6**.

McArthur, J.M. (1985). Francolite geochemistry–compositional controls during formation, diagenesis, metamorphism, and weathering. *Geochimica Cosmochimica Acta*, **49**, 23–35.

McClellan, G.H. & Lehr, J.R. (1969). Crystal chemical investigation of natural apatites. *American Mineralogist*, **54**, 1379–91.

McCrea, J.M. (1950). On the isotopic chemistry of carbonates and a paleotemperature scale. *Journal of Chemistry and Physics*, **18**, 849–57.

McNutt, M.K. (1984). Lithospheric flexure and thermal anomalies. *Journal of Geophysical Research*, **89**, 1180–94.

McNutt, M.K. & Menard, H.W. (1978). Lithospheric flexure and uplifted atolls. *Journal of Geophysical Research*, **83**, 1206–12.

Menard, H.W. (1973). Depth anomalies and the bobbing motion of drifting islands. *Journal of Geophysical Res*, **78**, 5128–37.

Montaggioni, L.F., Richard, G., Bourrouilh-Le Jan, F., Gabrie, C., Humbert, L., Monteforte, M., Naim, O., Payri, C. & Salvat, B. (1985). Geology and marine biology of Makatea, an uplifted atoll, Tuamotu Archipelago, central Pacific Ocean. *Journal of Coastal Research*, **1(2)**, 165–71.

Murata, K., Friedman, J.I. & Gulbrandsen, R.A. (1972). Geochemistry of carbonate rocks in Phosphoria and related formations of the western phosphate field. US Geological Survey Professional Paper, **800–D**, 103–10.

Niaussat, P.M. (1978). *Lelagon et l'Atoll de Clipperton*. Academie des Sciences d'Outre-Mer, Paris.

Owen, L. (1923). Notes on the phosphate deposit of Ocean Island; with remarks on the phosphates of the equatorial belt of the Pacific Ocean. *Quarterly Journal of the Geological Society of London*, **79**, 1–15.

Power, D.F. (1905). Phosphate deposit on Ocean and Pleasant Islands. *Transactions of the Australian Institute of Mining Engineering*, **10**, 213–32.

Power, D.F. (1925). Phosphate deposits of the Pacific. *Economic Geology*, **20**, 266–81.

Redfield, A.C. (1958). The biological control of chemical factors in the environment. *American Scientist*, **46**, 206–26.

Richards, F.A. (1965). Anoxic basins and fjords. In *Chemical Oceanography*, 1, ed. J.P. Riley & G. Skurrow, pp. 611–45. Academic Press, New York.

Riggs, S.R. (1984). Paleoceanographic model of Neogene phosphorite deposition. *Science*, **223**, 123–31.

Roe, K.K. & Burnett, W.C. (1985). Uranium geochemistry and dating of Pacific island apatite. *Geochimica Cosmochimica Acta*, **49**, 1581–92.

Rogers, K.A., Easton, A.J. & Downes, C.J. (1982). The chemistry of carbonate rocks on Niue Island, South Pacific. *Journal of Geology*, **90**, 645–62.

Schlanger, S.O. (1965). Dolomite–evaporite relations on Pacific islands. *Scientific Report of Tohoker University*, **37(2)**, 15–29.

Schlanger, S.O. & Tracey, J.I., Jr. (1970). Dolomitization related to recent emergence of Jarvis Island, Southern Line Island. *Geological Society of America Annual Meeting*, **2**, 676.

Sears, S.O. & Lucia, F.J. (1980). Dolomitization of northern Michigan Niagara reefs by brine refluxion and freshwater–seawater mixing. In *Concepts and Models of Dolomitization*, ed. D.H. Zenger, J.B. Dunham & R.L. Ethington, pp. 215–35. Society of Economic Palaeontologists and Mineralogists Special Publication No. 28, Los Angeles.

Sheldon, R.P. (1963). Physical stratigraphy and mineral resources of Permian rocks in western Wyoming. *US Geological Survey Professional Paper*, **313-B**, 273pp.

Sullivan, C.H. & Krueger, H.W. (1981). Carbon isotope analysis of separate chemical phases in modern and fossil bone. *Nature*, **292**, 333–5.

17

The submerged phosphate deposit of Mataiva Atoll, French Polynesia

A.M. ROSSFELDER

Abstract

The phosphate deposit discovered in 1976 in the shallow lagoon of Mataiva Atoll, in the Tuamotu Archipelago, French Polynesia, was extensively drilled and investigated, especially between 1976 and 1982. However the accent was on mining feasibility and too few scientific studies have been made of its genesis and history.

Its geological reserves are estimated at 20–25 million metric tons of high-grade phosphate, spread over some 5 km² on the west side of the shallow lagoon, along the present emersed atoll rim and under a Holocene carbonate fill averaging 7 m. The orebody lays over and within a karst corresponding to an antecedent rim. Disseminated phosphate grains in the Holocene overburden through most of the lagoon and a large apatite boulder at the Eastern end indicate that the original deposit was once more extensive.

It shares a pre-Holocene age and many physical and petrographic similarities with the major Pacific insular deposits of Nauru, Ocean Island (Banaba) and Makatea, which imply comparable origin if not history, the major difference being its recent submersion.

Its geological features generally support the classic view regarding the origin of these insular deposits, i.e. occurrence of nutrient-rich waters, transfer of phosphorus from water to plankton, from plankton to fish, from fish to birds, and reaction of guano on marine carbonate bioclasts, accompanied and followed by an intense diagenesis through chemical changes in the interstitial waters and bacterial activity. This deposit also underwent at least one episode of submersion and one of high-emersion, before being submerged again and covered by a bioclastic overburden during the last Holocene transgression.

Introduction

The phosphate deposit of Mataiva was discovered in September 1976 during a Pacific atoll exploration program which started in 1974 with a shakedown survey of Clipperton Atoll and was discontinued in 1984 after reconnaissance by drilling and/or seismic profiling of some 80 selected sites in the Pacific. The initial idea of this program was that former subaerial insular deposits of avian origin comparable to the known elevated-island deposits of Nauru, Ocean Island or Makatea could be found submerged in present-day atollian lagoons as a result of sea level changes or island subsidence. The discovery of Mataiva confirmed the exploration hypothesis, but remained unique.

Mataiva is located at 14° 54′ S latitude and 148° 40′ W longitude, 170 nautical miles north of Tahiti, at the western end of the Tuamotu Archipelago, in French Polynesia (Fig. 17.1). The Tuamotu, separated by a 4000–4500 m deep channel from the volcanic chain of the Society Islands, consists of a large elongated ridge with extensive shelf areas around the depths of 1400–1800 m down to 2800 m. This ridge, considered much older and more stable than the other archipelagos of this oceanic region, nevertheless presents along its western margin signs of continuous tectonism affecting a suite of atolls from Anaa in the south to Mataiva in the north as a result of an active deformation of the lithosphere moving over the distant hot-spot of the Society Islands near Mehetia, (McNutt & Menard, 1978; Pirazzoli & Montaggioni, 1985).

Mataiva (Fig. 17.2) is part of the municipality of Rangiroa. Its 150 inhabitants live in the village of Pupuehu in the west of the atoll along the boat channel, near the airstrip. The oval-shaped atoll, 10 km long and 6 km wide, is bounded by an emergent rim 0.5–1 km wide, extensively planted with coconuts. Its honeycomb lagoon is unique, consisting of some 100-odd basins separated by a network of shoals which make it difficult to cross by boat, except through intricated paths or via the deeper southern string of basins. The atoll economy is essentially based on the collection of copra and on the catch from fishtraps for export to Papeete.

After the initial discovery in 1976 in the course of the POLY-1 reconnaissance cruise, extensive evaluation programs, dubbed MTV-1 to MTV-5, were conducted from 1976–82 on the orebody; but they were done for mining assessment purposes with few accompanying scientific investigations, and, as of today, the origin and history of the Mataiva deposit cannot be reliably described. However, little additional fieldwork appears to be needed to complete the existing material and support the comprehensive study that this unusual deposit deserves. Meanwhile, the aim of this paper is to summarize the work done so far, its main results and shortcomings, and what we know and can assume about this deposit.

Reconnaissance and evaluation fieldwork

The insular phosphate deposits known in the Pacific Ocean are located in the subtropical and equatorial regions and are associated with a coral limestone basement. The largest

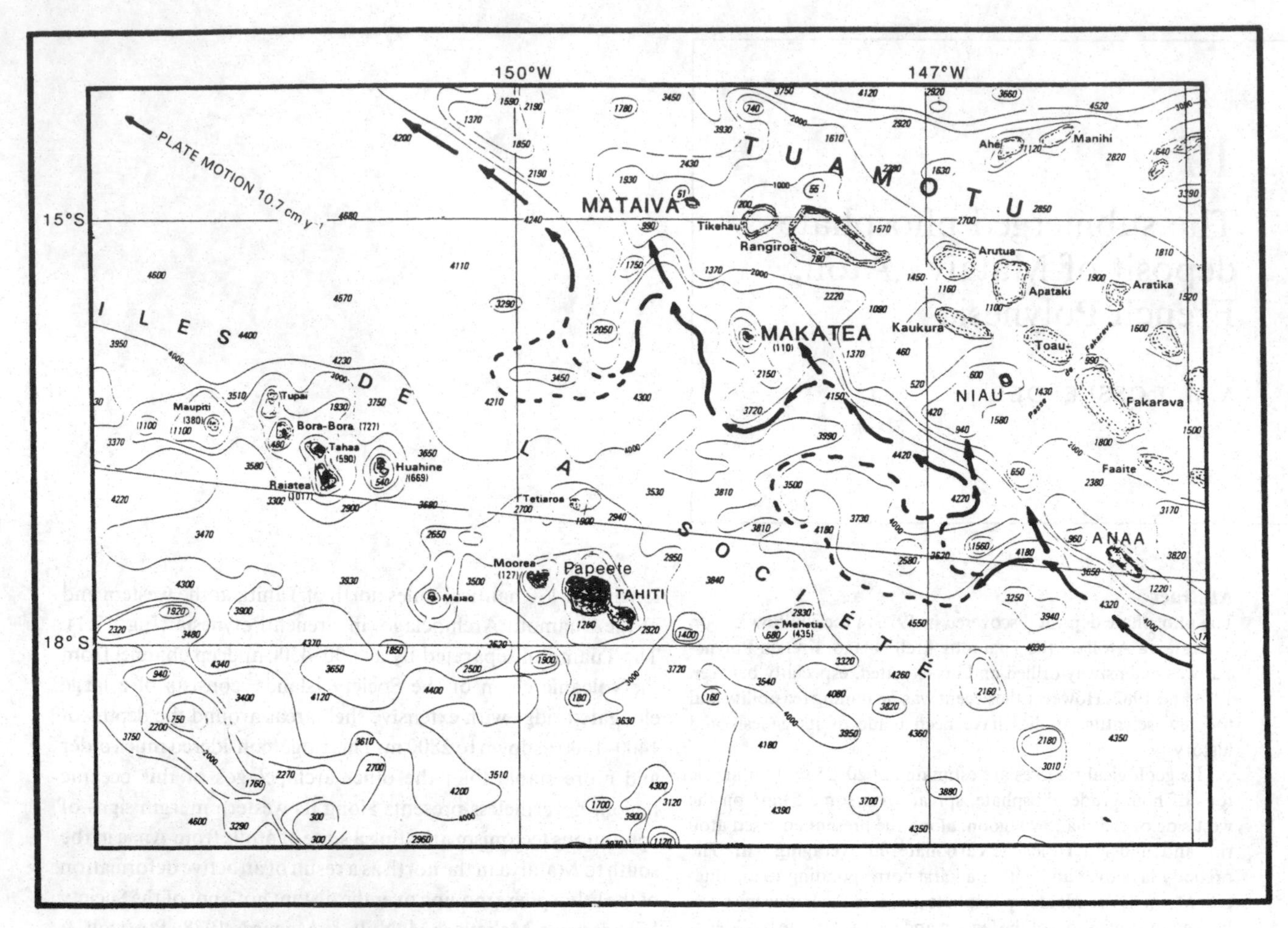

Fig. 17.1. Locator of the Northern Tuamotu, Tuamotu Channel and Society Islands (from SHOM 6607/Int 607; depths in m). Arrows represent the pattern of benthic circulation in the Tuamotu Channel.

deposits were mined at Nauru, Ocean Island (Banaba), and Makatea. Less important concentrations were also exploited at Angaur, Baker, Clipperton, Fais, Jarvis, Malden, Starbuck, the Datio Islands, and a few other minor sites (not to mention the contemporary guano deposits off Peru and Chile which are not associated with a coral substrate). With the exception of Nauru, now near exhaustion, and Bellona Island, which remains untouched, all these deposits have been mined-out, generally leaving no scientific studies and little information on their original state. The works of Elschner (1913) on Nauru and Obelliane (1963) on Makatea and the exhaustive review of Hutchinson (1950) represent the essential archives.

An exploration joint venture was initiated in 1972 under the sponsorship of UNOCAL mining division to assess whether similar phosphate orebodies might exist at shallow depths in atoll lagoons as a result of the submergence of avian phosphate deposits upon eustatic changes of the sea level or local tectonism. This joint venture was formalized in 1975 after an encouraging test survey of Clipperton Atoll. Soon afterwards, a Groupement d'Intérêt Economique, RARO MOANA GIE, was established in French Polynesia to carry on the exploration activities of the group in this territory, while similar programs were conducted in other Pacific jurisdictions. The criteria for target selection were simple: primarily, a shallow and closed lagoon in the intertropical strip, and secondarily, other factors such as reported phosphatic occurrences on the atoll rim, isolation and small size of the site favorizing bird concentration, and zones of known or suspected high oceanic productivity.

From 1976–80, the reconnaissance programs of RARO MOANA, named POLY-I, II and III, covered 24 selected atolls, 21 in the Tuamotu and three in the Society Islands, yielding a discovery at Mataiva, an uneconomical prospect at Puka-Puka and a still unconfirmed prospect at Niau. This general reconnaissance phase was performed through a preliminary site selection by a literature review and an aerial reconnaissance, followed by surveys of the targeted lagoons with sub-bottom profiling and/or rotary drilling.

High resolution, shallow-penetration seismic reflection was implemented with a 3.5 kHz transducer mounted on a lightweight aluminum boat about 3.5 m long capable of crossing the shallow passes and coral shoals. It was generally performed in advance of drilling when data were lacking on the depth and nature of the lagoon floor, eliminating those with excessive depth or heavy living coral formations, and also in conjunction with drilling to extend the ground-truth data and locate the next best drill site.

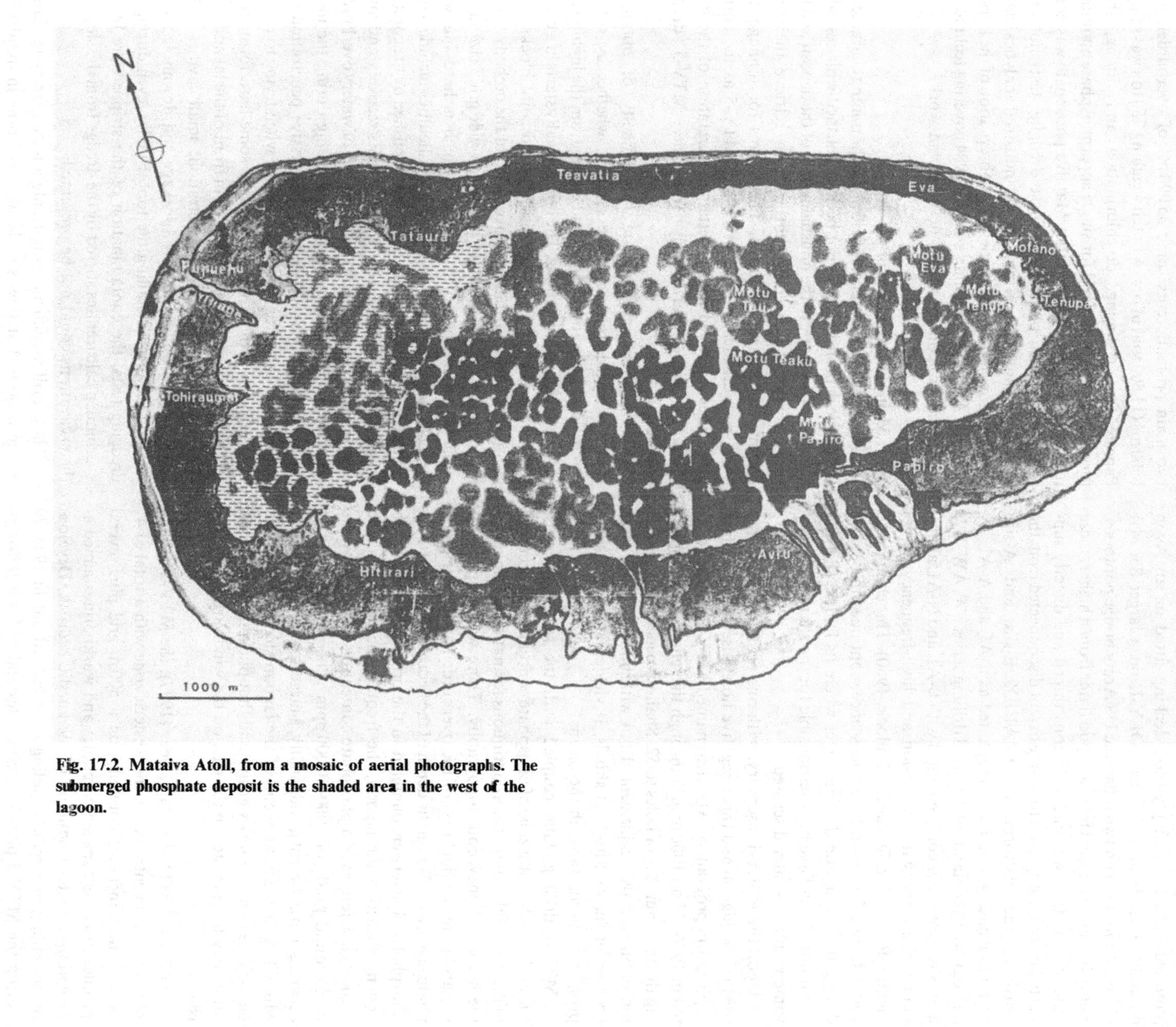

Fig. 17.2. Mataiva Atoll, from a mosaic of aerial photographs. The submerged phosphate deposit is the shaded area in the west of the lagoon.

Drilling was done with a portable rotary JKS-Winkie initially mounted on a 8 × 12 ft plywood decking fitted on an Avon inflatable river boat and, after POLY-I, on a larger 8 × 24 ft platform mounted on two inflatable US Army bridge pontoons. The drill was suspended from a tripod (later from a higher, more rigid tetrapod); it was manually constrained and directly supported by the casing, thus to a large extent disconnected from the heave of the platform. GEOMAREX E-size and A-size aluminum casings and rods (designated as E-AL and A-AL, by analogy with the standard diamond drilling series EW, AW, BW and NW) were later complemented with B-AL and N-AL aluminum casings and with percussion and high-frequency implements, (Rossfelder, Daguise & Pollock, 1980). This equipment, very effective for general reconnaissance, continued to be used during the evaluation drilling stages where its limited size and capabilities may have been responsible for the shortcomings of some results, as later discussed.

Upon the POLY-I discovery, with one positive hole on the west of the lagoon and three negative holes elsewhere, the first exploration program on Mataiva, named MTV-1, was conducted in 1976–77 on 1000 m grid through the whole lagoon bringing the total number of holes to 52. Shallow-penetration seismic reflection was also implemented but with limited effectiveness because of the multitude of shoals. It yielded a series of short profiles scattered through the basins.

MTV-1 drilling data, coupled with the seismic records, showed that the main zone of phosphate concentration occurred in the west of the lagoon as a continuous formation not linked to the surficial honeycomb topography. There was a significant variability in the thickness of the overburden and of the phosphate layer, as well as in the depth of the bedrock. Thin scattered phosphatized layers encountered in the drillholes in the north and northeast of the lagoon, the frequent occurrence in the lagoon calcarenites of a phosphate content distinctly above the background grade of 1000–10 000 ppm generally observed in other lagoon sediments, and, finally, a large apatite boulder near Motu Eva at the eastern end of the lagoon suggested that the deposit was once more extensive and that phosphatized formations may have existed in the past on the windward side of the island.

Following MTV-1 in same year (1977), the MTV-2 drilling phase was concentrated on the western zone with a total of 47 holes. These included coreholes on a 500 m grid plus several profiles of closely spaced coreholes and washborings aimed at providing a better resolution of the karstic bedrock. Development drilling and bulk sampling were carried out in 1978–79 during the MTV-3 phase (61 coreholes and 312 washborings, plus a 3.5 ton bulk sample lifted) and the MTV-4 phase (49 coreholes and 219 washborings, plus a 22 ton bulk sample lifted). Initial mining tests were also implemented during MTV-4 with a backhoe and suction equipment mounted on a barge. Further drilling was done during the MTV-5 dredging tests in 1981–82, bringing the total number of holes drilled in Mataiva lagoon from the initial reconnaissance to prefeasibility steps to more than 700.

The MTV-5 dredging tests, conducted with a prototype suction dredge designed to mine the phosphate layer above the top of the karst and to penetrate and cleanup the potholes within the karst, excavated a total area of 3900 m^2 and removed from the two main prismatic cuts some 35 000 m^3 of overburden and 6500 m^3 (11 500 ton) of ore. A batch weighing 750 ton was sent to Papeete for de-sliming, de-chlorination and drying, then forwarded to New Zealand for single superphosphate manufacture trials, which concluded that Mataiva phosphate was comparable to Nauru phosphate. These dredging tests not only provided mining engineering data and industrial analytical data, but also allowed the direct underwater observation of the karstic bedrock. A mining feasibility study in 1984 covering techniques and economics concluded this development work.

Main characteristics of the deposit

In spite of this extensive fieldwork, the mining characteristics of the deposit have still not been established with desirable accuracy. This is due to the unusual nature of the deposit and the shortcomings of the equipment used to retrieve the clean documented samples required. The figures for tonnage and grade rose drastically from MTV-1 to MTV-5 as the coring procedures improved and a better interpretation of the data could be made, particularly from the MTV-5 dredge excavations.

These corrections illustrate the difficulty of obtaining uncontaminated, representative samples, whether by rotary, percussion or vibration boring and coring from the highly variable waterlogged formations. The hole typically starts in a coarse calcarenitic lagoonal fill with disseminated coral blocks, then encounters an apatite hardpan over most of the orebody, enters into the loose phosphate material with its wide grain-size range, and ends in the potholed bedrock, which is hard but with frequent cavities. Examples of drilling difficulties include: coral clumps or pieces of hardpan tended to be pushed by the coretool into the softer phosphate layer, reducing its recovery and minimizing its thickness; a significant intrusion of watery overburden occurred along the outer wall of the casing during the incremental coring steps as a result of successive penetration and retraction of the corer acting as a piston; when trying to rotary-drill and core with a diamond bit into some hard formations, coral or apatite rubble accumulated in front of the bit and acted like ball-bearings impeding penetration; small cavities in the bedrock also resulted in core breakage and downhole rubble clogging the bit and stalling the tool. Finally, even during the dredging tests, the contamination of the stripped ore by the overburden calcarenites stirred by the dredge from the slopes of the excavations could not be prevented.

These drilling difficulties not only affected characterization of the deposit, but they should also be borne in mind when dealing with the existing phosphate samples from Mataiva. All have been contaminated to some extent by overburden carbonate particles. Contamination is generally heavy for the samples from the drillholes, although much less for the samples from the MTV-5 dredge excavations.

Table 17.1 presents both the orebody parameters which were retained for the 1984 mining feasibility study and also the figures now considered to be closer to reality and more in line with the phosphate extracted from nearby Makatea. In terms of phosphate grade and quality, Mataiva appears similar to the other major elevated-island deposits of the Pacific. Its reserves are in

Table 17.1. *Characteristics of the Mataiva phosphate deposit*

	Retained (1984)	Probable (1987)
Area of deposit (km^2)	5.0	5.0
Average water depth (m)	3.0	3.0
Average overburden thickness (m)	7.3	7.0
Average phosphate thickness (m)	2.33	2.6[a]
P_2O_5 grade (%)	37.5	38.0
In situ *density*	1.70	1.80
Geological reserves (MT)		
Proven	15.8	18.7[a]
Probable	2.9	3.4
Possible	1.3	1.5
Total resource	20.0	23.6[b]
At 35.3% P_2O_5	21.2	25.4[c]
Recoverable resource (MT)	14.8	17.8[d]

[a]Excluding about 1 km^2 of deposit area where the phosphate thickness is < 1 m.
[b]Difference between 'Retained' and 'Probable' due to different average phosphate thickness and density.
[c]Assuming that 35.3% grade is commercially acceptable therefore allowing some carbonate contamination.
[d]Assuming that the mining recovery rates of 60–80% observed during the dredge tests will be improved to an average of 75% during mining and deducting 5% for processing losses, i.e. 70% overall recovery.

the same range as Ocean-Banaba (about 25 MT) and ahead of Makatea (16 MT) but well behind Nauru (in the order of 100 MT).

Geological features

From bedrock to sea level, the summarized sequence of the formations is as follows (Fig. 17.3).

Karstic bedrock

At least in its upper part intersected by exploration drillholes, the karstic bedrock which underlies Mataiva lagoon appears as a hard yellowish coral limestone, dolomitized (4–17% Mg), with vugs and cavities commonly lined with recrystallized white carbonates. It is assumed to be of Miocene–Pleistocene age (Pirazzoli & Montaggioni, 1986).

The MTV-5 dredge excavations made direct observation of the karst possible through diving after the removal of the overburden and phosphate. The karst, with an amplitude averaging 1.5 m but ranging widely (1–6 m), bears a definite resemblance to the *karrenfeld* of the highlands of Nauru and, more strikingly, with the *Pot-Holes* area of Makatea, showing a caldron-type erosion rather than the maze of pinnacles, walls, and deep and narrow corridors which characterize the lowlands of Nauru. Some of these cavities, more or less circular and overlapping, with quasi-vertical walls, may be quite deep. At Hole 6, about 800 m southeast of the pass, where the first and largest dredge excavation was later located, the drill went down for 50 m into a narrow chimney filled with phosphate, increasingly soft and watery at depth, without reaching a hard bottom. From a set of holes placed around it, the diameter of this chimney was estimated at no more than 2 m and probably rapidly tapering down to approximately 1 m.

From the drillholes performed elsewhere in the lagoon, the paleosurface of the old reefal bedrock appears smoother than under the deposit. While this may indicate that the deep erosion of the karst in the west of the lagoon is related to the genesis and paragenesis of the overlying phosphate formation, it is nevertheless probable that a primary karst similar to the *feo* of the Austral Islands or to the *makatea* of the Cook Islands existed before the deposition of the phosphatic material.

The ancient reefal bedrock outcrops with a slight dip along the outer western and northwestern shores of the island. Its paleosurface follows the shape of an atoll (Fig. 17.4) with an elongated east–west lagoon, over 30 m deep below the present mean sea level, located at the center of the present lagoon and open to the sea through a pass under the present spillway of Papiro and possibly a second one under Hitirari (Fig. 17.4).

The bedrock under the deposit is part of this antecedent atoll rim. It dips gently at a 1–2° angle eastward from the present lagoon shoreline then at a steeper 5° angle around the 16–18 m contour below mean sea level towards the ancient lagoon. In this deeper southeastern fringe of the deposit, the lagoonal sands frequently found in karst pockets under the main orebody form a continuous layer over which the phosphate formation thins out in a spill-over fashion into the old lagoon. These loose to semi-consolidated lagoonal calcarenites are rich in foraminifera and occasional coral rubble, principally *Pocillopora*, and may reach a thickness of several meters.

In the western part of the deposit the old reef directly underlies the phosphate, but when moving eastward toward the spill-over zone, the transition from bedrock to phosphate is generally marked by an eluvium 0.1–1.0 m thick, of sand and fragments from the weathering of the old reef.

Phosphate formation

The phosphate formation, laying, as noted, directly over the karst on its shallower western part and on the karstic eluvium or the lagoonal sands toward the east, consists of gravel and sand in a mud matrix with disseminated clumps and boulders. Unlike Nauru where the tops of the pinnacles generally coincide with the top of the deposit, about half of the Mataiva orebody lies above the karst, forming an unobstructed horizon. However, pinnacles occasionally protrude through the top of the deposit into the overburden.

In a preliminary thin-section examination of the phosphate material, LeCouteur (1979) noted that it consists of subangular grains, rounded grains and ovules, ooliths and polynucleated pellets, phosphatized fossil fragments of gastropods, foraminifera and suspected bryozoa, pre-existing phosphatized sand recemented by collophane, and large grains themselves composed of aggregated phosphate grains. All this indicated an intense reworking of the phosphate components and movement of the phosphatizing solutions. He also noted that the sections, with a pale- to red-brown color, occasionally contained dark carbonaceous colorations likely attributable to carbon. If so, this indicated a biogenic influence in the phosphatization process and a reducing environment, while later cementation occurred under oxidizing conditions. A more recent

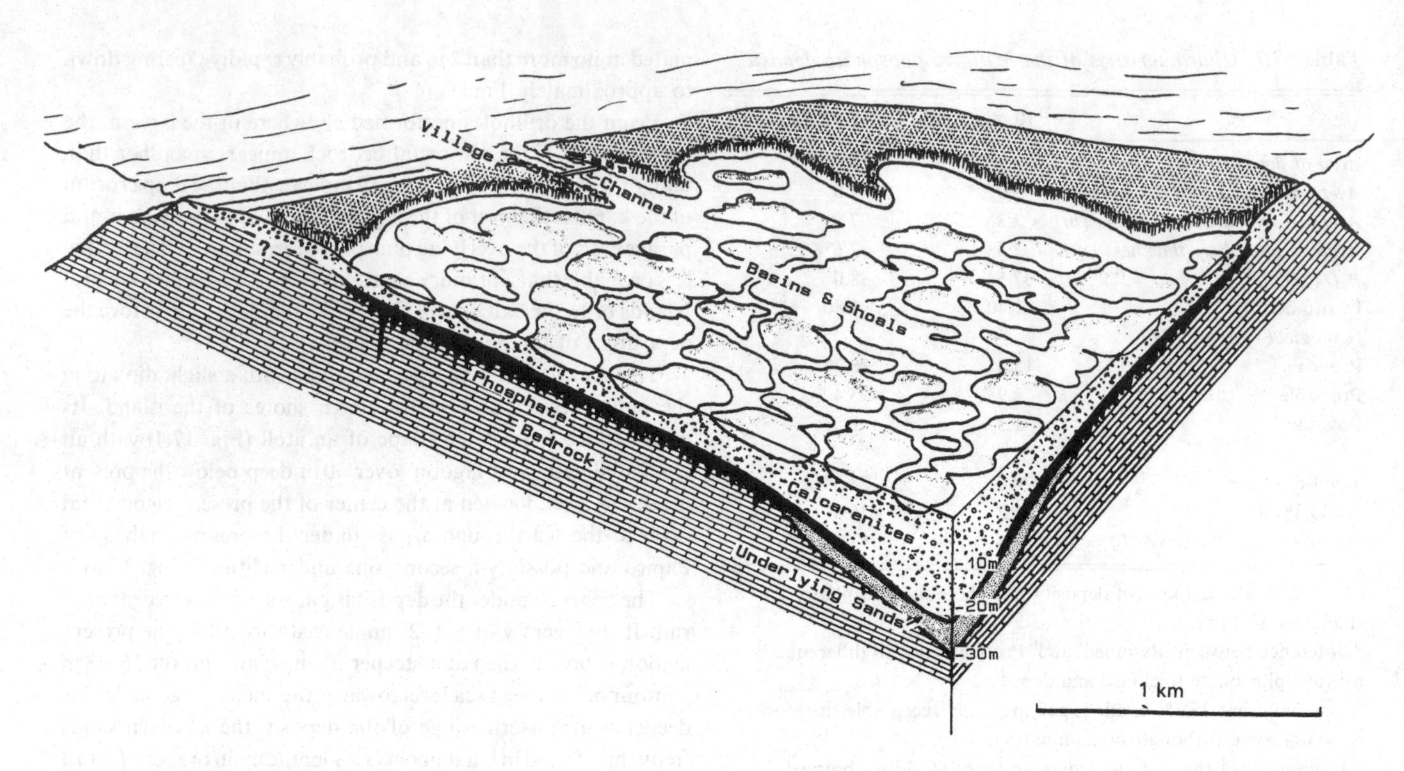

Fig. 17.3. General structure of the deposit.

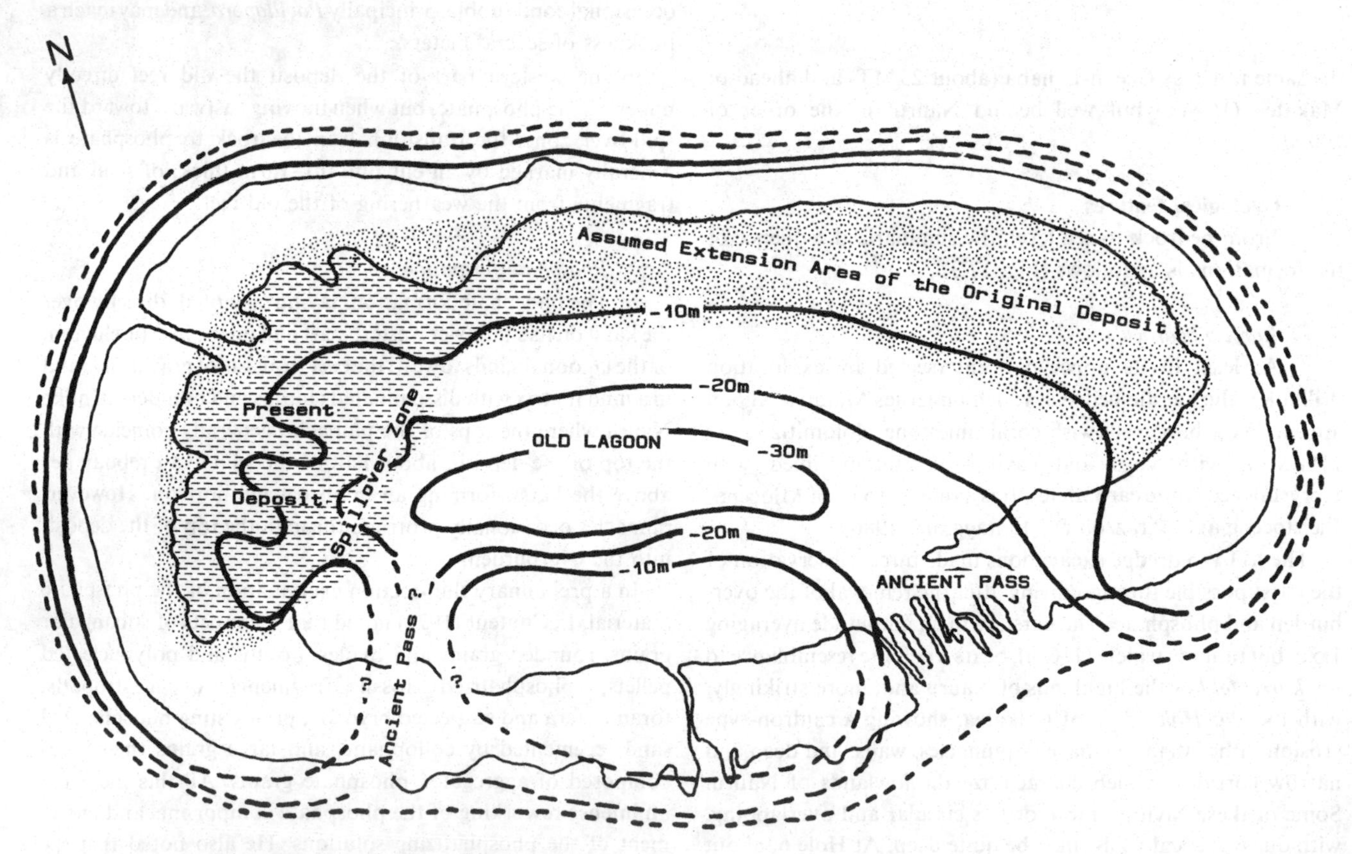

Fig. 17.4. General paleosurface of the limestone bedrock and shape of the antecedent atoll at the assumed time of phosphate deposition (depths from present sea level).

Table 17.2. *Analysis of Mataiva phosphate (weight %)*

Main components (from FMRA–1982)		Other components (from IFDC–1981)	
Al	0.05 (Al_2O_3 0.09)	Cd	50 ppm
CaO	53.1	Cl	250 ppm
$CaCO^3$	7.7	K_2O	0.01
F	3.1	MgO	0.23
Fe	0.10 (Fe_2O_3 0.15)	Na_2O	0.23
H_2O	1.7	S(total)	0.13
P_2O_5	37.1	SiO	0.13

petrographic study by D. Giot (unpublished data) confirms the extensive diagenetic processes which have affected the phosphatized material and casts some light on their sequence.

About half of the formation is covered by a discontinuous apatite hardpan a few centimeters to some 20 cm thick. The transition phosphate overburden is otherwise almost invariably marked by a layer of phosphatic gravel from sand–pebble-size, with the largest elements generally ochre-brown in color and frequently exhibiting a dark patina: an indication, in conjunction with the hardpan, that the deposit was at this time emergent and undisturbed. An analysis by the Fertilizer Manufacturers Research Association (FMRA) of the MTV-5 bulk sample shipped to New Zealand provided the data presented in the first column of Table 17.2, completed in the second column by an excerpt from an earlier analysis of the MTV-4 bulk sample by the International Fertilizer Development Center (IFDC), Alabama.

It is estimated that one-third–one-half of the $CaCO_3$ measured in the FMRA analysis results from contamination by overburden particles during excavation. Also noticeable are the exceptionally low Fe and Al concentrations which readily exclude a volcanic source for the phosphate.

Roe & Burnett (1985) dated samples from the Mataiva phosphate deposit from top to bottom at 130 000 y BP, 210 000 y BP and 300 000 y BP. These figures, being in normal sequence, may be construed as expressing successive phases of phosphatization; however, such phases are not reflected in any perceptible facies discontinuity through the orebody. Alternately, they could be explained by some contamination of the samples by the overburden during drilling; but the samples were carefully selected and manually cleaned of carbonate particles. Consequently, the most plausible explanation for this vertical variation of the age of the deposit is a *per-descensum* rejuvenation of the U/Th ratios by subsequent episodes of weathering and/or submersion. If so, these dates relate to diagenetic events and the protore of Mataiva is older than 300 000 y, and possibly much older, in line with what is now generally assumed for Nauru and Banaba and with the observations of Obelliane (1963) who concluded that the Makatea deposit may have originated with the settling of seabird colonies at the first emersion of the island in the Tertiary and continued into the Pliocene. The time of this emersion and initial phosphogenesis, which Obelliane thought to be Late Eocene or Early Miocene appears now to be Mid- or Late Miocene (Montaggioni, 1985).

The boundary of the phosphate orebody toward the west coincides quite closely with the present shoreline of the lagoon, except, possibly, for a small overlap on the Temiromiro peninsula, west of Tataura. However, the outcrops of the reefal bedrock on the western and northern seashore of the island present some scattered patches of apatite crust, also noted on land as fragments in the rubble from past diggings of water wells (M. Lasserre, pers. comm.). This indicates that some phosphatization also occurred in the past between the main deposit and the sea. Unfortunately, because of the villagers' concerns for their land and coconut groves no holes were drilled in this area. Such boreholes could have documented the stratigraphy of the atoll rim.

Overburden

The interface between the phosphate formation and the overburden is smooth and undulating. Within a -11 m–-17 m range below mean sea level, this interface is commonly marked above the apatite hardpan or gravel by a layer of brown peat, 10–50 cm thick. Traces of this peat were also unexpectedly encountered by the drill within the phosphate formation itself, however they were later shown by dredge tests to be roots descending into the deposit. This peat layer indicates an emersion stage in a wet climate and probably represents a mangrove swamp at the beginning of the last submersion of the lagoon.

Immediately above the phosphate or the peat layer, a clastic formation of coral rubble, mostly tubular *Acropora* debris in a greenish fetid phosphatized ooze, 0.5–2 m thick and containing about 5% P_2O_5, is commonly present. The original environment of this green layer is illustrated by live thickets of *Acropora* to the northeast of the lagoon.

The bulk of the principally aragonitic overburden consists of white calcarenite and calcilutite with clumps, slabs or large scattered blocks of coral, principally *Porites*. Minute phosphate grains and ooliths occasionally occur in this upper bioclastic formation through the entire lagoon area.

Roe & Burnett (1985) measured a U/Th date of 53 000 y BP, ± 2000 y, on a white phosphatized limestone sample (A4) from the base of the overburden. By contrast, Pirazzoli & Montaggioni (1986) showed the overburden to be entirely Holocene with its base estimated from C^{14} data at 6000 and 6500 y BP, dates comparable to those of Delesalle (1985). However, Roe & Burnett worked with a well documented sample from the transition zone, while the other authors utilized less reliable materials from the body of the overburden. If so, these data imply a submersion of the deposit ending sometime before 50 000–55 000 y BP, then an emergence gap until approximately 6500 y BP, followed by the final submersion of the lagoon and the deposition of the bulk of the clastic overburden.

The basins which form the present morphology of the lagoon have an average depth of 5–8 m, reaching 13 m in the south. They are filled with a fine white carbonate ooze settling from the cloudy waters. Their shape is generally lobated, and diving observations show that they frequently correspond to clusters of smaller circular pools. *Acropora* rubble commonly occupies the side of the pools. The partitioning shoals are covered by wheel-shaped *Porites lobata*. There is no correspondence between the morphology of the lagoon and the underlaying phosphate deposit or its bedrock; but *Porites* slabs appear more frequent under the shoals than under the basins.

A bathymetric survey conducted in 1981 to the 500–600 m

isobath on the western flank of the atoll showed a bench of jagged living reef, 50–60 m wide, sloping to a depth of about 20 m, where some terracing occurs. This correlates relatively well with the change of grade of the old reef bedrock inside the lagoon. The slope then plunges rapidly and regularly at a 35–40° angle.

Interpretation of the features of the deposit

The location of the deposit on the ancient rim of the atoll (Figs. 17.2 and 17.4) with only a limited spilling of the phosphate over the ancient lagoonal sands strongly suggests subaerial deposition of the original phosphatic material (the *protore*), and consequently the role of seabird colonies in line with the views of Elschner (1913) for Nauru, Barrie (1967) for Christmas, Obelliane (1963) for Makatea, and the reviews of Hutchinson (1950) and Stoddart & Scoffin (1983). This location also reinforces the image of such colonies established on the inside of the rim under the shelter of its crest, as they can be presently observed, for example, in Malden Island in the equatorial Pacific.

This avian-controlled phosphatization implies the existence of phosphate-rich waters with a teeming marine life to supply the food, the emersion of the depositional site to provide the habitat, and a dry climate to ensure the conservation of the guano. These conditions may have been satisfied during an ice age as suggested by Aharon & Veeh (1984). Also the presence of the Tuamotu Channel, over 4000 m deep, draining a deep circulation of highly productive subpolar waters toward the intertropical strip (Fig. 17.1), further enhanced by tradewind upwellings on the leeside of the Tuamotu platform, may have played a part in the presence there of the three known phosphatized islands of this region, namely Makatea, Mataiva and Niau. In this regard the behavioral trait of nest-site excretion conducive to guano formation is most prominent in certain cold water species of Pelicanidae typified by the ganaye birds of Peru, of subantarctic origin, now lagging along cold upwellings under the tropics.

The phosphate-rich low-pH excreta attacked not only the coral and foraminiferal sand, rubble and other bioclasts which may have covered the atoll rim at this time, but also the hard bedrock itself which appears, as already noted, more deeply etched under the deposit than elsewhere in the lagoon. This phosphate-related karstification may have taken place at the time of initial deposition, but may also correspond to a secondary process of concentration and weathering during a high-elevation period. In fact, from the observation of low-grade, recent phosphate occurrences in other Pacific atolls, we tend to visualize the protore not as a massive guano cap built in a single event over a few thousand years and leaching the underlaying carbonates, but as a layering of marine carbonates at various degrees of phosphatization gradually and intermittently enriched by avian excreta, reworked by bacteria and other organisms (such as the myriads of crabs of Clipperton Island), and concentrated by weathering.

The petrographic observations of LeCouteur & Giot (pers. comm.) indicate an intense diagenesis of the protore in a state of submergence, while the +50 m chimney of Hole 6 suggests a time of high emergence. It is not clear at this point which of these two episodes came first and there is a strong probability that more than one of each took place. We know, however, if we follow Pirazzoli & Montaggioni (1985), that Mataiva may have been in a stage of tectonic uplifting since the Mid-Pliocene when, moving WNW at a rate of 10.7 cm y^{-1}, it entered the flexuring zone of the Society Islands hot-spot somewhere south of the present location of Anaa. If so, any submergence after this event was chiefly of eustatic origin. Furthermore, assuming that the site was submerged until then, this event and Roe & Burnett's (1985) dating would place the time of avian phosphogenesis between 4.0 My and 0.3 My BP. We also know that the phosphate orebody was emergent and undisturbed during the hardpan episode and had already matured in its present state 6000 y ago; that it then underwent a final submersion marked by the green layer phase in very shallow and calm waters behind the protection of a new atoll rim, before being buried under the present lagoon fill which bear witness of passing hurricanes. This probably was the time when other phosphatized formations in the northern and eastern parts of the island were subjected to destruction and dispersion.

Conclusions

The interpretation of insular phosphate deposits is often subject to controversy for purely semantic reasons. Much of this controversy could be avoided if a clear distinction is made between the source of phosphorus, the pathways for its transfer and concentration in current depositional settings, the source of the calcium, the nature of the substrate, the mechanism of its phosphatization, and the subsequent processes of concentration, redissolution and reprecipitation – all part of the paragenesis resulting in the eventual orebody.

Mataiva's primary source of phosphorus can either be said to be volcanic, oceanic, planktonic, nektonic or avian, depending on how far back we want to go into the phosphorus cycle. The nature of the deposit can equally be termed subaerial or submarine depending on what diagenetic episode appears the most significant. But, as far as the formation of the protore is concerned, the macrofeatures of the Mataiva deposit, along with our observations of ancient and modern insular phosphate occurrences in the course of the 12 y exploration program during which it was discovered, do not suggest any other mechanism than the process of avian concentration and deposition which is generally accepted for this type of deposit.

Such an avian pathway of phosphogenesis is powerfully illustrated by the 40 m guano caps built in a few thousand years over the Chincha Islands. However, in the case of Mataiva and similar ancient phosphate-islands, we envision it as a slower process of limestone phosphatization of varying intensity with simultaneous, subsequent or alternate reworking and concentration by organisms, weathering, and vadose waters, all occurring within an oceanic environment and ecosystem different than those of today and under the sway of the ice ages.

The discovery of the Mataiva phosphate deposit confirmed the premises of our Pacific atoll exploration program, but the find remained unique. Mataiva shares with, at least, Makatea, Nauru and Christmas Island a pre-Holocene age and a history of submersion and emersion, but also a tectonic uplifting which has worked against all normal causes of island subsidence such as core settlement and plate cooling, and which placed it in a position accessible to discovery and mining. This may be why, in the final analysis, so few of these deposits were found and may ultimately be accounted for. Other deposits similar to Mataiva probably exist elsewhere in the Pacific and Indian Ocean, but are likely to be deeply buried in lagoons and out of economic reach, for the reason that they are much older and, therefore, more affected by island subsidence than originally anticipated.

Acknowledgments

The Pacific atoll exploration program during which Mataiva was discovered would not have been possible without the initial thrust from William R. Moran, vice president and general manager of the mining division of Unocal, and, later, without his continuous confidence and friendly support. I am equally grateful in this regard to Robert A. Spencer, exploration manager, Cominco. To belie a common view among marine miners, this program conducted through some 12 island jurisdictions could not have maintained a steady course without the restrained but efficient legal touch of Robert L. Humphrey, deputy general counsel of Unocal. This article relies heavily on the fieldwork and reports of Max Lasserre, senior geologist, Sogerem–Pechiney, who assisted the author during the early exploration phases on Mataiva before taking charge of the development drilling phases and dredging tests. Grant Gibson, senior geologist, Cominco, actively and efficiently advised and supported the field operations and the accompanying and subsequent mining studies. Finally, I wish to particularly thank Robert F. Dill, William C. Burnett and Joshua I. Tracey, Jr, for their patient and constructive review of this paper.

References

Aharon, P. & Veeh, H.H. (1984). Isotope studies of insular phosphate explain atoll phosphatization. *Nature*, **309**, 614–17.

Barrie, J. (1967). The geology of Christmas island. *Bureau of Mineral Resources, Geology and Geophysics*, **1967/37**.

Delesalle, B. (1985). Mataiva Atoll. Tuamotu archipelago. In *5th International Coral Reef Congress*, Tahiti, 1, ed. B. Delesalle, R. Galzin & B. Salvat, pp. 269–322. Muséum National d'Histoire Naturelle, EPHE, Antenne De Tahiti.

Elschner, C. (1913). *Corallogene Phosphat-Inseln Austral-Oceaniens und ihre Produkte*. Max Schmidt, Lubeck, 118 pp.

Hutchinson, G.E. (1950). Survey of contemporary knowledge of biogeochemistry. 1: The biogeochemistry of vertebrate excretion. *Bulletin of the American Museum of Natural History*, **96**, 554pp.

LeCouteur, P.C. (1979). Preliminary Report on Mataiva Phosphate Samples. Mataiva Nodules. Cominco Internal Memo. (unpublished).

McNutt, M. & Menard, H.W. (1978). Lithosphere flexure and uplifted atolls. *Journal of Geophysical Research*, **83(B3)**, 1206–12.

Montaggioni, L.F. (1985). Makatea Island, Tuamotu Archipelago. *5th International Coral Reef Congress*, Tahiti, vol. 1, ed. B. Delesalle, R. Galzin & B. Salvat, pp. 103–58. Muséum National d'Histoire Naturelle, EPHE, Antenne De Tahiti.

Obellianne, J.M. (1963). Le gisement de phosphate tricalcique de Makatea. *Sciences de la Terre*, **IX, 1962–63(1)**, 5–60.

Pirazzoli, P.A. & Montaggioni, L.F. (1985). Lithospheric deformation in French Polynesia (Pacific Ocean) as deduced from Quaternary shorelines. *5th International Coral Reef Congress*, Tahiti, vol. 3, 195–200. Muséum National d'Histoire Naturelle EPHE, Antenne De Tahiti.

Pirazzoli, P.A. & Montaggioni, L.F. (1986). Late Holocene sea-level changes in the northwest Tuamotu Islands, French Polynesia. *Quaternary Research*, **25**, 350–68.

Roe, K.K. & Burnett, W.C. (1985). Uranium geochemistry and dating of Pacific Islands apatite. *Geochemica Cosmochemica Acta*, **49**, 1581–92.

Rossfelder, A.M., Daguise D. & Pollock, R.J. (1980). Drilling and coring systems for shallow water exploration. *Offshore Technical Conference*, **OTC 3876**, 217–21.

Stoddart, D.R. & Scoffin, T.P. (1983). Phosphate rocks on coral reef islands. In *Chemical Sediments and Geomorphology, Precipitates and Residua in the Near-Surface Environment*, ed. A.S. Goudie & K. Pye, pp. 369–40. Academic Press, London.

PART 3

The Neogene environment

18

Paleoceanographic and paleoclimatic controls of the temporal and geographic distribution of Upper Cenozoic continental margin phosphorites

S. R. RIGGS AND R. P. SHELDON

Abstract
Phosphorite was formed during all major sea-level trangressions during the 67 million years of Cenozoic history. However, some periods were more important than others with respect to producing large volumes of phosphorites and preserving them in the geologic column. During the Upper Cretaceous, Paleocene and Eocene major episodes of phosphogenesis occurred within Tethys, a major east–west ocean, which produced extensive phosphorites throughout the Middle East, Mediterranean, and northern South American regions. By Middle Eocene, this circumglobal ocean had been destroyed by plate tectonic processes. During the rest of the Cenozoic, the north–south Pacific and Atlantic Oceans dominated global circulation patterns and phosphogenesis shifted to the bounding continental margins. Phosphogenesis took place on small and local scales throughout the Upper Cenozoic; however, specific episodes of phosphogenesis have occurred on much larger scales that reflect global paleoclimatic and paleoceanographic events. These Upper Cenozoic episodes of phosphogenesis have the following general relationships: 1) they were closely linked with development of polar glaciation that began during the Upper Oligocene; 2) they occurred during early to mid-stages of transgression associated with second-order eustatic sea-level fluctuations; 3) they reached maximum development during the TB2 second-order sea-level cycle (upper Lower–Middle Miocene); 4) they were controlled on a regional basis by inter-relationships between the geologic setting, paleoceanographic processes such as boundary current dynamics, and the third- and fourth-order sea-level fluctuations; and 5) they occur extensively in the subsurface of modern continental shelf and slope environments with local to regional occurrences of the updip portions on adjacent coastal plains.

Introduction

Modern understanding of phosphorite sedimentation began with the work of Kazakov (1937), who proposed a paleoceanographic model for phosphorite deposition in Eastern Boundary Current upwelling areas and used it to interpret ancient phosphorites in Russia. Some geochemical aspects of Kazakov's model have been discounted, but some of the geochemical and paleoceanographic aspects form the basis for much research in the modern oceans, as well as the geologic study of ancient phosphorite deposits. It has now been demonstrated that the authigenic–diagenetic formation of marine phosphorites occurs in low- to mid-latitudes in shallow continental margin sediment environments characterized by low rates of siliciclastic and biogenic sedimentation. Large supplies of nutrient phosphorus and nitrogen, derived from cold upwelling currents, are essential for production and sedimentation of large volumes of organic matter. Phosphorus from the organic matter is then concentrated by various diagenetic mechanisms, including bacterial and interstitial pore-water processes, either at or just below the sediment–water interface. These processes lead to the primary formation and growth of phosphate grains which may remain where they formed or be transported as clastic particles within their environment of formation. During subsequent periods of time, some primary phosphate grains may be physically reworked into younger sediment units in response to a different set of environmental conditions and depositional processes.

The present understanding of modern marine phosphogenesis has been developed by a multidisciplinary approach guided by the Kazakov model. Application of the modern examples of phosphogenesis to understanding Upper Cenozoic phosphorites has been highly successful because of similarities between the geographic, climatic, and oceanographic conditions. This paper examines the temporal and spatial distribution of known Upper Cenozoic phosphorite deposits.

Depositional framework for Cenozoic sediments

The depositional patterns of Cenozoic sediments on continental margins reflect cyclical deposition on at least three different time scales. These time scales represent second-order (greater than 10 my duration), third-order (1–10 my duration), and fourth-order (100 000–1 my duration) cycles of sea-level fluctuation (Vail, Mitchum & Thompson, 1977). The resulting continental margin rock record is characterized by changing sediment patterns and interbedded cyclic sediment units deposited in direct response to changing global tectonic, paleoclimatological and paleoceanographic conditions that are driving eustatic sea-level cyclicity.

The global cycles of relative sea level during the Cenozoic (Vail & Mitchum, 1979; Haq, Hardenbol & Vail, 1987) display a general change in sea level. The curve in Figure 18.1 depicts the second- and third-order sea level during the Paleogene to be at relatively higher levels than during the Neogene. After a major regression in the Upper Oligocene, Neogene and Quaternary sea

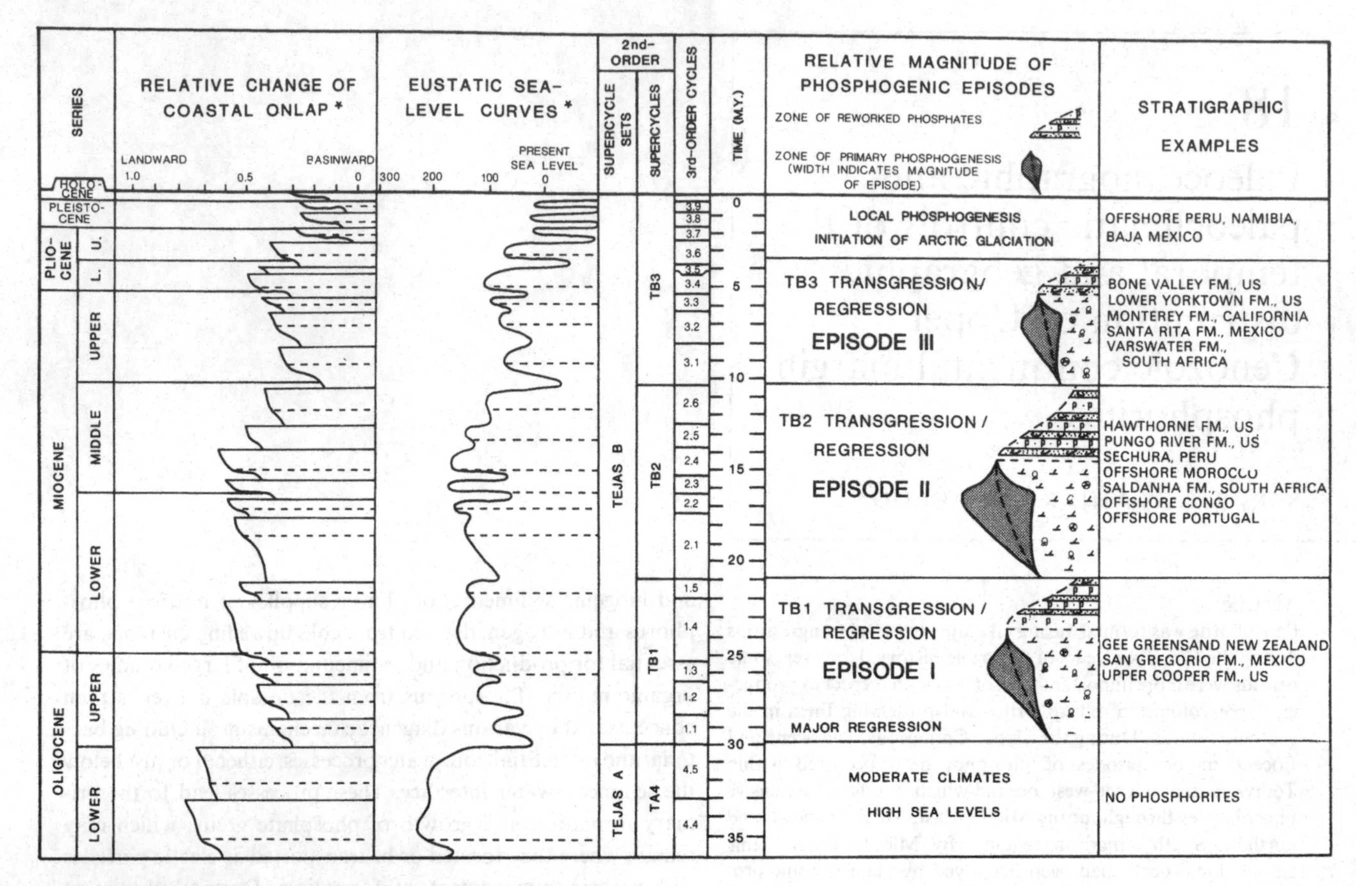

Fig. 18.1. Comparison of Upper Cenozoic phosphorite depositional episodes with the second- and third-order curves of coastal onlap and eustatic sea-level events of Haq *et al.* (1987). Interpretation of the pattern of deposition of phosphorites and associated sediment sequences and their relationship to the sea-level curves is based upon the model developed by Riggs *et al.*, Stephen W. Snyder, Hine & Riggs, Riggs & Mallette, and Scott W. Snyder (this volume) and is outlined in the summary of this paper.

levels were generally lower and declining. Concomitant changes in lithofacies patterns on the continental margin of southeastern United States (Fig. 18.2) were attributed to a major paleoclimatic change from nonglacial-dominated to one that was dominated by repeated cycles of glaciation and deglaciation (Riggs, 1984 and 1988).

The Neogene represents a transition zone into the well established period of Quaternary glaciation characterized by both Antarctic and Arctic glaciation. Increasing faunal and floral evidence suggests that global climates were becoming less stable with declining temperatures by the end of the Oligocene in Europe (Nilsson & Persson, 1983) and the United States (Dorf, 1964). Based upon oxygen-isotope data, Matthews & Poore (1980) and Matthews (1984) argued that glacio-eustatic sea-level fluctuations have been an important process in marine sedimentation at least since the Eocene. Recent oxygen isotopic studies of foraminifera from deep-sea sediments (Miller & Fairbanks, 1983; Keigwin & Keller, 1984; Miller *et al.*, 1985; Keigwin & Corliss, 1986) suggest that an Antarctic ice-cap was well developed by the Upper Oligocene. They believe that the opening of Drake Passage during the Lower Miocene established the deep circumpolar current, led to thermal isolation of Antarctica, and increased the global cooling processes. Keller & Barron (1983) recognized eight distinct deep-sea dissolution hiatuses within the Neogene. These hiatuses represent periods of cooling and intensification of bottom-water circulation, which they believe represent unstable climatic conditions characterized by cooling at high latitudes or increased glaciation. Furthermore, they found strong correlations between the Neogene deep-sea hiatal surfaces with onshore unconformities produced by eustatic lowering of sea level and severe basinward shifts of coastal onlap. Both sets of unconformities are interpreted to reflect periods of intensified polar glaciation. Woodruff, Savin & Douglas (1981) and Woodruff (1985) found that by the Middle Miocene (between 16 and 13 my), the climate had become extremely unstable with oxygen-isotope records suggesting variations similar to those of the Pleistocene with periodicities on the order of 100 000 years. Woodruff (1985) found that changes in distribution of deep-sea benthic foraminifera, as well as evolutionary originations and extinctions, were concomitant with paleoclimatic and paleoceanographic changes presumably related to Antarctic glacier expansion and cooling of the deep Pacific Ocean between 16 and 13 my. The Antarctic ice sheet continued to experience major expansion episodes between 13 to 11.5 my (Keigwin, 1979;

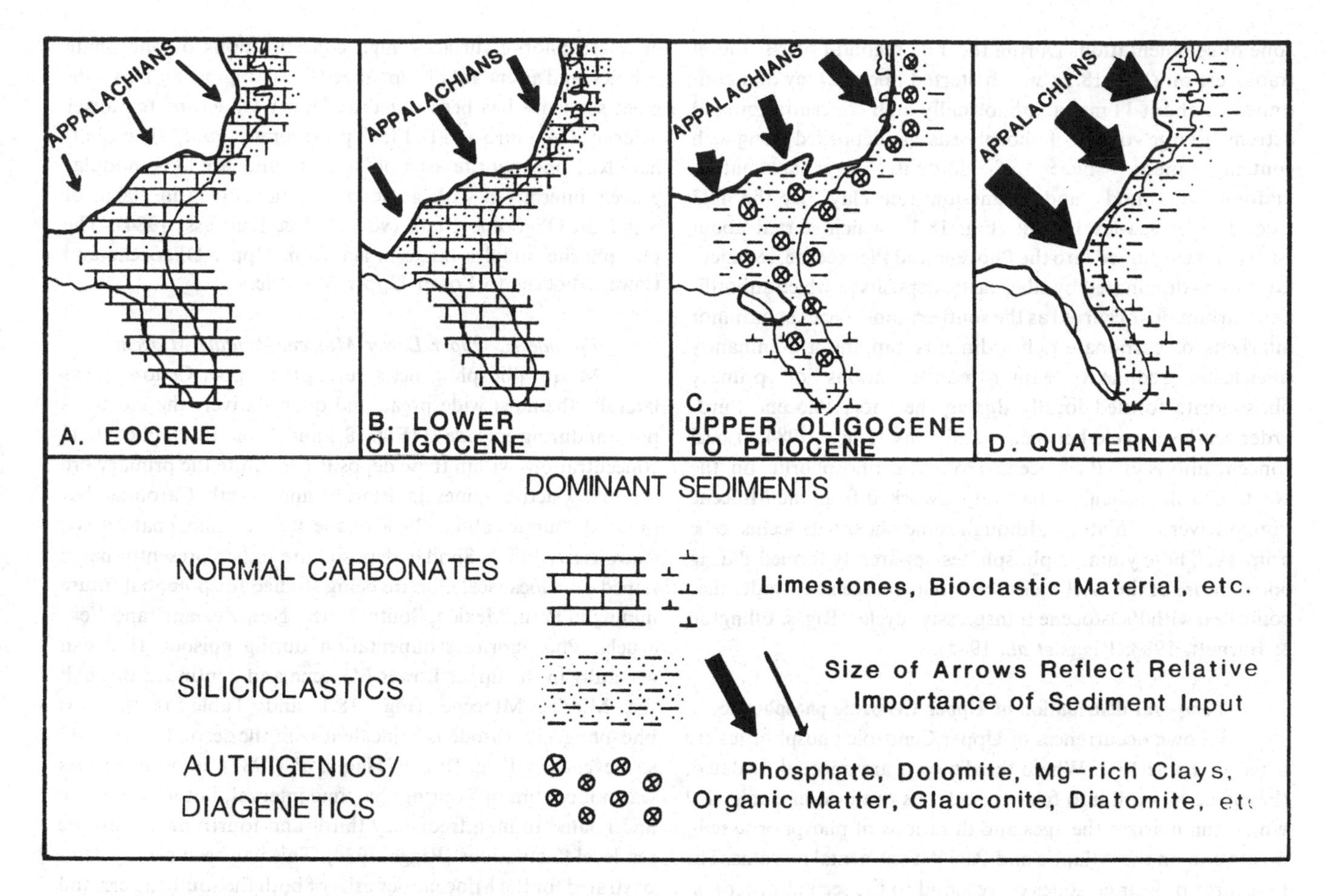

Fig. 18.2. Four-part map series of the southeastern US continental margin showing changing patterns of dominant sedimentation from the Eocene through the Quaternary (modified from Riggs, 1988).

Woodruff *et al.*, 1981; Keller & Barron, 1983) and again between 10 to 8 my (Woodruff, 1985).

Riggs (1986) found that major episodes of phosphogenesis and associated authigenic sediments do not form within extensive sequences of uniform sediments. Rather, most phosphogenic sequences occur in the transition zone between distinctly different lithologic groups of sediments (i.e. in the transition zone between thicker and more uniform sections of carbonate and siliciclastic or volcanoclastic sequences). Throughout the geologic column, this transition zone occurs in concert with major changes in sea level and is apparent in large first- and second-order sea-level fluctuations; it is also apparent on smaller time frameworks associated with local facies and unconformities resulting from third- and fourth-order fluctuations. The authors believe that major episodes of phosphorite sedimentation reflect changes in tectonism, paleogeography, or paleoclimate that affected and modified global oceanographic processes.

Most north–south Atlantic and Pacific continental margin sediment sequences display distinctive and changing patterns of sedimentation between Lower and Upper Cenozoic sediments, that suggest major paleoclimatic and paleoceanographic changes were taking place. For example, sediment sequences deposited through the Cenozoic of southeastern United States display the following changing patterns through time (Fig. 18.2).

Continental-shelf sediments deposited during the periods of relatively high sea level in the Eocene and Lower Oligocene (Fig. 18.1) were dominated by carbonate deposits characteristic of subtropical to tropical-bank deposition. Within the Eocene section, siliciclastic sediments were almost nonexistent in the southern part of the shelf and gradually increased in concentration northward, where they formed as thin siliciclastic interbeds and as major diluents within many carbonate facies. In addition to this south–north and carbonate–siliciclastic gradient, there is a general upsection increase in total siliciclastic sediment and a respective decrease in carbonate sediments throughout the Eocene and Oligocene deposits. During the Upper Oligocene and continuing through the Neogene and Quaternary, siliciclastic sediments became the dominant lithology with carbonates being subordinant; however, the same latitudinal and shoreward relationships still existed between the two lithofacies.

On the southeastern United States continental shelf, a major phosphogenic episode, with an associated suite of authigenic sediments, occurred in the transition zone between the Eocene and Oligocene limestones and siliciclastic sediment of Pliocene and Quaternary ages. This phosphogenic transition period began on a local basis during the Upper Oligocene, when phosphatic sediments formed in the Carolinas. This change coincided with the second-order TB1 sea-level event which began about 30 my (Fig. 18.1) and represents the beginning of the transition zone of sedimentation. During the TB2 second-order sea-level transgression (Fig. 18.1), which started about 21 my and continued to about 11 my, an abnormally high concentration and extremely large volume of phosphorus was deposited along with contemporaneous facies of dolomite, organic matter, diatomaceous muds, and magnesium-rich clays. By the TB3 second-order sea-level event (Fig. 18.1), which started about 10 my and continued into the Pliocene and Pleistocene, sedimentation was dominated by siliciclastic deposits in which subordinate carbonates occurred as the southernmost facies or as minor interbeds of carbonate-rich sediments capping a dominantly siliciclastic sediment. Minor concentrations of primary phosphorite formed locally during the short Pliocene third-order sea-level cycle between 5 and 4 my (Riggs, 1984). Local concentrations of Pleistocene–Holocene phosphorite on the North Carolina shelf are partially reworked from the Miocene Pungo River Formation, although some phosphate seems to be primary. These younger phosphates apparently formed during one or more of the brief Pleistocene phosphogenic episodes that coincided with Pleistocene transgressive cycles (Riggs, Ellington & Burnett, 1983; Riggs *et al.*, 1985).

Temporal distribution of Upper Cenozoic phosphorites

Known occurrences of Upper Cenozoic phosphorites are listed in Table 18.1. Where the deposits are adequately dated, they generally fall into four episodes as outlined in Table 18.2 which summarizes the ages and durations of phosphorite sedimentation on the Atlantic and Pacific continental margins. The first three of four episodes correspond to the second-order sea-level cycle of Haq *et al.* (1987) as outlined in Figure 18.1. Phosphorite deposits formed during episode II and associated with the TB2 second-order transgression are laterally and quantitatively the most extensive deposits within the Upper Cenozoic. During this episode, anomalous concentrations of phosphorite formed contemporaneously throughout a major portion of the world's continental margins at latitudes lower than 45° (Sheldon, 1964; Cook & McElhinny, 1979; Baturin, 1982).

Episode I: Upper Oligocene–Lower Lower Miocene

There is growing evidence that the Upper Cenozoic phosphogenic episode began in Upper Oligocene time. The Cooper Formation of South Carolina consists of two units, a lower limestone unit of Eocene age and an upper marl unit of Upper Oligocene age. The latter unit contains phosphatic sands and pebbles in a dolomite matrix (Force *et al.*, 1978). This is the first transition-zone sediment representing the extensive and multiple phosphogenic episodes occurring in the Upper Cenozoic section of southeastern United States (Riggs, 1987). The phosphorite of the San Gregorio Formation at San Juan de la Costa in Baja California, Mexico, previously regarded as Miocene in age (Salas, 1978; Ojeda Rivera, 1979; Galli-Oliver *et al.*, Chapter 10, this volume), has now been proven to be of Upper Oligocene age by diatom biostratigraphy (Kim & Barron, 1986; Kim, 1987) and by radiometric dating of overlying volcanic rocks (Hausback, 1984; McLean, Barron & Hausback, 1984).

In Oceania, phosphogenesis began in the Upper Oligocene. The Gee Greensand on the South Island of New Zealand is a glauconite sand that contains several phosphatized hardground or microsphorite surfaces, high concentrations of phosphate pebbles, and abundant phosphatized fossil fragments. This sediment sequence has been dated as Upper Oligocene to Lower Miocene (M. Laird and B. Field, pers. comm., 1986). The Chatham Rise phosphorites east of New Zealand, are loose nodular gravels intermixed with glauconitic sandy muds in 400 m of water on Oligocene chalk (von Rad & Kudrass, 1984). The phosphorite nodules range in age from Upper Oligocene and Lower Miocene to Lower Upper Miocene.

Episode II: Upper Lower Miocene–Middle Miocene

Marine phosphogenesis during the Upper Cenozoic was laterally the most widespread and quantitatively the most important during episode II (Fig. 18.1 and Table 18.2). Phosphate concentrations within these deposits constitute the primary ore for many active mines in Florida and North Carolina that presently supply about 25% of the world's phosphate needs (Stowasser, 1985). Similar deposits are either presently being mined on a local scale, or are being studied for potential future mining in Peru, Mexico, South Africa, New Zealand, and Venezuela. Phosphorite sedimentation during episode II began generally in the upper Lower Miocene and continued through the Middle Miocene (Fig. 18.1 and Table 18.2). This phosphogenic episode is coincident with the second-order TB2 sea-level event (Fig. 18.1) of Haq *et al.* (1987). Phosphogenesis was not continuous during this time interval, but was episodic and related to high frequency third- and fourth-order eustatic sea-level fluctuations (Riggs, 1984). This has been clearly demonstrated for the Miocene deposits of both the southeastern and southwestern United States (Riggs *et al.*, 1985; Garrison, Kastner & Reimers, Chapter 23, this volume; Riggs & Mallette, Chapter 31, this volume; Riggs *et al.*, Chapter 29, this volume; Stephen W. Snyder, Hine & Riggs, Chapter 30, this volume).

Episode III: Upper Miocene–Pliocene

Phosphorites of episode III are well known, but are not as extensive as episode II Miocene phosphorites. In North Carolina, the Lower Yorktown Formation is a phosphorite unit that was deposited during the Lower Pliocene transgression about 5–4 my (Scott W. Snyder, Mauger & Akers, 1983) and constitutes the uppermost cycle of the phosphatic Neogene sequence (Riggs, 1984; Riggs & Belknap, 1988). This Lower Yorktown phosphorite is thought to be equivalent to the Bone Valley Formation which constitutes a major portion of the ore section in the Central Florida Phosphate District (Riggs, 1979; Scott, Chapter 26, this volume).

A comparable, but far less extensive stratigraphic sequence occurs in South Africa (Hendey, 1981; Hendey & Dingle, 1989). The phosphatic Saldanha Formation was deposited during the Lower–Middle Miocene transgression. These sediments were partially eroded and reworked during regressions associated with higher frequency sea-level fluctuations (between 9.8 and 6.6 my) producing the Upper Gravel Member of the Saldanha Formation. The Lower Pliocene transgression led to the deposition of up to 20 m of very shallow-water phosphorite sands of the Varswater Formation, the main ore bed in the Varswater phosphate mine. The erosion surface between the Saldanha and Varswater Formations was correlated with the Upper Miocene

Table 18.1. *Upper Cenozoic marine phosphorites of the world*

Location	References
Episode I: Upper Oligocene – Lower Lower Miocene	
North America	
Mexico, Baja California Sur San Gregorio Fm.	Hausback (1984); Kim (1987); Galli-Oliver *et al.*, (Chptr. 10);
United States, South Carolina Cooper Fm.	Force *et al.* (1978)
South America	
Argentina, Santa Cruz San Julian Mbr. of Patagonia Fm.	H. Leanza (pers. comm., 1988)
Oceania	
New Zealand South Island, Gee Greensand	M. Laird & B. Fields (pers. comm., 1986)
Episode II: Upper Lower Miocene – Middle Miocene	
North America	
Cuba, Habana Province	Pokryshkin (1967 & 1969); Ilyin *et al.*, (Chptr. 9)
United States, Southeastern Province North Carolina, Pungo River Fm. South Carolina, Georgia, and Florida, Hawthorn Fm.	Woolsey (1976); Riggs (1984); Riggs *et al.* (1985, 1988, Chptr. 29); Scott (Chptr. 26); Snyder (1988)
United States, Blake Plateau Offshore	Manheim *et al.* (1980)
United States, California Monterey Fm.	Pisciotto & Garrison (1981); Roberts (1989); Garrison *et al.* (Chptr. 23)
United States, California offshore Borderlands	Dietz *et al.* (1942); Roberts (1989)
South America	
Venezuela, Capadave Fm.	Rodriguez (1989)
Brazil, Pernambuco and Bahia Offshore	Murray *et al.* (1891); Harrington *et al.* (1966)
Argentina, Santa Cruz Monte Leon Mbr. of Patagonia Fm.	H. Leanza (pers. comm., 1988)
Argentina, Tierra del Fuego	H. Leanza (pers. comm., 1988)
Argentina, Malvinas Plateau	Murray *et al.* (1891)
Peru, Sechura Basin	Cheney *et al.* (1980)
Peru, Pisco Basin	Allen & Dunbar (1988)
Chile, Caleta Herradura Fm.	Valdebenito (1989)
Africa	
Morocco, Offshore	Summerhayes *et al.* (1972 and Chptr. 14)
Gabon–Congo	Giresse & Cornen (1976); Giresse (1980); Slansky (1980)
South Africa, Saldanha Fm.	Hendey (1981); Hendey & Dingle (1989)
South Africa, Offshore	Birch (1979 and Chptr. 13)
Europe	
Italy, Salentino Peninsula, Pietra Leccese Fm.	Tarulli & Marcucci (1928); Melidoro & Zezza (1969); Notholt & Highley (1979)
Southeastern Sicily	Tedesco (1966); Di Grande *et al.*, (1979); Notholt & Highley (1979); Pedley & Bennett, 1985; Carbone *et al.*, (1987)
Spain, Northern Offshore	Lucas *et al.* (1978)
Portugal, Offshore	Gaspar (1982)
Asia	
Siberia, Sea of Okhotsk, Sea of Japan, and Sakhalin Peninsula	Oksengorn (1980); Bersenev *et al.* (Chptr. 15)
Philippines, Negros	Notholt (1968)
Indonesia, eastern Java	Hehuwat (1975)
Oceania	
New Zealand	
Otago, Clarenden Hill	Watters (1968)
Chatham Rise, Offshore	Cullen (1980)
Campbell Plateau, Offshore	Glasby & Summerhayes (1975)
Australia, southern Victoria	Baker (1945); Carter (1978); Cook & O'Brien (Chptr. 8)
Episode III: Upper Miocene – Pliocene	
North America	
United States, Southeastern Province	
North Carolina, Lower Yorktown Fm.	Snyder *et al.* (1983); Riggs (1984)
Florida, Bone Valley Fm.	Altschuler *et al.* (1964); Riggs (1979); Scott (Chptr. 26)

Table 18.1 *(Contd.)*

United States, Southwestern Province	
California, Sisquoq and Pancho Rico Fm.	Gower & Madsen (1964)
Africa	
South Africa, Varswater Fm.	Hendey (1981); Hendey *et al.* (1989)
Europe	
Italy, Salentino Peninsula	Galdieri (1918); Notholt & Highley (1979)
Asia	
Indonesia, eastern Java	Hehuwat (1975)
Philippines, Negros	Notholt (1968)
Oceania	
Australia, Eastern Continental Margin	S.W. Snyder, (pers. comm., 1988); Cook & O'Brien (Chptr. 8)
Episode IV: Quaternary	
North America	
United States, Southeastern Province Offshore	Gorsline (1963); Pevear & Pilkey (1966); Pilkey & Luternauer (1967); Riggs *et al.* (1983 and 1985); Riggs (1984)
Mexico, Baja California Coastal Region and Offshore	D'Anglejan (1967); Jahnke *et al.* (1983)
South America	
Chile–Peru, Offshore	Baturin & Petelin (1972); Burnett (1977); Baturin (1982)
Africa	
Namibia Offshore	Bremner (1980 and Chptr. 12); Baturin (1982)
Europe	
France, Mediterranean Sea, Offshore	Froget (1972)
Asia	
Arabian Peninsula, Offshore	Gevork'yan *et al.* (1970b)
India, Andaman Island, Offshore	Notholt (1968)
India, Southwestern Continental Shelf	Baturin (1982)
Oceania	
Australia, Eastern Continental Margin	Marshall & Cook (1980); O'Brien & Veeh (1980); Cook & Marshall (1981); Cook & O'Brien (Chptr. 8)

Table 18.2. *Summary of the phosphogenic episodes of the Upper Cenozoic*

Phosphogenic episode	Second-order sea-level cycle[a]	Duration (million years)	Geologic period	Time and extent of maximum phosphogenesis	
IV		3 my	Quaternary	3 my–present;	Regional
III	TB3	7 my	Pliocene–Upper Miocene	10–4 my	Regional
II	TB2	11 my	Middle Miocene–Upper Lower Miocene	20–14 my	Global
I	TB1	9 my	Lower Lower Miocene–Upper Oligocene	30–25 my	Regional

[a] *data from Haq* et al., *1987*

Messinian regression (Hendey, 1981). This Upper Miocene regression may also be correlatable with the major unconformity separating phosphorites of the Pungo River and Hawthorn Formations in southeastern United States from those in the overlying Lower Yorktown and Bone Valley Formations, respectively.

In California, the Upper Miocene–Lower Pliocene Sisquoq Formation and the Lower Pliocene Pancho Rico Formation are locally phosphatic units (Gower & Madsen, 1964). Minor occurrences of Pliocene phosphorite have also been reported in Italy and Indonesia (Table 18.1).

Episode IV: Quaternary

Phosphogenesis continued sporadically through the Quaternary (Table 18.1) producing phosphorites that have limited regional extent. As compared to the Miocene sediments, the amount of phosphate deposited in Quaternary sediments is minor. Some phosphate on continental margins is primary and formed during one or more brief Quaternary phosphogenic episodes coinciding with fourth-order sea-level cycles resulting from rapid glaciation and deglaciation. Known shelf–slope environments of Modern and Quaternary phosphogenesis include Peru–Chile (Burnett, 1977), Baja California, Mexico (Jahnke *et al.*, 1983), Namibia (Baturin, 1982), and East Australia (O'Brien & Veeh, 1980; O'Brien *et al.*, 1981).

In many shelf areas, phosphate occurring in the Pleistocene and Holocene sediments is largely reworked from underlying units of Tertiary age. Such deposits include those offshore of North Carolina (Luternauer & Pilkey, 1967; Pilkey & Luternauer, 1967; Riggs *et al.*, 1985), South Carolina, Georgia, and Florida (Gorsline, 1963; Pevear & Pilkey, 1966; Birdsall, 1978). An extensive belt of phosphatic Quaternary sediments occurs on the shelf of northwest Africa. Phosphate in these surface sediments was derived by erosion from underlying phosphatic units that range in age from the Cretaceous through the Miocene with no evidence for modern phosphate formation (Nutter, 1969; Summerhayes, 1970; Summerhayes, Nutter & Tooms, 1972; Bee, 1973; McArthur, 1974; Summerhayes & McArthur, Chapter 14, this volume). Similar deposits also occur in central-west Africa (Giresse, 1980; Barusseau & Giresse, 1987) and southwest Africa (Birch, 1979, Chapter 13, this volume). The reworked Quaternary phosphate is only a minor occurrence, except in areas adjacent to older Tertiary units that are major phosphorites. In fact, it has been demonstrated (Riggs *et al.*, 1985; Riggs, 1987) that the distribution of phosphate in Quaternary shelf sediments closely reflects the distribution within underlying sediment units and that the process of reworking significantly dilutes the concentration of phosphate in the surficial sediments.

Spatial distribution of Upper Cenozoic phosphorites

Most known Upper Cenozoic phosphate deposits occur on the emerged coastal plain and are geologically well known. In most cases, these emerged deposits are only the updip limit of much larger sections that extend seaward beyond the coastal plain and constitute large portions of the stratigraphic section underlying our modern continental shelves (Riggs, 1987). During the Upper Cenozoic, rapid sea-level cyclicity repeatedly moved the depositional and erosional processes across the shelf–slope system producing abnormally thick sections which led to major continental margin accretion and progradation, particularly during the Miocene. Locally, Tertiary sediments are still exposed on the seafloor such as in Onslow Bay, North Carolina (Fig. 18.3). However, they are generally buried below covers of Pleistocene and Holocene surface sediments such as in the Aurora Basin, North Carolina and Jacksonville Basin, Florida (Fig. 18.3). Thus, thicker and more extensive sequences of Upper Tertiary sediments probably occur on most Modern continental shelves than occur on the adjacent coastal plains. However, these sediments are poorly known due to the extensive surficial cover and the fact that the shallow subsurface Neogene geology of most of the world's shelves is poorly known.

Table 18.3 summarizes the known distribution of episode II (upper Lower–Middle Miocene) sediments containing major phosphorites along the modern continental margins. The distribution in Table 18.3 is based on known occurrences, many of which are minor and poorly known. If the previous paragraph is correct, this distribution is limited by lack of subsurface information on many continental margins. Consequently, future work could delineate significant phosphorite-bearing sediment sequences, particularly of Miocene age. For example, the Brazilian margin south of the Amazon, and the Gulf of Guinea on the West African margin could contain major stratigraphic sequences with extensive Miocene phosphorites in the modern shelf and slope environments.

If the Upper Cenozoic phosphorites are integral responses to common processes associated with oceanographic episodes of global extent, there should be large-scale temporal and geological similarities among the deposits. Details of the sediment patterns within each deposit should be dependent upon the local geologic setting interacting with regional oceanographic and climatic conditions. Each of the known deposits do contain similar patterns of phosphorite sedimentation suggesting that they are products of global events and changing paleoceanographic conditions through time.

Paleoceanographic setting

The distribution of Upper Cenozoic phosphorite deposits is plotted on Figure 18.4, using the Middle Miocene paleogeographic reconstruction of Denham & Scotese (1988). This allows classification of the paleoceanographic position of each deposit (Table 18.4), generally following the approach of Parrish (1982).

The majority of Upper Oligocene to Quaternary phosphorites listed in Table 18.1 occurred on continental shelves influenced by Eastern Boundary paleocurrents (EBP). These include the continental shelves of the North and South Pacific and Atlantic Oceans (Fig. 18.4). At least one major phosphogenic province occurred on continental shelves associated with Western Boundary paleocurrents (WBP) (Fig. 18.4). This includes the Upper Oligocene–Quaternary phosphorites of southeastern United States and Cuba. Perhaps, Miocene phosphorites of Negros, Philippines were associated with the WBP of the North Pacific, and Miocene–Quaternary phosphorites of the east Australian shelf were associated with the WBP of the South Pacific. Both of the latter regions would be excellent prospects for future continental margin drilling programs to delineate the potentially greater subsurface extent of larger sedimentologic provinces.

Other phosphorites of Table 18.1 occur too far south to be explained by paleo-boundary currents associated with major oceanic gyral circulation. These include the New Zealand phosphorites on the southeastern side of South Island, Chatham Rise, and Campbell Plateau; the Australian phosphorites in southern Victoria; the Agulhas Bank phosphorites off the southern coast of South Africa; and the southeastern Atlantic occurrences on the Malvinas Plateau and southeastern coast of Argentina. Rather, they all lie along the northern path of the deep Antarctic circumpolar current, which is thought to have developed with the opening of Drake Passage and thermal isolation of Antarctica in the Lower Miocene (Keller & Barron, 1983). These deposits could be associated with topographic upwelling of that paleocurrent as it crossed and interacted with the shelf and slope environments.

Minor occurrences of phosphorite have been found in the northern portions of the Indian Ocean (Table 18.1). Quaternary phosphorite occurs offshore of North Andaman Island in the Bay of Bengal (Notholt, 1968), phosphorite concretions have been found at a depth of 600 m on the continental slope off the southeastern Arabian Peninsula (Gevork'yan & Chugunnyy, 1970a; Baturin, 1982), and Holocene phosphatic sediments oc-

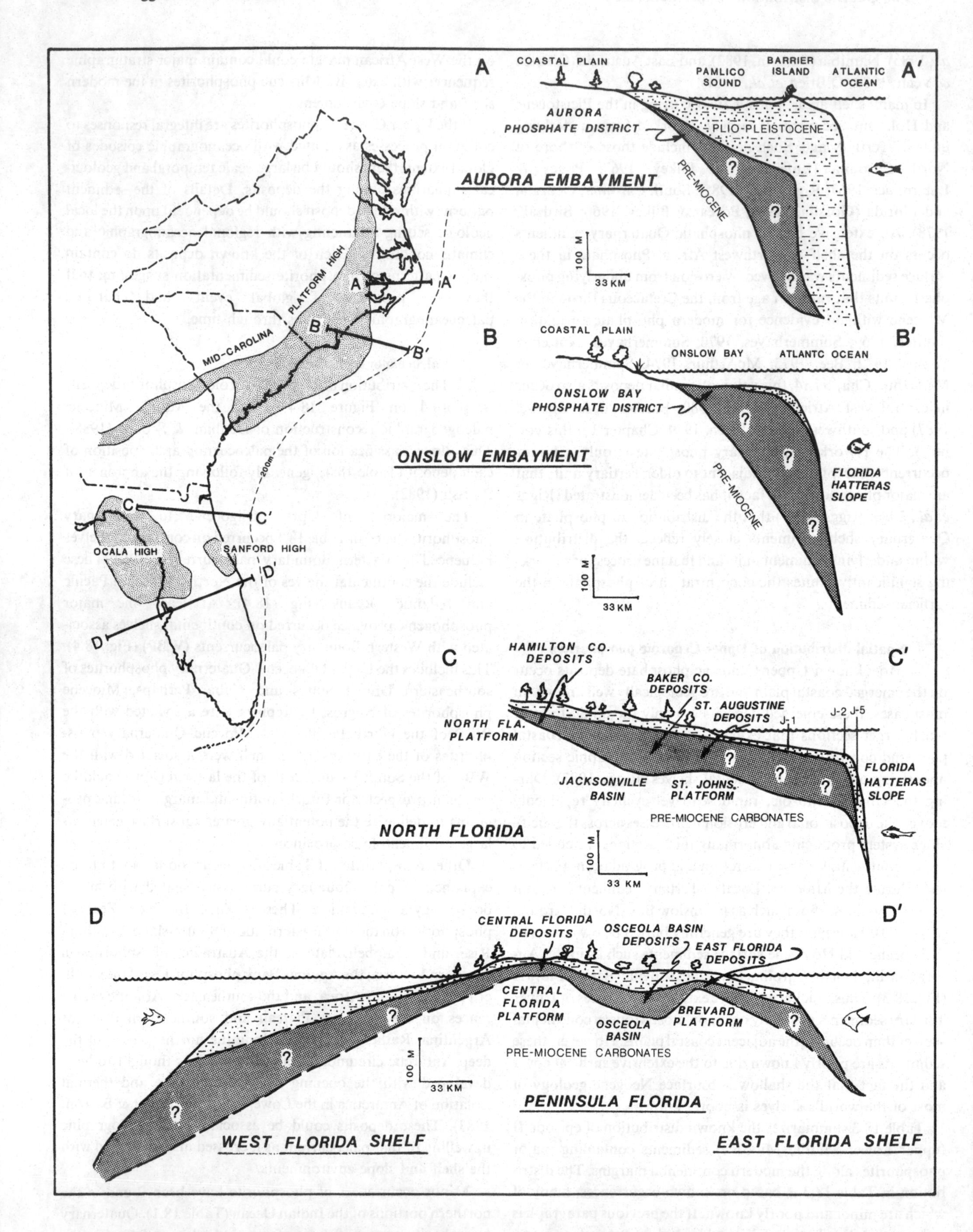

Fig. 18.3. Schematic cross-sections of the southeastern US continental margin. These sections show the distribution of Miocene sediments on the well-drilled coastal plain versus the distribution on the shelf and slope, which is based largely on seismic data with only rare drillhole information. Notice how the well known and extensively-drilled portion represents only the updip-most portion of a much more extensive Miocene section on the submerged continental shelf and slope.

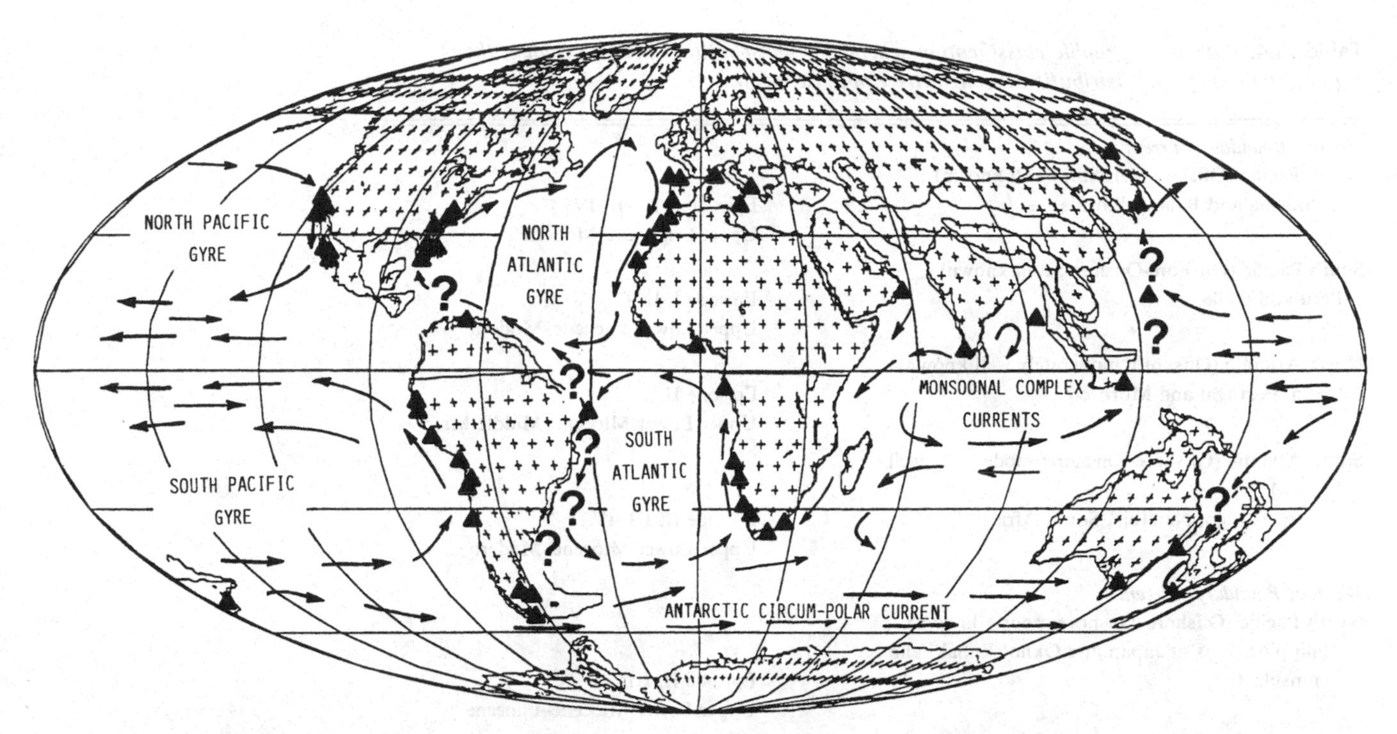

Fig. 18.4. Map showing the distribution of Oligocene–Quaternary phosphorites. Wind-driven surficial circulation is shown by arrows, with major ocean currents identified. Phosphate localities (triangles) are from Table 18.1 and Baturin (1982). Areas with question marks are continental margins with appropriate Western Boundary Currents to have produced topographic upwelling, phosphorites, and associated sediments during the various global episodes of the Upper Cenozoic. If such deposits formed as the model suggests, they probably occur within the shallow subsurface on the continental shelves and have not yet been discovered. Base map is from the Middle Miocene of Denham & Scotese (1988).

Table 18.3. *Regional distribution of known upper Lower–Middle Miocene sediments (Episode II) containing phosphorites along the modern continental margins*

Continental margin	Region
East Atlantic	Portugal, northwest Africa through South Africa, and Agulhas Bank
West Atlantic	North Carolina through Florida, Cuba, Venezuela, Brazil, and south Argentina
East Pacific	California through Baja California Mexico, and Peru through Chile
West Pacific	Sakalin Island, Sea of Japan, Indonesia, Chatham Rise east of New Zealand, and East Australian shelf

cur off of the southwestern tip of the Indian Peninsula (Baturin, 1982). Studies of the western Indian continental shelf by Schott (1970) and Choudhuri & Banerji (1968) have shown that the latter phosphorites are not widespread, but occur with laminated, organic-rich sediment containing radiolaria and diatoms within the oxygen-minimum zone. All of these occurrences may be associated with monsoonal upwelling (Parrish, 1982) resulting from the complex Indian Ocean currents driven by the monsoonal wind system. Phosphatic sediments of possible Quaternary age have been found on the shallow shelf offshore of Socotra Island near the Gulf of Aden (Gevork'yan & Chugunnyy, 1970b), but they appear to be erosion products from Mesozoic and Tertiary rocks on Socotra Island and not primary marine phosphorite.

Minor phosphorite occurrences are found in semi-restricted seas in several parts of the world (Tables 18.1 and 18.4). Miocene phosphorite occurs in the Sea of Japan (Bersenev *et al.*, Chapter 15, this volume) and onshore and offshore of Sakhalin Peninsula in the Sea of Okhotsk (Oksengorn, 1980; Bersenev *et al.*, Chapter 15, this volume). In eastern Java, just south of the Java Sea, Miocene and Pliocene phosphatic shales contain thin beds of phosphorite (Hehuwat, 1975). In the Mediterranean, Miocene and Pliocene phosphorite occurrences have been found in southeastern Italy and Sicily (Galdieri, 1918; Tarulli & Marcucci, 1928; Tedesco, 1966; Melidoro & Zezza, 1969; Di Grande, Lo Giudice & Battaglio, 1979; Notholt & Highley, 1979; Pedley & Bennett, 1985; Carbone *et al.*, 1987), and Quaternary phosphorite occurs offshore of Marseilles, France (Froget, 1972). The paleoceanographic conditions responsible for these minor shelf and slope deposits within restricted seas are not obvious. Further work needs to be done with these intriguing phosphorite occurrences to determine their spatial and temporal limits, petrology and geochemistry, and paleoceanographic setting.

Paleoclimatic setting

The genetic link between upwelling ocean currents, zonal winds, polar glaciation and large temperature gradients between low and high latitudes (Schopf, 1980; Parrish, 1982) suggests that periods of intense upwelling producing marine phosphorites should correlate with periods of polar glaciation

Table 18.4. *Paleoceanographic classification scheme for phosphorite deposits including their regional and temporal distribution through the Upper Cenozoic*

Eastern Boundary Current	
North Pacific (Offshore–Onshore; well known)	
California and Baja California	Episode I, II, III, IV Upper Oligocene–Modern
South Pacific (Offshore-Onshore; well known)	
Peru and Chile	Episode I, II, III, IV Upper Lower Miocene–Modern
North Atlantic (Offshore; moderately well known)	
Spain–Portugal and Morocco	Episode II Upper Lower Miocene–Middle Miocene
South Atlantic (Offshore–Onshore; moderately well known)	
Gabon–Congo, Namibia, South Africa	Episode II, III, IV Upper Lower Miocene–Modern
Western Boundary Current	
North Pacific (Offshore–Onshore; poorly known)	
Philippine, Seas of Japan and Okhotsk, Sakhalin Peninsula	Episode II, III Upper Lower Miocene–Pliocene
South Pacific (Offshore; poorly known)	
Australia, eastern continental shelf	Episode II, III, IV Upper Lower Miocene–Middle Miocene
North Atlantic (Offshore–Onshore; well known)	
Southeastern United States, Cuba, Venezuela	Episode I, II, III, IV Upper Oligocene–Modern
South Atlantic (Offshore–Onshore; very poorly known)	Episode?
Brazil, Argentina	Upper Tertiary
Indian Ocean monsoonal upwelling (Offshore; very poorly known)	
Arabian Penn., SW India, Andaman Island	Episode IV Quaternary
Antarctic circumpolar current (Offshore–Onshore; moderately well known)	
New Zealand, Victoria Australia, and southern Argentina and South Africa	Episode I, II Upper Oligocene–Middle Miocene
Restricted sea (Offshore–Onshore; very poorly known)	
Indonesia–Java Sea; USSR–Sakhalin Island, Okhotsk Sea; Japan–Sea of Japan; Italy and France–Mediterranean Sea	Episode II, III, IV Upper Lower Miocene–Quaternary

(Fig. 18.5). This has been demonstrated for Phanerozoic phosphorites in general (Sheldon, 1980 and 1987). Vail *et al.* (1977) believe that some second- and many third-order global sea-level fluctuations, particularly during the Upper Cenozoic, are the direct products of polar glaciation. If this is correct, a correlation should exist between phosphorites and sea-level fluctuation. This has been shown to be the case for the Neogene phosphorites of North Carolina (Stephen W. Snyder, 1982; Riggs, 1984; Riggs *et al.*, 1985; Stephen W. Snyder *et al.*, Chapter 30, this volume) and possibly for the Quaternary phosphorites of the Peru and Chile continental shelf (Burnett, 1979). Riggs *et al.*, Stephen W. Snyder *et al.*, Riggs & Mallette, and Scott W. Snyder (Chapters 29, 30, 31, 32, all in this volume) demonstrate that the Miocene Pungo River Formation consists of at least 18 depositional sequences that formed in response to fourth-order eustatic glacial and deglacial sea-level cycles that had an average duration of about 200 ky per cycle. Each of these 18 sea-level events contains, or has the potential for containing, a major phosphorite facies depending upon the regional conditions and detailed geological framework at that particular time.

Correlation of the known record of polar glaciation and temporal distribution of Oligocene–Holocene continental margin phosphorites to the sea-level curve of Haq *et al.* (1987) is shown in Figure 18.5. Problems exist in this correlation. Episodes of Antarctic and mid-latitude southern hemisphere glaciation are incompletely established and in dispute, particularly for the Oligocene (Mercer, 1983). In addition to effects of glaciation, sea-level curves contain the effects of marine sedimentation and volcanism, variations of sea-floor spreading (Hallam, 1984), and effects of ocean-basin enlargement by continent–continent collision (Pitman & Kominz, 1981) such as the Oligocene subduction of northern India beneath Asia (Powell,

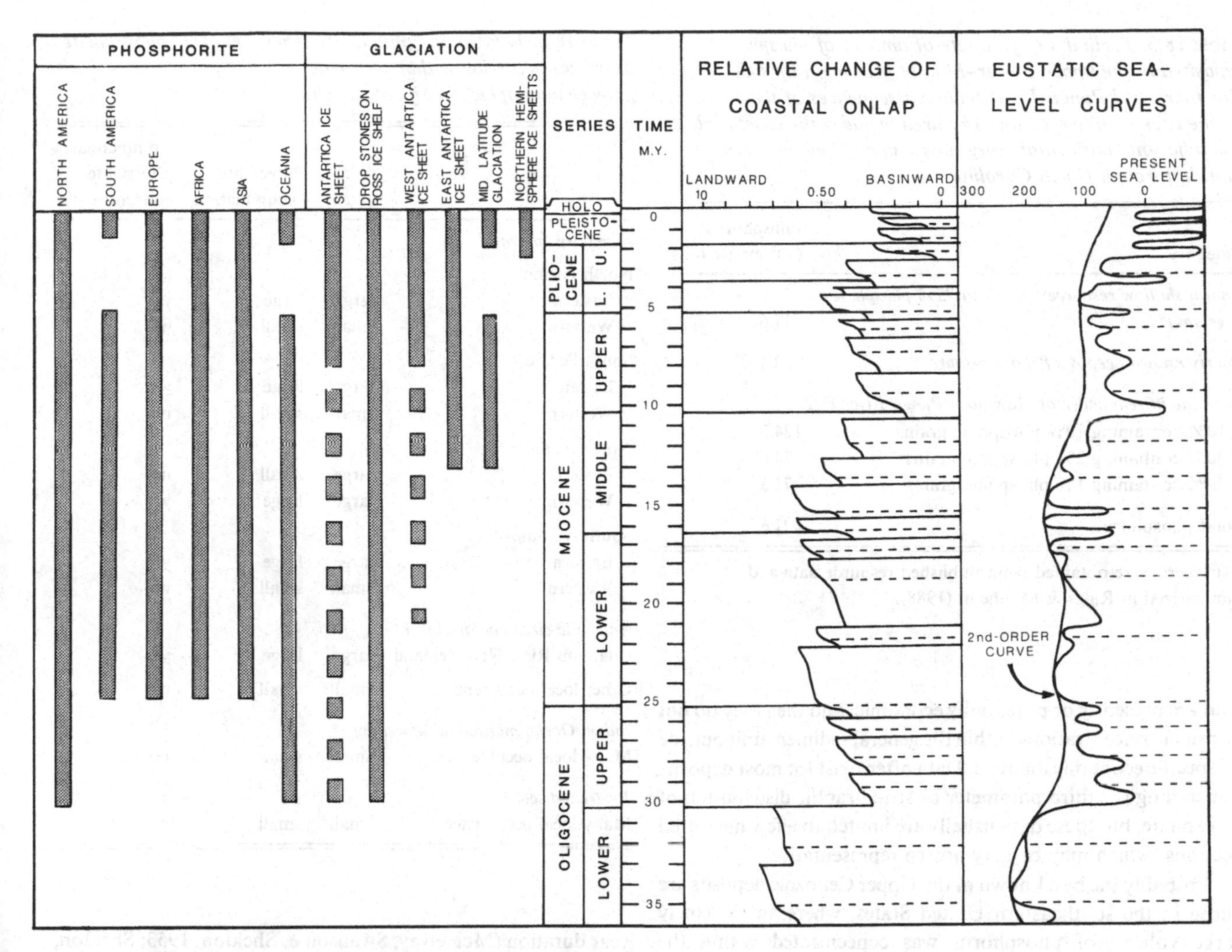

Fig. 18.5. Comparison of Upper Cenozoic phosphogenesis, glaciation, and sea-level variations. Glaciation data is from Mercer (1983) and coastal onlap and eustatic sea-level curves are from Haq *et al.* (1987). Glacial distribution is represented by dashed lines where in dispute.

1979). Upper Cenozoic phosphorites are difficult to date due to: 1) paucity of fossils; 2) lack of geochemical age dating techniques for phosphorites too old for the uranium disequilibrium method (about 800 000 y); and 3) poorly known geologic understanding of many deposits due to lack of detailed studies. Thus, the correlations shown on Figure 18.1 should be regarded as tentative. In general, the correlation shows a coincidence of the phosphogenic episodes from Oligocene to Quaternary to the general drop in sea level beginning in the Upper Oligocene and initiation of polar glaciation.

Phosphorite formation occurs in shelf and slope environments associated with the upper and lower boundaries of the oxygen-minimum zone formed by strong upwelling and associated organic productivity (Burnett, Veeh & Soutar, 1980). Lithofacies and biofacies relationships in major Neogene phosphorites support these conditions and environments of formation (Govean & Garrison, 1981; Riggs, 1984). At times of sea-level lowstands, zonal winds are strong and upwelling intense (Sarnthein, 1981); however, the ocean is withdrawn below the outer edge of shallow continental shelves, eliminating shelf environments favorable to phosphorite sedimentation (Arthur & Jenkyns, 1981). At times of sea-level highstands, during interglacials, zonal winds decrease and upwelling becomes less intense (Sarnthein, 1981); consequently, the broad, shallow-water shelf environments are well oxygenated with the associated boundary currents eroding the shelf–slope margin (Stephen W. Snyder *et al.*, Chapter 30, this volume).

Magnitude of phosphorite provinces

In spite of recent advances (Froelich *et al.*, 1983), the basic understanding of the phosphorus budget is limited. In order to resolve problems associated with episodicity of phosphorites in time and space, problems of phosphorus supply both within the ocean reservoir and in continental margin sediments must first be solved. Evaluation of the magnitude of phosphorus tied up in continental margin sediments through the Upper Cenozoic should be assessed on the basis of parameters such as aerial extent, total phosphate, and stratigraphic variations of phosphate content for each deposit. Detailed studies of the first two magnitude parameters are generally limited to the few deposits

Table 18.5. *Preliminary estimate of amount of phosphorus deposited in the upper Lower–Middle Miocene (Episode II) Hawthorn and Pungo River sediment sequences of the southeastern United States. The area includes the coastal plain and adjacent continental margin extending from southern Florida through North Carolina*

Category	Phosphorus (10^9 metric tons)
Known shallow resources (25% to 75% phosphate grains)[a]	17.6
Poorly known deep & offshore resources	132.0
Estimate in remainder of Hawthorn/Pungo River Fm.	
10% containing 10% phosphate grains	124.2
30% containing 2% phosphate grains	74.5
59% containing 1% phosphate grains	73.3
Total phosphorus	421.6

[a]Numbers were obtained from published resource data and summarized in Riggs & Manheim (1988).

that are presently or potentially economic, and then they do not consider concentrations within the general sediment unit outside of specific economic districts. Data often exist for most deposits concerning the third parameter or stratigraphic distribution of phosphate, but these data usually are limited to a few measured sections, which may or may not be representative.

Probably the best known of the Upper Cenozoic deposits are those in the southeastern United States, where an extremely large volume of phosphorus was concentrated within the stratigraphic section. The Miocene Pungo River and Hawthorn Formations occur over a very large area (approximately 1000 km in length, 200 km in width, and range from 0 to 500 m in thickness). Specific lithofacies represent periods of major phosphate formation and accumulation (25–75% phosphate grains); these areas represent the economic and potentially economic phosphate deposits of southeastern US. Much larger portions of these formations contain marginal (10%) and submarginal (3%) concentrations of phosphate (10–25% and 2–10% phosphate grains, respectively). All of the remaining lithofacies within the Pungo River and Hawthorn units contain minor concentrations of phosphate grains (approximately 1%). In North Carolina, these phosphatic units were deposited during multiple depositional events that represented about three million years of depositional history beginning about 18–19 million years ago and continuing through to about 12 million years ago (Scott W. Snyder, 1988; Riggs *et al.*, Chapter 29, this volume). Table 18.5 summarizes a preliminary, and very conservative, estimate of the amount of phosphorus concentrated within the Pungo River and Hawthorn Formations during the Miocene.

The estimated 422×10^9 metric tons of phosphorus deposited during three million years between 18 and 12 million years ago (Riggs *et al.*, Chapter 29, this volume) represents a very significant concentration. The southeastern US phosphate province is of a similar order of magnitude as that calculated for the Permian Phosphoria Formation deposited over a 2–6 million year duration (McKelvey, Swanson & Sheldon, 1953; Sheldon, 1981).

Table 18.6. *Relative magnitude of Upper Cenozoic phosphorite provinces occurring within each category of the paleoceanographic classification scheme*

	Areal extent	Phosphate accumulation	Commercial phosphate concentration
Boundary currents			
North Pacific			
Eastern	large	large	yes
Western	small	small	no
South Pacific			
Eastern	large	large	yes
Western	small	small	no
North Atlantic			
Eastern	large	small	no
Western	large	large	yes
South Atlantic			
Eastern	large	large	yes
Western	small	small	no
Antarctic circumpolar current			
Chatham Rise, New Zealand	large	large	yes
Other local occurrences	small	small	no
Indian Ocean monsoonal upwelling			
Many local occurrences	small	small	no
Restricted sea			
Many local occurrences	small	small	no

It is interesting now to speculate on the size of the global upper Lower- to Middle Miocene phosphogenic episode (II). The following considerations would suggest that this Miocene event was extremely large scale: 1) numerous phosphate deposits were formed during episode II (Table 18.1); 2) most of these other deposits are poorly explored and known (Table 18.4); 3) most of them have only a small known extent in their thin updip sections that occur on the coastal plains; 4) those that do crop out on the adjacent continental shelves have only been superficially investigated; and 5) most important is the fact that much of the Miocene section remains unknown and buried in the subsurface below the world's continental margins (Riggs, 1987).

In addition, the Upper Miocene and Pliocene (episode III) are characterized by a less extensive sequence of phosphatic sediments that represented a shorter duration of deposition. Local and generally minor concentrations of phosphate often occur within the Quaternary units (episode IV) on modern continental shelves. A general classification of magnitude of phosphorite paleoceanographic provinces is presented in Table 18.6. Enough data are available to allow crude estimates of hypothetical magnitudes, but the necessary analyses have not been carried out to quantify it beyond a few general terms as follows. Deposits described as *large* are of regional or basinal extent, contain sediments with major concentrations of phosphate that exceed a billion tons of resource, and have known beds of potentially economic grade and thickness. Deposits described as *small* have local occurrences and distributions, contain sediments with minor concentrations of phosphate, and do not have known beds of potentially economic grade and thickness.

Several important points can be made from Table 18.6. The EBC phosphate provinces, with the exception of the north Atlantic, are large in all respects. The only known large WBC phosphate province is in the north Atlantic. This WBC is the Gulf Stream, an unusually large, high-energy current system formed by merging the northern and southern equatorial currents, that dominates the continental margin processes along the southeastern United States (Pietrafesa, Chapter 1, this volume; Popenoe, Chapter 28, this volume; Riggs & Mallette, Chapter 31, this volume; Stephen W. Snyder *et al.*, Chapter 30, this volume).

It is interesting to speculate on the potential for major phosphorite deposits associated with the WBC in the south Atlantic. Only minor marine Upper Cenozoic sediments occur on the continental margin of eastern South America (Fig. 18.4 and Table 18.1). These include submarine occurrences off the eastern hump of Brazil and on the Malvinas Plateau off southern Argentina. Murray & Renard (1891) reported that HMS *Challenger* stations within these areas contained glauconitic pebbles and phosphatic concretions in blue muds. Harrington *et al.* (1966) described dredged phosphatic nodules off the coasts of Bahia and Sao Paulo. Also, several stratigraphic occurrences have now been described along the Atlantic coastal plain of Argentina (Leanza *et al.*, 1981 and 1989; H. Leanza, pers. comm., 1988). If the models are correct, there should be three areas of Upper Cenozoic sediments containing phosphorites and associated sediments in the subsurface of the continental shelf and slope environments. These three areas are indicated on Figure 18.4 and include the following: 1) the first area extends from the eastern hump of Brazil, southwest along the Brazilian coast to Argentina; 2) the second area extends from the eastern hump of Brazil, northwest towards Belem and southeast of the influence of Amazon River delta; and 3) the third area occurs within the Caribbean Sea north of Venezuela and west of the Windward Islands and the Orinoco River delta.

Similar deposits could exist in comparable continental margin situations of the western Pacific. The Kuroshio and East Australian WBC are similar to their counterpart currents in the Atlantic Ocean. The Kuroshio Current, a very strong component of the North Pacific Gyre, interacts with the Asian archipelagos of the Marianas, Japan, and Kuril Islands. Due to the sedimentation processes associated with active tectonic plate boundaries and subduction, there is low probability for the major development of extensive continental margin phosphorites and associated authigenic sediments. However, the presence of minor phosphorites on some of the archipelagos and within the restricted seas west of the archipelagos (Table 18.1 and Fig. 18.4) suggests that more work needs to be done in the Philippine Sea, Sea of Japan, and Sea of Okhotsk and associated land areas.

Another potential area of Neogene phosphogenesis is the northeastern Australian margin. The target area would be Miocene sediments occurring stratigraphically below the southern end of the modern Great Barrier Reef system and situated southward of Miocene carbonate sediment facies and northward of iron-rich and glauconite-dominated sediment systems of southeastern Australia. The latter iron-rich phosphorites occur in Neogene sediments on the outer shelf and upper slope between 29°S and 32°S latitude (Marshall & Cook, 1980; Cook & Marshall, 1981; O'Brien, 1983; Cook & O'Brien, Chapter 8, this volume) and southward in the Neogene basins of the southern coast of Victoria (Baker, 1945; Carter, 1978; Cook & O'Brien, Chapter 8, this volume).

Summary: model of continental margin phosphogenesis

A model, summarized in Figure 18.1, attempts to explain the relationship of phosphorite sedimentation to continental margin environments of deposition dominated by similar paleoclimatic and paleoceanographic conditions. This model has been developed for the well studied and well dated deposits on the Carolina margin (Riggs, 1984; Riggs & Mallette, Chapter 30, this volume; Riggs *et al.*, Chapter 29, this volume; Scott W. Snyder, Chapter 32, this volume; Stephen W. Snyder *et al.*, Chapter 30, this volume) and should be applicable to similar deposits on other Upper Cenozoic continental margins. Interpretation of changing patterns of seismic, faunal and floral, and lithologic depositional facies demonstrate that primary phosphorite sedimentation begins in the early stages of post-glacial transgression. Phosphorites and associated authigenic–diagenetic sediments begin to form on the continental slope in association with the oxygen-minimum zone and as organic matter begins to accumulate. Such phosphorites generally have low concentrations due to maximum levels of diluent siliciclastic sedimentation as the continental margins rapidly prograde during sea-level lowstands (Riggs & Belknap, 1988; Stephen W. Snyder *et al.*, Chapter 30, this volume). Conditions for major phosphorite accumulation are optimal during the mid-stages of the transgression: 1) zonal winds are still strong causing major upwelling to persist; 2) the boundary current has migrated landward with increased interaction with the continental margin topography intensifying the upwelling; 3) as the transgression progresses, continental shelf environments expand; and 4) dilution from land-derived siliciclastics rapidly decreases as the depositional environments deepen. During maximum transgression, zonal winds have diminished and boundary currents decrease in intensity and override the continental margin where they actively erode the previously deposited sediments (Pietrafesa, Chapter 1, this volume; Popenoe, Chapter 28, this volume; Stephen W. Snyder *et al.*, Chapter 30, this volume). Phosphorites that occur within the highstand lithofacies are often secondary deposits in which the phosphate grains have been eroded and redeposited from the primary phosphorites formed within earlier stage lithofacies. The following regression continues to rework these deposits which are then modified by late diagenetic and erosional processes during the subsequent sea-level lowstand.

Growing knowledge of Upper Cenozoic phosphogenesis constantly improves the models for interpretation of older phosphorites. This is particularly true for Jurassic, Permian and Lower Carboniferous phosphorites which were also deposited at times of polar glaciation (Sheldon, 1980). However, it is more difficult to apply the paleoceanographic and paleoclimatic models of Upper Cenozoic phosphogenesis to extensive Upper Cretaceous and Paleogene deposits of the Tethys phosphate provinces and to many Proterozoic and Cambrian phosphorites.

References

Allen, M.R. & Dunbar, R.B. (1988). Phosphatic sediments in the Pisco Basin. In *Cenozoic Geology of the Pisco Basin*, ed. R.B. Dunbar & P.A. Baker, pp. 109–26. Rice University, Houston, Texas.

Altschuler, Z.S., Cathcart, J.B. & Young, E.J. (1964). *Geology and Geochemistry of the Bone Valley Formation and its Phosphate Deposits, West Central Florida.* Guidebook Field Trip No. 6, Geological Society of America Annual Meeting, Miami Beach, 68pp.

Arthur, M.Z. & Jenkyns, H.C. (1981). Phosphorites and paleoceanography. *Oceanologica Acta*, **4 SP**, 83–96.

Baker, G. (1945). Phosphate deposit near Princetown, Victoria, Australia. *Journal of Sedimentary Petrology*, **15**, 88–92.

Barusseau, J.P. & Giresse, P. (1987). Some mineral resources of the west African continental shelves related to Holocene shorelines: phosphorite (Gabon, Congo), glauconite (Congo), and ilmenite (Senegal, Mauritania). In *Marine Minerals*, ed. P.G. Teleki, M.R. Dobson, J.R. Moore, & U. von Stackelberg, pp. 133–53. NATO Advanced Research Workshop, D. Reidel Publishing Co., Dordricht, Holland.

Baturin, G.N. (1982). *Phosphorites on the Sea Floor*. Elsevier, Amsterdam, Oxford and New York, 343pp.

Baturin, G.N. & Petelin, V.P. (1972). Phosphorite concretions on the shelf off Chile. *Litologiya i Poleznye Iskopaemye*, **3**, 3–10.

Bee, A.G. (1973). The marine geochemistry and geology of the Atlantic continental shelf of central Morocco. Unpublished Ph.D. thesis. Imperial College, University of London, 263pp.

Birch, G.F. (1979). Phosphatic rocks on the western margin of South Africa. *Journal of Sedimentary Petrology*, **49**, 93–110.

Birdsall, B.C. (1978). Eastern Gulf of Mexico, continental shelf phosphorite deposits. Unpublished M.Sc. thesis. University South Florida, St Petersburg.

Bremner, J.M. (1980). Concretionary phosphorite from SW Africa. *Journal of the Geological Society of London*, **137**, 773–86.

Burnett, W.C. (1977). Geochemistry and origin of phosphorite deposits from off Peru and Chile. *Geological Society of America Bulletin*, **88**, 813–23.

Burnett, W.C. (1979). Oceanic phosphate deposits. In *Fertilizer Mineral Potential in Asia and the Pacific*, ed. R.P. Sheldon & W.C. Burnett, pp. 119–44. East–West Center, Honolulu, Hawaii.

Burnett, W.C. (1980). Apatite–glauconite associations off Peru and Chile: paleo-oceanographic implications. *Journal of Geological Society of London*, **137**, 757–64.

Burnett, W.C., Veeh, H.H. & Soutar, A. (1980). U-series, oceanographic and sedimentary evidence in support of contemporary formation of phosphate nodules off Peru. In *Marine Phosphorites*, ed. Y.K. Bentor, pp. 61–77. Society of Economic Paleontologists and Mineralogists Spec. Pub. 29, Tulsa.

Carbone, S., Grasso, M., Lentini, F. & Pedley, H.M. (1987). The distribution and palaeoenvironment of early Miocene phosphorites of southeast Sicily and their relationships with the Maltese phosphorites. *Palaeogeography, Palaeoclimatology, Palaeoecology*, **58**, 35–53.

Carter, A.N. (1978). Phosphatic nodule beds in Victoria and the Miocene–Pliocene eustatic event. *Nature*, **276**, 258–9.

Cheney, T.M., McClellan, G.H. & Montgomery, E.S. (1980). Sechura phosphate deposits, their stratigraphy, origin, and composition. *Economic Geology*, **74**, 232–59.

Choudhuri, R. & Banerji, K.C. (1968). Some studies on shelf sediments from the western coast off Bombay and Ratnagiri. *Technology*, **5**, 192–203.

Cook, P.J. & Marshall, J.F. (1981). Geochemistry of iron and phosphorus-rich nodules from the east Australian continental shelf. *Marine Geology*, **41**, 205–21.

Cook, P.J. & McElhinny, M.W. (1979). A reevaluation of the spatial and temporal distribution of sedimentary phosphate deposits in the light of plate tectonics. *Economic Geology*, **74**, 315–30.

Cullen, D.J. (1980). Distribution, composition and age of submarine phosphorites on Chatham Rise, east of New Zealand. In *Marine Phosphorites*, ed. Y.K. Bentor, pp. 139–48. Society of Economic Paleontologists and Mineralogists Spec. Pub. No. 29, Tulsa.

D'Anglejan, B.F. (1967). Origin of marine phosphorites off Baja California, Mexico. *Marine Geology*, **5**, 15–44.

Denham, C.R. & Scotese, C.R. (1988). Terra Moblis: plate tectonics for the Macintosh. Earth in Motion Technologies, Austin, Texas.

Dietz, R.S., Emery, K.O. & Shepard, F.P. (1942). Phosphorite deposits on the sea floor off southern California. *Geological Society of America Bulletin*, **53**, 815–48.

Di Grande, A., Lo Giudice, A. & Battaglio, M. (1979). Dati geo-mineralogice preliminari sur livelli Miocenice a fosfate di Scicli-Donnalucata (Sicilia SE). *Bolletino della Societa Geologica Italiana*, **97**, 383–90.

Dorf, E. (1964). The use of fossil plants in paleoclimatic interpretations. In *Problems in Paleoclimatology*, ed. A.E.M. Narin, pp. 13–31, 46–48. Interscience, New York.

Force, D.R., Gohn, G.S., Force, L.M. & Higgins, B.B. (1978). Uranium and phosphate resources in the Cooper Formation of the Charleston region, South Carolina. *South Carolina Geological Survey, Geologic Notes*, **22**, 17–31.

Froelich, P.N., Kim, K.H., Jahnke, R., Burnett, W.C., Soutar, A. & Deakin, M. (1983). Pore water fluoride in Peru continental margin sediments: uptake from seawater. *Geochimica Cosmochimica Acta*, **47**, 1605–12.

Froget, C. (1972). Sur la presence de couches phosphatees pleistocene en Mediterranee nord-occidental (sud de Marseille, Bouches-du-Rhone). *Société Géologique de France*, **1**, 34–5.

Galdieri, A. (1918). Sullafosforite di Leuca. *Atti. R. Ist. Incoragg. Napoli*, **Ser. 6, 10**, 10pp.

Gaspar, L. (1982). Fosforites de margem continental Portugeusa – alguns aspectos geoquimicos: *Bull. Soc. Geol. Port. Lisboas*, **23**, 79–90.

Gevork'yan, V.K. & Chugunnyy, Y.G. (1970a). A new find of accumulations of phosphate concretions in the Indian Ocean. *Academy of Science USSR, Doklady, Earth Science Section*, **187**, 230–2.

Gevork'yan, V.K. & Chugunnyy, Y.G. (1970b). Phosphorite nodules in the bottom sediments of the Gulf of Aden. *Oceanology*, **10**, 233–41.

Giresse, P. (1980). The Maastrichtian phosphate sequence of the Congo. In *Marine Phosphorites*, ed. Y.K. Bentor, pp. 193–207. Society of Economic Paleontologists and Mineralogists Spec. Pub. No. 29, Tulsa.

Giresse, P. & Cornen, G. (1976). Distribution, nature et origine des phosphates miocenes et eocenes sous-marins des plateformes du Congo et du Gabon: *Bulletin du BRGM (deuxieme serie)* **Section IV, 1–1976**, 5–15.

Glasby, G.P. & Summerhayes, C.P. (1975). Sequential deposition of authigenic marine minerals around New Zealand: paleo-environmental significance. *New Zealand Journal of Geology and Geophysics*, **18**, 477–90.

Gorsline, D.S. (1963). Bottom sediments of the Atlantic shelf and slope of the southern United States. *Journal of Geology*, **71**, 422–40.

Govean, F.M. & Garrison, R.E. (1981). Significance of laminated and massive diatomites in the upper part of the Monterey Formation, California. In *The Monterey Formation and Related Siliceous Rocks of California*, ed. R.E. Garrison & R.G. Douglas, pp. 181–98. Pacific Section, Society of Economic Paleontologists and Mineralogists, Los Angeles, California.

Gower, H.D. & Madsen, B.M. (1964). The occurrence of phosphate rock in California. *US Geological Survey Professional Paper*, **501–D**, D79–85.

Hallam, A. (1984). Pre-Quaternary sea-level changes. *Annual Review of Earth and Planetary Science*, **12**, 205–43.

Haq, B.V., Hardenbol, J. & Vail, P.R. (1987). Chronology of fluctuating sea levels since the Triassic. *Science*, **235**, 1156–67.

Harrington, J.F., Ward, D.E. & McKelvey, V.E. (1966). Sources of fertilizer minerals in South America – a preliminary study. *US Geological Survey Bulletin, 1240.*

Hausback, B.P. (1984). Cenozoic volcanic and tectonic evolution of Baja California Sur, Mexico. In *Geology of the Baja California Peninsula*, ed. V.A. Frizzell, pp. 219–36. Society of Economic Paleontologists and Mineralogists, Pacific Section, Special Publication 39, Los Angeles.

Hehuwat, F. (1975). Marine phosphorite deposits in Indonesia. Committee for Co-Ordination of Joint Prospecting for Mineral Resources in Asian Offshore Areas (CCOP), Proceedings of the Twelfth Session, pp. 282–6.

Hendey, Q.B. (1981). Geological succession at Langebaanweg, Cape Province, and global events of the Late Tertiary. *South African Journal of Science*, **77**, 33–8.

Hendey, Q.B. & Dingle, R.V. (1989). Onshore sedimentary phosphate deposits in south-western Africa. In *Phosphates of the World Vol. II: Phosphate Rock Resources*, ed. A.J.G. Notholt, R.P. Sheldon & D.F. Davidson. Cambridge University Press, Cambridge, England.

Jahnke, R., Emerson, S., Roe, K.K. & Burnett, W.C. (1983). The present day formation of apatite in Mexican continental margin sediments. *Geochimica Cosmochimica Acta*, **47**, 259–66.

Kazakov, A.V. (1937). The phosphorite facies and the genesis of phosphorites. *Scientific Institute of Fertilizers and Insecto-Fungicides Transactions*, **142**, 95–113.

Keigwin, L.D. (1979). Late Cenozoic stable isotope stratigraphy and paleoceanography of DSDP site from nearby Pacific Ocean and Caribbean Sea cores. *Geology*, **6**, 630–4.

Keigwin, L.D. & Corliss, B.H. (1986). Stable isotopes in middle late Eocene to Oligocene foraminifera. *Geological Society of America Bulletin*, **97**, 335–45.

Keigwin, L.D. & Keller, G. (1984). Middle Oligocene cooling from Equatorial Pacific DSDP site 77B. *Geology*, **12**, 16–19.

Keller, G. & Barron, J.A. (1983). Paleoceanographic implications of Miocene deep-sea hiatuses. *Geological Society of America Bulletin*, **94**, 590–613.

Kim, W.H. (1987). Biostratigraphy and depositional history of the San Gregorio and Isidro Formations, Baja California Sur, Mexico. Unpublished Ph.D thesis. Stanford University, Stanford, California, 206pp.

Kim, W.H. & Barron, J.A. (1986). Diatom biostratigraphy of the upper Oligocene to lowermost Miocene San Gregorio Formation, Baja California Sur, Mexico. *Diatom Research*, **1**, 169–87.

Leanza, H.A., Spiegelman, A.T. & Hugo, C.A. (1981). Manifestaciones fosfaticas en la Formacion Patagonia: su genesis y relacion con el vulcanismo piroclastico siliceo. *Rev. AMPS*, **11**, 1–12.

Leanza, H.A., Spiegelman, A.T., Hugo, C.A., Mastandrea, O.O. & Oblitas, C.J. (1989). Phanerozoic sedimentary phosphatic rocks of Argentina. In *Phosphate Rock Resources*, A.J.G. Notholt, R.P. Sheldon & D.F. Davidson, eds. vol 2, Chpt. 24. Cambridge University Press, Cambridge, England.

Lucas, J., Prevot, L. & Lamboy, M. (1978). Phosphorites of the continental margin of northern Spain: chemistry, mineralogy, genesis. *Oceanologica Acta*, **1**, 55–72.

Luternauer, J.L. & Pilkey, O.H. (1967). Phosphorite grains: their application to the interpretation of North Carolina shelf sedimentation. *Marine Geology*, **5**, 315–20.

Manheim, F.T., Pratt, R.M. & McFarlin, P.F. (1980). Composition and origin of phosphorite deposits of the Blake Plateau. In *Marine Phosphorites*, ed. Y.K. Bentor, pp. 117–37. Society of Economic Paleontologists and Mineralogists, Special Publication No. 29, Tulsa.

Marshall, J.F. & Cook, P.J. (1980). Petrology of iron- and phosphorus-rich nodules from the eastern Australian continental shelf. *Journal Geological Society London*, **137**, 765–71.

Matthews, R.K. (1984). Oxygen-isotope record of ice-volume history; 100 million years of glacio-eustatic sea-level fluctuation. In *Interregional Unconformities and Hydrocarbon Accumulation*, ed. J.S. Schlee, pp. 97–107. American Association of Petroleum Geologists, Memoir 36.

Matthews, R.K. & Poore, R.Z. (1980). Tertiary delta ^{18}O record and glacio-eustatic sea level fluctuations. *Geology*, **8**, 501–4.

McArthur, J.M. (1974). The geochemistry of phosphorite from the continental margin off Morocco. Unpublished Ph.D thesis. Imperial College, University of London, 200pp.

McKelvey, V.E., Swanson, R.W. & Sheldon, R.P. (1953). The Permian phosphorite deposits of western United States. *19th International Geological Congress, Algiers, Comptes rendus*, **sec. 11**, 45–64.

McLean, H., Barron, J.A. & Hausback, B.P. (1984). The San Gregorio Formation of Baja California, Mexico, is late Oligocene. *Society of Economic Paleontologists and Mineralogists, Pacific Section*, Abstracts with Programs, Los Angeles.

Melidoro, G. & Zezza, F. (1969). Miocene phosphate deposits in the Salentine Peninsula, Apulia. *Geol. Appl. Idrogeol.*, **3**, 1–14.

Mercer, J.H. (1983). Cenozoic glaciation in the southern hemisphere. *Annual Review of Earth and Planetary Science*, **11**, 99–132.

Miller, K.G., Aubry, M.P., Khan, M.J., Melillo, A.J., Kent, D.V. & Berggren, W.A. (1985). Oligocene–Miocene biostratigraphy, magnetostratigraphy, and isotopic stratigraphy of the western North Atlantic. *Geology*, **13**, 257–61.

Miller, K.G. & Fairbanks, R.G. (1983). Evidence for Oligocene–Middle Miocene abyssal circulation changes in the western North Atlantic. *Nature*, **306**, 250–3.

Murray, J. & Renard, A.F. (1891). Scientific results, HMS *Challenger*. Deep-sea deposits, pp. 391–400. London.

Nilsson, S.T. & Persson, S. (1983). Tree-pollen spectra in the Stockholm region (Sweden). *International Palynological Conference*, **5**, 288.

Notholt, A.J.G. (1968). The stratigraphical and geographical distribution of phosphate deposits in the ECAFE region. In *Proceedings of the Seminar on Sources of Mineral Raw Materials for the Fertilizer Industry in Asia and the Far East*, pp. 57–73. United Nations Mineral Resources Development Series, No. 32, Bangkok.

Notholt, A.J.G. & Highley, D.E. (1979). *Dossier on phosphate*. Commission of the European Communities, Institute of Geological Science, London, 234pp.

Nutter, A.H. (1969). The origin and distribution of phosphate in marine sediments from the Moroccan and Portuguese continental margins. Unpublished D.I.C. thesis. Imperial College, University of London, 158pp.

O'Brien, G.W. (1983). Geochemistry and origin of phosphatic nodules and sediments from the East Australian continental margin. Unpublished Ph.D. thesis. Flinders University of South Australia, Adelaide, 431pp.

O'Brien, G.W., Harris, J.R., Milnes, A.R. & Veeh, H.H. (1981). Bacterial origin of east Australian continental margin phosphorites. *Nature*, **294**, 442–4.

O'Brien, G.W. & Veeh, H.H. (1980). Holocene phosphorite on the East Australian continental margin. *Nature*, **288**, 690–2.

Ojeda Rivera, J. (1979). Resumen de datos estratigraficos y estructurales de la Formacion Monterrey que aflora in el area de San Hilaria, Baja California Sur. *Revista Geomimet*, **no. 100**, 24pp.

Oksengorn, F.S. (1980). The first find of phosphorites on the Sea of Okhotsk shelf. *Oceanology*, **19**, 556–8.

Parrish, J.T. (1982). Upwelling and petroleum source beds, with reference to Paleozoic. *American Association of Petroleum Geologists*, **66**, 750–74.

Pedley, H.M. & Bennett, S.M. (1985). Phosphorites, hardgrounds and syndepositional solution subsidence: a palaeoenvironmental model from the Miocene of the Maltese Islands. *Sedimentary Geology*, **45**, 1–34.

Pevear, D.R. & Pilkey, O.H. (1966). Phosphorite in Georgia shelf sediments. *Geological Society of America Bulletin*, **77**, 849–58.

Pilkey, O.H. & Luternauer, J.L. (1967). A North Carolina shelf phosphate deposit of possible commercial interest. *Southeastern Geology*, **8**, 33–51.

Pisciotto, K.A. & Garrison, R.E. (1981). Lithofacies and depositional environments of the Monterey Formation, California. In *The Monterey Formation and Related Siliceous Rocks of California*, ed. R.E. Garrison & R.G. Douglas, pp. 97–122. Pacific Section, Society of Economic Paleontologists and Mineralogists, Los Angeles, California.

Pitman, W.C. & Kominz, M.A. (1981). New eustatic sea level curve from oceanographic data. *American Association of Petroleum Geologists Bulletin*, **65**, 973.

Pokryshkin, V. (1967). Areas de prospeccion y estudio de fosforitas en la Republica de Cuba. *Revista Tecnologica*, **2**, 3–16.

Pokryshkin, V. (1969). Phosphorites of Cuba. *Industry of Raw Chemical Materials*, **2**, 17–23.
Powell, C.M.A. (1979). *A Speculative Tectonic History of Pakistan and Surroundings: Some Constraints from the Indian Ocean*, pp. 5–24. Geological Survey of Pakistan, Quetta.
Riggs, S.R. (1979). Phosphorite sedimentation in Florida – a model phosphogenic system. *Economic Geology*, **74**, 285–314.
Riggs, S.R. (1984). Paleoceanographic model of Neogene phosphorite deposition, US Atlantic continental margin. *Science*, **223**, 123–31.
Riggs, S.R. (1986). Phosphogenesis and its relationship to exploration for Proterozoic and Cambrian phosphorites. In *Phosphates of the World, Vol. I, Proterozoic and Cambrian Phosphorites*, ed. P.J. Cook & J.H. Shergold, pp. 352–68. Cambridge University Press, Cambridge, England.
Riggs, S.R. (1987). Model of Tertiary phosphorites on the world's continental margins. In *Marine Minerals*, ed. P.G. Teleki, M.R. Dobson, J.R. Moore & von Stackelberg, pp. 99–118. NATO Advanced Research Workshop, D. Reidel Publishing, Dortrecht, Holland.
Riggs, S.R. (1988). Model for changing patterns of Cenozoic sedimentation on the southeastern United States Coastal Plain and Continental Shelf. In *Paleoenvironments: Offshore Atlantic US Margin*, by Schlee, J.S., Manspeizer, W., & Riggs, S.R., In *The Atlantic Continental Margin*, vol. 1–2, ed. R.E. Sheridan & J.A. Grow, pp. 376–85. Geological Society of America, The Geology of North America, Boulder, Colorado.
Riggs, S.R. & Belknap, D.F. (1988). Upper Cenozoic processes and environments of continental margin sedimentation: eastern United States. In *The Atlantic Continental Margin*, vol 1–2, ed. R.E. Sheridan & J.A. Grow, pp. 131–76. Geological Society of America, The Geology of North America.
Riggs, S.R., Ellington, M.D. Burnett, W.C. (1983). Geologic history of the Pleistocene–Holocene phosphorites on the North Carolina continental margin. *Geological Society of America Abstracts with Programs*, **15**, 105.
Riggs, S.R. & Manheim, F.T. (1988). Mineral resources of the US Atlantic continental margin. In *The Atlantic Continental Margin*, vol. 1–2, ed. R.E. Sheridan & J.A. Grow, pp. 501–20. Geological Society of America, The Geology of North America.
R ggs, S.R., Snyder, Stephen W., Hine, A.C., Snyder, Scott W., Ellington, M.D. Mallette, P.M. (1985). Geologic framework of phosphate resources in Onslow Bay, North Carolina continental shelf. *Economic Geology*, **80**, 716–38.
Roberts, A.E. (1989). Geology and resources of Miocene Coast Ranges and Cenozoic OCS phosphate deposits of California, USA. In *Phosphates of the World, vol. 2, Phosphate Rock Resources*, ed. A.J.G. Notholt, R.P. Sheldon & D.F. Davidson. Cambridge University Press, Cambridge, England.
Rodriguez, S.E. (1989). Phosphorite deposits of Venezuela. In *Phosphates of the World, vol. 2, Phosphate Rock Resources*, ed. A.J.G. Notholt, R.P. Sheldon & D.F. Davidson. Cambridge University Press, Cambridge, England.
Salas, G.P. (1978). Sedimentary phosphate deposits in Baja California, Mexico. *Society of Mining Engineers of AIME*, **Preprint No. 78-H-75**, 14pp.
Sarnthein, M. (1981). The aeolomarine sediment record of winds controlling upwelling. In *Program of Coastal Upwelling: Its Sediment Record*. Advanced Research Institute, Sept. 1–4, 1981, Dortrecht, Holland.
Schopf, T.J.M. (1980). *Paleoceanography*. Harvard University Press, Cambridge, Massachusetts, 341pp.
Schott, V.W. (1970). Geologische Untersuchungen au sedimenten des Indisch-Pakistanischen kontinental randes (Arabisches Meer). *Geol. Rundsch.*, **60**, 264–75.
Sheldon, R.P. (1964). Paleolatitudinal and paleogeographic distribution of phosphorites. *US Geological Survey Professional Paper*, **501–C**, C106–13.
Sheldon, R.P. (1980). Episodicity of phosphate deposition and deep ocean circulation – a hypothesis. In *Marine Phosphorites*, ed. Y.K. Bentor, pp. 239–47. Society of Economic Paleontologists and Mineralogists, Special Publication No. 29.
Sheldon, R.P. (1981). Ancient marine phosphorites. *Annual Review of Earth and Planetary Science*, **9**, 251–84.
Sheldon, R.P. (1987). Association of phosphatic and siliceous marine sedimentary deposits. In *Siliceous Sedimentary Rock-Hosted Ores and Petroleum*, ed. J.R. Hein, pp. 58–80. Van Nostrand Reinhold, New York.
Slansky, M. (1980). Ancient upwelling models: Upper Cretaceous and Eocene phosphorite deposits around West Africa. In *Fertilizer Mineral Potential in Asia and the Pacific*, ed. R.P. Sheldon & W.C. Burnett, pp. 145–58. East–West Center, Honolulu, Hawaii.
Snyder, Scott W. (ed.) (1988). *Micropaleontology of Miocene Sediments in the Shallow Subsurface of Onslow Bay, North Carolina Continental Shelf*. Cushman Foundation for Foraminiferal Research, Special Publication No. 25, Washington, DC., 189pp.
Snyder, Scott W., Mauger, L. & Akers, W.H. (1983). Planktonic foraminifera and biostratigraphy of the Yorktown Formation, Lee Creek Mine. In *Geology and Paleontology of the Lee Creek Mine, North Carolina*, ed. C.E. Ray, pp. 455–82. Smithsonian Contributions to Paleobiology, No. 53, Washington, DC.
Snyder, Stephen W. (1982). Seismic stratigraphy within the Miocene Carolina Phosphogenic Province: chronostratigraphy, paleotopographic controls, sea-level, cyclicity, Gulf Stream dynamics, and the resulting depositional framework. Unpublished M.Sc. thesis. University of North Carolina, Chapel Hill, NC, 183pp.
Stowasser, W.F. (1985). Phosphate rock. In *Mineral Facts and Problems*. US Department of Interior, Bureau of Mines, US Govt. Printing Office, Washington, DC.
Summerhayes, C.P. (1970). Phosphate deposits on the northwest African continental shelf and slope. Unpublished Ph.D thesis. Imperial College, University of London, 282pp.
Summerhayes, C.P., Nutter, A.H. & Tooms, J.S. (1972). The distribution and origin of phosphate in sediments off northwest Africa. *Sedimentary Geology*, **8**, 3–28.
Tarulli, G. & Marcucci, A. (1928). Minerali fosfatici nella provincia di lecce. *Annali di Chimica Applicata*, **18**, 40–6.
Tedesco, C. (1966). Recerche sulle mineralizzazioni fosfatiche della Sicilia orientale ed esame comparativo dei metodi di pprospezione: *Rivista Mineraila Sicilia*, **17**, 3–17.
Vail, P.R. & Mitchum, R.M. (1979). Global cycles of relative changes of sea level from seismic stratigraphy. In *Geological and Geophysical Investigations of Continental Margins*, ed. J.S. Watkins, L. Montadert, & P.W. Dickerson, pp. 469–72. American Association of Petroleum Geologists Memoir 29.
Vail, P.R., Mitchum, R.M. & Thompson, S. (1977). Seismic stratigraphy and global changes of sea level; Part 4: global cycles of relative changes of sea level. In *Seismic Stratigraphy – Applications to Hydrocarbon Exploration*, ed. C.E. Payton, pp. 83–97. American Association of Petroleum Geologists, Memoir 26.
Valdebenito, M. (1989). Phosphate deposits of the Caleta Herradura Formation, Mejillones Peninsula, Chile. In *Phosphates of the World, vol. 2, Phosphate Rock Resources*, ed. A.J.G. Notholt, R.P. Sheldon & D.F. Davidson, Cambridge University Press, Cambridge, England.
Von Rad, U. & Kudrass, J.R. (eds.) (1984). Geology of the Chatham Rise Phosphorite Deposits East of New Zealand: Results of a Prospection Cruise with R/V Sonne (1981). *Geologisches Jahrbuch Reihe D*, **Heft 65**, 252pp.
Watters, W.A. (1968). Phosphorite and apatite occurrences and possible reserves in New Zealand and outlying islands. In *Proceedings of the Seminar on Sources of Mineral Raw Materials for the Fertilizer Industry in Asia and the Far East*, pp. 144–51. United Nations Mineral Resources Development Series No. 32, Bangkok.
Woodruff, F. (1985). Changes in Miocene deep-sea benthic foraminiferal distribution in the Pacific Ocean; relationship to paleoceanography. In *The Miocene Ocean; Paleoceanography and Biogeography*, ed. J.P. Kennett, pp. 131–76. Geological Society of America, Memoir 163.
Woodruff, F., Savin, S.M. & Douglas, R.G. (1981). Miocene stable isotope record; a detailed deep Pacific Ocean study and its paleoclimatic implications. *Science*, **122**, 665–8.
Woolsey, R.J. (1976). Neogene stratigraphy of the Georgia coast and inner continental shelf. Unpublished Ph.D. thesis. University of Georgia, Athens, 222pp.

19

Paleoceanographic and paleoclimatic setting of the Miocene phosphogenic episode

J.T. PARRISH

Abstract
The combination of paleogeography and climate that occurred in the Miocene was unique in the Phanerozoic and resulted in a disproportionately high number of phosphate deposits at mid-latitudes, especially along east-facing coasts, relative to earlier intervals.

Introduction

The Miocene phosphogenic episode is the latest of many such episodes in earth history. The purpose of this paper is to compare the paleogeographic and general paleoclimatic setting of Miocene phosphogenesis with that of other periods.

Although there are some similarities between the Miocene phosphogenic episode and earlier ones, the Miocene was unique in many ways. Miocene geography was very similar to the present, particularly regarding the positions of the continents (Fig. 19.1). The major, climatically significant differences were: a slightly higher sea level than at present, resulting in less exposed land area; a lower Tibetan Plateau, which was just forming during the Miocene; and a Mediterranean Sea that was broader, largely because sea level was higher (Ziegler, Scotese & Barrett, 1983). This sea, however, was narrower than the western Tethyan seaway that occupied the same region earlier in the Phanerozoic. Although the Miocene had less land area than at present, more area was exposed than at any time since the Early Jurassic (Fig. 19.2). Consequently, the total albedo (reflectivity) of the earth's surface was greater. The latitudinal distribution of land and sea in the Miocene was such that more solar radiation would have been reflected back to space during the Miocene than at any time in the Phanerozoic since the Cambrian and Late Pennsylvanian, and the climate would have been cooler (Fig. 19.3).

Methods

Three types of information were used in this study – shelf area, number of phosphate deposits, and a latitudinal classification of upwelling zones. The data were compiled in the course of work on the global distribution of phosphates and their relation to upwelling zones predicted from global climate models (Parrish, 1982 and 1984; Parrish & Curtis, 1982; Parrish, Ziegler & Humphreville, 1983).

Shelf area

Shelf area was measured with a planimeter directly from Mollweide equal-area projections of the paleogeographic maps of Scotese *et al.* (1979) and Ziegler *et al.* (1983); these measurements are reported in Parrish (1985). The paleogeographic maps were constructed for selected geologic ages through the Phanerozoic. The criteria by which the ages were originally chosen included availability of data and general paleogeographic interest. For example, the Westphalian C-D of the Pennsylvanian was chosen because that was the time of maximum coal formation in North America.

The times covered by the paleogeographic maps, and, thus, the quantitative measurements of shelf area, are not necessarily the same times during which phosphate was deposited. Therefore, shelf area for each map interval may not be accurate for the precise time of phosphate deposition. However, the paleogeographic maps do tend to be for times of high sea-level stand, and phosphate tended to be deposited at such times (e.g. Sheldon, 1980; Arthur & Jenkyns, 1981). The measurements, therefore, probably are good first-order approximations for the phosphogenic episodes. Major changes in shelf area occur with the first-order sea-level curve (Vail, Mitchum & Thompson, 1977), and these changes are well represented in the curve that is based solely on the selected paleogeographic maps (Fig. 19.2).

Phosphate deposits were not normalized by time. Normalizing the number of deposits by the length of the geologic period in which they were formed would introduce substantially more error than exists in the raw data because phosphate deposition is 'bunched' in time. For example, a Permian phosphate map will be, in effect, a Late Permian phosphate map. Moreover, worldwide phosphogenic episodes are commonly so poorly dated on the absolute time scale that attempts to normalize for the length of each individual episode also is not productive and, in addition, would force artificial methods of dealing with phosphate deposits that were not formed during one of the

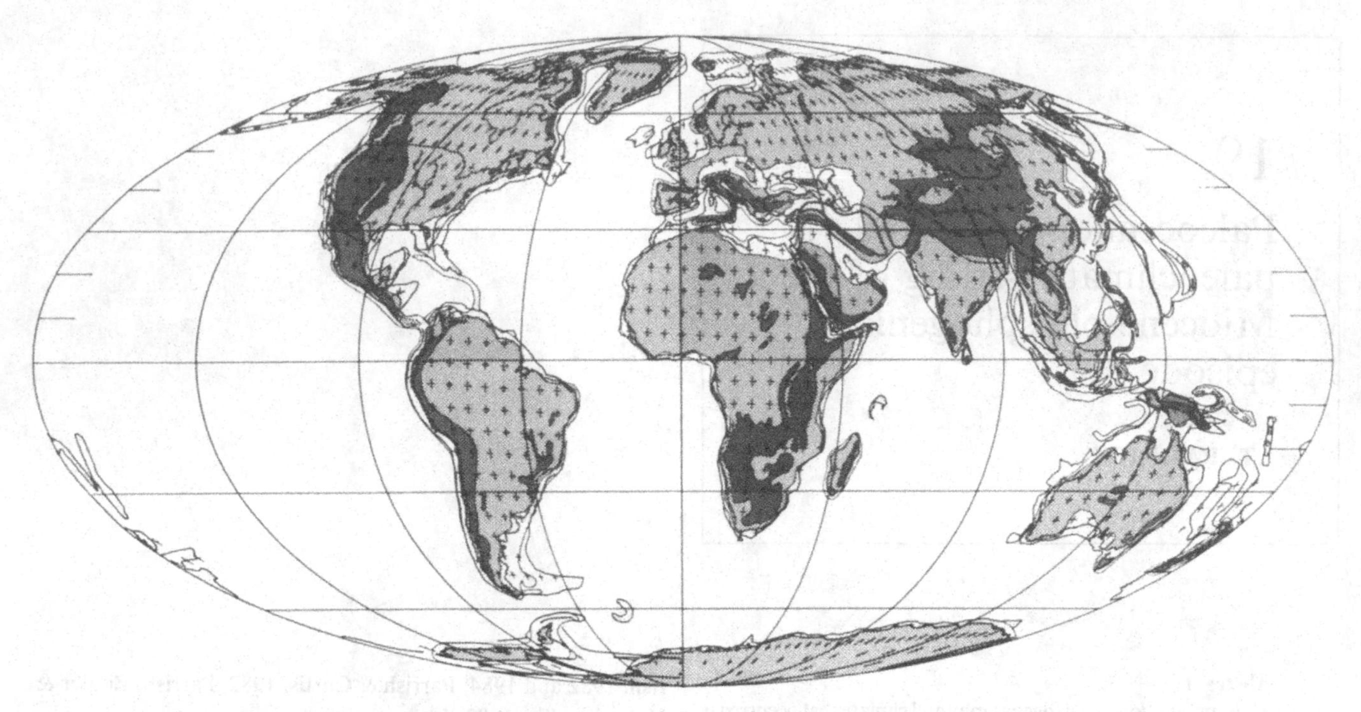

Fig. 19.1. Miocene paleogeography (from Ziegler *et al.*, 1983). Shading: dark – highlands; medium – lowlands; light – continental shelf.

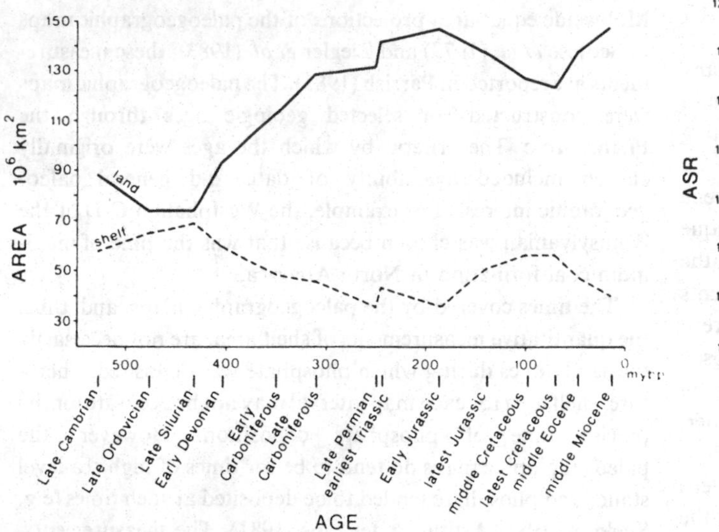

Fig. 19.2. Area of land and continental shelf through time (from Parrish, 1985). Stage names represent the times for which paleogeographic reconstructions were available (Scotese *et al.*, 1979; Ziegler *et al.*, 1983) and from which the measurements were made.

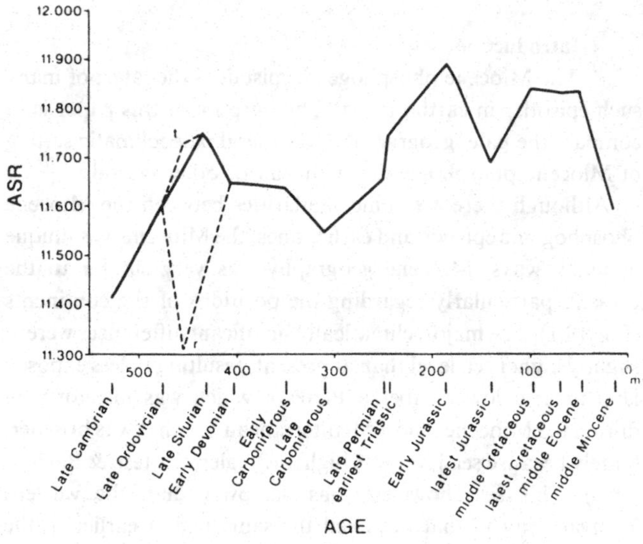

Fig. 19.3. Absorbed solar radiation through time (from Parrish, 1985). This curve includes assumptions about the distributions of a number of paleogeographic features, such as deserts, that would have affected albedo and also includes the solar evolution model of Sagan & Mullen (1972) for solar luminosity. See Parrish (1985) for details.

phosphogenic episodes. Short-term climatic changes that might have resulted in systematic geographic shifts of phosphate deposition will be obscured by the method used here.

Phosphate deposits

Data on Paleozoic phosphate deposits were reported in Parrish *et al.* (1983 and 1986). Those and the data presented here were collected from sources in the literature and, whenever possible, age and distribution were cross-checked from several references. Certain sources were exceptionally helpful. These are Vinogradov (1967 and 1968), Bushinskii (1969), and Mew (1980). References and data on Mesozoic and Cenozoic phosphates are listed in Table 19.1; Cambrian, Mesozoic, and Cenozoic data used in this study are updated from those summarized in Parrish (1984). The latitudinal distribution of Cambrian phosphate deposits was determined using the restored position of

South China (Lin, Fuller & Zhang, 1985; see Parrish *et al.*, 1986). References cited in Table 19.1 are representative of the literature on each deposit but, for many deposits, much more literature is available.

The raw data consist of phosphate localities plotted on a 0.1° latitude–longitude grid. For each time interval, all localities within a 5° latitude–longitude square were counted as one data point. This method has the advantage of giving statistical weight to very large deposits without overwhelming the smaller ones. Occurrences that cover more than one 5°-square in at least one dimension were counted as two or more data points, except at latitudes above 65° N and S (present coordinates), where occurrences within 5° latitude and 10° longitude were counted as one (four data points affected). In Table 19.1, the coordinates given for each data point mark the mid-point of the group of localities included in that data point, rounded to the nearest whole degree. The mid-point coordinates may be closer than 5° when, for example, a group of localities spanned 7° latitude. In such a case, the localities are divided into two data points, but the mid-points will be only 2.5° apart.

Estimates of phosphate volume or tonnage were not used for several reasons. Foremost is that the presence of phosphate is more significant than its abundance with regard to geographic distribution. This is an important point for a study of the type presented here. Almost all estimates of volume and tonnage are based on deposits that are already recognized as economic. However, if the distribution of phosphate is related to the distribution of upwelling, for example (McKelvey, 1959 and 1967; Freas & Eckstrom, 1968; and many others), the presence of phosphate, economic or not, is significant. An excellent example of such an occurrence is the phosphate in the Cretaceous chalks of England. This phosphate is non-economic, for a variety of reasons, yet its existence supports the conclusion that the chalks were the product of upwelling (Jarvis, 1980b). Moreover, undisputed upwelling deposits have varying amounts of phosphate. Compare, for example, the Monterey Formation, with no economic deposits (the Cuyama deposit is in the Santa Margarita Formation), and the Phosphoria Formation, with 126 million tonnes in the Montpelier district alone (Mew, 1980). The significance of phosphate as an upwelling indicator is the same for both deposits. Even if estimates of volume and tonnage were to prove useful in a study of this type, such estimates are not reported in the literature for most deposits. Moreover, the thicknesses and precise geographic extent of many deposits also are not reported, so that independent estimates cannot be made.

Igneous phosphorites (e.g. Notholt, 1979) and those that are clearly non-marine (e.g. Love, 1964; Pipiringos, Chisholm & Kepferle, 1965; Tsykin, 1980) were not included. Deposits that could not be located within 5° latitude–longitude or were not dated to at least the geologic period also were rejected.

Latitudinal classification of upwelling zones

The distribution of phosphate deposits and shelf area were divided into three latitudinal regions for the purpose of comparing the geographic settings of phosphate deposition among the different phosphogenic periods. This latitudinal classification of upwelling zones is genetic and derived from the distribution of upwelling zones of different types (Parrish, 1982). Thus, the latitudinal classification is not arbitrary, although peculiarities of geography may result in an arbitrary assignment of a few deposits to one or another type of upwelling.

Equatorial upwelling, 10° N–10° S. Upwelling in the equatorial region is driven by the equatorial easterlies (Sverdrup, Johnson & Fleming, 1942) or by monsoonal winds, as off Somalia (e.g. Swallow, 1980, and accompanying papers). The high productivity associated with these two types of upwelling generally is limited to the area between 10° N and 10° S latitude. Equatorial upwelling occurs in the open ocean today and does not, therefore, result in phosphate formation. However, in the past, high sea level could have permitted such upwelling over the continental shelf and consequent formation and preservation of phosphate.

Mid-latitude upwelling, 10–40° N & S. Mid-latitude upwelling occurs in eastern and western boundary currents, for example, off California and in the Gulf Stream. The latter is dynamic upwelling (McKelvey, 1967) and commonly is very productive (e.g. Yentsch, 1974). Like open-ocean upwelling, Gulf Stream-type upwelling commonly has been overlooked in models for ancient upwelling indicators because productivity in the Gulf Stream today occurs over deep water and is not reflected in the underlying sediment (see Parrish, 1982, for discussion). Upwelling associated with eastern and western boundary currents occurs mostly between 10° and 40° N and S. Mid-latitude upwelling can also occur along north-facing, east–west coastlines in the equatorial side of the region (Fukuoka, Ballester & Cervignon, 1968) and along south-facing, east–west coastlines in the poleward side (Parrish, 1982). West-coast and equatorial upwelling are contiguous off western South America today, but this is the result of an unusual geography and thus the boundary between the two types of upwelling at 10° latitude can be expected to be valid for most times.

High-latitude upwelling, 40°–90°N & S. High-latitude upwelling, above 40° N and S, generally occurs in open-ocean divergences at about 55–60° N and S, although coastal upwelling also may have occurred in the ancient Arctic Ocean during some intervals (Parrish & Curtis, 1982). In certain geographic settings, high-latitude upwelling in the divergences and west-coast upwelling might be contiguous but, as with the mergence of equatorial and west-coast upwelling, this is unlikely to have been common in the past.

Phosphate deposits

Along with more exposed land area, the continents had less shelf area during the Miocene than at any time since the Late Permian and Early Jurassic (Fig. 19.2). Phosphate distribution is indirectly dependent on shelf area because, during times of high sea-level stand (and greater shelf area), highly productive ocean currents can be brought up onto the shelf, where the organic matter produced in them is more likely to reach the sea bottom and create the conditions favorable for phosphate formation (e.g. Burnett, 1980; Riggs, 1984). Thus, the shelf-area curve may be regarded as a predictor for phosphate distribution; that is, shelf area (which is a reflection of sea level) and phosphate

Table 19.1.

Plate	Present lat.–long.	Age	Stratigraphic unit	Phosphate morphology; associated lithologies	References
Triassic					
Northern Pangaea (Laurasia)					
USA	69° N 145° W (comp.)	l	Shublik Fm.	Nod, pel, shell, hdgr; organic-rich ls	105, 106
Canada	50° N 114° W (comp.)	e–l	Spray River, Sulphur Mtn., Whitehorse Fms.	Black phos sh, peb, oo, pel; zs, ss	35, 92
USSR	65° N 57° E (comp.)	e–l		Nod; pyrite, red ss, sh	32, 148
	69° N 63° E	e–l		Nod; ss, sh	32, 148
	65° N 129° E (comp.)	l		Nod; ss, zs	148
	71° N 127° E	e,l		Nod, bed, oo, shell; ss, zs, clyst, siderite	140, 148
	74° N 118° E (comp.)	l		Nod; ss, zs, clyst	148
	75° N 140° E	e–l		Nod; black clyst, ls	69
Svalbard	77° N 16° E	undiff.	Liddalen, Botneheia Kapp Toscana Fms.	Nod, pel; ss, zs, sh, glauc	46
Southern Pangaea (Gondwana)					
Pakistan	26° N 67° E	l	Shirinab Fm.	Nod; zs, chert	136
	34° N 73° E	undiff	Galdanian Fm.	– dolomitic zs, ferruginous sh	16
Australia	21° S 122° E	undiff		–	50
	34° S 151° E (comp.)	e–m	Narrabeen Gr.	Pel, bed, nod; ss, sh	87
New Zealand	43° S 171° E	undiff	Fingers Fm.	Peb	12
Jurassic (Hettangian–Bathonian)					
Laurasia					
Canada	50° N 115° W (comp.)	Sin, Baj	Nordegg, Rock Creek Mbrs, Fernie Gr.	Phos beds; black sh, ls, chert	35, 44, 75, 92, 130
UK	53° N 1° W	e	Cephalopod Ls Mbr., Upper Lias; Lower Lias; Dogger Fm.	Oo, nod; ml, ss, siderite, goethite, pyrite, glauc	45, 61, 62
USSR	63° N 117° E (comp.)	e–m		Nod, cem, oo; zs, clyst	140, 148
	71° N 127° E (comp.)	e–m		Nod; silty ss, clyst	148
Svalbard	77° N 16°E	undiff	Brentskardhaugen Bed	Nod, pel; ss, zs, sh, glauc	46
Gondwana					
Peru	9° S 77° W	Sin	Aramachay Fm.	Bed, pel; organic-rich sh, chert, ss, glauc	79, 133
Pakistan	25° N 71° E	e		Phos ls	1
Jurassic (Callovian–Volgian)					
Laurasia					
Mexico	24° N 101° W	l	La Caja, La Casita Fms.	Nod, pel; chert, zs, ls	5, 92, 120, 126
France, Belgium	48° N 5° E (comp.)	undiff	Causes du Quercy, Gard ls	Nodules; ls	50, 92
Portugal	37° N 9° W	Kimm		Peb; ls	110
Poland	50° N 19° E (comp.)	Call		Nod; glauc, zs, ml	92, 141
USSR	49° N 57° E (comp.)	l		Nod; ml, ss	50, 148
	52° N 36° E (comp.)	l		Nod; ss, sandy clyst	148
	54° N 48° E (comp.)	l		Nod; ml	50, 92, 148

Table 19.1. *(Contd.)*

Plate	Present lat.–long.	Age	Stratigraphic unit	Phosphate morphology; associated lithologies	References
	57° N 38° E (comp.)	l		Nod; glauc ss, clyst	50, 92, 145, 148
	59° N 52° E (comp.)	l		Nod; hdgr; ml, glauc ss	92, 148
	58° N 42° E (comp.)	l		Nod; ml	148
	70° N 125° E (comp.)	l		Nod, peb, hdgr; biogenic silica, glauc ss, zs	17, 140, 148
	72° N 100° E (comp.)	l		Nod; ss, clyst	148
	73° N 115° E (comp.)	l		Nod; ss, zs, clyst	148
Gondwana					
Peru	11° S 75° W	l	Sincos Shale	Bed; black sh, cherty ls, chert	58, 126
Pakistan	34° N 72° E (comp.)	l	Chichali Fm.	Nod; sh, glauc ss	24, 136
India	30° N 79° E	l	Spiti Sh; Tal Series	Nod, phos sh; organic-rich sh, chert	103
Australia	29° S 115° E	undiff	Colalura Ss	Nod; ss	50, 92
Cretaceous (Berriasian–Cenomanian)					
Laurasia					
Canada	50° N 100° W	undiff	Ashville Fm.	Phos fish remains; sh	92
	68° N 137° W (comp.)	Alb	Rapid Creek Fm.	Pel, phos ironstone nod; sh, silty ms	119, 153, 154
Mexico	26° N 100° W (comp.)	m	Sierra Minas Viejas limestone	Nod; ls, limonite	92
UK	51° N 2° W (comp.)	Alb, Ceno	Lower Chalk, Upper Greensand	Nod, hdgr; ss, chalk, glauc	90, 92
France, Belgium	47° N 6° E (comp.)	Alb, Ceno		Nod; ss, chalk, glauc	27, 48, 92
Portugal	38° N 8° W	undiff		—	92
Spain	37° N 1° W	Alb		– glauc, carbonate	39, 92
Poland	51° N 22° E (comp.)	Alb, Ceno		Nod; clayey, calcareous, glauc ss	142, 144
	54° N 15° E (comp.)	Alb, Ceno		Nod; clayey, calcareous, glauc ss	142, 144
USSR	41° N 56° E (comp.)	Alb, Ceno		Nod; ss, clyst, zs	148
	43° N 65° E	undiff		Phos ss and sh	5
	45° N 46° E (comp.)	Apt–Alb		Nod; zs, glauc	56, 148
	48° N 43° E (comp.)	Alb, Ceno		Nod; sandy clyst, ss	148
	49° N 27° E (comp.)	e, Ceno		– clayey sh, ls, ss, ml, glauc	77*, 148
	51° N 38° E (comp.)	Val, Ceno		Nod; ss, ml, glauc	77, 92, 148
	52° N 33° E (comp.)	Val, Alb, Ceno		Nod; ss, sandy clyst, ls, ml	77, 148
	52° N 43° E (comp.)	Val, Alb, Ceno		Nod; ss, clyst	92, 148
	52° N 47° E (comp.)	Val, Ceno		Nod; ss, clyst	92, 148
	53° N 28° E (comp.)	Alb, Ceno		Nod; zs, sandy ls	148
	54° N 23° E (comp.)	Ceno		Nod; zs, sandy clyst	148

Table 19.1. *(Contd.)*

Plate	Present lat.–long.	Age	Stratigraphic unit	Phosphate morphology; associated lithologies	References
	57° N 43° E (comp.)	Val		Nod; ss	148
	57° N 39° E (comp.)	Val		Nod; ss, clyst	145, 148
	60° N 51° E (comp.)	Val	Vyatka–Kama deposits	Nod; glauc ss	92, 129, 145, 148
	65° N 55° E (comp.)	Val		Nod; ss, sandy clyst	148
	71° N 127° E	Val		Nod, hdgr; ss, ms	140
Gondwana					
Colombia	5° N 74° W (comp.)	Ceno	Capacho Fm.	Pel, phos sh; organic-rich sh, chert, ls, zs, glauc	58, 86
Brazil	8° S 34° W (comp.)	Alb–Ceno		Phos ls	91
Italy	46° N 11° E (comp.)	Apt, Alb	'Croûte phosphatée'	Hdgr; bituminous ls, glauc, dolomite	28, 73
Pakistan	33° N 71° E (comp.)	e	Chichali Fm.	Nod; sh, sandy sh, glauc	24, 136
India	13° N 80° E (comp.)	Alb–Ceno	Uttatur Stage	Nod; clyst	92, 98
Australia	12° S 130° E	Ceno	Bathurst Island Fm.	Nod, pel; ss, ms, zs, limonite	63
Cretaceous (Turonian–Maastrichtian)					
North America					
USA	45° N 104° W (comp.)	Tur	Niobrara Fm.	Nod, shell; ml, pyrite	139
Eurasia					
UK, France	51° N 2° E	Sen	Chalk	Pel, phos intraclasts, nod, oo, hdgr; chalk, glauc ss, ml	36, 70, 71, 72, 92
Germany (GDR)	52° N 11° E	Con		Nod; ss, glauc	137
Poland	51° N 22° E (comp.)	Tur, Sant–Maas		Nod, hdgr; chalk	92, 97, 143, 144
Bulgaria	43° N 25° E	Camp		–ls	152
Greece	39° N 23° E	l		Bed; ls, chert	67
Turkey	37° N 31° E (comp.)	l		Nod, bed; ls, chert	14
USSR	41° N 55° E (comp.)	Tur–Camp		Nod; ls, ml	148
	42° N 60° E (comp.)	Tur–Camp		Nod; ls, ss, ml	148
	49° N 38° E (comp.)	Con, Maas		Nod; ls, sandy chalk	148
	49° N 44° E (comp.)	Tur–Sant		Nod; ls, siliceous ml	148
	50° N 51° E (comp.)	Tur		Nod; ml, sandy chalk	148
	50° N 57° E	Sant–Camp		Nod, hdgr; ss, sh	78, 92
	52° N 47° eE (comp.)	Tur–Sant		Nod; ml, siliceous ml, sandy chalk	148
	53° N 43° E (comp.)	Tur–Camp		Nod; ls, ss, siliceous ml, sandy clyst	148
	53° N 26° E (comp.)	Tur		Nod; ls	148
	53° N 32° E (comp.)	Sant, Maas		Nod; ls, ml, chalk	148

Table 19.1. *(Contd.)*

Plate	Present lat.–long.	Age	Stratigraphic unit	Phosphate morphology; associated lithologies	References
	56° N 63° E (comp.)	Sant		Nod; silty, clayey ss	148
	61° N 87° E	Maas		Nod; clayey ss	148
	70° N 83° E	Sant		Nod; zs, clyst	148
South America					
Colombia, Ecuador	3° N 75° W (comp.)	Con or Sant	Lower Guadalupe, Napo Fms.	Pel; ss	31, 92, 151
Venezuela, Colombia	8° N 72° W (comp.)	Tur–Camp	La Morita Sh., Colóns Quevedo, La Luna, Guadalupe Fms.	Pel, fish remains; glauc, porcellanite, ml, dark sh	25, 31, 58, 86, 92
Brazil	8° S 35° W (comp.)	Maas	Gramame, Beberibe Fms.	Oo, nod, shell; argillaceous ls, ss	15, 58, 85, 92
Africa					
Algeria, Tunisia	35° N 8° E (comp.)	Maas	El Haria Fm.	–ml	92
Egypt	25° N 32° E (comp.)	Camp–Maas	'Nubian Sandstone', Phosphate Gr.	Bed, pel, nod, phos black sh; chert, clyst, ss, ms, glauc ss	53, 92, 108
Morocco	32° N 8° W (comp.)	Maas		Pel; chert, ls, ml	21, 92, 108
West Sahara	26° N 14° E	Maas		–clyst, ml	92
Congo, Gabon, Angola (Cabinda)	3° S 10° E (comp.)	Maas	Halle Series	Bed; ls	54, 55, 92
Turkey	36° N 44° E	Maas	Germav equivalents	Bed, oo; glauc	14
Turkey, Syria	36° N 39° E (comp.)	l	Germav, Karabogaz, Karababa Fms.	Bed, oo, pel, nod; ls, ml, chert, glauc, dolomite	6, 7, 14, 60, 80, 95, 108, 127
Iran	31° N 49° E	Camp	basal Gurpi Fm.	Oo, nod, bed, pel; ml, glauc	123, 124
Israel, Lebanon, Jordan, Iraq, Saudi Arabia	32° N 36° E (comp.)	Sant–Maas	Mishash, Ghareb, Aruma, Sayyarim Fms.	Nod, bed, pel, oo; gypsum, chalk, chert, organic-rich sh, iron, halite, glauc, bone, porcellanite	3, 13, 24, 89, 92, 96, 108, 111, 146
India					
Pakistan	30° N 69° E (comp.)	l	Kawagarh, Mughal Kot Fms.	Nod; siliceous sh, ls, glauc, oolitic hematite	1, 24, 68, 136
India	13° N 80° E	Sen	Ariyalur Stage	Nod; sandy sh	92, 98
	26° N 71° E	l	Fatehgarh deposit	Pel, phos ms, shell, oo; ss	94, 103
Australia					
Australia	32° S 116° E	l?	Ascot Beds	Nod (reworked from underlying Cretaceous)	11
Paleogene					
North America					
USA	29° N 82° W	Olig	Weaverville Fm.	Peb, pel; ss, tuff	81, 92
	32° N 96° W	Paleo	Wills Point Fm.	Nod; glauc clyst	92
Eurasia					
UK	52° N 1° E (comp.)	Eo	London Clay	Nod, pel; clyst, dolomite, glauc	8, 9
Poland	52° N 22° E (comp.)	Eo, Olig?		Nod; hdgr; black clyst, ss, ls, glauc zs	97, 143
USSR	32° N 68° E	Eo		Nod; organic-rich sh, pyrite	100
	40° N 68° E (comp.)	Eo	Alai Strata	Nod, bones, pel, teeth; clyst, ml, ls, chert	20, 64, 101, 108, 147
	45° N 34° E (comp.)	Paleo–Eo		Shell, nod; ml, ss, glauc	76
	45° N 60° E	Olig		'fucoids' (nod)	129

Table 19.1. *(Contd.)*

Plate	Present lat.–long.	Age	Stratigraphic unit	Phosphate morphology; associated lithologies	References
	50° N 40° E (comp.)	Eo		Nod; ss, clyst, ml	77, 147
	50° N 45° E (comp.)	Paleo–Eo		Nod; ss, sandy clyst, ml	147
	51° N 27° E (comp.)	Eo		Nod; clyst, sandy ml, ss, glauc	77, 147
	51° N 33° E (comp.)	Eo		Nod; ss, glauc	77, 147
	53° N 83° E	undiff		Bed; siliceous, dolomitic ls	26
Turkey	37° N 32° E	Eo–Mio		Bed; clastics	14
Africa					
Morocco	32° N 8° W (comp.)	Eo		Pel, shell; ls, glauc	65, 88, 132
	32° N 2° W (comp.)	Eo		Pel, nod; ls, chert	21, 95, 108
West Sahara	26° N 14° W (comp.)	Paleo–Eo?		Phos ss, nod; clyst, ml, glauc	92, 108, 132, 138
Algeria, Tunisia	33° N 9° E (comp.)	Paleo–Eo	Metlaoui Series	Nod, pel; ls, organic-rich ml, ss, gypsum, dolomite, glauc	49, 92, 95, 102, 108
Algeria	36° N 6° E (comp.)	Eo		Bed; ls, ml, chert	57, 108
Libya	31° N 14° E (comp.)	Eo		Nod, bed; ml, glauc ls	57
Iran	30° N 51° E (comp.)	Paleo–Eo	Pabdeh Fm.	Bed, pel, nod; ml, bituminous sh, glauc ls	92, 123, 124
Iraq, Turkey, Syria	35° N 39° E (comp.)	Eo	Becirman Fm.	Oo, nod, pel; chalk, glauc, dolomite	6, 7, 14, 92, 108
Saudi Arabia, Iraq	32° N 39° E (comp.)	Paleo–Eo	Hibr, Damman, Umm Er Radhuma Fms.	Hdgr, pel, oo; chert, ls, sh	2, 3, 4, 37, 89, 92, 146
Iraq	35° N 45° E	Paleo	Sinjar Fm.	Bed; ls	82
Senegal, Mauritania	15° N 16° W (comp.)	Eo–Olig		Nod, bed; organic-rich ml, chert	22, 24, 92, 108
Benin, Nigeria, Togo	6° N 2° E (comp.)	Eo	Lama Series	Oo, bed, pel; ls, clyst, ml	92, 99, 108
Mali	16° N 0°	Eo		Nod; ls, ss, sh	24
Central African Republic	6° N 23° E	undiff		Bed; ls, clyst, pyrite	92
Angola (Cabinda)	5° S 12° E (comp.)	Paleo–Eo		Nod; ls, siliceous zs	55
Southwest Africa	27° S 15° E (comp.)	Eo		–diatomite, ss, ms, glauc	131
South Africa	34° S 23° E (comp.)	Paleo–Eo		Peb, cem; glauc ss	42
	35° S 18° E (comp.)	Paleo–Eo		Peb, cem; glauc ss	42
India					
Pakistan	25° N 66° E	Paleo	'Jakkar Gr.'	Ferruginous nod, phos sh; calcareous sh, ss, ls	24, 136
	33° N 73° E (comp.)	Paleo–Olig	Badhrar, Patak Fms.	Nod, pel; dolomite, black sh, ls, chert, iron	1, 136
India	31° N 80° E	Paleo–Eo		Pel, nod; sh, zs, ls, pyrite	33
Atlantic Ocean					
Annan Seamount	9° N 21° W	Paleo or Eo		Phos ls	74
Oceania					
New Zealand	44° S 171° E (comp.)	Olig	Milburn Ls	–ls, glauc ss	150

Table 19.1. *(Contd.)*

Plate	Present lat.–long.	Age	Stratigraphic unit	Phosphate morphology; associated lithologies	References
	44° S 176° W	Paleo	Takatika Grit	Nod; zs	150
Neogene (Miocene only)					
North America					
USA	28° N 77° W (comp.)		Blake Plateau, Miami Terrace	Pel, nod, hdgr; ls, zs, clyst	83, 84, 93, 114
	28° N 82° W (comp.)		Hawthorne Group	Nod, pel; ss, dolomite, ls, chert	30, 112, 113, 125
	31° N 82° W (comp.)		Hawthorne Gr. equivalent	Pel, peb; ml, clyst, ss	30, 84, 92, 107, 112
	35° N 77° W (comp.)		Yorktown, Pungo River Fms.	Pel, bone, nod; clyst, ls, ss, ms	30, 92, 112, 115
	35° N 120° W (comp.)		Monterey, Santa Margarita Fms.	Pel, nod, phos sh, oo; diatomite, organic-rich sh, glauc	19, 24, 40, 52, 109, 118, 121, 128
Mexico	26° N 112° W (comp.)		Monterrey, Salada, San Isidro, Tortuga Fms.	Bed, nod, oo; ss, diatomite, dolomite, gypsum	47, 59, 116, 117, 122
Cuba	23° N 83° W			–	10
South America					
Venezuela	7° N 70° W (comp.)		Riecito, La Molina Fms.	Cem; ls, chert, black sh	58, 92
Peru	6° S 81° W (comp.)		Sechura Fm.	Pel; diatomite, ss, tuff	34, 58, 92
	14° S 76° W		Pasco Fm.	Nod; diatomite	92
Brazil	3° S 38° W (comp.)			Nod, hdgr; ls	91
Eurasia					
UK	52° N 2° E			Nod; clyst, ss	9
Italy	37° N 15° E			Nod, cem; ls, glauc	41
USSR	53° N 107° E (comp.)			–	147
	50° N 143° E			–	10
Pakistan	34° N 73° E		Murree Fm.	Bed; ss, sh, zs	136
Indonesia	7° S 113° E (comp.)		Rembang Fm.	Nod; glauc ss	43
Japan	38° N 137° E (comp.)		Nanao Gr.	Nod; ss, ms, diatomite	66, 92, 98
Philippines	10° N 121° E		Talave Limestone	Oo; ls	92
Africa					
South Africa	35° S 23° E (comp.)		Agulhas Group	Nod; glauc, ls	42, 51, 104
	33° S 18° E (comp.)		Saldanha Fm.	Pel, shell; ls, ss, zs, glauc, goethite	18, 42, 51, 104, 134, 135
	28° S 14° E (comp.)			Pel, phos ls; ls, glauc, goethite	18, 23
	19° S 14° E (comp.)			Pel, phos ls; ls, glauc, goethite, ss	18, 23, 131
Gabon, Congo, Angola (Cabinda)	5° S 12° E (comp.)			Pel; zs, ss	55
Australia and Oceania					
Australia	38° S 142° E (comp.)			Nod; calcareous clyst, ss	29

Table 19.1. *(Contd.)*

Plate	Present lat.–long.	Age	Stratigraphic unit	Phosphate morphology; associated lithologies	References
	38° S 147° E (comp.)			Nod	29
	30° S 153° E (comp.)			Nod	149
New Zealand	43° S 180° (comp.)			Nod; glauc, ls	38
	46° S 170° E		Clarendon Ss	Phos ss, bed; ss	150

*The nodules in the Cenomanian glauconitic sandstones are thought by Knyazev *et al.* (1974) to have been derived from Proterozoic beds that crop out to the east.

Abbreviations

Age:

Alb	Albian	Sant	Santonian
Apt	Aptian	Sen	Senonian
Baj	Bajocian	Sin	Sinemurian
Call	Callovian	Tur	Turonian
Camp	Campanian	Val	Valanginian
Ceno	Cenomanian		
Con	Coniacian	l	late
Eo	Eocene	m	middle
Kimm	Kimmeridgian	e	early
Maas	Maastrichtian	undiff	series or stage not known or not reported
Olig	Oligocene	,	signifies 'and'
Paleo	Paleocene	–	signifies 'to'

Phosphate morphology; associated lithologies:

bed	beds, laminae, lenses	clyst	claystone
cem	phosphate cement	glauc	glauconite (-itic)
hdgr	hardground, crust	ls	limestone
nod	nodules, concretions	ml	marlstone
oo	ooids	ms	mudstone
peb	pebbles, cobbles	sh	shale
pel	pellets	ss	sandstone
phos	phosphatic	zs	siltstone
shell	phosphatized shells, casts, molds		
–	morphology not reported		

Other:

comp. composite data point (see text)

deposition might be expected to increase and decrease together. This is not to say that sea level is the only control on phosphate deposition. Nevertheless, it is a useful point of reference for examining the differences among the phosphogenic episodes.

Equatorial region

A major setting for phosphate deposition is along the equator, which in the eastern Pacific today is a zone of high biologic productivity. Figure 19.4*a* shows that the number of phosphate deposits in the equatorial region through the Phanerozoic is loosely related to equatorial shelf area ($r = 0.58$). The anomalies are the Permian and early Tertiary, when proportionately more phosphate deposits were formed in the equatorial zone than would be predicted from the shelf-area curve (Figs. 19.4*a*, 19.5).

High-latitude region

Although phosphate deposits at latitudes greater than 40° are relatively rare, they exhibit the same pattern as equatorial phosphate deposits, generally tracking the trends in shelf area through time (Fig. 19.4*b*), particularly for the Mississippian onward ($r = 0.86$). The anomalies are in the early Paleozoic, when, depending on how the curve is interpreted, either the Cambrian had proportionately more high-latitude deposition or the Ordovician through Devonian had less than would be predicted from the shelf-area curve (Fig. 19.5). This pattern is

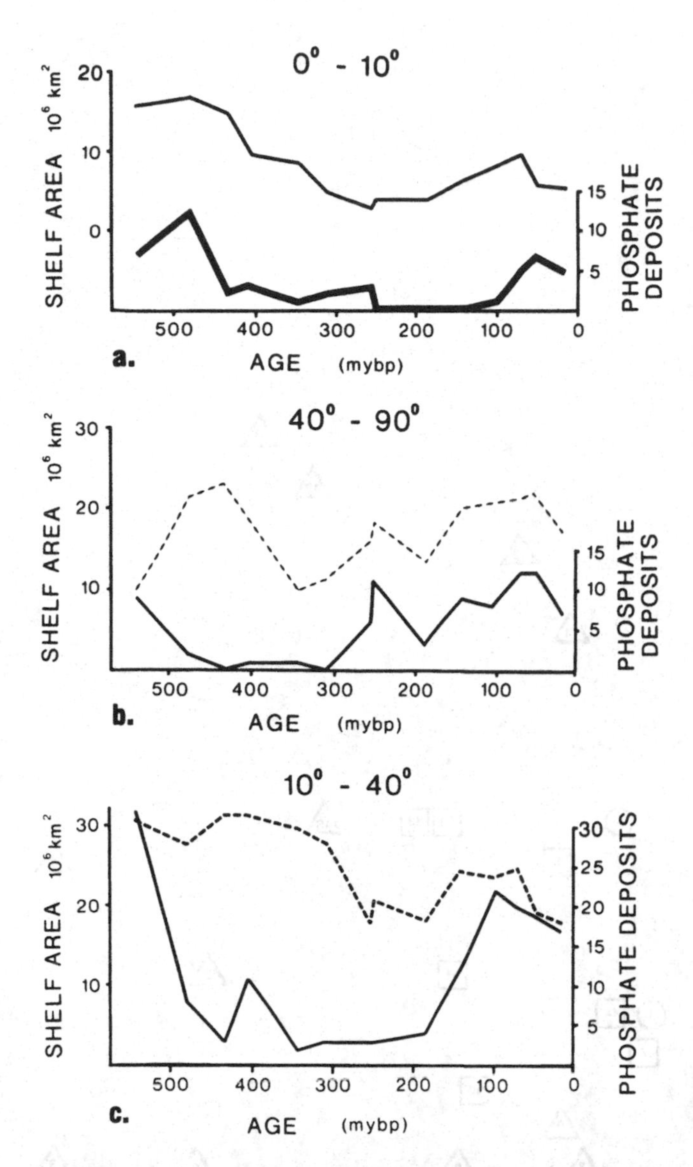

Fig. 19.4. Shelf area and phosphate deposits by latitude through time (Parrish *et al.*, 1983; Parrish, 1984). (a) Equatorial regions. Lines: thin – shelf area; thick – phosphate deposits. (b) High-latitude region. Lines: dotted – shelf area; solid – phosphate deposits. (c) Mid-latitude region. Lines: dashed – shelf area; solid – phosphate deposits.

probably an artifact of the small amount of continental crust that lay in high latitudes in the early Paleozoic (Parrish, 1985).

Mid-latitude region

The major setting for phosphate deposition is at mid-latitudes, either along west coasts or, less commonly, along east coasts. The greatest variation in the shelf area–phosphate relationship through the Phanerozoic is in the mid-latitudes (Figs. 19.4*c*, 19.5). Indeed, there is no correlation ($r = 0.09$) between shelf area and phosphate in this region. More mid-latitude deposition took place in the Cambrian, Cretaceous, and Tertiary, and possibly also in the Devonian and Permian, than predicted by the shelf-area curve, whereas less phosphate deposition than expected occurred in mid-latitudes during the Silurian and Carboniferous (Fig. 19.5).

Phosphate by latitude through time is summarized in Figure 19.6*a*. During most periods, mid-latitude phosphate deposition was dominant, entirely consistent with long-held concepts of phosphate deposition that warm waters are required (Gulbrandsen, 1969; Sheldon, 1981), and that most phosphate forms in west coast upwellings (e.g. Freas & Eckstrom, 1968). High-latitude deposition was relatively more important only in the Triassic and equatorial deposition was most important in the Ordovician. Figure 19.6*b* is a summary of shelf area in each province. Comparison of Figures 19.6*a* and 19.6*b* shows that the Miocene was unusual in that a high proportion of mid-latitude phosphogenesis occurred at that time despite the relative paucity of mid-latitude shelf area. By contrast, the Miocene had a relatively high proportion of shelf in the high-latitude and equatorial regions.

Miocene paleogeography and paleoclimatology

The major features of Miocene paleogeography compared to the rest of the Phanerozoic were low sea level, a low but rising Tibetan Plateau, and a wide Mediterranean. These features would be expected to have affected Miocene phosphate deposition in a number of ways. Although sea level was low in the Miocene, it was high enough to bring western boundary currents, with their high productivity (Yentsch, 1974), over the shelves (Riggs, 1984). This would have tended to increase mid-latitude phosphate deposition. This effect of high sea level does not necessarily apply to the earlier Phanerozoic because the presence of strong western boundary currents is dependent on factors other than sea level. The Mediterranean Sea may have been wide enough in the Miocene to produce its own atmospheric circulation cell and accompanying persistent winds (Parrish & Curtis, 1982), as did western Tethys earlier. By contrast, modern Mediterranean atmospheric circulation is dependent on processes outside the region, such as the North Atlantic subtropical high pressure cell and the Asian monsoon. Persistent winds would have enhanced the possibility of sustained wind-driven upwelling in the Mediterranean and further added to mid-latitude phosphate deposition. Finally, the development of the monsoon, which would have countered the establishment of a circulation cell over the Mediterranean, also would have affected equatorial phosphate deposition by starting up the Somalian upwelling and shutting down any west coast upwelling that might have occurred in western Australia. Miocene phosphates may be present in East Africa (Gevork'yan & Chugunnyy, 1969), and Miocene bedded chert and organic-rich rock are present; bedded chert in particular is indicative of upwelling (Hein & Parrish, 1987).

Discussion

Parrish (1984) found that atmospheric circulation models (Parrish & Curtis, 1982) were not effective for predicting the Miocene phosphate deposits, although the same predictions were quite successful in explaining the distribution of Miocene cherts. The failure resulted from a bias in the models, which concentrated solely on wind-driven upwelling and therefore could not predict the east coast phosphate deposits, which were formed in dynamic, not wind-driven, upwelling (e.g. Riggs, 1984). Such deposits were more abundant in the Miocene than at any other time in the Phanerozoic. The other phosphogenic episode for which the upwelling models failed was the Ordovi-

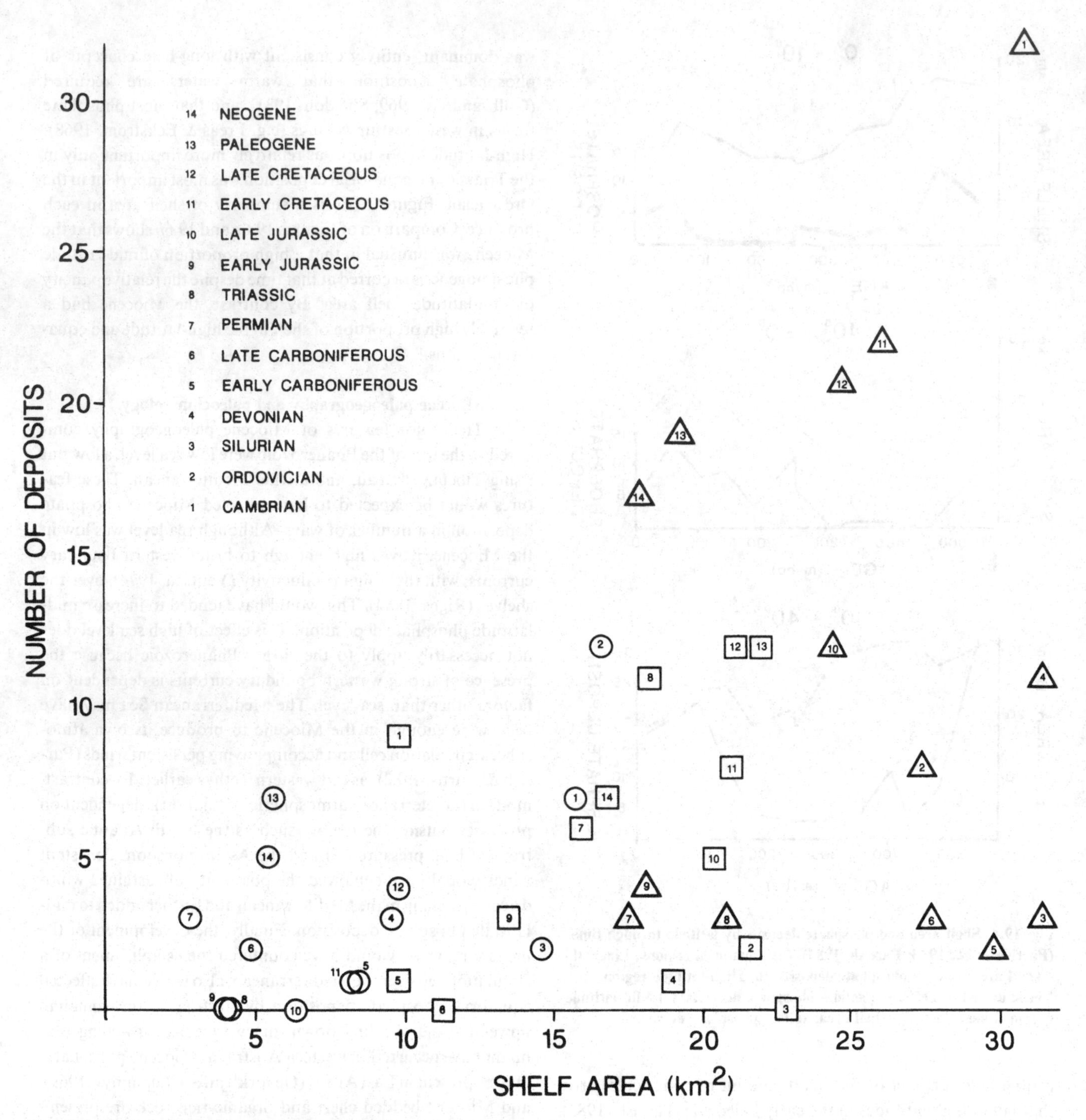

Fig. 19.5. Number of phosphate deposits versus shelf area for each upwelling region and time interval: circles – equatorial; triangles – mid-latitude; squares – high-latitude.

cian, which had an unusually high number of equatorial phosphate deposits (Fig. 19.5). Equatorial upwelling today occurs only in the open ocean, and Parrish (1982) was conservative in predicting equatorial upwelling on the continental shelves. Thus, it is apparent that the effectiveness of this type of upwelling on equatorial continental shelves was underestimated.

Not only was east-coast upwelling more important in the Miocene than at any other time, but west-coast upwelling also appeared to be more effective for generating phosphate. The only comparable times for the importance of phosphate deposits on west-facing coasts were the Cambrian and Permian (Parrish *et al.*, 1983). These, like the Miocene, also were cold phosphogenic periods (Fig. 19.3).

The quasi-cyclic distribution of phosphate through time (Cook & McElhinny, 1979) has engendered a number of efforts to tie phosphogenic episodes to a single mechanism (as opposed to phosphogenesis, for which upwelling still predominates in most discussions). For example, Fischer & Arthur (1977) and Slansky (1980) stressed warm global temperatures, Cook & McElhinny (1979) continental positions, Valeyev, Fayzullin & Yazmir (1979) tectonics and associated magmatism, Sheldon (1980) continental glaciation and global cooling, and Arthur &

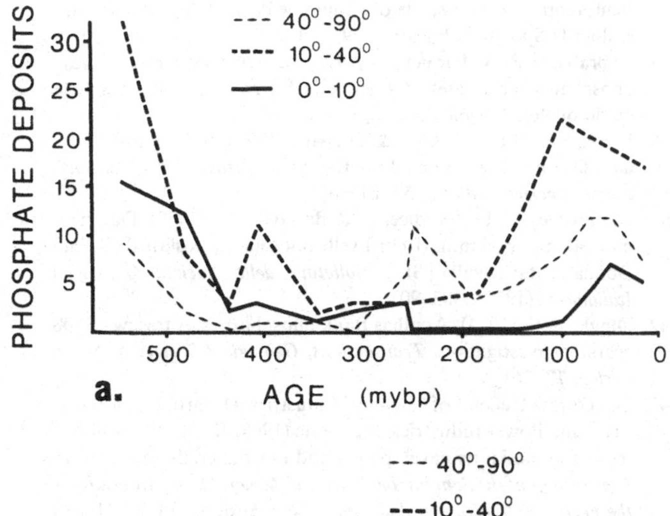

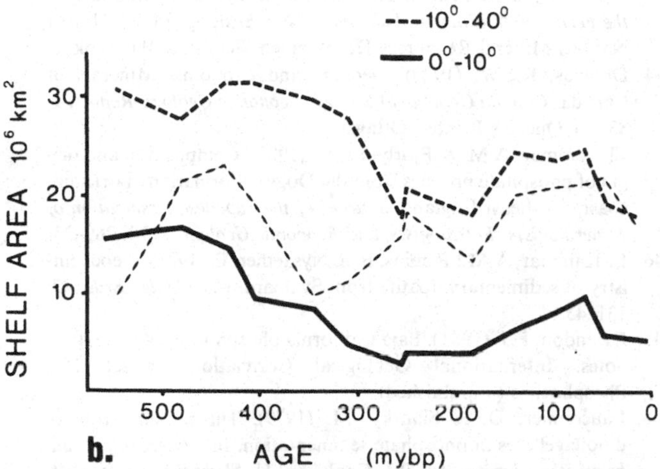

Fig. 19.6. Phosphate deposits and shelf area by latitude through time: (a) phosphate deposits; (b) shelf area.

Jenkyns (1981) transgression coupled with warm global temperatures. All of these workers, however, indicated that none of the proposed mechanisms works perfectly. Conversely, nearly all the proposed mechanisms work for some of the phosphogenic episodes. It seems reasonable to suggest, therefore, that each phosphogenic episode was the result of a unique combination of favorable conditions.

Data presented here suggest that cool temperatures (Sheldon, 1980) may have been important for increasing the efficiency of phosphate generation in west coast upwellings. Thus, the major factors accounting for the Miocene phosphogenic episode appear to be cool temperatures coupled with continental positions and sea level favorable to the flow of strong western boundary currents over shelves. This combination of climate and geography was unique in the Phanerozoic, as was the pattern of phosphate deposition.

Acknowledgments

Thanks are due to W.C. Burnett for encouraging me to undertake this study and to the other members of IGCP 156 for continuing inspiration. W.C. Burnett, E.K. Maughan, and S.M. Savin made many helpful suggestions for the improvement of the manuscript, but the author is solely responsible for its contents.

References

Numbered references are those cited in Table 19.1.

1. Ahmed, W. (1977). A short note on geological investigations for phosphate rocks in Sind and the adjoining areas. In *Phosphate Rock in the CENTO Region (Iran, Pakistan and Turkey)*, ed. A.J.G. Notholt, pp. 63–8. Central Treaty Organization, Ankara.
2. Al-Bassam, K.S. (1975). The mineralogy, geochemistry, and genesis of the Akashat phosphorite deposit, western Iraq. *Journal of the Geological Society of Iraq*, **9**, 1–33.
3. Al-Bassam, K.S., Al-Dahan, A.A. & Jamil, A.K. (1983). Campanian–Maastrichtian phosphorites of Iraq: petrology, geochemistry, and genesis. *Mineralium Deposita*, **18**, 215–33.
4. Al-Bassam, K.S. & Hagopian, D. (1983). Lower Eocene phosphorites of the Western Desert, Iraq. *Sedimentary Geology*, **33**, 295–316.
5. Altschuler, Z.S. (1980). The geochemistry of trace elements in marine phosphorites. Part 1. Characteristic abundances and enrichment. In *Marine Phosphorites-Geochemistry, Occurrence, Genesis*, ed. Y.K. Bentor, pp. 19–30. Society of Economic Paleontologists and Mineralogists Special Publication 29, Tulsa.

Arthur, M.A. & Jenkyns, H.C. (1981). Phosphorites and paleoceanography. *Oceanologica Acta*, **4 Supplement**, 83–96.

6. Atfeh, M.S. (1966). Phosphatic deposits in Syria and Safaga district, Egypt (discussion). *Economic Geology*, **61**, 1142–53.
7. Ayan, M. (1970). Fosfat yataklarinin teşekkülü ve aranmasi. *Türkiye Jeoloji Kurumu Bülteni*, **8(2)**, 49–82.
8. Balson, P.S. (1980). The origin and evolution of Tertiary phosphorites from eastern England. *Journal of the Geological Society of London*, **137**, 723–9.
9. Balson, P.S. (1984). Tertiary phosphorites in the area of the Southern Bight, North Sea: Origin, evolution and stratigraphic correlation. *International Geological Correlation Programme, Project 156, Newsletter*, **15**, 28.
10. Baturin, G.N. & Ilyin, A.V. (1985). Presentation at the 1985 Field Workshop and Seminar, International Geological Correlation Programme, Project 156, Phosphorites, May 1985, Tallahassee, Florida.
11. Baxter, J.L. & Hamilton, R. (1981). The Yoganup Formation and Ascot Beds as possible facies equivalents. *Western Australia Department of Mines Annual Report*, **1980**, 94–5.
13. Bein, A. & Amit, O. (1982). Depositional environments of the Senonian chert, phosphorite, and oil shale sequence in Israel as deducted from their organic matter composition. *Sedimentology*, **29**, 81–90.
14. Berker, K. (1977). Phosphate exploration and occurrences in Turkey. In *Phosphate Rock in the CENTO Region (Iran, Pakistan and Turkey)*, ed. A.J.G. Notholt, pp. 98–125. Central Treaty Organization, Ankara.
15. Beurlen, H. & Cassedanne, J.P. (1981). The Brazilian mineral resources. *Earth-Science Reviews*, **17**, 177–206.
16. Bhatti, N.A. (1977). Phosphorite deposits of the Kakul area, Hazara northwest Frontier Province in Pakistan. In *Phosphate Rock in the CENTO Region (Iran, Pakistan and Turkey)*, ed. A.J.G. Notholt, pp. 69–98. Central Treaty Organization, Ankara.
17. Bidzhiev, R.A., Koroleva, N.M. & Solev'eva, N.A. (1968). Phosphorites of the Volzhian Stage in the north of the Verkhoyansk Trough. *Lithology and Mineral Resources*, **1968(2)**, 165–73.
18. Birch, G.F. (1979). The nature and origin of mixed apatite–glauconite pellets from the continental shelf off South Africa. *Marine Geology*, **30**, 313–34.
19. Blake, G.H. (1981). Biostratigraphic relationship of Neogene benthic foraminifera from the southern California outer continental borderland to the Monterey Formation. In *The Monterey Formation and Related Siliceous Rocks of California*, ed. R.E. Garrison & R.G. Douglas, pp. 1–14. Society of Economic Paleontologists and Mineralogists, Pacific Section, Special Publication, Los Angeles.

20. Boiko, V.S., Kudryashev, N.S., Zhuravlev, Yu. P. & Shabanina, N.V. (1983). Lithology of middle Eocene phosphorite-bearing rocks of the central Kyzylkums. *Lithology and Mineral Resources*, **17**, 535–44.
21. Boujo, A. (1980). Un modéle de sédimentation phosphatée contrôlée par le fond: le golfe septentrional marocain (Oulad Abdon–Ganntour). In *Géologie Comparée des Gisements de Phosphates et de Pétrole*, pp. 129–40. Documents du Bureau de Recherches Géologiques et Minières 24, Orleans, France.
22. Boujo, A., Faye, B., Giot, D., Lucas, J., Manivit, H., Monciardini, C. & Prévôt, L. (1980). The early Eocene of the Lake of Guiers (western Senegal) – Reflections on some characteristics of phosphate sedimentation in Senegal. In *Marine Phosphorites – Geochemistry, Occurrence, Genesis*, ed. Y.K. Bentor, pp. 207–13. Society of Economic Paleontologists and Mineralogists Special Publication 29, Tulsa.
23. Bremner, J.M. (1978). Trace element concentrations in authigenic minerals from South West Africa. *Abstracts, 10th International Congress on Sedimentology*, Jerusalem 1978, **1**, 87–8.
24. British Sulphur Corporation (1971). *World Survey of Phosphate Deposits*, 3rd edn, ed. J.D. Waller. British Sulphur Corporation, London.
25. Bürgl, H. & Botero, G. (1967). Las capas fosfaticas de la Cordillera oriental. *Boletin Geologico*, **15**, 7–44.
Burnett, W.C. (1980). Apatite–glauconite associations off Peru and Chile: palaeo-oceanographic implications. *Journal of the Geological Society of London*, **137**, 757–64.
26. Bushinskii, G.I. (1969). *Old Phosphorites of Asia and Their Genesis*. Trudy Geolicheskii Institut, Akademia Nauk SSSR, 149 (translated by Israel Program for Scientific Translations, Jerusalem).
27. Busnardo, R., Enay, R., Latreille, G. & Rouquet, P. (1966). Le crétacé moyen détritique a céphalopodes près de Poncin (Jura méridional). *Travaux des Laboratoires de Géologie de la Faculté des Sciences de Lyon*, n.s., **13**, 205–28.
28. Caron, M., Rioult, M. & Royant, G. (1971). Position stratigraphique de la 'Croûte phosphatée' et des 'Calcschistes planctoniques' dans le versant méridional de l'Armetta (Alpes ligures). *Comptes Rendus Academie Sciences*, **Ser. D, 272**, 704–6.
29. Carter, A.N. (1978). Phosphatic nodule beds in Victoria and the late Miocene–Pliocene eustatic event. *Nature*, **276**, 258–9.
30. Cathcart, J.B. (1980). World phosphate reserves and resources. In *The Role of Phosphorous in Agriculture*, ed. F.E. Khasawneh, E.C. Sample & E.J. Kamprath, pp. 19–42. American Society of Agronomists, Madison, Wisconsin.
31. Cathcart, J.B. & Zambrano, F. (1967). Roca fosfatica en Colombia. *Boletin Geologico*, **15**, 65–162.
32. Chalyshev, V.I. (1968). The phosphorite assemblage of the Permian and Triassic deposits of the northern pre-Ural downwarp. *Lithology and Mineral Resources*, **1968(2)**, 174–83.
33. Chaudhri, R.S. & Gupta, G.D. (1977). Petrography and genesis of Nahan Phosphorite, Himachal Pradesh–northwest Himalayas. *Journal of the Geological Society of India*, **18**, 570–5.
34. Cheney, T.M., McClellan, G.H. & Montgomery, E.S. (1979). Sechura phosphate deposits, their stratigraphy, origin and composition. *Economic Geology*, **74**, 232–59.
35. Christie, R.L. (1978). Sedimentary phosphate deposits – an interim review. *Canada Geological Survey Paper*, **78–20**, 9pp.
Cook, P.J. & McElhinny, M.W. (1979). A reevaluation of the spatial and temporal distribution of sedimentary phosphate deposits in the light of plate tectonics. *Economic Geology*, **74**, 315–30.
36. Coulombeau, C. (1979). Une méthodologie originale de reconnaissance des dépôts phosphatées de la craie sénonienne de Picardie (France). *Chronique de la Recherche Minière*, **47(449)**, 5–24.
37. Čtyroký, P. & Karim, S.A. (1971). Stratigraphy and paleontology of the Umm er Radhuma Formation in the Akashat phosphate deposit, Ga'ara area, W. Iraq. *Journal of the Geological Society of Iraq*, **4**, 59–72.
38. Cullen, D.J. (1980). Distribution, composition and age of submarine phosphorites on Chatham Rise, east of New Zealand. In *Marine Phosphorites – Geochemistry, Occurrence, Genesis*, ed. Y.K. Bentor, pp. 139–48. Society of Economic Paleontologists and Mineralogists Special Publication 29, Tulsa.
39. Debrabant, P. & Paquet, J. (1975). L'association glauconites–phosphates carbonates (Albien de la Sierra de Espana, Espagne meridionale). *Chemical Geology*, **15**, 61–75.
40. Dietz, R.S., Emery, K.O. & Shepard, F.P. (1942). Phosphorite deposits on the sea floor off Southern California. *Geological Society of America Bulletin*, **53**, 815–48.
41. Di Grande, A., Lo Giudice, A. & Battaglia, M. (1979). Dati geomineralogici preliminari cui livelli miocenici a fosfati di Scicli–Donnalucata (Sicilia SE). *Bolletino della Societa Geologica Italiana*, **97(1978)**, 383–90.
42. Dingle, R.V. (1974). Agulhas Bank phosphorites: A review of 100 years of investigation. *Transactions, Geological Society of South Africa*, **77**, 261–4.
43. Directorate General of Chemical Industries, Department of Basic, Light and Power Industries, Indonesia (1968). The fertilizer industry in Indonesia: its development and raw materials situation. In *Proceedings of the Seminar on Sources of Mineral Raw Materials for the Fertilizer Industry in Asia and the Far East*, pp. 84–92. United Nations Mineral Resources Development Series 32, Bangkok.
44. Douglas, R.J.W. (1970). Geology and Economic Minerals of Canada. *Canada Geological Survey, Economic Geology Report*, **1**, 838pp. Queen's Printer, Ottawa.
45. El-Kammar, A.M. & Bjorlykke, K. (1983). Composition and origin of phosphate nodules from the Dogger Formation, Yorkshire coast, northeast England. *Journal of the Japanese Association of Mineralogists, Petrologists, and Economic Geologists*, **78**, 269–73.
46. El-Kammar, A.M., Robins, B. & Nysaether, E. (1983). Geochemistry of sedimentary apatite from Svalbard. *Chemie der Erde*, **42**, 131–43.
47. Escandon, F.J. (1981). Baja California phosphorite field trip, field notes. International Geological Correlation Project 156, Phosphorites (unpublished).
48. Fauconnier, D. & Slansky, M. (1979). The possible role of dinoflagellates in phosphate sedimentation. In *Proterozoic–Cambrian Phosphorites*, ed. P.J. Cook, & J.H. Shergold, pp. 93–101. First International Field Workshop and Seminar, International Geological Correlation Programme, Project 156, Canberra, Australia.
Fischer, A.G. & Arthur, M.A. (1977). Secular variations in the pelagic realm. In *Deep-Water Carbonate Environments*, ed. H.E. Cook & P. Enos, pp. 19–50. Society of Economic Paleontologists and Mineralogists Special Publication 25, Los Angeles.
49. Fournié, D. (1980). Phosphates et pétroles en Tunisie. In *Géologie Comparée des Gisements de Phosphates et de Pétrole*, pp. 157–66. Documents du Bureau de Recherches Géologiques et Minières 24, Orleans, France.
Frakes, L.A. (1979). *Climates Throughout Geologic Time*. Elsevier, New York.
50. Freas, D.H. & Eckstrom, C.L. (1968). Areas of potential upwelling and phosphorite deposition during Tertiary, Mesozoic, and Late Paleozoic time. *Proceedings of the Seminar of Sources of Mineral Raw Materials for the Fertilizer Industry in Asia and the Far East*, pp. 228–38. United Nations Mineral Resources Development Series 32, Bangkok.
Fukuoka, J., Ballester, A. & Cervignon, G. (1968). An analysis of hydrographical condition in the Caribbean Sea (III) – especially about upwelling and sinking. *Studies in Oceanography*, **47**, 145–9.
51. Fuller, A.O. (1979). Phosphate occurrences on the western and southern coastal areas and continental shelves of southern Africa. *Economic Geology*, **74**, 221–31.
52. Galliher, E.W. (1931). Collophane from Miocene brown shales of California. *American Association of Petroleum Geologists Bulletin*, **15**, 257–69.
53. Garrison, R.E., Glenn, C.R., Snavely, P.D. & Mansour, S.E.A. (1979). Sedimentology and origin of Upper Cretaceous phosphorite deposits at Abu Tartur, western desert, Egypt. *Annals of the Geological Survey of Egypt*, **9**, 261–81.
Gevork'yan, V. Kh. & Chugunnyy, Yu. G. (1969). A new find of

accumulations of phosphate concretions in the Indian Ocean. *Doklady Akademia Nauk SSSR*, **187**, 230–2.
54. Giresse, P. (1980). The Maastrichtian phosphate sequence of the Congo. In *Marine Phosphorites – Geochemistry, Occurrence, Genesis*, ed. Y.K. Bentor, pp. 193–205. Society of Economic Paleontologists and Mineralogists Special Publication 29, Tulsa.
55. Giresse, P. & Cornen, G. (1976). Distribution, nature et origine des phosphates miocènes et éocènes sousmarins des plate-formes du Congo et du Gabon. *Bureau de Recherches Géologiques et Minières Bulletin*, **4(1)**, 5–15.
56. Gorbunova, L.I. (1960). Glauconite from the Lower Cretaceous of Daghestan. *Doklady Akademia Nauk SSSR*, **130(4)**, 846–9.
57. Goudarzi, G.H. (1970). Geology and mineral resources of Libya–a reconnaissance. *US Geological Survey Professional Paper*, **660**, 104pp.
Gulbrandsen, R.A. (1969). Physical and chemical factors in the formation of marine apatite. *Economic Geology*, **64**, 365–82.
58. Harrington, J.F., Ward, D.E. & McKelvey, V.E. (1966). Sources of fertilizer minerals in South America – a preliminary study. *US Geological Survey Bulletin*, **1240**, 66pp.
59. Harris, R. (1981). Rofomex moves Mexico toward phosphate self-sufficiency. *Engineering and Mining Journal*, **July 1981**, 54–8.
60. Hassan, M. & Al-Maleh, A.K. (1976). La répartition de l'uranium dans les phosphates du Sénonien dans le Nord-Ouest Syrien. *Comptes Rendus Academie Sciences*, **Ser. D, 282**, 811–14.
Hein, J.R. & Parrish, J.T. (1987). Distribution of siliceous deposits in space and time. In *Siliceous Sedimentary Rock-Hosted Ores and Petroleum*, ed. J.R. Hein, pp. 10–57. Van Nostrand Reinhold, New York.
61. Hewitt, R.A. (1980). Microstructural contrasts between some sedimentary francolites. *Journal of the Geological Society of London*, **137**, 661–7.
62. Horton, A., Ivimey-Cook, H.C., Harrison, R.K. & Young, B.R. (1980). Phosphatic ooids in the Upper Lias (Lower Jurassic) of central England. *Journal of the Geological Society of London*, **137**, 731–40.
63. Hughes, R.J. (1978). The geology and mineral occurrences of Bathurst Island, Melville Island, and Coburg Peninsula, Northern Territory. *Bureau of Mineral Resources Bulletin of Geology and Geophysics*, **177**, 72pp.
64. Ibaydullaev, N. (1968). O raspredelenii redkikh elementov v razlichnykh tipakh fosfatizirovannykh kostnykh ostatkov paleogenovykh otlozheniy Kyzylkumov. *Uzbekskii Geologischeskii Zhurnal*, **1968(4)**, 44–8.
65. Ilyin, A.V. (1978). Sea floor spreading in the Atlantic and accumulation of phosphate. *Doklady Akademia Nauk SSSR*, **240**, 108–10.
66. Ilyin, A.V. (1985). Presentation at the 1985 Field Workshop and Seminar, International Geological Correlation Programme, Project 156, Phosphorites, May 1985, Tallahassee, Florida.
67. IGCP (1979). Phosphate rocks: selected bibliography and deposits of interest (Greece). *International Geological Correlation Programme, Project 156 Newsletter*, **5**, p. 28.
68. IGCP (1983). Phosphate investigations in Pakistan. *International Geological Correlation Programme, Project 156 Newsletter*, **13**, p. 15.
69. IGCP (1986). Triassic phosphorites of the New Siberian Islands, Arctic Ocean, USSR. *International Geological Correlation Programme, Project 156 Newsletter*, **17**, pp. 20–1.
70. Jarvis, I. (1980a). Geochemistry of phosphatic chalks and hardgrounds from the Santonian to early Campanian (Cretaceous) of northern France. *Journal of the Geological Society of London*, **137**, 705–21.
71. Jarvis, I. (1980b). The initiation of phosphatic chalk sedimentation – the Senonian (Cretaceous) of the Anglo-Paris Basin. In *Marine Phosphorites – Geochemistry, Occurrences, Genesis*, ed. Y.K. Bentor, pp. 167–92. Society of Economic Paleontologists and Mineralogists Special Publication 29, Tulsa.
72. Jarvis, I. & Woodroof, P. (1981). The phosphatic chalks and hardgrounds of Boxford and Winterbourne Berkshire-two tectonically controlled facies in the late Coniacian to early Campanian (Cretaceous) of southern England. *Geological Magazine*, **118**, 175–87.
73. Jenkyns, H.C. (1980). Cretaceous anoxic events: from continents to oceans. *Journal of the Geological Society of London*, **137**, 171–88.
74. Jones, E.J.W. & Goddard, D.A. (1979). Deep-sea phosphorite of Tertiary age from Annan Seamount, eastern equatorial Atlantic. *Deep-Sea Research*, **26**, 1363–79.
75. Killin, A.F. (1976). Phosphate. *Energy, Mines and Resources Canada, Mineral Bulletin*, **MR 160**, 18pp.
76. Kirikilitsa, S.I., Marchenko, Ye. Ya. & Vasenko, V.I. (1978). Phosphate concentration in Cretaceous–Paleogene boundary beds of the Crimea. *Doklady Akademia Nauk SSSR*, **241**, 48–50.
Knyazev, G.I., Vigdergauz, L.M. & Lazarenko, Yu. N. (1974). Some regularities of phosphorites distribution in the Dniester area (in Russian with English summary). *Dopovidi, Akademia Nauk Ukraïns'koi RSR*, **Seriya B, 2**, 113–17.
77. Kovalenko, D.II. & Semenov, V.G. (1964). *Fosforiti Ukraïni*, 179pp. Akademia Nauk Ukraïns'koi RSR, Kiev.
78. Lazur, O.G. (1976). Zakonomernosti razmeshcheniya i usloviya formirovaniya aktyubinskikh zhelvakovykh fosforitov. In *Ekzogennyye Poleznyye Iskopayemyye*, ed. V.K. Chaykovskiy, pp. 86–128. Izdatel'stvo Nauka, Moscow.
Lin, J.L., Fuller, M. & Zhang, W.Y. (1985). Preliminary Phanerozoic polar wander paths for the North and South China blocks, *Nature*, **313**, 444–9.
79. Loughman, D.L. & Hallam, A. (1982). A facies analysis of the Pucará Group (Norian to Toarcian carbonates, organic-rich shale, and phosphate) of central and northern Peru. *Sedimentary Geology*, **32**, 161–94.
Love, J.D. (1964). Uraniferous phosphatic lake beds of Eocene age in intermontane basins of Wyoming and Utah. *US Geological Survey Professional Paper*, **474–E**, E1–E66.
80. Lucas, J., Prévôt, L., Ataman, G. & Gündoğdu, N. (1980). Mineralogical and geochemical studies of the phosphatic formations in southeastern Turkey (Mazidaği-Mardin). In *Marine Phosphorites – Geochemistry, Occurrences, Genesis*, ed. Y.K. Bentor, pp. 149–52. Society of Economic Paleontologists and Mineralogists Special Publication 29, Tulsa.
81. Lydon, P.A. (1964). Unusual phosphatic rock – New deposit near Hyampom opens economic possibilities. *California Division of Mines and Geology Mineral Information Service*, **175(5)**, 65–74.
82. Mallick, K.A. & Al-Fadhli, I. (1980). Phosphorites in Sinjar Formation, Sulaimaniah area, Iraq, and its tectonic significance. *Abstracts of the 26th International Geological Congress*, **2**, p. 493.
83. Manheim, F.T., Popenoe, P., Siapno, W. & Lane, C. (1982). Manganese–phosphorite deposits of the Blake Plateau. In *Marine Mineral Deposits – New Research Results and Economic Prospects*, ed. P. Halbach & P. Winter, pp. 9–45. Verlag Glückauf GmbH, Essen.
84. Manheim, F.T., Pratt, R.M. & McFarlin, P.F. (1980). Composition and origin of phosphorite deposits of the Blake Plateau. In *Marine Phosphorites – Geochemistry, Occurrences, Genesis*, ed. Y.K. Bentor, pp. 117–37. Society of Economic Paleontologists and Mineralogists Special Publication 29, Tulsa.
85. Martins, L.R. & Coutinho, P.N. (1981). The Brazilian continental margin. *Earth-Science Reviews*, **17**, 87–108.
86. Maughan, E.K., Zambrano, F., Mojica, P. Abozaglo, J., Pachón, F. & Durán, R. (1979). Paleontologic and stratigraphic relations of phosphate beds in Upper Cretaceous rocks of the Cordillera Oriental, Colombia. *US Geological Survey Open-File Report*, **79–1525**, 97pp.
87. Mayne, S.J., Nicholas, E., Bigg-Wither, A.L., Rasidi, J.S. & Raine, M.J. (1974). Geology of the Sydney Basin: a review, *Bureau of Mineral Resources Bulletin of Geology and Geophysics*, **149**, 229pp.
88. McArthur, J.M. (1980). Post-depositional alteration of the carbonate-fluorapatite phase of Moroccan phosphates. In *Marine Phosphorites – Geochemistry, Occurrences, Genesis*, ed. Y.K. Bentor, pp. 53–60. Society of Economic Paleontologists and Mineralogists Special Publication 29, Tulsa.
McKelvey, V.E. (1959). Relation of marine upwelling waters to

phosphorite and oil. *Geological Society of America Bulletin*, **70**, 1783–4.

McKelvey, V.E. (1967). Phosphate deposits. *US Geological Survey Bulletin*, **1252-D**, D1–D21.

89. Meissner, C.R., Jr. & Ankary, A. (1972). Phosphorite deposits in the Sirhan–Turayf Basin, Kingdom of Saudi Arabia. In *Mineral Resources Report of Investigations (Kingdom of Saudi Arabia, Ministry of Petroleum and Mineral Resources)*, **2**, 27pp.
90. Melville, R.V. & Freshney, E.C. (1982). *The Hampshire Basin and Adjoining Areas*, 146pp. Institute of Geological Sciences, Her Majesty's Stationery Office, London.
91. Menor, E.A., Costa, M.P. deA. & Guazelli, W. (1979). Depósitos de fosfato. *Série Projeto REMAC, Reconhecimento Global de Margem Continental Brasileira*, **10**, 51–69.
92. Mew, M.C., ed. (1980). *World Survey of Phosphate Deposits*, 4th edn, 238pp. British Sulphur Corporation, London.
93. Mullins, H.T. & Neumann, A.C. (1979). Geology of the Miami terrace and its paleo-oceanographic implications. *Marine Geologist*, **30**, 205–32.
94. Nath, M. (1967). Phosphate deposits in Rajasthan. *Indian Minerals*, **21(2)**, 82–101.
95. Nathan, Y. & Nielson, H. (1980). Sulfur isotopes in phosphorites. In *Marine Phosphorites – Geochemistry, Occurrences, Genesis*, ed. Y.K. Bentor, pp. 73–8. Society of Economic Paleontologists and Mineralogists Special Publication 29, Tulsa.
96. Nathan, Y., Shiloni, Y., Roded, R., Gal, I. & Deutsch, Y. (1979). The geochemistry of the northern and central Negev phosphorites (southern Israel). *Geological Survey of Israel Bulletin*, **73**, 41pp.
97. Niedermayer, R.-O. & Schomburg, J. (1984). Phosphoritic nodules from the Late Cretaceous of Mielnik (Poland) and some aspects of the genesis of phosphorites. *Chemie der Erde*, **43**, 139–48.
98. Notholt, A.J.G. (1968). The stratigraphic and geographical distribution of phosphate deposits in the ECAFE region. In *Proceedings of the Seminar on Sources of Mineral Raw Materials for the Fertilizer Industry in Asia and the Far East*, pp. 57–73. United Nations Mineral Resources Development Series 32, Bangkok.

Notholt, A.J.G. (1979). The economic geology and development of igneous phosphate deposits in Europe and the USSR. *Economic Geology*, **74**, 339–50.

99. Notholt, A.J.G. (1980). Phosphatic and glauconitic sediments. *Journal of the Geological Society of London*, **137**, 657–9.
100. Oleinik, V.V. (1969). Formation conditions of Suzak (Lower Eocene) phosphorites of the Tadzhik Depression. *Lithology and Mineral Resources*, **1969(3)**, 289–96.
101. Oleinik, V.V. (1971). Phosphorites of Alai Strata (middle Eocene) of the Gissar–Turkestan mountain region. *Lithology and Mineral Resources*, **6**, 353–8.
102. Oussedik, M., Ousmer, N. & Belkhedim, M. (1980). Les minéralizations phosphatées éocènes en Algérie, et le gisement de phosphate du Djebel Onk. *Géologie Comparée des Gisements de Phosphates et de Pétrole*, pp. 141–54. Documents de Bureau de Recherches Géologiques et Minières 24, Orleans, France.
103. Pant, A., Dayal, B., Jain, S.C. & Chakravarty, T.K. (1979). Status of phosphorite investigation in Uttar Pradesh, India, and approach for future work. In *Proterozoic–Cambrian Phosphorites*. First International Field Workshop and Seminar, International Geological Correlation Programme, Project 156, ed. P.J. Cook & J.H. Shergold, p. 35–6. Bureau of Mineral Resources, Canberra, Australia.
104. Parker, R.S. & Siesser, W.G. (1972). Petrology and origin of some phosphorites from the South African continental margin. *Journal of Sedimentary Petrology*, **42**, 434–40.

Parrish, J.T. (1982). Upwelling and petroleum source beds, with reference to the Paleozoic. *American Association of Petroleum Geologists Bulletin*, **66**, 750–74.

Parrish, J.T. (1984). Phanerozoic paleoceanography and phosphorites. In *International Geological Correlation Programme, Project 156, Phosphorites*, **2**, pp. 221–4. 5th Field Seminar and Workshop, Kunming, China.

Parrish, J.T. (1985). Latitudinal distribution of land and shelf and absorbed solar radiation during the Phanerozoic. *US Geological Survey Open-File Report*, **85-31**, 21pp.

105. Parrish, J.T. (1987). Lithology, geochemistry, and depositional environment of the Shublik Formation (Triassic), northern Alaska. In *Alaskan North Slope Geology*, ed. I.L. Tailleur & P. Weimer, vol. 1, pp. 391–6. Pacific Section, Society of Economic Paleontologists and Mineralogists and Alaska Geological Society.

Parrish, J.T. & Curtis, R.L. (1982). Atmospheric circulation, upwelling, and organic-rich rocks in the Mesozoic and Cenozoic Eras. *Palaeogeography, Palaeoclimatology, Palaeoecology*, **40**, 31–66.

Parrish, J.T., Ziegler, A.M. & Humphreville, R.G. (1983). Upwelling in the Paleozoic Era. In *Coastal Upwelling: Its Sediment Record*, Part B., ed. J. Thiede & E. Suess, pp. 553–78. Plenum Press, New York.

Parrish, J.T., Ziegler, A.M., Scotese, C.R., Humphreville, R.G. & Kirschvink, J.L. (1986). Early Cambrian palaeogeography, palaeoceanography, and phosphorites. In *Phosphate Deposits of the World. Vol. 1. Proterozoic and Cambrian Phosphorites*, ed. P.J. Cook & J.H. Shergold, pp. 280–94. Cambridge University Press, Cambridge.

106. Patton, W.W., Jr. & Matzko, J.J. (1959). Phosphate deposits in northern Alaska. *US Geological Survey Professional Paper*, **302-A**, 1–17.
107. Pevear, D.R. & Pilley, O.H. (1966). Phosphorite in Georgia continental shelf sediments. *Geological Society of America Bulletin*, **77**, 849–58.

Pipiringos, G.N., Chisholm, A.W. & Kepferle, R.C. (1965). Geology and uranium deposits in the Cave Hills area, Harding County, South Dakota. *US Geological Survey Professional Paper*, **476-A**, A1–A64.

108. Pokryshkin, V.I., Boiko, V.S. & Il'yashenko, V.Ya. (1978). Distribution patterns of granular phosphorite deposits in the Arabian–African province and Central Asia. *Lithology and Mineral Resources*, **13(6)**, 733–47.
109. Poore, R.Z., McDougall, K., Barron, J.A., Brabb, E.E. & Kling, S.A. (1981). Microfossil biostratigraphy and biochronology of the type Relizian and Luisian stages of California. In *The Monterey Formation and Related Siliceous Rocks of California*, ed. R.E. Garrison & R.G. Douglas, pp. 15–42. Society of Economic Paleontologists and Mineralogists, Pacific Section, Special Publication, Los Angeles.
110. Pratsch, J.-C. (1958). Stratigraphische–tektonische Untersuchungen im Mesozoikum von Algarve (Südportugal). *Beihefte zum Geologischen Jahrbuch*, **30**, 123pp.
111. Reeves, M.J. & Saadi, T.A.K. (1971). Factors controlling the deposition of some phosphate bearing strata from Jordan. *Economic Geology*, **66**, 451–65.
112. Riggs, S.R. (1979). Petrology of the Tertiary phosphorite system of Florida. *Economic Geology*, **74**, 195–220.
113. Riggs, S.R. (1980). Intraclast and pellet phosphorite sedimentation in the Miocene of Florida. *Journal of the Geological Society of London*, **137**, 741–8.
114. Riggs, S.R. (1984). Paleoceanographic model of Neogene phosphorite deposition, US Atlantic continental margin. *Science*, **223**, 123–31.
115. Riggs, S.R., Hine, A.C., Snyder, S.W., Lewis, D.W., Ellington, M.D. & Stewart, T.L. (1982). Phosphate exploration and resource potential on the North Carolina continental shelf. *Offshore Technology Conference Paper*, **4295**, 737–8.
116. Rivera, J.O. (1979). Resumen de datos estratigráficos y estructurales de la Formación Monterrey que aflora en el área de San Hilario, Baja California Sur. Consejo de Recursos Minerales. México, *Revisto Geomimet*, **100**, 24pp.
117. Rivera, J.O. (1981). General geology and phosphate deposits of southern Baja California, Mexico. Notes prepared for the Baja California field trip, International Geological Correlation Programme, Project 156, Field Workshop and Seminar, La Paz, Mexico, 1981, 11pp.
118. Roberts, A.E. & Vercoutere, T.L. (1986). Geology and geochemis-

try of the Upper Miocene phosphate deposit near New Cuyama, Santa Barbara County, California. *US Geological Survey Bulletin*, **1635**, 89pp.

119. Robertson, B.T. (1982). Occurrence of epigenetic phosphate minerals in a phosphatic iron-formation, Yukon Territory. *Canadian Mineralogist*, **20**, 177–87.
120. Rogers, C.L., DeCserna, Z., Tavera, E. & Ulloa, S. (1956). General geology and phosphate deposits of Concepcion del Oro District, Zacatecas, Mexico. *US Geological Survey Bulletin*, **1037-A**, 102pp.
121. Rowell, H.C. (1981). Diatom biostratigraphy of the Monterey Formation, Palos Verdes Hills, California. In *The Monterey Formation and Related Siliceous Rocks of California*, ed. R.E. Garrison & R.G. Douglas, pp. 55–70. Pacific Section, Society of Economic Paleontologists and Mineralogists, Special Publication, Los Angeles.

Sagan, C. & Mullen, G. (1972). Earth and Mars: evolution of atmospheres and surface temperatures. *Science*, **177**, 52–6.

122. Salas, G.P. (1978). Sedimentary phosphate deposits in Baja California, Mexico. *American Institute of Mining Engineers Annual Meeting*, Denver.
123. Samimi Namin, M. & Ghasemipour, R. (1977). Phosphate deposits in Iran. In *Phosphate Rock in the CENTO Region*, ed. A.J.G. Notholt, pp. 1–41. Central Treaty Organization, Ankara.
124. Samimi Namin, M., Movahhed Aval, M., Ghasemipur, R. & Sharifi Norian, M. (1966). Cretaceous and Tertiary phosphate. Recent phosphate discoveries in Iran. *Geological Survey of Iran Report*, **10**, 58–79.

Scotese, C.R., Bambach, R.K., Barton, C., Van der Voo, R. & Ziegler, A.M. (1979). Paleozoic base maps. *Journal of Geology*, **87**, 217–77.

125. Scott, T.M. & MacGill, P.L. (1981). The Hawthorn Formation of central Florida. *Florida Department of Natural Resources, Report of Investigation*, **91**, 1–35.
126. Sheldon, R.P. (1964a). Paleolatitudinal and paleogeographic distribution of phosphorite. *US Geological Survey Professional Paper*, **501-C**, C106–C113.
127. Sheldon, R.P. (1964b). Exploration for phosphorite in Turkey – a case history. *Economic Geology*, **59**, 1159–75.

Sheldon, R.P. (1980). Episodicity of phosphate deposition and deep ocean circulation – an hypothesis. In *Marine Phosphorites – Geochemistry, Occurrence, Genesis*, ed. Y.K. Bentor, pp. 239–47. Society of Economic Paleontologists and Mineralogists Special Publication 29, Tulsa.

Sheldon, R.P. (1981). Ancient marine phosphorites. *Annual Reviews of Earth and Planetary Science*, **9**, 184–251.

Slansky, M. (1980). Localisation des gisements de phosphates dans les bassins sédimentaires. In *Géologie Comparée des Gisements de Phosphates et de Pétrole*, pp. 13–21. Documents du Bureau de Recherches Géologiques et Minières 24, Orleans, France.

128. Smith, P.B. (1968). Paleoenvironment of the phosphate-bearing Monterey Shale in the Salinas Valley, California. In *Proceedings of the Seminar on Sources of Mineral Raw Materials for the Fertilizer Industry in Asia and the Far East*, pp. 172–8. United Nations Mineral Resources Development Series 32, Bangkok.
129. Starkov, N.P. (1966). Nature of the friable fucoids of the Vyatka–Kama phosphorites. *Lithology and Mineral Resources*, **1966(3)**, 371–5.
130. Stott, D.F. (1983). Late Jurassic–Early Cretaceous foredeeps of northeastern British Columbia. *Transactions of the Royal Society of Canada*, **Series IV, 21**, 143–53.
131. Summerhayes, C.P., Birch, G.F., Rogers, J. & Dingle, R.V. (1973). Phosphate sediments off South-western Africa. *Nature*, **243**, 509–11.
132. Summerhayes, C.P., Nutter, A.H. & Tooms, J.S. (1972). The distribution and origin of phosphate in sediments off northwest Africa. *Sedimentary Geology*, **8**, 3–28.

Swallow, J.C. (1980). The Indian Ocean experiment: introduction. *Science*, **209**, 588.

Sverdrup, H.U., Johnson, M.W. & Fleming, R.H. (1942). *The Oceans: Their Physics, Chemistry, and General Biology*. Prentice-Hall, Englewood Cliffs, NJ.

133. Szekely, T.S. & Grose, L.T. (1972). Stratigraphy of the carbonate, black shale, and phosphate of the Pucará Group (Upper Triassic–Lower Jurassic), central Andes, Peru. *Geological Society of America Bulletin*, **83**, 407–28.
134. Tankard, A.J. (1974). Varswater Formation of the Langevaanweg–Saldanha area, Cape Province. *Transactions, Geological Society of South Africa*, **77**, 265–83.
135. Tankard, A.J. (1975). The marine Neogene Saldanha Formation. *Transactions, Geological Society of South Africa*, **78**, 257–64.
136. Tayyab Ali, S. (1977). Phosphate deposits of Pakistan. In *Phosphate Rock in the CENTO Region (Iran, Pakistan, and Turkey)*, ed. A.J.G. Notholt, pp. 42–62. Central Treaty Organization, Ankara.
137. Tiwari, R.N. & Roy, N.R. (1974). Sedimentpetrologische Untersuchungen an oberkretazischen Sandsteinen der Subherzynen Kreidemulde. *Freiberger Forschungschefte*, **C301**, 27–135.
138. Tooms, J.S. & Summerhayes, C.P. (1968). Phosphatic rocks from the north-west African continental shelf. *Nature*, **218**, 1241–2.
139. Tourtelot, H.A. & Cobban, W.A. (1968). Stratigraphic significance and petrology of phosphate nodules at base of Niobrara Formation, east flank of Black Hills, South Dakota. *US Geological Survey Professional Paper*, **594-L**, 22pp.

Tsykin, R.A. (1980). Phosphorus in the weathered crust and bauxite-containing deposits in the central part of the Chadobets. *Geology and Geophysics*, **21(3)**, 61–6.

140. Tuchkov, I.I. (1966). Phosphorites on the lower reaches of the Lena River. *Lithology and Mineral Resources*, **4**, 103–18.
141. Uberna, J. (1980). Les phosphorites dans le sédiments calloviens. In La Géologie et les gîtes mineraux en Pologne, ed. R. Osika. *Biuletyn Instytut Geologsiczny*, **251**, 442.
142. Uberna, J. (1980). Les phosphorites. In La Géologie et les gîtes mineraux en Pologne, ed. R. Osika. *Biuletyn Instytut Geologsiczny*, **251**, 470–3.
143. Uberna, J. (1981). Upper Eocene phosphate-bearing deposits in northern and eastern Poland. *Bulletin de l'Académie Polonaise des Sciences*, **29**, 81–90.
144. Uberna, J., Cieśliński, S., Błaszkiewicz, A., Jaskowiak, M. & Krassowska, A. (1971). Kredowe osady fosforytonośne i fosforyty w Polsce. *Biuletyn Instytut Geologsiczny*, **246**, 135–62.
145. USSR Delegation (1968). Geological characteristics of commercial deposits of main kinds of mineral raw material for fertilizers in the USSR. In *Proceedings of the Seminar on Sources of Mineral Raw Materials for the Fertilizer Industry in Asia and the Far East*, pp. 100–9. United Nations Mineral Resources Development Series 32, Bangkok.

Vail, P.R., Mitchum, Jr., R.M. & Thompson, III, S. (1977). Seismic stratigraphy and global changes of sea level, part 4. Global cycles of relative changes of sea level. In *Seismic Stratigraphy – Applications to Hydrocarbon Exploration*, ed. C.E. Payton, pp. 83–97. American Association of Petroleum Geologists Memoir 26.

Valeyev, R.N., Fayzullin, R.M. & Yazmir, M.M. (1979). Global rifting and deposition of phosphate ores. *Doklady Academia Nauk SSSR*, **249**, 56–8.

146. Vejlupek, M. (1979). Geologie a nerostné suroviny Iráku. *Geologický Průzkum*, **21**, 176–9.
147. Vinogradov, A.P., ed. (1967). *Atlas of the Lithological–Paleogeographical Maps of the USSR, Vol. 4, Paleogene, Neogene and Quaternary*. Ministerstvo Geologia, Akademia Nauk SSSR, Moscow.
148. Vinogradov, A.P., ed. (1968). *Atlas of the Lithological–Paleogeographical Maps of the USSR, Vol. 3, Triassic, Jurassic, Cretaceous*. Ministerstvo Geologia, Akademia Nauk SSSR, Moscow.
149. Von der Borch, C.C. (1970). Phosphatic concretions and nodules from the upper continental slope, northern New South Wales. *Journal of the Geological Society of Australia*, **16**, 755–9.
150. Watters, W.A. (1968). Phosphorite and apatite occurrences and possible reserves in New Zealand and outlying islands. In *Proceedings of the Seminar on Sources of Mineral Raw Materials for the*

Fertilizer Industry in Asia and the Far East, pp. 144–51. United Nations Mineral Resources Development Series 32, Bangkok.

151. Wilkinson, A.F. (1982). Exploration for phosphate in Ecuador. *Transactions, Institution of Mining and Metallurgy*, **Sec. B, 91**, 130–45.

Yentsch, C.S. (1974). The influence of geostrophy on primary production. *Tethys*, **6**, 111–18.

152. Yolkichev, N. (1978). Bivalvii probivachi v kampanskite i mastrikhtskite sedimenti mezhdu selata Debovo i Muselievo, Plevensko. *Godizhnik na Sophiiskiya Universitet, Geologo-Geographski Phacultet, Kniga 1, Geologiya*, **70**, 87–91.

153. Young, F.G. (1977). The mid-Cretaceous flysch and phosphatic ironstone sequence, northern Richardson Mountains, Yukon Territory. *Canada Geological Survey Paper*, **77-1C**, 67–74.

154. Young, F.G., Myhr, D.W. & Yorath, C.J. (1976). Geology of the Beaufort–MacKenzie Basin. *Canada Geological Survey Paper*, **76–11**, 65p.

Ziegler, A.M., Scotese, C.R. & Barrett, S.F. (1983). Mesozoic and Cenozoic paleogeographic maps. In *Tidal Friction and the Earth's Rotation*, II, ed. P. Brosche & J. Sündermann, pp. 240–52. Springer-Verlag, Berlin.

20

Isotopic evidence for temperature and productivity in the Tertiary oceans

S.M. SAVIN AND F. WOODRUFF

Abstract

Among the most striking conclusions of the study of oxygen isotope ratios of foraminifera from Tertiary sediments has been that during the early Middle Miocene, the time during which a large Antarctic ice-cap became permanently established, tropical regions warmed and latitudinal temperature gradients were intensified. This Middle Miocene intensification of the latitudinal temperature gradient may have been associated with more vigorous thermohaline circulation and enhanced upwelling, and may therefore be related to the major episodes of phosphogenesis that occurred during that time. There is some indication that the mean global temperature may also have increased at that time.

The interpretation of the foraminiferal carbon-isotope record is less firmly grounded than is the oxygen-isotope record, but rapid progress is now being made in the understanding of carbon isotopes. Oceanwide or global changes in the $\delta^{13}C$ of dissolved marine HCO_3^- during Tertiary time are now well established, and these must reflect changes in the global cycling of carbon. Miocene Pacific benthic foraminiferal $\delta^{13}C$ values are well correlated with marine onlap–offlap. The differences among $\delta^{13}C$ values of separated benthic species from single sediment samples co-vary with biological productivity in the overlying surface waters. This has not yet been developed as a quantitative indicator of paleoproductivity, but the future for the development of this approach is promising.

Introduction

Oxygen isotope ratios of the tests of planktic and benthic foraminifera in deep-sea sediments have been used for over 30 years to reconstruct ancient ocean temperatures and the extent of continental glaciation in the past. The pace of this research has grown increasingly during the 20 years since the commencement of the Deep Sea Drilling Project (DSDP). The isotopic data have provided one of the most important underpinnings of our understanding of the repetitive climatic fluctuations during the Quaternary, and have been the most reliable quantitative indicator of the thermal history of the oceans during the Tertiary period. Carbon isotope ratios of marine calcites are routinely obtained during the measurement of their $^{18}O/^{16}O$ ratios. Until recently, the significance of these carbon isotope ratios remained largely enigmatic. However, it has become apparent in the past few years that the $^{13}C/^{12}C$ ratios of benthic foraminiferal calcite respond to global carbon cycling, to bottom water circulation patterns, and to local fluctuations in biological productivity.

In this paper, we briefly review the temperature history of the oceans during Tertiary time. We then present new arguments that average global ocean temperatures rose during the Miocene, concurrently with the growth of a major Antarctic ice sheet. We conclude with an optimistic assessment of the potential for the use of carbon isotope ratios of benthic foraminifera as indicators of biological productivity in the ancient oceans.

Review of Tertiary marine temperatures

The thermal history of the oceans during the past 100 million years has been reviewed by Hecht (1976), Savin (1977 and 1983), Haq (1984), Savin & Douglas (1985), Miller, Fairbanks and Mountain (1987), and others. The major features of the isotopic record of Tertiary planktic and benthic foraminifera from low and intermediate latitudes are now reasonably well documented, although some ambiguities remain when the isotopic data are translated into bottom-, and especially, surface-water temperatures. The greatest obstacle to the interpretation of the isotopic data is the uncertainty in the $^{18}O/^{16}O$ ratio of sea water as a function of time and location. In the case of Tertiary benthic foraminifera this primarily reflects the uncertainty in the extent of continental glaciation and its effect on the isotopic composition of sea water. An additional uncertainty in the interpretation of planktic oxygen isotope data derives from the difficulty in assessing regional variations in evaporation and precipitation in the past, and their effects on the $^{18}O/^{16}O$ ratio of surface waters.

The oxygen isotope ratios of most species of benthic foraminifera are indicative of water temperature at the sediment–water interface. It is now known that most species of benthic foraminifera deposit calcite that is slightly out of oxygen isotopic equilibrium with the water in which they grow (Shackleton, 1974; Woodruff, Savin & Douglas, 1980; Belanger, Curry & Matthews, 1981; Graham *et al.*, 1981, and others). The amounts by which the oxygen isotope ratios of a species depart from equilibrium values seem to vary relatively little. As a result, it is possible to adjust measured $^{18}O/^{16}O$ ratios of calcite tests to approximate the values they would have had if the foraminifera had deposited the calcite in isotopic equilibrium with bottom water. While this practice can introduce some errors, these errors

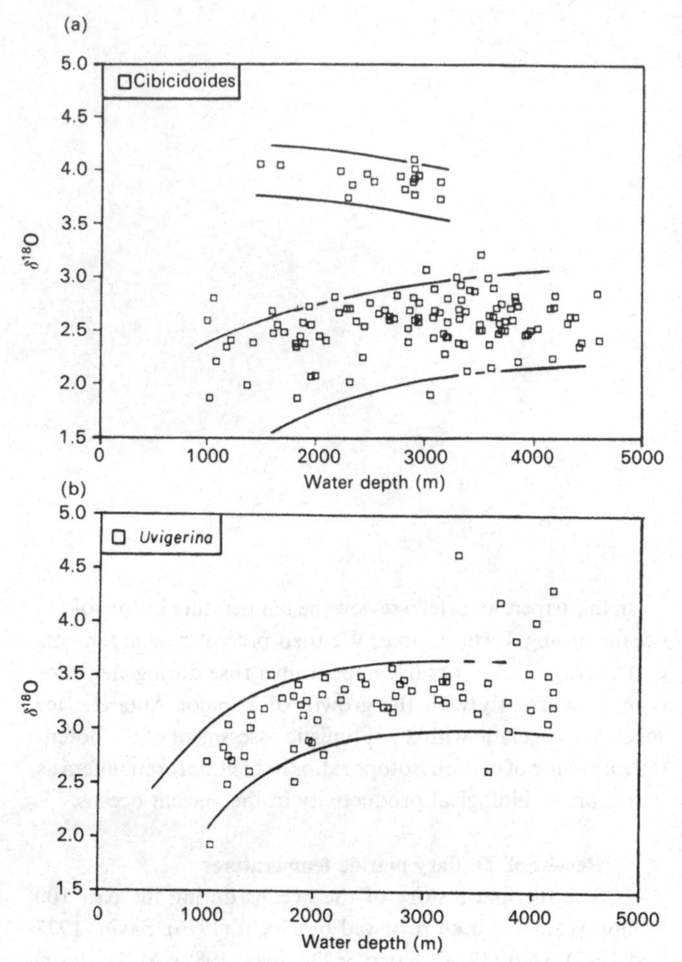

Fig. 20.1 Published $\delta^{18}O$ values of Holocene benthic foraminifera plotted as a function of the depth of the water at the sampling sites. Data and their sources are listed in Appendix 20.1: (a). *Cibicidoides*; (b). *Uvigerina*

are minor compared to the major temperature trends discussed in this paper.

A potentially more serious obstacle to the interpretation of isotopic compositions of old (especially Paleogene and older, and especially those from deeply buried sections) foraminifera is the uncertain extent of diagenesis and the uncertainty of its effect on the measured isotope ratios. Although in some cases alteration of foraminifera is obvious through examination with the binocular microscope, this is not a satisfactory technique for screening out all diagenetically altered samples. Scanning electron microscopy (SEM), a better indicator of the state of sample preservation (Barrera, Keller & Savin, 1985), has not been applied to many of the Early Tertiary sample sets that have been isotopically analyzed. Chemical analysis (Graham *et al.*, 1982; Delaney, 1983) and Sr isotope analysis (Elderfield *et al.*, 1982) as indicators of diagenesis have been applied to even fewer. Barrera (1987) has shown that some types of diagenetic alteration of benthic foraminifera are not accompanied by large shifts in isotope composition, although this may not be true for planktics.

Consideration of the variability of $\delta^{18}O$ values of geographically widely distributed Holocene benthic foraminifera gives some indication of the variability that might be expected in the $\delta^{18}O$ values of foraminifera from a narrowly defined stratigraphic interval in the past. Figure 20.1 is a compilation of 160-$\delta^{18}O$ values of globally distributed Holocene benthic foraminifera compiled from literature sources (approximately 56% from Duplessy *et al.*, 1984 and the remainder from other sources), plotted against depth of the water from which they were taken. All data are tabulated in Appendix 20.1. Figure 20.1*a* shows data for *Cibicidoides* and Figure 20.1*b* shows data for *Uvigerina*.

Most of the data for each data set fall within a range of about 0.6 mil^{-1}, with some indication of slightly lower $\delta^{18}O$ values at shallower depths, reflecting the slightly warmer waters. The especially ^{18}O-rich values for some of the *Cibicidoides* are, with one exception, from the Norwegian Sea. These must reflect growth in waters more ^{18}O-rich than those of deep waters in most of the ocean basins. Figure 20.1 suggests that in any narrowly defined time interval in the past when the deep waters of the ocean formed, as today, by the sinking of cold dense waters at high latitudes, a synoptic compilation of $\delta^{18}O$ values of geographically widely distributed benthic foraminifera would in most cases consist of values that fall within a range of approximately 0.6 mil^{-1}.

Figure 20.2 is a compilation of $\delta^{18}O$ values of published Tertiary benthic foraminifera from a number of Atlantic and Pacific DSDP sites. One of the most striking features of the figure is the similarity of the isotopic records from most of the sites. As was the case for the Holocene benthic foraminifera, the range of $\delta^{18}O$ values of the Tertiary benthic foraminifera for each narrowly defined time interval is consistently about 0.6 mil^{-1} or smaller. This, by itself, serves as evidence that the original $\delta^{18}O$ values in these samples have not, contrary to the suggestion of Killingly (1983), been grossly altered by diagenesis. It is improbable that diagenetic alteration of the isotope ratios of benthic foraminiferal calcite from a number of sites, with various sedimentation rates and geothermal gradients, would result in such uniformity of oxygen isotope ratios.

The isotopic record indicates that during the Tertiary there has been a net cooling of the deep waters of the ocean, but that the cooling has not been uniform. There have been episodes of warming as well as periods of relative temperature stability. During ice-free times the mean $\delta^{18}O$ value of sea water, and hence of the deep water of the ocean, was approximately -1.08 mil^{-1} (Shackleton, 1967). At present it is -0.18 mil^{-1}, and during times of maximum Pleistocene ice cover it was approximately $+1.2-1.6$ mil^{-1} (Savin & Yeh, 1981). The temperature scales on Figure 20.2 have been calculated using the modern and ice-free $\delta^{18}O$ values of seawater. As summarized by Savin & Douglas (1985) the oxygen isotope data have been commonly interpreted as indicating that the world was largely ice-free during the Paleocene, Eocene, and parts of the Oligocene and Early Miocene. During parts of the Oligocene and from the Middle Miocene to the present, there has been significant, but generally not well defined, amounts of continental ice. During most of that time, therefore, deep-water temperatures lay somewhere between those indicated by the ice-free temperature scale and those indicated by the modern temperature scale. The sharply defined increase in $\delta^{18}O$ values in latest Early and early Middle Miocene time reflects a combination of deep-water cooling and continental ice formation. Pleistocene deep-water tem-

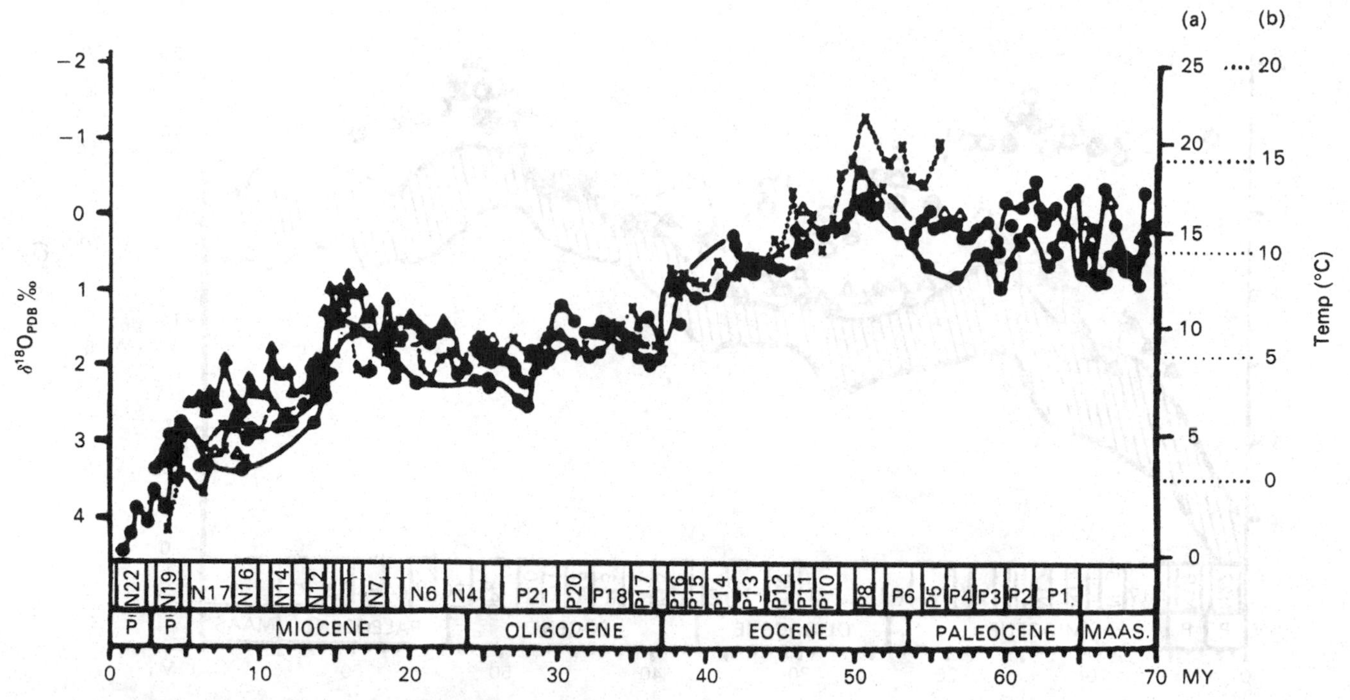

Fig. 20.2. Benthic foraminiferal $\delta^{18}O$ time series for the Tertiary for several sites in the Atlantic and Pacific Oceans. Data for DSDP Sites 20, 21, 44, 47, 94, 95, 167, 171, 277, 279, 281, 289, 305, 327, 329, 356, 357, 366, and 384 are shown. Temperature scales are shown for: (a) the modern world; and (b) the ice-free world. Data sources are Douglas & Savin (1971, 1973, 1975 and unpublished); Shackleton & Kennett (1975); Boersma & Shackleton (1977, 1978); Boersma *et al.* (1979); and Woodruff *et al.* (1981).

peratures were similar to or greater than those indicated by the modern temperature scale.

Representative oxygen isotope records for planktic foraminifera from high and low latitudes are shown in Figure 20.3, as is an envelope drawn about the benthic isotope data of Figure 20.2. During Paleogene and Early Miocene time, the major trends of the benthic isotope curve are repeated in the planktic curves, indicating that the oceans warmed and cooled simultaneously at both high and low latitudes. However, during the early Middle Miocene episode of cooling of bottom waters and middle and high latitude surface waters, $\delta^{18}O$ values of shallow-dwelling tropical planktic foraminifera either decreased (at some sites) or remained constant (at others). Accepting the interpretation that the Middle Miocene increase in benthic foraminiferal $\delta^{18}O$ values reflected, in part, a major phase of growth of the Antarctic ice-cap (and hence an increase in the mean $\delta^{18}O$ value of the ocean), the oxygen isotope data imply that the Middle Miocene was a time in which tropical surface temperatures must have warmed while high latitudes cooled and ice-caps grew. This Middle Miocene intensification of the latitudinal temperature gradient may have been associated with more vigorous thermohaline circulation and enhanced upwelling, and may therefore be related to the major episodes of phosphogenesis that occurred during that time. The increase in the latitudinal temperature gradient during the Middle Miocene is the most marked of a number of such increases in gradient during Tertiary time. These have been carefully documented by Keigwin & Corliss (1986), Kennett (1986) and Murphy & Kennett (1986). They have shown increases in the latitudinal temperature gradient in Early Oligocene, Middle Oligocene, early Middle Miocene and Late Pliocene intervals (Fig. 20.4).

An alternative interpretation of the Tertiary isotope data has been proposed by Matthews & Poore (1980). In that interpretation, continental ice volume was significant through much of Tertiary time, and tropical surface temperatures were constant. If so, the benthic isotope record throughout Tertiary time would reflect a combination of deep-water cooling and ice formation, and the Middle Miocene increase in benthic $\delta^{18}O$ values would primarily reflect high-latitude cooling. It would not indicate, primarily, an episode of massive Antarctic ice formation as suggested by Savin, Douglas & Stehli (1975), Shackleton & Kennett (1975) and others. We do not include a discussion of the relative merits of the two interpretations of the isotope data in this paper. However, in either the more conventional interpretation or the interpretation of Matthews & Poore, the latitudinal temperature gradient inferred from the difference between $\delta^{18}O$ values of tropical planktic foraminifera and those of benthic foraminifera is the same. Furthermore, as pointed out by Savin & Douglas (1985), the effect on the isotopic composition of seawater of an incremental increase in continental ice volume is relatively small in the early stages of ice-cap growth, and is large for an incremental increase of the same magnitude during the later stages of ice-cap growth.

Global temperature changes during the Miocene

The paleoceanography of the Miocene oceans is better known than that of any other Tertiary epoch. It was the focus of the recently completed Cenozoic Paleoceanography Project

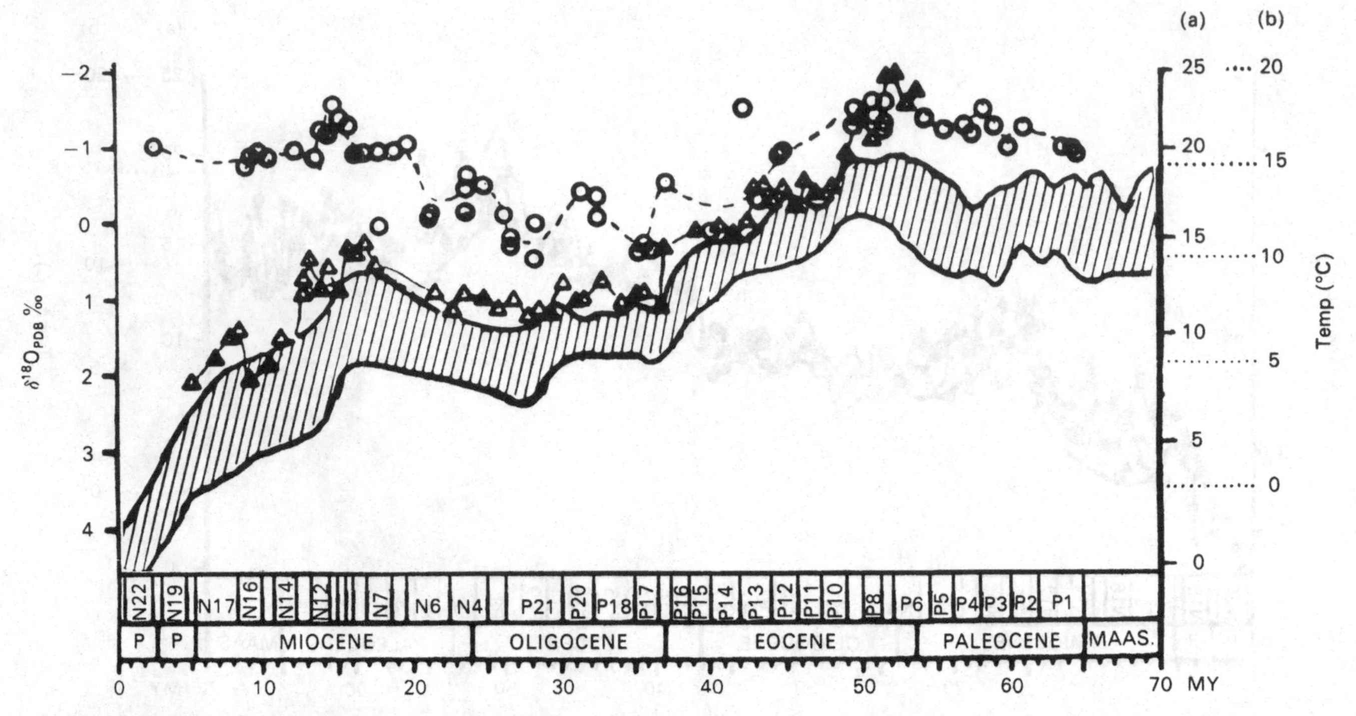

Fig. 20.3. Tertiary planktic foraminiferal $\delta^{18}O$ time series for the tropical (circles) and southern mid-latitude (triangles) Pacific Ocean. Shown for reference is an envelope drawn about the benthic data in Figure 20.2 (shaded region). Temperature scales are shown for: (a) the modern world; and (b) the ice-free world. Data sources are Savin *et al.* (1975, unpublished) and Shackleton & Kennett (1975).

(CENOP), many of the results of which are included in Kennett (1985). As part of the CENOP project, synoptic biogeographic and oxygen isotopic maps of planktic foraminiferal data were compiled for three Miocene time intervals: 22 Ma (foraminiferal zone N4B) in the Early Miocene when benthic foraminiferal $\delta^{18}O$ values indicated cool bottom waters; 16 Ma (foraminiferal zone N8) in the latest Early Miocene when benthic $\delta^{18}O$ values indicated the warmest bottom waters of the entire Miocene; and 8 Ma (foraminiferal zone N17), when high latitude regions were cold and continental ice cover was extensive.

Synoptic maps of $\delta^{18}O$ value of near-surface planktic foraminifera for the three times are shown in Figure 20.5. The isotopic data on these maps, from Savin *et al.* (1985), are an indication of the distribution of surface temperatures during the three time intervals, although reliable estimates of absolute temperature require accurate estimates of the mean $\delta^{18}O$ value of seawater for each interval. In comparing the three time intervals, the expansion with time, from Early–Late Miocene, of the area of the ocean for which low (i.e. warm) $\delta^{18}O$ values are obtained is apparent. This may be interpreted as reflecting a warming and an expansion of the tropics. No adjustment was made to the isotopic data in Figure 20.5 for changes in continental ice volume. If an adjustment for ice volume were made, the warming and expansion of the tropics in the Late Miocene synoptic map would be even more apparent. While the synoptic isotopic data indicate that the low latitudes warmed with time during the Miocene epoch, the small amount of data for intermediate latitudes show little temperature change, and the benthic foraminiferal isotopic data indicate that the high latitudes cooled.

In view of those observations, it is instructive to consider the change in average global (or at least average ocean) temperature during Miocene time. Conceptually, this can be done with the Miocene synoptic data by dividing the world into latitudinal bands (in this case, 10° bands), and determining the average temperature of each band. Then,

$$\text{Average temperature} = \frac{\sum_i \text{Area of band}_i \times \text{Average temperature of band}_i}{\sum_i \text{Area of band}_i} \quad (20.1)$$

where i indexes the 10° bands. In practice, because the available data are $\delta^{18}O$ values for shallow-dwelling planktic foraminifera and benthic foraminifera, it is more appropriate to calculate

$$\text{Average } \delta^{18}O = \frac{\sum_i \text{Area of band}_i \times \text{Average } \delta^{18}O \text{ of band}_i}{\sum_i \text{Area of band}_i} \quad (20.2)$$

Because the earth is a sphere, the area of low latitude 10° bands is much larger than that of bands at high latitudes (Fig. 20.6). The low latitudes are therefore more important than the high latitudes in determining the average temperature or average $\delta^{18}O$.

The results of the calculation are shown in Figure 20.7, in which an average $\delta^{18}O$ value for each of the three Miocene intervals has been calculated using the data from each of the three Miocene synoptic isotope maps. They show that the average ocean temperature increased (warmed) about 1.5°C (corresponding to a change in foraminiferal $\delta^{18}O$ of about 0.4 mil^{-1}) between 22 Ma and 16 Ma. The average $\delta^{18}O$ of shallow-dwell-

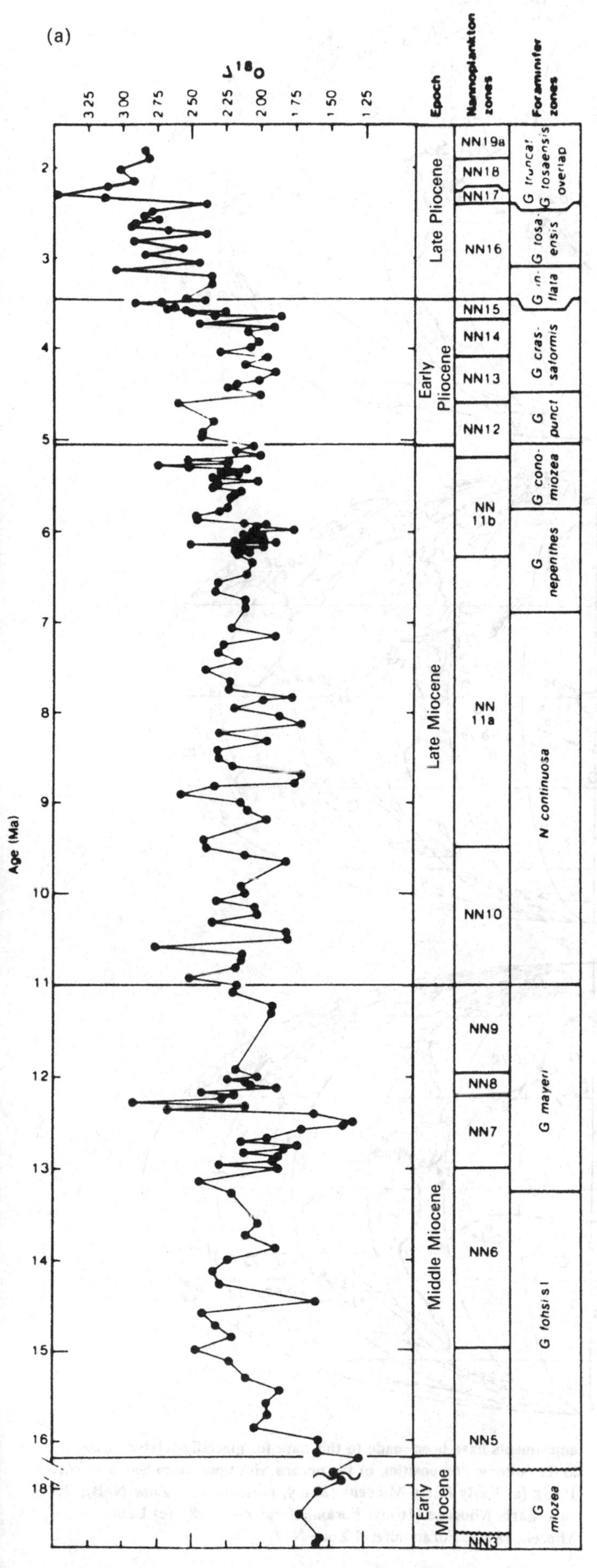

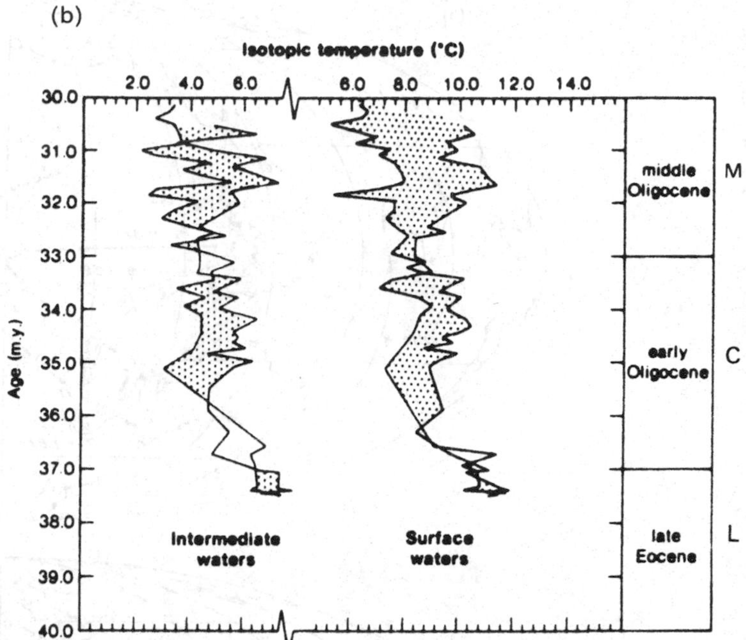

Fig. 20.4. (a) Difference between δ¹⁸O values of Miocene and Pliocene benthic and planktic foraminifera from DSDP Sites 590A and 590B (water depth 1300 m). This difference increases as the surface–bottom temperature gradient at the site steepens. Note the marked increases in earlier Middle Miocene time and between the Early and Late Pliocene (from Kennett, 1986). (b) Difference between δ¹⁸O values of planktic foraminifera from Sites 277 (present position 52°13′ S, 166°12′ W) and 593 (present position 40°31′ S, 167°41′ E). Shaded region indicates times when the more southerly site was cooler than the more northerly. The data show the latitudinal temperature gradient in this region to have been essentially non-existent in the Late Eocene and to have increased sharply in the Early Oligocene and again in the Middle Oligocene (from Murphy & Kennett, 1986).

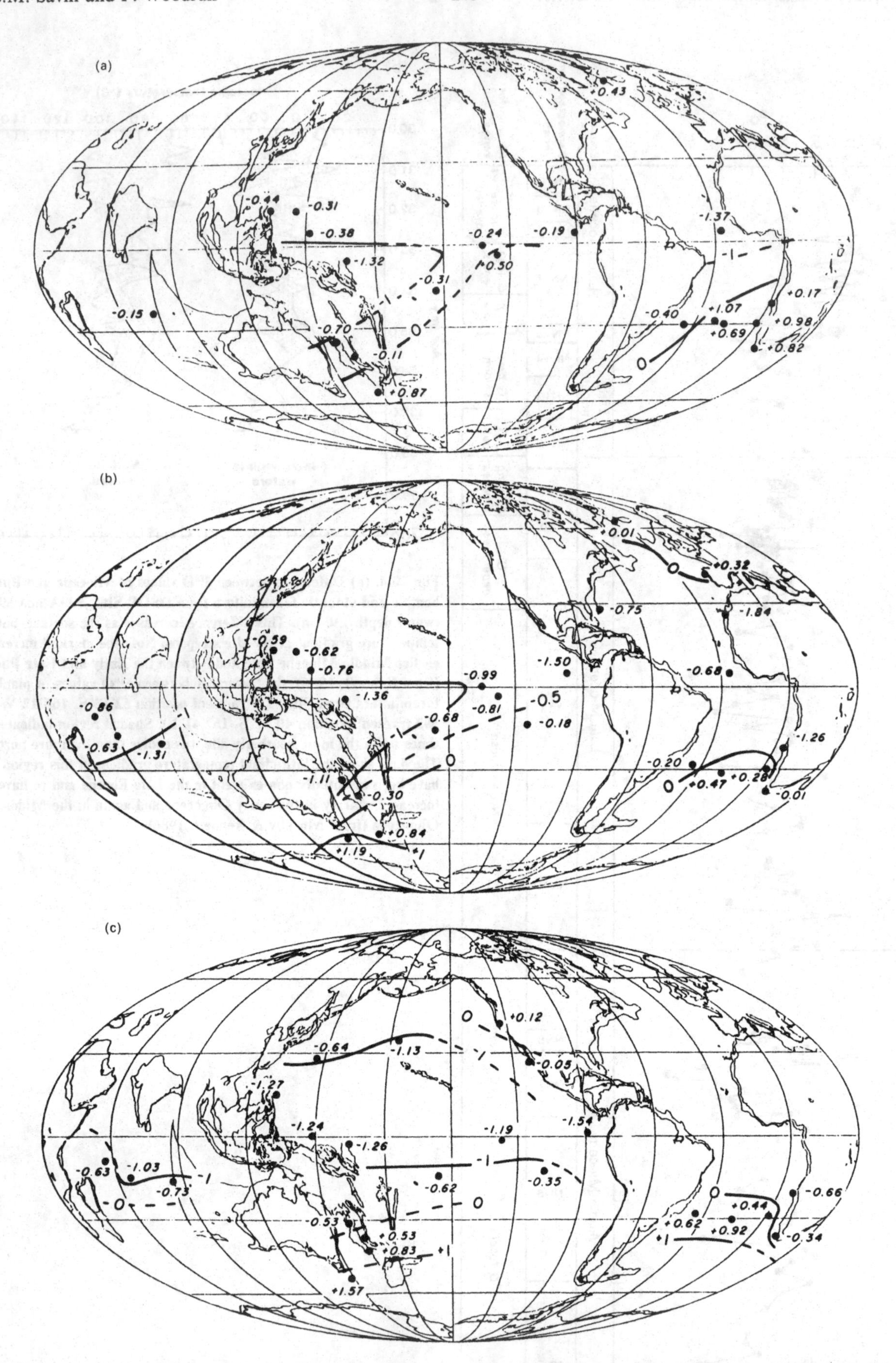

Fig. 20.5. Isotopic time slices for the Miocene, indicating the $\delta^{18}O$ values of shallow-dwelling planktic foraminifera from each of several sites. Although regional variations in the $^{18}O/^{16}O$ ratio of seawater have some effect on the patterns seen, lower (less positive or more negative) $\delta^{18}O$ values primarily indicate warmer temperatures. No adjustments have been made to this data for glacially-related changes in the isotopic composition of the oceans with time (from Savin *et al.*, 1985): (a) Early Early Miocene (22 my, Foraminiferal Zone N4B); (b) Late Early Miocene (16 my, Foraminiferal Zone N8); (c) Late Miocene (8 my, Foraminiferal Zone N17)

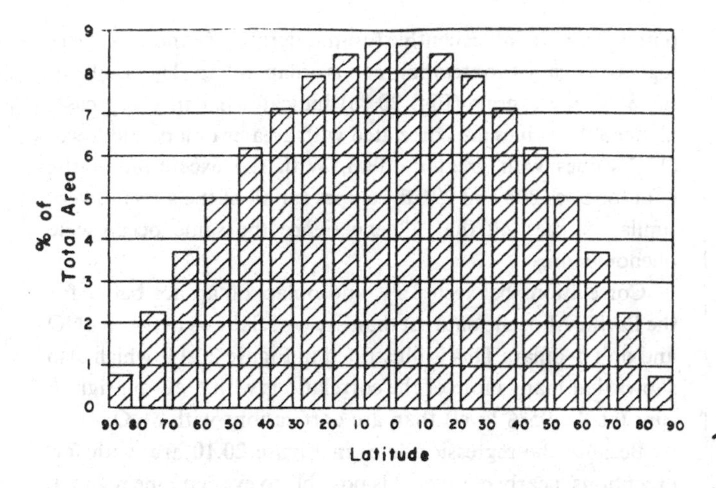

Fig. 20.6. Histogram showing the proportion of the area of the earth in each 10° band of latitude.

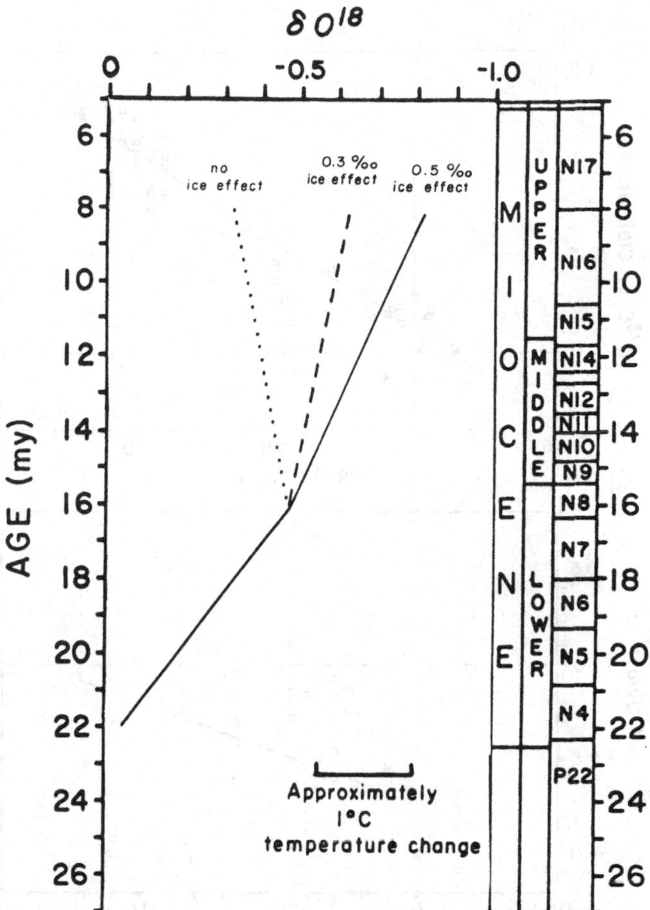

Fig. 20.7. Calculated mean $\delta^{18}O$ values of the oceans at 22 Ma, 16 Ma, and 8 Ma. The mean values were determined as described in the text, using the isotopic data from the maps in Figure 20.5. Geographic positions were backtracked for plate motion. Between 22 and 16 Ma there was probably little or no glacially-related change in the $^{18}O/^{16}O$ ratio of seawater. The dotted line between 16 and 8 Ma is calculated from the data with no adjustments for a glacial effect. The dashed line was drawn to include a 0.3 mil^{-1} glacial effect, and the solid line was drawn to include a change in the $\delta^{18}O$ value of seawater of 0.5 mil^{-1} between 16 and 8 Ma due to the growth of continental ice.

ing planktic foraminiferal calcite increased by about 0.3 mil^{-1} between 16 Ma, just before the onset of massive Antarctic ice-cap growth, and 8 Ma (shown by the dotted line in Fig. 20.7). That would correspond to a cooling of surface waters of about 1.5°C if the $\delta^{18}O$ value of seawater remained constant over that time interval. However, if the $\delta^{18}O$ of seawater increased by as little as 0.3 mil^{-1} (shown by the dashed line in Fig. 20.7) as the result of the growth of the Antarctic ice-cap, the data imply essentially no change in average ocean temperature during the period which saw the growth of a massive, permanent Antarctic ice-cap. If the increase in the $^{18}O/^{16}O$ of seawater during the 16–18 Ma interval was 0.5 mil^{-1}, as shown by the solid line in Figure 20.7 and as proposed by Savin *et al.* (1985), the average ocean temperature would have increased (warmed) by about 1.5°C.

The growth of continental ice during the Miocene epoch is usually regarded as having occurred in response to global cooling. However, we conclude from the discussion above that there was no global refrigeration at the time of onset of Antarctic ice-cap growth. Rather, glaciation probably occurred as the result of a decrease in the efficiency of equator-to-pole heat transport.

Carbon isotopes, carbon cycling and productivity

The magnitude of the fractionation of oxygen isotopes between $CaCO_3$ and H_2O is sensitive to temperature in the climatic temperature range, while the fractionation of carbon isotopes between $CaCO_3$ and HCO_3^- shows little variation with temperature. As a result, the $\delta^{18}O$ values of sedimentary carbonates vary both with the temperature and the $^{18}O/^{16}O$ ratio of the water in which they form, but the $\delta^{13}C$ values of the carbonates primarily reflect the $^{13}C/^{12}C$ ratio of the dissolved HCO_3^- from which they precipitate. This was clearly demonstrated by Duplessy *et al.* (1984) for a globally distributed Holocene benthic foraminiferal data set. They showed that the $\delta^{13}C$ values of both *Cibicidoides* and *Uvigerina* are well correlated with the $\delta^{13}C$ values of the dissolved inorganic carbon (T_{CO_2}) in the deep water overlying the sediment at the sites in which they grew. We have enlarged the foraminiferal data set of Duplessy *et al.* (1984) (see Appendix 20.1), and show the results of the correlations of the $S^{13}C$ values of the two benthic genera with $\delta^{13}C$ of T_{CO_2} in Figure 20.8. Carbon isotope ratios of T_{CO_2} were taken from the data of Craig (1970), Kroopnick, Weiss & Craig (1972), Kroopnick (1974), and Kroopnick (1985). The correlations with both species are good, although for reasons discussed below, the correlation between *Cibicidoides* and T_{CO_2} is better than that between *Uvigerina* and T_{CO_2}. The difference between $\delta^{13}C$ value of *Cibicidoides* and that of *Uvigerina* in any single site has been typically attributed to differences in the 'vital effects' of the two species, i.e. the biological processes that cause the isotopic composition of the biogenic calcite to differ from the thermodynamic equilibrium value. However, as discussed below, additional factors may be important in determining the intergeneric isotope difference.

Because the $\delta^{13}C$ values of a benthic foraminiferal species primarily reflect the $\delta^{13}C$ value of T_{CO_2}, benthic isotope records have been used to infer variations with time in the $^{13}C/^{12}C$ ratio of dissolved inorganic carbon in the oceans. It has been recognized for some time that at least some fluctuations in both the Miocene and the Pleistocene benthic foraminiferal $\delta^{13}C$ records are global or at least ocean-wide, and therefore reflect ocean-

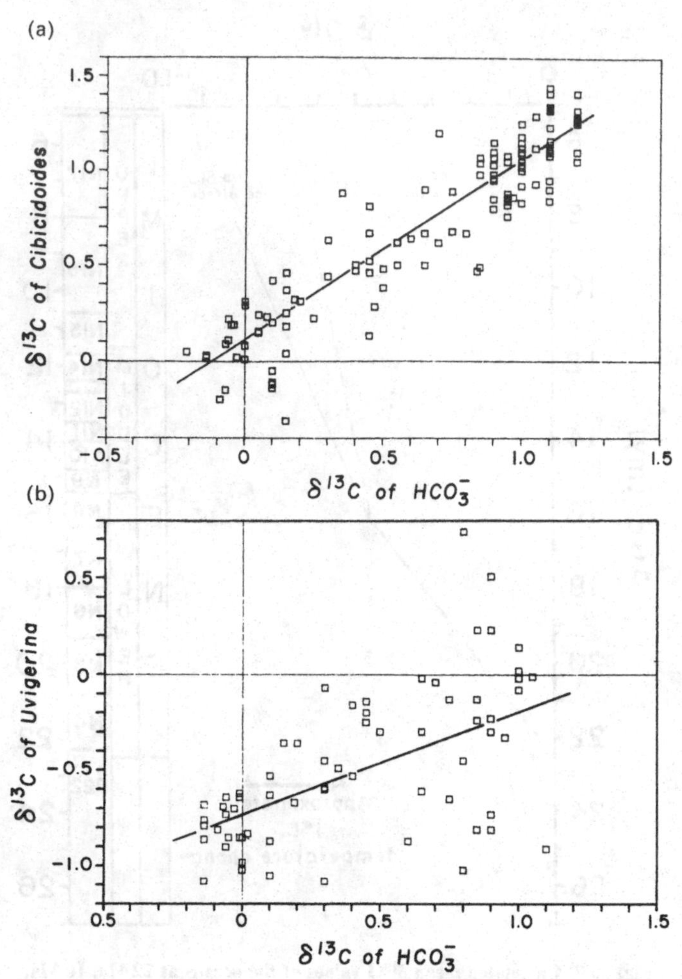

Fig. 20.8. $\delta^{13}C$ values of Holocene benthic foraminifera plotted against the $\delta^{13}C$ values of dissolved HCO_3^- in deep seawater at each of the sampling sites. Data, listed in Appendix 20.1 were taken from Duplessy *et al.* (1984) and a variety of other sources noted: (a) *Cibicidoides*. The regression line has a slope of 0.94 and a correlation coefficient of 0.92; (b) *Uvigerina*. The regression line has a slope of 0.55 and a correlation coefficient of 0.58.

wide or global changes in the $^{13}C/^{12}C$ ratio of dissolved HCO_3^- (Haq *et al.*, 1980; Loutit, Kennett & Savin, 1983; Berger, 1985; and others). The average $\delta^{13}C$ value of dissolved deep water marine HCO_3^- at any time reflects the fluxes of ^{13}C-enriched inorganic carbon and ^{13}C-depleted organic carbon into and out of the oceans (the carbon cycling effect). The $^{13}C/^{12}C$ ratio of the dissolved HCO_3^- in the deep water at any site reflects not only the average value for the ocean but an additional effect due to watermass aging. That is, as deep waters travel further from the surface regions in which they were generated, the oxidation of organic matter within the water mass causes the dissolved oxygen content to decrease, the HCO_3^- content to increase, and the $\delta^{13}C$ value of the HCO_3^- to decrease (Kroopnick, 1985).

Woodruff & Savin (1985) examined the $\delta^{13}C$ values of Miocene Pacific benthic foraminifera, and showed (Fig. 20.9) that they were well correlated with the extent of coastal onlap–offlap as determined from the curve of Vail & Hardenbol (1977). An arbitrary scale ranging from 0 (maximum Miocene offlap) to 17 (maximum Miocene onlap) was assigned to the curve of Vail & Hardenbol. For each of the 12 sites for which three or more data points were available, benthic foraminiferal $\delta^{18}O$ and $\delta^{13}C$ were regressed against contemporaneous onlap–offlap. The results of the $\delta^{13}C$ regressions (Fig. 20.10) indicate that in every case, higher $\delta^{13}C$ values are correlated with greater onlap, and lower $\delta^{13}C$ values with greater offlap. With the exception of the data for Sites 208 and 296 the slopes for all of the sites are very similar, suggesting that all experience the same ocean-wide phenomenon.

Correlations between $\delta^{13}C$ and onlap–offlap are better for the Pacific Miocene data set than are correlations between $\delta^{18}O$ and onlap–offlap. This is illustrated in Figure 20.11, which also shows that many more of the correlations with $\delta^{13}C$ are significant (at the 95% level) than are correlations with $\delta^{18}O$.

Because the regression lines in Figure 20.10 are, with few exceptions, nearly parallel, it is possible to evaluate the regional differences among benthic $\delta^{13}C$ values by comparing the $\delta^{13}C$ values of the regression lines at any chosen condition of onlap–offlap. For convenience, we have taken the intercept value on Figure 20.10, corresponding to maximum offlap (onlap–offlap = 0).

The total range of $\delta^{13}C$ intercept values in Figure 20.10 is approximately 1.2 mil^{-1}. This compares with the total range of $\delta^{13}C$ values of modern dissolved HCO_3^- of, with one exception, about 0.4 mil^{-1}, estimated for the sample sites using the reconstructions of Kroopnick (Fig. 20.12). It is possible that the greater range of Miocene $\delta^{13}C$ intercept values than of modern dissolved HCO_3^- $\delta^{13}C$ values reflects changes in ocean circulation and productivity patterns between Miocene time and the present. As argued below, however, it is much more likely that the Miocene $\delta^{13}C$ intercept values reflect a third factor (related to biological productivity) in addition to the carbon cycling effect and the watermass aging effect already discussed.

McCorkle, Emerson & Quay (1985) showed that the dissolved HCO_3^- in pore waters of the upper few centimeters of sediment was typically depleted in ^{13}C relative to that in the overlying seawater. The magnitude of the depletion increases with depth within the sediment and with the rain of organic matter to the bottom from the overlying surface waters (Fig. 20.13). In addition, they calculated that higher values of dissolved O_2 in the bottom water overlying the sediment should also lead to increasingly negative $\delta^{13}C$ values in the dissolved HCO_3^-. These effects are caused by *in situ* oxidation of low-^{13}C organic matter buried within the sediment. The greater the concentration of organic matter buried within the sediment when it is deposited, the greater the amount of ^{13}C-depleted HCO_3^- that can be produced.

The amount of organic matter found in the Miocene Pacific DSDP sediments (0–0.5% organic carbon) is far lower than the amount in modern Pacific sediments other than red clays (0.5–4%, Heath, Moore & Dauphin, 1977) because most of the organic carbon originally deposited has been degraded. Since the deep waters of the open ocean are sufficiently well oxygenated to permit the oxidation of organic matter, the residual amount preserved in Miocene sediments must be at least approximately related to the amount initially deposited during the time of sedimentation. Figure 20.14 is a graph in which the average content of organic matter in each of the Pacific Miocene sedimentary sections (taken from DSDP Initial Reports) is plotted

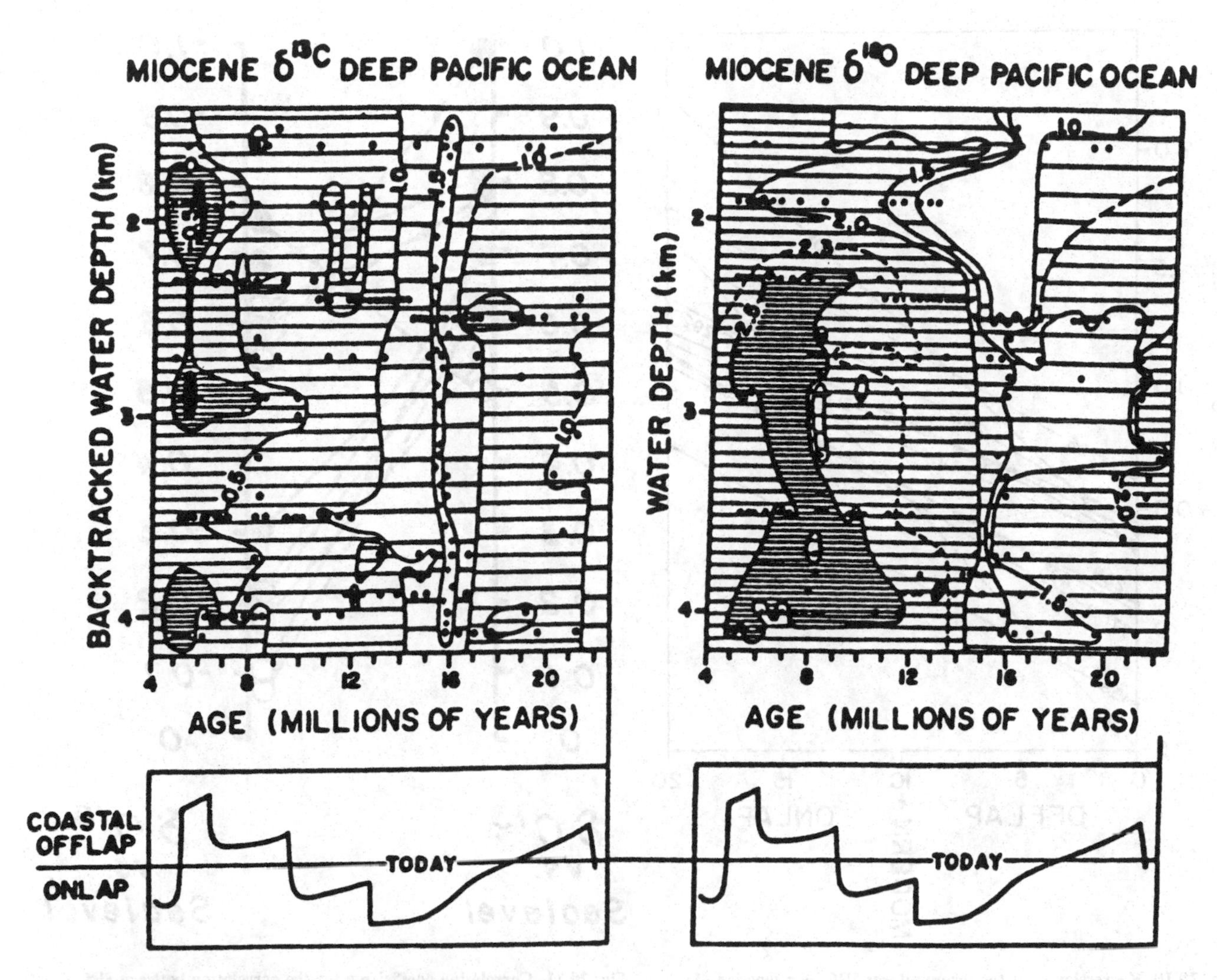

Fig. 20.9. Isotopic compositions of Miocene Pacific benthic foraminifera contoured on axes of (backtracked) water depth and sample age. Shown for reference, under both the oxygen and carbon isotope plots is the onlap–offlap curve of Vail & Hardenbol (1977). Note that for any chosen time interval there is a high degree of uniformity of the $^{13}C/^{12}C$ ratios of the Pacific samples. In part from Woodruff & Savin (1985).

against the $\delta^{13}C$ intercept for that site. The correlation is good, and shows that the $\delta^{13}C$ values of the Miocene benthic foraminifera from sediments with higher concentrations of residual organic matter have more negative $\delta^{13}C$ values than do those from sediments with lower concentrations of residual organic matter. This is almost certainly because the $^{13}C/^{12}C$ of the HCO_3^- in the water in which the foraminifera in the organic-rich sediments grew was lowered by mixing with low-^{13}C organically-derived HCO_3^-. This interpretation is consistent with the conclusions of Sarnthein *et al.* (1984) that the $\delta^{13}C$ values of Pleistocene benthic foraminifera from the eastern equatorial Atlantic were correlated with the flux of organic matter to the sediment.

The results of McCorkle *et al.* (1985) suggest that after any 'vital' or disequilibrium effects are accounted for, infaunal benthic foraminifera should have lower $\delta^{13}C$ values than epifaunal taxa, and that the difference between the $\delta^{13}C$ of an epifaunal species and an infaunal species from the same sample should be greater in sediments in which greater amounts of organic matter have been oxidized. More oxidation of organic matter might result from a greater rain of organic matter from surface waters, higher concentrations of dissolved O_2 in the seawater overlying the sediment, or both.

Corliss (1985) determined the depth within the sediment column at which living individuals of several taxa of benthic foraminifera were concentrated in two western North Atlantic box cores and found that different taxa preferentially inhabited different sediment depths, with some individuals living as deep as 10 or 15 cm. Living *Cibicidoides* were concentrated in the upper centimeter of the sediment column. Only one living specimen of *Uvigerina* was found in this study, and it was also in the upper centimeter of the sediment. In more recent work on the distribution of living benthic foraminifera on the eastern North American continental shelf, Corliss (pers. comm.) has found that living *Cibicidoides* and *Uvigerina* consistently occur within the uppermost centimeter of the sediment column. Based on an

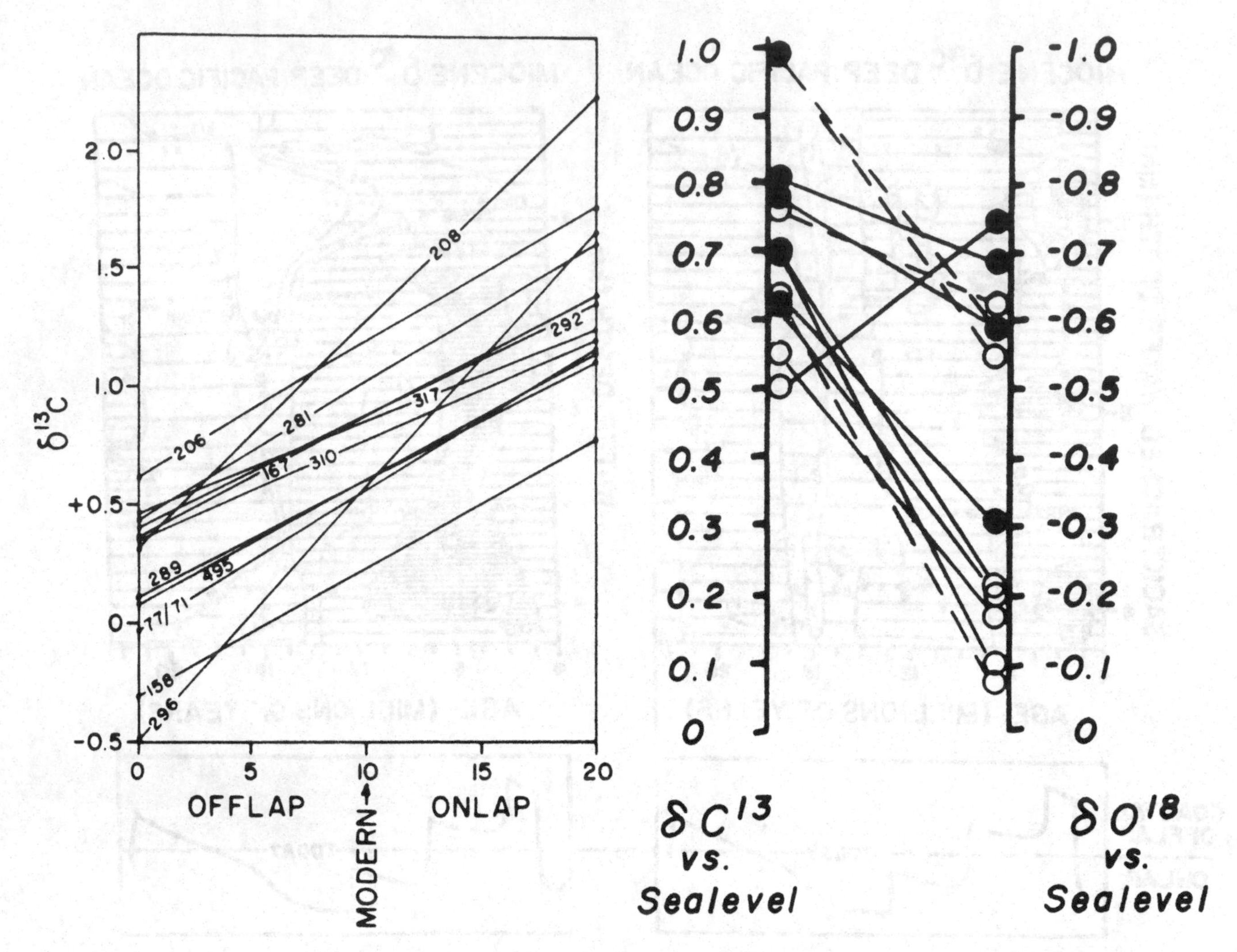

Fig. 20.10. Regression lines for foraminiferal $\delta^{13}C$ as a function of onlap–offlap for each of the Miocene Pacific sites studied. An arbitrary linear scale for onlap–offlap was assigned to the curve of Vail & Hardenbol (1977). Note that the regression lines for all but two of the sites are almost parallel. The extrapolated value of $\delta^{13}C$ for each site when onlap–offlap = 0 is termed the $\delta^{13}C$ intercept (from Woodruff & Savin, 1985).

Fig. 20.11. Correlation coefficients for the correlation between $\delta^{13}C$ and onlap–offlap are compared with those for the correlation between $\delta^{18}O$ and onlap–offlap. Lines connect carbon and oxygen data for the same sites. Solid circles are those for which the correlations have been determined to be highly significant. Note that not only are the correlations with carbon better than those with oxygen, but a greater proportion of the correlations with carbon are highly significant.

interpretation of their different test morphologies, he concluded that *Cibicidoides* has an epifaunal habitat whereas *Uvigerina* is infaunal within the upper centimeter.

We have examined the difference between the $\delta^{13}C$ value of *Cibicidoides* and that of T_{CO_2}, and compared that difference at each sample locality to the local biological productivity, as read from the map of Lieth (1975). We have similarly treated the carbon isotope data for *Uvigerina*. It is clear from the results, shown in Figure 20.15, that the $\delta^{13}C$ values of *Uvigerina* depend both upon the $\delta^{13}C$ value of T_{CO_2} and upon the biological productivity in the overlying surface waters, whereas those of *Cibicidoides* show no dependency upon biological productivity.

We interpret the results in Figure 20.15 as being consistent with the pore water $\delta^{13}C$ measurements of McCorkle *et al.* (1985). That is, regions of high biological productivity are generally ones in which the rain of organic matter to the sediment is greater and in which the organic content is higher. Carbon isotope ratios of infaunal *Uvigerina* should reflect the oxidation of organic matter in the pore waters in those regions to a greater extent than in zones of low biological productivity. This is observed.

The results of Figure 20.15 suggest the possibility of using the difference between the $\delta^{13}C$ value of *Cibicidoides* and that of *Uvigerina* as a quantitative indicator of paleoproductivity. We have attempted that with the Holocene data set, but the results have so far not been promising (Fig. 20.16). Among the reasons why this may be true are: (a) isotopic analyses for paired *Cibicidoides* and *Uvigerina* were available for only a small number of the Holocene sites, in part because *Uvigerina* may prefer to live in areas with a greater rain of organic carbon, both in the Recent (Miller & Lohmann, 1982; Lutze & Coulbourn, 1984) and during the Miocene (Woodruff, 1985); (b) among the small number of sites, the range of variation of productivity as read from the maps of Lieth is small; (c) the resolution with which the productivity data could be read is not high, and there may also be problems with the quality of the productivity data in some parts

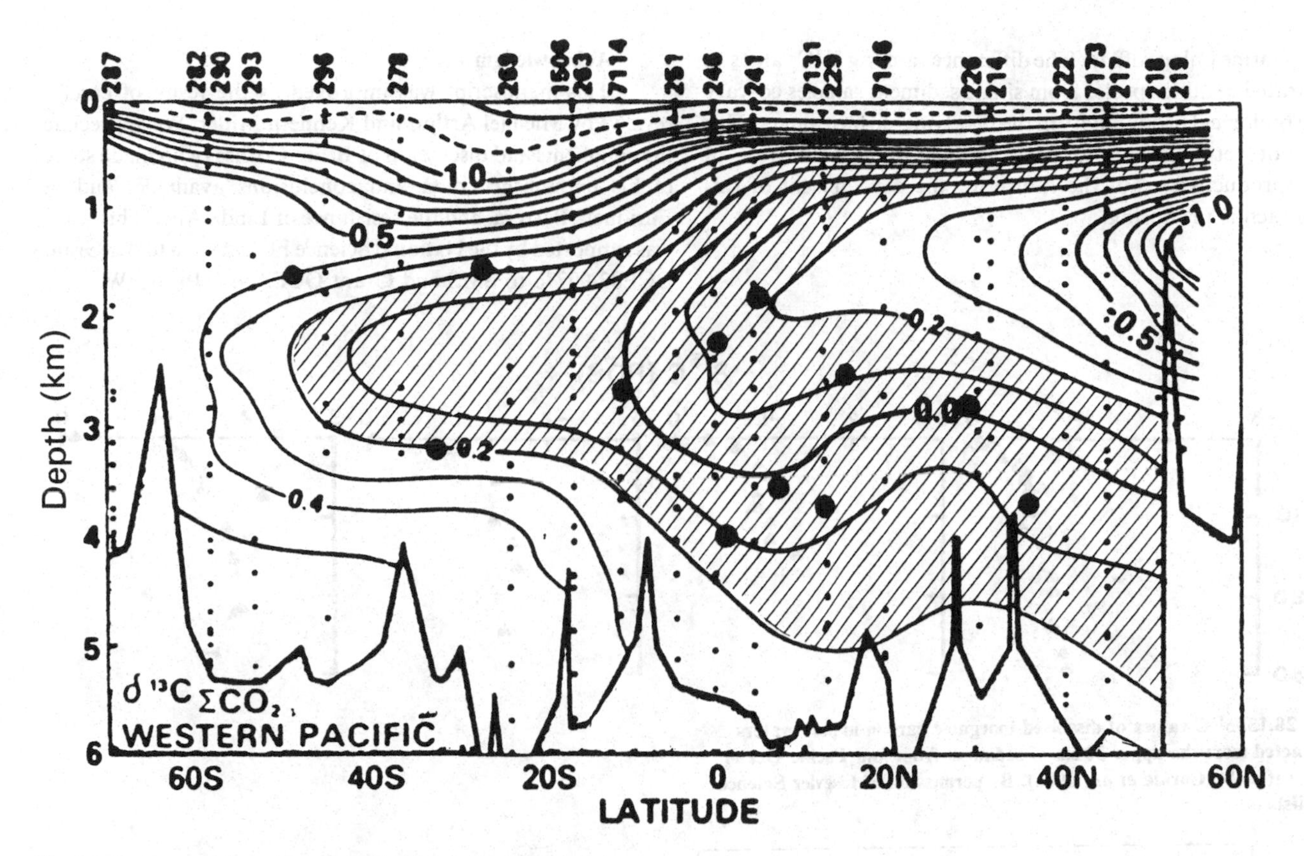

Fig. 20.12. Contoured cross-section of $\delta^{13}C$ values of dissolved inorganic carbon in the western Pacific Ocean (from Kroopnick, 1985). Dots show the projections of the locations of the Miocene benthic samples onto this cross-section. With one exception, all the dots plot within the shaded area, which indicates a range of 0.4 mil^{-1} $\delta^{13}C$.

of the world; (d) the rain of organic matter at the sediment–water interface, which we propose as a key factor in determining the $^{13}C/^{12}C$ ratio of *Uvigerina*, is a function not only of the surface productivity but of the oxidation of organic matter within the water column (Suess, 1980), and we did not consider oxidation of organic matter in the water column; (e) no attempt was made to take into account the dissolved O_2 of bottom waters, which according to McCorkle *et al.* (1985) may be an important factor in controlling pore water $\delta^{13}C$; and (f) *Uvigerina*, a shallow infaunal species, lives in the zone in which the $\delta^{13}C$ value of dissolved pore water HCO_3^- has the greatest decrease with depth, and hence a small vertical shift in the habitat of *Uvigerina* might cause a large change in the $\delta^{13}C$ of its calcite.

It is still possible that additional study of Holocene benthic foraminifera may permit the development of a paleoproductivity indicator based upon carbon isotope differences among benthic species, and we are continuing to investigate this. Downcore studies of Pleistocene sediments from the eastern Atlantic do indicate a correlation between the accumulation rate of organic carbon and the difference between $\delta^{13}C$ values of paired *Cibicidoides* and *Uvigerina* (Zahn, Winn & Sarnthein 1986). In a downcore study of Late Pliocene sediments from the eastern North Atlantic, Loubere (1987) concluded from paired analyses of *Cibicidoides* and *Uvigerina* that the $\delta^{13}C$ values of the benthic species may have been influenced by surface water processes in addition to the $^{13}C/^{12}C$ ratio of dissolved HCO_3^-.

Conclusions

The oxygen isotope ratios of planktic and benthic foraminifera are proven tools for the interpretation of surface- and deep-water temperature patterns in ancient oceans. Among the most striking conclusions of the isotopic study of foraminifera from Tertiary deep-sea sediments has been that during the Middle Miocene, at the same time that a large Antarctic ice-cap became permanently established, tropical regions were warming and latitudinal temperature gradients were intensified. This Middle Miocene intensification of the latitudinal temperature gradient may have been associated with more vigorous thermohaline circulation and enhanced upwelling, and may therefore be related to the major episodes of phosphogenesis that occurred during that time. There is some indication that the mean global temperature may have increased at that time.

The interpretation of the foraminiferal carbon isotope record is less firmly grounded than is the oxygen isotope record, but rapid progress is now being made in understanding of the carbon isotopes. Oceanwide or global changes in the $\delta^{13}C$ of dissolved marine HCO_3^- during Tertiary time are now well established, and these must reflect changes in the global cycling of carbon. Miocene Pacific benthic foraminiferal $\delta^{13}C$ values are well correlated with marine onlap–offlap. The differences among $\delta^{13}C$ values of separated benthic species from single sediment samples co-vary with biological productivity in the overlying surface waters. This has not yet been developed as a quantitative indicator of paleoproductivity, but the future for the development of this approach is promising.

Acknowledgments

This manuscript was improved as the result of careful reviews by Michael Arthur and Kenneth Miller. We appreciate the comments and discussion of Bruce Corliss (who made some of his unpublished data and conclusions available) and of Enriqueta Barrera, and the assistance of Linda Abel. This work was supported by the National Science Foundation under Grant OCE83-09776 to SMS and Grant OCE85-17110 to FW.

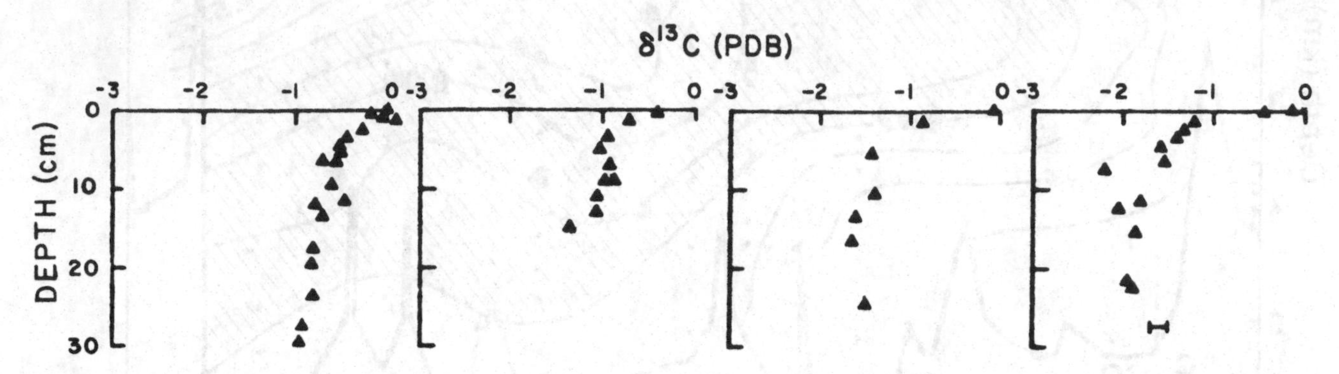

Fig. 20.13. $\delta^{13}C$ **values of dissolved inorganic carbon in pore waters extracted from the upper 30 cm of sediment from four Pacific Ocean sites (after McCorkle** ***et al.*****, 1985). By permission of Elsevier Science Publishers.**

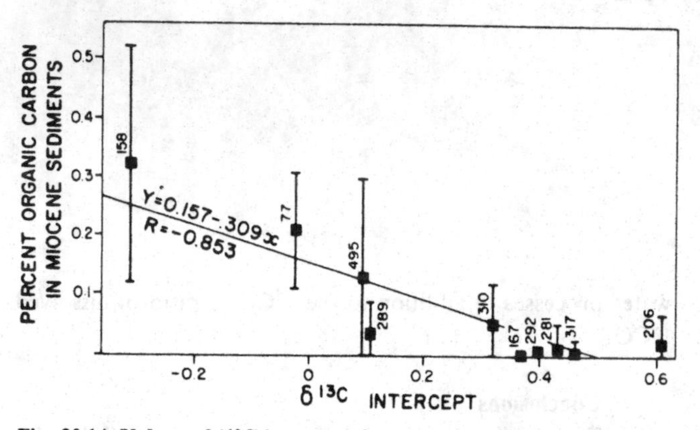

Fig. 20.14. Values of $\delta^{13}C$ **intercepts for each of the Miocene benthic foraminiferal data sets (see text and caption to Fig. 20.10) plotted against the average organic matter content in each of the Miocene sequences (from Woodruff & Savin, 1985).**

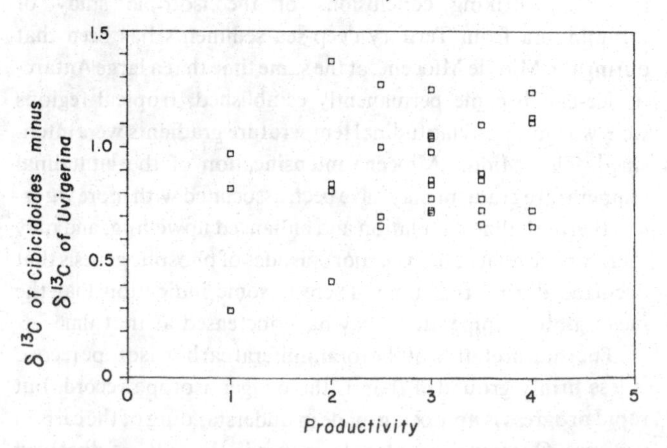

Fig. 20.16. Difference between $\delta^{13}C$ **values of all paired samples of** ***Cibicidoides*** **and** ***Uvigerina*** **in Appendix 20.1 plotted against the productivity of the overlying surface waters. See caption to figure 20.15 for explanation of productivity index.**

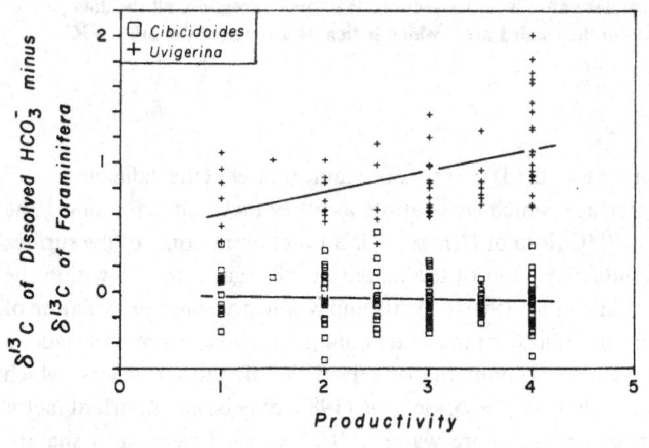

Fig. 20.15. Difference between the $\delta^{13}C$ **value of dissolved inorganic carbon at each of the Holocene benthic foraminiferal sample sites listed in Appendix 20.1 and the** $\delta^{13}C$ **value of the foraminifera from that site, plotted against the biological productivity in the surface waters overlying the site. The slope of the regression line through the data for** ***Uvigerina*** **(crosses) is significantly different from zero, but that through the** ***Cibicidoides*** **data (squares) is not. Productivity data read from the map of Lieth (1975) is indexed as follows: 1 = 0–50 g C(m²y)⁻¹; 2 = 50–100 g C(m²y)⁻¹; 3 = 100–200 g C(m²y)⁻¹; 4 = > 200 g C(m²y)⁻¹.**

Appendix 20.1. *Compilation of published isotopic data for Holocene benthic foraminifera*

Region	Core	Latitude (deg. min.)	Longitude (deg. min)	Water depth (m)	Data source	*Cibicidoides* $\delta^{18}O$	*Cibicidoides* $\delta^{13}C$	*Uvigerina* $\delta^{18}O$	*Uvigerina* $\delta^{13}C$	$\delta^{13}C$ of dissolved HCO_3^-	Surface productivity	*Cibicidoides* $\delta^{13}C$ − *Uvigerina* $\delta^{13}C$	$\delta^{13}C$ of HCO_3^- *Cibicidoides* $\delta^{13}C$	$\delta^{13}C$ of HCO_3^- *Uvigerina* $\delta^{13}C$
Arctic Ocean	T3-67-11	79 35 N	172 30 W	2810	*a*	3.82	1.12			1.10			−0.02	
Caribbean Sea	DSDP 502B	11 29 N	79 23 W	3051	*m*	1.90	0.86			0.97	1.5		0.11	
Gulf of Mexico	A64-9 (25)	22 54 N	90 48 W	1829	*m*	1.87	1.07	2.52	0.23	0.85	2	0.84	−0.22	0.62
Gulf of Mexico	A64-9 (29)	20 52 N	90 55 W	3599	*m*			2.63	−0.02	1.00	2			1.02
Indian Ocean	AII-15-765	32 1 S	49 56 E	3698	*m*	2.50	0.47			0.40	2		−0.07	
Indian Ocean	MD73-025	43 49 S	51 19 E	3284	*a*	3.00	0.28			0.47	4		0.19	
Indian Ocean	MD76-125	8 35 N	75 20 E	1878	*a*	2.58	0.02	3.23	−0.70	−0.03	3	0.72	−0.05	0.67
Indian Ocean	MD76-127	12 5 N	73 54 E	1610	*a*	2.48	0.22	3.51	−0.64	−0.06	3	0.86	−0.28	0.58
Indian Ocean	MD76-128	13 8 N	73 19 E	1712	*a*	2.49	0.09	3.29	−0.69	−0.07	3	0.78	−0.16	0.62
Indian Ocean	MD76-129	15 0 N	72 20 E	1954	*a*	2.56	0.19			−0.04	3		−0.23	
Indian Ocean	MD76-132	17 0 N	71 31 E	1430	*a*			3.13	−0.90	−0.06	3			0.84
Indian Ocean	MD76-135	14 27 N	50 31 E	1895	*a*	2.38	−0.15			−0.07	3		0.08	
Indian Ocean	MD76-136	12 52 N	46 49 E	1649	*a*	2.56	0.05			−0.21	3		−0.26	
Indian Ocean	MD76-191	7 30 N	76 43 E	1254	*a*			2.71	−0.64	−0.01	2.5			0.63
Indian Ocean	MD76-194	10 28 N	75 14 E	1222	*a*			2.75	−0.62	−0.01	3			0.61
Indian Ocean	MD76-195	11 30 N	74 32 E	1426	*a*	2.39	0.11	3.01	−0.73	−0.06	3	0.84	−0.17	0.67
Indian Ocean	MD77-200	16 33 N	67 54 E	2910	*a*	2.60	0.15			0.05	2		−0.10	
Indian Ocean	MD77-203	20 42 N	59 34 E	2442	*a*	2.54	−0.20			−0.09	2		0.11	
Indian Ocean	MD79-254	17 53 S	38 40 E	1934	*a*	2.73	0.67	3.27	−0.14	0.45	2	0.81	−0.22	0.59
Indian Ocean	MD79-256	19 35 S	37 2 E	1222	*a*	2.41	0.68	3.04	−0.13	0.75	2	0.81	0.07	0.88
Indian Ocean	RC11-120	43 31 S	79 52 E	3913	*m*	2.46	0.52			0.45	4		−0.07	
Indian Ocean	RC11-120	43 31 S	79 52 E	3135	*a*	2.67	0.46	3.29	−0.20	0.45	4	0.66	−0.01	0.65
Indian Ocean	RC11-147	19 4 S	112 45 E	1953	*m*	2.07	0.63	2.9	−0.07	0.30	2.5	0.70	−0.33	0.37
Indian Ocean	RC12-339	9 8 N	90 2 E	3010	*a*	3.07	0.23			0.08	1		−0.15	
Indian Ocean	RC17-85	31 5 S	57 25 E	3325	*m*	2.93	0.22			0.25	2		0.03	
Indian Ocean	V19-188	6 52 N	60 40 E	3356	*m*	2.37	0.46	3.31	−0.36	0.15	1	0.82	−0.31	0.51
Indian Ocean	V28-345	17 40 S	117 57 E	1904	*a*			3.13	−0.67	0.19	4			0.86
Indian Ocean	V29-29	5 7 N	77 35 E	2673	*a*	2.63	0.20			0.10	2.5		−0.10	
Indian Ocean	V34-88	16 31 N	59 32 E	2100	*a*			3.49	−0.85	−0.01	1			0.84
Northeast Atlantic Ocean	12301-5	27 3 N	15 3 W	2896	*j*	2.81	1.08			1.10	4		0.02	
Northeast Atlantic Ocean	12308-2	26 38 N	15 3 W	2090	*j*	2.41	1.15			0.90	4		−0.25	
Northeast Atlantic Ocean	12309-1	26 50 N	15 6 W	2849	*j*	2.52	1.14			1.10	4		−0.04	
Northeast Atlantic Ocean	12310-1	23 29 N	18 43 W	3075	*j*	2.62	1.16			1.10	4		−0.06	
Northeast Atlantic Ocean	12326-2	23 2 N	17 24 W	1024	*j*	1.87	1.20	2.70	−0.04	0.7	4	1.24	−0.50	0.74
Northeast Atlantic Ocean	12327-2	23 7 N	17 44 W	2032	*j*	2.45	0.98	3.09	−0.23	0.90	4		−0.08	
Northeast Atlantic Ocean	12328-1	21 8 N	18 34 W	2822	*j*			3 47	−0.91	1.10	4			2.01
Northeast Atlantic Ocean	12329-2	19 22 N	19 55 W	3314	*j*	2.61	0.85			0.90	4		0.05	
Northeast Atlantic Ocean	13238-1	14 8 N	17 52 W	1983	*j*			2.88	−1.02	0.80	4			1.82

Appendix 20.1. (*Contd.*)

Region	Core	Latitude (deg. min.)	Longitude (deg. min)	Water depth (m)	Data source	*Cibicidoides* $\delta^{18}O$	*Cibicidoides* $\delta^{13}C$	*Uvigerina* $\delta^{18}O$	*Uvigerina* $\delta^{13}C$	$\delta^{13}C$ of dissolved HCO_3^-	Surface productivity	*Cibicidoides* $\delta^{13}C$ − *Uvigerina* $\delta^{13}C$	$\delta^{13}C$ of HCO_3^- − *Cibicidoides* $\delta^{13}C$	$\delta^{13}C$ of HCO_3^- − *Uvigerina* $\delta^{13}C$
Northeast Atlantic Ocean	13280-1	18 43 N	16 57 W	1020	*j*			2.70	−0.87	0.60	4			1.47
Northeast Atlantic Ocean	13289-1	18 4 N	18 0 W	2470	*j*			3.43	−0.73	0.90	4			1.63
Northeast Atlantic Ocean	13290-1	18 3 N	18 4 W	2890	*j*			3.37	−0.81	0.90	4			1.71
Northeast Atlantic Ocean	13532-2	20 58 N	17 52 W	1418	*j*			2.61	−0.65	0.75	4			1.40
Northeast Atlantic Ocean	13533-1	20 59 N	18 1 W	2093	*j*			2.77	−0.81	0.85	4			1.66
Northeast Atlantic Ocean	15634-1	28 15 N	13 23 W	1215	*j*			2.49	−0.45	0.80	3.5			1.25
Northeast Atlantic Ocean	15641-2	32 25 N	9 51 W	1060	*j*			1.93	0.74	0.80	3.5			0.06
Northeast Atlantic Ocean	15651-4	33 11 N	9 49 W	3798	*j*	2.60	0.95			1.10	4		0.15	
Northeast Atlantic Ocean	15669-2	34 53 N	7 49 W	1997	*j*	2.08	1.04			0.85	3.5		−0.19	
Northeast Atlantic Ocean	15672-2	34 51 N	8 7 W	2430	*j*	2.25	0.90			1.10	3		0.20	
Northeast Atlantic Ocean	15676-2	34 45 N	8 50 W	3502	*j*	2.56	1.10			1.10	3		0.00	
Northeast Atlantic Ocean	15677-1	34 6 N	10 49 W	4380	*j*	2.63	0.88			0.95	3		0.07	
Northeast Atlantic Ocean	15678-1	33 28 N	12 45 W	4305	*j*	2.57	0.85			0.95	3		0.10	
Northeast Atlantic Ocean	16032-1	21 14 N	17 59 W	1176	*j*			2.83	−0.02	0.65	4			0.67
Northeast Atlantic Ocean	AII-31-14-14	11 32 N	42 43 W	3846	*m*	2.21	0.47			0.84	1		0.37	
Northeast Atlantic Ocean	CH115-11-10	9 16 N	19 26 W	4168	*m*	2.24	0.89			0.75	4		−0.14	
Northeast Atlantic Ocean	CH69-69	43 51 N	4 30 W	2075	*a*			3.33	0.14	1.00	3			0.86
Northeast Atlantic Ocean	CH71-07	4 23 N	20 52 W	3083	*a,d*	2.89	1.08			0.95	3		−0.13	
Northeast Atlantic Ocean	CH73-139C	54 38 N	16 21 W	2209	*a*	2.67	0.93			1.05	3		0.12	
Northeast Atlantic Ocean	DSDP 552A	56 3 N	23 13 W	2311	*a*	2.71	1.29			1.05	3.5		−0.24	
Northeast Atlantic Ocean	EN 23,LPC-01	34 43 N	61 24 W	4333	*m*	2.63	0.76			0.95	2		0.19	
Northeast Atlantic Ocean	EN 24,LPC-02	34 59 N	61 35 W	4606	*m*	2.41	0.82			0.95	2		0.13	
Northeast Atlantic Ocean	KW-31	3 31 N	5 34 E	1181	*a,f*	2.35	0.64			0.60	2		−0.04	
Northeast Atlantic Ocean	M12-310-4	23 30 N	18 43 W	3080	*a*	2.43	1.10			0.90	3		−0.20	
Northeast Atlantic Ocean	M12-392	25 10 N	16 50 W	2573	*a*	2.66	0.96	3.25	0.23	0.90	3.5	0.73	−0.06	0.67
Northeast Atlantic Ocean	M13-519	5 40 N	19 51 W	2862	*a,e*	2.69	1.05			0.95	3		−0.10	
Northeast Atlantic Ocean	M8-158	37 44 N	9 43 W	1819	*a*	2.35	0.95	2.82	−0.30	0.90	3	1.25	−0.05	1.20
Northeast Atlantic Ocean	NO75-08	45 42 N	31 22 W	3454	*a*	2.87	1.15			1.00	3		−0.15	
Northeast Atlantic Ocean	NO79-28	45 38 N	22 45 W	3625	*a*	2.90	0.92			1.00	3		0.08	
Northeast Atlantic Ocean	SL58/79/8	42 50 N	23 4 W	3520	*a*	2.50	1.05			1.00	2.5		−0.05	
Northeast Atlantic Ocean	SU8132	42 6 N	9 47 W	2280	*a*	2.71	1.00	3.38	0.02	1.00	3	0.98	0.00	0.98
Northeast Atlantic Ocean	V22-196	13 50 N	18 58 W	3728	*a*	2.50	0.85			0.95	4		0.10	
Northeast Atlantic Ocean	V22-197	14 10 N	18 35 W	3167	*a*	2.47	0.49	3.46	−0.24	0.85	4	0.73	0.36	1.09
Northeast Atlantic Ocean	V22-198	14 35 N	19 40 W	1082	*a*	2.21	0.50			0.55	2.5		0.05	
Northeast Atlantic Ocean	V29-179	44 0 N	24 32 W	3331	*a*	2.79	1.10			1.00	2.5		−0.10	
Northeast Atlantic Ocean	V30-97	41 0 N	32 56 W	3371	*a*	2.68	1.09			1.00	2		−0.09	
Norwegian Sea	CH77-07	66 36 N	10 31 W	1487	*a*	4.06	1.10			1.20	3.5		0.10	
Norwegian Sea	K-11	71 47 N	1 36 E	2900	*a*	3.92	1.27			1.20	4		−0.07	

Norwegian Sea	KN54-6/70	63 49 N	0 50 E	2217	g	3.99	0.84			1.10	2.5		0.26	
Norwegian Sea	V27-44	65 1 N	6 13 W	2772	g	3.94	1.44			1.10	4		−0.34	
Norwegian Sea	V27-60	72 11 N	8 34 E	2525	a	3.89	1.26			1.20	4		−0.06	
Norwegian Sea	V27-82	74 59 N	10 48 W	2895	g	4.10	1.41			1.20	2		−0.21	
Norwegian Sea	V27-86	66 36 N	1 7 E	2900	a,b	4.01	1.32			1.10	4		−0.22	
Norwegian Sea	V27-86	66 36 N	1 7 E	2900	a	3.77	1.34			1.10	4		−0.24	
Norwegian Sea	V28-25	76 49 N	1 20 W	3136	g	3.73	1.27			1.20	2		−0.08	
Norwegian Sea	V28-28	73 29 N	0 50 W	2288	g	3.74	1.25			1.20	3.5		−0.05	
Norwegian Sea	V28-48	73 20 N	12 29 E	1668	g	4.05	1.05			1.20	4			
Norwegian Sea	V28-49	74 29 N	11 40 E	2327	g	3.86	1.32			1.20	3		−0.12	
Norwegian Sea	V28-55	65 31 N	0 12 E	2886	g	3.89	1.33			1.10	4		−0.23	
Norwegian Sea	V28-56	68 2 N	6 7 W	2941	a	3.95	1.41			1.10	4		−0.31	
Norwegian Sea	V28-56	68 2 N	6 7 W	2941	g	3.95	1.23			1.10	4		−0.13	
Norwegian Sea	V29-210	66 44 N	6 44 W	2460	g	3.96	1.16			1.10	4		−0.06	
Norwegian Sea	V29-215	75 55 N	5 7 W	3139	g	3.89	1.28			1.20	2.5		−0.07	
Northwest Atlantic Ocean	AII 94-99-24	55 58 N	29 30 W	2848	m	2.39	1.02			1.00	4		−0.02	
Northwest Atlantic Ocean	AII-31-12-12	10 20 N	41 18 W	3182	m	2.28	1.07			0.90	1		−0.17	
Northwest Atlantic Ocean	CH82-11	42 0 N	52 0 W	3209	a,c	2.44	1.18			1.00	4		−0.18	
Northwest Atlantic Ocean	CH96-4-4	29 53 N	41 19 W	3608	m	2.15	0.83			1.00	1		0.17	
Northwest Atlantic Ocean	D114	41 45 N	51 56 W	3732	a	2.59	1.12			1.00	4		−0.12	
Northwest Atlantic Ocean	GS7205-86	29 12 N	44 14 W	3335	a			3.42	−0.08	1.00	1			1.08
Northwest Atlantic Ocean	HU75-58	62 46 N	59 22 W	1057	a	2.81	1.27			1.20	3		−0.07	
Northwest Atlantic Ocean	ORCH75-03	10 3 N	57 32 W	3410	a	2.88	0.84			0.95	1		0.11	
Northwest Atlantic Ocean	ORCH75-04	10 1 N	56 1 W	3820	a	2.82	0.86			0.95	1		0.09	
Northwest Atlantic Ocean	RC10-288	35 32 N	73 25 W	3678	a	2.57	1.03			1.00	3		−0.03	
Northwest Atlantic Ocean	TR 121-37	37 25 N	25 54 W	2310	m	2.02	1.25			1.00	2		−0.25	
Northwest Atlantic Ocean	TR141-1	61 32 N	28 38 W	1550	a			3.19	−0.13	0.85	4			0.98
Northwest Atlantic Ocean	V23-6	42 29 N	61 48 W	2246	a			3.28	−0.33	0.95	4			1.28
Northwest Atlantic Ocean	V28-14	64 47 N	29 34 W	1855	a	2.45	1.12	3.42	−0.01	1.05	4	1.13	−0.07	1.06
Pacific Ocean	22	3 11 N	101 51 W	3206	k	2.80	−0.11	3.51	−0.53	0.10	2	0.42	0.21	0.63
Pacific Ocean	35	2 37 N	107 1 W	3698	k	2.75	−0.14	4.20	−0.87	0.10	3	0.73	0.24	0.97
Pacific Ocean	57	1 28 N	113 46 W	3847	k	2.73	−0.05	3.84	−1.05	0.10	2.5	1.00	0.15	1.15
Pacific Ocean	67	1 59 N	117 44 W	4185	k			4.32	−0.87	0.10	3			0.97
Pacific Ocean	76	2 29 N	121 42 W	4583	k	2.85	−0.31			0.15	2.5		0.46	
Pacific Ocean	AII-54-9-5	9 31 S	94 13 W	3960	m	2.50	0.04			0.15	3		0.11	
Pacific Ocean	DSDP 503A	4 4 N	95 38 W	3672	m	2.47	−0.12			0.10	3		0.22	
Pacific Ocean	DSDP 506D	0 36 N	86 6 W	2707	a,i			3.58	−0.68	−0.14	2			0.54
Pacific Ocean	DSDP 508	0 32 N	86 5 W	2783	a			3.43	−0.79	−0.14	3			0.65
Pacific Ocean	ERDC112	1 37 S	159 14 E	2169	a,h	2.82	0.25			0.15	1		−0.10	
Pacific Ocean	ERDC123	0 1 S	160 25 E	2948	a,h	2.76	0.32			0.18	1		−0.14	
Pacific Ocean	ERDC135	0 53 N	161 0 E	3509	l	3.21								
Pacific Ocean	ERDC92	2 14 S	157 0 E	1598	l	2.69								
Pacific Ocean	MANOP C	1 3 N	138 56 W	4440	m	2.39	0.37			0.15	2.5		−0.22	
Pacific Ocean	PLDS66	0 57 N	104 6 W	3496	l	2.50		3.16						
Pacific Ocean	RC10-65	0 41 N	108 37 W	3588	a	2.64	0.42	3.45	−0.63	0.10	3	1.05	−0.32	0.73

Appendix 20.1. (*Contd.*)

Region	Core	Latitude (deg. min.)	Longitude (deg. min)	Water depth (m)	Data source	*Cibicidoides* $\delta^{18}O$	*Cibicidoides* $\delta^{13}C$	*Uvigerina* $\delta^{18}O$	*Uvigerina* $\delta^{13}C$	$\delta^{13}C$ of dissolved HCO_3^-	Surface productivity	*Cibicidoides* $\delta^{13}C$ − *Uvigerina* $\delta^{13}C$	$\delta^{13}C$ of HCO_3^- *Cibicidoides* $\delta^{13}C$	$\delta^{13}C$ of HCO_3^- *Uvigerina* $\delta^{13}C$
Pacific Ocean	RC10-65	0 41 N	108 37 W	3588	m	2.37	0.24			0.05	2.5		−0.19	
Pacific Ocean	RC11-210	1 49 N	140 3 W	4420	m	2.35	0.18			0.15	2.5		−0.03	
Pacific Ocean	RC13-113	1 39 S	103 38 W	3195	m	2.44	0.16			0.05	2.5		−0.11	
Pacific Ocean	RC15-52	29 14 S	85 54 W	3780	a	2.71	0.44	3.00	−0.45	0.30	3.5	0.89	−0.14	0.75
Pacific Ocean	RC15-61	40 37 S	77 12 W	3771	a			3.27	−1.08	0.30	3			1.38
Pacific Ocean	RC15-65	53 4 S	78 57 W	3200	a			3.26	−0.60	0.30	4			0.90
Pacific Ocean	V17-42	3 32 N	81 11 W	1814	m	2.38	0.08			0.00	3.5		−0.08	
Pacific Ocean	V17-42	3 32 N	81 11 W	1814	a			3.31	−0.76	−0.14	3			0.62
Pacific Ocean	V18-68	54 33 S	77 51 W	3982	a			3.20	−0.59	0.30	4			0.89
Pacific Ocean	V19-27	0 28 S	82 4 W	1373	m	1.99	0.31	2.75	−0.36	0.20	3.5	0.67	−0.11	
Pacific Ocean	V19-28	2 22 S	84 39 W	2720	a	2.61	0.19	3.17	−0.85	−0.05	3	1.04	−0.24	0.80
Pacific Ocean	V19-29	3 35 S	83 56 W	2673	a	2.60	0.02	3.21	−1.08	−0.14	3.5	1.10	−0.16	0.94
Pacific Ocean	V19-30	3 23 S	83 21 W	3091	a	2.69	0.03	3.39	−0.86	−0.14	3.5	0.89	−0.17	0.72
Pacific Ocean	V21-146	37 41 N	163 2 E	3968	a			3.54	−1.02	0.00	1.5			1.02
Pacific Ocean	V24-109	0 26 N	158 48 E	2367	a	2.59	0.46			0.15	1		−0.31	
Pacific Ocean	V28-304	28 32 N	134 8 E	2942	a	2.58	0.29		−0.98	0.00	2.5	1.27	−0.29	0.98
Pacific Ocean	V32-128	36 28 N	177 10 E	3623	a	2.63	0.31			0.00	1		−0.31	
Pacific Ocean	Y69-10-2	41 16 N	126 24 W	2743	a	2.83	0.01	3.32	−0.85	0.00	3.5	0.86	−0.01	0.85
Pacific Ocean	Y71-6-12	16 26 S	77 33 W	2734	a				−0.83	0.02	3.5			0.85
Red Sea	MD77-202	19 13 N	60 41 E	2427	a			3.50	−0.81	−0.09	1			0.72
Southeast Atlantic Ocean	BT-4	4 20 S	10 26 E	1000	a	2.60	0.38			0.50	2		0.12	
Southeast Atlantic Ocean	RC12-294	37 16 S	10 6 W	3308	a	2.70	0.81			0.45	2		−0.36	
Southeast Atlantic Ocean	RC12-294	37 16 S	10 6 W	3308	m	2.38	0.88	4.64	−0.49	0.35	2	1.37	−0.53	0.84
Southeast Atlantic Ocean	RC13-205	2 17 S	5 11 E	3731	a	2.59	0.98			0.85	1		−0.13	
Southeast Atlantic Ocean	RC13-228	22 20 S	11 12 E	3204	a	2.58	0.50	3.46	−0.61	0.65	4	1.11	0.15	1.26
Southeast Atlantic Ocean	RC13-229	25 30 S	11 18 E	4191	a			3.22	−0.53	0.40	4			0.93
Southeast Atlantic Ocean	RC13-229	25 30 S	11 19 E	4191	m	2.83	0.48	3.37	−0.3	0.50	3.5	0.78	0.02	0.8
Southeast Atlantic Ocean	RC13-253	46 36 S	7 38 E	2494	a	2.76	0.62			0.55	4		−0.07	
Southeast Atlantic Ocean	V22-174	10 4 S	12 49 W	2630	a	2.69	0.80	3.21	0.51	0.90	1	0.29	0.10	0.39
Southeast Atlantic Ocean	V22-182	0 33 S	17 16 W	3937	a	2.47	0.67			0.80	2		0.13	
Southwest Atlantic Ocean	CH115-140-90	30 51 S	38 22 W	3384	m	2.12	0.90			0.65	2		−0.25	
Southwest Atlantic Ocean	CH115-61	30 16 S	39 6 W	4181	a,g	2.72	0.13			0.45	2		0.32	
Southwest Atlantic Ocean	CH115-88	30 55 S	38 5 W	2941	a,g	2.62	0.68			0.75	2		0.07	
Southwest Atlantic Ocean	CH115-91	30 50 S	38 26 W	3576	a,g	2.99	0.62			0.55	2		−0.07	
Southwest Atlantic Ocean	CH99-19-14	8 15 S	17 40 W	4029	m	2.52	0.67	4.01	−0.3	0.65	1	0.97	−0.02	0.95
Southwest Atlantic Ocean	CH99-21-15	8 10 S	15 27 W	3652	m	2.71	0.62			0.70	1		0.08	

Southwest Atlantic Ocean	RC12-267	38 41 S	25 47 W	4144	[a]			3.52	−0.25	0.45	2			0.70
Southwest Atlantic Ocean	RC12-267	38 41 S	25 47 W	4144	[m]	2.71	0.50	3.07	−0.16	0.40	2.5	0.66	−0.1	0.56
Southwest Atlantic Ocean	V25-59	1 22 N	39 29 W	3824	[a]	2.78	1.03			0.90	1		−0.13	

Data sources:

[a] Data was taken from Duplessy, Chenouard & Reyss (1974); where indicated, original sources used by Duplessy *et al.* are shown.
[b] Streeter *et al.* (1982)
[c] Boyle and Keigwin (1982)
[d] Duplessy *et al.* (1974)
[e] Sarnthein *et al.* (1984)
[f] Pastouret *et al.* (1978)
[g] Belanger *et al.* (1981)
[h] Vincent, Killingley & Berger (1981a)
[i] Lalou *et al.* (1983)
[j] Ganssen (1983)
[k] Woodruff *et al.* (1980)
[l] Vincent, Killingley & Berger (1981b)
[m] Graham *et al.* (1981)

Productivity data are read from the map of Lieth (1975); values in gC $(m^2\ y)^{-1}$ are; 1 = 0–50; 2 = 50–100; 3 = 100–200; 4 = 200; where site locations were near a contour on Lieth's map, an intermediate value (e.g. 2.5 or 3.5) was read.

Dissolved HCO_3^- $\delta^{13}C$ values were taken as follows: for samples included in the report of Duplessy *et al.* (1984) the HCO_3^- values reported in that paper were used; they were taken from the series of papers by Kroopnick and co-workers listed in the text. For other samples, $\delta^{13}C$ values were estimated from the cross-sections of Kroopnick (1985).

References

Barrera, E. (1987). 'Isotopic Paleo-Temperatures: I. Effect of Diagenesis; II. Late Cretaceous Temperatures'. Ph.D. thesis, Case Western Reserve University, Cleveland, Ohio.

Barrera, E., Keller, G. & Savin, S.M. (1985). Evolution of the Miocene ocean in the eastern North Pacific as inferred from oxygen and carbon isotopic ratios of foraminifera. In *The Miocene Ocean: Paleoceanography and Biogeography*, ed. J.P. Kennett, pp. 83–102. Geological Society of America Memoir 163.

Belanger, P.E., Curry, W.B. & Matthews, R.K. (1981). Core-top evaluation of benthic foraminiferal isotopic ratios for paleoceanographic interpretation. *Palaeogeography, Palaeoclimatology, Palaeoecology*, **33**, 205–20.

Berger, W.H. (1985). CO_2 increase and climate prediction: Clues from deep sea carbonates. *Episodes*, **8**, 163–8.

Boersma, A. & Shackleton, N.J. (1977). Tertiary oxygen and carbon isotope stratigraphy, Site 357 (mid-latitude south Atlantic). In *Initial Reports of the Deep Sea Drilling Project*, vol. 39, ed. P.R. Supko, A. Perch-Nielsen, *et al.*, pp. 911–24. US Government Printing Office, Washington, DC.

Boersma, A. & Shackleton, N.J. (1978). Oxygen and carbon isotope record through the Oligocene, DSDP Site 366, equatorial Atlantic. In *Initial Reports of the Deep Sea Drilling Project*, vol. 41, ed. Y. Lancelot, E. Seibold, *et al.*, pp. 957–62. US Government Printing Office, Washington, DC.

Boersma, A., Shackleton, N.J., Hall, M. & Given, Q. (1979). Carbon and oxygen isotope records at DSDP Site 384 (North Atlantic) and some Paleocene paleotemperatures and carbon isotope variations in the Atlantic Ocean. In *Initial Reports of the Deep Sea Drilling Project*, vol. 43, ed. B.E. Tucholke, P.R. Vogt *et al.*, pp. 695–717. US Government Printing Office, Washington, DC.

Boyle, E. & Keigwin, L.D. Jr. (1982). Deep circulation of the North Atlantic over the last 200 000 years: Geochemical evidence. *Science*, **218**, 784–7.

Corliss, B.H. (1985). Microhabitats of benthic foraminifera within deep sea sediments. *Nature*, **314**, 435–8.

Craig, H. (1970). Abyssal carbon-13 in the South Pacific. *Journal of Geophysical Research*, **75**, 691–5.

Delaney, M.L. (1983). 'Foraminiferal Trace Elements: Uptake, Diagenesis and 100 M.Y. Paleochemical History.' Unpublished Ph.D. thesis, Woods Hole Oceanographic Institution and Massachusetts Institute of Technology, WHOI-84-2.

Douglas, R.G. & Savin, S.M. (1973). Oxygen and carbon isotope analyses of Cretaceous and Tertiary foraminifera from the central north Pacific. In *Initial Reports of the Deep Sea Drilling Project*, vol. 17, ed. E.L. Winterer, J.I. Ewing, *et al.*, pp. 591–605. US Government Printing Office, Washington, DC.

Douglas, R.G. & Savin, S.M. (1975). Oxygen and carbon isotope analyses of Tertiary and Cretaceous microfossils from Shatsky Rise and other sites in the north Pacific Ocean. In *Initial Reports of the Deep Sea Drilling Project*, vol. 32, ed. R.L. Larson, R. Moberly *et al.*, pp. 509–20. US Government Printing Office, Washington, DC.

Duplessy, J.-C., Chenouard, L. & Reyss, J.L. (1974). Paleotemperatures isotopique de l'Atlantique Equatorial. In *Variation du Climat au Cours du Pleistocene*, Colloques Internationaux du Centre de la Recherche Scientifique, No. 219, pp. 251–8. CNRS, Paris.

Duplessy, J.-C., Shackleton, N.J., Matthews, R.K., Prell, W., Ruddiman, W.F., Caralp, M. & Hendy, C.H. (1984). ^{13}C record of benthic foraminifera in the last interglacial ocean: Implications for the carbon cycle and the global deep water circulation. *Quaternary Research*, **21**, 225–43.

Elderfield, H., Gieskes, J.M., Baker, P.A., Oldfield, R.K., Hawkesworth, C.J. & Miller, R. (1982). $^{87}Sr/^{86}Sr$ and $^{18}O/^{16}O$ ratios, interstitial water chemistry and diagenesis in deep-sea carbonate sediments of the Ontong Java Plateau. *Geochimica Cosmochimica Acta*, **46**, 2259–68.

Ganssen, G. (1983). Dokumentation von kustennaham Auftrieb anhand stabiler Isotope in rezenten Foraminiferen vor Nordwest-Afrika. *Meteor Forschungsergebnisse*, **Part C**, 1–46.

Graham, D.W., Bender, M.L., Williams, D.F. & Keigwin, L.D., Jr. (1982). Strontium–calcium ratios in Cenozoic planktonic foraminifera. *Geochimica Cosmochimica Acta*, **46**, 1281–92.

Graham, D.W., Corliss, B., Bender, M.L. & Keigwin, L.D. Jr. (1981). Carbon and oxygen isotopic disequilibria of Recent deep-sea benthic foraminifera. *Marine Micropaleontology*, **6**, 483–97.

Haq, B.U. (1984). Paleoceanography: A synoptic overview of 200 million years of ocean history. In *Marine Geology and Oceanography of Arabian Sea and Coastal Pakistan*, ed. B.U. Haq & J.D. Milliman, pp. 201–31. Van Nostrand Reinhold Co., New York.

Haq, B.U., Worsley, T.R., Burckle, L.H., Douglas, R.G., Keigwin, L.D., Jr., Opdyke, N.D., Savin, S.M., Sommer, M.A., II, Vincent, E. & Woodruff, E. (1980). Late Miocene marine carbon isotopic shift and synchroneity of some biostratigraphic events. *Geology*, **8**, 427–31.

Heath, G.R., Moore, T.C. & Dauphin, J.P. (1977). Organic carbon in deep-sea sediments. In *The Fate of Fossil Fuel CO_2 in the Oceans*, ed. N.R. Anderson & A. Malahoff, pp. 605–25. Plenum Publishing, New York.

Hecht, A.D. (1976). The oxygen isotopic record of foraminifera in deep-sea sediment. In *Foraminifera*, vol. 2, ed. R.H. Hedley & C.G. Adams, pp. 1–43. Academic Press, London.

Keigwin, L.D. & Corliss, B.H. (1986). Stable isotopes in late middle Eocene to Oligocene foraminifera. *Geological Society of American Bulletin*, **97**, 335–45.

Kennett, J.P., ed. (1985). *The Miocene Ocean: Paleoceanography and Biogeography*. Geological Society of America Memoir 163.

Kennett, J.P. (1986). Miocene to early Pliocene oxygen and carbon isotope stratigraphy in the southwest Pacific, Deep Sea Drilling Project Leg 90. In *Initial Reports of the Deep Sea Drilling Project*, vol. 90, pt. 2, ed. J.P. Kennett, C.C. von der Borch, *et al.*, pp. 1383–411. US Government Printing Office, Washington, DC.

Killingley, J.S. (1983). Effects of diagenetic recrystallization on $^{18}O/^{16}O$ values of deep sea sediments. *Nature*, **301**, 594–7.

Kroopnick, P. (1974). The dissolved O_2–CO_2–^{13}C system in the eastern equatorial Pacific. *Deep-Sea Research*, **21**, 211–22.

Kroopnick, P. (1985). The distribution of ^{13}C of CO_2 in the world oceans. *Deep-Sea Research*, **32**, 57–84.

Kroopnick, P., Weiss, R.F. & Craig, H. (1972). Total CO_2, ^{13}C, and dissolved ^{18}O at Geosecs II in the North Atlantic. *Earth and Planetary Science Letters*, **16**, 103–10.

Lalou, C., Brichet, E., Leclaire, H. & Duplessy, J.-C. (1983). Uranium series disequilibrium and isotope stratigraphy in hydrothermal mounds samples from DSDP sites 506–509, leg 70, and site 424, leg 54: An attempt at chronology. In *Initial Reports of the Deep Sea Drilling Project*, vol. 70, ed. J. Honnorez, R.P. Von Herzen, *et al.*, pp. 303–14. US Government Printing Office, Washington, DC.

Lieth, H. (1975). Historical survey of primary productivity research. In *Primary Productivity of the Biosphere*, ed. H. Lieth & R.H. Whittaker, pp. 7–16. Springer-Verlag, New York.

Loubere, P. (1987). Late Pliocene variations in the carbon isotope values of North Atlantic benthic foraminifera: biotic control of the isotopic record? *Marine Geology*, **76**, 45–56.

Loutit, T.S., Kennett, J.P. & Savin, S.M. (1983). Miocene equatorial and southwest Pacific paleoceanography from stable isotope evidence. *Marine Micropaleontology*, **8**, 215–33.

Lutze, G.F. & Coulbourn, W.T. (1984). Recent benthic foraminifera from the continental margin of northeast Africa: Community structure and distribution. *Marine Micropaleontology*, **8**, 361–401.

Matthews, R.K. & Poore, R.Z. (1980). Tertiary $\delta^{18}O$ record and glacioeustatic sealevel fluctuations. *Geology*, **8**, 501–4.

McCorkle, D.C., Emerson, S.R. & Quay, P.D. (1985). Stable carbon isotopes in marine pore waters. *Earth and Planetary Science Letters*, **74**, 13–26.

Miller, K.G., Fairbanks, R.G. & Mountain, G.S. (1987). Tertiary oxygen isotope synthesis, sea-level history, and continental margin erosion. *Paleoceanography*, **2**, 1–19.

Miller, K.G. & Lohmann, G.P. (1982). Environmental distribution of Recent benthic foraminifera on the northeast United States continental slope. *Geological Society of America Bulletin*, **93**, 200–6.

Murphy, M.G. & Kennett, J.P. (1986). Development of latitudinal thermal gradients during the Oligocene: oxygen isotope evidence from the southwest Pacific. In *Initial Reports of the Deep Sea Drilling Project*, vol. 90, pt. 2, ed. J.P. Kennett, C.C. von der Borch, *et al.*, pp. 1347–60. US Government Printing Office, Washington, DC.

Pastouret, L., Chamley, H., Delibrias, G., Duplessy, J.C. & Thiede, J. (1978). Late Quaternary climatic changes in western tropical Africa deduced from deep-sea sedimentation off the Niger delta. *Oceanologica Acta*, **1**, 2217–32.

Sarnthein, M., Erlenkeuser, H., von Grafenstein, R. & Schroeder, C. (1984). Stable isotope stratigraphy for the last 750 000 years: Meteor core 13-519 from the eastern equatorial Atlantic. *Meteor Forschungsergebnisse*, **Part C, 38**, 9–24.

Savin, S.M. (1977). The history of the Earth's surface temperature during the past 100 million years. In *Annual Reviews of Earth and Planetary Science*, vol. 5, ed. F.A. Donath, F.G. Stehli & G.W. Wetherill, pp. 319–55. Annual Reviews Inc., Palo Alto.

Savin, S.M. (1983). Stable isotopes in climatic reconstructions. In *Climate in Earth History* (Studies in Geophysics), ed. W.H. Berger & J.C. Crowell, pp. 164–71. National Academy Press, Washington, DC.

Savin, S.M., Abel, L., Barrrera, E., Hodell, D., Keller, G., Kennett, J.P., Killingley, J., Murphy, M & Vincent, E. (1985). The evolution of Miocene surface and near surface marine temperatures: Oxygen isotopic evidence. In *The Miocene Ocean: Paleoceanography and Biogeography*, ed. J.P. Kennett, pp. 49–82. Geological Society of America Memoir 163.

Savin, S.M. & Douglas, R.G. (1985). Sea level, climate, and the Central American land bridge. In *The Great American Biotic Interchange*, ed. F.G. Stehli & S.D. Webb, pp. 303–24. Plenum Press, New York.

Savin, S.M., Douglas, R.G. & Stehli, F.G. (1975). Tertiary marine paleotemperatures. *Geological Society of America Bulletin*, **86**, 1499–510.

Savin, S.M. & Yeh, H.-W. (1981). Stable isotopes in ocean sediments. In *The Sea, Vol. 7., The Oceanic Lithosphere*, ed. C. Emiliani, pp. 1521–54. Wiley Interscience, New York.

Shackleton, N.J. (1967). 'The measurement of paleotemperatures in the Quaternary era.' Unpublished Ph.D. thesis, Cambridge University, Cambridge.

Shackleton, N.J. (1974). Attainment of isotope equilibrium between ocean water and the benthonic genus *Uvigerina*. In *Variation du Climat au Cours du Pleistocene*, Colloques Internationaux du Centre de la Recherche Scientifique, No. 219, pp. 203–9. CNRS, Paris.

Shackleton, N.J. & Kennett, J.P. (1975). Paleotemperature history of the Cenozoic and the initiation of Antarctic glaciation: Oxygen and carbon isotope analyses in DSDP Sites 277, 279 and 281. In *Initial Reports of the Deep Sea Drilling Project*, vol. 90, ed. J.P. Kennett, R.E. Houtz, *et al.*, pp. 743–55. US Government Printing Office, Washington, DC.

Streeter, S.S., Belanger, P.E., Kellogg, T.B. & Duplessy, J.C. (1982). Late Pleistocene paleo-oceanography of the Norwegian–Greenland Sea; benthic foraminiferal evidence. *Quaternary Research*, **18**, 72–90.

Streeter, S.S. & Shackleton, N.J. (1979). Paleocirculation of the deep North Atlantic: 150 000 year record of benthic foraminifera and oxygen-18. *Science*, **203**, 168–71.

Suess, E. (1980). Particulate organic carbon flux in the oceans – surface productivity and oxygen utilization. *Nature*, **288**, 260–3.

Vail, P.R. & Hardenbol, J. (1977). Sea-level changes during the Tertiary. *Oceanus*, **22**, 71–9.

Vincent, E., Killingley, J.S. & Berger, W.H. (1981a). Stable isotopes in benthic foraminifera from Ontong–Java plateau, box cores ERDC 112 and 123. *Palaeogeography, Palaeoclimatology, Palaeoecology*, **33**, 221–30.

Vincent, E., Killingley, J.S. & Berger, W.H. (1981b). Stable isotope composition of benthic foraminifera from the equatorial Pacific. *Nature*, **289**, 639–43.

Woodruff, F. (1985). Changes in Miocene deep-sea benthic foraminiferal distribution in the Pacific Ocean: Relationship to paleoceanography. In *The Miocene Ocean: Paleoceanography and Biogeography*, ed. J.P. Kennett, pp. 131–75. Geological Society of America Memoir 163.

Woodruff, F. & Savin, S.M. (1985). $\delta^{13}C$ values of Miocene Pacific benthic foraminifera: Correlations with sealevel and biological productivity. *Geology*, **13**, 119–22.

Woodruff, F., Savin, S.M. & Douglas, R.G. (1980). Biological fractionation of oxygen and carbon isotopes by Recent benthic foraminifera. *Marine Micropaleontology*, **5**, 3–11.

Woodruff, F., Savin, S.M. & Douglas, R.G. (1981). Miocene stable isotope record: a detailed deep Pacific Ocean study and its paleoclimatic implications. *Science*, **212**, 665–8.

Zahn, R., Winn, K. & Sarnthein, M. (1986). Benthic foraminiferal $\delta^{13}C$ and accumulation rates of organic carbon: *Uvigerina peregrina* group and *Cibicidoides wuellerstorfi*. *Paleoceanography*, **1**, 27–42.

21

Climate model evidence for variable continental precipitation and its significance for phosphorite formation

E.J. BARRON AND L.A. FRAKES

Abstract

The source of phosphate and the sites of phosphogenesis are potential problems in the explanation of major phosphate deposits. A variety of factors suggest that changes in climate and geography have bearing on these problems. A series of Cenozoic climate model simulations and Mid-Cretaceous climate model sensitivity studies indicate that precipitation is strongly influenced by geography and climatic warmth. In particular, the evolution of the Tethyan ocean modulates northern hemisphere continental precipitation in the model simulations. Geography and land-sea thermal contrasts may focus precipitation on continental margins, and changes in continental configuration in relation to the earth's general circulation may result in substantial variation in continental precipitation. Further, planetary warming tends to enhance the hydrologic cycle. The model-generated precipitation fields for different geography and different climates are sufficiently different to produce episodic periods of extreme continental weathering. Continental weathering may provide a source of phosphorus and consequently these weathering episodes may have bearing on the source, timing and location of major phosphate deposits.

Introduction

The ultimate sources of energy which drive the sedimentary system are from the earth's interior, manifested in plate tectonic processes, and from the solar input at the top of the atmosphere. Consequently, hypotheses and arguments concerning geologic problems, including the origin of Neogene sedimentary phosphorites, frequently involve geographic and climatic change (e.g. Cook & McElhinny, 1979; Arthur & Jenkyns, 1981; Sheldon, 1981). The above authors and others have presented a number of important 'factors' which define the problem:

Modern phosphate deposits are associated with regions of upwelling (Burnett, 1977; Sheldon, 1981) and upwelling environments have become a generally accepted analog for ancient phosphate accumulation. For many ancient deposits, the case for upwelling is plausible or even well demonstrated (e.g. Riggs, 1984). However, the upwelling model for ancient phosphate accumulation is directed largely toward explanations of the sites of deposition. The times of widespread accumulation and formation of major deposits appear to be episodic. Clearly, upwelling does not explain these variations in the geologic record, since there is little reason to postulate coincident episodic upwelling.

Major phosphate deposits contain considerable phosphate which may be difficult to explain by invoking changes in saturation or stripping of the marine reservoir. Some evidence suggests that phosphate concentrations in the ocean have not varied substantially. Today, $\delta^{13}C$ values and phosphate concentrations are well correlated in the same modern deep water masses. Given assumptions on the constancy of the Redfield ratio and the $\delta^{13}C$ values measured from the Cenozoic for total dissolved carbon in the deep sea, Cenozoic phosphate concentrations in the deep sea may not have varied by more than a factor of two (Arthur & Jenkyns, 1981). Arthur & Jenkyns (1981) question whether the time scale of deposition of major phosphate deposits and the size of the deposits can be achieved with reasonable oceanic mixing rates or with a reasonable efficiency for removal of phosphate given the evidence for a relatively constant phosphate concentration. These arguments are indicative of a source problem, and suggest that phosphate supply to the oceans must be an important factor.

Further, Sheldon (1964), Cook & McElhinny (1979) and Arthur & Jenkyns (1981) note that major phosphate deposits are associated with low latitudes (near tropics to 40° latitude) and with sea-level transgressions. In many cases, sea-level transgressions are associated with planetary warming and a more active hydrologic cycle. These three factors have been related to phosphogenesis in several ways: warm, humid climates may promote increased continental weathering and hence provide a source of phosphorus; higher sea level may result in estuarine traps of detritus and may provide low sedimentation shelf sites of phosphate formation; and low latitudes and open oceans may provide the opportunity for conducive oceanic circulation (e.g. upwelling).

The purpose of this contribution is to examine the results of Mesozoic–Cenozoic atmospheric simulations, using a General Circulation Model of the atmosphere, for relationships between geography and climate evolution, and to determine the potential of climatic change as a factor in explaining episodic phosphogenesis. Changes in continental precipitation as a function of geography and global temperatures are the most dramatic examples of climatic variability in the model simulations, and continental precipitation may be episodic. If these changes can be related to continental weathering and high organic productivity, then the results presented here may have bearing on the phosphorus supply to the oceans and the temporal and spatial problems associated with phosphogenesis.

Model description

The climate model referred to in this study is a version of the National Center for Atmospheric Research 'Community Climate Model' (CCM), a spectral general circulation model of the atmosphere. The CCM evolved from the model described by Bourke, *et al.*, (1977), and McAvaney, Bourke & Puri (1978) based on the hydrodynamic and thermodynamic laws as applied to the atmosphere. A new radiation–cloudiness formulation (Ramanathan *et al.*, 1983) and a more realistic hydrology treatment (Washington & Williamson, 1977) are the principle changes to the original model. The CCM includes nine levels in the vertical, an approximate 4.5° latitude resolution and a 7.5° longitude resolution.

The simulations described here employ an energy balance ocean formulation, sometimes referred to as a swamp ocean, which acts as a moisture source but does not include heat storage, transport or diffusion. Ocean surface temperatures are calculated based on the surface energy balance, and sea ice forms if the ocean temperature falls below −2°C (271°K). Experiments with this type of energy balance ocean are usually restricted to simulations with annual average insolation. One of the major advantages of this type of formulation is that the model reaches equilibrium relatively quickly and hence multiple simulations may be completed. The deficiences (lack of realistic oceans and lack of an annual cycle) suggest that this type of model should be used to examine primarily first-order effects.

Comparison of the present day model climatology with observations (Washington & Meehl, 1983) indicates that the model satisfactorily simulates many climatic characteristics, including the observed annual mean temperature structure and the zonal wind distribution.

Description of sensitivity experiments

The results presented here draw on a number of simulations including: (a) geographic sensitivity experiments designed to examine the role of continental positions, sea level and topography in climatic change based on mid-Cretaceous geography (Barron & Washington, 1984); (b) high atmospheric CO_2 simulations for the mid-Cretaceous (Barron & Washington, 1985); (c) January and July fixed insolation experiments with mid-Cretaceous geography (Barron & Washington, 1982); and (d) a Mesozoic–Cenozoic series of 'realistic' experiments based on Albian–Cenomanian, Campanian, Paleocene, Eocene, Miocene and Present day geography (Barron, 1985). The geographic boundary conditions are given in Figure 21.1.

In the mid-Cretaceous experiments, the continental positions and paleogeography are derived from Barron *et al.* (1981) with the exception that the paleogeography of Antarctica has been modified to be similar to that of Tarling (1978). Based on estimates of global sea level and continental size (Southam & Hay, 1981), the average elevation of the mid-Cretaceous continents is assumed to have been approximately 150 m less than the present day average of 744 m. The distributions of topographic highs is that reported by Ziegler, Scotese & Barrett (1982), with the exception that the Greenland elevation was increased in association with continental rifting at approximately 95 my ago.

The geographic sensitivity experiments for simulation (a) are based on geographic extremes using mid-Cretaceous continental positions. These experiments include mountain and no mountain cases and flooded and non-flooded continents. Approximately 20% of the present day area of continental terrain is covered by shallow seas in the flooded case.

The relative positions and paleolatitudes of the Tertiary continents are well constrained by seafloor-spreading magnetic anomalies and paleomagnetic pole positions. Based on seafloor-spreading anomaly time scales, the Cenozoic maps were determined for 60, 40 and 20 my ago. Paleocene, Eocene and Miocene paleogeography from Barron *et al.* (1981) were superimposed on the paleocontinental maps. Topography is specified following Ziegler *et al.* (1982). These data were then transformed into the 4.5° (latitude) by 7.5° (longitude) grid to conform to the model resolution. Permanent ice-caps were specified only in the modern control simulation.

Typical of model simulations with an energy balance ocean, the experiments were conducted for 300 model days, with the exception (150 d) of sensitivity experiments in simulation (a). The model simulations are near equilibrium at 50 d and exhibit almost no change beyond 200 d. The results in each case are based on an average of the last 100 d of model simulation time.

Cenozoic trends

The Paleocene, Eocene, Miocene and present day simulations indicate little systematic trend in surface temperature (Figs. 21.2, 21.3). For example, the southeastern USA is in the range of 17–27°C (290–300°K) in each simulation. The only major change in surface temperature occurs between the Miocene and the present day and this change is due to the specification of permanent polar ice in the present day simulation. These simulations do not address the timing of glaciation, only the fact that glaciation itself is a major factor in the global cooling. As discussed by Barron (1985), large-scale changes in geography do not seem to be a major explanation of the Cenozoic global cooling trend.

Interestingly, Barron & Washington (1985) note that one of the best correlated geographic variables with global warmth is continental flooding. Yet, climate model studies designed to test climate sensitivity to continental flooding indicate very little sensitivity (Barron & Washington, 1984). The lack of model sensitivity to continental flooding and the lack of a model-generated Cenozoic global cooling trend are enigmatic. Geography is well correlated with climatic change but this geography yields little model sensitivity. The explanation of the Cenozoic global cooling trend may lie in changes in atmospheric carbon dioxide concentration (Berner, Lasaga & Garrels, 1983; Barron & Washington, 1985). The work of Berner *et al.* (1983) suggests that rapid seafloor spreading and associated increase in volcanism would yield higher atmospheric CO_2 levels because an

20 MILLION YEARS

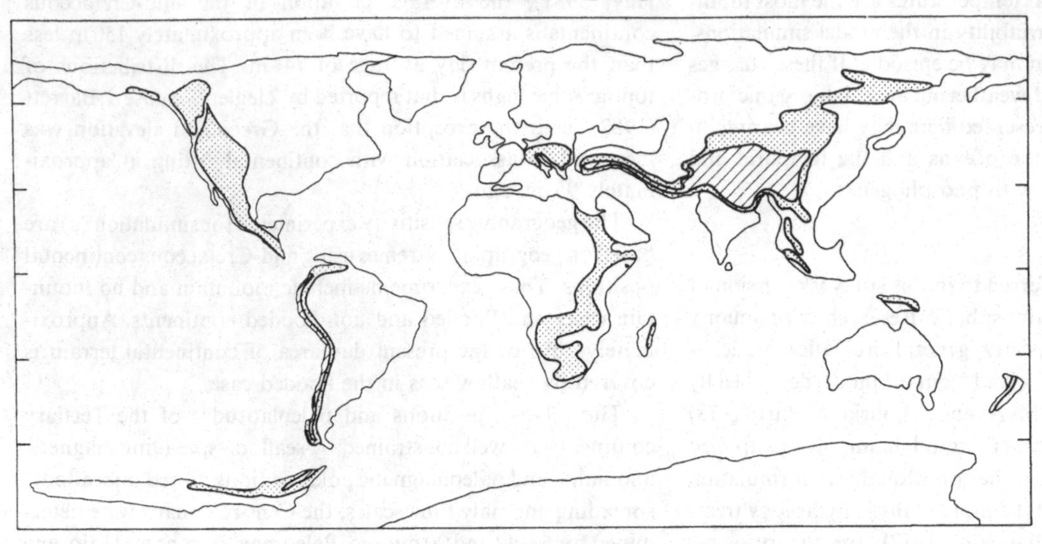

40 MILLION YEARS

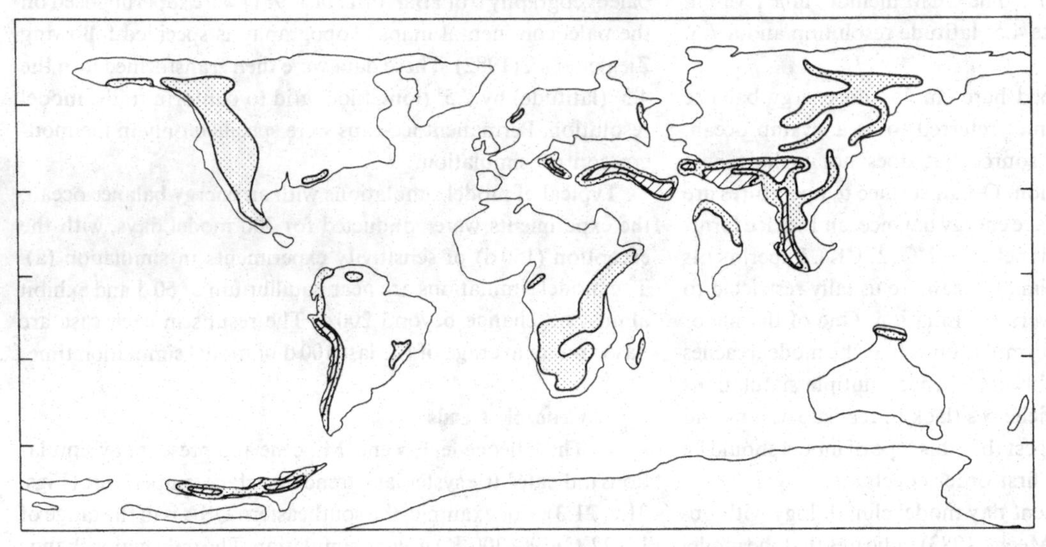

60 MILLION YEARS

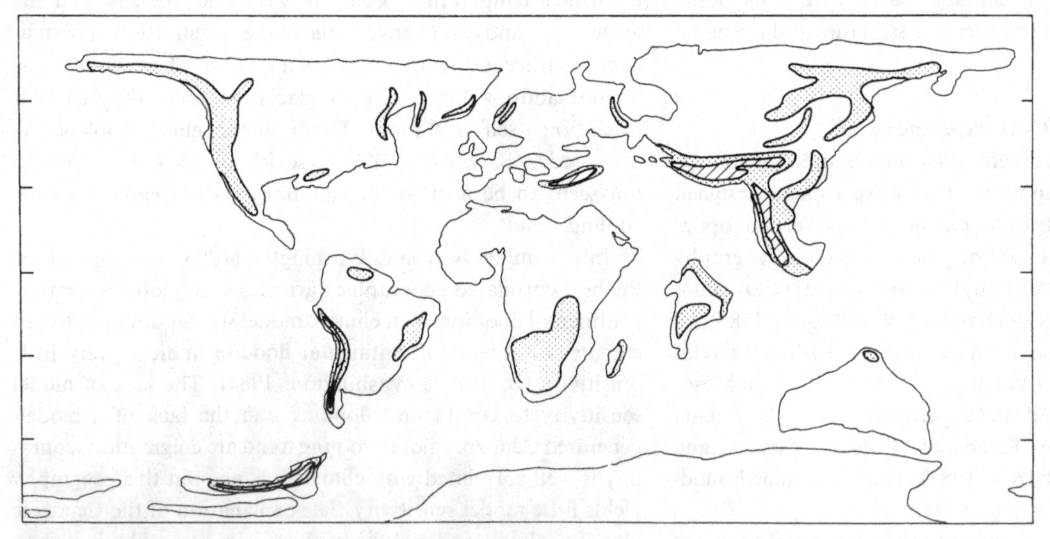

Fig. 21.1. Cenozoic paleogeography plotted on plate tectonic reconstructions for 60, 40 and 20 million years ago (from Barron, 1985). Light stipple – moderate topography; lined shading – high topography.

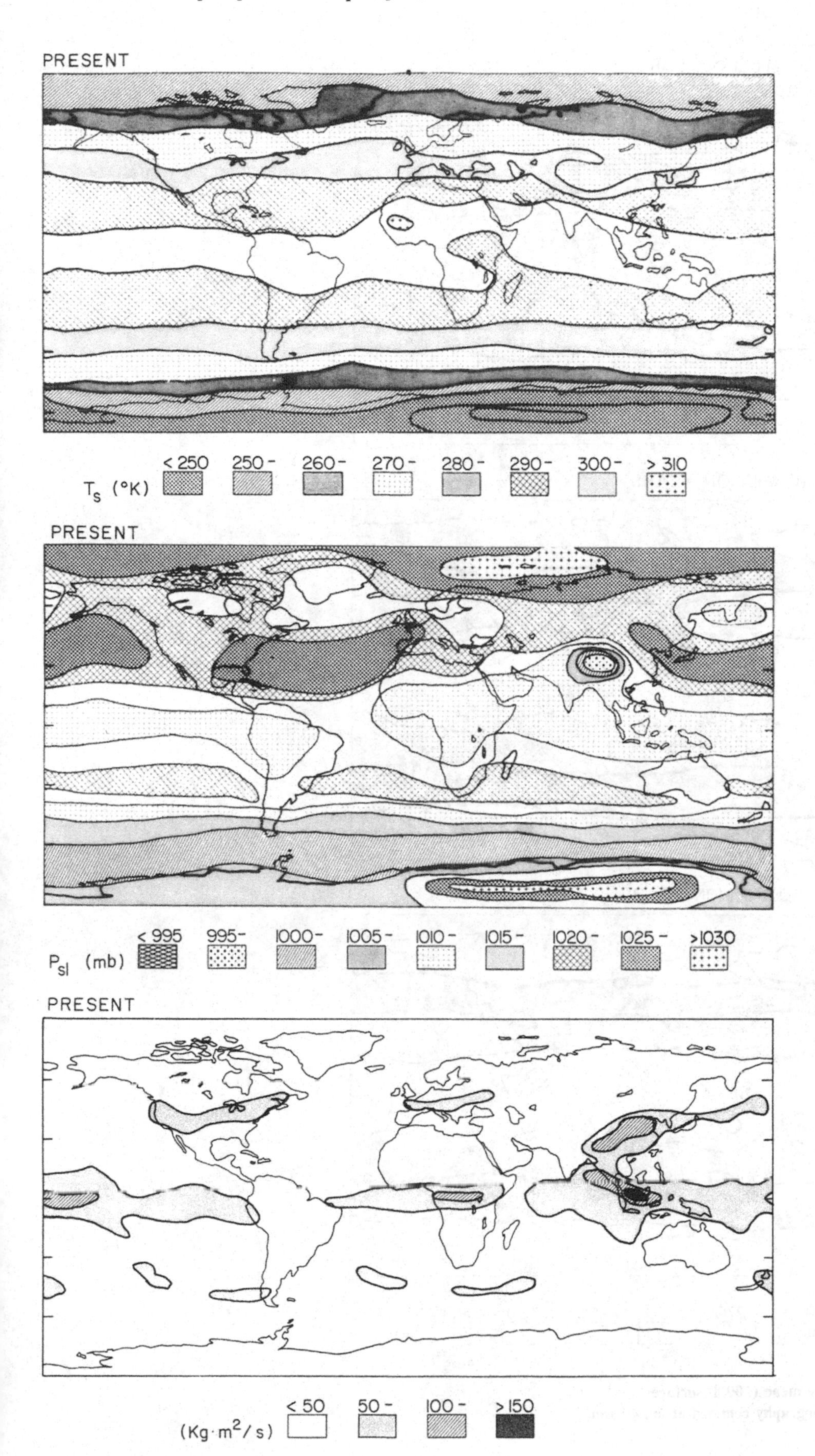

Fig. 21.2. Climate model-generated time mean (100 d) surface temperature (°K), sea-level pressure (mb) and precipitation $(kg(m^{-2}s)^{-1}$ or $mm\,s^{-1})$ for the present day control simulation.

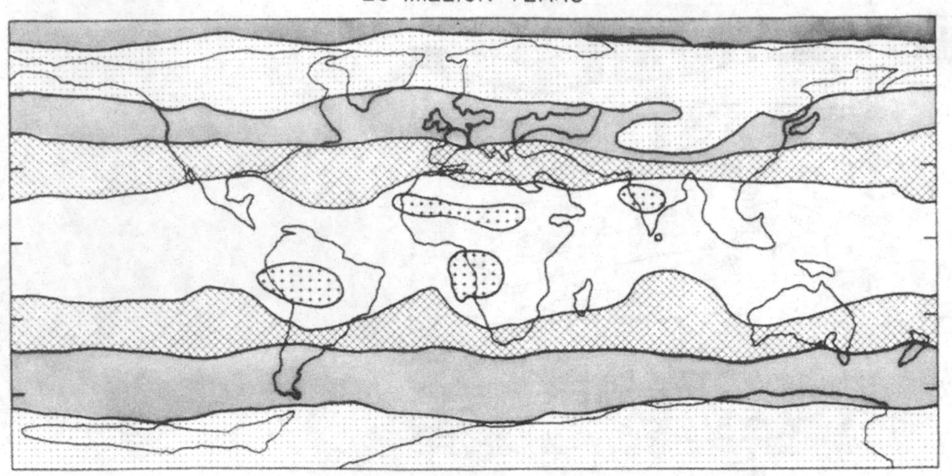

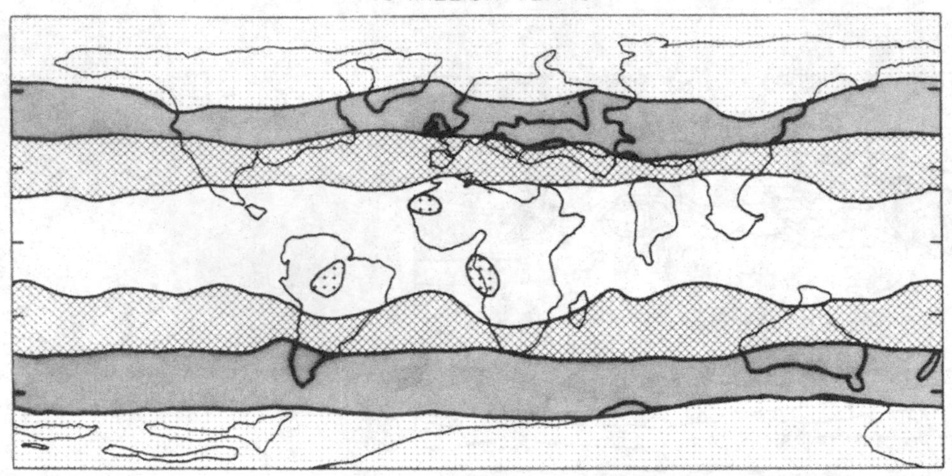

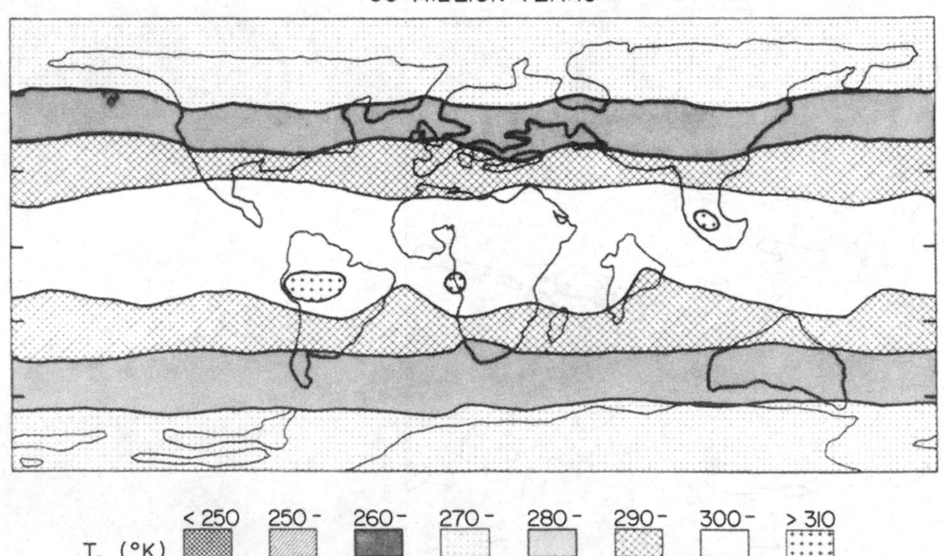

Fig. 21.3. Climate model-generated time mean (100 d) surface temperatures (°K) for Cenozoic paleogeography centered at 60, 40 and 20 million years ago.

increased volcanic input of CO_2 over several million years cannot be compensated by removal of atmospheric CO_2 by continental weathering. Since continental flooding is hypothesized to be a response to rapid seafloor spreading (e.g. Pitman, 1978), then the Berner *et al.* (1983) hypothesis may explain why continental flooding is correlated with global warmth and why the climate model sensitivity studies do not match observations of the global cooling trend.

The results reported here test the role of geography in climate change, but if the above hypothesis is correct, much of the observed climatic change during the Cenozoic may be the product of atmospheric CO_2 variation. This is a limitation in examining the model results, described here in reference to phosphogenesis. However, the proposed correlation between transgressions, atmospheric CO_2, and warming, and the model experiments designed to investigate the sensitivity to increased CO_2 may yield insights into the importance of this limitation for investigating sites and episodicity of phosphate accumulation.

The model-generated predictions of sea-level pressure for the Paleocene, Eocene, Miocene and for the present day (Figs. 21.2, 21.4) indicate some trends, including decreased zonality with time and increased pressure contrasts during more modern times. A large part of these changes are interpreted to be a function of increasing topographic expression. Note, however, that the predicted pressures for the southeastern USA for the Miocene are different in value but are little different in terms of the type of circulation from other time periods. Consequently wind patterns, to the extent that geography is a controlling factor, are likely to be similar (see Barron & Washington, 1982 for a discussion of possible temperature relationships).

Precipitation patterns illustrate large systematic changes (Figs. 21.2, 21.5) and in this regard are in contrast to the predicted temperature and pressure fields. Note that the Paleocene pattern in the northern hemisphere is of high precipitation on all the Northern Tethyan margins including the United States. With the decline in the extent and zonality of Tethys through the Cenozoic and the replacement of this subtropical zonal ocean by distinct oceanic basins, the well defined mid-latitude belt of high precipitation along the northern Tethyan margin declines through the Cenozoic. Precipitation is a truly 'geographic dependent' variable which exhibits large variations in the model simulations. The importance of Tethyan-controlled mid-latitude continental precipitation is supported further by mid-Cretaceous (the time period in which Tethys was best developed) sensitivity experiments.

Mid-Cretaceous sensitivity experiments

The Mid-Cretaceous sensitivity experiments of interest take three forms: (1) a January and July simulation; (2) a 'flooded' and 'non-flooded' simulation; and (3) a high CO_2 (4 × present) simulation. Figure 21.6 illustrates results from an earlier gridpoint version of a General Circulation Model with specified fixed January and fixed July solar insolation and energy balance ocean temperatures which were not allowed to cool below 10°C (an attempt to incorporate some data from oxygen isotopes). The most remarkable aspect of these two simulations involves northern hemisphere precipitation. The Intertropical Convergence Zone (ITCZ) penetrates a few degrees into the southern hemisphere during the southern summer. However, the ITCZ extends much farther poleward during the northern hemisphere summer, to a position directly over Tethys. The southern margin of Tethys is associated with high precipitation and a migrating ITCZ. Further, the northern hemisphere mid-latitude high precipitation belt is over the northern hemisphere continents in both winter *and* summer. The mechanisms for these regimes are described by Barron, Arthur & Kauffman (1985).

The northern hemisphere continental precipitation in summer occurs in association with continental heating, low pressures and rising motion. The positions of the continents in mid-latitudes helps to control the location of the mid-latitude low pressure belt. In winter, the simulation is characterized by cool northern hemisphere continents and warmer oceans. This temperature contrast along the northern margin of the Tethyan ocean controls the position of a mid-latitude storm track. The large thermal contrast with a cool continent and warm ocean is an ideal situation for high evaporation over the ocean and high precipitation on the continental margin. Importantly, the two seasonal simulations indicate the importance of a zonal Tethys in influencing precipitation patterns with the unique (?) situation of geography promoting regions of mid-latitude high precipitation in both summer and winter, and a strong tropical–subtropical precipitation regime associated with migration of the ITCZ on the southern margin of the zonal ocean.

In addition, the flooded and non-flooded mid-Cretaceous experiments described by Barron & Washington (1984) indicate that the influence of Tethys depends on its size and zonality. In other words, during times of extensive flooding, the precipitation of the northern Tethyan continental margin and the southern margin is enhanced. When Tethys is smaller and less zonal, the ocean has less influence on the characteristics of the general circulation and on the location of precipitation. The subtropical ocean is also the major evaporative source for mid-latitude and tropical precipitation.

Greater enhancement of mid-latitude and tropical precipitation apparently occurs with planetary warming. Barron & Washington (1985) show that, in present day, present day with 4 × present CO_2 concentration, mid-Cretaceous and mid-Cretaceous with 4 × present CO_2 simulations, globally-averaged precipitation increased with globally-averaged temperature (Fig. 21.7). An examination of the regions of increased precipitation (Fig. 21.8) shows that the areas of high precipitation in the Cretaceous simulation (tropics and the northern continental margin of Tethys) are accentuated by incorporating higher atmospheric CO_2 in the simulation. As a note of caution this relationship may not apply to other extreme geographic configurations.

A general hypothesis

The model simulations described above suggest that the evolution of Tethys exerts a major influence on continental precipitation. The following points are significant:

1) With the Cenozoic decline of the zonal Tethys into discrete ocean basins, model-predicted tropical and mid-latitude northern hemisphere precipitation declined. In contrast, the general pattern of atmospheric circulation changes much less as a function of Cenozoic geography.

20 MILLION YEARS

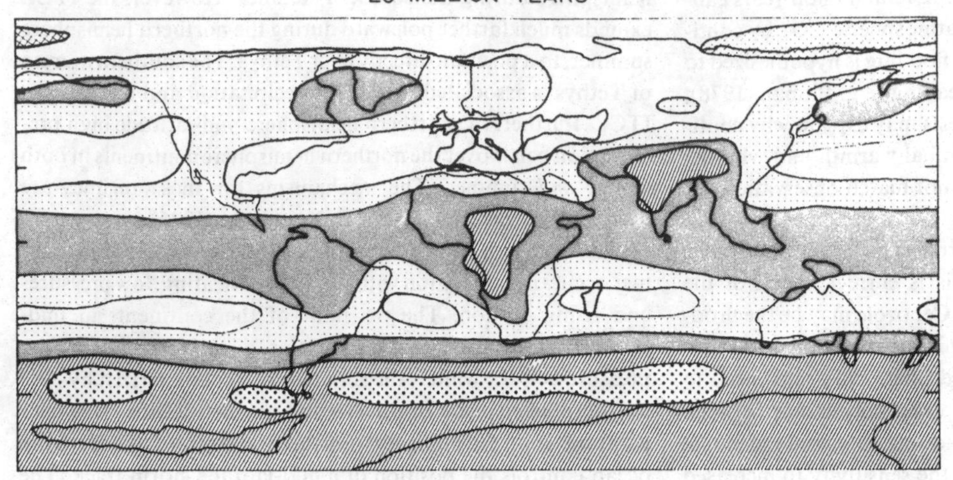

40 MILLION YEARS

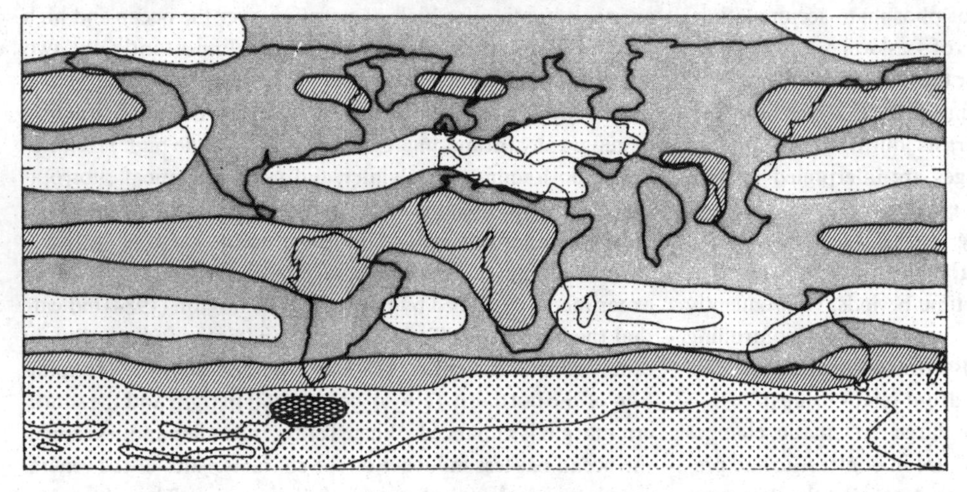

60 MILLION YEARS

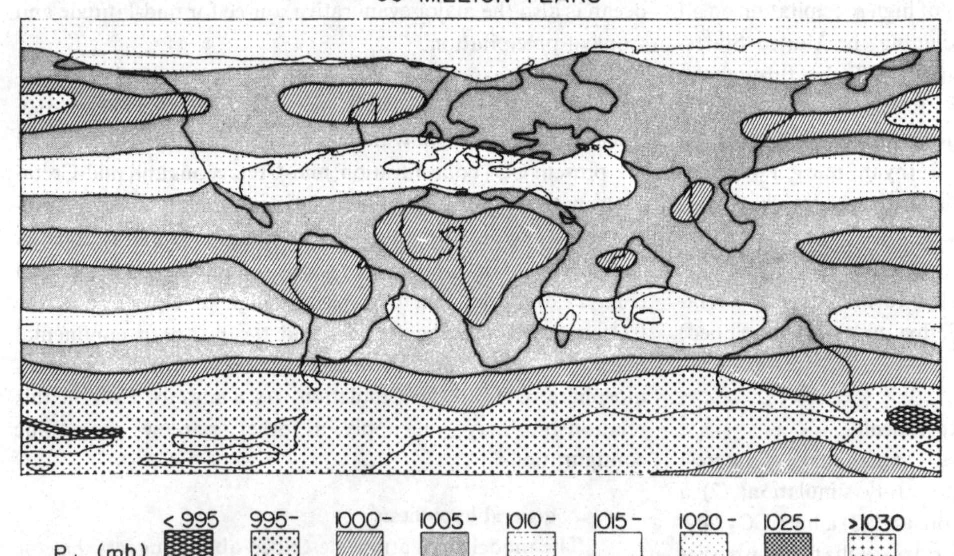

P_{sl} (mb) < 995 | 995- | 1000- | 1005- | 1010- | 1015- | 1020- | 1025- | >1030

Fig. 21.4. Climate model-generated time mean (100 d) sea-level pressures (mb) for Cenozoic paleogeography centered at 60, 40 and 20 million years ago.

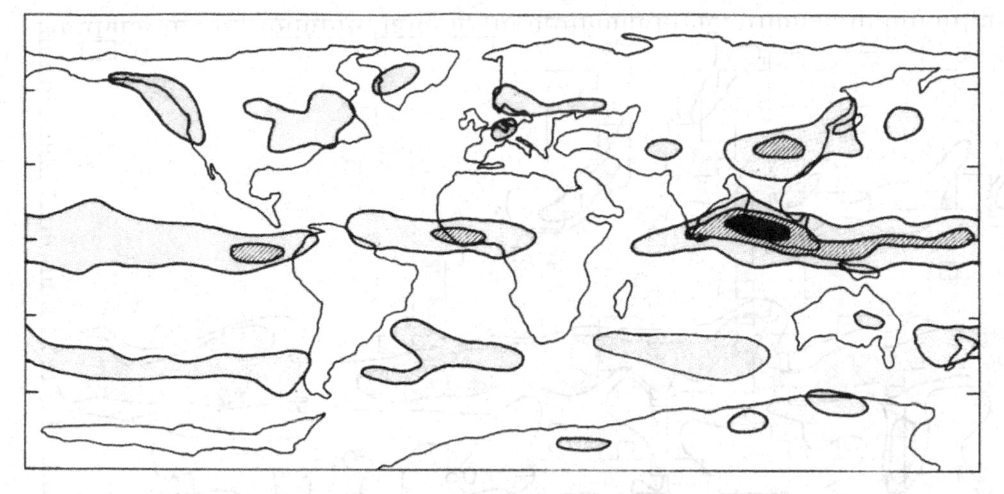

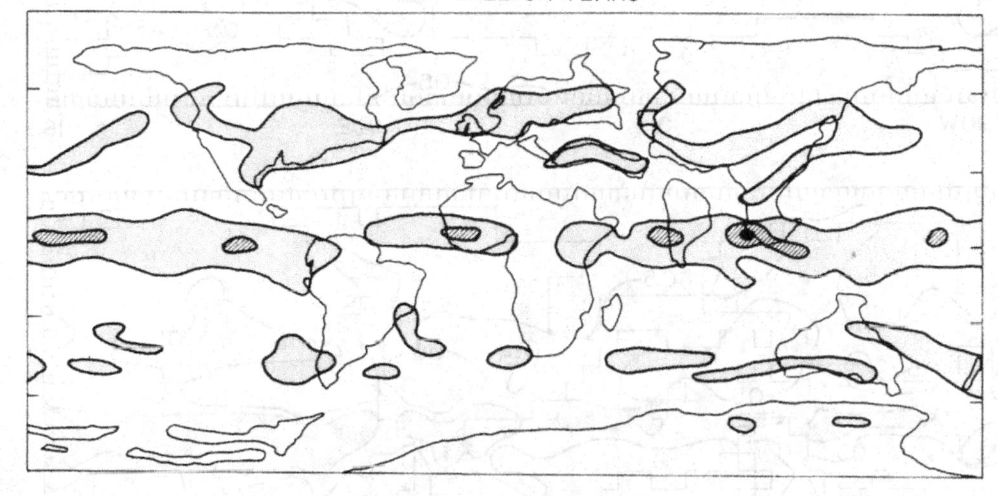

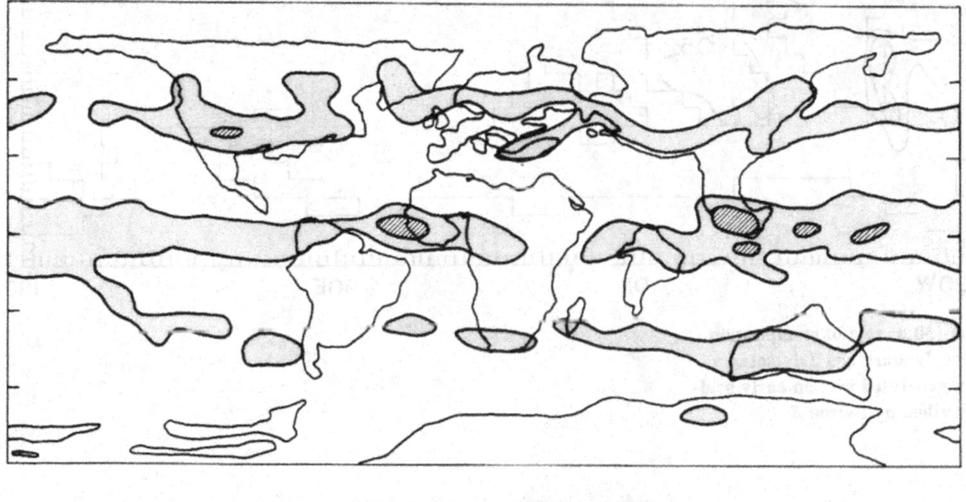

(Kg·m²/s) < 50 | 50 - | 100 - | > 150

Fig. 21.5. Climate model-generated time mean (100 d) precipitation (kg(m²s)$^{-1}$ or mm s^{-1}) for Cenozoic paleogeography centered at 60, 40 and 20 million years ago.

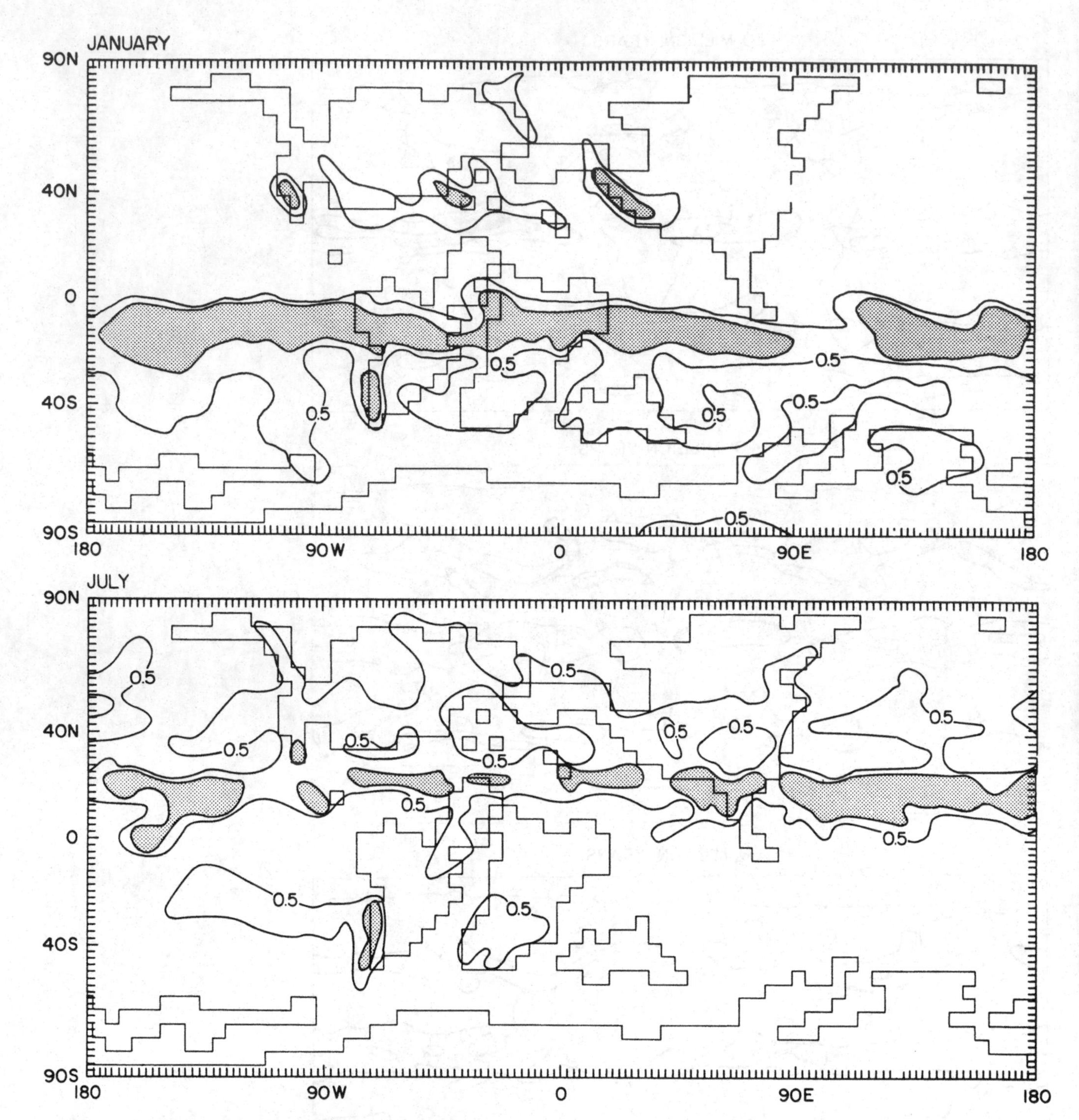

Fig. 21.6. A comparison of the time mean (30 d) rate of precipitation contoured at 0.5 cm d^{-1} and 1.5 cm d^{-1} for January and July solar insolation and mid-Cretaceous geography generated with an early grid-point General Circulation Model and described by Barron & Washington (1982).

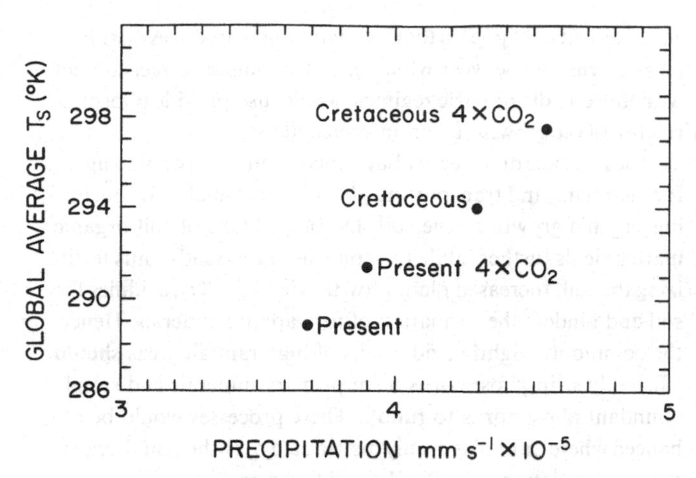

Fig. 21.7. The relationship between the globally-averaged rate of precipitation (mm s^{-1}) and globally-averaged surface temperature (°K) following Barron & Washington (1985).

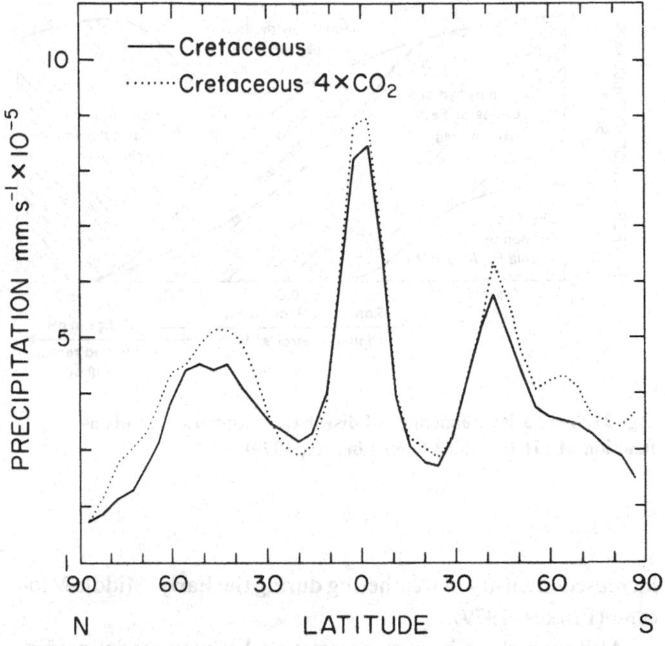

Fig. 21.8. The latitudinal sensitivity of precipitation rate (mm s^{-1}) for a quadrupling of atmospheric CO_2 with Cretaceous geography, following Barron & Washington (1985).

2) A well defined (mid-Cretaceous) Tethys produced both winter and summer amplification of the mid-latitude continental precipitation on its northern margin.
3) In as much as continental flooding influences the size and zonality of Tethys, continental flooding accentuates the mid-latitude continental precipitation on the northern Tethyan margin and precipitation in the ITCZ on the southern Tethyan margin.
4) Higher atmospheric CO_2, an important proposed explanation of Cretaceous warmth, increases the model-generated rate of precipitation on the northern margin of the Tethys and in the tropics.
5) At least for Cenozoic geographies, global warmth is likely to be correlated to increased continental precipitation rates (specifically on the northern margin of Tethys).

For the Cretaceous and the Cenozoic, the model simulations define a strong link between continental flooding, higher atmospheric CO_2, global warmth and continental precipitation. Large-scale continental flooding in response to an increase in seafloor-spreading rates may be associated with higher CO_2 levels in the atmosphere. Consequently, times of continental flooding would be correlated with times of global warmth. The specific characteristics of geography and global warmth produce model-simulated high precipitation along the northern Tethyan continental margin. The size of Tethys is an important factor in this regard, and the size and zonality are partly a function of continental flooding. Global warmth accentuates the zone of tropical and mid-latitude geographically-controlled high precipitation. If the model results are reasonable, then time periods of continental flooding and a well developed Tethys should be correlated with time periods of high precipitation and of extensive continental weathering in North America, much of Asia and in North Africa and northern South America. These model studies also provide cases where basin geometry and size and continental configuration greatly influence the patterns of high precipitation on continents which may occur during either warmer or cooler climatic conditions.

Continental weathering

The model simulations described above are sensitivity experiments designed to develop general physically-consistent concepts or as snapshots for particular fixed geographic configurations. As such these simulations do not provide a 'time series' of climates for comparison with geographic changes or with observations of weathering. However, the model results for the mid-Cretaceous are very different from the present day and if these are reasonable, then the Cretaceous northern hemisphere continents (and other tropical regions of high precipitation) should have evidence of extensive precipitation and weathering.

The Cretaceous is characterized by abundant laterites and widespread bauxites over large areas of North America, Europe and the Middle East (Khain & Ronov, 1960; Hallam, 1985). For example, Cretaceous karst bauxites and kaolinite-rich clays occur throughout southern Europe. The largest kaolin resource in the USA is a Cretaceous deposit in central Georgia and South Carolina. Further, many Deep Sea Drilling Project cores in the North Atlantic are characterized by abundant kaolinite and smectites of Cretaceous age, indicative of warm and humid conditions everywhere surrounding the North Atlantic (Chamley, 1979). The Cretaceous is also a time of extensive northern hemisphere coal formation (Beeson, 1984).

At least for the Cretaceous, the model simulations and the geologic data provide a consistent picture of high regional precipitation and associated extensive weathering. The Middle–Late Cretaceous also includes examples of major phosphate deposits in Russia, the Middle East, North Africa, Europe and northern South America (Sheldon, 1986) within the model-generated locations of high precipitation. Interestingly, the major Miocene phosphate deposits of the southeastern USA and other Miocene deposits are also apparently correlated with

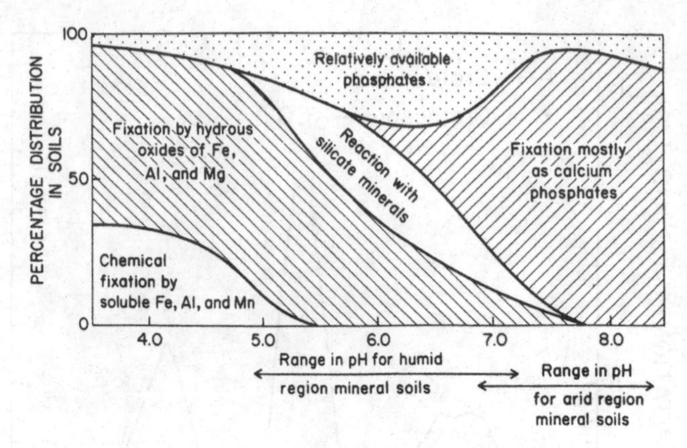

Fig. 21.9. Relative abundance of dissolved phosphorus in soils as a function of pH (modified from Lindsay, 1979).

increased intensity of weathering during the Early–Middle Miocene (Frakes, 1979).

Although there is some correlation between model predictions and weathering sequences, the weathering of bedrock, formation of regoliths and dispersal of dissolved solids in runoff are highly variable processes depending on, in addition to the parameters of climate, the chemistry of the bedrock and the groundwater, local and regional topography, and the history of the weathering process. The effects of climate change on the phosphorus cycle thus are disentangled only with difficulty from the effects of other variables. In view of the relationships shown earlier, however, we feel that the role of chemical weathering as a consequence of increased precipitation – increased flooding – increased atmospheric CO_2 concentrations – specific geographic configurations – deserves investigation.

The natural cycle of phosphorus (Burnett & Oas, 1979; Froelich *et al.*, 1982) has been greatly altered by man's activities but reasonable estimates have been made of pre-man annual input from runoff. However, the generation of natural variations in flux to the oceans has not been addressed. As a starting point, we suggest that global variations in chemical weathering rates have occurred as a result of climate change and, consequently, the supply of phosphorus varied.

Weathering of primary phosphorus minerals in soils appears to be strongly influenced by pH (Lindsay, 1979; Cole & Stewart, 1983). This is seen in the relative solubilities of calcium phosphates, such that a critical value between about pH6 and 7 yields substantial amounts of phosphorus to solution. At lower pH there is increased formation of insoluble aluminum- and iron-phosphate minerals, while higher pH results in non-alteration or reprecipitation of insoluble calcium- and fluoride-phosphate within the soil (Fig. 21.9). The activities of other species (Fe, Mg, F, Ca, S) no doubt bear on these processes but have not yet been fully assessed.

Being a primary nutrient together with carbon and nitrogen, phosphorus is also intimately involved in the biologic cycle. In soil environments, microbial activity delivers phosphorus for utilization in organic compounds and to combinations with labile organic matter. It is in the latter form, as well as in true solution, that phosphorus is carried from the soil to ocean environments. It appears that a system which produces dissolved phosphorus (pH between 6 and 7), and has limited direct contact with more acidic or basic regimes, would also provide phosphorus for plant growth, as an intermediate stage.

The presence of water in abundance ensures not only an agent for dissolving and transporting phosphorus but also for nurturing organic growth in the soil. Decomposition of soil organic matter yields further labile phosphorus compounds, thus fertilizing the soil. Increased plant growth, yielding CO_2, acidifies the soil and hinders the formation of new apatite minerals. Hence, the commonly slightly acidic soils of high rainfall areas should efficiently strip phosphorus from primary minerals and supply abundant phosphorus to runoff. These processes would be enhanced where rainfall is evenly distributed over the year, because seasonal variation of soil pH would be small.

The present study has shown substantial rainfall over the northern and southern margins of the Tethys in the Middle Cretaceous and the Palaeocene, and the foregoing discussion on weathering suggests that the abundance of phosphorites on the margins of Tethys may be a result of such regional climates. This relationship indicates in many cases an accumulation of phosphorite in areas not far removed from the source area – that is, high phosphorus concentrations may have occurred in limited nearshore areas, not necessarily throughout the world ocean. Estuaries, embayments and shallow parts of the continental shelf may have been effective traps for phosphorus during this interval and possibly at other times in other areas where rainfall was abundant year-round.

It is worth noting as well that during continental flooding, removal of phosphorus from soils in the form of labile organic matter and organic acids would have been enhanced through soil erosion. Any solid material containing phosphorus would have been concentrated in nearshore environments (Arthur & Jenkyns, 1981).

Summary and discussion

A number of previous arguments indicate that a problem in understanding major phosphate deposits and episodic phosphogenesis is the phosphate source. The widespread nature of phosphatization during past time intervals may also argue for environmental control of phosphate deposition in addition to upwelling zones. The plausibility of the existence of physical conditions suitable for upwelling (e.g. likely wind direction, current configurations for dynamic upwelling and bathymetric-current conditions for topographically-induced upwelling) may not be a sufficient criterion for explaining the distribution of all sites. These two factors indicate that a variety of approaches and hypotheses should be examined in order to explain episodic phosphogenesis.

The rate of phosphate supply to the oceans must ultimately depend on continental weathering. Of all the climate model parameters, continental precipitation is apparently the most easily varied. The climate model simulations described here suggest that continental precipitation is a function of geographical configuration and also a function of global warmth. Variations in geography and continental flooding may therefore provide episodic periods of extreme continental weathering. The mechanisms described here are plausible; however, a direct correlation between specific geographies, continental flooding episodes and weathering in specific regions through the Cenozoic has not been demonstrated. The model studies are limited snapshots or sensitivity experiments which may not reflect real situations. Only the conditions for such an occurrence have been described based on model simulations.

The model simulations specifically focus on the size and zonality of Tethys which is partly a function of continental flooding and the association between continental flooding, higher atmospheric CO_2, global warmth and increased precipitation. Geography sets the stage for the location of high precipitation regions, and global warmth accentuates these regions. A zonal Tethys is one specific condition for promoting geographically-controlled precipitation, but other types of continental configurations may have had a similar influence during earth history.

The model simulations described here suggest that continental precipitation is a major variable in earth history, with the potential to influence the sedimentary record. This hypothesis has a bearing on the phosphate problem if episodes of extensive continental precipitation can be related to phosphorus supply to the oceans. The conditions for geographically-controlled continental precipitation may be important for phosphate genesis in two ways: (1) The periods of intensive continental weathering may provide the source of phosphate necessary for the formation of major phosphate deposits; and (2) high runoff may influence continental margin productivity. However, a number of problems exist.

First, enhanced continental precipitation and weathering may be only one of the important controls on phosphogenesis or on the record of major phosphate deposits. Sediment supply, circulation and other factors may have an equal or greater role.

Second, we lack good correlations between the important data elements; specifically, well-mapped global climate, weathering sequences and circulation patterns in concert with the phosphate record.

Finally, the model simulations are flawed in at least three ways: (1) the model ocean is highly simplified and a more realistic simulation may yield greater insight; (2) the models used here lack a seasonal cycle; and (3) the Cenozoic model simulations do not produce a global cooling trend, leaving open the possibility that other factors necessary to produce circulation changes have not been incorporated in the simulations.

The value of the hypothesis described here is not as an answer to the phosphate problem, but rather that it provides directions for future research of significance for greater understanding of the sedimentary record and, potentially, the phosphate problem. A more detailed record of continental weathering of both a spatial and temporal nature should be obtained. The geographic configurations during episodes of phosphogenesis should be investigated in greater detail with respect to geographic control of precipitation patterns. The exceptions to the associations between continental flooding, global warmth, geographically-controlled precipitation and occurrence of phosphates must be examined to determine whether the conditions described here are of primary interest or are simply some of the variables in a complex interplay of many factors. The model-generated variability in precipitation indicates that, at a minimum, precipitation, runoff and episodic weathering must be considered as an integral factor in explaining major phosphate deposits.

Acknowledgments

The authors greatly appreciate the efforts of Stan Riggs and William Burnett in organizing the stimulating conference on Neogene Phosphates that prompted this contribution. We greatly appreciate critical reviews by R.P. Sheldon, A.M. Ziegler and S.R. Riggs. Eric Barron acknowledges support from the National Science Foundation, grant ATM-8521209.

References

Arthur, M.A. & Jenkyns, H.C. (1981). Phosphorites and paleoceanography. *Oceanologica Acta*, **4 Supplement**, 83–96.

Barron, E.J. (1985). Explanations of the Tertiary global cooling trend. *Palaeogeography, Palaeoclimatology, Palaeoecology*, **50**, 45–61.

Barron, E.J., Arthur, M.A. & Kauffman, E.G. (1985). Cretaceous rhythmic bedding sequences: a plausible link between orbital variations and climate. *Earth Planetary Science Letters*, **72**, 327–40.

Barron, E.J., Harrison, C.G.A., Sloan, J.L. & Hay, W.W. (1981). Paleogeography, 180 million years ago to the present. *Eclogae Geologicae Helvetiae*, **74**, 443–70.

Barron, E.J. & Washington, W.M. (1982). Cretaceous climate: a comparison of atmospheric simulations with the geologic record. *Palaeogeography, Palaeoclimatology, Palaeoecology*, **40**, 103–33.

Barron, E.J. & Washington, W.M. (1984). The role of geographic variables in explaining paleoclimates: results from Cretaceous climate model sensitivity studies. *Journal of Geophysical Research*, **89**, 1267–79.

Barron, E.J. & Washington, W.M. (1985). Warm Cretaceous climates: high atmospheric CO_2 as a plausible mechanism. In *The Carbon Cycle and Atmospheric CO_2: Natural Variations Archean to Present*, ed. E.T. Sundquist & W.S. Broecker, pp. 546–53. American Geophysical Union, Washington, DC.

Beeson, D.C. (1984). 'The relative significance of tectonics, sea level fluctuations, and paleoclimate to Cretaceous coal distribution in North America.' National Center for Atmospheric Research Cooperative Thesis No. 83. National Center for Atmospheric Research, Boulder, Colorado.

Berner, R.A., Lasaga, A.C. & Garrels, R.M. (1983). The carbonate–silicate geochemical cycle and its effect on atmospheric carbon dioxide over the last 100 million years. *American Journal of Science*, **283**, 641–83.

Bourke, W., McAvaney, B., Puri, K. & Thurling, R. (1977). Global modeling of atmospheric flow by spectral methods. In *Methods in Computational Physics, General Circulation Models of the Atmosphere*, ed. J. Chang, pp. 267–324. Academic Press, New York.

Burnett, W.C. (1977). Geochemistry and origin of phosphorite deposits from off Peru and Chile. *Geological Society of America Bulletin*, **88**, 813–23.

Burnett, W.C. & Oas, T.G. (1979). Environment of deposition of marine phosphate deposits off Peru and Chile. In *Proterozoic–Cambrian Phosphorites*, ed. P.J. Cook & J.H. Shergold, pp. 54–6, Australian National University Press, Canberra.

Chamley, H. (1979). North Atlantic clay sedimentation and paleoenvironment since the Late Jurassic. In *Deep Drilling Results from the Atlantic Ocean: Continental Margins and Paleoenvironment*, ed. M. Talwani, W. Hay & W.B.T. Ryan, pp. 342–61. American Geophysical Union, Washington, DC.

Cole, C.V. & Stewart, J.W.B. (1983). Impact of acid deposition on P cycling. *Environmental and Experimental Botany*, **23**, 235–41.

Cook, P.J. & McElhinny, M.W. (1979). A re-evaluation of the spatial and temporal distribution of sedimentary phosphate deposits in light of plate tectonics. *Economic Geology*, **74**, 315–30.

Frakes, L. (1979). *Climates Throughout Geologic Time*. Elsevier, Amsterdam.

Froelich, P.N., Bender, M.L., Leudtke, N.A., Heath, G.R. & DeVries,

T. (1982). The marine phosphorus cycle. *American Journal of Science*, **282**, 474–511.

Hallam, A. (1985). A review of Mesozoic climates. *Journal of the Geological Society*, **142**, 433–45.

Khain, V.E. & Ronov, A.B. (1960). World paleogeography and lithological associations of the Mesozoic Era. *Report of the 21st International Geological Congress*, **12**, 152–64.

Lindsay, W.L. (1979). *Chemical Equilibria in Soils*. John Wiley, New York.

McAvaney, B.J., Bourke, W. & Puri, K. (1978). A global spectral model for simulation of the general circulation. *Journal of the Atmospheric Sciences*, **35**, 1557–82.

Pitman, W. (1978). Relationship between eustasy and stratigraphic sequences of passive margins. *Geological Society of America Bulletin*, **89**, 1389–403.

Ramanathan, V., Pitcher, E.J., Malone, R.C. & Blackman, M.L. (1983). The response of a spectral general circulation model to refinements in radiative processes. *Journal of the Atmospheric Sciences*, **40**, 605–30.

Riggs, S.R. (1984). Paleoceanographic model of Neogene phosphorite deposition, US Atlantic continental margin. *Science*, **223**, 123–31.

Sheldon, R.P. (1964). Paleolatitudinal and paleogeographic distribution of phosphorite. *US Geological Survey Professional Paper*, **501-C**, C106–13.

Sheldon, R.P. (1981). Ancient marine phosphorites. *Annual Review of Earth and Planetary Sciences*, **9**, 251–84.

Sheldon, R.P. (1986). *Opening and closing of the Circumglobal Tethys Seaway, and Late Cretaceous–Early Tertiary Tethyan Phosphorite Deposition*. International Geological Correlation Program Project 156 Workshop, March, 1986. Caracas, Venezuela.

Southam, J.R. & Hay, W.W. (1981). Global sedimentary mass balance and sea level changes. In *The Sea*, vol. 7, ed. C. Emiliani, pp. 1617–84. Wiley Interscience, New York.

Tarling, D.H. (1978). The geological–geophysical framework of ice ages. In *Climatic Change*, ed. J. Gribben, pp. 3–24, Cambridge University Press, Cambridge.

Washington, W.M. & Meehl, G.A. (1983). General circulation model experiments on the climatic effects due to a doubling and quadrupling of carbon dioxide concentration. *Journal of Geophysical Research*, **88**, 6600–10.

Washington, W.M. & Williamson, D.L. (1977). A description of the NCAR global circulation models. In *Methods in Computational Physics, General Circulation Models of the Atmosphere*, ed. J. Chang, pp. 111–72. Academic Press, New York.

Ziegler, A.M., Scotese, C.R. & Barrett, S.F. (1982). Mesozoic and Cenozoic paleogeographic maps. In *Tidal Friction and the Earth's Rotation*, ed. P. Brosche & J. Sudermann, pp. 240–52. Springer-Verlag, New York.

22

Neogene geochemical cycles: implications concerning phosphogenesis

L.R. KUMP

Abstract

Isotopic and geologic indicators of paleoenvironments record significant excursions in the climate, depositional setting, and global biogeochemical cycles of the Neogene. The stable-isotope age curves of carbon and sulfur peak in the Middle Miocene after 40 my of increase, and then fall to today's values. Based on an understanding of the coupled operation of the cycles of carbon and sulfur and their effect on atmospheric oxygen, it appears that oxygen levels were significantly higher in the Miocene than today. This conclusion is supported by the results of a quantitative model of the global biogeochemical cycle which considers not only these isotope curves but also the records of global mean seafloor-spreading rates and land area, the subduction and metamorphism of sediments, and the exchange of calcium and magnesium at mid-ocean ridges. Notable changes in continental weathering rates, depositional settings for sediments, and global climate are indicated by the isotopic records of strontium and oxygen and the eustatic sea-level curve for the Neogene, and support the notion that the Neogene was the culmination of a number of important Tertiary trends in global geochemical cycling. It is speculated that relationships may exist among higher atmospheric oxygen levels, increased aerobic decay and fire on land and phosphogenesis.

Introduction

Several isotopic and geologic records of the Cenozoic highlight the Miocene as a time of minima or maxima, of turnarounds and excursions in values. A representative case is the isotopic record of marine sulfate, preserved in evaporites (Fig. 22.1A; Lindh, 1983). Throughout the Paleogene and early Neogene, $\delta^{34}S_{ocean}$ (the standardized ratio of $^{34}S/^{32}S$ in the ocean) increased, rising from an early value of 17‰ in the Paleocene to a peak value of 22‰. This pattern may be related to the dearth of Tertiary and abundance of Miocene evaporites (Holser, 1984). According to Ronov (1982), the Miocene evaporite mass exceeds that of the entire Paleogene. The implications for global ocean chemistry will be discussed below.

Another significant Miocene maximum was achieved in the isotopic record of marine inorganic carbon ($\delta^{13}C$), preserved in $CaCO_3$ deposits (Fig. 22.1B). Lindh's (1983) $\delta^{13}C$ curve displays a less impressive peak in the Miocene than does that of Shackleton (1985), whose data show a peak that is the culmination of a slow increase in $\delta^{13}C$ throughout the Tertiary. Both curves drop sharply from the Miocene to the Recent.

Figure 22.1C records the oxygen isotopic composition of marine planktonic and benthic foraminifera (cf. Shackleton & Kennett, 1975; Berger, Vincent & Thierstein, 1981; Vincent & Berger, 1985). The Miocene displays a re-initiation in the Cenozoic trends towards increasing $\delta^{18}O$ which was abandoned during the Oligocene.

The global rate of seafloor spreading as a function of time has been calculated based on the age distribution of the seafloor and on analyses of the paleo-poles of plate rotations (cf. Southam & Hay, 1977; Pitman, 1978; Kominz, 1984). The global mean spreading rate curve of Kominz (1984; Fig. 22.1D) shows a large drop during the Early Eocene, fluctuating rates during the Miocene, and a gradual rise to the present value from the Late Miocene minimum.

Vail & Hardenbol (1979) have produced a detailed Tertiary eustatic sea-level curve largely based on the stratigraphic interpretation of seismic data. As shown in Figure 22.1E, the Paleogene was a period of high global sea level that oscillated about a slowly decreasing mean of around 150–200 m above the present level. The most dramatic Cenozoic sea-level event occurred in the Late Oligocene when sea level fell more than 220 m in a few million years or less. Sea level then rose to a small Early Miocene maximum, fell slightly, then rose steadily to a major Middle Miocene maximum of about 120 m above present at a time when there was extensive deposition of phosphorites in Florida and North Carolina (Riggs, 1984). The Late Miocene was a time of continual fall in sea level. A sharp spike in the sea-level curve appears in the Early Pliocene which coincides with another phosphorite event (Riggs, 1984); the dramatic Pleistocene swings in sea level are not shown in this plot.

Finally, Figure 22.1F (Palmer & Elderfield, 1985) shows the isotopic ratio of radiogenic $^{87}Sr/^{86}Sr$ in carbonate rocks, a measure of the relative importance of continental weathering rates over isotopic exchange during marine hydrothermal activity. During the Paleogene, this ratio slowly increased; after and perhaps as a result of the tremendous sea-level fall of the Late Oligocene it began to increase rapidly, assuming a new, larger, steady value in the Late Miocene. In the post-Miocene Cenozoic, $^{87}Sr/^{86}Sr$ began to increase again. Today's ratio is the highest achieved during the entire Cenozoic.

In summary, the Middle Miocene was a time in which peaks were observed in $\delta^{34}S$, $\delta^{13}C$, $^{87}Sr/^{86}Sr$, eustatic sea level and

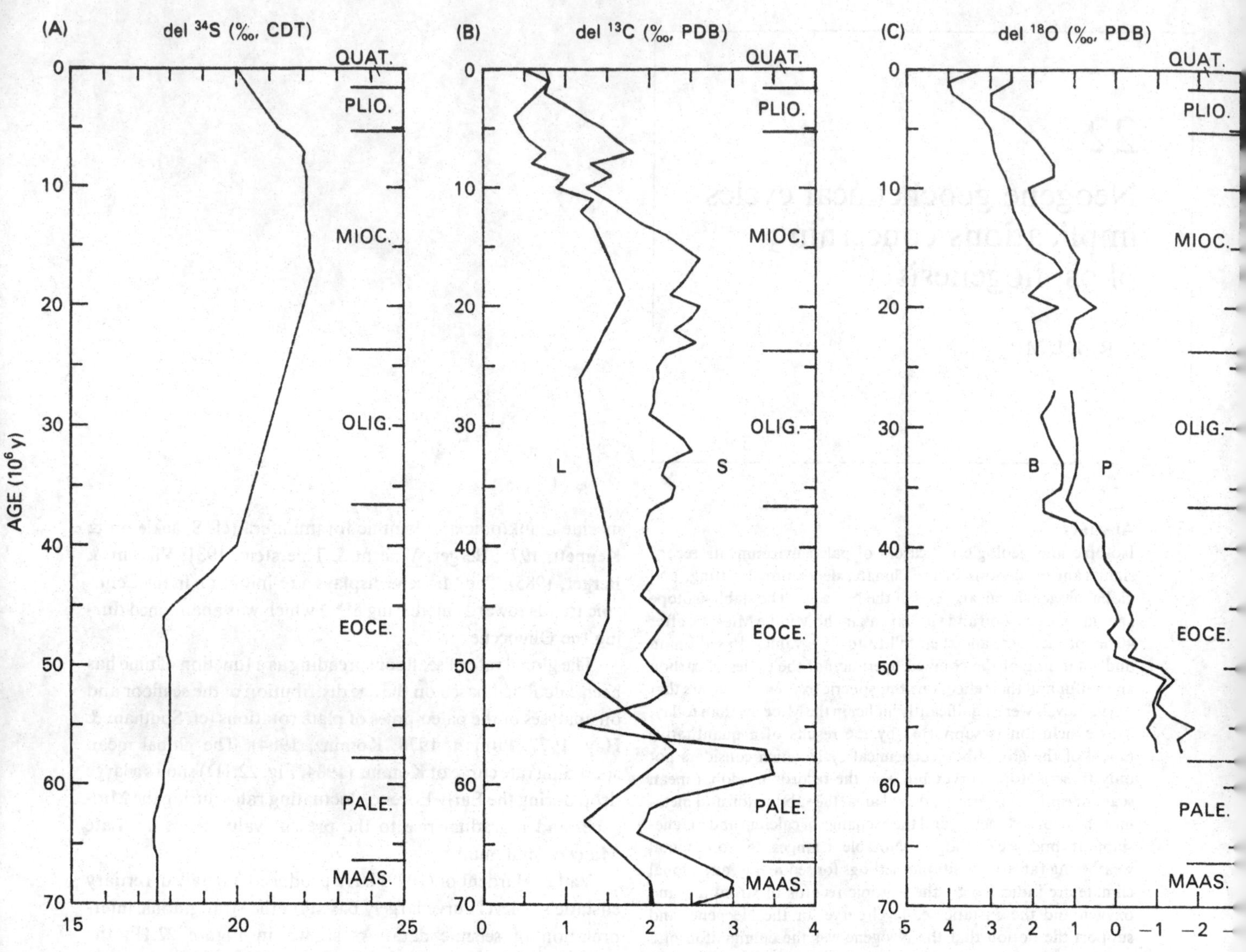

Fig. 22.1. The isotopic and geologic records of change in the global geochemical cycle during the Cenozoic. (A) The isotopic ratio of $^{34}S/^{32}S$ in marine evaporites, expressed as ‰ relative to the Canyon Diablo Troilite (CDT) standard (Lindh, 1983). (B) The isotopic ratio of $^{13}C/^{12}C$ in marine carbonates, expressed as ‰ relative to the Peedee Belemnite (PDB) standard (L, Lindh, 1983; S, Shackleton, 1985). (C) The isotopic ratio of $^{18}O/^{16}O$ in marine carbonates, expressed as ‰ relative to PDB (P, planktonic, B, benthic foraminifera; Shackleton & Kennett, 1975; graph taken from Vincent & Berger, 1985). (D) The global mean oceanic spreading rate (cm y^{-1}) (Kominz, 1984). (E) Eustatic sea level (m) relative to today (Vail & Hardenbol, 1979) (F) The isotopic ratio of $^{87}Sr/^{86}Sr$ in marine carbonate rocks (Palmer & Elderfield, 1985).

mean, global seafloor-spreading rate, and in which a trend towards increasing $\delta^{18}O$ was initiated.

Although it is tempting to relate these curves intuitively to changes in global geochemical cycling, productivity and climate, the intricate interdependence of the various processes involved necessitates the use of quantitative models to unravel the responses of the global system to the changes indicated by the curves.

This paper will present a discussion of the feedbacks that may exist in the global geochemical cycle that have a bearing on climate and phosphogenesis. First, the coupled carbon–sulfur cycle is described. The feedbacks in this system control atmospheric oxygen levels and influence atmospheric carbon dioxide. The other controls on carbon dioxide are then presented in the form of a global feedback model. Finally, the implications for interpretations of the Neogene world are addressed, and model calculations of atmospheric carbon dioxide and oxygen levels in the Cenozoic are presented.

Carbon–sulfur cycling

With the exception of iron, an element whose global, geochemical redox cycle is rather poorly understood, the control of the redox state of the earth's surface rests entirely in the global cycles of carbon and sulfur (cf. Holland, 1973; Garrels & Perry, 1974; Schidlowski, Junge & Pietrek, 1977). Sedimentary carbon and sulfur occur in both an oxidized and reduced state, and these states are generally treated as separate reservoirs in the sedimentary rock mass (Fig. 22.2).

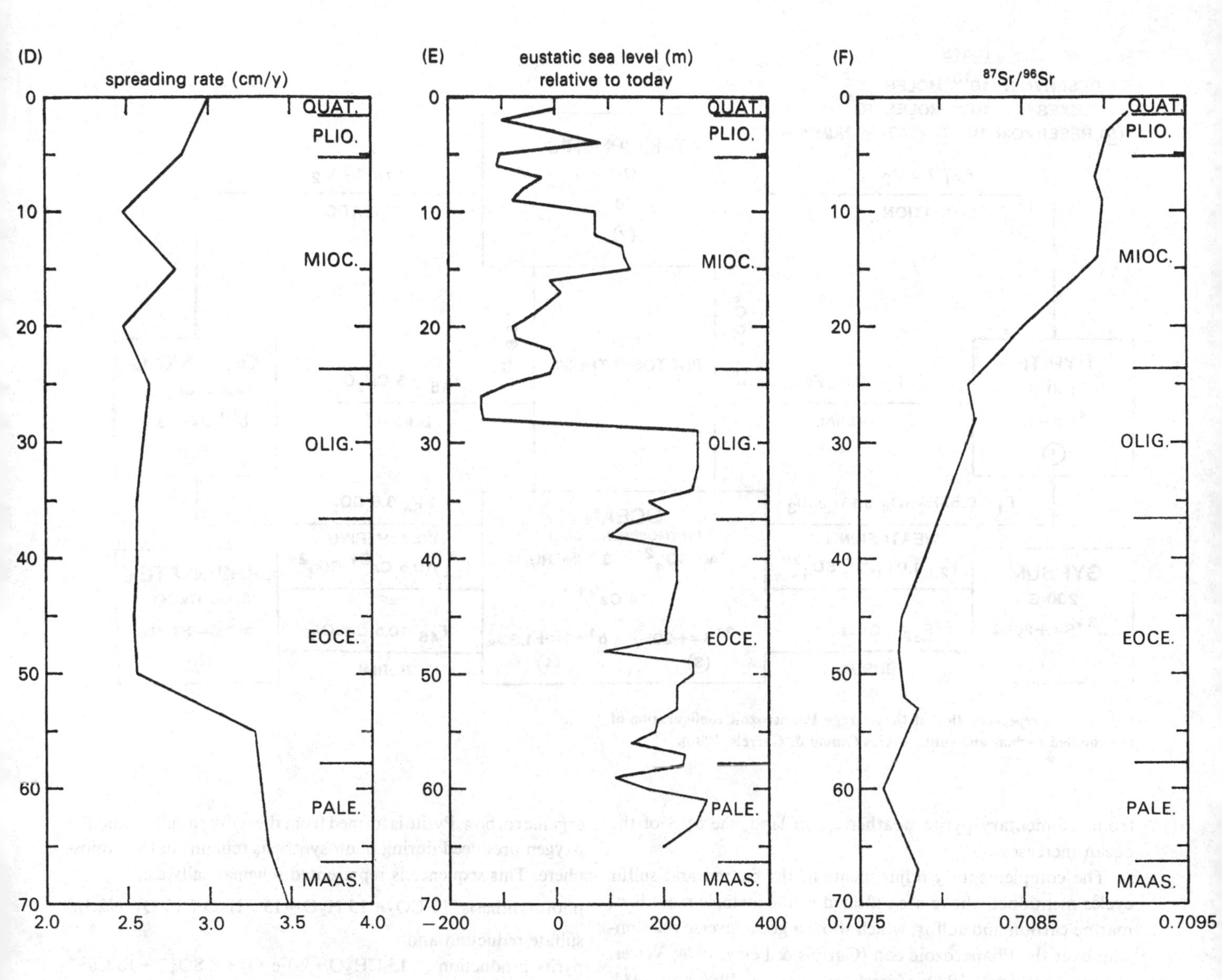

Carbon in its oxidized form is found in sedimentary carbonates, especially limestone and dolomite. The reduction of carbon occurs during photosynthesis, and the accumulation of buried organic carbon comprises the reduced carbon reservoir. The ocean, atmosphere, and biosphere contain very little carbon in comparison.

The global reservoirs of oxidized sulfur include evaporite deposits of gypsum and anhydrite, and the rather substantial quantity of dissolved sulfate in the ocean. Bacteria reduce oceanic, porewater sulfate to sulfide in anoxic sediments, and through combination with reduced iron, sedimentary pyrite is formed.

Global environmental changes affect the partitioning of the total amounts of cycling carbon and sulfur between their reduced and oxidized reservoirs. These changes have acted over time scales of tens of millions of years to produce complementary adjustments in the carbon and sulfur cycles; periods of high burial rates of organic carbon, and thus transfer of sedimentary carbon from the oxidized to reduced reservoir, have, in general, coincided with periods of high precipitation rates of evaporitic, oxidized sulfur, and thus transfer of sedimentary sulfur from the reduced to oxidized reservoirs, and vice versa.

Organic carbon in recent marine sediments is approximately 25‰ more enriched in ^{12}C over ^{13}C than is the inorganic, oxidized carbon in seawater (Deines, 1980). In contrast, carbonates are deposited with essentially the same $\delta^{13}C$ as seawater. Thus, the burial of organic carbon in sediments provides a sink for ^{12}C, and the ocean becomes isotopically 'heavier' (more enriched in ^{13}C). Sedimentary organic carbon weathering on land provides a source of ^{12}C which tends to 'lighten' the ocean; the imbalance between burial and weathering of organic carbon determines the net effect on ocean carbon-isotope composition.

An analogous situation occurs in the sulfur cycle, with pyrite sulfur playing the role of organic carbon, and gypsum (and anhydrite) that of carbonate. The bacterial reduction of oceanic sulfate to sulfide occurs with a fractionation of about 35‰ (i.e. $\delta^{34}S_{ocean} = \delta^{34}S_{pyrite} - 35‰$), while the precipitation of evaporite sulfur leads to a relatively insignificant fractionation of sulfur isotopes (Holser & Kaplan, 1966; Claypool *et al.*, 1980). If the depositional rate of pyrite exceeds the supply of light sulfur

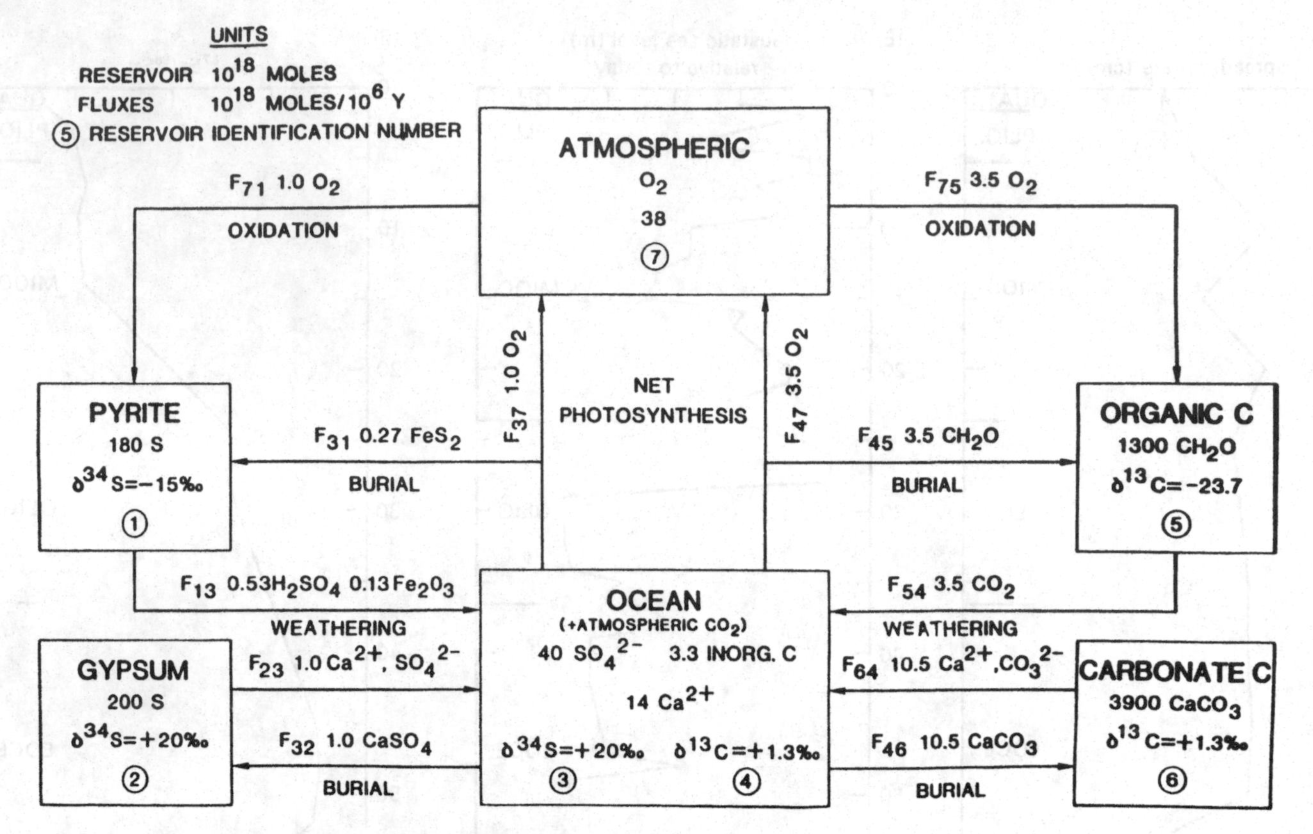

Fig. 22.2. A representation of the average Phanerozoic configuration of the coupled carbon and sulfur cycles (Kump & Garrels, 1986).

from sedimentary pyrite weathering on land, the $\delta^{34}S$ of the ocean increases.

The complementary adjustments in the carbon and sulfur cycles mentioned above are reflected in the isotopic records of marine carbon and sulfur, which show a good inverse relationship over the Phanerozoic eon (Garrels & Perry, 1974; Veizer, Holzer & Wilgus, 1980; Garrels & Lerman, 1981 and 1984; Berner & Raiswell, 1983; Lindh, 1983; Holland, 1984).

An important consequence of the operation of the carbon and sulfur cycles is the regulation of atmospheric oxygen levels (Kump & Garrels, 1986 and references therein). The burial of organic carbon represents an excess of photosynthesis over respiration and decay, and thus the production of oxygen:

$$CO_2 + H_2O \xrightarrow{h\nu} CH_2O + O_2 \qquad (22.1)$$

The weathering of organic carbon (CH_2O) is an oxidation reaction, and is merely the reverse of equation (22.1). Growth of the sedimentary CH_2O reservoir therefore leads to the net production and increase of atmospheric oxygen, and vice versa.

In a less apparent fashion, the burial of pyrite sulfur also releases oxygen to the atmosphere. The process begins with photosynthesis (eqn (22.1)), which liberates oxygen. During aerobic respiration an equivalent amount of oxygen is used to oxidize the organic matter recently photosynthesized, and there is no net effect on atmospheric oxygen. However, if the organic matter is deposited in anoxic sediments, a portion of it is consumed by anaerobic bacteria that utilize SO_4^{2-} ions rather than dissolved oxygen as the electron acceptor in oxidation of the organic carbon. Pyrite is formed from the reduced sulfur, and the oxygen produced during photosynthesis remains in the atmosphere. This sequence is represented schematically as:

photosynthesis $15\ CO_2 + 15\ H_2O \rightarrow 15\ CH_2O + 15\ O_2$ (22.1*a*)

sulfate reduction and pyrite production

$$15\ CH_2O + 2\ Fe_2O_3 + 8\ SO_4^{2-} + 16\ Ca^{2+} + 16\ HCO_3^- \rightarrow 4\ FeS_2 + 16\ CaCO_3 + 15\ CO_2 + 23\ H_2O \qquad (22.2)$$

net reaction

$$8\ SO_4^{2-} + 16\ Ca^{2+} + 2\ Fe_2O_3 + 16\ HCO_3^- \rightarrow 4\ FeS_2 + 16\ CaCO_3 + 8\ H_2O + 15\ O_2 \qquad (22.3)$$

The weathering of sedimentary pyrite consumes oxygen according to the reverse of equation 22.3. As with growth of the organic carbon reservoir, growth of the sedimentary pyrite reservoir (implying an excess of deposition over weathering) indicates an increase in atmospheric oxygen. The sum of the contributions from the growth of the reduced carbon and sulfur reservoirs determines the change of atmospheric oxygen according to:

$$\Delta M(O_2) = \Delta M(CH_2O) + (2 \cdot 15/8 \cdot \Delta M(FeS_2)) \qquad (22.4)$$

where $M(x)$ is the number of moles in the x reservoir.

The inverse trends in the long-term records of carbon and sulfur show that atmospheric oxygen fluctuations have been moderated to some extent by the compensatory operation of the carbon and sulfur cycles (Holland, 1973; Garrels & Perry, 1974; Schidlowski *et al.*, 1977; Veizer *et al.*, 1980; Garrels & Lerman, 1981 and 1984; Schidlowski & Junge, 1981; Berner & Raiswell, 1983; Lindh, 1983; Holland, 1984; Kump & Garrels, 1986). Indeed, this must be so, for given the relatively short residence

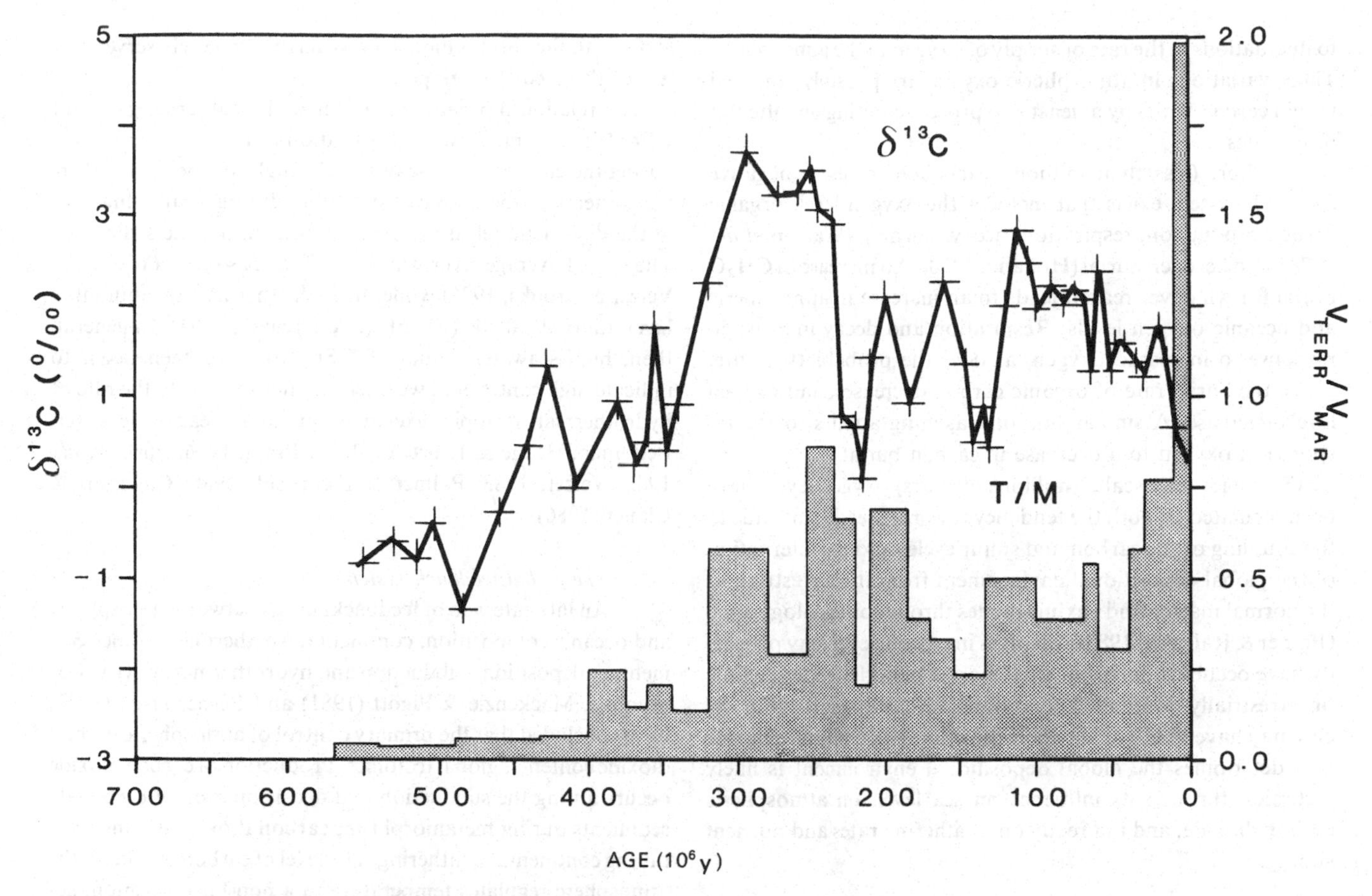

Fig. 22.3. Demonstrated relationship between $\delta^{13}C$ (+, Lindh, 1983) and the ratio of the volumes of terrigenous and marine sediments deposited in a given period (bars, Ronov, 1982) as suggested by R.M. Garrels (pers. comm.).

time of atmospheric oxygen (about 10 my) if the carbon and sulfur cycles were not roughly compensatory over the long term, oxygen levels would fluctuate rapidly between low and high values intolerable to complex life.

Organic carbon burial

The burial rate of organic carbon over millions of years is essentially regulated by the rate of nutrient supply (primarily of phosphorus) from rock weathering and the efficiency of utilization of these nutrients. The global depositional rate of organic carbon, designated as F_c^g, can be resolved into the component supplied by terrestrial CH_2O production, F_c^t, which may be buried on land or beneath the sea, and its marine component, F_c^m:

$$F_c^g = F_c^t + F_c^m \tag{22.5}$$

The global phosphorus weathering–supply rate, F_p, and its removal rate with CH_2O burial can be written:

$$F_p = F_c^t (P/C)_{org,t} + F_c^m (P/C)_{org,m} \tag{22.6}$$

where $(P/C)_{org,t}$ is the phosphorus: carbon ratio in buried terrestrial organic matter, and $(P/C)_{org,m}$ is the phosphorus: carbon ratio in buried marine organic matter.

Since the terrestrial phosphorus: carbon ratio is much smaller than the marine ratio (cf. Sholkovitz, 1973; Froelich *et al.*, 1982), an increase in F_c^t relative to F_c^m under constant nutrient supply would lead to an increase in F_c^g according to equations (22.5) and (22.6). Conversely, an increase in the removal of nutrients during marine burial, F_c^m, could well result in a decrease in F_c^t and F_c^g under constant F_p. Only with an increase in F_p could an increase in F_c^m lead to an increase in F_c^g.

It is suggested then that a significant part of the variation in the global depositional rate of organic carbon, and thus in the carbon isotopic composition of the ocean, is due to change in the ratio of the supply of terrestrial to marine organic matter to the sediments. If this is so one should be able to observe two phenomena in the geologic record: (1) a correlation between the carbon isotopic record and the ratio of terrestrial to marine sediment deposition; and (2) a lack of correlation prior to the rise of land plants.

Figure 22.3 (R.M. Garrels, pers. comm., 1986) plots both the $\delta^{13}C$ age curve (Lindh, 1983) and the terrestrial: marine ratio for the Phanerozoic (data from Ronov, 1982). Both of the phenomena described above can be observed in this figure, especially during the Paleozoic, which suggests that indeed the global depositional rate of organic carbon is controlled at least in part by the source of organic carbon. Terrestrial organic carbon, with its large C:P ratio, utilizes the available nutrients more efficiently, thus allowing a larger global burial flux of organic carbon.

Variation in the source of CH_2O may then be expected to lead to fluctuations in the rate of supply of oxygen to the atmosphere. These variations in atmospheric oxygen are possibly confined within certain limits by at least two processes acting on different time scales.

On short (less than million year) time scales, a negative feedback system exists that includes the oxygen level, organic carbon production, respiration, decay, burning (Watson *et al.*, 1978), and kerogen burial (Holland, 1978). An increase in CH_2O burial for whatever reason leads to an increase in atmospheric and oceanic oxygen levels. Respiration and decay increase in response to increased oxygen, as does the probability of fire; hence the burial rate of organic carbon decreases, and oxygen levels decrease. A similar line of reasoning applies to the response of oxygen to a decrease in carbon burial.

On longer time scales (millions of years) oxygen levels have been regulated by both the tendency towards steady state due to the coupling of the carbon and sulfur cycles and the alternation of the global depositional environment from the terrestrial, to the normal marine and euxinic modes throughout geologic time (Berner & Raiswell, 1983). Changes in oxygen level may primarily have occurred during normal marine periods; under euxinic or terrestrially-dominated conditions the carbon and sulfur cycles may have been roughly compensatory. The ultimate factor that determines the global depositional environment is likely tectonics, through its influence on sea level, on atmospheric carbon dioxide, and indirectly on weathering rates and nutrient supply.

Tectonic, climate and the global geochemical cycle

Climate and weathering

The global geochemical cycling models of Berner *et al.* (1983), Lasaga, Berner & Garrels (1985) and the extension presented here do not utilize all of the historic records presented in Figure 22.1. In particular, variations in global mean temperatures and continental weathering rates, reflected in the isotopic records of oxygen and strontium, respectively (e.g. Veizer, 1983), are not part of the model. Thus these records can be used as independent checks of the model results if the reasons for their secular trends are well understood.

The response of $\delta^{18}O$ to global temperature changes arises from the temperature dependence of the isotope fractionation of oxygen (Urey, 1947). Other factors influence the sedimentary carbonate $\delta^{18}O$, inluding changes in oceanic $\delta^{18}O$ due to continental ice buildup or melting (cf. Emiliani, 1955; Shackleton & Opdyke, 1973), biological 'vital' effects on the fractionation factor (Urey *et al.*, 1951), and isotopic exchange during carbonate diagenesis (see summary in Bathurst, 1975). If these factors are isolated from consideration, it is possible to assign temperatures corresponding to the $\delta^{18}O$ values of Figure 22.1F, with more negative values (to the right) indicating warmer global temperatures.

Higher temperatures are assumed to lead to increases in weathering rates as a result of the stimulation of microbial activity in soils; aerobic bacteria oxidize soil organic carbon to carbon dioxide which is then consumed during rock weathering. In the Berner *et al.* (1983) model the global mean temperature is calculated from the mass of atmospheric carbon dioxide via a 'greenhouse' equation. The obvious alternative would be to use $\delta^{18}O$; with the current model, however, $\delta^{18}O$ merely serves as a test of the predicted temperature.

The relationship between continental weathering rate and $^{87}Sr/^{86}Sr$ is complicated by variations in isotopic exchange during the circulation of seawater through seafloor hydrothermal systems, in the source of strontium during weathering, and in the diagenetic release of strontium from marine sediments. The global average river water $^{87}Sr/^{86}Sr$ is ~ 0.711 (Wadleigh, Veizer & Brooks, 1985) while the hydrothermal strontium flux has a ratio of ~ 0.704 (Elderfield & Greaves, 1981). In general, then, high seawater values of $^{87}Sr/^{86}Sr$ have been taken to indicate high continental weathering rates relative to the rate of hydrothermal isotopic exchange at mid-ocean ridges (cf. Peterman, Hedge & Tourtelot, 1970; Brass, 1976; Burke *et al.*, 1982; Veizer, 1983; Palmer & Elderfield, 1985; Chadhuri & Clauer, 1986).

The global feedback system

An intricate web of feedbacks exists between atmospheric and oceanic composition, continental weathering, marine sedimentary deposition, subduction and hydrothermal activity. For example, Mackenzie & Pigott (1981) and Berner *et al.* (1983) have concluded that the primary control of atmospheric carbon dioxide content is global tectonics. Production of carbon dioxide occurs during the subduction and decarbonation of carbonate sediments during metamorphism; carbon dioxide is consumed during continental weathering. The level of carbon dioxide in the atmosphere regulates temperature in a non-linear fashion, according to the 'greenhouse' principle. Temperature in turn affects weathering rates; thus an increase in carbon dioxide leads to a global warming and increased consumption of carbon dioxide during weathering.

However, if increases in carbon dioxide are largely due to increases in the rates of subduction and decarbonation, and if in turn, these are a function of oceanic-spreading rates, the net effect of increased weathering rates due to high carbon dioxide and thus high temperature may be moderated by the rise of sea level and decrease in land area (especially on the emergent continental shelves) associated with fast spreading rates (Pitman, 1978), though there may be a time lag (Heller & Angevine, 1985). In addition, an increase in weathering rates on land may lead to greater nutrient supply and organic carbon burial rates, which would tend to draw down atmospheric carbon dioxide via its consumption during photosynthesis. The weathering of organic carbon produces carbon dioxide and this may tend to offset the depletion due to weathering of other crustal materials. It becomes apparent that at this level of complexity, intuition may fail, and numerical box models have been developed to deal quantitatively with the system of fluxes (e.g. Berner *et al.*, 1983; Lasaga *et al.*, 1985). In the next section some results from an example of this type of model will be presented.

Tentative interpretations of the Miocene world

From a global modeling perspective the Neogene provides an ideal set of conditions under which a geochemical cycling model can be tested. It is a time during which many of the expected long-term relationships, such as the inverse relationship between $\delta^{13}C$ and $\delta^{34}S$, are not observed. The Miocene epoch also exhibits an interesting correspondence between several of the age curves in Figure 22.1. It appears, for example, that the eustatic sea-level rise during the period of 20–15 my BP, when many of the world's major phosphorites formed, was associated with an increase in the global mean spreading rate. Increased spreading may have led to an increase in the carbon dioxide content of the atmosphere and in continental weathering rates, as suggested by the strontium-isotope record. The global organic carbon and pyrite burial rates were likely very high during this period (given the peaks in their isotope records), indicating an increase in the nutrient supply rate or an increase in the proportion of terrestrial organic carbon burial.

From 15–10 my BP the trends were reversed. Sea level fell as did the rate of seafloor spreading. $\delta^{13}C$ dropped sharply, but, within the resolution of the record, $\delta^{34}S$ was maintained at a high level. A global cooling step occurred over this period (Berger *et al.*, 1981), according to the $\delta^{18}O$ record. Little change in continental weathering rates is indicated by the Sr-isotope curve, however.

The Late Miocene patterns in many ways reflect a miniature event of the type that occurred during the Early Miocene. Sea level rose and fell again, as spreading rates increased; global temperatures and CH_2O burial rates also peaked during this period. There is little evidence of change in the sulfur or strontium cycles during this period.

The remainder of the Neogene record suggests a continual cooling of the global climate, significant drops in organic carbon and pyrite burial rates, and a general increase in continental weathering rates, all perhaps associated with the onset of widespread glaciation.

For the Neogene, at least, there is a good, direct correspondence between the carbon- and oxygen-isotope curves. As more negative $\delta^{18}O$ values and more positive $\delta^{13}C$ values coincide, the suggestion is that warm climates were contemporaneous with times of increased organic carbon burial, either through increased nutrient supply or increased burial of terrestrial organic carbon.

The paucity of Paleogene and abundance of Neogene evaporites may have great implications for the oceanic chemistry of Na, Cl, Ca and SO_4, and indirectly, for the nutrient chemistry of the ocean during the Cenozoic. In the absence of the globally-significant deposition of evaporites during the first 40 million years of the Cenozoic, an important removal mechanism for these elements was not active, and their concentrations in the ocean and ocean salinity conceivably increased. What effect this change in ocean chemistry would have had on phosphorite solubility in the ocean has yet to be adequately modeled. The rough correspondence between periods of major evaporites (Early Cambrian, Permo-Triassic, Cretaceous, and Miocene) and phosphorites suggests a causal link between the two, although other important factors, especially the development of suitable basins for evaporite deposition, and the facies relationship between phosphorites and evaporites, may explain the co-occurrence of these rock types (Hite, 1978). As the Neogene evaporites were deposited ocean salinity perhaps dropped, and the major ions assumed their present concentrations.

The above interpretations are provided to demonstrate the interrelationships between the observed isotopic and geologic records and the conditions they are believed to record. They are in no way definitive, and alternate interpretations are indeed possible. As an example, Palmer & Elderfield (1985) argue that the Cenozoic increase in $^{87}Sr/^{86}Sr$ was a result of a change in the isotopic ratio of the continental source rather than an overall increase in weathering rates. Work in progress by R.B. Stallard and R.J. Murnane at Princeton University on an alternative indicator of the relative importance of continental weathering rates, the Ge:Si ratio in deep-sea sediments, may help resolve this issue (Murnane, Mortlock & Stallard, 1984).

The carbon and sulfur isotopic records are consistent with the proposition that oxygen levels were higher in the Miocene than today. The sharp drop in $\delta^{34}S$ and $\delta^{13}C$ from the Miocene to the present reasonably represents a significant reduction in the rate of O_2 production from lowered CH_2O and FeS_2 burial, relative to its consumption during weathering, i.e. a significant drop in atmospheric O_2. Numerical models have been developed to gauge the effect of these isotopic changes, and they do indeed predict higher O_2 levels during the Miocene (Shackleton, 1985; Kump & Garrels, 1986).

We have recently incorporated the isotope balance equations into the Berner, Lasaga & Garrels (BLAG) carbonate–silicate geochemical model (Berner *et al.* 1983), relying heavily on the work of Lasaga *et al.* (1985) but modifying their treatment by assuming that the weathering rate of the reduced sedimentary components (CH_2O and FeS_2) is proportional both to the reservoir size and to the amount of oxygen in the atmosphere (Kump & Garrels, 1986).

Figures 22.4 and 22.5 show the calculated burial rates of CH_2O and FeS_2 for the last 100 my. These rates are certainly higher in the earlier Neogene than today and show distinct peaks in the Miocene. However, earlier Tertiary and Cretaceous events perhaps are more spectacular.

As a consequence of the high burial rates of reduced sediments, oxygen production was high in the Miocene, and the Miocene displays a Cenozoic peak in the atmospheric oxygen mass (Fig. 22.6). This result was shown by Shackleton (1985) as well. It is also consistent with a dramatic increase in charcoal production which occurred during the Cenozoic (Herring, 1985). As a CO_2-consuming process the increased burial rate of organic carbon tended to counteract the increased production of carbon dioxide due to the higher seafloor-spreading rates of the Middle Miocene (Fig. 22.7).

An implication of low carbon dioxide is that weathering rates, which were generally on the rise throughout much of the Cenozoic due to an increase in land area, may have been somewhat supressed during much of the Miocene. This is indicated by Figure 22.8, which shows the weathering rate of the $MgSiO_3$ reservoir as a function of time. Because of the tremendous size of this reservoir it is virtually unaffected by fluxes to and from it, so that its weathering rate in the model is responsive only to land area and P_{CO_2}. Thus Figure 22.8 provides a good measure of global trends in weathering rates. The predicted general increase in this rate over the last 40 million years is substantiated by the Sr-isotope record (Fig. 22.1F) which displays a similar trend.

An important but difficult question to address is how increased atmospheric oxygen in the Miocene may have affected the phosphorus cycle. The ideas presented here are highly specu-

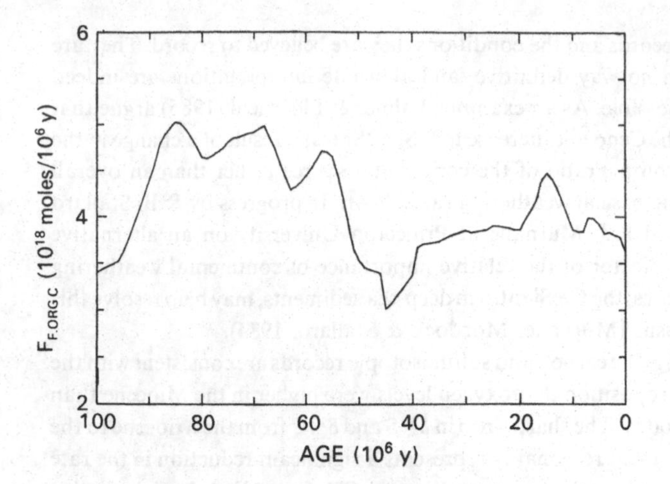

Fig. 22.4. Model calculation of the burial rate of organic carbon.

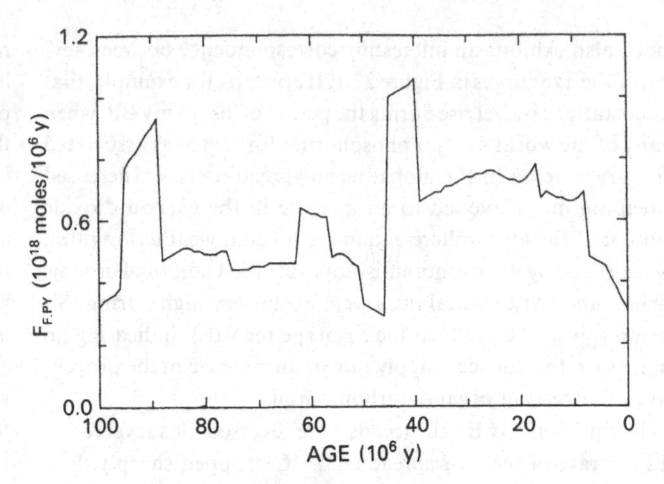

Fig. 22.5. Model calculation of the burial rate of pyrite sulfur.

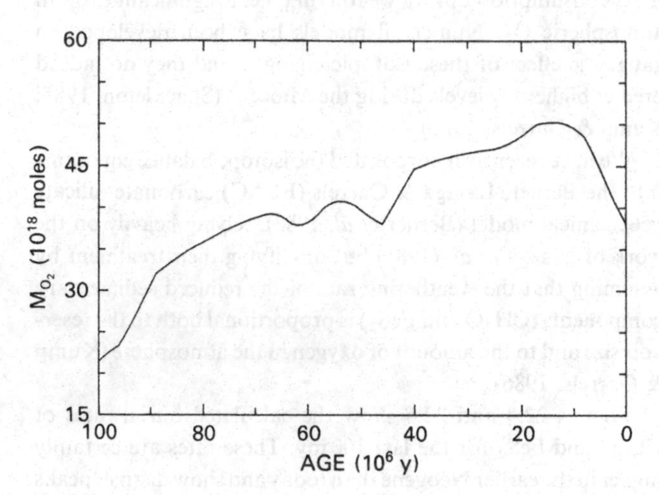

Fig. 22.6. Model calculation of the mass of atmospheric O_2.

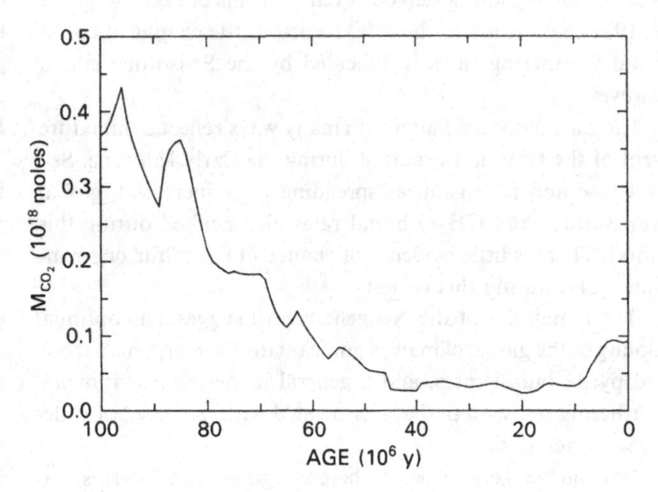

Fig. 22.7. Model calculation of the mass of atmospheric CO_2.

lative and should be considered as such. The model results are only dangerously interpolated within the resolution of the data, which is roughly 5–10 million years, and only longer term trends contain truly significant information. The question is, then, if oxygen *was* higher in the Miocene, how might the phosphorus cycle have responded?

Giant phosphorite deposits require a localized source of phosphorus. Upwelling seems to be the currently preferred mechanism, while a river discharge mechanism is out of favor (but see Barron & Frakes, Chapter 21, this volume). However, there are potential feedbacks between oxygen and phosphorus that might favor the latter mechanism. High oxygen levels in the Miocene would have stimulated the aerobic degradation of organic matter in soils. Phosphorus fluxes from soils to the hydrologic cycle could have increased substantially as a result, increasing stream loadings. This mechanism would thus transfer phosphorus from terrestrial to marine ecosystems, and support high productivity at discharge sites.

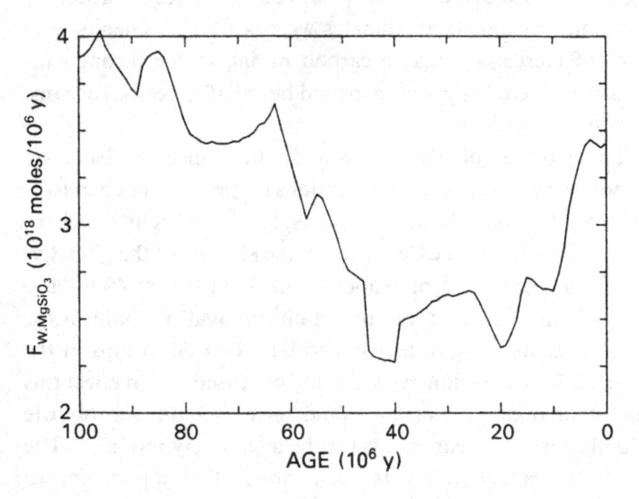

Fig. 22.8. Model calculation of the rate of weathering of magnesium silicate rocks.

Another way that phosphorus may have bypassed terrestrial ecosystems is by fire. Today's oxygen level is critically near the point (~25% O_2) at which moist tropical forests would likely ignite following a lightning strike, even if it were accompanied by rain. The model calculations give a Miocene peak of 52 × 10^{18} mol O_2 in the atmosphere, or an increase to approximately 27% O_2 (by volume). If correct, the probability of fire was high, and may have limited the development of terrestrial ecosystems. The Tertiary expansion of fire-resistant vegetation (Herring, 1985; Wolfe, 1985) may be related to the increase in oxygen and in the probability of fire. Might the replacement of the high biomass, deciduous forests of the earlier Tertiary by low biomass later-Tertiary vegetation have allowed increased transfer of phosphorus from land to sea?

Summary

A discussion of the geological records of the carbon, sulfur, strontium and oxygen isotopic composition of seawater, and of the changes of global spreading rate and eustatic sea level has revealed a number of interesting excursions centered on the periods of Neogene phosphogenesis. Interpretations based on these records should be made only after careful consideration of the interdependent causes of variation in the records. In general, numerical box models are used to sort out the feedbacks involved and make calculations.

A global sedimentary redox model has been developed and added to the BLAG geochemical model. The model predicts that the Miocene was a period of high burial rates of organic carbon and pyrite sulfur and of levels of atmospheric oxygen higher and carbon dioxide lower than today.

Phosphorus is an important player in the global geochemical cycle, yet it has not been incorporated into comprehensive models of the exogenic system. The global depositional rate of CH_2O is for the most part controlled by the supply of phosphorus and its distribution between terrestrial and marine ecosystems. High global burial rates are more likely to occur when the majority of the organic carbon is synthesized on land, where C:P ratios are high. Only under extremely high supply rates of phosphorus to the sea will increased global burial rates of CH_2O be due to marine deposition.

Acknowledgments

This work is largely the result of collaboration with the late Professor Robert M. Garrels. Support was provided by a fellowship from the Poynter Fund of the University of South Florida, and by the Earth System Science Center of the Pennsylvania State University.

References

Bathurst, R.G.C. (1975). *Carbonate Sediments and their Diagenesis.* Elsevier, Amsterdam.

Berger, W.H., Vincent, E., & Thierstein, H.R. (1981). The deep-sea record: major steps in Cenozoic ocean evolution. Society of Economic Paleontologists and Mineralogists Special Publication 32, p. 489–504.

Berner, R.A., Lasaga, A.C. & Garrels, R.M. (1983). The carbonate–silicate geochemical cycle and its effect on atmospheric carbon dioxide over the past 100 million years. *American Journal of Science*, **283**, 641–83.

Berner, R.A. & Raiswell, R. (1983). Burial of organic carbon and pyrite sulfur in sediments over Phanerozoic time: a new theory. *Geochimica Cosmochimica Acta*, **47**, 855–62.

Brass, G.W. (1976). The variation of the marine $^{87}Sr/^{86}Sr$ ratio during Phanerozoic time: interpretation using a flux model. *Geochimica Cosmochimica Acta*, **40**, 721–30.

Burke, W.H., Denison, R.E., Hetherington, E.A., Koepnick, R.B., Nelson, H.F. & Otto, J.B. (1982). Variation of seawater $^{87}Sr/^{86}Sr$ throughout Phanerozoic time. *Geology*, **10**, 516–19.

Chadhuri, S. & Clauer, N. (1986). Fluctuations of isotopic composition of strontium in seawater during the Phanerozoic Eon. *Chemical Geology*, **59**, 293–303.

Claypool, G.E., Holser, W.T., Kaplan, I.R., Sakai, H. & Zak, I. (1980). The age curves of sulfur and oxygen isotopes in marine sulfate and their mutual interpretation. *Chemical Geology*, **28**, 199–259.

Deines, P. (1980). The isotopic composition of reduced organic carbon. In *Handbook of Environmental Isotope Geochemistry*, vol. 1, ed. P. Fritz & J.C. Fontes, pp. 329–406. Elsevier, Amsterdam.

Elderfield, H. & Greaves, M.J. (1981). Strontium isotope geochemistry of Icelandic geothermal systems and implications for seawater chemistry. *Geochimica Cosmochimica Acta*, **45**, 2201–12.

Emiliani, C. (1955). Pleistocene temperatures. *Journal of Geology*, **63**, 2202–12.

Froelich, P.N., Bender, M.L., Luedtke, N.A., Heath, G.R. & DeVries, T. (1982). The marine phosphorus cycle. *American Journal of Science*, **282**, 474–511.

Garrels, R.M. & Lerman, A. (1981). Phanerozoic cycles of sedimentary carbon and sulfur. *Proceedings of the National Academy of Science*, **78**, 4652–6.

Garrels, R.M. & Lerman, A. (1984). Coupling of sedimentary sulfur and carbon cycles – an improved model. *American Journal of Science*, **284**, 989–1007.

Garrels, R.M. & Perry, E.A., Jr. (1974). Cycling of carbon, sulfur, and oxygen through geologic time. In *The Sea*, vol. 5, ed. E.D. Goldberg, pp. 303–36. Wiley Interscience, New York.

Heller, P.L. & Angevine, C.L. (1985). Sea-level cycles during the growth of Atlantic-type oceans. *Earth Planetary Science Letters*, **75**, 417–26.

Herring, J.R. (1985). Charcoal fluxes into sediments of the north Pacific Ocean: the Cenozoic record of burning. In *The Carbon Cycle and Atmospheric CO_2: Natural Variations Archean to Present*, ed. E.T. Sundquist & W.S. Broecker, pp. 419–42. American Geophysical Union, Washington, DC.

Hite, R.J. (1978). Possible genetic relationships between evaporites, phosphorites and iron-rich sediments: *Mountain Geologist*, **14**, 97–107.

Holland, H.D. (1973). Systematics of the isotopic composition of sulfur in the oceans during the Phanerozoic and its implications for atmospheric oxygen. *Geochimica Cosmochimica Acta*, **37**, 2605–16.

Holland, H.D. (1978). *The Chemistry of the Atmosphere and Oceans.* John Wiley & Sons, New York.

Holland, H.D. (1984). *The Chemical Evolution of the Atmosphere and Oceans.* Princeton University Press, New Jersey.

Holser, W.T. (1984). Gradual and abrupt shifts in ocean chemistry during Phanerozoic time. In *Patterns of Change in Earth Evolution*, ed. H.D. Holland & A.F. Trendall, pp. 123–43. Dahlem Konferenzen, Springer-Verlag, Berlin.

Holser, W.T. & Kaplan, I.R. (1966). Isotope geochemistry of sedimentary sulfates. *Chemical Geology*, **1**, 93–135.

Kominz, M.A. (1984). Oceanic ridge volumes and sea-level change – an error analysis. In *Interregional Unconformities and Hydrocarbon Accumulation*, ed. J.S. Schlee, pp. 108–27. American Association of Petroleum Geologists Memoir 36.

Kump, L.R. & Garrels, R.M. (1986). Modeling atmospheric O_2 in the global sedimentary redox cycle. *American Journal of Science*, **286**, 337–60.

Lasaga, A.C., Berner, R.A. & Garrels, R.M. (1985). A geochemical model of atmospheric CO_2 fluctuations over the last 100 million years. In *The Carbon Cycle and Atmospheric CO_2: Natural Variations Archean to Present*, ed. E.T. Sundquist & W.S. Broecker, pp. 397–411. American Geophysical Union, Washington, DC.

Lindh, T.B. (1983). 'Temporal variations in ^{13}C, ^{34}S and global sedimentation during the Phanerozoic.' Unpublished M.Sc. thesis, University of Miami.

Mackenzie, F.T. & Pigott, J.D. (1981). Tectonic controls of Phanerozoic sedimentary rock cycling. *Journal of the Geological Society of London*, **138**, 183–96.

Murnane, R.J., Mortlock, R.A. & Stallard, R.F. (1984). Ge/Si ratios in biogenic opal and ambient waters. *Transactions of the American Geophysical Union*, **65**, 950.

Palmer, M.R. & Elderfield, H. (1985). Sr isotope composition of sea water over the past 75 Myr. *Nature*, **314**, 526–8.

Peterman, Z.E., Hedge, C.E. & Tourtelot, H.A. (1970). Isotopic composition of strontium in sea water throughout Phanerozoic time. *Geochimica Cosmochimica Acta*, **34**, 105–20.

Pitman, W.C. (1978). Relationship between eustacy and stratigraphic sequences of passive margins. *Geological Society of America Bulletin*, **89**, 1289–403.

Riggs, S.R. (1984). Paleoceanographic model of Neogene phosphorite deposition, US Atlantic continental margin. *Science*, **223**, 123–31.

Ronov, A.B. (1982). The earth's sedimentary shell (quantitative patterns of its structure, compositions, and evolution). *International Geology Review*, **24**, 1313–63.

Schidlowski, M., Junge, C.E. & Pietrek, H. (1977). Sulfur isotope variations in marine sulfate evaporites and the Phanerozoic oxygen budget. *Journal of Geophysical Research*, **82**, 2557–65.

Schidlowski, M. & Junge, C.E. (1981). Coupling among the terrestrial sulfur, carbon and oxygen cycles: numerical modeling based on revised Phanerozoic carbon isotope record: *Geochimica Cosmochimica Acta*, **45**, 589–94.

Shackleton, N.J. (1985). Oceanic carbon isotope constraints on oxygen and carbon dioxide in the Cenozoic atmosphere. In *The Carbon Cycle and Atmospheric CO_2: Natural Variations Archean to Present*, ed. E.T. Sundquist & W.S. Broecker, pp. 412–19. American Geophysical Union, Washington, DC.

Shackleton, N.J. & Kennett, J.P. (1975). Paleotemperature history of the Cenozoic and the initiation of Antarctic glaciation: oxygen and carbon isotope analyses in DSDP sites 277, 279 and 281. *Initial Report Deep Sea Drilling Project*, **29**, 734–56.

Shackleton, N.J. & Opdyke, N.D. (1973). Oxygen isotope and paleomagnetic stratigraphy of equatorial Pacific core V28-238: oxygen isotope temperatures and ice volumes on a 10^5 year and 10^6 year scale. *Quaternary Research*, **3**, 39–55.

Sholkovitz, E. (1973). Interstitial water chemistry of Santa Barbara Basin sediments. *Geochimica Cosmochimica Acta*, **37**, 2043–73.

Southam, J.R. & Hay, W.W. (1977). Time scales and dynamic models of deep-sea sedimentation. *Journal of Geophysical Research*, **82**, 3825–42.

Urey, H.C. (1947). The thermodynamic properties of isotopic substances. *Journal of the Chemical Society*, **1947**, 562–81.

Urey, H.C., Lowenstam, H.A., Epstein, S. McKinney, C.R. (1951). Measurements of paleotemperatures and temperatures of the Upper Cretaceous of England, Denmark, and southeastern US. *Geological Society of America Bulletin*, **62**, 399–416.

Vail, P.R. & Hardenbol, J. (1979). Sea-level changes during the Tertiary. *Oceanus*, **22**, 71–9.

Veizer, J. (1983). Trace elements and isotopes in sedimentary carbonates. In Carbonates: *Mineralogy and Chemistry*, vol. 11, ed. R.J. Reeder, pp. 265–99. Mineralogical Society of America, Reviews in Mineralogy, Washington, DC.

Veizer, J., Holser, W.T. & Wilgus, C.K. (1980). Correlation of $^{13}C/^{12}C$ and $^{34}S/^{32}S$ secular variations. *Geochimica Cosmochimica Acta*, **44**, 579–87.

Vincent, E. & Berger, W.H. (1985). Carbon dioxide and polar cooling in the Miocene: The Monterey Hypothesis. In *The Carbon Cycle and Atmospheric CO_2: Natural Variations Archean to Present*, ed. E.T. Sundquist & W. Broecker, pp. 455–68. American Geophysical Union, Washington, DC.

Wadleigh, M.A., Veizer, J. & Brooks, C. (1985). Strontium and its isotopes in Canadian Rivers: fluxes and global implications. *Geochimica Cosmochimica Acta*, **49**, 1727–36.

Watson, A., Lovelock, J.E. & Margulis, L. (1978). Methanogenesis, fires and the regulation of atmospheric oxygen. *BioSystems*, **10**, 293–8.

Wolfe, J.A. (1985). Distribution of major vegetational types during the Tertiary. In *The Carbon Cycle and Atmospheric CO_2: Natural Variations Archean to Present*, ed. E.T. Sundquist & W. Brocker, pp. 357–76. American Geophysical Union, Washington, DC.

PART 4

Neogene phosphorites of California and the southeastern USA

23

Miocene phosphogenesis in California

R.E. GARRISON, M. KASTNER AND
C.E. REIMERS

Abstract

Miocene phosphatic rocks in California occur in two major facies within small Neogene basins associated with the San Andreas transform system. The *phosphatic marlstone facies* consists of coccolith–diatom–foraminiferal sediments slowly deposited in low-oxygen environments; authigenic carbonate fluorapatite replaced carbonate components and is present as small nodules, peloids and laminae; fossilized bacterial mats are common. The *pelletal–oolitic phosphorite facies* consists of phosphatic sandstones deposited on shelves and banks and in basinal turbidites. Phosphatization in these deposits was complex; it occurred in environments which were relatively well-oxygenated and in which siliceous (diatomaceous) sedimentation dominated, and it may have commonly involved cyanobacterial mediation. In some instances, sedimentary reworking concentrated sand-size phosphate grains, but in others phosphate peloids and ooids may have formed *en masse* on the seafloor without significant reworking.

Phosphatic marlstones formed mainly between 15.5 and 13.2 my, coincident with a high sea-level stand, coccolith-rich sedimentation beneath warm water masses, widespread low oxygen conditions, and a global heavy carbon excursion. The marked diachroneity of this facies along the California margin was probably due to preferential preservation of calcareous sediments and phosphatization in persistently anoxic basins.

Pelletal–oolitic phosphorites formed mainly between 14 and 8 my, during the change from warm seas to cool, highly fertile, diatom-rich seas along the northeastern Pacific margin. Increasingly vigorous oceanic circulation during this time interval favored more widespread oxygenated depositional environments.

The specific form of Miocene phosphogenesis in California thus resulted from the interplay of global climatic events and local oceanographic and tectonic conditions. Overall, however, the propensity toward phosphogenesis in California appears to have been created by the onset of Antarctic glaciation in Middle Miocene time; this accords well with Sheldon's (1980) proposal that global Miocene phosphogenesis resulted from phosphorus withdrawal from the deep-ocean phosphorus sink by trade-wind upwelling at the time of transition from high-level, warm oceans to low-level, cold oceans.

Introduction

Among discussions of the episodicity of phosphogenesis in geologic time, the Miocene epoch has received considerable attention (e.g. Cook & McElhinney, 1979; Sheldon, 1980; Arthur & Jenkyns, 1981; Riggs, 1984). This article reviews the attributes and chronostratigraphy of Miocene phosphatic rocks in California and attempts to establish relationships among the formation of phosphatic facies, Miocene paleoceanographic trends, and local tectonic–oceanographic conditions. What emerges is a portrayal of the complexities which existed within a single phosphogenic province like Miocene California – complexities due to the interplay of global and local events.

These complexities produced intricate stratigraphic relationships which have become encoded in a plethora of local stratigraphic terminology. In so far as possible, we have sought to use as few local stratigraphic names as practical. The main Miocene phosphate-bearing unit is the Monterey Formation; since the late 1930s, Kleinpell's (1938 and 1980) provincial stages have been used for age dating and correlation within the Miocene of California and are herein utilized in discussions of chronostratigraphy.

Geologic framework

Miocene units, including phosphatic facies, occur chiefly in a series of Neogene basins in the western half of California (Fig. 23.1). These are extensional, continental borderland basins formed in response to strike-slip faulting, beginning in Late Oligocene time, along the San Andreas transform system (Blake *et al.*, 1978; Crouch, 1979; Howell *et al.*, 1980; Graham, 1987). Rapid subsidence, at rates up to 500 m per million years, during the Miocene typically led to the formation of small basins, up to 2000 m deep, in which the Monterey Formation and related units were deposited. The manner in which these basins became filled depended, in part, on their geographic position. The most offshore and distal basins (e.g. Santa Barbara and Santa Maria Basins) experienced long intervals of sediment starvation during which pelagic–hemipelagic sediments accumulated at relatively slow rates. Proximal basins (e.g. San Joaquin Basin), in contrast, were usually more rapidly filled with either detrital-rich hemipelagic sediments or with turbidite fan deposits or both. Some Neogene basins were of intermediate character, among them the Los Angeles, Cuyama, Salinas and La Honda Basins; the Miocene lithologic fill of these basins typically comprises hemipelagic deposits intercalated with sandstones of shelfal or turbiditic character.

In addition to tectonism, paleoceanographic events also

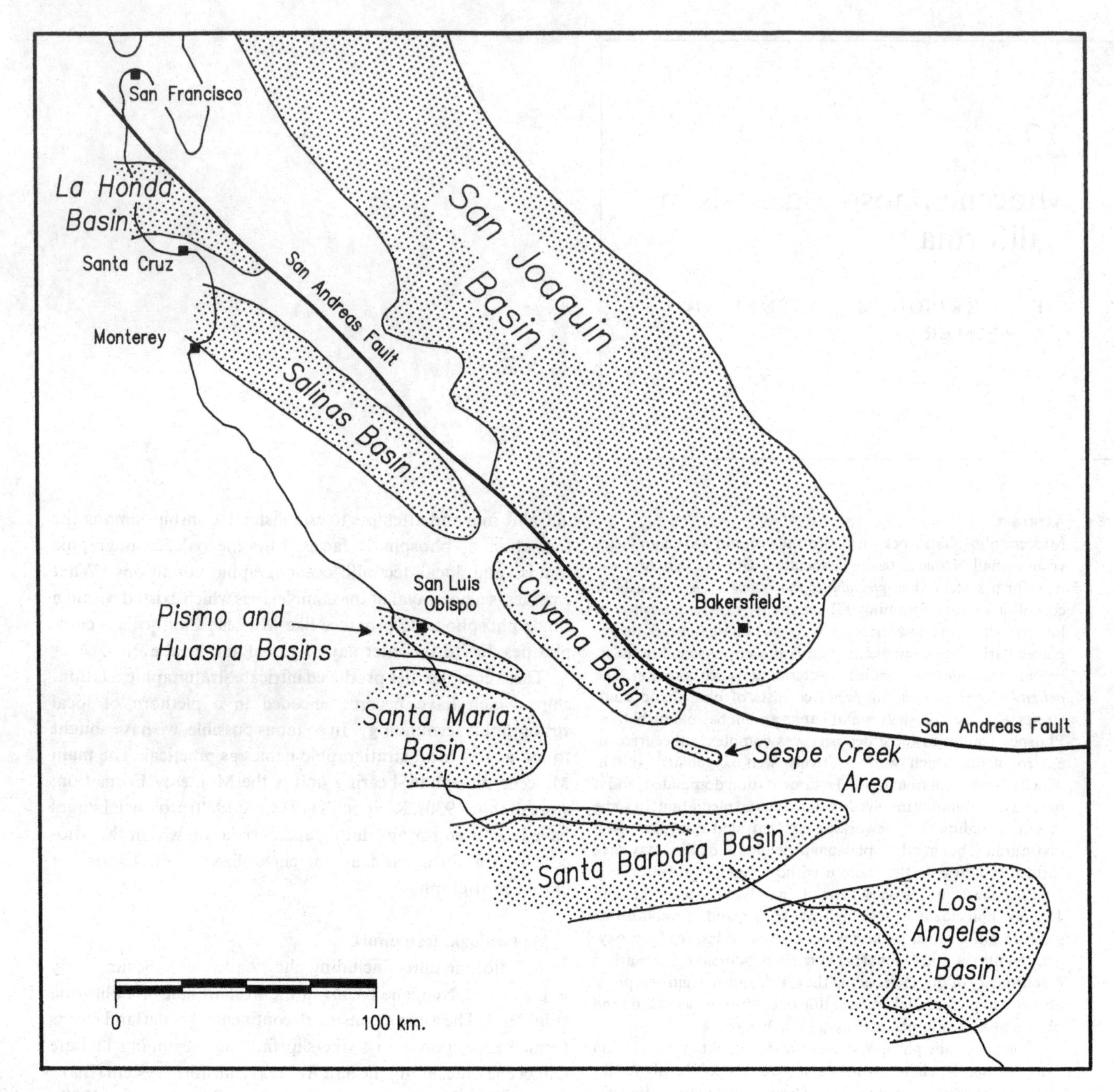

Fig. 23.1. Principal Upper Cenozoic basins of central and southern California.

markedly affected the character of the Miocene basin fill. In Miocene pelagic and hemipelagic sediments, the most notable effect was a change from calcareous–siliceous sedimentation in the lower part of the Monterey Formation to dominantly siliceous sedimentation in the upper part. This change appears to record the transition from generally weak coastal upwelling to very intense upwelling that occurred in Middle Miocene time. High nutrient levels associated with intense upwelling led to plankton communities dominated by diatoms during late Middle and Late Miocene time; the resulting sediments are now represented by the distinctive diatomites, porcelanites and cherts of the upper Monterey. Prior to intensification of upwelling, Miocene plankton communities were comprised of mixtures of diatoms and coccolithophorids, so that the lower Monterey, particularly in distal basins, is a siliceous limestone or marlstone. In some basins, particularly the more distal basins, a distinctive phosphatic marlstone facies occurs between the calcareous–siliceous and siliceous facies. These patterns have led some workers (e.g. Pisciotto, 1978; Pisciotto & Garrison, 1981) to recognize three facies in the Monterey Formation and related units: a lower calcareous or calcareous–siliceous facies, a middle phosphatic facies and an upper siliceous facies (cf. Fig. 23.8). More detailed subdivisions have been recognized in some basins (e.g. Isaacs, 1980 and 1983) and, as documented later, local preservational effects complicate the simple threefold subdivision noted above.

Periodic anoxic conditions, particularly during Middle Miocene time, also left an imprint on the lithic fill of some basins. These intervals of anoxia produced widespread laminated sediments (in the siliceous, phosphatic, and calcareous–siliceous facies), and led to the preservation of organic matter rich in carbon and phosphorus and to the development of important petroleum source beds.

The sea-level lowerings of Late Miocene time (Haq, Hardenbol & Vail, 1987), combined with tectonic uplift in Late Miocene–Pliocene time, led to marked increases in supply of detrital sediment and elevated sedimentation rates. In proximal basins, this resulted in deposition of shallow-water clastic sediments above the Monterey. In more distal basins, it led to deposition of hemipelagic sediments which were mixtures of diatoms and abundant fine-grained detritus (e.g. the Sisquoc Formation of Santa Barbara and Santa Maria Basins).

Miocene phosphatic rocks

Previous reviews of Miocene phosphatic rocks in California Neogene basins were by Gower & Madsen (1964) Dickert (1966 and 1971), Roberts (1981), and Garrison, Kastner & Kolodny (1987). Although several types of phosphatic rocks are present (ranging, for example, from nodular phosphorites to glaucophosphorites), two facies dominate: phosphatic marlstones and pelletal–oolitic phosphorites.

Phosphatic marlstone facies

Phosphatic marlstones are laminated, commonly fissile, organic-rich rocks which contain peloids, small nodules and laminae of light-colored and friable phosphate in the form of carbonate fluorapatite (Figs. 23.2 and 23.3). They are best developed in Middle Miocene sequences of distal basins, particularly in the western part of the Santa Barbara Basin, where Isaacs (1983) has assigned them to the Carbonaceous Marl member of the Monterey Formation, and in the Santa Maria Basin, where Pisciotto (1978; see also Pisciotto & Garrison, 1981) placed them in the informal phosphatic facies of the Monterey Formation.

The original sediments of this facies were coccolith–diatom–foraminiferal muds and oozes containing varying amounts of detrital grains. Table 23.1 shows the compositional range for rocks of this facies in the western Santa Barbara Basin.

Several characteristics indicate that phosphatic marlstones were deposited in low oxygen environments which may have varied from anoxic to suboxic, in the sense of Demaison & Moore (1980). Intensively burrowed rocks are not present, and most marlstones are banded (see Fig. 23.2) and have laminations that are either regular and continuous or irregular and discontinuous (Isaacs, 1981; Mertz, 1984). In the Salinas Basin, continuously and discontinuously laminated pockets of phosphatic marlstones are interbedded with massive rocks, suggesting variable but generally low-oxygen bottom waters which inhibited a deeply burrowing infauna (Mertz, 1984). Phosphatic marlstones typically contain benthic foraminiferal assemblages, most notably the buliminids, indicative of low-oxygen environments (Ingle, 1967 and 1981). Most phosphatic marlstones have high Total Organic Carbon (TOC) values. Phosphatic marlstones in the western Santa Barbara Basin average 8.0% TOC (Isaacs, 1980 and 1981), with maximum values of nearly 30%. Laminated marlstones in the Salinas Basin range from $<1.0\%$ to 7.0% TOC (Mertz, 1984). Bacterial mats indicative of anoxic to suboxic environments (Williams & Reimers, 1983; Williams, 1984) are particularly common in this facies and may be a major component of the organic matter (Mertz, 1984; Williams, 1984; Reimers, *et al.*, Chapter 24, this volume).

Thin section and scanning electron microscope observations show that apatite in these rocks replaced carbonate-rich host sediment (particularly foraminiferal tests and micritic matrix) and was precipitated in void space such as foraminiferal chambers. Sand-size phosphatic peloids are particularly common and may be in large part phosphatized fecal pellets and foraminiferal tests (Fig. 23.4). Garrison *et al.* (1987) have demonstrated phosphatization of bacterial cells, and Reimers *et al.* (Chapter 24, this volume) document the phosphatization of bacterial filaments. Compactional bending of phosphatized laminae around earlier formed phosphate nodules (Fig. 23.3) suggests several phases of phosphatization, including very early diagenetic events near the seafloor.

Based on regional stratigraphic and structural considerations, Garrison *et al.* (1987) concluded that phosphatic marlstones developed on outer shelves, slopes, basin floors, and submerged offshore ridges that were bathed by low-oxygen water masses. As noted previously, they are especially prominent in slowly deposited middle Miocene sections of distal basins but absent or poorly developed in the more proximal basins such as the San Joaquin, Cuyama and Salinas Basins. Significantly, however, they are very well developed in the Salinas Basin on top of a sediment-starved Middle Miocene offshore ridge, termed the 'Lockwood High' by Graham (1976; cf. Fig. 23.3 of this paper). This suggests that, in addition to low-oxygen environments, a further prerequisite for development of this facies was very slow sedimentation, with comparatively little detrital input. A possible reason may be found in the observations of Froelich *et al.* (1983) that authigenic apatite formation in modern Peruvian shelf and slope sediments is apparently linked to fluorine availability. Because the major source of fluorine is seawater fluorine which diffuses downward into sediment, apatite precipitation is limited to the upper few centimeters in slowly deposited sediments. More rapidly deposited sediments would be too quickly buried beneath the zone of fluorine diffusion for significant apatite to form.

Pelletal–oolitic phosphorite facies

Pelletal–oolitic phosphorites are phosphatic sandstones which may contain, in addition to peloids and coated phosphatic grains, silt- to fine sand-size detrital grains, glauconite peloids and phosphatic micronodules, intraclasts, and fish fragments (Figs. 23.5 and 23.6). As documented in the following sections, three distinct varieties are present.

Turibiditic phosphorites These rocks occur as thin (10–30 cm thick), scattered layers interbedded with deep-water, hemipelagic marlstones and porcelanites of basinal origin. Most such layers have sharp erosional basal contacts; many also contain rip-up clasts of the underlying hemipelagic sediment and display graded bedding. Typically, they are well sorted rocks,

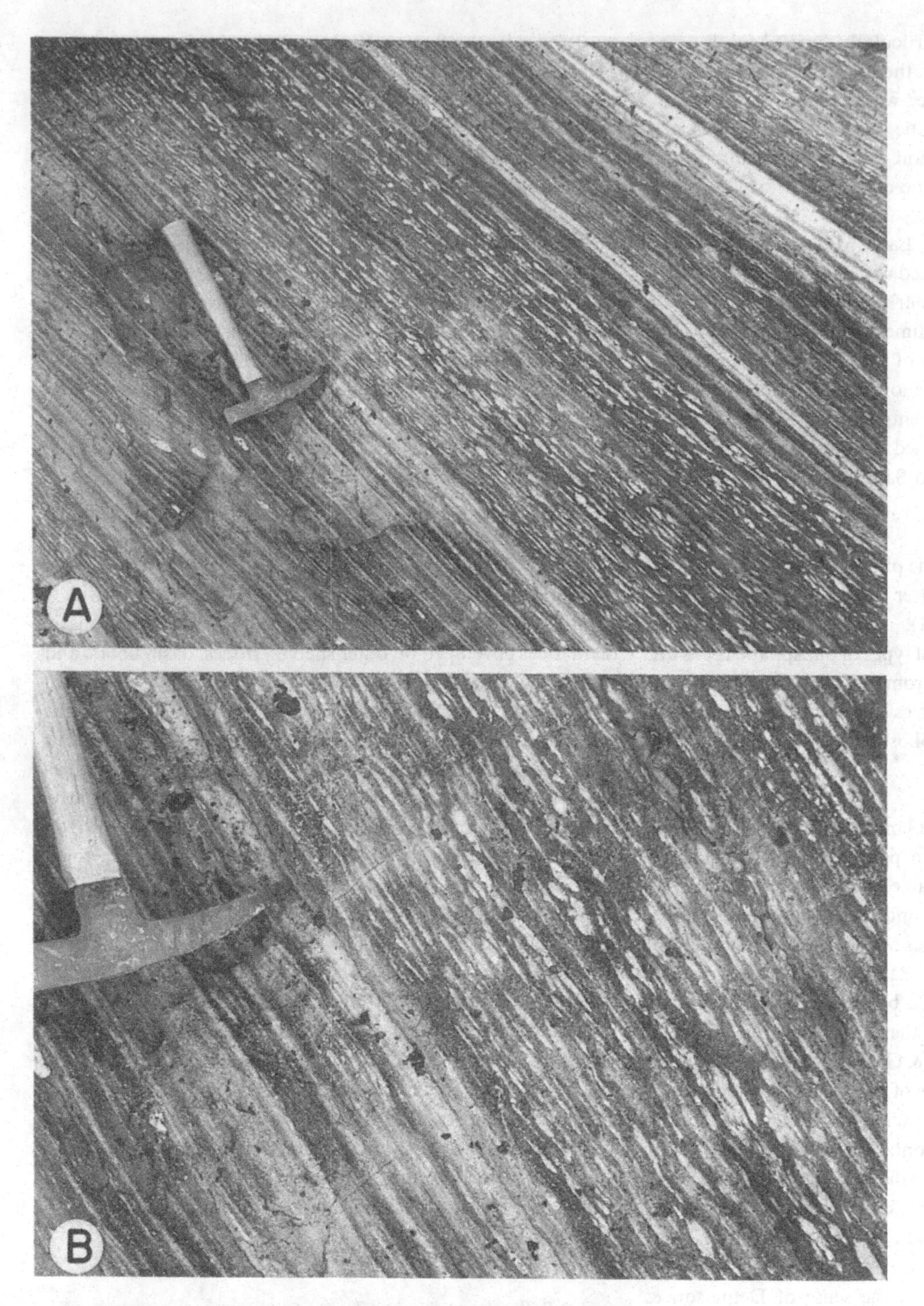

Fig. 23.2. Phosphatic marlstone facies, Naples Beach section, western Santa Barbara Basin. (a) Dark, laminated, organic-rich marlstone contains layers and small nodules of light-colored carbonate fluorapatite; (b) Close-up of above view showing phosphatized layers and dense concentrations of elongate phosphate nodules.

with medium sand-size phosphatic grains along with a finer grained admixture of detrital quartz and feldspar grains (Fig. 23.5). Some also contain displaced shallow-water, benthic foraminifera and glauconite peloids. Phosphatic components include coated grains, peloids without apparent nuclei, and sparse fish bones. The presence of coated grains with multiple nuclei and off-center nuclei (Fig. 23.5) indicates the coated grains are not simple ooids and suggests they may be instead phosphatized cyanobacterial coatings (cf. Soudry & Champetier, 1983; Garrison *et al.*, 1987).

Turbiditic phosphorites were products of redeposition of phosphatic grains, formed originally on shelves and submerged banks, into deep-water basins, with estimated depths of hundreds to as much as 2000 m (Graham, 1976; Ingle, 1981; Mertz, 1984). They are most common in the Cuyama and Salinas Basins, especially the latter. More detailed discussions of their

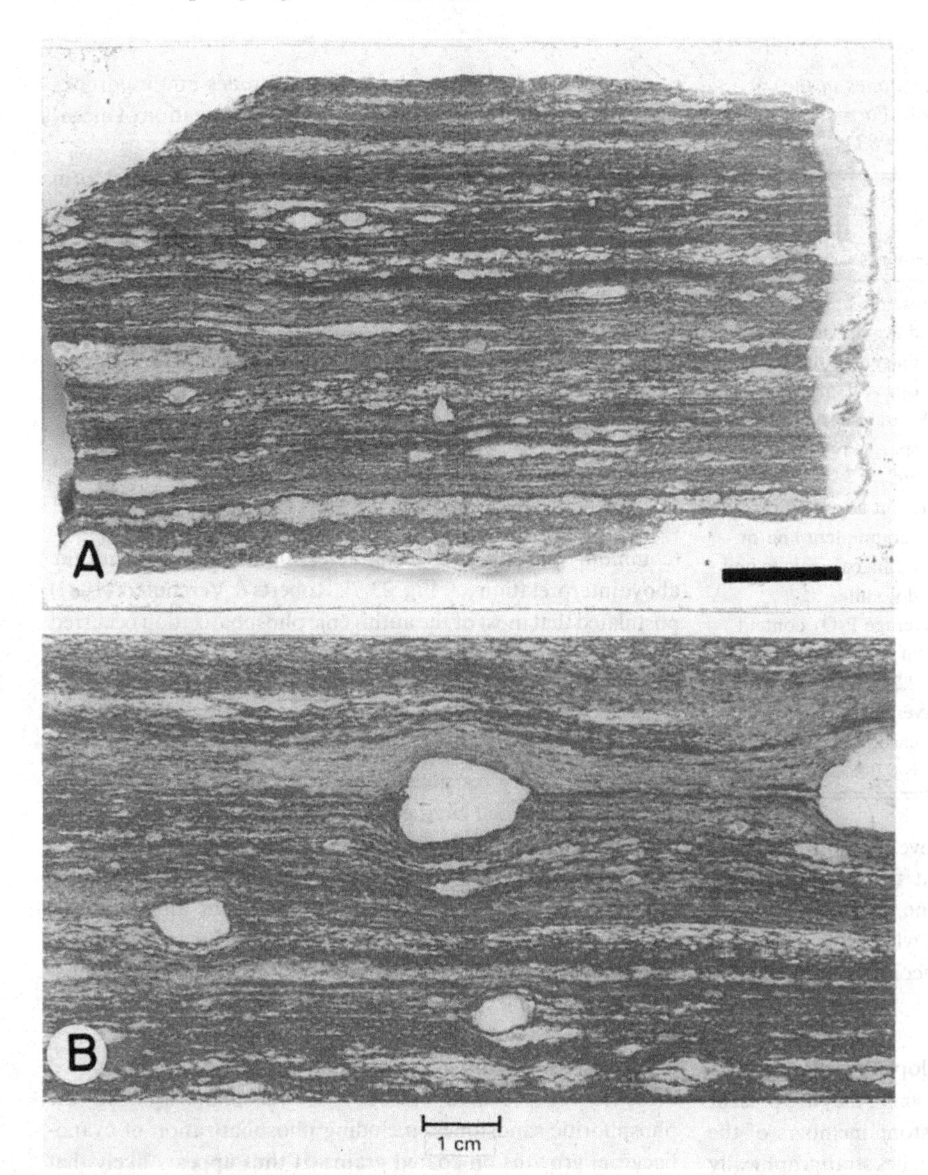

Fig. 23.3. Phosphatic marlstone facies deposited on top of the 'Lockwood High', a Middle Miocene offshore bank in the Salinas Basin. Photographs of sawed surfaces: (a) Light-colored peloids, small nodules and discontinuous layers of finely-crystalline carbonate fluorapatite. Note how phosphate layers are formed by coalescence of peloids and nodules. Scale bar is 1 cm long. (b) Detailed view of highly-phosphatized marlstone. Note bending of phosphatized laminae around some of the phosphate nodules which must have been hard at the time of compaction. Scale bar is 1 cm long.

stratigraphy, sedimentology and petrology are in Dickert (1966 and 1971), Graham (1976 and 1979), Younse (1979), Mertz (1984), and Garrison *et al.* (1987).

Banktop phosphorites These phosphorites occur in thin, condensed sequences which were deposited on top of submerged offshore banks. The best documented example is within the Salinas Basin where a meter or less of phosphoritic sandstone was deposited on the 'Lockwood High', a submerged Middle Miocene bank with an estimated water depth of about 200 m (Garrison, Stanley & Horan, 1979; Graham, 1979; Mertz, 1984). The 'Lockwood High' was apparently an upfaulted block of crystalline basement rock. In the southern Santa Maria Basin, Middle Miocene volcanic rocks formed a somewhat similar submerged bank on which phosphoritic sandstones and breccias accumulated. In both examples, glauconite occurs together with phosphatic grains, and the textural evidence indicates that phosphatization followed glauconization, and several phases of phosphatization occurred. Coated phosphatic grains, ranging from medium sand- to pebble-size, are abundant and closely resemble phosphatized bacterial or cyanobacterial coatings described by other workers (cf. Soudry & Champetier, 1983).

Because these banktop phosphorites lack a macrofauna and are generally well sorted, Garrison *et al.* (1987) speculated that they formed on banktops which were shallow enough to be winnowed by at least occasional currents (perhaps storm-in-

Table 23.1. *Composition of phosphatic marlstones in the carbonaceous marl Member of the Monterey Formation, western Santa Barbara Basin (from Isaacs, 1983)*

Component	Mean abundance (weight %)	Range (weight %)	Remarks
Biogenic silica	16	3–90	Present as opal-A diatom frustules or as diagenetically-derived opal-CT or quartz.
Detrital minerals	23	5–65	Detrital quartz, feldspar, clay minerals, etc.
Carbonate minerals	42	5–75	Present as coccoliths, foraminifera and/or authigenic calcite and dolomite.
Apatite	6	0–20	Average P_2O_5 content of these rocks is 2.2%.
Organic matter	13	2–24	Average total organic carbon in these rocks is 8.0%.

duced currents), but deep enough to prevent colonization by macrofauna. They further suggested that these banktops lay near the edge of the oxygen minimum zone, in a setting analogous to the present Peru–Chile outer shelf where Burnett (1980) has documented the co-generation of glauconite and phosphate in mildly reducing environments.

Shelfal phosphorites Well developed shelfal phosphorites occur only in the Cuyama Basin, where they are present in the upper and lower phosphatic mudstone members of the Santa Margarita Formation. The latter unit lies stratigraphically above the Monterey Formation as it is defined in the Cuyama Basin, but the Santa Margarita is correlative with the upper part of the Monterey in other basins and consists mainly of siliceous mudrocks. Shelfal phosphorites are present also in the Santa Margarita Formation of the Sespe Creek area (Fig. 23.1), a region which is probably the southern extension of the Cuyama Basin.

Shelfal phosphorites are phosphatic sandstones which contain mixtures of silt- to fine sand-size detrital grains and phosphatic peloids, micronodules, intraclasts, coated grains and fish fragments (Fig. 23.6). Some are well sorted, others have a muddy matrix. They occur as relatively thin (0.2–2.0 m) layers interbedded with massive siliceous mudrocks (Fig. 23.7). The latter rocks were originally diatom-rich muds which contained small amounts of phosphatic material (peloids, fish bones, etc.), but burial diagenesis has largely destroyed the diatom frustules and converted the muds into opal-CT mudrocks. In contrast to the low-oxygen, relatively deep-water environments in which the Miocene phosphatic marlstones formed, deposition of both the phosphoritic sandstones and siliceous mudrocks apparently occurred in comparatively shallow and oxygenated shelfal environments inhabited by a flourishing benthic community, including infaunal elements (Lagoe, 1985; Roberts & Vercoutere, 1987). Consequently, both sandstones and mudrocks were thoroughly burrowed, with only rare preservation of mechanical sedimentary structures.

Shelfal phosphorites have P_2O_5 contents of 5–25% and form only about 5–10% of the total thickness of the two phosphatic mudstone members of the Santa Margarita Formation (Fedawa & Hovland, 1981). The siliceous mudrocks, the main lithology of the two members, average $<1\%$ P_2O_5.

Previous workers have interpreted the phosphorite beds in both the Cuyama and Sespe Creek areas as products of current reworking and winnowing of diatomaceous muds which contained dispersed fish debris and authigenic phosphate grains (Thor, 1977 and 1978; Roberts & Vercoutere, 1987). Most phosphorite beds have sharp, erosional basal contacts and many contain rip-up clasts of mudrock and reworked bivalve remains (commonly phosphatized), characteristics compatible with the above interpretation (cf. Fig. 23.7). Roberts & Vercoutere (1987) postulated that most of the authigenic phosphatization occurred in organic-rich diatomaceous muds deposited in anoxic shelfal environments which promoted early diagenetic formation of apatite. As noted above, however, these sediments were bioturbated, thus it seems more likely that, if anoxic conditions existed, they were restricted to more localized burial environments (such as shell cavities or micropores in fecal pellets) in organic-rich diatomaceous sediments which were relatively rapidly deposited (cf. Demaison & Moore, 1980). In the scenario proposed by Roberts & Vercoutere (1987), the alternation of phosphoritic sandstones and siliceous (diatomaceous) mudrocks records minor transgressions and regressions, phosphoritic sandstones being products of winnowing and reworking during regressions.

Roberts & Vercoutere (1987) and Garrison *et al.* (1987) noted evidence for repeated episodes of phosphatization in the phosphoritic sandstones, including phosphatization of cyanobacterial growths on coated grains. It thus appears likely that formation of these sandstones was complex and may have involved repeated episodes of reworking, burrowing and phosphatization, including some phosphatization directly on the seafloor of individual grains. It is noteworthy that many of the phosphoritic sandstones in the Cuyama area are poorly sorted, which argues against extensive winnowing. Recent studies along the Peruvian margin by W.C. Burnett and his colleagues (Baker & Burnett, 1988; Burnett *et al.*, 1988) suggest that somewhat poorly sorted pellatal phosphorite layers of Holocene age were formed *in situ* rather than concentrated during reworking. Radiometric age dating of individual Peruvian pellets, including coated grains that look identical to those in the Cuyama deposits, indicate they formed very rapidly, on time scales of a few years (Burnett *et al.*, 1988). These modern outer shelf-upper slope phosphorite layers may, as they become better known, turn out to be fairly precise analogs of the shelfal phosphorites in California.

Chronostratigraphic distribution of Miocene phosphatic facies

Figure 23.8 portrays the chronostratigraphic distribution of major Miocene phosphatic facies in California. As noted before, the benthic foraminiferal stages of Kleinpell (1938 and

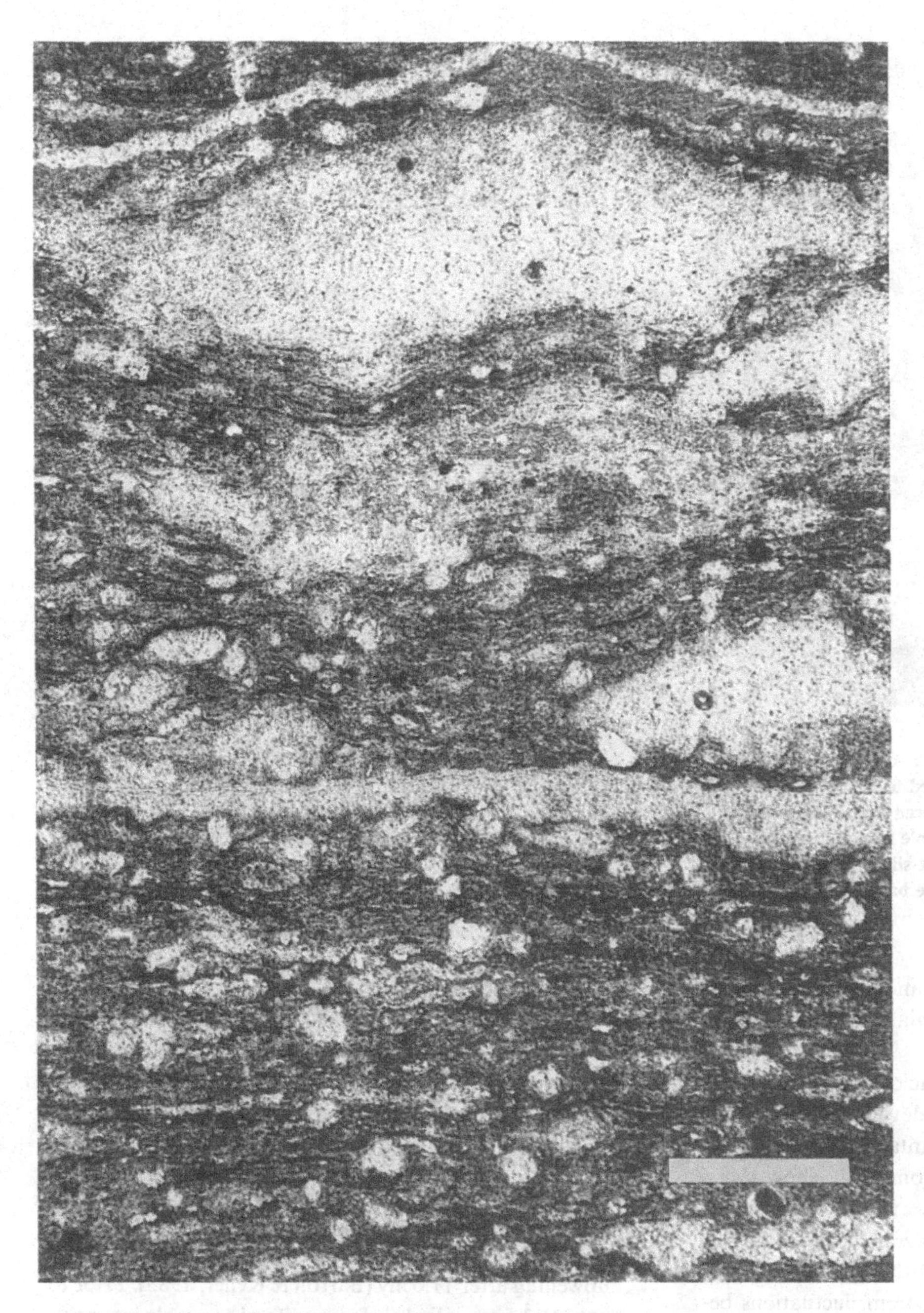

Fig. 23.4. Photomicrograph of phosphatic marlstone from the Santa Maria Basin. Light-colored peloids, micronodules and laminae are carbonate fluorapatite. Dark-colored matrix is organic-rich micrite with abundant coccoliths and possible bacterial mats. Many of the peloids are phosphatized foraminiferal tests. Scale bar is 500 μm.

1980) have formed the basis for age dating and correlation of Miocene sedimentary rocks in California since the late 1930s. More recently, this dating has been supplemented and refined through use of other microfossils, mainly planktonic foraminifera and diatoms (Barron & Keller, 1983). Diatoms are especially useful, allowing, in favorable cases, resolution of time on the order of 100 000 y (Barron, Keller & Dunn, 1985).

Phosphatic marlstone facies

Phosphatic marlstones are most prominent in Middle Miocene parts of the Monterey Formation within the more distal basins. Compilations by Garrison *et al.* (1987) suggest this facies was best developed between about 15.5 and 13.2 my, corresponding to the Luisian and Early Mohnian stages of Kleinpell (1938 and 1980). In the Santa Maria Basin, phosphatic marlstones and shales are irregularly distributed in rocks as old as 17.5 my and as young as about 7.5–7.0 my (Woodring & Bramlette, 1950; Pisciotto, 1978). In the Santa Barbara Basin, well developed phosphatic marlstones (= the Carbonaceous Marl Member of Isaacs, 1983) are as young as about 8.5 my (Lagoe, 1985 and in press; Arends & Blake, 1986; Barron, 1986b). In the Los Angeles Basin, the base of this facies is about 14.2 my (Conrad & Ehlig, 1983), and in the Salinas Basin the base of this facies is about 17.5 my (Mertz, 1984). Phosphatic

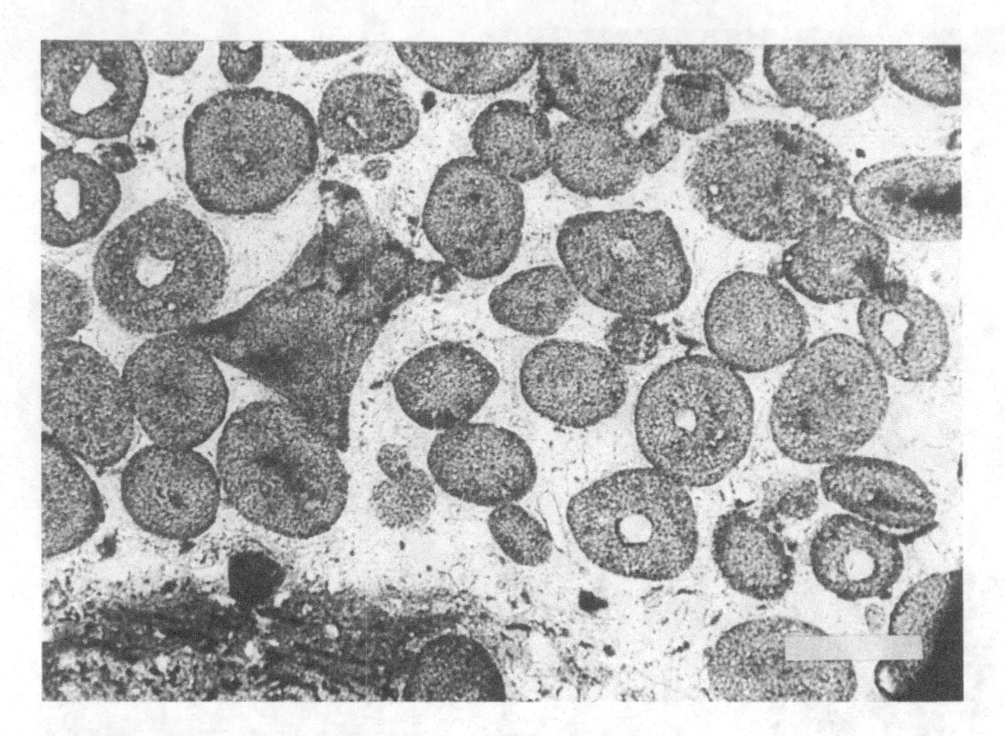

Fig. 23.5. Photomicrograph of turbiditic pelletal–oolitic phosphorite from the Salinas Basin. A diagenetically-altered fish vertebra is at left center. Several peloids have more-or-less centered nuclei composed of detrital quartz, others have off-center or multiple nuclei. Light-colored material between the grains is a mixture of silt-size detrital quartz and very finely crystalline phosphatic cement. Scale bar is 500 μm.

marlstones of Early Miocene age (Saucesian stage) are present in the La Honda Basin in northern California (the Lambert Shale of Clark, 1981).

The entire chronologic extent of this facies is thus from about 23 to 7.0 my, but its most pronounced development occurred between 15.5 and about 13.2 my. In the Santa Barbara and Santa Maria Basins, phosphatic marlstones commonly show cyclic interbedding with siliceous pelagic rocks (diatomites, porcelanites and cherts) in the transition zone between the calcareous and overlying siliceous parts of the Monterey. These cycles appear to record relatively short-term fluctuations between warm, coccolith-rich and cool, diatom-dominated water masses (cf. Barron & Keller, 1983).

Pelletal–oolitic phosphorite facies

Compared to the phosphatic marlstone facies, the age range of this facies is less well constrained. According to Dickert (1966), sandstones with phosphatic pellets of possible latest Oligocene to earliest Miocene age are present in the Salinas and western San Joaquin Basins. Turbiditic phosphorites of late Early–early Middle Miocene (Relizian and Luisian) age occur sporadically in the Salinas and Cuyama Basins (Graham, 1976; Younse, 1979; Mertz, 1984; Lagoe, 1985) and suggest the presence of at least some phosphorite-generating shelves or banks during this time span. The most extensive occurrences of this facies, however, appear to be in rocks of Middle–early Late Miocene (Mohnian) age in shelfal regions of the Cuyama Basin and the Sespe Creek area. Lagoe's (1985) correlations suggest the maximal development of this facies was between about 14 and 8 my, particularly during the earlier part of this interval.

Miocene climatic–oceanic events

As shown in Figure 23.8, major Miocene events and trends included:

1) The formation of the Antarctic ice-cap between about 16.5 and 13.5 my, followed by expanded glaciation and marked cooling of deep ocean waters (Savin *et al.*, 1981; Woodruff, Savin & Douglas, 1981).
2) In the northeastern Pacific, increasing upwelling between about 12.5 and 11.0 my, followed by generally very strong upwelling after 11.0 my (Barron & Keller, 1983). Prior to about 13.5 my, planktic foraminiferal assemblages in this region suggest subtropical influences, but between 13.5 and 12.5 my, cold California Current assemblages became dominant and persisted during the remainder of Miocene time (Barron & Keller, 1983).
3) A generally high stand of sea level during late Early and Middle Miocene time, between about 20.0 and 10.0 my, according to Haq et al., (1987). This was followed by fluctuating but generally lower sea-level stands, during Late Miocene time.
4) A pronounced positive excursion of $\delta^{13}C$ signals in shells of both benthic and planktic foraminifera between 17.5 and 13.5 my (Vincent & Berger, 1985) and a negative shift in the $S^{13}C$ values of deep-sea benthic foraminifera shells at about 6 my (Keigwin, 1979; Woodruff & Savin, 1985).

The interval between 15 and 10 my was a period of transition between the non-glacial world of Early Miocene time and the onset of intense glaciation (accompanied by marked lowering of sea level) of Late Miocene time. Events in the northeastern

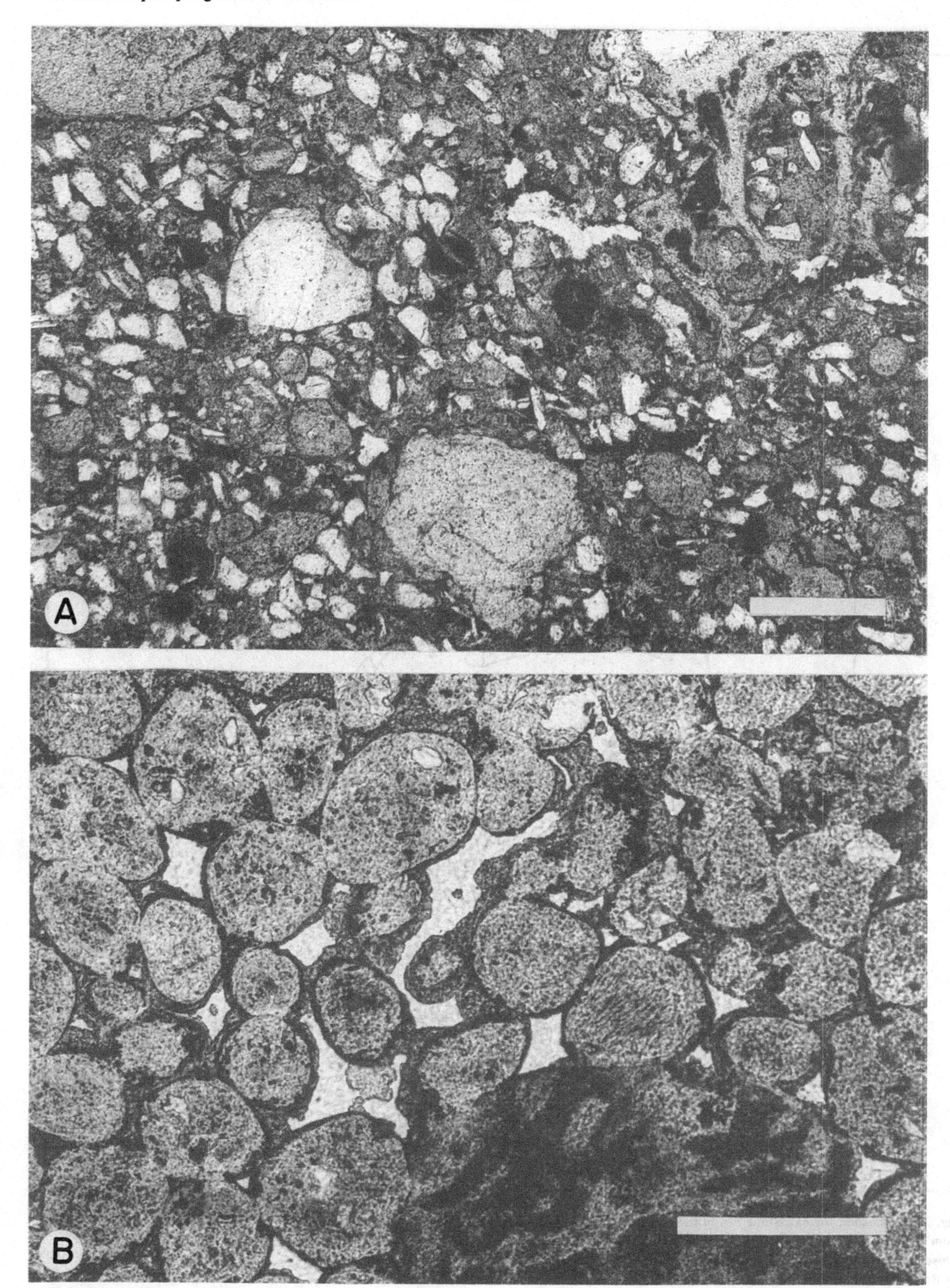

Fig. 23.6. Photomicrographs of shelfal pelletal–oolitic phosphorites. Scale bars are 500 μm; (a) Mixture of phosphatic peloids, fish debris (upper right), angular silt-size detrital grains of quartz and feldspar, and an argillaceous–siliceous matrix (Cuyama Basin). (b) Phosphatic peloids and ooids; most of the latter have multiple, off-center nuclei, suggesting they were accreted, perhaps by cyanobacterial growth. Note rims of dark phosphatic cement showing an uneven distribution ('meniscus cement') and suggesting cementation in the vadose zone (see Garrison *et al.*, 1987) (Sespe Creek area).

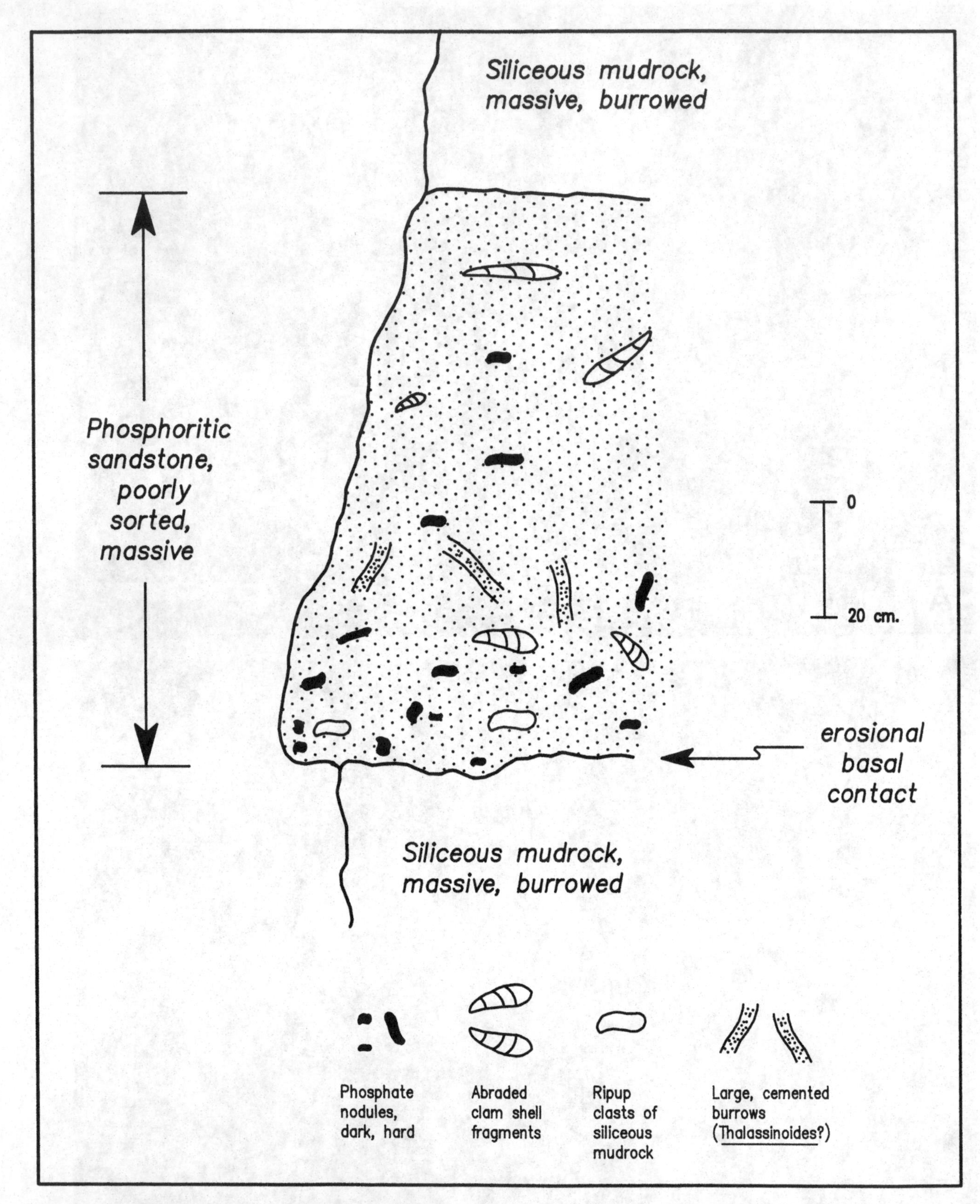

Fig. 23.7. Phosphoritic sandstone bed in the Cuyama Basin; sketch from a field photograph. The siliceous mudrocks above and below the phosphorite were originally diatomaceous muds, but the opal-A diatom frustules were converted to opal-CT by burial diagenesis.

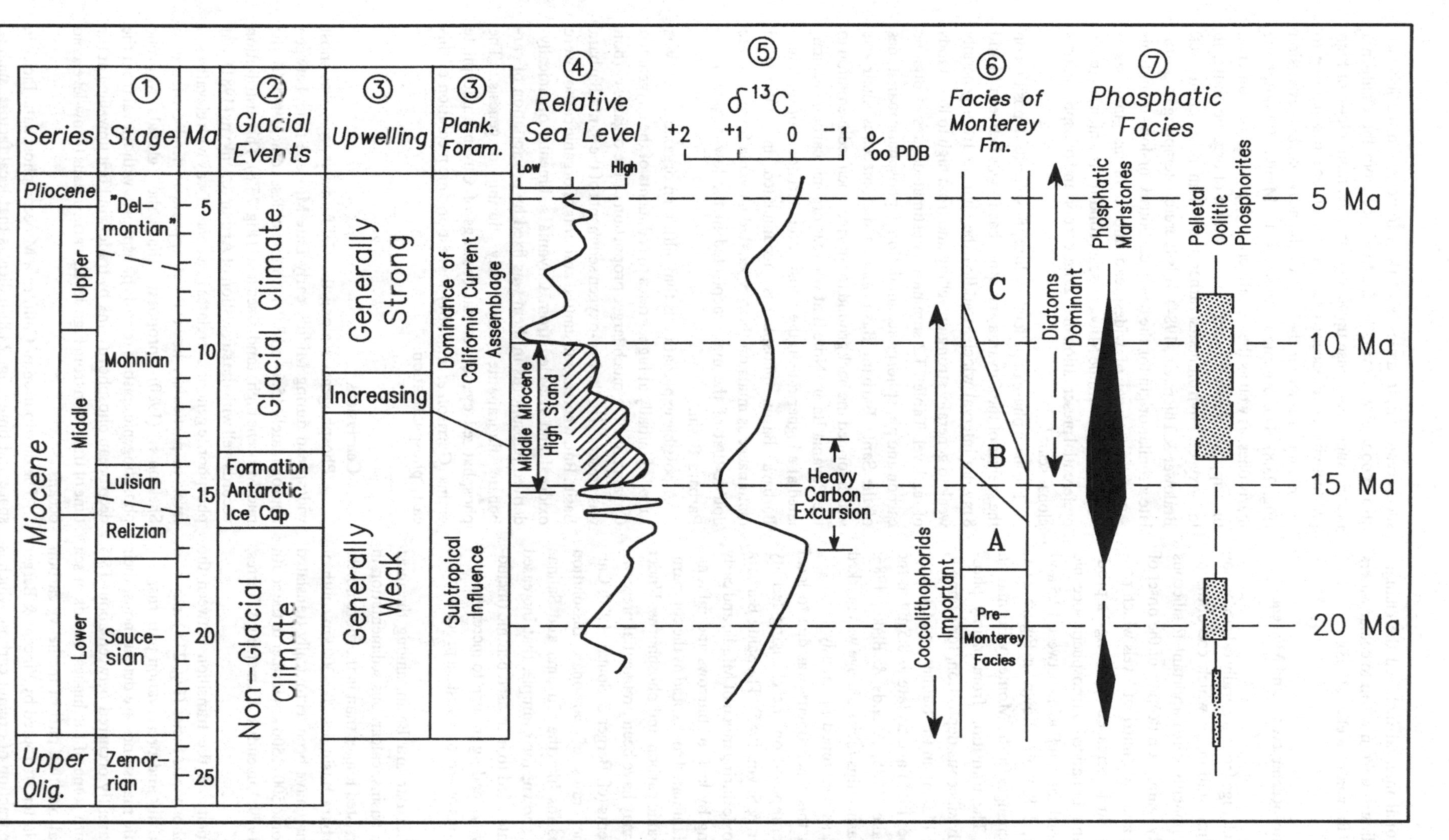

Fig. 23.8. Correlation of Miocene facies in California with major paleoceanographic events. Column 1 shows the California Miocene stages of Kleinpell (1938, 1980) as calibrated by Barron (1986a,b). Column 2 is after Woodruff *et al.* (1981). Column 3, refers to the northeastern Pacific, after Barron & Keller (1983). Column 4 is the short-term eustatic sea-level curve of Haq, Hardenbol & Vail (1987). Column 5 after Savin *et al.* (1981). Column 6 shows the generalized facies of the Monterey Formation (Pisciotto, 1978; Pisciotto & Garrison, 1981) correlated according to various sources noted in the text; Facies A is the calcareous or calcareous–siliceous facies; Facies B is the phosphatic facies; and Facies C is the siliceous facies. Column 7 after various sources quoted in the text. The width of the bars represents approximate abundance of the two phosphatic facies, and dotted lines indicate that the facies is sporadically present.

Pacific during this transitional interval included the changes from weak to strong upwelling and from warm to cooler waters between 13.5 and 11.0 my (Barron & Keller, 1983).

Correlation of climatic–oceanic events and Miocene facies in California

As noted earlier, the change from generally calcareous or calcareous–phosphatic sedimentation, in which coccoliths or calcareous–phosphatic components were important, to siliceous sedimentation was probably mainly a reflection of the onset of vigorous coastal upwelling and high nutrient levels which promoted diatom productivity. Such increased upwelling – a product of climatic cooling and enhanced atmospheric–oceanic circulation – began in the northeastern Pacific between 12.5 and 11.0 my (Barron & Keller, 1983). As shown in Figure 23.8, however, the major facies changes in the Monterey Formation were highly diachronous. The transition from mainly phosphatic marlstones to mainly siliceous deposition, for example, occurred as early as about 14.0 my in the Salinas and Cuyama Basins (Mertz, 1984; Lagoe, 1985) and as late as 8.5 my in the Santa Barbara Basin (Blake, 1985; Arends & Blake, 1986; Barron, 1986b). In some basins, this facies change was marked by carbonate–silica cycles, as mentioned previously.

We may speculate that this diachroneity was due to local preservational effects superimposed on the changing fertility patterns (cf. Barrera, Keller, & Savin, 1985). The Santa Barbara Basin, for example, was anoxic during much of Middle and early Late Miocene time, judging by lack of burrows and infaunal remains, the abundance of laminated rocks, and by the comparatively high amounts of organic carbon and phosphorus. Under these circumstances, there may have been preferential preservation of carbonate components (cf. Berger & Soutar, 1970; Carlos, 1985), whereas higher rates of carbonate dissolution occurred in less anoxic basins like the Cuyama and Salinas Basins. Carlos (1985) has shown, for example, that increased oxygen levels in bottom waters led to increased benthic (including infaunal) organic activity, leading in turn to increased biological breakage of calcareous microfossil tests, a process which promotes dissolution.

In addition, in the Santa Barbara Basin, among the most southerly of the California basins, calcareous sedimentation in warm waters may have persisted longer than in the more northerly basins. The Santa Barbara Basin lies 200–300 km south of the Salinas Basin, and it may have been tectonically translated from an original position some 500–1500 km farther to the south (Kamerling & Luyendyk, 1979; Luyendyk, Kamerling & Terres, 1980).

Similarly, the diachroneity of the transition between the calcareous–siliceous and phosphatic facies (Fig. 23.8) may be a reflection of local productivity and preservation patterns.

Although the phosphatic marlstones are diachronous, their optimal development apparently occurred between about 15.5 and 13.2 my, coincident with some of the highest stands of sea level in the Miocene and also with a part of the heavy carbon excursion (Fig. 23.8). The latter attributed by Vincent & Berger (1985) to an initial rapid extraction of organic carbon, which is depleted in ^{13}C, from the ocean–atmosphere system, followed by continued extraction and preservation of organic carbon into such ocean-margin deposits as the Monterey Formation (see also Woodruff & Savin, 1985). Evidence from the California margin suggests intense upwelling, and the main phase of high productivity diatomaceous sedimentation began some one to two million years after the peak of the heavy carbon excursion (Fig. 23.8). The organic richness of the Monterey phosphatic marlstones, as well as their relatively high phosphorus contents, thus appear to reflect mainly preservation of organic matter in low-oxygen environments rather than high productivity (cf. Bralower & Thierstein, 1984). It is therefore tempting to postulate a similar origin for the heavy carbon excursion – could this reflect an interval of widespread anoxia in marginal basins beneath warm Middle Miocene seas, conditions which led to the preferential preservation of organic carbon and phosphorus on a global scale?

The main phase of pelletal–oolitic phosphorite deposition on the shelves of the Cuyama and other basins was between 14 and 8 my, an interval which straddled the changes from generally weak to generally strong upwelling and from high to low stands of sea level. Table 23.2 summarizes estimated rates of organic carbon and phosphorus accumulation in phosphatic marlstones of the Santa Barbara Basin and in diatomaceous mudrocks which formed the background for normal shelf sedimentation in the Cuyama Basin. Note that whereas organic carbon accumulated at a higher rate in the phosphatic marlstones of the Santa Barbara Basin, phosphorus accumulated in bioturbated diatomaceous muds on the oxygenated Cuyama Shelf at a rate double that of the marls deposited in the low-oxygen Santa Barbara Basin.

A possible explanation is that, although organic matter was supplied initially at higher rates to sediments of the oxygenated Cuyama Shelf, a much larger proportion of the organic carbon was preserved in the low-oxygen sediments of the Santa Barbara Basin. But, whereas substantial amounts of organic carbon were oxidized and recycled from Cuyama sediments, organically-derived phosphorus in them was fixed by precipitation of carbonate fluorapatite very early during diagenesis. The phosphatized cyanobacterial coatings of Cuyama phosphate grains (Garrison *et al.*, 1987) suggest organic mediation of this early phosphatization.

Conclusions

Phosphogenesis along the California margin was most widespread during Middle–early Late Miocene time, between approximately 15.5 and 8.0 my, more or less coincident with the mid-Miocene high stand of sea level (Fig. 23.8). This distribution accords well with the suggestion of Arthur & Jenkyns (1981) that phosphorite genesis correlates in a general way with elevated sea level; in particular, however, it appears to lend credence to Sheldon's (1980) proposal that the global Miocene phosphogenic episode records phosphorus withdrawal from the deep ocean phosphorus sink by trade-wind belt upwelling at the time of transition from high-level, warm oceans to low-level cold oceans.

The distribution of California Miocene phosphatic facies in space and time (Fig. 23.8) indicates that local factors, such as tectonism and persistence of anoxic basins, became superimposed on global climatic–oceanographic trends in determin-

Table 23.2. *Accumulation rates for organically-derived carbon and phosphorus in the Santa Barbara and Cuyama basins (from Garrison, Kastner & Kolodny, 1987)*

Stratigraphic unit and basin	Accumulation Rates: Decompacted sedimentation rate (m my^{-1})	Organic carbon (10^{-5}mol $cm^{-2}y^{-1}$)	Phosphorus (10^{-6}mol $cm^{-2}y^{-1}$)
Phosphatic marlstones (carbonaceous marl Member of Isaacs, 1983), Monterey Formation, Santa Barbara Basin	50	3.7	1.7
Siliceous mudrocks, (diatom muds), upper phosphatic mudstone Member, Santa Margarita Formation, shelfal region of the Cuyama Basin	275	2.6	3.4

ing the nature of the rock record in individual basins. Riggs (1984) has demonstrated similar patterns in the Neogene phosphorite province of the US Atlantic continental margin, and it appears likely that most phosphogenic provinces record similar kinds of complexities, whose recognition and interpretation depend upon precise chronostratigraphy and careful sedimentology.

Acknowledgments

Financial support for various aspects of this work came through grants from the National Science Foundation (EAR85-19113) and the Petroleum Research Fund (PRF 13953-AC), and acknowledgment is made to donors of these funds. Research was also aided by grants from the Faculty Research Committee of U.C. Santa Cruz and the Committee on Pacific Rim Research of the University of California. For discussions and generous access to unpublished data. We are very grateful to M. Arthur, J. Barron, C. Blake, W. Burnett, R. Douglas, M. Lagoe, Y. Nathan and R. Sheldon. The manuscript benefitted from reviews by R. Moberly and S. Savin. This paper is a contribution to Project 156-Phosphorites, of the International Geological Correlation Project.

References

Arends, R.G. & Blake, G.H. (1986). Biostratigraphy and paleoecology of the Naples Bluff coastal section based on diatoms and benthic foraminifera. In *Siliceous Microfossil and Microplankton of the Monterey Formation and Modern Analogs*, ed. R.E. Casey & J.A. Barron, pp. 121–35. Society of Economic Paleontologists and Mineralogists, Pacific Section Book 45, Los Angeles.

Arthur, M.A. & Jenkyns, H.C. (1981). Phosphorites and paleoceanography. *Oceanologica Acta*, **4 Supplement**, 83–96.

Baker, K.B. & Burnett, W.C. (1988). Distribution, texture and composition of modern phosphate pellets in Peru shelf muds. *Marine Geology*, **80**, 195–213.

Barrera, E., Keller, G. & Savin, S.M. (1985). Evolution of the Miocene ocean in the eastern North Pacific as inferred from oxygen and carbon isotopic ratios of foraminifera. In *The Miocene Ocean: Paleoceanography and Biogeography*, ed. J.P. Kennett, pp. 83–102. Geological Society of America Memoir 163.

Barron, J.A. (1986a). Paleoceanographic and tectonic controls on deposition of the Monterey Formation and related siliceous rocks in California. *Palaeogeography, Palaeoclimatology, Palaeoecology*, **53**, 27–45.

Barron, J.A. (1986b). Updated diatom biostratigraphy for the Monterey Formation of California. In *Siliceous Microfossil and Microplankton of the Monterey Formation and Modern Analogs*, ed. R.E. Casey & J.A. Barron, pp. 105–19. Society of Economic Paleontologists and Mineralogists, Pacific Section, Book 45, Los Angeles.

Barron, J.A. & Keller, G. (1983). Paleotemperature oscillations in the middle and late Miocene of the northeastern Pacific. *Micropaleontology*, **29**, 150–81.

Barron, J.A., Keller, G. & Dunn, D.A. (1985). A multiple microfossil biochronology for the Miocene. In *The Miocene Ocean: Paleoceanography and Biogeography*, ed. J.P. Kennett, pp. 21–36. Geological Society of America Memoir 163.

Berger, W.H. & Soutar, A. (1970). Preservation of plankton shells in an anaerobic basin off California. *Geological Society of America Bulletin*, **81**, 275–82.

Blake, G.H. (1985). 'The faunal response of California continental margin benthic foraminifera to the oceanographic and depositional events of the Neogene.' Unpublished Ph.D. thesis, University of Southern California.

Blake, M.C., Jr., Campbell, R.H., Dibblee, T.W., Jr., Howell, D.G., Nilsen, T.H., Normark, W.R., Vedder, J.C. & Silver, E.A. (1978). Neogene basin formation in relation to plate-tectonic evolution of San Andreas fault system, California. *American Association of Petroleum Geologists Bulletin*, **62**, 344–72.

Bralower, T.J. & Thierstein, H.R. (1984). Low productivity and slow deep-water circulation in mid-Cretaceous oceans. *Geology*, **12**, 614–18.

Burnett, W.C. (1980). Apatite–glauconite associations off Peru and Chile: palaeo-oceanographic implications. *Journal of Geological Society London*, **137**, 757–64.

Burnett, W.C., Baker, K.B., Chin, P., McCabe, W. & Ditchburn, R. (1988). Uranium-series and AMS ^{14}C studies of modern phosphatic pellets from Peru shelf muds: *Marine Geology*, **80**, 215–30.

Carlos, A.P. (1985). 'Comparative study of paleoenvironmental factors on the preservation of calcareous microfossils in the Miocene Monterey Formation, Upper Newport Bay, Newport Beach, California'. Unpublished M.Sc. thesis, University of Southern California, Los Angeles.

Clark, J.C. (1981). Stratigraphy, paleontology, and geology of the central Santa Cruz Mountains, California Coastal Ranges. *US Geological Survey Professional Paper*, **1168**, 51pp.

Conrad, C.L. & Ehlig, P.L. (1983). The Monterey Formation of the Palos Verdes Peninsula, California – an example of sedimentation in a tectonically active basin within the California Continental Borderland. In *Cenozoic Marine Sedimentation, Pacific Margin, USA*, ed. D.L. Larue, & R.J. Steel, pp. 103–16. Society of Economic Paleontologists and Mineralogists, Pacific Section, Special Publication, Los Angeles.

Cook, P.J. & McElhinny, M.W. (1979). A reevaluation of the spatial and temporal distribution of sedimentary phosphate deposits in light of plate tectonics. *Economic Geology*, **74**, 315–30.

Crouch, J.K. (1979). Neogene tectonic evolution of the California continental borderland and western Transverse Ranges. *Geological Society of America Bulletin*, **90**, 338–45.

Demaison, G.J. & Moore, G.T. (1980). Anoxic environments and oil source bed genesis. *American Association of Petroleum Geologists Bulletin*, **64**, 1179–209.

Dickert, P.F. (1966). Tertiary phosphatic facies of the Coast Ranges. *California Division of Mines Geological Bulletin*, **190**, 289–304.

Dickert, P.F. (1971). 'Neogene phosphatic facies in California.' Unpublished Ph.D. thesis, Stanford University, Stanford.

Fedawa, W.T. & Hovland, R.D. (1981). Phosphate resources of the Upper Miocene phosphate deposit near New Cuyama, Santa Barbara County, California. In *Geology and Petrology of the Upper Miocene Phosphate Deposit Near New Cuyama, Santa Barbara County, California*, ed. A.E. Roberts & T.L. Vercoutere, pp. 68–74, US Geological Survey Open-File Report **81–1037**.

Froelich, P.N., Kim, K.H., Jahnke, R., Burnett, W.C., Soutar, A. & Deakin, M. (1983). Pore water fluoride in Peru continental margin sediments: uptake from seawater. *Geochimica Cosmochimica Acta*, **47**, 1605–12.

Garrison, R.E., Kastner, M. & Kolodny, Y., (1987). Phosphorites and phosphatic rocks in the Monterey Formation and related Miocene units, coastal California. In *Cenozoic Basin Development in Coastal California, Rubey Volume VI*, ed. R.V. Ingersoll, & W.G. Ernst, pp. 348–81. Prentice-Hall, Englewood Cliffs, NJ.

Garrison, R.E., Stanley, R.G. & Horan, L.J. (1979). Middle Miocene sedimentation on the southwestern edge of the Lockwood High, Monterey County, California. In *Tertiary and Quaternary Geology of the Salinas Valley and Santa Lucia Range, Monterey County, California*, ed. S.A. Graham, pp. 51–65. Society of Economic Paleontologists and Mineralogists, Pacific Section Field Guide #4, Los Angeles.

Gower, H.D. & Madsen, B.M. (1964). The occurrence of phosphate rock in California: *US Geological Survey Professional Paper*, **501-D**, 79–85.

Graham, S.A. (1976). 'Tertiary sedimentary tectonics of the central Salinian block of California'. Unpublished Ph.D. thesis, Stanford University, Stanford.

Graham, S.A. (1979). Tertiary stratigraphy and depositional environments near Indians Ranch, Monterey County, California. In *Tertiary and Quaternary Geology of the Salinas Valley and Santa Lucia Range, Monterey County, California*, ed. S.A. Graham, pp. 3–12. Society of Economic Paleontologists and Mineralogists, Pacific Section, Pacific Coast Paleogeography Field Guide #4, Los Angeles.

Graham, S.A. (1987). Tectonic controls on petroleum occurrence in central California. In *Cenozoic Basin Development in Coastal California, Rubey Volume VI*, ed. R.V. Ingersoll & W.G. Ernst pp. 47–63, Prentice-Hall, Englewood Cliffs, NJ.

Haq, B.V., Hardenbol, J. & Vail, P.R. (1987). Chronology of fluctuating sea levels since the Triassic (250 Million years ago to present). *Science*, **235**, 1156–67.

Howell, D.G., Crouch, J.K., Greene, H.G., McCulloch, D.S. & Vedder, J.G. (1980). Basin development along the late Mesozoic and Cainozoic California margin: a plate tectonic margin of subduction, oblique subduction and transform tectonics. In *Sedimentation in Oblique Slip Mobile Zone's*, ed. P.F. Ballance & H.G. Reading, pp. 43–62. International Association of Sedimentologists, Special Publication Number 4, Los Angeles, California.

Ingle, J.C., Jr. (1967). Foraminiferal biofacies variation and the Miocene-Pliocene boundary in southern California. *Bulletin of American Paleontology*, **52(236)**, 210–394.

Ingle, J.C., Jr. (1981). Origin of Neogene diatomites around the north Pacific rim. In *The Monterey Formation and Related Siliceous Rocks of California*, ed. R.E. Garrison & R.G. Douglas, pp. 159–80. Society of Economic Paleontologists and Mineralogists, Pacific Section, Los Angeles.

Isaacs, C.M. (1980). 'Diagenesis in the Monterey Formation examined laterally along the coast near Santa Barbara, California.' Unpublished Ph.D. thesis, Stanford University, Stanford.

Isaacs, C.M. (1981). Lithostratigraphy of the Monterey Formation, Goleta to Point Conception, Santa Barbara Coast, California. In *Guide to the Monterey Formation in the California Coastal Area, Ventura to San Luis Obispo*, ed. C.M. Isaacs, pp. 9–24. Pacific Section, AAPG Publication, vol. 52.

Isaacs, C.M. (1983). Hemipelagic deposits in a Miocene basin, California: toward a model of lithologic variation and sequence. In *Marine Sedimentation on the Pacific Margin, USA*, ed. D.K. Larue & R.J. Steel, pp. 117–32. Society of Economic Paleontologists and Mineralogists Publication, Pacific Section, Los Angeles.

Kamerling, M.J. & Luyendyk, B.P. (1979). Tectonic rotations of the Santa Monica Mountains region, western Transverse Ranges, California, suggested by paleomagnetic vectors. *Geological Society of America Bulletin*, **90**, 331–7.

Keigwin, L.D. (1979). Late Cenozoic stable isotope stratigraphy and paleoceanography of DSDP sites from the east equatorial and central North Pacific Ocean. *Earth and Plantetary Science Letters*, **45**, 361–82.

Kleinpell, R.M. (1938). *Miocene Stratigraphy of California*. American Association of Petroleum Geologists, Tulsa.

Kleinpell, R.M. (1980). *Miocene Stratigraphy of California Revisited*. American Association of Petroleum Geologists, Studies in Geology 11, 182pp.

Lagoe, M.B. (1985). Depositional environments in the Monterey Formation, Cuyama Basin, California. *American Association of Petroleum Geologists Bulletin*, **96**, 1296–312.

Lagoe, M.B. (1987). Middle Cenozoic basin development, Cuyama Basin, California. In *Cenozoic Basin Development in Coastal California, Rubey Volume VI*, ed. R.V. Ingersoll & W.G. Ernst, pp. 172–205. Prentice-Hall, Englewood Cliffs, NJ.

Luyendyk, B.P., Kammerling, M.J. & Terres, R. (1980). Geometric model for Neogene crustal rotations in southern California. *Geological Society of America Bulletin*, **91**, 211–17.

Mertz, K.A., Jr. (1984). 'Origin and depositional history of the Sandholdt Member, Miocene Monterey Formation, Santa Lucia Range, California.' Unpublished Ph.D. thesis, University of California, Santa Cruz.

Pisciotto, K.A. (1978). 'Basinal sedimentary facies and diagenetic aspects of the Monterey shale, California.' Unpublished Ph.D. thesis, University of California, Santa Cruz.

Pisciotto, K.A. & Garrison, R.E. (1981). Lithofacies and depositional environments of the Monterey Formation, California. In *The Monterey Formation and Related Siliceous Rocks of California*, ed. R.E. Garrison & R.G. Douglas, pp. 97–122. Society of Economic Paleontologists and Mineralogists Publication, Pacific Section.

Riggs, S. (1984). Paleoceanographic model of Neogene phosphorite deposition, US Atlantic continental margin. *Science*, **223**, 123–31.

Roberts, A.E. (1981). Phosphatic rock localities in California. US Geological Survey Open-File Report, **79–1466**, 64pp.

Roberts, A.E. & Vercoutere, T.L. (1987). Geology and petrology of the upper Miocene phosphate deposit near New Cuyama, Santa Barbara County, California. *US Geological Survey Bulletin*, **B1635**, 61pp.

Savin, S.M., Douglas, R.G., Keller, G., Kilingley, J.S., Shaughnessy, L., Sommer, M.A., Vincent, E. & Woodruff, F. (1981). Miocene benthic foraminiferal isotope records: a synthesis. *Marine Micropaleontology*, **6**, 423–50.

Sheldon, R.P. (1980). Episodicity of phosphate deposition and deep ocean circulation – an hypothesis. In *Marine Phosphorites – Geochemistry, Occurrence, Genesis*, ed. Y.K. Bentor, pp. 239–47. Society of Economic Paleontologists and Mineralogists Special Publication No. 29, Tulsa.

Soudry, D. & Champetier, Y. (1983). Microbial processes in the Negev phosphorites (southern Israel). *Sedimentology*, **30**, 411–23.

Thor, D.R. (1977). 'Depositional environments and paleogeographic setting of the Santa Margarita Formation, Ventura County, California.' Unpublished M.Sc. thesis, California State University, Northridge.

Thor, D.R. (1978). Depositional environments and paleogeographic setting of the Santa Margarita Formation, Ventura County, California. In *Depositional Environments of Tertiary Rocks along Sespe Creek, Ventura County, California*, ed. A.E. Fritsche, pp. 42–59. Society of Economic Paleontologists and Mineralogists, Pacific Section, Pacific Coast Paleogeography Field Guide 3, Los Angeles.

Vail, P.R. & Hardenbol, J. (1979). Sea-level changes during the Tertiary Oceans. *Oceanus*, **22**, 71–9.

Vincent, E. & Berger, W.H. (1985). Carbon dioxide and polar cooling in the Miocene: the Monterey hypothesis. In *Natural Variations in Carbon Dioxide and the Carbon Cycle*, ed. E.T. Sundquist & W.S. Broecker, pp. 455–68. American Geophysical Union Monography 32.

Williams, L.A. & Reimers, C. (1983). Role of bacterial mats in oxygen-deficient marine basins and coastal upwelling regimes: preliminary report. *Geology*, **11**, 267–9.

Williams, T.A. (1984). Subtidal stromatolites in Monterey Formation and other organic rich rocks as suggested source contributors to petroleum formation. *American Association of Petroleum Geologists Bulletin*, **68**, 1879–93.

Woodring, W.P. & Bramlette, M.N. (1950). Geology and paleontology of the Santa Maria district, California. *US Geological Survey Professional Paper*, **222**, 185pp.

Woodruff, F. & Savin, S.M. (1985). $\delta^{13}C$ values of Miocene Pacific benthic foraminifera: correlations with sea level and biological productivity. *Geology*, **13**, 119–22.

Woodruff, F., Savin, S. & Douglas, R.G. (1981). Miocene stable isotope record: a detailed Pacific Ocean study and its paleoclimatic implications. *Science*, **212**, 665–8.

Younse, G.A. (1979). 'The stratigraphy and phosphoritic rocks of the Robinson Canyon–Laureles Grade area, Monterey County, California.' Unpublished M.Sc. thesis, San Jose State University, San Jose, California.

24

The role of bacterial mats in phosphate mineralization with particular reference to the Monterey Formation

C.E. REIMERS, M. KASTNER AND R.E. GARRISON

Abstract

Organic-rich shales sampled from the Naples Beach section of the Monterey Formation, California, contain authigenic phosphorite phases which apparently formed within bacterial mat laminations. Light and scanning electron microscope observations show that the fossilized remains of these mats are similar to communities of filamentous sulfur-oxidizing bacteria found today at slope-depth, sediment, oxic–anoxic interfaces. The crystal size and habits of the phosphatic minerals resemble contemporary precipitates in contact with pore water solutions.

Mineralogical and chemical analyses of the most phosphatic layers show that some are a nearly pure carbonate fluorapatite (or francolite). These layers are lightly colored and are sandwiched between black siliceous muds with $<0.2\%$ $CaCO_3$, and organic carbon contents that exceed 25%. In other phosphatic intervals of the Naples Beach section, calcite is a major diluent of light and dark layers.

The C:P ratio of a single modern bacterial mat sample shows that these benthic communities are more enriched in phosphorus than planktonic organic matter. Thus, the generally believed premise that authigenic phosphorite formation results from organic matter diagenesis in near-surface anoxic sediments is here modified by the contention that such transformations are more highly favored in sediments supporting massive microbial communities.

Introduction

Phosphorites occur in a wide variety of forms in modern, organic-rich, diatomaceous sediments, but only thin laminae and lenses of unconsolidated phosphatic material in sediments from off the coasts of Namibia and Peru–Chile have been shown to be wholly contemporaneous (Baturin, Merkulova & Chalov 1972; Veeh, Burnett & Soutar, 1973; Veeh, Calvert & Price, 1974). To explain the origin of these deposits, Price & Calvert (1978) have argued that phosphorus must be precipitated diagenetically from surrounding pore waters. Similar arguments have been made by Burnett (1977), Burnett, Beers & Roe (1982), Froelich *et al.* (1983), and Jahnke *et al.* (1983) to reconcile pore-water and solid phase phosphorus profiles, and/or growth rates of Holocene–Late Pleistocene phosphatic nodules located near the sediment–water interface. The source of the dissolved phosphorus has been presumed by these authors to be sedimented planktonic organic phosphorus. Suess (1981), however, suggested dissolution of fish remains as an important phosphorus source.

The objective of this paper is to explore the possibility that in specific present and past continental margin settings the benthic community is and was a more immediate source of phosphorus for concurrent phosphate mineralization. Such a benthic community is dominated by colorless sulfur-oxidizing bacteria belonging to the family Beggiatoaceae. These organisms are rich in organic forms of phosphorus, and are taxonomically subdivided by morphological criteria into three genera: *Beggiatoa*, *Thioploca*, and *Vitreoscilla* (Wiessner, 1981). They have been found as extensive layers of intertwining white filaments, or 'mats', at non-bioturbated, sediment, oxic–anoxic interfaces in phosphorite source regions off Peru–Chile (*Thioploca*, spp., Gallardo, 1977; Rosenberg *et al.*, 1983), off southwest Africa (unidentified spp., Copenhagen, 1934; Gallardo, 1977), in the Gulf of California (*Beggiatoa* spp., Williams & Reimers, 1983) and in the Santa Barbara Basin (*Beggiatoa* spp., Soutar & Crill, 1977). Since these organisms derive energy from reduced sulfur compounds, they also thrive near hydrothermal vents (Jannasch, 1984).

We suggest that an active bacterial role in phosphatization may also be feasible because: the Beggiatoaceae deposit intercellular polyphosphate granules (volutin) (Strohl & Larkin, 1978; Wiessner, 1981); they are associated with many other varieties of chemosynthetic bacteria (whose activities are unknown); and, both mat communities and modern phosphatic minerals are frequently associated with and are contributors to laminated sediment deposits. In ancient phosphatic sediments, laminae may have been obliterated by later diagenetic processes; but we and coworkers have observed that in some, Miocene, laminated, phosphatic shales, filamentous fossils and thin carbonate fluorapatite layers or lenses are preserved together with little apparent alteration (Williams & Reimers, 1983; Mertz, 1984; Garrison, Kastner & Kolodny, 1987). These findings are based on light and scanning electron microscopy observations of samples from the Monterey Formation, California. Analogous studies have been undertaken by Soudry & Champetier (1983) and O'Brien *et al.* (1981) to verify a bacterial influence at origin for Late Cretaceous, shallow marine phosphorites in the Negev, Israel, as well as for Recent phosphorites from the low productivity continental margin of eastern Australia. Because the ages and environments of formation of these deposits are clearly different from those of the Miocene and Recent 'western margin' phosphorites, we believe the geologic record must preserve many examples of microbial phosphatization processes.

Background and geologic setting

The suggestion that there may be a cause-and-effect relationship between bacterial mats and authigenic phosphorite formation was first made by L. Williams (Williams & Reimers, 1983; Williams, 1984). The relationship was proposed because filamentous microfossils and phosphorite layers and nodules were abundant in finely laminated, subsurface, siliceous sediments from the Miocene, Monterey Formation in the western San Joaquin Basin of California (Graham & Williams, 1985). The filamentous microfossils were presumed to be remnants of Beggiatoaceae mats because the laminated rocks contained benthic foraminifera indicative of upper- to mid-bathyal water depths (150–1500 m), and stratigraphical, mineralogical and organic geochemical evidence suggested that deposition had occurred in a dysaerobic slope environment. These criteria distinguished these microfossils from the remains of photosynthetic cyanobacteria (Williams, 1984).

The Monterey phosphatic lithofacies which Williams sampled occurs in most of the Neogene basins in California, and in many localities appears to be marked by mat-like, thin, organic-rich laminations (Mertz, 1984; Garrison *et al.*, 1987). Further characteristics of this lithofacies are a relatively low rate of accumulation and cyclic sedimentation of exceedingly pure, carbonate fluorapatite layers. A schematic stratigraphic diagram for one well-exposed coastal section at Naples Beach, Santa Barbara County, demonstrates the correlation between low sedimentation rates and highly organic-rich, phosphatic units (Fig. 24.1). The correlations between age, thickness, and lithofacies in Figure 24.1 are a synthesis of numerous stratigraphical, paleontological and paleomagnetical studies compiled by Isaacs (1981a, b and 1983), Blake (1985), Arends & Blake (1986), and Barron (1986). The paleoenvironment of the phosphatic facies represents a high productivity offshore region with an intense oxygen minimum, and a periodically sediment-starved continental slope or distal basin floor. This is one of the environments in which dense bacterial mats occur today. The samples studied in detail are from two distinct zones, one pervasively laminated, the other partially laminated (Figs. 24.1 and 24.2). Sedimentation rates are generally lower in the vicinity of the well laminated zone. This may be due in part to numerous hiatuses which are marked by pebbly phosphatic layers and conglomerates (Blake, 1985).

Modern bacterial mats

Living bacterial mats are composed of microaerophilic organisms which grow in a horizontal layer a few millimeters thick. At a sediment–water interface, mat formation tends to be favored where pore water gradients of O_2 and H_2S overlap. This helps to promote rapid biological oxidation of the sulfide, and apparently confines the oxidation of sulfide to within the mat (Jørgensen & Revsbech, 1983). pH is lowered within a mat, as a consequence of the H_2S oxidation and other metabolic reactions (Jørgensen & Revsbech, 1983). The exact nature of these reactions, particularly the type of carbon metabolism, is debated by microbiologists (see Strohl, 1989, for a summary).

To gain a better understanding of mat organisms in modern marine environments, in August 1985 we sampled freshly recovered box-cored sediments from the floor of the Santa Barbara Basin (water depth $\sim$600 m). The surface of the sediments was covered with a patchy layer of filaments presented in Figure 24.3. These photographs, taken under transmitted light, show mostly 5–7 μm diameter *Beggiatoa*, and a few other sulfur bacteria which have been identified as *Thiothrix* and *Thiobacillus ferrooxidans* by H. Jannasch (pers. comm., 1985). The *Beggiatoa* contain visible inclusions of elemental sulfur (Fig. 24.3). They also may include polyphosphate (volutin) and poly-β-hydroxybutyric acid (PHB) granules which are characteristic constituents, but less refractile in light (Strohl, 1989).

Elemental analyses of the Santa Barbara *Beggiatoa* mat revealed that these organisms are relatively enriched in phosphorus (Table 24.1). The sample analyzed was pipetted from the core surface, frozen, and later freeze-dried. It was not a pure bacteria sample, but contained some detritus. Ignited and untreated splits of dried mat were acid treated, and acid extracts were analyzed colorimetrically for total and inorganic phosphorus, respectively (Andersen, 1976). Total carbon and carbonate carbon were determined as liberated CO_2 after ignition (total C) and acid digestion (carbonate C), using a Coulometrics carbon analyzer. The resulting organic C:P mole ratio of 68, listed in Table 24.1, is significantly lower than the planktonic Redfield ratio of 106. We assume this is because *Beggiatoa* cell walls contain high concentrations of phospholipids (Wiessner, 1981).

The bacterial mats that we observed on box core surfaces from the floor of the Santa Barbara Basin are known to be dynamic and often intermittent sediment–interface features (Jørgensen, 1977a; Soutar & Crill, 1977). Mat intermittence is most probably caused by up and down movements of mat bacteria, in the sediment, in response to fluctuations in sedimentation or redox conditions (Jørgensen, 1977a). As new mats reform and grow, postmortem degradation will usually obliterate most of the cellular structure of earlier populations. This tends to leave only an organically-bound sediment fabric for fossilization (Reimers, 1982). Early degradation is primarily due to sulfate-reducing organisms which are abundant in organic-rich sediments over several to tens of centimeters (Jørgensen, 1977b). As discussed by Burnett (1977), Suess (1981), and Mertz (1984), sulfate reduction is accompanied by the release and build up of inorganic PO_4^{-3} in pore waters. At the sediment surface the assimilation of phosphorus into the cell constituents of new-generation mat organisms could remove phosphate from adjacent pore waters, partially capping the phosphate flux out of the sediment.

Laminated phosphorites in the Monterey Formation

Occurrence, petrography and textures

The phosphatic lithofacies of California's Monterey Formation contains prime examples of phosphatic laminae which seem to have originated as early authigenic precipitates. At Naples Beach the thickest and purest phosphatic laminae are located just below and within a zone of condensed sedimentation in which about 15 m of section represents the time interval from about 14 to 9 my (Fig. 24.1) (Blake, 1985; Arends & Blake, 1986; Barron, 1986). This condensed zone occurs within the Carbonaceous Marl member described by Isaacs (1981a). As discussed

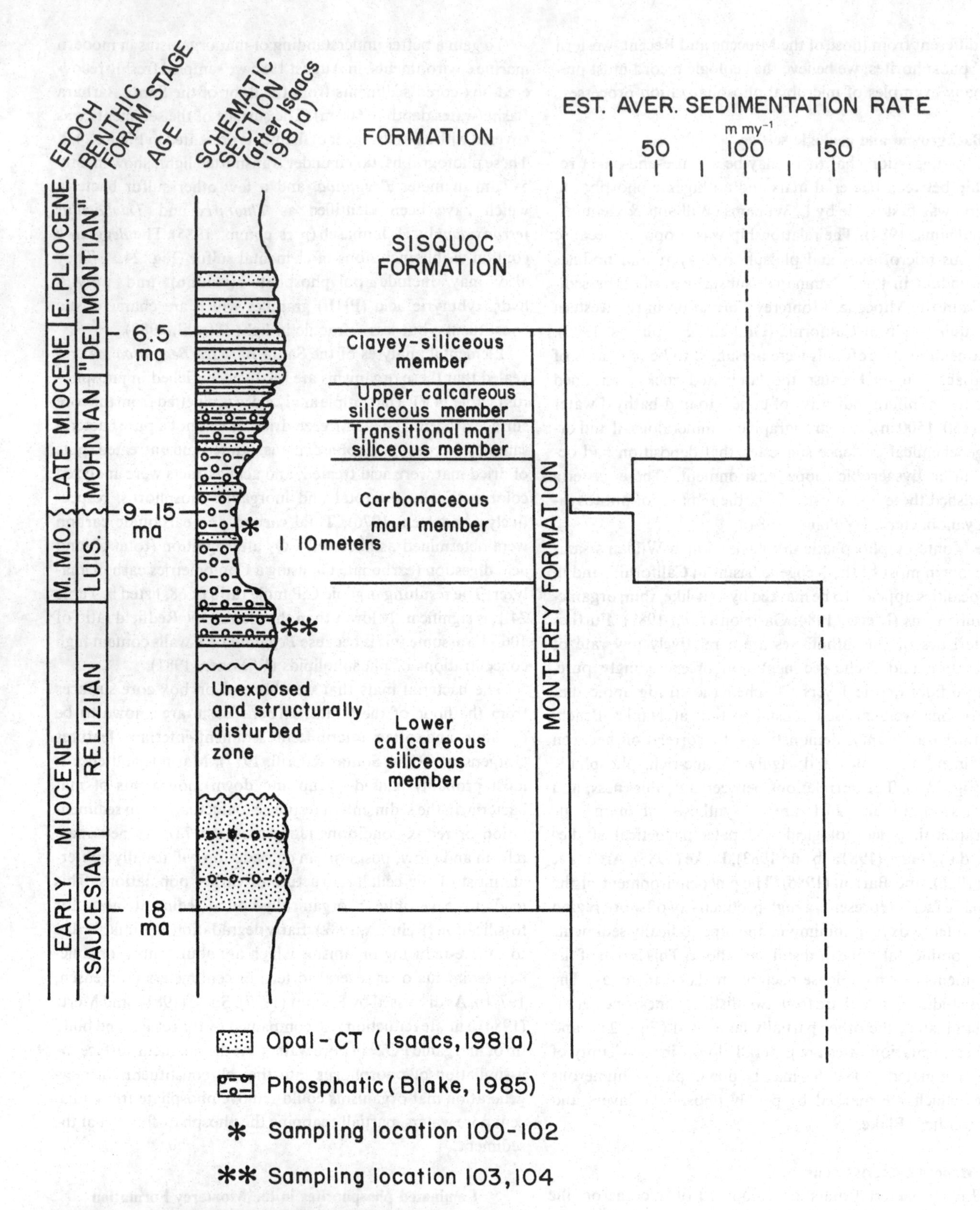

Fig. 24.1. Generalized lithostratigraphic column of the Miocene sequence at Naples Beach, Santa Barbara County, California. Modified from Isaacs (1981a, 1983), with additional stratigraphic data from Blake (1985) and Barron (1986). Stage names are from Kleinpell (1938, 1980). The unconformity between the Middle and Late Miocene is based on biostratigraphic data in Blake (1985) and Barron (1986). The estimated average sedimentation rates are compacted rates.

A

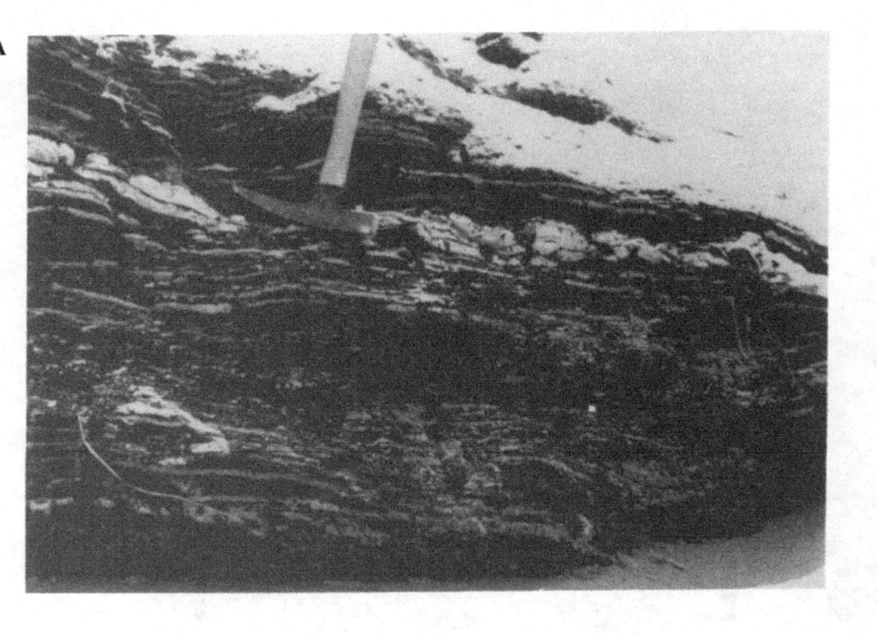

B

C

D

Fig. 24.2. (A) An outcrop of laminated phosphatic- and organic-rich shale within the carbonaceous marl member at Naples Beach. Samples 100–102 were collected from this exposure. (B) Representative sample (101) from the carbonaceous marl member. (C) A view of phosphatic- and organic-rich shales within the lower calcareous–siliceous member at Naples Beach. (D) Representative sample (104) from the calcareous–siliceous member.

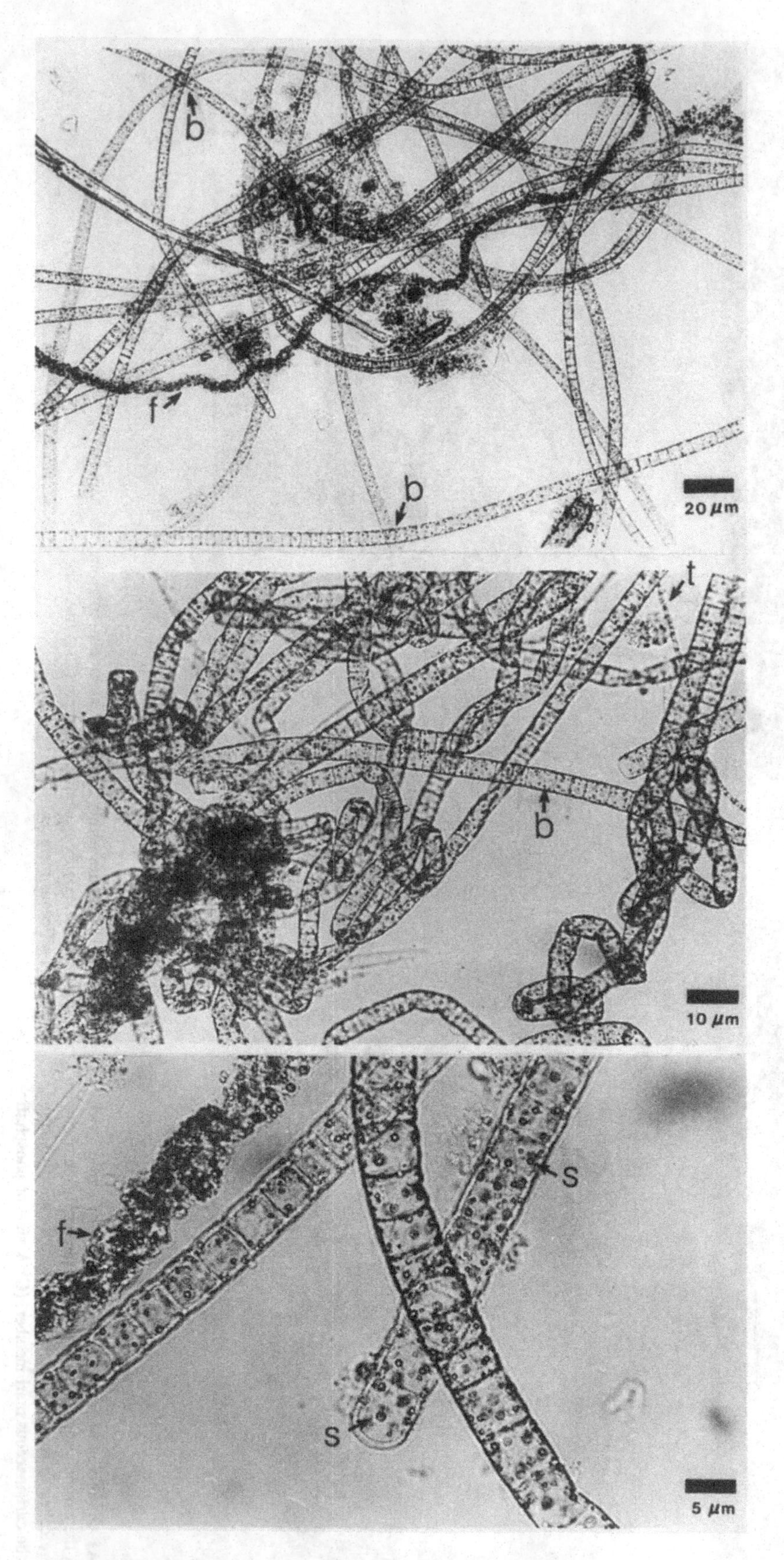

Fig. 24.3. Photographs taken under transmitted light of modern mat microorganisms from the Santa Barbara Basin; b – *Beggiatoa*; f – *Thiobacillus ferrooxidans*; t – Thiothrix; and s – elemental sulfur.

Table 24.1. *Analyses of a Fresh Beggiatoa Mat from the Santa Barbara Basin*

Analytical parameter	Weight % of salt-free sample	Weight ratio	Mole ratio
Total carbon	6.68, 6.58 (run in duplicate)		
Inorganic carbon	0.54		
Total phosphorus	0.28		
Inorganic phosphorus	0.05		
% Ash	34.3		
Total C: Total P		24	61
Org C: Org P		26	68

Organic fractions are calculated as the difference between total and inorganic fractions.

earlier, we sampled it in one area shown in close-up views in Figure 24.2A and B. These samples, 100–102, were later separated into their discrete 'light' and 'dark' layers. A second set of samples, 103 and 104, was taken in a more irregularly laminated, phosphatic interval within the lower Calcareous–Siliceous member of the section (Figs. 24.1 and 24.2C, D). Since the layering of these samples was thin and discontinuous, individual light and dark horizons were combined into light and dark subsamples, respectively. The only other striking visual difference between the samples from the two areas of the section is that in thin sections one sees many calcitic and phosphatized foraminifera tests in both light and dark layers in samples 103 and 104, but not in samples 100–102.

Figure 24.4 displays photomicrographs of the kind of light and dark layering that is characteristic of the Naples Beach samples. Many of the light layers are composed of a nearly homogeneous groundmass of submicroscopic particles (Fig. 24.4A). There are, however, many mixed layers in the section, in which an organic mesh wraps over and around phosphatic particles and peloids (Fig. 24.4B). The structure is analogous to the cyanobacterial lamination described by Soudry & Champetier (1983) in the Negev, Israel, phosphorites. It is also reminiscent of what Price & Calvert (1978) describe as the contemporary phosphorite laminae in the diatomaceous oozes off Namibia. The phosphatic particles observed within these mixed layers appear mostly as agglutinated grains.

Scanning electron microscope (SEM) observations and standard X-ray diffraction-powder studies of different light, dark and mixed layers confirm striking differences between them (Fig. 24.5; Table 24.2). Thick (i.e. 1–10 mm) light layers are composed of a porous mass of apatite crystals, often radially aggregated, that are generally 1–3 μm in length and $<1\ \mu$m in width (Fig. 24.5A–C). The crystals are subhedral to euhedral and have textured surfaces. In appearance, they are very similar to experimentally precipitated carbonate fluorapatite crystals synthesized in seawater on calcite and apatite seed crystals (Gulbrandsen, Roberson & Neil 1984). They also resemble francolite formed interstitially in Peru margin muds (Burnett, 1977).

In contrast to the light layers, dark layers appear to be composed primarily of the spongy, structureless organic matter and the biogenic opaline silica (mainly diatoms and sponge spicules) that together form the matrix of most modern coastal upwelling sediments (Fig. 24.5D, E). With the exception of some detrital silicates and pyrite (Fig. 24.5F; Table 24.2), non-biogenic crystalline material is rare in these layers. Porosities are high, particles are small, and on a submicroscopic scale particles do not appear preferentially oriented parallel to the bedding planes. Occasionally, by SEM, one sees filamentous structures draping over and amongst the debris.

The identification of fossilized bacteria filaments is constrained by preservation and based on morphological similarities between modern and fossilized forms. In Figure 24.6, a modern Santa Barbara *Beggiatoa* filament is compared to fossilized examples from the Carbonaceous Marl member at Naples Beach (where bacterial fossil preservation was exceptionally good). The modern example was critical point dried after gluteraldehyde fixation and an exchange series replacement of interstitial water with distilled water–ethanol and ethanol–freon. The Naples Beach examples were observed on freshly exposed surfaces of mixed but predominantly light, and dark layers of sample 101. By Energy Dispersive X-ray (EDX) analysis, it appeared the fossils in Figure 24.6B were phosphatized. However, they may still have been organic, since the cellular structure and intertwining character of Beggiatoaceae filaments were retained. As seen in the background area in Figure 24.6B, apatitic crystals are less distinctly developed and are bound by an organic coating throughout such mixed layers. The diameters of the fossilized filaments in Figure 24.6B are smaller than the ones in Figure 24.6C and the Santa Barbara modern analog. This is not unusual because modern *Beggiatoa* range from $<1\ \mu$m to $>60\ \mu$m in diameter, and different size classes often exist in a single mat community (Jørgensen, 1977a).

Chemical composition

The organic matter content and acid-soluble constituents of 12 subsamples of Naples Beach shales are reported in the lower half of Table 24.2. Organic carbon and carbonate carbon were determined by the same coulometric method used for analyzing the *Beggiatoa* mat sample. Soluble phases were extracted overnight at room temperature in 1N HCl. In several light layers, the soluble fraction constituted $>90\%$ of the sample. Inorganic phosphorus was determined, from the acid leachate, colorimetrically; and calcium, magnesium and iron were measured by standard atomic absorption spectroscopy methods. We also attempted to measure organic phosphorus concentrations using the ignition method of Andersen (1976). In dark layers, total phosphorus after ignition exceeded inorganic phosphorus by 200–600 ppm which was less than the reproducibility of the method. In light layers, total phosphorus equaled inorganic phosphorus ± 400 ppm. Accordingly, in all cases analyzed, organic phosphorus comprises $<0.1\%$ of the samples.

The compositions of the acid-soluble fractions of the various subsamples indicate that the light layers of samples 100–102 are composed of almost pure carbonate fluorapatite mixed with some organic matter. All the calcium is presumed to be from apatite as no calcite was detected by X-ray diffraction (Table 24.2). Weight percent $CaO:P_2O_5$ ratios are between 1.38 and 1.52 compared to 1.50 in pure average francolite (McClellan, 1980; Thomson *et al.*, 1984). CO_2 contents, however, range from

A

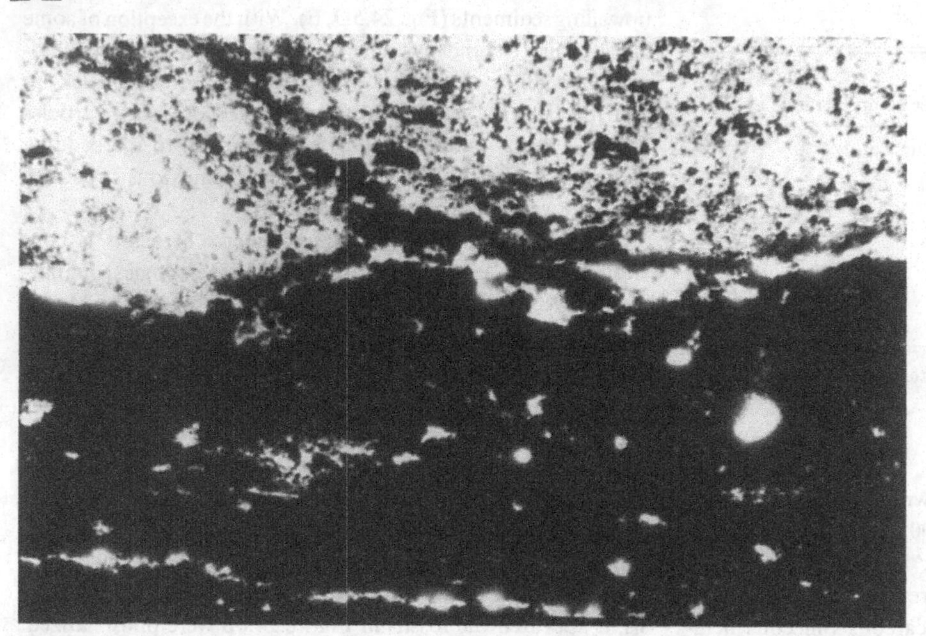

B

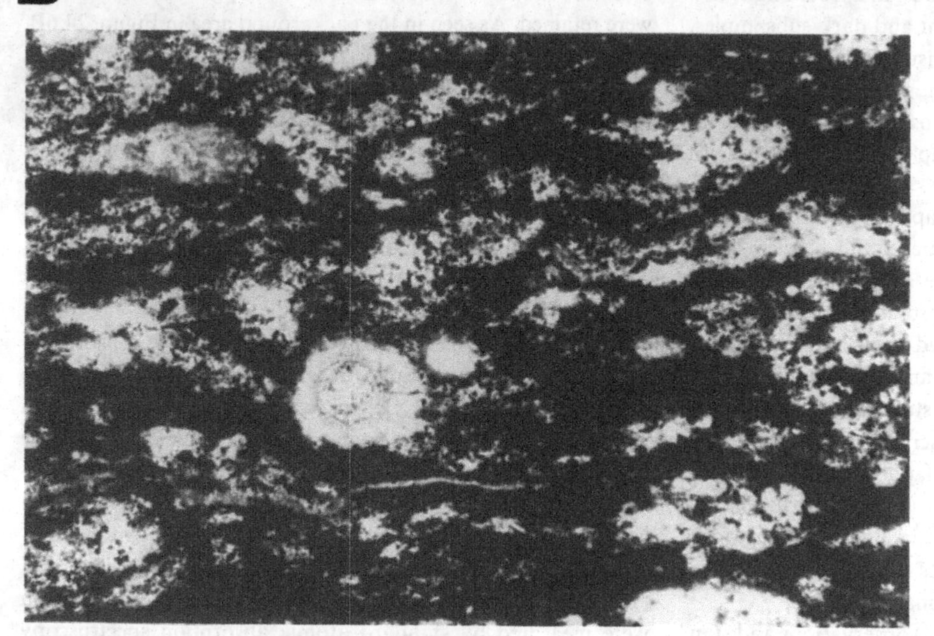

300 µm

Fig. 24.4. Typical photomicrographs of phosphatic organic-rich shales, Naples Beach, showing: (A) sharp transition between a thick light and dark layer (sample 101); and (B) a mixed light and dark layer (sample 100). Note phosphatic halo around a partially replaced foraminifer in the mixed layer. Calcareous microfossils are rare in samples 100–102, but abundant in samples 103–104.

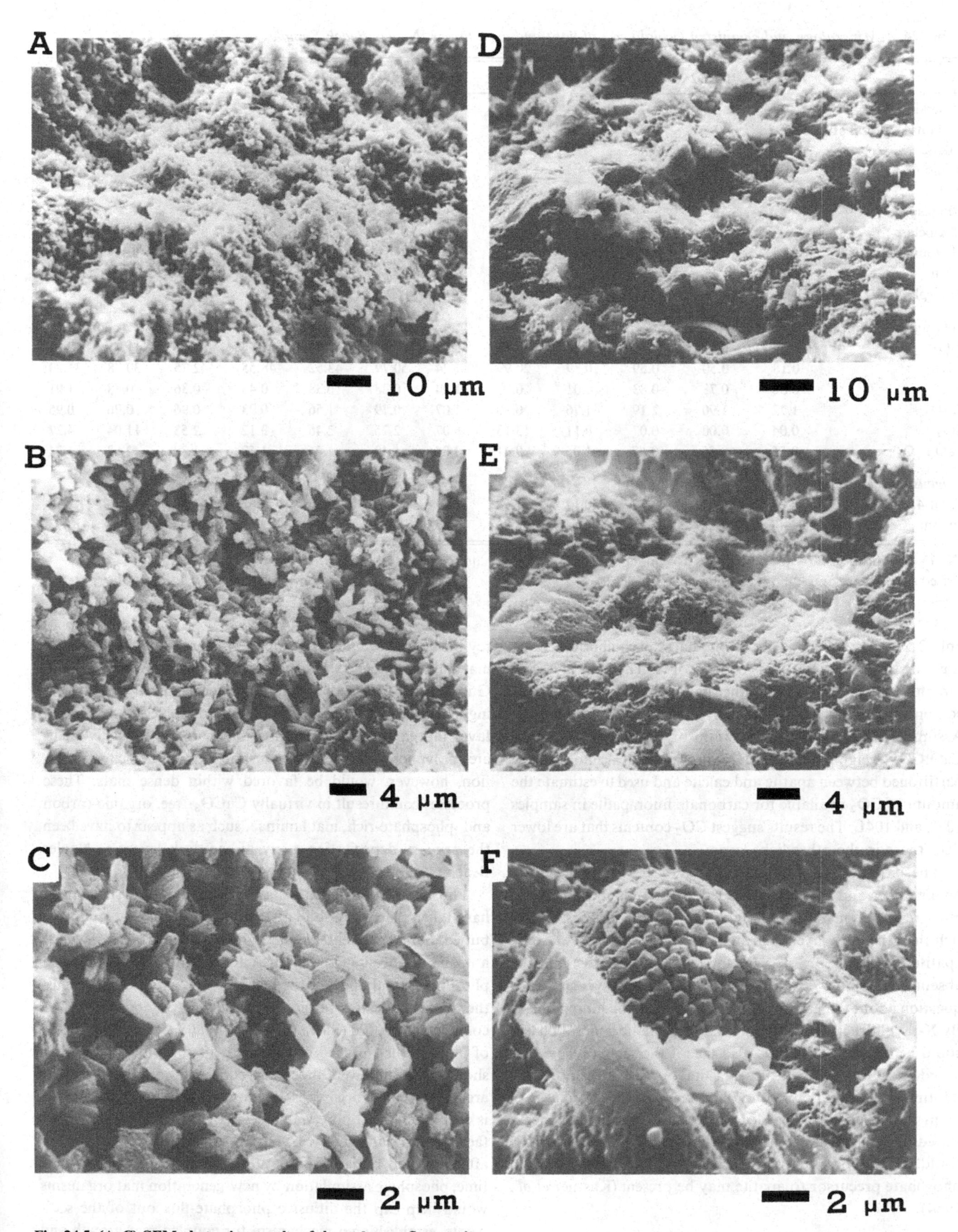

Fig. 24.5. (A–C) SEM photomicrographs of the carbonate fluorapatite crystals which comprise the groundmass of most thick (1–10 mm) light layers in the Naples Beach samples studied. (D,E) Typical amorphous organic and opaline material which together make up the matrix of Naples Beach dark layers. (F) Framboidal pyrite within a dark layer.

Table 24.2. *Mineralogy and chemistry (weight %) of light and dark layers, Naples Beach Samples*

	100 D1	100 D2	101 D1	102 D2	103 D	104 D	100L2	100L3	101 L1	102 L1	103 L	104 L
Mineralogy[a]												
Carbonate-fluorapatite	0	0	0	0	+	+	++	++	++	++	++	++
Calcite	0	0	0	0	++	+	0	0	0	0	++	+
Dolomite	0	0	0	0	0	0	0	0	0	0	0	0
Quartz	+	+	+	+	+	+	0	0	0	0	+	+
k-Feldspar	+	+	+	+	+	+	0	0	0	0	+	+
Plagioclase	+	+	+	+	+	+	0	0	0	0	+	+
Clay minerals	+	+	+	+	+	+	0	0	0	0	+	+
Pyrite	+	+	+	+	+	+	0	0	0	0	0	0
Amorphous material[b]	++	++	++	++	++	++	0	0	0	0	+	+
Acid-soluble constituents												
P_2O_5	0.33	0.89	0.88	0.37	2.23	3.09	34.14	30.25	31.85	30.93	11.71	17.96
CaO	0.18	0.50	0.59	0.39	20.99	11.24	50.79	43.52	48.55	42.75	30.78	31.20
MgO	0.86	0.73	0.83	1.01	0.98	1.24	0.38	0.38	0.45	0.36	0.88	1.01
Fe_2O_3	1.72	1.90	2.19	1.76	0.80	1.07	0.79	1.56	0.93	0.96	0.76	0.95
CO_2	0.04	0.00	0.07	0.11	13.13	6.07	2.75	2.46	3.12	2.53	11.04	4.27
$CaO:P_2O_5$	0.55	0.56	0.67	1.05	9.41	3.84	1.49	1.44	1.52	1.38	2.63	1.74
Combustible constituents												
Total 450° C	39.03	40.77	42.37	40.12	21.78	33.93	5.40	9.98	7.35	7.38	12.36	17.12
Organic Carbon	24.48	23.25	26.21	24.14	8.83	11.65	2.46	5.61	3.27	3.94	6.17	2.93

[a]XRD identifications: 0 = not detectable, + = minor component, ++ = major component.
[b]based on presence and intensities of hump at 22–26° 2θ.

only 2.5 to 3.1%. This is about one-half the equilibrium value suggested by Jahnke (1984) for francolite. The light layers of samples 103 and 104, on the other hand, have higher bulk CO_2 contents because they are diluted by varying amounts of calcite. Assuming a $CaO:P_2O_5$ weight ratio of 1.50 in apatite and a $CaO:CO_2$ weight ratio of 1.27 in calcite, calcium can be partitioned between apatite and calcite and used to estimate the amount of CO_2 available for carbonate fluorapatite in samples 103L and 104L. The results suggest CO_2 contents that are lower than those in the other light layers.

The dark layers of samples 103 and 104 contain calcite, opaline silica and an assemblage of detrital components that are also present in their adjacent light layers. They are less organic-rich than the dark layers in samples 100–102 and contain some apatite (Table 24.2). The very low $CaO:P_2O_5$ ratios, and the absence of soluble CO_2, in the 100–102 dark layers raises a question about the nature of their phosphate-containing phase. By X-ray diffraction techniques, no phosphatic mineral phases could be detected. This was also true if the samples were first ashed in a low temperature plasma furnace to remove the diluting organic matter. Without further work the only clue may be in the observation that leachable iron is relatively concentrated in these layers. This suggests that the phosphate phase could be iron bearing. Alternatively, an amorphous calcium phosphate precursor to apatite may be present (Kastner *et al.*, 1984).

Discussion

A microbial role in phosphate mineralization

Bacterial mats and authigenic carbonate fluorapatite are commonly associated in Recent upwelling systems and Monterey Formation sediments. It thus seems probable that bacterial mat activity influences phosphate mineralization. According to Gulbrandsen (1969), the interstitial microenvironment of a living mat is probably too low in pH and dissolved phosphate to favor inorganic apatite precipitation there. Biologically-mediated polyphosphate accumulation and inorganic calcite dissolution, however, would be favored within dense mats. These processes could result in virtually $CaCO_3$-free, organic-carbon and -phosphate-rich, mat laminae, such as appear to have been the primary deposits for many of the dark layers at Naples Beach.

Phosphatization of filaments in the Naples Beach shales must have taken place very soon after mat organisms died and were buried. Buried mats enter a sulfate-reducing environment, where a rise in pH occurs (Jørgensen & Revsbech, 1983), and where phosphate should be desorbed from iron oxyhydroxides. Since the molar C:P ratios of biogenous detrital organic matter are commonly 180–400 (Suess & Müller, 1980), the decomposition of phosphorus-enriched mat constituents of a lower C:P ratio should raise dissolved phosphate concentrations to levels that are higher than in organic-rich sediments without mats. Thus, it is conceivable that pore waters near the sediment–water interface become saturated with respect to carbonate fluorapatite after repeated burial of massive microbial mats. At the same time, phosphate assimilation by new-generation mat organisms would help cap the diffusive phosphate-flux out of the sediments, establishing a mechanism for coupled mat growth and phosphatization.

The foregoing scenario describes a relatively steady-state depositional environment (i.e. no drastic changes in sediment influx or redox conditions). However, for individual laminae or

lens of carbonate fluorapatite to develop, other factors may be necessary to initiate and continue confined phosphate mineral accumulations. The presence of nucleation sites has often been regarded as an important prerequisite for early phosphate mineral precipitation. Although bacteria cells have been suggested as sites (O'Brien *et al.*, 1981; Garrison *et al.*, 1987), it would seem from experimental and field observations that calcite is the favored seed material, followed by siliceous skeletal fragments, feldspars or fish debris (Gulbrandsen *et al.*, 1984; Burnett, 1977). In light of these data, we suggest that sediment layers in mat-associated sediments that are buried a few millimeters to centimeters, with some remaining calcite, are probably the most reactive, but not the only possible, layers for early phosphatization. Once phosphatization is initiated, renewed precipitation may occur at greater depths in the sediments due to a resupply of dissolved phosphate by diffusion. This suggestion is based on the nature of pore-water phosphate by diffusion. This suggestion is based on the nature of pore-water phosphate profiles in mat-covered sediments from off Peru (Jahnke & Soutar, pers. comm.). It also agrees with interpretations of pore-water fluoride distributions (Froelich *et al.*, 1983), and with signs of calcite replacement in some phosphatic laminae within the Monterey Formation.

Paleoenvironmental and late diagenetic conditions leading to the laminated phosphatic shales of the Monterey Formation

Given that the lenses or laminae of contemporary apatite observed within Recent, coastal upwelling sediments do not occur in a regular alternating pattern of relatively thick phosphorite laminae and black mud, like those observed at Naples Beach, the shelf-slope paleoenvironment of the Middle Miocene must have been more favorable for cyclic phosphatization than are similar Recent depositional environments, and/or later diagenetic alteration of the Monterey Formation sediments produced or enhanced a cyclic signal. Globally, Middle Miocene time was an epoch of increasing atmospheric–oceanic circulation and generally high but variable sea-level stands (Romaine & Lombari, 1985). These conditions isolated distal basins and slopes from continent-derived sediment supplies but favored coastal upwelling and an intense oxygen minimum zone (Garrison *et al.*, 1987). We have interpreted the organic- and phosphorus-rich zones of the Naples Beach section as having formed on such a sediment-starved slope under the waters of the paleo-California current. As evidence, average compacted sedimentation rates for the Carbonaceous Marl member (Fig. 24.1) are 30 to nearly 0 $m(my)^{-1}$, which when decompacted represent roughly one-tenth the average rate in the present day Peru–Chile mud lens (Reimers & Suess, 1983). If mid-Miocene rates were relatively unsteady, it is reasonable to assume that bacterial mats were most optimally developed, and widespread during periods with little non-biogenic detrital input

Fig. 24.6. SEM photomicrographs of (A) a modern *Beggiatoa* filament; (B) fossilized filaments from a mixed, but predominantly light layer of sample 101; and (C) from a dark layer. The variation in the filament diameters is not surprizing since modern *Beggiatoa* range from $<1\,\mu m$ to $>60\,\mu m$ diameter.

and low bottom-water oxygen. Then, depending on their rate of decay and other, difficult to reconstruct, microenvironmental conditions, some bacteria quickly became fossilized within discrete, very organic-rich layers (Soudry & Champetier, 1983). Continued early diagenesis of these and surrounding layers must have remobilized all or nearly all forms of buried organic-phosphorus, gradually releasing phosphate into pore waters for the growth of phosphatic phases. We suggest that the initial phosphatic precipitates were largely amorphous precipitates, similar to the magnesium and calcium phosphate amorphous phases experimentally synthesized from seawater as precursors for low CO_2-content carbonate fluorapatites (Gulbrandsen *et al.*, 1984). Then during later stages of diagenesis (the timing of which is unclear), advanced phosphorus remobilization and crystallization of carbonate fluorapatite erased all traces of early phosphatic phases and the primary sediment fabric from what are now the thickest light layers exposed at Naples Beach. In other layers, perhaps those with initially little calcite and/or high concentrations of mat and detrital organic matter, apatite crystallization was incomplete or totally inhibited.

If organic matter was the deterrent for the formation of crystalline carbonate fluorapatite in the dark layers from the Naples Beach section, it could mean that alternating patterns of phosphorus and organic matter-rich layers in many sections of the Monterey Formation were ultimately produced by fluctuating paleoenvironmental conditions affecting the burial rate and preservation of organic matter. Early effects of this mechanism were suggested by Jahnke *et al.* (1983) to be detectable in modern environments. More specifically, they proposed that organic matter slows down the formation of apatite by blocking surface nucleation sites. Their example was a single ~1 cm layer of authigenic apatite in a core of Mexican continental muds. Its position could only be explained by the low organic content of the layer relative to surrounding sediment (Jahnke *et al.*, 1983).

Conclusions

The microcrystalline phosphorites examined in this study are the end-products of incipient phosphatization processes that may have been coupled to the growth and burial of massive microbial communities. These communities develop and become prominent in dysaerobic coastal upwelling environments. The filamentous bacteria are able to assimilate large quantities of phosphorus, and their growth and burial appear to catalyze phosphate mineralization near the sediment–water interface. An analysis of a single sample of a modern bacterial mat yielded a molar C.P ratio 3–6 times lower than ratios in typical detrital planktonic organic matter, due to phosphorus enrichment. Anoxic diagenesis of postmortem bacterial mats is proposed as a mechanism that could have contributed to many organic-rich phosphatic deposits. Well known specific examples could include, in addition to the Monterey Formation, the Permian Phosphoria Formation of the Rocky mountains and Upper Cretaceous units in northern South America and in the Middle East (Kolodny, 1981). Filamentous bacterial fossils exist in rocks since the Early Precambrian (Awramik, Schoff & Walter, 1983; Walsh & Lowe, 1985), but in most cases their structures must have been obliterated by episodes of reworking or diagenesis.

Acknowledgments

This work was initiated under an award from the University-wide Energy Research Group of the University of California. Later support was provided by grants from TEXACO, USA and by NSF grant# EAR85-19113. Ms. Gretchen Andersen assisted in the analytical work.

References

Andersen, J.M. (1976). An ignition method for determination of total phosphorus in lake sediments. *Water Resources*, **10**, 329–31.

Arends, R.G. & Blake, G.H. (1986). Biostratigraphy and paleoecology of the Naples Bluff coastal section based on diatoms and benthic foraminifera. In *Siliceous Microfossil and Microplankton of the Monterey Formation and Modern Analogues*, ed. R.E. Casey & J.A. Barron, pp. 121–35. Society of Economic Paleontologists and Mineralogists, Pacific Section, Los Angeles. Book 45.

Awramik, S.M., Schoff, J.W. & Walter, M.R. (1983). Filamentous fossil bacteria from the Archean of Western Australia. *Precambrian Resources*, **20**, 357–74.

Barron, J.A. (1986). Paleoceanographic and tectonic controls on deposition of the Monterey Formation and related siliceous rocks in California. *Palaeogeography, Palaeoclimatology, Palaeoecology*, **53**, 27–45.

Baturin, G.N., Merkulova, K.I. & Chalov, P.I. (1972). Radiometric evidence for recent formation of phosphatic nodules in marine shelf sediments. *Marine Geology*, **13**, M37–41.

Blake, G.H. (1985). 'The faunal response of California continental margin benthic foraminifera to the oceanographic and depositional events of the Neogene.' Unpublished Ph.D. thesis, University of Southern California.

Burnett, W.C. (1977). Geochemistry and origin of phosphorite deposits from off Peru and Chile. *Geological Society of America Bulletin*, **88**, 813–23.

Burnett, W.C., Beers, M.J. & Roe, K.K. (1982). Growth rates of phosphate nodules from the continental margin off Peru. *Science*, **215**, 1616–18.

Copenhagen, W.J. (1934). Occurrence of sulphides in certain areas of the sea bottom on the South African coast. Fish. Mar. Biol. Sur. Div. Union South Africa, Report, No. 3, 1–18.

Froelich, P.N., Kim, K.H., Jahnke, R., Burnett, W.C., Soutar, A. & Deakin, M. (1983). Pore water fluoride in Peru continental margin sediments: uptake from seawater. *Geochimica Cosmochima Acta*, **47**, 1605–12.

Gallardo, V.A. (1977). Large benthic microbial communities in sulfide biota under Peru–Chile subsurface countercurrent. *Nature*, **268**, 331–2.

Garrison, R.E., Kastner, M. & Kolodny, Y. (1987). Phosphorites and phosphatic rocks in the Monterey Formation and related Miocene units, coastal California. In *Cenozoic Basin Development in Coastal California*, ed. R.V. Ingersoll & W.G. Ernst, Prentice-Hall, Inc., Inglewood, NJ.

Graham, S.A. & Williams, L.A. (1985). Tectonic, depositional, and diagenetic history of Monterey Formation (Miocene), Central San Joaquin basin, California. *American Association of Petroleum Geologists Bulletin*, **69**, 385–411.

Gulbrandsen, R.A. (1969). Physical and chemical factors in the formation of marine apatite. *Economic Geology*, **64**, 365–82.

Gulbrandsen, R.A., Roberson, C.E. & Neil, S.T. (1984). Time and the crystallization of apatite in seawater. *Geochimica Cosmochimica Acta*, **48**, 213–18.

Isaacs, C.M. (1981a). Field trip guide for the Monterey Formation. Santa Barbara Coast, California. In *Guide to the Monterey Formation in the California Coastal Area, Ventura to San Luis Obispo*, ed. C.M. Isaacs, pp. 55–72. Pacific Section American Association of Petroleum Geologists.

Isaacs, C.M. (1981b). Lithostratigraphy of the Monterey Formation, Goleta to Point Conception, Santa Barbara Coast, California. In *Guide to the Monterey Formation in the California Coastal Area, Ventura to San Luis Obispo*, ed. C.M. Isaacs, pp. 55–72. Pacific Section

American Association of Petroleum Geologists.
Isaacs, C.M. (1983). Hemipelagic deposits in a Miocene basin, California: toward a model of lithologic variation and sequence, In *Marine Sedimentation On the Pacific Margin*, ed. D.K. Larre & R.J. Steel, pp. 117–32. Society of Economic Paleontologists and Mineralogists, Pacific Section, Los Angeles.
Jahnke, R.A. (1984). The synthesis and solubility of carbonate fluorapatite. *American Journal of Science*, **284**, 58–77.
Jahnke, R.A., Emerson, S.R., Roe, K.K. & Burnett, W.C. (1983). The present day formation of apatite in Mexican continental margin sediments. *Geochimica Cosmochimica Acta*, **47**, 259–66.
Jannasch, H.W. (1984). Chemosynthetic microbial mats of deep-sea hydrothermal vents. In *Microbial Mats: stromatolites*, pp. 121–31. Alan R. Liss, Inc, New York.
Jørgensen, B.B. (1977a). Distribution of colorless sulfur bacteria (*Beggiatoa* spp.) in a coastal marine sediment. *Marine Biology*, **41**, 19–28.
Jørgensen, B.B. (1977b). The sulfur cycle of a coastal marine sediment (Limfjorden, Denmark). *Limnology and Oceanography*, **22**, 814–32.
Jørgensen, B.B. & Revsbech, N.P. (1983). Colorless sulfur bacteria, *Beggiatoa* spp. and *Thiovulum* spp., in O_2 and H_2S microgradients. *Applied Environmental Microbiology*, **45**, 1261–70.
Kastner, M., Mertz, K., Hollander, D. & Garrison, R. (1984). The association of dolomite–phosphorite–chert: causes and possible diagenetic sequences. In *Dolomites of the Monterey Formation and Other Organic-rich Units*, ed. R.E. Garrison, M. Kastner & D.H. Zenger, pp. 75–86. Society of Economic Palaeontologists and Mineralogists Pacific Section, Los Angeles.
Kleinpell, R.M. (1938). *Miocene Stratigraphy of California*. American Association of Petroleum Geologists, Tulsa.
Kleinpell, R.M. (1980). *Miocene Stratigraphy of California Revisited*. American Association of Petroleum Geologists, Studies in Geology, **11**, Tulsa.
Kolodny, Y. (1981). Phosphorites. In *The Sea*, vol. 7, ed. J. Emiliani, pp. 981–1023. John Wiley & Sons, New York.
McClellan, G.H. (1980). Mineralogy of carbonate fluorapatites. *Journal of the Geological Society*, London, **137**, 675–81.
Mertz, K.A., Jr. (1984). Diagenetic aspects, Sandholt member, Miocene Monterey Formation, Santa Lucia Mountains, California: implications for depositional and burial environments. In *Dolomites of the Monterey Formation and Other Organic-rich Units*, ed. R.E. Garrison, M. Kastner & D.H. Zenger, pp. 49–73. Pacific Section, Society of Economic Palaeontologists and Mineralogists Special Publication 41, Los Angeles.
O'Brien, G.W., Harris, J.R., Milnes, A.R., & Veeh, H.H. (1981). Bacterial origin of East Australian continental margin phosphorites. *Nature*, **292**, 442–4.
Price, N.B. & Calvert, S.E. (1978). The geochemistry of phosphorites from the Namibian Shelf. *Chemical Geology*, **23**, 151–70.
Reimers, C.E. (1982). Organic matter in anoxic sediments off central Peru: relations of porosity, microbial decomposition and deformation properties. *Marine Geology*, **46**, 175–97.
Reimers, C.E. & Suess, E. (1983). Spatial and temporal patterns of organic matter accumulation on the Peru continental margin. In *Coastal Upwelling Its Sediment Record*, Pt. B, ed. J. Thiede & E. Suess, pp. 311–45. Plenum, New York.
Romaine, K. & Lombari, G. (1985). Evolution of Pacific circulation in the Miocene: radiolarian evidence from DSDP site 289. In *The Miocene Ocean: Paleoceanography and Biogeography*, ed. J.P. Kennett, pp. 273–90. Geological Society of America Memoir 163.
Rosenberg, R., Arntz, W.E., de Flores, E.C., Flores, L.A., Carbajal, G., Finger, I. & Tarazona, J. (1983). Benthos biomass and oxygen deficiency in the upwelling system off Peru. *Journal of Marine Research*, **41**, 263–79.
Soudry, D. & Champetier, Y. (1983). Microbial processes in the Negev phosphorites (southern Israel). *Sedimentology*, **30**, 411–24.
Soutar, A. & Crill, P.A. (1977). Sedimentation and climatic patterns in the Santa Barbara Basin during 19th and 20th centuries. *Geological Society of America Bulletin*, **88**, 1161–72.
Strohl, W.R. (1989) *Beggiatoa*. In *Bergey's Manual of Systematic Bacteriology*, vol. 3. Williams & Wilkins, New York.
Strohl, W.R. & Larkin, J.M. (1978). Enumeration, isolation, and characterization of *Beggiatoa* from freshwater sediments. *Applied Environmental Microbiology*, **36**, 755–70.
Suess, E. (1981). Phosphate regeneration from sediments of the Peru continental margin by dissolution of fish debris. *Geochimica Cosmochimica Acta*, **45**, 577–88.
Suess, E. & Müller, P.J. (1980). Productivity, sedimentation rate and sedimentary organic matter in the oceans II. – Elemental fractionation. In *Biogeochimie de la Matiere Organique a L'Interface Eau-Sediment marin*, pp. 17–26. Elsevier, Amsterdam.
Thomson, J., Calvert, S.E., Mukherjee, S., Burnett, W.C. & Bremner, J.M. (1984). Further studies of the nature, composition and ages of contemporary phosphorite from the Namibian Shelf. *Earth and Planetary Science Letters*, **69**, 341–53.
Veeh, H.H., Burnett, W.C. & Soutar, A. (1973). Contemporary phosphorites on the continental margin of Peru. *Science*, **181**, 844–5.
Veeh, H.H., Calvert, S.E. & Price, N.B. (1974). Accumulation of uranium in sediments and phosphorites on the south west African shelf. *Marine Chemistry*, **2**, 189–202.
Walsh, M.M. & Lowe, D.R. (1985). Filamentous microfossils from the 3,500-Myr-old Onverwacht Group, Barberton Mountain Land, South Africa. *Nature*, **314**, 530–2.
Wiessner, W. (1981). The family Beggiatoaceae. In *The Prokaryotes*, ed. M.P. Starr, H. Stolp, H.G. Truper, A. Balows & H.G. Schegel, pp. 380–9. Springer-Verlag, Berlin.
Williams, L.A. (1984). Subtidal stromatolites in Monterey Formation and other organic-rich rocks as suggested source contributors to Petroleum Formation. *American Association of Petroleum Geologists*, **68**, 1879–93.
Williams, L.A. & Reimers, C.E. (1983). Role of bacterial mats in oxygen-deficient marine basins and coastal upwelling regimes: preliminary report. *Geology*, **11**, 267–9.

25

Coupled changes of oxygen isotopes in PO_4^{3-} and CO_3^{2-} in apatite, with emphasis on the Monterey Formation, California

M. KASTNER, R.E. GARRISON,
Y. KOLODNY, C.E. REIMERS AND
A. SHEMESH

Abstract

Apatites from Monterey Formation phosphatic marls and pelletal–nodular phosphorites from four basins were analyzed for oxygen isotopes of the PO_4^{3-} ion, for oxygen and carbon isotopes of the CO_3^{2-} ion, and for the amount of structural CO_3^{2-}. Values of $\delta^{18}O_{PO_4^{3-}}$ range between 14.9 and 20.8‰ and of $\delta^{18}O_{CO_3^{2-}}$ between 19.5 and 32.2‰ (SMOW). Most of the $\delta^{18}O_{PO_4^{3-}}$ values of these apatites are significantly lower than those of Miocene fish bones from the Monterey and related formations, which have a more restricted range of values (19.9–21.5‰).

The large ranges of $\delta^{18}O_{PO_4^{3-}}$ and $\delta^{18}O_{CO_3^{2-}}$ values in Monterey Formation apatites encompass most of the variations in both isotopes of sedimentary apatites of all ages. Oxygen isotopes of the two ions in apatites change proportionally. The strong correlation between them, $R = 0.89$ for Middle Miocene Monterey Formation apatites, and $R = 0.91$ for sedimentary apatites of all ages, suggests that (1) both ions re-equilibrated with the environmental (diagenetic or epigenetic) water oxygen, and (2) the isotopic re-equilibration of the oxygen in the PO_4^{3-} ion was enzymatically catalyzed. No petrologic criteria are known at present to distinguish re-equilibrated from non-re-equilibrated apatites. Only the latter could be used for paleoceanographic studies.

The amount of structural CO_3^{2-} in the Monterey Formation apatites, ranging from 1.7 to 4.5% CO_2, appears to be controlled by the carbonate-ion concentration in the environmental water and by temperature. Consequently, apatites with $< 6 \pm 1\%$ CO_2 are not necessarily weathered. The environmental carbonate-ion concentration most probably also controls the formation of diagenetic apatite versus carbonate.

The $\delta^{13}C$ values of most Monterey Formation apatites are negative. In organic-rich water-dominated systems, phosphatized carbonates acquire negative $\delta^{13}C$ values from the pore-water dissolved carbonate. Only in (carbonate) rock-dominated systems could void-filling apatites be distinguished from phosphatized carbonates by their negative $\delta^{13}C$ values.

Introduction

The Miocene Monterey Formation, the product of a high productivity upwelling system, long known for its distinctive siliceous deposits, also contains a variety of calcareous and phosphatic rocks (Garrison, Kastner & Kolodny, 1987; Garrison *et al.*, this volume). The deposition of these sediments occurred in a number of relatively small, fault-bounded and sediment-starved basins (Fig. 25.1) which formed in late Oligocene and Miocene time in response to tensional stresses associated with offsets along the San Andreas and related faults.

Contemporaneous with the development of these basins were several paleoceanographic events, summarized in Garrison *et al.*, (this volume). The most important of these were: a generally high sea-level stand during much of Middle Miocene time, about 16 to 10 my ago, and increasing coastal upwelling, beginning along the California margin at about 12 my and extending through the remainder of the Miocene. The latter trend, associated with increases in Antarctic glaciation and intensified atmospheric–oceanic circulation (Ingle, 1981), led to the distinctive siliceous (diatomaceous) facies of the upper Monterey. The middle and lower parts of the Monterey, in contrast, have important calcareous and phosphatic rocks. Phosphatogenesis was apparently most widespread during the Middle Miocene high sea-level stand and global cooling event, the time of the onset of the modern ocean–atmosphere system.

Fig. 25.1. Location map of major Neogene basins of central and southern California; the stippled line shows the present coastline.

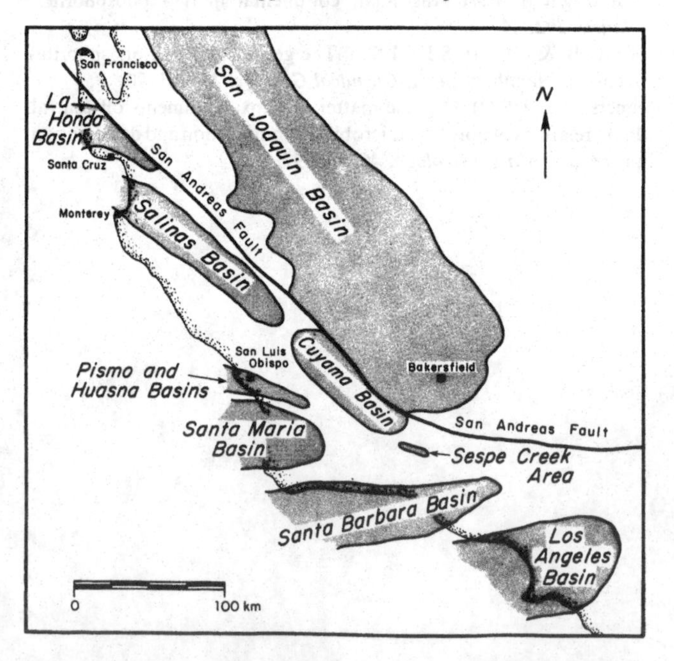

Table 25.1. *Stratigraphic subdivision of Miocene rocks established by different workers in the coastal basins of Central California. Modified from Pisciotto (1981) and Isaacs (1983)*

Epoch	Stage	Formation	Age	GENERALIZED LITHOLOGY (After Isaacs (1980))	SANTA BARBARA BASIN: Isaacs (1983) (informal)	SANTA MARIA BASIN: Woodring & Others (1943); Woodring & Bramlette (1950)	SANTA MARIA BASIN: Pisciotto (1978) (informal)	SALINAS BASIN: Graham (1976)	CUYAMA BASIN: Roberts & Vercoutere (1986)
E. PLIOCENE	'DELMONTIAN'	SISQUOC FORMATION							
LATE MIOCENE	'DELMONTIAN' / MOHNIAN	MONTEREY FORMATION	5.5 my	SILICEOUS ROCKS: Diatomites, changing down-section to rhythmically-bedded, laminated and massive chert, porcellanite and siliceous mudrocks (variable carbonate)	MONTEREY FORMATION: Clayey–siliceous Member	MONTEREY FORMATION: Upper Member	MONTEREY FORMATION: Siliceous facies	MONTEREY FORMATION: Hames Member	Santa Margarita Formation
LATE MIOCENE	MOHNIAN	MONTEREY FORMATION	8 my	PHOSPHATIC ROCKS: Phosphatic, organic-rich shale and mudstone with cyclic alternations with siliceous rocks near top (variable carbonate)	Upper calcareous–siliceous Member; Transitional marl–siliceous Member	Middle Member	Siliceous facies; Phosphatic facies	Hames Member	Santa Margarita Formation
M. MIOCENE	MOHNIAN / LUISIAN	MONTEREY FORMATION	11 my	PHOSPHATIC ROCKS (continued)	Carbonaceous marl Member	Lower Member	Phosphatic facies	Hames Member; Sandholdt Member	Santa Margarita Formation; Monterey Formation
EARLY MIOCENE	RELIZIAN / SAUCESIAN	MONTEREY FORMATION	15 my – 18 my	CALCAREOUS ROCKS: Foraminifer and coccolith shale and mudstone with abundant biogenic and authigenic carbonate	Lower calcareous–siliceous Member	Point Sal Formation	PT. SAL FM.: Calcareous facies	Sandholdt Member	Monterey Formation

Various subdivisions of the Monterey and related formations, based on these lithologic variations, are summarized in Table 25.1.

Phosphate occurs in the Monterey and related units in several distinct forms (Garrison *et al.*, 1987; Garrison *et al.*, this volume). The most important ones are:

(1) *Phosphatic marls* – The marls are commonly organic-rich, and many are laminated unburrowed sediments deposited on the outer shelves, slopes and floors of low-oxygen basins. They contain small lenses and pelloids of generally light colored and friable carbonate fluorapatite.
(2) *Nodular phosphorites* – These are the products of complex cycles of phosphatization, exhumation, reburial rephosphatization, etc. They occur in thin beds which commonly mark surfaces of hiatus.
(3) *Pelletal and pelletal oolitic phosphorites* – These are best developed in decimeter-thick beds of shelf units (e.g. the Santa Margarita Formation of the Cuyama Basin), as thin banktop deposits, or as turbidite beds within basinal sequences of the Monterey. They formed in a more oxygenated environment than the phosphatic marls.

The deposition of the Monterey Formation phosphatic marl facies, mostly 16.0 to 13.5 my ago, coincides with a global oceanic $\delta^{13}C$ excursion to heavy values, the 'Monterey C isotope excursion' of Vincent & Berger (1985). The $\delta^{13}C$ excursion denotes excess extraction of organic matter and phosphate by deposition in oxygen-poor continental margin sediments, producing the youngest global phosphatogenic period (e.g. Ingle, 1981; Suess & Thiede, 1983; Riggs, 1984; Garrison *et al.*, 1987; Garrison *et al.*, this volume).

On the basis of pore-water profiles of dissolved phosphate and fluoride, in continental margin phosphatic sediments beneath highly productive upwelling regions, e.g. in the Peruvian and Mexican continental margins, Froelich *et al.* (1983) and Jahnke *et al.* (1983) concluded that sedimentary apatite forms at or near the sediment–water interface. If indeed oxygen isotopes of the phosphate ion in sedimentary apatite (carbonate fluorapatite) were a reliable recorder of paleoenvironments, particularly of temperatures, as argued by Kolodny, Luz & Navon (1983) and Shemesh, Kolodny & Luz (1983), then the apatites of the Monterey Formation phosphatic rocks should contain critical information about seafloor temperatures, hence about important Middle Miocene tectonic, oceanographic, and climatic events.

Preliminary work on the oxygen isotopes of the phosphate ion ($\delta^{18}O_P$) of Monterey apatites from the Sandholdt Member in the Salinas Basin (Fig. 25.1; Table 25.1), recorded light $\delta^{18}O_P$ values, between 14.5 and 18.5‰ (SMOW) (Kastner *et al.*, 1984; Mertz, 1984a, b). Such light values suggest apatite formation or diagenetic isotopic exchange, either at moderate to elevated temperatures or in the presence of ^{18}O-depleted water. Assuming $\delta^{18}O_{H_2O} = -1$‰, $\delta^{18}O_P$ of 14.5–18.5‰ correspond to approximately 28–45°C, using the Longinelli & Nuti (1973a) phosphate-water isotopic temperature scale, 45–80°C using the Karhu & Epstein (1986) temperature scale, and to 34–66°C using the Shemesh, Kolodny & Luz (1988) scale.

The phosphate-water isotopic temperature linear relationship of Longinelli & Nuti (1973a), based on shell phosphate from environments with narrow temperature ranges (4–27°C), is:

$$t = 111.4 - 4.3(\delta^{18}O_{PO_4} - \delta^{18}O_{H_2O}) \qquad (25.1)$$

This temperature scale was independently confirmed by Kolodny *et al.* (1983).

The two alternative phosphate-water isotopic temperature scales are: the Karhu & Epstein (1986) equation, based on shell phosphate at low temperatures and on phosphate from a metamorphic pegmatite at high temperatures (~350°C), is:

$$\Delta^{18}O(PO_4 - H_2O) = 2.17(10^6 \cdot T^{-2}) - 1.94 \qquad (25.2)$$

The Shemesh *et al.* (1988) equation, also primarily an empirical equation, based on fish apatites at low temperatures, on the Karhu & Epstein (1986) metamorphic phosphate and on one experimental point (510°C) at high temperatures is:

$$\Delta^{18}O(PO_4 - H_2O) = 2.12(10^6 \cdot T^{-2}) - 2.98 \qquad (25.3)$$

As discussed by Karhu & Epstein (1986) and Shemesh *et al.* (in press), none of these equations is a rigorously calibrated phosphate-water temperature scale; therefore the calculated temperatures are correct only in a relative sense.

The above derived temperatures from the Longinelli & Nuti (1973a) equation are high, and those from the equations by Karhu & Epstein (1986) and by Shemesh *et al.* (in press) are even higher. These results are difficult to reconcile with the petrographic observations of Mertz (1984a) that these apatites formed during early diagenesis. Higher temperatures could have been obtained only deeper down. Mertz (1984a) reported maximum burial depth of ~2400 m and maximum diagenetic temperatures of 80–105°C for the Sandholdt Member, Salinas Basin. Oxygen isotopes of associated dolomites also suggest similar formation temperatures, between 20 and 80°C (Kastner *et al.*, 1984; Mertz, 1984a, b). If the high temperatures, calculated from the phosphate oxygen isotopes of these apatites, are approximately correct, they signify the burial diagenetic temperatures and not the Middle Miocene seafloor temperatures. This implies that the apatites either formed at greater burial depths, in conflict with the conclusion of Froelich *et al.* (1983) that sedimentary apatites form at or near the sediment–water interface, or that they recrystallized at depth, in conflict with the assumption of Kolodny *et al.* (1983) and Shemesh *et al.* (1983) that the phosphate ion is practically inert to isotopic exchange in inorganic post-depositional reactions.

In an attempt to resolve these apparently incongruous problems, thoroughly characterized Monterey Formation apatites from four Neogene coastal basins of Central California, with distinct tectonic and sedimentologic histories, were chosen for this study (Fig. 25.1, Tables 25.1 and 25.2). They were analyzed for $\delta^{18}O_{PO_4}$ ($\delta^{18}O_P$) in order to record paleotemperatures, for oxygen isotopes of the structural CO_3^{2-} ($\delta^{18}O_C$) to test the reliability of the carbonate ion for paleotemperature studies, and for carbon isotopes of the structural CO_3^{2-} ($\delta^{13}C_C$) for information about their mode of formation by phosphatization of precursor carbonate or by void filling, as suggested by Benmore, Coleman & McArthur (1983), and McArthur, Coleman & Bremner (1980) and McArthur *et al.* (1986). The $\delta^{18}O_P$ values of these apatites are compared with the $\delta^{18}O_P$ values of well preserved Miocene marine fish from the Monterey and related

Table 25.2. *Phosphatic rocks, Monterey Formation and related units*

Sample	Location	Neogene basin	Lithostratigraphic Member of Monterey Formation or Santa Margarita Formation (Cuyama Basin)	Age[a]	Lithology	Sedimentologic setting
MO-5	Naples Beach	Santa Barbara	Carbonaceous marl Member	Middle Miocene 12 ± 2 Ma	Dark, redeposited phosphate nodules, condensed zone	Slope
MO-4	Haskel Beach near Goleta	Santa Barbara	Carbonaceous marl Member	Middle Miocene 14–15 Ma	Light-colored phosphate nodules and pelloids in laminated, organic-rich marls	Slope
MO-3	Wood Canyon	Santa Barbara	Carbonaceous marl Member	Middle Miocene 14–15 Ma	Light-colored phosphate nodules and pelloids in laminated, organic-rich marls	Slope
MR-36A	North of Mussel Rock	Santa Maria	Contact between Monterey and Sisquoc Formations	Late Miocene 5–6 Ma	Dark, redeposited phosphate nodules	Slope?
MO-6	Lions Head	Santa Maria	Lower Member	Middle Miocene 14–16 Ma	Light brown, redeposited phosphate nodules or crusts	Slope or basin floor
MO-1	Lions Head	Santa Maria	Lower Member	Middle Miocene 14–16 Ma	Light-colored phosphate nodules and pelloids in laminated, organic-rich marls	Slope or basin floor
Tmt-93	Near Mussel Rock	Santa Maria	Lower Member	Middle Miocene 14–16 Ma	Light-colored phosphate nodules and pelloids in laminated, organic-rich marls	Basin floor?
SH-75-02	Carmel Valley	Salinas	Hames Member	Late Middle Miocene 11–13 Ma	Pelletal phosphoritic sandstone	Basinal turbidite
Tmt-63	Carmel Valley	Salinas	Hames Member	Middle–Late Miocene 10–14 Ma	Pelletal phosphoritic sandstone	Basinal turbidite
V-9a	Vaqueros Canyon	Salinas	Sandholdt Member	Late Middle Miocene ~14 Ma	Light-colored phosphate nodules and pelloids in laminated, organic-rich marls	Slope
As-22	Arroyo Seco	Salinas	Sandholdt Member	Middle Miocene 14.5–15 Ma	Light-colored phosphate nodules and pelloids in laminated, organic-rich marls	Slope
As-15	Arroyo Seco	Salinas	Sandholdt Member	Middle Miocene ~15 Ma	Light-colored phosphate nodules and pelloids in laminated, organic-rich marls	Slope

Table 25.2. *(Contd.)*

Sample	Location	Neogene basin	Lithostratigraphic Member of Monterey Formation or Santa Margarita Formation (Cuyama Basin)	Age[a]	Lithology	Sedimentologic setting
MO-2	Jolon area	Salinas	Sandholdt Member	Middle Miocene 14–16 Ma	Light-colored phosphate nodules and pelloids in laminated, organic-rich marls	Slope?
J-6	Sam Jones Canyon	Salinas	Sandholdt Member	Late early–early Middle Miocene 16–16.5 Ma	Light-colored phosphate nodules and pelloids in laminated, organic-rich marls	Slope
I-3	Indians Ranch	Salinas	Sandholdt Member	Late Early–early Middle Miocene 16–16.5 Ma	Pelletal phosphorite	Slope
I-12	Indians Ranch	Salinas	Sandholdt Member	Late Early Miocene 16.5 Ma	Pelletal phosphorite	Slope
Sa-Ma-2	Cuyama phosphorite deposit	Cuyama	Upper phosphatic mudstone Member[b]	Late Middle–early Late Miocene 10–13 Ma	Pelletal phosphoritic sandstone	Winnowed shelf
Sa-Ma-6	Cuyama phosphorite deposit	Cuyama	Upper phosphatic mudstone member [b]	Late Middle–early Late Miocene 10–13 Ma	Pelletal phosphoritic sandstone	Winnowed shelf
Tmt-66b	Indian Creek	Cuyama	Whiterock bluff shale Member	Late Middle Miocene ~13 Ma	Pelletal phosphoritic sandstone	Basinal turbidite

[a]Ages based on following references: Woodring & Bramlette (1950); Graham (1976); Pisciotto (1981); Isaacs (1983); Lagoe (1984); Roberts & Vercoutere (1986).
[b]Santa Margarita Formation

Table 25.3. *Oxygen isotope values of marine fish from the Miocene of California*

				Temperature (°C)[a]		
Sample description	Location	LACM #	$\delta^{18}O$ (‰ SMOW)	Longinelli & Nuti (1973a)	Karhu & Epstein (1986)	New equation of this Paper
Myliobatid (ray) tooth	Orange County	1945/11367	21.1	16.4	27.3	13.7
Clupeid from Sisquoc, Formation	Santa Barbara County	6589/12119	21.1	16.4	27.3	13.7
Clupeid from Modelo Formation	Santa Monica Mountains	5435	20.9	17.2	28.5	15.0
Clupeid from Puente Formation	San Gabriel Valley, Los Angeles Co.	1937	19.9	21.5	35.1	22.2
Fish from Monterey Formation	Lompoc quarry	—	21.5	14.7	24.8	11.0

[a]Assuming $\delta^{18}O_{H_2O} = -1‰$

formations (Table 25.3). These fish $\delta^{13}O_P$ values should yield the closest estimates to the depositional temperatures of the Monterey basins.

Experimental methods

Samples of oxygen and carbon isotope analyses of apatites were carefully chosen from four Neogene Basins (Table 25.1) to represent different tectonic and paleoceanographic conditions; and petrographic and mineralogic analyses were performed by microscopy and X-ray diffraction (XRD), respectively.

Oxygen isotopes of phosphate

The chemical procedures for phosphate purification and $BiPO_4$ precipitation developed by Tudge (1960) were followed. Oxygen was liberated by reacting the $BiPO_4$ with BrF_5 at 150°C, converted to CO_2 and analyzed mass spectrometrically. Details of the procedures followed are those described in Kolodny *et al.* (1983) and Shemesh *et al.* (1988). Several of the samples were run in duplicate. The standard deviation did not exceed 0.3‰. Except for the $\delta^{18}O_P$ analyses of apatite which were carried out at the Department of Geology, Hebrew University, all other analyses were done at Scripps Institution of Oceanography.

Oxygen and carbon isotopes of structural carbonate

Prior to isotope analysis, guided by XRD data, samples without or with small amounts of calcite were treated with triammonium citrate solution (Silverman, Fuyat & Weiser, 1952), which selectively dissolves calcite. Samples with moderate amounts of calcite were treated twice. Complete removal of calcite was checked by XRD and microscopy. As triammonium citrate solution does not remove dolomite, samples MO-6 and Tmt-66b (Tables 25.2 and 25.3) could not be analyzed for oxygen and carbon isotopes of structural CO_3^{2-}.

Oxygen and carbon isotopes were analyzed mass spectrometrically following the phosphoric acid method of McCrea (1950). Each sample was run at least in duplicate. 65% of the samples were run in triplicate. The precision was within 0.15 for $\delta^{18}O$ and 0.10 for $\delta^{13}C$, except for samples Sa-Ma-2 and Sa-Ma-6 from Cuyama Basin (Table 25.2) where precisions were within 0.50 and 0.30, respectively. All results are reported in the δ‰ notation with respect to SMOW standard for oxygen and with respect to PDB standard for carbon.

Temperatures of formation were calculated from $\delta^{18}O_{PO_4^{-3}}$($\delta^{18}O_P$) values, using the phosphate-water temperature equations determined by Longinelli & Nuti (1973a), Karhu & Epstein (1986), and by Shemesh *et al.* (1988) for $\delta^{18}O$ of seawater of −1‰ SMOW. A new empirical equation and the corresponding calculated temperatures of formation are also presented in this paper (Table 25.4).

Amount of structural CO_3^{2-} in apatite

The content of CO_3^{2-} was determined by two methods, the XRD method by Gulbrandsen (1970) and by CO_2-coulometry. Precision by CO_2-coulometry is within 3% of the values reported in Table 25.4.

Results and discussion

Oxygen isotopes of the PO_4^{3-} ion

The $\delta^{18}O$ values in phosphate of five fossil fish from the marine Miocene sequences of California, and their translation into temperatures of formation, are summarized in Table 25.3. Except for the Karhu & Epstein (1986) equation, the Table 25.3 data indicate that temperatures of the upper water mass varied between 11 and 25°C. These values are in good agreement with previous estimates of water temperatures of the Miocene Central California coastal basins. Based on planktonic foraminifera, surface-water temperatures fluctuated between 10 and 20°C, possibly with short excursions up to a maximum of 25°C (Ingle, 1981). And benthic foraminiferal assemblages suggest that during the climatic optimum, prior to the late Middle Miocene rapid increase of Antarctic ice buildup, bottom-water temperatures were 9 to 15°C. In late Middle Miocene they dropped to perhaps 5–7°C (J. Ingle, pers. comm.). A surface–bottom temperature differential of nearly 10°C persisted.

The $\delta^{18}O$ values in phosphate of the pelloid and nodular apatites from the Monterey and related formations, summarized in Table 25.4, differ sharply from those of the fossil fish. Further-

Table 25.4. *Oxygen and carbon isotopic data, CO_2 values, and calculated formation temperatures in apatites from the Monterey and related Formations*

		$\delta^{18}O$(‰ SMOW)		$\delta^{13}C$(‰ PDB)	Weight %CO_2		Temperature (°C)[b] calculated from $\delta^{18}O_{PO_4}$ values according to phosphate–water equations of			
Location	Sample	PO_4	CO_3	CO_3	CO_2-Coulometer	$\Delta 2\theta$°(004-410)[a]	Longinelli & Nuti (1973a)	Shemesh, *et al.* (1988)	Karhu & Epstein (1986)	New equation of this paper
Santa Barbara basin	MO-5	20.2	30.1	0.3	3.6	4.0	20	23	33	20
	MO-4	20.8	30.8	−1.0	3.4	3.2	18	19	29	16
	MO-3	17.7	27.1	−4.3	2.7	3.1	31	40	51	40
Santa Maria basin	MR-36-A	19.6	32.2	−1.8	4.4	3.3	23	27	37	25
	MO-6[c]	18.3	—	—	—	—	28	35	47	35
	MO-1	15.4	21.9	−4.6	1.7	2.3	41	58	71	63
	Tmt-93[c]	15.8	—	—	—	1.9	39	54	67	59
Salinas basin	SH-75-02	17.3	22.0	1.0	1.7	3.1	33	43	54	44
	Tmt-63	15.8	20.3	−1.3	2.0	3.5	39	54	67	59
	V-9a[d]	17.3	22.7	3.1	—	—	33	43	54	44
	AS-22[d]	16.6	20.2	−2.6	—	—	36	48	60	50
	AS-15[d]	16.5	19.9	−3.0	—	—	36	49	61	51
	MO-2[d]	15.6	22.4	−3.4	2.4	—	40	56	69	61
	J-6[d]	17.9	22.8	—	—	—	30	38	50	38
	I-3[d]	15.7	19.5	1.5	—	—	40	55	68	60
	I-12[d]	14.9	20.4	−1.9	—	—	43	62	76	68
Cuyama basin	Sa-Ma-[e]	20.4	26.9	−7.9	2.6	3.6	19	22	32	19
	SaMa-6[e]	20.6	26.8	−7.7	4.5	3.6	19	21	37	17
	Tmt-66b	19.6	26.9	−4.3	2.8	4.1	23	27	37	25

[a] Cuk_{α} radiation.
[b] Assuming $\delta^{18}{}_{H_2O} = -1$‰.
[c] Isotopic composition of CO_3 not determined because sample contains dolomite.
[d] $\delta^{18}O_{PO_4}$ from Kastner *et al.* (1984) and Mertz (1984b).
[e] Standard deviation of CO_3 isotopic analyses were unusually low, 0.50‰ for $\delta^{18}O$ and 0.30‰ for $\delta^{13}C$; most likely due to inhomogeneities in these samples. Standard deviation of all other CO_3 isotopic analyses were 0.15‰ for $\delta^{18}O$ and 0.10‰ for $\delta^{13}C$.

more, unlike other Neogene phosphorites with $\delta^{18}O_P$ values of 19–25‰ (SMOW) (Shemesh *et al.*, 1983 and 1988), the $\delta^{18}O_P$ values of Monterey Formation apatites from the four different basins, described in Tables 25.2 and 25.4, range between 15 and 21‰. In fact, these values overlap the Mesozoic phosphorites with $\delta^{18}O_P$ of 16–20‰ (Shemesh *et al.*, 1983 in press) and the Ordovician–Pennsylvanian conodonts in North America with $\delta^{18}O_P$ of 15–19‰ (Luz, Kolodny & Kovach, 1984). During Early–Middle Miocene time seawater isotopic composition was near −1‰ (SMOW); hence, changes in isotopic composition of seawater cannot be invoked to explain the Monterey oxygen isotopic record.

Of all the phosphorite samples studied the only $\delta^{18}O_P$ values which might reflect the middle Miocene maximum surface water temperatures are those of the pelletal phosphoritic sandstones of the Cuyama Basin (Table 25.4). But these apatites, among the youngest analyzed in this study, formed between 10–13 my ago toward the end of the major mid-Miocene cooling event, and even these rather low calculated temperatures are too high for the bottom water temperatures, the zone in which phosphorites are supposed to form. Similarly, to explain the relatively low formation temperature of 18–25°C, of the two light-colored nodular-pelloidal *slope* apatite samples (MO-5, MO-4, Table 25.4) from the oxygen-minimum zone of the paleo-Santa Barbara Basin, formation in isotopic equilibrium with Middle Miocene warm *surface* waters would have to be invoked.

The oxygen isotopes of all the apatite samples from the distal Santa Maria Basin and the proximal Salinas Basin, as well as of sample MO-3 from the Santa Barbara Basin, definitely do not reflect the known paleoceanographic conditions (Table 25.4). The only possibilities to explain the $\delta^{18}O_P$ values of the Monterey Formation apatites are that they represent either diagenetic isotopic re-equilibration or diagenetic formation values:

(1) at moderate to elevated temperatures
(2) with isotopically changed interstitial waters
(3) or combinations of (1) plus (2)

Geochemical mass balance considerations support early formation, near the sediment–water interface, of phosphoritic apatites, with almost contemporaneous and/or subsequent reworking and mechanical enrichment (e.g. Baturin, 1971; Froelich *et al.*, 1982 and 1983, and references therein). This, however, does not exclude some apatite formation at greater burial depths during later diagenesis.

Sedimentological, petrographical, and textural observations suggest early formation of most of the important apatite occurrences in the Monterey Formation (Pisciotto, 1978; Mertz, 1984b). Hence, diagenetic isotopic re-equilibration of the PO_4^{3-} oxygen must have occurred.

Tudge (1960), Longinelli & Nuti (1973), and Kolodny *et al.* (1983) have demonstrated convincingly that the oxygen in orthophosphate resists isotopic exchange during inorganic reactions, such as dissolution and reprecipitation, at least up to 80°C. Thus, isotopic exchange with meteoric water in the low-temperature weathering environment is excluded. Isotopic exchange of water-oxygen with inorganic phosphate (Pi), $O_{Pi} \rightleftharpoons O_{HOH}$, is, however, rapid in biochemical reactions due to enzymatic catalysis (e.g. Boyer *et al.*, 1977; Faller & Elgavish, 1984), and the fractionation is temperature-dependent (Kolodny *et al.*, 1983). The $\delta^{18}O_P$ should thus faithfully record temperatures of equilibration between water and phosphate in the last enzymatically catalyzed reaction. Hence, on the basis of only stable isotope values in apatite, its origin and diagenetic history might be obscured.

The existence of continuous bacterial degradation reactions of organic matter, enhancing burial diagenesis, has been documented extensively using the carbon isotopes of diagenetic carbonates, of dissolved organic matter, and of organic gases (e.g. Rosenfeld & Silverman, 1959; Claypool & Kaplan, 1974; Nissenbaum, Presley & Kaplan, 1977; Garrison, Kastner & Zenger, 1984, and references therein).

Irrespective of the phosphate-oxygen isotopes of organic matter in the sediments, which are unrelated to the water temperature of the original environment (Longinelli, Bartelloni & Cortecci, 1976), interstitial water-dissolved phosphate, if generated solely by enzymatic catalytic bacterial decay of the organic matter, should be in isotopic equilibrium with the interstitial water at the diagenetic temperature. Apatite which forms inorganically from this dissolved phosphate, either by void filling or by phosphatization of precursor carbonates, will thus record the *in situ* diagenetic temperature of formation. This should also apply to dissolved biogenic apatite, such as fish debris, irrespective of its initial $\delta^{18}O_P$ value; if the dissolution and/or reprecipitation were enzymatically (bacterially?) controlled, their newly acquired $\delta^{18}O_P$ values must be in equilibrium with the temperature and isotopic composition of the fluid at depth. The $\delta^{18}O_P$ values of the Monterey Formation apatites suggest that at greater burial depths, at elevated temperatures, most of the early-formed apatites were involved in enzymatically-catalyzed bacterial reactions which prompted oxygen isotopic re-equilibration. Such moderate to deep burial diagenetic apatites will henceforth be designated as 'burial' apatites, to distinguish them from 'oceanic' apatites. Oxygen isotopes of 'burial' apatites record burial diagenetic environments, and those of 'oceanic' apatites record paleoceanographic environments. Petrographical or geochemical criteria for discriminating between these two types of apatites are not available. Caution is thus necessary in using $\delta^{18}O_P$ for paleoceanographic interpretations. Only recently has the role of bacterially-mediated reactions in apatite formation received serious attention (e.g. Soudry & Champetier, 1983; Lucas & Prévôt, 1985; Soudry, 1987; Reimers, Kastner & Garrison, Chapter 24, this volume).

At moderate to deep burial, interstitial water–oxygen isotopic composition changes; at first the water generally becomes isotopically lighter than seawater, and with depth it becomes heavier. The depth and temperature at which the transition from lighter to heavier than seawater oxygen isotopes takes place depends mainly on the water–rock mass ratio, the geothermal gradient, and the extent of water–rock interactions. Within the first few hundred meters of burial, a high water–rock mass ratio environment, interstitial water–oxygen isotopes vary only slightly, generally by no more than 1–2‰ (e.g. Lawrence, Gieskes & Broecker, 1975). It is, however, imperative to emphasize that deeply buried isotopically re-equilibrated 'burial' apatites record diagenetic environments and not just diagenetic temperatures.

It is hence interesting to note that the $\delta^{18}O_P$ values of the apatites from Cuyama and Santa Barbara Basins are consistently heavier than those of the apatites from the Santa Maria and Salinas Basins (Table 25.4).

	$\delta_{18}O_P$(‰, SMOW)
Cuyama and Santa Barbara Basins	17.7–20.8
Santa Maria and Salinas Basins	14.9–18.3

Estimated maximum burial depths for these basins are: 1100 m for Cuyama, < 2000 m for Santa Barbara, > 2500 m for Santa Maria and Salinas (Woodring & Bramlette, 1950; Graham, 1976; Pisciotto, 1978 and 1981; Isaacs, 1980 and 1983; Lagoe, 1984; Mertz, 1984a; Roberts & Vercoutere, 1986). Moreover, estimates of geothermal gradients suggest the possibility of higher gradients in the latter two basins (Pisciotto, 1978 and 1981; Mertz, 1984a, b).

As this paper concentrates on the Monterey and related apatites which formed during the Miocene high sea-level stand, sample MR-36-A was not included in the above tabulation. Its origin is unclear, the apatite occurs as dark to black nodules in a 5–6 my ago phosphatic conglomerate, which marks the contact between the Monterey and Sisquoc Formations in the Mussel Rock section, Santa Maria Basin. Petrographic analysis indicates multiple phosphatization episodes.

Structural CO_3^{2-} in apatite

Evidence for carbonate ion substitution in the apatite (francolite) structure was first provided by Altschuler, Cisney & Barlow (1952) by careful X-ray diffraction analysis. The controversy over the structural site (or sites) of the planar CO_3^{2-} ion was summarized by McClellan (1980).

Substitution of the PO_4^{3-} tetrahedron by a distorted $(CO_3 \cdot F)^{3-}$ tetrahedron:

$$PO_4^{3-} \rightleftharpoons (CO_3 \cdot F)^{3-}$$

originally proposed by Borneman-Starynkevitch & Belov (1940), was confirmed by Bacquet *et al.* (1980) by the ESR method. Chemical analyses of carbonate fluorapatites, however, indicate F-deficiency relative to CO_3^{2-}. Hence, additional probable substitutions were suggested: $PO_4^{3-} \rightleftharpoons (CO_3 \cdot OH)^{3-}$ (McClellan, 1980), or for just charge balance, $2Na^+ \rightleftharpoons Ca^{2+}$, or unoccupied Ca^{2+} positions (Gulbrandsen, 1966).

Weight percent structural CO_3^{2-} in apatites, analyzed mostly by the X-ray diffraction method of Gulbrandsen (1970) and reported as $\%CO_2$, range between 1 and 6.5% (Gulbrandsen, 1970; McArthur, 1978; McClellan, 1980; McClellan & Saavedra, 1986).

Carbonate fluorapatite is the stable apatite phase in a high-carbonate solution such as seawater (Jahnke, 1984, and references therein). He then concluded that all marine apatites forming near the sediment–water interface should contain a rather similar amount of CO_2, about 6% by weight, a value previously suggested as a criterion for non-weathered apatite by McArthur (1978).

McArthur (1978), McClellan (1980), and McClellan & Saavedra (1986) further suggested that the observed large variations of CO_2 content in apatite and the trend of decreasing CO_2 with increasing age indicate that CO_3^{2-} is relatively easily exchangeable during weathering, diagenesis, and metamorphism. It was concluded by them that the amount of CO_2 in apatite does not indicate its origin; it only aids in distinguishing between weathered and unweathered samples. Gulbrandsen (1970) suggested temperature and differences in depositional environments as causes for the observed CO_2 variations in apatites. Loss of structural CO_2 by experimental apatite heating and by the natural thermal metamorphism of the Hatrurim Formation ('Mottled Zone') was documented by Lehr *et al.* (1967) and by Matthews & Nathan (1977), respectively.

In Monterey Formation apatites, the amount of structural carbonate does not seem to vary randomly. In the 'colder' Cuyama and Santa Barbara Basins, it ranges between 2.6–4.5 weight % CO_2, and in the 'warmer' Santa Maria and Salinas Basins % CO_2 is lower, ranging between 1.7–2.8% (Table 25.3). Our data (unpublished) of % CO_2 in apatites from subsurface samples from the Santa Maria Basin fall within these ranges. Are all Monterey Formation apatites weathered to various degrees, or do these low to medium carbonate values instead contain important information about the depositional or diagenetic environments of these apatites? To understand the geochemical significance of % CO_2 in apatites, the oxygen isotopes of the CO_3^{2-} ion in these apatites were determined.

Oxygen isotopes of structural CO_3^{2-}, with emphasis on its relationship with $\delta^{18}O_P$

Similar to the time trend of structural carbonate content (McArthur, 1978; McClellan, 1980; McClellan & Saavedra, 1986), an overall weathering trend of decreasing $\delta^{18}O_{CO_3^{2-}}$ ($\delta^{18}O_C$) values with increasing age was suggested by McArthur *et al.* (1980 and 1986) and Benmore *et al.* (1983). Shemesh *et al.* (1983) originally supported the view that unlike ^{18}O in the phosphate ion of apatite, the ^{18}O in the carbonate ion is rather easily exchangeable. But in a recent paper by Shemesh *et al.* (1988) in which they observed a good correlation between $\delta^{18}O_P$ and $\delta^{18}O_C$, they abandoned the weathering interpretation.

Apatite is the only anhydrous marine mineral which contains oxygen in two different ions, PO_4^{3-} and CO_3^{2-}, which can easily be analyzed separately. If the oxygens of the two ions equilibrate with the surrounding water, the $\Delta^{18}O_{C-P} = \delta^{18}O_C - \delta^{18}O_P$ value would make it possible to calculate the temperature at which equilibrium took place independent of the oxygen isotopes of the water. Furthermore, a knowledge of the $\Delta^{18}O_{C-P}$ temperature would then allow calculations of the $\delta^{18}O$ value of the environmental water. Oxygen isotopes of a coexisting mineral pair is probably a less reliable temperature indicator than of ion pairs in the same mineral, because in the former simultaneous equilibration must be assumed.

Such an approach for marine apatites was previously attempted by Shemesh *et al.* (1983), from which they concluded that for paleoceanography, except for 'young' apatites, the $\delta^{18}O_C$ values are not as reliable as the $\delta^{18}O_P$ ones. However, in their later paper (Shemesh *et al.*, 1988) they observed a high correlation between the oxygen isotopes in the two structural sites in apatites of all ages, and thus concluded that both $\delta^{18}O_P$ and $\delta^{18}O_C$ reflect the environmental conditions of apatite formation. A different approach was taken by Karhu & Epstein (1986). In an attempt to determine oceanic paleotemperatures and possible variations in the $\delta^{18}O$ value of

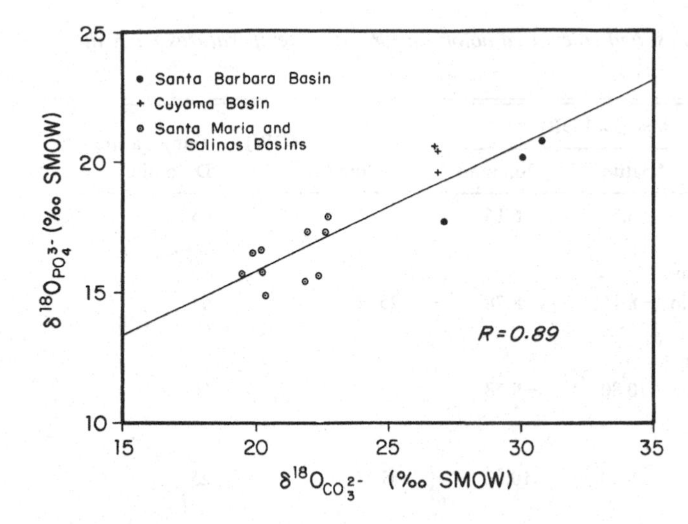

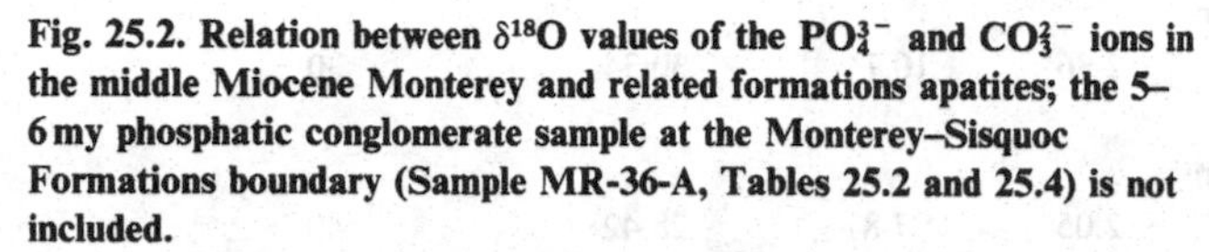

Fig. 25.2. Relation between $\delta^{18}O$ values of the PO_4^{3-} and CO_3^{2-} ions in the middle Miocene Monterey and related formations apatites; the 5–6 my phosphatic conglomerate sample at the Monterey–Sisquoc Formations boundary (Sample MR-36-A, Tables 25.2 and 25.4) is not included.

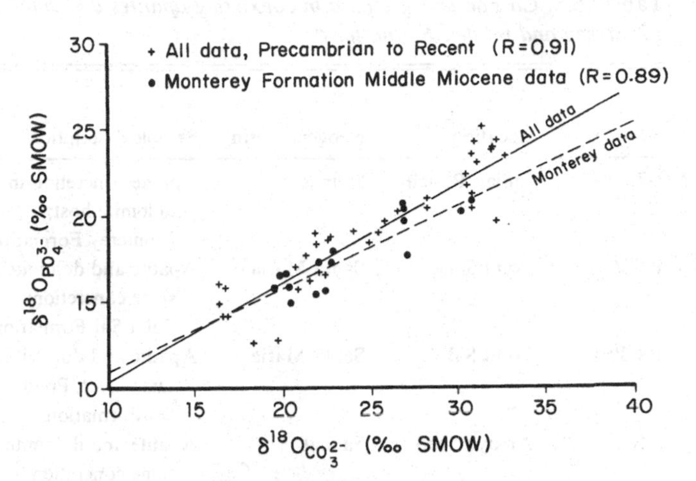

Fig. 25.3. Comparison of the relation between $\delta^{18}O$ values of the PO_4^{3-} and CO_3^{2-} ions: (a) in Miocene Monterey and related formations apatites (R = 0.89), and (b) Precambrian–Recent apatites from the major phosphatogenic provinces (R = 0.91) (from Shemesh *et al.* (1988)).

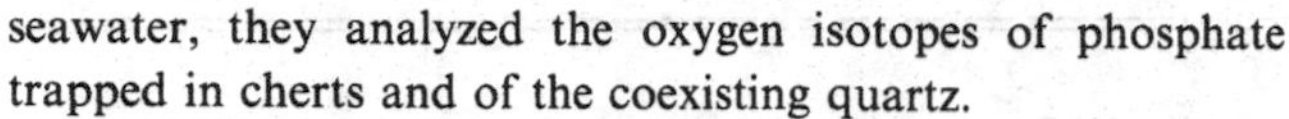

seawater, they analyzed the oxygen isotopes of phosphate trapped in cherts and of the coexisting quartz.

If, as previously assumed, the $\delta^{18}O_C$ in apatites can be used to determine the temperature of diagenesis, whereas the $\delta^{18}O_P$ would indicate the temperature of the original apatite formation, the $\delta^{18}O_C$ versus $\delta^{18}O_P$ values should show no linear relationship. To test this assumption, apatites from the well characterized Middle Miocene Monterey and related formations (Table 25.2), which were analyzed for $\delta^{18}O_P$, were also analyzed for $\delta^{18}O_C$. The results are summarized in Table 25.4, and a $\delta^{18}O_P$ versus $\delta^{18}O_C$ plot is presented in Figure 25.2.

The $\delta^{18}O_C$ (SMOW) values of the Monterey Formation apatites are: 26.8–26.9‰ in the Cuyama Basin; 27.1–30.8‰ in the Santa Barbara Basin; and 19.5–22.8‰ in the Santa Maria and Salinas Basins. These data show that not only do the $\delta^{18}O_C$ values have a trend similar to the $\delta^{18}O_P$ values, both being lighter in the warmer than in the colder basins, but that these two changes are proportional, with a correlation as high as R = 0.89 (Fig. 25.2). Such a high correlation surely cannot be coincidental.

In discussing the $\delta^{18}O_P$ values of these apatites, arguments for diagenetic temperatures rather than Miocene oceanic bottom- or surface-water temperatures were presented. It was concluded that the phosphate ion in most Monterey Formation apatites has re-equilibrated isotopically with the interstitial water oxygen during diagenesis. The excellent correlation between the $\delta^{18}O_P$ and $\delta^{18}O_C$ values, seen in Figure 25.2, indicates that oxygen in the phosphate and carbonate ions of these apatites were introduced simultaneously, the changes in the oxygen isotopes of both ions being proportional. During burial diagenesis, changes in both the $\delta^{18}O$ values of the water and in temperature occur. Recently, a similar conclusion reached by Shemesh *et al.* (1988) was based on analyses of Precambrian–Recent apatites from all major phosphatogenic provinces. A δ–δ space plot comparing the Monterey and related formations data with those of Shemesh *et al.* (1988) is shown in Figure 25.3. The slopes of the two regression lines are most similar, the slightly lower slope of the Monterey regression line is insignificant.

Having rigorously calibrated phosphate-water and carbonate-water temperature scales is a prerequisite for any quantitative analysis of the diagenetic conditions recorded in our data. Unfortunately, no such temperature scale exists for phosphate-water, only three non-equivalent empirical temperature scales exist (eqns (25.1)–(25.3)). Therefore precise temperature and $\delta^{18}O_{H_2O}$ estimates from ΔO_{C-P} will not be attempted here. The following discussion is based on the adoption of the experimental calcite-water isotopic temperature scale of O'Neil, Clayton & Mayeda (1969), implying equivalence of calcite and apatite CO_3^{2-}, as discussed in Kolodny & Kaplan (1970). Accordingly, unlike both the original Longinelli & Nuti (1973b) isotopic phosphate-water temperature scale and its subsequent reconfirmation by Kolodny *et al.* (1983), which are based on isotopic data from a small temperature range (~20°C), as a first approximation, the observed span of $\delta^{18}O_C$ values of the apatites studied here (Table 25.4) suggests a significantly larger range of apatite formation temperatures, on the order of 50–60°C.

A new empirical isotopic phosphate-water temperature relationship is herewith proposed:

$$\Delta^{18}O(PO_4 - H_2O) = 1.73(10^6 \cdot T^{-2}) + 1.07 \qquad (25.4)$$

It was derived by utilizing the excellent correlation between the $\delta^{18}O_C$ and $\delta^{18}O_P$ seen in Figure 25.2 and by assuming that the $\delta^{18}O_{H_2O}$ of the environmental water maintained its original seawater value. Although this is our weakest assumption, modeling of oxygen isotope values of Monterey Formation cherts and dolomites (unpublished, and P. Knauth, pers. comm.) suggests relatively moderate *increases* in the $\delta^{18}O_{H_2O}$ values with depth, on the order of 2 to possibly 3‰ in the first 500–800 m of the sediment column. Hence, the temperature slope might be slightly steeper than indicated in Table 25.4, but the estimated temperatures are correct in a relative sense, and are similar to the temperatures derived from the Shemesh *et al.* (1988) equation.

Table 25.5. *Carbon isotopic data in coexisting apatites and dolomites, and calculated dolomite formation temperatures from the Monterey and related Formations*[a]

Sample	Location	Neogene basin	Sample description	$\delta^{13}C$(‰ PDB)		$\delta^{18}O$(‰ SMOW)	Temperature° C[b]
				Apatite	Dolomite	Dolomite	Dolomite
I-7	Indian Ranch	Salinas	Apatite concretion in dolomite host; Monterey Formation	−6.51	−6.16	24.43	65
PS-30I	Point Sal	Santa Maria	Apatite and dolomite in same concretion; Point Sal Formation	−8.15	−8.78	25.06	61
PS-30II	Point Sal	Santa Maria	Apatite and dolomite in same layer; Point Sal Formation	−10.80	−9.58	26.77	50
AS-1	Arroyo Seco	Salinas	Apatite and dolomite in same concretion; Monterey Formation	6.93	10.62	31.46	25
AS-3	Arroyo Seco	Salinas	Apatite concretion in dolomite host; Monterey Formation	−2.86	10.14	30.34	30
AS-5	Arroyo Seco	Salinas	Apatite concretion in dolomite host; Monterey Formation	−2.05	7.87	28.42	40

[a]Data from Kastner *et al.* (1984), and Mertz *et al.* (1984b).
[b]Calculated using the dolomite–water equation of Fritz & Smith (1970), assuming $\delta^{18}O_{H_2O} = -1$‰.

The δ–δ space plot (Fig. 25.3) of the Monterey data which comprise 16.5–10 my, but mostly just 25 my, 16.5–14 my ago, overlap the data from the entire range of sedimentary apatites, from Precambrian to Recent. It is therefore proposed that both data sets actually represent a diagenetic rather than an oceanic paleotemperature record.

In conclusion, we propose the following hypothesis regarding the amount of structural carbonate in apatites and their $\delta^{18}O_P$ and $\delta^{18}O_C$ records. The amount of CO_2 in apatite is controlled by the carbonate ion concentration in the environmental (diagenetic) water and by temperature; the higher the temperature, the lower the % CO_2 in the apatite. Even at low temperatures, apatites which formed near the sediment–seawater interface should contain structural carbonate in different amounts than apatites which formed between >0.5 and <50 m burial depth within a sulfate-reduction zone, an environment of higher carbonate alkalinity. The thermodynamic stability of apatite requires drastically increased phosphate concentrations as carbonate substitution in apatite increases (Jahnke, 1984). Therefore, eventually with burial, when the carbonate-ion concentration exceeds a certain as yet undetermined threshold, carbonate fluorapatite would become thermodynamically unstable, and only authigenic carbonates, calcite, and/or dolomite would form. In surficial weathering environments, controlled by inorganic geochemical processes, oxygen in the CO_3^{2-} ion may be more easily exchangeable than that in PO_4^{3-}. During burial diagenesis of organic-rich sediments, in which bacterial activity persists to great depth and elevated temperatures, the oxygen in both apatite ions re-equilibrates with the oxygen of the environmental (diagenetic) water. The establishment of an experimental phosphate-water isotopic temperature equation will improve the accuracy of the existing $\Delta^{18}O_{C-P}$ values and thus provide a powerful tool for recording diagenetic temperatures and synchronous isotopic evolution of the diagenetic pore fluids.

Carbon isotopes of CO_3^{2-}

In support of the proposition that microbial activity played an important role in diagenesis of the Monterey organic-rich phosphatic sediments, are the $\delta^{13}C$ values of associated dolomites. They vary significantly from negative to positive values, suggesting formation in the sulfate-reduction or methanogenesis zones (e.g. Claypool & Kaplan, 1974). The data in Table 25.5 were chosen to represent the range of Monterey dolomite $\delta^{13}C$ values reported in Kastner *et al.* (1984) and Mertz (1984b). The $\delta^{13}C$ values of some of the coexisting apatites and dolomites are almost identical, but other apatite–dolomite pairs have exceedingly different $\delta^{13}C$ values. We do not at present have a simple explanation for the latter case. Most significantly, the temperatures indicated by the $\delta^{18}O$ values of the dolomites (Table 25.5) are in excellent agreement with the temperatures indicated by the $\delta^{18}O_P$ of Monterey apatites (Table 25.4).

No trend in $\delta^{13}C$ values of the structural carbonate is seen in the apatites studied here (Table 25.4). Most values lie between −1 and −8‰ (PDB); the others are 'typical' marine $\delta^{13}C$ values of 0 ± 1‰. McArthur *et al.* (1980 and 1986) and Benmore *et al.* (1983) proposed using the $\delta^{13}C$ values of apatites as a discriminating indicator for the mode of formation. In their view, negative carbon isotopes, with less than −2‰ (PDB), indicate void filling apatite formation in organic-rich sediments with active anaerobic bacterial degradation of organic matter. They believe that phosphatized carbonates, on the contrary, retain the $\delta^{13}C$ signature of the precursor and have $\delta^{13}C$ values of 0 ± 1‰. This is correct only where the system is (carbonate) rock-dominated. In organic-rich water-dominated systems, even phosphatized carbonates might acquire negative $\delta^{13}C$ values. Thus negative $\delta^{13}C$ values of CO_3^{2-} in apatites should be used as a genetic indicator only with great caution.

In the Monterey Formation, phosphorites of both apatite types, void-filling and phosphatized carbonate, exist.

Conclusions

1. Unlike other Cenozoic apatites, which have $\delta^{18}O_P$ values of 19–25‰ (SMOW), those of the Miocene Monterey Formation apatites, from two proximal and two distal basins range between 15 and 21‰; these $\delta^{18}O_P$ values span a large temperature range.
2. The strong linear correlation between oxygen isotopes of coexisting PO_4^{3-} and CO_3^{2-} ions in the Monterey and related Formations apatites is not coincidental. It indicates a common genesis, and the difference between the two oxygen isotopes, $\Delta^{18}O_{C-P}$, permits estimates of environmental temperature and $\delta^{18}O$ of the diagenetic fluid.
3. The clear difference between the high $\delta^{18}O_P$ values of the Miocene marine fish from California (19.9–21.5‰), and the low $\delta^{18}O$ values of the Monterey pelloidal and nodular apatites (14.9–20.8‰), suggests that the apatite $\delta^{18}O_P$ values signify diagenetic isotopic re-equilibration.
4. Isotopic exchange of interstitial water oxygen with inorganic phosphate seems to be controlled by bacterial enzymatic catalytic reactions which are pervasive in organic-rich sediments.
5. The strong diagenetic imprint on the young Monterey Formation phosphorites, demonstrated here, casts serious doubts on the validity of interpreting the variations in the $\delta^{18}O_P$ values of pelloidal or nodular apatites of different ages as reflecting either oceanic paleotemperatures or variations in seawater $\delta^{18}O$. This conclusion should not, however, disqualify all $\delta^{18}O_P$ results for paleoceanographic studies. The presently available results of conodents, fish (and other fossils) apatites support the validity of these $\delta^{18}O_P$ values as approximate *depositional* indicators. Also, pelloidal and nodular phosphorites which did not undergo deep burial diagenesis, such as most Mesozoic phosphorites, might be reliable depositional recorders.
6. Petrographic or geochemical criteria which could enable us to discriminate between apatites that formed near the sediment–water interface and retained their original depositional oxygen isotopic record, the 'oceanic' apatites, and those which subsequently re-equilibrated isotopically with water oxygen in a diagenetic new environment, the 'burial' apatites, do not exist at present.
7. Carbonate-ion concentration in the environmental water and *in situ* temperature are important factors controlling the amount of structural carbonate in apatite. A range of carbonate values in apatite should thus be expected.
8. In water-dominated systems, phosphatized carbonates might acquire from anaerobic decay of organic matter negative $\delta^{13}C$ values which are similar to void-filling apatite.

Acknowledgments

We greatly appreciate the review by Y.K. Bentor, and we benefited from discussions with B. Luz, R.A. Jahnke and J. Schuffert. The fish specimens were most generously contributed by J.D. Stewart from the Los Angeles County Museum of Natural History. This research was supported mainly by NSF Grant EAR85-19113 and partially by grants from Texaco USA, from ACS/PRF #15262-AC2, and the Univ. of CA #UER-101-R2. The conscientious analytical work by G. Emanuel, D. Introne and R. Nissan, and the help by T. Shaw in figure preparation are acknowledged.

References

Altschuler, Z.S., Cisney, E.A. & Barlow, I.H. (1952). X-ray evidence of the nature of carbonate-apatite. *Geological Society of America Bulletin.* **63**, 1230–1.

Bacquet, G., Vo Quang, T., Bonel, G. & Vignoles, M. (1980). Résonance paramagnétique électronique du centre F dans les fluorapatites carbonatées de type B. *Journal of Solid State Chemistry*, **33**, 189–95.

Baturin, G.N. (1971). States of phosphorite formation on the ocean floor. *Nature*, **232**, 61–2.

Benmore, R.A., Coleman, M.L. & McArthur, J.M. (1983). Origin of sedimentary francolite from its sulphur and carbon isotope composition. *Nature*, **302**, 516–18.

Borneman-Starynkevitch, I.D. & Belov, N.V. (1940). Isomorphic substitution in carbonate-apatite. *Comptes Rendus Doklody Academic Sciences URSS*, **XXVI, 18**, 804–6.

Boyer, P.D., deMeis, L., Carvalho, M.G.C. & Hackney, D.D. (1977). Dynamic reversal of enzyme carboxyl group phosphorylation as the basis of the oxygen exchange catalyzed by sacroplasmic reticulum Adenosine Triphosphase. *Biochemistry*, **16**, 136–40.

Claypool, G.E. & Kaplan I.R. (1974) The origin and distribution of methane in marine sediments. In: *Natural Gases in Marine Sediments*, ed. I.R. Kaplan, pp. 99–139. Plenum Press, NY.

Cook, P.J. & McElhinny, M.W. (1979). A reevaluation of the spatial and temporal distribution of sedimentary phosphate deposits in the light of plate tectonics. *Economic Geology*, **74**, 315–30.

Faller, L.D. & Elgavish, G.A. (1984). Catalysis of oxygen-18 exchange between inorganic phosphate and water by the gastric H, K-ATPase. *Biochemistry*, **23**, 6584–90.

Fritz, P. & Smith, D.G.W. (1970). The isotopic composition of secondary dolomites. *Geochimica Cosmochimica Acta*, **34**, 1161–73.

Froelich, P.N., Bender, M.C., Luedtke, N.A., Heath, G.R. & DeVries, T. (1982). The marine phosphorus cycle. *American Journal of Science*, **282**, 474–511.

Froelich, P.N., Kim, K.H., Jahnke, R., Burnett, W.C., Soutar, A. & Deakin, M. (1983). Pore water fluoride in Peru continental margin sediments: uptake from seawater. *Geochimica Cosmochimica Acta*, **47**, 1605–12.

Garrison, R.E., Kastner, M. & Kolodny, Y. (1987). Phosphorites and phosphatic rocks in the Monterey Formation and related Miocene units, Coastal California. In *Cenozoic Basin Development of Coastal California*, Rubey Vol. VI, eds. R.V. Ingersoll, & W.G. Ernst, pp. 349–81, Prentice-Hall, Englewood Cliffs, NJ.

Garrison, R.E., Kastner, M. & Zenger, D.H. (1984). *Dolomites of the Monterey Formation and Other Organic-Rich Units.* Pacific Section, Society of Economic Paleontologists and Mineralogists, 215pp.

Graham, S.A. (1976). 'Tertiary sedimentary tectonics of the central Salinian block of California'. Ph.D. thesis, Stanford University, 510pp.

Gulbrandsen, R.A. (1966). Chemical composition of phosphorites of the Phosphoria formation. *Geochimica Cosmochimica Acta*, **30**, 769–78.

Gulbrandsen, R.A. (1970). Relation of carbon dioxide content of apatite of the Phosphoria Formation to regional facies. *US Geological Survey Professional Paper*, **700-B**, B9–B13.

Ingle, J.C. (1981). Origin of Neogene diatomites around the north Pacific Rim. In: *The Monterey Formation and Related Siliceous Rocks of*

California, ed. R.E. Garrison, R.G. Douglas, K.E. Pisciotto, C.M. Isaacs & J.C. Ingle, pp. 159–79, Pacific Section, Society of Economic Paleontologists and Mineralogists.

Isaacs, C.M. (1980). 'Diagenesis in the Monterey Formation examined laterally along the coast near Santa Barbara, California'. Ph.D. thesis, Stanford University, 329pp.

Isaacs, C.M. (1983). Compositional variation and sequence in the Miocene Monterey Formation, Santa Barbara coastal area, California. In *Cenozoic Marine Sedimentation, Pacific Margin, USA*, ed. D.K. Larue & R.J. Steel, pp. 117–32, Pacific Section, Society of Economic Paleontologists and Mineralogists.

Jahnke, R.A., (1984). The synthesis and solubility of carbonate fluorapatite. *American Journal of Science*, **284**, 58–78.

Jahnke, R.A., Emerson, S.E., Roe, K.K. & Burnett, W.C. (1983). The present day formation of apatite in Mexican continental margin sediments. *Geochimica Cosmochimica Acta*, **47**, 259–66.

Karhu, J. & Epstein, S., (1986). The implication of oxygen isotope records in coexisting cherts and phosphates. *Geochimica Cosmochimica Acta*, **50**, 1745–56.

Kastner, M., Mertz, K., Hollander, D. & Garrison, R. (1984). The association of dolomite-phosphorite-chert: causes and possible diagenetic sequences. In *Dolomites of the Monterey Formation and other Organic-Rich Units*, ed R.E. Garrison, M. Kastner & D.H. Zenger, pp. 75–86. Pacific Section, Society of Economic Paleontologists and Mineralogists.

Kolodny, Y. & Kaplan, I.R. (1970). Carbon and oxygen isotopes in apatite CO_2 and co-existing calcite from sedimentary phosphorite. *Journal of Sedimentary Petrology*, **40**, 954–9.

Kolodny, Y., Luz, B. & Navon, O. (1983). Oxygen isotope variations in phosphate of biogenic apatite, I. Fish bone apatite – rechecking the rules of the game. *Earth Planetary Science Letters*, **64**, 398–404.

Lagoe, M.B. (1984). Paleogeography of Monterey Formation, Cuyama basin, California. *American Association of Petroleum Geologists Bulletin*, **68**, 610–27.

Lawrence, J.R., Gieskes, J.M. & Broecker, W.S. (1975). Oxygen isotope and cation composition of DSDP pore waters and the alteration of Layer II basalts. *Earth Planetary Science Letters*, **27**, 1–10.

Lehr, J.R., McClellan, G.H., Smith, J.P. & Frazier, A.W. (1967). Characterization of apatites in commercial phosphate rocks. In *Colloque International sur les Phosphates Mineraux Solides*, vol. 2, pp. 29–44. Societé Chemique de France, Toulouse.

Longinelli, A., Bartelloni, M. & Cortecci, G. (1976). The isotopic cycle of oceanic phosphate, I. *Earth Planetary Science Letters*, **32**, 389–92.

Longinelli, A. & Nuti, S. (1973a). Revised phosphate-water isotopic temperature scale. *Earth Planetary Science Letters*, **19**, 373–6.

Longinelli, A. & Nuti, S. (1973b). Oxygen isotope measurements of phosphate from fish teeth and bones. *Earth Planetary Science Letters*, **20**, 337–40.

Lucas, J. & Prévôt, L. (1985). The synthesis of apatite by bacterial activity: mechanism. *Sciences de la Géologie Mémoire*, **77**, 83–92.

Luz, B., Kolodny, Y. & Kovach, J. (1984). Oxygen isotope variations in phosphate of biogenic apatites, III. Conodonts. *Earth Planetary Science Letters*, **69**, 255–62.

Matthews, A. & Nathan, Y. (1977). The decarbonation of carbonate fluorapatite (francolite). *American Mineralogist*, **62**, 565–73.

McArthur, J.M. (1978). Systematic variations in the contents of Na, Sr, CO_3 and SO_4 in marine carbonate-fluorapatite and their relation to weathering. *Chemical Geology*, **21**, 89–112.

McArthur, J.M., Benmore, R.A., Coleman, M.L., Soldi, C., Yeh, H.W. & O'Brien, G.W. (1986). Stable isotope characterisation of francolite formation. *Earth Planetary Science Letters*, **77**, 20–34.

McArthur, J.M., Coleman, M.L. & Bremner, J.M. (1980). Carbon and oxygen isotopic composition of structural carbonate in sedimentary francolite. *Journal of the Geological Society, London*, **137**, 669–73.

McClellan, G.H. (1980). Mineralogy of carbonate fluorapatite. *Journal of the Geological Society, London*, **137**, 675–81.

McClellan, G.H. & Saavedra, F.N. (1989). Chemical and mineral characteristics of some Cambrian and Precambrian phosphorites. In *Phosphate Deposits of the World*, vol. 1, Proterozoic and Cambrian Phosphorites, ed. P.J. Cook & J.H. Shergold, pp. 244–67. Cambridge University Press, Cambridge. (in press).

McCrea, J.M. (1950). On the isotopic chemistry of carbonate and paleotemperature scale. *Journal of Chemistry and Physics*, **18**, 849–57.

Mertz, K.A. (1984a). 'Origin and depositional history of the Sandholdt Member, Miocene Monterey Formation, Santa Lucia Range, California'. Ph.D. thesis, University of California, Santa Cruz, 295pp.

Mertz, K.A. (1984b). Diagenetic aspects, Sandholdt Member, Miocene Monterey Formation, Santa Lucia Mountains, California: implications for depositional and burial environments. In *Dolomites of the Monterey Formation and Other Organic-Rich Units*, ed. R.E. Garrison, M. Kastner & D.H. Zenger, pp. 49–73. Pacific Section, Society of Economic Paleontologists and Mineralogists.

Nissenbaum, A., Presley, B.J. & Kaplan, I.R. (1972). Early diagenesis in a reducing fjord, Saanich Inlet, British Columbia, I. Chemical and isotopic changes in major components of interstitial water. *Geochimica Cosmochimica Acta*, **36**, 1007–27.

O'Neil, J.R., Clayton, R.N. & Mayeda, T.K. (1969). Oxygen isotope fractionation in divalent metal carbonates. *Journal of Chemistry and Physics*, **51**, 5547–58.

Pisciotto, K.A. (1978). 'Basinal sedimentary facies and diagenetic aspects of the Monterey Shale, California'. Ph.D. thesis, University of California, Santa Cruz, 450pp.

Pisciotto, K.A. (1981). Notes on Monterey rocks near Santa Maria, California. In *Guide to the Monterey Formation in the California Coastal Area, Ventura to San Luis Obispo*, ed. C.M. Isaacs, pp. 73–81. Pacific Section, American Association of Petroleum Geologists *52*.

Riggs, S.R. (1984). Paleoceanographic model of Neogene phosphorite deposition, U.S. Atlantic continental margin. *Science*, **223**, 123–31.

Roberts, A.E. & Vercoutere, T.L. (1986). Geology and petrology of the upper Miocene phosphate deposit near New Cuyama, Santa Barbara County, California. *US Geological Survey Bulletin*, **B-1635**, 61pp.

Rosenfeld, W.D. & Silverman, S.R. (1959). Carbon isotope fractionation in bacterial production of methane. *Science*, **130**, 1658–9.

Shemesh, A., Kolodny, Y. & Luz, B. (1983). Oxygen isotope variations in phosphate in biogenic apatites, II. Phosphorite rocks. *Earth Planetary Science Letters*, **64**, 405–16.

Shemesh, A., Kolodny, Y. & Luz, B. (1988). Isotope geochemistry of oxygen and carbon in phosphate and carbonate of phosphorite francolite. *Geochimica Cosmochimica Acta*, **52**, 2565–72.

Silverman, S.R., Fuyat, R.K. & Weiser, J.D. (1952). Quantitative determination of calcite associated with carbonate-bearing apatite. *American Mineralogist*, **37**, 211–22.

Soudry, D. (1987). Ultra-fine structures and genesis of the Campanian Negeve high-grade phosphorites (southern Israel). *Sedimentology*, **34**, 641–60.

Soudry, D. & Champetier, Y. (1983). Microbial processes in the Negev phosphorites (southern Israel). *Sedimentology*, **30**, 411–23.

Suess, E. & Thiede, J. (eds.) (1983). *Coastal Upwelling, Its Sedimentary Record, Part A: Responses of the Sedimentary Regime to Present Coastal Upwelling*. Plenum Press, New York, 640pp.

Tudge, A.P. (1960). A method of analysis of oxygen isotopes in orthophosphate – its use in the measurement of paleotemperatures. *Geochimica Cosmochimica Acta*, **18**, 81–93.

Vincent, E. & Berger, W.H. (1985). Carbon dioxide and polar cooling in the Miocene: The Monterey hypothesis. In *Natural Evolution in Carbon Dioxide and the Carbon Cycle*, ed. E.T. Sundquist & W.S. Broecker, pp. 455–68. American Geophysical Union, Monograph 32, Washington, DC.

Woodring, W.P. & Bramlette, M.N. (1950). Geology and paleontology of the Santa Maria district, California. *US Geological Survey Professional Paper*, **222**, 185pp.

26

The lithostratigraphy of the Hawthorn Group of peninsular Florida

T.M. SCOTT

Abstract

The Hawthorn Group has been a problematic unit since it was named by Dall & Harris (1892). It is a complex unit consisting of interbedded and intermixed carbonate and siliciclastic sediments containing variable concentrations of phosphate. Scott (1988) upgraded the Hawthorn to group status in Florida and delineated its component formations.

The Hawthorn Group in northern Florida has been subdivided, in ascending order, into the Penney Farms, Marks Head, Coosawhatchie and Statenville Formations. These units range in age from Early Miocene (Aquitanian) to Middle Miocene (mid Serravalian) (Huddlestun, pers. comm., 1983). Lithologically, the Hawthorn Group in northern Florida is made up of a basal carbonate with interbedded siliciclastics (Penney Farms), a complexly interbedded siliciclastic–carbonate sequence (Marks Head), a siliclastic unit with variable concentrations of carbonate in the matrix and individual beds (Coosawhatchie), and a crossbedded, predominantly siliciclastic unit (Statenville). Phosphate grains are present throughout most of these sediments, varying in concentration from absent to as much as 60% of the sediment.

In southern Florida, the Hawthorn Group is comprised of, in ascending order, the Arcadia and Peace River Formations. The Tampa Formation or Limestone of former usage is included as a lower member of the Arcadia Formation due to the Tampa's limited areal extent, lithologic similarities and lateral relationship with the undifferentiated Arcadia. Similarly, the Bone Valley Formation of former usage is incorporated as a member in the Peace River Formation. The southern Florida Hawthorn Group sediments range in age from Early Miocene (Aquitanian) to Early Pliocene (Zanclean) (Hunter, pers. comm., 1985). Lithologically, the Arcadia Formation is composed of carbonate with varying amounts of included and interbedded siliciclastics which are most prevalent in the Nocatee Member (basal Arcadia). The Peace River Formation is predominantly a siliciclastic unit with some interbedded carbonates. The carbonates are often dolostone. Phosphate grains are virtually ubiquitous in the south Florida Hawthorn sediments with the exception of the Tampa Member where it is often absent.

The Hawthorn sediments of peninsular Florida reflect a series of sea-level events and phosphogenic episodes. These sediments are characterized as being deposited in inner shelf, nearshore environments. Erosion and reworking of pre-existing sediments played an important role in the development of the Miocene sediment packages.

Introduction

The sediments of the Hawthorn Group represent a dramatic change in sedimentation in Florida. From the end of the Mesozoic to the beginning of the Neogene, carbonate sediments with a very minor siliciclastic component dominated the depositional environments of peninsular Florida. During the Early Miocene, siliciclastics, derived from the southern Appalachians, filled the Gulf Trough and encroached into the carbonate-producing environments of peninsular Florida. The carbonate-producing environments were gradually pushed further south as siliciclastic deposition increased. During the Miocene, phosphogenesis became an important depositional–diagenetic process. Phosphate grains occur within the Hawthorn sediments in amounts ranging from trace to 60%. Palygorskite, sepiolite, smectite and minor illite are the clay minerals associated with the Hawthorn sediments.

The lithostratigraphy of the Hawthorn Group throughout peninsular Florida provides the regional framework for attempting to understand the complex nature of the phosphate-bearing Miocene sediments.

There have been numerous investigations of the Hawthorn sediments during the last century. Scott (1988) provides a lengthy discussion of these previous investigations and as such they will not be discussed in this paper.

Structure

The Hawthorn Group is present throughout much of peninsular Florida (Figs. 26.1, 26.2) except on top of the Ocala Platform and the Sanford High where it is missing due to erosion or non-deposition. The main structures which controlled the deposition and present distribution of the Hawthorn Group are shown on Figure 26.3. These features have been considered structural in origin; however, there is considerable discussion concerning their origin, whether they are depositional or structural. Nonetheless, the features were topographically expressed on the surface underlying the Hawthorn Group. The two major positive features were the Ocala Platform with the Central Florida Platform and the Sanford High with the St Johns and

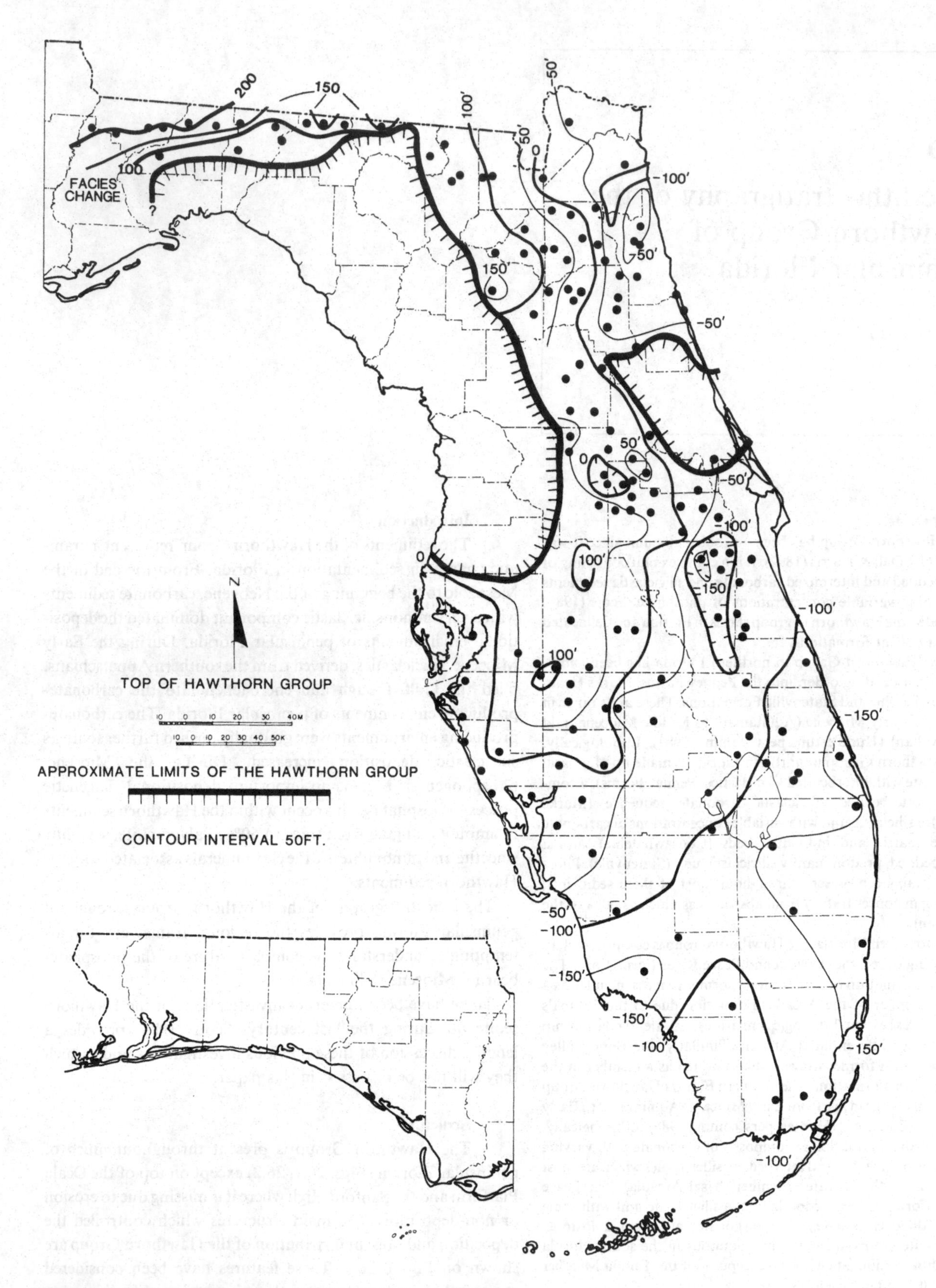

Fig. 26.1. Top of Hawthorn Group in Florida based on cores and well cuttings.

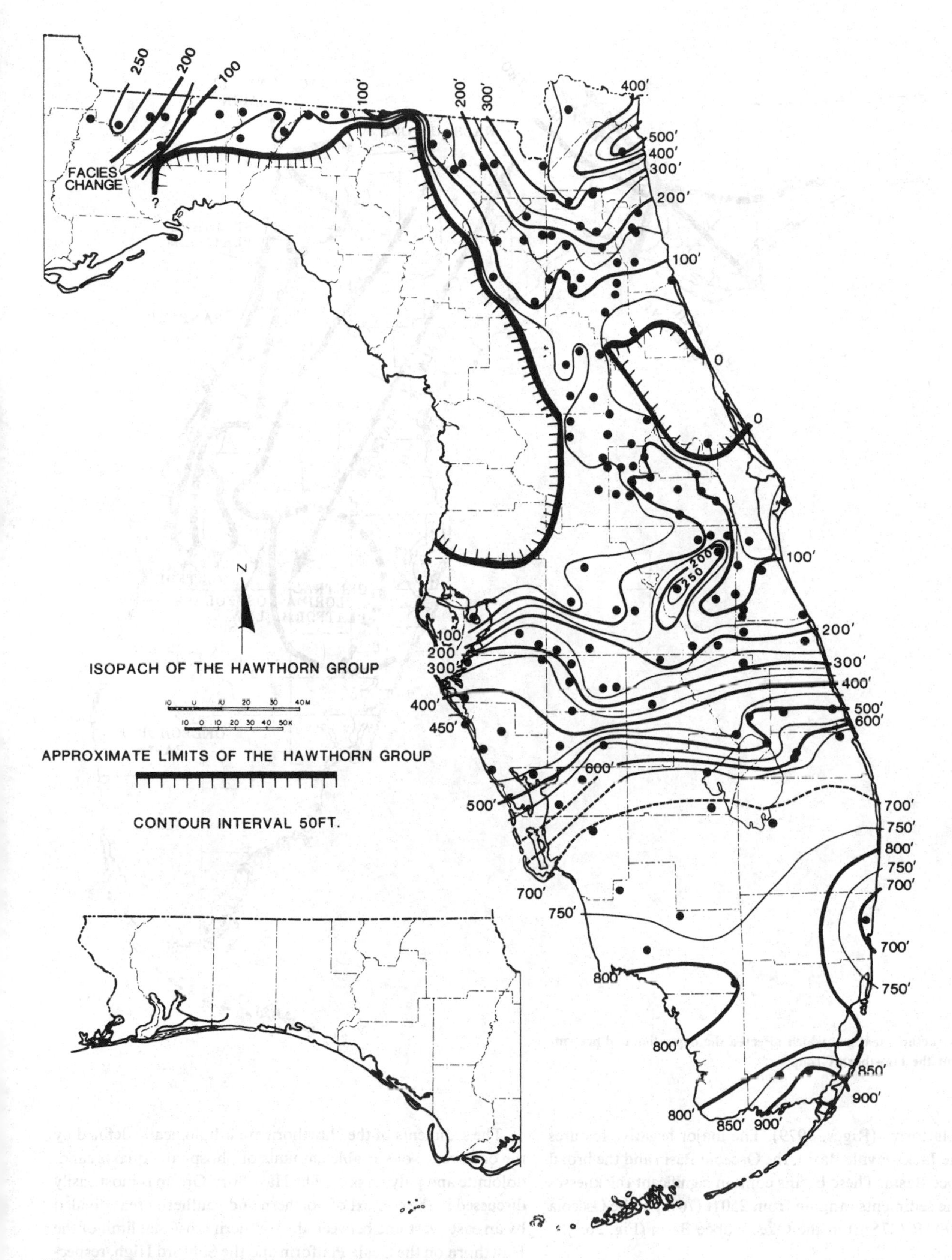

Fig. 26.2. Isopach of the Hawthorn Group in Florida based on cores and well cuttings.

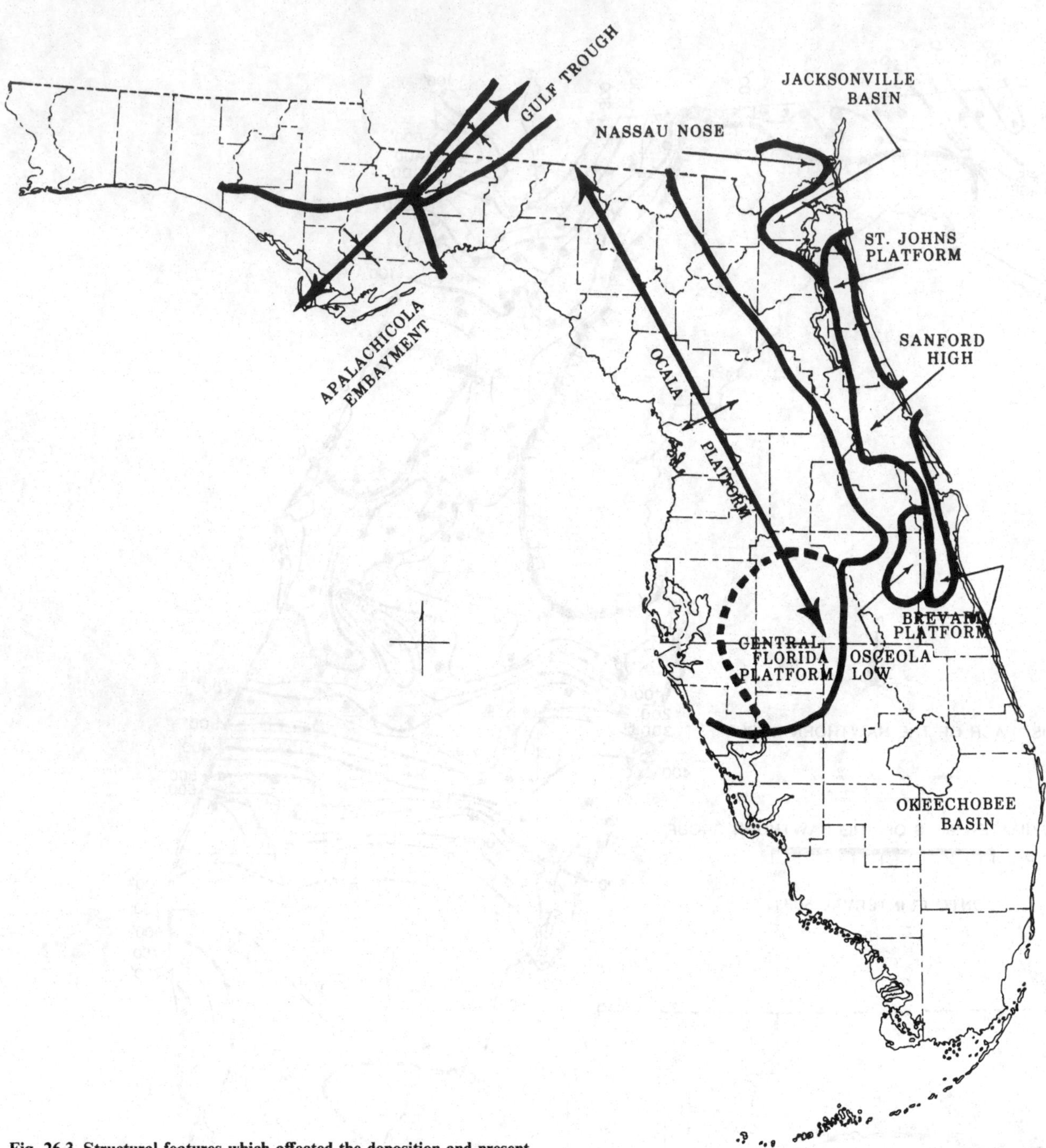

Fig. 26.3. Structural features which affected the deposition and present occurrence of the Hawthorn Group.

Brevard platforms (Riggs, 1979). The major negative features include the Jacksonville Basin, the Osceola Basin and the broad Okeechobee Basin. These basins contain significant thicknesses of Miocene sediments ranging from 250 ft (76 m) in the Osceola Basin to 900 ft (275 m) in the Okeechobee Basin (Fig. 26.2).

Lithostratigraphy

Previous investigations considered the Hawthorn sediments as a single formation which displayed great variability throughout Florida. The resulting confusion is evident in the literature. Elevating the Hawthorn to group status eliminates much of the confusion.

The sediments of the Hawthorn are lithologically defined by the occurrence of variable amounts of phosphate, quartz sand, dolomite and palygorskite. The Hawthorn Group is most easily discussed in the context of northern and southern areas divided by an east–west line between the southern erosional limit of the Hawthorn on the Ocala Platform and the Sanford High, respectively (see Fig. 26.4). Lateral and vertical variability is most pronounced in north Florida.

North Florida

The Hawthorn Group in north Florida contains significantly higher percentages of siliciclastic sediments than in south

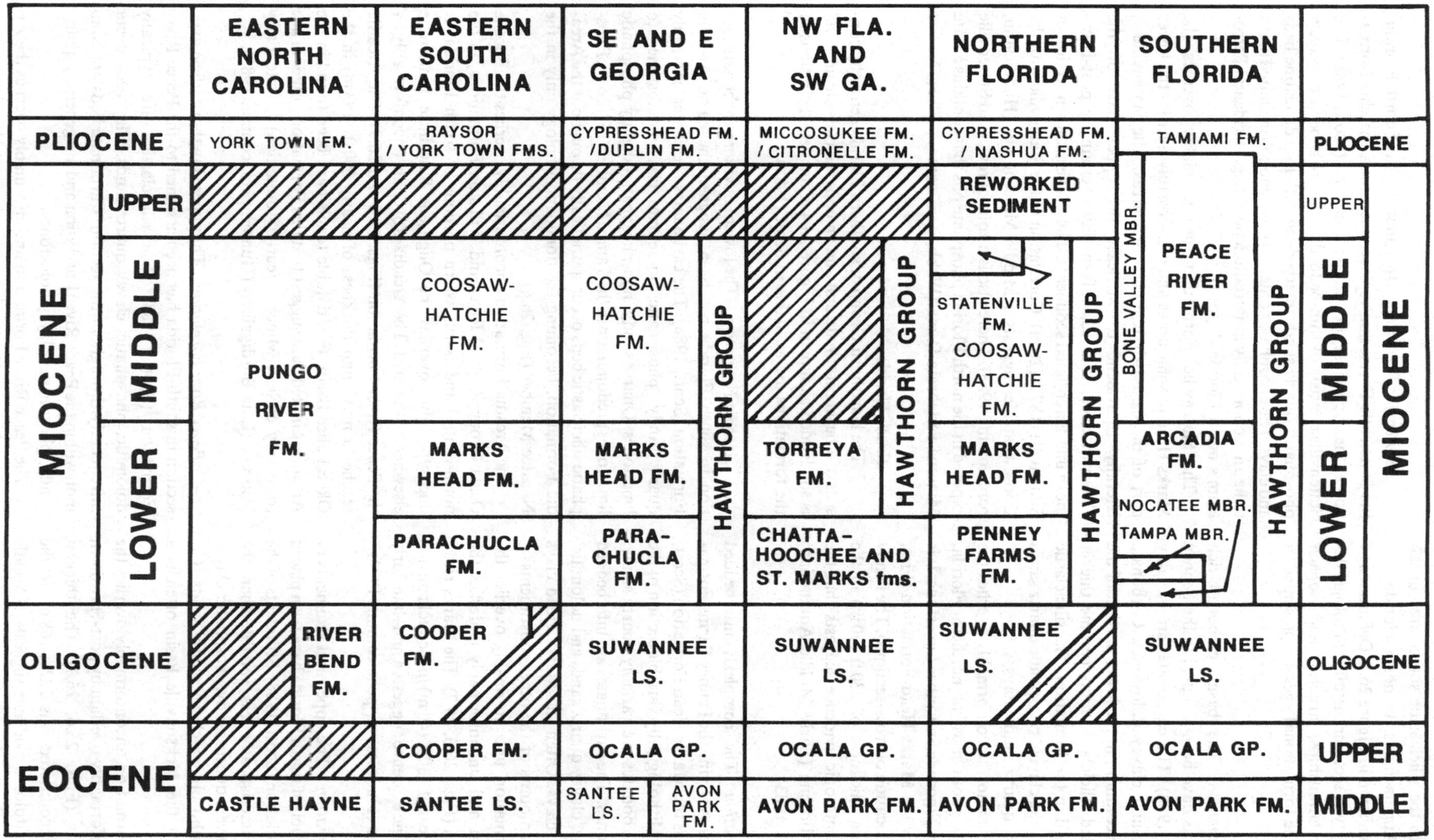

		EASTERN NORTH CAROLINA	EASTERN SOUTH CAROLINA	SE AND E GEORGIA	NW FLA. AND SW GA.	NORTHERN FLORIDA	SOUTHERN FLORIDA		
PLIOCENE		YORK TOWN FM.	RAYSOR / YORK TOWN FMS.	CYPRESSHEAD FM. /DUPLIN FM.	MICCOSUKEE FM. / CITRONELLE FM.	CYPRESSHEAD FM. / NASHUA FM.	TAMIAMI FM.	PLIOCENE	
MIOCENE	UPPER					REWORKED SEDIMENT	HAWTHORN GROUP: BONE VALLEY MBR.; PEACE RIVER FM.	UPPER	MIOCENE
MIOCENE	MIDDLE	PUNGO RIVER FM.	COOSAW-HATCHIE FM.	HAWTHORN GROUP: COOSAW-HATCHIE FM.	HAWTHORN GROUP	HAWTHORN GROUP: STATENVILLE FM. COOSAW-HATCHIE FM.	HAWTHORN GROUP: BONE VALLEY MBR.; PEACE RIVER FM.	MIDDLE	MIOCENE
MIOCENE	LOWER	PUNGO RIVER FM.	MARKS HEAD FM.	HAWTHORN GROUP: MARKS HEAD FM.	HAWTHORN GROUP: TORREYA FM.	HAWTHORN GROUP: MARKS HEAD FM.	HAWTHORN GROUP: ARCADIA FM. NOCATEE MBR. TAMPA MBR.	LOWER	MIOCENE
MIOCENE	LOWER	PUNGO RIVER FM.	PARACHUCLA FM.	HAWTHORN GROUP: PARA-CHUCLA FM.	CHATTA-HOOCHEE AND ST. MARKS fms.	HAWTHORN GROUP: PENNEY FARMS FM.	HAWTHORN GROUP: ARCADIA FM. NOCATEE MBR. TAMPA MBR.	LOWER	MIOCENE
OLIGOCENE		RIVER BEND FM.	COOPER FM.	SUWANNEE LS.	SUWANNEE LS.	SUWANNEE LS.	SUWANNEE LS.	OLIGOCENE	
EOCENE	UPPER		COOPER FM.	OCALA GP.	OCALA GP.	OCALA GP.	OCALA GP.	UPPER	
EOCENE	MIDDLE	CASTLE HAYNE	SANTEE LS.	SANTEE LS. / AVON PARK FM.	AVON PARK FM.	AVON PARK FM.	AVON PARK FM.	MIDDLE	

Fig. 26.4. Generalized stratigraphic correlations from Florida to North Carolina (modified from unpublished Gulf Coast COSUNA chart and Atlantic Coastal Plain chart, American Association of Petroleum Geologists, 1983).

Florida since it is closer to the siliciclastic source area in the Piedmont and southern Appalachians. In north Florida, the Hawthorn is subdivided as shown in Figure 26.5. Definitions of these formations including type sections, detailed lithologic descriptions, stratigraphic relationships, structure and isopach maps and discussions of the formational ages are presented in Scott (1988).

Penney Farms Formation The basal, carbonate-rich Hawthorn Group sediments have been assigned to the Penney Farms Formation (Scott, 1988). The carbonates are variably quartz sandy, phosphatic and clayey dolostones. Carbonate beds are thicker and more common in the lower portion of the Penney Farms. Interbedded siliciclastic units become thicker and more abundant upward in the formation. The siliciclastic sediments consist of dolomitic, clayey, phosphatic quartz sands and dolomitic, phosphatic, quartz sandy clays.

The Penney Farms Formation unconformably overlies the Eocene Ocala Group throughout most of northern Florida. Occasionally it overlies the Oligocene Suwannee Limestone. It is overlain unconformably by the Marks Head Formation (Figs. 26.6, 26.7; Fig. 26.5 shows location of cross-sections). The Penny Farms reaches a maximum thickness of 230 ft (70 m) in the Jacksonville Basin. Palaeontologic evidence suggests that the Penney Farms is Early Miocene (Early–Middle Aquitanian) (Huddlestun, pers. comm., 1983).

Marks Head Formation The complexly interbedded siliciclastics and carbonates overlying the Penney Farms Formation have been referred to as the Marks Head Formation (Scott, 1988). This unit is the most lithologically variable section of the Hawthorn Group. The carbonates are variably quartz sandy, clayey, phosphatic dolostones. The carbonates are interbedded with dolomitic, phosphatic, clayey quartz sands and dolomitic, phosphatic, quartz sandy clays. Often the Marks Head clays may contain only minor amounts of the other constituents.

The Marks Head Formation unconformably overlies the Penney Farms Formation and unconformably underlies the Coosawhatchie Formation (Figs. 26.6, 26.7). The Marks Head reaches a maximum thickness of 130 ft (40 m) in the Jacksonville Basin. Limited paleontologic evidence suggests a mid–late Early Miocene (Burdigalian) age (Huddlestun, pers. comm., 1983).

Coosawhatchie Formation The uppermost sediments of the Hawthorn Group in much of north Florida have been placed in the Coosawhatchie Formation (Scott, 1988). Sediments of the Coosawhatchie are characteristically less variable than the Marks Head, consisting of very quartz sandy, phosphatic, clayey dolostones to dolomitic, phosphatic, clayey quartz sands. The base of this formation in the Jacksonville Basin often is a dolomitic, silty clay.

The Coosawhatchie Formation unconformably overlies the Marks Head and is unconformably overlain by post-Hawthorn undifferentiated sediments (Figs. 26.6, 26.7). The thickest Coosawhatchie section encountered was 222 ft (68 m) in the Jacksonville Basin. Paleontologic evidence suggests a Middle Miocene (Early Serravalian) age (Huddlestun, pers. comm., 1983).

Statenville Formation In a limited area of north Florida, the Statenville Formation replaces the Coosawhatchie Formation as the uppermost Hawthorn Group unit (Scott, 1988). The Statenville consists of thin-bedded, often cross-bedded, clayey, phosphatic, quartz sands interbedded with thin dolostone beds and clays. The phosphate content is great enough in limited areas to be an economically important deposit (e.g. Occidental Petroleum's north Florida deposit).

The Statenville Formation unconformably overlies the Marks Head in some areas and conformably overlies the lower part of the Coosawhatchie in other areas. The Statenville is laterally equivalent to at least the upper part of the Coosawhatchie. It is unconformably overlain by post-Hawthorn undifferentiated sediments. The thickest known section of Statenville is 87 ft (26 m). Correlation of the Statenville with the Coosawhatchie suggests a Middle Miocene age (Huddlestun, pers. comm., 1983). There also is a zone of reworked Statenville exposed in the north Florida phosphate pits that contains a Late Miocene vertebrate fauna (Webb, pers. comm., 1985).

South Florida

The Hawthorn Group of southern Florida generally consists of a basal carbonate unit and an upper siliciclastic unit and is subdivided as shown in Figure 26.4. Complete definitions of the formations are available in Scott (1988).

Arcadia Formation The lower Hawthorn carbonate section in south Florida has been assigned the name Arcadia Formation (Scott, 1988). The carbonates are characteristically quartz sandy, phosphatic, sometimes clayey dolostones to limestones. Quartz sandy, non-phosphatic–slightly phosphatic limestones predominate in the Tampa Member (Fig. 26.4). Phosphatic siliciclastic beds occur sporadically throughout the Arcadia Formation, becoming the dominant lithology only in the Nocatee Member (Fig. 26.4).

The Arcadia Formation unconformably overlies the Eocene Ocala Group, Crystal River and Williston Formations, in the north-central and northeastern portions of south Florida. It unconformably overlies the Oligocene Suwannee Limestone elsewhere in this area. The Arcadia is overlain unconformably by the Peace River Formation (Figs. 26.8 and 26.9). The Arcadia reaches a maximum thickness of nearly 600 ft (183 m) in the Okeechobee Basin. Few datable fossils have been found in the Arcadia. Limited data suggest that this Formation ranges in age from very earliest Miocene (early Aquitanian) to late Early Miocene (Late Burdigalian) (Hunter, pers. comm., 1985).

Peace River Formation The upper Hawthorn siliciclastic section in south Florida has been assigned to the Peace River Formation (Scott, 1988). The siliciclastics are typically dolomitic, phosphatic, clayey quartz sands. Clay beds occur infrequently throughout the section. Carbonate beds are common within the Peace River Formation and are generally quartz sandy, phosphatic, clayey dolostones.

The Peace River Formation unconformably overlies the Arcadia and is unconformably overlain by post-Hawthorn undifferentiated sediments (Figs. 26.8 and 26.9). The Peace River

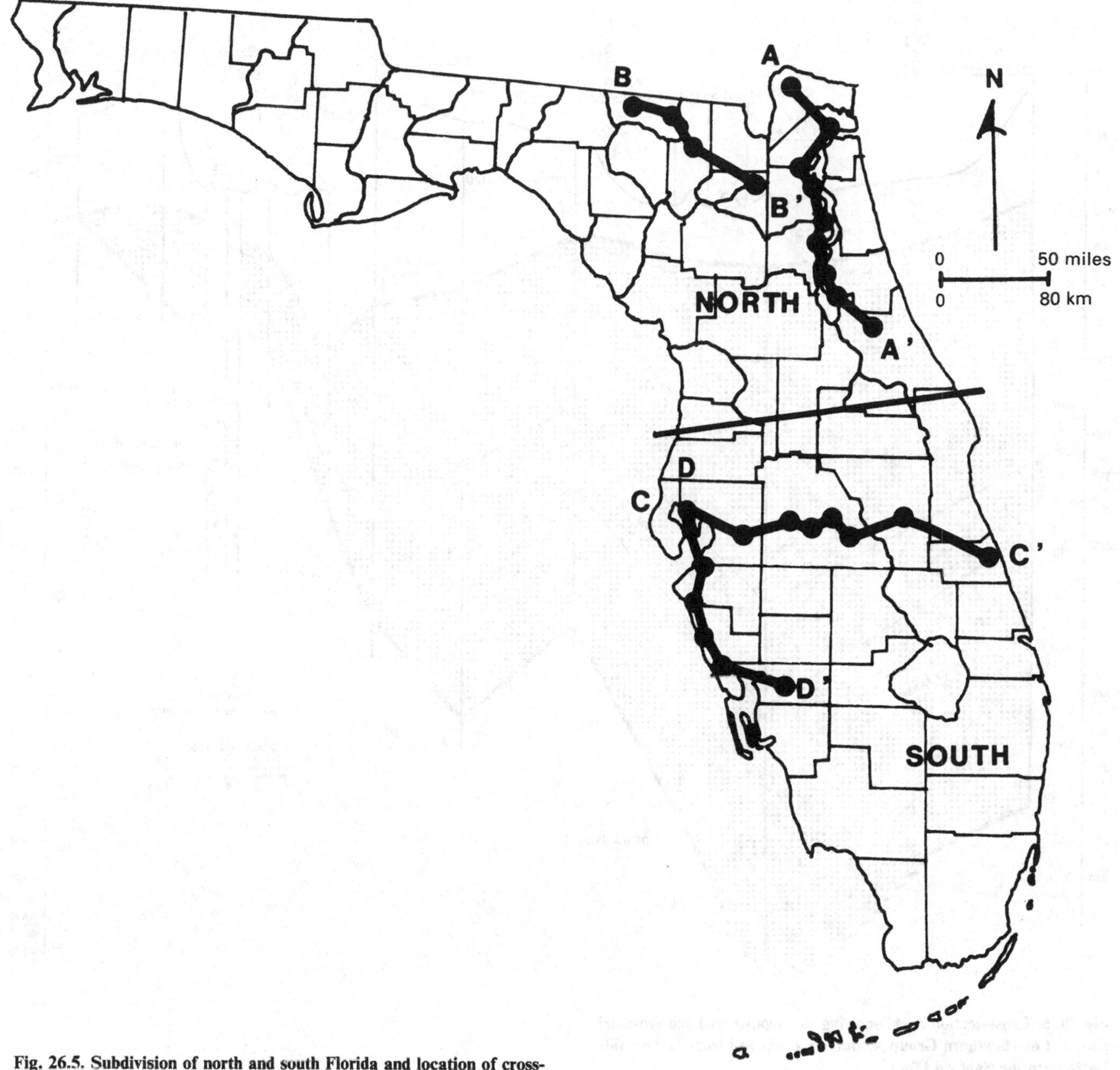

Fig. 26.5. Subdivision of north and south Florida and location of cross-sections.

Formation reaches a maximum known thickness of 650 ft (198 m) in the Okeechobee Basin. Faunal evidence suggests that the Peace River Formation ranges in age from Middle Miocene (Serravalian) to no younger than earliest Pliocene (earliest Zanclean) (Hunter, pers. comm., 1985).

Scott (1988) reduced the status of the former Bone Valley Formation to Member. The status reduction was due to the limited areal extent of this unit and the lithologic relationship and demonstrated time equivalence to the undifferentiated Peace River Formation. Lithologically, the most important factor for separating the Bone Valley from the remainder of the Peace River Formation is the occurrence of phosphate gravel in the Bone Valley Member. Phosphate gravel and sand are mixed with quartz sand and clay in proportions that vary from bed to bed vertically, and within beds laterally. This unit is the main phosphate-producing horizon in the Central Florida Phosphate District.

The boundaries of the Bone Valley Member are complex. In some areas of the district it unconformably overlies the Arcadia Formation while in other areas the Bone Valley conformably to unconformably overlies the undifferentiated Peace River Formation. The Bone Valley Member is unconformably overlain by post-Hawthorn undifferentiated sediments (Figs. 26.8 and 26.9). The maximum recognized thickness for this unit is approximately 50 ft (15 m). The age of the Bone Valley Member is derived entirely from vertebrate fossils collected from this until MacFadden & Webb (1982) and Webb & Crissinger (1983) suggest that the Member ranges in age from as old as latest Early Miocene (latest Burdigalian) to as young as Early Pliocene (Zanclean).

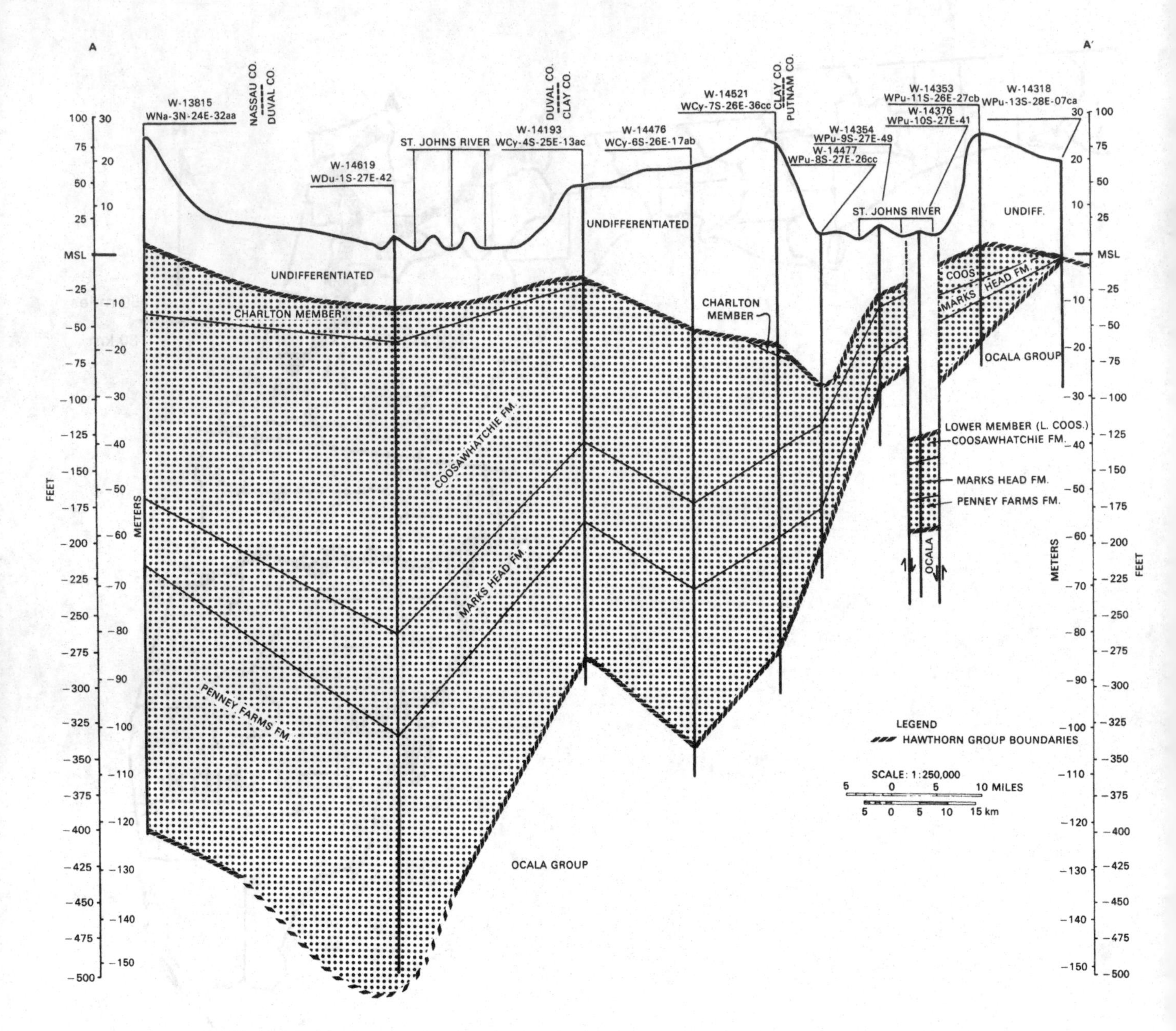

Fig. 26.6. Cross-section A–A′ showing the depositional and erosional pinchout of Hawthorn Group sediments southward from Jacksonville Basin onto the Sanford High.

Geologic history

During the Miocene, peninsular Florida was subjected to numerous major and minor sea-level fluctuations and periodically subaerially exposed. The sediments of the basal Hawthorn Group are the first record of Miocene transgressions upon the eroded, karstic, limestone surface of Florida. These sediments, carbonates and siliciclastics of the Penney Farms and lower Arcadia Formations, contain significant but not economically important amounts of apatite, indicating the Miocene phosphogenic episode had begun.

Carbonate sediments dominated deposition in the Early Miocene, although they contained a significant siliciclastic component. The siliciclastics present in the earliest Miocene represent the first recorded Cenozoic influx of siliciclastics onto the carbonate bank of peninsular Florida. This major shift in sedimentation may have been the result of eperiogenic uplift in the Appalachians, possibly coupled with a climatic change. The dramatic influx of siliciclastic sediments filled the Gulf Trough which had effectively isolated the carbonate bank from siliciclastic input from at least Early Paleocene through Late Oligocene. Siliciclastic deposition was very important in northern Florida in the Lower Miocene but did not dominate in southernmost Florida until possibly as late as early Late Miocene (Scott, 1983). Phosphate occurs as discrete apatite grains and as rims on and replacement of carbonate intraclasts. Sediments deposited during the early part of the Lower Miocene (Early Aquitanian) probably did not completely cover the Ocala Platform or the Sanford High.

Sediments deposited in the mid-Lower Miocene (Late Aquitanian–early Burdigalian) have not been recognized in

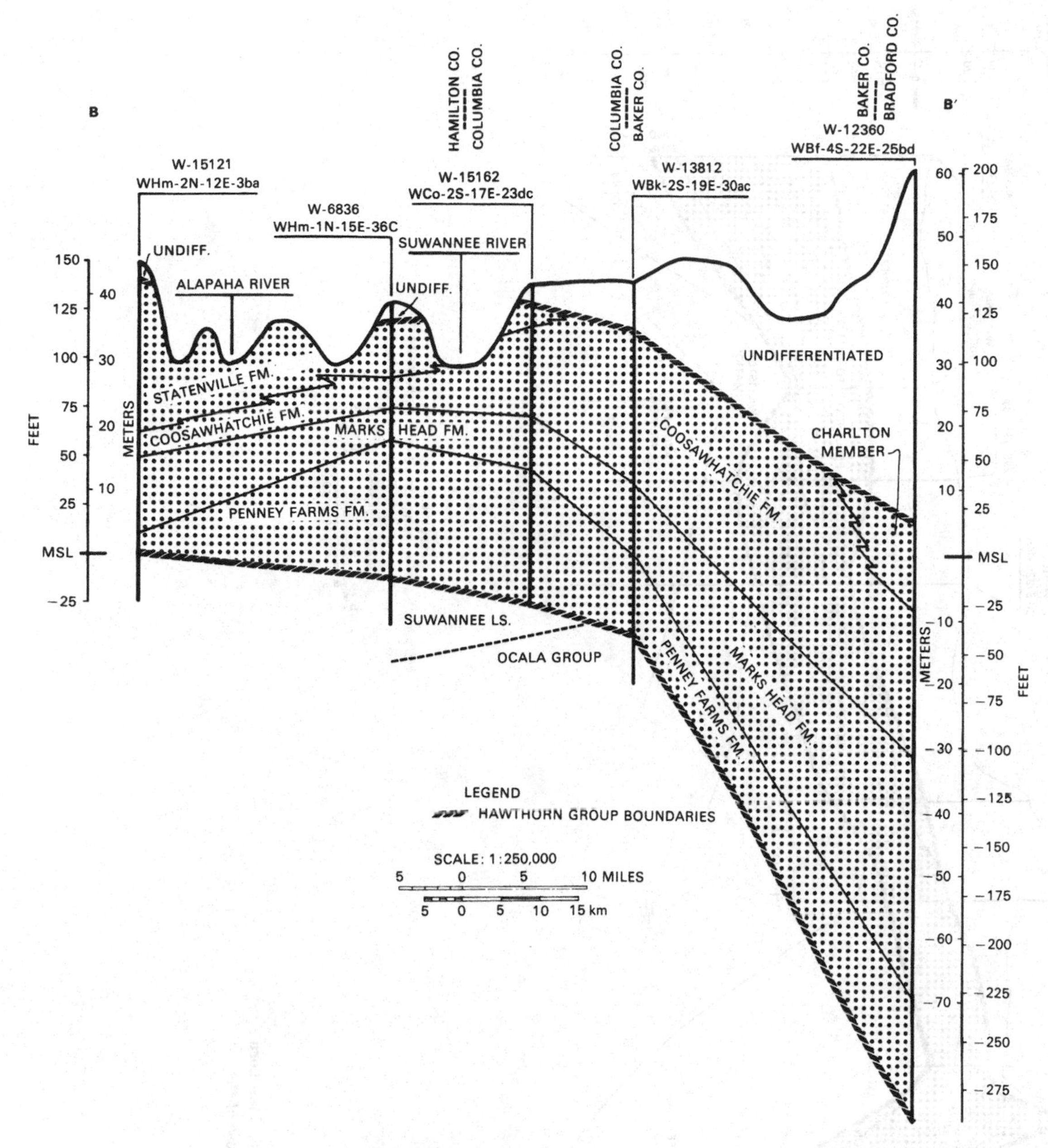

Fig. 26.7. Cross-section B–B′ showing the relationship of the Statenville Formation to the remainder of the Hawthorn Group in north Florida.

Florida. If sediments of this age are to be found onshore, it may be in the thick Arcadia Formation section in south Florida's Okeechobee Basin. It is not known whether the absence of these sediments is due to non-deposition or erosion, but it is helpful to note that they are not recognized in Georgia, South Carolina or North Carolina.

Lower Miocene (Mid-Burdigalian) sediments of the Marks Head and upper Arcadia formations were deposited on the mid-Lower Miocene unconformity. Siliciclastic sedimentation dominated in northern Florida while carbonate deposition, mixed with variable amounts of siliciclastics, continued in southern Florida. Once again apatite is present, but not in economically important quantities. In north Florida, the Lower Miocene sediments were probably not deposited over the Ocala Platform or the Sanford High. However, the present distribution and thickness of these sediments suggest that more of the structures were covered than earlier in the Early Miocene.

Middle Miocene sediments disconformably overlie the Lower Miocene throughout much of the State. The unconformity is often marked by a bored, phosphatized, well-lithified carbonate bed or a rubble resulting from the rip-up of this bed. Deposition of the mid-Miocene sediments may have begun as early as very early Middle Miocene (Langhian) (Snyder, pers. comm., 1985) to mid-Middle Miocene (Early Serravalian) (Huddlestun, pers. comm., 1985). The deposition of the Middle Miocene Hawthorn sediments covered the entire peninsula (Scott, 1988). There, sediments were later removed by erosion from the crest and upper flanks of the Ocala Platform,

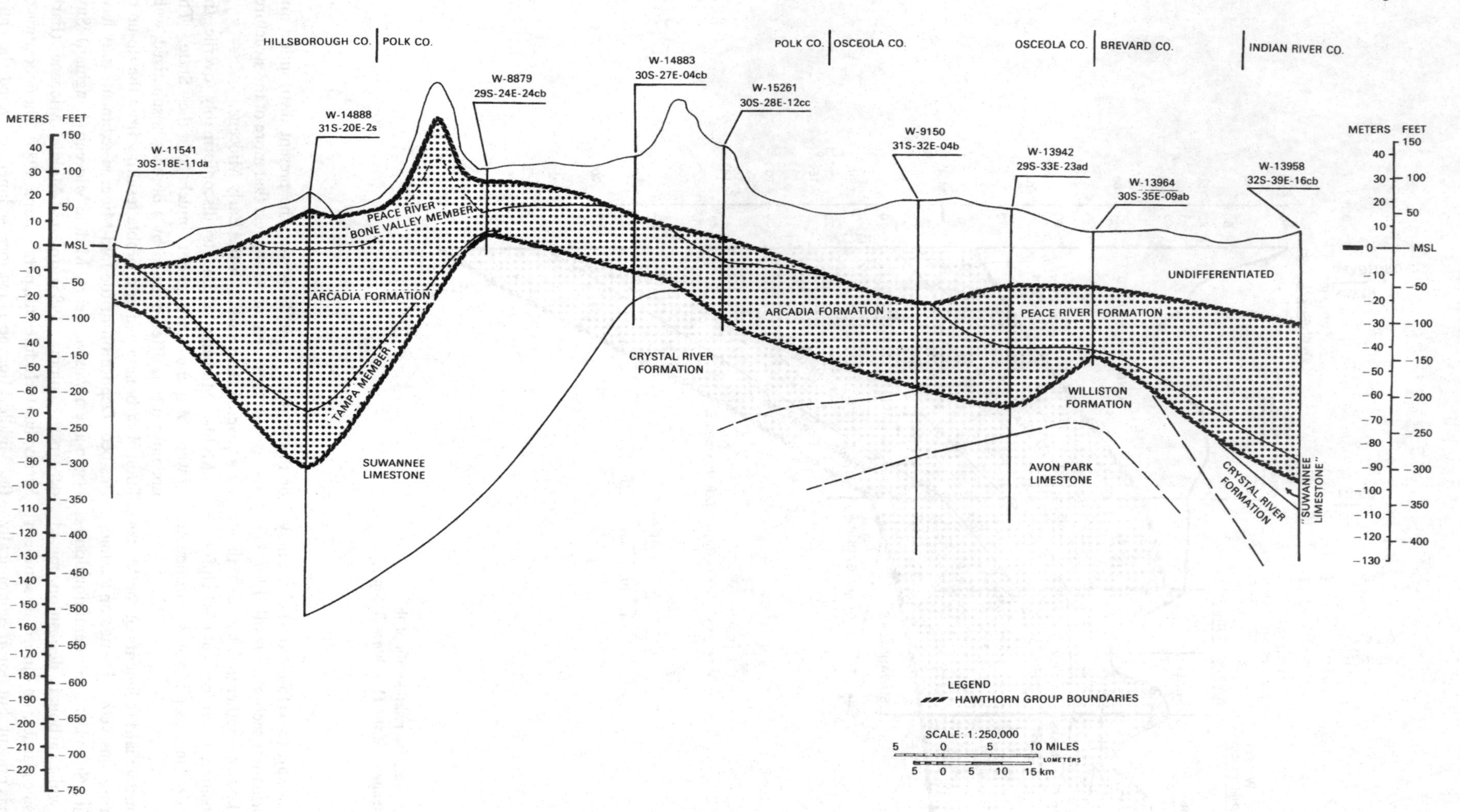

Fig. 26.8. Cross-section C–C′ showing the southern Florida Hawthorn Group sediments from west to east across the southern nose of the Ocala Platform to the Brevard Platform.

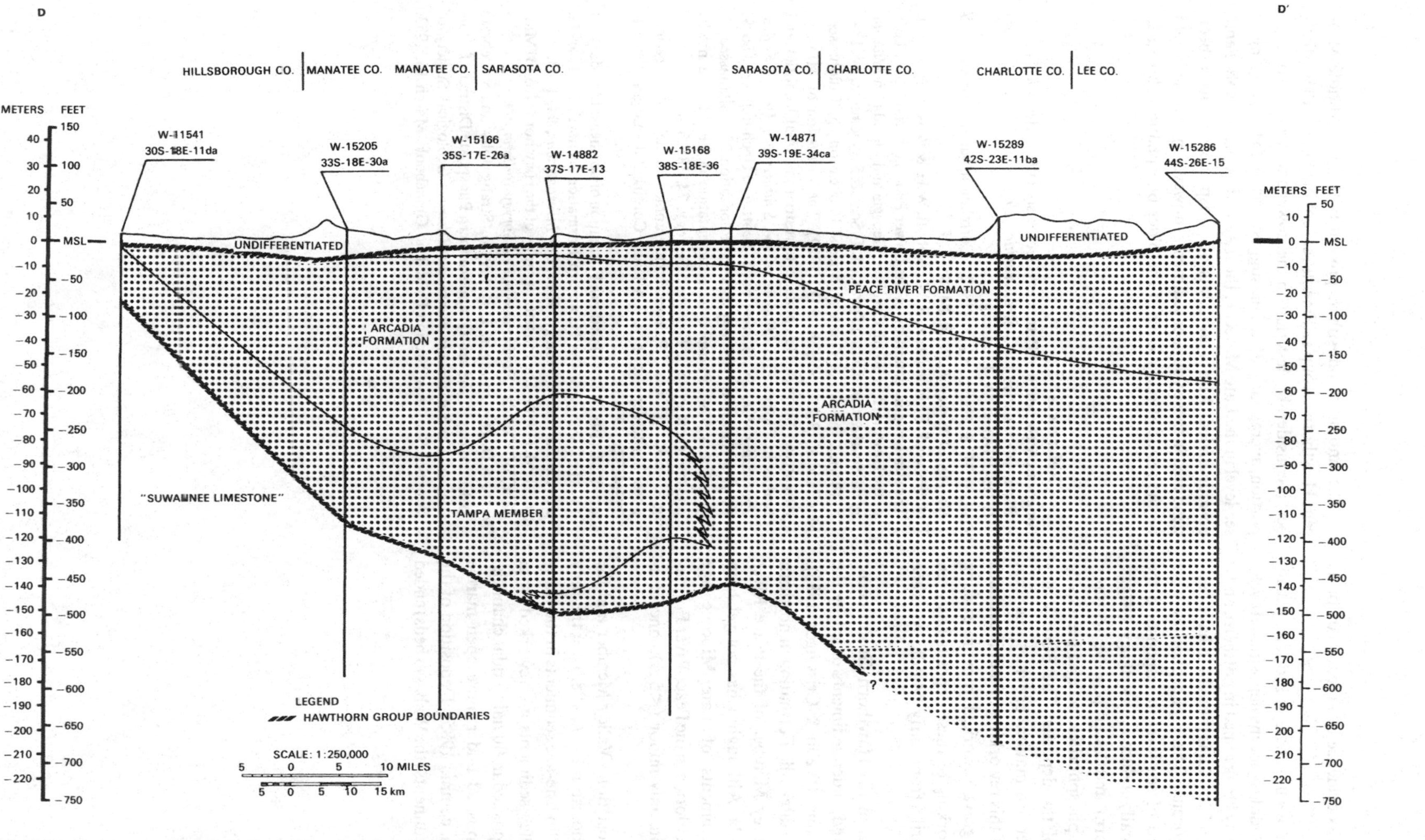

Fig. 26.9. Cross-section D–D′ showing thickening of the Hawthorn Group southward into the Okeechobee Basin.

leaving only isolated outliers as evidence of at least part of its former extent.

The greatest accumulation of phosphate in Florida appears to have occurred during this Middle Miocene depositional cycle. Much of the phosphate currently being mined in northern, and a portion of that mined in central Florida, is thought to have been deposited in the Middle Miocene. These phosphorites were deposited during a high stand of sea level in response to the impingement of topographically-induced upwellings (Riggs, 1984) onto the Ocala Platform and lesser positive features. Subsequent reworking of the phosphorites has also been quite important in development of the deposits. The Bone Valley Member of the Peace River Formation provides the best example of the importance of this reworking in that it contains large quantities of phosphate gravel reworked from pre-existing phosphate deposits. The reworking of these deposits has been multicyclic, occurring primarily from early Middle Miocene through Early Pliocene.

Upper Miocene sediments in the Hawthorn Group occur primarily as zones of reworked, older sediments dated by the occurrence of vertebrate remains (Webb & Crissinger, 1983). These occur at the top of the Statenville Formation in northern Florida and in the Bone Valley Member of the Peace River Formation in southern Florida. Although it has not yet been fully documented, other sediments of Late Miocene age (Tortonian) occur in southern Florida in the Peace River Formation. These sediments contain only minor (< 5%) concentrations of phosphate grains.

Sediments of the uppermost Bone Valley Member and the informal Wabasso beds (Huddlestun *et al.*, 1982) are the only documented occurrences of Pliocene-age sediments in the Hawthorn Group. The Bone Valley sediments are reworked from older Hawthorn beds and deposited in fluvial to tidal channels. The Wabasso beds were deposited under more open marine conditions (Huddleston, pers. comm., 1985). Deposition of the youngest Hawthorn Group sediments (the Wabasso beds) ended in late Early Pliocene.

The amount of phosphate present in the post-Middle Miocene Hawthorn Group sediments (except the Bone Valley Member) is considerably less than in the Lower and Middle Miocene portion, averaging < 3%. This suggests that with the lowering of sea level in the Late Miocene, the major Miocene phosphogenic episode ended. Minor amounts of phosphate may have been deposited in these younger sediments; however, much of it could be the result of reworking of the older phosphatic sediments.

References

American Association of Petroleum Geologists. (1983). *Atlantic Coastal Plain, Correlation of Stratigraphic Units of North America* (COSUNA) ed. A. Lindberg.

Dall, W.H. & Harris, G.D. (1892). Correlation paper–Neocene. *US Geological Survey Bulletin*, **84**, 85–158.

Huddlestun, P.F., Hoenstine, R.W., Abbott, W.H. & Wosley, R. (1982). The stratigraphic definition of the Lower Pliocene Indian River beds of the Hawthorn in South Carolina, Georgia and Florida. In *Miocene of the Southeastern United States*, ed. T. Scott & S. Upchurch, pp. 184–5. Florida Bureau of Geology, Special Publication 25, Tallahassee.

MacFadden, B.J. & Webb, S.D. (1982). The succession of Miocene Arikareean through Hemphillian terrestrial mammal localities and faunas in Florida. In *Miocene of the Southeastern United States, Proceeding of the Symposium*, ed. T. Scott & S. Upchurch, pp. 186–9. Florida Bureau of Geology, Special Publication 25, Tallahassee.

Riggs, S.R. (1979). Phosphorite sedimentation in Florida – a model phosphogenic system. *Economic Geology*, **74**, 185–314.

Riggs, S.R. (1984). Paleoceanonographic model of Neogene phosphorite deposition, US Atlantic Continental Margin. *Science*, **223(4632)**, 123–31.

Scott, T.M. (1983). The Hawthorn Formation of northeast Florida: Part I – The geology of the Hawthorn Formation of northeast Florida. *Florida Bureau of Geology, Report of Investigations*, **94**, 1–43.

Scott, T.M. (1988). The lithostratigraphy of the Hawthorn Group (Miocene) of Florida. *Florida Geological Survey Bulletin*, **59**.

Webb, S.D. & Crissinger, D.B. (1983). Stratigraphy and vertebrate paleontology of the central and southern Phosphate District of Florida. In *Central Florida Phosphate District*, Geological Society of America Southeast Section Field Trip Guidebook, March 16, 1983.

27

Clay mineralogy of the phosphorites of the southeastern United States

G.H. McCLELLAN AND S.J. VAN KAUWENBERGH

Abstract
Common clay minerals associated with the phosphorite deposits of the southeastern United States (SEUS) include illite, kaolinite, and smectite. Palygorskite and sepiolite are rather uncommon clay minerals found in these sediments. The significance of the association of these uncommon clays with phosphorites is unclear. Although these clays and phosphorites may develop in related environments, their formation may not be contemporaneous. New data are presented on the specific mineralogic and chemical characteristics of the clays in these deposits. These data and those of previous studies have been combined to develop possible modes of formation for these clay minerals. Global occurrences and associations of phosphorites, clays, and zeolites are related on a temporal and spatial basis.

Introduction

The Miocene phosphorite deposits of the southeastern United States (SEUS) extend along the Atlantic Coastal Plain from North Carolina to the center of the Florida Peninsula. The identified recoverable resources of these deposits are estimated at approximately seven billion tons of phosphate rock (Cathcart, Sheldon & Gulbrandsen, 1984). The sedimentology and mineralogy of these Miocene deposits have been the subject of a number of investigations. The general impetus of these studies has been to gain a better understanding of the important economic aspects of the deposits of clays and phosphates that occur within these sediments. Supporting academic studies have addressed the question of their formation in geologic models as well as regional stratigraphy. Because the clays and phosphates are each important industrial minerals in their own right, previous work has been generally concentrated on their individual geologic problems.

Smectites and palygorskite–sepiolite have been identified as common clay minerals in these sediments. Smectite is used in this paper as a general term to designate expanding clay minerals (Bailey *et al.*, 1971). Based on this nomenclature, montmorillonite is considered a smectite species. To maintain uniformity in usage, references to montmorillonite have been changed to smectite when the original author used the name as a group designation. Palygorskite has been used throughout this chapter in the place of attapulgite. These names are considered synonymous for the purposes of this presentation.

A recent review on the mineralogy of palygorskite–sepiolite compiled by Singer & Galan (1984) contains contributions from a number of authors on the various occurrences and uses of these unusual clay minerals. Velde (1985) provides an analysis of the possible physico-chemical environments in which these fibrous-textured, chain-structure clay minerals occur. Readers are referred to these texts and their extensive bibliographies for the details on the properties of these and related clay minerals.

During the Paleocene, Eocene, and Oligocene epochs, the Gulf Coast Geosyncline was filled with a thick wedge of clastics deposited in sedimentary environments such as deltas, barrier islands, bays, marshes, and shelf areas (Murray, 1961). During the same time thinner beds of clastic sediments were deposited in the Atlantic Coastal Province, Georgia, and Florida. These sediments became progressively calcareous to the south. Shallow-water limestones predominated in southern Georgia and Florida and grade eastward into shallow marine and coastal lagoon environments on the continental shelf (JOIDES, 1965).

During the Miocene epoch, southern Florida developed limestone deposits nearly 300 m thick; northward into southern Georgia, the Miocene deposit becomes thinner, more sandy, and contains calcareous silts and oozes on the continental shelf. The Atlantic Coastal Province contains up to 30 m of Miocene sands and clays.

Regional and local correlations of the Miocene deposits are not well established. Freas (1968) and Freas & Riggs (1968) have identified various features that may have influenced sediment deposition or formation in the region. The work by Weaver & Beck (1977) on the clay mineralogy and the work by Riggs (1984) on the phosphate deposition are particularly noteworthy since they provide models of the geologic processes that may have been operative.

Riggs (1984) broadly divided the region into the Carolina and Florida phosphogenic provinces. A prominent feature in the paleogeographic reconstructions is the Gulf Trough (Patterson & Herrick, 1971), which is a remnant of the Suwannee Channel (Jordan, 1954) that connected the Atlantic Ocean with the ancestral Gulf of Mexico. This feature divides the region into the two phosphogenic provinces. The Florida province is characterized by clastic sediments and is an area of major commercial palygorskite clay deposits.

Phosphate is relatively abundant throughout the SEUS in

sediments of Miocene age. In North Carolina, phosphate occurs on the Atlantic Coastal Plain and offshore on the continental shelf. The Florida phosphate is generally thought to have been deposited in the Early–Middle Miocene epoch and reworked into younger sediments. The Georgia and South Carolina phosphates were deposited during the same general time as the Florida deposits and also show evidence of reworking.

Previous studies of the mineralogy and significance of clay minerals associated with the main phosphate-producing areas of the SEUS have not been integrated. The most comprehensive effort is the work of Weaver & Beck (1977). This was a broad regional study of the clay minerals with a subordinate emphasis on the associated phosphate deposits. Their sample base was limited outside of Georgia and the area of commercial 'fuller's earth' deposits in Florida (Weaver & Beck (1977) for sample location details). Few samples were studied from the commercial phosphate areas of peninsular Florida; limited sampling occurred in southeastern South Carolina, and no samples are reported for North Carolina.

According to Weaver & Beck (1977), smectite is the dominant clay mineral in the region although palygorskite–sepiolite, detrital illite, and kaolinite are relatively abundant and opal-cristobalite may be locally important. Glauconite is abundant in the Atlantic Coastal Plain sediments beginning in northern North Carolina and continuing northward. Palygorskite with some sepiolite can be found over all the region south of North Carolina. Clinoptilolite is a common but minor component of the North Carolina deposits and also occurs in the Miocene of South Carolina (Heron & Johnson, 1966). Zeolites (clinoptilolite and heulandite) are also present in the continental shelf carbonates of the Atlantic shelf (Gibson, 1970).

Rooney & Kerr (1967) described smectite as the most common clay mineral of the North Carolina phosphate deposit; they had reported the occurrence of clinoptilolite in the clay fraction in an earlier publication (Rooney & Kerr, 1964). Later work by Siedlecki (1983) and Lyle (1984) confirmed that smectite was the main clay mineral present with illite as a ubiquitous associate. Rooney & Kerr (1967) and Lyle (1984) used cation saturation, glycolation, and heating to determine that the smectite present was montmorillonite rather than beidellite, saponite, or nontronite. Rooney & Kerr (1967) report that clinoptilolite occurred in all but one of the samples taken from the Pungo River formation in the area of the commercial deposits. The zeolite is reported to be restricted to the Burdigalian depositional sequence of the Pungo River formation in Onslow Bay (Lyle, 1984).

In South Carolina, Heron & Johnson (1966) reported smectite as the principal clay mineral from the Coosawhatchie clay (Hawthorn formation) with small amounts of kaolinite and illite (up to 10%). The clay-size minerals in the underlying, more typical Hawthorn units are smectite, palygorskite, and sepiolite. Clinoptilolite is reported to occur sparingly in one Hawthorn sample; this is the only literature citation where palygorskite and clinoptilolite are reported to occur together in the SEUS.

McClellan (1962) and Reik (1982) presented data on the clay mineralogy of various geological sections in northern and peninsular Florida. In general, they confirm smectite as the ubiquitous clay mineral in the Miocene sediments of Florida. Kaolinite is common as a minor constituent in many sections and can be very abundant, particularly at the top of altered sections. In northern Florida and southern Georgia, sepiolite can be the dominant clay phase (i.e. > 50%) in certain areas or beds (Weaver & Beck, 1977); it occurs as a minor component in association with palygorskite in the basal portions of some of the least altered sections in central Florida (Van Kauwenbergh & McClellan, 1985).

Mineralogy and chemistry

In the phosphatic sediments of the SEUS, smectite and palygorskite–sepiolite are the most abundant clay minerals. The presence of palygorskite in the Florida deposits and its absence in North Carolina deposits are an indication that the depositional environments and post-depositional histories of the areas may have been different. The presence of the magnesium-rich clay minerals (palygorskite–sepiolite) in the SEUS phosphorite deposits is often emphasized as an unusual feature although these clays are much less abundant than the associated smectites. The zeolite, clinoptilolite, is a minor but potentially important diagnostic or sedimentological indicator in the North Carolina deposit and also has been reported in deep cores in the southern extension of the Florida deposit (Cathcart, pers. comm., 1985). Unfortunately, its very limited abundance (2–5% of the sediments in which it occurs) has discouraged detailed study. These clay and zeolite minerals may be sensitive indicators of the depositional environments and diagenesis in these sediments. Although vertical and lateral variability in the clay mineral assemblage in Florida is apparent, the data from North Carolina indicate a rather homogeneous composition.

The general vertical sequence of clays in the commercial phosphate zones of the central Florida phosphate deposits is, from top to base, kaolinite→smectite→smectite + palygorskite (Altschuler, Cathcart & Young, 1964). A series of X-ray patterns is given in Figure 27.1 through a producing section in central Florida. Because this is a producing section, the aluminophosphate and kaolinite zones at the top of the section have been excluded. The smectite present may represent original detrital material (Weaver & Beck, 1977), the result of diagenesis, or a mixture of unaltered and altered material. The kaolinite is thought to be a mixture of detrital sources (Weaver & Beck, 1977) and/or derived from the alteration of montmorillonite (Altschuler, Dwornik & Kramer, 1963). There are no data available to indicate the relative proportions of kaolinite originating from the two possible sources. Similarly, the illite in these sediments is considered detrital (Weaver & Wampler, 1972) and only a minor, although persistent, phase.

Palygorskite–sepiolite

Barron & Frost (1985) have provided details on the structure of these chain or ribbon magnesium silicates. They describe them as 2:1 phyllosilicates that contain two-dimensional tetrahedral sheets in which the individual tetrahedra are linked with their neighbors by sharing three corners. Talc and pyrophyllite would be the respective magnesium and aluminium end-members of this group at zero interlayer charge. Palygorskite and sepiolite differ from the 2:1 sheet structures because they exhibit a 2:1 chain or ribbon structure (Bradley,

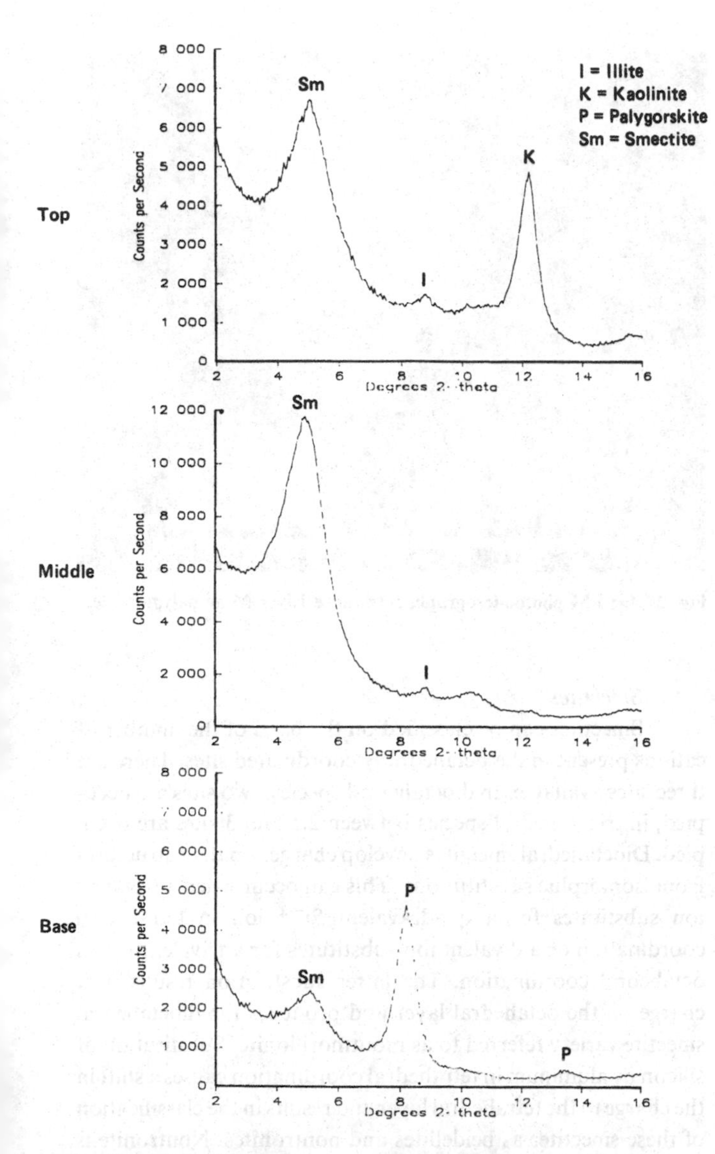

Fig. 27.1. X-ray diffraction patterns of glycolated clay fractions through a producing section in central Florida.

1940; Preisinger, 1957). Palygorskite has five octahedral positions whereas sepiolite has eight octahedral positions; silica tetrahedra occur on either side of the octahedral sheet. These structures can be considered as two or three amphibolic chains (inosilicates), respectively. The amphibolic chains alternate in a checkerboard pattern leaving open channels of fixed dimensions running parallel to the chain length. These channels contain water molecules to complete the coordination of the octahedral sheet, cations, and zeolitic water. Ideal structural formulas are $Mg_2Al_2Si_8O_{20}(OH)_2(OH_2)_4 \cdot 4H_2O$ for palygorskite (50% Al in the octahedral sheet) and $Mg_8Si_{12}O_{30}(OH)_4(OH_2)_4 \cdot 8H_2O$ for sepiolite. Tetrahedral substitutions are relatively unimportant in both minerals.

Huggins, Denny & Shell (1962) were able to distinguish long-fiber and short-fiber varieties of palygorskite although the materials were apparently structurally identical. Weaver & Pollard (1973) report the short-fiber variety (with fibers averaging approximately 1 μm in length) is the dominant type from the commercial beds near Attapulgus, Georgia; these authors speculate that there may be chemical differences between the varieties, with the short-fiber variety containing more iron. Weaver & Beck (1977) interpreted the long-fiber variety to be the result of a more evaporitic environment.

Two nearly pure samples of palygorskite were studied in some detail in the present study. One is from the Clear Springs mine (Strom & Upchurch, 1985) in central Florida, and the other is from the Occidental Suwannee River Mine in northern Florida. Chemical analyses of the samples (Table 27.1) are similar to those for palygorskites summarized by Weaver & Pollard (1973) and give excellent palygorskite X-ray diffraction patterns indicating a very minor content of intimately associated smectite (Fig. 27.2). Scanning electron microscope studies show that these samples are the long-fiber variety (Figs. 27.3 and 27.4) on the basis of the fibers that are 2 μm or longer.

Gard & Follett (1968) demonstrated the existence of two palygorskite symmetries using electron diffraction. Christ *et al.* (1969) examined five palygorskite samples by X-ray diffraction and reported that some samples exhibit orthorhombic symmetry while others appear to be monoclinic. These authors reported that the powder data of a sample from Attapulgus, Georgia, can readily be indexed on the basis of orthorhombic symmetry. Weaver & Beck (1977) interpreted their palygorskite X-ray data to indicate that palygorskite has orthorhombic symmetry because of a lack of separation of the (121) reflection. The two samples reported here (Fig. 27.2) show a clear separation of the (121) and $(12\bar{1})$ reflections near 20° 2-theta, indicating these samples are both monoclinic. It is likely that the variation in symmetry of the palygorskite samples results from variations in chemical composition. Unfortunately, an adequate data base has not been developed to establish this relationship.

Chemical analyses reported in the literature for palygorskites are quite variable. In part, these differences are the result of variations in analytical techniques and the amounts of impurities that these samples contain. The data in Table 27.1 are for palygorskite samples that contain only minor amounts of smectite. When converted to structural formulas, these two samples contain 0.18 or less tetrahedral aluminum per eight tetrahedral positions. This is in the lower range of values for

Table 27.1. *Chemical analyses of Florida palygorskites*

	Weight %	
	Clear Springs Mine	Suwannee River Mine
Al_2O_3	9.76	10.32
CaO	0.00	0.02
Fe_2O_3	3.46	4.94
H_2O^{+a}	12.60	12.93
K_2O	0.57	0.39
MgO	12.15	10.51
MnO	0.03	0.02
Na_2O	0.18	0.37
SiO_2	60.04	58.65
TiO_2	0.41	0.28
Total	99.20	98.43
Octahedral Al/Mg	0.59	0.71

[a] H_2O^+ = water retained above 105° C.

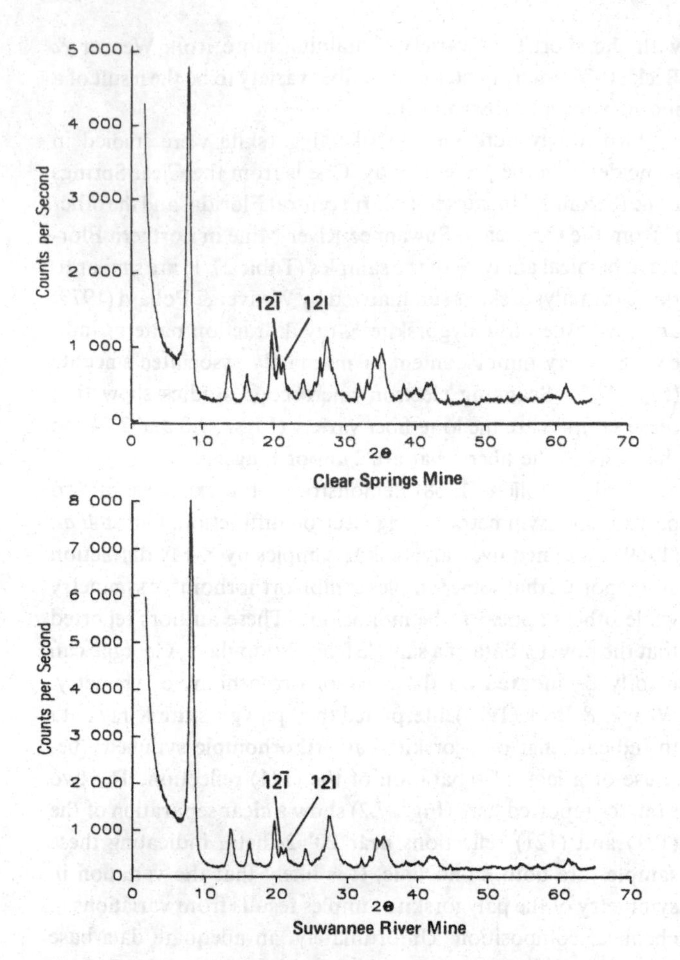

Fig. 27.2. X-ray diffraction patterns of Florida palygorskites.

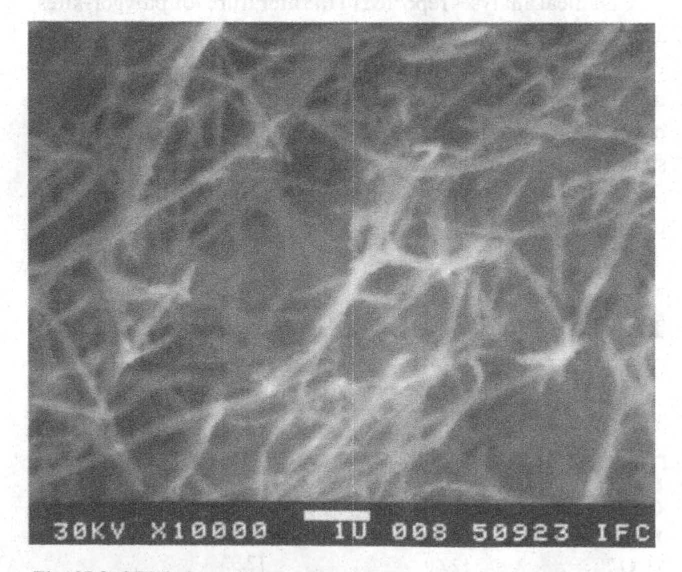

Fig. 27.3. SEM photomicrograph, Clear Springs Mine palygorskite.

palygorskite. Aluminum fills about 36% of the octahedral sites while magnesium fills about 48%; the remaining octahedral sites probably are filled by iron. The aluminum:magnesium ratio in these two samples is about 0.75, which is among the lowest values reported for palygorskite by Weaver & Pollard (1973). The ratio of divalent to trivalent cations in the octahedral layer is approximately 1.

Fig. 27.4. SEM photomicrograph, Suwannee River Mine palygorskite.

Smectites

Smectites can be classified on the basis of the number of cations present in the octahedrally coordinated sites. There are three sites available. In dioctahedral species, two sites are occupied; in trioctahedral species between 2.5 and 3 sites are occupied. Dioctahedral smectites develop charges on their structures from isomorphic substitutions. This can occur when a trivalent ion substitutes for a quadravalent Si^{+4} ion in tetrahedral coordination or a divalent ion substitutes for a trivalent ion in octahedral coordination. The latter substitution results in a charge on the octahedral layer and produces the dioctahedral smectite variety referred to as montmorillonite. Substitution of silicon by aluminum in tetrahedral coordination causes a shift in the charge to the tetrahedral layer and results in the classification of these smectites as beidellites and nontronites. Nontronite is the dioctahedral smectite with significant octahedrally coordinated iron and exhibits a d-spacing of the (060) reflection at approximately 0.1520 nm or slightly greater. The lowest Fe_2O_3 concentration known to the authors that still exhibits the (060) characteristic of nontronite is 15.22% wt (Gruner, 1935).

Brusewitz (1975) refined the lithium saturation–heating–glycolation tests originally proposed by Greene-Kelley (1953) for differentiating smectites by X-ray diffraction into the procedures used in this work. XRD methods are often selected as the single procedure to determine smectite varieties because they are reasonably uncomplicated, fast, and effective in identifying mixtures. Data are presented for several techniques, including chemical, XRD, and thermal methods. The results of the various methods should reinforce one another in the identification of the variety or varieties present.

Surprisingly little work has been done on the detailed mineralogy of the smectites in the SEUS phosphorite deposits. In 1963, Altschuler *et al.* described the smectites in the 'unaltered' Bone Valley deposits as nontronitic montmorillonites on the basis of a chemical analysis that showed that they contained more than 6% iron oxides. Other authors have followed this terminology or have used 'iron-rich montmorillonites' in descriptions of the clays (Bromwell, 1982; Strom & Upchurch, 1985). Van

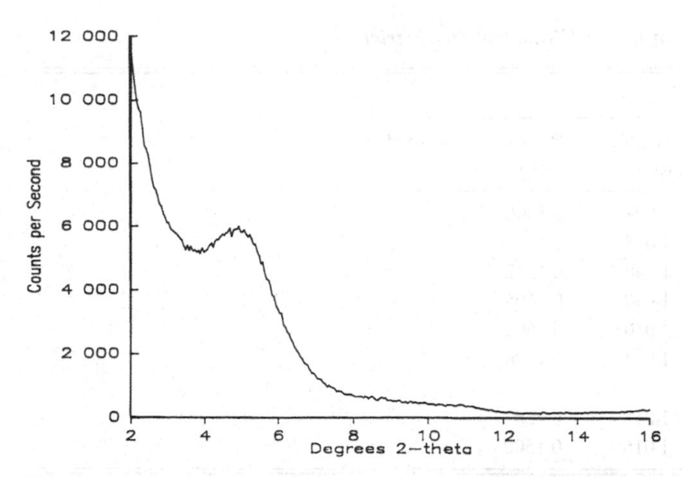

Fig. 27.5. X-ray diffraction pattern of a Li-saturated, heated, and glycolated beidellite sample (No. 42–84), base of Four Corners Mine.

Kauwenbergh & McClellan (1985) reported that the smectites in a group of samples from central Florida belonged to the beidellite–nontronite series. A calculation of the structural formula using the chemical data of Altschuler *et al.* (1963) indicates that the charge on the structure results from tetrahedral aluminum substitution and that the molar magnesium:iron ratio is about 1. This agrees with the data of Weaver & Beck (1977) for the structural compositions of smectites from the Miocene of Florida and Georgia. The data indicate these smectites have most of their charge originating in the tetrahedral layer and thus are part of the beidellite–nontronite series. Van Kauwenbergh & McClellan (1985) presented data for the (060) reflections that indicated their samples contained smectites that are essentially dioctrahedral beidellites rather than nontronites. Wissa, Fuleihan & Ingra (1982) also concluded that the smectites in Florida slimes are not montmorillonite, nontronite, or nontronite-like; their data are consistent with the interpretation of the smectite as beidellite.

Stevensite, a trioctahedral magnesium smectite, has been reported in association with African palygorskite–sepiolite deposits (Millot, 1970; Trauth, 1977). Weaver (1984) states that the smectite intimately admixed with palygorskite in southwest Georgia may be stevensite rather than a dioctahedral smectite. This interpretation assumes a fixed composition for palygorskite and calculates the stevensite-like composition by difference. The smectite intimately associated with the palygorskite–sepiolite may be a high-magnesium variety, but there is no direct supporting evidence at the present time.

After treatment using the procedures described by Brusewitz (1975), similar results have been obtained on smectite samples from phosphate-producing horizons and formations at several locations in Florida. An example is given of an essentially monomineralic, lithium-saturated smectite from the base of the producing zone at the Four Corners Mine that expands to about 1.7 nm upon glycolation after heat treatment (Fig. 27.5). This is the typical X-ray diffraction response for beidellite and nontronite, in contrast with montmorillonite which does not expand upon glycolation after heating.

Measurements of the (060) reflection have been performed on several kaolinite-free smectite samples from Florida. The results (Table 27.2) show an average value of 0.1502 nm, which basically agrees with the 0.1495 nm value reported by Altschuler *et al.* (1963). Brindley (1980) and Weir & Greene-Kelly (1962) report an (060) value of 0.1498 nm for beidellite. Schneiderhöhn (1964), Carroll (1970), and Chen (1977) indicate that nontronites have (060) values of 0.1522–0.1525 nm. Thus, the X-ray data indicate that the smectites studied have the smaller (060) values characteristic of beidellites rather than nontronites.

When the chemical data for the smectites (Table 27.3) are converted to structural formulas using the methods of Marshall (1949), Ross & Hendricks (1945) and Weaver & Pollard (1973), the results indicate most of the charge occurs in the tetrahedral layer. This is the characteristic that separates beidellite fom montmorillonite (Greene-Kelly, 1955; Weir & Greene-Kelly, 1962). The charge calculation of Sample 26–84 (Lonesome Mine, Bone Valley) is nearly equally divided; however, it reacted positively to the beidellite test. These structural formulas and layer charge calculations are not inerrant, but they may be useful data transformations when used in conjunction with other data. For a more complete discussion of the errors and limitations in the interpretation of such chemical data and structural formulas, see Schultz (1969). The levels of MgO in the samples suggest that smectites are members of a biedellite–montmorillonite series rather than a biedellite–nontronite series.

The presence of rather high iron contents in various smectite samples from the SEUS complicates the interpretation of the chemical data. Field and microscopic evidence indicate that part of the iron occurs as discrete clay-sized particles and surface coatings. Although routine chemical extraction of clay minerals should be avoided, Samples 26–84 and 42–84 (Table 27.3) were extracted with sodium dithionite (Mehra & Jackson, 1960). A reduction of 40–50% in the Fe_2O_3 content of these samples resulted in analyses that are very similar to the average montmorillonite–beidellite of Weaver & Pollard (1973). XRD analyses before and after extractions indicated no significant change in the smectite. Analysis of the supernatant liquid from the extractions indicated that minimal preferential stripping of the octahedral or tetrahedral layers occurred.

Trauth & Lucas (1967) reported that the substitution of iron for aluminum in beidellite altered the thermal behavior of the mineral by decreasing the magnitude of the 850°C endothermic reaction. Nontronite dehydroxylates at approximately 500°C while beidellites undergo dehydration at 550°C. Thermal analyses of the central Florida smectites all indicate endotherms at approximately 550°C and 850°C. Schultz (1969) classified montmorillonites and beidellites into seven groups based on their composition, the amount and distribution of their layer charge, thermal behavior, and properties which are revealed by Li^+ and K^+ treatments. Among these groups are non-ideal montmorillonites and non-ideal beidellites. Both of these groups can be distinguished from their ideal counterparts by their characteristic dehydroxylation endotherms (550–590°C = non-ideal types versus 665–730°C for all other types). Based on the thermal and other forms of data, the Florida smectites may be more appropriately termed non-ideal beidellites.

The smectites from the North Carolina phosphorites cannot be clearly classified because all the samples studied were complex mixtures of smectites, illite, sepiolite, clinoptilolite, and possibly

Table 27.2. *Characteristic d-spacings of some smectite samples from the SEUS phosphate districts*

Mine, Sample Number	Formation	d-spacing (nm) Air dried (001)	Glycolated (001)	Heated $(001)^a$	Randomly oriented $(060)^b$
Florida Lonesome 26–84	Bone Valley	1.466	1.736	1.008	0.1500
Phosphoria 30a–84	Hawthorn	1.493	1.703	1.015	0.1502
Hookers Prairie 38–84	Hawthorn	1.508	1.726	1.040	0.1502
Four Corners 42–84	Hawthorn	1.454	1.730	1.030	0.1505
Kingsford 45–84	Hawthorn	1.472	1.714	1.010	0.1502
Hookers Prairie FHPC	Hawthorn	1.543	1.755	1.015	0.1498
North Carolina Lee Creek					
Ore Zone C	Pungo River	1.540	1.716	1.015	0.1501
Ore Zone B	Pungo River	1.448	1.722	1.016	0.1502

[a]Heated at 350° C for 24 h.
[b]No kaolinite detectable by XRD.

Table 27.3. *Chemical and layer charge analyses of Florida smectite samples*[a]

Sample Number / Location and Formation	Weight % 26–84[b] Lonesome Mine Bone Valley	SC1[c] Silver Creek Mine Bone Valley	FHPC Hookers Prairie Mine Hawthorn	42–84[b] Four Corners Mine Hawthorn
Al_2O_3	21.54	22.65	22.89	20.95
CaO	0.75	1.47	0.67	0.27
FeO		0.11		
Fe_2O_3	3.13	6.29	6.48	4.75
H_2O^+	11.21[b]	7.34	10.87[d]	11.54[d]
K_2O	1.64	0.74	1.32	1.41
MgO	3.28	3.62	2.59	3.90
MnO	0.03	—	0.02	0.04
Na_2O	0.21	0.18	0.87	0.06
SiO_2	57.25	56.91	53.38	55.46
TiO_2	0.95	0.65	0.76	0.64
Total	99.99	99.96	99.85	99.02
Octahedral charge	−0.28	0.00	+0.04	+0.05
Tetrahedral charge	−0.27	−0.58	−0.70	−0.39
Total charge	−0.55	−0.58	−0.66	−0.34

[a]Analyses corrected for identified apatite, dolomite, crandallite, and wavellite after chemical determination of P_2O_5, CaO, MgO, CO_2, and Al_2O_3.
[b]Sodium dithionite extracted, original Fe_2O_3 analyses, 26–84 = 5.1 wt %; 42–84 = 10.6 wt %.
[c]From Altschuler, Dwornik & Kramer (1963).
[d]H_2O = water present above 105° C.

opal C-T. Samples from a core of the ore zone C (Fig. 27.6) and a section in a working pit that were treated using the Brusewitz techniques (1975) indicate the clay mineral fractions are mixtures of beidellite, illite, and possibly montmorillonite. Measurements of the (060) reflection of several North Carolina smectites averaged 0.1501 nm, which is in the range of montmorillonite and beidellite (Table 27.2). Additional sampling and study will be needed to precisely determine the detailed mineralogical composition of these sediments.

Other minerals

There are several other minerals that occur in the clay-size fractions of the Florida and North Carolina phosphate ores. Of these, sepiolite and clinoptilolite, which occur as minor constituents, are of special interest. In Florida, the sepiolite occurs with palygorskite in the lower portions of the least altered sediments. In North Carolina, Lyle (1984) reports sepiolite in the Burdigalian (late Early Miocene) and Langhian (early Middle Miocene) portions of the section. These results have been confirmed by this study, which shows sepiolite in intimate physical association (Figs. 27.7 [XRD pattern] and 27.8 and 27.9 [SEM photo]) with smectites in ore zone C. It is interesting to note that sepiolite occurs without palygorskite in these sediments. In fact, palygorskite has not been reported north of the Coosawhatchie district in South Carolina (Heron & Johnson, 1966). This sepiolite may be authigenic and derived from the smectite.

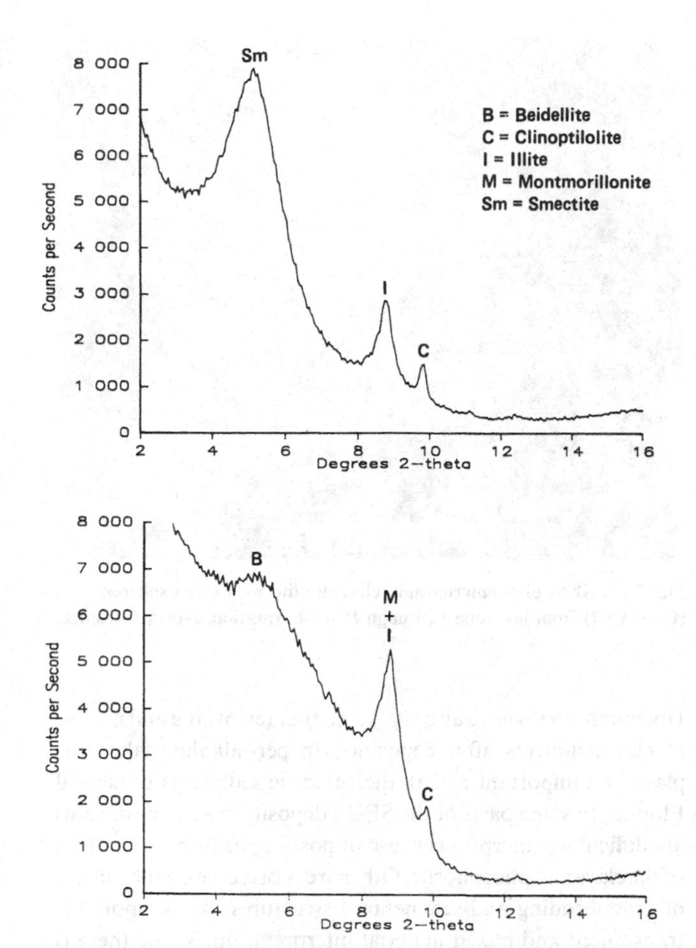

Fig. 27.6. X-ray diffraction patterns of the clay fraction of ore zone B, Pungo River Formation, North Carolina. The upper pattern is a glycolated slide while the lower pattern has been Li-saturated, heated and glycolated.

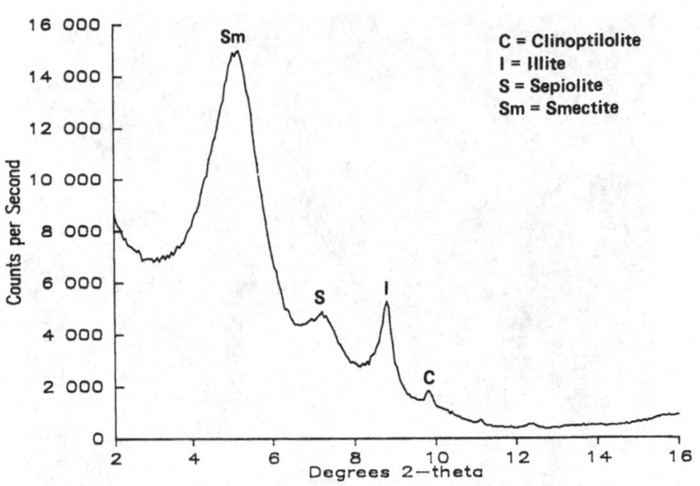

Fig. 27.7. X-ray diffraction pattern of the clay fraction of ore zone C, Pungo River Formation, North Carolina.

Clinoptilolite was detected as a minor component (Figs. 27.6 and 27.7) of the minus 2- and minus 10-μm fractions of Units B and C in the North Carolina deposits. Cathcart (pers. comm., 1985) reports one or two occurrences of clinoptilolite at depth in central Florida. This zeolite is interpreted as having an authigenic origin in many of its sedimentary occurrences. In North Carolina, clinoptilolite occurs as small prismatic crystals (Fig. 27.9) in intimate association with small silica spheres.

Francolite and aluminophosphates also are common constituents of the clay-size fraction, particularly in the weathered zones of the Florida districts. The francolite in the clay fraction may be due to recycling and reprecipitation (Altschuler, 1965). Concentrations of up to 10.6% wt P_2O_5 (approximately 28% francolite) have been noted in the clay fractions from the tops of sections (IFDC, unpublished data).

McClellan (1980) has shown that the mineralogy of the carbonate–fluorapatites (francolites) varies systematically with geologic time. Because francolites are subjected to weathering, heat, pressure, and other diagenetic variables, they are altered to less carbonate-substituted forms. These changes are rather subtle and the basic apatite structure is preserved even though carbonate and fluorine are lost. A range of variation can be found between high carbonate-substituted forms to low carbonate-substituted forms in a single deposit (Van Kauwenbergh & McClellan, 1985). In Florida, palygorskite most often occurs with the least altered francolite, generally in the lowermost beds.

Origin of the clays

Palygorskite occurs in a wide variety of sedimentary environments: continental, continental–lacustrine, peri-marine, and marine (for a review, see Singer, 1979). The salinities of associated waters range from fresh to hypersaline. It has been cited as forming from precursor clays, volcanic glasses, or precipitation from fluids. To the authors' knowledge there has been no successful synthesis of palygorskite in the laboratory.

In 1959 Jeanette *et al.* and Slansky, Camez & Millot reported that phosphates and palygorskite occur in separate alternating beds in Morocco and Benin. Smectites are reported to occur with both palygorskite and phosphate. These results were interpreted to indicate that palygorskite requires very calm conditions for its deposition while phosphate deposition occurs under agitation. This may be true, but the ubiquitous presence of smectites with phosphate sediments suggests that factors other than physical agitation play an important role. Smectites seem to be tolerant to a range of geochemical and physical conditions that occur with both the deposition or formation of phosphate and palygorskite. A combination of data from Millot (1970), Salvan (1985), and IFDC (unpublished) shows that palygorskite deposits often precede phosphate deposition in several deposits in sedimentary basins along the west coast of Africa (Fig. 27.10). In north Florida the palygorskite is believed to have been deposited prior to the main phosphorite deposits (Weaver & Beck, 1977). In central Florida, palygorskite beds occur below or within the phosphorites.

Millot (1963 and 1970) reviewed work that he and his colleagues did during the 1950s and 1960s to develop the concept of neoformation (i.e. neogenesis, authigenesis) and explain the development of various clay mineral suites that showed little or no evidence of detrital origin or formation from precursor minerals. The chemical formation model used alumina activity to explain thick seaward successions of smectite, palygorskite, and sepiolite that occur in the tertiary basins of west Africa. Isphording (1973 and 1984) favors a similar model to describe

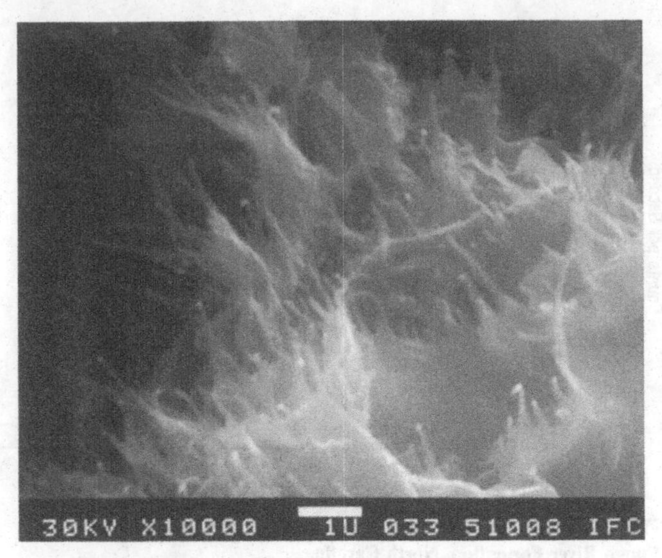

Fig. 27.8. SEM photograph, sepiolite on smectite in a sample from the ore zone C, Pungo River Formation, North Carolina.

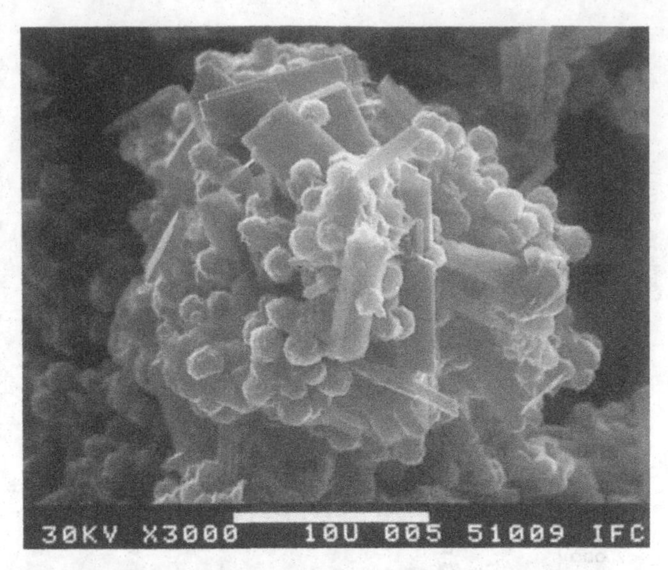

Fig. 27.9. SEM photomicrograph, clinoptilolite with silica spheres (Opal-CT?) from ore zone B, Pungo River Formation, North Carolina.

the origin of the palygorskite-containing deposits of the Yucatan Peninsula. A strong argument supported by both Millot (1970) and Isphording (1973 and 1984) for neoformation is the lack of associated detrital sediments and the frequent association with other chemical sediments such as carbonates, cherts, and phosphates. Since the idea of neoformation was proposed, the pro- and anti-neoformationists (and all those in between) have argued about the origins of palygorskite in the clay mineral literature. Although there is no consensus of opinion on this subject, neoformation is a widely accepted model for palygorskite formation in soils and peri-marine environments. Singer (1979) specified that palygorskite forms under alkaline conditions with high silica and magnesium activities but low aluminum activity. He considers that palygorskite is created by neoformation rather than by diagenesis.

Palygorskite is commonly associated with smectites. In some occurrences, palygorskite is believed to weather to smectite under the influence of acidic groundwater (McClellan, 1964). Smectites also form from palygorskite in calcareous soils (Barshad, *et al.*, 1956) and under alkalinic conditions in laboratory-scale tests (Golden *et al.*, 1985). In other cases the smectite may have been transformed to palygorskite by the addition of magnesium (Loughnan, 1959; Altschuler *et al.*, 1964; Weaver and Beck, 1977). Huertas, Linares & Martin-Vivaldi (1970) and Decarreau, Sautereau & Steinberg (1975) have proposed beidellites as the precursor smectite. Couture (1977 and 1978) presented evidence that palygorskite can form authigenically in deep-sea sediments from combinations of volcanogenic material, zeolites, smectites, and biogenic silica. Galan & Ferrero (1982) have proposed the formation of palygorskite from illite. It is not certain that all these transformations occur although field, petrographic, and mineralogical studies have been interpreted to support these conclusions.

Weaver & Beck (1977) suggest that the palygorskite in northwest Florida and southwest Georgia was deposited prior to phosphate and that the subsequent mixing of these components was the result (in part) of erosion and redeposition. Strom & Upchurch (1985) have also suggested that reworking and mixing of clay sediments after deposition in peri-alkaline lakes has played an important role in the Miocene sediments of central Florida. In some parts of the SEUS deposit, these relationships are difficult to interpret because of post-depositional alteration of the clays and phosphorite. Other areas preserve clear evidence of cross-bedding and sedimentary structures that support the transported and mixed material interpretation. While there is evidence that the phosphatic and palygorskite-rich sediments have been reworked (a process that might explain their admixtures and association), this would have required an environment that would have preserved these minerals. Among the criteria for such an environment are neutral–alkaline pH, presence of components that buffer the system (i.e. carbonates), arid conditions, and an abundance of magnesium and silica in solution. Millot (1970) has questioned the possibility that palygorskite can be reworked and redeposited.

The abundance of palygorskite in nearshore marine environments may indicate an association of these occurrences with epeiric seas. Callen (1984) reports a crude correspondence of these occurrences with the global transgression curves of Jenkyns (1980). Riggs (1984) has established an analogous correlation between increases in sea level and phosphate deposition on the southern Atlantic coast of the USA. The cyclic model of phosphorite formation combines fluctuating sea levels, changing climatic conditions, and multiple transgressive–regressive stages during the Miocene epoch and younger times. These cycles are characterized by systematic changes from terrigenous sediments with cold-water fauna to carbonate deposition with subtropical fauna in multiple vertical sequences. Weaver (1984) interpreted his palygorskite depositional data as cyclic; the palygorskite in west Florida and southwest Georgia was deposited during the regressive phases of two cycles of lagoonal and tidal environments.

According to Callen (1984), palygorskite–sepiolite minerals can be formed by three mechanisms: (1) chemical sedimentation or diagenesis in epicontinental seas and lakes; (2) hydrothermal

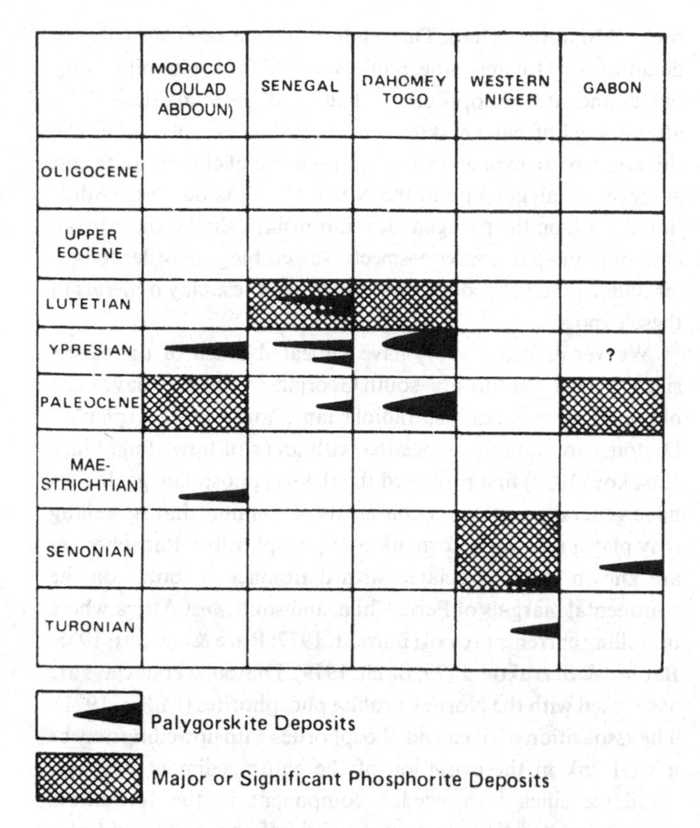

Fig. 27.10. Palygorskite and phosphorite relationships in north and west Africa (Millot, 1970; Salvan, 1985; and IFDC, unpublished data.)

alteration of basaltic glass or volcanic sediments in fore-arc basins or on oceanic rises; and (3) direct crystallization in calcareous soils. Palygorskite formed in shallow seas may be a paleoclimatic indicator of a Mediterranean to semiarid climate that occurs at low latitudes (below about 40°). The distribution of palygorskite deposits with age may indicate drier climatic conditions or regions in the geological record although the distribution may have been modified by ocean currents or continental effects.

Thus climate, temperature, and cyclic deposition seem to be factors that may affect the distribution of the non-detrital mineral components of the system (in this case both clay minerals and phosphates). The chain-structure clay minerals (palygorskite and sepiolite) may be indicative of warmer temperatures; their occurrence in the SEUS is restricted to areas south of central North Carolina and east of Alabama. Alt (1974) suggests that both phosphorite and the palygorskite–smectite assemblage in the SEUS can be reasonably accounted for by using a model of the southeastern coast during Miocene time in which cold upwellings brought seawater rich in phosphate and silica into restricted basins along an arid coast. Hite (1978) has suggested even more extreme conditions that result in a direct association of phosphorites and evaporites as a result of interaction between phosphorus-rich continental brines and cold oceanic upwellings. Although semiarid or seasonally arid conditions may have existed in the southeastern United States during the Miocene, they may have lacked the duration and intensity to form characteristic desert deposits such as sand dunes and evaporite formations. Force & Garnar (1985) extend Alt's arid climate scenario and interpret portions of the Trail Ridge of Florida as aeolian sand dunes although they consider the deposit as much younger in age than Miocene, as proposed by Alt. Thus, there is evidence in the sediment associations in the SEUS that supports the interpretation of palygorskite occurrences as paleoclimatic indicators of arid or semiarid conditions as suggested by Callen (1984) and others.

Weaver & Beck (1977) present several arguments that relate the origin of palygorskite–sepiolite to the occurrence of dolomite and phosphate in the SEUS Miocene sediments. These authors conclude that the peri-marine environment provides the restricted brackish and schizohaline conditions required to penecontemporaneously form palygorskite and the high-calcium, magnesium-deficient dolomites that occur in northern Florida and southern Georgia. Altschuler *et al.* (1963) suggested dolomitization on a regional scale was the result of incorporation of magnesium released from smectites as part of the overall diagenetic scheme. The alteration of francolite that occurs in the sediments also can release magnesium for dolomitization during diagenesis (Van Kauwenbergh & McClellan, 1985). Weaver & Beck (1977) noted the presence of palygorskite cutans on preexisting dolomite grains from southern Georgia. Such a relationship has been also noted in this study (Fig. 27.4). The formation of a portion of these clays may be linked to the dissolution of dolomite, high magnesium-calcite, or magnesium-rich clays higher in sections. These mineral associations need considerable additional study to resolve their relationships and paragenesis.

Thermodynamic calculations show that palygorskite and stevensite could form from an unspecified montmorillonite with only slight modifications of normal seawater conditions (Weaver, 1984). Trauth (1977) proposed the formation of palygorskite and sepiolite from smectite by the progressive alteration of Wyoming-type (smectite Al, Fe) and Cheto-type smectites (smectite Al, Mg) to aluminum-saponite, magnesium-saponite, and ultimately stevensite.

The origins of the smectites of the SEUS phosphate deposits are a matter of speculation. Weaver & Beck (1977) refer to the smectites as detrital material and suggest a volcanic origin (p. 186). Grim (1933) first proposed that the clays of the SEUS might be the result of alteration of volcanic debris based on the identification of isotropic fragments as glass shards and the presence of beds of smectite in the sediments. Mansfield (1940), Gremillion (1965), and Heron & Johnson (1966) all support the interpretation that these clays, or a portion thereof, were derived from volcanogenic material. Kerr (1937), Espenshade & Spencer (1963), and McClellan (1964) could not find any evidence of volcanic material in these sediments. Heron & Johnson (1966) believe the absence of glass shards to be inconclusive; vitreous volcanic material would probably be destroyed or altered by subsequent weathering or diagenetic alteration. Based on province studies using heavy and clay mineral suites from the western, central, and eastern Gulf Coast Miocene sediments, Isphording (1973) has argued against a volcanic origin for these clays. Just as Altschuler (1965) has interpreted portions of the phosphorite deposits in central Florida to have been multiply physically and chemically recycled, these smectites may have been similarly recycled, which would mask their original character.

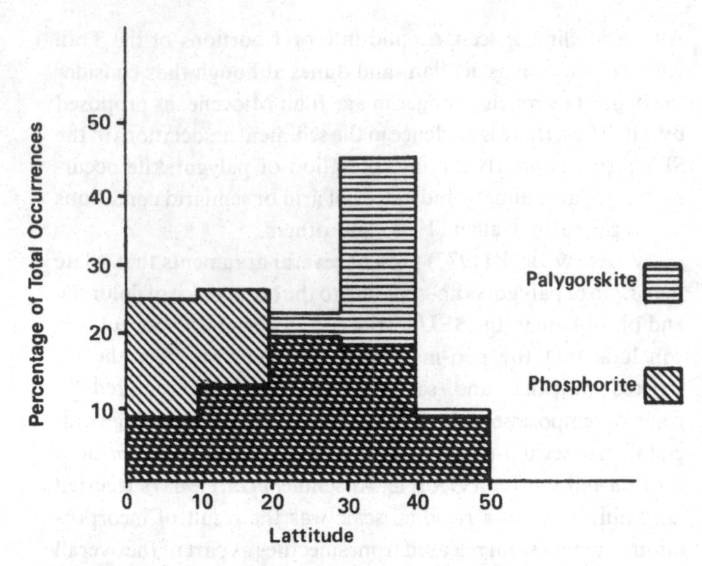

Fig. 27.11. Paleolatitudinal distribution of palygorskite and phosphorite land occurrences (from Cook & McElhinney, 1979; Callen, 1984).

Because the smectites in the sediments of SEUS usually occur in intimate physical mixtures with other clays and minerals, precise identification of the varieties has been difficult or impossible. The result has been that neither the palygorskite precursor nor stevensite has been definitely identified. Additional work on these details is essential in establishing similarities or differences in these materials and the clay minerals associated with other phosphate deposits.

Rooney & Kerr (1967) interpreted the occurrence of clinoptilolite in North Carolina as evidence for a volcanic origin for the deposits. Nathan & Soudry (1982) interpret clinoptilolite not as an indicator of a volcanic origin, but only an environment with an abundant source of silica. Clinoptilolite is commonly described as an authigenic mineral in sediments that occur in shallow water as well as in the deep sea. Nathan & Flexer (1977) report an antipathetic relationship between palygorskite and clinoptilolite in some Middle Eocene chalks and limestones. Weaver (1984) interprets this antipathy as indicating fresher water for the occurrence of palygorskite and more saline conditions for the occurrence of clinoptilolite. Thus phosphorites, smectites, and palygorskite may have a paragenetic association with clinoptilolite.

Nathan & Flexer (1977) suggest that clinoptilolite can form from the diagenesis of opal and hydrated alumina or Al-smectite; the presence of active silica (amorphous or opal C-T) is a necessary condition for the formation of these authigenic minerals. As can be seen in Figure 27.9, the clinoptilolite (and clays) in North Carolina is often associated with small silica spheres that may be opal C-T.

Golden *et al.* (1985) report that zeolites are formed by the alteration of palygorskite under alkaline conditions and are a common alteration product of clay minerals in alkali soil systems. Baldar & Whitlig (1968) report that phillipsite occurs rather than analcime when the Si:Al ratio is low. The zeolites that form at low temperatures in sedimentary rocks are more siliceous and alkali-rich than the same types found in igneous rocks (Mumpton, 1960). This relationship has been most clearly demonstrated for analcime, phillipsite, and the heulandite group (which includes clinoptilolite). If the pedogenic alkaline alteration model of palygorskite was applied to marine sediments, this could be an explanation of the presence of clinoptilolite and absence of palygorskite in the North Carolina deposits. Additional work on the paragenetic relationships should include the clinoptilolite–palygorskite–smectite assemblage in order to better understand the geochemical history of these clay minerals in these deposits.

Weaver & Beck (1977) have indicated much of the excess silica in the north Florida–south Georgia system may have been obtained from diatoms, radiolarians, and sponge spicules. Diatoms are notably associated with areas of upwelling. Since Kasakov (1937) first proposed the theory, phosphate geologists have generally reached a consensus of opinion that upwelling may play a role in the formation of phosphorites. Phosphorites are known to be associated with diatomaceous oozes on the continental margins of Peru, Chile, and southwest Africa where upwelling currents prevail (Burnett, 1977; Price & Calvert, 1978; Baturin & Bezrukov, 1979; Birch, 1979). Diatomaceous clays are associated with the North Carolina phosphorites (Miller, 1971). The association of silica and phosphorites with upwelling may be a vital link in the evolution of the entire sediment package. Available silica is a needed component in the formation, diagenetic alteration, and paragenesis of the zeolite and clay components.

Distribution of palygorskite and phosphorite

The general acceptance of plate tectonics has resulted in the reinterpretation of many structural and sedimentary features and has facilitated the logical understanding of many geological processes. Plate tectonics has been used to develop sedimentological models that explain why certain deposits occur at specific sites and what their relationship to associated sediments and minerals is.

Some interesting observations can be made when paleogeographic data from disparate sources are combined. Callen (1984) has compiled data on palygorskite occurrences at 10° paleolatitudinal intervals. Compilation of data for continental deposits indicates a concentration of occurrences between 10° and 40° and especially between 30° and 40°. Sheldon (1964) pointed out that the paleolatitudes of ancient and young phosphorites both fall within 40° of the equator. Cook & McElhinny (1979) used additional data to confirm the earlier observation of Sheldon and concluded that all phosphate deposits considered occur at low latitudes, with a clear maximum within 20° of the paleoequator. Their plot of major deposits shows a clear peak only between 10° and 20° from the equator.

When the paleogeographic data of Callen (1984) are plotted with those of Cook & McElhinny (1979) (Fig. 27.11), 54% of the data overlap although the maxima for each occurrence are separated by 20° of latitude. This may indicate that while phosphates and palygorskite can be part of the same macrosedimentological system, they may not be deposited in immediate juxtaposition. For example, palygorskite might be deposited at a more northerly latitude in a basin while phosphate was forming at a more southerly latitude; changes in conditions (i.e. sea level rising or falling, paleoclimate, changing patterns of ocean currents, etc.) could cause lateral migration in the basins, resulting in mixtures of phosphorite and palygorskite.

In addition to a favorable latitude, the physical environment must be conducive to formation and preservation of both types of deposits. These favorable conditions seem to be best met by epi-continental seas and interplay between marine and non-marine environments of deposition that may occur along continental boundaries. The relative position of deposition of phosphate and palygorskite is very speculative because of the various uncertainties involved (i.e. paleolatitudinal reconstructions, possibility of reworking, etc.). Perhaps it is best to say that both materials may be deposited under similar physical and latitudinal conditions, and their mutual association can be expected.

Callen (1984) also summarized the temporal distribution of palygorskite–sepiolite. The greatest number of continental deposits occur in the Cambrian, Devonian–Carboniferous, Permian–Triassic, Late Cretaceous–Eocene, Late Oligocene with major deposition in Middle Miocene and numerous occurrences in Pliocene–Pleistocene. When the temporal data of Callen are plotted with those of Cook & McElhinny (1979) (Fig. 27.12), the occurrences of phosphates and palygorskite since the Cretaceous are strikingly similar. The pre-Cretaceous correlations may be influenced by increasing uncertainty in the paleogeographic reconstructions or systematic variations in clay mineralogy with time. Studies of clay mineral distribution with age (Weaver, 1967; Dunoyer de Segonzac, 1970) have shown the apparent abundance of smectites and kaolinite decreases with increasing geologic age. At least two interpretations are possible: conditions favoring the formation of smectites and kaolinite developed during more recent geologic history, or these minerals were metastable and disappeared from the older sediments during burial. The second interpretation (i.e. disappearance of metastable clay minerals) has been supported by data from deep boreholes. Although the data base is smaller, similar arguments could be made for palygorskite distribution with age. The palygorskites of ancient deposits may now be represented by chlorites or other magnesium-rich species.

When the Deep Sea Drilling Project (DSDP) data compiled by Nathan & Flexer (1977) for clinoptilolite are compared with Callen's DSDP data (1984) for palygorskite (Fig. 27.13), it is clear that the variations in abundance of these two minerals through the Cretaceous show a remarkable similarity with maxima for both minerals in the Cretaceous, Eocene, and Miocene. Unfortunately, the lack of compiled data on land deposits of clinoptilolite does not allow a comparison of clinoptilolite with the major land occurrences of palygorskite and phosphorites.

Thus, it is apparent that there are spatial and temporal similarities between palygorskite and phosphorites, particularly in deposits of younger geologic ages. The data are not sufficiently precise to say whether these two materials are pene-contemporaneous. The data may only indicate that they can form in similar or associated environments. The palygorskite beds in Florida and some west African basins are generally below and may precede the main development of the principal phosphate deposits (Fig. 27.10). This clay mineral may be part of the sedimentary sequence associated with phosphate development. Although there is an apparent association, it is not ubiquitous; there are sedimentary palygorskite deposits that are not associated with phosphorites and vice versa.

Conclusions

As the results presented have shown, there are unquestionable associations of the palygorskite–smectite assemblage with phosphorites. This relationship is particularly clear for deposits younger than the Cretaceous period. As previously mentioned, the frequency of palygorskite occurrences is much less in rocks of older geologic ages (see Fig. 27.12). The significance of the palygorskite–phosphorite association is unclear. It seems that they occur in related sedimentary environments but may not be contemporaneous. The high silica and magnesium environment that is favorable for the deposition or formation of palygorskite may be antagonistic to the deposition or formation of phosphorite. Magnesium has been reported to inhibit apatite precipitation in seawater and synthetic chemical systems (Martens & Harris, 1970; LeGeros *et al.*, 1967). Thus the peri-marine environment of Weaver (1984) or the peri-alkaline lakes of Strom & Upchurch (1985) may have been the alkaline-, silica-, and magnesium-rich locales that occurred in the nearshore, or perhaps onshore, environments during a transgressive sequence or onlap cycle. These models seem to be compatible with the model of phosphogenesis (Riggs, 1984) and could account for the physical separation of the palygorskite depositional zones from those of phosphate deposition. Obviously, transgressive sequences based on these models would result in the palygorskite occurring below the main phosphorite deposition in an undisturbed section. However, dynamics of transgressive–regressive sequences often can result in reworking of such deposits and the mixing of clays, phosphates, and other minerals from various sources.

While the magnesium-rich double chain clay minerals are emphasized as an unusual feature of the SEUS phosphorite sediments, smectites are volumetrically more important, just as they are volumetrically more important in other types of sediments since the beginning of Cenozoic time (Weaver, 1967). Mineralogically, they present a difficult problem because they usually occur as intimate mixtures with other clays. Based on the various literature reports, these deposits might contain a number of types of smectites of various derivations: a precursor material to authigenic palygorskite and other smectites, a trioctahedral magnesium smectite associated with palygorskite, and authigenic smectites that may result from the alteration of palygorskite by acidic groundwater. Neoformed smectites also need to be considered.

The clay mineralogy of the phosphates in the SEUS is rather similar to that reported for other post-Cretaceous phosphorites in Africa and the Middle East. Unfortunately, the efforts of most authors have concentrated on only one aspect of this multicomponent problem (either the clays or phosphorite). Much remains to be done along the lines of the integrated mineralogic-basin analysis of Weaver & Beck (1977) to relate the occurrences of these mineral assemblages.

The paleogeographic data show significant similarities in the occurrences of phosphorites and palygorskites at low latitudes.

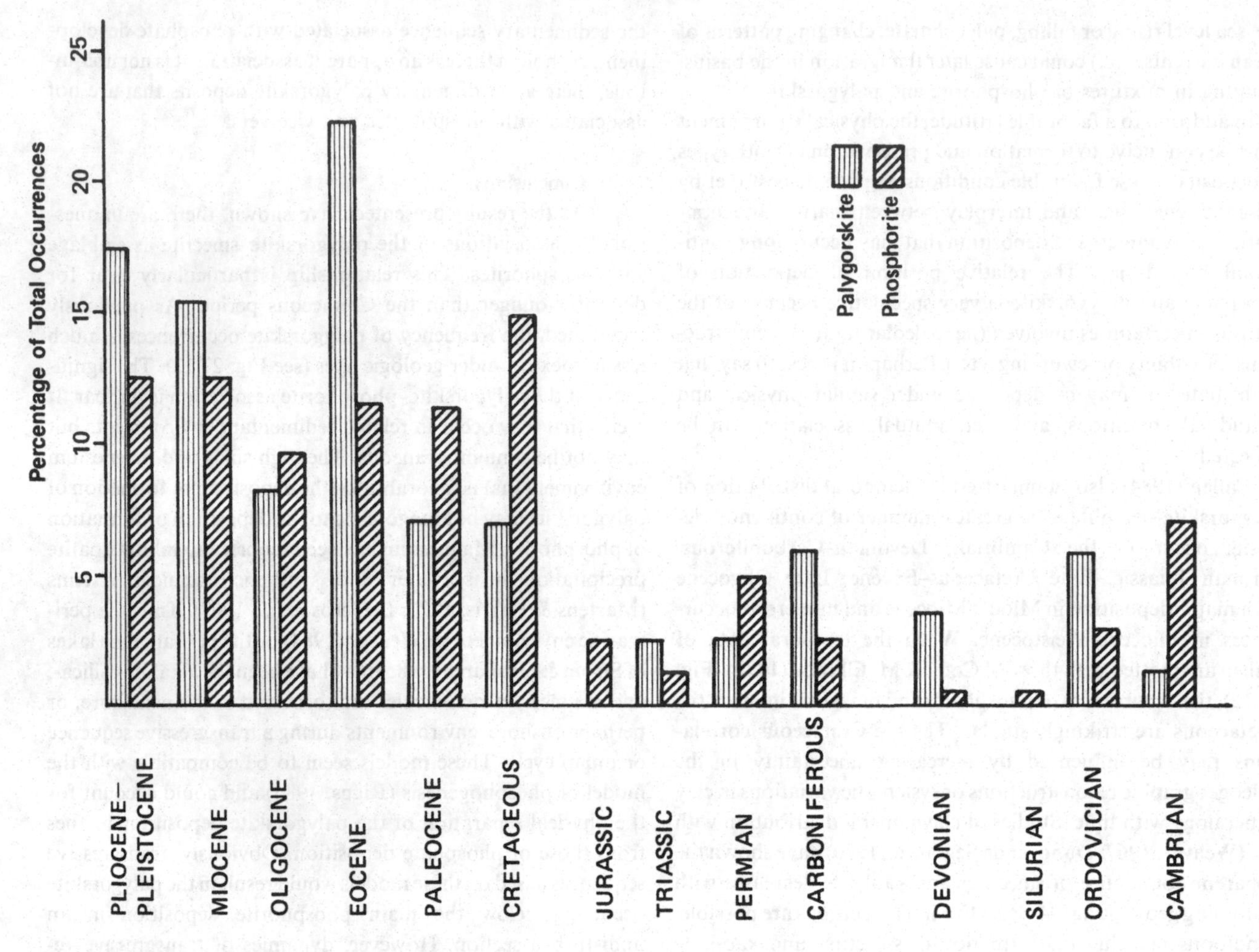

Fig. 27.12. Variation in palygorskite and phosphorite land occurrences with geologic time (from Cook & McElhinney, 1979; Callen, 1984).

Palygorskite may be an indicator of semiarid or seasonally arid climates where geochemical weathering could provide influxes of solubilized silica, magnesium, and calcium into restricted onshore or nearshore basins. These soluble constituents could provide the ingredients necessary to form new minerals or alter precursors that might be in the basins. Transgressive sequences and chemically rich upwelling ocean currents also could create conditions favorable for the formation of both the clay mineral suites and phosphorites. This interpretation fits with proposed models and might explain the occurrence of palygorskite below phosphorite in various locations. Regressive sequences would result in a reversal of this relationship.

Post-depositional alteration (including very early diagenesis) of the total sediment sequence (carbonates, clays, phosphates, and nonclay silicates) has not been adequately investigated. The North Carolina sediments show little evidence of alteration and occur as a freshwater confining bed in the Lee Creek area. The sediments in Florida have had multiple intervals of alteration with rising and falling sea and groundwater levels that may have caused schizohaline conditions with vadose and phreatic zones acting on the sediments. This complicated scenario has not been comprehensively studied for a multicomponent system although Weaver & Beck (1977) have touched on many of the individual components.

Preliminary data on precise mineral identifications, chemical analyses, and quantitative determinations have begun to be collected. The paragenetic sequences and stability relationships have not been established for many mineral groups or suites in these sediments. Thermodynamic studies by Nriagu (1976), Vieillard (1978) and Vieillard, Tardy & Nahon (1979) broadly define the system with an emphasis on aluminophosphate formation. Kinetic factors also need to be considered in this scenario. Future work on these aspects may allow the refinement of existing models and might lead to the development of new models. Although much work has been done on these sediments, significant new data are needed to develop an acceptable general model or models for the system.

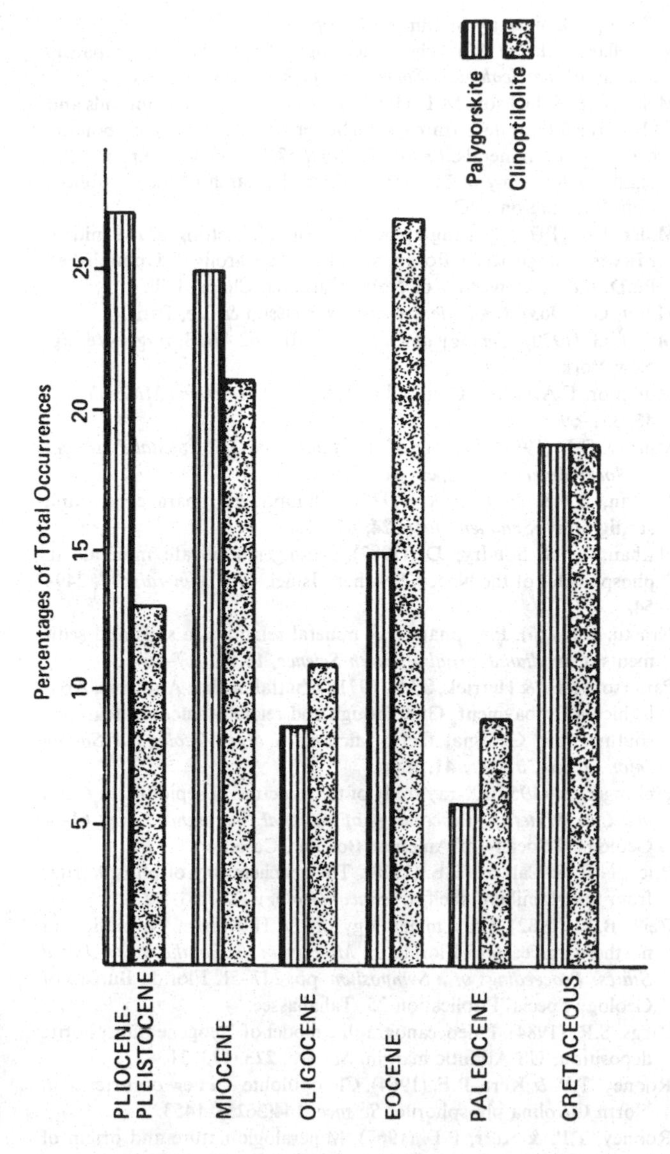

Fig. 27.13. Variation in palygorskite and clinoptilolite occurrences in DSDP cores (from Nathan & Flexer, 1977; Callen, 1984).

References

Alt, D. (1974). Arid climatic control of Miocene sedimentation and origin of modern drainage, southeastern United States. In *Post-Miocene Stratigraphy, Central and South Atlantic*, ed. R.Q. Oaks, pp. 21–9. Utah State Press, Salt Lake City.

Altschuler, Z.S. (1965). Precipitation and recycling of phosphate in the Florida land-pebble phosphate deposits. *US Geological Survey Professional Paper*, **525-B**, B91–5.

Altschuler, Z.S., Cathcart, J.B. & Young, E.J. (1964). *Geology and Geochemistry of the Bone Valley Formation and its Deposits, West Central Florida*. Field Guide, Geological Society of America Annual Convention, Miami Beach.

Altschuler, Z.S., Dwornik, E.J. & Kramer, H. (1963). Transformation of montmorillonite to kaolinite during weathering. *Science*, **141**, 148–52.

Bailey, S.W., Brindley, G.W., Johns, W.D., Martin, R.T. & Ross, M. (1971). Summary of national and international recommendations on clay mineral nomenclature. *Clays and Clay Minerals*, **19**, 129–32.

Baldar, N.A. & Whitlig, L.D. (1968). Occurrence and synthesis of soil zeolites. *Soil Science Society of America Proceedings*, **32**, 235–8.

Barron, P.F. & Frost, R.L. (1985). Solid state ^{29}Si NMR examination of the 2:1 ribbon magnesium silicates, sepiolite and palygorskite. *American Mineralogist*, **70**, 758–66.

Barshad, I., Halevy, E., Gold, H.A. & Hagin, J. (1956). Clay minerals in some limestone soils of Israel. *Soil Science*, **81**, 423–37.

Baturin, G.N. & Bezrukov, P.L. (1979). Phosphorites on the sea floor and their origin. *Marine Geology*, **31**, 317–32.

Birch, G.F. (1979). Phosphorite pellets and rock from the western continental margin and adjacent coastal terrace of South Africa. *Marine Geology*, **33**, 91–116.

Bradley, W.F. (1940). The structural scheme of attapulgite. *American Mineralogist*, **25**, 405–13.

Brindley, G.W. (1980). Order–disorder in clay mineral structures. In *Crystal Structures of Clay Minerals and Their Identification*. Springer-Verlag, New York.

Bromwell, L.G. (1982). *Physico-Chemical Properties of Florida Phosphatic Clays*. Florida Institute of Phosphate Research, Publication No. 02-003-020, Bartow.

Brusewitz, A.M.B. (1975). Studies on the Li test to distinguish between beidellite and montmorillonite. In *Proceedings International Clay Conference*, pp. 419–28.

Burnett, W.C. (1977). Geochemistry and origin of phosphorite deposits from off Peru and Chile. *Bulletin of the Geologic Society of America*, **88**, 813–23.

Callen, R.A. (1984). Clays of the palygorskite–sepiolite group: depositional environment, age and distribution in palygorskite–sepiolite. In *Developments in Sedimentology 37*, ed. A. Singer & E. Galan, pp. 1–38. Elsevier, New York.

Carroll, D. (1970). *Clay Minerals: A Guide to their X-ray Identification*. Geological Society of America Special Paper 126, Boulder, Colorado, 80pp.

Cathcart, J.B., Sheldon, R.P. & Gulbrandsen, R.A. (1984). Phosphate rock resources of the United States. *US Geological Survey Circular*, **88**, 48pp.

Chen, P.Y. (1977). Table of key lines in X-ray powder diffraction patterns of minerals in clays and associated rocks. *Indiana Geological Survey Occasional Paper*, **No. 21**, 67pp.

Christ, C.L., Hathaway, J.C., Hostetler, P.B. & Shepard, C.L. (1969). Palygorskite: new X-ray data. *America Mineralogist*, **85**, 198–205.

Cook, P.J. & McElhinny, M.W. (1979). A reevaluation of the spatial and temporal distribution of sedimentary phosphate deposits in the light of plate tectonics. *Economic Geology*, **74**, 315–30.

Couture, R.A. (1977). Composition and origin of palygorskite-rich and montmorillonite-rich zeolite-containing sediments from the Pacific Ocean. *Chemical Geology*, **19**, 113–30.

Couture, R.A. (1978). Miocene of the SE United States: a model for chemical sedimentation in a peri-marine environment–comments. *Sedimentary Geology*, **21**, 149–57.

Decarreau, A., Sautereau, J.P. & Steinberg, M. (1975). Genęse des Mineraux Argileaux du Barontonieu Moyer du Bassin de Paris. *Bulletin de la Societe Francaise de Mineralogie et de Cristallographie*, **98**, 142–51.

Dunoyer de Segonzac, G. (1970). The transformation of clay minerals during diagenesis and low grade metamorphism: a review. *Sedimentology*, **53**, 281–346.

Espenshade, G. & Spencer, C. (1963). Geology of phosphate deposits of northern peninsular Florida. *US Geological Survey Bulletin*, **1118**, 115.

Force, E. & Garnar, T. (1985). High-angle aeolian crossbedding in Trail Ridge, Florida. *Industrial Minerals*, **August**, 55–9.

Freas, D.H. (1968). Exploration for Florida phosphate deposits. In *Proceedings of the Seminar on Sources of Mineral Raw Materials for the Fertilizer Industry in Asia and the Far East*, United Nations Mineral Resources Development Series No. 32, pp. 187–99, Bangkok.

Freas, D.H. & Riggs, S.R. (1968). Environments of phosphorite deposition in the central Florida phosphate district. In *4th Forum on Geology of Industrial Minerals*, Austin, Texas, pp. 117–28.

Galan, E. & Ferrero, A. (1982). Palygorskite–sepiolite clays of Lebrija Southern Spain. *Clays and Clay Minerals*, **30**, 191–9.

Gard, J.A. & Follett, E.A.C. (1968). A structural scheme for palygorskite. *Clay Minerals*, **7**, 367–9.

Gibson, T.G. (1970). Late Mesozoic–Cenozoic tectonic aspects of the Atlantic Coastal margin. *Geologic Society of America Bulletin*, **81**, 1813–22.

Golden, D.C., Dixon, J.B., Shadfan, H. & Kippenberger (1985). Palygorskite and sepiolite alteration to smectite under alkaline conditions. *Clays and Clay Minerals*, **33(1)**, 44–50.

Greene-Kelly, R. (1953). The identification of montmorillonoids in clays. *Journal of Soil Science*, **4**, 233–7.

Greene-Kelly, R. (1955). Dehydration of the montmorillonite minerals. *Mineralogic Magazine*, **30**, 604.

Gremillion, L. (1965). 'The origin of attapulgite in the Miocene strata of Florida and Georgia'. Unpublished Ph.D. thesis, Department of Geology, Florida State University, Tallahassee.

Grim, R.E. (1933). Petrography of the fuller's earth deposits, Olmstead, Illinois, with a brief study of some non-Illinois earths. *Economic Geology*, **29**, 344–63.

Grim, R.E. (1968). *Clay Mineralogy*, 2nd edn. McGraw-Hill, New York.

Gruner, J.W. (1935). The structural relationship of nontronites and montmorillonite. *American Mineralogist*, **20(7)**, 475.

Heron, S.P. & Johnson, H.S. (1966). Clay mineralogy, stratigraphy, and structural setting of the Hawthorn Formation, Coosawhatchie District, South Carolina. *Southeastern Geology*, **7**, 51–63.

Hite, R. (1978). Possible genetic relationships between evaporites, phosphorites and iron-rich sediments. *The Mountain Geologist*, **14(3)**, 97–107.

Huertas, F., Linares, J. & Martin-Vivaldi, J.L. (1970). Clay mineral geochemistry in basic sedimentary environments. *Reun Hispano-Belga de Minerals de la Arcilla*, pp. 211–14. CSIC, Madrid.

Huggins, C.W., Denny, M.V. & Shell, H.R. (1962). Properties of palygorskite, an asbestiform mineral. *US Bureau of Mines Report of Investigations*, **6071**, 8pp.

Isphording, W.C. (1973). Discussion of the occurrence and origin of sedimentary palygorskite–sepiolite deposits. *Clays and Clay Minerals*, **21**, 391–401.

Isphording, W.C. (1984). The clays of Yucatan, Mexico. In *Developments in Sedimentology 37*, ed. A. Singer & E. Galan, pp. 59–74. Elsevier, New York.

Jeannette, A., Monition, A., Oertli, I. & Salvan, H. (1959). Premiers resultats de l'étude des argiles de la série phosphatée du Bassin du Khouribga (Maroc). *20th International Geologic Congress*, Mexico, 1956, pp. 53–62.

Jenkyns, H.C. (1980). Cretaceous anoxic events. *Journal of the Geological Society, London*, **137**, 171–88.

JOIDES (1965). Ocean drilling on the continental margin. *Science*, **150**, 709–16.

Jordan, L. (1954). Oil possibilities in Florida. *Oil and Gas Journal*, **53**, 370–5.

Kasakov, A.V. (1937). The phosphorite facies and genesis of phosphorites. *Scientific Institute of Fertilizers and Insecto-fungicides Transactions*, **142**, pp. 95–113.

Kerr, P. (1937). Attapulgus clay. *American Mineralogist*, **22**, 548.

LeGeros, R., Trautz, O., LeGeros, P. & Klein, E. (1967). Apatite crystallites: effects of carbonate on morphology. *Science*, **155**, 1409–11.

Loughnan, F.C. (1959). Further remarks on the occurrence of palygorskite at Red Banks Plains, Queensland. *Proceedings of the Royal Society of Queensland*, **71**, 43–50.

Lyle, M.E. (1984). 'Clay mineralogy of the Pungo River Formation, Onslow Bay, North Carolina Continental Shelf'. Unpublished M.Sc. thesis, East Carolina University, North Carolina, 129pp.

Mansfield, G. (1940). Clay investigations in the southern states, 1934–1935: Introduction. *US Geological Survey Bulletin*, **901**, 9.

Marshall, C.E. (1949). *The Colloid Chemistry of the Silicate Minerals*. Academic Press, New York.

Martens, C. & Harris, R. (1970). Inhibition of apatite precipitation in the marine environment by magnesium ions. *Geochimica Cosmochimica Acta*, **34**, 621–5.

McClellan, G.H. (1962). 'Identification of clay minerals from the Hawthorn Formation from the Devil's Mill Hopper, Alachua County, Florida'. Unpublished M.Sc. thesis, University of Florida, Tallahassee, 119pp.

McClellan, G.H. (1964). 'Petrology of attapulgus clay in north Florida and southwest Georgia'. Unpublished Ph.D. thesis, Department of Geology, University of Illinois, 127pp.

McClellan, G.H. (1980). The mineralogy of carbonate fluorapatite. *Journal of the Geological Society, London*, **137(6)**, 675–81.

Mehra, O.P. & Jackson, M.L. (1960). Iron oxide removal from soils and clays by a dithionite–citrate system buffered with sodium bicarbonate, clays and clay minerals. *Proceedings of the 7th Conference*, pp. 317–27, National Academy of Science–National Research Council Publication, Washington, DC.

Miller, J.A. (1971). 'Stratigraphic and structural setting of the middle Miocene Pungo River Formation of North Carolina'. Unpublished Ph.D. thesis, University of North Carolina, Chapel Hill.

Millot, G. (1963). *Geologie des Argiles*. Masson & Cie, Paris.

Millot, G. (1970). *Geology of Clays*, pp. 201, 262–76. Springer-Verlag, New York.

Mumpton, F.A. (1960). Clinoptilolite redefined. *American Mineralogist*, **45**, 351–69.

Murray, G.E. (1961). *Geology of the Atlantic and Gulf Coastal Provinces of North America*. Harper, New York.

Nathan, Y. & Flexer, A. (1977). Clinoptilolite, paragenesis and stratigraphy. *Sedimentology*, **24**, 845–55.

Nathan, Y. & Soudry, D. (1982). Authigenic silicate minerals in phosphorites of the Negev, Southern Israel. *Clay Minerals*, **17**, 249–54.

Nriagu, J. (1976). Phosphate–clay mineral relations in soils and sediments. *Canadian Journal of Earth Science*, **13(6)**, 717–36.

Patterson, S.H. & Herrick, S.M. (1971). Chattahoochee Anticline, Apalachicola Embayment, Gulf Trough and related structural features, southwestern Georgia, fact or fiction. *Georgia Geological Survey Information Circular*, **41**, 16pp.

Preisinger, A. (1957). X-ray study of the structure of sepiolite. In *Clays and Clay Minerals, Proceedings of National Conference 6*, pp. 61–7. Geological Society of America, Boulder, Colorado.

Price, N.B. & Calvert, S.E. (1978). The geochemistry of phosphorites from the Namibian shelf. *Chemical Geology*, **23**, 151–70.

Reik, B.A. (1982). Clay mineralogy of the Hawthorn Formation in northern and eastern Florida. In *Miocene of the Southeastern United States: Proceedings of a Symposium*, pp. 247–51. Florida Bureau of Geology Special Publication 25, Tallahassee.

Riggs, S.R. (1984). Paleoceanographic model of Neogene phosphorite deposition, US Atlantic margin. *Science*, **223**, 123–31.

Rooney, T.P. & Kerr, P.F. (1964). Clinoptilolite – a new occurrence in North Carolina phosphorite. *Science*, **144(3625)**, 1453.

Rooney, T.P. & Kerr, P.F. (1967). Mineralogic nature and origin of phosphorite, Beaufort County, North Carolina. *Geological Society of America Bulletin*, **78**, 731–48.

Ross, C.S. & Hendricks, S.B. (1945). Minerals of the montmorillonite group. *US Geological Survey Professional Paper*, **205-B**. 79pp.

Salvan, H. (1985). Particulantés de répartition stratigraphique des dépôts phosphatés des la Mésogée et de la bordure Atlantique du continent Africain. In *Sciences Géologiques*, eds. J. Lucas & L. Prévôt, pp. 93–8. Memoir No. 77.

Schneiderhöhn, P. (1964). Nontronit von Hohen Hagen und chloropal von Meenser Steinberg bei Göttingen. *Tschermaks Mineralogische und Petrographische Mitteilungen*, **10(1–4)**, 385–99.

Schultz, L.G. (1969). Lithium and potassium absorption, dehydration temperature and structural water content of aluminous smectites. *Clays and Clay Minerals*, **17**, 115–49.

Sheldon, R.P. (1964). Paleolatitudinal and paleogeographic distribution of phosphate. *US Geological Survey Professional Paper*, **501-C**, C106–13.

Siedlecki, M. (1983). 'Trace element analysis of the clay silt fraction of the Pungo River Formation, North Carolina'. Unpublished M.Sc. thesis, North Carolina State University.

Singer, A. (1979). Palygorskite in sediments: detrital, diagenetic or neoformed – a critical review. *Geologische Rundschau*, **68**, 996–1008.

Singer, A. & Galan, E. (1984). Palygorskite–sepiolite; occurrences, genesis, and uses. In *Developments in Sedimentology 37*. Elsevier, New York.

Slansky, M., Camez, T. & Millot, G. (1959). Sédimentation argileuse et phosphatée au Dahomey. *Bulletin de la Société Géologique de France*,

1, 150–5.
Strom, R.N. & Upchurch, S.B. (1985). Palygorskite distribution and silicification in the phosphatic sediments of central Florida. In *Guidebook Eighth International Field Workshop and Symposium*, International Geological Correlations Program Project 156 (Phosphorites), pp. 118–26.
Trauth, N. (1977). Argiles évaporitiques dans la sédimentation carbonatée continentale et épicontinentale tertiaire. *Sciences Géologiques*, Mémoire No. 49, 195pp.
Trauth, N. & Lucas, J. (1967). Thermal methods in the study of clay minerals. *Bulletin du Groupe Francais des Argiles*, **19(2)**, 11–24.
Van Kauwenbergh, S.J. & McClellan, G.H. (1985). Variations in the mineralogy of the Florida phosphate deposits. In *Florida Land-Pebble Phosphate District Field Guidebook*, ed. J.B. Cathcart & T.M. Scott, pp. 38–67. Geological Society of America.
Velde, B. (1985). Clay minerals: A physico-chemical explanation of their occurrence. In *Developments in Sedimentology 40*. Elsevier, New York, 427pp.
Vieillard, P. (1978). Géochimie des phosphate. Etude thermodynamique, application à la genise et à l'alteration des apatites. *Sciences Geologiques*, Memoir No. 51, University of Strasbourg, Strasbourg.
Vieillard, P., Tardy, Y. & Nahon, D. (1979). Stability fields of clays and aluminum phosphates: parageneses in lateritic weathering of argillaceous phosphatic sediments. *American Mineralogist*, **64**, 626–34.
Weaver, C.E. (1967). The significance of clay minerals in sediments. In *Fundamental Aspects of Petroleum Geochemistry*, ed. B. Nagy & U. Colombo, pp. 37–75. Elsevier, Amsterdam.
Weaver, C.E. (1984). Origin and geologic implications of the palygorskite in the southeastern United States in palygorskite–sepiolite. In *Developments in Sedimentology 37*, ed. A. Singer & E. Galan, pp. 39–58. Elsevier, New York.
Weaver, C.E. & Beck, K.C. (1977). Miocene of the southeastern United States: a model for chemical sedimentation in a peri-marine environment. In *Developments in Sedimentology 22*. Elsevier, New York, 234pp.
Weaver, C.E. & Pollard, L.D. (1973). The chemistry of clay minerals. In *Developments in Sedimentology 15*. Elsevier, New York, 213pp.
Weaver, C.E. & Wampler, J.M. (1972). The illite–phosphate association. *Geochimica Cosmochimica Acta*, **36**, 1–13.
Weir, A.H. & Greene-Kelly, R. (1962). Beidellite. *American Mineralogist*, **47**, 137–46.
Wissa, A.E.Z., Fuleihan, N.F. & Ingra, T.S. (1982). Evolution of phosphatic clay disposal and reclamation methods. In *Mineralogy of Phosphatic Clays*, vol. 2. Florida Institute of Phosphate Research, Bartow, 55p.

28

Paleoceanography and paleogeography of the Miocene of the southeastern United States

P. POPENOE

Abstract
Coastal plain stratigraphy, modern concepts of margin subsidence, cycles of relative sea-level change, and seismic stratigraphy are used to reconstruct the paleoceanographic and paleogeographic conditions of Miocene deposition over the southeastern United States. An intimate relationship between changing eustatic sea level and shifts in paleocirculation controlled Miocene deposition. The Ocala High, and the St Johns and Brevard Platforms of Florida formed as a result of vertical carbonate accretion in the Late Cretaceous and Eocene when the Suwannee Current isolated the Florida peninsula from siliciclastic input. During this time of high sea level, the Gulf Stream flowed across southern Georgia and southern South Carolina, cutting major basins which later became Miocene depocenters. Basins of Miocene deposition were also formed in the Early Oligocene in eastern North Carolina by marine currents that flowed through the Gulf Trough. All of the basins were modified by exposure and subaerial erosion in the Late Oligocene and Early Miocene.

Transgressive seas in the earliest Miocene (Aquitanian) inundated much of the lower coastal plain widely depositing marine sediments of the Arcadia and Penney Farms Formations across Florida, and the Parachucla, Edisto, and Belgrade Formations across the coastal plain of Georgia and the Carolinas. Following an Early Burdigalian regression during which the Florida Platform, coastal plain, and shelf were subaerially exposed, seas again inundated the coastal plain in the late Burdigalian, Langhian, and Serravallian. Sea levels were low in the Tortonian when most of the coastal plain and the continental shelf was exposed to subaerial erosion and when the earlier deposits were subaerially reworked.

The Gulf Stream flowed as the Suwannee Current across southern Georgia in the Late Cretaceous, Eocene, and Early Oligocene, but flowed through the Straits of Florida for the entire Neogene. During the Aquitanian and Late Burdigalian, Langhian and Early Serravallian, the Gulf Stream was pressed tightly against the Florida–Hatteras Slope. During the Early Burdigalian and Tortonian regressions, it was forced offshore across the central Blake Plateau and around the Charleston Bump. Reconstructions of Early Burdigalian, Late Burdigalian, and Langhian paleoceanography and paleogeography are presented that are consistent with both the marine and land facies.

Introduction

The Florida Current flows through the Straits of Florida with a surface velocity of up to six knots and a transport volume of more than 70 times the combined flow of all the land rivers of the world (Pratt, 1966). Over the Blake Plateau the Florida Current is joined by the Antilles Current (Fig. 28.1) where the volume flow more than doubles (Worthington, 1976) and the current is known as the Gulf Stream. Off the southeastern United States the voluminous and rapid flow of the Gulf Stream is pushed against the Florida–Hatteras Slope as a result of the large-scale response of the North Atlantic Gyre to wind stress. Here, the deep, narrow, fast-moving current controls its channel, that is, the volume of material transported into any segment of the channel is the same as that transported out (Pratt, 1966). Where flow is confined in the northern Straits of Florida and over the inner Blake Plateau almost no sedimentation occurs and the bottom is scoured down to resistant, indurated strata of Cretaceous, Paleocene, Eocene, and Oligocene age that is armored by phosphatic pavements and manganese rinds (Manheim, Pratt, & McFarlin, 1980, Manheim *et al.*, 1982). On the northern Blake Plateau the bottom shallows over a pre-Gulf Stream feature, the Charleston Bump (Fig. 28.2), and intensified bottom currents have scoured large erosional holes and troughs that may be kilometers across and over 200 m deep (Fig. 28.3). The Gulf Stream forms a barrier to the eastward transport of sediments from the continent, sweeping sediments northward, so that the outer part of the continental shelf – the Blake Plateau – has developed into a deep-water, sediment-starved area. Only a thin conformable drape of hemipelagic ooze rained down from biologic production in the overlying water column is deposited on the outer Blake Plateau. In contrast, on the landward side of the current, a shallow-water shelf, – the Florida–Hatteras Shelf (Fig. 28.2) – has prograded out to the current, developed from both terrigenous and bioclastic deposition.

The US Geological Survey, in a cooperative program with the US Bureau of Land Management, began collecting seismic-reflection data over the southeastern US continental margin in 1976. The program was related to environmental and resource evaluation studies preceding petroleum exploration and possible development in this area. As part of these studies single-channel high-resolution seismic-reflection records, using chiefly airgun arrays, sparker, and 3.5 kHz signal sources, were obtained across the shelf, the Blake Plateau, and the continental slope, on 20 km-spaced dip lines, between northern Florida and Virginia

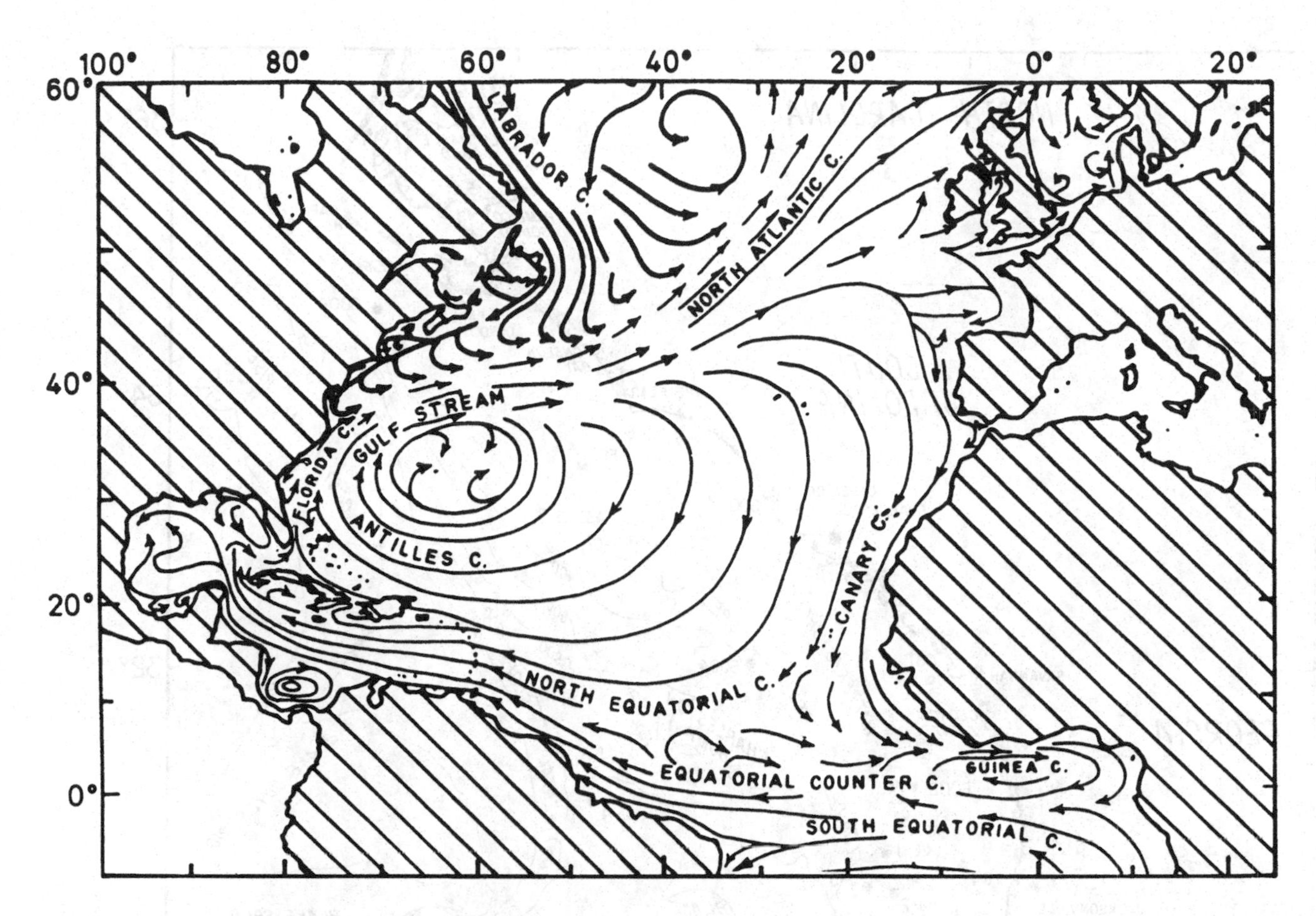

Fig. 28.1. Surface currents of the north Atlantic Ocean (modified from Emery & Uchupi, 1984).

(Fig. 28.4) (Bailey & Grow, 1980a, b; Paull & Dillon, 1980; Popenoe, 1980, 1983a, b; Popenoe & Meyer, 1983; Popenoe & Ward, 1983). Preliminary seismic–stratigraphic analyses have been completed on these data, and some of the results have been published (Paull & Dillon, 1980; Paull *et al.*, 1980; Pinet & Popenoe, 1985a, b; Popenoe, 1985). Analyses of these records have yielded both a history of Gulf Stream circulation and of eustatic change. Rugged unconformities cut by the bottom currents of the Gulf Stream in the past were traceable on the records. Seismic stratigraphic analyses show that the Gulf Stream – cut unconformities shift in position both laterally and vertically, and that these shifts are essentially coincident with the major shifts of eustatic sea level (Fig. 28.5) proposed by Vail, Mitchum & Thompson (1977). Concomitant shifts of the flanking facies, that is, the clastic shelf facies west of the unconformity, and the deep-water pelagic facies east of the unconformity, have also been mapped and are in agreement with the above concept.

This paper uses Paleogene and Neogene Gulf Stream positions, seismic and litho-stratigraphic facies, isopach and structure contour maps derived from both interpretation of seismic data and published material, modern biostratigraphic dating of depositional sequences and unconformities (Fig. 28.6), and analyses of the depositional paleoenvironment of these sediments to define the regional paleoenvironment of the Miocene. Miocene sediments settled into structures of both oceanographic and subaerial origin that were created by eustatic change and related shifts in sedimentation and ocean currents in Late Cretaceous, Eocene and Oligocene time. Phosphate deposition in the Miocene appears to conform to the Riggs model of phosphogenesis.

The Riggs model of phosphogenesis

The relationship of phosphorite deposition to primary and secondary structure of the margin has been discussed by Riggs (1979 and 1984) and Riggs *et al.*, (1985). In these papers a model has been developed wherein topographic upwelling is induced by bottom interference of primary margin structures to Gulf Stream flow. During the middle stages of transgression there is both high density stratification of the shelf and increased upwelling due to interaction of the current with both bottom and shelf bathymetry. During middle stages upwelled waters migrate farther into the embayed shelf system where phosphate deposition is controlled by topographic features. Optimum phosphogenesis occurs on shallow-water platforms and the adjacent entrapment basins that project onto the continental shelf, chiefly in the transition zone between the siliciclastic and carbonate facies. Maximum transgression culminates with warming shelf waters, increased carbonate deposition, and decreased phosphogenesis.

The primary margin structures

Figure 28.7 shows depth contours on the crystalline and metamorphic basement surface (post-rift unconformity)

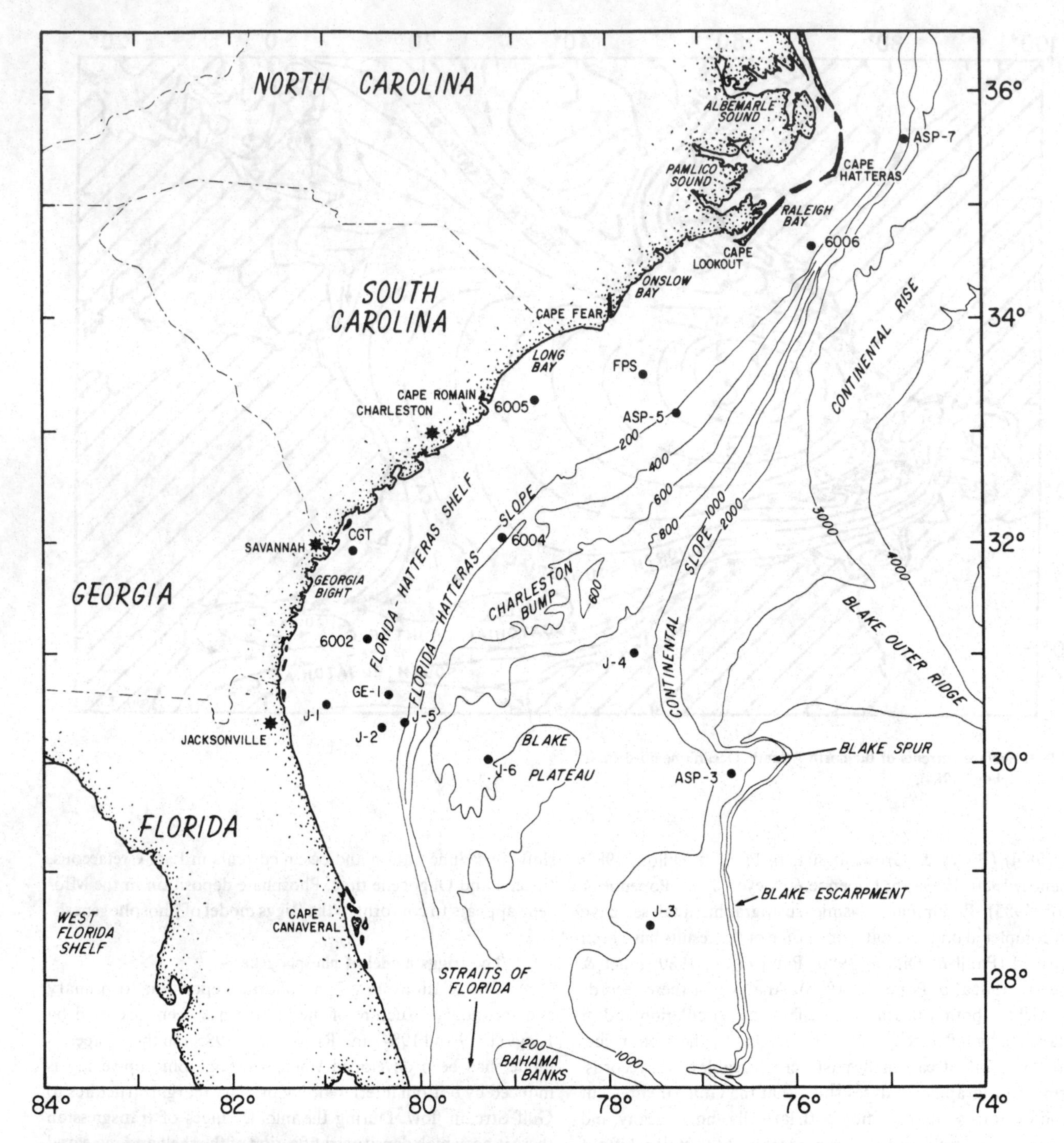

Fig. 28.2. Physiographic features of the US southeastern Atlantic continental margin and locations of offshore drill holes. CGT – Coast Guard Tower well (McCollum & Herrick, 1964); J1–J6 – JOIDES drillholes (JOIDES, 1965); GE-1 – Continental Offshore Stratigraphic Test well GE-1 (Scholle, 1979); 6002, 6004, 6005 – USGS Atlantic Margin Coring Project drillholes (Hathaway *et al.*, 1979); ASP-3, ASP-5, ASP-7 – Atlantic Stratigraphic Project core holes (Poag, 1978).

underlying the southeastern US continental margin. The contours define the primary structures of the southeastern margin, that is, the structures that resulted from subsidence of the continental edge following the separation of Africa and North America. The two major offshore basins, the Carolina Trough and Blake Plateau Basin (Klitgord & Behrendt, 1979), lie normal to the direction of ocean opening and overlie the transitional crust that formed at the edge of the rifted continent. This thinned, stretched, heated, and intruded crust subsided exponentially from thermal contraction and later sediment loading following continental separation (Sleep, 1971; Falvey, 1974; Watts, 1981; Hutchinson *et al.*, 1982). These basins are now filled with sedi-

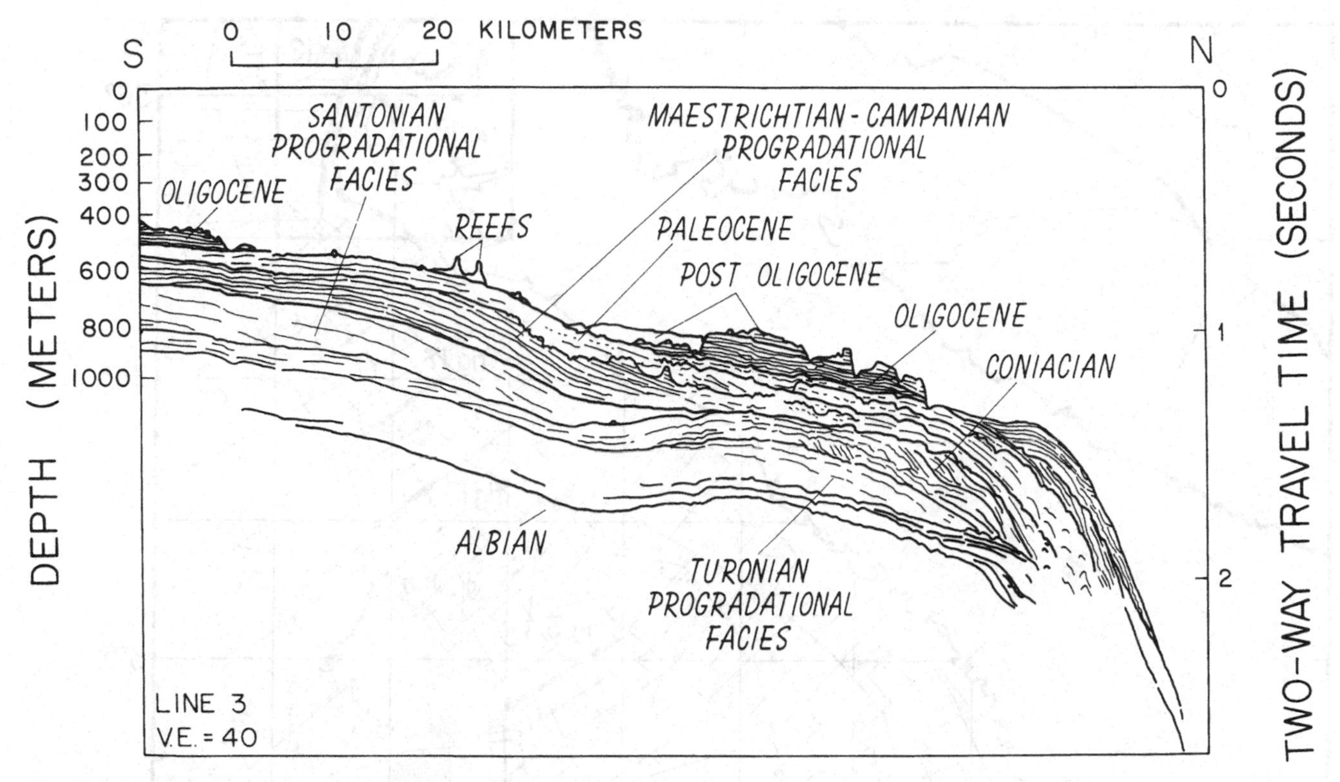

Fig. 28.3. Line drawing of high-resolution seismic-reflection profile across the southern flank of the Charleston Bump showing the extensively eroded bottom. The location of the profile is shown on Figure 28.4. Note the dissected blanket of flat-lying Middle Oligocene sediments deposited across the Charleston Bump while Gulf Stream was flowing across the southern Blake Plateau.

ment so that they are not evident in the bathymetry of the ocean floor.

The Blake Plateau Basin is offset from the Carolina Trough (Fig. 28.7) reflecting an original offset in the breakup boundary (Dillon *et al.*, 1979; Klitgord & Behrendt, 1979; Dillon & Popenoe, 1988). The offset occurs across the extension of the Blake Spur Oceanic Fracture Zone, which juxtaposes continental crust to the north with transitional crust to the south. The northern Blake Plateau, underlain by continental crust of the Carolina Platform corner, has been dominantly positive relative to the southern Plateau, underlain by transitional crust of the Blake Plateau Basin. The effect of the different rate of subsidence across the fracture zone causes a series of stacked facies boundaries in the overlying sedimentary column (Pinet & Popenoe, 1985a). The differing subsidence rate between the northern and southern Plateau is also evident in the bathymetry (Fig. 28.2) as the Charleston Bump (Brooks & Bane, 1978; Bane & Brooks, 1979; Chao & Janowitz, 1979; Legeckis, 1979). The shallower northern Blake Plateau overlies the Carolina Platform corner where older, more indurated rocks, are near the surface. The corner of the Carolina Platform, the Charleston Bump, has been excavated by the Gulf Stream, and is notable for its effect on Gulf Stream flow throughout the Tertiary.

The structural high of the basement near Cape Fear, the Cape Fear Arch (Fig. 28.7), is the landward effect of the offset in the offshore crustal boundaries. Because the margin has been shaped by erosional and depositional processes as well as subsidence, the shoreline diagonally crosses the corner of the Carolina Trough toward Cape Hatteras. South of Cape Fear the coastline both approaches the Blake Plateau Basin and overlies the Southeast Georgia Embayment. The coast is farthest from the offshore basins at Cape Fear and the crystalline basement is shallowest in this area.

Landward of the basins the surface of the basement is relatively flat, sloping gently upward from a basement hinge zone toward the Fall Line where crystalline and metamorphic rocks crop out at the surface. These 'platform' areas overlie normal-thickness continental crust whose edges have been pulled down by the subsiding offshore basins. The platforms are dominantly positive and tectonically stable. They formed the broad positive ramps for Miocene transgression.

The Florida and Carolina Platforms are separated by the Southeast and Southwest Georgia Embayments, and the linking Suwannee Saddle. These embayments, which show a continuing history of slightly more subsidence than the adjoining platforms, are developed over old Triassic rift basins (Popenoe & Zietz, 1977; Nelson *et al.*, 1985). The South Florida Embayment, on the other hand, is developed over rift stage crust that was fragmented in the opening of the Gulf of Mexico (Klitgord, Popenoe, & Schouten, 1984) and has been dominantly negative throughout its history. These negative areas were influential in channelling currents across Georgia during the Paleogene.

Although the primary margin structures played the dominant role in sediment distribution in the Mesozoic, changes in ocean

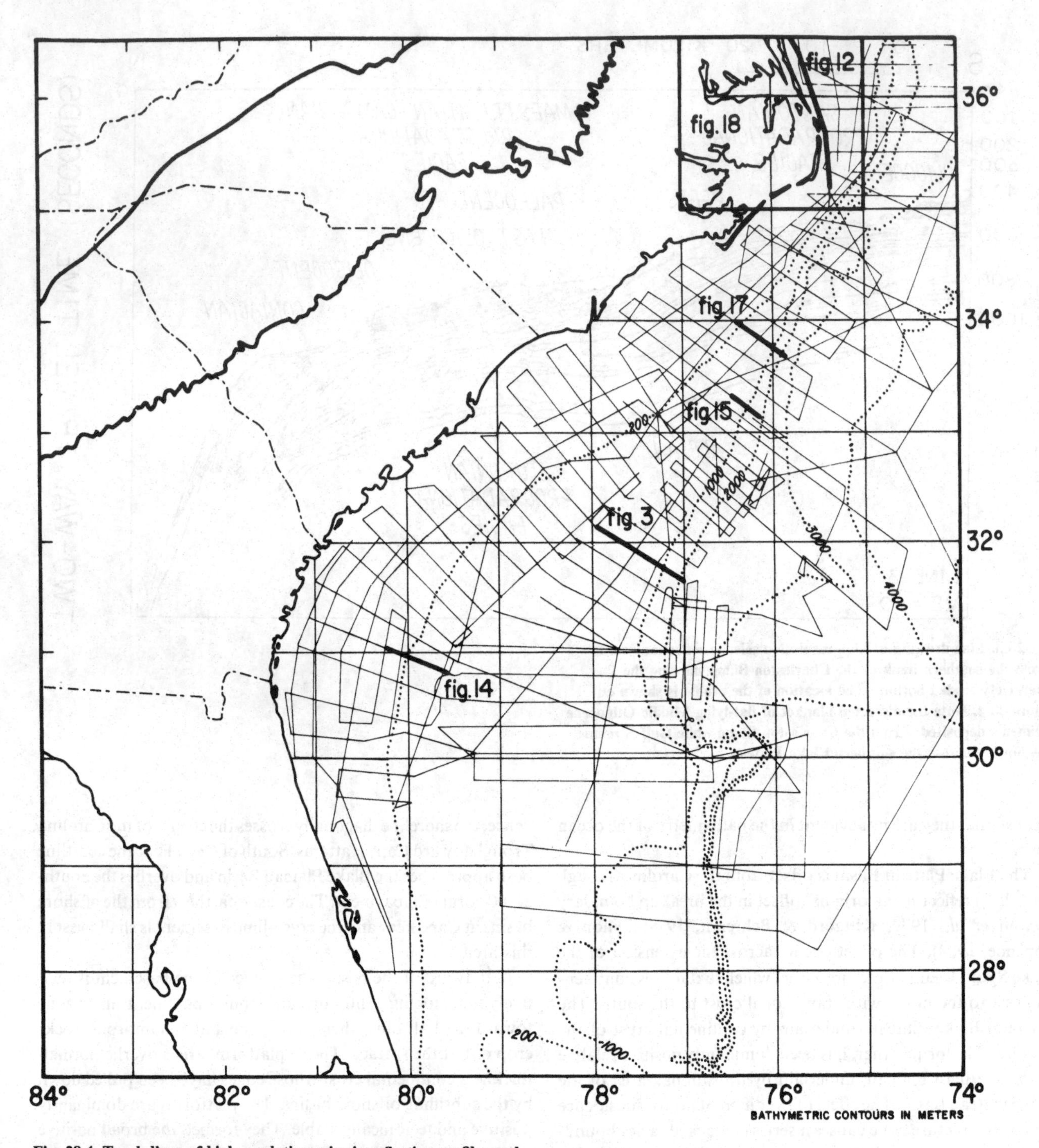

Fig. 28.4. Track lines of high-resolution seismic-reflection profiles and location of figures discussed in text.

current structures and sea level were far more important than crustal adjustments in controlling the location of depocenters in the Tertiary. The primary structures played a more subtle role in Tertiary deposition by influencing the path taken by ocean currents and by interfering with flow, thus controlling the location of areas of major topographically-induced upwelling (Riggs, 1984; Riggs *et al.*, 1985).

Crustal adjustments along the southeastern margin do not appear to have influenced Miocene sedimentation greatly. Minor crustal adjustments of the margin may have occurred on the platforms, but these were of small amplitude and epeirogenic in extent; as a result their effects are difficult to separate by seismic stratigraphy from the overriding larger and more local effects of depositional and erosional processes. Seismic reflection records offshore clearly show that the secondary basins of the margin were created by depositional and erosional, rather than tectonic, processes.

The secondary structures influencing Miocene deposition

The Suwannee Strait and Gulf Trough

The secondary structures that influenced Miocene deposition were largely formed by oceanographic conditions of the Late Cretaceous and Paleogene. These periods were dominated by higher eustatic sea levels than present, which innundated both

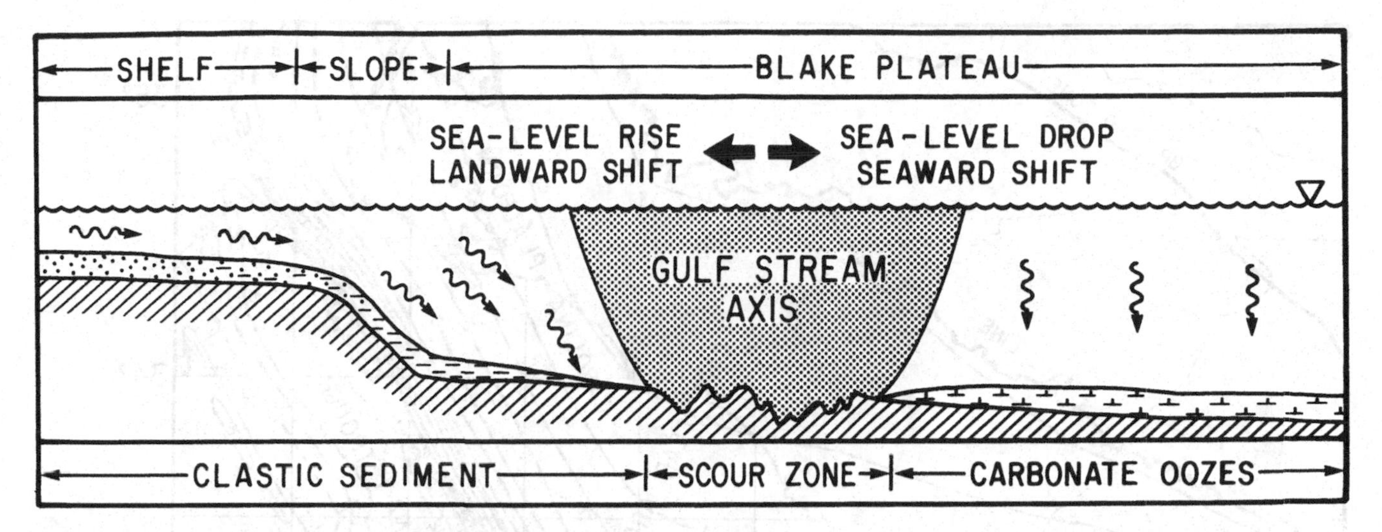

Fig. 28.5. Diagram of unconformity cut by bottom currents of the Gulf Stream and flanking facies; the Florida–Hatteras Shelf to its west, and the deep-water hemipelagic drape of sediments to its east. All three of these signatures of Gulf Stream presence are mapped and utilized to construct paleo-Gulf Stream positions. All facies shift laterally and vertically with eustatic change.

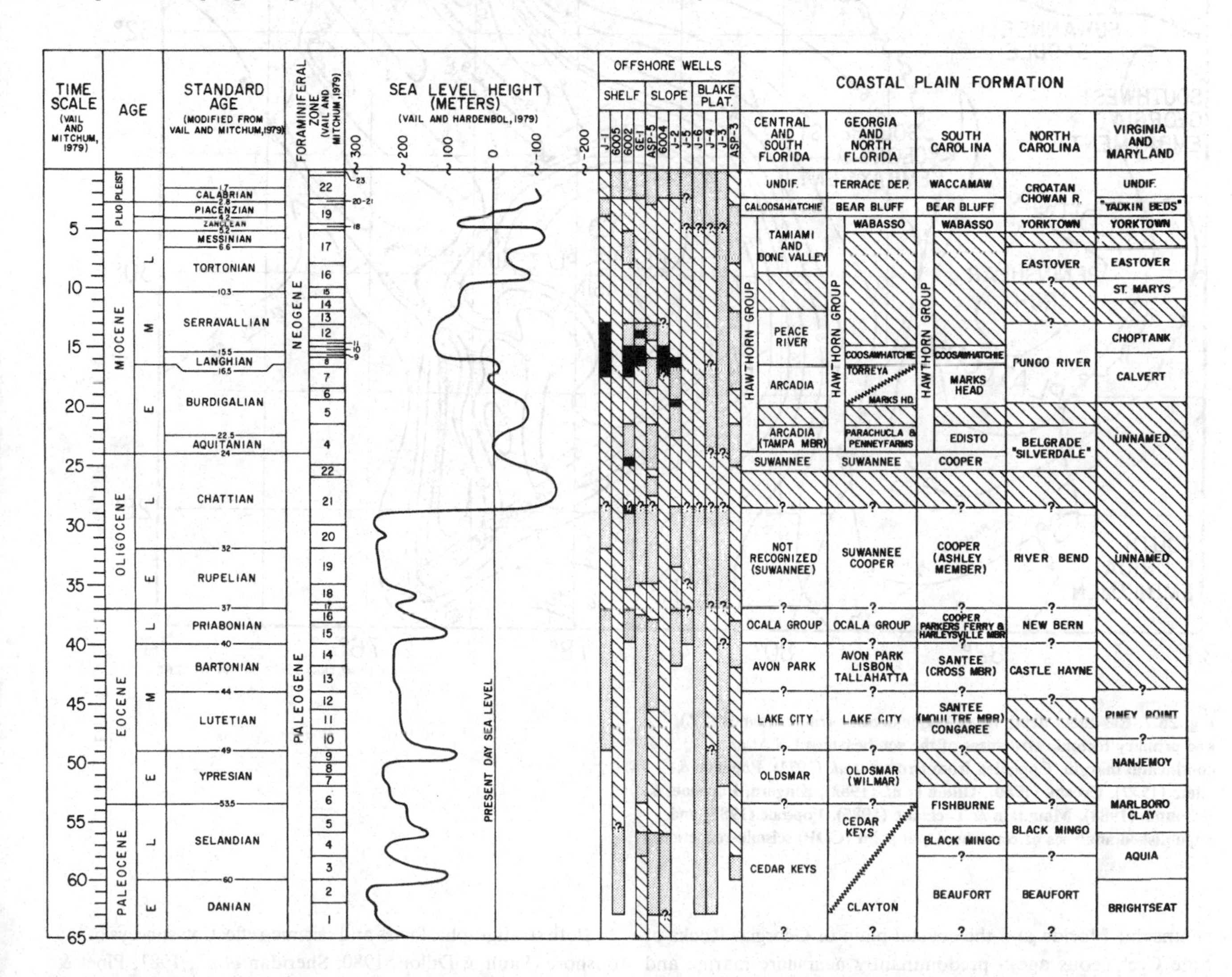

Fig. 28.6. Sea-level height and depositonal sequences of the southeastern Atlantic Coastal Plain and continental margin. Boundaries are queried where depositional sequences are not tied to specific foraminiferal zones. Zones of notable phosphorite are shown in black (phosphorite content has not been studied in the ASP-5 well). Time scale, age, standard age, and foraminiferal zone from Vail & Mitchum (1979); sea-level height from Vail & Hardenbol (1979). Offshore wells from JOIDES (1965), Poag (1978 and 1985) Hathaway *et al.* (1979), Poag & Hall (1979), and Popenoe (1985). Stratigraphy chiefly after Chen (1965), Hazel *et al.* (1977), Blackwelder & Ward (1979), Ward *et al.* (1979), Ward & Blackwelder (1980), Blackwelder (1981a,b); Gibson (1982), Huddlestun (1982 and 1984), Snyder, Scott W. *et al.* (1982), Hall (1983), Scott (1985a,b).

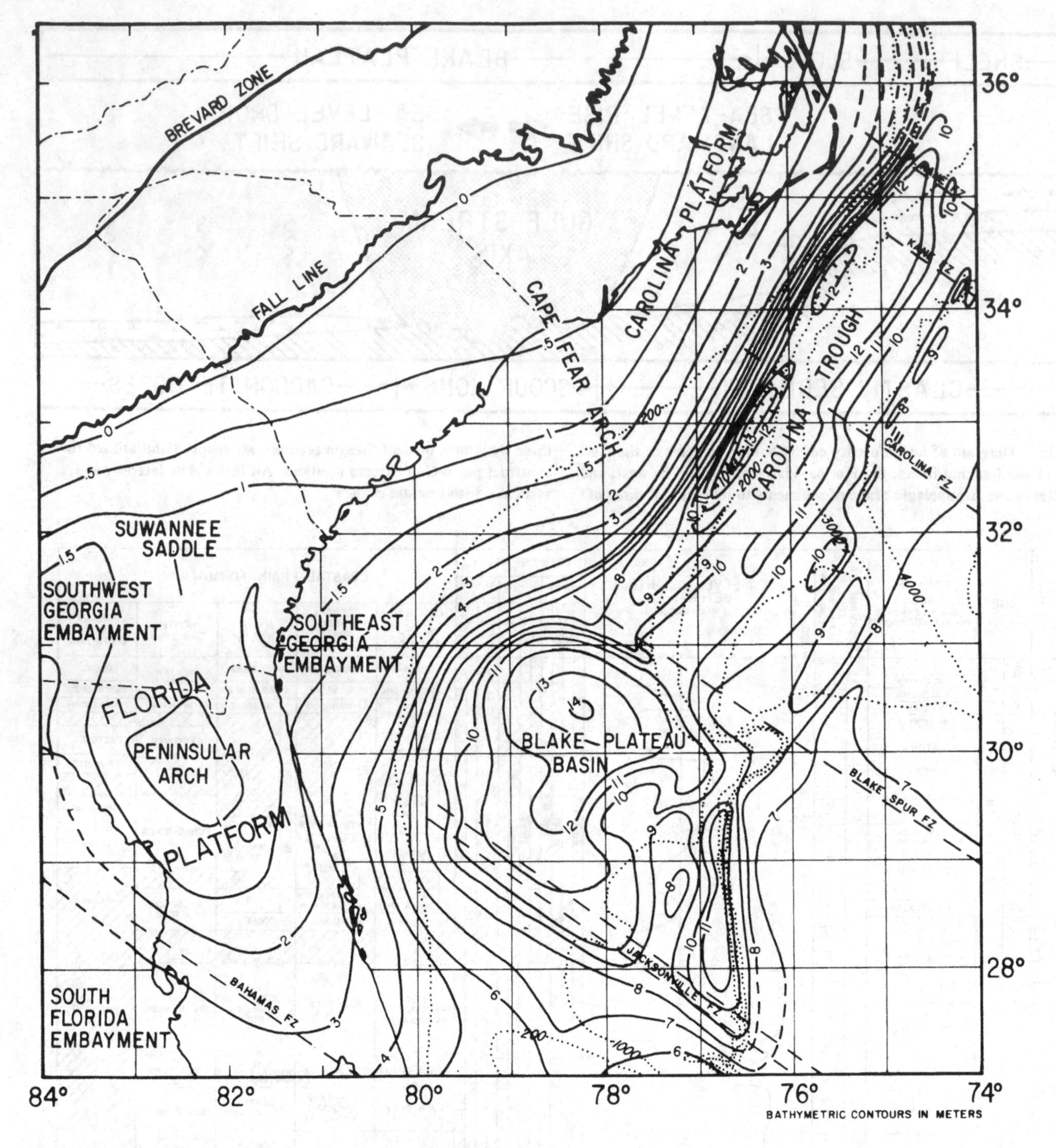

Fig. 28.7. Basement depth (km), major oceanic fracture zones (FZ), and primary tectonic structures of the southeastern US Atlantic continental margin. Compiled from Brown *et al.* (1972), Popenoe & Zietz (1977), Crosby (1980), Dillon *et al.* (1982), Klitgord, Popenoe, & Schouten (1984), Mountain & Tucholke (1985), Popenoe (1985) and unpublished analyses of common-depth-point (CDP) seismic-reflection data.

peninsular Florida and the coastal plain of Georgia. Rocks of Late Cretaceous age – predominantly nearshore marine and continental clastics – crop out continuously along the Fall Line from Alabama to North Carolina. In the Southeast and Southwest Georgia Embayments they grade into deeper-water facies, and southward into shallow-water marine limestones over Florida (Chen, 1965; Applin & Applin, 1967).

Both stratigraphic facies and seismic-reflection surveys of the offshore (Paull & Dillon, 1980; Sheridan *et al.*, 1981; Pinet & Popenoe, 1985b; Dillon & Popenoe, 1988) indicate that during the Late Cretaceous and Paleocene most surface flow from the Gulf of Mexico to the Atlantic was across the present coastal plain of southern Georgia through a channel (Fig. 28.8) known as the Suwannee Strait (Hull, 1962; Chen, 1965; Pinet &

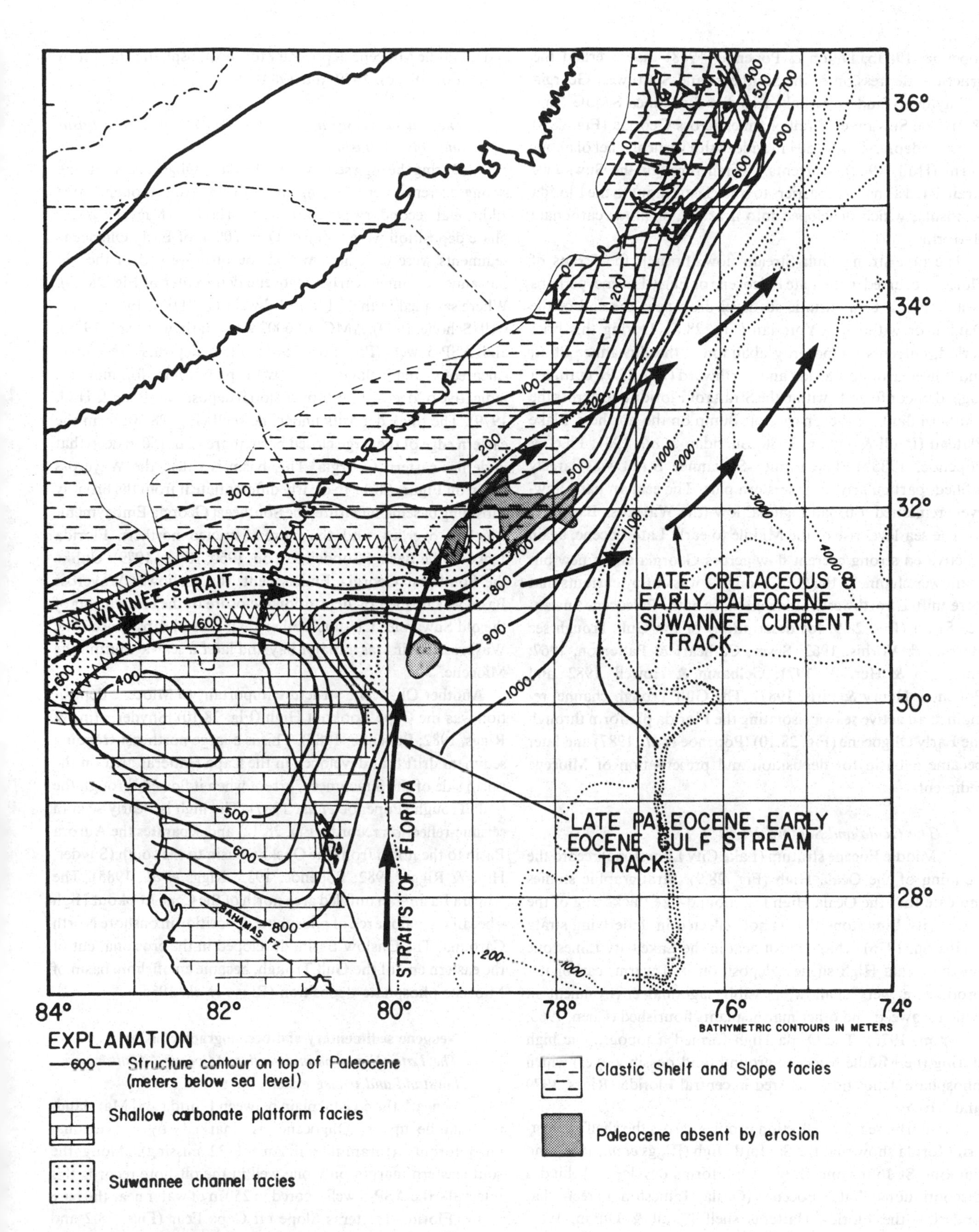

Fig. 28.8. Structure contour map of the top of the Paleocene deposits, facies of the Paleocene, track of the Early Paleocene Suwannee Current through the Suwannee Strait, and the initial track of the Gulf Stream through the Straits of Florida (after Dillon & Popenoe, 1988).

Popenoe, 1985b; Dillon & Popenoe, 1988) that overlaid the structural depression of the Southeast and Southwest Georgia Embayments and were linked by the Suwannee Saddle (Fig. 28.7). The Suwannee Strait channel across Georgia (Fig. 28.8) formed a depression 40–50 km wide with a vertical relief of about 140 m (Hull, 1962). Currents flowing through the Suwannee Strait acted as natural barriers to siliciclastic input to the Florida peninsula, which developed into a broad, shallow, carbonate platform.

The first strong Gulf Stream flow through the Straits of Florida occurred in the Late Paleocene or Early Eocene, perhaps as a response to a eustatic sea-level drop near the end of the Danian or within the Ypresian (Fig. 28.6). During this brief period, currents temporarily abandoned the Suwannee Strait and flowed around Florida and northward (Fig. 28.9), cutting a rugged unconformity within the Straits of Florida and across the southern flank of the Charleston Bump on the northern Blake Plateau (Paull & Dillon, 1980; Sheridan *et al.*, 1981; Pinet & Popenoe, 1985b). The former Suwannee Strait was largely infilled, particularly in its western part. The eastern part, however, remained a broad regional low (the Waycross Basin).

The sea-level rise of the Middle to early Late Eocene, again reactivated strong current flow across Georgia and a new but shallower channel – the Gulf Trough – was cut by currents that were shifted northwestward from those of the Paleocene Suwannee Strait (Fig. 28.9; for descriptions of the Gulf Trough see Herrick & Vorhis, 1963; Sever, Cathcart & Patterson, 1967; Patterson & Herrick, 1971; Gelbaum & Howell, 1982; and Popenoe, Henry & Idris, 1987). The Gulf Trough channel remained an active seaway isolating the Florida platform through the Early Oligocene (Fig. 28.10) (Popenoe *et al.*, 1987) and later became a basin for deposition and preservation of Miocene sediments.

The Ocala and Sanford Highs

Middle Eocene stratum (Lake City Limestone) record the building of the Ocala High (Fig. 28.9). Stratigraphic studies indicate that the Ocala High is a mound-like thickening of the Lake City Limestone that is not reflected in underlying strata (Winston, 1976). Thin beds of peat in the Lake City Limestone on the Ocala High suggest deposition in a warm, carbonate-productive, very shallow, possibly lagoonal environment in which seaweed and other marine plants flourished (Chen, 1965; Winston, 1976). The Ocala High formed a topographic high during the Middle Miocene around (and possibly over) which phosphate deposition occurred in central Florida (Riggs, 1979 and 1984).

Similarly, seismic-reflection profiles across the shelf off eastern Florida show that the Sanford High (Riggs *et al.*, 1985) and flanking St Johns and Brevard Platforms developed behind a discontinuous Late Eocene (Ocala Limestone) reef that underlies the Florida–Hatteras shelf (Paull & Dillon, 1980; Sheridan *et al.*, 1981; Popenoe, Kohout & Manheim, 1984). The Sanford High (Riggs, 1979) is separated from the Ocala High by a lower area that includes the Jacksonville Basin, Kissimmee Saddle, and Osceola Basin (Goodell & Yon, 1960; Riggs, 1979 and 1984). This depression formed an Early Miocene estuary and a Middle Miocene depocenter for the phosphatic Hawthorn Formation (Riggs, 1979 and 1984).

The Waycross Basin, Cape Lookout High, Aurora Basin and Onslow Basin

During the high sea levels of the Early Oligocene, while the strong currents were flowing through the Gulf Trough, three additional secondary structures important to Miocene phosphate deposition were formed. Over 200 m of Early Oligocene sediments were deposited within the offshore end of the old Suwannee Channel (early Oligocene depocenter of Fig. 28.10). Where sampled in the J-1, and J-2 wells (JOIDES, 1965), GE-1 well (Scholle, 1979), AMCOR 6002 well (Hathaway *et al.*, 1979), and ASP-5 well (Popenoe, 1985), these deposits are chiefly calcareous hemipelagic ooze with planktonic foraminifera indicative of deep-water (upper slope) deposition (Poag & Hall, 1979). The thick deposits under the shelf (Fig. 28.10) form the eastern edge of the large closed basin more than 150 m deep that underlies eastern Georgia. This basin is called the Waycross Basin on Figure 28.11 to clearly differentiate it from the broadly subsiding basement feature, the Southeast Georgia Embayment, which it overlies, and the more restricted northern Florida Jacksonville Basin (Goodell & Yon, 1960; Riggs, 1979). Unlike the Southeast Georgia Embayment, the shallower Waycross Basin was formed by depositional closure of the eastern end of the old Suwannee Channel rather than crustal subsidence. The Waycross Basin was a major bay and later a seaway during the Miocene.

Another Oligocene structure important to Miocene deposition was the Cape Lookout High (Fig. 28.10) (Snyder, Hine & Riggs, 1982; Popenoe, 1985). This is a large northeast-trending sediment drift that developed in the Cape Hatteras area on the north side of the Suwannee Current when it flowed through the Gulf Trough (Popenoe *et al.*, 1987). The high is clearly seen in seismic-reflection records (Fig. 28.12) and separates the Aurora Basin to the north from the Onslow Basin to the south (Snyder, Hine & Riggs, 1982; Popenoe, 1985; Riggs *et al.*, 1985). The Aurora Basin – an unfilled area north of the Cape Lookout High – became a major area of phosphate deposition in onshore North Carolina. The Onslow Basin, developed in the erosional cut of the eastern end of the Gulf Trough, became an offshore basin of Miocene phosphate deposition (Riggs *et al.*, 1985).

Neogene sedimentary and oceanographic history

The Late Oligocene and earliest Miocene (Aquitanian) lowstand and transgression

Beneath the coastal plain between Florida and Maryland, the middle–upper Oligocene is marked by a regional unconformity (foraminiferal zone P-22 missing). Along the southeastern margin only one well in the offshore records this interval – the ASP-5 well – cored in 250 m of water near the base of the Florida–Hatteras Slope off Cape Fear (Figs. 28.2 and 28.6). In this well the P-22 foraminiferal zone sediments are calcarenites (Poag, in Popenoe, 1985) that are part of a broad depocenter of flat-lying Oligocene strata extending across the northern Blake Plateau, in the lee of the Charleston Bump off North Carolina (Middle Oligocene depocenter; Fig. 28.10). This

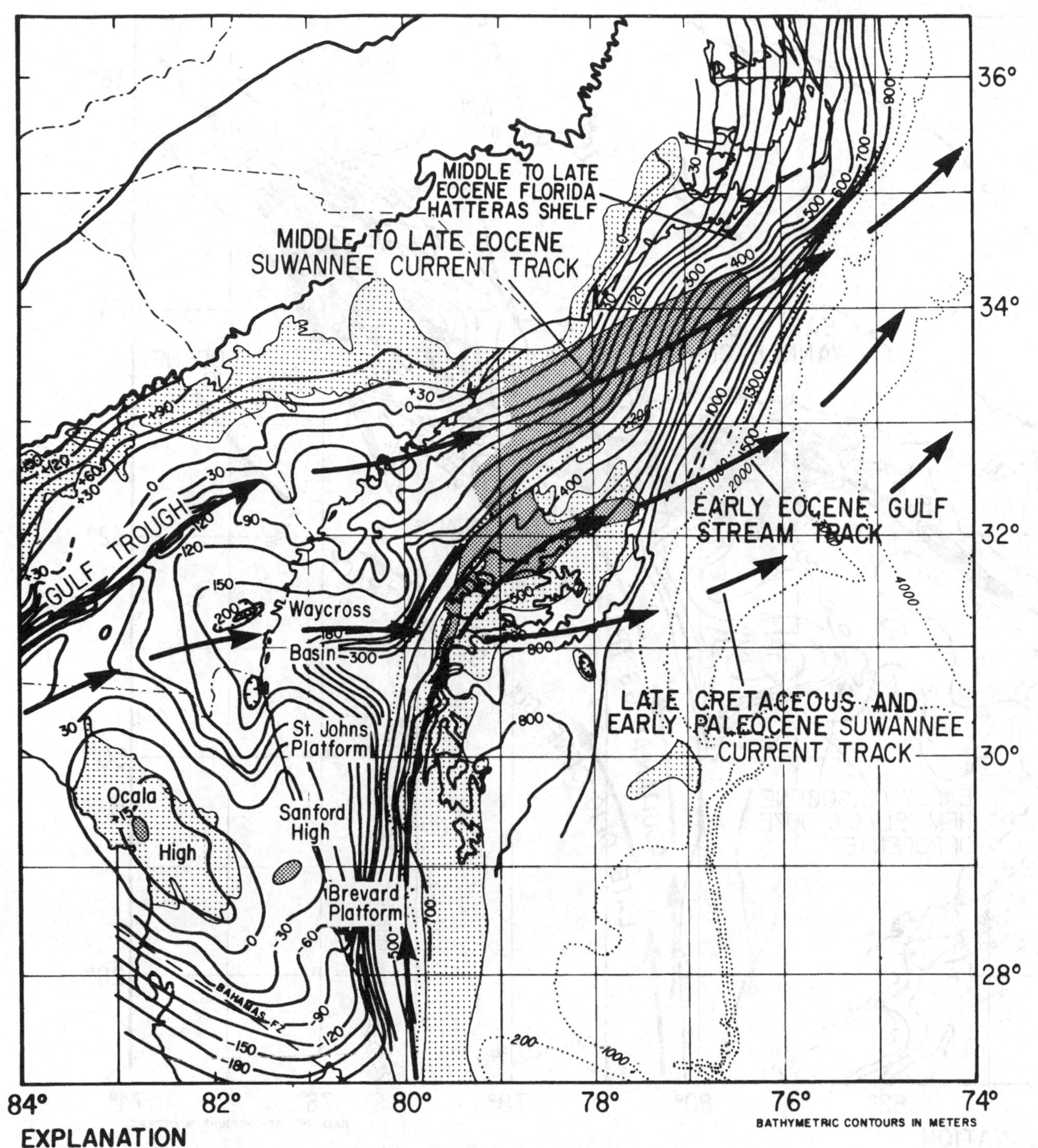

Fig. 28.9. Structure contour map of the top of the Eocene (m below sea level), showing major topographic and bathymetric features on this surface. Also shown are the Suwannee Current track for the Late Cretaceous–Early Paleocene and the Middle–Late Eocene, and the Gulf Stream track for the Early Eocene. Compiled, in part, from Cooper & Stringfield (1950), Malde (1959), Chen (1965), Brown *et al.* (1972), Meisburger & Field (1975), Weaver & Beck (1977), Paull & Dillon (1980), Gelbaum & Howell (1982); Manheim *et al.* (1982) and Popenoe (1985).

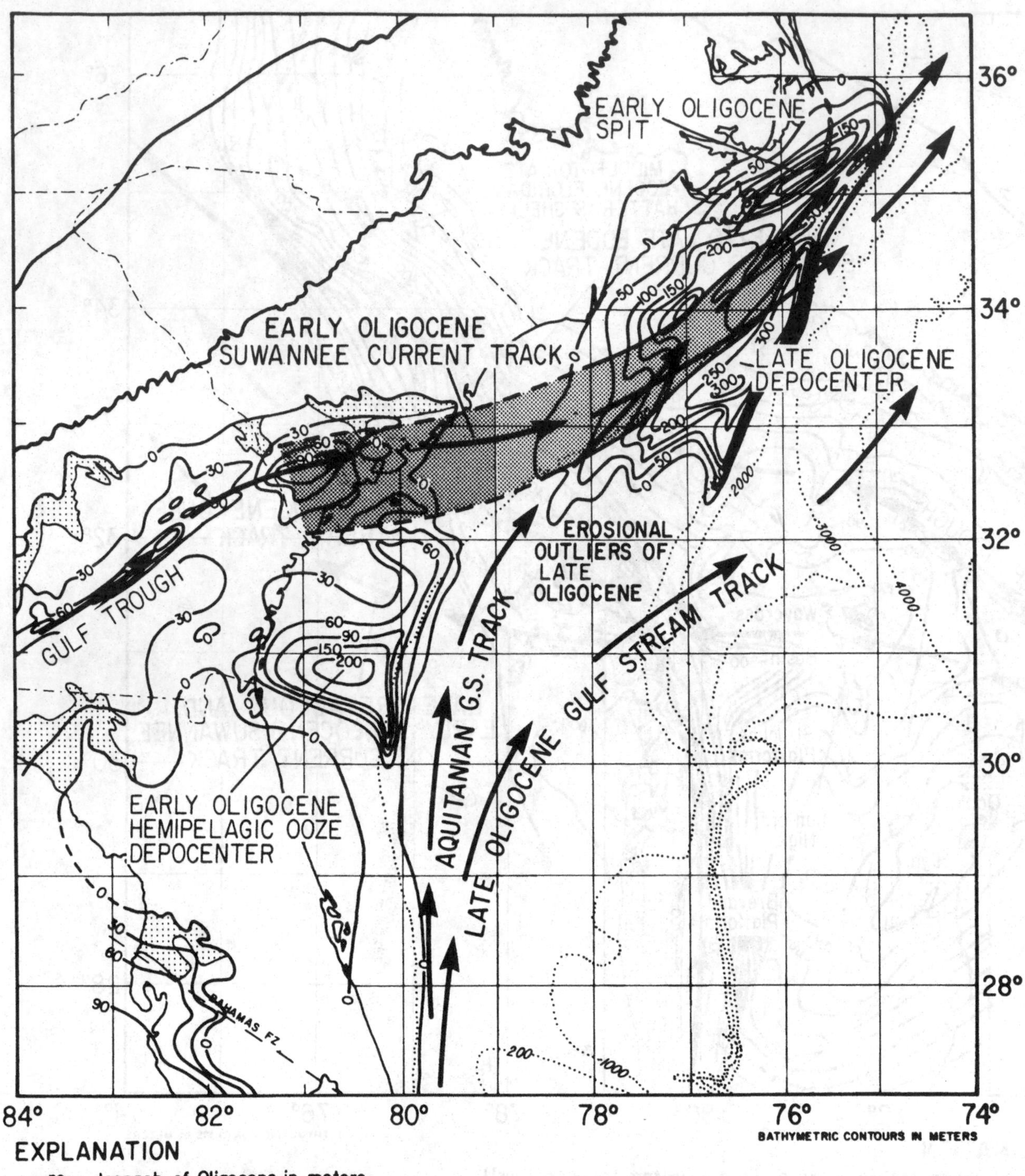

Fig. 28.10. The thicknes of Oligocene sediments (m). Also shown are Early Oligocene Suwannee Current and Late Oligocene and Early Miocene (Aquitanian) Gulf Stream tracks. Compiled in part from Toulmin (1952), Brown *et al.* (1972), Cramer (1974), Edsall (1978), Paull & Dillon (1980), Gelbaum & Howell (1982), Idris (1983), Pinet & Popenoe (1985a) and Popenoe (1985).

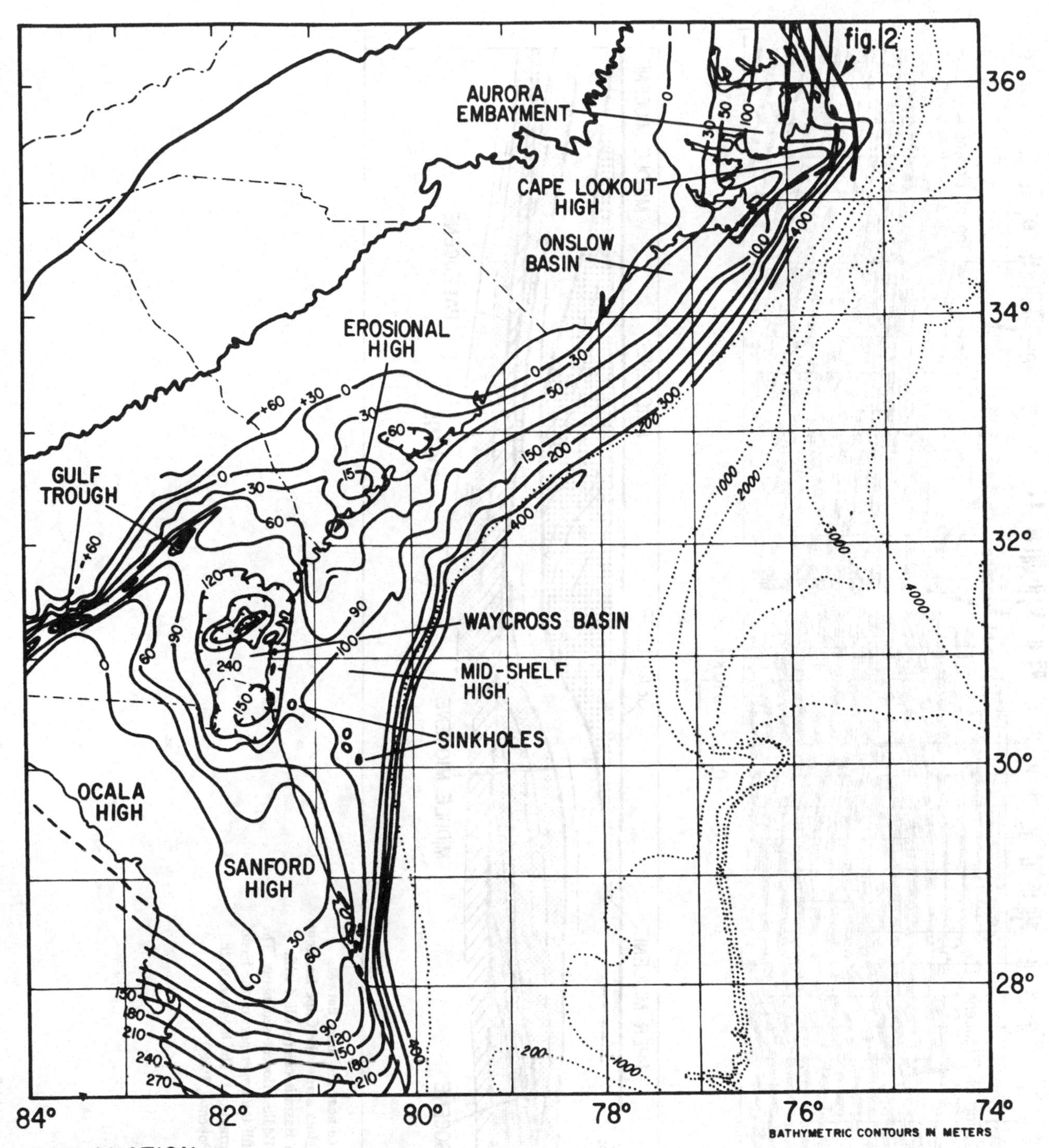

EXPLANATION

—30— Elevation (below sea level) of the base of the Miocene

Fig. 28.11. Structure contour map of the base of the Miocene sediments showing major topographic and bathymetric features. Compiled, in part, from Cooper & Stringfield (1950), Herrick (1961), Siple (1969), Miller (1971), Brown *et al.* (1972), Cramer (1974), Gelbaum & Howell (1982), Idris (1983) and Popenoe (1985).

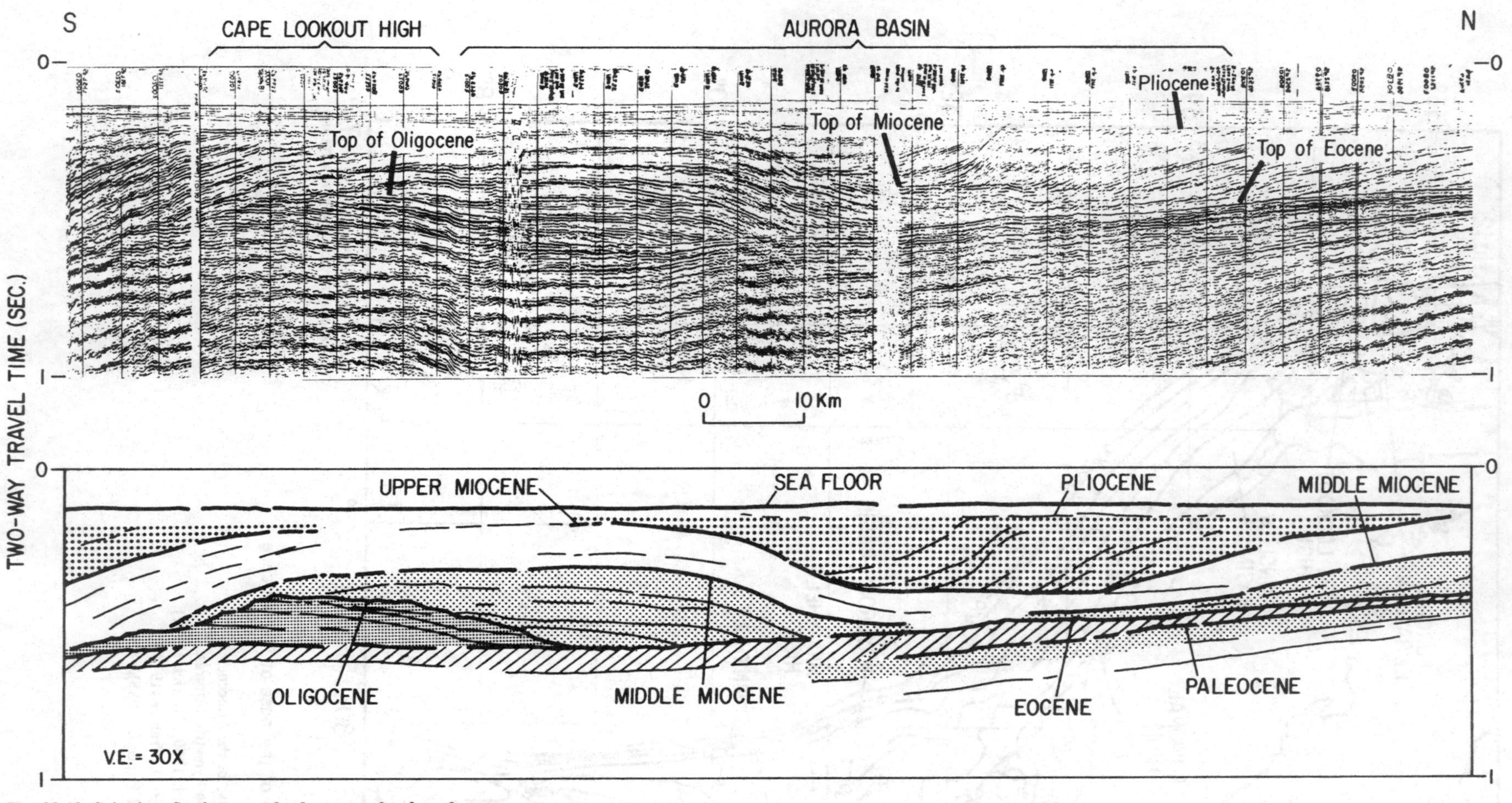

Fig. 28.12. Seismic-reflection record taken near the shore between Cape Hatteras, North Carolina and Virginia across the Cape Lookout High and the Aurora Basin. The Aurora Basin is chiefly filled with northward- and southward-prograding Middle Miocene sediments. The steeply southward-prograding beds that fill the upper part of the basin are Early Pliocene in age (Yorktown Formation). The location of the profile is shown on Figure 28.11 (after Popenoe, 1985).

suggests subaerial exposure of the coastal plain and a downdip shift of the Florida–Hatteras Shelf to across the northern Blake Plateau.

Seismic profiles across the Straits of Florida and the southern Blake Plateau show an absence of Late Oligocene strata and a rugged unconformity, indicating that the Gulf Stream was flowing strongly through the Straits and across the southern Blake Plateau, south of the Charleston Bump in Late Oligocene time (Pinet, Popenoe & Nelligan, 1981). The offshore shift of sedimentation to across the northern Blake Plateau and the southward shift of the Gulf Stream to deeper water suggest a major sea-level fall. This is in agreement with a global fall of sea level to 150 m below present (Fig. 28.6) during that time, proposed by Vail & Hardenbol (1979).

The transgression of sea level of the Late Oligocene, culminating in the Aquitanian, can be traced both offshore and onshore. Offshore, the sea-level rise caused the Gulf Stream to progressively straighten its course and transgress the Charleston Bump on the northern Blake Plateau, cutting into previously-deposited Middle Oligocene sediments (Popenoe, 1985). During the maximum transgression of the Aquitanian, the Gulf Stream cut a rugged linear unconformity near the base of the present Florida–Hatteras Slope off North Carolina (Aquitanian Gulf Stream track; Fig. 28.10).

Onshore, as seas flooded the coastal plain, the Suwannee Limestone and later the Parachucla (Huddlestun, 1982) and Penney Farms Formations (Scott, Chapter 26, this volume), were deposited within the Gulf Trough across Georgia, and the upper Cooper Formation and Edisto Formations were deposited across South Carolina. Across much of Georgia and north Florida, the Parachucla is a siliciclastic deposit of probable deltaic origin, indicating deltaic retreat of the river systems with rising seas into developing bays and estuaries in the Gulf Trough area.

Late Chattian and Aquitanian seas were high enough to inundate much of low-lying Florida where the Penney Farms and Tampa Member of the Arcadia Formations (Scott, 1985a, b) were deposited. If these seas inundated the Ocala or Sanford Highs, the evidence was destroyed by erosion in the Early Miocene (Burdigalian) lowstand. However, remnants are found as high as 45 m above present sea level in northern Florida (T.M. Scott, pers. comm., 1985).

The Gulf Stream appears to have been confined to the Straits of Florida for the entire Neogene. The Florida–Hatteras Shelf was well developed along the Straits of Florida and off Georgia and South Carolina. It confined the current, even during the high sea levels of the Middle Miocene. Neogene seas were not as high as those of the Paleogene and were largely lower than present. Neogene deposition is distinctly more siliceous than that of the Paleogene signaling the end of the Suwannee Straits as a barrier to terrigenous input to southern Georgia and the Florida peninsula (Riggs, 1979; Riggs *et al.*, 1985).

The Early Miocene (Burdigalian) lowstand and transgression

The evidence for a major regression during the Early Burdigalian is clear. In Georgia, Florida, South Carolina, and North Carolina, sediments that contain N.5 foraminifera (Blow, 1969) are missing onshore (P.F. Huddlestun, pers. comm., 1985). In the offshore, the J-2 well (Figs. 28.2 and 28.6), drilled on the continental slope off Georgia, penetrated a relatively complete section of very coarse-grained quartzose sand and sandy silt of deltaic character and Early Burdigalian age (Poag & Hall, 1979). The sediments at the well can be traced by seismic stratigraphy to a 200 m thick pod of sediments that underlies the slope north of the J-2 well near 31° N latitude, 80° W longitude (Fig. 28.13). Seismic-reflection records (Fig. 28.14) show that this pod is built of obliquely downlapping clinoforms whose morphologic character and position suggest that a delta of the ancestral Savannah River, then far offshore, dumped sediment over the Florida–Hatteras Slope. Later Burdigalian sediments onlap the shelf, and thin updip over a high in the middle shelf which formed over the thick depocenter of Lower and Middle Oligocene strata off Georgia (high on Oligocene top; Fig. 28.13). On the crest of the high, Burdigalian age sediments are absent (Poag & Hall, 1979) in the GE-1 well (Figs. 28.2 and 28.6).

The Early Burdigalian lowstand sequence is even better expressed off North Carolina. Seismic stratigraphic data indicate that the lowest Burdigalian sediments occur over the northern Blake Plateau, north of the Charleston Bump. At the edge of the northern plateau, the sequence is expressed on seismic-reflection records as a mass of transparent–hyperbolic returns (Fig. 28.15) whose form is characteristic of a euphotic-zone, shallow-water reef. This facies has not been sampled but can be seismically traced along the edge of the plateau for over 120 km. Where it outcrops, it forms a resistant ledge. Sediments behind this facies grade shoreward into a sequence of coarsely-layered, high-amplitude, parallel reflectors that pinch out updip against the underlying Oligocene strata in about the position of the present Florida–Hatteras Slope (Popenoe, 1985).

Although lower Burdigalian age sediments are not present across the south flank of the Charleston Bump, there is evidence that a thin layer of these sediments may have once been present. The morphology of erosional outliers of Lower Miocene sediments preserved near the crest of the Charleston Bump (Fig. 28.16) suggests that these deposits were once relatively thick and extensive over part of the south flank of the Bump. The phosphorite pavement deposits of the Blake Plateau may have originated as lag gravels washed from these deposits in the later Burdigalian and Langhian as the Gulf Stream overrode the earlier deposits due to deepening water (Manheim *et al.*, 1982).

Both a hiatus of Burdigalian strata in the Straits of Florida and across the southern Blake Plateau, and the presence of facies of Lower Miocene strata to the east and west of this unconformity, indicate that the Gulf Stream was flowing from the Straits of Florida across the southern Blake Plateau during the Early Burdigalian (Pinet & Popenoe, 1982).

The sea-level rise in the Late Burdigalian produced dramatic effects on the north side of the Charleston Bump off North Carolina. As a result of deepening water, the Gulf Stream progressively straightened its course, overriding the northern Blake Plateau. In doing so, it stripped sediments from the south flank of the Charleston Bump and deposited them in the lee of the bump as a northward-prograding downstream sediment drift (see arched beds on Fig. 28.12). These mounded beds were later completely overridden by the Gulf Stream which stripped most (> 200 m) of the upper part of them away (Middle Miocene unconformity; Figs. 28.15 and 28.17). In the latest Burdigalian

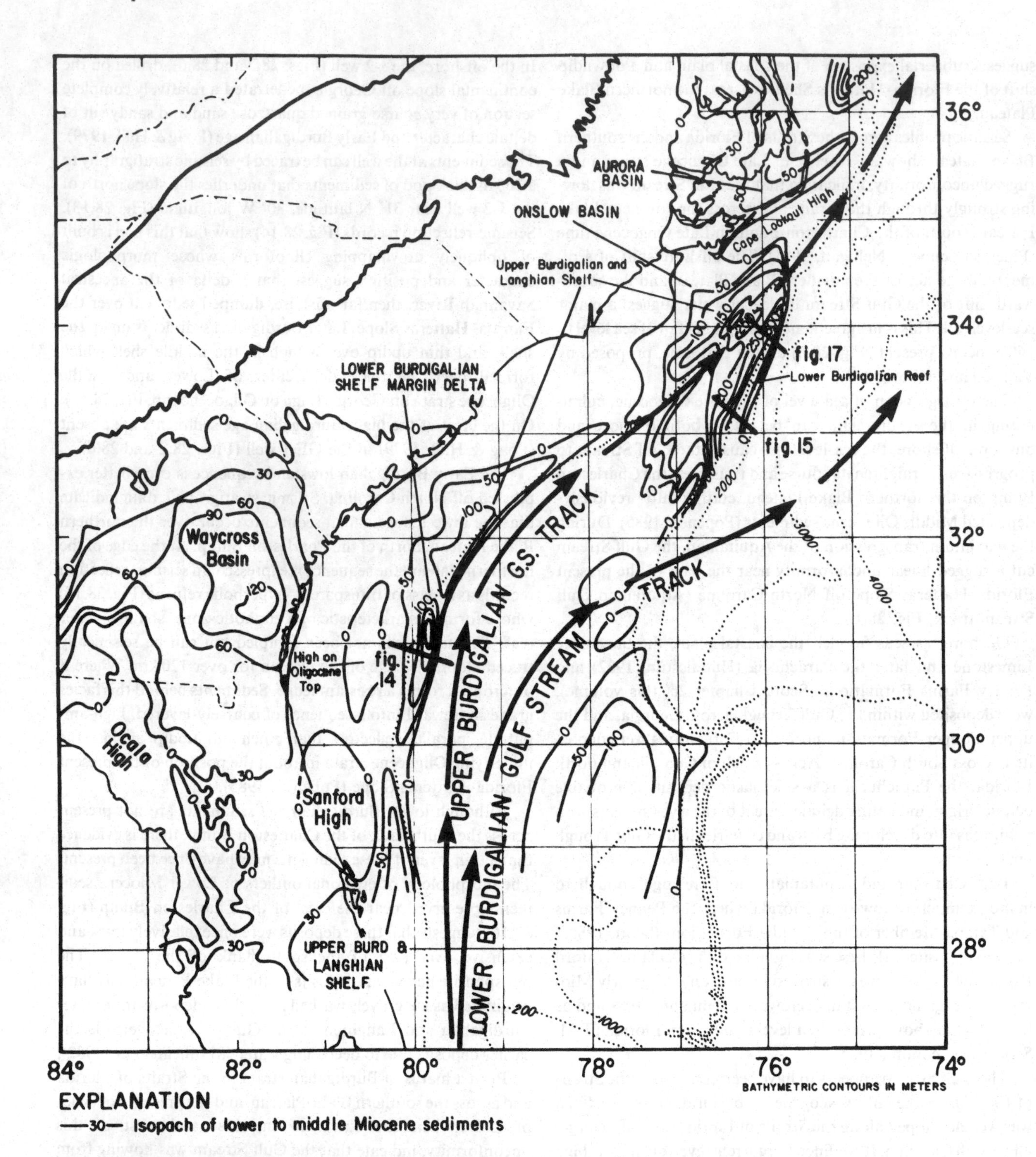

Fig. 28.13. Thickness of Burdigalian and Langhian sediments (m) and Gulf Stream tracks of the Burdigalian. Compiled, in part, from Weaver & Beck (1977), Riggs (1979), and Popenoe (1985).

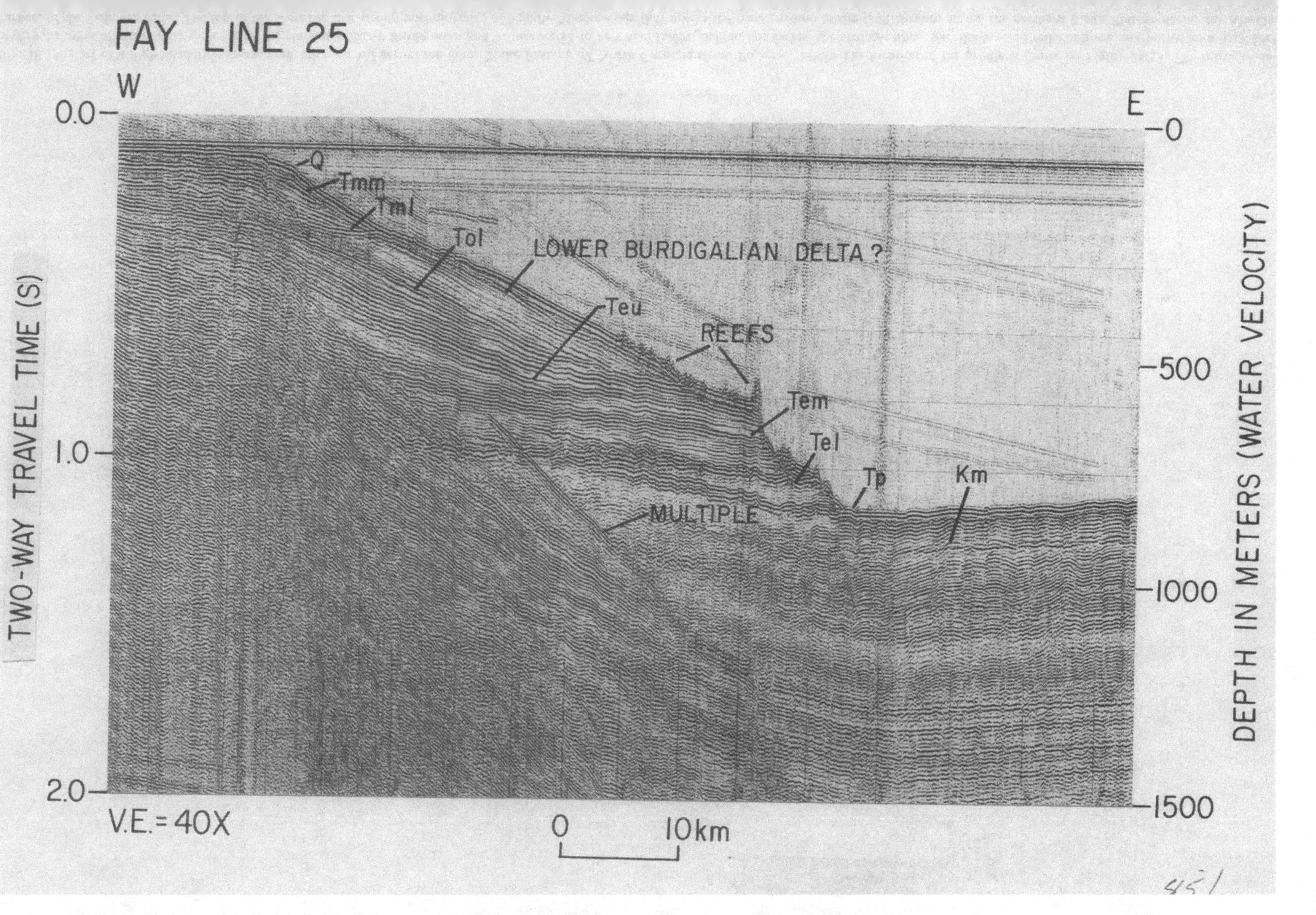

Fig. 28.14. High-resolution seismic-reflection profile across the Florida–Hatteras Slope off Georgia. Tops of depositional sequences are marked. The location of the profile is shown on Figure 28.4. From its seismic character and limited distribution (Fig. 28.13), the progradational sequence of Early Miocene age (Tml) is interpreted to be deltaic; probably the ancestral Savannah River. Note the truncation of the Middle and Upper Eocene (Tem, Teu) and Lower Oligocene (Tol) strata near the base of Florida–Hatteras Slope. These beds were deposited out across the Straits of Florida and Blake Plateau when currents were in the Gulf Trough. They have since been dissected by Gulf Stream currents discharging from the Straits of Florida. Km – Upper Cretaceous; Tp – Paleocene, Tmm – Middle Miocene, Q – Quaternary. Spike-like reflectors near the base of the slope are deep-water reefs that flourish in areas of Gulf Stream turbulence. Similar reef build-ups are present in buried sequences associated with paleo-Gulf Stream erosional tracks.

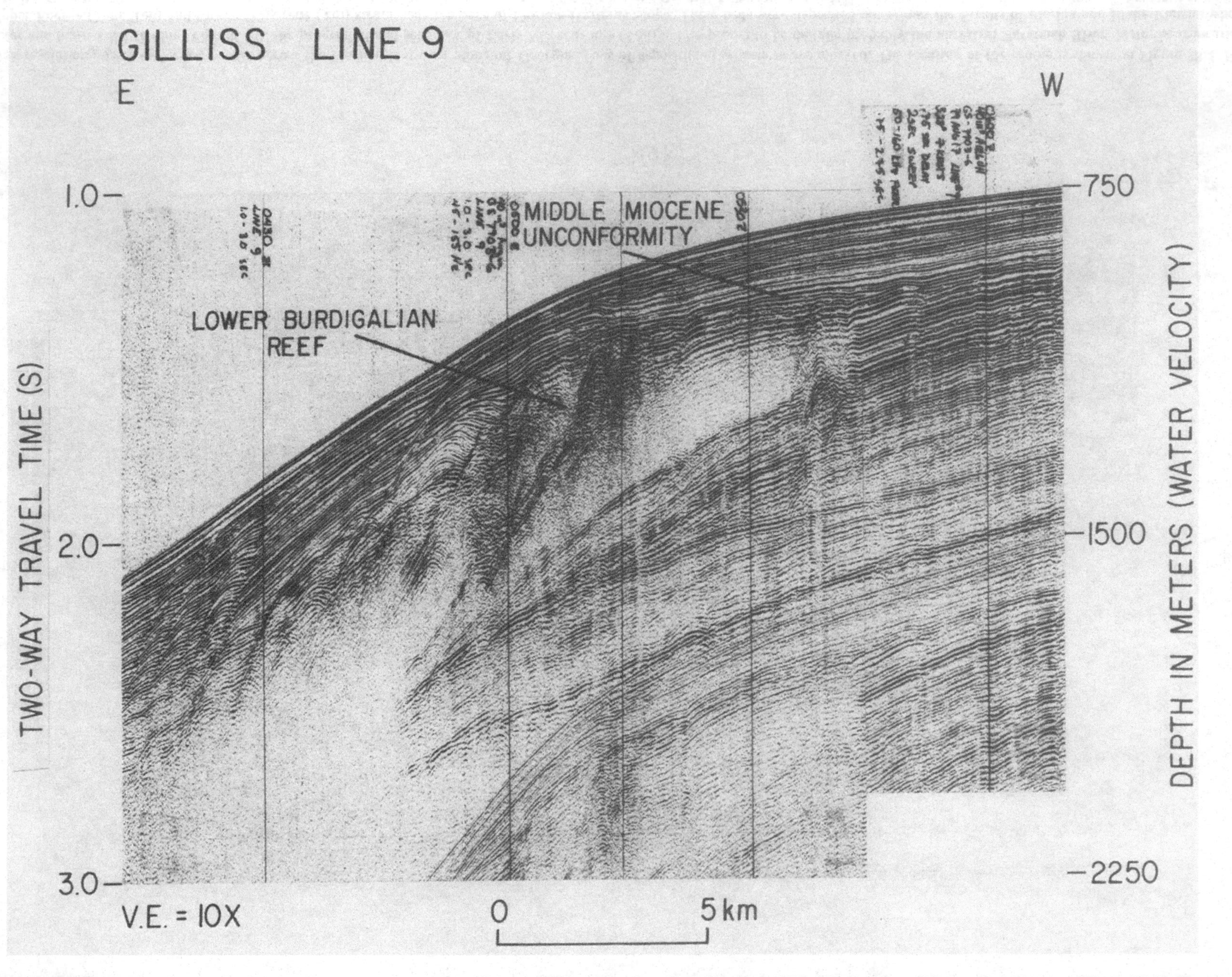

Fig. 28.15. Part of a high-resolution seismic-reflection record across the outer Blake Plateau off North Carolina (from Popenoe, 1985). The location of the profile is shown on Figure 28.13. The transparent–hyperbolic mass of strata near the edge of the plateau is Early Burdigalian and is interpreted to be a reef facies. Behind this facies, the strong-return, parallel-layered reflectors are interpreted as a back-reef carbonate platform sequence. The top of the sequence is a strong unconformity of Middle Miocene age that marks the transgression of the Gulf Stream across the northern Blake Plateau during the Middle

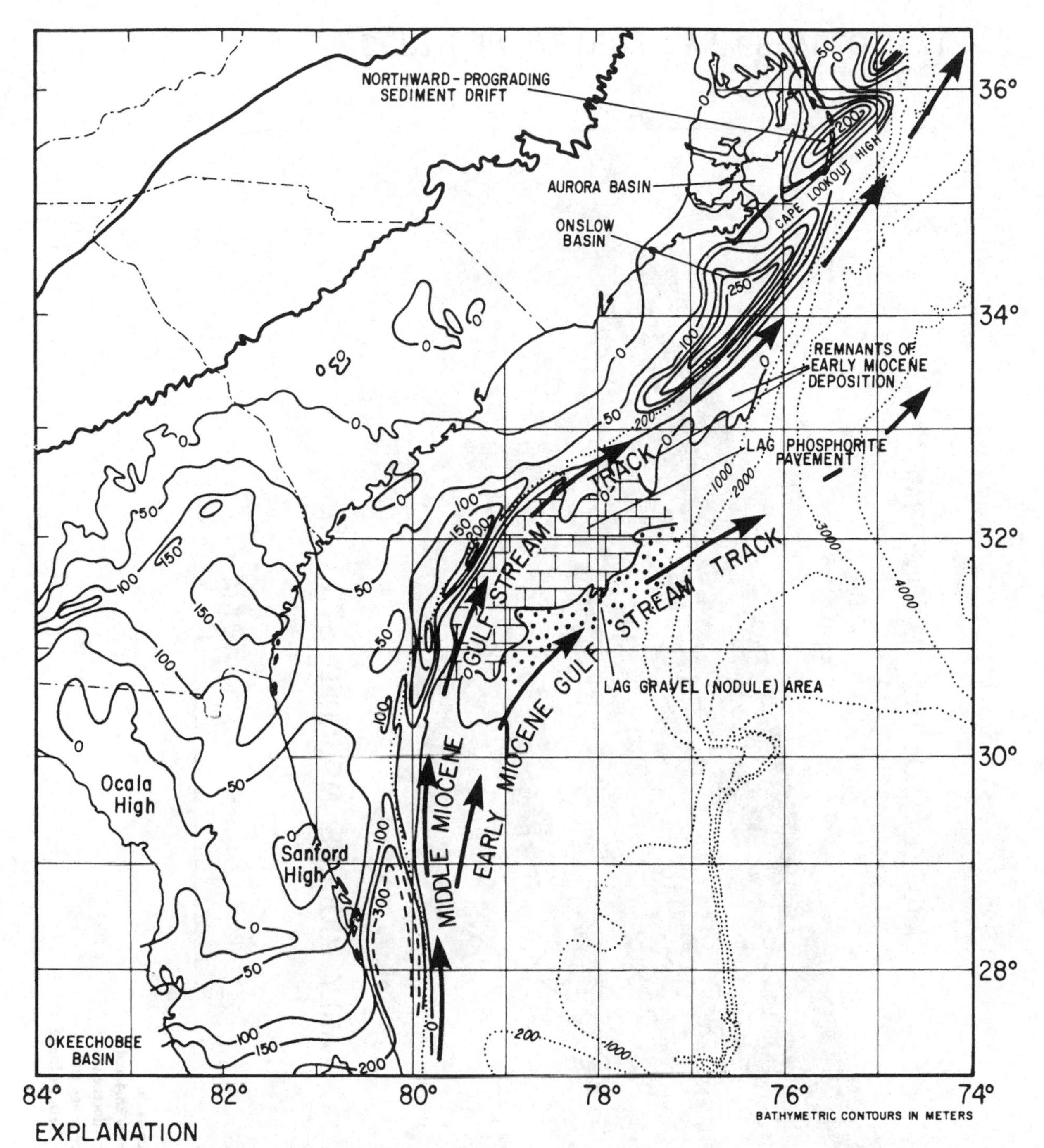

Fig. 28.16. The thickness of Miocene sediments across the southeastern US continental margin (m). This isopach map does not include the Early Miocene depocenter on the northern Blake Plateau. Note the buildouts of the Florida–Hatteras Shelf off North Carolina and off Cape Canaveral, Florida. Compiled, in part, from Vernon (1951), Riggs (1979), Popenoe (1984) and Miller (1986).

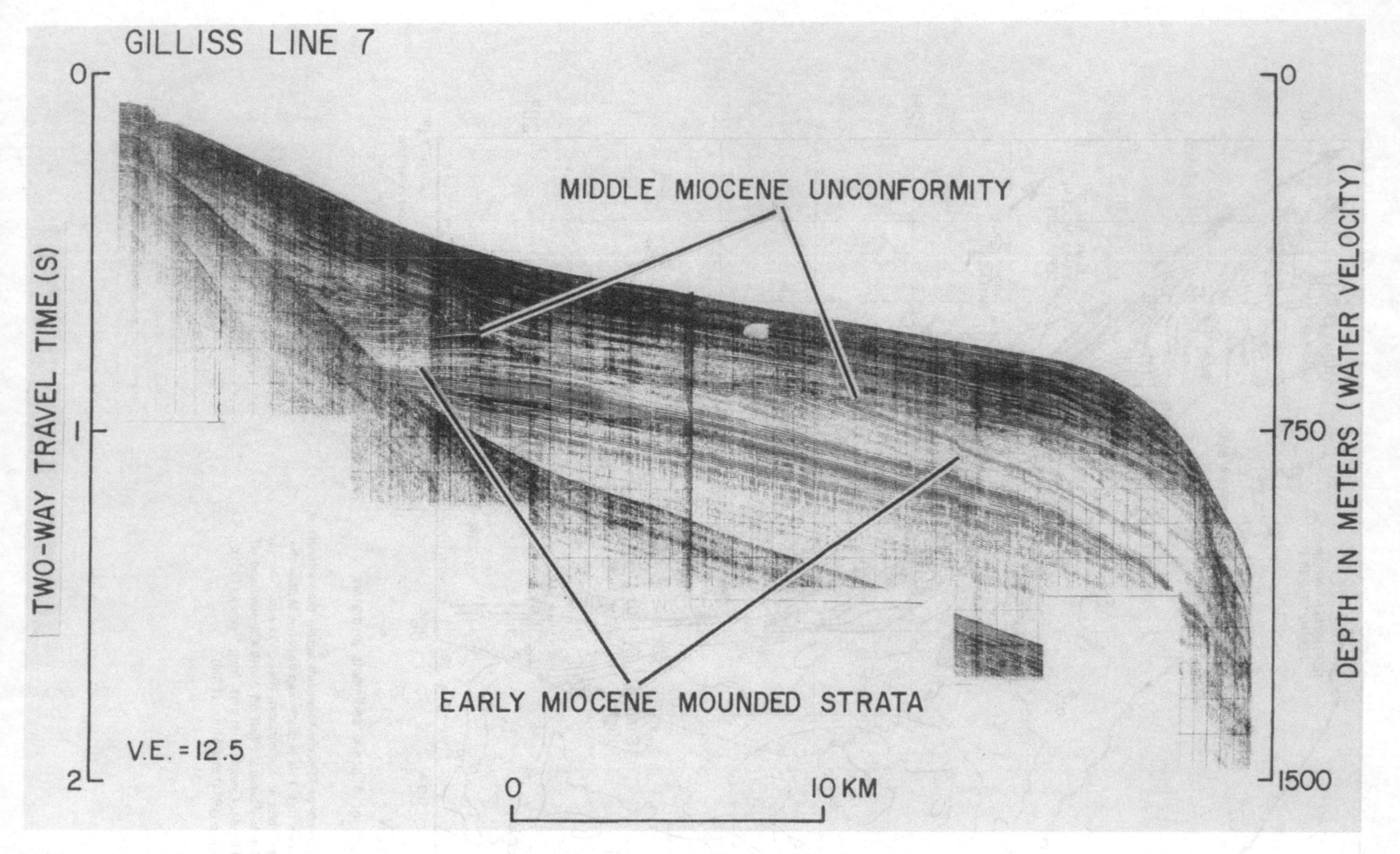

Fig. 28.17. High-resolution seismic-reflection profile across the northern Blake Plateau. The location is shown on Figures 28.4 and 28.14. Mounded beds prograde north and are of Early Miocene (Late Burdigalian) age. The unconformity is of Middle Miocene age and marks the transgression of the Gulf Stream across the northern Blake Plateau (after Dillion & Popenoe, 1988).

and Langhian, the Gulf Stream assumed a position very near its present location on the inner Blake Plateau. The Florida–Hatteras Shelf prograded out to this position on its landward side, filling most of the Onslow Basin and a basin to the south (Late Burdigalian and Langhian shelf; Fig. 28.13) along the shelf edge off southern South Carolina.

The Middle Miocene (Late Burdigalian, Langhian and Serravallian) highstand in Georgia and north Florida

The Miocene stratigraphy of the coastal plain of Georgia and northern Florida is summarized in Figure 28.6. The Hawthorn Group of Georgia is divided (Huddlestun, 1982) into the Parachucla Formation (Tampa Formation of Weaver & Beck, 1977) of Aquitanian age (foraminiferal zone N.4), the Marks Head Formation of eastern Georgia and equivalent Torreya Formation of the panhandle, both of Burdigalian age (zones N.6–N.7) and the Coosawhatchie Formation of Langhian and Serravallian age (zone N.8–N.11). The Hawthorn Group also includes the Wabasso Formation of the Pliocene (Zanclean) (Fig. 28.6). All marine facies in Georgia grade updip into prograding, fluvial, clastic facies of the Altamaha Formation, which outcrops extensively northwest of the Gulf Trough (Huddlestun, 1982).

Aquitanian deposition was widespread across Georgia, the Carolinas, and Florida (Fig. 28.6) as discussed earlier. Early Burdigalian sedimentation was mainly within the Gulf Trough and Waycross Basin lows, where deposits above the Parachucla Formation are chiefly poorly-sorted sands, silts, and black waxy clays that represent fluvial and lake deposits (Altamaha Formation of Abbott & Huddlestun, 1980). These deposits grade eastward into fine- to coarse-grained sands with scattered beds of pebbles deposited as shallow-water deltas. The eastern, downdip facies contains phosphatic sands, clays, and interbedded dolomites (Marks Head Formation of Huddlestun, 1982) of estuarine or restricted marine character. The Waycross Basin and Gulf Trough lows were largely filled by deltaic deposits by the end of the Burdigalian.

The Middle Miocene deposits (Coosawhatchie Formation of Langhian and lower Serravallian age) of Georgia are mainly marine clays rich in phosphate, diatoms, and sponge spicules. These beds are chiefly preserved in two areas – a shallow elongate basin in the area of the Gulf Trough, and a shore-parallel depression on the east flank of the Waycross Basin (the Ridgeland Trough and connecting Jacksonville Basin on Fig. 28.21). Although Weaver & Beck (1977) suggest that deposition in the Gulf Trough occurred in an almost enclosed elongate basin whose eastern end was probably a shallow littoral zone where sand and phosphate were reworked but where deposition was minor, the equivalent beds in the Ridgeland Trough contain facies that represent deltaic–lacustrine, littoral, and neritic environments (Abbott, 1974). These indicate that shelf-depth seas inundated most of eastern Georgia during the Middle Miocene, although only remnants of this deposition remain.

The Miocene deposits in northern Florida have been discussed by Miller (1982), and Scott (Chapter 23, this volume). In northern Florida the thickest Miocene deposits are situated on the southern flank of the Waycross Basin and the Jacksonville Basin (Fig. 28.16). Precise age dating of these deposits has not been done because foraminifera are scarce (J.A. Miller, pers. comm., 1985). Two environments of deposition are present: a shallow marine–deltaic environment characterized by sand-sized phosphate pellets and pebble sized phosphate mixed with coarse clastics and molluscan limestone; and, a more open-marine environment characterized by pelletal phosphate and fine- to medium-quartz sands interbedded with thick, olive-green diatomaceous clays. The source of pebbles and coarse clastics was to the southwest – eroded from older deposits on the Ocala High (Miller, 1982) and the Sanford High. Farther west, on the flanks of the Florida Platform in the eastern panhandle, the Lower Miocene strata are a mixture of both calcareous beds and phosphatic and diatomaceous sand and clay containing brackish-water foraminifera (Torreya Formation) (P.F. Huddlestun, pers. comm., 1985). The presence of several terrestrial vertebrate fossil layers within this formation (Hunter & Huddlestun, 1982) indicate that the Late Burdigalian transgression of Florida, similar to that of North Carolina (Riggs *et al.*, 1982, 1985, Chapter 31, this volume) was punctuated by minor regressions.

The Miocene stratigraphy of central Florida

In central Florida the Hawthorn Group is divided into the lithostratigraphic units of the Arcadia, Peace River (Noralyn), and Bone Valley Formations (Freas & Riggs, 1968; Riggs, 1979; Scott, 1985a, b, Chapter 23, this volume). The basal dolomitic unit – the Arcadia Formation – which interfingers with and underlies the more sandy and phosphatic Peace River Formation, is of Middle–Late Langhian age (Webb & Crissinger, 1983). This is the period indicated by offshore data as the time of maximum water depths, as well as the age of the richest phosphorites in the J-1, 6002, COST GE-1, and 6004 wells of the offshore (Fig. 28.6). In onshore Florida, the Arcadia Formation contains a vast resource of low-grade pelletal phosphate (Riggs, 1979). The top of Langhian deposition in the Central and Southern Districts is marked by an unconformity and a hiatus along which calcareous and silty clay beds containing terrestrial fossils occur. The presence of solution channels and sinkholes on this surface attests to a period of minor regression and subaerial exposure in the Early Serravallian (Webb & Crissinger, 1983).

The Peace River Formation is a complex shallow-water coastal and nearshore shelf facies. It is discontinuous and thin in the Central District, but grades downdip in the Southern District into a predominantly clastic unit containing a heterogenous mixture of shallow marine, littoral, lagoonal, and fluvial sediments that show considerable evidence of reworking and mixing of previous material (Hall, 1983). This formation contains the most concentrated phosphate deposits within central and southern Florida.

The Bone Valley Formation (Bone Valley Member of the Peace River Formation of Scott, Chapter 21, this volume) is predominantly a terrigenous sand and clay that occurs as a complex set of fluvial channel deposits in upslope areas adjacent to the Ocala High and grades downdip into more extensive and uniform estuarine and shallow marine facies of the Peace River Formation (Riggs, 1979). The lower Bone Valley is of latest Serravallian age (Late Clarendonian), and the upper Bone Valley is earliest Zanclean (latest Hempillian) age (Webb &

Crissinger, 1983). Most of the Serravallian is represented by an unconformity.

The facies in Florida and offshore indicate that the highest sea levels of the Miocene occurred during the Langhian, when the upper marine facies of the Arcadia Formation were deposited high across central Florida onto the Ocala High. As sea level dropped in the Serravallian, the predominantly clastic and shallow-marine Peace River Formation was deposited. The basal Bone Valley Formation represents reworking during minor rises of the sea in the Late Serravallian and Early Tortonian. The marine component in the upper part of the Bone Valley Formation may have been deposited in the Late Tortonian and Zanclean (see Fig. 28.6).

Well data and interpretation of seismic-reflection records over the northern Florida shelf (Popenoe *et al.*, 1984) indicate that the Sanford High was not inundated until the latest Burdigalian and Langhian interval (see the J-1 well, Fig. 28.6). The latest Burdigalian beds, both erosionally and depositionally thin, updip (southward) from the J-1 well toward the Sanford High and pinch out entirely against Oligocene strata about 15 km due south of the well. Erosional remnants of these beds occur discontinuously farther south indicating that a more extensive distribution of these strata was once present. The Langhian and Serravallian sediments also pinch out southward over the Sanford High. However, thick remnants of Middle Miocene beds are preserved in sinkhole depressions near the crest of the Sanford High indicating that the high was inundated by shallow seas during the Middle Miocene and covered by a fairly thick layer (>20 m) of marine sediments. Most of these sediments were removed during the Late Miocene lowstand (Tortonian and Messinian), when the platform areas were exposed to subaerial erosion. T.M. Scott (pers. comm., 1986) classifies the sediments as Coosawhatchie Formation over the St Johns Platform, and Peace River Formation over much of the Brevard Platform.

Outer slope planktonic foraminifera in the Middle Miocene sediments at the COST GE-1 well (Poag & Hall, 1979) and the facies and distribution of Middle Miocene sediments in Florida (Riggs, 1979) suggest that Middle Miocene seas inundated all of southeastern Georgia and Florida during the Langhian and Early Serravallian. However, much of the accompanying deposition was removed by Late Miocene subaerial erosion. One major depocenter in the offshore (Fig. 28.16) was the eastern Waycross Basin, where the mid-shelf high was covered by 50–100 m of Early Serravallian-age shell hash and quartz sands (Rhodehammel, 1979) that were deposited in slope depths (Poag & Hall, 1979).

The Miocene of North Carolina

The Onslow Basin and other depressions along the continental shelf were filled in the Middle Miocene with prograded shelf deposits (Fig. 28.16), which produced a relatively smooth-topped shelf.

The Aurora Basin off North Carolina contains a thick wedge of northward-prograding Middle Miocene sediments on the north flank of the Cape Lookout High (Fig. 28.12). The obliquely downlapping character and strong seismic return of these beds indicates they are probably shelf sands that were swept across the Cape Lookout High by north-flowing high-energy Gulf Stream eddy currents, probably during the Late Burdigalian and Langhian. Late Serravallian-age sediments in the Aurora Basin were deposited more uniformly in lower energy conditions.

The Aurora Basin also contains southward-prograding Middle Miocene beds on its north flank (Fig. 28.12) indicating that the basin was at the merge-point of the northward-flowing Gulf Stream and southward-flowing shelf currents, similar to the Cape Hatteras area today. Sediments on the north flank are dominantly clastic marine muds (Brown, Miller & Swain, 1972) similar to the Chesapeake Group of Maryland and Virginia. Those of the south flank of the embayment are related to the more calcareous Pungo River Formation indicating a southward derivation.

The offshore and onshore data give a consistent picture of Late Burdigalian and Middle Miocene geography. Late Burdigalian seas progressively inundated the Gulf Trough and Waycross Basin into which the deltas of rivers were retreating with the rise. The Florida Platform was dissected by deep drainage valleys that were cut during Late Oligocene and Early Miocene subaerial exposure, and its surface was riddled with sinkholes, sinkhole lakes, and other karst features (Weaver & Beck, 1977; Riggs, 1979 and 1984). The river valleys (Jacksonville Basin; see Goodell & Yon, 1960; Riggs, 1979 and 1984) became large embayments that extended southward from the Waycross Basin. The divides between embayments were eventually eroded and flooded by marine seas in the Langhian, followed by a minor regression and reworking of sediment in the Serravallian.

The Late Miocene (Tortonian and Messinian)

The major sea-level regression that marks the Late Miocene began in the Late Serravallian and Early Tortonian. The forcing of the Gulf Stream offshore during this regression is apparent in the sedimentary record of the northern Blake Plateau. During the Langhian and Early Serravallian, only erosional cutting occurred on the northern plateau under the Gulf Stream that was overriding the Charleston Bump and pressed against the Florida–Hatteras Slope. During the Late Serravallian, deposition occurred downstream of the Charleston Bump signaling increased bottom friction and diversion of the Gulf Stream offshore by shoaling water.

Over most of the Georgia and Florida coastal plain, the Late Miocene is expressed as an erosional hiatus (Fig. 28.6). The AMCOR 6002 well on the mid-shelf off Georgia (Fig. 28.2) cored 10 m of olive silty clay topped by dark grayish-brown phosphatic sand containing Late Serravallian and Late Miocene diatoms (Abbott, in Hathaway *et al.*, 1979). This lens of Late Miocene sediments occurs only on the west flank of the mid-shelf high. Its top is a conspicuous unconformity. C.W. Poag's (pers. comm., 1985) identification of N.17 foraminifera in the core samples indicates that at least part of these strata are upper Tortonian in age and equivalent to the Eastover Formation of North Carolina, and part of the Peace River and Bone Valley Formation of central Florida (Scott, Chapter 21, this volume).

The extent of the Late Miocene (Early Tortonian) regression is probably best expressed on the Miami and Portales Terraces, which are drowned carbonate platforms off Miami Florida, and the Florida Keys. Rock samples dredged from the upper surface of the Miami Terrace at a depth of 200–375 m are early Middle Miocene in age ($\sim$15 my BP) (Mullins & Neumann, 1979). The surface of the terrace is riddled with sinkholes and karst features (Mullins & Neumann, 1979; Freeman-Lynde, Popenoe & Meyer, 1982), suggesting that it has been subaerially exposed. Analyses of phosphatized limestones recovered from the Portales Terrace at greater than 300 m depth indicate that during the Late Miocene the upper part of the terrace underwent freshwater diagenesis (Burnett & Gomberg, 1977). Mullins & Neumann (1979) calculated that, after considering regional tectonic subsidence, a eustatic lowering of sea level of 100 m or less could have subaerially exposed the terrace top.

A transgression to about 50 m above present sea level in the Late Tortonian resulted in deposition of the Eastover Formation, which is centered chiefly in Virginia (Ward & Blackwelder, 1980). In North Carolina, within the Aurora Basin a seismic stratigraphic sequence is present that is possibly equivalent to the Eastover Formation (Popenoe, 1985). This sequence includes both sigmoidally-thinned beds on the northern Blake Plateau and the conformably-draped beds within the center of the Aurora Basin that were included by Miller (1971) and Brown *et al.*, (1972) as the upper part of the Pungo River Formation. Within the Aurora Basin the beds are chiefly clays; their depositional character suggests that in the lower seas of the Late Miocene, the basin was isolated from both southward-flowing shelf- and northward-flowing Gulf Stream circulation (Fig. 28.18). Later deposition (Pliocene) is characterized by increasingly higher-energy deposits as sea level rose, and finally oblique progradations occurred as southward-flowing shelf currents flowed diagonally across the divide that formerly isolated the embayment. Blackwelder (1981a) states that the top of the Eastover Formation is a pronounced unconformity in which [subaerial] erosion took place prior to deposition of the lower Pliocene Yorktown Formation. The Aurora Basin was filled in Early Pliocene (Zanclean) time by shallow-water-derived sands that were swept into the basin by southward-flowing shelf currents (see the Pliocene sequence of Fig. 28.12) (Popenoe, 1985).

Late Tortonian-age (N.17) sediments have also been recognized in the Coostawhatchie Formation of the Alum Bluff Group of panhandle Florida (Huddlestun, 1984) and in the lower part of the Tamiami Formation in south Florida (Peck & Wise, 1982), indicating that the Late Miocene transgression was widespread. This transgression is not shown on the Vail & Hardenbol (1979) sea-level curve (Fig. 28.6).

Miocene paleogeographic summary

Based on the above discussions and on the geologic facies of the offshore, three reconstructions of time slices of Miocene paleogeography have been created. These time slices, derived from seismic-reflection, lithofacies, and biostratigraphic data, indicate what conditions may have been during the Early Burdigalian, Middle–Late Burdigalian, and Langhian. The Late Miocene paleogeography of the coastal plain was probably similar to present geography. Essentially, all of the relict topography of the margin, such as the Gulf Trough and Waycross Basin created in the Eocene and Oligocene, was filled by Late Miocene time. Only the outer end of the Aurora Basin survived through the Pliocene. Sea-level transgressions and regressions of the Late Miocene exposed or inundated a continental shelf or coastal plain with little topography.

The Early Miocene reconstruction (Fig. 28.19) represents early Burdigalian (N.6) time. The lack of N.5 foraminifera in coastal plain sediments suggests that the Waycross Basin may have been an inland sea. By N.6 time, however, marine circulation to the Atlantic was established across the mid-shelf high, and brackish-water planktonic foraminifera of Atlantic affinities occur within sediments in the Waycross Basin of eastern Georgia (Huddlestun, pers. comm., 1985). During N.6 time, Florida became a large island that was isolated from the continent by the seaway of the Gulf Trough and the Waycross Basin. The island was etched by deep drainage basins, sinkholes, sinkhole lakes, and other karst features developed by limestone dissolution during subaerial exposure. These basins subsequently accumulated thick deposits of Late Burdigalian phosphorites (Riggs, 1979).

In the offshore, the Gulf Stream flowed through the Straits of Florida and along the Late Eocene (Ocala) carbonate bank edge located just east of Cape Canaveral. Over the Blake Plateau, its flow was south of the Charleston Bump and around the higher, probably reef-fringed northern Blake Plateau. Upwelling was probably occurring across the present phosphate pavement area on the south flank of the Charleston Bump, with related phosphorite deposition, north of where the currents were diverted by the shoal bathymetry of the northern Blake Plateau. The Cape Lookout High was subaerially exposed and the Aurora Basin was a small bay into the coast.

The Middle–Late Burdigalian reconstruction (Fig. 28.20) represents foraminiferal zone N.7 time. Continued transgression caused the Waycross Basin to develop into an open seaway marine system. The seaway was open to marine flushing through the Waycross Basin and the Gulf Trough. Over Florida, long, deep open embayments extended up flooded valleys between the Ocala Upland and St Johns and Brevard Platforms. As water on the West Florida Shelf deepened, Gulf Stream currents were probably able to circulate across southern Florida from the Gulf of Mexico to the Atlantic within the deeper Okeechobee Basin.

In Georgia and southern South Carolina, the deltas of the ancient Savannah, Ogeechee, Oconee, and Ocmulgee Rivers that had retreated during rising sea level now dropped their clastic loads into the Gulf Trough and northwestern margin of the Waycross Basin, filling the proximal flank of this large seaway with estuarine arkosic sands and gravel (Altamaha Formation). The east flank of the Waycross Basin received only finer clastics. Muddy phosphatic sands, clays, and interbedded limestones (St Marks Formation) were deposited behind the peninsula that overlaid the mid-shelf high off South Carolina and northeastern Georgia. Based on the extent of the mid-shelf high, this peninsula was larger than the present Delmarva Peninsula of Delaware, Maryland, and Virginia, and was created in part by inherited topographic relief from Oligocene deposition and in part by longshore sediment drift.

In the offshore, the Gulf Stream migrated westward with the transgression and climbed the south flank of the Charleston Bump, where it eroded and winnowed previously deposited

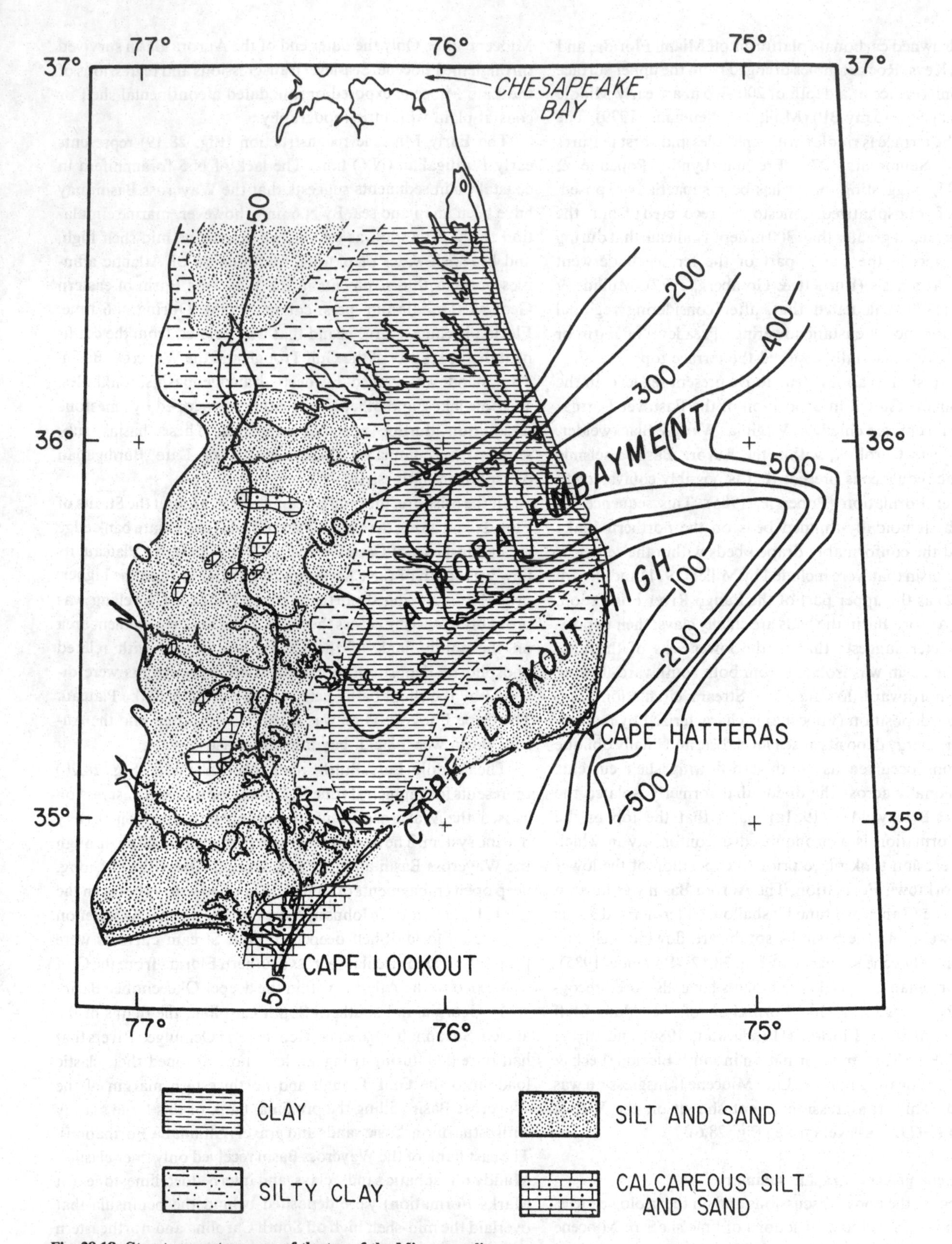

Fig. 28.18. Structure contour map of the top of the Miocene sediments in the Aurora Basin and facies of the Pungo River Formation as mapped by Miller (1971). The clay facies in the center of the basin is probably chiefly of Late Miocene age, deposited when the basin was isolated from shelf circulation. Calcareous silt and sand, and silt and sand facies represent older, erosionally beveled, Early and Middle Miocene strata (after Popenoe, 1985).

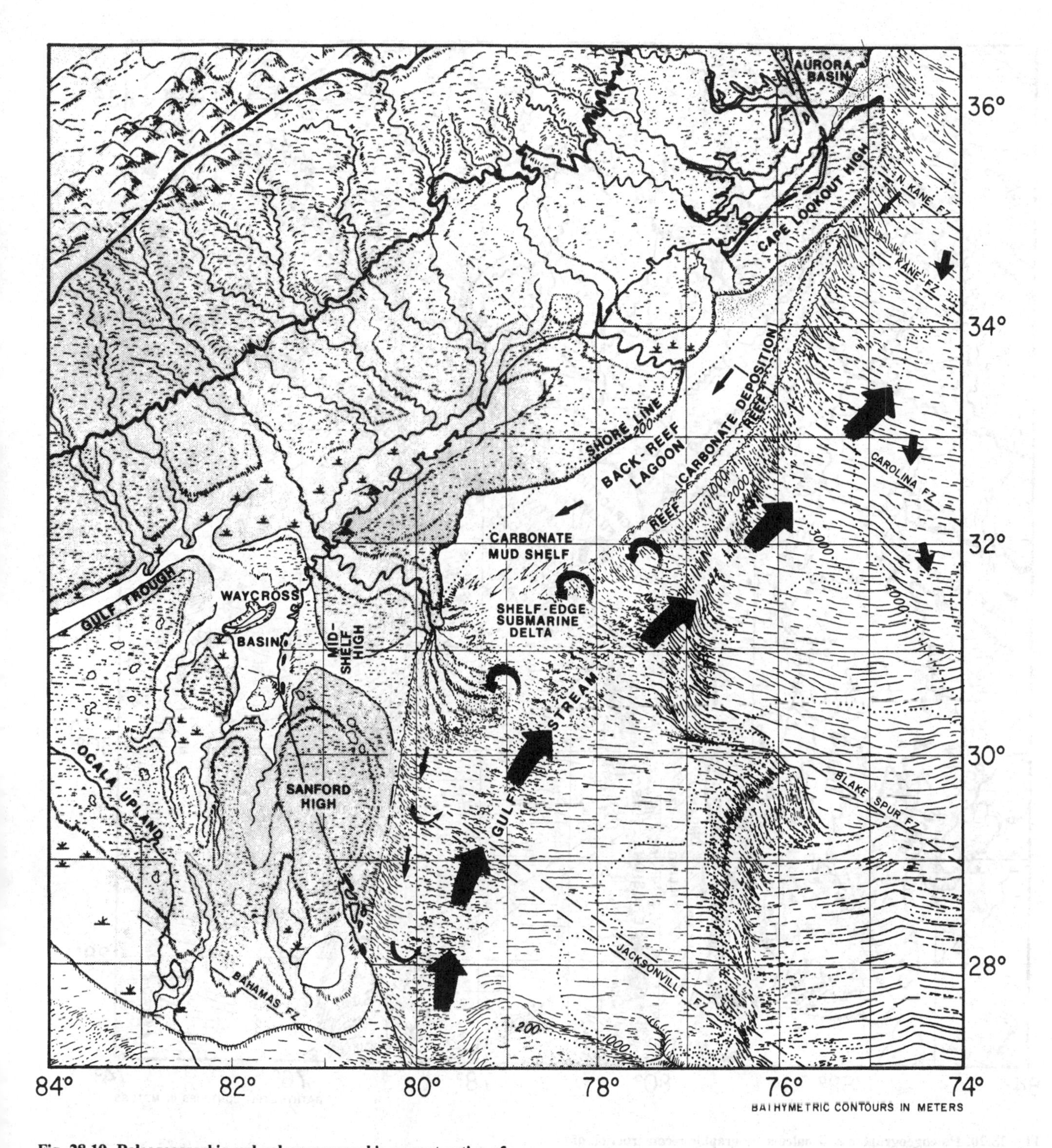

Fig. 28.19. Paleogeographic and paleoceanographic reconstruction of the Early Burdigalian (N.6 time).

lower Burdigalian carbonate muds and left only a few sediment outliers and lag phosphate gravels. These muds were deposited in the lee of the Charleston Bump, as a large northward-prograding mounded drift. Upwelling induced by bottom topography was probably occurring off South Carolina and southern North Carolina. Muds swept off the northern plateau and over the continental slope, became entrained in the southward-flowing upper abyssal current, the Western Boundary Undercurrent, and deposited, as current velocities dropped, along the crest of the Blake Outer Ridge.

The Cape Lookout High in the early part of the transgression was then either a long peninsula or a shoal, isolating the Aurora Basin from Gulf Stream currents. Later in the Burdigalian, Gulf Stream eddies swept across the shoal, prograding sediments into the southern part of the embayment.

The Middle Miocene reconstruction (Fig. 28.21) represents the maximum sea-level advance of Late Burdigalian and Langhian time (foraminifera zones N.8–N.9). The sea had inundated most of Georgia and Florida, and created open marine conditions on the flanks of the Ocala High and over the Sanford

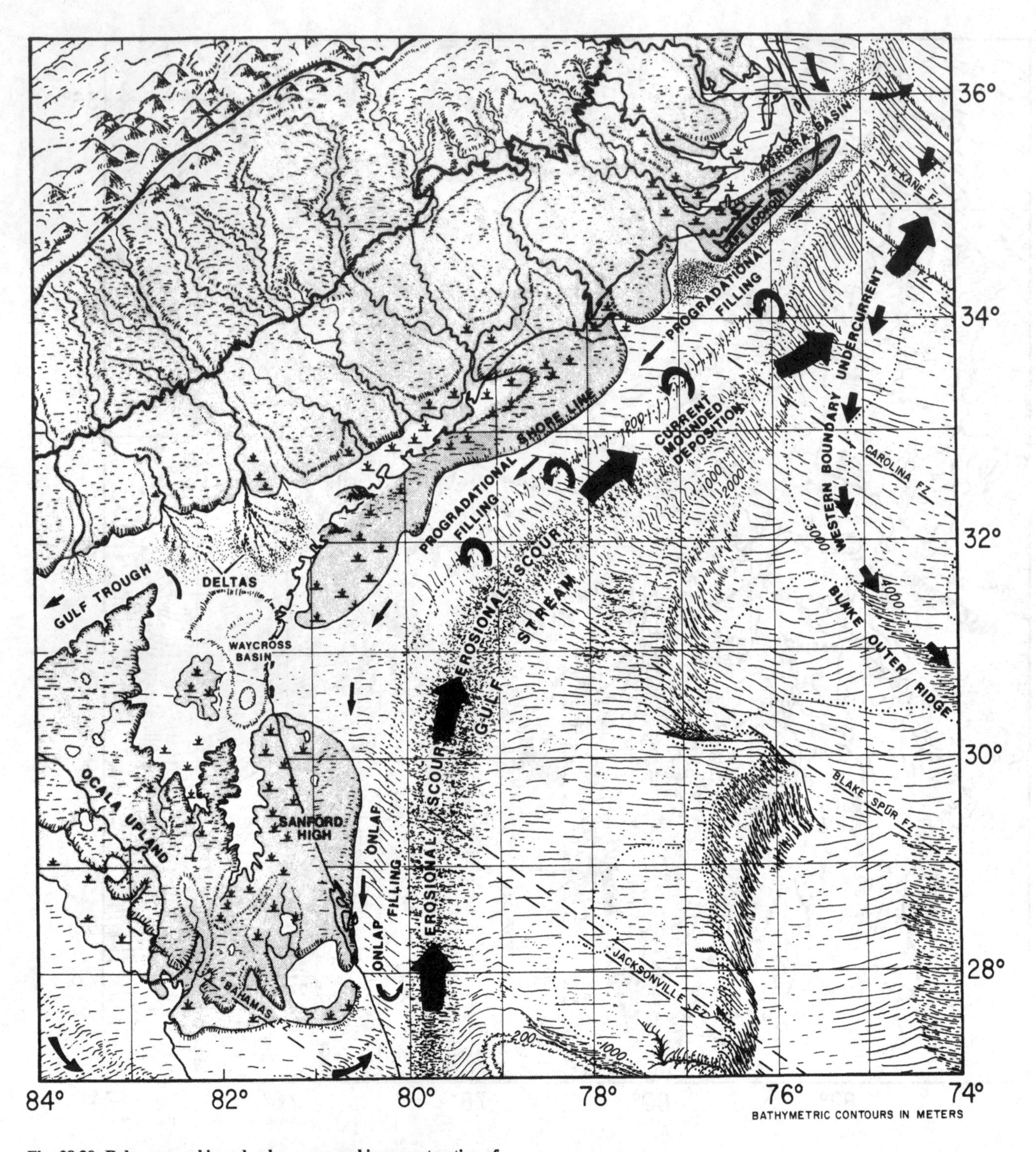

Fig. 28.20. Paleogeographic and paleoceanographic reconstruction of the Middle Burdigalian (N.7 time).

High. The highest parts of the Ocala High may not have been inundated; however, the inundation of the Sanford High is indicated by remnants of Middle Miocene marine strata preserved in offshore limestone collapse features (Popenoe *et al.*, 1984). The thickness and morphology of these beds indicates that the platform was once extensively covered by a thick blanket of Middle Miocene strata that was eroded away during the Late Miocene regression.

Strong Gulf Stream currents probably crossed southern Florida during the Middle Miocene through the Okeechobee Basin, which may have interacted with bathymetric features causing upwelling, as proposed by Riggs (1984). The strongest evidence for these currents is the major outbuilding of the Florida – Hatteras Shelf off Cape Canaveral – probably built from sediments swept off the platform. Both drillholes along the shoreline and seismic-reflection records show this outbuilding, however a control well is needed offshore of Cape Canaveral to determine if the major post-Eocene outbuilding is of Middle Miocene age.

In northern Florida and southeastern Georgia, the Waycross

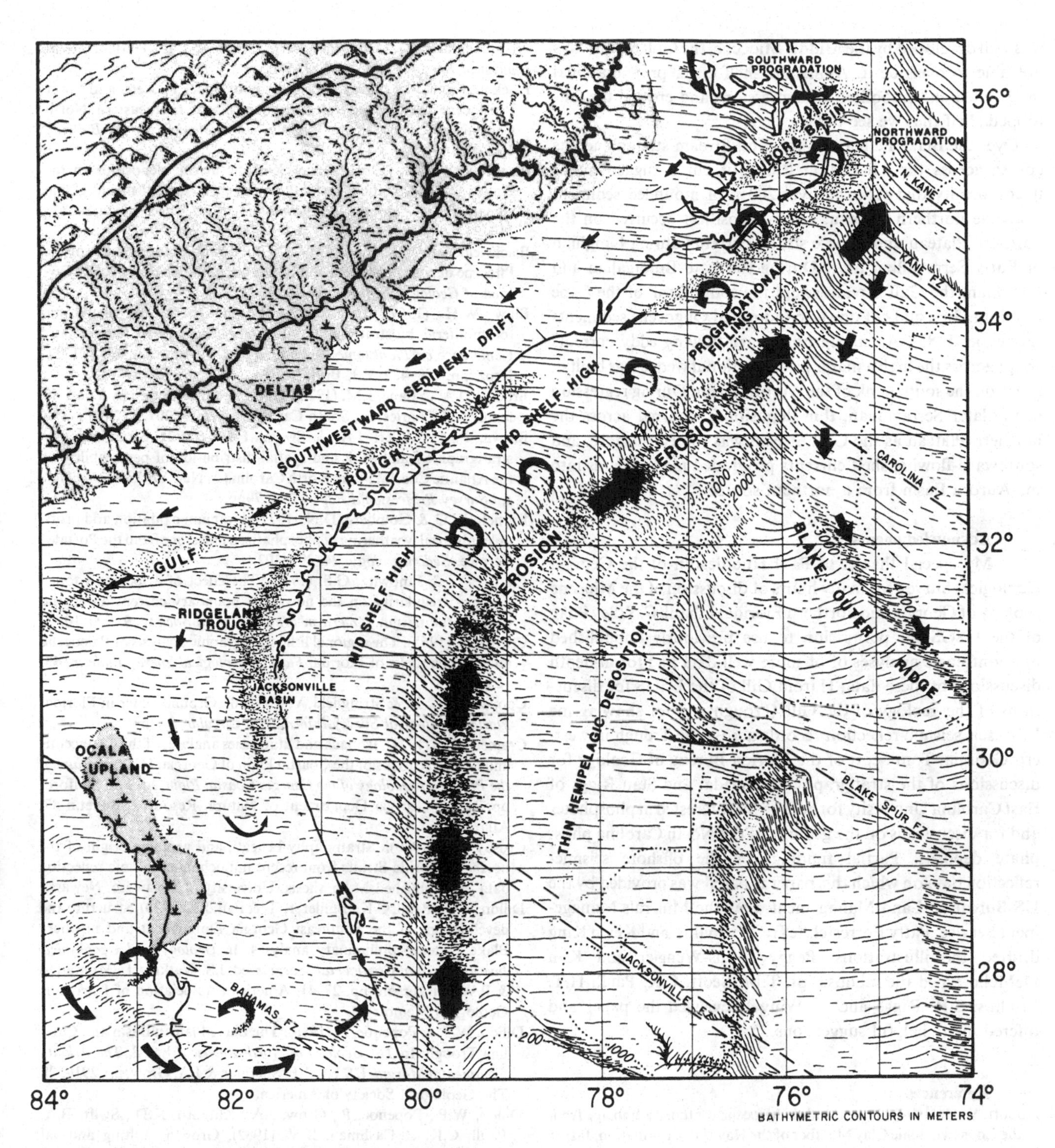

Fig. 28.21. Paleogeography and paleoceanography of the Langhian (N.8–N.9 time).

Basin and the Gulf Trough had been almost completely filled by sediment in the Early Miocene. Only minor depressions remained on the eastern flank of the Waycross Basin (the Ridgeland Trough and Jacksonville Basin) and over the former Gulf Trough. These became the basins in which the upper Langhian and Serravallian deposits – the Coosawhatchie Formation – were preserved. Deposits in most other areas were stripped away during the subaerial exposure of the Early Tortonian. During the early phases of the Serravallian regression, and during each minor sea-level fluctuation, the primary phosphate deposits of the Burdigalian and Langhian were reworked and enriched in the nearshore environment and become preserved as the Peace River Formation (Scott, 1985a). Reworking by marine seas probably occurred again in the sea-level advance of the Late Tortonian and Zanclean.

In the offshore, the Middle Miocene Gulf Stream was pressed tightly against the Florida–Hatteras Slope where it eroded an unconformity and probably induced topographic upwelling. Ekman transport would have carried these upwelled waters high onto the Florida platform. Because the Florida–Hatteras Shelf was well developed by the Middle Miocene, the Gulf Stream was not able to override it. Along the shelf edge, progradational sands filled and smoothed all depressions and created the flat-topped shelf that we see today.

Over the Blake Plateau, as the Gulf Stream straightened its course across the Charleston Bump, it cut a rugged major unconformity that stripped over 200 m of mounded sediment from the northern Plateau. Little deposition occurred on the northern Plateau under these strong currents during Langhian or Early Serravallian time. During the Late Burdigalian and Langhian, the Gulf Stream swept the south flank of the Cape Lookout High and cut down to indurated Oligocene sediments (Popenoe, 1985). As strong northward-flowing eddy currents swept across the high a major progradational wedge was built in its lee on the south flank of the Aurora Basin. During regression in the later Serravallian, thin deposition occurred across the northern Plateau as the Gulf Stream was forced offshore, and southward-flowing shelf currents prograded marine muds into the Aurora Basin from a northern source.

Acknowledgments

My special thanks to Paul Huddlestun of the Georgia Geological Survey for many hours of discussion of the Miocene geology of Georgia and northern Florida. I thank C. Wiley Poag of the USGS for permission to use previously unpublished palaeontologic analyses of offshore well data and for in-depth discussions of these data; Harold Gill of the USGS for discussions of the geology of the Gulf Trough and for showing me Vibroseis seismic-reflection records across the Trough in western Georgia; Tom Scott of the Florida Bureau of Geology for discussions of the stratigraphy of Florida; and Stan Riggs of East Carolina University for getting me interested in phosphates and for discussions on the geology of the North Carolina phosphate deposits. Partial funding for the offshore seismic-reflection data on which this paper is based was provided by the US Bureau of Land Management, now the Minerals Management Service. Patty Forrestel, Jeffrey Zwinakis, and Kevin King drafted the illustrations. Peggy Mons-Wengler and Kim DeMello typed the manuscript. T.M. Scott, C.K. Paull, H.T. Mullins, S.R. Riggs and E. Winget reviewed the paper and offered many helpful suggestions.

References

Abbott, W.H. (1974). Lower middle Miocene diatom assemblage from the Coosawhatchie Clay Member of the Hawthorn Formation, Jasper County, South Carolina. *South Carolina Division of Geology Geologic Notes*, **18**, 46–53.

Abbott, W.H. & Huddlestun, P.F. (1980). The Miocene of South Carolina. In *Excursions in Southeastern Geology*, vol. 1, ed. R.W. Frey, p. 110. Geological Society of America, Boulder.

Applin, P.L. & Applin, E.R. (1967). The Gulf Series in the subsurface of northern Florida and southern Georgia. *US Geological Survey Professional Paper*, **524-G**, 34pp.

Bailey, N.G. & Grow, J.A. (1980a). Single-channel seismic-reflection profiles collected over the Atlantic US continental shelf, slope, and rise east of Cape Hatteras. *US Geological Survey Open File Report*, **80–510**, 4pp.

Bailey, N.G. & Grow, J.A. (1980b). Single-channel seismic-reflection profiles from the Blake Plateau and Blake Outer Ridge. *US Geological Survey Open File Report*, **80–652**, 4pp.

Bane, J.M. & Brooks, D.A. (1979). Gulf Stream meanders along the continental margin from the Florida Straits to Cape Hatteras. *Geophysical Research Letters*, **6**, 280–2.

Blackwelder, B.W. (1981a). Stratigraphy of upper Pliocene and lower Pleistocene marine and estuarine deposits of northeastern North Carolina and southeastern Virginia. *US Geological Survey Bulletin*, **1502-B**, 16pp.

Blackwelder, B.W. (1981b). Late Cenozoic marine deposition in the United States Atlantic Coastal Plain related to tectonism and global climate. *Paleogeography, Paleoclimatology, and Paleoecology*, **34**, 87–113.

Blackwelder, B.W. & Ward, L.W. (1979). Stratigraphic revision of the Pliocene deposits of North and South Carolina. *South Carolina Division of Geology Geologic Notes*, **23**, 33.

Blow, W.H. (1969). Late middle Eocene to Recent planktonic foraminiferal biostratigraphy. In *Proceedings, First International Conference on Planktonic Microfossils*, ed. R. Bronnimann & H.H. Renz, pp. 199–422. E.J. Brill, Leiden.

Brooks, D.A. & Bane, J.M. (1978). Gulf Stream deflection by a bottom feature off Charleston, South Carolina. *Science*, **201**, 1225–6.

Brown, P.M., Miller, J.A. & Swain, F.M. (1972). Structural and stratigraphic framework and spatial distribution of permeability of the Atlantic Coastal Plain, North Carolina to New York. *US Geological Survey Professional Paper*, **796**, 79pp.

Burnett, W.C. & Gomberg, D.N. (1977). Uranium oxidation and probable subaerial weathering of phosphatized limestone from the Portales Terrace. *Sedimentology*, **24**, 291–302.

Chao, S-Y. & Janowitz, G.S. (1979). The effect of a localized topographic irregularity on the flow of a boundary current along the continental margin. *American Meteorological Society*, **9**, 900–10.

Chen, C.S. (1965). The regional lithostratigraphic analysis of Paleocene and Eocene rocks of Florida. *Florida Geological Survey Bulletin*, **45**, 105pp.

Cooper, H.H., Jr. & Stringfield, V.T. (1950). Ground water in Florida. *Florida Geological Survey Information Circular*, **3**, 7pp.

Cramer, H.R. (1974). Isopach and lithofacies analyses of the Cretaceous and Cenozoic rocks of the Coastal Plain of Georgia. In *Symposium of the Petroleum Geology of the Georgia Coastal Plain*, ed. L.P. Stafford, pp. 21–44. Georgia Department of Natural Resources Bulletin 87, Atlanta.

Crosby, J.T. (1980). 'Stratigraphy beneath and geologic origin of the northern Florida Straits from recent multichannel seismic reflection data'. Unpublished M.Sc. thesis, University of Delaware, Newark.

Dillon, W.P., Paull, C.K., Buffler, R.T. & Fail, J.P. (1979). Structure and development of the Southeast Georgia Embayment and northern Blake Plateau: preliminary analyses. In *Geology and Geophysical Investigations of Continental Margins*, ed. J.S. Watkins, L. Montadert & P.W. Dickerson, pp. 27–41. American Association of Petroleum Geologists Memoir 29.

Dillon, W.P. & Popenoe, P. (1988). The Blake Plateau Basin and Carolina Trough. In *The Geology of North America, V.1–2, The Atlantic Continental Margin, US*, ed. R.E. Sheridan & J.A. Grow, pp. 291–328. The Geological Society of America.

Dillon, W.P., Popenoe, P., Grow, J.A., Klitgord, K.D., Swift, B.A., Paull, C.K. & Cashman, K.V. (1982). Growth faulting and salt diapirism, their relationships and control in the Carolina Trough, eastern North America. In *Studies in Continental Margin Geology*, ed. J.S. Watkins & C.L. Drake, pp. 21–46. American Association of Petroleum Geologists Memoir 34.

Edsall, D.W. (1978). Southeast Georgia Embayment high-resolution seismic-reflection survey. *US Geological Survey Open File Report*, **78–800**, 90pp.

Emery, K.O. & Uchupi, E. (1984). *The Geology of the Atlantic Ocean*. Springer Verlag, New York.

Falvey, D.A. (1974). The development of continental margins in plate tectonic theory. *Australian Petroleum Exploration Association Journal*, **14**, 95–106.

Freas, D.H. & Riggs, S.R. (1968). Environments of phosphorite deposition in the Central Florida Phosphate District. In *Proceedings, Fourth Forum on Geology of Industrial Minerals*, ed. L.F. Brown, Jr., pp. 117–28. University of Texas, Austin.

Freeman-Lynde, R.P., Popenoe, P. & Meyer, F.W. (1982). Seismic stratigraphy of the western Straits of Florida (abstract) *Geological Society of America, Abstracts with Programs*, **14(1&2)**, 18.

Gelbaum, C. & Howell, J. (1982). The geohydrology of the Gulf Trough. In *Proceedings, Second Symposium on the Geology of the Southeastern Coastal Plain*, ed. D.D. Arden, B.F. Beck & E. Morrow, pp. 140–53. Georgia Geological Survey Information Circular 53, Atlanta.

Gibson, T.G., 1982. Depositional framework and paleoenvironments of Miocene strata from North Carolina to Maryland. In *Miocene of the Southeastern United States*, ed. T.M. Scott & S.B. Upchurch, pp. 1–34. Florida Bureau of Geology Special Publication 25.

Goodell, H.G. & Yon, F.W., Jr. (1960). The regional lithostratigraphy of the post Eocene rocks of Florida. In *Southeast Geological Society Guidebook*, Field Trip 9, pp. 75–113. Southeast Geological Society, Durham, NC.

Hall, R.B. (1983). General geology and stratigraphy of the Southern Extension of the Central Florida Phosphate District. In *The Central Florida Phosphate District*, pp. 1–25. Geological Society of America, Southeastern Section, Field Trip Guidebook, Tampa.

Hathaway, J.C., Poag, C.W., Valentine, P.C., Miller, R.E., Schultz, D.M., Manheim, F.T., Kohout, F.A., Bothner, M.H. & Sangrey, D.A. (1979). US Geological Survey core drilling on the Atlantic shelf. *Science*, **206**, 515–27.

Hazel, J.E., Bybell, L.M., Christopher, R.A., Frederiksen, N.O., May, F.E., McLean, D.M., Poore, R.Z., Smith, C.C., Sohl, N.F., Valentine, P.C. & Witmer, R.J. (1977). Biostratigraphy of the deep corehole (Clubhouse Crossroads Corehole 1) near Charleston, South Carolina. *US Geological Survey Professional Paper*, **1028**, 71–89.

Herrick, S.M. (1961). Well logs of the Coastal Plain of Georgia. *Georgia Geological Survey Bulletin*, **70**, 462pp.

Herrick, S.M. & Vorhis, R.S. (1963). Subsurface geology of the Georgia Coastal Plain. *Georgia Geological Survey Information Circular*, **25**, 80pp.

Huddlestun, P.F. (1982). The stratigraphic subdivision of the Hawthorn Group in Georgia. In *Miocene of the Southeastern United States*, ed. T.M. Scott & S.B. Upchurch, pp. 183–4. Florida Bureau of Geology Special Publication 25.

Huddlestun, P.F. (1984). 'The Neogene Stratigraphy of the central Florida panhandle'. Unpublished Ph.D. thesis, Florida State University, Tallahassee.

Hull, J.P.D., Jr. (1962). Cretaceous Suwannee Strait, Georgia and Florida. *American Association of Petroleum Geologists Bulletin*, **46**, 118–22.

Hunter, M.E. & Huddlestun, P.F. (1982). The biostratigraphy of the Torreya Formation of Florida. In *Miocene of the Southeastern United States*, ed. T.M. Scott & S.B. Upchurch, pp. 211–33. Florida Bureau of Geology Special Publication 25.

Hutchinson, D.R., Grow, J.A., Klitgord, K.D. & Swift, B.A. (1982). Deep structure and evolution of the Carolina Trough. In *Studies in Continental Margin Geology*, ed. J.S. Watkins & C.L. Drake, p. 1292. American Association of Petroleum Geologists Memoir 34.

Idris, F.M. (1983). 'Cenozoic stratigraphy and structure of the South Carolina lower coastal plain and continental shelf'. Unpublished Ph.D. thesis, University of Georgia, Athens.

JOIDES, 1965. Ocean drilling on the continental margin. *Science*, **150(3697)**, 709–16.

Klitgord, K.D. & Behrendt, J.C. (1979). Basin structure of the US Atlantic Margin. In *Geological and Geophysical Investigations of Continental Margins*, ed. J.S. Watkins, L. Montadert & P.W. Dickerson, pp. 85–112. American Association of Petroleum Geologists Memoir 29.

Klitgord, K.D., Popenoe, P. & Schouten, H. (1984). Florida: Jurassic transform plate boundary. *Journal of Geophysical Research*, **89**, 7753–72.

Legeckis, R.V. (1979). Satellite observations of the influence of bottom topography on the seaward deflection of the Gulf Stream off Charleston, South Carolina. *Journal of Physical Oceanography*, **9**, 483–97.

Malde, H.E. (1959). Geology of the Charleston phosphate area, South Carolina. *US Geological Survey Bulletin*, **1079**, 105pp.

Manheim, F.T., Popenoe, P., Siapno, W. & Lane, C. (1982). Manganese–phosphorite deposits of the Blake Plateau. In *Marine Mineral Deposits, Proceedings of the Clausthaler Workshop*, September, 1982, ed. P. Halback & P. Winter, pp. 9–44. Verlag Gluckauf, Essen W. Germany.

Manheim. F.T., Pratt, R.M. & McFarlin, P.F. (1980). Composition and origin of phosphorite deposits of the Blake Plateau. *Society of Economic Paleontologists and Mineralogists Special Publication* 29. p. 117–137.

McCollum, M.J. & Herrick, S.M. (1964). Offshore extension of the upper Eocene to Recent stratigraphic sequence in southeastern Georgia. *US Geological Survey Professional Paper*, **501-C**, C61–3.

Meisburger, E.P. & Field, M.E. (1975). Geomorphology, shallow structure, and sediments of the Florida inner continental shelf, Cape Canaveral to Georgia. US Army Corps of Engineers, Coastal Engineering Research Center Technical Memorandum 54, Jacksonville, 119pp.

Miller, J.A. (1971). 'Stratigraphic and structural setting of the middle Miocene Pungo River Formation of North Carolina'. Unpublished Ph.D. thesis, University of North Carolina, Chapel Hill.

Miller, J.A. (1982). Structural and sedimentary setting of phosphorite deposits in North Carolina and in northern Florida. In *Miocene of the Southeastern United States*, ed. T.M. Scott & S.B. Upchurch, pp. 162–82. Florida Bureau of Geology Special Publication 25.

Miller, J.A. (1986). Hydrogeologic framework of the Floridan aquifer system in Florida and parts of Georgia, Alabama, and South Carolina. *US Geological Survey Professional Paper*, **1403-B**, 91pp.

Mountain, G.S. & Tucholke, B.E. (1985). Mesozoic and Cenozoic geology of the US Atlantic continental slope and rise. In *Geologic Evolution of the United States Atlantic Margin*, ed. C.W. Poag, pp. 293–341. Van Nostrand Reinhold, New York.

Mullins, H.T. & Neumann, C.A. (1979). Geology of the Miami Terrace and its paleogeographic implications. *Marine Geology*, **30**, 205–32.

Nelson, K.D., Arnow, J.A., McBride, J.H., Willemin, J.H., Huang, J., Zheng, J., Oliver, J.E., Brown, L.D. & Kaufman, S. (1985). New COCORP profiling in the southeastern United States Part 1. Late Paleozoic suture and Mesozoic rift basin. *Geology*, **13**, 714–17.

Patterson, S.H. & Herrick, S.M. (1971). Chattahoochee Anticline, Appalachicola Embayment, Gulf Trough, and related structural features, southwestern Georgia, fact or fiction. *Geological Survey of Georgia Information Circular*, **41**, 16pp.

Paull, C.K. & Dillon, W.P. (1980). Structure, stratigraphy, and geologic history of the Florida–Hatteras Shelf and inner Blake Plateau. *American Association of Petroleum Geologists Bulletin*, **64**, 339–58.

Paull, C.K., Popenoe, P., Dillon, W.P. & McCarthy, S.M. (1980). Geologic subcrop map of the Florida–Hatteras Shelf, Slope, and inner Blake Plateau. *US Geological Survey Miscellaneous Field Investigations Map*, **MF-1171**, scale 1:500 000.

Peck, D.M. & Wise, S.W. (1982). Upper Miocene–Pliocene planktic and benthic Foraminifers from Lee County, Florida. In *Miocene of the Southeastern United States*, ed. T.M. Scott & S.B. Upchurch, p. 300. Florida Bureau of Geology Special Publication 25.

Pinet, P.R. & Popenoe, P. (1982). Blake Plateau; control of Miocene sedimentation patterns by large-scale shifts of the Gulf Stream axis. *Geology*, **10**, 57 9.

Pinet, P.R. & Popenoe, P. (1985a). Shallow seismic stratigraphy and post-Albian geologic history of the northern and central Blake Plateau. *Geological Society of America Bulletin*, **96**, 627–38.

Pinet, P.R. & Popenoe, P. (1985b). A scenario of Mesozoic–Cenozoic ocean circulation over the Blake Plateau and its environs. *Geological Society of America Bulletin*, **96**, 618–26.

Pinet, P.R., Popenoe, P. & Nelligan, D.F. (1981). Reconstruction of Cenozoic flow patterns over the Blake Plateau. *Geology*, **9**, 266–70.

Poag, C.W. (1978). Stratigraphy of the Atlantic continental shelf and slope of the United States. *Earth and Planetary Sciences*, **6**, 251–80.

Poag, C.W. (1985). Benthic foraminifera as indicators of potential petroleum sources. In *Habitat of Oil and Gas in the Gulf Coast*, ed. B.F. Perkins & G.B. Martin, pp. 275–83. Proceedings, Gulf Coast Section, SEPM, Fourth Annual Conference, Austin.

Poag, C.W. & Hall, R.E. (1979). Foraminiferal biostratigraphy, paleoecology, and sediment accumulation rates. In *Geological Studies of the COST GE-1 Well, United States Outer Continental Shelf Area*, ed. P.A. Scholle, pp. 49–63. US Geological Survey Circular 800.

Popenoe, P. (1980). Single-channel seismic-reflection profiles collected on the northern Blake Plateau, 29 September to 19 October, 1978. *US Geological Survey, Open-File Report*, **80–1265**, 4pp.

Popenoe, P. (1983a). High-resolution seismic reflection profiles collected August 4–28, 1979, between Cape Hatteras and Cape Fear, North Carolina, and off Georgia and northern Florida. *US Geological Survey Open-File Report*, **83–512**, 4pp.

Popenoe, P. (1983b). Description of a high-resolution seismic reflection profile collected between Chesapeake Bay, Virginia and Cape Romain, South Carolina, Sept. 26-Oct. 2, 1981. *US Geological Survey Open-File Report*, **83–514**, 4pp.

Popenoe, P. (1984). Summary geologic report for the South Atlantic Outer Continental Shelf Planning Area (a supplement to *US Geological Survey Report*, **83–186**). *US Geological Survey Open File Report*, **84–476**, 17pp.

Popenoe, P. (1985). Cenozoic depositional and structural history of the North Carolina margin from seismic stratigraphic analyses. In *Stratigraphy and Depositional History of the US Atlantic Margin*, ed. C.W. Poag, pp. 125–87. Van Nostrand Reinhold, Stroudsburg.

Popenoe, P., Henry, V.J., & Idris, F.M., 1987. Gulf Trough – the Atlantic connection. *Geology*, **15**, 327–32.

Popenoe, P., Kohout, F.A. & Manheim, F.T. (1984). Seismic-reflection studies of sinkholes and limestone dissolution features on the northeastern Florida shelf. In *Sinkholes, their Geology, Engineering, and Environmental Impact*, ed. B.F. Beck, pp. 43–57. A.A. Balkema, Boston.

Popenoe, P. & Meyer, F.W. (1983). Description of high-resolution seismic reflection data collected from the shelf, slope, and upper rise between Cape Hatteras, North Carolina, and Norfolk, Virginia, and Vero Beach to Miami, Florida. *US Geological Survey Open-File Report*, **83–515**, 4pp.

Popenoe, P. & Ward, L.W. (1983). Description of high-resolution seismic reflection data collected in Albemarle and Croatan Sounds, North Carolina. *US Geological Survey Open-File Report*, **83–513**, 3pp.

Popenoe, P. & Zietz, I. (1977). The nature of the geophysical basement beneath the Coastal Plain of South Carolina and northeastern Georgia. *US Geological Survey Professional Paper*, **1028**, 119–37.

Pratt, R.M. (1966). The Gulf Stream as a graded river. *Limnology and Oceanography*, **11**, 60–7.

Rhodehammel, E.C. (1979). Lithologic descriptions. In *Geological Studies of the COST GE-1 Well, United States Outer Continental Shelf Area*, ed. P.A. Scholle, pp. 24–36. Geological Survey Circular **800**.

Riggs, S.R. (1979). Phosphorite sedimentation in Florida – A model phosphogenic system. *Economic Geology*, **74**, 285–314.

Riggs, S.R. (1984). Paleoceanographic model of Neogene phosphorite deposition, US Atlantic continental margin. *Science*, **223**, 123–31.

Riggs, S.R., Lewis, D.W., Scarborough, A.K. & Snyder, S.W. (1982). Cyclic deposition and Neogene phosphorites in the Aurora area, North Carolina, and their possible relationship to global sea-level fluctuations. *Southeastern Geology*, **23(4)**, 189–204.

Riggs, S.R., Snyder, S.W.P., Hine, A.C., Snyder, S.W., Ellington, M.D. & Malette, P.M. (1985). Geologic framework of phosphorite resources in Onslow Bay, North Carolina continental shelf. *Economic Geology*, **80**, 716–38.

Scholle, P.A., ed. (1979). Geological studies of the COST GE-1 well, United States Outer Continental Shelf area: *US Geological Survey Circular*, **800**, 114pp.

Scott, T.M. (1985a). The geology of the Florida peninsula and its relationship to economic phosphate deposits. In *International Geological Correlation Program Project 156 – Phosphorites. Eighth International Field Workshop and Symposium Guidebook*, ed. S. Snyder, pp. 97–113, Greenville, North Carolina.

Scott, T.M. (1985b). The lithostratigraphy of the central Florida Phosphate District and its southern extension. In *Florida Land Pebble Phosphate District. Geological Society of America Annual Meeting Guidebook*, ed. J.B. Cathcart & T.M. Scott, pp. 28–37. Geological Society of America, Tallahassee.

Sever, C.W., Cathcart, J.B. & Patterson, S.H. (1967). *South Georgia Minerals Program–Project Report 7, Phosphate Deposits of South-Central Georgia and North-Central Peninsular Florida*. Georgia Department of Mines and Mining Geology, Atlanta.

Sheridan, R.E., Crosby, J.T., Bryan, G.M. & Stoffa, P.L. (1981). Stratigraphy and structure of the southern Blake Plateau, northern Florida Straits and northern Bahama platform from multichannel seismic-reflection data. *American Association of Petroleum Geologists Bulletin*, **65**, 2571–93.

Siple, G.E. (1969). Salt water encroachment of Tertiary limestones along coastal South Carolina. *South Carolina Division of Geology Geologic Notes*, **13(2)**, 51–65.

Sleep, N.H. (1971). Thermal effects of the formation of the Atlantic continental margins by continental breakup. *Geophysical Journal of the Royal Astronomical Society*, **24**, 325–50.

Snyder, S.W., Hine, A.C. & Riggs, S.R. (1982). Miocene seismic stratigraphy, structural framework, and sea-level cyclicity, North Carolina continental shelf. *Southeastern Geology*, **23**, 247–66.

Snyder, S.W., Riggs, S.R., Katrosh, R., Lewis, D.W. & Scarborough, K.A. (1982). Synthesis of phosphatic sediment–faunal relationships within the Pungo River Formation. Environmental implications. *Southeastern Geology*, **23**, 233–45.

Toulmin, L.D. (1952). Sedimentary volumes in Gulf Coastal Plain of the United States and Mexico. *Geological Society of America Bulletin*, **63**, 1165–76.

Vail, P.R. & Hardenbol, J. (1979). Sea level changes during the Tertiary. *Oceanus*, **22**, 71–9.

Vail, P.R. & Mitchum, R.M., Jr. (1979). Global cycles of relative changes of sea level from seismic stratigraphy. In *Geological and Geophysical Investigations of Continental Margins*, ed. J.S. Watkins, L. Montadert & P.W. Dickerson, pp. 469–72. American Association of Petroleum Geologists Memoir 29.

Vail, P.R., Mitchum, R.M., Jr. & Thompson, S., III (1977). Global cycles of relative changes of sea level. In *Seismic Stratigraphy – Applications to Hydrocarbon Exploration*, ed. C.E. Payton, pp. 83–97. American Association of Petroleum Geologists Memoir 26.

Vernon, R.O. (1951). Geology of Citrus and Levy counties, Florida. *Florida Geological Survey Bulletin*, **33**, 256pp.

Ward, L.W. & Blackwelder, B.W. (1980). Stratigraphic revision of upper Miocene and lower Pliocene beds of the Chesapeake Group, middle Atlantic Coastal Plain. *US Geological Survey Bulletin*, **1482-D**, 59pp.

Ward, L.W., Blackwelder, B.W., Gohn, G.S. & Poore, R.Z. (1979). Stratigraphic revision of Eocene, Oligocene, and lower Miocene formations of South Carolina. *South Carolina Division of Geology Geological Notes*, **23(1)**, 2–32.

Watts, A.B. (1981). The US Atlantic continental margin: subsidence history, crustal structure, and thermal evolution. In *Geology of Passive Continental Margins*, ed. A.W. Bally *et al.*, pp. 2-1–2-75. American Association of Petroleum Geologists Education Course Note Series No. 19.

Webb, S.D. & Crissinger, D.B. (1983). Stratigraphy and vertebrate paleontology of the Central and Southern Phosphate Districts of Florida. In *The Central Florida Phosphate District, Southeastern Section Meeting Guidebook*, pp. 28–72. Geological Society of America.

Weaver, C.E. & Beck, K.C. (1977). Miocene of the SE United States: a model for chemical sedimentation in a peri-marine environment. *Sedimentary Geology*, **17**, 234pp.

Winston, G.O. (1976). Florida's Ocala Uplift is not an uplift. *American Association of Petroleum Geologists Bulletin*, **60**, 992–4.

Worthington, L.V. (1976). *On the North Atlantic Circulation*. Johns Hopkins University Press, Baltimore.

29

Stratigraphic framework for cyclical deposition of Miocene sediments in the Carolina Phosphogenic Province

S.R. RIGGS, STEPHEN W. SNYDER,
SCOTT W. SNYDER AND A.C. HINE

Abstract

This paper, PART I of a four-part series, presents the general geologic setting for the Miocene phosphorites and associated sediments of the Pungo River Formation on the Carolina continental margin. It is a chronostratigraphic overview of at least 18 fourth-order seismic sequences interpreted to be the product of high-frequency (< 200 ky), high-amplitude (> 50 m), sea-level fluctuations during the Miocene. Subsequent papers of the series (Chapters 30–32) describe in detail the sedimentological responses, patterns of environmental change, and complex history of deposition and erosion along the Carolina margin through these 18 or more sea-level events.

Introduction

Many Upper Cenozoic strata along the southeastern United States continental margin contain some phosphate (Altschuler *et al.*, 1964; Cathcart, 1968a, b; Miller, 1971 and 1982; Brown, Miller & Swain, 1972; Riggs, 1979 and 1984; Riggs & Belknap, 1988; Schlee, Manspeizer & Riggs, 1988; Scott, Chapter 26, this volume). Minor concentrations of phosphate occur in the Late Oligocene Cooper Marl, in several correlative Late Miocene and Pliocene units including the Bone Valley, Tamiami, Duplin, and Yorktown Formations, and locally in Pleistocene sediments.

Major concentrations of phosphate occur over broad regions of the southeastern United States in sediments of Early–Middle Miocene age. Regionally, the Hawthorn and Pungo River Formations are approximately contemporaneous Miocene units that have abnormally high concentrations of sedimentary phosphate. Hawthorn sediments and their stratigraphic equivalents extend from southern Florida into southern South Carolina (Fig. 29.1). In northern South Carolina and southern North Carolina, these sediments crop out on the continental shelf in Long and Onslow Bays and extend northward across the North Carolina Coastal Plain where they are called the Pungo River Formation.

The seaward displacement of Miocene sediments in northern South Carolina and southern North Carolina resulted from deposition around the Carolina Platform (Fig. 29.2), an Early Mesozoic syn-rift, tectonic feature composed of continental crust (Klitgord & Behrendt, 1979). The topographically highest portion of this basement feature has historically been called the Cape Fear Arch; more recent literature refers to it as the Mid-Carolina Platform High (Figs. 29.1 and 29.2). The Miocene section in North Carolina occurs as a seaward thickening sedimentary wedge (Fig. 29.3) deposited off the northeastern flank of the Carolina Platform as it drops into the Carolina Trough, a syn-rift basin filled with over 8 km of Mesozoic and Cenozoic sediments (Dillon *et al.*, 1979; Klitgord & Behrendt, 1979; Popenoe, 1985; Schlee, Dillon & Grow, 1979). The Carolina Platform is a first-order structural feature that controlled the deposition and concentric distribution pattern of apparent offlapping Tertiary units (Stephen W. Snyder, 1982; Stephen W. Snyder, Hine & Riggs, 1982 and 1983; Riggs *et al.*, 1985; Riggs & Belknap, 1988), as well as the formation of Miocene phosphate in the Carolina Phosphogenic Province (Fig. 29.1) (Riggs, 1979, 1981 and 1984).

The distribution of phosphate-rich sediments within the Miocene section is neither laterally nor vertically continuous or uniform. The spatial distribution of phosphate deposition (Fig. 29.1) was controlled regionally by the tectonic setting and first-order structural framework, and locally by second-order structural and topographic features. Phosphogenic episodes were temporally controlled by cyclical sea-level oscillations, changing current patterns, and fluctuations in chemical environments of deposition along the continental margin (Riggs, 1979 and 1984; Riggs *et al.*, 1982 and 1985; Stephen W. Snyder, 1982; Stephen W. Snyder *et al.*, 1982 and 1983; Riggs & Belknap, 1988). Miocene phosphorites formed in specific depositional environments within restricted portions of the continental margin of southeastern United States. Major phosphate deposits that occur on the emerged coastal plain are geologically well known; however, they represent only the updip limit of a much larger Miocene section that extends seaward beyond the coastal plain to form a thick sediment wedge underlying most of the modern continental shelf and upper slope (Fig. 29.3).

Carolina Phosphogenic Province

In North Carolina, Miocene sediments were deposited on a large, open continental shelf–slope system off the east flank of the Carolina Platform. The Cape Lookout High (Fig. 29.4) is an east–west oriented, Late Oligocene and Miocene,

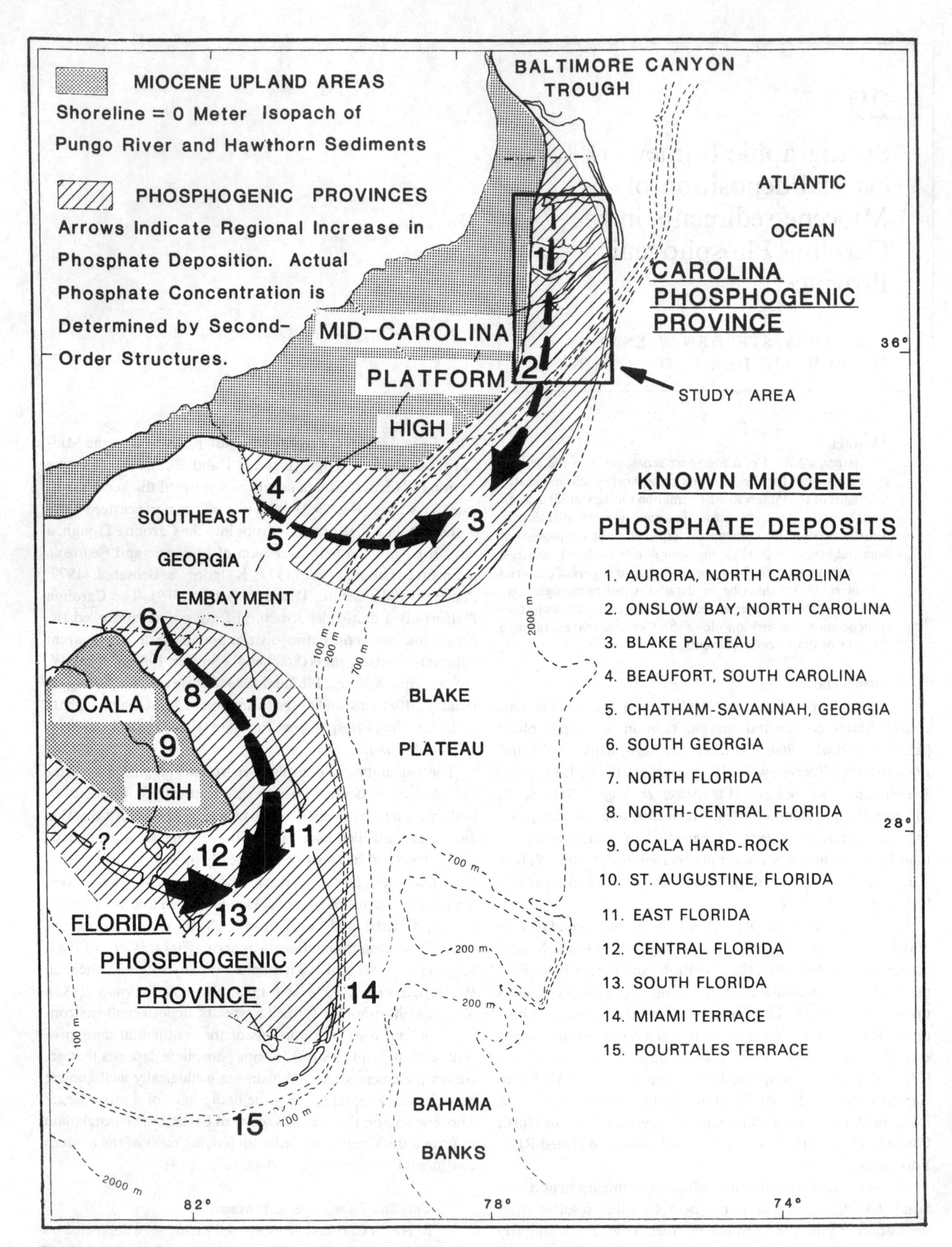

Fig. 29.1. Map of the southeastern United States showing: (a) major first-order structural features which controlled Miocene phosphate sedimentation; (b) the regional distribution of cumulative phosphate formed during the Miocene; and (c) the location of major phosphate deposits (modified from Riggs, 1984).

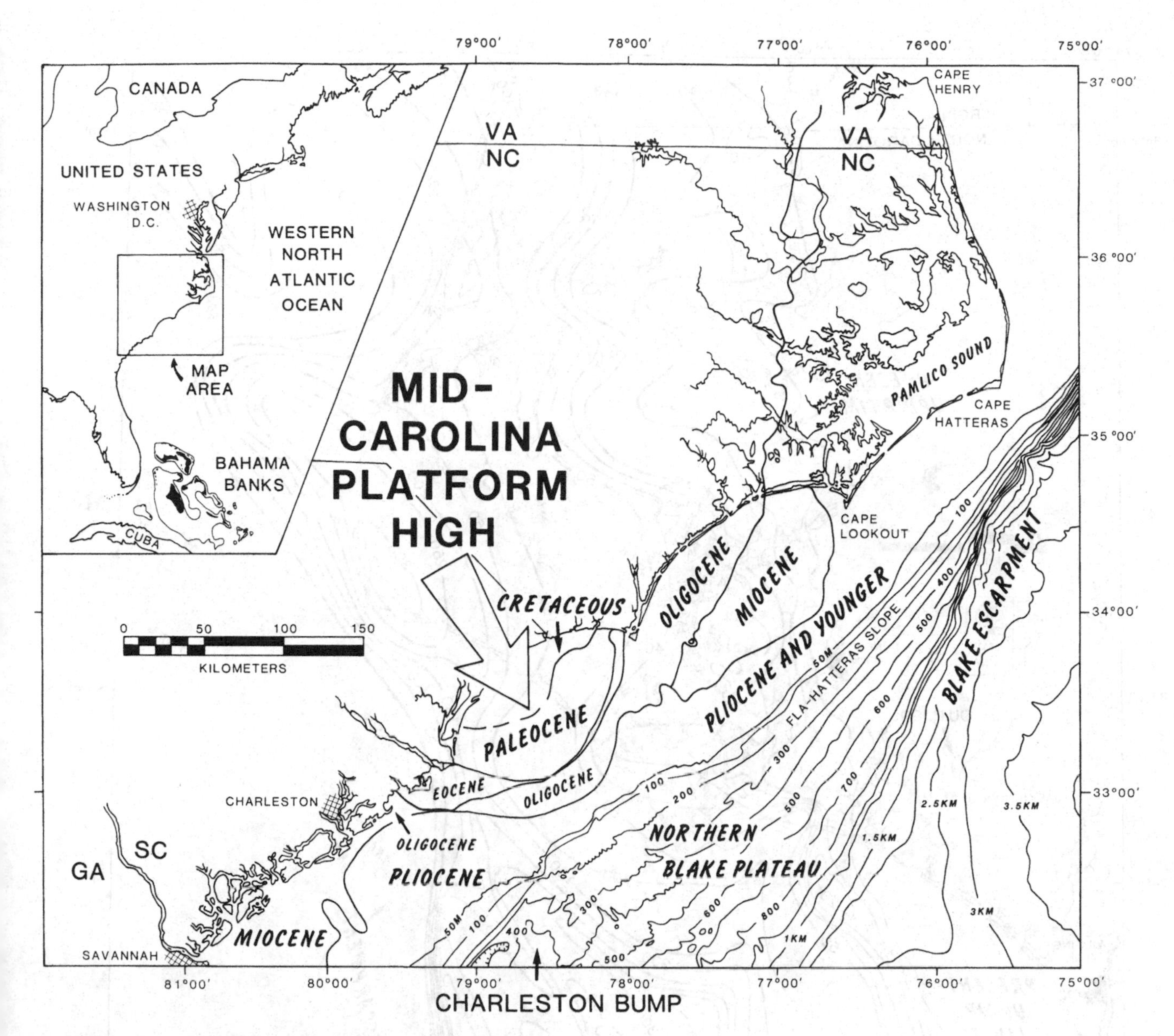

Fig. 29.2. Distribution of outcropping Cenozoic and Cretaceous sequences within the continental shelf of the Carolinas and Georgia as mapped from chronostratigraphic correlations via the seismic-reflection studies of Stephen W. Snyder (1982), Idris (1983) and Popenoe (1985) (modified from Stephen W. Snyder, 1982). Note that the Miocene sequences only crop out on the seafloor across the limbs of the Mid-Carolina Platform High and are partially missing from the nose of this first-order paleotopographic feature; the remaining Miocene sediment has been buried by Pliocene and Quaternary shelf-margin sediments.

paleotopographic high that subdivided the North Carolina Miocene shelf into two embayments (Riggs *et al.*, 1982 and 1985; Stephen W. Snyder, 1982; Stephen W. Snyder *et al.*, 1982 and 1983; Riggs, 1984; Popenoe, 1985; Riggs & Belknap, 1988). North of Cape Lookout High is the Aurora Embayment, which underlies most of the outer coastal plain of northeastern North Carolina and contains a major producing phosphate mine near Aurora. South of Cape Lookout High is Onslow Embayment, which underlies the modern continental shelf in Onslow Bay and contains two recently discovered phosphate deposits in the northeastern Onslow Bay and Frying Pan areas (Riggs *et al.*, 1985).

The Miocene Pungo River Formation, which includes major sedimentary phosphorites, underlies the north-central coastal plain of North Carolina (Fig. 29.3). Many workers, including Kimrey (1964 and 1965), Gibson (1967), and Rooney & Kerr (1967) have studied the formation since Brown (1958) first described subsurface phosphorites from Beaufort County, NC. More recent paleontologic, sedimentologic, and stratigraphic studies have increased our understanding of phosphatic sedi-

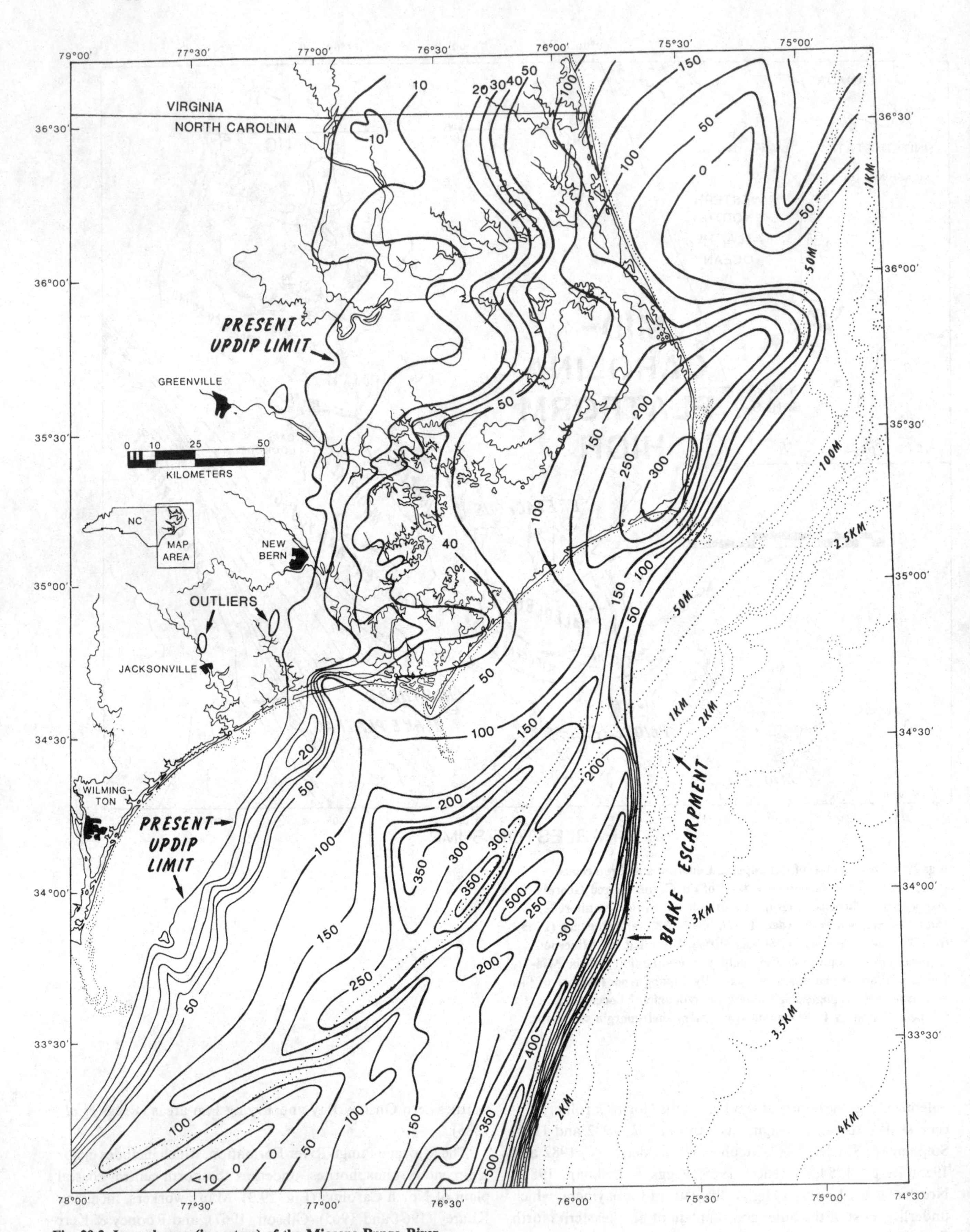

Fig. 29.3. Isopach map (in meters) of the Miocene Pungo River Formation in the North Carolina continental margin. This map was produced from seismic-reflection data of Stephen W. Snyder (1982) and Popenoe (1985) and drill hole data of Miller (1971 and 1982) (modified from Riggs *et al.*, 1985). Notice how the formation thickens seaward to between 300 and 500 m and is abruptly terminated.

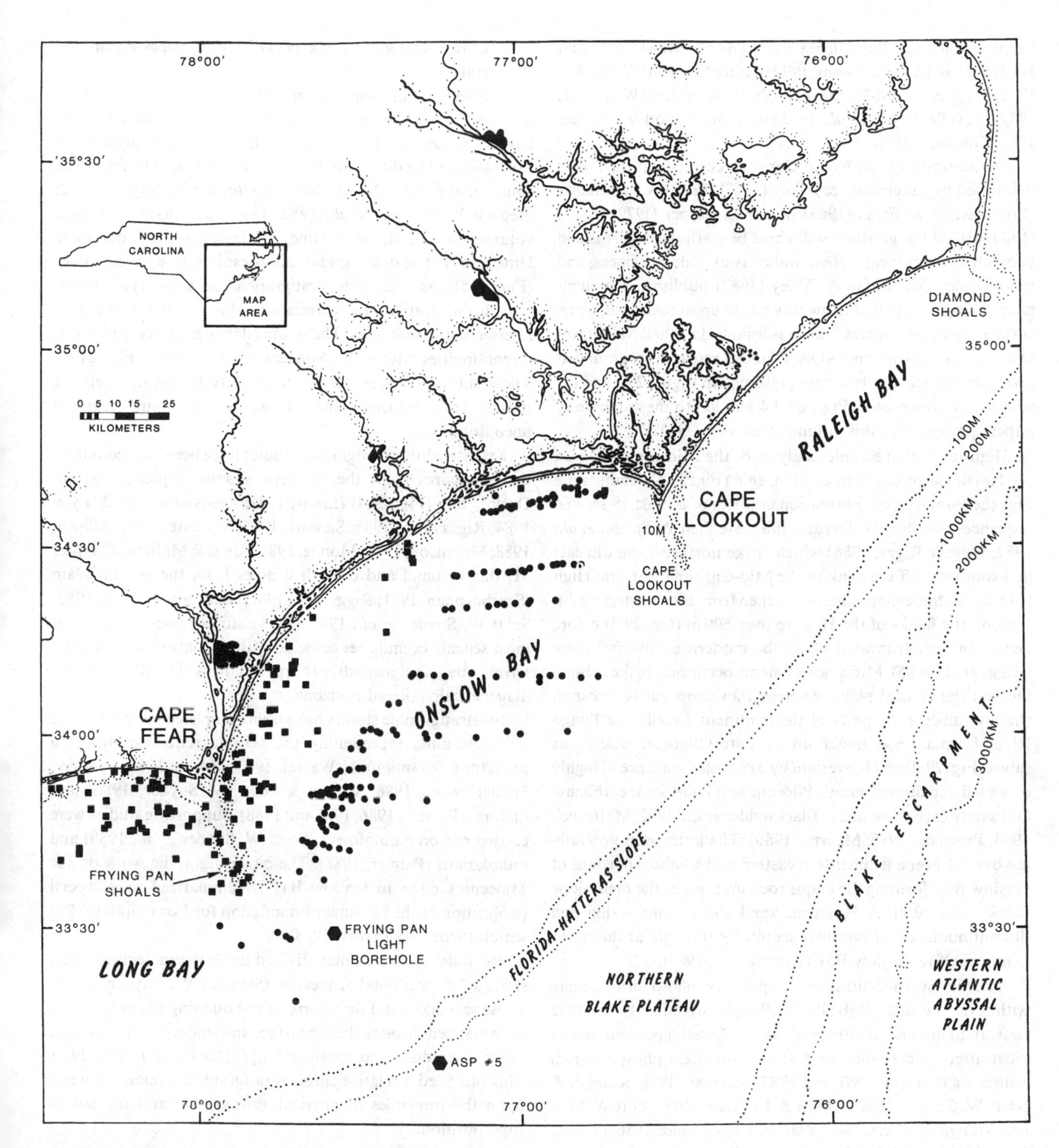

Fig. 29.4. Location of continental shelf vibracores and ocean drill sites used for lithologic and biostratigraphic support data in identifying sequences within the seismic-reflection data (modified from Hine & Riggs, 1986). Alignment of vibracores resulted from selecting sites along key seismic-reflection profiles.

ments throughout the Aurora Embayment (Miller, 1971 and 1982; Brown, Miller & Swain, 1972; Katrosh & Scott W. Snyder, 1982; Riggs *et al.*, 1982; Scarborough, Riggs & Scott W. Snyder, 1982; Scott W. Snyder *et al.*, 1982 and 1986; Abbott & Ernissee, 1983; Gibson, 1983).

Phosphorites of the North Carolina continental shelf were first noted by Luternauer & Pilkey (1967), Pilkey & Luternauer (1967), Riggs & Freas (1967), and Meisburger (1979). Steele (1980) drilled Pungo River sediments beneath the barrier island between the continental shelf and coastal plain provinces, and Blackwelder, MacIntyre & Pilkey (1982) published a preliminary geologic map of Onslow Bay based upon rock dredge and surface sediment samples. This was followed by the development of an extensive NSF- and NOAA-funded research program that produced the seismic data base presented in Figure 29.5 and the vibracores presented in Figure 29.4 and led to the sequence of papers presented in this volume (Chapters 29–32).

High-resolution seismic analyses of the Miocene section on the North Carolina continental margin (Fig. 29.5) demonstrate that the Pungo River Formation consists of at least 18 seismic sequences (Stephen W. Snyder, 1982; Stephen W. Snyder *et al.*, 1982; Hine & Riggs, 1986) which strike northeast and dip east and southeast off the flank of the Mid-Carolina Platform High (Fig. 27.2). Miocene sediments thicken from their western updip limit on the flanks of the High to over 500 m (Fig. 29.3) before being abruptly truncated along the modern continental slope (Riggs *et al.*, 1985). Miocene sediments occur only in the subsurface on the coastal plain, whereas they crop out or occur in shallow subcrop on parts of the continental shelf. The Pungo River Formation is underlain by Late Oligocene calcareous sands (Fig. 29.2) and is overlain by a complex sequence of highly dissected and discontinuous Pliocene and Pleistocene carbonate and sandy carbonate units (Blackwelder *et al.*, 1982; Matteucci, 1984; Popenoe, 1985; Mearns, 1986). The latter units generally lap over Miocene sediments in eastern and southern portions of Onslow Bay, forming the major rock units along the shelf–slope break (Fig. 29.2). A Holocene sand sheet forms a thin and discontinuous cover on older sediments throughout the continental shelf in Onslow Bay (Riggs *et al.*, 1985).

The major concentration of primary phosphorite occurs within various units of the Pungo River Formation, both on the coastal plain and continental shelf. Based upon numerous biostratigraphic studies carried out on these phosphate-rich sediments (Gibson, 1967 and 1983; Katrosh, 1981; Katrosh & Scott W. Snyder, 1982; Abbot & Ernissee, 1983; Scott W. Snyder, Mauger & Akers, 1983; Waters, 1983; Moore, 1986; Powers, 1986, 1987 and 1988; Waters & Scott W. Snyder, 1986; Palmer, 1988; Scott W. Snyder, 1988), the general age assignment is late Early–late Middle Miocene. Lesser concentrations of primary phosphorite occur in the lower Yorktown Formation on the coastal plain (Riggs *et al.*, 1982; Riggs, 1984). This unit is Pliocene in age (Gibson, 1983; Hazel, 1983; Scott W. Snyder *et al.*, 1983). Throughout Onslow Bay, local concentrations of phosphate occur in Pleistocene and Holocene sediments that are largely reworked from underlying Miocene units (Luternauer & Pilkey, 1967; Pilkey & Luternauer, 1967; Blackwelder *et al.*, 1982; Riggs *et al.*, 1985).

Chronostratigraphy of cyclical Miocene depositional units

Stratigraphic basis for chronostratigraphy

A detailed seismic stratigraphy was developed for the Upper Cenozoic section, including the Miocene Pungo River Formation, extending from the lower coastal plain to the upper continental slope of North Carolina (Stephen W. Snyder, 1982; Stephen W. Snyder *et al.*, 1982, 1983 and Chapter 30, this volume; Matteucci, 1984; Hine & Riggs, 1986; Matteucci & Hine, 1987). The dense grid of high-resolution seismic profiles (Fig. 29.5) was reduced to stratigraphic line drawings to facilitate chronostratigraphic correlations through the network of intersecting seismic lines. The seismic data depict many physical discontinuities within the Miocene stratal record (Fig. 27.6). These subsurface erosional surfaces were traced through the mosaic of reduced line drawings to map regional unconformities.

Extensive lithostratigraphic studies have been carried out on the vibracores from the Miocene seismic sequences within Onslow Bay (Fig. 29.4) (Lewis, 1981; Lewis *et al.*, 1982; Lyle, 1984; Riggs *et al.*, 1985; Stewart, 1985; Mallette, 1986; Allison, 1988; Mallinson, 1988; Moretz, 1988; Riggs & Mallette, Chapter 31, this volume) and deep drill holes from the coastal plain (Scarborough, 1981; Riggs *et al.*, 1982; Scarborough *et al.*, 1982; Scott W. Snyder *et al.*, 1986). These studies demonstrate that most seismic boundaries coincide with unconformities characterized by hardgrounds, abrupt changes in lithology, or diagenetically-altered horizons.

Biostratigraphic studies have been completed on most of the sediment units representing the seismic sequences utilizing planktonic foraminifera (Waters, 1983; Moore, 1986; Scott W. Snyder *et al.*, 1986; Waters & Scott W. Snyder, 1986) and diatoms (Powers, 1986, 1987 and 1988). Supporting studies were carried out on nannofossils (Scott W. Snyder *et al.*, 1988) and radiolarians (Palmer, 1988). The biostratigraphic work on the Miocene section in Onslow Bay is summarized in a special publication of the Cushman Foundation for Foraminiferal Research (Scott W. Snyder, 1988).

Regional unconformities divided the Miocene section into a series of depositional sequences (Stephen W. Snyder, 1982). These sequences and their correlative bounding unconformities were mapped through the subsurface and stacked in chronologic order to produce a composite stratigraphic column (Fig. 29.7). This provided a relative chronostratigraphic framework based upon the principles of physical stratigraphy and the law of superposition.

Miocene chronostratigraphy for North Carolina

Within the Pungo River Formation of Onslow Bay, three seismic sections and the associated unconformities have been mapped across Onslow Bay (Fig. 29.6) and biostratigraphically resolved. These include the mid-Burdigalian Frying Pan Sequence (FPS), the Langhian Onslow Bay Sequence (OBS), and the Serravallian Bogue Banks Sequence (BBS). Biostratigraphic age assignments for the three third-order sections and associated unconformities are summarized in Figure 29.7. These sections can be biostratigraphically correlated with standard European

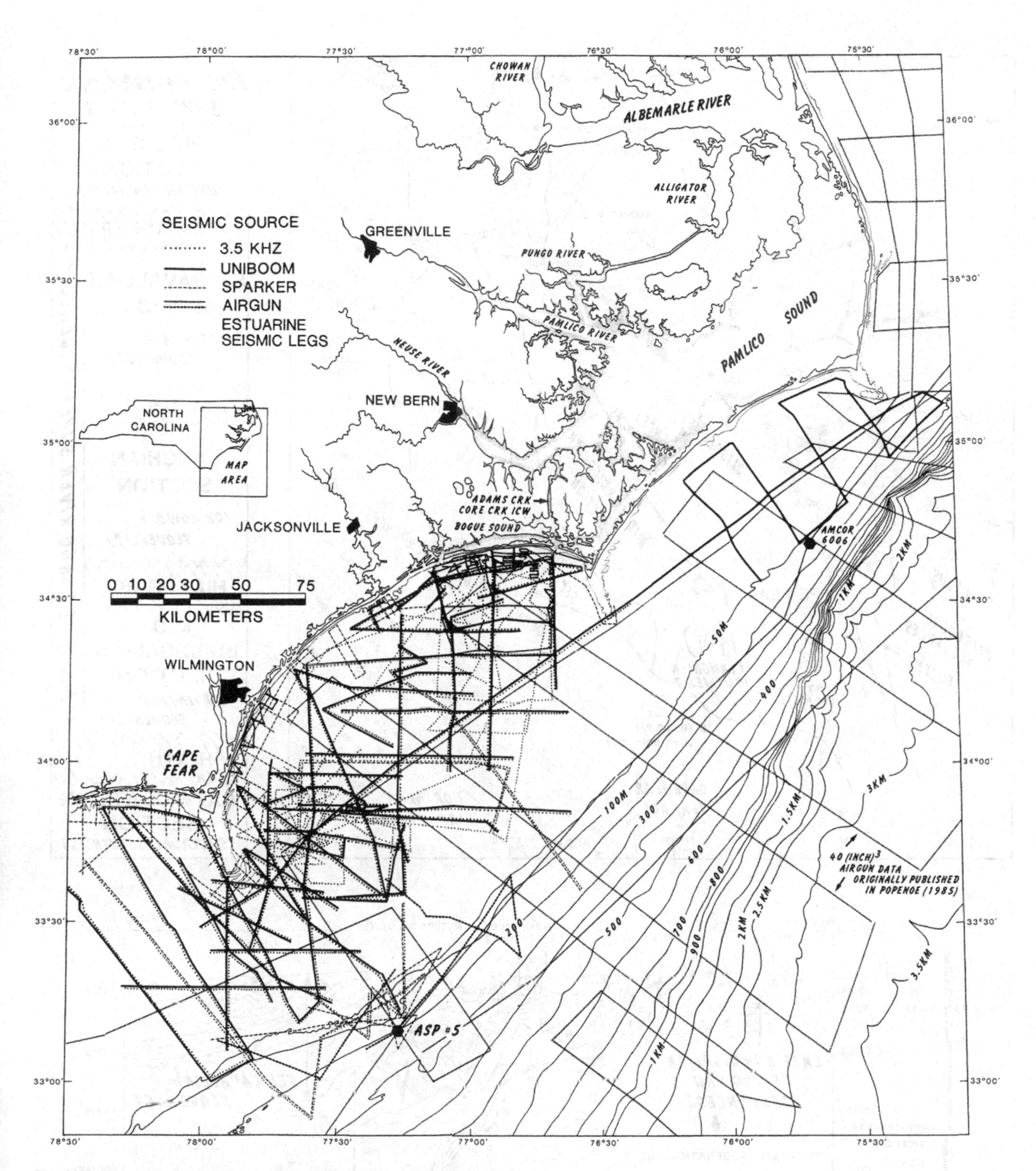

Fig. 29.5. Location of seismic-reflection profiles used in this study for mapping the Miocene subsurface stratigraphy of the North Carolina continental margin (from Hine & Riggs, 1986). The estuarine seismic lines included sparker and boomer profiles, and were used to tie the offshore stratigraphy to coastal plain stratigraphy known from drill holes and quarries.

(A)

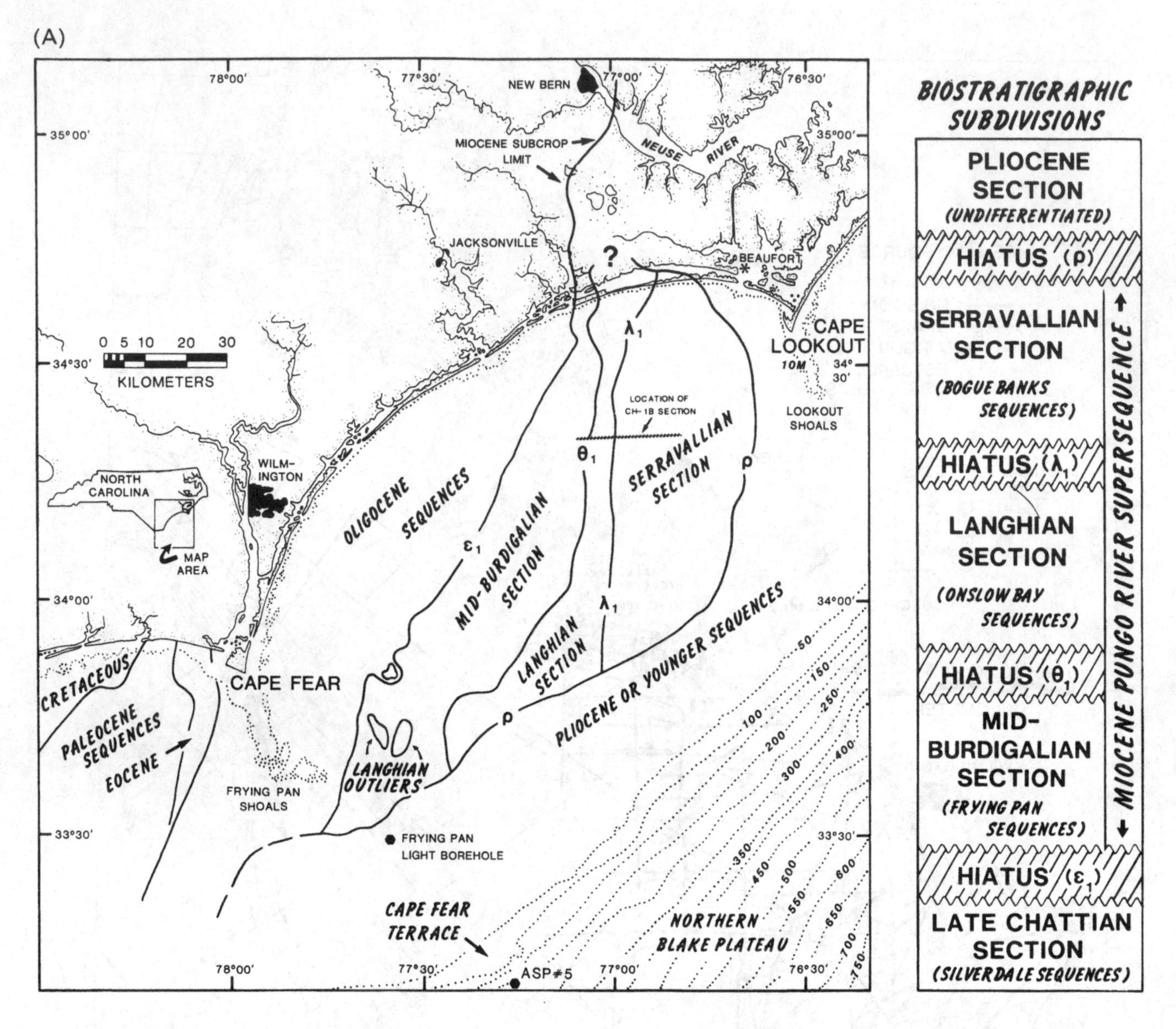

(B)

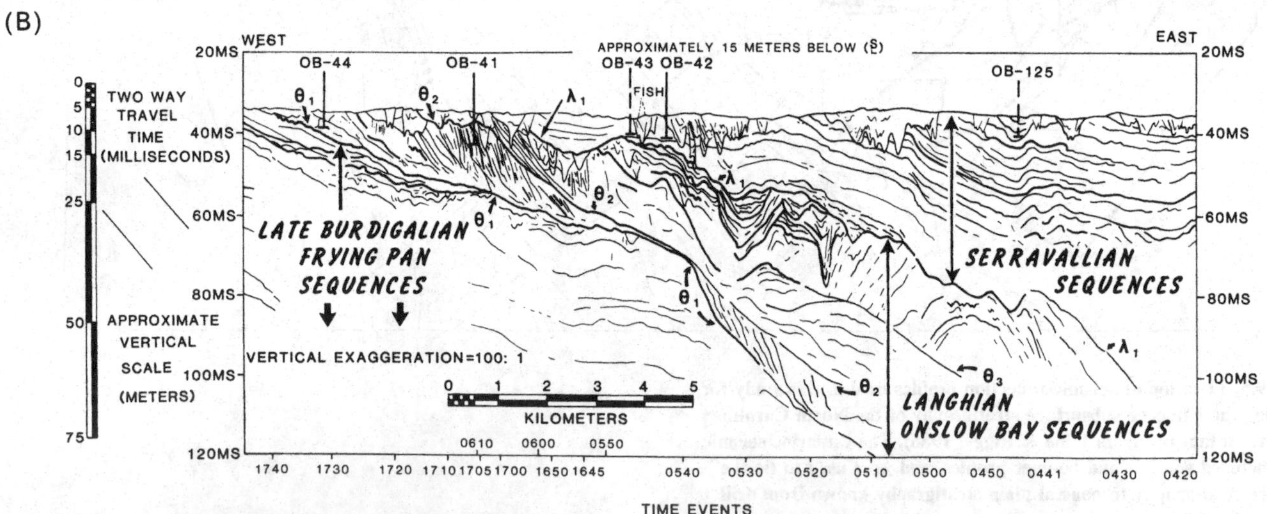

Fig. 29.6. (A) Distribution of the three third-order Miocene sections and associated unconformities which crop out on the North Carolina continental shelf of Onslow Bay (modified from Stephen W. Snyder, 1982). (B) Reduced statigraphic line-drawing of seismic line CH–1B showing the biochronostratigraphic sections illustrated in (A) (modified from Stephen W. Snyder, 1982). Notice that the three biochronostratigraphic sections illustrated in (A) each consist of several unconformity-bound sequences that are not resolvable by biostratigraphic studies, but are easily recognized in seismic data via lateral termination of reflectors in onlap, downlap, or erosional relationships.

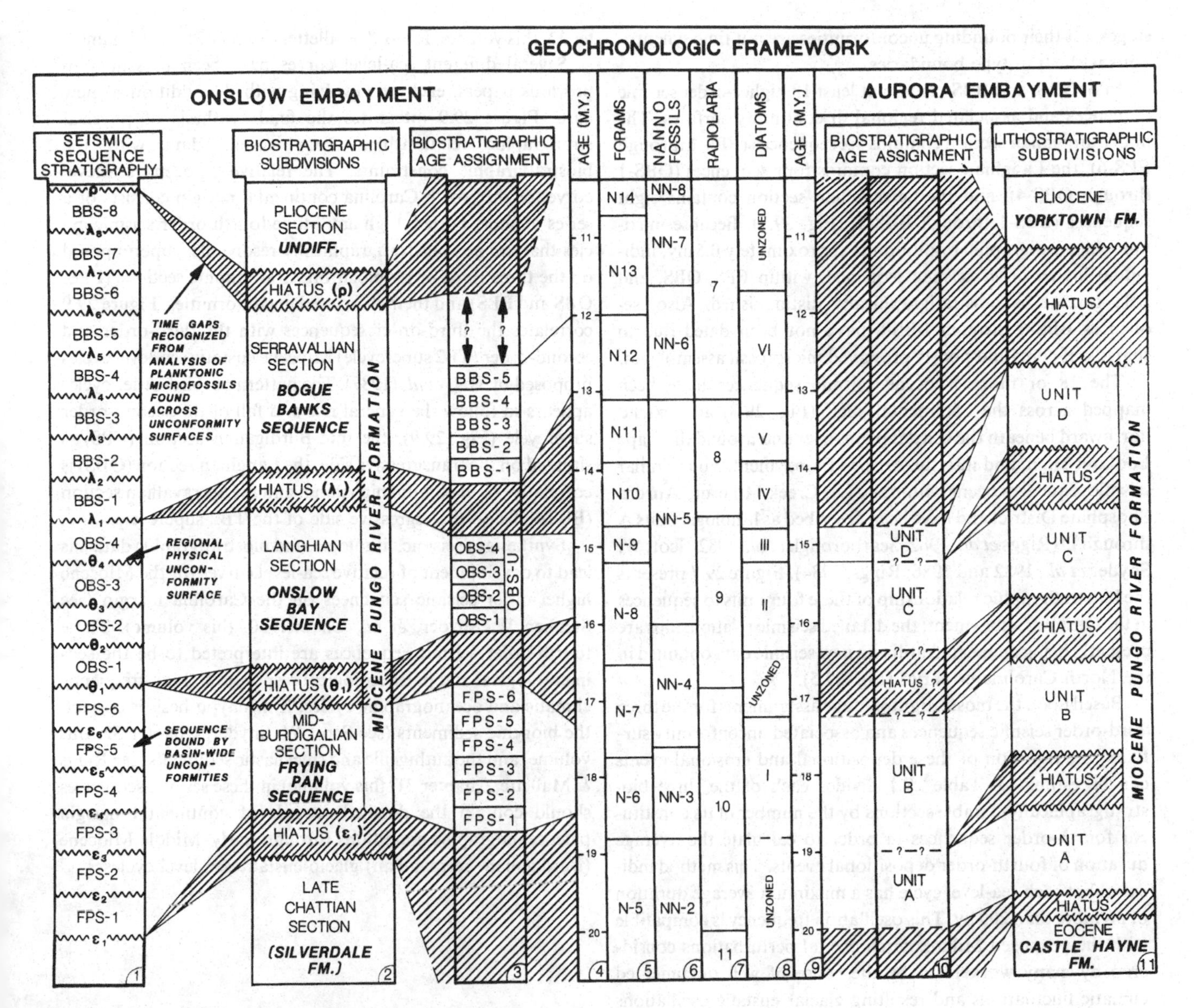

Fig. 29.7. Three stratigraphic columns depicting the relationships between sequence stratigraphy, biostratigraphy, and geochronology of the composite Miocene chronostratigraphic framework of North Carolina. (A) Relative chronostratigraphic framework generated by identification of physical unconformities from seismic-reflection profiles (from Stephen W. Snyder, 1982). (B) Biostratigraphic sections for Onslow Bay and associated time–stratigraphic unconformities identified using planktonic microfossils (from Scott W. Snyder, 1988). (C) Biostratigraphic sections for Aurora Phosphate District and associated time–stratigraphic unconformities identified using diatoms (from Powers, 1987). (D) Geochronology of Miocene sequences and unconformities using microfossils to correlate to global planktonic biochrons and thereby place the sequences on the calibrated time scale of Haq *et al.* (1987) (from Stephen W. Snyder, 1982; Scott W. Snyder, 1988).

stages but their bounding unconformities are not time synchronous with stratotype boundaries.

FPS, OBS, and BBS contain at least 18 higher-order seismic sequences and associated regional discontinuity surfaces. The mid-Burdigalian section contains six sequences (FPS-1 through FPS-6), the Langhian section contains four sequences (OBS-1 through OBS-4), and the Serravallian section contains eight sequences (BBS-1 through BBS-8) (Fig. 29.7). Because maximum biostratigraphic resolution is approximately 0.5 my, individual higher-order seismic sequences within FPS, OBS, and BBS cannot be biostratigraphically distinguished. Also, sequences BBS-6 through BBS-8 have not been dated due to sparsity of cores and poorly preserved microfossil assemblages.

The 18 or more Miocene seismic sequences have been mapped across the continental shelf (Fig. 29.8) and extend northward beneath the coastal plain, over and around the Cape Lookout High, and into the Aurora Embayment. Four similar seismic sequences occur within the Lee Creek Mine and Aurora Phosphate District and have been described as lithologic units A through D (Riggs *et al.*, 1982; Scarborough *et al.*, 1982; Scott W. Snyder *et al.*, 1982 and 1986; Riggs, 1984). Figure 29.7 presents the biostratigraphic relationship of these four units to sequences in the Onslow Embayment; the detailed seismic relationships are presently being delineated utilizing new seismic data obtained in the North Carolina estuaries (Fig. 29.5).

Based upon the biostratigraphic age assignments for the three third-order seismic sequences and associated unconformity surfaces, the duration of these depositional and erosional events can be estimated. Table 29.1 divides each of the three biostratigraphically-datable sections by the number of its constituent fourth-order sequences in order to calculate the average duration of fourth-order depositional events. This method indicates that each sea-level cycle has a maximum average duration of no more than 200 ky. This oscillation frequency is compatible with the proposed Milankovitch orbital perturbations considered by many workers to be the cause of well documented climatic fluctuations and resulting glacial–eustatic oscillations of the Late Quaternary sediment record (Shackleton & Opdyke, 1973 and 1976; Hays, Imbrie & Shackleton, 1976; Imbrie & Imbrie, 1980; Imbrie, 1985; Prell *et al.*, 1986).

Miocene sea-level curve for North Carolina

Biostratigraphic data summarized in Figure 29.7 are plotted against the sequence stratigraphy and eustatic sea-level curves of Haq, Hardenbol & Vail (1987) in Figure 29.9. Thus, deposition of phosphorites associated with the Pungo River Formation began about 18.4 million years ago (late Early Miocene) and continued cyclically until at least 12 million years ago (late Middle Miocene) (Scott W. Snyder, 1988). Since changing sea level is the only known mechanism capable of generating regional unconformities stretching across an entire continental shelf and into the emerged coastal plain, the Miocene depositional record and associated unconformities are interpreted to be products of at least 18 high-frequency (< 200 ky), high-amplitude (> 50 m) glacio-eustatic sea-level fluctuations (Stephen W. Snyder, 1982; Stephen W. Snyder *et al.*, 1983, 1986 and Chapter 30, this volume; Riggs, 1984; Riggs *et al.*, 1985; Riggs & Hine, 1986; Riggs & Belknap, 1988; Scott W. Snyder, 1988 and Chapter 32, this volume; Riggs & Mallette, Chapter 31, this volume).

Several different sea-level curves have been presented in previous papers, each one evolving with the addition of new data. Figure 29.9 integrates the final synthesis of physical stratigraphic relationships from seismic data with the biostratigraphic constraints. The relative Miocene sea-level curve for the North Carolina continental margin consists of a series of short-pulsed, high amplitude fourth-order seismic cycles that are not biostratigraphically resolvable, superimposed on the three third-order biochronostratigraphic sections (FPS, OBS and BBS) and their associated unconformities. Figure 29.9 correlates the third-order sequences with the third-order and second-order (TB2 supercycle) Miocene eustatic sea-level cycles proposed by Haq *et al.* (1987). The pattern of third-order cycles appears to follow the general rise and fall of the second-order supercycle (Fig. 29.9); the mid-Burdigalian section (FPS) is situated on the transgressive side, the Langhian section (OBS) is correlative to maximum highstand, and the Serravallian section (BBS) falls on the regressive side of the TB2 supercycle.

Synthesis of seismic, lithologic, and biostratigraphic data has lead to development of relative sea-level curves for the Miocene higher-order seismic sequences on the Carolina margin (see Stephen W. Snyder, *et al.*, Chapter 30, this volume). These fourth-order seismic sequences are interpreted to be the sedimentological and paleoecological responses to high-frequency climatic and oceanographic cycles. If this hypothesis is correct, the biogenic sediments (see Scott W. Snyder, Chapter 32, this volume) and the authigenic and siliciclastic sediments (see Riggs & Mallette, Chapter 31, this volume) in these seismic sequences should contain the detailed records of continental margin paleoceanographic conditions through Early–Middle Miocene (Burdigalian–Serravallian) glacio-eustatic sea-level cycles.

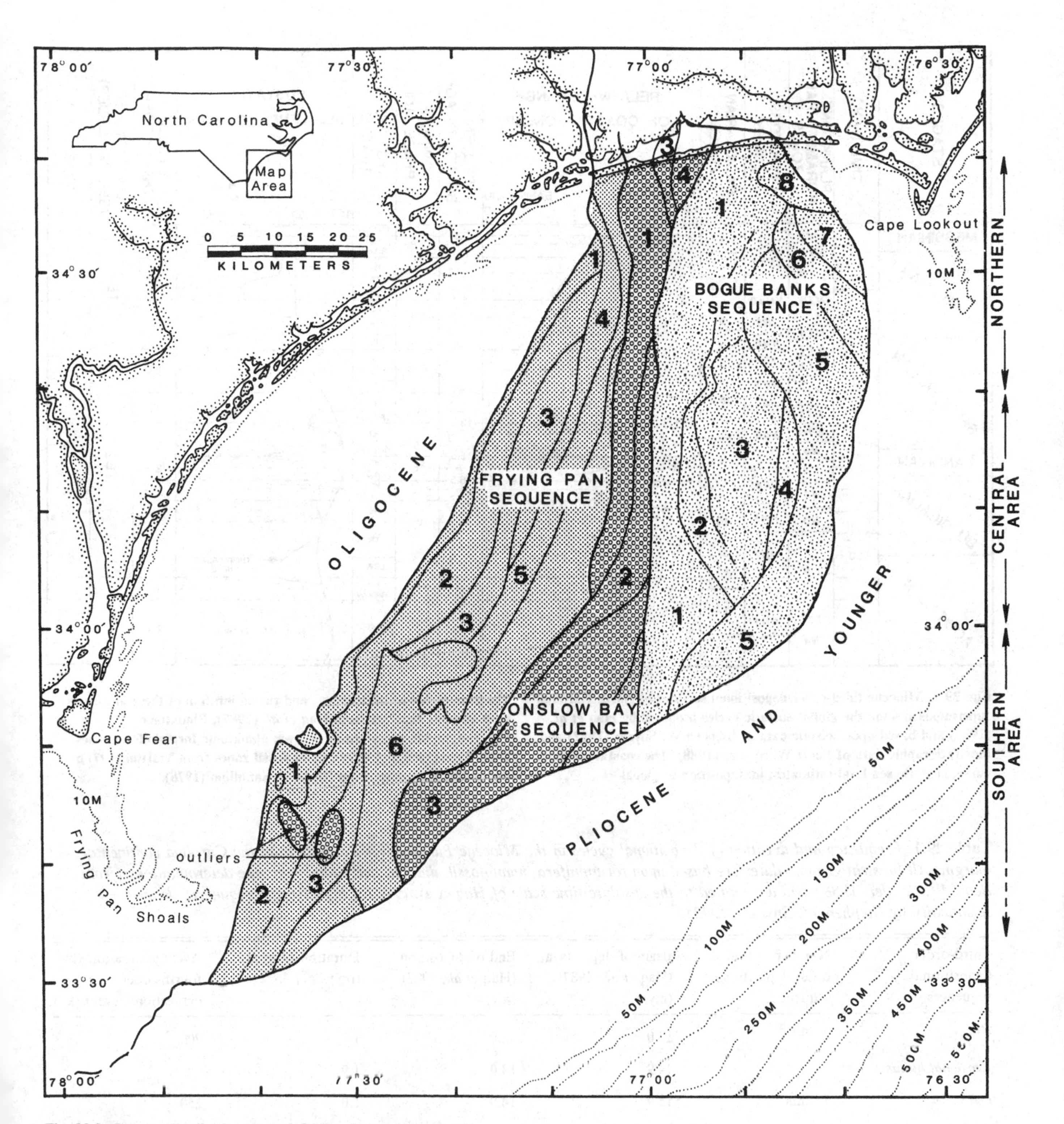

Fig. 29.8. Outcrop distribution map of Onslow Bay, North Carolina showing 16 (of the 18 or more) fourth-order seismic sequences of the Miocene Pungo River Formation (modified from Stephen W. Snyder, 1982 by Mallette, 1986). Additional fourth-order seismic sequences have been mapped in the subsurface but do not crop out. Also, shown on the map are the three third-order seismic, biostratigraphic, and depositional sections referred to as the Frying Pan (FPS), Onslow Bay (OBS), and Bogue Banks Sections (BBS) of Mid-Burdigalian, Langhian, and Serravallian ages, respectively.

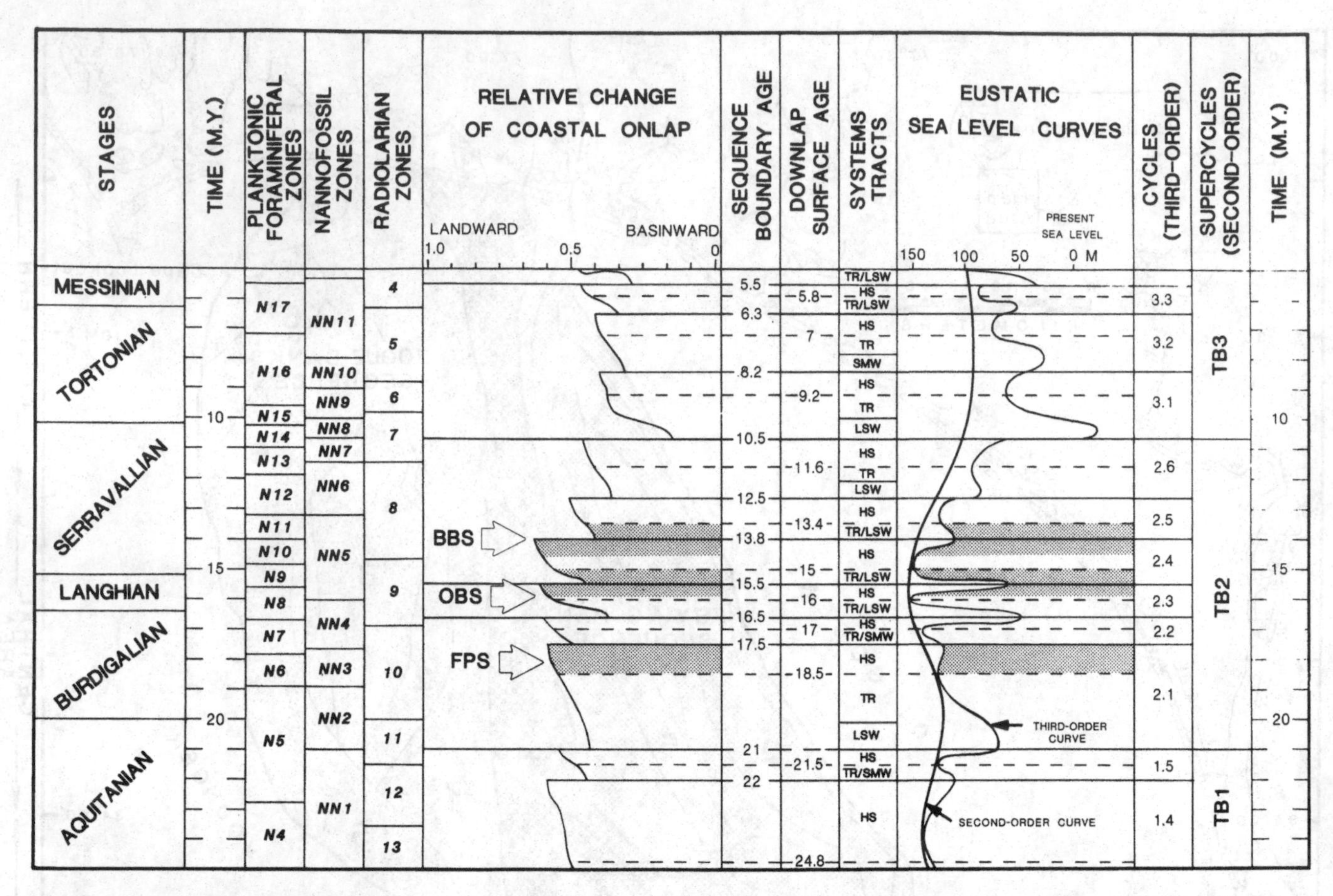

Fig. 29.9. Miocene third-order depositional sections of North Carolina superimposed upon the global eustatic cycles proposed by Haq *et al.* (1987) and based upon seismic data of Stephen W. Snyder (1982) and biostratigraphic data of Scott W. Snyder (1988). The coastal onlap curve, eustatic sea-level estimates, juxtaposition of global biostratigraphic zonal schemes, and the calibration of these biochrons with absolute time are from Haq *et al.* (1987). Planktonic biostratigraphic zones are as follows: planktonic foraminiferal zones from Blow (1969 and 1979); nannofossil zones from Martini (1971); and radiolarian zones from Riedel & Sanfillipo (1978).

Table 29.1. *Frequency and duration of depositional cycles in the Miocene Pungo River Formation on the Carolina continental margin. All biostratigraphic dates are based upon foraminifera, nannofossil, diatom, and radiolarian age determinations (from Scott W. Snyder, 1988) and are related to the absolute time scale of Haq* et al. *(1987); higher-order sequences defined seismically by Stephen W. Snyder (1982).*

Third-order depositional sequences	Number of fourth-order depositional sequences	Begin of deposition (Haq *et al.*, 1987) (my)	End of deposition (Haq *et al.*, 1987) (my)	Duration of event (my)	Average duration (ky) fourth-order depositional event (ky)
BBS	5 units[a]	13.0	12.0	1.0	200
Erosional hiatus		14.9	13.0	1.9	
OBS	4 units	15.9	14.9	1.0	250
Erosional hiatus		17.4	15.9	1.5	
FPS	6 units	18.4	17.4	1.0	167
Pungo River Formation[a]					
Total Time		18.4	12.0	6.4	
Total duration of known erosional events				3.4	
Total duration of fourth-order deposition events[b]				3.0	
Average duration of fourth-order depositional event[b]					200

[a]The 3 uppermost BBS units have not been biostratigraphically dated.
[b]These 15 depositional units include 14 erosional hiatuses, 12 of which separate fourth-order sea-level events, but which cannot be biostratigraphically resolved.

Acknowledgments

The research results presented in Chapters 29 through 32 were supported by 1) National Science Foundation grants OCE-798949, OCE-8118164, OCE-8400383, and OCE-8609161 from 1979 through 1987 and 2) University of North Carolina Sea Grant College grants R/AO-3 and R/AO-4 from 1982 through 1986. Appreciation is expressed to 1) the many members of the Geology Department at East Carolina University, Greenville, NC and the Marine Science Program at the University of South Florida, St Petersburg, Florida who supplied help both at sea and in the laboratory; 2) Texasgulf Inc and the former North Carolina Phosphate Corp who supplied samples and analytical capabilities; and 3) the ongoing programs associated with the International Geological Correlation Program (IGCP) 156 on phosphorites sponsored by UNESCO and IUGS.

References

Abbott, W.H. & Ernissee, J.J. (1983). Biostratigraphy and paleoecology of a diatomaceous clay unit in the Miocene Pungo River Formation of Beaufort County, North Carolina. In *Geology and Paleontology of the Lee Creek Mine, North Carolina, Part I*, ed. C.E. Ray, pp. 287–354. Smithsonian Contributions to Paleobiology, No. 53, Washington, DC.

Allison, M.A. (1988). 'Mineralogy and sedimentology of the clay-sized fraction, Miocene Pungo River Formation, North Carolina continental margin'. Unpublished M.Sc thesis, East Carolina University, Greenville, NC, 112pp.

Altschuler, Z.S., Cathcart, J.B. & Young, E.J. (1964). *The Geology and Geochemistry of the Bone Valley Formation and its Phosphate Deposits, West Central Florida.* Geological Society of America Annual Meeting, Guidebook Field Trip 6, Miami Beach, 68pp.

Blackwelder, B.W., MacIntyre, I.G. & Pilkey, O.H. (1982). Geology of the continental shelf, Onslow Bay, North Carolina, as revealed by submarine outcops. *American Association of Petroleum Geologists Bulletin*, **66**, 44–56.

Blow, W.H. (1969). Late middle Eocene to Recent planktonic foraminiferal biostratigraphy. In *Proceedings of the First International Conference on Planktonic Microfossils* vol. 1, ed. P. Bronniman & H.H. Renz, pp. 199–421. E.J. Brill, Leiden, The Netherlands.

Blow, W.H. (1979). *The Cainozoic Globigerinida.* E.J. Brill, Leiden, The Netherlands, 1413pp.

Brown, P.M. (1958). The relation of phosphorites to ground water in Beaufort County, North Carolina. *Economic Geology*, **53**, 85–101.

Brown, P.M., Miller, J.A. & Swain, F.M. (1972). Structural and stratigraphic framework, and spatial distribution of permeability of the Atlantic Coastal Plain, North Carolina to New York. *US Geological Survey Professional Paper*, **796**, 79pp.

Cathcart, J.B. (1968a). Florida-type phosphate deposits of the United States; origin and techniques for prospecting. In *Seminar on Sources of Mineral Raw Materials for the Fertilizer Industry in Asia and the Far East*, pp. 178–86. Proceedings of the United Nations ECAFE, Mineral Resources Development Series 32, Bangkok.

Cathcart, J.B. (1968b). Phosphate in the Atlantic and Gulf Coastal plains. In *Proceedings of the 4th Forum of Geology of Industrial Minerals*, ed. L.F. Brown, pp. 23–4. University of Texas Press, Austin.

Dillon, W.P., Paull, C.K., Buffler, R.T. & Fail, J.P. (1979). Structure and development of the Southeast Georgia Embayment and northern Blake Plateau; preliminary analysis. In *Geologic and Geophysical Investigations of Continental Margins*, ed. J.S. Watkins, L. Montadert & P.W. Dickerson, pp. 27–46. American Association of Petroleum Geologists Memoir 29.

Gibson, T.C. (1967). Stratigraphy and paleoenvironment of the phosphatic Miocene strata of North Carolina. *Geological Society of America Bulletin*, **78**, 631–50.

Gibson, T.C. (1983). Stratigraphy of Miocene through Lower Pleistocene strata of the US Central Atlantic Coastal Plain. In *Geology and Paleontology of the Lee Creek Mine, North Carolina*, ed. C.E. Ray, pp. 35–80. Smithsonian Contributions to Paleobiology, No. 53, Washington, DC.

Haq, B.U., Hardenbol, J. & Vail, P.R. (1987). Chronology of fluctuating sea levels since the Triassic. *Science*, **235**, 1156–67.

Hays, J.D., Imbrie, J. & Shackleton, N.J. (1976). Variations in the earth's orbit: pacemaker of the ice ages. *Science*, **194**, 1121–32.

Hazel, J.E. (1983). Age and correlation of the Yorktown (Pliocene) and Croatan (Pliocene and Pleistocene) Formations at the Lee Creek Mine. In *Geology and Paleontology of the Lee Creek Mine, North Carolina*, ed. C.E. Ray, pp. 81–200. Smithsonian Contributions to Paleobiology, No. 53, Washington, DC.

Hine, A.C. & Riggs, S.R. (1986). Geological framework, Cenozoic history, and modern processes of sedimentation on the North Carolina continental margin. In *SEPM Field Guidebooks, Southeastern US Third Annual Midyear Meeting*, ed. D.A. Textoris, pp. 129–94. Society of Economic Paleontologists and Mineralogists, Tulsa.

Idris, F.M. (1983). 'Cenozoic stratigraphy and structure of the South Carolina lower coastal plain and continental shelf.' Unpublished Ph.D. thesis, University of Georgia, Athens, 125pp.

Imbrie, J. (1985). A theoretical framework for the Pleistocene ice ages. *Journal of the Geological Society of London*, **142**, 417–32.

Imbrie, J. & Imbrie, J.Z. (1980). Modeling the climatic response to orbital variations. *Science*, **207**, 943–53.

Katrosh, M.R. (1981). 'Foraminiferal paleoecology and biostratigraphy of the Yorktown and Pungo River Formations: Beaufort, Pamlico, Craven and Carteret Counties, North Carolina.' Unpublished M.Sc. thesis, East Carolina University, Greenville, NC, 198pp.

Katrosh, M.R. & Snyder, Scott W. (1982). Diagnostic foraminifera and paleoecology of the Pungo River Formation, central coastal plain of North Carolina. *Southeastern Geology*, **23**, 217–32.

Kimrey, J.O. (1964). The Pungo River Formation, a new name for middle Miocene phosphorites in Beaufort County, North Carolina. *Southeastern Geology*, **5**, 195–205.

Kimrey, J.O. (1965). Description of the Pungo River Formation in Beaufort County, North Carolina. *North Carolina Department of Conservation and Development, Division of Mineral Resources Bulletin*, **79**, 132pp.

Klitgord, K.D. & Behrendt, J.C. (1979). Basin structure of the United States Atlantic continental margin. In *Geologic and Geophysical Investigations of Continental Margins*, ed. J.S. Watkins, L. Montadert & P.W. Dickerson, pp. 85–112. American Association of Petroleum Geologists Memoir 29.

Lewis, D.W. (1981). 'Preliminary stratigraphy of the Pungo River Formation of the Atlantic continental shelf, Onslow Bay, North Carolina'. Unpublished M.Sc. thesis, East Carolina University, Greenville, NC, 75pp.

Lewis, D.W., Riggs, S.R., Hine, A.C., Snyder, Stephen W., Snyder, Scott W. & Waters, V.J. (1982). Preliminary stratigraphic report on the Pungo River Formation of the Atlantic continental shelf, Onslow Bay, North Carolina. In *Miocene of the Southeastern United States*, ed. T.M. Scott & S.B. Upchurch, pp. 122–37. Southeastern Geological Society and Florida Bureau of Geology Special Publication No. 25.

Luternauer, J.L. & Pilkey, O.H. (1967). Phosphorite grains: their application to the interpretation of North Carolina shelf sedimentation. *Marine Geology*, **5**, 315–20.

Lyle, M. (1984). 'Clay mineralogy of the Pungo River Formation, Onslow Bay, North Carolina continental shelf'. Unpublished M.Sc thesis, East Carolina University, Greenville, NC, 129pp.

Mallette, P.M. (1986). 'A lithostratigraphic analysis of cyclical phosphorite sedimentation, Miocene Pungo River Formation, North Carolina continental shelf'. Unpublished M.Sc. thesis, East Carolina University, Greenville, NC, 186pp.

Mallinson, D.J. (1988). 'Distribution and petrology of glauconitic sediments in the Miocene Pungo River Formation, North Carolina continental margin'. Unpublished M.Sc. thesis, East Carolina University, Greenville, NC, 125pp.

Martini, E. (1971). Standard Tertiary and Quaternary calcareous nannoplankton zonation. In *Proceedings of the Second Planktonic Conference*, v.2, pp. 739–85.Edizioni Tecnoscienza, Rome.

Matteucci, T.D. (1984). 'High-resolution seismic stratigraphy of the North Carolina continental margin – the Cape Fear region: sea-level cyclicity, paleobathymetry, and Gulf Stream dynamics'. Unpublished M.Sc. thesis, University of South Florida, St Petersburg, 100pp.

Matteucci, T.D. & Hine, A.C. (1987). Evolution of the Cape Fear Terrace: a complex interaction between the Gulf Stream and a paleoshelf edge delta. *Marine Geology*, **77**, 1–22.

Mearns, D.L. (1986). 'Continental shelf hardbottoms in Onslow Bay, North Carolina: their geology, biological erosion, and resistance to Hurricane Diana, September 11–13, 1984'. Unpublished M.Sc. thesis, University of South Florida, St Petersburg, 150pp.

Meisburger, E.P. (1979). Reconnaissance geology of the inner continental shelf, Cape Fear region, NC. *US Army Corps of Engineers, Coastal Engineering Research Center, Ft Belvoir, Va., Technical Report*, **TP79-3**, 135pp.

Miller, J.A. (1971). 'Stratigraphic and structural setting of the middle Miocene Pungo River Formation of North Carolina'. Unpublished Ph.D. thesis, University of North Carolina, Chapel Hill, 82pp.

Miller, J.A. (1982). Stratigraphy, structure, and phosphate deposits of the Pungo River Formation of North Carolina. *North Carolina Department of Natural Resources and Community Development, Geological Survey Bulletin*, **87**, 32pp.

Moore, T.L. (1986). 'Foraminiferal biostratigraphy and paleoecology of the Miocene Pungo River Formation, central Onslow Embayment, North Carolina continental margin'. Unpublished M.Sc. thesis, East Carolina University, Greenville, NC, 160pp.

Moretz, L. (1988). 'Diagenesis of benthic foraminifera in the Miocene Pungo River Formation of Onslow Bay, North Carolina continental shelf'. Unpublished M.Sc. thesis, East Carolina University, Greenville, NC, 92pp.

Palmer, A.A. (1988). Radiolarians from the Miocene Pungo River Formation of Onslow Bay, North Carolina continental shelf. In *Micropaleontology of Miocene Sediments in the Shallow Subsurface of Onslow Bay, North Carolina Continental Shelf*, ed. Scott W. Snyder, pp. 163–78. Cushman Foundation for Foraminiferal Research, Special Publication No. 25, Washington, DC.

Pilkey, O.H. & Luternauer, J.L. (1967). A North Carolina shelf phosphate deposit of possible commercial interest. *Southeastern Geology*, **8**, 33–51.

Popenoe, P. (1985). Cenozoic depositional and structural history of the North Carolina margin from seismic–stratigraphic analyses. In *Geologic Evolution of the United States Atlantic Margin*, ed. C.W. Poag, pp. 125–88. Van Nostrand Reinhold Co, New York.

Powers, E.R. (1986). Biostratigraphic correlation of Miocene phosphorites on the North Carolina continental margin using diatoms and silicoflagellates. *Society of Economic Paleontologists and Mineralogists, Annual Midyear Meeting Abstracts*, **3**, 91.

Powers, E.R. (1987). 'Diatom biostratigraphy and paleoecology of the Miocene Pungo River Formation, North Carolina continental margin'. Unpublished M.Sc. thesis, East Carolina University, Greenville, NC, 225pp.

Powers, E.R. (1988). Diatom biostratigraphy and paleoecology of the Miocene Pungo River Formation, Onslow Bay, North Carolina continental shelf. In *Micropaleontology of Miocene Sediments in the Shallow Subsurface of Onslow Bay, North Carolina Continental Shelf*, ed. Scott W. Snyder, pp. 97–162. Cushman Foundation for Foraminiferal Research, Special Publication No. 25.

Prell, W.L., Imbrie, J., Martinson, D.G., Morley, J.J., Pisias, N.G., Shackleton, N.J. & Streeter, H.E. (1986). Graphic correlation of oxygen isotope stratigraphy application to the late Quaternary. *Paleoceanography*, **1**, 137–62.

Riedel, W.R. & Sanfillipo, A. (1978). Stratigraphy and evolution of tropical Cenozoic radiolarians. *Micropaleontology*, **24**, 61–96.

Riggs, S.R. (1979). Phosphorite sedimentation in Florida – a model phosphogenic system. *Economic Geology*, **74**, 285–314.

Riggs, S.R. (1981). Relation of Miocene phosphorite sedimentation to structure in the Atlantic continental margin, southeastern United States. *American Association of Petroleum Geologists Bulletin*, **65**, 1669.

Riggs, S.R. (1984). Paleoceanographic model of Neogene phosphorite deposition, US Atlantic continental margin. *Science*, **223**, 123–31.

Riggs, S.R. & Belknap, D.F. (1988). Upper Cenozoic processes and environments of continental margin sedimentation. In *The Geology of North America, Volume 1–2, The Atlantic Continental Margin, US*, ed. R.E. Sheridan & J.A. Grow, pp. 131–76. Geological Society of America.

Riggs, S.R. & Freas, D.H. (1967). Submerged shoreline features on the shelf near Cape Fear, North Carolina. *Geological Society of America Special Paper*, **115**.

Riggs, S.R., Lewis, D.W., Scarborough, A.K. & Snyder, Scott W. (1982). Cyclic deposition of Neogene phosphorites in the Aurora area, North Carolina, and their possible relationship to global sea-level fluctuations. *Southeastern Geology*, **23**, 189–204.

Riggs, S.R., Snyder, Stephen W., Hine, A.C., Snyder, Scott W., Ellington, M.D. & Mallette, P.M. (1985). Geologic framework of phosphate resources in Onslow Bay, North Carolina continental shelf. *Economic Geology*, **80**, 716–38.

Rooney, T.P. & Kerr, P.F. (1967). Mineralogic nature and origin of phosphorite, Beaufort County, North Carolina. *Geological Society of America Bulletin*, **78**, 731–48.

Scarborough, A.K. (1981). 'Stratigraphy and petrology of the Pungo River Formation, central coastal plain of North Carolina'. Unpublished M.Sc. thesis, East Carolina University, Greenville, NC, 78pp.

Scarborough, A.K., Riggs, S.R. & Snyder, Scott W. (1982). Stratigraphy and petrology of the Pungo River Formation, central coastal plain, North Carolina. *Southeastern Geology*, **23**, 205–16.

Schlee, J.S., Dillon, W.P. & Grow, J.A. (1979). Structure of the continental slope off the eastern United States. In *Geology of Continental Slopes*, ed. L.J. Doyle, pp. 95–118. Society of Economic Paleontologists and Mineralogists Special Paper 27, Tulsa.

Schlee, J.S., Manspeizer, W. & Riggs, S.R. (1988). Paleoenvironments: offshore Atlantic US margin. In *The Geology of North America, Volume 1–2, The Atlantic Continental Margin, US*, ed. R.E. Sheridan & J.A. Grow, pp. 365–85. Geological Society of America.

Shackleton, N.J. & Opdyke, N.C. (1973). Oxygen isotope and paleomagnetic stratigraphy of equatorial Pacific core V28–V238: oxygen isotope temperatures and ice volumes on a 10^5 and 10^6 year scale. *Quaternary Research*, **3**, 39–55.

Shackleton, N.J. & Opdyke, N.C. (1976). Oxygen isotope and paleomagnetic stratigraphy of Pacific core V28–V239 late Pliocene to latest Pleistocene. In *Investigation of Late Quaternary Paleoceanography and Paleoclimatology*, ed. R.M. Cline & J.D. Hays, pp. 449–64. Geological Society of America Memoir 145.

Snyder, Scott W., ed. (1988). *Micropaleontology of Miocene Sediments in the Shallow Subsurface of Onslow Bay, North Carolina Continental Shelf*. Cushman Foundation for Foraminiferal Research, Special Publication No. 25, 189pp.

Snyder, Scott W., Crowson, R.A., Riggs, S.R. & Mallette, P.M. (1986). Geology of the Aurora Phosphate District. In *SEPM Field Guidebooks, Southeastern US Third Annual Midyear Meeting*, ed. D.A. Textoris, pp. 345–57. Society of Economic Paleontologists and Mineralogists.

Snyder, Scott W., Mauger, L.L. & Akers, W.H. (1983). Planktonic foraminifera and biostratigraphy of the Yorktown Formation, Lee Creek Mine. In *Geology and Paleontology of the Lee Creek Mine, North Carolina*, ed. C.E. Ray, pp. 455–82. Smithsonian Contributions to Paleobiology, No. 53, Washington, DC.

Snyder, Scott W., Riggs, S.R., Katrosh, M.R., Lewis, D.W. & Scarborough, A.K. (1982). Synthesis of phosphatic sediment–faunal relationships within the Pungo River Formation: paleoenvironmental implications. *Southeastern Geology*, **23**, 233–46.

Snyder, Scott W., Steinmetz, J.C., Waters, V.J. & Moore, T.L. (1988). Occurrence and biostratigraphy of planktonic foraminifera and calcareous nannofossils in Pungo River Formation sediments from Onslow Bay, North Carolina continental shelf. In *Micropaleontology of Miocene Sediments in the Shallow Subsurface of Onslow Bay, North Carolina Continental Shelf*, ed. Scott W. Snyder, pp. 15–42. Cushman Foundation for Foraminiferal Research, Special Publication No. 25.

Snyder, Stephen W. (1982). 'Seismic stratigraphy within the Miocene Carolina Phosphogenic Province: chronostratigraphy, paleotopographic controls, sea-level cyclicity, Gulf Stream dynamics, and the resulting depositional framework'. Unpublished M.Sc. thesis, University of North Carolina, Chapel Hill, NC, 183pp.

Snyder, Stephen W., Hine, A.C. & Riggs, S.R. (1982). Miocene seismic stratigraphy, structural framework and sea-level cyclicity: North Carolina Continental Shelf. *Southeastern Geology*, **23**, 247–66.

Snyder, Stephen W., Hine, A.C. & Riggs, S.R. (1983). 3-D stratigraphic modeling from high-resolution seismic reflection data: an example from the North Carolina continental shelf. *American Association of Petroleum Geologists Bulletin*, **67**, 549–50.

Snyder, Stephen W., Hine, A.C., Riggs, S.R. & Snyder, Scott W. (1986). Miocene unconformities, chronostratigraphy, and sea level cyclicity: fine-tuning early Neogene relative coastal onlap curve for North Carolina continental margin. *American Association of Petroleum Geologists Bulletin*, **70**, 651.

Steele, G.A. (1980). 'Stratigraphy and depositional history of Bogue Banks, North Carolina'. Unpublished M.Sc. thesis, Duke University, Durham, NC, 201pp.

Stewart, T.L. (1985). 'Carbonate petrology and sedimentology of the Miocene Pungo River Formation, Onslow Bay, North Carolina continental shelf'. Unpublished M.Sc. thesis, East Carolina University, Greenville, NC, 184pp.

Waters, V.J. (1983). 'Foraminiferal paleoecology and biostratigraphy of the Pungo River Formation, southern Onslow Bay, North Carolina continental shelf'. Unpublished M.Sc. thesis, East Carolina University, Greenville, NC, 186pp.

Waters, V.J. & Snyder, Scott W. (1986). Planktonic foraminiferal biostratigraphy of the Pungo River Formation, southern Onslow Bay, North Carolina continental shelf. *Journal of Foraminiferal Research*, **16**, 9–23.

30

The seismic stratigraphic record of shifting Gulf Stream flow paths in response to Miocene glacio-eustacy: implications for phosphogenesis along the North Carolina continental margin

STEPHEN W. SNYDER, A.C. HINE AND S.R. RIGGS

Abstract

Miocene deposition along the North Carolina continental margin occurred in response to a global, second-order, high stand of sea level (TB2). However, the depositional evolution of this continental margin was completely regulated by multiple, higher-frequency (100–300 ky), high-amplitude (50–200 m), sea-level fluctuations. Physical stratal relationships indicate the ancestral western boundary current (Gulf Stream precursor) traversed a lateral path tens of kilometers wide in response to each of these higher-frequency cycles of sea-level change. Moreover, the geometrics portrayed in regional Miocene outcrop–subcrop maps suggest the shifting flow-track positions also dictated the evolution and decay of local paleotopographic controls, the development of shelf embayments and basins, and also significantly influenced the sedimentary infill histories of these local depocenters.

Along the North Carolina continental margin, secular variations in location and flow configuration of the Miocene western boundary current appear to have been determined by an interplay between the position of sea level relative to the elevation of major bathymetric highs (obstructions). Important Miocene bathymetric barriers included the Charleston Bump Complex (CBC) and the Cape Lookout High (CLH). During relatively low sea levels, major boundary current frontal events initiated by the CBC resulted in steering the Gulf Stream seaward of the North Carolina margin. This stimulated shelf-margin growth and sediment drift-building episodes. Conversely, when sea levels were relatively high, the Gulf Stream bypassed barriers within the CBC and scoured the North Carolina shelf margin before being deflected eastward by the CLH farther downstream. Repeated scour associated with these higher stands of sea level eventually excavated a large shelf-margin basin (outer Onslow Bay Basin).

Persistent upwelling episodes resulted when the Gulf Stream had to steer seaward of topographic obstructions. Topographic steering originates from instabilities set-up via horizontal shear between streamlines in a boundary current as it swiftly flows past stationary bathymetric features. Common by-products of this type of frictional interaction are the initiation of topographic waves and the subsequent shedding of cyclonic eddies. The latter are characterized as shallow, warm-cored, frontal filaments separated from the main axis of the Gulf Stream by upwelled slope waters. Miocene flow-track maps show that the ancestral western boundary current had to steer around the Cape Lookout High several times. The sites of predicted upwelling episodes resulting from these deflection events appear to coincide with the organic-rich depositional regimes which ultimately led to phosphate mineralization within the Aurora Embayment. Although the link between the Cape Lookout High and the Aurora Phosphate District is now established, more work is needed to better define the relative timing between topographic steering, upwelling, and phosphogenesis within this system.

Repetitive lateral shifts by the Gulf Stream in response to multiple sea-level changes provided the foundation for the cyclicity observed in Miocene depositional and erosional patterns of North Carolina. More importantly, however, it was the relative position of sea level with respect to major bathymetric obstructions which controlled the position, flow dynamics, and therefore the influence of the Miocene boundary current on sedimentary patterns found in the Pungo River Formation.

Introduction

The goal of this paper is to demonstrate that the evolution of the North Carolina continental margin was strongly influenced by multiple landward–seaward shifts in the position of the Gulf Stream during the Miocene. Each lateral shift by this boundary current was a direct response to short-pulsed, high-amplitude, sea-level fluctuations (Stephen W. Snyder, 1982 and 1989; Stephen W. Snyder *et al.*, 1986b; Riggs *et al.*, Chapter 29, this volume).

Documenting that the ancestral Gulf Stream physically interacted with the North Carolina shelf margin during the Miocene represents a significant step forward in understanding the origin of the phosphate-rich sequences found within the Pungo River Formation by Brown (1958) and others (Kimrey, 1964; Gibson, 1967 and 1982; Miller, 1971 and 1982; Brown, Miller & Swain, 1972; Steele, 1980; Lewis *et al.*, 1982; Riggs *et al.*, 1982, 1985 and Chapter 29, this volume; Mallette, 1986; Riggs & Mallette, Chapter 31, this volume). The significance of establishing this relationship arises from three facts: (1) Pungo River phosphorites formed on an east-facing continental margin; (2) upwelling conditions are usually required to induce primary phosphogenic episodes; and (3) within eastern pericontinental settings, upwelling events are chiefly generated via frictional interactions between a strong boundary current and topographic highs.

Boundary currents, upwelling and phosphogenesis

The steering of a boundary current around bathymetric obstructions is known as topographic steering. It creates instabilities within the flow structure of the boundary current (Fofonoff, 1954, 1962 and 1981; Holton, 1972; Pedlosky, 1987; Veronis, 1981; Pond & Piccard, 1983; Hendershot, 1987). Downstream from obstructions, the boundary current adjusts to compensate for these instabilities in order to maintain coherent flow in a steady-state system. Adjustments may include the initiation of topographic waves and the shedding of meso-scale eddies (Lee, 1975; Legeckis, 1976 and 1979; Lee & Mayer, 1977; Maul, 1977 and 1978; Brooks & Bane, 1978 and 1981; Maul *et al.*, 1978; Pietrafesa, Atkinson & Blanton, 1978; Bane & Brooks, 1979; Chao & Janowitz, 1979; Vukovitch *et al.*, 1979a, b; Hulbert & Thompson, 1980; Vukovitch & Crissman, 1980; Bane, Brooks & Lorenson, 1981; Bane, 1983; Lee & Atkinson, 1983; Lee & Waddel, 1983; Olson, Brown & Emmerson, 1983; Singer *et al.*, 1983; Vukovitch & Maul, 1985; Auer, 1987; Leaman, Molinari & Verltes, 1987).

Some eddies travel westward, up the slope, and onto the outer continental shelf. These westward-propagating gyres are characterized by counter-clockwise or cyclonic circulation. Ekman pumping within the center of cyclonic gyres (on the northern hemisphere) commonly results in a slow but continuous vertical flux of nutrients residing near or below the thermocline (Janowitz & Pietrafesa, 1980 and 1982; Lee, Atkinson & Legeckis, 1981; Yodder *et al.*, 1981a, b, 1983; Brooks & Bane, 1983; Paluszkiewicz *et al.*, 1983; McClain, Pietrafesa & Yodder, 1984 and 1985; McClain & Atkinson, 1985; Vukovitch, 1986; Zantopp *et al.* 1987). This type of upwelling is described as topographically-induced. It is the only known modern mechanism for generating annual upwelling events on east-facing continental margins (Bane, 1983; Lee & Waddell, 1983; Pietrafesa, 1983; Singer *et al.*, 1983; Atkinson, Menzel & Bush, 1985; Auer, 1987).

Upwelling conditions favor the burial of organic carbon (Mueller & Suess, 1979). Thus, they also favor interstitial water chemistries and diagenetic reactions leading to phosphate mineralization (Burnett, 1977). Rising nutrients enter the euphotic zone and are fixed in organic carbon complexes via photosynthesis (e.g. carbohydrates, proteins, and lipids). Some organic matter produced in the water column eventually settles to the seafloor where it begins to decay. Heterotrophic bacteria oxidize the carbon, a process which consumes oxygen to produce bicarbonate at seawater pH. This process continues to deplete dissolved oxygen concentrations in the near bottom waters until dysoxic–anoxic seafloor conditions prevail. Once aerobic bacteria can no longer tolerate the low levels of oxygen, much of the particulate organic carbon (POC) raining to the seafloor is buried instead of remineralized at the surface (Aller, 1980). The buried POC becomes a key ingredient for anaerobic bacterial use in catalyzing authigenic precipitation of inorganic minerals within the sulfate reduction and methanogenic zones of the sediment column (Claypool & Kaplan, 1974; Berner, 1980; Baker & Burns, 1986; Compton, 1988). It is within these suboxic–anoxic zones of early diagenesis where the bulk of the phosphate minerals are formed (Burnett, 1977 and Chapter 5, this volume; McArthur *et al.*, 1986).

For these reasons, most investigators now consider intense or prolonged upwelling to be the foremost prerequisite for a primary phosphogenic event (Kazakov, 1937; Cook & McElhinny, 1979; Sheldon, 1981; Parrish, 1982 and Chapter 19, this volume). Because the Miocene phosphorites of North Carolina formed on an east-facing continental margin, and long-term upwelling is a prerequisite to phosphorite formation, boundary current upwelling must then be considered essential to explain both the genesis and distribution of phosphatic sequences found within the Pungo River Formation.

Links between boundary current upwelling and the occurrence of Miocene phosphorites in North Carolina have been previously inferred from a succession of characteristic vertical lithofacies changes (Riggs, 1984; Riggs *et al.*, 1985; Mallette, 1986; Riggs & Mallette, Chapter 31, this volume). However, direct evidence of boundary current activity influencing Miocene depositional patterns within the North Carolina continental margin has not been presented. In this paper, we attempt to fill this gap by using physical stratigraphic relationships to map relict scour paths of the ancestral Gulf Stream. These data depict frequent, large-scale, lateral shifts by the main axis of flow. Submarine erosional scars identified in the Miocene sequences also testify that the Gulf Stream frequently passed across the North Carolina continental margin. Thus, data and interpretations presented here complement extrapolations made solely from lithofacies patterns. They show that boundary current scour was indeed proximal to the local sites of phosphorite formation.

Miocene biochronostratigraphic sections: definition, distribution and anatomy

Previous Miocene stratigraphic studies of the North Carolina continental margin utilized a dense network of continuous seismic reflection data to discriminate between local and regional (cross-shelf) truncation surfaces (Stephen W. Snyder, 1982; Popenoe, 1985). Regional unconformities were used to construct a relative chronostratigraphic framework for the entire Miocene section of the North Carolina continental margin. That framework now consists of at least 18 discrete depositional sequences (Fig. 30.1). The time–stratigraphic framework provided by the sequence stratigraphy has been used as a basin-wide correlation tool. Physical stratigraphic, biostratigraphic, and lithostratigraphic data were integrated (Fig. 30.2), and a preliminary sea-level curve (Fig. 30.3) for the North Carolina continental margin was constructed (Stephen W. Snyder *et al.*, 1986a). The detailed history of these Miocene sea-level changes including estimates for highstand–lowstand positions, frequency of change, nature of hiatuses, as well as all supporting data, are presented elsewhere (Stephen W. Snyder, 1989). However, it is important to note here that the periodicity and amplitudes of the higher-frequency sea-level cycles preclude any other driving mechanism other than glacio-eustacy (see Table 29.1 in Riggs *et al.*, Chapter 29, this volume).

The grid of intersecting seismic profiles used to develop both the high-resolution chronostratigraphic framework (Figs. 30.1 and 30.2) and the local sea-level curve (Fig. 30.3) is illustrated in Figure 29.9 of Riggs *et al.* (Chapter 29, this volume). This data base also supplied subbottom stratigraphic control for mapping

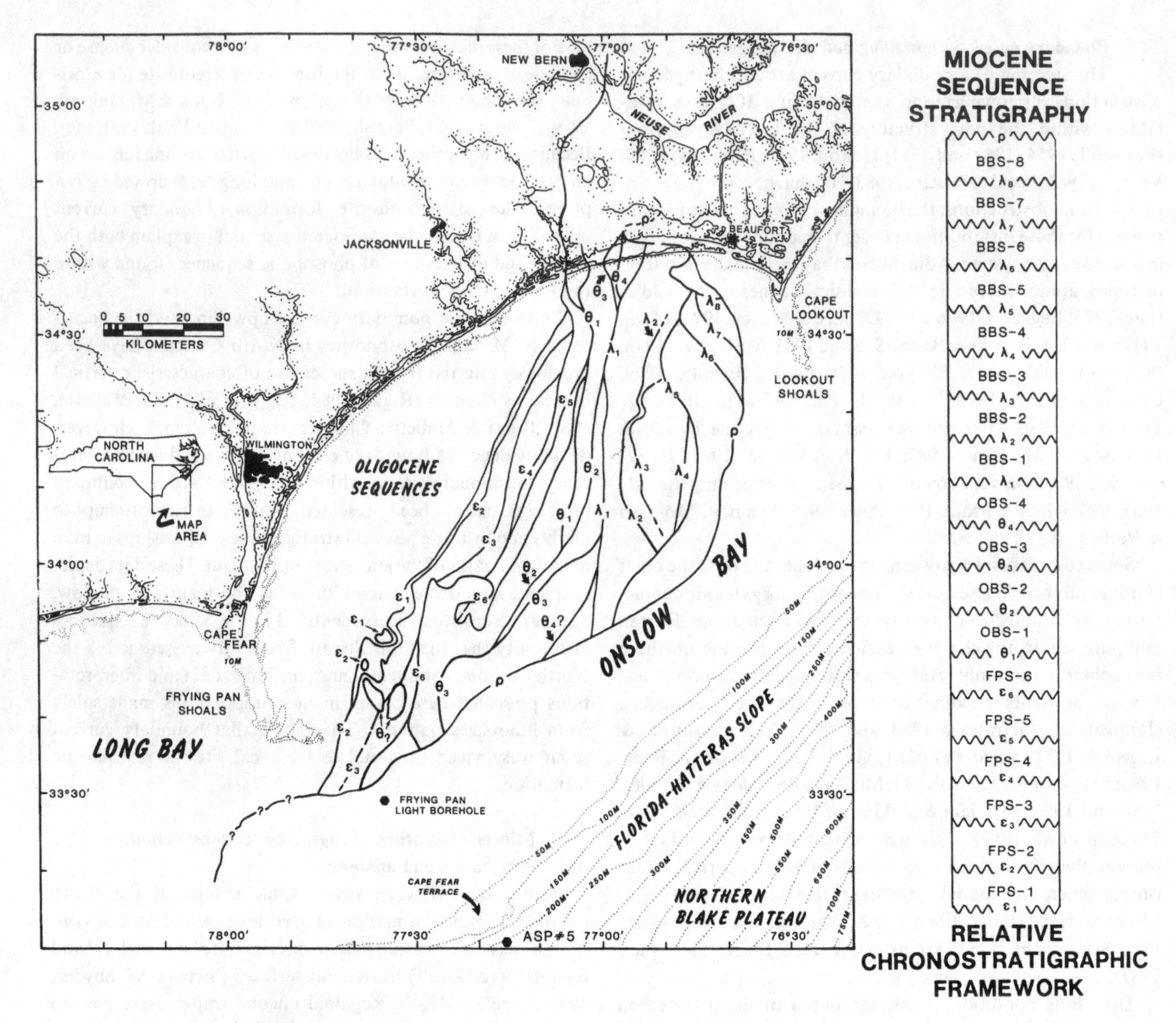

Fig. 30.1. Map shows the distribution of the Miocene sequences and correlative unconformities that crop out on the North Carolina continental shelf of Onslow Bay (after Snyder, Stephen W., 1982). Chronostratigraphic column (right) illustrates the relative chronology of the Miocene sequences and unconformities based on their superposition as identified in the subsurface of the North Carolina continental margin.

the spatial distribution and geometries of specific Miocene time–stratigraphic sections (Stephen W. Snyder, 1982; Stephen W. Snyder, Hine & Riggs, 1983; Stephen W. Snyder *et al.*, 1986a, b). Maps for the three major biochronostratigraphic sections are presented here. These sections were defined by mating the seismic and biostratigraphic data. They are referred to as: (1) the mid-Burdigalian Section (≃ 18–17 ma); (2) Langhian Section (≃ 16–15 ma); and (3) Serravallian Section (≃ 13–12 ma). Figure 30.2 shows that each of these stratal sections is separated by a major hiatus; yet, each section can be subdivided into 4 or more distinct depositional sequences based on the identification of basin-wide unconformities (Figs. 30.1 and 30.4). Each of the sequences represents a separate depositional episode associated with one of the individual glacio-eustatic cycles illustrated in Figure 30.3. The larger biochronostratigraphic sections may correspond to segments or even complete cycles of lower-frequency (3rd order) sea-level events (Scott W. Snyder, 1988).

Burdigalian Section

Definition The Burdigalian section is represented by six depositional sequences (Figs. 30.1 and 30.2). These sequences are referred to as the Frying Pan Sequences (FPS) because they crop out on the continental shelf adjacent to Frying Pan Shoals, a large Quaternary sand massif located seaward of the Cape Fear cuspate foreland. This section extends seaward in the subsurface. It thickens to over 90 m beneath the present shelf margin off Cape Fear (Fig. 30.5).

Planktic biostratigraphic studies indicate that the Frying Pan

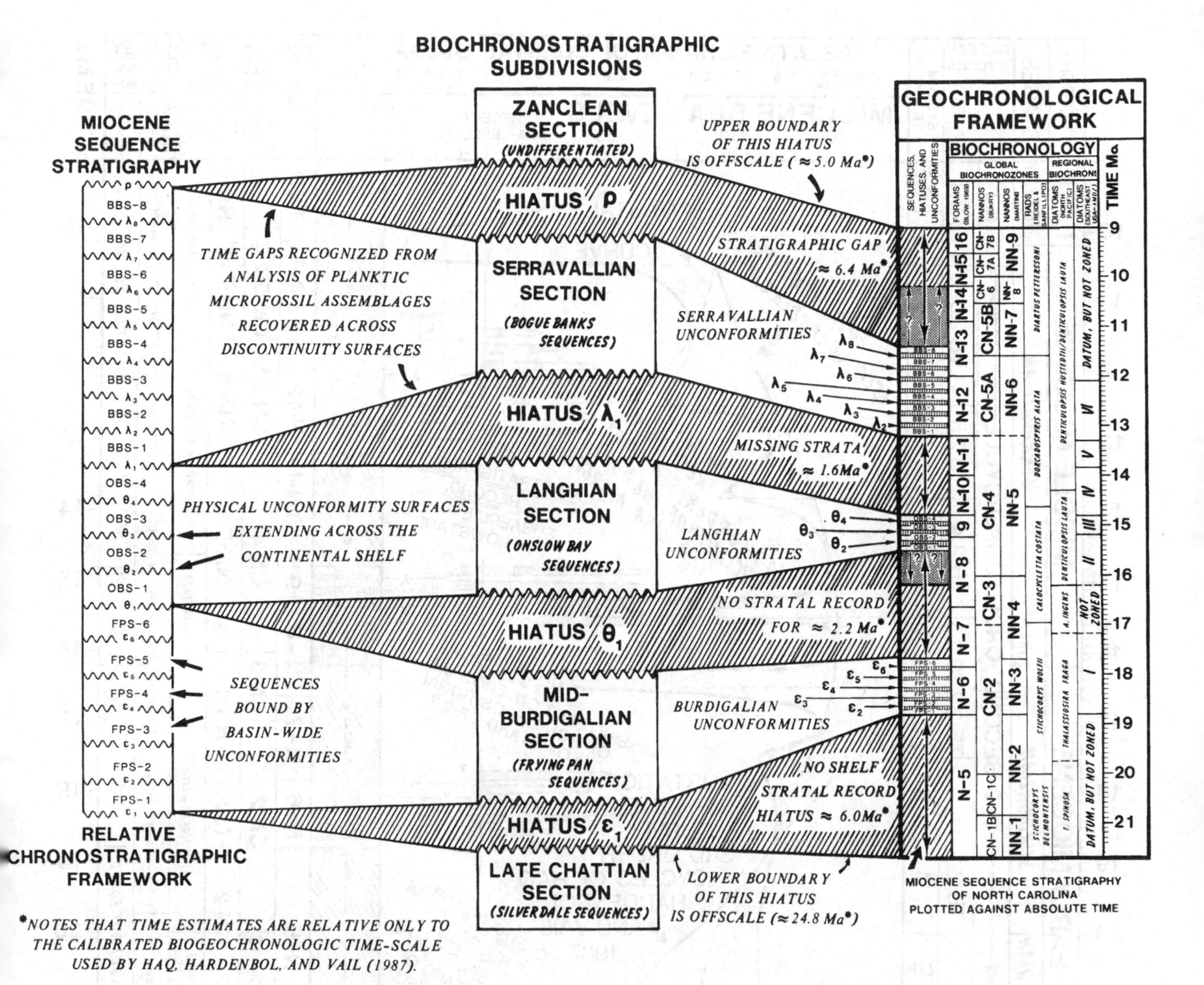

Fig. 30.2. Three stratigraphic columns depicting the relationships between the sequence stratigraphy, the biostratigraphy, and the geochronology of the composite chronostratigraphic framework for the Miocene section of Onslow Bay, North Carolina. (left column) Relative chronostratigraphic framework is based on identification of physical unconformities from seismic-reflection profiles. (middle column) Biostratigraphic sections and correlative time–stratigraphic gaps (hiatuses) identified using planktic microfossils based on Scott W. Snyder (1988). (right column) Geochronology of the Miocene sequences, unconformities, and hiatuses using the *in situ* microfossils to correlate to global planktonic biochrons and, thereby, hang the sequences on the calibrated time scale of Haq *et al.* (1987). Differences in absolute values of time from that shown in Riggs *et al.* (Chapter 29, this volume) are a by-product of precision below the resolution of modern biochrons.

Sequences and their bounding unconformities represent an absolute time frame spanning approximately 1.0 ma (Figs. 30.2 and 30.3; Scott W. Snyder, 1988). Deposition started at the N-5–N-6 zonal boundary of Blow (1969) and terminated near the NN-3–NN-4 nannoplankton zonal boundary of Martini (1971).

Distribution and lithologies Vibracores penetrating the Frying Pan Sequences in southwest Onslow Bay have recovered a variety of unlithified quartz sand and finer-grained lithofacies. They include muddy phosphorites, muddy phosphatic sands, and microfossiliferous very muddy sand (Lewis *et al.*, 1982; Riggs *et al.*, 1985; Mallette, 1986; Riggs & Mallette, Chapter 31, this volume). In general, FPS lithologies increase in siliciclastic sand content northward to central Onslow Bay, and contain abundant molluscan shells and barnacle plates in northern Onslow Bay (Steele, 1980; Lewis *et al.*, 1982; Mallette, 1986; Riggs & Mallette, Chapter 31, this volume). Analyses of the benthic foraminiferal assemblages depict a middle sublittoral (shelf) environment with deposition in < 100 m of water (Waters, 1983; Moore & Scott W. Snyder, 1985; Scott W. Snyder, Chapter 32, this volume).

In the subsurface, the Mid-Burdigalian shelf sequences form

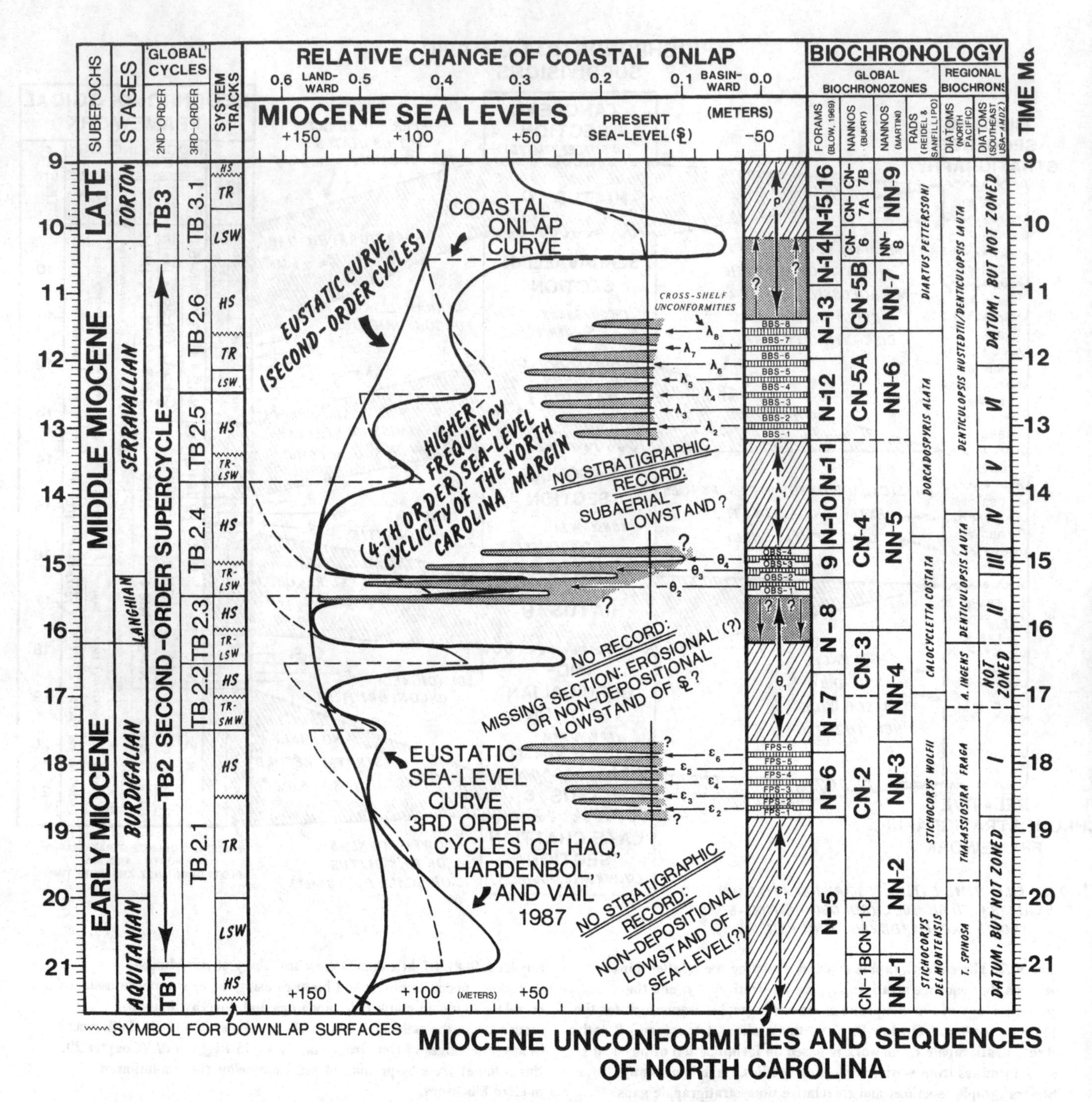

Fig. 30.3. Miocene sea-level curve depicting the North Carolina record of high-frequency, high-amplitude cycles, and the eustatic cycles proposed by Haq *et al.* (1987). The coastal onlap curve, eustatic sea-level estimates, juxtaposition of global biostratigraphic zonal schemes, as well as the calibration of these biochrons with absolute time are all from Haq *et al.* (1987). Planktic biostratigraphic zones are as follows: FORAMS (N zonal scheme) – Blow (1969); NANNOS (CN zonal scheme) – Okada & Bukry (1980), and Bukry (1981); NANNOS (NN zonal scheme) – Martini (1971); RADS (radiolarian zonal scheme) – Reidel & Sanfillipo (1978) and Sanfillipo *et al.* (1981) DIATOMS (North Pacific regional zonal scheme) – Fenner (1985) and Barron (1985a); DIATOMS (southeast USA continental margin regional zonal scheme) – Abbott (1978, 1980 and 1984). Figure modified from Stephen W. Snyder *et al.* (1986a).

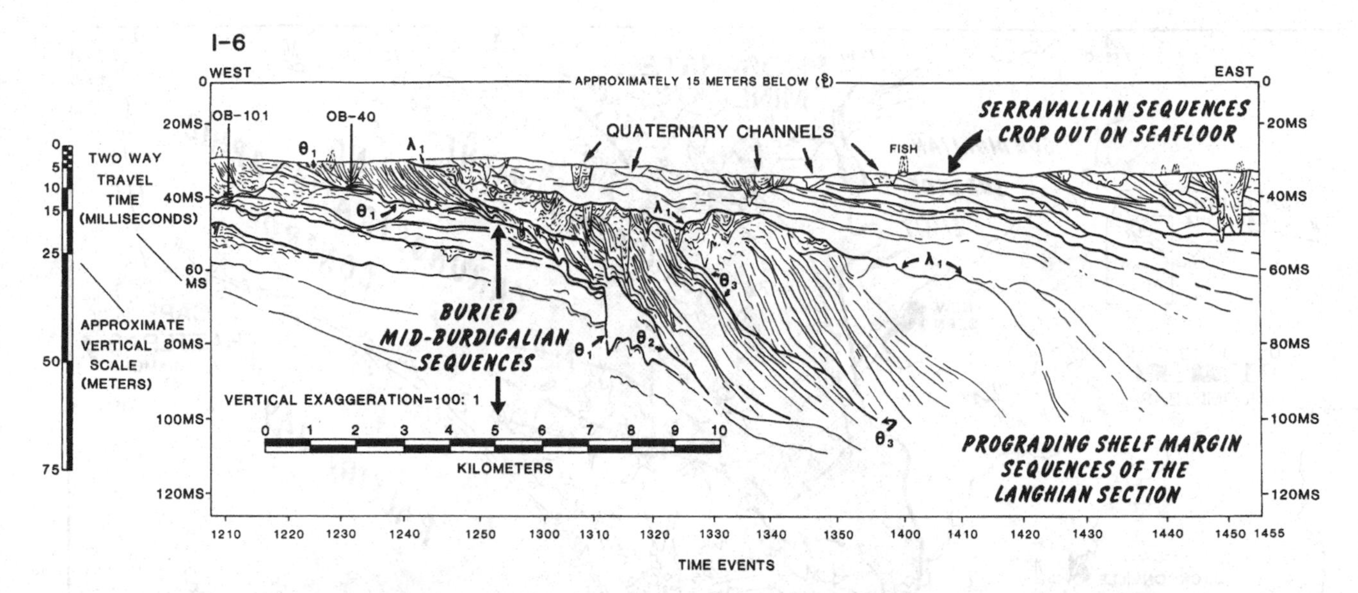

Fig. 30.4. Reduced stratigraphic line-drawing as interpreted from UNIBOOM seismic-reflection profile I-6. The parallel to tangential oblique-clinoforms represent the characteristic seismic facies of the Langhian Onslow Bay Sequences where they prograde across the subbottom escarpment defined by reflector θ_1. Strata horizons below θ_1 are Burdigalian in age, while those above reflector λ_1 are Serravallian in age. For more details on reflector nomenclature see Figure 30.1. Vertical lines labelled OB-101 and OB-40 represent the position and penetration limits of vibracores recovering Miocene sediments. The approximate vertical scale was calculated using seismic velocities (two-way travel time) of 1500 m s^{-1} in the water column and 1700 m s^{-1} in the subsurface. Horizontal scale and vertical exaggeration (100:1) were rectified and held constant via a digitizing program used to reduce the raw data to stratal line-drawings. Location of this seismic profile is shown in Figures 30.9 and 30.12.

a depositional wedge which thickens from its western edge to over 90 m beneath the present shelf margin of southeast Onslow Bay (Fig. 30.5). The wedge then thins abruptly beneath the modern shelf–slope break. No Burdigalian sequences were mapped in subcrop adjacent to Atlantic Slope Project well ASP-5 (Fig. 30.5). Seismic data show the absence of Burdigalian strata in this area is not the product of lateral thinnning across a paleotopographic high, but rather a consequence of severe erosional truncation (Fig. 30.6). Abrupt lateral thinning of the Burdigalian stratigraphic wedge also occurs in northeastern Onslow Bay west of Cape Lookout (Fig. 30.5). This pattern of thinning was caused by erosional truncation as well (Fig. 30.7).

Beneath the emerged coastal plain of eastern North Carolina, the Burdigalian section becomes thicker again as it enters a large, relict, Miocene shelf embayment referred to as the Aurora Embayment (Fig. 30.5). Here, the Burdigalian section is represented by the two lower lithostratigraphic units in the Pungo River deposits exposed in the quarry of Texasgulf Chemical Company (Aurora Lithostratigraphic Units ALU-A and ALU-B in Fig. 30.8). These two units, like all the Aurora Lithostratigraphic Units described and defined by Riggs *et al.* (1982), are characterized by the vertical succession of three lithofacies. In ascending order, they are: (1) a basal fine-grained siliciclastic facies; (2) a muddy phosphorite; and (3) a lithified carbonate caprock which may include borings from macro-infauna (endoliths).

Langhian section

Definition The Langhian section forms a north–south striking outcrop belt across central Onslow Bay (Fig. 30.1). It consists of four prograding shelf-margin sequences (Fig. 30.4) which have been designated the Onslow Bay Sequences (OBS).

Planktic fauna and flora from the Onslow Bay Sequences have been used to constrain the timing of deposition to the Langhian (Moore, 1986; Powers, 1986, 1987 and 1988; Scott W. Snyder *et al.*, 1988). These indices depict initial deposition started just above the NN-4–NN-5 boundary of Martini (1971), while the youngest Langhian sequence was deposited prior to the N-9–N-10 zonal boundary of Blow (1969).

Distribution and lithologies As illustrated in Figure 30.9, the Langhian section thickens eastward from its updip limit in central Onslow Bay to 250 m beneath the northern Blake Plateau, and to 350 m along the Blake Escarpment at 24° 30′ N latitude. The Langhian section is thicker along its updip limit in Onslow Bay than along its subcrop limit at the western margin of the Aurora Embayment (Fig. 30.9). This pattern is primarily due to a subbottom erosional escarpment which directly controls the infill pattern of Langhian sequences in central Onslow Bay (Fig. 30.4). This is the same escarpment responsible for the truncated Burdigalian subcrop in northeast Onslow Bay (Figs. 30.5 and 30.7).

North of Onslow Bay, the Langhian section thins and pinches

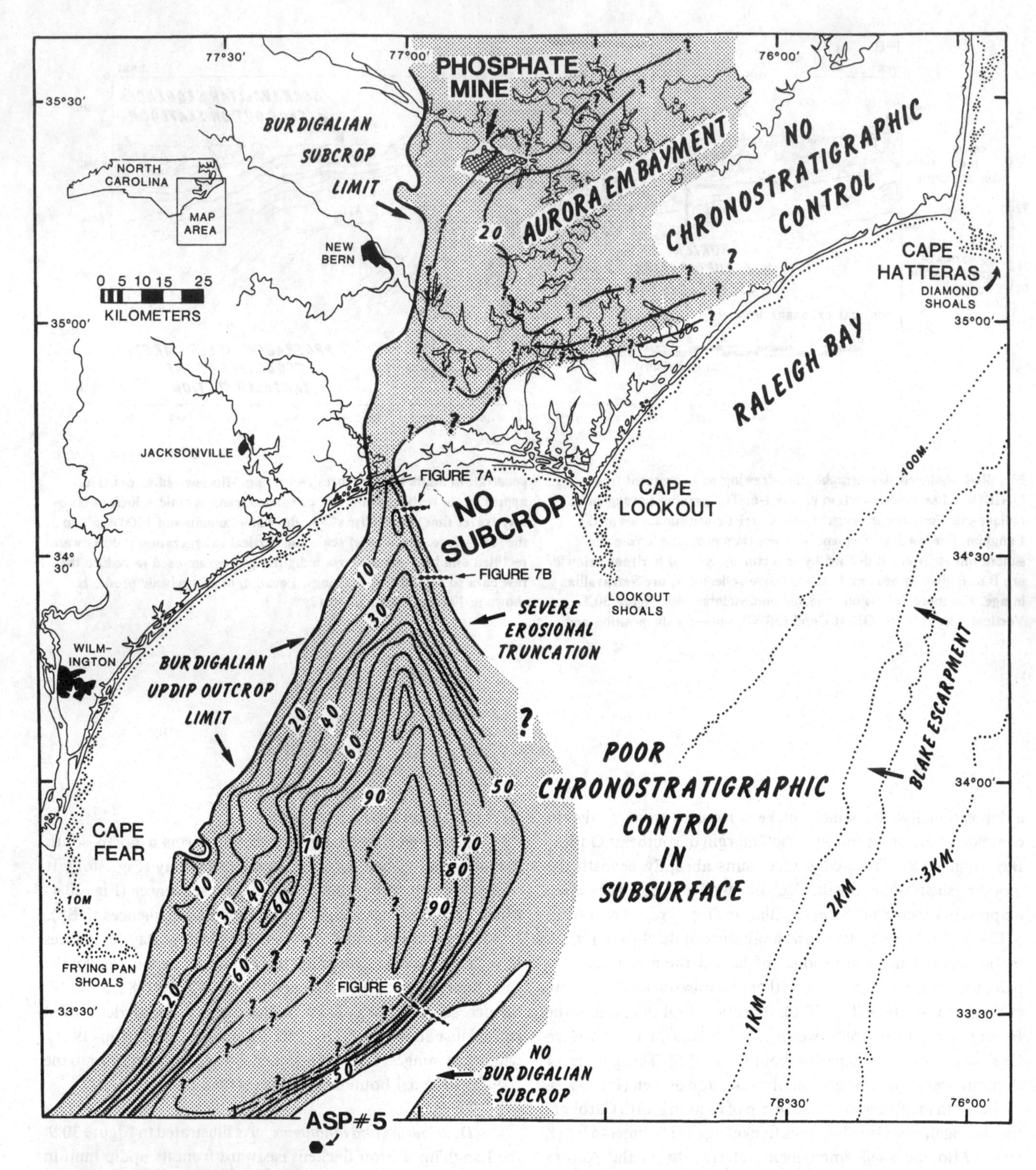

Fig. 30.5. Isopach map of the Mid-Burdigalian section of North Carolina based on interpretations from the network of seismic-reflection data illustrated in Figure 29.4 (Riggs *et al.*, Chapter 29, this volume). Contours are in meters assuming a constant subsurface velocity of 1700 m s^{-1} two-way travel time. Poor chronostratigraphic control along the shelf margin is a product of the section dipping below the penetration limitations of the seismic tool used on the shelf of Onslow Bay. This precluded direct correlations to the air-gun records of Popenoe (1985). No chronostratigraphic control west of Cape Hatteras is the product of penetration limitations of the seismic source used in Pamlico Sound. Contour map modified from Stephen W. Snyder *et al.* (1986a,b).

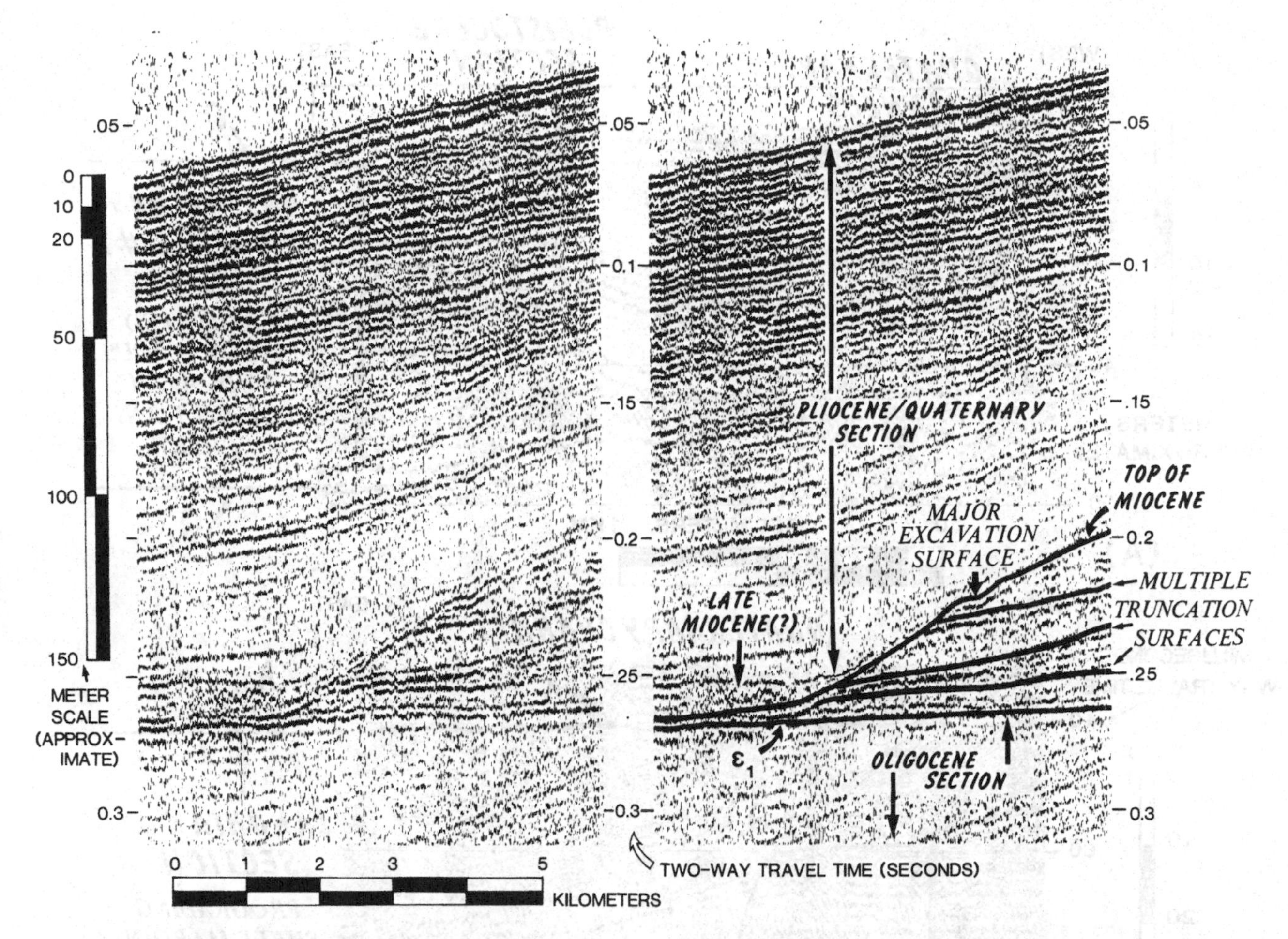

Fig. 30.6. Raw (left) and interpreted (right) seismic record of multiple Miocene truncation surfaces. Most of the Miocene subcrop in the proximity of Atlantic slope core-site ASP-#5 was stripped away when this surface was cut (e.g. no or thin subcrops shown in Figs. 30.5, 30.9, and 30.12). The major cut probably evolved during the 2nd-order lowstand between global supercycles TB-2 and TB-3 (Tortonian through Messinian Stages). The multiple truncation surfaces found on the landward side (northwest) of this unconformity were cut during the Burdigalian, Langhian, and Serravallian Stages. This seismic section is located in Figures 30.5, 30.9 and 30.12. Two-way subsurface travel time is given in seconds. Approximate vertical scale is given in meters assuming a constant velocity of 1700 m s^{-1}.

out (northward) across an antecedent high beneath the Cape Lookout area. It thickens slightly again to the north within the relict shelf embayment of Aurora (Fig. 30.9). In this embayment, the Langhian section is represented by Aurora lithostratigraphic units ALU-C and ALU-D of Riggs *et al.* (1982) (Fig. 30.8). Correlation between the Aurora Embayment and Onslow Bay is primarily based on calcareous plankton and siliceous floral indices (Katrosh & Scott W. Snyder, 1982; Powers, 1986 and 1987; Scott W. Snyder, 1988). Direct ties via the estuarine seismic lines were inhibited by gas-charged sediments and stratal pinch-out patterns.

Vibracores penetrating the shelf-margin sequences in central Onslow Bay depict a series of interbedded lithic sections. Clean quartz sands (0.5–3 m thick) alternate with 10–30 cm thick mud lenses composed primarily of silt-sized dolomite or 'dolosilt' (Mallette, 1986; Riggs & Mallette, Chapter 31, this volume). This interbedded facies grades northward into a carbonate facies characterized by non-reefal skeletal sands (barnacle, molluscan, and bryozoan macrofossil shell hash) with varying amounts of quartz sand and lime mud (Steele, 1980; Lewis *et al.*, 1982; Scarborough, Riggs & Lewis, 1982; Mallette, 1986; Riggs & Mallette, Chapter 31, this volume).

Interpretations of benthic foraminiferal assemblages indicate a middle–outer sublittoral paleoenvironment of deposition (Katrosh & Scott W. Snyder, 1982; Moore & Scott W. Snyder, 1985; Moore, 1986; Scott W. Snyder, Chapter 32, this volume). This is consistent with the shelf-margin setting inferred from the seismic facies which are characterized by steep, prograding, parallel oblique and tangential clinoforms (Fig. 30.4).

The Langhian shelf-margin sequences of central Onslow Bay grade downdip (eastward) into a thicker sequence of laminated slope-front fill sediments (Figs. 30.9, 30.10 and 30.11). This

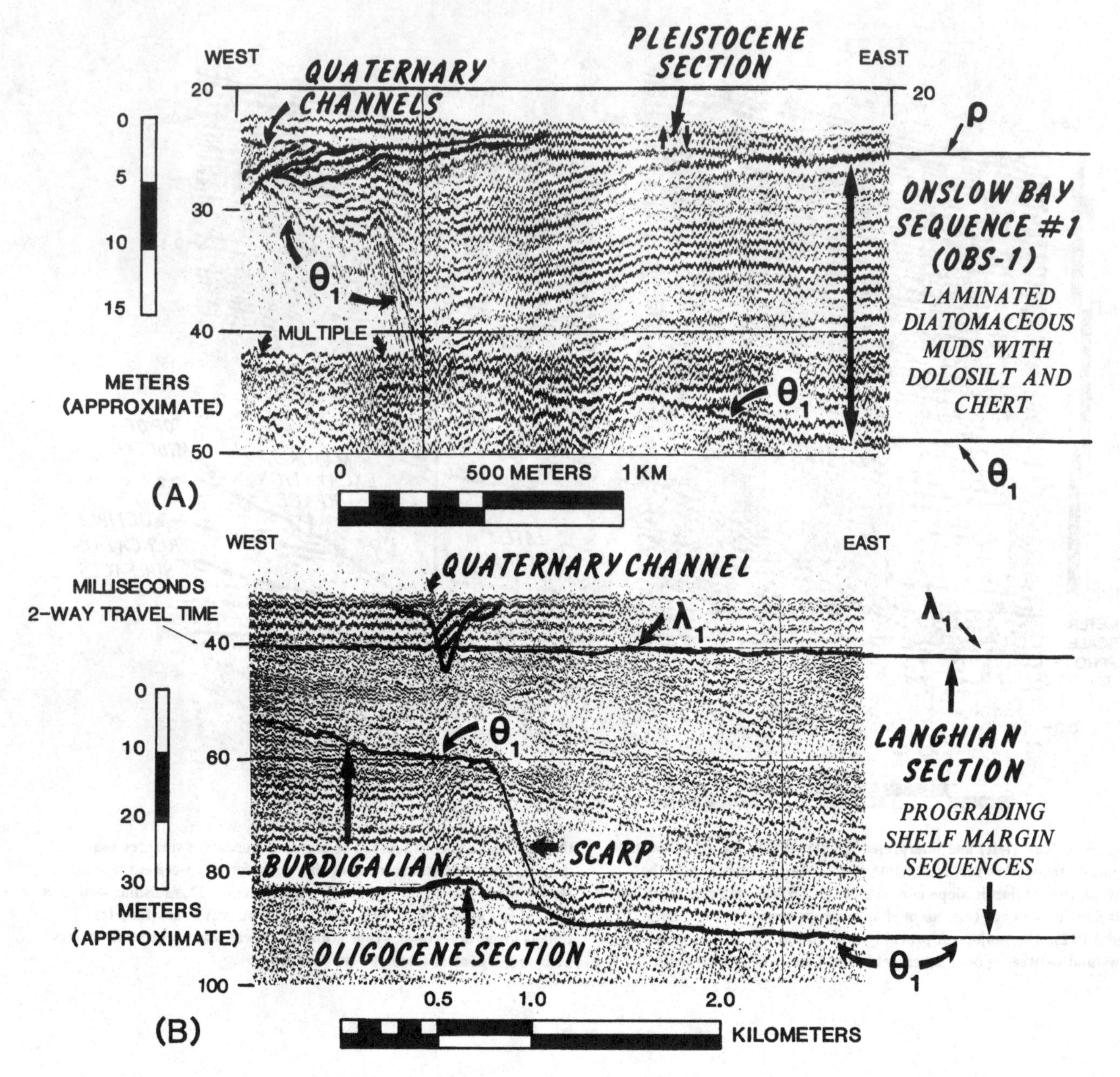

Fig. 30.7. Two seismic-reflection profiles across the same erosional escarpment. These subbottom features are responsible for the lack of a Burdigalian subcrop beneath northeastern Onslow Bay. Two-way subsurface travel time is given in milliseconds. Approximate vertical scale is given in meters assuming a constant velocity of 1700 m s^{-1} two-way travel time. Location of both seismic sections are shown in Figure 30.5.

sedimentary section reaches 250 m thick beneath the northern Blake Plateau and is interpreted to represent uniform microfossiliferous sandy muds with abundant organic matter and diatoms (Riggs & Mallette, Chapter 31, this volume). There are no major discontinuity surfaces correlative to unconformities separating the four Langhian shelf-margin sequences updip to the west. That is, unconformities θ_2, θ_3 and θ_4 in Figures 30.2, 30.3 and 30.4 have no correlative discontinuity surfaces downdip.

The absence of unconformities within the Langhian section underlying the Northern Blake Plateau indicates that, at least within this deeper-water paleoenvironmental setting, deposition of pelagic and hemipelagic sediments was relatively uninterrupted. This style of deposition produced a more complete Langhian record beneath the present northern Blake Plateau off North Carolina. Only partial records are preserved within the shelf-margin sequences to the west due to alternate exposure and flooding during successive sea-level cycles.

The eastern margin of the Langhian subcrop is a very sharp boundary which parallels the Blake Escarpment across most of the North Carolina continental margin (Fig. 30.9). This truncated boundary is interpreted to be a post-depositional scour

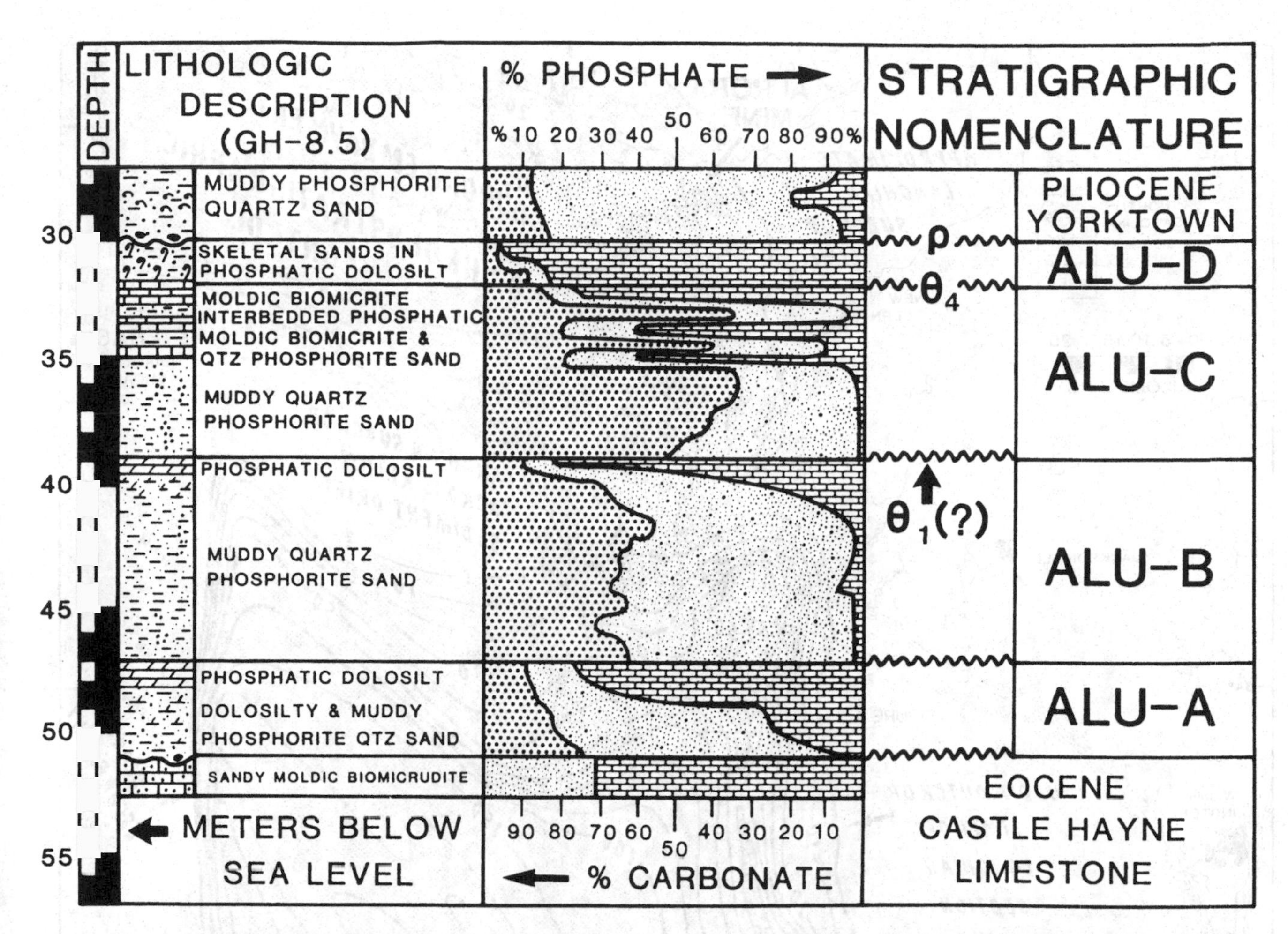

Fig. 30.8. Miocene lithostratigraphic units (ALU) defined from core and quarry sections within the Aurora Embayment. Modified from Riggs *et al.* (1982). Chronostratigraphic correlations with the sequences and unconformities defined in Onslow Bay are based on: (A) limited correlations through estuarine seismic lines; (B) lithologic correlations (across the Cape Lookout High) using drill, bore, and vibracore data from Steele (1980), Lewis *et al.* (1982) and Scarborough *et al.* (1982), and, planktonic microfossil assemblages as defined in Katrosh & Scott W. Snyder (1982), Powers (1987 and 1988), and Scott W. Snyder (1988). Location of the quarry (Aurora Mine) is shown in Figures 30.5, 30.9 and 30.12.

surface cut by the Western Boundary Undercurrent during the Late Neogene (Popenoe, 1985).

The western limit of the Langhian section and associated stratal thinning south of 33° 30′ N latitude (Fig. 30.9) resulted from multiple truncation events. The Langhian subcrop thins adjacent to drill site ASP-5, a trend which continues toward the northeast beneath the entire length of the present shelf–slope break. Note that this erosional surface has severed the 250 m isopach into two, separate, closed contours (Fig. 30.9).

Serravallian section

Definition The Serravallian section includes at least eight depositional sequences which crop out at inner–middle shelf depths across most of eastern Onslow Bay (Fig. 30.1). These sequences have been designated the Bogue Banks Sequences (BBS).

Biostratigraphic constraints on the timing of deposition for the Serravallian sequences are not very tight. In fact, only the lower five Bogue Banks Sequences have yielded useful planktonic indices. They indicate initiation of Serravallian deposition at the N-11–N-12 zonal boundary of Blow (1969). The upper limits are still undetermined because timing of deposition for BBS-8 is not firmly constrained (see Fig. 30.2; Scott W. Snyder, 1988).

Distribution and lithologies The Serravallian sequences underlie the entire outer-shelf environment of the present North Carolina continental margin. This section thins in three directions (Fig. 30.12): (1) westward to its updip outcrop limit in central Onslow Bay; (2) seaward to its eastern subcrop limit along the present 100 m isobath; and (3) northward across the paleotopographic Cape Lookout High.

The Serravallian section reaches a maximum thickness of about 200 m beneath eastern Onslow Bay (Fig. 30.12). Here, the

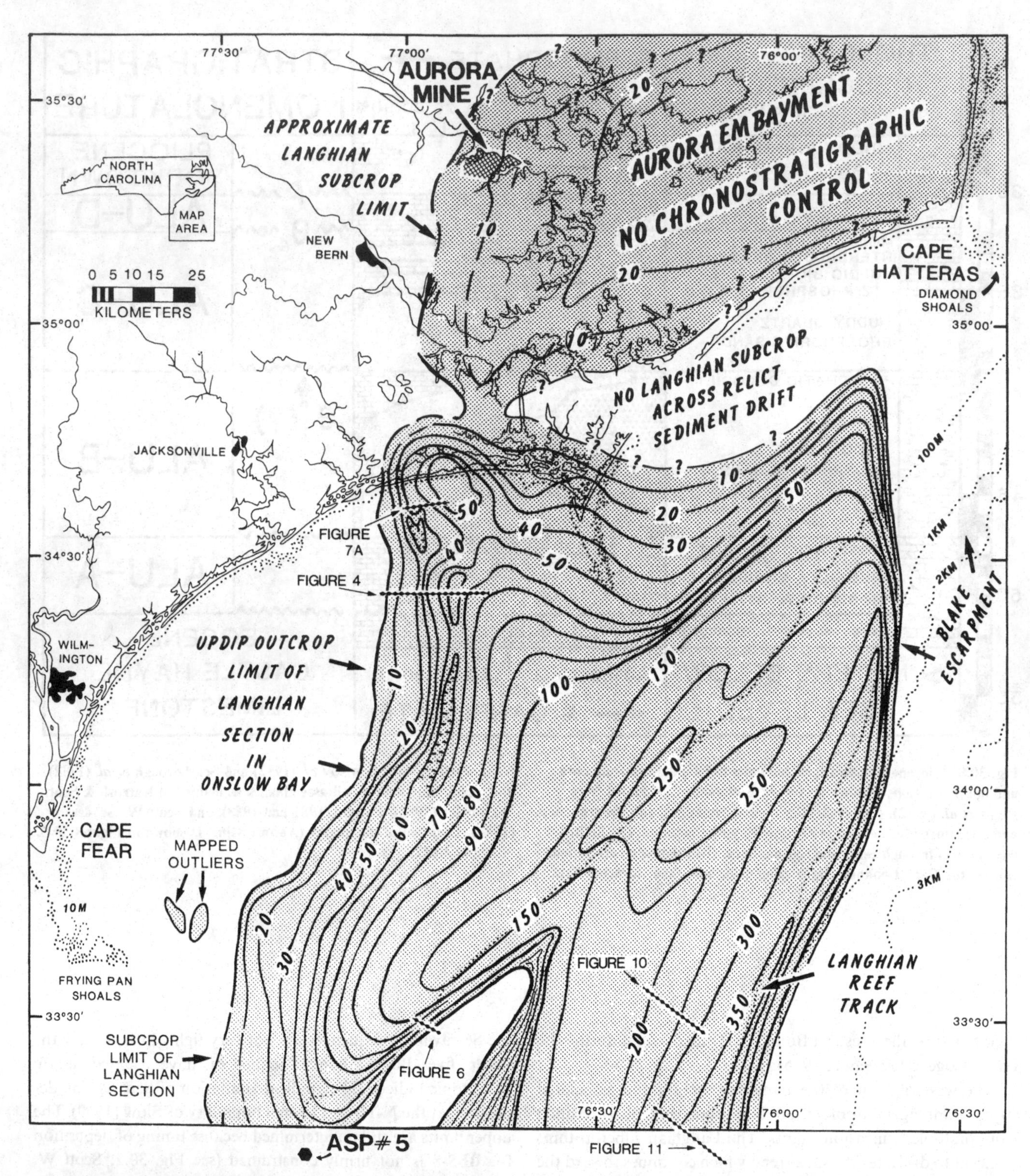

Fig. 30.9. Isopach map of the Langhian section of North Carolina based on interpretations from the network of seismic-reflection data shown in Figure 29.4 (Riggs *et al.*, Chapter 29, this volume). Contours are in meters assuming a constant subsurface velocity of 1700 m s^{-1} two-way travel-time. No chronostratigraphic control west of Cape Hatteras is due to penetration limitations of the seismic source used in Pamlico Sound. Also, the extremely thick stratigraphic section east of the present 100 m bathymetric contour (dotted line) may include some Late Burdigalian as well as Early Serravallian sediments. Drilling through this thick sedimentary prism is required before correlations to (and within) the seismic network of Popenoe (1985) can be confirmed and ages accurately assigned. Isopach map is modified from Stephen W. Snyder *et al.* (1986a,b).

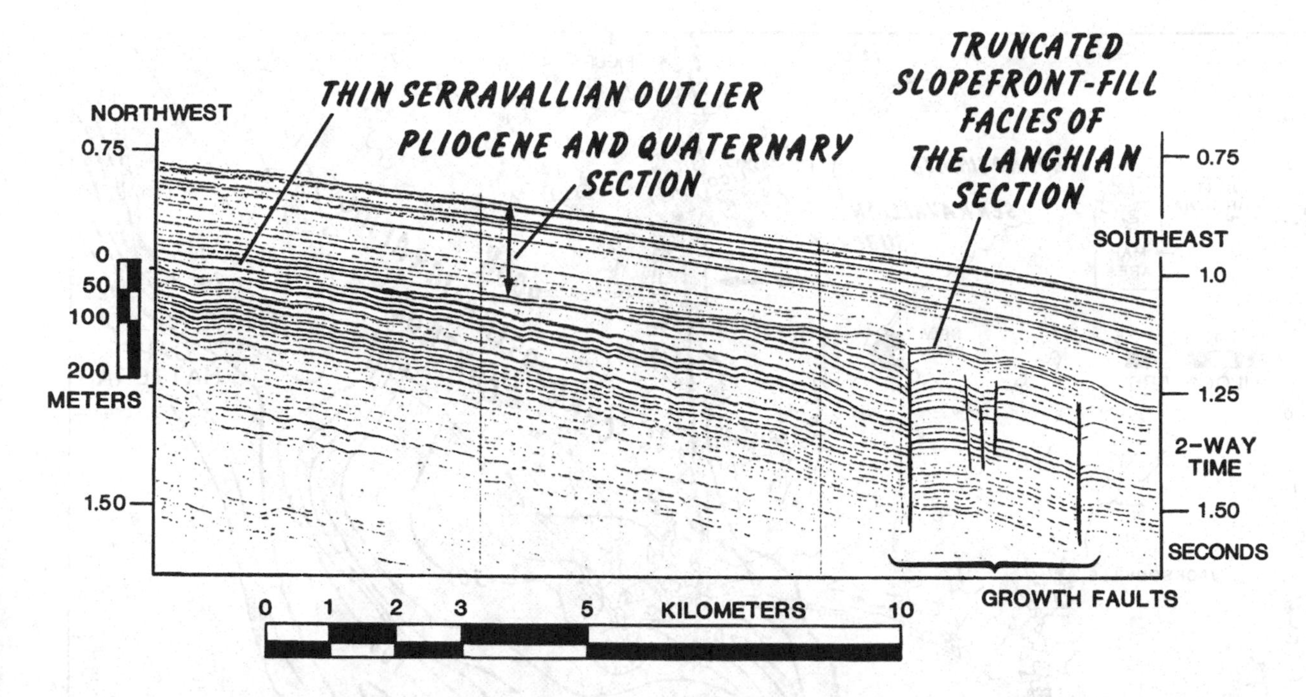

Fig. 30.10. Seismic section illustrating parallel, moderate–low amplitude, seismic reflectors which characterize the slope-front fill facies of the Langhian section beneath the present Northern Blake Plateau. Location of this seismic line is shown in Figure 30.9. Two-way subsurface travel time is given in seconds. Approximate vertical scale is given in meters assuming a constant velocity of 1700 m s^{-1}. Water depth varies from ≃575 m (left margin) to ≃850 m (right margin). This section is from the Seismic Library at Woods Hole Oceanograpic Institute courtesy of P. Popenoe and the US Geological Survey.

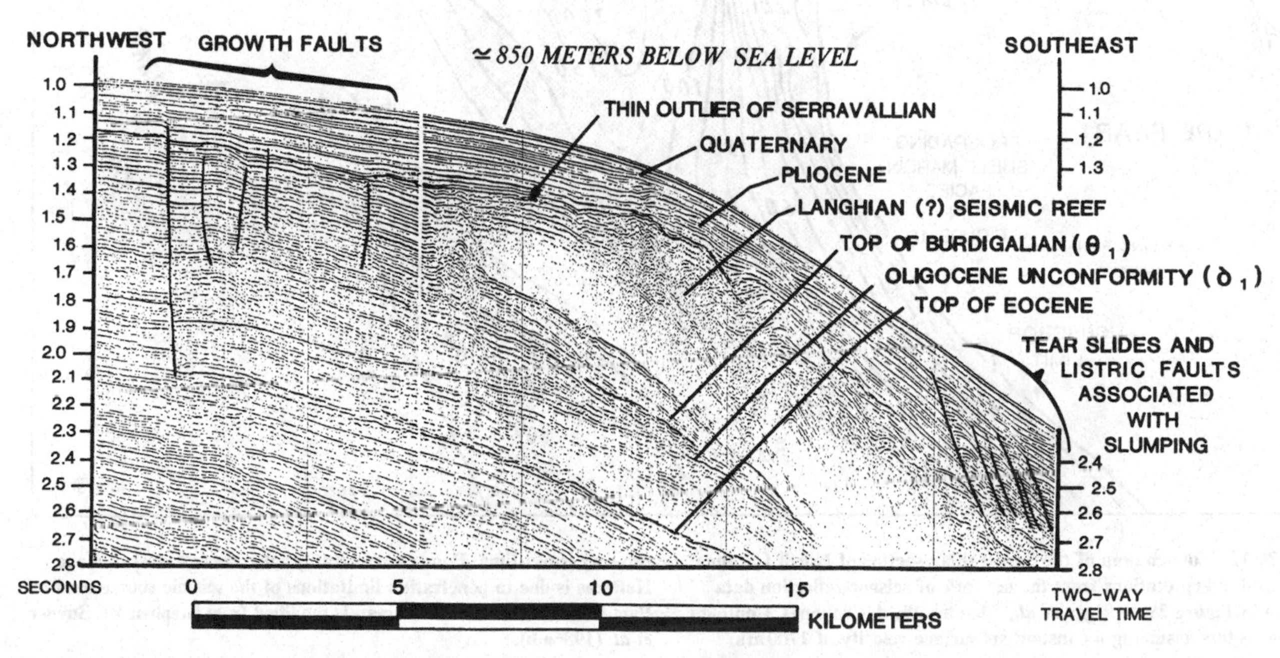

Fig. 30.11. Seismic section of buried Langhian(?) reef which helped contribute to the large-scale Middle Miocene build-up along the North Carolina continental margin. Beneath the present Northern Blake Plateau this accretionary wedge locally reached over 400 m in thickness. Location of this seismic line is shown in Figure 30.9. Two-way travel time is given in seconds. Data are from the Seismic Library at Woods Hole Oceanographic Institute courtesy of P. Popenoe and the US Geological Survey.

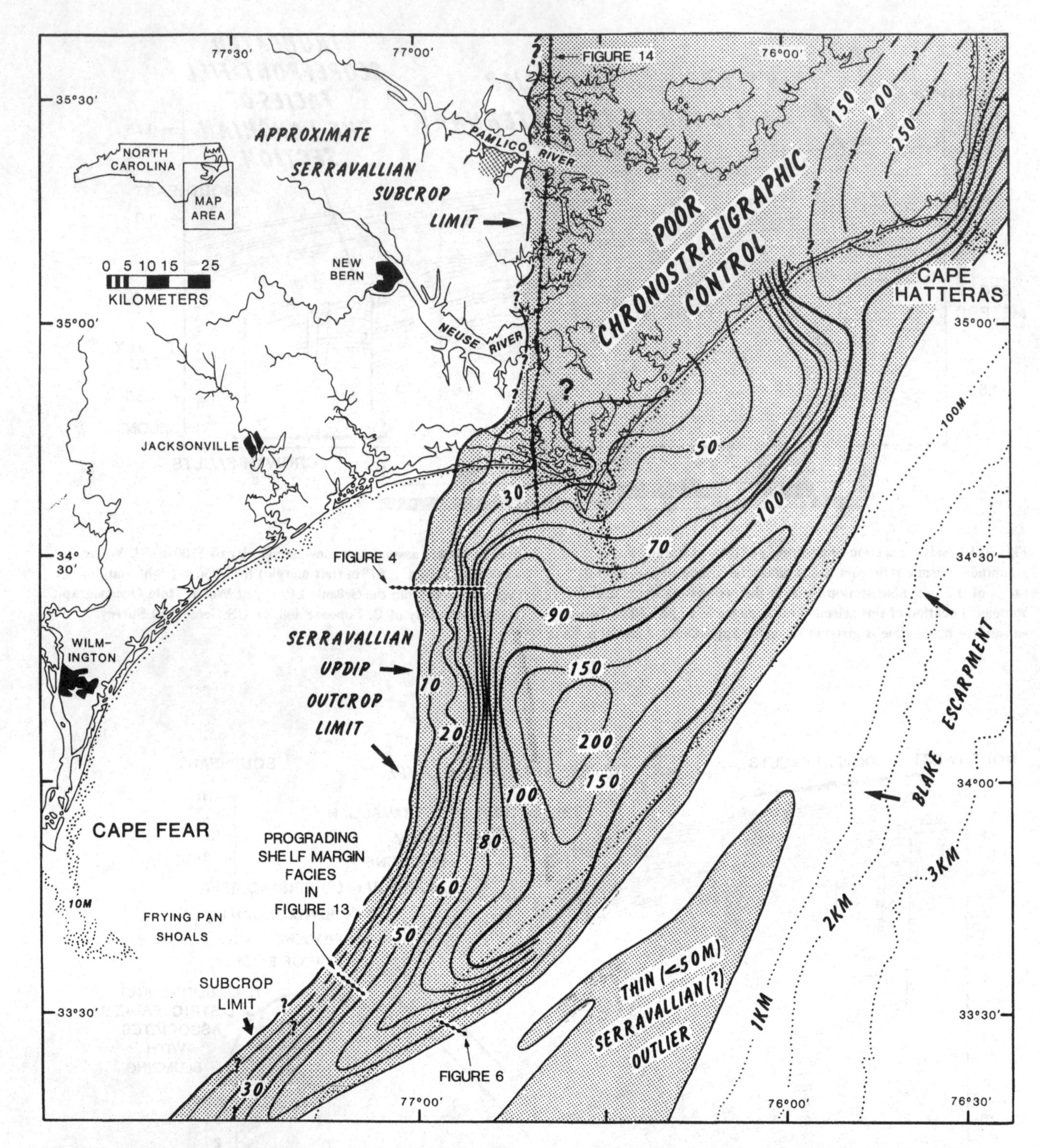

Fig. 30.12. Isopach map of the Serravallian section of North Carolina based on interpretations from the network of seismic-reflection data shown in Figure 29.4 (Riggs *et al.*, Chapter 29, this volume). Contours are in meters assuming a constant subsurface velocity of 1700 m s^{-1} two-way travel-time. Poor chronostratigraphic control west of Cape Hatteras is due to penetration limitations of the seismic source used in Pamlico Sound. The contour map is modified from Stephen W. Snyder *et al.* (1986a,b).

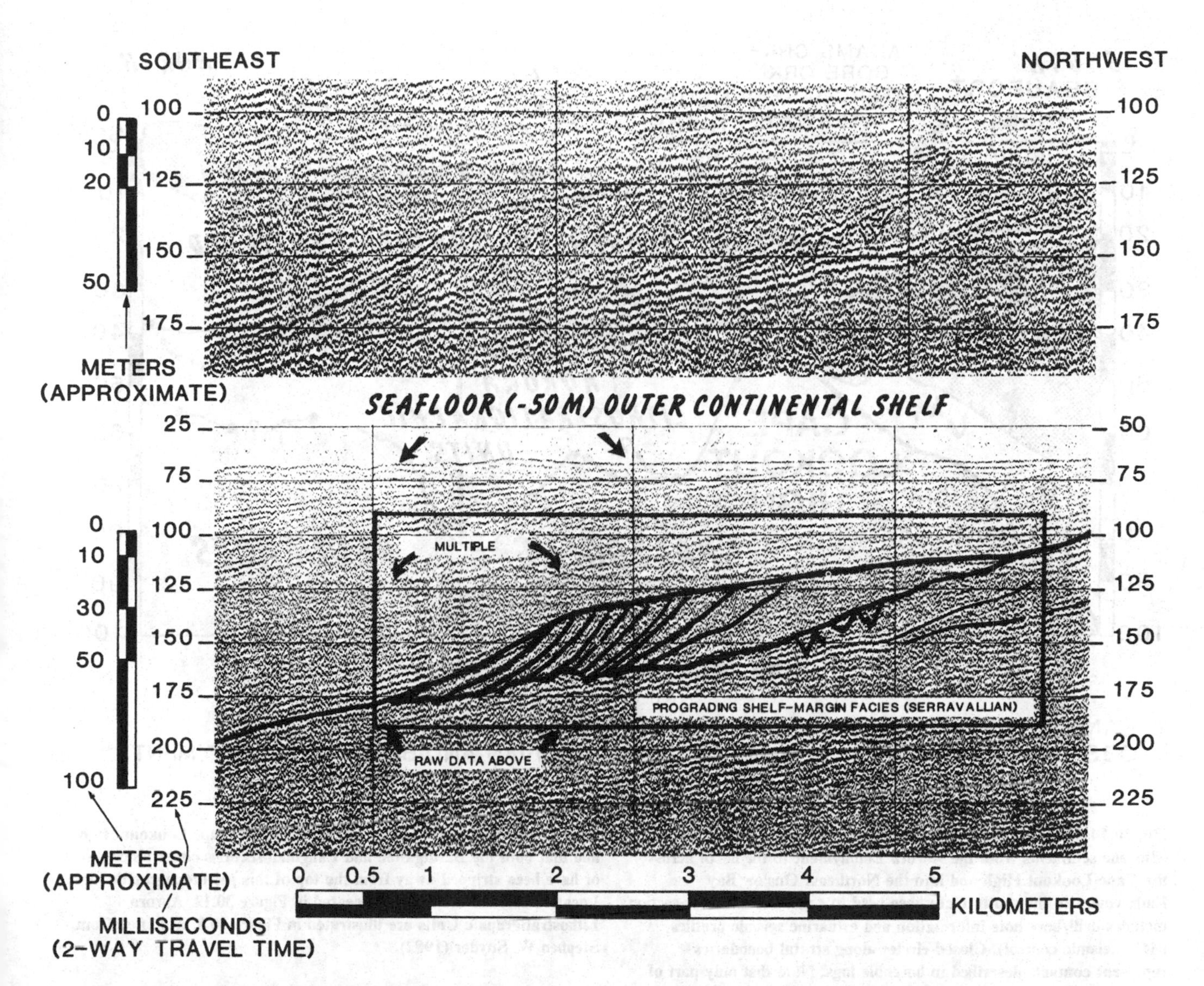

Fig. 30.13. Raw (upper panel) and interpreted (lower panel) seismic sections illustrating the distribution and seismic facies (prograding clinoforms) of Serravallian shelf-margin sequences buried beneath the present outer shelf and shelf margin of Onslow Bay. Two-way travel time given in milliseconds. Approximate vertical scale is given in meters assuming a constant subsurface velocity of 1700 m s^{-1} two-way travel time. Location of the seismic section is shown in Figure 30.12.

isopach lines are elongated in a NE–SW direction due, in part, to a regional erosional truncation surface which parallels the entire shelf–slope break. The small Serravallian outlier in the subcrop beneath the Northern Blake Plateau is a remnant of this erosional activity (Figs. 30.10, 30.11 and 30.12).

Muddy quartz sand recovered from vibracores penetrating the Serravallian outcrops in central and northeastern Onslow Bay are similar to the Burdigalian lithofacies cropping out in southwest Onslow Bay. They contain varying amounts of phosphorite, phosphatic sand, phosphatic muddy sand, skeletal sands (macrofossil hash), and dolomitic mud (Mallette, 1986; Riggs & Mallette, Chapter 31, this volume). The dolomitic mud lithofacies primarily consists of silt-sized rhombohedral dolomite, but also includes siliceous microfossils, and/or aggregates of clinoptilolite (Mallette, 1986). The benthic foraminiferal assemblages indicate a middle sublittoral paleoenvironment of deposition (Moore & Scott W. Snyder, 1985; Scott W. Snyder, Chapter 32, this volume). Some diatom species found within the mud lithofacies of sequences BSS-3 through BBS-5 suggest an outer sublittoral–upper bathayal depositional environment (Powers, pers. comm., 1986, 1987).

Seismic data penetrating the elongated subcrop beneath the outer shelf of southern Onslow Bay depict the Serravallian section as a series of prograding shelf-margin sequences (Fig. 30.13). The Serravallian shelf sequences of eastern Onslow Bay (muddy sands) apparently grade laterally (downdip and eastward) into prograding shelf-margin sequences beneath the present shelf–slope break. This lateral transition suggests that the 200 m thick Serravallian depositional wedge beneath eastern Onslow Bay represents the infill pattern associated with a major progradational phase of the North Carolina continental margin. This was probably the final phase of Miocene lateral accretion

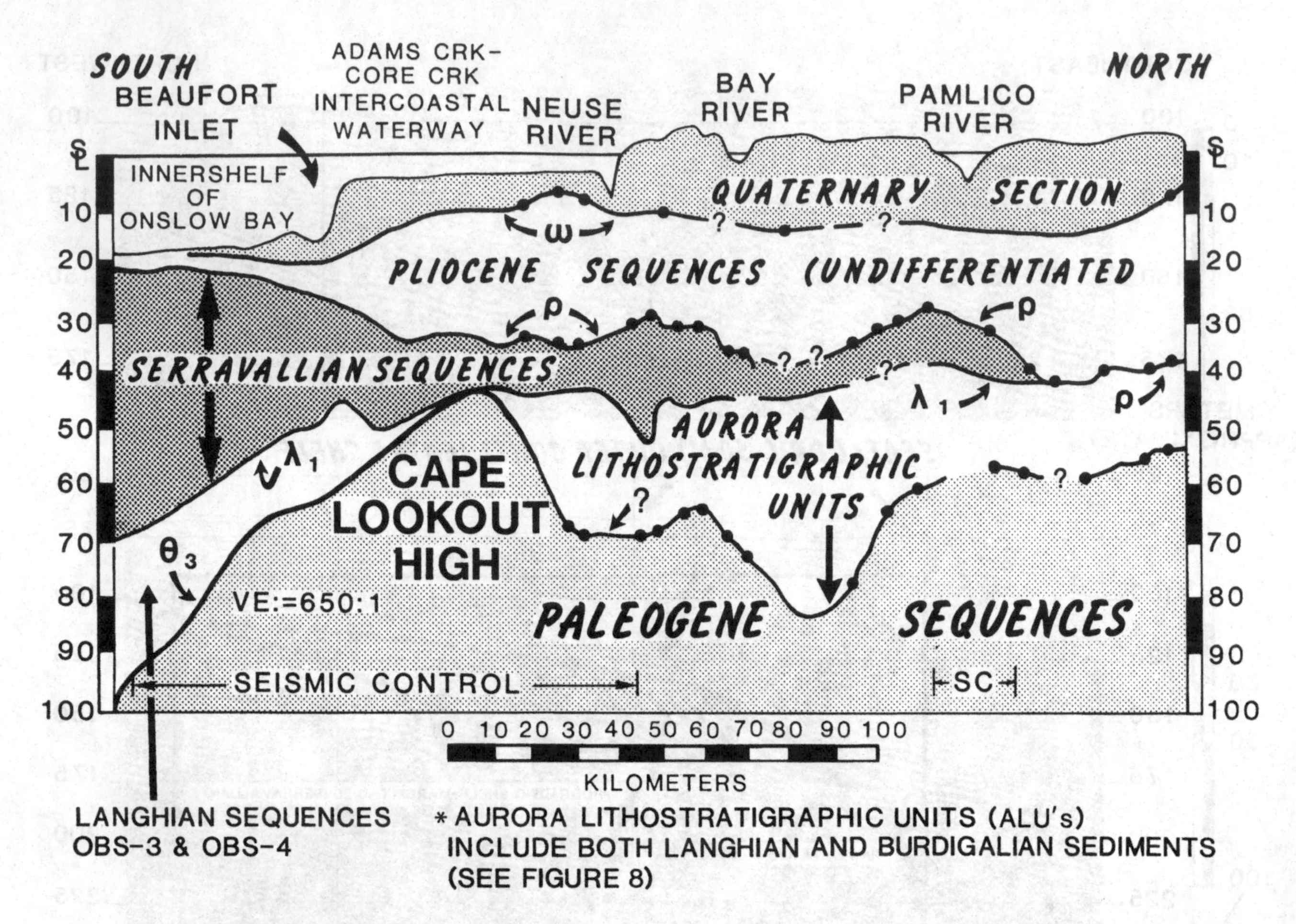

Fig. 30.14. Stratigraphic section delineating the distribution of Miocene sediments from the Aurora Embayment to the north across the Cape Lookout High and into the Northeast Onslow Bay Embayment to the south. Data base used to construct the cross-section includes drill/bore hole information and estuarine seismic profiles (SC = seismic control). Closed circles along stratal boundaries represent contacts described in borehole logs. Note that only part of the Serravallian section can be traced across the Cape Lookout High and that both the Burdigalian and Langhian sections either pinch-out or have been stripped away from the top of this paleotopographic high. Location of cross-section is delineated in Figure 30.12. Aurora Lithostratigraphic Units are illustrated in Figure 30.8. Modified from Stephen W. Snyder (1982).

along this continental margin. No Tortonian or Messinian sediments have yet been conclusively identified within the North Carolina margin.

North of Onslow Bay, the Serravallian section thins but does not pinch out across the persistent paleotopography of the Cape Lookout High (Fig. 30.12). A few of the Bogue Banks Sequences have been traced into the Aurora Embayment via estuarine seismic tie-lines (Fig. 30.14). Correlations between seismic data and adjacent drill holes suggest that the sediments are dominantly non-reefal macrofossil hash and quartz sand (Brown *et al.*, 1972; Scarborough *et al.*, 1982).

Further north, deposition of Burdigalian and Langhian sediments had filled the western half of the Aurora Embayment. This forced the locus of Serravallian deposition to shift eastward (Fig. 30.12). Consequently, no Serravallian sequences are exposed within the phosphate mine at Aurora (Figs. 30.8 and 30.12; Katrosh & Scott W. Snyder, 1982; Powers, 1987). However, a very thick section (>250 m) lies just to the north of Cape Hatteras (Fig. 30.12). Drill data from the Cape Hatteras cuspate foreland indicate that this thick Serravallian sedimentary lobe is an interbedded sequence of calcareous sands–muddy calcareous sands which are capped by a shelly limestone (Miller, 1971; Brown *et al.*, 1972).

Miocene sea levels and Gulf Stream flow patterns

Miocene excavation surfaces

Mapped spatial distributions and geometries of the Miocene chronostratigraphic sections define three subcrop features which are characteristic of major excavation surfaces. They are outlined below.

Truncation surfaces Several erosional surfaces dip sharply seaward to form escarpment-like topography in structure contour maps and seismic-reflection profiles. These surfaces have larger, thicker, and more complete stratigraphic sections preserved on the landward side of the erosional cut (Figs. 30.6 and 30.7). This testifies that excavation events began seaward of the escarpments and migrated landward while eroding the continental margin and exhuming a piece of the shelf-margin stratigraphic record.

Regional thinning Each of three Miocene sections displayed regional thinning in isopach maps. Some showed total removal of major time–stratigraphic sections across the crest and southern flanks of antecedent highs.

Linear excavation surfaces and severed subcrops Truncation surfaces commonly have linear or elongated geometries which parallel the present shelf–slope break (Figs. 30.5, 30.9, and 30.12). In some cases, these linear scour surfaces have simply resulted in the thinning of Miocene units along portions of the continental margin (elongated or severed isopachs in Fig. 30.9). Occasionally, erosion has isolated small stratigraphic sections from the main body of the subcrop (e.g. the thin outlier of Serravallian sediments lying beneath the Northern Blake Plateau in Fig. 30.12).

The seismic data demonstrate that most of the mapped anomalies in subcrop geometries are the direct consequence of major erosional unconformity surfaces such as those illustrated in Figures 30.6 and 30.7. These excavation features have a few consistent traits which strongly suggest they are products of submarine erosion by a boundary current. First, erosion started seaward and migrated landward. Second, most of the erosional swaths can be traced continuously across the continental margin as a narrow, linear band. Third, they usually parallel the present path of the Gulf Stream. However, they are buried erosional surfaces (Figs. 30.6 and 30.7), and therefore, they must be a consequence of ancestral rather than modern erosional processes.

Erosional scour and boundary current flow tracks
The characteristics outlined above suggest that ancestral flow paths of the western boundary current are responsible for inscribing most of the major submarine erosional scars found in the Miocene stratigraphy of North Carolina. No other mechanism has the potential to exhume such large linear tracks along the continental margin from a seaward approach. In addition, many elongated erosional swaths mapped in this study were shown by Stephen W. Snyder (1982) to be coincident with the northern extension of subbottom channels previously mapped in the central Blake Plateau area of the Southeast Atlantic Bight (Fig. 30.15). The channel networks mapped to the south have been interpreted as former boundary current flow paths as well (Pinet, Popenoe & Nelligan, 1981; Pinet & Popenoe, 1982, 1985a, b; Popenoe, 1985 and Chapter 28, this volume; Popenoe, Henry & Idris, 1987).

Lateral Gulf Stream shifts in response to sea-level changes Seismic data portray the scour routes of the ancestral western boundary current as consistently migrating westward from a more seaward position. Other seismic stratigraphic studies within the continental margin of the Southeast Atlantic Bight

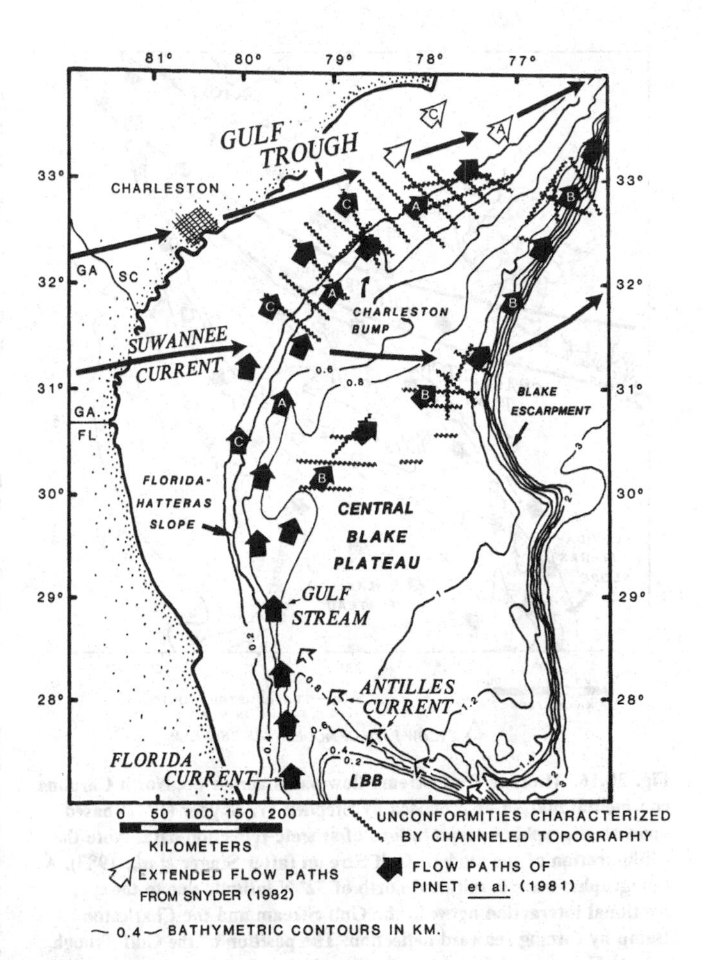

Fig. 30.15. Tertiary flow paths of the Western Boundary Current as proposed by Pinet *et al.* (1981), Pinet & Popenoe (1982, 1985a,b), Popenoe (1985 and this volume), and Popenoe *et al.* (1987). Data are from seismic stratigraphic studies within the continental margins of the Southeast Atlantic Bight and the emerged coastal plain of South Carolina and Georgia. The Swannee Channel was proposed by Pinet & Popenoe (1985a,b) as the flow track of the Western Boundary Current during the Late Cretaceous, a time when the Florida Platform was entirely submerged. The Gulf Trough is interpreted to have been intermittently active during the Late Eocene and Oligocene (Popenoe *et al.*, 1987; Popenoe, this volume). Modified from Snyder (1982).

have mapped similar Gulf Stream migration pathways (Pinet *et al.*, 1981; Stephen W. Snyder, 1982; Pinet & Popenoe, 1982, 1985a, b; Matteucci, 1984; Riggs, 1984; Popenoe, 1985 and Chapter 28, this volume). In these prior studies, landward directed shifts were interpreted as a dynamic response of the western boundary current to a rise in sea level. This apparent temporal relationship between sea-level change and shifting Gulf Stream flow paths is a very important concept; one that is relied on to explain sequence distribution and geometry, as well as to reconstruct the chronology of changing flow tracks for the Miocene Gulf Stream. Thus, it is important to develop the basis for this concept, and illustrate how the position and configuration of the Gulf Stream on the North Carolina continental margin was influenced by major frontal events that occurred in the upstream area.

The central Blake Plateau record The original idea that sea-level change resulted in lateral migrations of the western

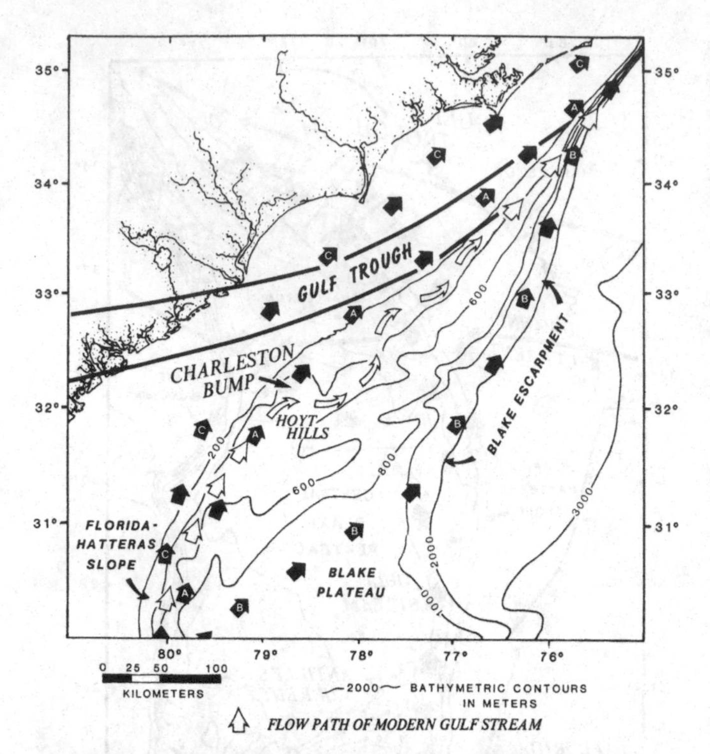

Fig. 30.16. Miocene Gulf Stream flow paths across the North Carolina continental margin as proposed by Stephen W. Snyder (1982) based upon stratigraphic interpretations of seismic-reflection data. Note the configuration of the modern Gulf Stream (after Singer *et al.*, 1983). A topographic wave is initiated north of 32° N latitude due to the frictional interaction between the Gulf Stream and the Charleston Bump by forcing seaward deflection. The position of the Gulf Trough, a Late Eocene and Oligocene flow track, is taken from Popenoe *et al.* (1987).

boundary current was first introduced from the Cenozoic stratal record of the central Blake Plateau (Pinet *et al.*, 1981; Pinet & Popenoe, 1982, 1985a, b; Popenoe, 1985 and Chapter 28, this volume). Gulf Stream scour was evidenced by large gullied swaths cut tens to hundreds of meters deep and tens of kilometers in width. A regional network of seismic-reflection data showed these gullied bands of Gulf Stream scour were elongated thalwegs (Pinet *e al.*, 1981). They were traced across the central Blake Plateau, and were mapped out to show the tracks extend over hundreds of kilometers. (Fig. 30.15).

Along one regional unconformity separating Paleocene and Eocene strata, two distinct gullied terranes were mapped. These two erosional swaths are separated by 200 km at their maximum divergence (flow paths A and B in Fig. 30.15). Utilizing the coastal onlap curve of Vail *et al.* (1977) and stratigraphic control from two adjacent drill sites, Pinet *et al.* (1981) suggested that the Gulf Stream shifted between these two extreme positions in response to a 3rd-order fluctuation in sea level. More specifically, during an early Eocene 3rd-order highstand, the western boundary current migrated landward to a position beneath the present Florida–Hatteras Slope (flow path A). When the water shoaled during the subsequent regression, the Gulf Stream shifted seaward and flowed across the central Blake Plateau (flow path B).

The second scour path (B) is approximately 500 m below the highstand scour track (A). Yet, the change in sea level as estimated by Vail & Hardenbol (1979) and Haq, Hardenbol & Vail (1987) was < 100 m. Why, then, was there such a severe shift in the position of the Gulf Stream? Apparently, the drop in sea level forced the Gulf Stream to interact with a large bathymetric protuberance known as the Charleston Bump (Figs. 30.15 and 30.16). The bump impeded the Gulf Stream's downstream path, deflecting its main axis seaward across the Blake Plateau.

Pinet & Popenoe (1982 and 1985a) and Popeneoe (1985 and Chapter 28, this volume) portrayed the Gulf Stream as shifting between extreme landward and seaward positions throughout the Tertiary, while inferring that each lateral migration was induced by 3rd-order fluctuations in global sea level. Specific Neogene flow paths are difficult to reconstruct due to poor preservation of time-equivalent stratigraphic sections (Pinet & Popenoe, 1982). Yet, where thin Neogene sections are preserved on the central Blake Plateau, they are always characterized by a complex mosaic of subbottom truncation surfaces (Pinet & Popenoe, 1985a; Popenoe, Chapter 28, this volume).

The Charleston Bump Although induced by changing sea levels, repeated shifting patterns between two extreme flow paths is interpreted to be caused by 'deflection' of Gulf Stream flow by the large bathymetric feature known as the Charleston Bump (Brooks & Bane, 1978; Pinet & Popenoe, 1982). This positive antecedent feature consists of a series of bathymetric hills and depressions. Locally, elevations change abruptly (± 100 m) and the bathymetry generally ranges between 300 and 500 m below present sea level (Figs. 30.15 and 30.16). The Charleston Bump, as used here, refers to the composite bathymetric feature and includes all topographic irregularities such as the Hoyt Hills and associated seachannels delineated in NOAA bathymetric charts (1:250 000 series).

This suite of seafloor mounds and valleys is cored by Cretaceous shelf margin sediments and rocks (Manhiem *et al.*, 1982; Pinet & Popenoe, 1985a). Its age implies that the Charleston Bump has been obstructing Gulf Stream flow during periods of low sea level throughout the Cenozoic. Over the past decade, satellite imagery and current-meter moorings have shown that seaward deflection of the modern Gulf Stream is most pronounced immediately downstream (to the north) of obstruction sites (Fig. 30.16; Brooks & Bane, 1978; Yodder *et al.*, 1981a, b, 1983; Lee & Atkinson, 1983; McClain & Atkinson, 1985). The North Carolina margin is located only a few hundred kilometers downstream from the Charleston Bump, and this bathymetric high existed during the Miocene. Therefore, boundary current steering (seaward) around this complex of hills must have been a critical factor in determining the position of the Miocene Gulf Stream along the North Carolina continental margin 20–10 million years ago.

Topographic steering, vorticity changes, and upwelling

Topographic steering is defined as the seaward steering of a boundary current as it approaches shallow bathymetric features. It originates from barotropic instabilities within the flow structure of a boundary current which are initiated by changes in water depth (Stommel, 1948 and 1965). This type of instability can be related to horizontal shear. As the boundary current flows along the bathymetric contours of a continental margin, it swiftly passes over small bathymetric irregularities. The friction produced as it moves past these stationary features creates shear zones between streamlines. This shearing flow then forces water parcels to begin spinning in a cyclonic fashion (counter-clockwise relative to a fixed point on the earth's surface). Vorticity is another name for this 'spin' and, by convention, cyclonic spin is referred to as positive vorticity.

Three other forces in addition to friction can induce water parcels to spin relative to a stationary point on the earth's surface. They are: (1) vertical shear derived from density-related (baroclinic) instabilities such as differences in dynamic topography, or steep isopycnal surfaces; (2) the curl of the wind stress (both local and regional); and (3) the Coriolis force generated by the continuous rotation of the earth around a central axis (Pedlosky, 1987). Spin induced by the earth's rotation is referred to as planetary vorticity. Because it is a function of latitude, the magnitude of spin initiated by planetary vorticity changes with north–south flow.

In steady-state flow, all relative vorticity changes must be balanced (i.e. positive spin must equal negative spin). This balance is required to conserve potential vorticity (Pedlosky, 1987). When the flow structure of a boundary current is seriously perturbed such that potential vorticity is no longer conserved, the current will either seek a path to balance the relative vorticity changes initiated by the perturbation, or break-up into an incoherent field of turbulence until all the energy of the boundary current system is dissipated via meso-scale mixing with the surrounding water masses. Observations demonstrate that the latter process does not commonly occur in today's ocean.

Using the argument that potential vorticity must always be conserved, theoretical studies have implied that most changes in the flow configuration of western boundary currents are achieved through significant relative vorticity changes (Stommel, 1948 and 1965; Fofonoff, 1954 and 1962; Pond and Piccard, 1983). Relative vorticity changes are, in turn, primarily initiated by frontal events such as topographic deflection (Veronis, 1981).

Within the subtropical gyre of the North Atlantic Basin, the Gulf Stream seeks some frictional interaction with the continental margin in order to balance planetary vorticity changes acquired as it flows northward. In other words, the western boundary current needs to rub against the continental margin in order to produce enough positive (+) spin to counter the negative (−) spin it receives from planetary rotation and resulting Coriolis forces. Significant perturbations are produced via the addition of excess positive relative vorticity when the boundary current feels bathymetric obstructions. In order to conserve potential vorticity, the boundary current must then steer to deeper waters (seaward around the obstruction). Eventually, changes in planetary vorticity force the current to steer back towards the shelf margin rather than continuing to flow off into deeper water. It once again requires some frictional interactions with the continental margin (positive spin) in order to counter the negative spin initiated by additional planetary vorticity changes. After turning landward, contact with the seafloor is made and the delicate vorticity balance is approached. More commonly, however, the boundary current overshoots this balance and continues to meander in a tireless effort to balance the vorticity argument. This process initiates topographic waves (e.g. the meanders illustrated to north of the Charleston Bump in Fig. 30.16).

Observations within the Southeast Atlantic Bight during the past three decades have shown that topographic waves are generated when the modern Gulf Stream passes over bathymetric hills within the Charleston Bump complex (Von Arx *et al.*, 1955; Webster, 1961; Brooks & Bane, 1979; Bane & Brooks, 1979; Bane *et al.*, 1981; Blanton *et al.*, 1981; Lee *et al.*, 1985; Yodder *et al.*, 1981a, b and 1983; Bane, 1983; Lee & Atkinson, 1983; Pietrafesa, 1983 and Chapter 1, this volume; McClain & Atkinson, 1985). Shedding of cyclonic eddies is also observed (McClain *et al.*, 1984). These gyres are born immediately downstream of deflection sites within the Charleston Bump Complex (Brooks & Bane, 1978; Legeckis, 1979; Olson *et al.*, 1983; Singer *et al.*, 1983). A shallow filament of Gulf Stream water is shed landward of the main axis of the boundary current, and it begins to translate westward, then southwestward, along the shelf margin. These warm-cored filaments are separated from the main axis of the Gulf Stream by cold, upwelled slope waters which are relatively enriched in nutrients (Lee *et al.*, 1981; Yodder *et al.*, 1981a, b and 1983; Brooks & Bane, 1983; Paluszkiewicz *et al.*, 1983; McClain *et al.*, 1984 and 1985; McClain & Atkinson, 1985; Vukovitch, 1986; Zantopp *et al.*, 1987).

Although the modern upwellings are episodic, they do occur throughout the year and in the same geographical area. Thus, persistent upwelling appears to be a significant by-product of the vorticity changes generated when a boundary current steers around large-scaled bathymetric obstructions.

Miocene Gulf Stream flow paths on the North Carolina continental margin: shifts in response to glacio-eustacy

Based upon physical models of boundary current flow patterns both from the modern oceanographic record and the record of flow paths over the Blake Plateau, a Miocene Gulf Stream flow path history can be developed for the North Carolina continental margin. As a general rule, when sea level was relatively low, the Gulf Stream occupied a path located east of the ancestral shelf margin (flow path B in Fig. 30.16). When sea level was higher, the Gulf Stream migrated westward to occupy a flow path similar to tracks A and C in Figure 30.16. The relative position between tracks A and C was dependent on the absolute height of the highstands. As outlined below, it was during these relatively higher sea-level stands when the Gulf Stream cut the erosional scars found in the Miocene section. It was also during the transgressive and highstand phases of several sea-level cycles when boundary current steering around major bathymetric features on the North Carolina continental margin gave rise to intense upwelling episodes.

Earliest Miocene

Prior to deposition of the Frying Pan Sequences, the North Carolina continental margin was subaerially exposed for about 6.5 Ma (Fig. 30.2). The evidence is the large time-stratigraphic hiatus found by biostratigraphic analyses across reflector/unconformity ϵ_1, as well as the identification of several buried channels of probable paleofluvial origin (Stephen W.

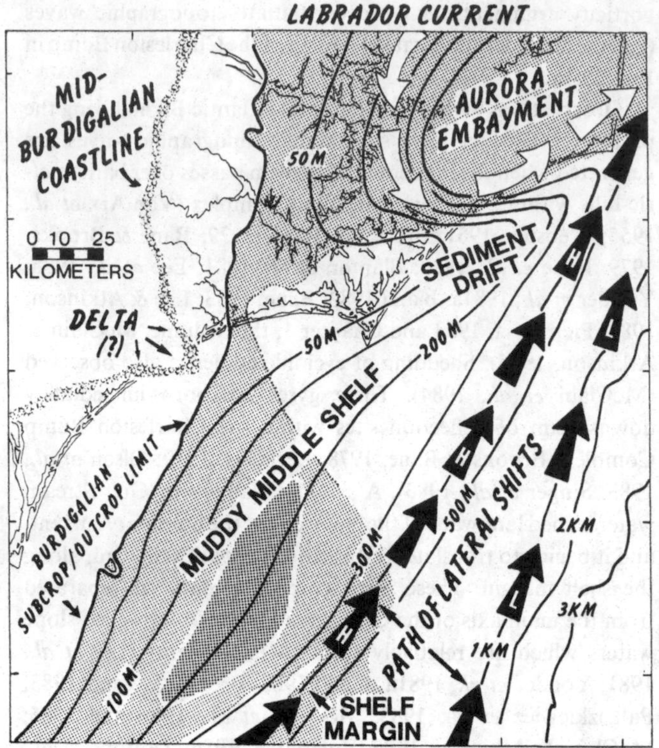

Fig. 30.17. Reconstruction of the North Carolina continental margin depicting a shoreline position and general flow-path configurations of the Gulf Stream for the Mid-Burdigalian Stage. The position of Delta front is based on Mallette (1986), and Riggs & Mallette (this volume). The two different paths represent flow tracks inferred for 4th-order sea-level highstands (path H) and lowstands (path L). Shifts between these positions were dictated by the glacioeustatic sea-level cycles shown in Figure 30.3. Light stipple depicts the present known subcrop distribution of the Early–Mid-Burdigalian section, while the dark stipple denotes areas of relatively thick Burdigalian sections. Contours are in meters below present sea level to reflector–unconformity ϵ_1 assuming a velocity of 1500 m s^{-1} in the water column and a constant subsurface velocity of 1700 m s^{-1}.

Snyder, 1982 and 1989). During this time, Gulf Stream flow was deflected seaward of the North Carolina slope. It probably occupied an eastern flow track half-way between paths A and B in Figure 30.16 as originally suggested by Pinet & Popenoe (1982).

Mid-Burdigalian

Gulf Stream flow paths The Gulf Stream migrated landward in response to initial Miocene submergence of the North Carolina continental margin during the Burdigalian. Sea levels were not high enough, however, to prevent partial deflection by some of the shallower hills within the Charleston Bump Complex. Consequently, during the early phases of Miocene flooding, the Gulf Stream crossed the North Carolina continental margin seaward of the shelf–slope break. This flow path can only be inferred because the shelf-margin facies of the Frying Pan Sequences were completely removed by subsequent Gulf Stream scour events (e.g. no distal subcrops in Fig. 30.5 due to the erosion shown in Fig. 30.6).

In the reconstruction presented in Figure 30.17, it is assumed that the Mid-Burdigalian boundary current flowed northeasterly, and that its westernmost flow track remained approximately 60–100 km east of the Burdigalian outcrops in southwest Onslow Bay (path H in Fig. 30.17). This position is based on flow path configurations required for renewed (Burdigalian) construction along an ancestral sediment drift system.

Depositional consequences Most of the continental margin underlying Onslow Bay represented a shelf during deposition of the Frying Pan Sequences. A constant supply of siliciclastic sediments was delivered to this shallow shelf environment from a point source located in northwest Onslow Bay (delta front in Fig. 30.17; also see Riggs & Mallette, Chapter 31, this volume). Note that even during the highstand flow tracks, the main axis of the ancestral Gulf Stream was located at least 50 km to the east at the time of their deposition. This suggests that the Frying Pan Sequences were deposited too far landward to be directly affected by Gulf Stream scour. Yet, these muddy shelf sediments were close enough to be influenced by eddies and associated boundary current upwelling if the Gulf Stream had interacted with a major bathymetric barrier immediately to the south.

As illustrated in Figure 30.17, a sediment drift began to prograde toward the northeast, with growth primarily occurring between the modern Capes Lookout and Hatteras. This drift-building episode significantly enlarged the paleotopographic feature known as the Cape Lookout High (Figs. 30.17 and 30.18). Sediments contributed from the siliciclastic point source inland as well as winnowed sands from the ancestral shelf margin off Onslow Bay were transported northeast. These siliciclastic sediments, in addition to some skeletal sediments shed from the top of the Cape Lookout High, were the chief sediment sources for building this depositional feature.

The Cape Lookout High sediment drift system had initially been established in the Oligocene via the confluence of the southward-flowing Labrador Current and the northward-flowing Gulf Stream (Stephen W. Snyder, 1982). Growth of the Oligocene precursor was arrested during the period of relatively low sea levels between latest Chattian and early Burdigalian (Fig. 30.3). Deposition along the Cape Lookout High did not resume until the early floodstages of the Middle Miocene supercycle (TB2), and construction was chiefly performed during the highstands of Burdigalian 4th-order sea-level cycles (path H in Fig. 30.14). The sum of these building episodes significantly changed the overall morphology of the North Carolina continental margin. A broad shallow shelf occupied the margin on the south flank of the sediment drift, and a more restricted embayment (Aurora Embayment) developed on the northern flank of this accreting sand massif. The Aurora Embayment became an important site of phosphate deposition during both the Burdigalian and Langhian Stages.

Late Burdigalian

Gulf Stream Elevated seas associated with the continued rise of the Miocene 2nd-order sea level further submerged the North Carolina margin. During highstands of the shorter-pulsed sea-level events, the latest mid-Burdigalian Gulf Stream was capable of overriding many of the bathymetric barriers in the Charleston Bump complex. Consequently, the boundary

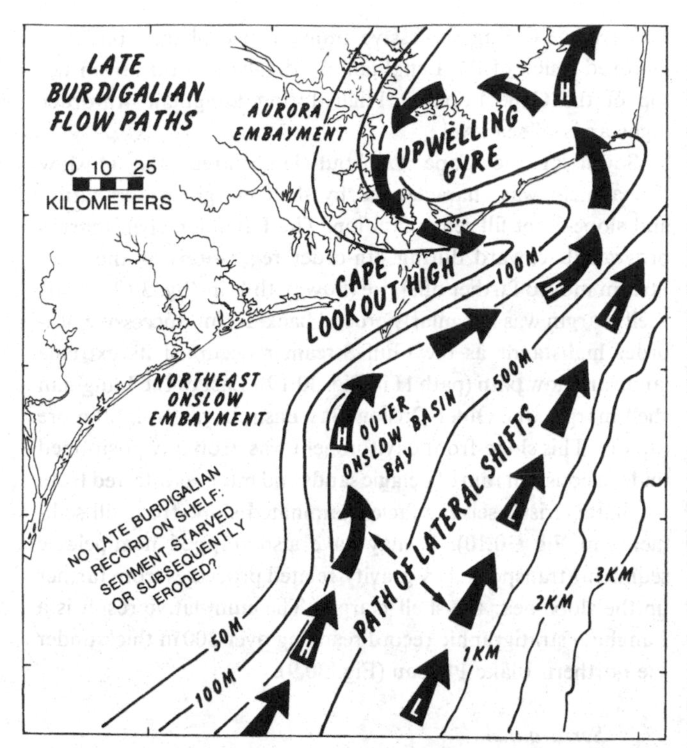

Fig. 30.18. Reconstruction of the North Carolina continental margin illustrating inferred highstand–lowstand positions of the ancestral Gulf Stream for the Mid- through Late Burdigalian Stage. Highstand positions allowed the Gulf Stream to excavate the southern flank of the Cape Lookout High which resulted in the carving of the Outer Onslow Bay basin and the White Oak Scarp. Perturbations associated with this erosional activity resulted in the generation of a relatively persistent upwelling gyre within the shelf embayment at Aurora. The cannibalistic nature of Gulf Stream scour left no record of short-pulsed sea-level fluctuations within the Onslow Bay area and, therefore, they can only be inferred. Contours represent the bathymetry of the continental margin as interpreted for the Late Burdigalian. They are in meters below present sea level.

current shifted landward and began to erode the continental margin south of Cape Lookout. Repeated excavations produced the channel deposits associated with FPS-6 while also carving an enormous notch on the southeast flank of the Cape Lookout High sediment drift (compare Figs. 30.17 and 30.18). The excavated depression resulting from these erosional events is known as the Outer Onslow Bay Basin (Popenoe, 1985), while the shallow shelf embayment located along the northwestern margin of this basin is referred to as the Northeast Onslow Embayment (Stephen W. Snyder, 1982; Stephen W. Snyder, Hine & Riggs, 1982; Riggs, 1984; Riggs *et al.*, 1985).

In eroding the Outer Onslow Bay Basin, the Gulf Stream demonstrated its cannibalistic nature as it repeatedly consumed the Late Burdigalian stratigraphic record on the north and western sides of the basin. It even exhumed the entire Burdigalian record in northern Onslow Bay (i.e. no Burdigalian subcrop in Fig. 30.5). The only stratigraphic record left from these severe erosional events is a linear, generally north–south-striking, subbottom escarpment (Fig. 30.4) extending over 100 km beneath the continental shelf of Onslow Bay. This erosional feature was originally designated the White Oak Lineament (Stephen W. Snyder, 1982; Stephen W. Snyder *et al.*, 1982), but is now commonly referred to as the White Oak Scarp (Mallette, 1986; Riggs & Mallette, Chapter 31, this volume). The erosional hiatus associated with the scarp-cutting event (reflector–unconformity θ_1 in Figure 30.4) represents at least 1.5 Ma of missing stratigraphic section (see Fig. 30.2 for time constraints).

Depositional consequences Erosion dominated the shallow shelf sea of the Northeast Onslow Embayment, while episodes of topographically-induced upwelling were probably established in the Aurora Embayment (Fig. 30.18). Both processes resulted from the configuration of the ancestral Gulf Stream which had to steer around the Cape Lookout High.

The geometric relationships between the Aurora Embayment and the Cape Lookout High are illustrated in Figure 30.18. Deflection sites within the southern flank of the sediment drift are approximately 100–150 km upstream from the restricted shelf embayment at Aurora. This is analogous to the geographic relationships between deflection sites in the Charleston Bump and the position of modern cyclonic eddies generated immediately downstream (Bane, 1983; McClain & Atkinson, 1985). Topographic steering initiated by the Cape Lookout High during the Mid-Late Burdigalian probably resulted in the persistent shedding of cyclonic eddies. By analogy, these upwelling gyres would have been centered in the deeper part of the Aurora Embayment. They generated the organic-rich depositional regimes which ultimately produced the phosphate found within some of the Mid-Burdigalian Aurora Lithostratigraphic Units (e.g. ALU-B in Fig. 30.8) as well as Late Burdigalian sections not yet sampled downdip (to the east) beneath Pamlico Sound.

The distal nose of the Cape Lookout High sediment drift had grown to nearly the position of modern Cape Hatteras. This rejuvenated growth was severely retarded during the Late Burdigalian because the flow configuration of the Gulf Stream generally eroded rather than contributed sediment to this part of the continental margin (flow path H in Fig. 30.18).

By the end of the Burdigalian the Gulf Stream had again resculptured the morphology of the North Carolina continental margin. This time it was almost entirely by virtue of its erosive capability. The northern shelf embayment (Aurora Embayment) was partially filled with organic-rich, fine-grained sediments deposited under upwelling conditions. The southern shelf embayment (Northeast Onslow) and adjacent deeper basin (Outer Onslow Bay Basin) continued to be dominated by Gulf Stream erosion. The Cape Lookout High located between these two depocenters represented the erosional remains of what was once a larger, active sediment drift.

Langhian

Gulf Stream flow paths The Gulf Stream climbed over shallower bathymetric barriers as the Miocene 2nd-order global sea-level cycle reached its maximum flood stage during the Langhian (Fig. 30.3). Along the North Carolina continental margin, the boundary current continued to migrate further landward. In fact, during the highstands of the shorter-pulsed Langhian sea-level cycles (Fig. 30.3), the western margin of the Gulf Stream may have actually breached the bathymetric barrier represented by the Cape Lookout High and crossed over into the Aurora Embayment (flow path H in Fig. 30.19).

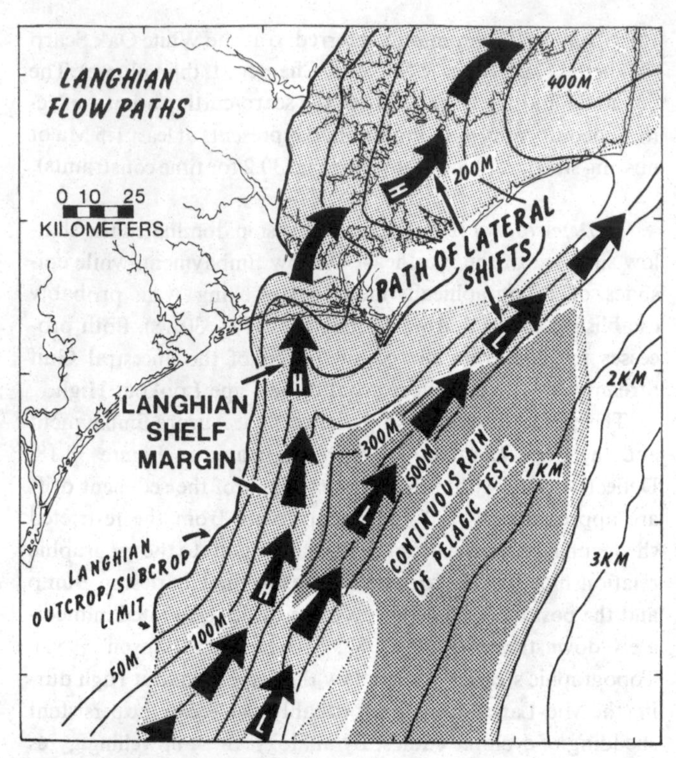

Fig. 30.19. Reconstruction of the North Carolina continental margin depicting the general flow path configurations of the Gulf Stream for the Langhian Stage. The two different flow paths represent the positions inferred for 4th-order sea-level highstands (path H) and lowstands (path L). Shifts between the two flow paths were dictated by the glacioeustatic sea-level cycles shown in Figure 30.3. Light stipple depicts the present subcrop distribution of the Langhian section, while the dark stipple denotes areas of relatively thick Langhian sections. Note that multiple migrations across the Cape Lookout High stripped away all Langhian and Burdigalian sediments such that no subcrop remains (compare with Figures 30.5 and 30.9). Contours are in meters below present sea level to reflector–unconformity θ_1 assuming a velocity of 1500 m s^{-1} in the water column and a constant subsurface velocity of 1700 m s^{-1}.

Depositional consequences Multiple shifts between flow path H and L in Figure 30.19 resulted in lateral scarping and stratigraphic thinning of sequences previously deposited across the crest of the Cape Lookout High. These erosional episodes help to explain the conspicuous lack of both Langhian and Burdigalian strata over much of this elevated area (Figs. 30.5, 30.9, 30.17 and 30.19). Sediments were eroded and reworked. Non-reefal macrofossils were transported off the top of this antecedent high. These sediments partially filled the northern margin of the Onslow Embayment and the southern margin of the Aurora Embayment. This resulted in sediments rich in molluscan shells and barnacle plate fragments being deposited as thin interbeds within the muddy, organic-rich sediments of the Aurora Embayment.

Topographically-induced upwelling events probably contributed to authigenic formation of the phosphorites in the Langhian section of the Aurora Embayment. The upwelling is again interpreted to be a by-product of friction rendered each time the Gulf Stream migrated westward across the Cape Lookout High. Thus, repeated Langhian boundary current migrations not only induced upwelling events in the Aurora Embayment during successive transgressive phases, but also removed much of the Langhian–Burdigalian section from the top of the Cape Lookout High during Langhian 4th-order highstands of sea level.

South of the Cape Lookout High area, the Onslow Embayment was characterized by shelf margin progradation and slope-front fill sedimentation. The Langhian shelf margin prograded seaward during 4th-order regressions as the Gulf Stream moved farther offshore (flow path L in Fig. 30.19). The shelf margin was irregularly eroded back during successive 4th-order highstands as the Gulf Stream reoccupied its extreme landward flow path (path H in Fig. 30.19). East of the Langhian shelf margin, the Outer Onslow Bay Basin began to infill more rapidly. This slope-front environment was probably positioned under a constant rain of pelagic sands and muds as inferred from its characteristic seismic facies (laminated slope-front fill sediments in Fig. 30.10). It may have also trapped hemipelagic sediments transported by gravity-related processes from farther up the slope near the shelf margin. The cumulative result is a Langhian stratigraphic record reaching over 300 m thick under the northern Blake Plateau (Fig. 30.9).

Serravallian

Gulf Stream flow paths By the Serravallian Stage, the Miocene 2nd-order sea-level cycle had entered its regressive phase along the North Carolina continental margin (Fig. 30.3). This caused the highstands of the Serravallian 4th-order sea-level cycle to be generally lower than the Langhian lowstand positions (compare Langhian lowstand positions with the Serravallian highstand positions in Figure 30.3). As a consequence of these relative sea-level positions, the path of lateral shifts moved seaward forcing the track of active Gulf Stream scour to migrate between the upper continental slope and the northern Blake Plateau (compare shifts between paths L and H in Fig. 30.20).

As the Gulf Stream was forced seaward, the shelf margin began to prograde, filling much of the remaining depression associated with the Outer Onslow Bay Basin (Figs. 30.12 and 30.20). The zone of submarine scour migrated westward during each Serravallian 4th-order transgression, exhuming segments of previously deposited Serravallian sediments. In the relatively deep-water environment of the northern Blake Plateau, the landward–seaward migrations caused stratigraphic thinning and eventually severed a portion of the subcrop from the main body of Serravallian sediments (outlier in Figs. 30.12 and 30.20). Underlying Langhian slope-front fill sequences were also eroded and thinned as evidenced by the segregation of the 250 m isopach into two separate ellipsoids in (Figure 30.9), and the truncated nature of Langhian slope-front fill strata in Figure 30.10.

Depositional consequences Sediments removed by Serravallian high-stand erosional events were carried along the slope and deposited at the distal nose of the sediment-drift system. The nose of the Cape Lookout High was now located northeast of Cape Hatteras as indicated by the 200 m isopach in Figure 30.12 and heavy stippled 'Hatteras Sediment Drift' in Figure 30.20. This rejuvenated growth was initiated as the Serravallian sea-level positions forced the Gulf Stream to re-

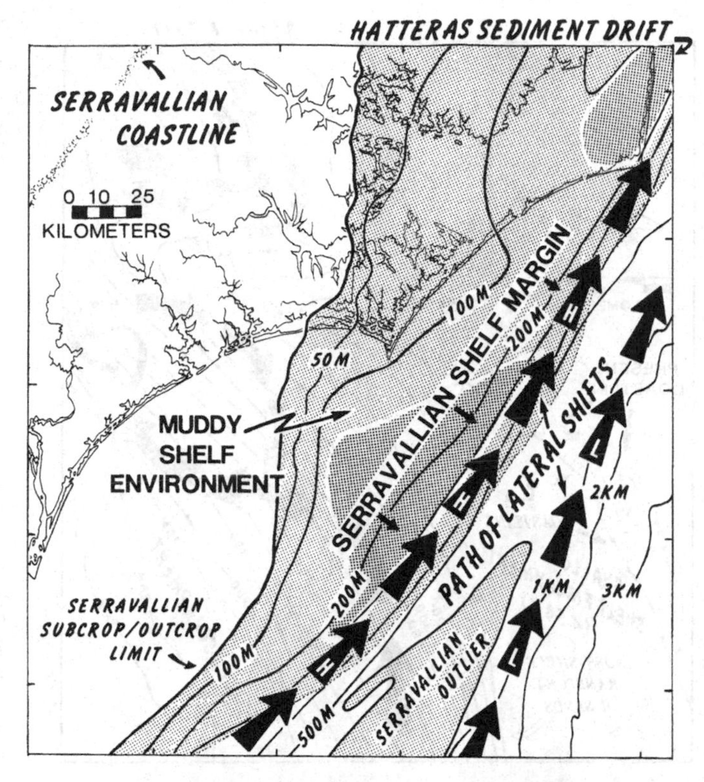

Fig. 30.20. Reconstruction of the North Carolina continental margin illustrating a shoreline position and general flow-path configurations of the Gulf Stream during the Serravallian Stage. The two different flow paths represent positions inferred for 4th-order sea-level highstands (path H) and lowstands (path L) based on subcrop geometries and physical stratal relationships. Shifts between the two flow-path positions were dictated by the glacioeustatic sea-level cycles shown in Figure 30.3. Light stipple depicts the present subcrop distribution of the Serravallian section, while the dark stipple denotes areas of relatively thick Serravallian sections. Contours are in meters below present sea level to reflector–unconformity λ_2 assuming a velocity of 1500 m s^{-1} in the water column and a constant subsurface velocity of 1700m s^{-1}. Note the Gulf Stream has returned to paths established during the Early–Mid-Burdigalian (compare with Fig. 30.18), and sediment drift building is rejuvenated near the modern cuspate foreland at Cape Hatteras.

establish the Early–Mid-Burdigalian locations (compare Figs. 30.12 and 30.20).

The similarity between Mid-Burdigalian and Serravallian flow-path positions is also reflected in lithologies of the Frying Pan and Bogue Banks Sequences. All of these shelf sequences consist mainly of siliciclastic muddy sands with variable concentrations of phosphate, dolosilts, and diatomaceous muds. Their correlative ancient shelf environments were also located too far landward of the shelf margin to be directly affected by Gulf Stream erosion. However, both depositional systems were close enough to the shelf margin to be influenced by upwellings associated with cyclonic spin-off eddies. The cyclonic eddies could have been shed from the main axis as the boundary current changed its configuration to accomodate bathymetric barriers 100–150 km to the south. Upwelling could have also been driven by baroclinic boundary current instabilities similar in mode and origin to the summer upwelling fronts observed today along the mid–outer continental shelf of eastern Florida (Paffenhofer *et al.*, 1987). More subsurface seismic data to the south of Onslow Bay are required to delineate the paleotopography and its relationship to Burdigalian and Serravallian upwelling events.

Relevance to Miocene Pungo River phosphorites

The Miocene Gulf Stream flow tracks engraved in the physical stratigraphic record demonstrate that flow across the North Carolina continental margin was periodically modified by frictional events. Studies of the modern Gulf Stream show upwelling of nutrient-rich slope waters commonly results from frictional interaction and resulting vorticity-related instabilities. At least one site initiated horizontal shear and seaward steering of the Miocene boundary current along North Carolina. This site, the southern flank of the Cape Lookout High, is located approximately 120 km upstream of a known phosphate resource (Aurora Phosphate District). Their geographic relationship is analogous to modern upwelling gyres shed from Gulf Stream via interaction with bathymetric obstructions in the Charleston Bump complex. Thus, topographically-induced boundary current upwelling is clearly a plausible mechanism for generating the Miocene phosphorites of North Carolina.

Boundary current upwelling is also geographically inferred from a series of preliminary paleoceanographic and paleoenvironmental reconstructions which step through one complete Middle Miocene glacio-eustatic sea-level cycle (Stephen W. Snyder *et al.*, 1984b and 1986b). These reconstructions are illustrated in Figures 30.21 through 30.25. They represent a first attempt at modeling the higher-frequency flow-path changes occurring in response to a single Miocene glacio-eustatic sea-level cycle. The reconstructions integrate the position of sea level (as defined in Fig. 30.3) with the continental margin morphology (defined by structure contour maps generated from the seismic data) to portray the paleotopography and paleobathymetry at five different stages through one complete sea-level cycle. The predicted temporal and spatial flow-path changes (Figs. 30.21–30.25) are now being compared with the lateral and vertical facies relationships emerging from the lithostratigraphic analyses of individual Miocene sequences (Riggs *et al.*, 1982 and 1985; Riggs, 1984; Mallette, 1986; Riggs & Mallette, Chapter 31, this volume).

Collectively, these data and interpretations imply that boundary current upwelling resulting from the shedding of cyclonic gyres is an important continental margin process. It not only explains the genesis of phosphorites on eastern continental margins, but also the variety of depositional regimes that coexisted across the Miocene continental margin at any one time, as well as their spatial changes through time. The ephemeral nature of boundary current upwelling observed in the reconstructions also helps explain the pattern and position of organic-rich sediment deposition. Oxygen consumed by heterotrophic bacteria to oxidize the organic carbon results in oxygen-deficient bottom conditions. These suboxic seafloor conditions were recorded by Miocene benthic foraminiferal communities (Scott W. Snyder *et al.*, 1982 and Chapter 32, this volume). Distributional patterns within the Miocene section show that benthic assemblages tolerant of suboxic conditions were restricted to specific geographic sites during short-lived upwelling episodes. Further, these benthic assemblages changed geographic location from one de-

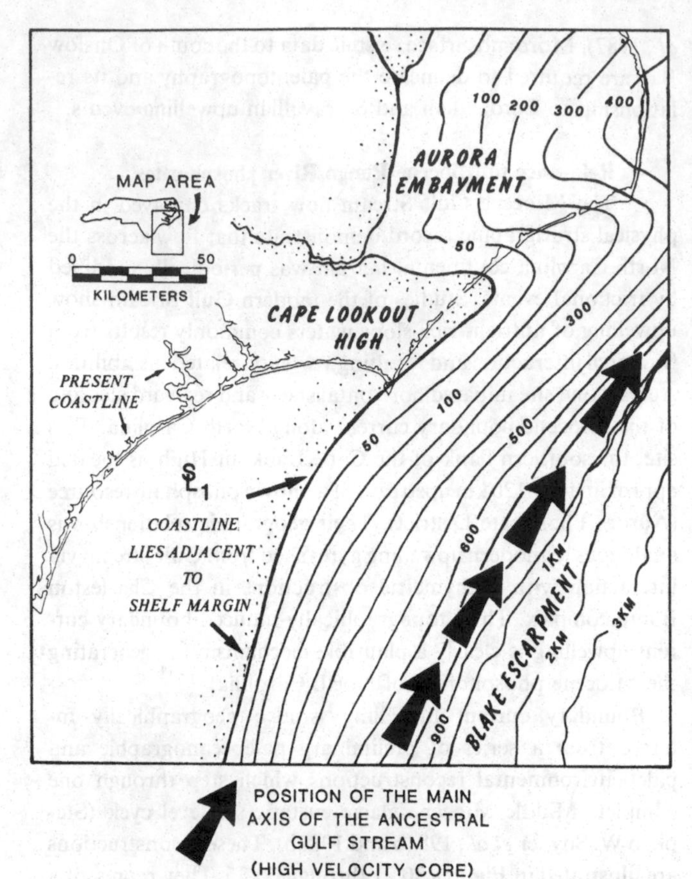

Fig. 30.21. Paleoenvironmental–paleoceanographic reconstruction of the North Carolina continental margin illustrating the inferred shoreline position and flow-path configuration of the Gulf Stream prior to deposition of Onslow Bay Sequence #1. Contours are in meters relative to the sea-level #1 position which is approximately 100 m below present sea-level. Seaward deflection of the boundary current by bathymetric hills to the south is inferred. This forced the Gulf Stream to occupy an extreme eastward flow track across the North Carolina continental margin. It promoted lateral growth via off-shelf sand transport and progradation of the shelf margin.

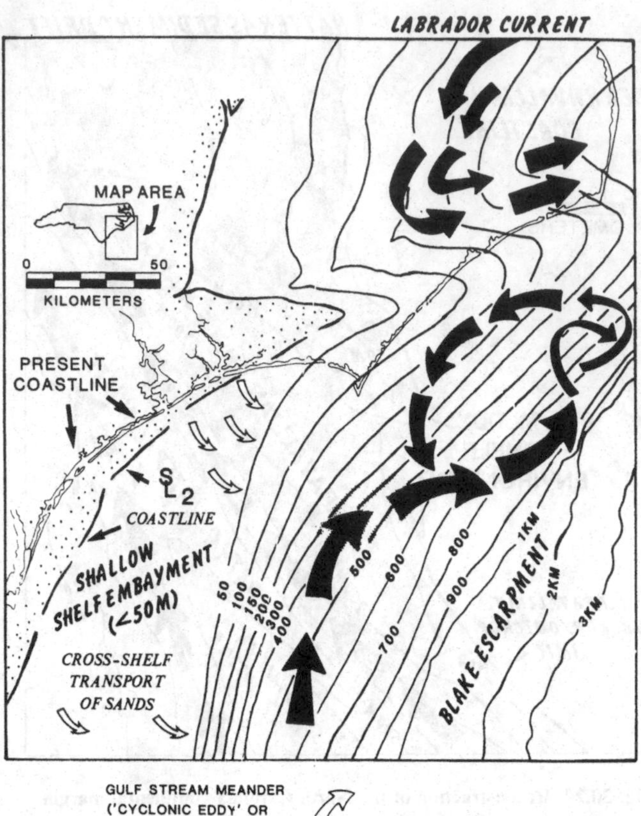

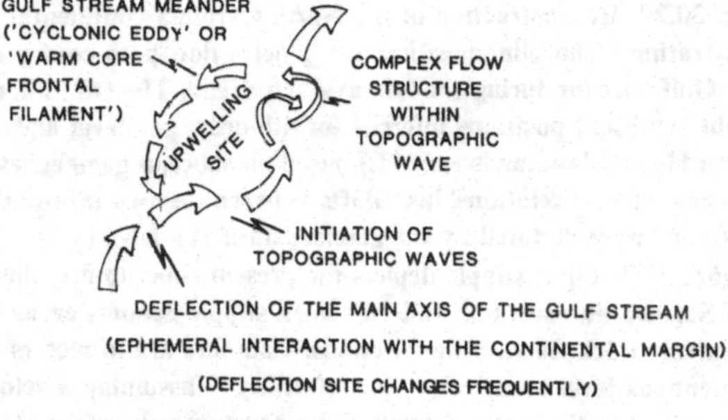

Fig. 30.22. Paleoenvironmental–paleoceanographic reconstruction of the North Carolina continental margin illustrating the inferred shoreline position and flow-path configuration of the Gulf Stream during an early transgressive phase of sea-level cycle OBS-1. Contours are in meters relative to sea-level position #2, or approximately equal to present sea level. Note that bypassing of some of the bathymetric barriers to the south has resulted in the Gulf Stream impinging on the North Carolina continental margin. As a consequence of this frictional interaction, ephemeral cyclonic eddies may have been generated and shed downstream from the site(s) of impact (stippled area).

positional sequence to the next (Scott W. Snyder, Chapter 32, this volume). Thus, they reflect the rapid shifting of upwelling sites with changing sea levels on a time scale consistent with glacio-eustatic fluctuations. All of these processes are implicit in the reconstructions (Figs. 30.21–30.25).

Conclusions

The signature of Gulf Stream scour is firmly inscribed on the physical stratigraphic record of the Miocene section along the North Carolina continental margin. Direct evidence includes: (1) large-scale erosional surfaces cut into pre-existing stratigraphic units; (2) preferential preservation of sediment units on the landward side of these excavation surfaces; (3) elongated, coast-parallel erosional zones of stratigraphic thinning; and (4) stratigraphic outliers severed from the main body of their equivalent time–stratigraphic sections.

Maps delineating the spatial distribution of specific time–stratigraphic sections further imply that sequence geometries and basin infill histories have been significantly influenced, if not completely molded, by multiple changes in the flow path of the western boundary current. These shifts were driven by short-pulsed (<300 ka), high-amplitude (50–200 m) sea-level fluctuations.

Reconstructions of Miocene flow-track positions show the migration pathways between glacial lowstands and interglacial highstands extended several tens of kilometers across the continental margin. These relatively wide migration pathways were a consequence of the high amplitude nature of the sea-level cycles. The position of the western boundary current at any one time during the Miocene appears to have been controlled by the position of sea level relative to the local bathymetry. Seafloor obstructions dynamically steered the boundary current seaward by inducing vorticity-related instabilities through frictional shear.

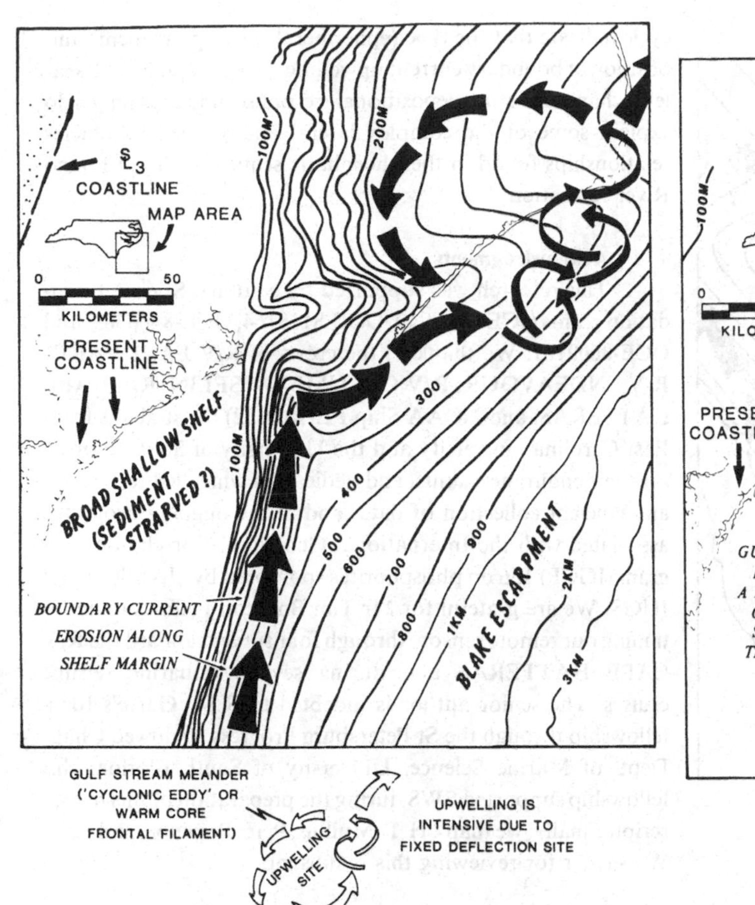

Fig. 30.23. Paleoenvironmental–paleoceanographic reconstruction of the North Carolina continental margin illustrating the inferred shoreline position and flow path configuration of the Gulf Stream during the Mid–Late transgressive phase of sea-level cycle OBS-1. Contours are in meters relative to sea-level position #3 which is approximately 100 m above present sea level. Note that the southern flank of the Cape Lookout High has become a persistent bathymetric barrier to the northward flowing Gulf Stream. Cyclonic eddies reside within the Aurora Embayment because the western boundary current is consistently forced seaward by the Cape Lookout High. Consequently, upwelling conditions prevailed within this shelf embayment and adjacent basin. The landward translation of the shoreline removed siliciclastic sources. The northeast Onslow Embayment became sediment-starved, severely restricting any lateral growth.

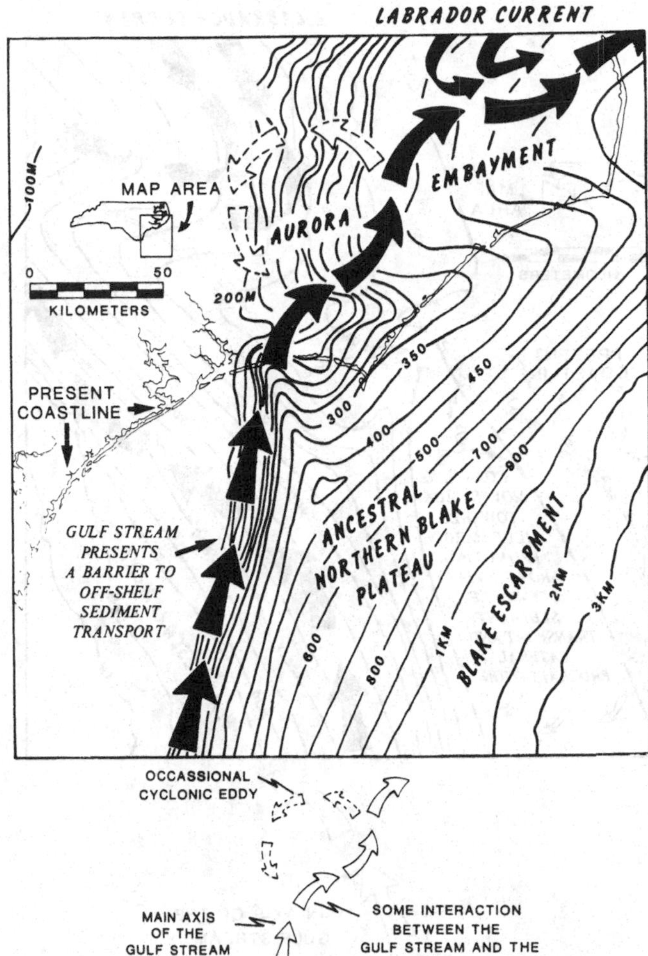

Fig. 30.24. Paleoenvironmental–paleoceanographic reconstruction of the North Carolina continental margin illustrating the inferred position and configuration of the Gulf Stream during the terminal transgressive phase of sea-level cycle OBS-1. Contours are in meters relative to sea-level position #4 which is approximately 150 meters above present sea level. Note that elevated sea levels facilitated a breach of the Cape Lookout High bathymetric barrier. Consequently, the Gulf Stream entered and flowed through the Aurora Embayment, bathing this deep-shelf environment in warm, subtropical waters. Friction generated as the Gulf Stream over-ran the top of this bathymetric hill may have induced additional but less intense upwelling events within the western margin of the Aurora Embayment.

Major Miocene deflection events occurred when the Gulf Stream interacted with the bathymetric hills of the Charleston Bump complex (Fig. 30.15) and the Cape Lookout High (Figs. 30.17 and 30.19). More specifically, Aquitanian and Early Burdigalian sea-level positions, as well as the glacial lowstands during the Mid-Burdigalian and the Serravallian, all forced the Gulf Stream to adjust its flow path to accommodate topographic obstructions within the Charleston Bump complex. As a result of these frontal events, the ancestral boundary current flowed seaward of the North Carolina shelf–slope break. Reoccupation of similar flow path positions during these relative low positions of sea level governed the growth patterns associated with renewed accretion along a major sediment drift system (Cape Lookout High). When eustatic changes resulted in relatively high sea levels, the Miocene boundary current was capable of overriding many of the barriers in the Charleston Bump complex. The boundary current then flowed against and sometimes across the North Carolina shelf margin, excavating major stratigraphic sections. This was particularly evident during the latest Mid-Burdigalian through Early Langhian, when the flow paths of this boundary current carved out an enormous embayment along the North Carolina shelf margin and upper slope (Outer Onslow Bay Basin). During the remainder of the Langhian Stage, laterally migrating flow paths controlled the initial infill history of this excavated basin. Offshelf sediment transport and shelf-margin progradational infilling developed

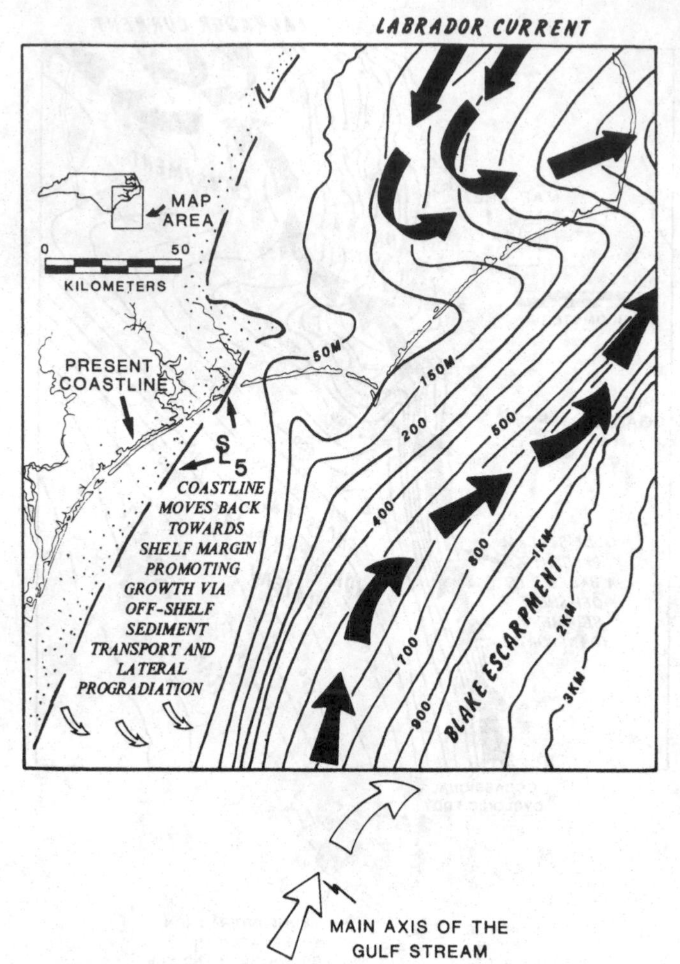

Fig. 30.25. Paleoenvironmental–paleoceanographic reconstruction of the North Carolina continental margin illustrating the inferred shoreline position and flow-path configuration of the Gulf Stream during the maximum regressive phase of sea-level cycle OBS-1 (i.e. termination of OBS-1 and initiation of OBS-2). Contours are in meters relative to sea-level position #5 which was approximately 50 meters *above* present sea level. Note that the Gulf Stream still rubs against the North Carolina continental margin even though the lower sea level has forced this boundary current to interact and became slightly deflected seaward by bathymetric barriers within the Charleston Bump Complex. This type of interaction may have led to the initiation of topographic waves and the subsequent shedding of ephemeral cyclonic gyres similar to those illustrated in Figure 30.22.

during lowstands. These shelf margin sequences were then partially cut back via boundary current erosion during succeeding highstands. Downstream from the outer Onslow Bay Basin, the Gulf Stream traversed the Cape Lookout High. Boundary current erosion during the terminal highstands of each Langhian sea-level cycle totally removed all Burdigalian and Langhian stratal sequences from the top of this relict sediment drift.

Relative sea-level positions during the Serravallian approximated Mid-Burdigalian positions as the Mid-Miocene supercycle (TB2) started into a regressive phase (Fig. 30.3). Consequently, the Serravallian Gulf Stream occupied flow patterns similar in position and configuration to the Mid-Burdigalian flow paths. Initial attempts at reconstructing higher-order flow path shifts through one complete sea-level cycle indicate that the type (ephermeral *versus* permanent) and position of boundary current upwelling change rapidly with sea-level changes. Future depositional–response modeling may help explain some of the complex lateral and vertical lithofacies relationships found in the phosphatic sequences of the Pungo River formation.

Acknowledgments

This research was supported by National Science Foundation grants OCE-7908949, OCE-8118164, OCE-8400383, and OCE-8609161. We thank 1) the crews of R/V EASTWARD, R/V ENDEAVOUR, R/V COLUMBUS ISELIN, R/V CAPE HATTERAS; and NOAA Ship PEIRCE, 2) the students from East Carolina University and the University of South Florida for their endurance, skills, and dedication which led to the safe and efficient collection of data; and 3) the ongoing programs associated with the International Geological Correlation Program (IGCP) 156 on phosphorites sponsored by UNESCO and IUGS. We are grateful for Mr Tim Boynton's efforts in maintaining our remote sensors through four surveys aboard the R/V CAPE HATTERAS and during several estuarine seismic cruises. The senior author is indebted to R.M. Garrels for a fellowship through the St Petersburg Progress Endowed Chair, Dept. of Marine Science, University of South Florida; this fellowship supported SWS during the preparation of this manuscript. Finally, we thank H.T. Mullins, Peter Popenoe, and Scott W. Snyder for reviewing this manuscript.

References

Abbott, W.H. (1978). Correlation and zonation of Miocene strata along the Atlantic margin on North America using diatoms and silicoflagellates. *Marine Micropaleontology*, **3**, 15–34.

Abbott, W.H. (1980). Diatoms and stratigraphically significant silicoflagellates from the Atlantic Margin Coring Project and other Atlantic Margin Sites. *Micropaleontology*, **26**, 49–80.

Abbott, W.H. (1984). Progress in the recognition of Neogene diatom datums along the US Atlantic Coast. *Palaeogeography, Palaeoclimatology, Palaeoecology*, **47**, 5–20.

Aller, R.C. (1980). Diagenetic processes near the sediment–water interface of Long Island Sound. I. Decomposition and nutrient element geochemistry (S,N,P). *Advances in Geophysics*, **22**, 237–350.

Atkinson, L.P., Menzel, D.W. & Bush, K.A. (1985). *Oceanography of the Southeastern US Continental Shelf.* Coastal and Estuarine Science Series, v. 2, American Geophysical Union, Washington, DC.

Auer, S.J. (1987). Five-year climatological survey of the Gulf Stream system and its associated rings. *Journal of Geophysical Research*, **92**, 709–26.

Baker, P.A. & Burns, S.J. (1986). Occurrence and formation of dolomite in organic-rich continental margin sediments. *American Association of Petroleum Geologists Bulletin*, **69**, 1917–30.

Bane, J.M., Jr. (1983). Initial observations of the subsurface structure and short-term variability of the seaward deflection of the Gulf Stream off Charleston, South Carolina. *Journal of Geophysical Research*, **88**, 4673–84.

Bane, J.M., Jr. & Brooks, D.A. (1979). Gulf Stream meanders along the continental margin from the Florida Straits to Cape Hatteras. *Geophysical Research Letters*, **6**, 280–2.

Bane, J.M., Jr., Brooks, D.A. & Lorenson, K.R. (1981). Synoptic observations of the three-dimensional structure and propagation of Gulf Stream meanders along the Carolina Continental Margin. *Journal of Geophysical Research*, **86**, 6411-25.

Barron, J.A. (1985a). Late Eocene to Holocene diatom biostratigraphy of the equatorial Pacific Ocean, Deep Sea Drilling Project Leg 85. In

Initial Reports DSDP, vol 85, ed. L. Mayer, F. Theyer *et al.*, pp. 413–56.
Barron, J.A. (1985b). Miocene to Holocene planktonic diatoms. In *Plankton Stratigraphy*, ed. H.M. Bolli, J.B. Saunders & K. Perch-Neilsen, pp. 763–809. Cambridge University Press, Cambridge, England.
Berner, R.A. (1980). *Early Diagenesis: A Theoretical Approach*. Princeton University Press, Princeton, NJ.
Blanton, J.L., Atkinson, L.P., Pietrafesa, L.J. & Lee, T.N. (1981). The intrusion of Gulf Stream water across the continental shelf due to topographically induced upwelling. *Deep-Sea Research*, **28A**, 339–405.
Blow, W.H. (1969). Late Middle Eocene to Recent planktonic foraminiferal biostratigraphy. In *Proceedings of the 1st International Conference on Planktonic Microfossils*, Geneva, 1967, vol. 1, ed. P. Bronniman & H.H. Renz, pp. 199–421. E.J. Brill, Leiden.
Brooks, D.A. & Bane, J.M., Jr. (1978). Gulf Stream deflection by a bottom feature off Charleston, South Carolina. *Science*, **201**, 1225–6.
Brooks, D.A. & Bane, J.M., Jr. (1981). Gulf Stream fluctuations and meanders over the Onslow Bay upper continental slope. *Journal of Physical Oceanography*, **11**, 247–56.
Brooks, D.A. & Bane, J.M., Jr. (1983). Gulf Stream meanders off North Carolina during winter and summer, 1979. *Journal of Geophysical Research*, **88**, 4633–50.
Brown, P.M. (1958). The relation of phosphorites to groundwater in Beaufort County, North Carolina. *Economic Geology*, **53**, 85–101.
Brown, P.M., Miller, J.A. & Swain, F.M. (1972). Structural and stratigraphic framework and spatial distribution of permeability of the Atlantic Coastal Plain, North Carolina to New York. *US Geological Survey Professional Paper*, **796**, 79pp.
Burky, D. (1981). Synthesis of silicoflagellate stratigraphy for Maastrichtian to Quaternary marine sediment. In *The Deep Sea Drilling Project: A Decade of Progress*, ed. J.E. Worme, R.G. Douglas & E.L. Winterer, pp. 433–44. Society of Economic Paleontologists and Mineralogists Special Publication 32, Tulsa.
Burnett, W.C. (1977). Geochemistry and origin of phosphorite deposits from off Peru and Chile. *Geological Society of America Bulletin*, **88**, 813–23.
Chao, S.Y. & Janowitz, G.S. (1979). The effect of a localized topographic irregularity on the flow of a boundary current along the continental margin. *Journal of Physical Oceanography*, **9**, 900–10.
Claypool, G.E. & Kaplan, I.R. (1974). The origin and distribution of methane in marine sediments. In *Natural Gases in Marine Sediments*, ed. I.R. Kaplan, pp. 99–139. Plenum, New York.
Compton, J.C. (1988). Degree of supersaturation and precipitation of organogenic dolomite. *Geology*, **16**, 318–21.
Cook, P.J. & McElhinny, M.W. (1979). A reevaluation of the spatial and temporal distribution of sedimentary phosphate deposits in the light of plate tectonics. *Economic Geology*, **74**, 315–30.
Fenner, J. (1985). Cretaceous to Oligocene planktic diatoms. In *Plankton Stratigraphy*, ed. H.M. Bolli, J.B. Saunders & K. Perch-Nielsen, pp. 713–62. Cambridge University Press, Cambridge, England.
Fofonoff, N.P. (1954). Steady flow in a frictionless homogeneous ocean. *Journal of Marine Research*, **13**, 254–62.
Fofonoff, N.P. (1962). Dynamics of ocean currents. In *The Sea, Vol. 1*, ed. M.N. Hill, pp. 323–95. Interscience Publishers, John Wiley & Sons, New York.
Fofonoff, N.P. (1981). The Gulf Stream System. In *Evolution of Physical Oceanography*, ed. B.A. Warren & C. Wunsch, pp. 11–139. MIT Press, Cambridge, MA.
Gibson, T.G. (1967). Stratigraphy and paleoenvironment of the phosphatic Miocene strata of North Carolina. *Geological Society of America Bulletin*, **78**, 631–49.
Gibson, T.G. (1982). Depositional framework and paleoenvironments of Miocene strata from North Carolina to Maryland. In *Miocene of the Southeastern US*, ed. T.M. Scott & S.B. Upchurch, pp. 1–22. Florida Bureau of Geology Special Publication No. 25.
Haq, B.V., Hardenbol, J. & Vail, P.R. (1987). Chronology of fluctuating sea levels since the Triassic. *Science*, **235**, 1156–67.
Hendershot, M.C. (1987). Single layer models of the general circulation. In *General Circulation of the Ocean*, ed. H.D.I. Abarbanel & W.R. Young, pp. 202–67. Springer-Verlag, New York.
Holton, J.R. (1972). Circulation and vorticity. In *Introduction to Dynamic Meteorology*, Chapter 5, International Geophysical Series, vol. 16, Academic Press, New York.
Hulbert, H.E. & Thompson, J.D. (1980). A numerical study of Loop Current intrusions and eddy shedding. *Journal of Physical Oceanography*, **10**, 1611–51.
Janowitz, G.S. & Pietrafesa, L.J. (1980). A model and observations of time dependent upwelling over the mid-shelf and slope. *Journal of Physical Oceanography*, **10**, 1574–83.
Janowitz, G.S. & Pietrafesa, L.J. (1982). The effects of a longshore variation in bottom topography on a boundary current or topographically induced upwelling. *Continental Shelf Research*, **1**, 123–41.
Katrosh, M.C. & Snyder, Scott W. (1982). Foraminifera of the Pungo River Formation, central coastal plain of North Carolina. *Southeastern Geology*, **23**, 217–23.
Kazakov, A.V. (1937). The phosphate facies and the genesis of phosphorites. *Scientific Institute of Fertilizers and Insecto-Fungicides Proceedings*, **142**, 95–113.
Kimrey, J.W. (1964). The Pungo River Formation, a new name for the middle Miocene phosphorites in Beaufort County, North Carolina. *Southeastern Geology*, **5**, 195–205.
Kimrey, J.W. (1965). Description of the Pungo River Formation in Beaufort County, North Carolina. *North Carolina Department of Conservation and Development Division of Mineral Resources, Bulletin*, **79**, 131pp.
Leaman, K.D., Molinari, R.L. & Vertes, P.S. (1987). Structure and variability of the Florida current at 27° N: April 1982–July 1984. *Journal of Physical Oceanography*, **17**, 565–83.
Lee, T.N. (1975). Florida current spin-off eddies. *Deep Sea Research*, **22**, 753–65.
Lee, T.N. & Atkinson, L.P. (1983). Low-frequency current and temperature variability from Gulf Stream frontal eddies and atmosphere forcing along the southeast US outer continental shelf. *Journal of Geophysical Research*, **88**, 4541–67.
Lee, T.N., Atkinson, L.P. & Legeckis, R. (1981). Observations of a Gulf Stream frontal eddy on the Georgia Continental Shelf, April 1977. *Deep Sea Research*, **28A**, 347–78.
Lee, T.N. & Mayer, D. (1977). Low-frequency current variability and spin-off eddies on the shelf off southeast Florida: *Journal of Marine Research*, **35**, 193–220.
Lee, T.N. & Waddell, E. (1983). On Gulf Stream variability and meanders over the Blake Plateau at 30° N. *Journal of Geophysical Research*, **88**, 4617–31.
Legeckis, R.V. (1976). The influence of bottom topography on the path of the Gulf Stream at latitude 31° N from NOAA's satellite imagery. *Transactions of the American Geophysical Union*, **57**, 260.
Legeckis, R.V. (1979). Satellite observations of the influence of bottom topography on the seaward deflection of the Gulf Stream off Charleston, South Carolina. *Journal of Physical Oceanography*, **9**, 483–97.
Lewis, D.W., Riggs, S.R. Snyder, Stephen W., Hine, A.C., Snyder, Scott W. & Waters, V.J. (1982). Preliminary stratigraphic report on the Pungo River Fm. in Onslow Bay, Continental Shelf, North Carolina. In *Miocene of the southeastern US*, ed. T.M. Scott & S.B. Upchurch, pp. 122–37. Florida Bureau of Geology Special Publication No. 25.
Mallette, P.M. (1986). 'Lithostratigraphic analysis of cyclical phosphorite sedimentation within the Miocene Pungo River Formation, North Carolina Continental Shelf'. Unpublished M.Sc. thesis, Department of Geology, East Carolina University, Greenville, NC.
Manheim, F.T., Popenoe, P., Siapno, W. & Lane, C. (1982). Manganese–phosphorite deposits of the Blake Plateau. In *Marine Mineral Deposits – New Research Results and Economic Prospects*, ed. P. Halbach & P. Winter, pp. 9–44. Gluckauf, Essen, West Germany.
Martini, E. (1971). Standard Tertiary and Quaternary calcareous nannoplankton zonation: *Proceedings II Planktonic Conference Rome, Italy*, pp. 739–85. Edizioni Tecnoscienza, Roma.
Martini, E. & Muller, C. (1976). Eocene to Pleistocene silicoflagellates from the Norwegian–Greenland Sea, DSDP Leg 38. In *Initial Reports of the Deep Sea Drilling Project*, vol. 38, pp. 857–95. US Government Printing Office, Washington DC.

Martini, E. & Worsley, T.R. (1970). Standard Neogene calcareous nannoplankton zonation. *Nature*, **215**, 289–90.

Matteucci, T.D. (1984). 'High-resolution seismic stratigraphy of the North Carolina continental margin – the Cape Fear region: sea-level cyclicity, paleobathymetry, and Gulf Stream dynamics'. Unpublished M.Sc. thesis, University of South Florida, St Petersburg.

Maul, G.A. (1977). The annual cycle of the Gulf Loop. Part I. Observations during a one-year time series. *Journal of Marine Research*, **35**, 219–47.

Maul, G.A. (1978). The 1972–73 cycle of the Gulf Loop Current. Part II: mass and salt balances of the basin. In *Cooperative Investigations of the Caribbean and Adjacent Regions*, ed. H.B. Stewart, pp. 567–619. FAO Fisheries Report 200, Rome.

Maul, G.A., DeWitt, P.W., Yanaway, A. & Baig, S.R. (1978). Geostationary satellite observations of Gulf Stream meanders: infrared measurements and time series analysis. *Journal of Geophysical Research*, **83**, 6123–35.

McArthur, J.M., Benmore, R.A., Coleman, M.L., Soldi, C., Yeh, H.-W. & O'Brien, G.W. (1986). Stable isotopic characterization of francolite formation. *Earth Planetary Science Letters*, **77**, 20–34.

McClain, C.R. & Atkinson, L.P. (1985). A note on the Charleston gyre. *Journal of Geophysical Research*, **90**, 857–61.

McClain, C.R., Pietrafesa, L.J. & Yodder, J.A. (1984). Observations of Gulf Stream-induced and wind-driven upwelling in the Georgia Bight using ocean color and infrared imagery. *Journal of Geophysical Research*, **89**, 3705–23.

McClain, C.R., Pietrafesa, L.J. & Yodder, J.A. (1985). Corrections to 'observations of Gulf Stream-induced and wind-driven upwelling in the Georgia Bight using ocean color and infrared imagery'. *Journal of Geophysical Research*, **90**, 15–18.

Miller, J.A. (1971). 'Stratigraphic and structural setting of the middle Miocene Pungo River Formation of North Carolina'. Unpublished Ph.D. thesis, University North Carolina, Chapel Hill, NC.

Miller, J.A. (1982). Structural and sedimentary setting of phosphorite deposits in North Carolina and in northern Florida. In *Miocene of the Southeastern US*, ed. T.M. Scott & S.B. Upchurch, pp. 162–82. Florida Bureau of Geology Special Publication No. 25.

Moore, T.L. (1986). 'Foraminiferal biostratigraphy and paleoecology of the Miocene Pungo River Formation, central Onslow embayment, North Carolina continental margin'. Unpublished M.Sc. thesis, East Carolina University, Greenville, North Carolina.

Moore, T.L. & Snyder, Scott W. (1985). Benthic foraminiferal paleoecology of Miocene deposits in central and northern Onslow Bay, North Carolina Continental Shelf. *Geological Society of America*, Abstract with Programs, **17**, 126.

Mueller, P.J. & Suess, E. (1979). Productivity, sedimentation rate, and sedimentary organic matter in the ocean. I. Organic carbon preservation. *Deep Sea Research*, **26**, 1347–62.

Okada, H. & Bukry, D. (1980). Supplementary modification and introduction of code numbers to the low-latitude coccolith biostratigraphic zonation (Bukry, 1973–1975). *Marine Micropaleontology*, **5**, 321–5.

Olson, D.B., Brown, O.B. & Emmerson, S.R. (1983). Gulf Stream frontal statistics from Florida Straits to Cape Hatteras derived from satellite and historical data. *Journal of Geophysical Research*, **88**, 4569–77.

Paffenhofer, G.A., Atkinson, L.P., Blanton, J.O., Lee, T.N., Pomeroy, L.R. & Yodder, J.A. (1987). Summer upwelling on the southeastern Continental Shelf of the USA during 1981. *Progress in Oceanography*, **19**, 221–30.

Palmer, A. (1988). Radiolarian biostratigraphy and paleoecology. In *Micropaleontology of the Miocene Pungo River Formation from Shallow Subsurface of Onslow Bay, North Carolina Continental Shelf. Journal of Foramininferal Research Special Publication*, **25**, 163–78.

Paluszkiewicz, T., Atkinson, L.P., Posmentier, E.S. & McClain, C.R. (1983). Observations of a loop current frontal eddy intrusion onto the West Florida Shelf. *Journal of Geophysical Research*, **88**, 9639–51.

Parrish, J.T. (1982) Upwelling and petroleum source beds, with reference to Paleozoic. *American Association of Petroleum Geologists Bulletin*, **66**, 750–74.

Pedlosky, J. (1987). *Geophysical Fluid Dynamics*, 2nd edn. Springer-Verlag, New York.

Pietrafesa, L.J. (1983). Shelfbreak circulation, fronts, and physical oceanography: east and west coast perspectives. In *Shelf Break: Critical Interface on Continental Margins*, ed. D.J. Starkey & G.T. Moore, pp. 233–50. Society of Economic Paleontologists and Mineralogists Special Publication 33, Tulsa.

Pietrafesa, L.J., Atkinson, L.P. & Blanton, J.O. (1978). Evidence for deflection of the Gulf Stream by the Charleston Rise. *Gulf Stream*, **4**, 3–7.

Pinet, P.R. & Popenoe, P. (1982). Blake Plateau: control of Miocene sedimentation patterns by large-scale shifts of the Gulf Stream axis. *Geology*, **10**, 257–9.

Pinet, P.R. & Popenoe, P. (1985a). A scenario of Mesozoic–Cenozoic ocean circulation over the Blake Plateau. *Geological Society of America Bulletin*, **96**, 618–26.

Pinet, P.R. & Popenoe, P. (1985b). Shallow seismic stratigraphy and post-Albian geologic history of the northern and central Blake Plateau. *Geological Society of America Bulletin*, **96**, 627–38.

Pinet, P.R., Popenoe, P., & Nelligan, D.F. (1981). Gulf Stream reconstruction of Cenozoic flow patterns over the Blake Plateau. *Geology*, **9**, 266–70.

Pond, S. & Piccard, G.L. (1983). *Introductory Dynamical Oceanography*, 2nd edn. Pergamon Press, Oxford.

Popenoe, P. (1985). Seismic stratigraphy and Tertiary development of the North Carolina continental margin. In *Geological Evolution of the US Atlantic margin*, ed. C.W. Poag, pp. 125–87. Van Nostrand Reinhold, New York.

Popenoe, P., Henry, V.J. & Idris, F.M. (1987). Gulf Trough – the Atlantic connection. *Geology*, **15**, 327–32.

Powers, E.R. (1986). Biostratigraphic correlation of Miocene phosphorites on the North Carolina continental margin using diatoms and silicoflagellates. *Society of Economic Paleontologists and Mineralogists, Abstracts with Programs*, **3**, 91.

Powers, E.R. (1987). 'Diatom biostratigraphy and paleoecology of the Miocene Pungo River formation, North Carolina Continental Margin'. Unpublished M.Sc. thesis, East Carolina University, Greenville, NC.

Powers, E.R. (1988). Diatom biostratigraphy and paleoecology. In *Micropaleontology of the Miocene Pungo River Formation from Shallow Substance of Onslow Bay, North Carolina Continental Shelf*, ed. S.W. Snyder. *Journal of Foraminiferal Research Special Publication*, **25**, 97–162.

Reidel, W.R. & Sanfillippo, A. (1978). Stratigraphy and evolution of tropical Cenozoic Radiolarians. *Micropaleontology*, **29**, 61–96.

Riggs, S.R. (1984). Paleoceanographic model of Neogene phosphorite deposition, US Atlantic continental margin. *Science*, **224**, 123–31.

Riggs, S.R., Lewis, D.W., Scarborough, A.K. & Snyder, S.W. (1982). Cyclic deposition of Neogene phosphorites in the Aurora area North Carolina and their possible relationship to global sea-level fluctuations. *Southeastern Geology*, **23**, 189–204.

Riggs, S.R., Snyder, Stephen W., Hine, A.C., Snyder, Scott W., Ellington, M. & Mallette, P.M. (1985). Geologic framework of phosphate resources in Onslow Bay, North Carolina Continental Shelf. *Economic Geology*, **80**, 716–38.

Sanfillipo, A., Westberg, M.J. & Reidel, W.R. (1981). Cenozoic radiolarians at Site 462, Deep Sea Drilling Project Leg 61, Western Tropical Pacific. In *Initial Report DSDP*, vol. 61, ed. J. Shambach & L.N. Stout, pp. 495–505. US Government Printing Office, Washington, DC.

Scarborough, A.K., Riggs, S.R. & Lewis, D.W. (1982). Stratigraphy and petrology of the Pungo River Formation, central coastal plain, North Carolina. *Southeastern Geology*, **23**, 205–15.

Sheldon, R.P. (1981). Ancient marine phosphorites: *Annual Review of Earth and Planetary Science*, **9**, 251–84.

Singer, J.J., Atkinson, L.P., Blanton, J.O. & Yodder, J.A. (1983). Cape Romain and the Charleston Bump: historical and recent hydrographic observations. *Journal of Geophysical Research*, **88**, 4685–97.

Snyder, Scott W., ed. (1988). *Micropaleontology of the Miocene Pungo*

River Formation from Shallow Subsurface of Onslow Bay, North Carolina Continental Shelf. Journal of Foramiferal Research Special Publication, **25**, 189pp.

Snyder, Scott W., Riggs, S.R., Katrosh, M.R., Lewis, D.W. & Scarborough, A.K. (1982). Synthesis of the phosphatic sediment–faunal relationships within the Pungo River Formation: paleoenvironmental implications. *Southeastern Geology*, **23**, 233–45.

Snyder, Scott W., Steinmetz, J.C., Waters, V.J. & Moore, T.L. (1988). Occurrence and biostratigraphy of planktonic foraminifera and calcareous nannofossils. *Journal of Foraminiferal Research Special Publication*, **#25**, 15–42.

Snyder, Stephen W. (1982). 'Seismic stratigraphy within the Miocene Carolina Phosphogenic Province: chronostratigraphy, paleotopographic controls, sea-level cyclicity, Gulf Stream dynamics, and the resulting depositional framework'. Unpublished M.Sc. thesis, Geology Department, University of North Carolina Chapel Hill, NC.

Snyder, Stephen W. (1989). 'Miocene sea-level cyclicity: resolution of frequency and estimates of amplitudes from the stratigraphic record of the Carolina Continental Margin'. Ph.D. thesis, Department of Marine Science, University of South Florida, St Petersburg.

Snyder, Stephen W., Hine, A.C. & Riggs, S.R. (1982). Miocene seismic stratigraphy, structural framework and sea-level cyclicity: North Carolina continental shelf. *Southeastern Geology*, **23**, 247–66.

Snyder, Stephen W., Hine, A.C. & Riggs, S.R. (1983). 3-D stratigraphic modeling from high-resolution seismic reflection data: an example from the North Carolina Continental Shelf. *American Association of Petroleum Geologists Bulletin*, **67**, 549–50.

Snyder, Stephen W., Hine, A.C. & Riggs, S.R. (1984a). Cyclic Miocene phosphogenic episodes within the North Carolina continental margin: a paleoceanographic model. *Society of Economic Paleontologists and Mineralogists Abstracts*, **1**, 76.

Snyder, Stephen, W., Hine, A.C. & Riggs, S.R. (1986b). Shifting sites of Miocene phosphorite deposition: the critical depth hypothesis. *Society of Economic Paleontologists and Mineralogists Abstracts*, **III**, 105.

Snyder, Stephen W., Riggs, S.R., Hine, A.C., & Snyder, Scott W. (1984b). Lithostratigraphic response to glacioeustatic sea-level cyclicity and concomittant Gulf Stream flow dynamics: Miocene stratigraphy, North Carolina continental margin. *Society of Economic Paleontologists and Mineralogists Abstracts*, **1**, 77.

Snyder, Stephen W., Hine A.C., Riggs, S.R. & Snyder, Scott W. (1986a). Miocene unconformities, chronostratigraphy and sea-level cyclicity: fine-tuning the Early Neogene relative coastal onlap curve for the North Carolina continental margin. *American Association of Petroleum Geologists Bulletin*, **70**, 651.

Steele, G.A. (1980). 'Stratigraphy and depositional history of Bogue Banks, North Carolina'. Unpublished M.Sc. thesis, Duke University, Durham, NC.

Stommel, H. (1948). The westward intensification of wind-driven ocean currents. *Transactions of the American Geophysical Union*, **29**, 202–6.

Stommel, H. (1965). *The Gulf Stream*. University of California Press, Berkeley.

Vail, P.R. & Hardenbol, J. (1979). Sea-level changes during the Tertiary. *Oceanus*, **22**, 71–9.

Vail, P.R., Mitchum, R.M. & Thompson, S. 1977. Seismic stratigraphy and global changes of sea level; Part 4: global cycles of relative changes of sea level. In *Seismic Stratigraphy – Application to Hydrocarbon Exploration*, ed. C.E. Payton, pp. 83–97. American Association of Petroleum Geologists Memoir.

Veronis, G. (1981). Dynamics of larger-scaled ocean circulation. In *Evolution of Physical Oceanography*, ed. B.A. Warren & C. Wunsch, pp. 140–83. MIT Press, Cambridge, Massachussetts.

Von Arx, W.S., Bumpus, D.F. & Richardson, W.S. (1955). On the fine structure of the Gulf Stream. *Deep Sea Research*, **3**, 46–55.

Vukovitch, F.M. (1986). Aspects of the behavior of cold perturbations in the eastern Gulf of Mexico: a case study. *Journal of Physical Oceanography*, **16**, 175–88.

Vukovitch, F.M. & Crissman, B.W. (1980). Some aspects of Gulf Stream western boundary eddies from satellite and *in situ* data. *Journal of Physical Oceanography*, **10**, 1792–813.

Vukovitch, F.M., Crissman, B.W., Bushnell, M. & King, W.J. (1979a). Some aspects of the oceanography of the Gulf of Mexico using satellite and *in situ* data. *Journal of Geophysical Research*, **84**, 7749–68.

Vukovitch, F.M., Crissman, B.W., Bushnell, M. & King, W.J. (1979b). Gulf Stream boundary eddies of the east coast of Florida. *Journal of Physical Oceanography*, **9**, 1214–22.

Vukovitch, F.M. & Maul, G.A. (1985). Cyclonic eddies in the eastern Gulf of Mexico. *Journal of Physical Oceanography*, **15**, 105–17.

Waters, V.J. (1983). 'Foraminiferal paleoecology and biostratigraphy of the Pungo River Formation, Southern Onslow Bay, North Carolina continental shelf'. Unpublished M.Sc. thesis, East Carolina University, Greenville, NC.

Waters, V.J. & Snyder, Scott W. (1986). Planktonic foraminiferal biostratigraphy of the Pungo River Formation, Southern Onslow Bay, North Carolina Continental Shelf. *Journal of Foraminiferal Research*, **16**, 9–23.

Webster, F.A. (1961). Description of Gulf Stream meanders off Onslow Bay, *Deep Sea Research*, **9**, 130–43.

Yodder, J.A., Atkinson, L.P., Lee, T.N., Kimm, H.H. & McClain, C.R. (1981a). Role of Gulf Stream frontal eddies in forming phytoplankton patches on the outer southeastern shelf. *Limnological Oceanography*, **26**, 1103–10.

Yodder, J.A., Atkinson, L.P., Lee, T.N., Kim, H.H. & McClain, C.R. (1981b). Role of Gulf Stream frontal plankton productivity and the distribution of fishes on the southeastern US continental shelf. *Science*, **214**, 352–3.

Yodder, J.A., Atkinson, L.P., Bishop, S.S., Hoffman, E.E. & Lee, T.N. (1983). Effect of upwelling on phytoplankton productivity on the outer southeastern US continental shelf. *Continental Shelf Research*, **1**, 385–404.

Zantopp, R.J., Leaman, K.D. & Lee, T.N. (1987). Florida current meanders: a close look in June–July 1984. *Journal of Physical Oceanography*, **17**, 584–95.

31

Patterns of phosphate deposition and lithofacies relationships within the Miocene Pungo River Formation, North Carolina continental margin

S.R. RIGGS AND P.M. MALLETTE

Abstract

From southern Florida through North Carolina, the Miocene section contains: abnormally high concentrations of phosphate; an associated suite of authigenic minerals (dolomite, silica polymorphs, clinoptilolite) and organic matter; regional patterns of phosphate, carbonate, and siliciclastic sedimentation; and cyclical lithologic patterns produced by alternating deposition and erosion (Riggs, 1984). Data from the Miocene Pungo River Formation in North Carolina suggest that paleoenvironmental conditions producing these phosphorite sediment sequences changed dramatically and periodically in response to complex interactions of both global and regional paleoceanographic events during the Miocene.

Introduction

Distribution of phosphate-rich sediments within the Pungo River Formation is not continuous or uniform either laterally or vertically. Spatial distribution of abundant phosphate was controlled by four interacting parameters (Riggs, 1979, 1980, 1981 and 1984; Stephen W. Snyder, 1982; Stephen W. Snyder, Hine & Riggs, 1982 and 1983; Stephen W. Snyder *et al.*, 1986): 1) Regional distribution was primarily determined by tectonic setting and first-order structural framework of the continental margin as indicated by the relationship of the Carolina Phosphogenic Province to the Carolina Platform (Figs. 29.1 and 29.2 in Riggs *et al.*, Chapter 29, this volume). 2) Distribution of phosphate, associated sediments, and petrographic and textural characteristics within the larger framework were controlled by smaller-scale, second-order structural and topographic elements within the continental margin sediment system. 3) Paleoceanographic cyclicity including sea-level oscillations, changing current patterns, and fluctuations in chemical parameters along the ocean margin, produced changing environmental conditions that determined the spatial distribution through time. 4) The ultimate concentration of phosphate components was also affected by local rate and volume of sedimentation, largely through dilution by nonphosphatic components. This paper considers the latter three parameters, which collectively determined the composition, distribution, and depositional patterns of Miocene lithofacies within the Carolina Phosphogenic Province.

Miocene sedimentation along the North Carolina continental margin was characterized by high-frequency, cyclical episodes of deposition alternating with episodes of non-deposition, erosion, and late diagenetic alteration that complicated the primary sediment patterns and altered the mineralogic components. Eighteen fourth-order seismic sequences (Figs 29.7 and 29.8 in Riggs *et al.*, Chapter 29, this volume) have been defined within the Miocene section on the Carolina margin (Stephen W. Snyder, 1982; Stephen W. Snyder *et al.*, 1982 and Chapter 30, this volume; Riggs *et al.*, 1985; Riggs & Belknap, 1988). Fourth-order seismic sequences were produced by transgressive and regressive sea-level events that repeatedly moved depositional and erosional processes across the shelf to produce the thick Miocene section (Fig. 29.3 in Riggs *et al.*, Chapter 29, this volume). This produced the high-degree of truncation of fourth-order seismic sequences that is apparent in the pattern of outcropping units on the continental shelf of Onslow Bay (Fig. 29.8 in Riggs *et al.*, Chapter 29, this volume).

Complex distributional patterns of siliciclastic, biogenic, and mixed authigenic–diagenetic sediments were initially determined by depositional controls of each sediment group. Subsequent episodes of destruction and modification associated with each sea-level cycle complicated previous depositional patterns (Stephen W. Snyder, 1982; Stephen W. Snyder *et al.*, 1982 and Chapter 30, this volume; Riggs, 1984; Hine & Stephen W. Snyder, 1985; Popenoe, 1985; Riggs & Belknap, 1988). Regressive and low-stand portions of each cycle were characterized by fluvial scouring and channeling with subaerial, late diagenetic alteration along the inner shelf; non-deposition, sediment bypass, and submarine late diagenetic alteration on the outer shelf; and contemporaneous siliciclastic deposition on the slope. The subsequent transgression resulted in shoreface retreat on the inner shelf and erosional processes by the Gulf Stream on the outer shelf.

Miocene lithostratigraphy

Lithologic components of the Pungo River Formation

Miocene sediments of the Carolina continental margin are complex mixtures of four major lithologic components (Table

Table 31.1. *Lithologic components of the Miocene Pungo River Formation along the North Carolina continental margin*

Sediment category	Mineral components	Relative abundance[a]
Siliciclastics		
Sand and Silt	Quartz	Major
	Feldspar	Minor
	Heavy minerals	Trace
Clays	Smectite	Major
	Fe-rich illites	Major
	Chamosite	Minor
	Sepiolite	Minor
Biogenics		
Macrofossil Allochems		Major
	Arthropods (dom. barnacles)	
	Bryozoans (assoc. with barnacles)	
	Echinoderms (dom. echinoids)	
	Mollusks (dom. bivalves)	
Microfossil Allochems		Major
	Foraminifera	
	Diatoms	
Organic Matter		Minor
Authigenics/Early diagenetics		
	Carbonate fluorapatite	Major
	Dolomite	Major
	Calcite	Minor
	Glauconite	Minor
Late diagenetics		
	Clinoptilolite	Major
	Calcite cements	Minor
	Dolomite	Minor
	Opal CT	Minor
	Microcrystalline quartz	Trace
	Pyrite	Trace
	Carbonate fluorapatite	Trace
	Glauconite	Trace

[a] Relative abundances are based upon the following average percentages: Major = > 10%; minor = 1–10%; Trace = < 1%. This does not mean that all units contain these abundances; specific depositional units may contain these concentrations.

31.1). Siliciclastics (group 1) are ubiquitous and volumetrically the most important component. Biogenics (group 2) and authigenics and early diagenetics (group 3) are locally abundant. Late diagenetics (group 4) are volumetrically least important, but are genetically the most complex and difficult to differentiate from the others. At any given point in time and space within the depositional system, any of these components may become the dominant sediment and constitute a major sediment lithofacies. Both depositional processes and sediment composition changed significantly through time; some of the resulting sediments were subjected to repeated episodes of erosion and late diagenetic alteration.

In order to sort out the many complex variables of the Pungo River Formation, the lithostratigraphy will be systematically discussed as follows. First, the four end-member lithologic components and resulting lithofacies will be defined. Second, the regional controls and resulting lateral distribution of lithofacies will be considered through time. Third, vertical lithofacies patterns that result from migrating sea level across the continental margin through time will be described. Fourth, complications from late diagenetic alteration will be superimposed upon the primary sediment patterns.

Siliciclastic sediments

Siliciclastic sediments consist of relatively stable silicates and oxides that represent the sedimentological framework of the Pungo River Formation. Sand- and silt-sized sediments consist primarily of angular–subangular, fresh, concoidally-fractured quartz grains with minor amounts of feldspar and heavy minerals. Clay minerals consist primarily of illite and smectite, with minor amounts of kaolinite, chamosite, palygorskite, and sepiolite (Lyle, 1984; Allison, 1988).

Siliciclastic sediments are derived from one of two sources: allocthonous siliciclastics are volumetrically the most important and were delivered into the depositional regime from landward point sources at the time of deposition; and less abundant autochthonous siliciclastic sediments are ubiquitous across the continental margin. Shoreface retreat, development of cape-shoal structures, and scour and channel cutting by oceanic-current systems such as the Gulf Stream, actively eroded and reworked previously deposited siliciclastic sediments.

Biogenic sediments

Biogenic sediments include skeletal material or organic matter that formed at or near the time of deposition, either within the water column or at or below the sediment–water interface. Planktonic forms reflect oceanographic conditions within the water column, benthic forms reflect conditions at or near the sediment–water interface, and infaunal forms reflect conditions within shallow interstitial waters. Like siliciclastic sediments, skeletal material may be eroded and reworked from previously deposited sediments. However, reworked biogenic material is more readily recognized because it is less stable and often shows physical and chemical evidence of reworking.

Biogenic sediments occur in minor concentrations throughout all lithofacies, often become major sediment components, and include calcareous macrofossils, calcareous and siliceous microfossils, and organic matter Calcareous macrofossils are generally disarticulated, abraded, gravel- and coarse sand-sized fragments of arthropods (barnacles), mollusks (bivalves and gastropods), and echinoderms (echinoid plates and spines). Calcareous mud is common and is interpreted to result from physical and biomechanical breakdown of pre-existing calcareous skeletal components. Calcareous microfossils include planktonic and benthic foraminifera, ostracods, and nannofossils. Siliceous microfossils, though less abundant than calcareous microfossils, occasionally constitute prominent portions of the biogenic sediment. Diatoms are locally abundant, particularly in muddy and clayey faces, while radiolarians, silicoflagellates, and sponge spicules occur in trace amounts. Organic matter is extremely variable in concentration ranging from 0.1% up to 6% TOC as follows:

Lithofacies Dominated By	*Range of TOC Values*
Siliciclastic sands or biogenic macrofossils	0.1–0.4%
Mixed authigenic/early diagenetic sediments and biogenic microfossils	0.4–1.0%
Phosphorite and dolomud facies	1.0–6.0%

Authigenic and early diagenetic sediments

Authigenic and early diagenetic sediments are those which formed at or near the time of deposition, either inorganically or organically, and at the sediment–water interface or in shallow interstitial waters. They reflect chemical, biological, and physical conditions of basal ocean waters and associated pore waters below the sediment–water interface at the time of their formation. Genesis of this group of mineral components is poorly known and it is very difficult to recognize reworked components.

The most common authigenic and early diagenetic sediments are phosphate, dolomite, and glauconite. Phosphate occurs mainly as sand-sized peloidal, intraclastic, and skeletal grains. Less frequently, it occurs as thin microsphorite laminae associated with hardgrounds and unconformities, or as phosphate-coated quartz grains derived from the erosion of microsphorites. Dolomite occurs primarily as isolated, very fine sand–silt-sized rhombs or as rhomb aggregates. Pungo River Formation dolomite is a non-ferroan, non-stoichiometric, and poorly ordered mineral interpreted to have formed as a primary authigenic mineral phase on the continental margin in cool marine pore waters and within the zone of sulfate reduction (Allen, 1985; Stewart, 1985). Trace amounts (locally up to a few percent) of light–dark green, silt- and sand-sized glauconite grains occur disseminated through some of the lithofacies.

Late diagenetic components

Late diagenetic components are those which form at some time after deposition in response to changed pore-water chemistry. They result either from alteration of pre-existing sediment components or from contribution of new chemicals to the sediments by interstitial fluids. Table 31.2 lists the specific types of alteration recognized within the Pungo River Formation. Due to depositional cyclicity in response to sea-level fluctuations, both submarine and subaerial diagenetic alteration are important and complex processes within the Formation. Recognition and removal of these complicating factors are prerequisite to understanding the genesis of authigenic and early diagenetic sediments.

Important late diagenetic components within the Pungo River Formation are listed in Table 31.1 and include zeolites, silica polymorphs, calcite cements, phosphate, dolomite, and pyrite. Clinoptilolite occurs as lath-shaped or bladed, silt- and sand-sized crystal aggregates that form as moldic infillings of foraminifera. Moldic aggregates and fine broken pieces of aggregates constitute up to 30% of the total sediment in some organic-rich, sandy muds. Opal-CT lepispheres and nodular cherts commonly occur disseminated through sediments and along unconformities that are laterally equivalent to diatom-rich facies, and in some carbonate-rich sediments. Calcite cements are abundant, indurating carbonate-rich sediments and occurring as crystal overgrowths on calcareous fossils throughout other lithofacies. Secondary phosphate occurs as internal and external molds of foraminiferal tests and various other fossil fragments. Sediments that contain external moldic phosphate grains with glossy and smooth surface textures, usually contain associated primary phosphate grains that have highly pitted surface textures. The latter grains have often been leached internally to produce an external 'eggshell' of the original grain. Minor amounts of dolomite occur as rhombic crystals growing on calcareous fossil fragments, often at the expense of the fossil. Trace amounts of pyrite occur within organic-rich mud facies and grow as framboidal aggregates on clinoptilolite crystals in foraminiferal tests.

Table 31.2. *Specific types of alteration recognized within each of the three categories of late diagenesis occurring in the Miocene Pungo River Formation on the North Carolina continental margin.*

Late Diagenetic Signatures
Modification of Microfossil Tests (Foraminifera and Diatoms)
A. Dissolution
B. Recrystallization
C. Mineralization
Modification of Authigenic/Early Diagenetic Phosphate Grains
A. Dissolution
B. Recrystallization
C. Rimming
D. Trace element profiles
Late Diagenetic Mineral Formation
A. Carbonate fluorapatite
B. Clinoptilolite
C. Calcite cements
D. Dolomite
E. Glauconite
F. Opal CT and other silica polymorphs

Lithofacies of the Pungo River Formation

Siliciclastic, biogenic, and authigenic–early diagenetic components represent primary sediment end-members, each associated with specific depositional controls. Ideally, each end-member can occur as a pure sediment. However, pure end-member sediments are rare within the Pungo River Formation; most sediments are mixtures of these components, as outlined in Figure 31.1. For example, all Pungo River lithofacies contain some siliciclastic sediment, even within depocenters of dominantly authigenic–early diagenetic and biogenic end members. Late diagenetic components were secondarily superimposed upon some Pungo River sediments sometime after primary sediment components were deposited. The next two sections of this paper discuss only the three primary lithologic components; secondary components associated with late diagenetic overprinting are considered in a later section.

Siliciclastic lithofacies

Although minor amounts of autochthonous siliciclastic material are ubiquitous, the greatest volume of siliciclastics within the Pungo River Formation are allochthonous. These

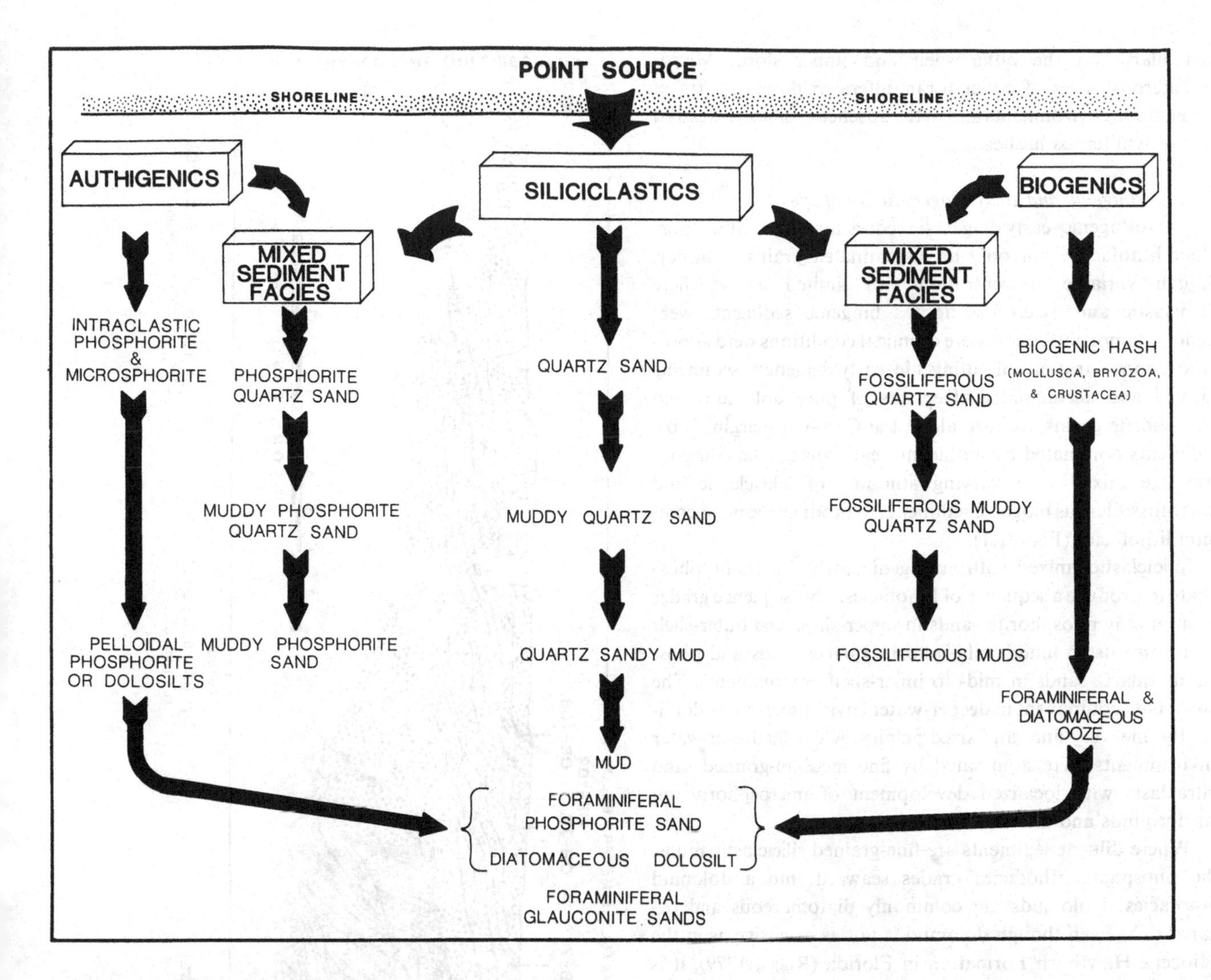

Fig. 31.1. Lithofacies common to the Pungo River Formation on the North Carolina continental margin.

point-source-derived siliciclastics were discharged onto the continental shelf and redistributed by normal marine shelf processes to produce the lithofacies diagramatically outlined in Figure 31.1. Consequently, siliciclastic lithofacies display strong regional controls with the location of point sources determining their distribution and textural characteristics. Also, siliciclastic sediments were the main diluent and are, therefore, an important factor in determining the relative concentrations of biogenic and authigenic–early diagenetic components.

In regions where point-source-derived siliciclastic sediments were discharged, allochthonous sediments form a nearly pure siliciclastic end-member with minor amounts of other sediment components present. A gradational series of facies results in which grain sizes decrease laterally away from the point source and seaward across the continental shelf (Fig. 31.1). Clean quartz sands were deposited either on the inner shelf in the vicinity of point-source discharges or off the shelf edge in response to major cross-shelf transport associated with systems such as the modern cape shoals (Matteuci, 1984; Matteuci & Hine, 1987). Muddy quartz sands were deposited across the shelf and laterally along the shelf away from point-source discharges. Quartz sandy mud and mud were deposited on the outer shelf and laterally along the shelf in areas remote to point-source discharges.

Biogenic lithofacies

In the absence of allochthonous siliciclastics and authigenic–early diagenetic sediments, the biogenic component dominates (Fig. 31.1). Biogenic hash consists of concentrated molluscan, bryozoan, and arthropod (primarily barnacles) fragments that are produced and deposited on the shelf. The richest biogenic hashes form on topographic highs and local hardgrounds where there is only minor dilution by siliciclastics due either to absence of a source area or to active winnowing of fines. Coarse biogenic sediments are often deposited as clinoform beds off the flanks of topographic highs.

Biogenic components are usually mixed with siliciclastic components throughout most shelf environments. Consequently, mixed macrofossiliferous and siliciclastic lithofacies, with varying concentrations of microfossils, represent a widespread lithofacies. Seaward across the shelf, macrofossil-dominated lithofacies grade into microfossil-dominated lithofacies, particularly on the outer shelf and upper slope. Microfossiliferous oozes of either foraminifera or diatoms form in upper-slope environments and in the absence of siliciclastics and macrofossiliferous hashes.

Authigenic and early diagenetic lithofacies

Authigenic–early diagenetic sediments occur within most other lithofacies, but only as disseminated grains in minor, though variable concentrations. In limited areas where siliciclastic and macrofossiliferous biogenic sediments were minor components, and where chemical conditions were appropriate, concentrations of authigenic–early diagenetic sediments formed and accumulated. Deposits of pure dolomuds and phosphorite grains are rare along the Carolina margin. Most sediments dominated by authigenic–early diagenetic components are mixed with varying amounts of siliciclastic and microfossiliferous biogenic sediments to produce the most common lithofacies (Fig. 31.1).

Siliciclastics mixed with authigenic–early diagenetic phosphate to produce a sequence of lithofacies. This sequence grades from muddy phosphorite sands in upper-slope and outer-shelf environments to muddy phosphorite quartz sands and phosphatic quartz sands in mid- to inner-shelf environments. The phosphate component in deeper-water environments was dominantly fine–very fine sand-sized peloids, while shallower-water environments were dominated by fine–medium-grained sand intraclasts with localized development of microsphorite on hardgrounds and unconformities.

Where diluent sediments are fine-grained siliciclastic muds, the phosphate lithofacies grades seaward into a dolomud lithofacies. Dolomuds are commonly diatomaceous and organic-rich. Even though dolomud is not as extensive as in the Miocene Hawthorn Formation in Florida (Riggs, 1979), it is probably the most extensive authigenic–early diagenetic lithofacies within the Pungo River Formation.

Many other sediment combinations occur as local facies, but they are not common and usually are not repeated in space and time. For example, glauconite forms internal molds of foraminiferal tests in fine-grained siliciclastic muds with little or no dolomite, phosphate, or diatom content.

Regional distribution of lithofacies in Onslow Bay

The 18 fourth-order seismic sequences that constitute the Pungo River Formation can be grouped into three depositional sections separated by major erosional unconformities (Figs. 29.6, 29.7 and 29.8 in Riggs *et al.*, Chapter 29, this volume). From oldest to youngest they are the Frying Pan (FPS), Onslow Bay (OBS), and the Bogue Banks Sections (BBS). Each has distinct regional lithofacies patterns. Differences between most successive fourth-order depositional units are not as distinctive as those between the three depositional sections. Changing depositional patterns through time reflect the evolution of the Miocene continental margin. The following discussion summa-

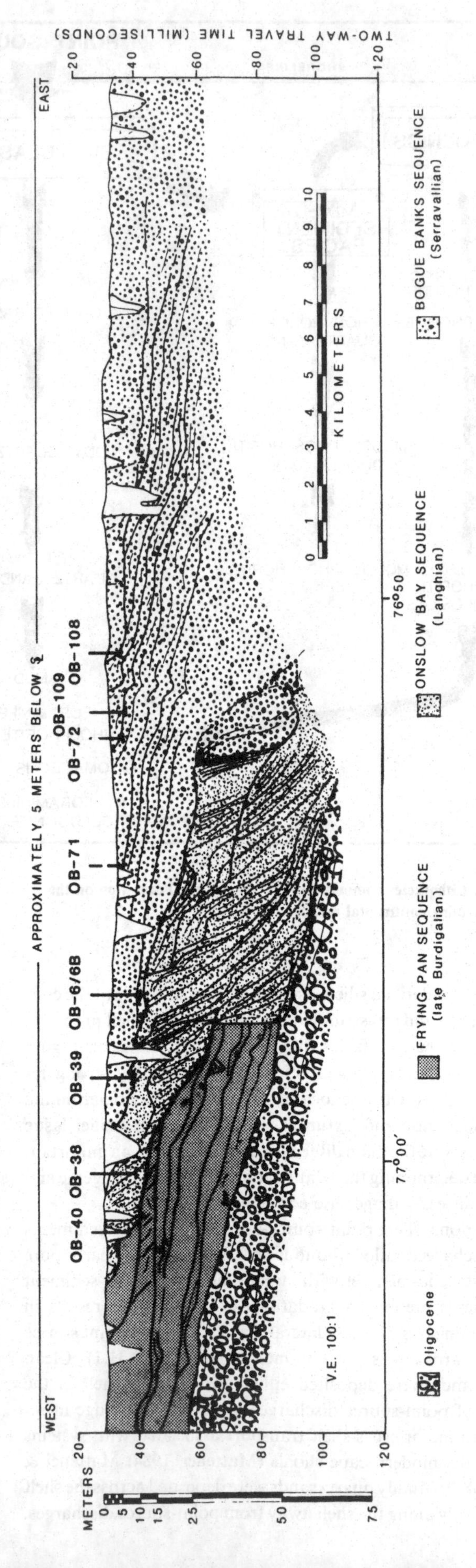

Fig. 31.2. Interpreted west to east uniboom seismic profile through northern Onslow Bay (22 m Profile) with vibracore locations superimposed (modified from Stephen W. Snyder in Riggs *et al.*, 1985).

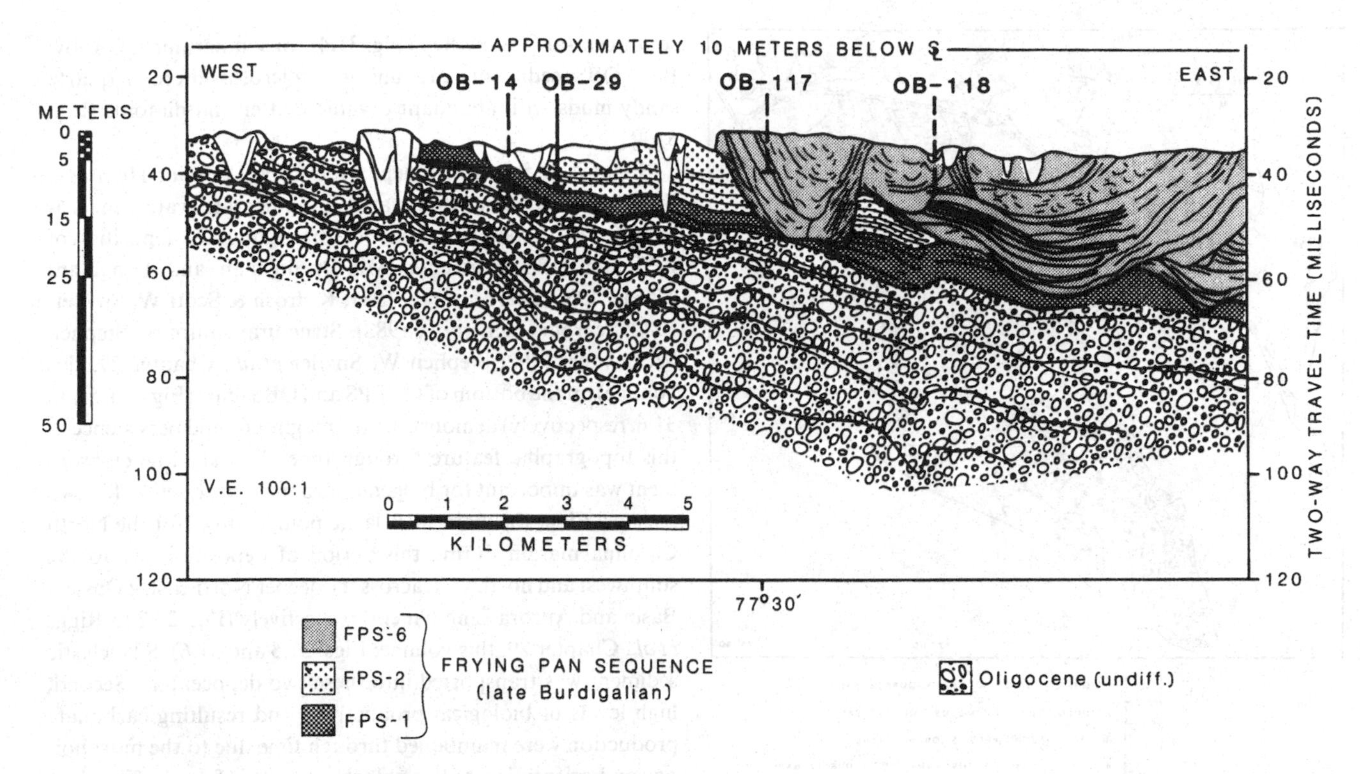

Fig. 31.3. Interpreted west to east uniboom seismic profile through southern Onslow Bay (EN-8C Profile) with vibracore locations superimposed (modified from Stephen W. Snyder in Riggs *et al.*, 1985).

rizes lateral lithofacies through each of these three depositional sections, as developed by Mallette (1986).

Frying Pan Section

The Frying Pan Section (FPS) of mid-Burdigalian age is composed of six fourth-order seismic sequences (Figs. 29.7 and 29.8 in Riggs *et al.*, Chapter 29, this volume). Seismic interpretations by Stephen W. Snyder (1982) suggest that a) the lower five units (FPS-1 through FPS-5) parallel one another and dip gently to the east and southeast (Fig. 31.2); b) these five units are abruptly truncated downdip at the 25 m high White Oak Scarp (Fig. 31.2); and c) the uppermost unit (FPS-6) cuts into the underlying units in the southern portion of Onslow Bay (Fig. 31.3). FPS-6 represents a very large erosional channel characterized by channel backfill geometry.

The Frying Pan Section is characterized by five major lithofacies (Fig. 31.4). Siliciclastic lithofacies are the most widespread throughout FPS. Sediments are coarsest in the north and fine southward, accompanied by increasing concentrations of biogenic, authigenic, and diagenetic components. In northern Onslow Bay, clean quartz sands and muddy quartz sands are interbedded with biogenic hash. Central Onslow Bay is characterized by uniform muddy quartz sands interbedded with quartz sandy muds. These lithofacies grade southward into muddy phosphorite sands overlain by microfossiliferous sandy muds with high concentrations of organic matter, dolosilt, and either foraminifera or clinoptilolite molds of foraminifera.

Textural data and regional distribution of lithofacies suggest a siliciclastic point source northwest of Onslow Embayment, while biogenic hash was derived from shoal environments associated with the Cape Lookout High to the northeast (Fig. 31.5). Southern Onslow Embayment, which lacked coarse siliciclastics, was characterized by chemical environments dominated by high nutrient levels and organic production and by low dissolved oxygen concentrations, as indicated by abundant authigenic–early diagenetic phosphate, dolomite, organic matter, and planktonic microfossils and by the species composition of the benthic fauna (Riggs, 1979 and 1984; Waters, 1983; Moore, 1986; Scott W. Snyder, 1988, Chapter 32, this volume). Central Onslow Embayment represented a mixed zone between siliciclastic and biogenic hash facies to the north and authigenic–early diagenetic and biogenic facies to the south (Fig. 31.5).

Depositional environments in southern Onslow Embayment were dominated by chemical sediments associated with upwelling from the beginning of FPS-1 deposition through FPS-5. FPS-6, a microfossiliferous muddy quartz sand, was deposited as sediment backfill under normal marine conditions in a very large and complex channel system (Figs. 31.3 and 31.4). Hine & Stephen W. Snyder (1985) interpreted this feature to be a fluvial channel solely on the basis of the seismic records. However, rich foraminiferal faunas with planktonic to benthic ratios greater than 1:1 in combination with deeper-water benthic species (Waters, 1983) suggest an open marine depositional regime. Based upon the lithology and associated microfauna, scale of the channel feature relative to known fluvial channels on the continental shelf (Fig. 31.3), and physical and temporal proximity to the ancient shelf-edge and White Oak Scarp at the end of FPS and beginning of OBS deposition. Riggs (1984) and Riggs *et al.* (1985) interpreted this feature to be a product of submarine erosion by currents associated with the Gulf Stream. During a

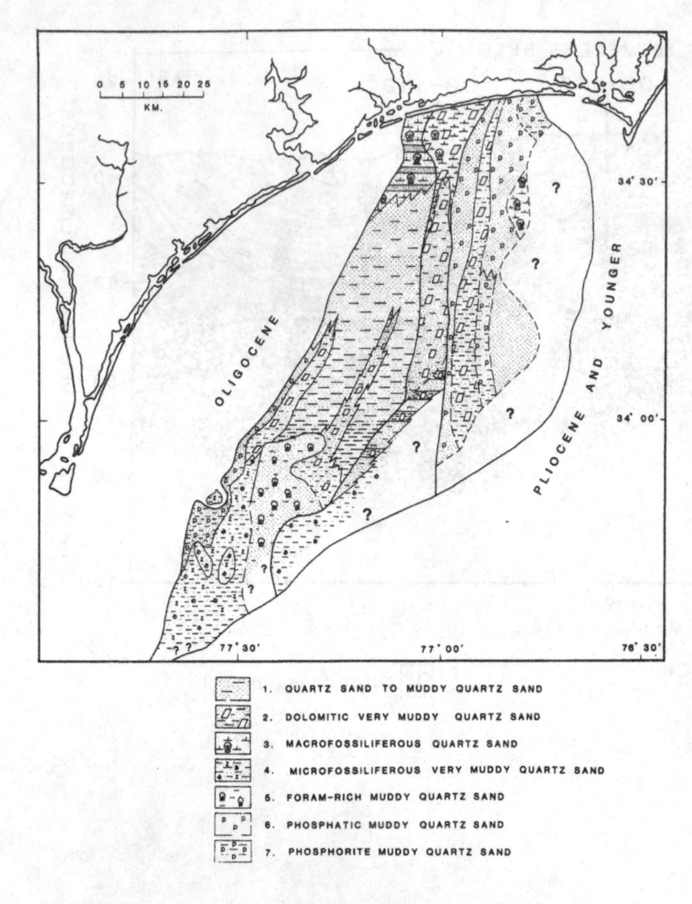

Fig. 31.4. Regional distribution of major lithofacies within the Frying Pan (FPS), Onslow Bay (OBS), and the Bogue Banks Sections (BBS) of the Miocene Pungo River Formation in Onslow Bay (modified from Mallette, 1986).

high stage of sea level, the Gulf Stream impinged upon the outer continental shelf, eroding the seaward side of units FPS-1 through FPS-5 to produce the White Oak Scarp (Fig. 31.2). Major channels were scoured along the outer continental shelf by currents related to the Gulf Stream. The reworked sediments were deposited as backfill (FPS-6) in channel features (Fig. 31.3), with a new suite of microfossils reflecting open-marine, well-oxygenated environments.

Onslow Bay Section

The Onslow Bay Section (OBS) of Langhian age consists of four fourth-order seismic sequences (Figs. 29.7 and 29.8 in Riggs *et al.*, Chapter 29, this volume) and is characterized by four major lithofacies (Fig. 31.4). Northern Onslow Bay is characterized by a macrofossiliferous biogenic hash lithofacies composed of barnacles, bryozoans, and mollusks, with variable amounts of quartz sand and disseminated phosphate (1–7%) and glauconite. This lithofacies occurs across and along the flanks of the Cape Lookout High (Steele, 1980; Lewis, 1981; Scarborough, 1981; Scarborough, Riggs & Scott W. Snyder, 1982; Mallette, 1986). The biogenic hash diminishes in size and abundance downslope and away from the axis of Cape Lookout High. Southward, biogenic hash grades into quartz sand interbedded with dolomud lithofacies (at a scale of 0.5–3 m), a lithofacies which extends through north-central, central and south-central Onslow Bay (Fig. 31.4). In southernmost Onslow Bay, OBS sediments are uniform microfossiliferous quartz sandy muds with abundant organic matter and diatoms (Fig. 31.4).

Figures 31.5 and 31.6 depict the Cape Lookout High as a shallow environment dominated by temperate water, macrofossiliferous carbonate production throughout the deposition of FPS and OBS sediments (Mid-Burdigalian and Langhian) (Steele, 1980; Scarborough, 1981; Katrosh & Scott W. Snyder, 1982; Scarborough, *et al.*, 1982). Structural contours (Stephen W. Snyder, 1982; Stephen W. Snyder *et al.*, Chapter 29, this volume) on the bottom of the FPS and OBS units (Figs. 31.5 and 31.6, respectively) demonstrate the magnitude and persistence of this topographic feature through time. This shallow environment was important for biogenic carbonate production for two reasons. First, the main siliciclastic point sources for the North Carolina margin during this period of deposition lay to the southwest and northwest, across the deeper Northeast & Onslow Basin and Aurora Embayment, respectively (Fig. 29.2 in Riggs *et al.*, Chapter 29, this volume; Figs. 31.5 and 31.6). Siliciclastic sediment was transported into these two depocenters. Second, high levels of biological productivity and resulting carbonate production were maintained through time due to the morphology and orientation of this paleotopographic feature. The shallow, cross-shelf orientation ensured that benthic environments remained well oxygenated, while also being supplied with generally high nutrient levels characteristic of the continental margin during this portion of the Miocene. As a result, thin temperate-water carbonate sediments accumulated along the axis of the High and were periodically deposited on the flanks as abraded hash in response to punctuated storm events.

Shallow-shelf areas to the north and southwest of Cape Lookout High supported extensive communities of calcareous benthic organisms and accumulated during periods of low siliciclastic sedimentation. Episodic and punctuated events of rapid deposition of siliciclastic sands repeatedly buried these benthic communities producing the sharply interbedded lithologies which characterize the gradational zone around the High. Biogenic hash contains barnacles, bivalves, echinoid spines and plates, and bryozoans, in decreasing order of abundance and is a mixture of whole, unabraded forms and highly abraded fragments. This suggests that biogenic horizons represent autochthonous sedimentation, while lack of fossil material in quartz sand interbeds suggests that they represent the allochthonous component.

Seismic interpretations (Stephen W. Snyder, 1982) suggest that units of the Onslow Bay Sequences (OBS) in northern and central Onslow Embayment were deposited as a series of prograding clinoforms over and east of the White Oak Scarp (Fig. 31.2). This scarp, with its origin linked to Gulf Stream erosional processes during prior sea-level highstands, is interpreted to represent the shelf edge at the time of deposition of OBS-1 (see Stephen W. Snyder, Chapter 30, this volume).

Siliciclastics probably dominated continental shelf sedimentation west of the scarp, but OBS sediments in that area were not preserved (Fig. 31.6). East of the scarp, OBS sediments consist of two lithologies interbedded on a scale of 0.5–3 m. Normal sedimentation in this upper-slope environment was slow accumula-

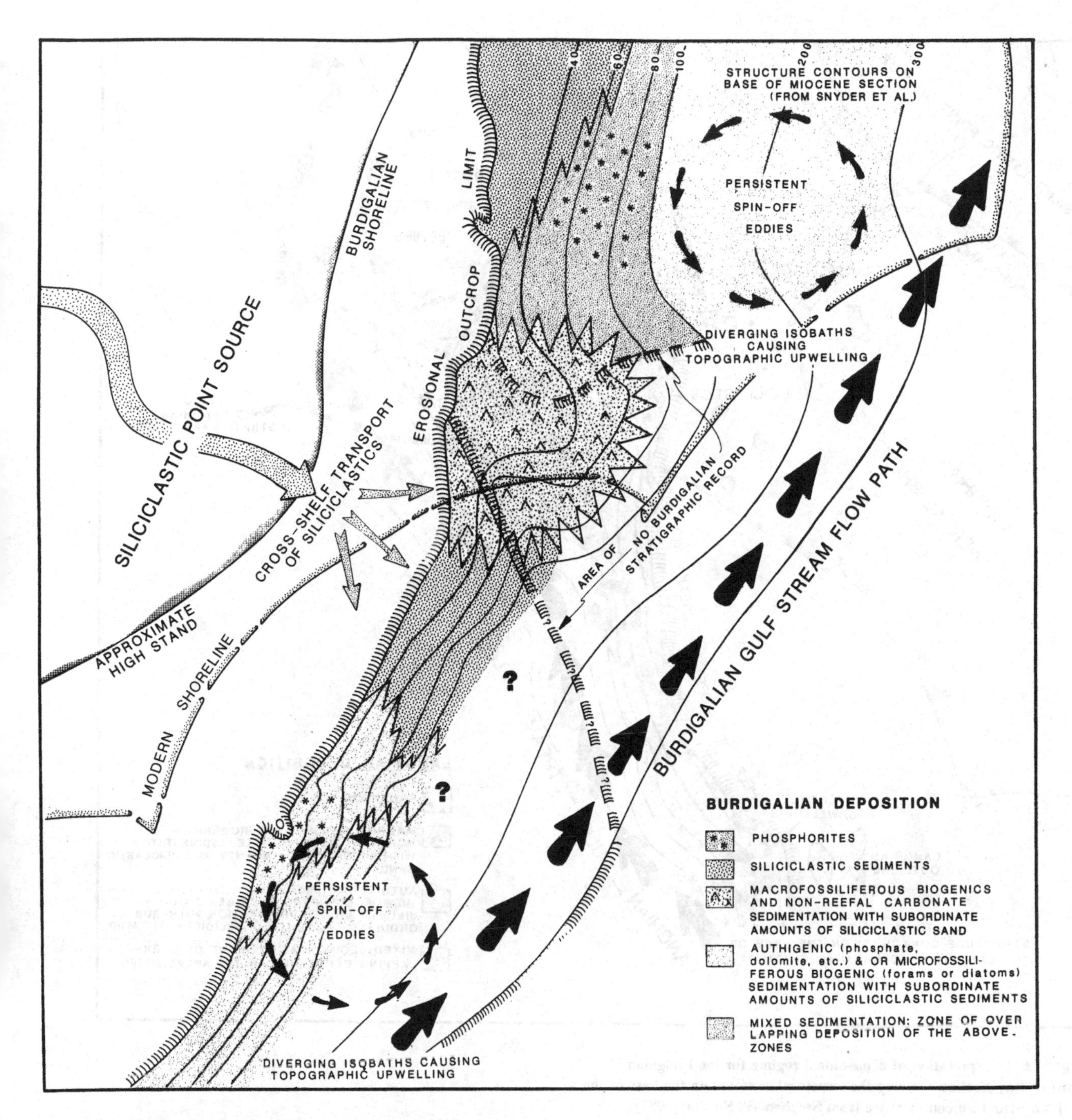

Fig. 31.5. Interpretation of depositional regime for the Mid-Burdigalian Frying Pan Section showing the structural contours on the base of the section. Structure contours are from Stephen W. Snyder (1982).

tion of fine-grained, organic-rich dolomuds. This background sedimentation was periodically interrupted by rapid deposition of shelf-derived, slightly-graded quartz sands. These episodic siliciclastic interbeds contributed most of the total sediment and resulted in greater shelf-slope progradation. The apparent toplap geometric pattern suggests that offshelf siliciclastic deposition repeatedly interrupted normal upper-slope authigenic sedimentation. Deposition was periodically interrupted by erosion resulting from ongoing transgressions and regressions and truncation of upper portions of previously deposited fourth-order seismic sequences (Fig. 31.2). Depositional patterns of OBS sediments are similar to those described for Pliocene–Pleistocene depositional units in the Cape Fear region of the Carolina continental margin (Matteucci, 1984; Matteucci & Hine, 1987).

In southernmost Onslow Bay, the interbedded quartz sands and dolomuds grade laterally into a uniform facies of microfossiliferous, quartz sandy mud (Fig. 31.4). Microfossils, both foraminifera and diatoms (up to 10%), in this organic-rich mud facies are associated with abundant dolosilt and

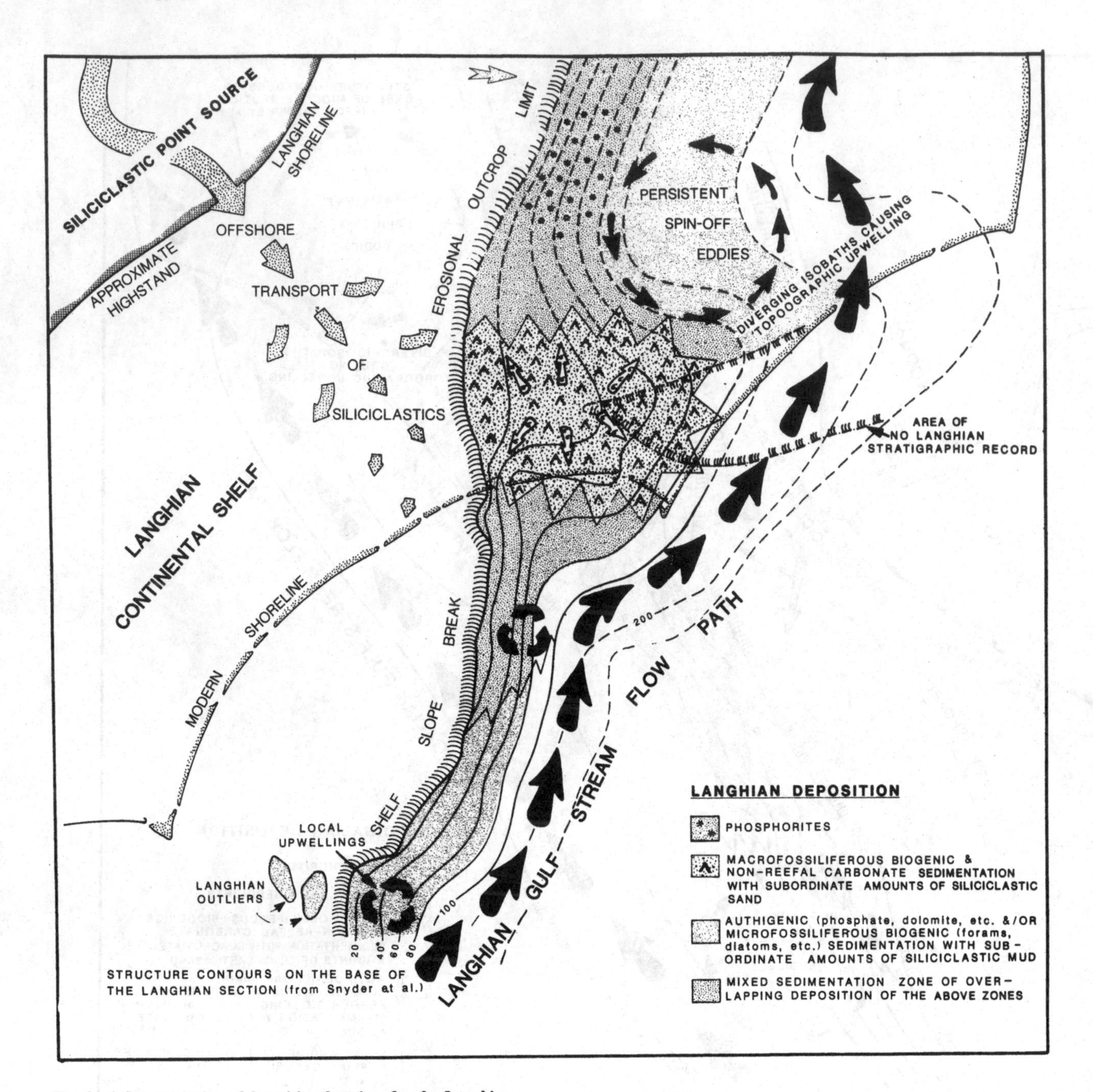

Fig. 31.6. Interpretation of depositional regime for the Langhian Onslow Bay Section showing the structural contours on the base of the section. Structure contours are from Stephen W. Snyder (1982).

clinoptilolite. Environments in southern Onslow Bay during deposition of OBS sediments were similar to those of FPS-1 through FPS-5 in the same region (Fig. 31.6). OBS sediments differ from FPS in having relatively small concentrations of phosphate and abundant diatoms. The depositional environment was characterized by low energy, lack of siliciclastic input, and by high nutrient levels and organic production in combination with low dissolved oxygen concentrations.

Bogue Banks Section

The Bogue Banks Section (BBS) of Serravallian age is composed of eight fourth-order depositional units (Figs 29.7 and 29.8 in Riggs *et al.*, Chapter 29, this volume) defined by seismic stratigraphy; however, only six units were vibracored. Consequently, lithostratigraphic data is limited to depositional units BBS-1 through BBS-5 and BBS-8; biostratigraphic data is limited to BBS-1 through BBS-5 because BBS-8 does not contain an age diagnostic microfossil assemblage. Seismic interpretations by Stephen W. Snyder (1982) indicate that reflectors of BBS sediments dip gently seaward with little disruption (Fig. 31.2).

BBS is characterized by five major lithofacies (Fig. 31.4). Relative to the FPS and OBS sections, siliciclastic sedimentation became much more important in the BBS section. BBS-1 through BBS-4 are dominated by siliciclastic sand and mud sedimentation laterally throughout Onslow Bay. Within each of

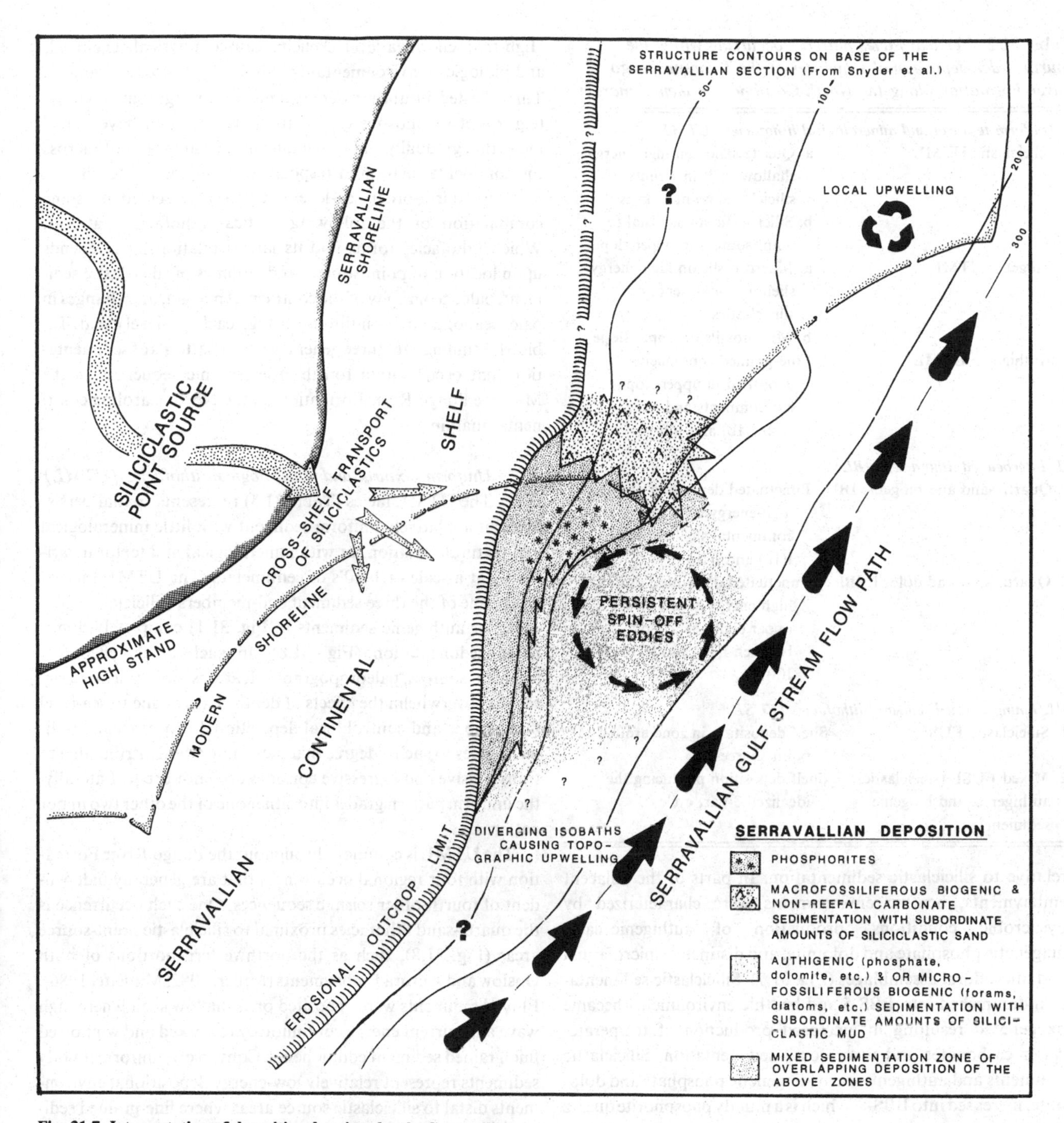

Fig. 31.7. Interpretation of depositional regime for the Serravallian Bogue Banks Section showing the structural contours on the base of the section. Structure contours are from Stephen W. Snyder (1982).

these units, variations occur through time producing a vertical pattern of fining upward sediment, with coarser siliciclastic sediments grading upward into muddy phosphatic and phosphorite sands and culminating in dolomuds. Within most units, the fining upward siliciclastics are increasingly diluted by authigenic sediments including phosphate (2–25%) in sandier facies, and by diatoms, chert, and dolosilt in muddier facies.

BBS-1 through BBS-4 do not contain the biogenic hash lithofacies that characterized northern Onslow Bay during deposition of the FPS and OBS sections. However, one core in unit BBS-5 is dominated by biogenic hash (Fig. 31.4). This suggests that siliciclastic sedimentation had decreased sufficiently for increased development and accumulation of a diverse benthic community of macroinvertebrates to produce biogenic hash. Relief associated with the Cape Lookout High was progressively diminished with time by deposition of FPS, OBS, and the lower portion of BBS; and the Northeast Onslow Basin along the south flank of the Cape Lookout High filled, causing seaward progradation of the southwest–northeast-trending continental margin (Fig. 31.7). During deposition of BBS-1 through BBS-4, the eastward nose of the Cape Lookout High was still a prominent topographic feature, but carbonate production diminished

Table 31.3. *Vertical facies patterns occurring within the fourth-order depositional sequences of the Miocene Pungo River Formation along the North Carolina continental margin*

I. Uniform textural and mineralogical lithofacies (UTML)	
1. Siliciclastic UTML	a. Quartz sands on high-energy shallow shelf in vicinity of siliciclastic point sources
	b. Siliciclastic muds distal to point sources in upper slope
2. Biogenic UTML	a. Macrofossils on high-energy shelf and in absence of siliciclastics
	b. Microfossils on upper slope
3. Authigenic UTML	Fine-grained components deposited in upper-slope environments and mixed with I(1b) and I(2b)
II. Interbedded lithofacies (IBL)	
1. Quartz sand and biogenic IBL	Punctuated deposition during high-energy events on continental shelf between I(1a) and I(1a)
2. Quartz sand and dolosilt IBL	Punctuated deposition during high-energy events on the upper continental slope between I(1b) and I(2b) or I(3)
III. Fining upward sequence lithofacies (FUSL)	
1. Siliciclastic FUSL	Shelf deposition in zone around point sources
2. Mixed FUSL (siliciclastic, authigenic, and biogenic sediment)	Shelf deposition producing the idealized 3-part cycle

relative to siliciclastic sedimentation. In parts of the adjacent embayments, benthic environments were characterized by dysaerobic conditions, production of authigenic–early diagenetic phosphate and dolomite, and distinctive microfaunal and microfloral assemblages (Fig. 31.7). Siliciclastic sedimentation diminished into BBS-5 and benthic environments became oxygenated resulting in increased production of temperate-water carbonates and macrofossil sedimentation. Siliciclastic sediments and authigenic–early diagenetic phosphate and dolomite, increased into BBS-8, which is a muddy phosphorite quartz sand.

Vertical lithofacies patterns

Lithologic facies within fourth-order seismic sequences are complex, with subtle but significant lateral and vertical changes (Lewis, 1981; Scarborough, 1981; Riggs *et al.*, 1982 and 1985; Stewart, 1985; Mallette, 1986). Paleotopography interacted with changing depth and position of the paleo-Gulf Stream to control; (a) sites and rates of deposition and erosion; (b) location and volume of point-source siliciclastic sedimentation; (c) intensity of topographic upwelling and downwelling; (d) formation of authigenic–early diagenetic (i.e. phosphate, dolomite, glauconite) and biogenic (i.e. carbonate and siliceous tests, organic matter) components; (e) offshelf sand transport; and (f) formation of carbonate hardgrounds and phosphatic crusts. High-frequency sea-level cyclicity caused physical, chemical, and biological environmental thresholds to be crossed rapidly. This resulted in abrupt depositional or stratigraphic changes (e.g. onset of phosphate formation) rather than 'layer cake' facies that gradually and systematically migrated laterally across the continental margin in response to changing sea levels.

Each fourth-order sea-level event is represented by some combination of the following vertical lithofacies patterns. Which lithofacies forms and its lateral relationships depends upon location of point sources and volumes of siliciclastic sediment, paleogeography of the continental margin, and changes in paleoceanographic conditions through each sea-level event. Table 31.3 summarizes three general vertical patterns of sedimentation that occur within fourth-order seismic sequences of the Miocene Pungo River Formation on the North Carolina continental margin.

Uniform textural and mineralogical lithofacies (UTML)

The UTML facies (Table 31.3) represents spatial persistence of a relatively uniform sediment with little mineralogical and textural variation, or with mineralogical and textural variability at a scale of 1–10's of centimeters. The UTML formed where one of the three sediment end-members (siliciclastic, biogenic, or authigenic sediments in Fig. 31.1) completely dominated sedimentation (Fig. 31.8). In such areas, effects of sediment sources, paleotopographic features, or physical energy regimes overwhelm the effects of depth changes due to sea-level fluctuations and control local depositional processes and sediment types to such a degree that facies migration attributable to transgressive and regressive episodes does not occur. Laterally, the uniform pattern grades into either one of the other two major patterns.

The UTML is common throughout the Pungo River Formation with four regional occurrences that are generally independent of fourth-order seismic sequences. One such occurrence is the quartz-sand lithofacies proximal to siliciclastic point-source areas (Fig. 31.8), such as the northwestern portions of both Onslow and Aurora Embayments (Miller, 1982; Mallette, 1986). Fluvial sediments were deposited on a shallow shelf where high wave and current energy continuously reworked and winnowed fine-grained sediment components. Conversely, uniform muddy sediments represent relatively low-energy depositional environments distal to siliciclastic source areas where fine-grained sediments accumulate throughout a depositional cycle (Fig. 31.8). Mud fractions of this UTML facies commonly contain significant concentrations of fine-grained authigenic and biogenic sediments. The muddy UTML facies is characteristic of sedimentation in deeper-water environments in the eastern and southern portions of Onslow and Aurora Embayments (Miller, 1982; Mallette, 1986).

Another distinct UTML lithofacies is composed of coarse biogenic sediments which formed in high-energy environments on the shallow shelf (Fig. 31.8). Coarse, macrofossiliferous biogenic hash formed primarily on and around a shallow topographic, carbonate-producing platform, the Cape Lookout High, and was diluted with varying amounts of siliciclastic sands (Steele, 1980; Scarborough, 1981; Miller, 1982; Scarborough *et al.*, 1982). Textural variations were controlled by local environ-

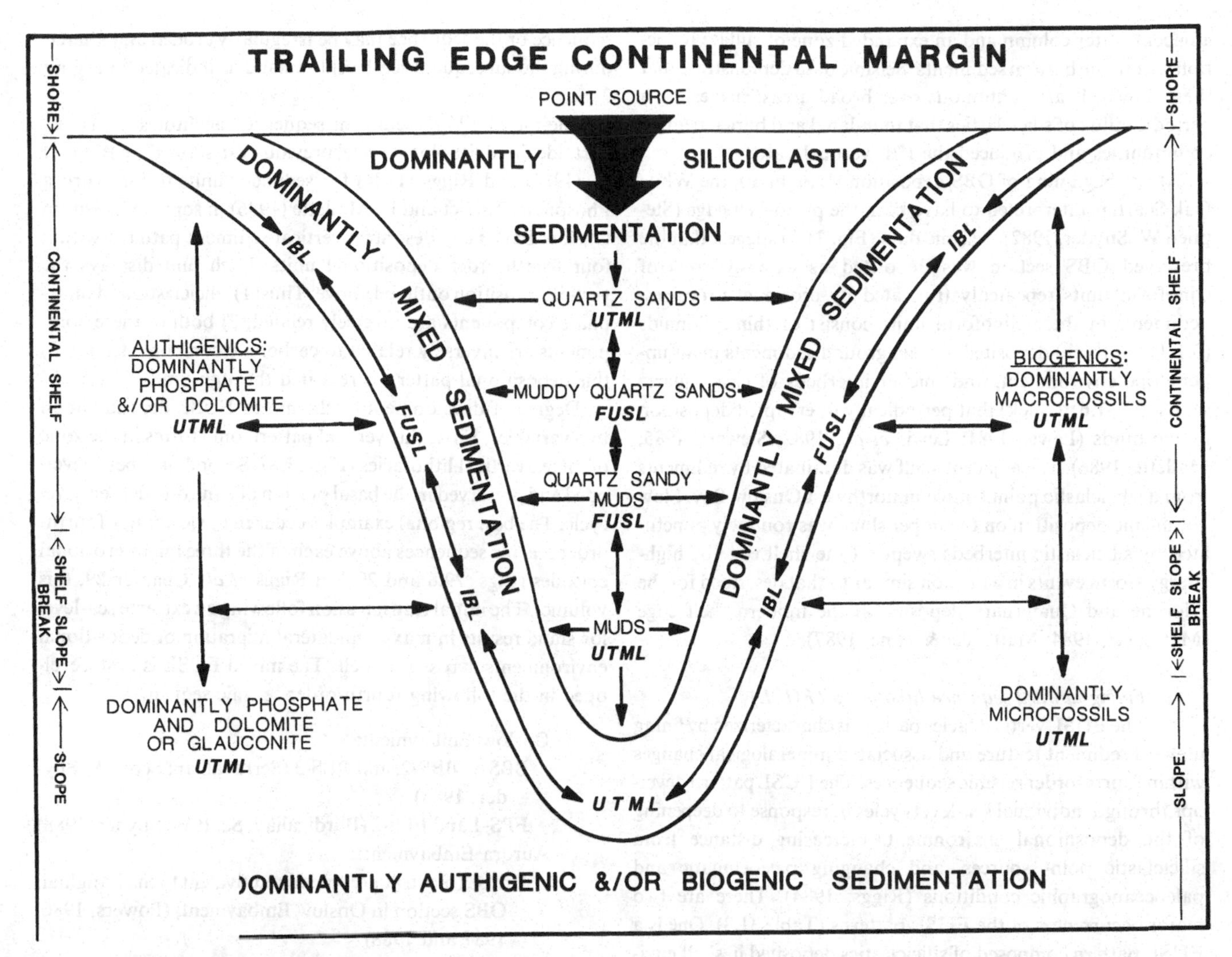

Fig. 31.8. Theoretical interpretation of distribution of vertical facies patterns for North Carolina continental margin.

mental conditions that influenced composition of macrofossil assemblages and subsequent degree of fragmentation. Carbonate shell material produced on Cape Lookout High extends into the adjacent Onslow and Aurora Embayments where it grades into the other vertical lithofacies patterns (Fig. 31.8).

The final occurrence of the UTML is associated with microfossiliferous muddy quartz sands of unit FPS-6, which infills a large and complex channel feature (Figs. 31.3 and 31.4) in southern Onslow Embayment (Waters, 1983; Mallette, 1986). Other lithofacies may be present in deeper portions of the FPS-6 channel system and longer cores might necessitate modification of this lithofacies category.

Interbedded lithofacies (IBL)

The IBL pattern consists of distinctly interbedded sequences of clean quartz sands with either biogenic hash or dolomuds (Table 31.3). Interbeds with generally sharp contacts are repeated at scales from 0.25 to 4 m. IBL patterns developed in areas transitional between two contiguous environments where uniform textural and mineralogical lithofacies (UTML) were being deposited (Fig. 31.8). The interbedded character resulted from sediment transfer between these two distinct environments during periodic, high-energy depositional events. This process is referred to as 'punctuated mixing' by Mount (1984).

IBL patterns generally have limited lateral extent, but are common within the Miocene section and have two general occurrences. First, they occur adjacent to topographic features on the continental shelf, such as the Cape Lookout High and smaller hardground platforms. Second, they occur along the continental slope below the shelf edge, such as the White Oak Scarp during the Langhian. Either quartz sand, biogenic hash, or dolomud may represent the background sediment regime operating within a depositional environment. Normal sedimentation is punctuated by episodes of high-energy transport of either siliciclastic sands or coarse biogenic hash to produce the vertical IBL pattern.

Deposition on Cape Lookout High was dominated by biogenic lithofacies (Fig. 31.4), whereas deposition to the south and north of the High was dominated by siliciclastics derived from major point sources. Although these areas are generally dominated by UTML vertical facies patterns, there are intermediate zones where vertical facies are mixed to produce alternating thin beds of biogenic hash (0.6–1 m thick) and thicker beds of clean quartz sand (1.5–4 m thick). In interbedded zones around the flanks of Cape Lookout High, a dolomite matrix occurs within the biogenic hash beds suggesting decreasing oxygen levels in the adjacent water column and an expanded zone of sulfate reduction within the bottom sediments. Seismic data demonstrate that these interbeds are continuous over broad areas, suggesting a periodic influx of siliciclastics that inundated and buried benthic communities and produced the IBL vertical patterns.

At the beginning of OBS deposition (Langhian), the White Oak Scarp is interpreted to have been the paleoshelf edge (Stephen W. Snyder, 1982). Seismic data (Fig. 31.2) suggest that the preserved OBS section was deposited as extensive sets of clinoform units repeatedly truncated by erosional processes. Sediments in these clinoform units consist of thin dolomuds (0.1–0.6 m thick) deposited as background sediments in an upper-slope environment, and thicker interbeds of clean quartz sands (0.3–1.6 m thick) that periodically interrupted deposition of the muds (Lewis, 1981; Lewis *et al.*, 1982; Stewart, 1985; Mallette, 1986). The adjacent shelf was dominated by sediments from a siliciclastic point source in northwest Onslow Bay. Slow authigenic deposition on the upper slope was routinely punctuated by siliciclastic interbeds swept off the shelf edge by high-energy storm events in a fashion similar to that described for the Pliocene and Quaternary deposits on the modern shelf edge (Matteucci, 1984; Matteucci & Hine, 1987).

Fining upward sequence lithofacies (FUSL)

The FUSL vertical facies pattern is characterized by fining upward sediment texture and associated mineralogical changes within fourth-order seismic sequences. The FUSL pattern develops through individual sea-level cycles in response to deepening of the depositional environment, increasing distance from siliciclastic point sources, and changing paleoclimatic and paleoceanographic conditions (Riggs, 1984). There are two major occurrences of the FUSL patterns (Table 31.3). One is a FUSL pattern composed of siliciclastics deposited in shelf environments away from the main point-source discharge area where the vertical UTML patterns form in shallow-water environments. The latter grade seaward into siliciclastic FUSL patterns (Fig. 31.8). The resulting FUSL pattern is characterized by decreasing siliciclastic grain size upward through the depositional unit, and it grades laterally into a second, mixed FUSL pattern, the 'ideal fining upward sequence' (Fig. 31.9).

The mixed FUSL (Table 31.3) forms in the environmental transition between siliciclastic and authigenic or biogenic lithofacies (Fig. 31.8). Within these areas, changes in sea level through time generate both an overall fining upward sequence and mineralogical evolution as summarized in Figure 31.9. The basal zone or transgressive facies is dominated by: a) siliciclastic muddy sands that fine upward; b) high but variable amounts of organic carbon and authigenic and early diagenetic phosphate; and c) a biogenic component consisting of siliceous or calcareous microfossils. A thin intermediate zone or condensed section consists of: a) high concentrations of organic carbon; b) authigenic or early diagenetic dolomite; and c) highly recrystallized and dissolved siliceous and calcareous microfossils. The upper zone or regressive facies is dominated by: a) siliciclastic muddy sands; b) low concentrations of organic carbon and reworked phosphorites; and c) increasing upward concentrations of macrofossils dominated by bivalves and barnacles. Hardgrounds may be developed on top of this sediment sequence or the sequence may be irregularly eroded and altered during the subsequent sea-level lowstand as indicated in Figure 31.9.

The mixed FUSL sediment sequence constitutes the three-part 'idealized lithologic cycle' previously described by Riggs *et al.* (1982) and Riggs (1984) for sediment units in the Aurora Phosphate District and by Mallette (1986) in some units within Onslow Bay. They described vertical sediment patterns within four fourth-order depositional units. Each unit displays the vertical transition outlined above. Thus: 1) siliciclastic and phosphate components are inversely related; 2) both of these components are inversely related to carbonate components; and 3) this depositional pattern is repeated through time.

Degree of development of the mixed FUSL depends upon two variables. First, this vertical pattern only forms in the zone of mixed vertical lithofacies (Fig. 31.8). Second, it is best developed and preserved in the basal portion of a third-order sea-level cycle. The best regional examples occur in the lower few fourth-order seismic sequences above each of the three major erosional episodes (Figs. 29.6 and 29.7 in Riggs *et al.*, Chapter 29, this volume). The initial transgression following an extreme sea-level lowstand results in maximum lateral migration of depositional environments across the shelf. The mixed FUSL is best developed in the following fourth-order seismic sequences.

Onslow Embayment:
- BBS-1, BBS-2, and BBS-3 (Serravallian; Scott W. Snyder, 1988)
- FPS-1 and FPS-2 (Burdigalian; Scott W. Snyder, 1988)

Aurora Embayment:
- Unit C (biostratigraphically equivalent to the Langhian OBS section in Onslow Embayment; (Powers, 1986, 1987 and 1988)
- Unit B (biostratigraphically equivalent to the Burdigalian FPS section in Onslow Embayment; (Powers, 1986, 1987 and 1988)

FPS-1 and FPS-2 in central Onslow Embayment are characterized by the total mixed FUSL pattern (Fig. 31.10). However, siliciclastics decrease southwestward away from their point source, until they are almost absent. In this situation, sediments in both the base and intermediate portions of the unit consist dominantly of authigenic–early diagenetic and biogenic components. This environmental setting represents the optimum situation for maximum development of phosphorites; the basal and middle portions of FPS-1 contain up to 22.9% P_2O_5 in the total sediment (Riggs *et al.*, 1985).

Sediment responses to multiple sea-level events

Cyclical depositional units

The Miocene section on the North Carolina continental margin consists of 18 seismic sequences (Figs. 29.7 and 29.8 in Riggs *et al.*, Chapter 29, this volume) defined by high-resolution seismic stratigraphy. Lithostratigraphic studies of sediments within the seismic sequences demonstrated that they represent discrete depositional units. Biostratigraphic analyses established the chronostratigraphic framework (Fig. 29.9 in Riggs *et al.*, Chapter 29, this volume) for the lower 15 fourth-order seismic sequences. Table 29.1 in Riggs *et al.* (Chapter 29, this volume)

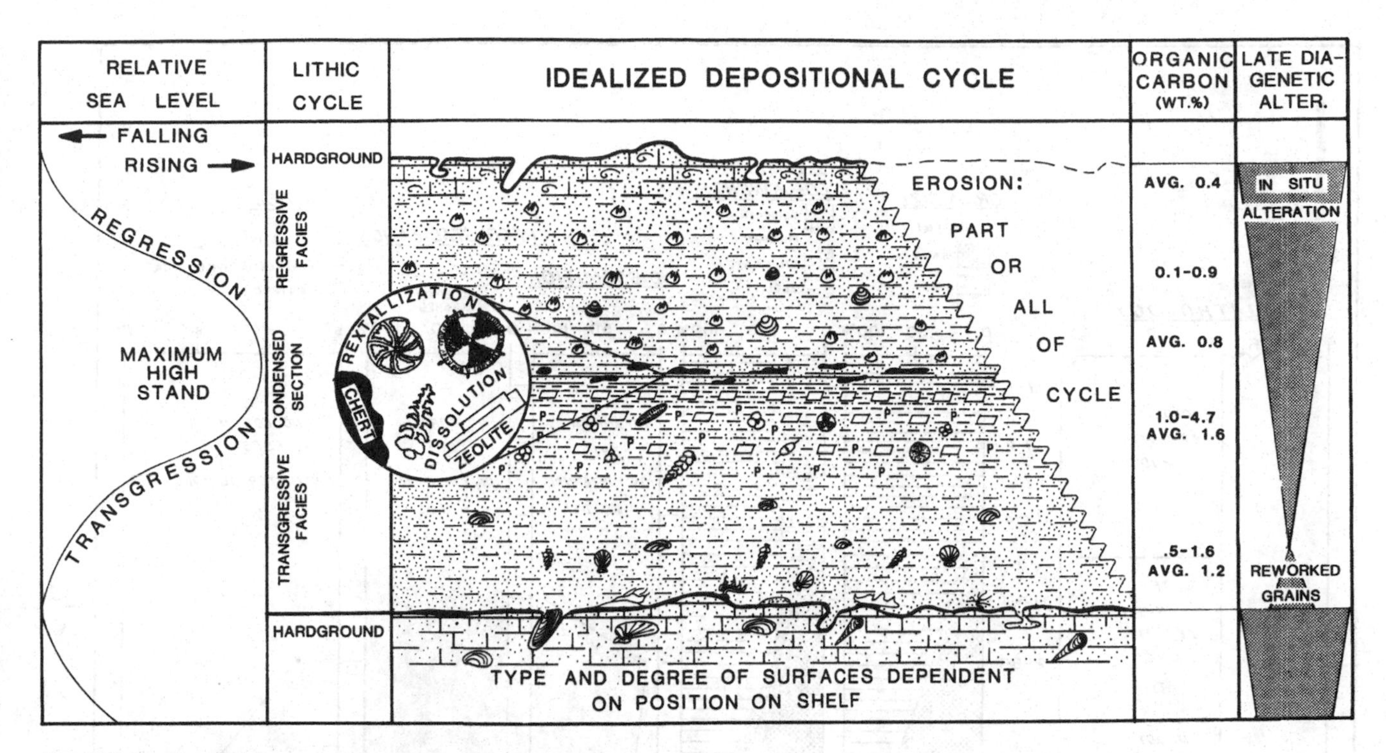

Fig. 31.9. Idealized lithologic cycle common to some of the Miocene depositional units within the Aurora and Onslow Embayments of the Carolina Phosphogenic Province. This figure shows the following features: (a) mixed fining upward sequence lithofacies (FUSL) through the various stages of one fourth-order sea-level cycle; (b) associated organic carbon content; (c) erosional potential of the deposited unit due to subsequent submarine and subaerial processes; and (d) late diagenetic alteration patterns resulting from submarine exposure and development of a condensed section during the maximum sea-level highstand and hardground development during the sea-level lowstand. The downward decrease in size of the pattern in the column entitled diagenetic alteration indicates decreasing degree of *in situ* late diagenesis downsection from the hiatal surfaces. Note the basal zone of diagenetically altered grains that have been reworked from the erosion of the preceding hardground surface.

summarizes the time framework for the depositional cycles within the Pungo River Formation based upon the biostratigraphic assignments. It is believed that these high-frequency sediment units were deposited in response to fourth-order paleoclimatic and paleoceanographic cycles with a maximum average duration of about 200 ky (Table 29.1 in Riggs *et al.*, Chapter 29, this volume), and that they are detailed records of glacio-eustatically controlled sea-level variations.

Thus, each fourth-order transgression and regression shifted the processes of sedimentation across the shelf–slope system producing changing patterns of deposition and erosion. Subaerial and submarine erosional events have been documented seismically and lithologically to occur on top of each of the 18 fourth-order seismic sequences. Seismic profiles (Figs. 31.2 and 31.3) and isopach and structure contour maps of seismic sequences (Stephen W. Snyder, 1982; Stephen W. Snyder *et al.*, 1982 and Chapter 30, this volume; Popenoe, 1985; Riggs, *et al.*, 1985; Hine & Riggs, 1986) demonstrate extensive fluvial channeling associated with the upper surface of many seismic sequences, and large submarine scours and erosional scarps attributed to lateral excursions of the Gulf Stream in response to sea-level fluctuation. Most sequence-bounding seismic reflectors coincide with unconformities characterized by hardgrounds, changes in lithology, or late diagenetic profiles. Biostratigraphy, though unable to resolve time gaps between fourth-order seismic sequences, has established the initiation and termination of Pungo River deposition and documented two significant time gaps within the Miocene section (Figs. 29.6, 29.7 and 29.9 and Table 29.1 in Riggs *et al.*, Chapter 29, this volume).

Late diagenetic overprinting

Based upon compositional and crystallographic characteristics of the carbonate fluorapatite, Pungo River phosphorites are among the world's least weathered and diagenetically altered anywhere (McClellan & Lehr, 1969; McClellan, 1980). In spite of the apparent pristine nature of the sediments, Stewart (1985), Mallette (1986), Powers (1987), and Moretz (1988) have demonstrated that there is a subtle but important late diagenetic overprint on the depositional record of the Pungo River Formation. This differs from the more typical lateritic weathering

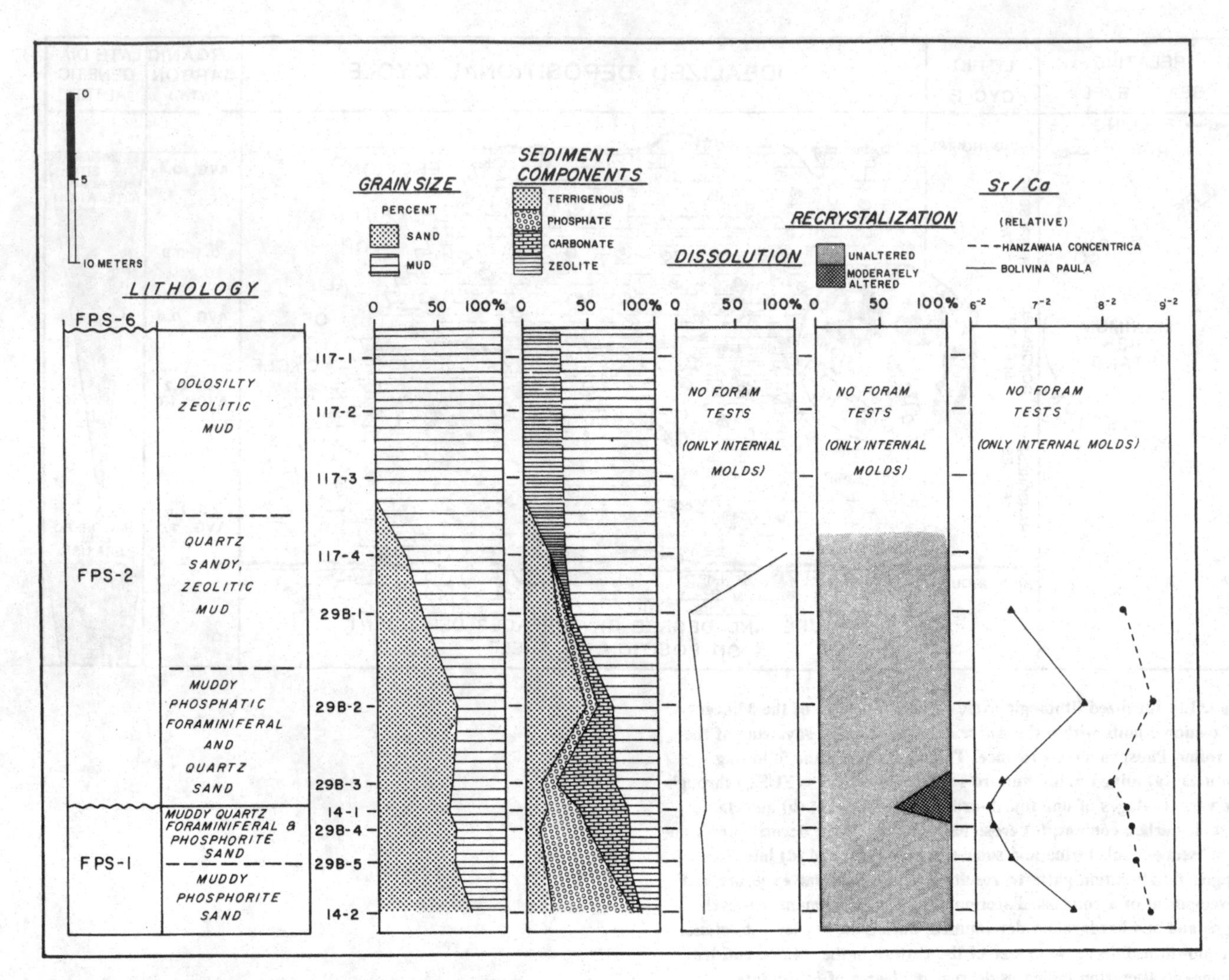

Fig. 31.10. Plot of decreasing benthic foraminifera diagenesis downsection from the hiatal surfaces of Units FPS-1 and FPS-2 as characterized by relative dissolution and recrystallization indices, and Sr/Ca ratios within the Pungo River Formation in southern Onslow Bay (modified from Mallette, 1986; Moretz, 1988). The depositional units display typical fining upward sequence lithofacies (FUSL) characteristic of the 'idealized lithologic cycle' of Riggs *et al.* (1982) and the lithology shows the inverse relationship between foram preservation and development of clinoptilolite.

profile that characterizes many phosphate deposits in Florida, Africa, Asia and the Middle East (Altschuler, 1974; Bentor, 1980; McClellan, 1980).

Due to sea-level cyclicity, deposition of each fourth-order seismic sequence in the Pungo River Formation was followed by a period of non-deposition and erosion during the subsequent sea-level low. These periods of subaerial and submarine exposure produced subtle late diagenetic overprints by modifying the sediment texture and mineralogy of many depositional units. For example, the fining upward sequence lithofacies (FUSL) outlined in Figure 31.9 is developed and preserved only on a local basis throughout the continental margin. Where it occurs, carbonate caprocks are generally indurated, moldic, and extensively bored and stained, indicating development of a hardground on the surface of non-deposition (Riggs, *et al.*, 1982; Riggs, 1984; Stewart, 1985; Forgang & Riggs, 1986; Mallette, 1986; Riggs & Belknap, 1988). The mineralogy and chemistry of underlying sediments show an alteration profile in which diagenesis decreases in intensity downward from the unconformity (Fig. 31.9). Wherever the upper portions of a fining upward sequence lithofacies have been truncated by erosion (Fig. 31.9), subtle alteration profiles within similar lithologies of successive depositional sequences are the only evidence for cyclic deposition and an intervening hiatus.

Three categories of late diagenetic signatures are recognized within the Pungo River Formation and include modification of microfossil tests, modification of phosphate grains, and new mineral formation. Table 31.2 lists specific types of alteration which occur within each category of late diagenesis. Figure 31.10 outlines the late diagenetic overprint of the three categories of alteration across two depositional units; intensity of alteration generally decreases downward from each hiatal surface. Because

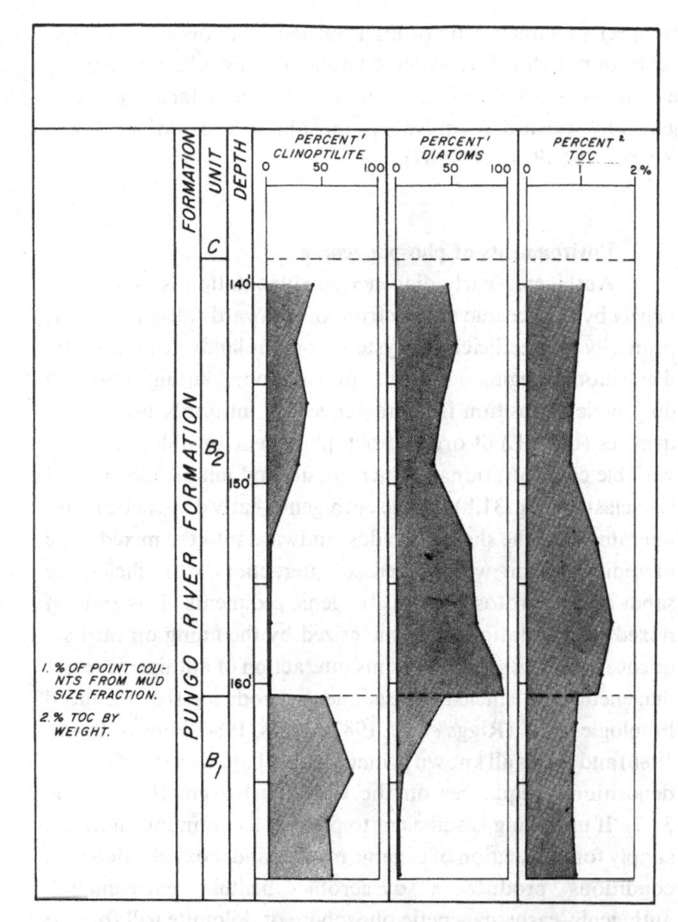

Fig. 31.11. Plot of concentration and degree of preservation of diatom frustules, abundance of late diagenetic clinoptilolite, and concentration of total organic carbon (unpublished data, E. Maxwell, Department of Marine Science, UNC, Chapel Hill) downward from the hiatal surface of 4th-order depositional Units B1 and B2 in the Pungo River Formation, core GH8.5R, Aurora Phosphate District.

microfossils (diatom frustules and foraminiferal tests) are readily altered by post-depositional processes and because they have well known primary mineralogy and chemistry, they are sensitive indicators of late diagenesis that can be used to map late diagenetic facies and to reflect alteration in more complex and less well known minerals (Mallette, 1986; Riggs & Scott W. Snyder, 1986; Powers, 1987; Moretz, 1988). For example, phosphate grain modification is considerably more difficult to recognize because most grains are complex sediment aggregates consisting of multiple components of variable mineralogical and chemical composition with little knowledge as to what the 'primary' characteristics were prior to late diagenesis (Riggs, 1979; Ellington, 1984).

Figure 31.11 presents examples of two types of late diagenetic signatures plotted through fourth-order depositional sequences of the Miocene Pungo River Formation. Concentrations of late diagenetic clinoptilolite, which occurs as internal molds of foraminiferal tests, are plotted against abundance and degree of preservation of diatom frustules from an unconformity, downward to least altered portions of the underlying depositional unit (Scott W. Snyder *et al.*, 1984; Powers, 1987). Perturbations on the idealized pattern (Fig. 31.9) reflect localized microenvironmental effects including: a) variability in primary deposition of specific sediment components (e.g. diatoms vs. foraminifera); b) change in mud content that controlled porosity and permeability (e.g. sand infilling of burrows in a sandy mud or carbonate lenses in a muddy quartz sand); and c) presence of other specific sediment components (e.g. organic matter or carbonates) which influenced the chemistry of local pore waters. Similar profiles occur in microtextures on phosphate grain surfaces, P_2O_5 concentration within specific phosphate grain types (Ellington, 1984; Mallette, 1986), and in preservation of foraminiferal tests (Fig. 31.8; Moretz, 1988).

Secondary calcite cements are important components (up to 10% of the total sediment) in many carbonate-rich caprocks associated with unconformities within the Pungo River Formation (Stewart, 1985; Forgang & Riggs, 1986; Mallette, 1986). Freshwater diagenesis is the primary agent of textural and mineralogical alteration; in the Aurora Phosphate District, all calcite cements in the carbonate caprocks of Units A through D are interpreted to be products of meteoric diagenesis. Co-occurring textures such as moldic porosity, mineralogical inversion of aragonite and high-Mg: low-Mg calcite, and dolomite dissolution also suggest freshwater diagenesis.

Miocene depositional patterns on the Carolina continental margin

The Miocene Pungo River Formation contains three depositional sections (Figs 29.6 and 29.7 in Riggs *et al.*, Chapter 29, this volume) Frying Pan Section (FPS) of mid-Burdigalian age, Onslow Bay Section (OBS) of Langhian age, and Bogue Banks Section (BBS) of Serravallian age. Complex depositional patterns within each of these sections are determined by two sets of factors. First, the resulting sediment unit is a product of depositional controls for siliciclastic, biogenic, and authigenic–early diagenetic sediments which include: a) paleotopographic features of the continental margin; b) siliciclastic point sources; and c) paleoceanographic conditions controlling benthic depositional environments. Second, the preservation of any given sediment unit is then dependent upon subsequent sea-level history which dictated: a) Gulf Stream erosion during high sea-level events (Stephen W. Snyder *et al.*, Chapter 30, this volume); b) fluvial erosion during low sea-level events; and c) associated late diagenetic overprinting.

Mallette (1986) found that regional lithofacies patterns varied only slightly from one fourth-order depositional sequence to another. The greatest differences occurred between the FPS, OBS, and BBS sections as deposition shifted from continental-shelf environments in FPS to shelf-edge and upper-slope environments in OBS and back to shelf environments in BBS (Fig. 31.4). The regional distribution of lithofacies within each of these three depositional sections in Onslow Embayment and southern Aurora Embayment is presented in Figures 31.5–31.7. A schematic summarization for the Miocene continental margin of North Carolina is presented in Figure 31.8.

The zero isopach line marking the western limit of the Pungo River Formation in Figures 31.5–31.7 is an erosional line and does not represent the ancient shoreline. Miocene sediments representing coastal depositional environments have not yet been recognized on the Carolina margin. Unit A in the Aurora Phosphate District contains medium-grained, clean siliciclastic sands interbedded with oyster bioherms and biogenic hash, both interbedded with authigenic–early diagenetic phosphorites and dolomuds (IBL). This unit is early Burdigalian (Fig. 29.7 and Table 29.1 in Riggs *et al.*, Chapter 29, this volume) and appears to be the oldest recognized portion of the Pungo River Formation (Powers, 1986, 1987 and 1988). Unit A is interpreted to represent deposition in inner-shelf environments along the leading edge of the Miocene second-order transgression.

Among the most important features controlling deposition of the Pungo River Formation are two areas of point source-derived siliciclastic sediments. During deposition of FPS sediments (Fig. 31.5), these source areas for northwestern Onslow and Aurora Embayments were close to the deposits preserved in the stratigraphic record. Texturally- and mineralogically-uniform siliciclastic quartz sands and muddy quartz sands are an important lithofacies (UTML) within the FPS section. The UTML generally grades seaward and laterally along the shelf into finer-grained siliciclastic sediments with fining upward textures (FUSL) and locally grades into interbedded lithofacies (IBL). Siliciclastics remain important within the OBS and BBS sections, but the influence of point sources is diminished (Figs. 31.6 and 31.7, respectively). This may reflect higher sea levels that moved the coastal system farther inland relative to the preserved depositional record. The UTML to the west was not preserved, and only the more seaward IBL and FUSL were preserved in the OBS and BBS sediments, respectively.

Occasionally siliciclastic sands reached the shelf edge such as during the deposition of the OBS section in Onslow Embayment (Fig. 31.6). Episodic dumping of these siliciclastics off the shelf edge interrupted normal upper slope biogenic or authigenic–early diagenetic sedimentation to produce an interbedded sequence of prograding clinoforms (IBL) (Matteucci, 1984; Mallette, 1986; Matteucci & Hine, 1987; Stephen W. Snyder *et al.*, Chapter 30, this volume).

Two types of biogenic sediment occurred within very different depositional environments on the Carolina continental margin (Fig. 31.8). On the shelf, macrofossiliferous biogenic hash accumulated as a mostly uniform, carbonate lithofacies (UTML) on the top and flanks of the shallow platform of Cape Lookout High. Siliciclastics were mixed with biogenic sediments on top of the High in response to high wave-energy conditions. Laterally off its northern and southern flanks, siliciclastics and biogenic sediments formed a zone of interbedded lithofacies (IBL). The lateral extent of both biogenic hash and the interbedded lithofacies away from the High varied through time (Figs. 31.5–31.7).

The second area of biogenic sedimentation occurred on the outer portion of the continental margin (Fig. 31.8). These sediments are generally characterized by uniform bedding (UTML) and are dominated by microfossils, either diatoms or foraminifera, but rarely both. Where siliciclastic muds were the major diluent and authigenic–early diagenetic sediments were scarce, foraminifera dominate the microfossil component and reflect oxygenated, open-marine depositional environments (Scott W. Snyder, Chapter 32, this volume). Where authigenic sediments were the major diluent, both forams and diatoms occur within the uniform sediments. Foraminiferal assemblages primarily coincide with the phosphate facies and are dominated by species characteristic of high-nutrient and low-oxygen concentrations (Scott W. Snyder, Chapter 32, this volume). Diatom assemblages occur primarily in the dolomud facies and are generally seaward of equivalent phosphate facies (Miller, 1982; Riggs, 1984; Powers, 1987).

Environments of phosphogenesis

Authigenic–early diagenetic sedimentation is driven primarily by the oceanic system from the seaward side and diluted primarily by the siliciclastic system from the landward side of the depositional regime. On the upper slope, authigenic–early diagenetic deposition is characterized by unformly bedded sequences (UTML) of organic-rich phosphate or dolomite, with variable concentrations of microfossils and minor fine-grained siliciclastics (Fig. 31.8). As the authigenic–early diagenetic facies migrates onto the shelf, it grades landward into the mixed zone of sedimentation with increased interaction with siliciclastic sands and macrofossiliferous biogenic sediments. This zone of mixed sedimentation is characterized by the fining upward sequence lithofacies (FUSL). This interaction of authigenic–early diagenetic and siliciclastic sediments produces the 'idealized lithologic cycle' (Riggs *et al.*, 1982; Riggs, 1984; Hine & Riggs, 1986) and forms all known primary phosphate-rich fourth-order depositional sequences on the Carolina margin (Figs. 31.5–31.7). If upwelling is sufficient to provide a continuous nutrient supply for production of organic matter, and if oxygen-depleted conditions produce a dysaerobic benthic environment, authigenic–early diagenetic phosphate or dolomite will form at or below the sediment–water interface. Grain morphology, isotopic and trace-element geochemistry, and petrology of specific phosphate grains will change as the energy and chemical conditions of the benthic environment and siliciclastic input change landward across the shelf.

Deposition and relative abundance of authigenic–early diagenetic sediments is inversely proportional to deposition of diluent sediments, mainly siliciclastics and to a lesser extent biogenics. This suggests that major phosphogenesis occurs in environments away from siliciclastic point sources. Within any given depositional environment on the continental margin, the grain size of siliciclastic sediments generally decreases and their diluent effect is reduced as sea-level rises and point-source areas retreat landward. Paleotopography of the continental margin also influences which environments are dominated by open-ocean circulation or upwelling and dysaerobic conditions. Seismic data demonstrate that these conditions migrate in response to interactions between paleotopography and the Gulf Stream as the water deepens through each transgressive event (Stephen W. Snyder, *et al.*, Chapter 30, this volume). Riggs (1984) and Stephen W. Snyder, *et al.*, (Chapter 30, this volume) suggest that Gulf Stream interaction leading to maximum topographic upwelling will occur on the continental margin during early–mid-stage transgression within each sea-level event. Consequently, maximum authigenic–early diagenetic sedimentation should occur in seaward environments during early-stage transgression, and in more landward environments during mid-stage transgression.

Along the Carolina continental margin, authigenic–early

Table 31.4. *Depositional characteristics of three phosphorite units that are interpreted to be fourth-order depositional sequences of the Pungo River Formation, Aurora Phosphate District*

Feature	Units A and B Burdigalian	Unit C Langhian
Benthic foram fauna (Snyder *et al.*, 1982; Snyder, Chptr. 32)	Low species diversity, stressed environment (low O_2 and high nutrient conditions)	High species diversity, normal environment (normal O_2 and nutrient conditions)
Planktonic foram fauna (Snyder *et al.*, 1982; Snyder, Chptr. 32)	Low species diversity and low populations	Abundant, normal marine assemblages or not preserved
Diatom composition (powers, 1987)	Abundant, well-preserved suite reflecting cold-water & upwelling conditions	Absent or poorly preserved
Phosphate grain types (Riggs, 1979; Scarborough, 1981; Ellington, 1984)	Abundant pelloids mixed with dominantly fine–medium sand intraclasts; deep-water, outer-shelf	Dominantly medium–coarse sand intraclasts; shallow-water, inner- to mid-shelf
Phosphate grain texture (Ellington, 1984)	Smooth and resinous	Dull, highly pitted, and fractured
Elemental composition of phosphate grains (Ellington, 1984)	High Al & Si (clay) & Fe (Pyrite) contents as inclusions & low P suggest minimal reworking & diagenetic alteration	Moderate Al (lower clay content), low Fe & high P suggest increased reworking & diagenetic alteration
Trace element composition of phosphate grains (Indorf, 1982; Ellington, 1984)	U = 76 ppm Unit A U = 63 ppm Unit A Cd = 15 ppm Units A & B Zn = 171 ppm Units A & B	U = 102 ppm Unit C Cd = 37 ppm Unit C Zn = 296 ppm Unit C
Carbonate caprock: macroinvertebrates (Riggs *et al.*, 1982; unpub. data)	Very low species diversity & low populations of macroinvertebrates	Very high species diversity & high populations of macroinvertebrates
Carbonate caprock: mineralogy (Riggs *et al.*, 1982; unpub. data)	Primary dolomite of a non-replacive origin (no calcite or aragonite precursor)	Primary calcite with minor dolomite of a replacive origin (with a calcite or aragonite precursor)

diagenetic phosphate sedimentation reached significant concentrations within the following units.

Stage	Onslow Embayment	Aurora Embayment
Serravallian	BBS-8	
	BBS-2	
	BBS-1	
Langhian		Unit C
Burdigalian	FPS-2	Unit B
	FPS-1	
		Unit A

Units FPS-1 and FPS-2 in central and southern Onslow Embayment show a gradual southerly increase in phosphate, organic-rich muds, and diatoms relative to siliciclastic components to the north (Fig. 31.4). Also, each unit is a fining upwards sequence (FUSL) which grades vertically into diatomaceous, dolosilty, siliciclastic muds (Fig. 31.10). This suggests sustained upwelling in the area of diverging bathymetric contours into Onslow Embayment (Fig. 31.5) during the initial Miocene transgression. Similar bathymetric divergence occurred on the north side of Cape Lookout High, producing major topographic upwelling into the Aurora Embayment as the Gulf Stream flowed past the High during deposition of Units A and B (Figs 31.5 and 31.6).

With minor variations, each of the four depositional sequences of the Pungo River Formation in the Aurora Phosphate District represents a FUSL (Riggs *et al.*, 1982; Riggs, 1984; Fig. 31.4). On a more detailed basis, each of the phosphorite sequences (Units A, B, and C) contains distinctive criteria which directly reflect the depositional environment for the phosphate component (i.e. primary *in situ* versus reworked phosphate sediments). Characteristics of these depositional sequences are summarized in Table 31.4. Unit A was deposited on the leading edge of the second-order Miocene transgression. Where preserved, Unit A grades upward from shallow-water, inner-shelf facies of interbedded quartz sands, biogenic hash, oyster bioherms, and phosphorite quartz sands to a more extensive, deeper-water facies. The upper portion of Unit A and Unit B are sedimentologically very similar and are interpreted to represent slow, *in situ* phosphate formation and deposition in environments characterized by nutrient-rich, oxygen-depleted conditions in outer-shelf environments. Phosphorites in Unit C were initially deposited in inner- to mid-shelf environments under similar chemical conditions, possibly as the shallow-water equivalent of Unit B. However, these primary deposits were subjected to weathering during subsequent erosion, mechanically reworked, and redeposited as the present Unit C under normal marine conditions in middle–outer-shelf environments (Scott W. Snyder, *et al.*, 1982).

Phosphate occurrences in BBS-1 and BBS-2 are also part of cyclic FUSLs that culminate in diatomaceous dolomuds or moldic limestones similar to Units A, B, and C in the Aurora Phosphate District. These poorly known Serravallian sequences appear to have formed in response to: (1) initiation of the Miocene second-order sea-level regression (Fig. 29.9 in Riggs *et al.*, Chapter 29, this volume); and (2) a change in orientation of the continental margin (Fig. 29.8 in Riggs *et al.*, Chapter 29, this volume). Due to rapid eastward shelf progradation during deposition of the OBS sediments, orientation of the shelf edge in northeastern Onslow Embayment changed from northeast–southwest to north–south by the time of BBS deposition (Stephen W. Snyder, *et al.*, Chapter 30, this volume) producing a major bathymetric nose in southern Onslow Embayment. Diverging isobaths downstream from this feature produced sustained upwelling and reinitiated authigenic–early diagenetic phosphate sedimentation in BBS-1 and BBS-2.

BBS-8 has a limited outcrop area (Fig. 29.8 in Riggs *et al.*, Chapter 29, this volume), limited seismic data, and only two core holes. Consequently, there are many unknowns about the distribution and pattern of sedimentation within this depositional sequence. Scott W. Snyder (Chapter 32, this volume) described a benthic foram faunal assemblage in BBS-8 characteristic of highly-oxygenated bottom conditions. On the basis of extensively abraded and leached intraclastic phosphate grains and the benthic foraminiferal fauna, he interprets the phosphate in BBS-8 to have been reworked and transported to its present site of accumulation in a fashion similar to Unit C in the Aurora Phosphate District.

Acknowledgments

The research presented in this paper resulted from National Science Foundation grants OCE-7908949, OCE-8118164, OCE-8400383, and OCE-8609161 and University of North Carolina Sea Grant College grants R/AO-3 and R/AO-4 from 1982 through 1986. Appreciation is expressed to: 1) the many members of the Geology Department at East Carolina University, Greenville, NC and the Marine Science Program at the University of South Florida, St Petersburg, Florida who supplied help both at sea and in the laboratory; 2) Texasgulf Inc and North Carolina Phosphate Corp who supplied samples and analytical capabilities; and 3) the ongoing programs associated with the International Geologic Correlation Program (IGCP) 156 on Phosphorites sponsored by UNESCO and IUGS.

References

Allen, M.R. (1985). 'The origin of dolomites in the Miocene phosphorites of the Pungo River Formation, North Carolina'. Unpublished M.Sc. thesis, Duke University, Durham, NC, 43pp.

Allison, M.A. (1988). 'Mineralogy and sedimentology of the clay-sized fraction, Miocene Pungo River Formation, North Carolina continental margin'. Unpublished M.Sc. thesis, East Carolina University, Greenville, NC, 112pp.

Altschuler, Z.S. (1974). The weathering of phosphate deposits – environmental and geochemical aspects. In *Environmental Phosphorus Handbook*, ed. E.J. Griffith, A. Beeton, J.M. Spencer & D.T. Mitchell, pp. 33–96. John Wiley, New York.

Bentor, Y.K. (1980). Phosphorites – the unresolved problems. In *Marine Phosphorites – Geochemistry, Occurrence, Genesis*, ed. Y.K. Bentor, pp. 3–18. Society of Economic Paleontologists and Mineralogists Special Publication 29, Tulsa.

Ellington, M.D. (1984). 'Major and trace element composition of phosphorites of the North Carolina continental margin'. Unpublished M.Sc. thesis, East Carolina University, Greenville, NC, 93pp.

Forgang, J.A. & Riggs, S.R. (1986). Stratigraphy and diagenesis of carbonate sediments in the Miocene Pungo River Formation, Aurora Phosphate District, North Carolina. *Society of Economic Paleontologists and Mineralogists, Abstracts*, **III**, 39.

Hine, A.C. & Riggs, S.R. (1986). Geological framework, Cenozoic history, and modern processes of sedimentation on the North Carolina continental margin. In *SEPM Field Guidebooks, Southeastern US Third Annual Midyear Meeting*, ed. D.A. Textoris, pp. 129–94. Society of Economic Paleontologists and Mineralogists.

Hine, A.C. & Snyder, Stephen W. (1985). Coastal lithosome preservation: evidence from the shoreface and inner continental shelf off Bogue Banks, North Carolina. *Marine Geology*, **63**, 307–30.

Indorf, M.S. (1982). 'Uranium–phosphorus determinations for selected phosphate grains from the Miocene Pungo River Formation, North Carolina'. Unpublished M.Sc. thesis, East Carolina University, Greenville, NC, 80pp.

Katrosh, M.R. & Snyder, Scott W. (1982). Diagnostic foraminifera and paleoecology of the Pungo River Formation, central coastal plain of North Carolina. *Southeastern Geology*, **23**, 217–32.

Lewis, D.W. (1981). 'Preliminary stratigraphy of the Pungo River Formation of the Atlantic continental shelf, Onslow Bay, North Carolina'. Unpublished M.Sc. thesis, East Carolina University, Greenville, NC, 75pp.

Lewis, D.W., Riggs, S.R., Hine, A.C., Snyder, Stephen W., Snyder, Scott W. & Waters, V.J. (1982). Preliminary stratigraphic report on the Pungo River Formation of the Atlantic continental shelf, Onslow Bay, North Carolina. In *Miocene of the Southeastern United States*, ed. T.M. Scott & S.B. Upchurch, pp. 122–37. Southeastern Geological Society and Florida Bureau of Geology Special Publication No. 25.

Lyle, M. (1984). 'Clay mineralogy of the Pungo River Formation, Onslow Bay, North Carolina continental shelf'. Unpublished M.Sc. thesis, East Carolina University, Greenville, NC, 129pp.

Mallette, P.M. (1986). 'A lithostratigraphic analysis of cyclical phosphorite sedimentation, Miocene Pungo River Formation, North Carolina continental shelf'. Unpublished M.Sc. thesis, East Carolina University, Greenville, NC, 186pp.

Matteucci, T.D. (1984). 'High-resolution seismic stratigraphy of the North Carolina continental margin – the Cape Fear region: sea-level cyclicity, paleobathymetry, and Gulf Stream dynamics'. Unpublished M.Sc. thesis, University of South Florida, St Petersburg, 100pp.

Matteucci, T.D. & Hine, A.C. (1987). Evolution of the Cape Fear Terrace: a complex interaction between the Gulf Stream and a paleo-shelf edge delta. *Marine Geology*, **77**, 1–22.

McClellan, G.H. (1980). Mineralogy of carbonate fluorapatites. *Journal of the Geological Society of London*, **137**, 675–82.

McClellan, G.H. & Lehr, J.R. (1969). Crystal chemical investigation of natural apatites. *Mineralogy*, **54**, 1374–91.

Miller, J.A. (1982). Stratigraphy, structure, and phosphate deposits of the Pungo River Formation of North Carolina. *North Carolina Department of Natural Resources and Community Development, Geological Survey Bulletin*, **87**, 32pp.

Moore, T.L. (1986). 'Foraminiferal biostratigraphy and paleoecology of the Miocene Pungo River Formation, central Onslow Embayment, North Carolina continental margin'. Unpublished M.Sc. thesis, East Carolina University, Greenville, NC, 160pp.

Moretz, L. (1988). 'Diagenesis of benthic foraminifera in the Miocene Pungo River Formation of Onslow Bay, North Carolina continental shelf'. Unpublished M.Sc. thesis, East Carolina University, Greenville, NC, 92pp.

Mount, J.F. (1984). Mixing of siliciclastic and carbonate sediments in shallow shelf environments. *Geology*, **12**, 432–5.

Popenoe, P. (1985). Cenozoic depositional and structural history of the North Carolina margin from seismic-stratigraphic analyses. In *Geological Evolution of the United States Atlantic Margin*, ed. C.W. Poag, pp. 125–88. Van Nostrand Reinhold Co., New York.

Powers, E.R. (1986). Biostratigraphic correlation of Miocene phosphorites on the North Carolina continental margin using diatoms and silicoflagellates. *Society of Economic Paleontologists and Mineralogists, Annual Midyear Meeting Abstracts*, **3**, 91.

Powers, E.R. (1987). 'Diatom biostratigraphy and paleoecology of the Miocene Pungo River Formation, North Carolina continental margin'. Unpublished M.Sc. thesis, East Carolina University, Greenville, NC, 225pp.

Powers, E.R. (1988). Diatom biostratigraphy and paleoecology of the Miocene Pungo River Formation, Onslow Bay, North Carolina continental shelf. In *Micropaleontology of Miocene Sediments in the Shallow Subsurface of Onslow Bay, North Carolina Continental Shelf*, ed. Scott W. Snyder, pp. 97–162. Cushman Foundation for Foraminiferal Research, Special Publication No. 25.

Riggs, S.R. (1979). Phosphorite sedimentation in Florida – a model phosphogenic system. *Economic Geology*, **74**, 285–314.

Riggs, S.R. (1980). Intraclast and pellet phosphorite sedimentation in the Miocene of Florida. *Journal of the Geological Society (London)*, **137**, 741–8.

Riggs, S.R. (1981). Relation of Miocene phosphorite sedimentation to structure in the Atlantic continental margin, southeastern United States. *American Association of Petroleum Geologists Bulletin*, **65**, 1669.

Riggs, S.R. (1984). Paleoceanographic model of Neogene phosphorite deposition, US Atlantic continental margin. *Science*, **223**, 123–31.

Riggs, S.R. & Belknap, D.F. (1988). Upper Cenozoic processes and environments of continental margin sedimentation. In *The Geology of North America, Volume I-2, The Atlantic Continental Margin, US*, ed. R.E. Sheridan & J.A. Grow, pp. 131–76. Geological Society of America.

Riggs, S.R., Lewis, D.W., Scarborough, A.K. & Snyder, Scott W. (1982). Cyclic deposition of Neogene phosphorites in the Aurora area, North Carolina, and their possible relationship to global sea-level fluctuations. *Southeastern Geology*, **23**, 189–204.

Riggs, S.R. & Snyder, Scott W. (1986). Patterns of cyclic sedimentation of the Upper Cenozoic section, North Carolina Coastal Plain. In *SEPM Field Guidebooks, Southeastern US Third Annual Midyear Meeting*, ed. D.A. Textoris, pp. 333–72. Society of Economic Paleontologists and Mineralogists.

Riggs, S.R., Snyder, Stephen W., Hine, A.C., Snyder, Scott W., Ellington, M.D. & Mallette, P.M. (1985). Geologic framework of phosphate resources in Onslow Bay, North Carolina continental shelf. *Economic Geology*, **80**, 716–38.

Scarborough, A.K. (1981). 'Stratigraphy and petrology of the Pungo River Formation, central coastal plain of North Carolina'. Unpublished M.Sc. thesis, East Carolina University, Greenville, NC, 78pp.

Scarborough, A.K., Riggs, S.R. & Snyder, Scott W. (1982). Stratigraphy and petrology of the Pungo River Formation, central coastal plain, North Carolina. *Southeastern Geology*, **23**, 205–16.

Snyder, Scott W., ed. (1988). *Micropaleontology of Miocene Sediments in the Shallow Subsurface of Onslow Bay, North Carolina Continental Shelf.* Cushman Foundation for Foraminiferal Research, Special Publication No. 25, 189pp.

Snyder, Scott W., Hale, W.R., Riggs, S.R., Spruill, R.K. & Waters, V.J. (1984). Occurrence of clinoptilolite as moldic fillings of foraminiferal tests in continental margin sediments. *Geological Society of America Abstracts with Programs*, **16 (6)**, 662.

Snyder, Scott W., Riggs, S.R., Katrosh, M.R., Lewis, D.W. & Scarborough, A.K. (1982). Synthesis of phosphatic sediment–faunal relationships within the Pungo River Formation: paleoenvironmental implications. *Southeastern Geology*, **23**, 233–46.

Snyder, Stephen W. (1982). 'Seismic stratigraphy within the Miocene Carolina Phosphogenic Province: chronostratigraphy, paleotopographic controls, sea-level cyclicity, Gulf Stream dynamics, and the resulting depositional framework'. Unpublished M.Sc. thesis, University of North Carolina, Chapel Hill, NC, 183pp.

Snyder, Stephen W., Hine, A.C. & Riggs, S.R. (1982). Miocene seismic stratigraphy, structural framework and sea-level cyclicity: North Carolina continental shelf. *Southeastern Geology*, **23**, 247–66.

Snyder, Stephen W., Hine, A.C. & Riggs, S.R. (1983). 3-D stratigraphic modeling from high-resolution seismic reflection data: an example from the North Carolina continental shelf. *American Association of Petroleum Geologists Bulletin*, **67**, 549–50.

Snyder, Stephen W., Hine, A.C., Riggs, S.R. & Snyder, Scott W. (1986). Shifting sites of Miocene phosphorite deposition: the critical depth hypothesis. *Society of Economic Paleontologists and Mineralogists, Abstracts*, **III**, 105.

Steele, G.A. (1980). 'Stratigraphy and depositional history of Bogue Banks, North Carolina'. Unpublished M.Sc. thesis, Duke University, Durham, NC, 201pp.

Stewart, T.L. (1985). 'Carbonate petrology and sedimentology of the Miocene Pungo River Formation, Onslow Bay, North Carolina continental shelf'. Unpublished M.Sc. thesis, East Carolina University, Greenville, NC, 184pp.

Waters, V.J. (1983). 'Foraminiferal paleoecology and biostratigraphy of the Pungo River Formation, southern Onslow Bay, North Carolina continental shelf'. Unpublished M.Sc. thesis, East Carolina University, Greenville, NC, 186pp.

32

Relationships between benthic foraminiferal assemblages and Neogene phosphatic sediments, North Carolina coastal plain and continental shelf

SCOTT W. SNYDER

Abstract

The Miocene Pungo River Formation and Pliocene Yorktown Formation record numerous episodes of cyclic sedimentation along the North Carolina continental margin. Pungo River sediments occur in the subsurface of the modern coastal plain and in outcrop–shallow subcrop across the modern continental shelf of Onslow Bay. Yorktown sediments crop out across the coastal plain but are best exposed in the quarry of Texasgulf Inc.

Sediment cycles are largely siliciclastic, but some include phosphorites and/or carbonate caprocks. Presumably, phosphate-rich sediments were deposited in nutrient-rich, oxygen-depleted environments, whereas 'cleaner' quartzitic sands accumulated in well-oxygenated conditions. Benthic foraminiferal assemblages are consistent with this interpretation. Phosphatic horizons contain assemblages that are numerically dominated by species of *Bolivina*, *Bulimina* and *Buliminella.* Muddy sediments with moderate phosphate content contain an abundance of species of *Florilus* and *Nonionella*. Such assemblages occur in modern seas where there are oxygen-minimum zones and sewage outfalls. Quartz and carbonate sands are predominated by species of *Cibicides*, *Valvulineria* and *Hanzawaia.* Extant representatives exist in similar sediment types where bottom waters are well oxygenated.

The relationship between foraminiferal assemblages and phosphate content is not entirely predictable because other environmental variables affect foraminifera. However, assemblages are generally sensitive indicators of bottom-water mass conditions (i.e. nutrient and oxygen levels).

Introduction

Most of the world's phosphorus resources occur as ancient, bedded sedimentary deposits of marine origin (Manheim & Gulbrandsen, 1979). Marine phosphorites represent a complex sediment system for which the precise environmental setting, locale relative to the sediment–water interface, and mechanism of phosphate formation are not fully understood. Discrete deposits may differ through space and time with regard to any or all of these factors.

Several important points regarding phosphate formation are essential to the thrust of this paper. Phosphate forms in suboxic environments (Froelich, *et al.*, 1988). Phosphate pellets form rapidly in the uppermost few centimeters of marine sediments (Burnett, Chapter 5, this volume). Because benthic foraminifera live on or within the upper few centimeters of marine sediments, they may provide information on conditions necessary to generate phosphate in this environment. Carbonate components may be destroyed during some phosphate-fixing processes, such as those associated with filamentous bacterial mats (Reimers, Chapter 24, this volume). However, some processes may not involve such destruction, thus enhancing chances that calcareous microfossils will be preserved.

Even where phosphate-fixing processes do not involve destruction of carbonate, calcareous microfossils may be destroyed by diagenetic processes. Consequently, rich calcareous microfossil assemblages are rarely found in phosphorites. The North Carolina phosphate deposits, being among the least altered phosphorites in the world (G.H. McClellan, pers. comm.), contain a rich and diverse calcareous microfauna that is numerically dominated by foraminifera. These foraminiferal assemblages and their relationships to enclosing sediments provide an opportunity to investigate the depositional environments in which the North Carolina phosphates formed and/or accumulated.

This paper summarizes the use of benthic foraminifera as indicators of ancient depositional environments within phosphatic Neogene sediments of North Carolina. More detailed faunal data are published elsewhere (Scott W. Snyder, Waters & Moore, 1988).

The regional geologic setting and seismic stratigraphy of Miocene deposits along the North Carolina continental margin are described by Riggs *et al.* (this volume) and Stephen W. Snyder *et al.* (this volume). Patterns of change in the Miocene benthic foraminiferal assemblages of Onslow Bay are herein related to their detailed seismic subdivision of the Pungo River Formation (three third-order stratigraphic sections which collectively comprise 18 fourth-order seismic sequences). Assemblage changes in Pungo River sediments of the Aurora area are related to cyclic sediment patterns described by Riggs *et al.* (1982). The overlying Yorktown Formation, formally named and described by Clark & Miller (1906), contains variable amounts of phosphate, most of which is concentrated in its lower portions. Yorktown strata crop out across northwestern portions of the coastal plain, but exposures are scattered and stratigraphically incomplete. The open-pit phosphate mine of

Texasgulf Inc. provides the most complete Yorktown exposure in North Carolina. Yorktown sediments at the mine have been interpreted to be of Pliocene age on the basis of ostracods (Hazel, 1983) and planktonic foraminifera (within Blow's Zones N18/N19 according to Scott W. Snyder, Mauger & Akers, 1983). Pliocene faunal–sediment relationships discussed herein are based on this exposure.

General depositional environments in the North Carolina Neogene

Modern benthic foraminiferal distribution patterns indicate that species are influenced by numerous, interrelated ecological factors. Although it is often impossible to isolate the effects of specific environmental variables, many workers have focused on a select few, particularly water depth.

Water depth directly affects benthic marine organisms via hydrostatic pressure, which becomes important only through depth intervals of great magnitude. Numerous depth-related factors (e.g. light intensity, salinity, temperature, food supply, dissolved oxygen) are much more variable through smaller bathymetric ranges. Though depth itself cannot be considered a limiting environmental factor, many modern species are distributed in patterns that parallel depth contours. Still, ancient water depth is difficult to determine, even in an approximate way (Raup & Stanley, 1978). Using data from modern benthic foraminifera, which often show a distinct depth zonation, it is sometimes possible to estimate at least the general paleobathymetric setting represented by fossil assemblages. In the Aurora district, the lower part of the Pungo River Formation was deposited in inner to middle sublittoral environments, while its upper portion accumulated in middle to outer sublittoral environments (Katrosh & Scott W. Snyder, 1982). In Onslow Bay, Pungo River sediments were deposited in a variety of environments, ranging from outer sublittoral to upper bathyal for the Frying Pan Section in southern Onslow Bay to middle sublittoral for the Onslow Bay and Bogue Banks Sections in central and northern Onslow Bay (Waters, 1983; Moore & Scott W. Snyder, 1985). The lower part of the Yorktown Formation in the Aurora district represents deposition in middle to outer sublittoral conditions followed by progressive shoaling to inner sublittoral environments (Mauger, 1979).

Ecology of selected modern benthic foraminiferal species

Within the context of the generalized depositional environments described above, focusing on environmental variables other than depth can provide more detailed interpretations. The physical and chemical characteristics of water masses, especially with regard to dissolved oxygen content, appear to be more important to foraminiferal distributions than changes in depth. Van der Zwaan (1982) argued that oxygen content of water is by far the most important denominator of benthic ecology. Where environmental parameters and benthic foraminiferal abundance patterns are statistically analyzed, one of the few parameters which almost always shows up is oxygen content. Van der Zwaan stated that oxygen almost never acts as a limiting factor by hampering metabolism. Reduced oxygen content may alter foraminiferal assemblages by excluding predators and by minimizing competition (Phleger & Soutar, 1973). However, its control appears to stem primarily from nutrient availability and the amount and quality of available food (Van der Zwaan, 1982). The following discussion focuses on oxygen content–nutrient availability as a primary factor in controlling foraminiferal species distributions. Tolerances noted for living species will later be used to infer environmental conditions associated with ancient assemblages.

Modern benthic foraminifera in oxygen minima

The eastern margin of the Pacific Ocean provides numerous examples of coastal upwelling systems and their associated oxygen-minimum zones. Despite differences in latitude, bathymetry, and ambient water mass composition, all such systems support benthic foraminiferal assemblages that are taxonomically similar. Though fewer in number, similar examples of relationships between oxygen-minimum–nutrient-enriched zones and benthic forminiferal assemblages have been noted in the Atlantic. Table 32.1 summarizes the predominant benthic foraminiferal species that characterize oxygen-minimum zones.

Perhaps the relative abundance of some species depends more upon organic content than upon dissolved oxygen content. For example, Corliss, Martinson & Keffer (1986) stated that the percentage of *Uvigerina peregrina* can vary independently of dissolved oxygen in deep water. They concurred with Miller & Lohman (1982), who found that the abundance of this species correlates instead with high amounts of organic carbon and fine-grained sediment. However, in sublittoral and upper bathyal environments oxygen-minimum zones and high productivity often occur in the same areas, such as those associated with coastal upwelling (Phleger & Soutar, 1973). The low diversity, high-predominance benthic foraminiferal assemblages that characterize such areas may be useful in recognizing ancient low oxygen, high-productivity environments.

Modern benthic foraminifera in polluted environments

Modern benthic foraminifera are also sensitive to environmental modification resulting from pollution of various types. Commonly, large-scale pollution of continental shelf environments results from sewage outfalls. Although effects vary according to the type and volume of effluent, most outfalls cause localized nutrient enrichment and correspondingly low dissolved oxygen concentrations. For example, phosphate concentrations associated with the Hyperion outfall system in Santa Monica Bay, California were reported to be 20-times those of nearby uncontaminated waters (Bandy, Ingle & Resig, 1965). Nutrient salts of nitrogen and silicon are also greater near outfalls. Assemblages beneath outfalls differ markedly from those in adjacent, unpolluted sediments. Differences in species composition and diversity are attributed to the effects of pollution because temperature and salinity values vary only slightly in many outfall areas (Bandy, Ingle & Resig, 1964a and 1965). The benthic foraminiferal assemblages associated with outfalls located in open-shelf areas are taxonomically similar to those which inhabit naturally occurring oxygen-minimum zones. Table 32.2 summarizes assemblages associated with a variety of open-shelf areas characterized by pollution.

Examples of benthic assemblages associated with polluted

Table 32.1. *Summary of modern benthic foraminiferal species that numerically dominate naturally occurring oxygen-minimum zones*

Location	Bathymetry (m)	Oxygen concentration (ml l^{-1})	Predominant species	Comments	Source of information
Pacific Ocean					
Off Point Sur, California	700–750	0.27	*Bolivina argentea, B. spissa, Uvigerina* spp.	Bolivinids account for 80% of living assemblage	Mullins *et al.*, (1985)
Santa Barbara Basin, California	500–600	0.10	*Globobulimina hoeglundi, Bolivina seminuda, Suggrunda eckisi, Nonionella stella*	Three most abundant species account for 80–86% of living assemblage	Phleger & Soutar (1973)
South of Punta Abreojos, Baja California, Mexico	75–100	0.10	*Bolivina* spp., *Uvigerina* spp, and *Bulimina* spp.		Phleger & Soutar (1973)
	530	0.10	*Bolivina seminuda*	*Bolivina seminuda* accounts for 71% of living assemblage	Phleger & Soutar (1973)
Off Callao, Peru	200	Described as 'low oxygen'	*Bolivina*, cf. *B. pacifica*	*Bolivina*, cf. *B. pacifica* accounts for 85% of living assemblage	Phleger & Soutar (1973)
Gulf of California	440–880	<0.20	*Bolivina* spp., *B. subadvena*		Streeter (1972)
Peru–Chile trench area	150–400	<1.00	*Bolivina rankini, B. interjuncta*	*Bolivina rankini* and *interjuncta* account for 78–93% of living assemblage	Ingle, Keller & Kolpack (1980)
	900–1700	Not specified	*Bolivina spissa, Uvigerina peregrina, Epistominella exigua, Eilohedra levicula, Bulimina striata*		Ingle *et al.* (1980)
Atlantic Ocean					
Off Daytona Beach, Florida	185	Not specified	*Bolivina subaenariensis*	*Bolivina subaenariensis* is major species in dense population	Sen Gupta, Lee & May (1981)
North Atlantic	2000–4000	<0.50	*Uvigerina peregrina*	Distribution congruent with disposition of dissolved oxygen	Streeter & Shackleton (1979)

waters along the Atlantic margin of the United States are limited primarily to nearshore, marginal marine, somewhat restricted environments. Unlike the open shelf, these environments are characterized by large seasonal fluctuations in temperature and salinity. Table 32.3 summarizes the benthic foraminiferal species that predominate in marginal marine outfall areas.

Clearly, species composition of marginal marine benthic foraminiferal assemblages is primarily controlled by factors other than organic enrichment and dissolved oxygen content. For example, formae of *Elphidium excavatum* respond to a combination of water temperature, salinity, proximity to shore, and range of annual climatic variation (Feyling-Hanssen, 1972; Miller, Scott & Medioli, 1982). Because species like *E. excavatum* thrive in both polluted and unpolluted waters, they are of limited use in assessing nutrient enrichment–oxygen depletion in ancient sediments.

Benthic foraminifera associated with ancient oxygen minima

Well documented examples of foraminiferal response to oxygen depletion are rare in the marine sedimentary record. Sapropels, dark organic-rich layers that mark basin stagnation and anoxic bottom-water conditions, provide opportunity to study foraminiferal changes associated with ancient events of oxygen depletion. Neogene strata of the Mediterranean contain numerous sapropels which have recently received attention.

Cita & Podenzani (1979) studied faunal changes immediately below and above sapropels and ash falls in Pleistocene sediments of the eastern Mediterranean. The onset of anoxic conditions was usually sudden, and was not preceded by detectable change in environmental conditions, as observable in composition of the benthic assemblage. However, for Sapropel S-1 (8000 years old), faunal density and diversity changes recorded progressve dete-

Table 32.2. *Summary of modern benthic foraminiferal species that numerically dominate polluted environments of the continental shelf and slope*

Location	Volume of discharge (10^6 gal d^{-1})	Type of effluent	Predominant species	Comments	Source of information
Orange County, California	25	Primary, chlorinated	*Buliminella elegantissima*, agglutinated species	*Buliminella elegantissima* accounts for 79% of living assemblage	Watkins (1961)
Laguna Beach, California	1	Primary, chlorinated	*Bolivina vaughni*, *Buliminella elegantissima*, *Nonionella miocenica*	*B. vaughni* has specificity for outfall, *B. elegantissima* as much as 50% of living assemblage, *N. miocenica* at periphery of outfall	Bandy *et al.* (1964a)
Los Angeles County, California	275	Untreated	*Buliminella elegantissima*, *B. marginata*, *B. denudata*, *Nonionella* spp.	*Nonionella* spp. at periphery of outfall	Bandy *et al.* (1964b)
Hyperion (Los Angeles City) California	264	Unchlorinated mixed primary–secondary	*Buliminella elegantissima*, *B. marginata*, *B. denudata*, agglutinated species, *Nonionella* spp.	*Nonionella* spp. at periphery of outfall	Bandy *et al.*, (1965)
Pacific margin of Panama	?	Not specified but sediments organic-rich	*Bulimina marginata*		Seiglie (1968)
Venezuelan shelf (Cariaco Gulf)	?	Not specified but sediments have 10% organic C	*Buliminella silviae*, *B. elegantissima*, *Nonionella opima*, *Florilus grateloupi*		Seiglie (1968)
Venezuelan shelf (Carupano Depression)	?	Not specified but sediments have 3.2% organic C	*Florilus atlanticus*, *Buliminella elegantissima*, *B. silviae*	*F. atlanticus* as much as 75% and *B. elegantissima* as much as 20% of assemblage	Seiglie (1968)
Venezuelan shelf (Coche Depression)	?	Not specified but sediments have 4.3% organic C	*Buliminella* spp.		Seiglie (1968)
Cienfuegos Bay, Cuba	?	Not specified	*Buliminella elegantissima*, *B. silviae*		Seiglie (1968)
Western shelf of Puerto Rico	?	Not specified	*Florilus grateloupi*, *F. pontoni*		Seiglie (1968)

Type and volume of discharge relate to time studies were done and do not indicate present conditions.

rioration of environmental conditions. Sediments immediately below the sapropel contain abundant (44%) *Bolivina*, the genus most tolerant of low oxygen levels in modern seas (Boltovskoy & Wright, 1976).

Mullineaux & Lohmann (1981) also studied the response of deep-water foraminifera to stagnating conditions implied by sapropel deposition in the eastern Mediterranean. They noted that '*Globobulimina affinis*, the taxon most closely associated with anoxic sediment in the cores, is found immediately below, above, and occasionally within a sapropel'. *Articulina tubulosa* and *Chilostomella mediterranensis* are also abundant near sapropelic sediments, but not in layers reflecting complete anoxia. Such assemblages are most similar to those presently living in deep, isolated basins.

Katz & Thunell (1984) quantitatively examined benthic foraminiferal response to oxygen depletion during the Middle Miocene–Early Pliocene of the eastern Mediterranean. They recognized such episodes by the deposition of sapropels, which they defined as organic-rich muds with more than 2% organic carbon. The sapropel assemblage is predominated by *Uvigerina mediterranea* and *Bolivina dilitata*, both of which are most abundant in modern sediments at depths having low oxygen content

Table 32.3. *Summary of modern benthic foraminiferal species that numerically dominate polluted environments in marginal marine, partially restricted environments*

Location	Volume of discharge	Type of effluent	Predominant species	Comments	Source of information
Moorehead City, North Carolina	Not applicable	Secondary	*Elphidium excavatum* (*clavatum*)	Sets of artificial ponds (one effluent and one control set) both predominated by *E. excavatum* (*clavatum*)	LeFurgey & St. Jean (1976)
Southern Chesapeake Bay (Chesapeake–Elizabeth outfall)	6×10^6 gal d^{-1}	Secondary chlorinated	Undifferentiated formae of *Elphidium excavatum*	*E. excavatum* predominated in both polluted and nearby unpolluted waters, averaged 81% of all living assemblages, and showed little correlation with distribution of organic carbon	Bates & Spencer (1979)
Chaleur Bay, New Brunswick, Canada	45 000 m^3 y^{-1}	Organics from municipal waste systems and paper mills	*Eggerella advena* and *Ammotium cassis* (1978); *Eggerella advena* and *Elphidium excavatum* (1980)	Dump site sampled one month and two years after dumping ceased (1978 and 1980, respectively)	Schafer (1982)

Type and volume of discharge relate to time studies were done and do not indicate present conditions.

relative to other parts of the Mediterranean basin.

Using quantitative analyses, Van der Zwaan (1983) concluded that ecological patterns of Miocene and Pliocene benthic species in the Mediterranean were determined to a great extent by oxygen content. Certain clusters are related to oxygen depletion, as recognized by the presence of finely laminated muds which, though not so organic-rich as sapropels, indicate the absence of benthic organisms responsible for bioturbation in well-oxygenated bottom conditions. Miocene–lower Pliocene benthic foraminiferal assemblages tolerant of oxygen deficiency at the seafloor are predominated by *Bulimina costata*, *B. elongata*, *Bolivina spathulata*, and *Uvigerina cylindrica gaudryinoides*. During the Late Pliocene, *Bulimina exilis* and species of *Globobulimina* and *Chilostomella* appear to have developed an increasing tolerance for strong oxygen deficiency.

Implications of ecological data for Neogene assemblages

Oxygen deficiency in marginal marine settings does not produce a distinctive foraminiferal assemblage because predominant species, such as *Elphidium excavatum*, are limited by other environmental variables. However, areas of oxygen depletion in open-shelf or more offshore environments support assemblages that differ markedly from adjacent assemblages and yet have a striking degree of taxonomic similarity to one another. Species of the genera *Bolivina*, *Bulimina*, *Globobulimina*, *Buliminella*, *Uvigerina*, *Nonionella* and *Florilus* recur in oxygen minimum zones and near sewage outfalls (Tables 32.1 and 32.2), and are associated with sapropels. The species may differ, but representatives of these genera are consistently predominant in oxygen-deficient bottom conditions. Van der Zwaan (1982) noted that different species are abundant in different water masses under apparently similar physical and chemical conditions. He attributed such faunal differences to biological interaction, where abundance of a given species is strongly dependent on competitive abilities of other species present in the association. It is also possible that some foraminifera have specific food requirements, so that nutrient composition could have a profound effect upon species composition. Hence, variation in species composition in no way diminishes the environmental significance of pronounced faunal similarity at the generic level. For example, Poag (1985) noted that '*Bolivina* is predominant in many upper slope oxygen-minima around the world as a result of its adaptation to oxygen depletion and nutrient enrichment . . .'.

Select species of *Bolivina*, *Bulimina*, *Buliminella*, and *Uvigerina* characterize conditions of intense oxygen deficiency, whereas those of *Nonionella* and *Florilus* thrive in peripheral zones with somewhat higher oxygen and lower nutrient levels. Faunal predominance by various combinations of species belonging to these genera should reliably indicate rather specific environmental conditions. According to Schafer & Cole (1976), when quantitatively establishing biofacies that reflect environmental conditions, fluctuations in the proportions of abundant

Fig. 32.1. (a) ***Bolivina calvertensis*** **Dorsey, side view; (b)** ***Bolivina paula*** **Cushman and Cahill, side view; (c)** ***Buliminella elegantissima*** **d'Orbigny, side view; (d)** ***Bulimina elongate*** **d'Orbigny, side view; (e)** ***Bolivinopsis fairhavensis*** **Gibson, side view; (f)** ***Uviqerina auberiana*** **d'Orbigny, side view; (g)** ***Uvigerina juncea*** **Cushman and Todd, side view; (h)** ***Uvigerina calvertensis*** **Cushman, side view; (i)** ***Uvigerina subperegrina*** **Cushman and Kleinpell, side view; (j–k)** ***Florilus grateloupi*** **d'Orbigny, (j) side view, (k) edge view; (l–m)** ***Florilus pizarrensis*** **Berry, (l) side view, (m) edge view; (n)** ***Nonionella miocenica*** **Cushman, side view; (o)** ***Eliphiduim excavatum*** **Terquem, side view. Each scale bar is 100 μm.**

species should yield about the same degree of resolution as is obtained from analyzing the entire assemblage. Phleger & Soutar (1973) concluded that fossil benthic assemblages of foraminifera in low-oxygen environments should exhibit good preservation, an abundance of small tests, and predominance by a few species. It also appears that the genera to which those species belong are predictable. Species whose abundance in Pungo River and Yorktown sediments implies oxygen deficiency are listed below (Fig. 32.1). Hereafter this assemblage is referred to as Association 1.

Bolivina calvertensis Dorsey
Bolivina paula Cushman and Cahill
Bulimina elongata d'Orbigny
Buliminella elegantissima (d'Orbigny)
Bolivinopsis fairhavenensis Gibson
Uvigerina auberiana d'Orbigny
Uvigerina calvertensis Cushman
Uvigerina juncea Cushman and Todd
Uvigerina subperegrina Cushman and Kleinpell
Florilus grateloupi (d'Orbigny)
Florilus pizarrensis (Berry)
Nonionella miocenica Cushman

The last three species suggest less intense oxygen deficiency.

The predominant taxa associated with well-oxygenated, more nutrient-depleted bottom conditions are generally quite distinct from those associated with oxygen depletion. This group of species maintains itself in stable marine environments inhabited by many species. The actual taxonomic composition will vary greatly, depending upon the interaction and influence of numerous environmental variables (temperature, salinity, substrate type, sedimentation rates, etc.). Detailed discussion of such assemblages is beyond the scope of this paper, and comment is limited to those few taxa that are abundant in the Pungo River and Yorktown Formations listed below (Fig. 32.2). For the purposes of our discussion this assemblage will be referred to as Association 2.

Cibicides americanus (Cushman)
Cibicides floridanus (Cushman)
Cibicides lobatulus (Walker and Jacob)
Hanzawaia concentrica (Cushman)
Valvulineria floridana Cushman
Valvulineria laevigata Phleger and Parker
Valvulineria olssoni Redmond
Valvulineria venezuelana Hedberg
Rosalina cavernata (Dorsey)

Ecological tolerances of the four extant species from this list indicate that the entire association thrives in normal marine, middle- to outer-sublittoral environments. *Cibicides floridanus* occurs at numerous, widespread localities across the middle and outer shelf of the Gulf of Mexico (Culver & Buzas, 1981). *Cibicides lobatulus* is generally confined to depths of < 200 m in cool waters along the Atlantic continental margin from Cape Hatteras to Newfoundland (Culver & Buzas, 1980). Because it is an attaching form that often conforms in shape to the particle serving as a substrate, its distribution may be influenced by grain size of the sediment. Along the Atlantic continental margin, *Hanzawaia concentrica* is largely confined to depths of 200 m or less (Culver & Buzas, 1980), is a predominant taxon across the shelf (Todd, 1979), and is particularly characteristic of the middle shelf (Schnitker, 1971). It is a common species in inner- to middle-shelf environments of the Gulf of Mexico (Poag, 1981). *Valvulineria laevigata* occurs sporadically across the outer shelf and slope of the Gulf of Mexico (Culver & Buzas, 1981). It is rare along the mid-Atlantic outer shelf (Schnitker, 1971).

Association 1 is predominant in low oxygen, high-nutrient conditions, while Association 2 thrives in more normal marine conditions that are well oxygenated and contain lesser amounts of nutrients. Predominance of either association in a fossil assemblage is presumed to indicate physico-chemical conditions similar to those that favor its numerical abundance in modern environments.

A third faunal association is marked by the conspicuous presence of *Siphogenerina lamellata* and *S. transversa*. Several species characteristic of Association 1 are also present, particularly those of the genus *Florilus*. This association is virtually absent in the Aurora area, and, though more persistent in Onslow Bay, it is always a relatively minor component of the total assemblage. The presence of Association 3 is noted, but cyclical faunal variations are defined on the basis of Association 1 versus Association 2.

Phosphatic sediment–benthic foraminiferal relationships

Sediment cyclicity in Neogene formations of the North Carolina Coastal Plain and continental shelf is associated with concomitant changes in benthic foraminiferal assemblages. The following discussion summarizes sediment and assemblage patterns for the Pungo River (Miocene) and Yorktown (Pliocene) Formations in the Aurora area and Onslow Bay.

Aurora area

Pungo River Formation (Miocene) The Pungo River Formation in the vicinity of Aurora, NC comprises four sediment cycles, the lower three of which are characterized by an admixture of phosphatic and terrigenous materials overlain by carbonate (Fig. 32.3). The uppermost cycle is predominantly carbonate. Analysis of 64 samples from five drill cores (Katrosh & Snyder, 1982; Snyder, unpublished data) reveals two distinct benthic foraminiferal assemblages within these Pungo River cycles.

One assemblage includes many taxa common to Association 1, as defined above. Predominant species are *Buliminella elegantissima* and *Elphidium excavatum*. Present in minor quantities are several species characteristic of Association 2. These include *Cibicides lobatulus* and *Hanzawaia concentrica*.

The two lower Pungo River cycles (units A and B of Riggs *et al.*, 1982) are moderately phosphatic (Fig. 32.3), but contain 15–50% pelletal phosphate grains, which is the highest concentration of this grain type near Aurora (Scarborough, Riggs & Scott W. Snyder, 1982). Macrofaunal skeletal remains are rare. Samples from these cycles are predominated by *Buliminella elegantissima* (up to 90% of the fauna) or *Elphidium excavatum* (up to 48% of the fauna). *B. elegantissima* thrives in high-nutrient–low-oxygen conditions in modern seas, whereas *E. excavatum* tolerates both high and low oxygen concentrations. The most persistent species of lesser abundance are: *Bulimina*

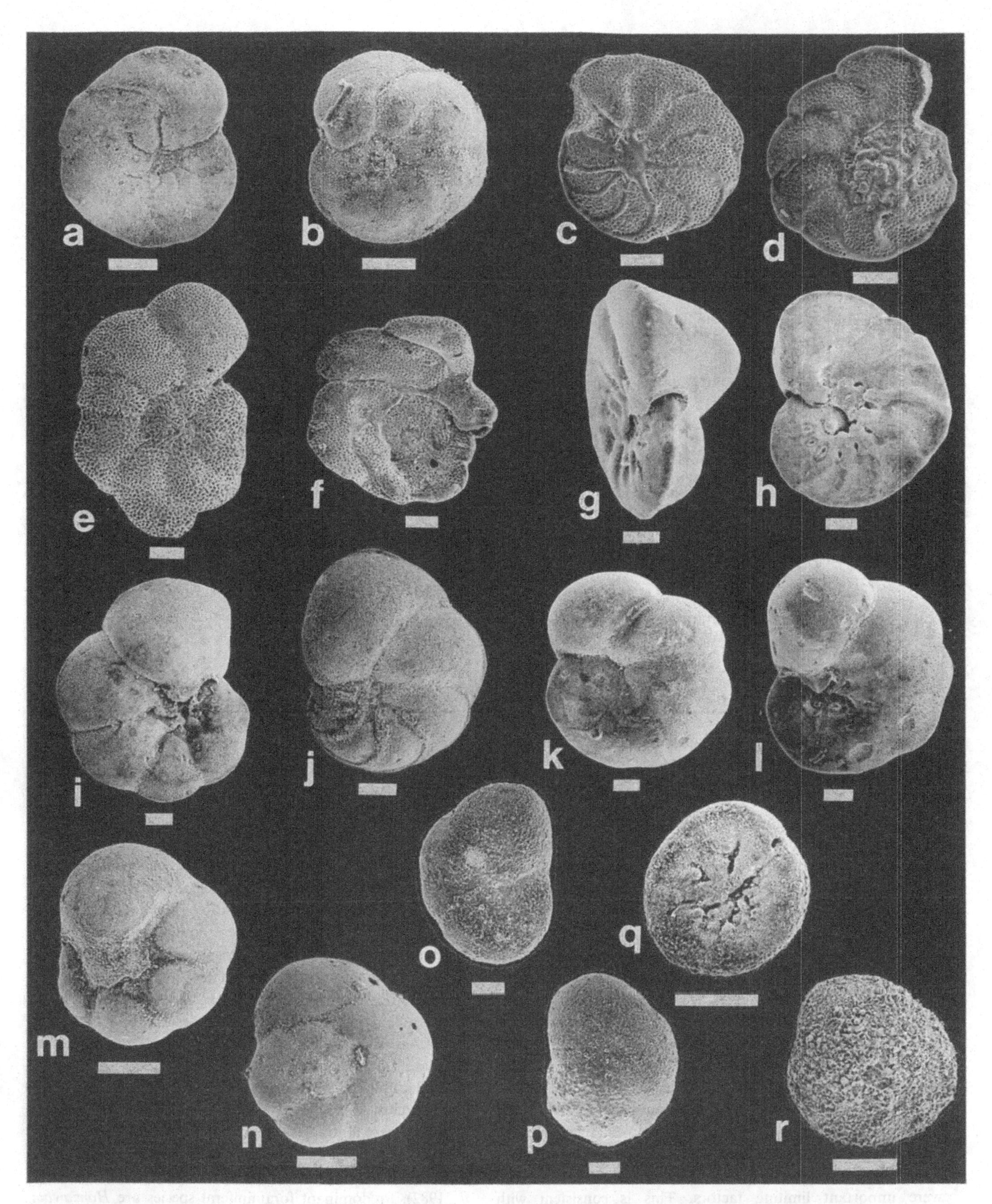

Fig. 32.2. (a–b) *Cibicides americanus* Cushman, (a) umbilical view, (b) spiral view; (c–d) *Cibicides floridanus* Cushman, (c) umbilical view, (d) spiral view; (e–f) *Cibicides lobatulus* Walker and Jacob, (e) umbilical view, (f) spiral view; (g–h) *Hanzawaia concentrica* Cushman, (g) oblique edge view, (h) spiral view; (i–j) *Valvulineria floridana* Cushman, (i) umbilical view, (j) spiral view; (k–l) *Valvulineria venezuelana* Hedberg, (k) spiral view, (l) umbilical view; (m–n) *Valvulineria laevigata* Phleger and Parker, (m) umbilical view, (n) spiral view; (o–p) *Valvulineria olssoni* Redmond, (o) umbilical view, (p) spiral view; (q–r) *Rosalina cavernata* Dorsey, (q) umbilical view, (r) spiral view. Each scale bar is 100 μm.

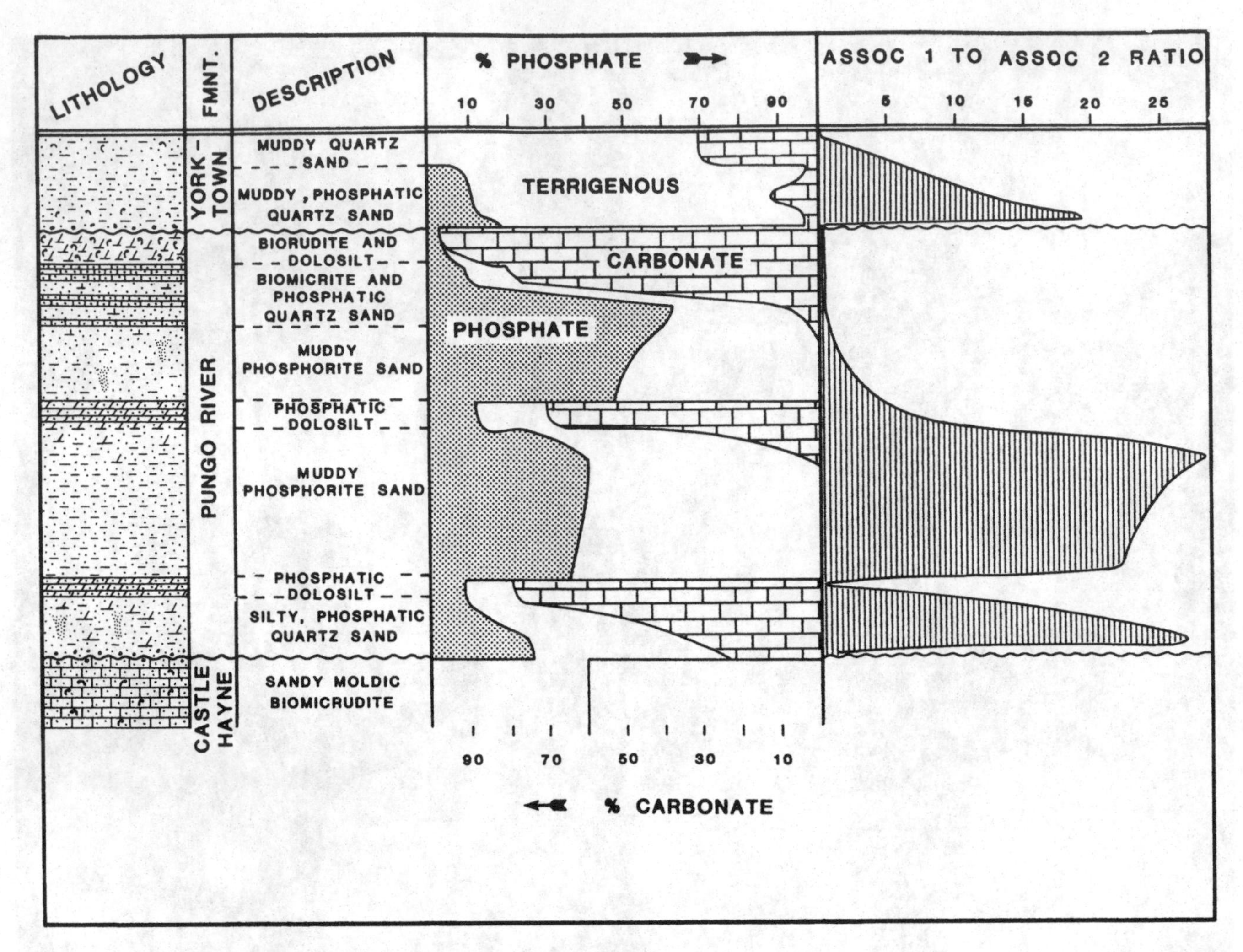

Fig. 32.3. Composite of Pungo River and lower Yorktown Formations in Aurora area: lithology, phosphate content and benthic foraminiferal faunal trends (modified from Riggs *et al.*, 1982).

elongata and *Bolivinopsis fairhavensis*, extinct forms which have no closely related modern analogues, but consistently occur in phosphatic sediments and with species of Association 1; and *Florilus pizarrensis* and *Nonionella miocenica*, both of which thrive under conditions of moderately elevated nutrient levels and associated reductions in dissolved oxygen.

Foraminifera are occasionally scarce within phosphatic portions of units A and B, but low diversity and high faunal predominance characterize even the richest, best preserved samples. Species diversity, calculated using the Shannon–Wiener Information Function [$H(S)$], averages only 1.4. The relative abundance of the most abundant species averages 51%. Planktonic specimens are rare, or often absent. These characters, along with the abundance of *Elphidium* suggest an inner-sublittoral environment with conditions unsuitable for many benthic species. High Association 1: Association 2 ratios (Fig. 32.3) suggest that nutrient enrichment and oxygen depletion were important limiting factors. This is consistent with upwelling of deeper water which was instrumental in phosphate formation. Scott W. Snyder *et al.* (1982) interpreted these phosphates as *in situ* deposits. Riggs (1984) discussed possible mechanisms for upwelling.

Carbonates which cap each of the two lower phosphorites are dolomitic and variably indurated. Foraminifera have often been destroyed during diagenesis, and those that do remain must be observed in thin section. Well preserved specimens from a few horizons are mostly forms belonging to Association 2. A largely moldic molluscan assemblage indicates warm water, normal marine conditions (J.G. Carter, pers. comm.).

The stratigraphically highest phosphorite (unit C of Riggs *et al.*, 1982) is also the richest in phosphate (Fig. 32.3). It is the only phosphate-rich deposit on the North Carolina Coastal Plain which is not predominated by species of Association 1. Although taxa of Association 1 are present in minor amounts, species of Association 2 are predominant.

Unlike units A and B, phosphate in unit C is composed almost exclusively of intraclastic grains; skeletal and pelletal grains are rare. Intraclasts represent dislodged fragments of orthochemical phosphate mud that were reworked and possibly transported before accumulating within unit C (Scarborough *et al.*, 1982). Predominant foraminiferal species are *Hanzawaia concentrica* (up to 67% of the fauna) and *Cibicides lobatulus* (up to 26% of the fauna). Species of lesser abundance which are consistently present include *Valvulineria floridana*, *V. olssoni*, *Florilus pizarrensis*, *Nonionella miocenica*, and *Ammonia beccarii*.

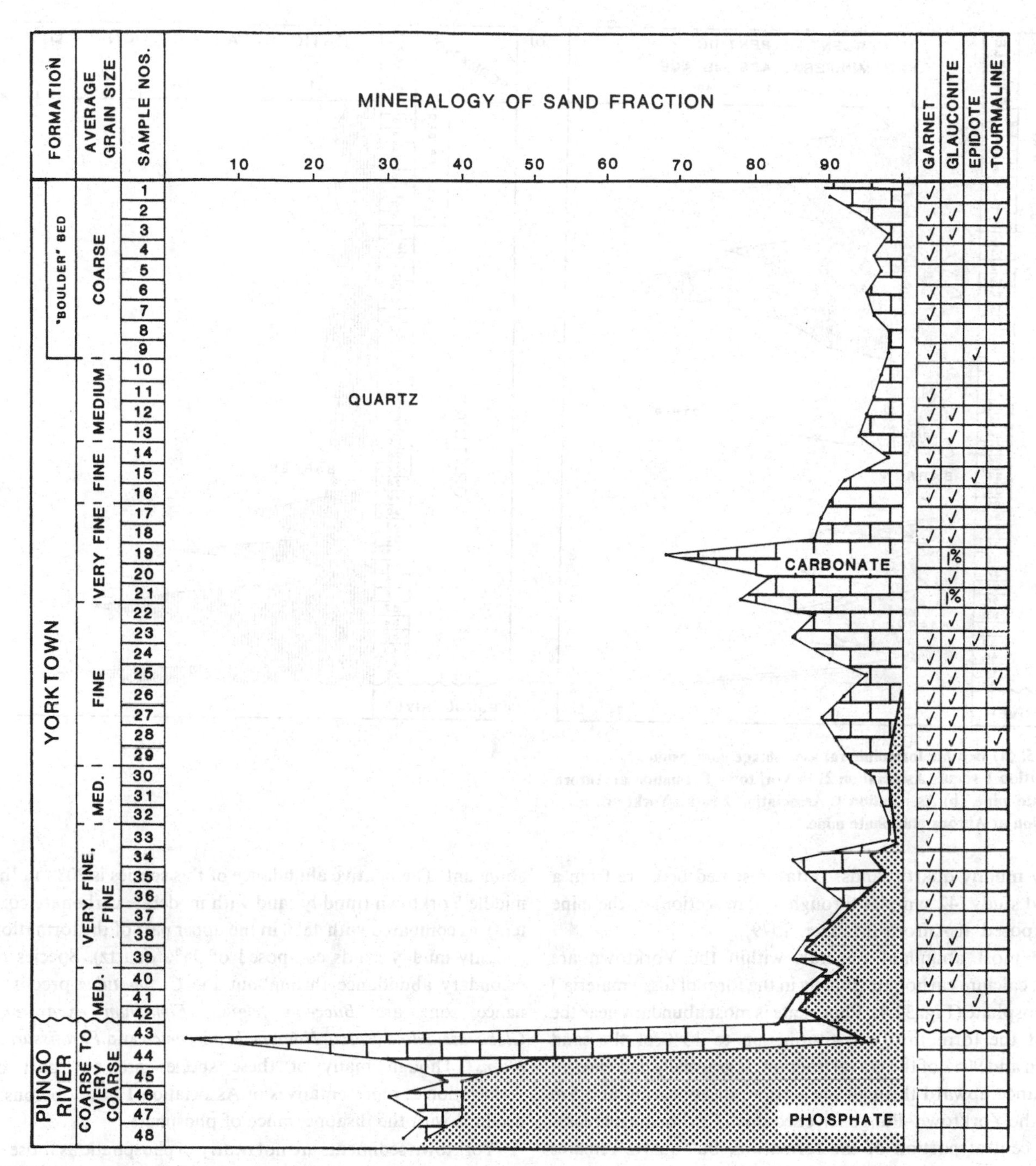

Fig. 32.4. Mineralogy of the sand fraction, Yorktown Formation at Aurora phosphate mine (modified from Mauger, 1979).

Foraminifera in unit C phosphorites are usually common and well preserved. Species diversity [mean value of $H(S)=2.3$] is higher than in the stratigraphically lower phosphates, and faunal predominance (mean = 33%) is lower. Planktonic specimens range from rare to common. These characteristics, along with known environmental tolerances of predominant species, suggest middle to outer sublittoral environments with well-oxygenated bottom conditions. The Associations 1:Association 2 ratio is markedly lower than in units A and B (Fig. 32.3). Scott W. Snyder *et al.* (1982) interpreted unit C phosphorites to have been transported from the site of original phosphate formation. Such reworking can, in the absence of terrigenous sediment input, enrich the phosphate content (Baturin, 1971). Constituent intraclasts merely accumulated as sedimentary particles on normally oxygenated bottoms which supported a type-2 association.

The carbonates which cap the unit C phosphorite, and those of the predominantly carbonate fourth cycle (unit D of Riggs *et al.*, 1982) consist of mostly calcite. Calcareous macro- and microfaunal elements are similar to those described for carbonates of the lower cycles.

Yorktown Formation (Pliocene) Yorktown strata unconformably overlie Pungo River deposits throughout the Aurora area. The Yorktown grades upward from phosphatic quartz sand into sandy blue-gray muds overlain by muddy–

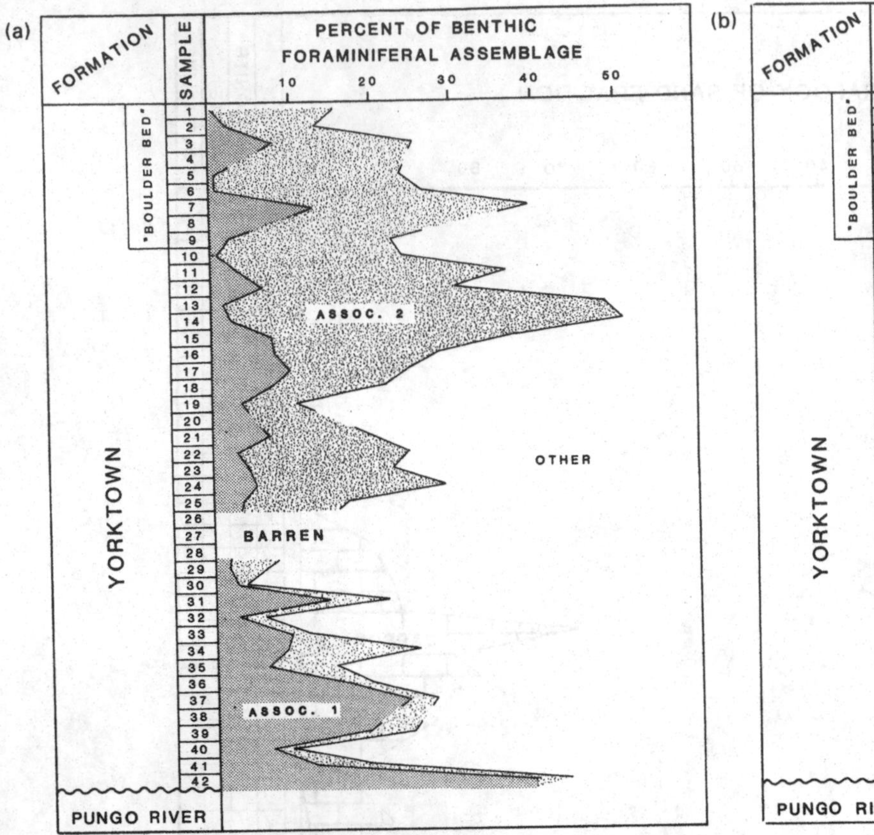

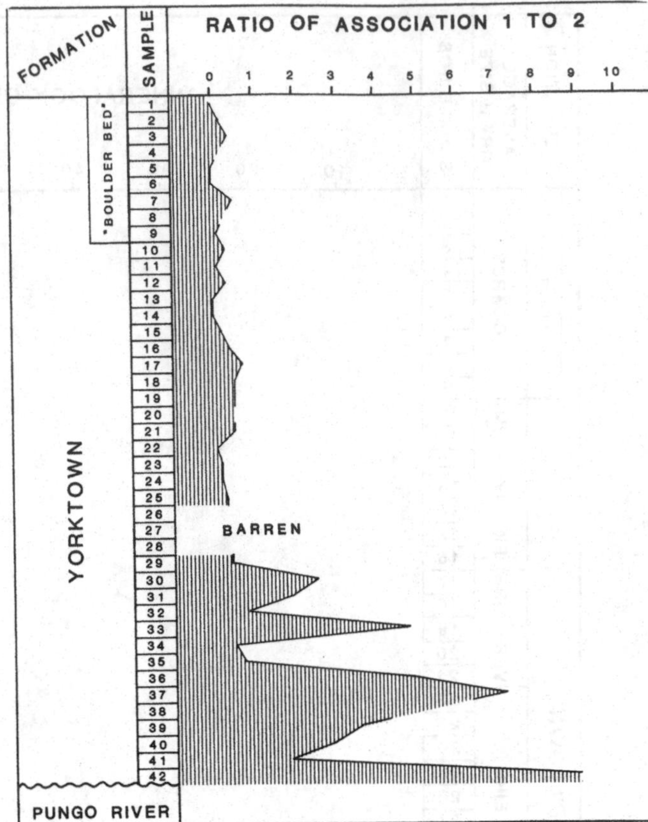

Fig. 32.5. (a) Benthic foraminiferal assemblage composition (Association 1 versus Association 2) in Yorktown Formation at Aurora phosphate mine. (b) Association 1: Association 2 ratio, Yorktown Formation at Aurora phosphate mine.

slightly muddy quartz sands. Data presented here are from a detailed study (42 samples through a 15 m section) of the mine face exposed at Aurora (Mauger, 1979).

The most abundant minerals within the Yorktown are quartz, calcium carbonate (mostly in the form of fossil material) and phosphate (Fig. 32.4). Phosphate is most abundant near the base of the formation, constituting up to 15% of the sand fraction and 90% of the gravel fraction. It gradually decreases in abundance upward through the section, disappearing at 5.5 m above the Yorktown–Pungo River contact. Carbonate increases in the central portion of the formation, but quartz remains predominant throughout (Fig. 32.4).

Of 157 benthic foraminiferal species which occur in the Yorktown, only 11 account for 10% or more of the assemblage in one or more samples. *Nonionella miocenica* is numerically dominant in the lower part of the formation and *Cibicides lobatulus* predominates in the upper part.

Nonionella miocenica predominates in the lowermost 4 m characterized by the presence of phosphate. Other abundant species are *Bulimina elongata*, *Buliminella elegantissima*, *Cassidulina laevigata*, *Elphidium excavatum*, *Florilus pizarrensis* and *Epistominella danvillensis*. Three of these species are members of Association 1. *E. excavatum*, although tolerant of a wide variety of conditions, could have tolerated conditions like those indicated by Association 1.

The overlying *Cibicides lobatulus* predominance zone is particularly well developed where clean quartz sands are most abundant. The relative abundance of this species is 20% in the middle Yorktown (muddy sand with moderate carbonate content), as compared with 48% in the upper part of the formation (slightly muddy sands composed of 95% quartz). Species of secondary abundance throughout the *C. lobatulus* predominance zone are *Buccella frigida*, *Elphidium excavatum*, *Globocassidulina crassa*, *Nonionella miocenica* and *Parafissurina bidens*. Though many of these species are not part of Association 2, representatives of Association 1 have obviously declined with the disappearance of phosphate.

Yorktown sediments are not nearly so phosphatic as those of the Pungo River Formation (Fig. 32.4). Hence, Associations 1 and 2, largely defined by trends within the Pungo River, compose a lesser percentage of the total benthic assemblage. Even so, percentages of these associations change systematically with changing phosphate content (Fig. 32.5*a*). Association 1 clearly thrived during deposition of the lower Yorktown, while Association 2 expanded later in the depositional sequence. Association 1:Association 2 ratios demonstrate the magnitude of change (Fig. 32.5*b*). Understandably, less dramatic changes in phosphate content are accompanied by less intense foraminiferal assemblage variations.

Phosphate in the lower Yorktown represents, at least in part, material reworked from the underlying Pungo River Formation. Certainly, phosphate gravel at the base of the Yorktown is derived from phosphatic pavements in the upper Pungo River. However, benthic foraminiferal trends suggest that conditions

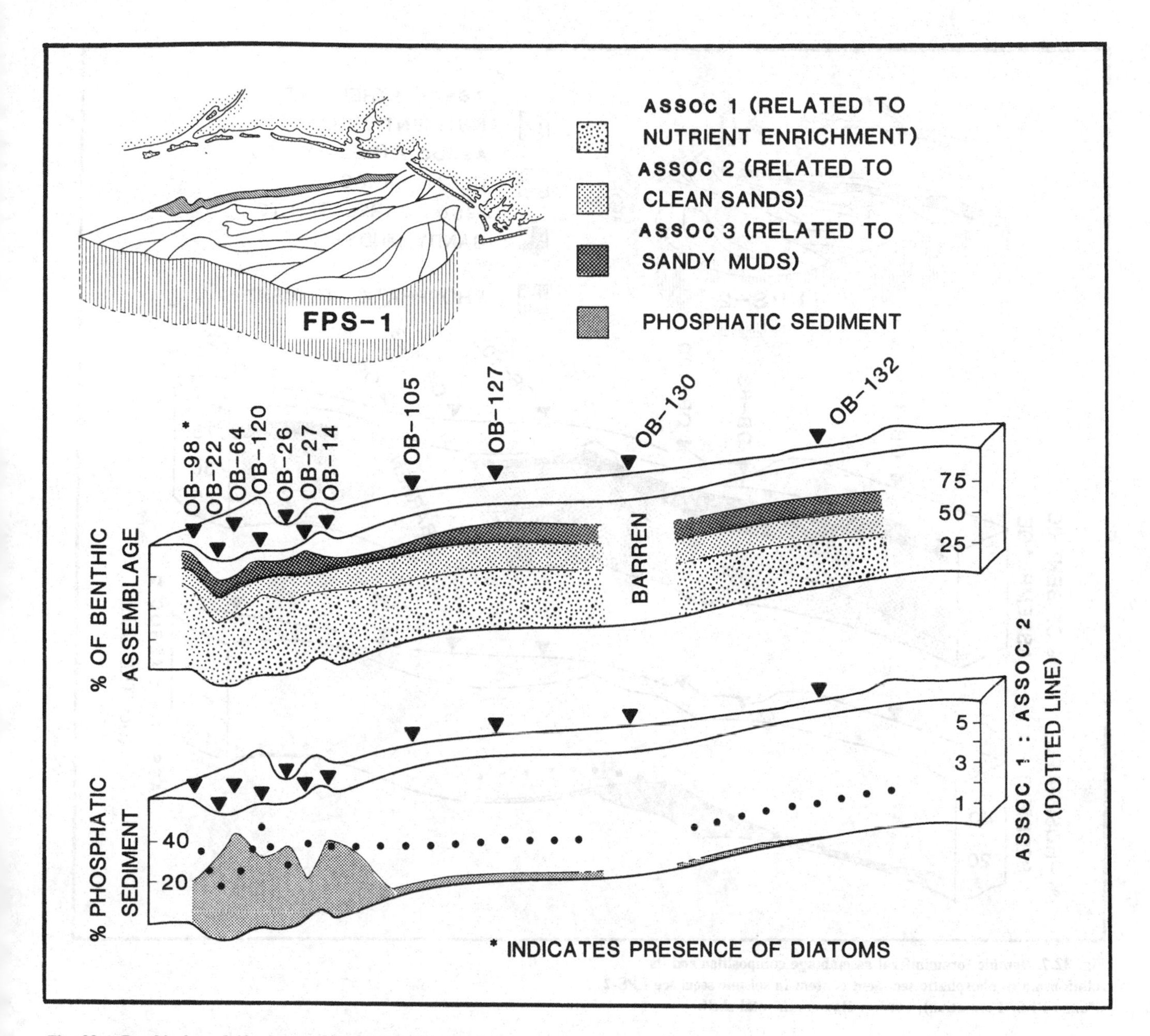

Fig. 32.6. Benthic foraminiferal assemblage composition and its relationship to phosphatic sediment content in seismic sequence FPS-1 (Pungo River Formation), Onslow Bay continental shelf.

of modest nutrient enrichment accompanied deposition of lower Yorktown sediments. Perhaps some of the phosphate resulted from a subdued or short-lived pulse of primary phosphate formation.

Onslow Bay

Pungo River Formation (Miocene) The vast majority of vibracores (maximum length 9 m) taken in Onslow Bay penetrate Pungo River sediments. So few samples are available from Pliocene material that the discussion of Onslow Bay will address faunal–sediment relationships only within the Miocene. The discussion that follows is based upon analyses of 155 samples from 55 vibracores. Though distributed throughout the area of Miocene outcrop–shallow subcrop, vibracores are most densely concentrated in southern and northeastern Onslow Bay. The volume of data from these regions precludes complete graphic presentation on some of the figures that follow. Representative cores were selected to graphically summarize data trends recognized in these most densely sampled regions.

Seismic sequences within the Pungo River Formation of Onslow Bay are discussed in stratigraphic order, from oldest to youngest. Not all sequences have been cored. Only those for which there are data on foraminiferal assemblages are discussed below. Figures are provided only for those sequences with sufficient vibracore coverage to reveal regional trends.

In southern Onslow Bay, seismic *sequence FPS-1* is a muddy, slightly quartzitic, foraminiferal phosphorite (50–75%) sand which grades upward into a muddy, foraminiferal, phosphatic (15–20%) quartz sand. Mean phosphatic sediment content for the entire sequence ranges from 25 to 40% (Fig. 32.6). *Bolivina paula* is the predominant species in all cores (relative abundance ranging from 27 to 45%). In descending order of abundance,

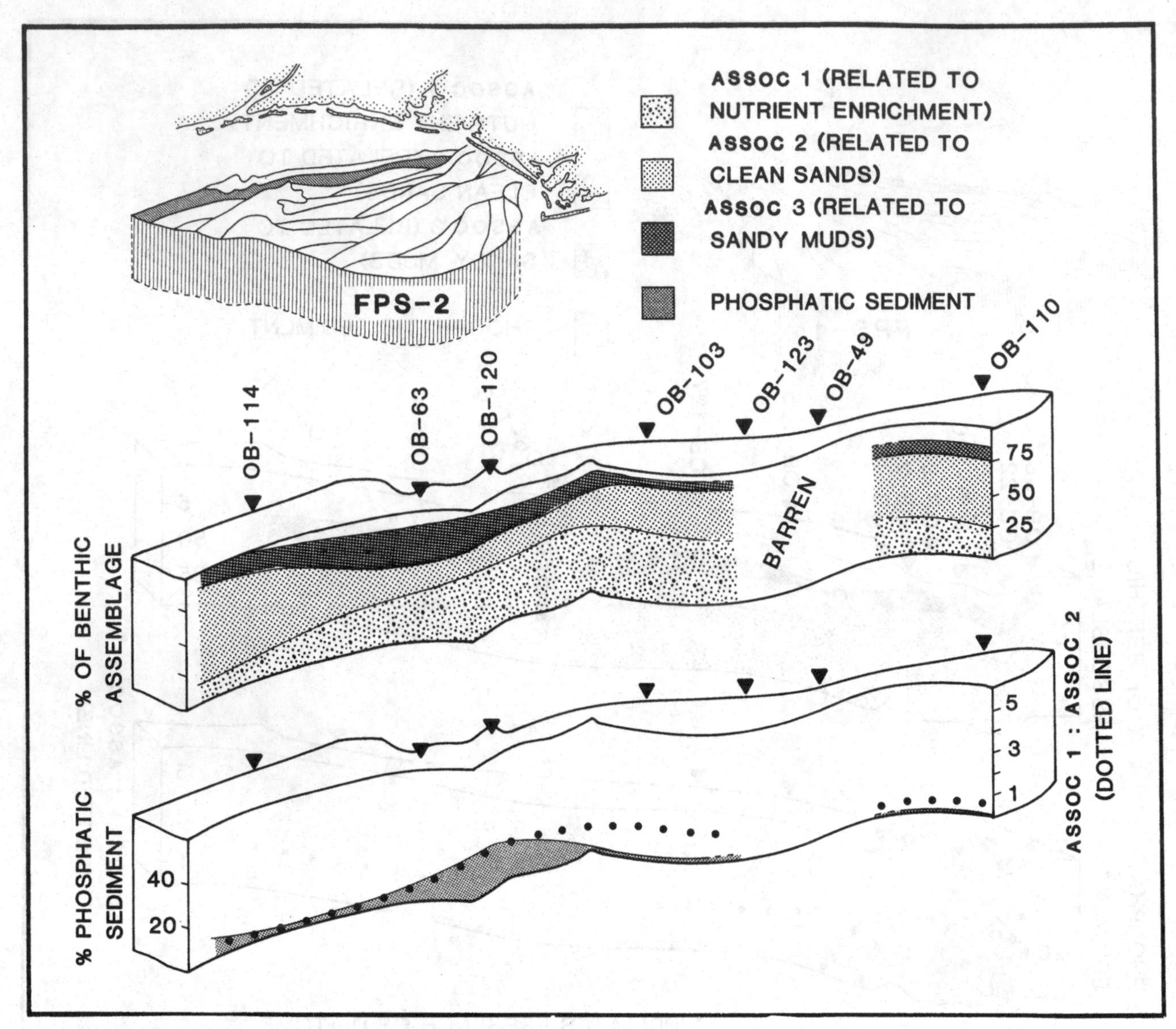

Fig. 32.7. Benthic foraminiferal assemblage composition and its relationship to phosphatic sediment content in seismic sequence FPS-2 (Pungo River Formation), Onslow Bay continental shelf.

subordinate species are *Hanzawaia concentrica*, *Bulimina elongata*, *Florilus pizarrensis*, *Buliminella elgantissima* and *Uvigerina auberiana*. Taxa of Association 1 are clearly predominant, and Association 1:2 ratios range from 3 to 5.

Northward within FPS-1, there is an abrupt decrease in phosphate content between vibracores 14 and 105 (Fig. 32.6). Phosphatic sediment content then decreases gradually from about 3% until it disappears at core 132. The sediment from 105 northward to 132 is a muddy, variably fossiliferous quartz sand. In contrast to the abrupt decrease in phosphate, the abundance of taxa belonging to Association 1 decreases gradually. The predominant species is either *Bolivina paula* or *Buliminella elegantissima* (15–38% of the benthic assemblage). Secondary species are *Florilus pizarrensis*, *Valvulineria floridana*, *Cibicides americanus* and *Uvigerina auberiana*. Taxa of Association 1 remain predominant, but Association 1:2 ratios decline northward to about 2 (Fig. 32.6). Perhaps the taxa of Association 1 decline gradually because the mud content and its associated organic enrichment remain high (Martens, unpublished data).

Throughout southern and central Onslow Bay, seismic *sequence FPS-2* is a moderately phosphatic (5–20%), foraminiferal, slightly quartzitic mud. In core 114, where phosphate content is low, *Hanzawaia concentrica* is predominant (55%) and the Association 1:2 ratio is <1 (Fig. 32.7). Where phosphate content increases (cores 63 and 120), *Bolivina paula* becomes predominant (24–27%) and other species of Association 1 increase in abundance. Association 1:2 ratios increase to about 2. Northward, as phosphate content diminishes, *Hanzawaia concentrica* and *Valvulineria floridana* become sufficiently abundant to reduce the ratio to <1. The sediment is a rather uniform muddy, fine quartz sand. Again, faunal change is more gradual than reduction in phosphate content (Fig. 32.7).

The outcrop–shallow subcrop pattern of seismic *sequence FPS-3* is discontinuous. The southern sector consists of fossiliferous (foraminifera and diatoms), slightly phosphatic (5%), quartzitic, dolosilt-rich mud. *Bolivina paula* (19% of assemblage) is the most abundant species, followed by *Hanzawaia concentrica*, *Florilus pizarrensis*, *Bulimina elongata* and

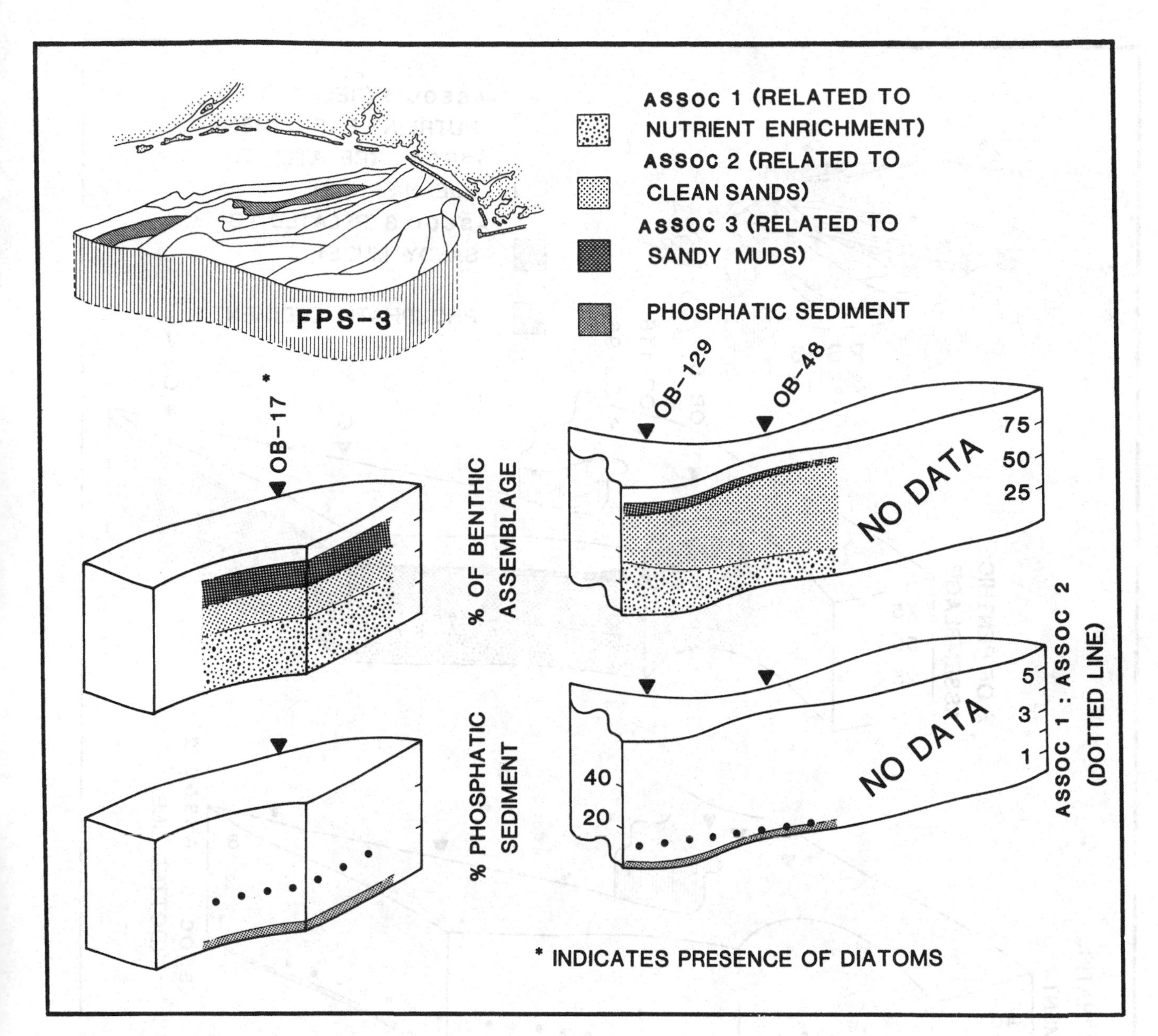

Fig. 32.8. Benthic foraminiferal assemblage composition and its relationship to phosphatic sediment content in seismic sequence FPS-3 (Pungo River Formation), Onslow Bay continental shelf.

Valvulineria floridana. In view of low phosphate content, the Association 1:2 ratio of about 2 is higher than might be expected (Fig. 32.8). High organic content of this muddy sediment may explain this relationship (Martens, unpublished data).

Sequence FPS-3 in the northern sector is a muddy quartz sand with minor phosphate (<3%) and dolosilt content. *Bolivina paula* (27% of assemblage) remains predominant at core 129, but taxa of Association 2 increase sufficiently to reduce the Association 1:2 ratio to about 1. At core 48, three species (*Cibicides americanus*, *Hanzawaia concentrica* and *Valvulineria floridana*) each compose 18% of the benthic assemblage. The increased abundance of these taxa reduces the Association 1:2 ratio to <1 (Fig. 32.8).

The irregular pattern of seismic *sequence FPF-6* sharply truncates older sequences. No cores are available from its southernmost portions. Its central part consists of slightly muddy, foraminiferal quartz sand with minor phosphate (<2%). *Hanzawaia concentrica* (28–42% of assemblage) is the predominant species in all cores. Secondary species are *Florilus pizzarensis*, *Cibicides floridanus*, *Valvulineria floridana* and *Bolivina paula*. Association 1:2 ratios are consistently low (<1) (Fig. 32.9).

To the north, FPS-6 becomes a fossiliferous (barnacle-rich), muddy, dolosilt-rich quartz sand with minor phosphate (<3%). Though *Bolivina paula* is predominant in core 50 (27% of assemblage) and shares predominance with *Valvulineria floridana* in core 45 (both 25% of assemblage), secondary species are largely those of Association 2 – *Cibicides americanus* and *Hanzawaia concentrica*. Consequently, the Association 1:2 ratio diminishes to <1 (Fig. 32.9).

Seismic *sequence OBS-1* occurs in central and northern Onslow Bay. In its southern portion, OBS-1 is a variably muddy

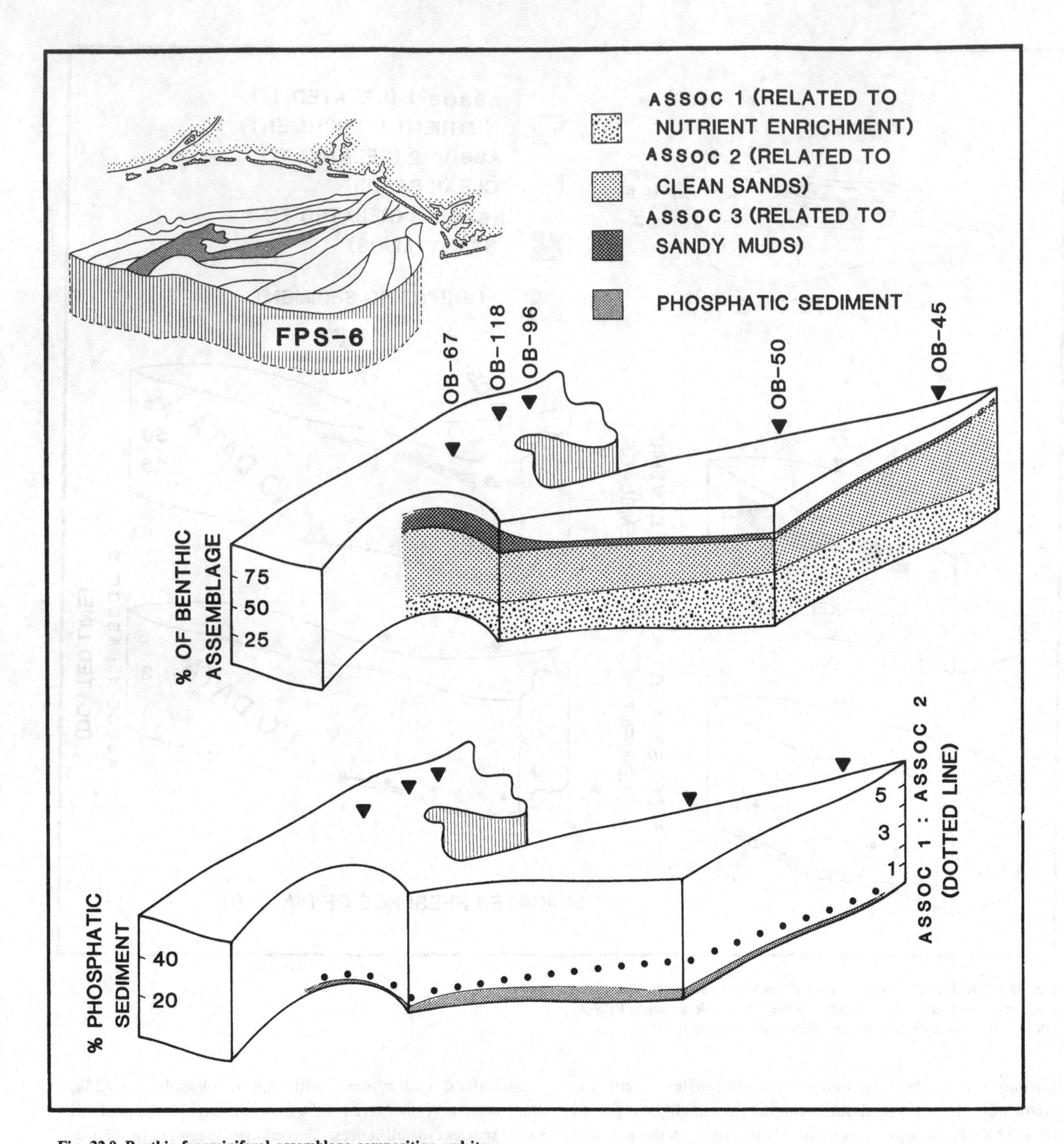

Fig. 32.9. Benthic foraminiferal assemblage composition and its relationship to phosphatic sediment content in seismic sequence FPS-6 (Pungo River Formation), Onslow Bay continental shelf.

quartz sand with minor dolosilt and phosphate (<2%) and no calcareous microfauna. Farther north (cores 34 and 35), this sequence is composed of fossiliferous (foraminifera and diatoms), quartzitic, dolosilt-rich muds interbedded with foraminiferal quartz sands containing minor phosphate (<2%).

In core 35, where *Cibicides lobatulus* accounts for 27% of the benthic assemblage, the Association 1:2 ratio is <1 (Fig. 32.10). *Valvulineria floridana*, *Bolivina paula*, *Buliminella elegantissima* and *Rosalina cavernata* are of secondary abundance. In core 34, marked by an increased abundance of diatoms, *Bolivina paula* and *Buliminella elegantissima* are most abundant (both 20% of benthic assemblage). Secondary species are *Cibicides lobatulus* and *Valvulineria floridana*. Though phosphate content is low, the Association 1:2 ratio approaches 3 (Fig. 32.10).

The only cores that penetrate *sequence OBS-2* are located in its northernmost portion. Sediments are composed of variably muddy quartz sand with minor dolosilt and phosphate (<2%). *Hanzawaia concentrica* (25%) and *Bolivina paula* (17%) predominate. Associated species, many nearly as abundant as the two mentioned above, include *Cibicides americanus*, *C.*

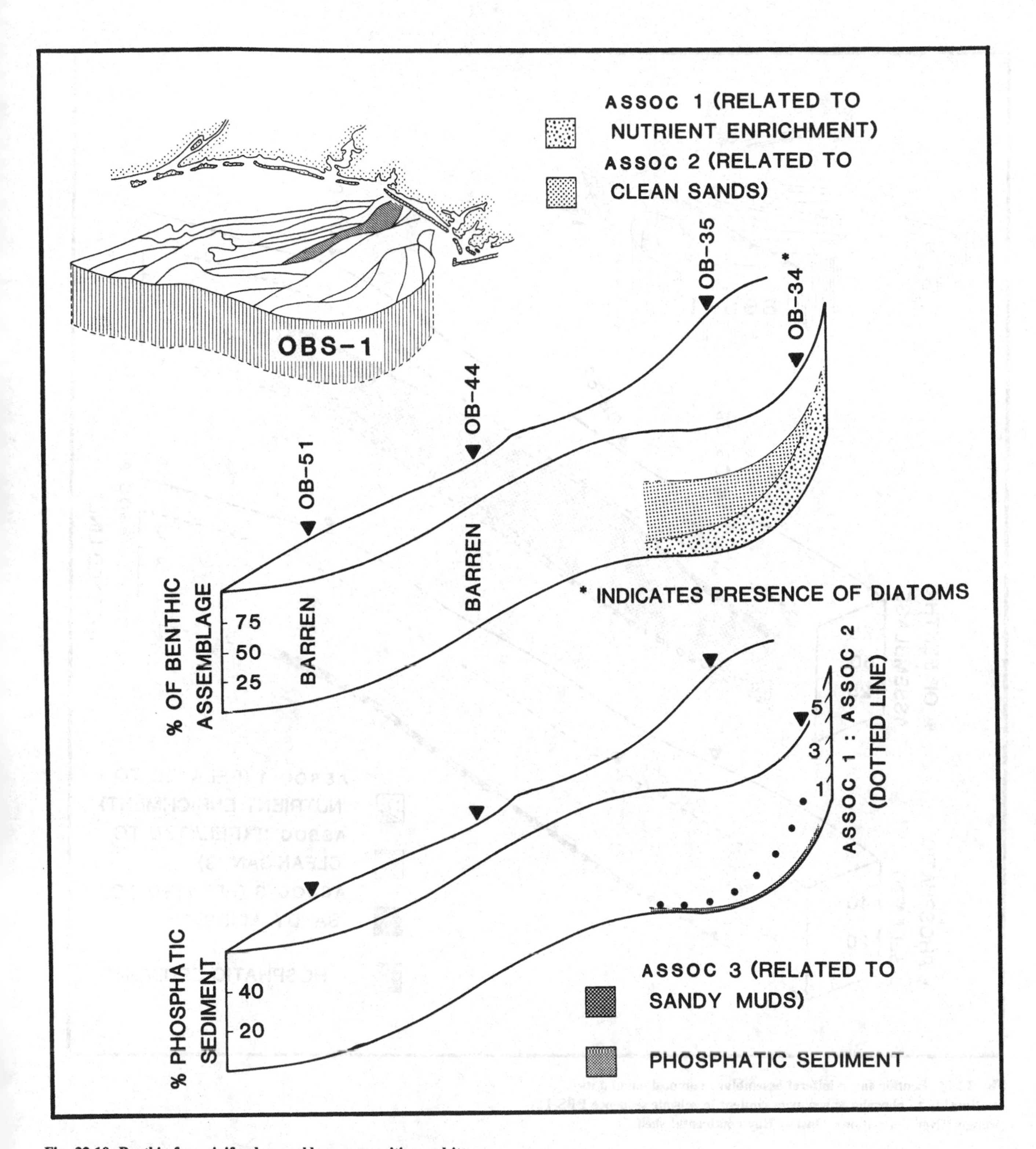

Fig. 32.10. Benthic foraminiferal assemblage composition and its relationship to phosphatic sediment content in seismic sequence OBS-1 (Pungo River Formation), Onslow Bay continental shelf.

lobatulus, *Bolivina calvertensis* and *Valvulineria floridana*. The Association 1:2 ratio is consistently <1.

Information about *sequence OBS-3* is based on only one core, located in the southern segment of its two-part outcrop–shallow subcrop pattern. Muddy, quartzitic calcareous sands in this core contain only trace amounts of phosphate. The foraminiferal assemblage is predominated by *Cibicides americanus* (48%), with *Valvulineria floridana*, *Cibicides lobatulus* and *Bolivina paula* following in that order. The Association 1:2 ratio is nearly zero.

Sequence OBS-4 also has a two-part outcrop–shallow subcrop pattern. The only data available come from two cores in its small northern segment. Sediments are muddy, shelly, slightly quartzitic, dolosilt-rich calcareous sands which contain only a

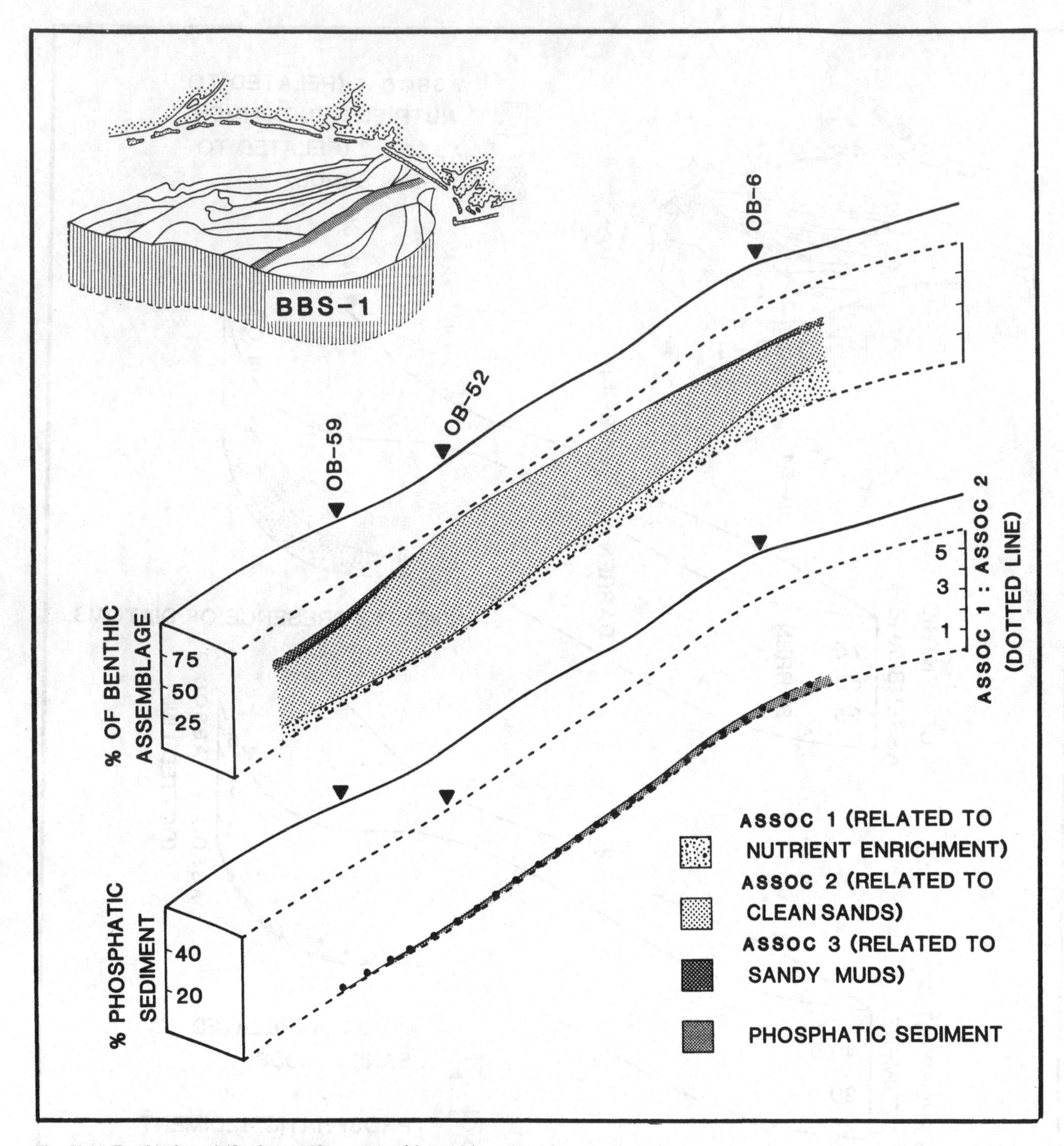

Fig. 32.11. Benthic foraminiferal assemblage composition and its relationship to phosphatic sediment content in seismic sequence BBS-1 (Pungo River Formation), Onslow Bay continental shelf.

trace of phosphate. The benthic foraminiferal assemblage is nearly identical in both cores. *Bolivinia paula* (24%) is most abundant, but species of Association 2 compose nearly all of the remaining assemblage. *Cibicides lobatulus*, *Rosalina cavernata* and *Cibicides americanus* are most abundant among the secondary species. Association 1:2 ratios are <1.

Sequence BBS-1 (Fig. 32.11) is composed of slightly muddy quartz sand with minor phosphate. In the north (core 6), *Cibicides floridanus* (23% of benthic assemblage) predominates, while *Valvulineria floridana* (12%) is most abundant in the south (core 59). Species of secondary abundance are *Hanzawaia concentrica*, *Cibicides americanus* and *Bolivina paula*. Association 1:2 ratios remain well below 1. The assemblage in core 52 is similar except that *Cibicides floridanus* accounts for 78% of the assemblage and taxa of Association 1 virtually disappear. This assemblage occurs in nearly pure quartz sands, and the Association 1:2 ratio is nearly zero (Fig. 32.11).

Sequence BBS-2 contains dramatic variations in both phosphate content and faunal composition (Fig. 32.12). The northernmost core (72) contains muddy, very fine quartz sand with

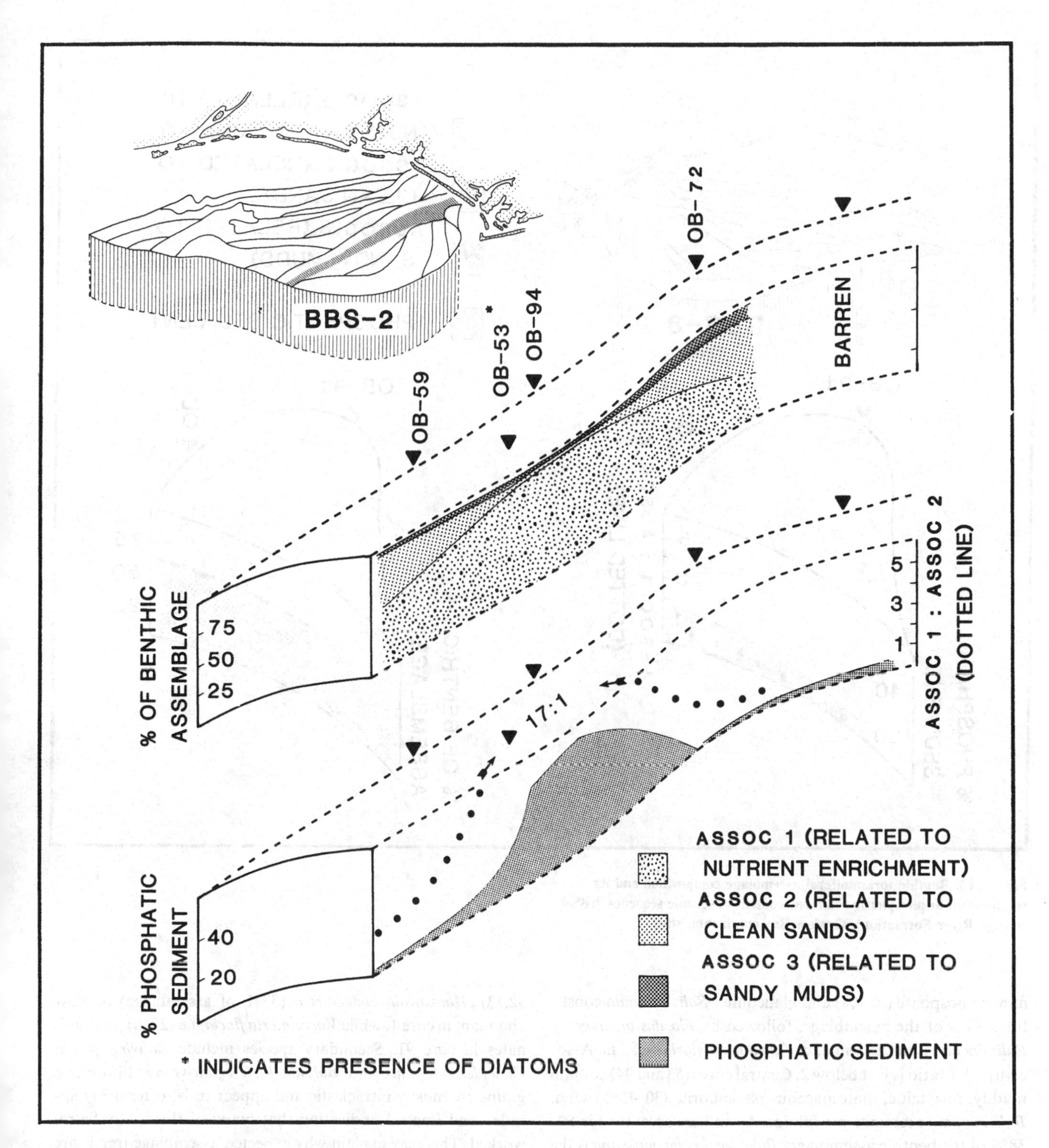

Fig. 32.12. Benthic foraminiferal assemblage composition and its relationship to phosphatic sediment content in seismic sequence BBS-2 (Pungo River Formation), Onslow Bay continental shelf.

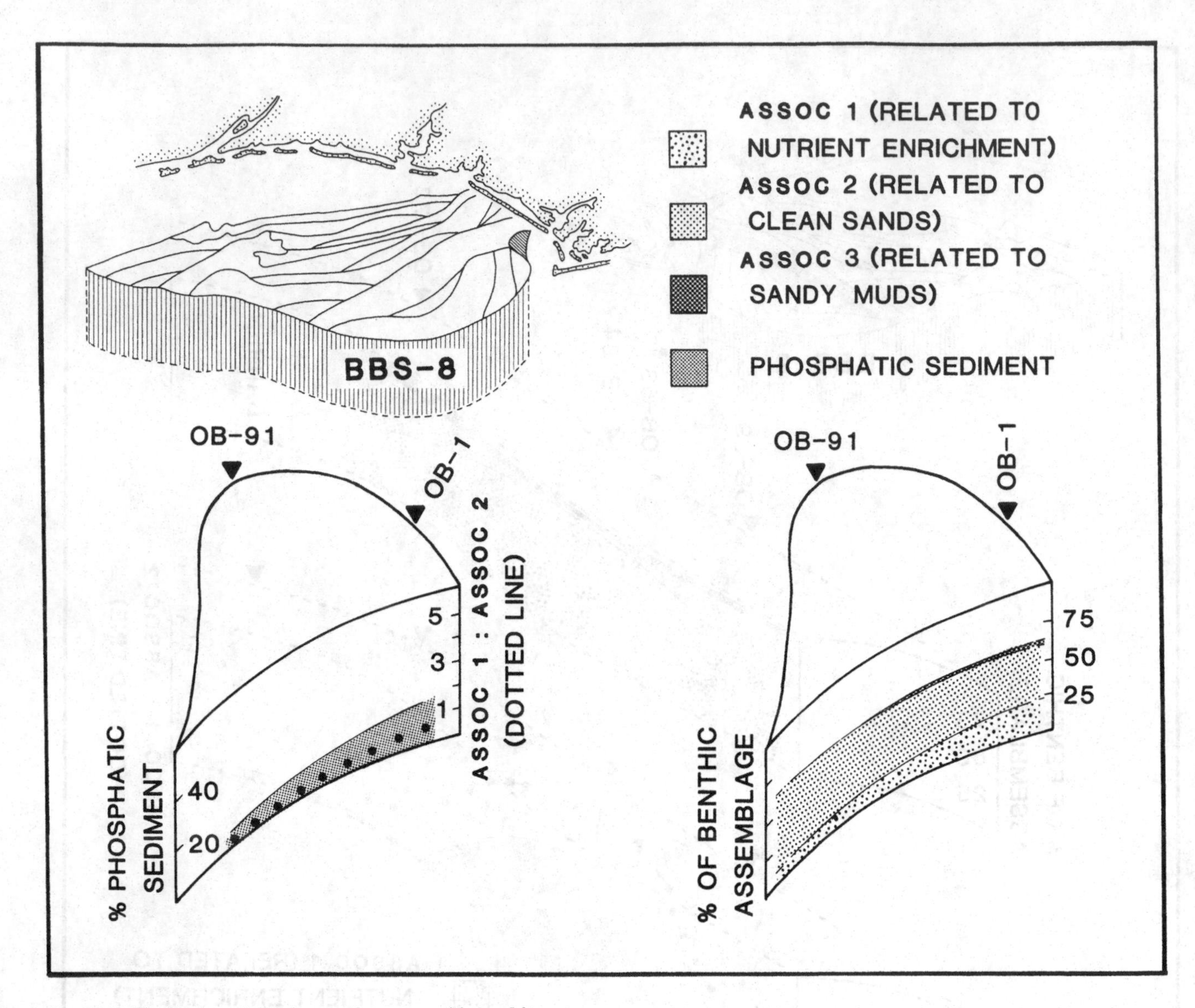

Fig. 32.13. Benthic foraminiferal assemblage composition and its relationship to phosphatic sediment content in seismic sequence BBS-8 (Pungo River Formation), Onslow Bay continental shelf.

minor phosphate (<4%) and glauconite. *Bolivina paula* constitutes 33% of the assemblage, followed by *Florilus pizarrensis*, *Buliminella elegantissima* and *Valvulineria floridana*. The Association 1:2 ratio is just below 2. Central cores (53 and 94) contain muddy, quartzitic, diatomaceous phosphorite (30–45%) sand. *Bolivina paula* remains predominant, but here accounts for 50–58% of the benthic assemblage. *Buliminella elegantissima* is the next most abundant species, and the Association 1:2 ratio attains a maximum value of 17 (the highest encountered in Onslow Bay and comparable to values in the lower Pungo River phosphorites at Aurora). The southernmost core (59) contains foraminiferal, quartzitic, dolosilt-rich mud with minor phosphate (<2%). *Bolivina paula*, though still the most abundant species, composes only 21% of the assemblage, and the Association 1:2 ratio declines to about 2 (Fig. 32.12).

Sediments of seismic *sequence BBS-8* consist of slightly muddy, phosphatic (10–20%) quartz sands. Despite rather high phosphate content, Association 1:2 ratios remain below 1 (Fig. 32.13). *Hanzawaia concentrica* (33% of assemblage) is most abundant in core 1, while *Valvulineria floridana* (27%) predominates in core 91. Secondary species include *Bolivina paula*, *Cibicides lobatulus* and *Buliminella elegantissima*. Phosphate grains are mostly intraclastic and appear to be extensively abraded and leached, indicating that many of them may be reworked. This may explain why expected assemblage trends are not observed in this sequence.

Summary and conclusions

Neogene phosphorites of North Carolina contain an unusually rich calcareous microfauna composed largely of foraminifera. As phosphate content within Pungo River and Yorktown sediments varied, so did the abundance of benthic foraminifera which today thrive in nutrient-enriched, oxygen-depleted bottom conditions. Where phosphate is abundant, species of *Bolivina*, *Bulimina*, *Buliminella*, *Uvigerina*, *Nonionella* and *Florilus* predominate benthic assemblages. These taxa, here designated as Association 1, thrive in modern upwelling zones and sewage outfalls, and they are associated with ancient conditions of oxygen depletion (sapropels). In non-phosphatic horizons, species of *Hanzawaia valvulineria*, *Cibicides* and *Rosalina* (Association 2 of this paper) are more abundant. Extant representatives of these taxa thrive in clean, high-oxygenated bottom conditions.

Changes in Neogene benthic foraminiferal faunal composition can be conveniently expressed as the Association 1:2 ratio, which generally varies directly with phosphatic sediment content. However, the relationship between this ratio and phosphate content is not entirely uniform or predictable. For example, high Association 1:2 ratios sometimes occur where Pungo River sediments have little phosphate (Onslow Bay cores 98 (FPS-1), 17 (FPS-3) and 34 (OBS-1)). In these particular cores foraminifera occur with abundant diatoms (Figs. 32.6, 32.8 and 32.10). While the presence of diatoms does not in itself document nutrient enrichment, the species composition of floras in these sequences indicates nearby marine upwelling (Powers, 1988). Some phosphate-poor sediments having a high Association 1:2 ratio lack diatoms (e.g. cores 127 and 132 in FPS-1). These cores, as well as most which contain diatoms, have a high mud content. Perhaps organic enrichment associated with these muds supports the predominantly type-1 association (Martens, unpublished data). Hence, the occasional abundance of Association 1 in the absence of phosphorites does not refute any relationship between them. It merely indicates that foraminifera may respond to different types of nutrient enrichment, some of which do not result in phosphate formation.

At two locations (unit C at Aurora and to a lesser degree sequence BBS-8 in Onslow Bay), phosphate-rich horizons in the Pungo River Formation contain mostly taxa belonging to Association 2. In both cases, phosphate appears to have been reworked and transported to its present site of accumulation. Hence, nutrient-rich conditions associated with its formation did not affect foraminiferal assemblages at these secondary sites of deposition.

Lack of relationship between faunal composition and phosphate content is the exception, not the rule. High Association 1:2 ratios most often coincide with phosphate-rich horizons (Pungo River units A and B at Aurora, lower Yorktown sediments at Aurora, Pungo River sequences FPS-1 and BBS-2 in Onslow Bay). Low ratios most often coincide with scarcity or absence of phosphate (Pungo River carbonates at Aurora; upper Yorktown sediments at Aurora; Pungo River sequences FPS-2, FPS-3 [part], FPS-6, OBS-2 through OBS-4, and BBS-1 in Onslow Bay).

The correlation of faunal changes with cyclic sedimentation does not prove causation. Conversely, repeated occurrences of the same pattern are most likely not fortuitous. Foraminiferal assemblage and sediment composition are probably responding to a common set of paleoceanographic conditions. In assessing the future of paleontology, Thomson (1985) stressed the importance of studying interconnections between organisms and environment during cyclical changes. Benthic foraminiferal – phosphatic sediment interconnections may aid greatly in understanding cyclical changes which controlled the formation of marine sedimentary phosphorites. Certainly, other Neogene sequences characterized by upwelling and phosphate formation should be investigated to determine whether similar relationships exist.

Acknowledgments

Research summarized here was supported by grants from the National Science Foundation (OCE-7908949, OCE-8110907, OCE-8118164 and OCE-8342777, S.R. Riggs and A.C. Hine principal investigators), the North Carolina Sea Grant College (NA83AA-D-00012/R/AO-3 and NA85AA-D-SGO22/R/AO-4, Scott W. Snyder and S.R. Riggs principal investigators) and the North Carolina Geological Survey Section. I thank C.W. Poag, V.J. Waters, S.W. Wise, R.P. Sheldon and one anonymous reviewer for their critical reading of the manuscript.

References

Bandy, O.L., Ingle, J.C. & Resig, J.M. (1964a). Foraminiferal trends, Laguna Beach outfall area, California. *Limnology and Oceanography*, **9(1)**, 112–23.

Bandy, O.L., Ingle, J.C. & Resig, J.M. (1964b). Foraminifera, Los Angeles County outfall area, California. *Limnology and Oceanography*, **9(1)**, 124–37.

Bandy, O.L., Ingle, J.C. & Resig, J.M. (1965). Foraminiferal trends, Hyperion outfall, California. *Limnology and Oceanography*, **10(3)**, 314–32.

Bates, J.M. & Spencer, R.S. (1979). Modification of foraminiferal trends by the Chesapeake–Elizabeth sewage outfall, Virginia Beach, Virginia. *Journal of Foraminiferal Research*, **9(2)**, 125–40.

Baturin, G.N. (1971). Stages of phosphorite formation on the ocean floor. *Natural and Physical Science*, **232**, 61–2.

Boltovskoy, E. & Wright, R.H. (1976). *Recent Foraminifera*. Junk, The Hague, Netherlands.

Cita, M.B. & Podenzani, M. (1979). Destructive effects of oxygen starvation and ash falls on benthic life: a pilot study. *Quaternary Research*, **13**, 230–41.

Clark, W.B. & Miller, B.L. (1906). A brief summary of the geology of the Virginia coastal plain. *Virginia Department of Agriculture and Immigration, Geological Series Bulletin*, **2**, 11–24.

Corliss, B.H., Martinson, D.G. & Keffer, T. (1986). Late Quaternary deep-ocean circulation. *Geological Society of America Bulletin*, **97**, 1106–21.

Culver, S.J. & Buzas, M.A. (1980). Distribution of Recent benthic foraminifera off the North American Atlantic coast. *Smithsonian Contributions to Marine Sciences*, **6**, 512pp.

Culver, S.J. & Buzas, M.A. (1981). Distribution of Recent benthic foraminifera in the Gulf of Mexico. *Smithsonian Contributions to Marine Sciences*, **8**, 898pp.

Feyling-Hanssen, R.W. (1972). The foraminifer *Elphidium excavatum* (Terquem) and its variant forms. *Micropaleontology*, **18(3)**, 337–54.

Froelich, P.N., Arthur, M., Burnett, W.C., Deakin, M., Hensley, V., Jahnke, R., Kaul, R., Kim, K.H., Roe, K., Soutar, A. & Vathakanan, C. (1988). Early diagenesis of organic matter in Peru continental margin sediments: phosphorite precipitation. *Marine Geology*, **80**, 309–46.

Hazel, J.E. (1983). Age and correlation of the Yorktown (Pliocene) and Croatan (Pliocene and Pleistocene) Formations at the Lee Creek mine. *Smithsonian Contributions to Paleobiology*, **53**, 81–199.

Ingle, J.C., Keller, G. & Kolpack, R.L. (1980). Benthic foraminiferal biofacies, sediments, and water masses of the southern Peru–Chile Trench area, southeastern Pacific Ocean. *Micropaleontology*, **26**, 113–50.

Katrosh, M.R. & Snyder, Scott W. (1982). Diagnostic foraminifera and paleoecology of the Pungo River Formation, central coastal plain of North Carolina. *Southeastern Geology*, **23(4)**, 217–31.

Katz, M.E. & Thunell, R.C. (1984). Benthic foraminiferal biofacies associated with middle Miocene to early Pliocene oxygen-deficient conditions in the eastern Mediterranean. *Journal of Foraminiferal Research*, **14(3)**, 187–202.

LeFurgey, A. & St. Jean, J. (1976). Foraminifera in brackish-water ponds designed for waste control and aquaculture studies in North Carolina. *Journal of Foraminiferal Research*, **6(4)**, 274–94.

Manheim, F.T. & Gulbrandsen, R.A. (1979). Marine phosphorites. In *Marine Minerals*, pp. 151–73. Mineralogical Society of America Short Course Notes, **6**.

Mauger, L.L. (1979). 'Benthonic foraminiferal paleoecology of the Yorktown Formation at Lee Creek mine, Beaufort County, North Carolina'. Unpublished M.Sc. thesis, East Carolina University, North Carolina.

Miller, A.L., Scott, D.B. & Medioli, F.S. (1982). *Elphidium excavatum* (Terquem): ecophenotypic versus subspecific variation. *Journal of Foraminiferal Research*, **12**, 116–44.

Miller, K.G. & Lohman, G.P. (1982). Environmental distribution of Recent benthic foraminifera on the northeast United States continental slope. *Geological Society of America Bulletin*, **93**, 200–6.

Moore, T.L. & Snyder, Scott W. (1985). Benthic foraminiferal paleoecology of Miocene deposits in central and northern Onslow Bay, North Carolina continental shelf. *Geological Society of America, Abstracts with Programs*, **17(2)**, 126.

Mullineaux, L.S. & Lohmann, G.P. (1981). Late Quaternary stagnations and recirculation of the eastern Mediterranean: changes in the deep water recorded by fossil benthic foraminifera. *Journal of Foraminiferal Research*, **11**, 20–39.

Mullins, H.T., Thompson, J.B., McDougall, K. & Vercoutere, T.L. (1985). Oxygen-minimum zone edge effects: evidence from the central California coastal upwelling system. *Geology*, **13(7)**, 491–4.

Phleger, F.B. & Soutar, A. (1973). Production of benthic foraminifera in three east Pacific oxygen minima. *Micropaleontology*, **19**, 110–15.

Poag, C.W. (1981). *Ecologic Atlas of Benthic Foraminifera of the Gulf of Mexico*. Marine Science International, Woods Hole, Massachusetts.

Poag, C.W. (1985). Benthic foraminifera as indicators of potential petroleum sources. *Proceedings of Fourth Annual Research Conference*, pp. 275–84. Gulf Coast Section, Society of Economic Paleontologists and Mineralogists.

Powers, E.R. (1988). Diatom biostratigraphy and paleoecology of the Miocene Pungo River Formation, Onslow Bay, North Carolina continental shelf. In *Micropaleontology of Miocene Deposits from the Shallow Subsurface of Onslow Bay, North Carolina Continental Margin*, ed. Scott W. Snyder. *Journal of Foraminiferal Research Special Publication*, **25**, 97–162.

Raup, D.M. & Stanley, S.M. (1978). *Principles of Paleontology*. W.H. Freeman, San Francisco.

Riggs, S.R. (1984). Paleoceanographic model of Neogene phosphorite deposition, US Atlantic continental margin. *Science*, **223**, 123–31.

Riggs, S.R., Lewis, D.W., Scarborough, A.K. & Snyder, Scott W. (1982). Cyclic deposition of the upper Tertiary phosphorites of the Aurora area, North Carolina, and their possible relationship to global sea level fluctuations. *Southeastern Geology*, **23(4)**, 189–204.

Scarborough, A.K., Riggs, S.R. & Snyder, Scott W. (1982). Stratigraphy and petrology of the Pungo River Formation, central coastal plain, North Carolina. *Southeastern Geology*, **23(4)**, 205–16.

Schafer, C.T. (1982). Foraminiferal colonization of an offshore dump site in Chaleur Bay, New Brunswick, Canada. *Journal of Foraminiferal Research*, **12(4)**, 317–26.

Schafer, C.T. & Cole, F.E. (1976). Foraminiferal distribution patterns in the Restigouche estuary. In *The First International Symposium on Benthic Foraminifera of Continental Margins, Part A, Ecology and Biology*, n.1, pp. 1–24. Maritime Sediments, Special Publication, Halifax, Nova Scotia.

Schnitker, D. (1971). Distribution of foraminifera on the North Carolina continental shelf. *Tulane Studies in Geology and Paleontology*, **8(4)**, 169–215.

Seiglie, G.A. (1968). Foraminiferal assemblages as indicators of high organic carbon content in sediments and polluted waters. *American Association of Petroleum Geologists Bulletin*, **52(11)**, 2231–41.

Sen Gupta, B.K., Lee, R.F. & May, M.S., III (1981). Upwelling and an unusual assemblage of benthic foraminifera on the northern Florida continental slope. *Journal of Paleontology*, **55(4)**, 853–7.

Snyder, Scott W., Mauger, L.L. & Akers, W.H. (1983). Planktonic foraminifera and biostratigraphy of the Yorktown Formation, Lee Creek mine. *Smithsonian Contributions to Paleobiology*, **53**, 455–82.

Snyder, Scott W., Riggs, S.R., Katrosh, M.R., Lewis, D.L. & Scarborough, A.K. (1982). Synthesis of phosphatic sediment–faunal relationships within the Pungo River Formation: paleoenvironmental implications. *Southeastern Geology*, **23**, 233–45.

Snyder, Scott W., Waters, V.J. & Moore, T.L. (1988). Benthic foraminifera and paleoecology of Miocene Pungo River sediments in Onslow Bay, North Carolina continental shelf. In *Micropaleontology of Miocene Deposits in the Shallow Subsurface of Onslow Bay, North Carolina Continental Shelf*, ed. Scott W. Snyder, *Journal of Foraminiferal Research Special Publication*, **25**, 43–96.

Streeter, S.S. (1972). Living benthonic foraminifera of the Gulf of California, a factor analysis of Phleger's (1964) data. *Micropaleontology*, **18**, 64–73.

Streeter, S.S. & Shackleton, N.J. (1979). Paleocirculation of the deep North Atlantic: 150 000-year record of benthic foraminifera and oxygen-18. *Science*, **203**, 168–71.

Thomson, S.T. (1985). Is paleontology going extinct? *American Scientist*, **73**, 570–2.

Todd, R. (1979). Depth occurrences of foraminifera along the southeastern United States. *Journal of Foraminiferal Research*, **9(4)**, 277–301.

Van der Zwaan, G.J. (1982). Paleoecology of late Miocene Mediterranean foraminifera. *Utrecht Micropaleontological Bulletins*, **25**, 1–202.

Van der Zwaan, G.J. (1983). Quantitative analyses and the reconstruction of benthic foraminiferal communities. *Utrecht Micropaleontological Bulletins*, **30**, 49–69.

Waters, V.J. (1983). 'Foraminiferal paleoecology and biostratigraphy of the Pungo River Formation, southern Onslow Bay, North Carolina continental shelf.' Unpublished M.Sc. thesis, East Carolina University, North Carolina.

Watkins, J.G. (1961). Foraminiferal ecology around the Orange County, California ocean sewer outfall. *Micropaleontology*, **7(2)**, 199–206.